TABLE OF ATOMIC WEIGHTS AND NUMBERS

Based on the 1977 Report of the Commission on Atomic Weights of the International Union of Pure and Applied Chemistry. Scaled to the relative atomic mass of carbon-12.

Element	Symbol	Atomic Number	Atomic Weight	
Actinium	Ac	89	227.0278	(e)
Aluminum	Al	13	26.98154	
Americium	Am	95	(243)	(f)
Antimony	Sb	51	121.75	(a)
Argon	Ar	18	39.948	(a, b, c)
Arsenic	As	33	74.9216	
Astatine	At	85	(210)	(f)
Barium	Ba	56	137.33	(c)
Berkelium	Bk	97	(247)	(f)
Beryllium	Be	4	9.01218	
Bismuth	Bi	83	208.9804	
Boron	B	5	10.81	(b, d)
Bromine	Br	35	79.904	
Cadmium	Cd	48	112.41	(c)
Calcium	Ca	20	40.08	(c)
Californium	Cf	98	(251)	(f)
Carbon	C	6	12.011	(b)
Cerium	Ce	58	140.12	(c)
Cesium	Cs	55	132.9054	
Chlorine	Cl	17	35.453	
Chromium	Cr	24	51.996	
Cobalt	Co	27	58.9332	
Copper	Cu	29	63.546	(a, b)
Curium	Cm	96	(247)	(f)
Dysprosium	Dy	66	162.50	(a)
Einsteinium	Es	99	(252)	(f)
Erbium	Er	68	167.26	(a)
Europium	Eu	63	151.96	(c)
Fermium	Fm	100	(257)	(f)
Fluorine	F	9	18.998403	
Francium	Fr	87	(223)	(f)
Gadolinium	Gd	64	157.25	(a, c)
Gallium	Ga	31	69.72	
Germanium	Ge	32	72.59	(a)
Gold	Au	79	196.9665	
Hafnium	Hf	72	178.49	(a)
Helium	He	2	4.00260	(c)
Holmium	Ho	67	164.9304	
Hydrogen	H	1	1.0079	(b)
Indium	In	49	114.82	(c)
Iodine	I	53	126.9045	
Iridium	Ir	77	192.22	(a)
Iron	Fe	26	55.847	(a)
Krypton	Kr	36	83.80	(c, d)
Lanthanum	La	57	138.9055	(a, c)
Lawrencium	Lr	103	(260)	(f)
Lead	Pb	82	207.2	(b, c)
Lithium	Li	3	6.941	(a, b, c, d)
Lutetium	Lu	71	174.967	(a)
Magnesium	Mg	12	24.305	(c)
Manganese	Mn	25	54.9380	
Mendelevium	Md	101	(258)	(f)
Mercury	Hg	80	200.59	(a)
Molybdenum	Mo	42	95.94	
Neodymium	Nd	60	144.24	(a, c)
Neon	Ne	10	20.179	(a, d)
Neptunium	Np	93	237.0482	(e)
Nickel	Ni	28	58.70	
Niobium	Nb	41	92.9064	
Nitrogen	N	7	14.0067	
Nobelium	No	102	(259)	(f)
Osmium	Os	76	190.2	(c)
Oxygen	O	8	15.9994	(a, b)
Palladium	Pd	46	106.4	(c)
Phosphorus	P	15	30.97376	
Platinum	Pt	78	195.09	(a)
Plutonium	Pu	94	(244)	(f)
Polonium	Po	84	(209)	(f)
Potassium	K	19	39.0983	(a)
Praseodymium	Pr	59	140.9077	
Promethium	Pm	61	(145)	(f)
Protactinium	Pa	91	231.0359	(e)
Radium	Ra	88	226.0254	(c, e)
Radon	Rn	86	(222)	(f)
Rhenium	Re	75	186.207	
Rhodium	Rh	45	102.9055	
Rubidium	Rb	37	85.4678	(a, c)
Ruthenium	Ru	44	101.07	(a, c)
Samarium	Sm	62	150.4	(c)
Scandium	Sc	21	44.9559	
Selenium	Se	34	78.96	(a)
Silicon	Si	14	28.0855	(a)
Silver	Ag	47	107.868	(c)
Sodium	Na	11	22.98977	
Strontium	Sr	38	87.62	(c)
Sulfur	S	16	32.06	(b)
Tantalum	Ta	73	180.9479	(a)
Technetium	Tc	43	(98)	(f)
Tellurium	Te	52	127.60	(a, c)
Terbium	Tb	65	158.9254	
Thallium	Tl	81	204.37	(a)
Thorium	Th	90	232.0381	(c, e)
Thulium	Tm	69	168.9342	
Tin	Sn	50	118.69	(a)
Titanium	Ti	22	47.90	(a)
Tungsten	W	74	183.85	(a)
(Unnilhexium)	(Unh)	106	(263)	(f, g)
(Unnilpentium)	(Unp)	105	(262)	(f, g)
(Unnilquadium)	(Unq)	104	(261)	(f, g)
Uranium	U	92	238.029	(c, d)
Vanadium	V	23	50.9415	(a)
Xenon	Xe	54	131.30	(c, d)
Ytterbium	Yb	70	173.04	(a)
Yttrium	Y	39	88.9059	
Zinc	Zn	30	65.38	
Zirconium	Zr	40	91.22	(c)

Except as noted in the footnotes that follow, the atomic weight values are good to ±1 unit in the last place.

(a) Precise to ±3 units in the last place.

(b) Atomic weight cannot be expressed more precisely because among normal terrestrial materials there are known variations in isotopic compositions.

(c) Geological samples of this element have been found with anomalous isotopic composition and atomic weights different from this value.

(d) Considerable variations from this atomic weight value can occur in commercial samples because of changes in isotopic composition.

(e) The atomic weight of the radioisotope of longest half-life.

(f) The mass number of the radioisotope of longest half-life.

(g) The official name and symbol has not been agreed to. Element 105 is unofficially called hahnium. Element 104 is called rutherfordium by American scientists and kurchatovium by Russian scientists.

61.10x

GENERAL CHEMISTRY
PRINCIPLES AND STRUCTURE

GENERAL CHEMISTRY 4/E

PRINCIPLES AND STRUCTURE
SI Version

JAMES E. BRADY

St. John's University
Jamaica, New York

GERARD E. HUMISTON

Skidmore College
Saratoga Springs, New York

**SI Version Prepared by
Henry Heikkinen**

University of Maryland at College Park

John Wiley & Sons

New York   Chichester   Brisbane   Toronto   Singapore

Cover Photograph: Computer Graphics Laboratory, University of California, San Francisco.

Cover: Molecular model of acetylcholine receptor ion channel (blue) with sodium ion (yellow) surrounded by water (red and white) traveling down the channel, which appeared on the cover of Chemical and Engineering News August 12, 1985.

Production supervised by Linda R. Indig

Cover designed by Karin Gerdes Kincheloe

Text designed by Judith Fletcher Getman/Joan Willens

Photo researched by John Schultz and Elyse Rieder

Photo editor: Stella Kupferberg

Illustrations by John Balbalis with the assistance of the Wiley Illustration Department

Manuscript edited by Patti Brecht under the supervision of Tamara Lee.

Library of Congress Cataloging-in-Publication Data

Brady, James E., 1938–
  General chemistry.
  1. Chemistry.   I. Humiston, Gerard E., 1939–
II. Heikkinen, Henry.   III. Title.
QD31.2.B7 1986   540   86-7786
ISBN 0-471-83841-1

Printed in the United States of America

10 9 8 7 6 5 4 3 2 1

PREFACE

Throughout the evolution of this book we have pursued two principal goals. First, we wished to make the book interesting and easily understood, thereby making it useful to the student. Second, we wished to keep the book up-to-date with respect to current trends in the teaching of chemistry, and thereby make it useful to the teacher. As we undertook preparation of this Fourth Edition these remained our goals, and as in previous editions we have kept what has proven valuable while responding to suggestions for changes and improvements. The level of the book and its general approach therefore remain unchanged; the text continues to be designed and written for the first year chemistry course for science majors.

In this edition we have continued to focus attention on improving the usefulness and readability of the text. Major sections have been rewritten to improve clarity, and in some places entire discussions have been reworked to make them easier to understand. Additional worked examples have been included, and in this edition each includes a title that describes the kind of problem being solved. To aid the student as well as the teacher, end-of-chapter exercises have been sorted and arranged by category. The more difficult problems are marked by asterisks. Answers to approximately half the Review Problems as well as to selected Review Questions are included in Appendix C at the back of the book.

Also new to this edition is the availability of a unique software package, THE CHEMISTRY TUTOR, which is keyed to the critically important first eight chapters of the book. This tutorial software was designed to complement discussions in the text and to improve student performance in such diverse areas as balancing equations, writing Lewis structures, predicting the products of metathesis reactions, and naming chemical compounds. To assist students who wish to avail themselves of this valuable supplement, we have indicated in the margin of the text at appropriate locations the titles of the programs that provide the necessary tutorial assistance. Additional details concerning this software supplement appear on a separate page following this preface.

In response to suggestions from users of the previous edition and from reviewers, we have made some changes in topic sequence. In doing so, there were two objectives. First, we wished to make it more apparent to students that chemistry is an experimental science, and that theory is developed to explain observed facts. Second, we wished to integrate additional descriptive chemistry with discussions of principles so instructors would find it easier to teach and students would appreciate that learning chemical facts is an important part of learning chemical principles. Toward these ends, former Chapter 9 (Periodic Table Revisited) and Chapter 14 (Acids and Bases) have been dismantled and the material in them redistributed principally among earlier chapters. For example, the general properties of metals, nonmetals, and metalloids now introduce the chapter that covers the periodic table; the molecular structures of the nonmetals are discussed in the chapter on modern theories of covalent bonding to illustrate how an atom's ability to form pi bonds can determine the complexity of molecules; and all of acid–base chemistry is now discussed together in Chapter 7. We have also added a new section at the beginning of the elementary bonding chapter that describes the properties of ionic and molecular compounds. This serves as a prelude to discussions of ionic and covalent bonding.

Among other significant changes is the division of the very long former Chapter 3 (Atomic Structure and the Periodic Table) into two chapters. The first discusses the periodic table and the experiments that led to the discovery of atomic structure. The next examines the electronic structures of atoms. We have also divided former Chapter 6 into two chapters. One (now Chapter 7) focuses on reaction types, including ionic, acid–base, and redox reactions. This chapter also presents an opportunity to introduce some related topics in descriptive chemistry. Here we include sections on metals as reducing agents and nonmetals, especially molecular oxygen, as oxidizing agents. This is followed by a second chapter (Chapter 8) that examines ionic reactions in solution in greater detail, including solution stoichiometry. Because the details of solution stoichiometry have now been postponed until Chapter 8, some aspects of this topic are introduced toward the end of Chapter 2.

There have been some other changes as well. The chapter on physical properties of solutions now includes a section that treats the properties of colloids. The discussion of metallurgy has been moved to Chapter 18. The chapter on nuclear chemistry has been moved forward to precede organic and biochemistry. A glossary has been added, and for each entry there is a reference to the section in the text where the term is discussed.

Despite these changes, features that have appealed to users of previous editions have been retained. As in the past, we assume no prior student background in chemistry, and mathematical sophistication is limited to simple algebra. Generally it is assumed that students who use this book will have a scientific calculator at their disposal, and there is a section in Appendix A that discusses the use of calculators. A table of common logarithms is still included, however, for the rare student willing to use it. New terms have been set in bold type and nearly all are included in the glossary.

Although there have been changes made in topic sequence, concepts continue to be presented in a logical sequence that permits early introduction of quantitative experiments in the laboratory. Chapters 1 and 2 introduce the concepts of atoms, molecules, and the mole. The next four chapters go on to explore the makeup of atoms and the kinds of bonds that they form with each other. In reworking these chapters we have woven in descriptive topics that serve as a foundation for theory.

Once atomic structure and chemical bonding have been discussed, we turn our attention in Chapters 7 and 8 to reaction types. We begin with ionic reactions, including aqueous acid–base chemistry. The acid–base concept is then extended to Brønsted–Lowry and Lewis acids and bases. Redox reactions are discussed next, and that leads to an examination of periodic trends in the ease of oxidation of metals and the tendencies of nonmetals to serve as oxidizing agents. The special role of oxygen as an oxidizing agent and the products formed in typical reactions are also included.

Chapters 9 and 10 consider the physical properties of the states of matter, and they are followed by a chapter dealing with the physical properties of mixtures, especially solutions.

The next three chapters cover the topics of thermodynamics, kinetics, and equilibrium. They are discussed in this sequence because they address the questions: "Is a reaction possible, how fast does it occur, and what is the system like when it reaches equilibrium?" In the chapter on thermodynamics, the discussions of the First Law and of entropy have been revised. The equilibrium chapter focuses on gaseous and heterogeneous systems, the way thermodynamics relates to equilibrium, and a thorough discussion of Le Châtelier's principle.

Chapters 15 and 16 treat ionic equilibria in detail. Chapter 15 deals with acid–base equilibria in aqueous solutions; Chapter 16 covers solubility and complex ion equilibria. In Chapter 15, the discussion of hydrolysis of salts has been totally revised in terms of Brønsted–Lowry acid–base chemistry. These chapters are followed by one on electrochemistry.

Chapters 18 through 21 are descriptive chemistry chapters. They provide a systematic discussion of the chemistry of the most important elements. Frequent reference to important and familiar chemicals illustrates how the elements and their compounds impact modern society. Next is a chapter on nuclear chemistry, and the book finishes with organic and biochemistry—topics that serve as an introduction to the second year course.

We have been gratified by the reception this text has received in its previous editions. It has been a source of great satisfaction that both students and teachers have found the book and its associated supplements useful. We hope that this edition will also please you, and we invite your comments and suggestions.

James E. Brady
Gerard E. Humiston

ACKNOWLEDGMENTS

It has always been a pleasure to thank those who have contributed in their own special way to the completion of a project such as this. First, we wish to express affectionate appreciation to our wives and children who have been patient with us and have been willing to put off pet projects around the house while we toiled to meet deadlines. We offer our special appreciation to the staff at Wiley for their careful work, cheerful spirit, and attention to detail, especially our Editor Dennis Sawicki for his guidance and encouragement; his assistants, Tomi Navedo and Blanca Ferreris; our Copy Editor, Patty Brecht; our Picture Editor, Stella Kupferberg; our Illustrator, John Balbalis; and our Designer, Karin Kincheloe. Special mention has to be made of our Production Supervisor, Linda Indig, who with delightful good humor and just a bit of arm-twisting has guided this book to completion. We are also grateful to our colleagues and students for their many constructive suggestions and helpful discussions, especially Drs. Ernest Birnbaum, Eugene Holleran, Neil Jespersen, William Pasfield, and Siao Sun. And finally, our special thanks go to the following colleagues who have helped shape this book by their thoughtful reviews and criticisms of the manuscript and their many valuable suggestions:

Carlo Alfare
Department of Chemistry
Mercer County Community College
Trenton, NJ 08690

Orville T. Beachley
Department of Chemistry
SUNY Buffalo
Buffalo, NY 14214

David Becker
Department of Chemistry
Oakland Community College
Farmington Hill, MI 48018

Jo A. Beran
Department of Chemistry
University of Colorado
Boulder, CO 80309

James J. Bohning
Department of Chemistry
Wilkes College
Wilkes-Barre, PA 18766

Luther K. Brice
Department of Chemistry
Virginia Polytechnic Institute &
State University
Blacksburg, VA 22901

Alice Corey
Department of Chemistry
Pasadena City College
Pasadena, CA 91106

Marcia D. Davies
Department of Chemistry
Creighton University
Omaha, NE 68178

Wade Freeman
Department of Chemistry
University of Illinois
Chicago, IL 60680

Graham P. Glass
Department of Chemistry
Rice University
Houston, TX 77251

Forrest C. Hentz Jr.
Department of Chemistry
North Carolina State University
Raleigh, NC 27650

Paul B. Kelter
Department of Chemistry
Manhattan College
Riverdale, NY 10471

Philip S. Lamprey
Department of Chemistry
University of Lowell
Lowell, MA 01854

E. R. Magnuson
Department of Chemistry
Milwaukee School of Engineering
Milwaukee, WI 53201

Lawrence C. Nathan
Department of Chemistry
University of Santa Cruz
Santa Cruz, CA 95053

John P. Oliver
Department of Chemistry
Wayne State University
Detroit, MI 48202

Larry Peck
Department of Chemistry
Texas A & M University
College Station, TX 77843-3255

Helen Place
Department of Chemistry
Washington State University
Pullman, WA 99164

David Price
Department of Chemistry
Glendale Community College
Glendale, AZ 85302

Lewis A. Radonovich
Department of Chemistry
University of North Dakota
Grand Forks, ND 58202

Nancy S. Rowan
Department of Chemistry
American University
Washington, DC 20016

Dennis P. Ryan
Department of Chemistry
Hofstra University
Hempstead, NY 11550

Morris B. Silverman
Department of Chemistry
Portland State University
Portland, OR 97207

Scott Sinex
Department of Chemistry
Prince George's Community College
Largo, MD 20772

Ronald Strothkamp
Department of Chemistry
Hofstra University
Hempstead, NY 11550

William Tucker
Department of Chemistry
North Carolina State University
Raleigh, NC 27650

R. D. Willet
Department of Chemistry
Washington State University
Pullman, WA 99164

J. E. B.
G. E. H.

PHOTO CREDITS

Figure 18.14: Lester V. Bergman and Associates, Inc.
Page 674: U.S. Thermite, Inc.
Figure 18.16: Peter Lerman.

CHAPTER 19
Opener: Courtesy of Environmental Protection Agency.
Page 685: Peter Lerman.
Page 692: Dick Breho/Associated Photographers, Courtesy of
    Calgon Corporation.
Page 693: Peter Lerman.
Figure 19.2: James Brady.
Figure 19.3: Courtesy of The Permuitt Company, Division of
    Sybron Company.
Page 697: Courtesy NASA.
Page 699: Peter Lerman.
Page 706: (top) Peter Lerman.
Figure 19.7: Peter Lerman.

CHAPTER 20
Opener: Arthur D'Arazien/The Image Bank.
Figure 20.1: Peter Lerman.
Figure 20.6: © Leo Touchet.
Figure 20.8: Mario Fantain/Photo Researchers.
Figure 20.9: Peter Lerman.
Page 728: Kollar/TVA.
Figure 20.10: Schmidt-Thomsen, Landesdenkmalamt, Westfalen-
    Lippe, Münster, Germany.

Figure 20.11: (a) © Erich Hartman/Magnum; (b) Ted Spiegel/
    Black Star.
Figure 20.12: Peter Lerman.
Page 746: The Dow Chemical Company.
Page 750: Kathy Bendo.

CHAPTER 21
Opener: Gene Ahrens.
Page 758: General Electric Company, Neal Park, Cleveland, Ohio.
Page 764: National Archives.
Page 767: Grant Heilman/Grant Heilman Photography.
Page 768: PAR/NYC.
Page 771: Alyeska Pipeline Service Company.
Page 771: Kathy Bendo.

CHAPTER 22
Opener: William Felger/Grant Heilman.
Figure 22.5: Courtesy Nuclear Equipment Chemical.
Page 819: U.S. Department of Energy
Figure 22.14: U.S. Department of Energy.

CHAPTER 23
Opener: Randy Taylor/Sygma.
Page 827: The Bettmann Archive.

CHAPTER 24
Opener: Alfred Owczarzak/Taurus Photos.

SUPPLEMENTS

A complete package of supplements to accompany this text is available to assist both the teacher and the student.

Study Guide to Accompany General Chemistry, Principles and Structure, Fourth Edition, by James E. Brady. This softcover book has been carefully structured to assist students in mastering concepts and developing problem-solving skills. It is keyed section by section to the text, and for each section there is a set of objectives, a brief review (sometimes with additional worked examples), a self-test with answers, and a list of the new terms (each carefully defined) that are introduced in that section.

The Chemistry Tutor, by Frank Rinehart. This software, available for the IBM PC and Apple II/II+/IIe computers, provides tutorial assistance for major topics that serve as a foundation in the learning of general chemistry. It is discussed more fully on a preceding page.

Laboratory Manual for General Chemistry, Principles and Structure, Third Edition, by Jo Beran and James E. Brady. This manual features a thorough techniques section with photographs that illustrate important apparatus and manipulations, and 47 experiments sequenced to follow the topical development of the text. For the teacher, an instructor's manual accompanies the laboratory manual.

Solutions Manual for General Chemistry, Principles and Structure, SI Version, by M. Larry Peck. This softcover supplement provides detailed solutions to all the numerical problems in the text, as well as answers to all the questions.

Problem Exercises for General Chemistry, Third Edition, by G. Gilbert Long and Forrest C. Hentz, Jr. Intended to bridge the gap between textbook-style exercises and those that students encounter on examinations, this softcover book features over 1000 problems and questions in multiple-choice format. Throughout the book students are taught to use the basic "tools of the chemical trade."

Lecture Outline for General Chemistry, Principles and Structure, by Ronald Ragsdale. This outline provides the student with a framework about which meaningful class notes can be built. Through its structure, it provides the instructor with more class time to cover difficult topics.

Test File for General Chemistry, Principles and Structure, by David Becker. This package of multiple choice questions is available from Wiley at no charge for teachers who adopt this book.

Microcomputerized Testing System for the IBM PC and Apple II/II+ /IIe. This testing system, with questions, allows the user to prepare multiple choice examinations. It can be obtained from Wiley without charge by instructors who adopt this book.

Transparency Masters Instructors who adopt this book may obtain from Wiley, without charge, a set of enlarged black and white line drawings that duplicate key figures in the text. They can be used to prepare transparencies.

CONTENTS

CHAPTER 1
INTRODUCTION   1

1.1 The Scientific Method   3
1.2 Measurement   5
1.3 Units of Measurement   11
1.4 Matter and Energy   17
1.5 Properties of Matter   21
1.6 Elements, Compounds, and
     Mixtures   25
1.7 Conservation of Mass and
     Definite Proportions   29
1.8 The Atomic Theory of
     Dalton   30
1.9 Atomic Weights   32
1.10 Symbols, Formulas, and
     Equations   33
Review Questions and Problems   35

CHAPTER 2
STOICHIOMETRY:
CHEMICAL ARITHMETIC   41

2.1 Chemical Symbols and For-
     mulas—Another Look   42
2.2 The Mole   42
2.3 Measuring Moles of Atoms   46
2.4 Measuring Moles of Com-
     pounds: Molecular Weights
     and Formula Weights   49
2.5 Percentage Composition   51
2.6 Chemical Formulas   52

2.7 Empirical Formulas   53
2.8 Molecular Formulas   56
2.9 Balancing Chemical Equa-
     tions   57
2.10 Calculations Based on
     Chemical Equations   58
2.11 Limiting Reactant Calcula-
     tions   61
2.12 Theoretical Yield and Per-
     centage Yield   63
2.13 Molar Concentration   64
2.14 Preparing Solutions by Di-
     lution   67
2.15 The Stoichiometry of Reac-
     tions in Solutions   70
Review Questions and Problems   73

CHAPTER 3
THE PERIODIC TABLE AND THE
MAKEUP OF ATOMS   81

3.1 Some Properties of the Ele-
     ments   82
3.2 The First Periodic Table   85
3.3 Atomic Numbers and the
     Modern Periodic Table   86
3.4 The Structures of Atoms   89
3.5 The Charge on the Electron   92
3.6 Positive Particles and the
     Mass Spectrometer   93
3.7 Radioactivity   96

3.8   The Nuclear Atom   96
3.9   The Neutron   97
3.10 Isotopes   98
Review Questions and Problems   99

CHAPTER 4
ELECTRONIC STRUCTURE AND
THE PERIODIC TABLE   103

4.1   Electromagnetic Radiation
      and Atomic Spectra   104
4.2   The Bohr Theory of the Hy-
      drogen Atom   110
4.3   The Wave Nature of Matter:
      Wave Mechanics   114
4.4   Electron Spin and the Pauli
      Exclusion Principle   119
4.5   The Electron Configurations
      of the Elements   120
4.6   The Periodic Table and Elec-
      tron Configurations   124
4.7   The Spatial Distribution of
      Electrons   128
4.8   The Variation of Properties
      with Atomic Structure   131
Review Questions and Problems   138

CHAPTER 5
CHEMICAL BONDING:
GENERAL CONCEPTS   142

5.1   Properties of Ionic and Mo-
      lecular Compounds   143
5.2   Lewis Symbols   145
5.3   The Ionic Bond   146
5.4   Factors That Influence the
      Formation of Ionic Com-
      pounds   151
5.5   The Covalent Bond   154
5.6   Drawing Lewis Structures   157
5.7   Bond Order and Some Bond
      Properties   160
5.8   Resonance   163
5.9   Coordinate Covalent Bonds   165
5.10 Polar Molecules and Elec-
      tronegativity   166

5.11 The Naming of Chemical
      Compounds   169
Review Questions and Problems   173

CHAPTER 6
COVALENT BONDING AND
MOLECULAR STRUCTURE   178

6.1   Molecular Shapes   179
6.2   Valence Shell Electron-Pair
      Repulsion Theory   182
6.3   Polarity of Molecules and
      Molecular Structure   190
6.4   Valence Bond Theory   194
6.5   Hybrid Orbitals   196
6.6   Multiple Bonds   205
6.7   Resonance   208
6.8   Single Bonds Versus Multi-
      ple Bonds: The Molecular
      Structures of the Elemental
      Nonmetals   209
6.9   Molecular Orbital Theory   213
Review Questions and Problems   219

CHAPTER 7
CHEMICAL REACTIONS AND
THE PERIODIC TABLE   222

7.1   Reactions in Solution   223
7.2   Acids and Bases in Aqueous
      Solutions   232
7.3   Brønsted–Lowry Acids and
      Bases   237
7.4   The Strengths of Acids and
      Bases: Periodic Trends   241
7.5   Lewis Acids and Bases   246
7.6   Oxidation-Reduction Reac-
      tions; Oxidation Numbers   248
7.7   Balancing Redox Equations
      Using Oxidation Numbers   251
7.8   Balancing Redox Equations
      by the Ion-Electron Method   253
7.9   Metals as Reducing Agents   259
7.10 Nonmetals as Oxidizing
      Agents   265
7.11 Molecular Oxygen as an
      Oxidizing Agent   267
Review Questions and Problems   270

CHAPTER 8
IONIC REACTIONS IN
SOLUTION—A CLOSER LOOK   275

8.1 Metathesis Reactions: Why
They Occur   276

8.2 The Preparation of Inorganic
Salts by Metathesis Reactions   283

8.3 Stoichiometry of Ionic Reactions   287

8.4 Chemical Analysis and Titrations   291

8.5 Oxidizing and Reducing
Agents in the Laboratory   295

Review Questions and Problems   300

CHAPTER 9
PROPERTIES OF GASES   305

9.1 Volume and Pressure   306
9.2 Boyle's Law   310
9.3 Charles' Law   313
9.4 Gay–Lussac's Law   315
9.5 The Combined Gas Law   316
9.6 Dalton's Law of Partial
Pressures   318
9.7 Chemical Reactions Between
Gases   321
9.8 The Ideal Gas Law   323
9.9 Graham's Law of Effusion   328
9.10 Kinetic Molecular Theory
and the Gas Laws   329
9.11 Real Gases   334
Review Questions and Problems   337

CHAPTER 10
STATES OF MATTER AND
INTERMOLECULAR FORCES   342

10.1 Comparing the Properties
of Gases, Liquids, and
Solids   343
10.2 Intermolecular Attractive
Forces   349
10.3 Heat of Vaporization   351
10.4 Vapor Pressures of Liquids   354

10.5 Boiling Point   360
10.6 Freezing Point   363
10.7 Crystalline Solids   364
10.8 Lattices   367
10.9 Types of Crystals   371
10.10 Liquid Crystals   374
10.11 Heating and Cooling
Curves; Changes of State   375
10.12 Vapor Pressures of Solids   377
10.13 Phase Diagrams   378
Review Questions and Problems   382

CHAPTER 11
PHYSICAL PROPERTIES OF
SOLUTIONS AND COLLOIDS   387

11.1 Kinds of Mixtures: Suspensions, Colloids, and Solutions   388
11.2 Types of Solutions   393
11.3 Concentration Units   394
11.4 The Solution Process in
Liquid Solutions   398
11.5 Heats of Solution   403
11.6 Solubility and Temperature   408
11.7 The Effect of Pressure on
Solubility   409
11.8 Vapor Pressures of Solutions   411
11.9 Fractional Distillation   416
11.10 Colligative Properties of Solutions   418
11.11 Osmotic Pressure   422
11.12 Solutions of Electrolytes   425
Review Questions and Problems   430

CHAPTER 12
CHEMICAL THERMODYNAMICS   435

12.1 Some Commonly Used
Terms   436
12.2 The First Law of Thermodynamics   438
12.3 Heats of Reaction: Thermochemistry   444
12.4 Hess's Law of Heat Summation   448

12.5 Standard States 451

12.6 Bond Energies 454

12.7 Energy, Entropy, and the Spontaneity of Chemical and Physical Changes 458

12.8 The Second Law of Thermodynamics 462

12.9 Free Energy and Useful Work 465

12.10 Standard Entropies and Free Energies 466

12.11 Free Energy and Equilibrium 469

Review Questions and Problems 471

CHAPTER 13
CHEMICAL KINETICS: THE STUDY OF THE RATES OF REACTIONS 477

13.1 Reaction Rates and Their Measurement 478

13.2 Rate Laws 481

13.3 Concentration and Time: Half-Lives 486

13.4 Collision Theory 489

13.5 Reaction Mechanisms 490

13.6 Effective Collisions 493

13.7 Transition State Theory 494

13.8 Effect of Temperature on Reaction Rates 497

13.9 Catalysts 501

13.10 Chain Reactions 503

Review Questions and Problems 505

CHAPTER 14
CHEMICAL EQUILIBRIUM 511

14.1 The Equilibrium Law for a Chemical Reaction 512

14.2 The Equilibrium Constant 514

14.3 Thermodynamics and Chemical Equilibrium 515

14.4 The Relationship Between K_p and K_c 518

14.5 Heterogeneous Equilibria 520

14.6 Le Châtelier's Principle and Chemical Equilibria 522

14.7 Equilibrium Calculations 525

Review Questions and Problems 532

CHAPTER 15
ACID–BASE EQUILIBRIA IN AQUEOUS SOLUTIONS 536

15.1 The Ionization of Water and the pH Concept 537

15.2 Dissociation of Weak Acids and Bases 543

15.3 Dissociation of Polyprotic Acids 550

15.4 Buffers 554

15.5 Hydrolysis of Salts 559

15.6 Acid–Base Titrations: The Equivalence Point 568

15.7 Acid–Base Indicators 572

Review Questions and Problems 574

CHAPTER 16
SOLUBILITY AND COMPLEX ION EQUILIBRIA 579

16.1 Solubility Product 580

16.2 The Common Ion Effect and Solubility 587

16.3 Complex Ions and Their Equilibria 590

16.4 Complex Ions and Solubility 591

Review Questions and Problems 594

CHAPTER 17
ELECTROCHEMISTRY 596

17.1 Metallic and Electrolytic Conduction 597

17.2 Electrolysis 599

17.3 Practical Applications of Electrolysis 602

17.4 Quantitative Aspects of Electrolysis 608

17.5 Galvanic Cells 611

17.6 Cell Potentials 613

17.7 Reduction Potentials 614

17.8 Spontaneity of Oxidation-Reduction Reactions 621

17.9 Thermodynamic Equilibrium Constants from Standard Cell Potentials 623

17.10 The Effects of Concentration on Cell Potentials: The Nernst Equation 625

17.11 Applications of the Nernst Equation 626

17.12 Practical Applications of Galvanic Cells 631

Review Questions and Problems 635

CHAPTER 18
METALS AND THEIR COMPOUNDS; THE REPRESENTATIVE METALS 640

18.1 Metallurgy 641

18.2 Trends in Metallic Behavior 650

18.3 Ionic-Covalent Character of Metal–Nonmetal Bonds 652

18.4 Colors of Metal Compounds 654

18.5 Group IA: The Alkali Metals 657

18.6 Group IIA: The Alkaline Earth Metals 665

18.7 Metals of Groups IIIA, IVA, and VA 671

Review Questions and Problems 678

CHAPTER 19
THE CHEMISTRY OF SELECTED NONMETALS PART I: HYDROGEN, CARBON, OXYGEN, AND NITROGEN 682

19.1 Hydrogen 683

19.2 Carbon 691

19.3 Oxygen 696

19.4 Nitrogen 703

Review Questions and Problems 713

CHAPTER 20
THE CHEMISTRY OF SELECTED NONMETALS PART II: PHOSPHORUS, SULFUR, THE HALOGENS, THE NOBLE GASES, AND SILICON 717

20.1 Phosphorus 718

20.2 Sulfur 724

20.3 The Halogens 733

20.4 Noble Gas Compounds 743

20.5 Silicon 745

Review Questions and Problems 751

CHAPTER 21
THE TRANSITION ELEMENTS 755

21.1 General Properties 756

21.2 Electronic Structure and Oxidation States 758

21.3 Atomic and Ionic Radii 761

21.4 Magnetism 762

21.5 Properties of Some Important Transition Metals 764

21.6 Coordination Compounds 772

21.7 Coordination Number and Structure 776

21.8 Naming Coordination Compounds 776

21.9 Isomerism and Coordination Compounds 779

21.10 Bonding in Coordination Compounds: Valence Bond Theory 784

21.11 Crystal Field Theory 788

Review Questions and Problems 793

CHAPTER 22
NUCLEAR CHEMISTRY 797

22.1 Spontaneous Radioactive Decay 798

22.2 Applications of Nuclear Reactions 804

22.3 Nuclear Stability 808

22.4 Nuclear Transformations 811

22.5 Extension of the Periodic Table 812

22.6 Nuclear Binding Energy   815
22.7 Fission, Fusion, and Nuclear Energy   816
Review Questions and Problems   822

CHAPTER 23
ORGANIC CHEMISTRY   826

23.1 Hydrocarbons   827
23.2 Isomers of Organic Compounds   831
23.3 Naming Organic Compounds   836
23.4 Cyclic Hydrocarbons   842
23.5 Aromatic Hydrocarbons   845
23.6 Hydrocarbon Derivatives   849
23.7 Halogen Derivatives   852
23.8 Organic Compounds that Contain Oxygen   853
23.9 Amines and Amides: Organic Derivatives of Ammonia   860
23.10 Polymers   862
Review Questions and Problems   864

CHAPTER 24
BIOCHEMISTRY   868

24.1 Proteins   869
24.2 Enzymes   875
24.3 Carbohydrates   878
24.4 Lipids   882
24.5 Nucleic Acids   887
24.6 Protein Synthesis   891
Review Questions and Problems   896

APPENDIX A

Mathematics for General Chemistry   A-1

APPENDIX B

Common Logarithms   A-12

APPENDIX C

Answers to Selected Questions and Problems   A-14

GLOSSARY   G-1

INDEX   I-1

GENERAL CHEMISTRY
PRINCIPLES AND STRUCTURE

CHAPTER 1

INTRODUCTION

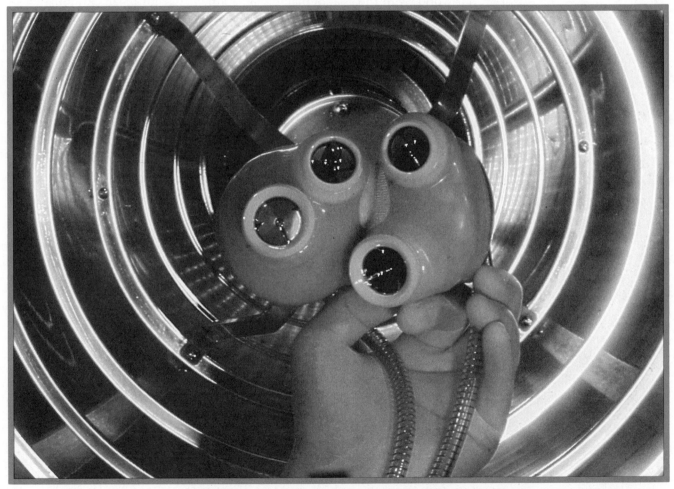

Chemistry is a subject that touches human lives in many ways, but none quite as dramatically as in modern medicine. The Jarvik 7 artificial heart shown in this photo owes its construction and life-saving functions to materials that are not normally found in nature, materials that are the fruits of chemical research.

Anyone who reads the newspapers or watches TV is aware of the impact that chemicals have had on our lives. All too often the news has been bad, as we hear about toxic waste dumps or chemicals in foods and the environment that are potential cancer producers. These are such terrifying things that it is difficult to appreciate that it is the same science—chemistry—that is responsible for the development and synthesis of drugs that can save our lives, or has led to the formulation of pesticides that have protected food crops and prevented people from starving to death. We take for granted the benefits that chemical research has brought, while our attention is drawn to the actions of relatively few careless or greedy chemical producers and users. Chemistry has become a social issue, and there are people who demand that all chemicals be banned. Through all of this, scientists who study chemicals have come to be viewed with some fear and suspicion.

Fears about the goals of science in general, and of chemistry in particular, stem from ignorance. All the branches of science seek a greater understanding of our world, with the goal of making life a little better for us all. And just look at how marvelously successful we've been. A good part of the clothing we wear, the automobiles we drive, and other products we encounter daily simply did not exist at the turn of the century, and would not exist today if it were not for chemical research. Medicines created in the laboratory have made us healthier and prolonged our lives. The realization that a living organism is a complex chemical "factory" has generated a strong interest in the life sciences, especially biology and medicine. As a result, the study of biochemistry has brought great advances in our knowledge of the nature of life. So as you begin your study of chemistry, be sure to maintain a balanced view of the subject and keep an open mind. It will make learning the subject a lot easier, and you may even find that you like it.

What chemistry is about

Those who would ban chemicals from the face of the earth would ban themselves as well. A chemical is anything that we can see or touch. All the objects around us, including our own bodies, are composed of substances that we call chemicals. An unpolluted spring spouts a liquid just packed full of a chemical—a chemical called water. Clean fresh air is filled with chemicals—chemicals called nitrogen and oxygen, and a little argon and water vapor. In fact, the only place you will not find chemicals is in a vacuum.

Chemistry is a branch of science that studies the behavior of chemicals. It is concerned with the compositions of materials and the way their properties are related to their compositions. And of course, chemistry is concerned with the way different substances interact with each other in chemical reactions to form new substances with new properties. In fact, it is the fantastic changes that take place during chemical reactions that make chemistry such an interesting science. Who would believe that a poisonous war gas, and a metal so sensitive to moisture that an explosion will result if a chunk of it is dropped into water, can be combined in a violent reaction to form simple table salt? Yet that is what happens when the metal sodium is dropped into a container full of the gas chlorine. To wonder how this can be, and to seek an explanation, is to study chemistry.

In this introductory chapter, we examine how science in general operates. You will learn about the materials and some of the concepts with which chemists and chemistry students work. You also will learn how the concept of the atom became firmly established, and you will begin to learn some of the jargon used by chemists. If you've had a previous course in chemistry, much of what is covered here will be familiar. Nevertheless, be sure you understand it all, be-

Webster's defines science as "a branch of study concerned with observation and classification of facts."

cause students who have mastered these fundamental concepts and learned to "speak the language" of chemistry have little difficulty later on.

1.1 THE SCIENTIFIC METHOD

Many of the most important advances in science, such as the discoveries of radioactivity by Henri Becquerel and penicillin by Alexander Fleming, have come about by accident. These discoveries were really only partly accidental, however, because the people involved had learned to think "scientifically" and were aware that they had observed something new and exciting.

Progress in chemistry, as well as in other sciences, is usually much less spectacular than the discoveries by Becquerel and Fleming. It is accomplished by many hours of careful work that follows a more or less systematic approach toward answering scientific questions. This approach is called the **scientific method.**

Scientists usually follow the scientific method as much by instinct as by design.

The scientific method is really nothing more than a formal statement of the steps that we follow as we logically approach any problem. Consider, for example, how a TV technician handles the repair of a disabled TV set. First, the defective component is located by observing the results of a series of tests. Then, the bad component is replaced, and finally the TV set is turned on to check that the proper repair had been made.

When we approach a problem in the sciences, we proceed in much the same way. The first step in the scientific method can be called **observation.** That is the purpose of the experiments that you, or any other scientist, perform in the laboratory. There nature is observed under controlled conditions so that the results of the experiments are reproducible. The bits of information you obtain are called **data** and can be classified as either qualitative or quantitative. **Qualitative** observations do not have numbers associated with them. An example is the observation that when sodium bicarbonate (baking soda) is added to acetic acid (vinegar), the mixture bubbles vigorously as the two chemicals react. However, if we measured the *amount* of sodium bicarbonate needed to react with a given quantity of acetic acid, we would be making a **quantitative** observation because it would result in numerical data. We will see that quantitative measurements are generally more useful to a scientist than qualitative observations because they provide more information.

The statement that "a gas is formed in the reaction of baking soda with something in vinegar" is a qualitative observation. No numerical data are involved.

It is natural for humans to seek order amid disorder. When confronted with massive quantities of information, we look for similarities and differences in an attempt to acquire a view of the "big picture." So it is in the sciences, too. As we collect data, we look for trends and seek generalizations that help us summarize the information, so that it is easier to comprehend.

In science, a general statement that summarizes facts that come from many experiments is called a **law,** or sometimes a **natural law.** In one sense, laws serve as a convenient means of storage for the vast quantities of data from which they are derived. But laws are really more useful than that. They also allow us to predict the outcome of yet untried experiments. For instance, chemists have found that when hydrogen gas and oxygen gas, at the same temperature and pressure, are allowed to react with each other to form water, two volumes of hydrogen are always needed to completely use up one volume of oxygen. This simple statement is a law dealing with the reaction of hydrogen with oxygen (see Figure 1.1). It allows us to predict that if we had five cubic metres of oxygen gas, ten cubic metres of hydrogen gas would be needed for a complete reaction. We can make this prediction with confidence, even if nobody had ever worked with these volumes of these gases, because the law that we applied was based on the

Figure 1.1

The fierce reaction of hydrogen with oxygen (from the air) led to the fiery destruction of the Hindenburg, a German airship filled with hydrogen, in Lakehurst, New Jersey, in 1937. Thirty people died in the disaster.

Physicians apply the scientific method when they use the results of lab tests and a patient's symptoms to diagnose a disease and follow its treatment by various medications.

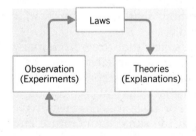

The scientific method is a cyclic process. Experiments suggest laws, which are explained by theories, which themselves suggest new experiments to be done, which produce new laws and new theories, and so on.

results of many reactions with varying amounts of hydrogen and oxygen.

Sometimes a law is expressed as a simple verbal statement, as in the law just discussed regarding the reaction of hydrogen and oxygen. On the other hand, it is often more useful to have a law stated in the form of an equation. For example, it is observed that the force of attraction between oppositely charged particles decreases as their distance of separation increases. This is more accurately stated by Coulomb's equation, or law,

$$F = k\frac{q_1 q_2}{r^2}$$

in which F is the force of attraction between the two oppositely charged particles, q_1 and q_2 are the charges on the particles, and r is their distance of separation. Laws quite commonly are expressed in equation form.

As we have noted, a law simply correlates large quantities of information. Laws in themselves do not explain why nature behaves as it does. Scientists, being human (despite what you may have heard to the contrary), are not satisfied with simple statements of fact and seek to explain their observations. Thus, the second step in the scientific method is to propose tentative explanations, or **hypotheses,** that may be tested by experiment. If they are not disproven by repeated experiments they develop into **theories.** Theories themselves always serve as a guide to new experiments and are constantly being tested. When a theory is proven incorrect by experiment, it must either be discarded in favor of a new one or, as is often the case, modified so that all the experimental observations may be accounted for. Science develops, then, through a constant interplay between theory and experiment.

As a final note, you should understand that theories can seldom be *proven* to be correct. Usually, the best we can do is fail to find an experiment that disproves a theory. Of course, this does not mean that such an experiment is not possible, so we can never be quite sure that a theory is true. There always should be a lingering doubt, and a scientist must always be careful not to confuse theory with experimental fact. Too many times in the past incorrect theories have been accepted as fact, and the progress of science slowed because of it.

1.2   MEASUREMENT

The great British physicist William Thomson (better known by his title, Lord Kelvin) once wrote, ''When you can measure what you are speaking about, and express it in numbers, you know something about it.'' This statement is true not only in physics but in all the sciences. No science can proceed very far without resorting to quantitative observations—the making of measurements.

In making a measurement, a scientist is concerned about two things. One, of course, is the *size* of the measured quantity. The other, which is equally important, is how *reliable* the measurement is. The actual numerical value of a measurement depends on the units used to express it, and we will discuss units in some detail in the next section. But first, let's examine the process of making a measurement, because it is *how* we measure something that determines how confident we are in the numerical value we obtain.

Significant figures

When you make a measurement, you obtain numbers by reading them from a scale of some sort on the measuring device, and because of this there is nearly always some limitation on the number of meaningful digits that can be obtained. As an illustration, let's consider measuring the length of a piece of wood with two different scales, as shown in Figure 1.2.

Using the scale in Figure 1.2a, we might estimate the length of the piece of wood to be 3.2 centimetres (symbolized 3.2 cm). Notice that to arrive at this number, we are forced to estimate the second digit; that is, we must decide whether the length lies closer either to 3.2 or 3.3 cm. Because we are making an estimate, some uncertainty exists in the second digit (the 2), and the third digit, for all practical purposes, is completely unknown. Therefore, for measurements made with the scale in Figure 1.2a, we are not justified in reporting numbers containing more than two figures.

Digits that are obtained as the result of a measurement are called **significant figures.** When a number is written to represent the result of a measurement, it is always assumed that, unless stated otherwise, only the rightmost digit is

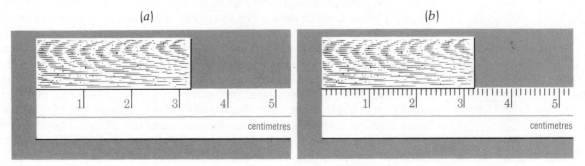

Figure 1.2

Measuring the length of a piece of wood with two different scales. (a) *Length = 3.2 cm* (b) *Length = 3.24 cm.*

uncertain. It is also assumed that any digits farther to the right are unknown. Thus, the measurement illustrated in Figure 1.2*a* yields a number with two significant figures.

In Figure 1.2*b*, we see the same piece of wood measured with a scale subdivided by additional graduations. Now we can observe that both the 3 and the 2 are known for sure, and we can try to estimate the third digit. One estimate of the length might be 3.24 cm, although some people might judge it to be either 3.23 or 3.25 cm. The measurement 3.24 cm contains three significant figures because the 3 and the 2 are known positively and only the 4 has some uncertainty. Digits to the right of the 4 cannot be estimated at all using the scale in Figure 1.2*b*, so none are written.

The importance of significant figures is that they indicate to us the reliability of our measurements. In the determination of the length of the piece of wood previously described, we saw two different values, obtained with two different measuring devices; our intuition tells us that we can place more confidence in the value with the greater number of significant figures. Laws and theories are derived from measured quantities, and our confidence in them is directly related to the quality of the data on which they are based.

EXAMPLE 1.1   THE MEANING OF SIGNIFICANT FIGURES

PROBLEM   The length of a room was measured to be 11.0 m using a tape measure graduated in tenths of a metre.

(a) How many significant figures are in this measurement?

(b) What would be wrong with reporting the length simply as 11 m?

SOLUTION   (a) There are three significant figures. Both digits to the left of the decimal point can be assumed to be known for sure, and only the zero to the right of the decimal point is uncertain. In other words, the length could lie between 10.9 m and 11.1 m.

(b) If the length is reported as 11 m, it is implied that the measurement is uncertain by at least ±1 m (i.e., the length lies between 10 m and 12 m). But we know the measurement is more reliable than that. If you have done the work to get that third significant figure, don't waste it by reporting only two.

Precision and accuracy

In discussing measured quantities, the words precision and accuracy are often used. The term **precision** refers to how closely two measurements of the same quantity come to each other. For example, repeating a measurement using the scale in Figure 1.2*a* could be expected to give values that differ from each other by about 0.1 cm. We say that lengths measured with this scale are *uncertain* by about 0.1 cm, which means that a given measured value could be 0.1 cm larger or 0.1 cm smaller. The way this is usually expressed is to say that the uncertainty is ±0.1 cm. If we make repeated measurements with the scale in Figure 1.2*b*, we could expect values that differ from each other by about 0.01 cm, so the uncertainty in measurements with this scale is about ±0.01 cm. The measurements made with the second scale have a *smaller degree of uncertainty*, and they are therefore considered to be more precise.

In general, the more significant figures there are in a measured quantity, the greater the precision of measurement. A reported value of 3.24 cm implies that if

The uncertainty in a measurement is an estimate of how much larger or smaller another measured value is likely to be.

The uncertainty in a measurement is sometimes reported along with the value of the measured quantity—for example, 3.2 ± 0.1 cm or 3.24 ± 0.01 cm.

the measurement were repeated, the result would be a value that differs from 3.24 cm by, at most, only a few hundredths of a centimetre. On the other hand, a reported value of 3.2 cm suggests that another measurement made with the same instrument would yield a value that might differ from 3.2 cm by a few tenths of a centimetre. Thus, the value 3.24 cm implies a smaller uncertainty (greater precision) than the value 3.2 cm.

The term **accuracy** refers to how close an experimental observation lies to the true value. Generally, a more precise measurement will also be a more accurate measurement. In our example above, the value 3.24 cm has greater precision than 3.2 cm and probably also lies closer to the true length (whatever that may be).

There are instances when a measurement may be precise, but not particularly accurate. The scale in Figure 1.3, for instance, is improperly marked. Failure to notice this error, and therefore failure to compensate for it, would yield measurements that would be wrong by 1 cm, even though three significant figures can be read from the scale. To avoid these kinds of problems, scientists **calibrate** their instruments. This involves adjusting the instrument to give correct values while the instrument is being used to measure accurately known standards.

Scientific notation and significant figures

Many of the measurements that we make in science give either very large or very small numbers. For example, astronomers have been able to determine, by various methods, the distances to nearby stars. These distances, which are enormous, have often been expressed in units of *light years*—the distance that light travels through space in one year. Expressed in these units, the brightest star in the heavens, Sirius, is 8.70 light years away. Expressed in more familiar units, however, this value is 82 400 000 000 000 km. To avoid having to write down large numbers of zeros like this, it is often convenient to express quantities as a *product* of a number lying between 1 and 10 multiplied by 10 raised to an appropriate power. This type of representation is called **scientific notation,** or sometimes **exponential notation,** and is discussed in detail in Appendix A at the back of this book. In scientific notation, the distance to Sirius is 8.24×10^{13} km.

If you are not familiar with scientific notation and exponential arithmetic, you should be sure to study Appendix A.

A bonus of scientific notation is that it allows us to remove any ambiguity with regard to the number of significant figures in a measurement. Normally, it is simple to count the number of significant figures in a measured value just by looking at it. A measured length of 23.6 m obviously contains three significant figures, and a measurement of 1.203 cm has four significant figures. But how about the numbers 0.002 15 m and 12.30 m? Do the zeros in these measured values count as significant figures?

The general rule is that a zero *does not* count as a significant figure if its purpose is merely to locate the position of the decimal point. Consider the measured value 0.002 15 m, without the zeros to the left of the 2 we wouldn't know how large the measurement is. If we rewrite the value in scientific notation, the zeros disappear

$$0.002\ 15 \text{ m} = 2.15 \times 10^{-3} \text{ m}$$

Figure 1.3

An improperly calibrated scale. All measurements made with this scale will be in error by 1 cm.

"Scientific" calculators allow you to enter numbers in scientific notation. If you have this kind of calculator, read Appendix A or the direction booklet for your calculator to be sure you can correctly enter these numbers.

which means that in normal decimal notation, they only were needed to tell us where the decimal point is. Therefore, we don't count them as significant figures. In general, *zeros that lie to the left of the first nonzero digit are not counted as significant figures.*

How about the zero in the measured value 12.30 m? This zero *is* counted as a significant figure, because it would not have been written unless the digit had been estimated to be a zero. This zero does not disappear when the number is written in scientific notation

$$12.30 \text{ m} = 1.230 \times 10^1 \text{ m}$$

This leads to another general statement: In a measured quantity, *a zero that lies to the right of the decimal point, and which also lies to the right of the first nonzero digit, counts as a significant figure.* If we put all this together, in the measurement below only the nonzero digits plus the zeros above the asterisks count as significant figures

$$0.0054070 \text{ m}$$
$$* \quad *$$

and the measurement therefore contains five significant figures.

Scientific notation is especially useful for removing ambiguities that arise in counting significant figures in large numbers. For example, suppose that the length of a field had been reported to be 1200 m. How many significant digits are in this value? If the measurement was made with a device that is reliable to the nearest metre, then the value is 1200 ±1 m, and the measurement has four significant figures. On the other hand, if the distance had simply been estimated at 1200 m with an uncertainty of 100 m, then the value is 1200 ± 100 m, and the measurement has only two significant figures. By just looking at the number 1200, however, we cannot tell whether the zeros are significant figures or whether they merely locate the decimal point. This is when scientific notation comes to the rescue. If we wish to express 1200 to four significant figures, we can write it as 1200×10^3, but if we wished only to show two significant figures, we could write it as 1.2×10^3.

EXAMPLE 1.2	USING SCIENTIFIC NOTATION TO EXPRESS SIGNIFICANT FIGURES
PROBLEM	Suppose the height of a building has been estimated to be 140 m, with an uncertainty of ±10 m. Express this measurement in scientific notation to the proper number of significant figures.
SOLUTION	With an uncertainty of 10 m, the actual height of the building could range from 130 to 150 m. The uncertainty is in the second digit, so the value has only two significant figures. In scientific notation, this can be expressed as

$$1.4 \times 10^2 \text{ m}$$

Significant figures and calculations

In almost all cases, the numbers that we obtain from measurements are used to calculate other quantities, and we must exercise care to report the calculated result in a way that neither overstates nor understates the level of confidence we have in it. This means that we must be careful to report the computed value to the proper number of significant figures. When an electronic calculator is used to perform the arithmetic, this is particularly important because calculators usually give answers with eight or ten digits. Most of these digits are meaningless in the sense that they shouldn't be counted as significant figures.

To see how problems can arise, suppose we wished to calculate the area of a rectangular carpet whose sides have been measured (using two different scales) to be 6.2 m and 7.00 m long. We know that the area is the product of these two numbers

$$\text{Area} = 6.2 \text{ m} \times 7.00 \text{ m} = 43.4 \text{ m}^2$$

(m^2 means "square metres"). How many significant figures are justified in the answer?

The length 6.2 m has two significant figures, which implies an uncertainty of about ±0.1 m. Suppose that an error of 0.1 m had been made, and that the scale of the scale should have read 6.3 m. How much would this error influence the result? Let's see by recalculating the area using 6.3 m instead of 6.2 m

$$\text{Area} = 6.3 \text{ m} \times 7.00 \text{ m} = 44.1 \text{ m}^2$$

Notice that this error in the measured length would cause a change in the second digit of the answer (it changes from a 3 to a 4). An uncertainty in the second digit of the length causes an uncertainty in the second digit of the answer.

The length of the other dimension, 7.00 m, has an implied uncertainty of about ±0.01 m. If it were this value that was in error, instead of the 6.2 m, how much would the calculated area change? Suppose this length should have been read as 7.01 m. The area would therefore be

$$\text{Area} = 6.2 \text{ m} \times 7.01 \text{ m} = 43.5 \text{ m}^2 \text{ (rounded)}$$

An error in the third digit of 7.00 m causes a change in the third digit of the answer, so if only the 7.00 m were in error, the answer would have its uncertainty in the third digit.

Since *either* the 6.2 or 7.00 m could be in error, we really cannot rely on our answer to have an uncertainty less than ±1 m^2, so we should round off[1] the area properly to 43 m^2. Notice that there are two significant figures in the answer, which is the same as the number of significant figures in the least precise factor in the calculation, 6.2 m.

The analysis that we have done leads to a general rule, which applies to calculations that involve either multiplication or division. *For multiplication and division, the product or quotient should not be written with more significant figures than are present in the least precisely known factor in the calculation.*

For addition and subtraction, the procedure used to determine the number of significant figures in an answer is slightly different. Here, the number we write as the result of a calculation is determined by the figure with the fewest number of decimal places. Thus, in the sum of these measured quantities,

$$
\begin{array}{r}
4.371 \\
\underline{302.5} \\
306.871
\end{array}
$$

[1]When we wish to round off a number at a certain point, we simply drop the digits that follow if the first of them is less than five. Thus 6.2317 rounds to 6.23, if we wish only two decimal places. If the first digit following the point of round off is greater than 5, or if it is 5 and other nonzero digits follow after that, we add 1 to the preceding digit. Thus 6.236 and 6.2351 both round off to 6.24. Finally, when the digit following the point of round off is 5, and no other nonzero digits follow the 5, then we drop the 5 if the preceding digit is even and we add 1 if it is odd. Thus 8.165 rounds to 8.16 and 8.175 rounds to 8.18.

we must round off the answer to 306.9. The reason for this is that the two digits that follow the 5 in the number 302.5 are completely unknown; that is, they could conceivably have any value from 0 to 9. As a result, the last two digits in the answer 306.871 must also be completely uncertain. If we follow our rules for writing numbers such that *only* those digits with real significance are included, we are not justified in writing these last two digits and must round off the answer to 306.9. For addition and subtraction, the rule is that *the absolute uncertainty in a sum or difference cannot be smaller than the largest absolute uncertainty in any of the terms in the calculation.* In the example above, the absolute uncertainty implied by the number 4.371 is ±0.001, whereas the uncertainty in 302.5 is implied to be ±0.1. According to our rule, the sum of these two cannot have an uncertainty less than ±0.1; therefore, the sum 306.871 must be rounded to 306.9 in order to suggest an uncertainty of ±0.1.

Exact numbers

In some calculations we employ numbers that come from definitions (such as 10 dm = 1 m) or that are the result of a direct count (such as the number of people in a room). These numbers are called **exact numbers** and contain no uncertainty (e.g., there are exactly 10 dm in 1 m, or the number of people in a room must be a whole number). When such quantities are used in computation, they may be considered to possess an infinite number of significant figures. Thus, the conversion of a measured length of 4.27 metres into decimetres would be accomplished as

$$4.27 \text{ m} \times \left(\frac{10 \text{ dm}}{1 \text{ m}} \right) = 42.7 \text{ dm}$$

3 significant figures

Notice that the number of significant figures in the product is determined by the number of significant figures in the *measured* length. Also, notice that in performing this calculation we have cancelled the units, metres. We will see more of this **factor–label method** of setting up arithmetic in future problems. The factor–label method is explained in detail in Appendix A at the back of this book.

Units undergo the same kinds of mathematical operations that numbers do. If you are unfamiliar with this concept, be sure to read Appendix A at the back of the book.

EXAMPLE 1.3	PERFORMING CALCULATIONS INVOLVING SIGNIFICANT FIGURES
PROBLEM	Perform the following computations and report the results to the proper number of significant figures. Assume all the numbers are the result of measurement. (a) 3.142 m ÷ 8.05 s; (b) 29.3 cm + 213.87 cm; (c) 144.3 m² + (2.54 m × 8.3 m)
SOLUTION	(a) Using a calculator, we get $$\frac{3.142 \text{ m}}{8.05 \text{ s}} = 0.390310559 \text{ m/s}$$ The number 3.142 contains four significant figures; the number 8.05 contains three significant figures. The result must be rounded off to 0.390 m/s because for division the answer should not contain more significant figures than are found in the factor with the fewest number of significant figures.

Sometimes, a calculator gives too few significant figures. If you multiply 0.500 by 6.00, a calculator may give 3 as the answer instead of 3.00.

(b) In performing this computation, note that the 7 must be added to a question mark.

$$
\begin{array}{r}
29.3?\ \text{cm} \\
+\,213.87\ \text{cm} \\
\hline
243.1?\ \text{cm}
\end{array}
$$

The only thing that can be said about the numerical value of the question mark in the answer is that it must be at least 7, and therefore we should round off the answer to

$$243.2\ \text{cm}$$

In this kind of situation, you may be tempted to make the error of writing a zero after the 3 in 29.3 (that is, 29.30). The digit following the 3 is unknown. If it were known to be a zero, a zero would have been written there!

(c) In mixed computations such as this, we perform multiplications and divisions before additions and subtractions.

$$2.54\ \text{m} \times 8.3\ \text{m} = 21.082\ \text{m}^2\ (\text{using a calculator})$$

Since 8.3 contains only two significant figures, the product must be rounded to 21. Now the appropriate addition is performed

$$
\begin{array}{r}
144.3\ \text{m}^2 \\
+\ \ 21.?\ \text{m}^2 \\
\hline
165.? = 165\ \text{m}^2
\end{array}
$$

1.3 UNITS OF MEASUREMENT

Units form an integral part of any measurement. For instance, to say that a chemical system has been heated for a time interval of "three" is meaningless unless a particular unit or units (seconds, minutes, hours, etc.) is associated with the number. Chemists have traditionally used metric-based units in their measurements. In 1960, the General Conference of Weights and Measures, an international body, adopted and recommended for worldwide use a modified version of the older metric system. The new system is called the **International System of Units** (abbreviated **SI** from the French *le Système International d'Unités*). As a beginning, the SI specifies a set of seven **base units,** which are given in Table 1.1.

Derived units

The SI base units are very precisely defined and serve as a foundation for all other units of measure. These other units, called **derived units,** are obtained from the base units by appropriate combinations, which depend on the dimensions of the measured quantities. For instance, if you wished to calculate the area of a rectangular carpet, you would multiply its length by its width. The *unit* for area is likewise the product of the *unit* for length by the *unit* for width. Since

Table 1.1
The seven SI base units

Physical Quantity	Name of Unit	Symbol
Mass	Kilogram	kg
Length	Metre[a]	m
Time	Second	s
Electric current	Ampere	A
Temperature	Kelvin	K
Luminous intensity	Candela	cd
Amount of substance	Mole	mol

[a]All English-speaking nations except the United States have adopted the -re spelling for metre. Both -re and -er forms have support within the United States, although the National Bureau of Standards currently favors the -er spelling.

length and width are both distances, for which the SI unit is the metre (m), the SI derived unit for area is m^2 (metre squared or square metre).

$$\text{length} \times \text{width} = \text{area}$$
$$\text{m} \quad \times \quad \text{m} \quad = m^2$$

Similarly, speed is a ratio of distance to time. It is computed as the distance traveled divided by the elapsed time. The SI derived unit for speed is therefore metre per second, or m/s.

Creating larger or smaller units

Often, either the base units or the derived units are of a size that makes them inconvenient for ordinary measurements. The SI unit for volume, for instance, is the cubic metre (m^3). For expressing volumes measured in a laboratory, however, the cubic metre is awkward. An ordinary glass of water, for example, is about $0.000\ 25\ m^3$. The SI solves this problem by modifying units with decimal factors and prefixes to obtain multiples or submultiples of them. Prefixes involving exponent multiples or submultiples of three are preferred in SI. Thus prefix factors representing, for example, 10^3 or 10^6 are favored in practice over prefixes representing, say, 10^1 or 10^2. A complete list of the SI prefixes is given in Table 1.2. Be sure to learn those that are printed in color because they are the ones you

$0.000\ 25\ m^3 = 250\ cm^3$

Table 1.2
The sixteen SI prefixes

Factor	Prefix	Symbol	Factor	Prefix	Symbol
10^{18}	exa	E	10^{-1}	deci	d
10^{15}	peta	P	10^{-2}	centi	c
10^{12}	tera	T	10^{-3}	milli	m
10^{9}	giga	G	10^{-6}	micro	μ
10^{6}	mega	M	10^{-9}	nano	n
10^{3}	kilo	k	10^{-12}	pico	p
10^{2}	hecto	h	10^{-15}	femto	f
10^{1}	deka	da	10^{-18}	atto	a

Table 1.3
Modifying the size of SI units with prefixes

Prefix	Multiplication Factor	Examples			Symbol
kilo-	$1000\ (10^3)$	1 kilometre	=	1000 metre $(10^3\ \text{m})$	km
		1 kilogram	=	1000 gram $(10^3\ \text{g})$	kg
deci	$1/10\ (10^{-1})$	1 decimetre	=	0.1 metre $(10^{-1}\ \text{m})$	dm
centi-	$1/100\ (10^{-2})$	1 centimetre	=	0.01 metre $(10^{-2}\ \text{m})$	cm
milli-	$1/1\ 000\ (10^{-3})$	1 millimetre	=	0.001 metre $(10^{-3}\ \text{m})$	mm
		1 millisecond	=	0.001 second $(10^{-3}\ \text{s})$	ms
		1 milligram	=	0.001 gram $(10^{-3}\ \text{g})$	mg
micro-	$1/1\ 000\ 000\ (10^{-6})$	1 micrometre	=	0.000 001 metre $(10^{-6}\ \text{m})$	μm
		1 microgram	=	0.000 001 gram $(10^{-6}\ \text{g})$	μg
nano-	$1/1\ 000\ 000\ 000\ (10^{-9})$	1 nanometre	=	0.000 000 001 metre $(10^{-9}\ \text{m})$	nm
		1 nanogram	=	0.000 000 001 gram $(10^{-9}\ \text{g})$	ng

will encounter most frequently in this course. Table 1.3 illustrates how these multipliers work and how the prefixes associated with them are used to name modified units. Examples 1.4 and 1.5 show how conversions from one unit to another are easily accomplished.

EXAMPLE 1.4 CONVERSIONS AMONG SI UNITS

PROBLEM A certain person is 172 cm tall. Express this height in millimetres.

SOLUTION This problem illustrates how unit conversions of this type can be easily accomplished using the factor–label method. The first step is to analyze the problem, which can be restated

$$172\ \text{cm} = ?\ \text{mm}$$

We need a way of relating the units cm to mm. Reviewing Table 1.2, we can't see a way of doing this directly, but we can write the relationships

$$1\ \text{cm} = 10^{-2}\ \text{m} \quad \text{or} \quad 100\ \text{cm} = 1\ \text{m}$$

$$1\ \text{mm} = 10^{-3}\ \text{m} \quad \text{or} \quad 1000\ \text{mm} = 1\ \text{m}$$

Notice that these provide a path from cm to mm. Centimetres can be converted to metres using the first relationship, and then metres can be changed to millimetres using the second. When we use them, we keep our eye on the units to be cancelled, because that controls how the factors are set up. Thus,

$$172\ \text{cm} \times \left(\frac{1\ \text{m}}{100\ \text{cm}}\right) = 1.72\ \text{m}$$

$$1.72\ \text{m} \times \left(\frac{1000\ \text{mm}}{1\ \text{m}}\right) = 1720\ \text{mm}$$

We can also "string together" the conversion factors and obtain the same net result.

$$172\ \text{cm} \times \left(\frac{1\ \text{m}}{100\ \text{cm}}\right) \times \left(\frac{1000\ \text{mm}}{1\ \text{m}}\right) = 1720\ \text{mm}$$

to metres

to millimetres

EXAMPLE 1.5 CONVERSIONS AMONG SI UNITS

PROBLEM Calculate the number of cubic centimetres in 0.255 cubic decimetres.

SOLUTION This time we need a relationship between cubic units, because our problem can be restated as

$$0.255 \text{ dm}^3 = ? \text{ cm}^3$$

Table 1.2 does not give us a way of converting dm^3 to cm^3 directly, but we saw that it does provide a means of relating dm to cm.

$$1 \text{ dm} = 10^{-1} \text{ m}$$
$$1 \text{ cm} = 10^{-2} \text{ m}$$

To obtain the relationships between the corresponding cubic units, we simply cube each side of the equation.

A review of arithmetic operations involving numbers expressed in powers of 10 is located in Appendix A.

$$(1 \text{ dm})^3 = (10^{-1} \text{ m})^3$$
$$1 \text{ dm}^3 = 10^{-3} \text{ m}^3$$

and

$$(1 \text{ cm})^3 = (10^{-2} \text{ m})^3$$
$$1 \text{ cm}^3 = 10^{-6} \text{ m}^3$$

These can now be used to construct conversion factors to solve the problem

$$0.255 \text{ dm}^3 \times \left(\frac{10^{-3} \text{ m}^3}{1 \text{ dm}^3} \right) \times \left(\frac{1 \text{ cm}^3}{10^{-6} \text{ m}^3} \right) = 255 \text{ cm}^3$$

Until recently, the older metric units were used in the scientific literature. A practicing scientist must therefore be familiar with both the old and the new units.

SI units have gained increasing acceptance within the international scientific community. However, the presence of older units in the scientific literature demands that we be aware of both the old and the new. SI units are used consistently throughout this book, but we will also identify important non-SI units at appropriate times.

Units for laboratory measurements

In chemistry, it is necessary to measure, on a routine basis, mass, length, volume, and temperature. The units that we ordinarily use to express the first three quantities are based on the **gram** (symbolized **g**), the **metre** (**m**), and the **cubic decimetre** (**dm³**), respectively. The gram itself is a conveniently sized unit for most laboratory measurements of mass (about which we will say more in Section 1.4). However, for most laboratory purposes we measure lengths in units of centimetres or millimetres.

$$1 \text{ cm} = 10^{-2} \text{ m}$$
$$1 \text{ mm} = 10^{-3} \text{ m}$$

It is often easier to remember these relationships if they are expressed as

$$1 \text{ m} = 100 \text{ cm} = 1000 \text{ mm}$$

$1 \text{ L} = 1 \text{ dm}^3$

Also, it is useful to remember that 1 cm = 10 mm.

The litre is defined by the SI as exactly 1000 cubic centimetres (cm^3), which is the same as one cubic decimetre (Figure 1.4).

$$1 \text{ litre} = 1000 \text{ cm}^3 = 1 \text{ dm}^3$$

Figure 1.4
One litre equals one cubic decimetre. 1 L = 1 dm³ = 1000 cm³ = 1000 mL.

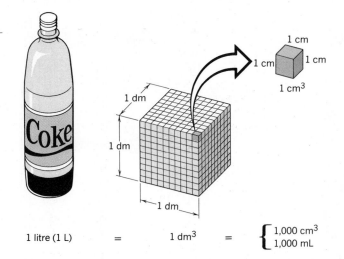

| 1 litre (1 L) | = | 1 dm³ | = | $\begin{cases} 1,000 \text{ cm}^3 \\ 1,000 \text{ mL} \end{cases}$ |

Sometimes, you may see the abbreviation cc used for cubic centimetre instead of cm³.

Since there are 1000 millilitres in 1 litre, the size of the millilitre and cubic centimetre are the same—they are identical.

$$1 \text{ mL} = 1 \text{ cm}^3$$

EXAMPLE 1.6   CONVERTING SI UNITS

PROBLEM   An object is moving at a speed of 14.2 cm s^{-1}. Express this speed in units of kilometres per hour (km h^{-1}).

SOLUTION   Let's use the factor–label method again. First, we list the relationships that we know among the units.

$$1 \text{ cm} = 10^{-2} \text{ m}$$

$$1 \text{ km} = 10^3 \text{ m}$$

$$60 \text{ s} = 1 \text{ min}$$

$$60 \text{ min} = 1 \text{ h}$$

Now, we use these relationships to construct conversion factors that enable us to cancel unwanted units. Our problem involves converting the units "centimetres" into "kilometres" and the units "seconds" into "hours."

$$14.2 \, \frac{\text{cm}}{\text{s}} = ? \, \frac{\text{km}}{\text{h}}$$

If you set up the problem so that you obtain the correct units for the answer, you can be confident your answer is also correct.

$$14.2 \, \frac{\text{cm}}{\text{s}} \times \left(\frac{10^{-2} \text{ m}}{1 \text{ cm}} \right) \times \left(\frac{1 \text{ km}}{10^3 \text{ m}} \right) \times \left(\frac{60 \text{ s}}{1 \text{ min}} \right) \times \left(\frac{60 \text{ min}}{1 \text{ h}} \right) = 0.511 \, \frac{\text{km}}{\text{h}}$$

Note that the answer has been rounded to give three significant figures.

In a formal sense, **temperature** is a quantity that determines the direction in which heat will flow spontaneously—heat always flows from something at a high temperature into something at a low temperature. This is such a common experience that the definition is hardly necessary, and you certainly recognize when something is hotter than something else. You also know that temperature is usually measured with a **thermometer** (Figure 1.5). A thermometer consists of

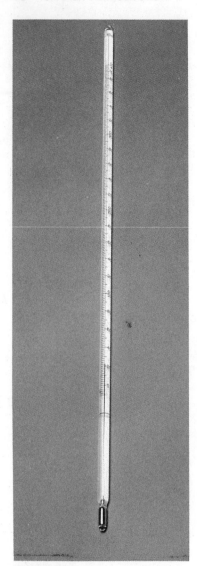

Figure 1.5

A typical laboratory thermometer.

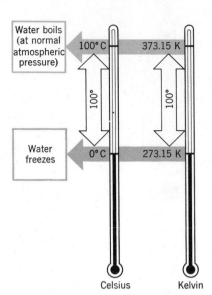

Figure 1.6

Comparison of temperature scales.

The symbol for the temperature unit is capitalized, but the name of the unit, the kelvin, is not.

a narrow capillary tube connected to a thin-walled bulb-shaped reservoir that is filled with some liquid (usually mercury). As the temperature of the bulb rises, the liquid expands and pushes its way up the capillary, and the height of the liquid in the capillary is directly proportional to the temperature.

To make the markings on the scale of a thermometer, two reference temperatures are chosen. The height of the mercury column is marked after the thermometer is brought to each temperature, and the distance between these two marks is then divided into some number of degree units, depending on the temperature scale used.

The reference temperatures used for the common temperature scales are the freezing point and boiling point of water. These are chosen simply for practical experimental reasons. At its freezing point (which is also its melting point), a pure substance such as water can exist as solid (ice) and liquid, both in contact with each other *at the same temperature*. If heat is added to a mixture of solid and liquid water, some of the solid melts, but the temperature doesn't change. If some heat is removed from the mixture in an attempt to cool it, some of the liquid freezes, but once again there is no temperature change. A mixture of the solid and liquid therefore maintains a *constant* temperature, and this allows the experimentalist plenty of time to accurately mark a thermometer immersed in the mixture. Similarly, a pure liquid will boil at a constant temperature that is independent of how fast heat is supplied to the liquid. Thus, the boiling point also serves as a convenient reference temperature.

The **Celsius scale** defines 0°C as the freezing point of water and 100°C as the boiling point of water, both measured under reproducible conditions. Today, SI defines the Celsius temperature scale in terms of its relation to the kelvin temperature scale (discussed below), even though the 100-degree interval between the freezing and boiling points of water is maintained.

The SI base temperature unit is the **kelvin (K)**. The size of the degree unit on the kelvin scale is the same as on the Celsius scale; the difference between the two is the location of the zero point. On the kelvin scale, water freezes at 273.15 K (note that the degree symbol ° is not used in the SI unit), so the relationship between the Celsius and kelvin temperature is[3]

$$K = °C + 273.15$$

[3] Important equations are set in colored type to help you identify them.

Often, we will only need to express these temperatures to the nearest whole degree, so we can use

$$K = °C + 273$$

Thermometers are never graduated directly in kelvins, so if you need to know the temperature in kelvins, you must measure the Celsius temperature and then calculate the corresponding kelvin temperature. Figure 1.6 illustrates the differences between these two temperature scales.

Zero kelvins is known as absolute zero—the lowest temperature possible. We will say more about it later in the book.

1.4   MATTER AND ENERGY

Now that you have learned a little about how science operates and the importance of measurements and their units, we turn our attention to the main subject of this book—chemistry. We begin by taking a closer look at the kinds of things with which chemistry is concerned.

Matter

All the materials that interest chemists—in fact, all the things we can see or touch or feel—are examples of *matter*, whether they be books, pencils, telephones, hamburgers, or people. **Matter** is defined as anything that takes up space and has mass. In setting down this definition, we are very careful to specify the term *mass* rather than *weight*, even though we often use the terms as if they were interchangeable. Mass and weight are not really the same. The **mass** of a body is a measure of its resistance to a change in velocity. A Ping-Pong ball moving at 30 km h^{-1} (30 kilometres per hour), for example, is easily deflected by a soft breeze, but a cement truck moving at the same speed is not. Quite clearly, the mass of the cement truck is considerably greater than that of the Ping-Pong ball. The term **weight** refers to the force with which an object of a certain mass is attracted by gravity to the earth or to some other body that it may be near, such as the moon. Force and mass are related to each other by Newton's equation (Newton's law).

$$F = ma$$

where F = force, m = mass, and a = acceleration. In order to accelerate a body, a force must be applied to it. When an object is dropped, it accelerates because of the gravitational attraction of the earth. An object resting on the earth or moon exerts a force (its weight, W) that is equal to its mass, m, multiplied by the acceleration due to gravity, g, that is,

$$W = mg$$

For example, at the earth's surface, g = 9.81 m s^{-2}. Thus a one-kilogram mass would have a weight (or experience a force downward at the earth's surface) of

$$W = mg = 1.00 \text{ kg} \times \left(\frac{9.81 \text{ m}}{s^2} \right) = 9.81 \frac{\text{kg m}}{s^2} = 9.81 \text{ kg m s}^{-2}$$

For convenience, the derived SI unit of force or weight, possessing units of kg m s^{-2}, is defined as the newton (N). A one-kilogram mass has a weight of 9.81 N at the earth's surface. You can experience the magnitude of a one-newton force by placing a large lemon (or any object with slightly more than 100 g mass) in your hand; the "push" downward represents a one-newton force. On the moon the gravitational acceleration is only about one-sixth of that on earth, so

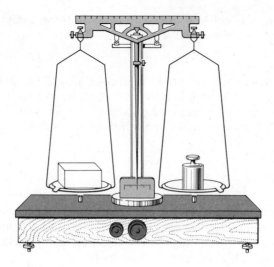

Figure 1.7

A traditional two-pan balance. Known masses are added to the right pan until the pointer is centered. The contents of each pan then have the same weight and therefore also possess the same mass.

No one ever talks about "massing" an object.

the same object weighs only one-sixth as much on the moon as on the earth, even though its mass is the same in both locations. Even on the earth the value of *g* has been found to vary slightly from place to place. This means that if very precise measurements are made, an object's weight also varies slightly according to its location on the earth. Because of this, we specify an object's mass instead of its weight when we wish to report the quantity of matter in the object.

The measurement of mass (a process, oddly enough, called *weighing*) is actually performed by comparing the weights of two objects, one of known mass, the other of unknown mass. The apparatus used for this is called a **balance**. Figure 1.7 is a drawing of a traditional two-pan balance. The object to be weighed is placed on the left pan of the balance and objects of known mass are added to the other until the pointer comes to the center of the scale. At this point the contents of both pans weigh the same, and since they both experience the same gravitational acceleration, both pans contain equal masses. In chemistry, as mentioned previously, we generally measure mass in grams.

Most balances found in labs today function on the same principle as their older cousin, even though their outward appearances may seem quite different. The most modern electronic balances, like the one shown in Figure 1.8, operate on a somewhat different principle. They are rugged, fast, and convenient, and make the measurement of mass a routine laboratory operation.

Energy

When chemical changes occur, they are almost always accompanied by either an absorption or release of energy. These energy changes tell us a great deal about the nature of the chemicals that are reacting, and equally important to us, certain chemical reactions provide the energy that our bodies and our society need in order to function. Therefore, an understanding of what energy is and how it can be transferred from one object to another is as important to those who study chemistry as an understanding of matter itself.

Energy itself is a difficult concept to comprehend, because energy is so different from matter. You can't put energy in a bottle to examine it. All you can do is examine the effects of energy on objects. **Energy** is usually defined as the capacity to do work or to make things happen. When an object has energy, it can

Figure 1.8

A modern electronic balance, such as the one shown here, permits rapid, precise, and accurate measurement of mass.

affect other objects by doing work on them. A moving car possesses energy because it can do work on another car by moving it some distance in a collision. Coal and oil possess energy because they can be burned and the heat that is liberated can be harnessed to do work. Because energy can be transferred from one object to another as work, the *units* of energy and work are the same.

An object can possess energy in two ways: as kinetic energy and as potential energy. **Kinetic energy (K.E.)** is associated with motion and is equal to one-half of an object's mass, *m*, multiplied by its velocity, *v*, squared.

$$\text{K.E.} = \tfrac{1}{2}mv^2$$

Doubling the speed of an auto increases its kinetic energy by a factor of four.

Thus, we see that the amount of work a moving body can do depends on both its mass and its velocity. For example, a truck moving at 30 km h^{-1} can do more work on the rear end of a car than a bicycle moving at the same speed. We also know that a truck moving at 80 km h^{-1} can do more work on a car than one traveling at only 5 km h^{-1}.

Potential energy (P.E.) represents energy a body possesses because of the attractive or repulsive forces it experiences with other objects. If there are no attractive or repulsive forces, the body does not possess potential energy.

An object has potential energy *only* if it experiences attractive or repulsive forces.

To see how potential energy is affected by these forces, consider two balls connected by a coiled spring as shown in Figure 1.9. When the balls are pulled apart, the spring is stretched and the energy spent in separating the balls is

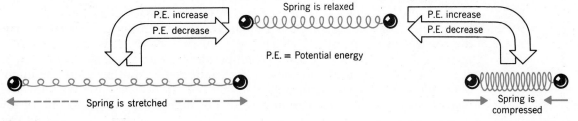

Figure 1.9

Potential energy exists between objects that either attract or   *repel each other. When the spring is either stretched or compressed, the P.E. of the two balls increases.*

stored in the stretched spring. We say, therefore, that the potential energy of the two balls has increased. This energy is released and converted to kinetic energy when the balls are allowed to move toward one another. In this case, there was an attractive force (the stretched spring) between the two objects.

With this same apparatus, energy can also be stored by compressing the spring, thereby producing a repulsive force between the two balls. This stored energy is released as kinetic energy when the balls are permitted to move away from one another.

The kinds of changes in potential energy that accompany changes in the relative position of objects that either attract or repel one another are very important to remember. They will help us analyze many physical and chemical changes in the chapters ahead.

As we have indicated, chemical substances contain energy that can be released through chemical reactions. For example, when wood burns, it reacts with oxygen from the air. As the products of combustion are formed, rather sizable quantities of heat are released. In this case, the wood and the oxygen are sources of potential energy, sometimes called "chemical energy," and the reaction causes this energy to be released into the surroundings as heat. The quantity of energy released or absorbed in a chemical reaction depends on the amounts of materials that react. The burning of a match, for example, releases only a very small quantity of energy, but a large bonfire produces much more.

When a change (either chemical or physical) releases energy to the surroundings, it is called an **exothermic** change. Changes that absorb energy are termed **endothermic**. It is usually easy to recognize whether a process is exothermic or endothermic; an exothermic change raises the temperature of its surroundings, and an endothermic change causes its surroundings to become cool.

The total quantity of energy that an object possesses is equal to the sum of its kinetic energy and its potential energy. In an isolated system, such as we believe our universe to be, the total quantity of energy is constant. This leads to a very important physical law, called the **law of conservation of energy,** which states that *energy can be neither created nor destroyed. It only can be changed from one kind of energy to another.*

Units of energy

Energy is transmitted from one body to another in a variety of ways, for example, as light, sound, electricity, and/or heat. These various forms of energy can be converted from one to another and therefore are ultimately equivalent. The SI unit for energy is the **joule (J),** which is defined as 1 kg m^2/s^2 in terms of the SI base units. It represents the kinetic energy possessed by an object with a mass of 2.0 kg traveling at 1 m/s.[4] In referring to energies involved in chemical reactions, it is useful to use the term **kilojoule (kJ).** One kilojoule is equal to 1000 joules (1 kJ = 1000 J).

All forms of energy ultimately are transformed into heat, and when chemists measure energy, it is usually in the form of heat. For example, we might measure the heat liberated during the combustion of natural gas, which is composed principally of the compound methane. The measurement of heat involves the concept of temperature, which we discussed earlier. It is important to remember that heat is not the same as temperature. Heat *is* energy, while temperature is a measure of the intensity of heat or hotness. Another useful definition of temperature is that it defines the *direction* and *rate* of heat flow. We know that heat always flows by itself from a region of high temperature to one of low

---

[4] Remember, K.E. = $\frac{1}{2}mv^2$. In this case, K.E. = $\frac{1}{2}$(2 kg)(1 m/s)^2, or K.E. = 1 kg m^2/s^2 = 1 J.

temperature, as when a hot cup of coffee becomes cool. We also know that the rate of heat transfer depends on the difference in temperature between two objects. If we want to warm something slowly, we place it in contact with an object just slightly warmer; if we wish to heat it quickly, we place it in contact with something very hot.

The original definition of the unit of heat energy called the **calorie** (symbolized **cal**) relied on the concept of temperature. The calorie was defined as the quantity of heat needed to raise the temperature of 1 gram of water, initially at 15°C, by 1°C.[5] The **kilocalorie (kcal)**, like the kilojoule, is a more appropriately sized unit for dealing with energy changes in chemical reactions. It has also been the unit used in nutrition for reporting the energy content of foods—the Calorie (written with a capital C) is the same as a kilocalorie. Thus, if we read that a serving of mashed potatoes contains 230 Cal, we are being told that 230 kcal of energy are liberated when the body metabolizes this food.

Until recently, nearly all the chemical scientific literature used the calorie (or kilocalorie) in reporting energy changes. With the introduction of SI units, the joule (or kilojoule) is now preferred and the calorie has been redefined in terms of this SI unit. The calorie and kilocalorie are defined exactly by the relationships

$$1 \text{ cal} = 4.184 \text{ J}$$

$$1 \text{ kcal} = 4.184 \text{ kJ}$$

Thus in terms of SI, one Calorie of food energy is more properly expressed as 4.184 kJ.

1.5 PROPERTIES OF MATTER

Objects are recognized by their characteristics, or **properties.** For example, you recognize your chemistry book by the way it looks and by the printing on its cover. But how could you tell what material was used to construct the objects in Figure 1.10? Their shiny surfaces suggest that they are metallic, and if you brought a magnet near them, you would feel the magnet pull at the objects. This tells you that they might be made of iron. If you left them overnight in the rain, they would begin to rust, so now you would be even more confident in saying that the objects are composed of iron.

Luster, color, magnetism, and the tendency to corrode are just some of the many properties that we use to recognize and classify different samples of matter. These properties themselves can be divided into two broad categories: **extensive properties,** which depend on the size of a sample of matter; and **intensive properties,** which are independent of the sample size. Of the two, intensive properties are the more useful, because a substance will exhibit the same intensive property regardless of how much of it we examine.

Examples of extensive properties are mass and volume—as the amount of a substance increases, its mass and volume also increase. Some examples of intensive properties are melting point and boiling point. Another intensive property is **density,** which is defined as the ratio of an object's mass to its volume.

In chemistry, we use the word **substance** to mean the chemical material of which an object is composed. An ice cube, for example, is composed of the *substance* water.

$$\text{density} = \frac{\text{mass}}{\text{volume}}$$

---

[5] It is necessary to specify the temperature of the water because the quantity of heat required to raise the temperature of 1 g of water by 1°C varies slightly with the temperature of the water. For example, at 25°C (room temperature) it takes 0.998 cal to raise the temperature of 1 g of water by 1°C.

Figure 1.10

What properties could you use to determine what these objects are composed of?

A sample of a liquid can be identified by its density—an intensive property—but not by either its mass or volume alone.

Liquid water, for instance, has a density of 1 g cm^{-3}. This means that if we had 1 g of water, it would occupy a volume of 1 cm^3. If we had 20 g of water, it would occupy a volume of 20 cm^3, but the *ratio* of the water's mass to its volume is the same: 20 g/20 cm^3 = 1 g/1 cm^3 = 1 g cm^{-3}. Notice that we have created an intensive property by taking the ratio of two extensive ones. Later in our discussion of chemistry, we will encounter quite a few other intensive properties defined in a similar way.

Normally, when a substance is heated or cooled, its volume expands or contracts. This means that the object's mass is packed into either a larger or smaller volume, so the density also changes with temperature. Therefore, for very accurate work, the temperature corresponding to a reported density must be specified. For example, at 25.0°C (room temperature) the density of water is 0.9970 g cm^{-3}, while at 35.0°C its density is 0.9956 g cm^{-3}. (For most purposes, we can take the density of water to be 1.00 g cm^{-3}.) The following examples show how density is calculated and how it can be used.

EXAMPLE 1.7   CALCULATING DENSITY

PROBLEM   An aluminum bar was weighed and found to have a mass of 14.2 g. Its volume was measured to be 5.26 cm^3. What is the density of aluminum?

SOLUTION   To calculate the density, d, we simply take the ratio of the object's mass to its volume.

$$d = \frac{14.2 \text{ g}}{5.26 \text{ cm}^3}$$

$$= 2.70 \text{ g cm}^{-3}$$

We could also express the density of aluminum as 2.70 g mL^{-1}. (Why?)

EXAMPLE 1.8   USING DENSITY IN CALCULATIONS

PROBLEM   A certain copper penny has a mass of 3.14 g. The density of copper is 8.96 g cm^{-3}. What is the volume of the penny?

SOLUTION   Density provides a relationship between an object's mass and its volume. It tells us, in this instance, that if we have 1.00 cm³ of copper, its mass is 8.96 g. It also tells us that if we have 8.96 g of copper, its volume is 1.00 cm³. We can use the density as a conversion factor in two ways

$$\frac{8.96 \text{ g copper}}{1.00 \text{ cm}^3} \quad \text{or} \quad \frac{1.00 \text{ cm}^3}{8.96 \text{ g copper}}$$

Since we are given the mass of copper, we must multiply by the second factor to eliminate the units *grams.*

$$3.14 \text{ g copper} \times \left(\frac{1.00 \text{ cm}^3}{8.96 \text{ g copper}}\right) = 0.350 \text{ cm}^3$$

The volume of the penny is 0.350 cm³ (or 0.350 mL).

A property closely related to density is **specific gravity** (often abbreviated *sp.gr.*), or **relative density,** which is defined as the ratio of the density of a substance to the density of water.

For precise work, the temperature of the substance and the temperature of the water to which it is compared must both be specified

$$\text{sp. gr.} = \frac{d_{\text{substance}}}{d_{\text{water}}} = \text{relative density}$$

Its usefulness is that it allows us to compute the density of the substance in various units simply by multiplying its specific gravity by the density of water expressed in those units. In this way two tables—one containing the specific gravities of substances and the other containing the density of water in a variety of units—take the place of the many tables that would otherwise be needed to express the densities of substances in all those different units.

EXAMPLE 1.9   FINDING DENSITY FROM SPECIFIC GRAVITY

PROBLEM   Hexane, a solvent used for rubber cement, has a specific gravity of 0.668. Water is the density of hexane in g cm⁻³ and in kg m⁻³. Water has a density of 1.00 g cm⁻³ or 1.00×10^3 kg m⁻³.

SOLUTION   By definition

$$\text{sp. gr. hexane} = \frac{d_{\text{hexane}}}{d_{\text{water}}}$$

Therefore

$$(\text{sp. gr. hexane}) \times d_{\text{water}} = d_{\text{hexane}}$$

In units of grams per cubic centimetre

$$d_{\text{hexane}} = (0.668) \times (1.00 \text{ g/cm}^3) = 0.668 \text{ g cm}^{-3}$$

Notice that in units of g cm⁻³ the numerical values of density and specific gravity are the same. In units of kilograms per cubic metre,

$$d_{\text{hexane}} = (0.668) \times (1.00 \times 10^3 \text{ kg/m}^3) = 6.68 \times 10^3 \text{ kg m}^{-3}$$

Specific heat capacity

An intensive property of matter associated with energy is **specific heat capacity** (sometimes termed **specific heat**), the quantity of heat required to raise the temperature of 1 g of a substance by 1°C. The specific heat capacity of water is 4.184 J g^{-1} °C^{-1}. Most other substances have smaller specific heats. Iron, for example, has a specific heat of only 0.452 J g^{-1} °C^{-1}. This means that it takes less heat to raise the temperature of 1 g of iron by 1°C than it does to cause the same temperature change for a gram of water. It also means that a given quantity of heat will raise the temperature of 1 g of iron more than it will raise the temperature of 1 g of water.

The large specific heat capacity of water is responsible for the moderating effect the oceans have on weather, since they cool very slowly in winter and warm up slowly in the summer. Air moving over the oceans in winter never gets very cold, and in the summer the air never gets very hot.

EXAMPLE 1.10   CALCULATIONS INVOLVING SPECIFIC HEAT

PROBLEM   How many joules are required to raise the temperature of a 7.5-cm iron nail with a mass of 7.05 g from room temperature (25°C) to 100°C? The specific heat capacity of iron is 0.452 J g^{-1} °C^{-1}.

SOLUTION   To solve this problem, we must multiply the specific heat capacity by mass (g) and the temperature change (°C) to eliminate these units and obtain joules as the units of the answer.

specific heat capacity × mass × temperature change = heat energy

$$\left(\frac{J}{g\,°C}\right) \times (g) \times (°C) = J$$

Using our data (the temperature change is 75°C), we get

$$\left(\frac{0.452\ J}{g\,°C}\right) \times (7.05\ g) \times (75°C) = 240\ J\ \text{(rounded)}$$

Notice that the answer has been rounded to two significant figures. Do you know why? If not, review the significant figure rules for multiplication.

Physical and chemical properties

In speaking of the properties of substances, we also distinguish between physical properties and chemical properties. A **physical property** can be specified without reference to any other substance. Density, specific heat, color, magnetism, mass, and volume are all examples of physical properties. A **chemical property,** on the other hand, states some interaction between chemical substances. When iron is exposed to oxygen and water, it corrodes and produces a new substance called iron oxide—rust. This is a chemical property of iron. We also say, for instance, that sodium is very reactive toward water. Reactivity is a chemical property that refers to the tendency of a substance to undergo a particular chemical reaction. However, to say simply that a substance is very reactive, without specifying "with what" or under what conditions, is not particularly helpful. Sodium, for example, is very reactive with water but quite unreactive toward the gas helium.

1.6 ELEMENTS, COMPOUNDS, AND MIXTURES

The three words that form the title to this section lie very close to the heart of chemistry, because we work with elements, compounds, and mixtures in the laboratory. We must therefore understand what they are and how to distinguish among them.

Elements are the simplest forms of matter that can exist under conditions that we find in a chemical laboratory; they thus are the simplest forms of matter with which the chemist deals directly. Elements serve as the building blocks for all of the more complex substances that we encounter, from common table salt to extremely complex proteins. All are composed of a limited set of elements. At present, there are 108 known elements, but only a much smaller number will be of real interest to us.

Elements combine with each other to form compounds. A **compound** is characterized by having its constituent elements always present in the same proportions. For example, you probably know that water is composed of two elements: hydrogen and oxygen. All samples of pure water contain these two elements combined in the proportion of one part by mass hydrogen to eight parts by mass oxygen (for example, 1.0 g of hydrogen to 8.0 g of oxygen). Also, when hydrogen is allowed to react with oxygen to produce water, the relative quantities of hydrogen and oxygen that combine are always the same. Thus, whenever 1.0 g of hydrogen reacts, it is always observed that only 8.0 g of oxygen are consumed, even if more than that quantity of oxygen is available.

Mixtures differ from elements and compounds in that they may be of variable composition. (As a result, they are not considered to be pure substances.) A solution of sodium chloride (table salt) in water is a mixture of two substances. We know that by dissolving varying quantities of salt in water (or a bowl of soup), we can obtain solutions with a wide range of compositions. Most materials found in nature or prepared in the laboratory are not pure but instead are mixtures.

In nature, there are very few materials that even approach being pure compounds—nearly everything is a mixture.

Mixtures can be described as being either **homogeneous** or **heterogeneous**. A *homogeneous mixture is called a* **solution** *and has uniform properties throughout.* If we were to sample any portion of a sodium chloride solution, we would find that it has the same properties (e.g., composition) as any other portion of the solution; we say that it consists of a single **phase.** Thus, we define a phase as any part of a system that has uniform properties and composition.

A heterogeneous mixture, such as oil and water, is not uniform (Figure 1.11). If we were to sample one portion of the mixture, it would have the properties of water, while some other part of the mixture would have the properties of oil. This mixture consists of two phases: the oil and the water. If we shook the

Figure 1.11

Oil and water—a heterogeneous mixture. (a) *Before mixing.* (b) *After mixing.*

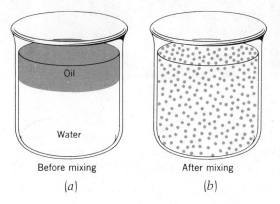

Before mixing
(a)

After mixing
(b)

mixture so that the oil was dispersed throughout the water as small droplets (as in a salad dressing), all the oil droplets taken together would still constitute only a single phase, since the oil in one droplet has the same properties as the oil in another. If we added an ice cube to this "brew," we would then have three phases: the ice (a solid), the water (a liquid), and the oil (another liquid). In all these examples, we can detect the presence of two or more phases because a boundary exists between them.

A useful feature of pure substances is that they undergo **phase changes** (e.g., solid to liquid or liquid to gas) at constant temperature. Ice, for instance, melts at a temperature of 0°C, a temperature that remains constant while the water undergoes the change from solid to liquid. When mixtures undergo phase changes, they generally do so over a range of temperatures. This phenomenon provides us with one experimental test to determine when we have obtained a pure substance.

There is another way that mixtures differ from compounds and elements. When a mixture is prepared, the chemical properties (and often, the physical properties) of the components do not change, but when elements are combined to form a compound, very profound changes occur in both chemical and physical properties. For example, copper and sulfur are two elements. Copper, of course, is a reddish-colored metal, a good conductor of electricity, and is relatively resistant to corrosion. Sulfur is a yellow nonmetallic substance, shown in powdered form in Figure 1.12. A mixture of sulfur and copper is easily prepared, but in the mixture we can still see traces of the properties of copper and the properties of sulfur. The formation of the mixture has involved a **physical process**—a process that has not altered the chemical characteristics of the components.

If the mixture of copper and sulfur is heated, a **chemical reaction,** or **chemical change,** takes place. The copper and sulfur combine to form a compound, and this is accompanied by dramatic changes in the properties of the substances. After the reaction is over, we can't find anything that has the properties of copper, and we can't find anything that has the properties of sulfur. In their place is a new substance, called copper(II) sulfide, that has new properties (Fig-

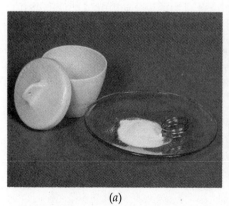 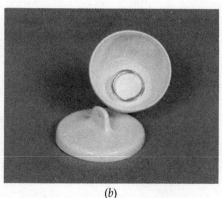 

(a) (b) (c)

Figure 1.12

(a) *Alongside a crucible and its cover we see a coil of red-colored copper wire and yellow powdered sulfur.* (b) *When mixed in the crucible, the copper and sulfur retain their individual properties.* (c) *When the mixture of copper and sulfur is heated, a reaction takes place and a new substance called copper(II) sulfide is formed. The copper(II) sulfide has the same shape as the coiled copper wire from which it was formed. Notice that its properties differ from both the copper and sulfur.*

Figure 1.13

Classification of matter.

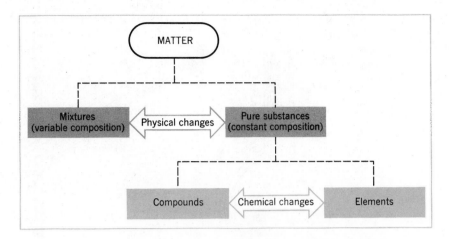

ure 1.12*c*). It doesn't conduct electricity; it doesn't have the color of either copper or sulfur; it has a density that is different from both sulfur and copper; and its chemical properties are entirely different, too. Such changes are what characterize chemical reactions. The relationship among elements, compounds, and mixtures is summarized in Figure 1.13.

Separating elements, compounds, and mixtures

Elements are usually defined as the simplest substances ever produced in chemical reactions.

Since elements form compounds by chemical reactions, decomposing compounds into their elements also involves chemical reactions, and they also are accompanied by drastic changes in properties. The nature of these reactions are as varied as the properties of the elements themselves, so few generalizations are possible. You will learn how some of the elements are obtained from their compounds later in this book.

Mixtures can be separated by physical processes in which the chemical properties of the components are not altered. Even so, separating mixtures is not always an easy job, and a variety of methods have been devised by chemists for this purpose. For example, a mixture of salt in water can be left to evaporate, and the departure of the water leaves the salt behind as a solid. If we wished to recover the water as well, we could boil the mixture in an apparatus similar to the one in Figure 1.14 and collect the water after it has condensed from the steam. This process is called **distillation.** It is one method that is used to obtain drinking water from sea water (the desalination of sea water).

Adsorption means sticking to a surface.
Absorption means soaking up like a sponge.

Another method of separating mixtures, called **chromatography,** makes use of the different tendencies that substances have for being adsorbed onto the surface of certain solids. For example, in thin-layer chromatography (Figure 1.15), a small spot of a solution containing several components is placed onto one end of a glass plate that is thinly coated with a material such as silica gel. A suitable solvent is then allowed to creep up the coating from a reservoir. As the solvent flows past the spot, the different components tend to be lifted from the silica gel surface with different degrees of ease. This causes the components of the mixture to move through the silica gel at different rates, with the more strongly adsorbed components moving more slowly. The result is a separation of the original spot into a set of spots, each containing (we hope) one component. This technique is widely used today by chemists who synthesize new compounds.

Figure 1.14

Simple distillation apparatus. The sodium chloride solution is boiled in the flask and the steam is converted to the liquid in the condenser.

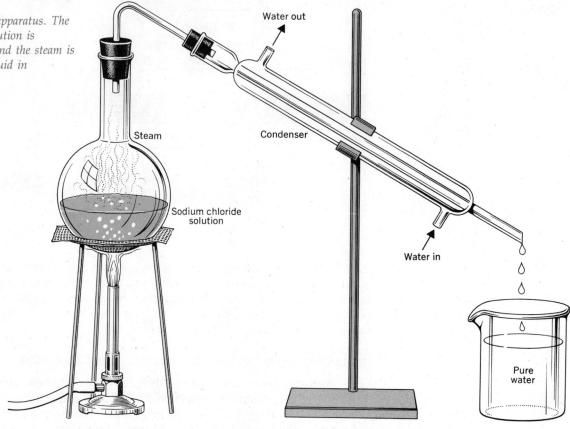

Water out

Steam

Condenser

Sodium chloride solution

Water in

Pure water

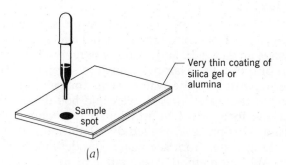

Very thin coating of silica gel or alumina

Sample spot

(a)

Figure 1.15

Thin-layer chromatography (TLC). (a) A solution containing a mixture is spotted near one end of a glass plate coated with silica gel or alumina. (b) The plate is placed with the end nearest the spot in a tray of solvent that rises through the coating by capillary action. The mixture is separated, with substances least strongly adsorbed by the coating moving farthest.

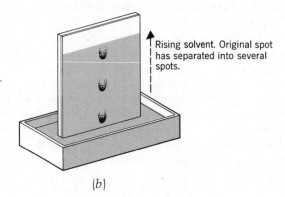

Rising solvent. Original spot has separated into several spots.

(b)

1.7 CONSERVATION OF MASS AND DEFINITE PROPORTIONS

The early history of chemistry was marked, not surprisingly, by incorrect theories about what occurred during chemical reactions. It had long been observed, for example, that when the combustion of wood took place, the resulting ash was very light and fluffy. Metals also changed their appearance when heated in air. The resulting material was less dense than the original metal and thus appeared lighter. These observations led to the conclusion that "something," which the German chemists Becher and Stahl called **phlogiston,** was lost by substances when they burned. Even when it was pointed out that metals gained mass when they were heated in air, the theory was salvaged by concluding that phlogiston simply had negative mass!

The reluctance to abandon the crumbling phlogiston theory demonstrates a general human phenomenon. New theories are difficult to come by, and old ones become so thoroughly entrenched that it is often tempting to try to shore up a sagging theory rather than to dream up a new one that will do a better job of explaining all the observed facts.

It was Antoine Lavoisier (1743–1794),[6] a French chemist, who finally laid the phlogiston theory to rest and set chemistry on the proper course again. He demonstrated by his experiments that the combustion process actually occurred by the reaction of substances with oxygen. He also showed, *through careful measurements,* that if a reaction is carried out in a closed container, so that none of the products of reaction escape, the total mass of all substances present after the reaction has occurred is the same as before the reaction began. These observations form the basis of the **law of conservation of mass,** which states that *mass is neither created nor destroyed in a chemical reaction.*[7] This is one of the most important principles in chemistry, even today.

The work of Lavoisier clearly demonstrated the importance of careful measurement. After his book, *Traité Elementaire de Chimie,* appeared in 1789, many chemists were inspired to investigate the quantitative aspects of chemical reactions. These investigations led to another important chemical law, the law of definite proportions.

The **law of definite proportions** (also called the **law of definite composition**) states that, *in a pure chemical substance, the elements are always present in definite proportions by mass.* In the substance water, for instance, the ratio of the mass of hydrogen to the mass of oxygen is always 1/8, regardless of the source of the water. Thus, if 9.0 g of water are decomposed, 1.0 g of hydrogen and 8.0 g of oxygen are always obtained. If 18.0 g of water are broken down, 2.0 g of hydrogen and 16.0 g of oxygen are produced. Furthermore, if 2.0 g of hydrogen are mixed with 8.0 g of oxygen and the mixture ignited, 9.0 g of water are formed and 1.0 g of hydrogen remains unreacted. In the water that is formed, the hydrogen-to-oxygen mass ratio once again is 1/8. Thus, varying the amounts of hydrogen and oxygen that are present during the reaction does not alter the composition of the water produced.

The definition of *compound* is a statement of the law of definite proportions.

[6] Lavoisier was a victim of the French Revolution. He went to the guillotine on May 8, 1794.
[7] Einstein showed that there is a relationship between mass and energy, $E = mc^2$, where c is the speed of light. Energy changes that take place during chemical reactions therefore are also accompanied by mass changes; however, the changes in mass are far too small to be detected experimentally. For example, the energy change associated with the reaction of 2 g of hydrogen with 16 g of oxygen happens to be equivalent to a mass change of approximately 10^{-9} g. Sensitive analytical balances can only detect mass differences of 10^{-6} to 10^{-7} g. Consequently, as far as the average chemist is concerned, there are no observable mass increases or decreases accompanying chemical reactions.

1.8 THE ATOMIC THEORY OF DALTON

The real father of modern chemistry could well be considered the Englishman John Dalton (1766–1844), who proposed his atomic theory of matter around 1803. The concept of the atom (from the Greek *atomos*, meaning indivisible) did not originate with Dalton. The Greek philosophers Leucippus and Democritus suggested, as early as 400 to 500 B.C., that matter cannot be forever divided into smaller and smaller parts and that ultimately particles would be encountered that would be indivisible. These early proposals, however, were not based on the results of experiments and were little more than exercises in thought. Dalton's theory was different because it was based on the laws of conservation of mass and definite proportions, laws that had been derived from many direct observations.

The theory Dalton proposed can be expressed by the following postulates:

1. Matter is composed of indivisible particles called atoms.
2. All atoms of a given element have the same properties (e.g., size, shape, and mass), which differ from the properties of all other elements.
3. A chemical reaction merely consists of a reshuffling of atoms from one set of combinations to another. The individual atoms themselves, however, remain intact.

The test of any theory, of course, is how well it explains already existing fact and whether it can predict as yet undiscovered laws. Dalton's theory proved successful on both counts.

First, it accounts for the law of conservation of mass. If a chemical reaction does nothing more than redistribute atoms and no atoms are lost from the system, it follows that the total mass must remain constant when the reaction occurs.

Second, it explains the law of definite proportions. To see this, imagine a substance formed from two elements, say, A and B, in which each molecule of the substance is composed of one atom of A and one atom of B. We define a **molecule** as *a group of atoms bound tightly enough together that they behave as, and can be recognized as, a single particle* (just as a car is composed of many parts held together tightly enough so that we identify them as *a car*). Let's also suppose that the mass of an atom of A is twice the mass of an atom of B. Then, in one molecule of this substance, twice as much mass is contributed by A as by B and the ratio of the mass of A to the mass of B in this molecule is 2/1. If we take a large collection of these molecules, we will always have equal numbers of A and B atoms; therefore, regardless of the size of the sample, we still always have a mass ratio (A to B) of 2/1. Also, if we were to react A and B together to form this compound, each atom of A would combine with only one atom of B. If we were to mix 100 atoms of A with 110 atoms of B, after the reaction was over, we would be left with 100 molecules of AB and 10 atoms of unreacted B. The substance AB would still have a mass ratio, A to B, of 2 to 1.

Third, Dalton's theory predicted the **law of multiple proportions.** This law can be stated as follows: *Suppose that we have samples of two different compounds formed by the same two elements. If the mass of one of the elements is the same in the two samples, the masses of the other element are in a ratio of small whole numbers.* Actually, this may seem more complicated than it really is. Let's consider the two compounds formed by carbon and oxygen. In one of them (carbon monoxide), we find 1.33 g of oxygen combined with 1.00 g of carbon, while in the second (carbon dioxide) there are 2.66 g of oxygen combined with 1.00 g of carbon. If we examine the ratio of the masses of oxygen (1.33 g/2.66 g), that combines with a fixed mass of carbon (1.00 g), we observe a ratio of small whole numbers

The smallest particle of carbon dioxide is a molecule of carbon dioxide, which consists of one atom of carbon and two atoms of oxygen.

$$\frac{1.33}{2.66} = \frac{1.33/1.33}{2.66/1.33} = \frac{1}{2}$$

This is consistent with the atomic theory if we consider that carbon monoxide contains *one* atom of carbon and *one* of oxygen, whereas carbon dioxide contains *one* atom of carbon and *two* atoms of oxygen. Because carbon dioxide has twice as many oxygen atoms bound to a carbon atom as does carbon monoxide, the mass of oxygen in a molecule of carbon dioxide *must* be twice the mass of oxygen in a molecule of carbon monoxide. (See Figure 1.16.)

EXAMPLE 1.11	**DEMONSTRATING THE LAW OF MULTIPLE PROPORTIONS**
PROBLEM	Nitrogen forms several different compounds with oxygen. In one (called laughing gas), it is observed that 2.62 g of nitrogen are combined with 1.50 g of oxygen. In another (a major air pollutant), 0.655 g of nitrogen is combined with 1.50 g of oxygen. Show that these data demonstrate the law of multiple proportions.
SOLUTION	In both cases, we are dealing with a mass of nitrogen that combines with 1.50 g of oxygen. If these data do fit the law of multiple proportions, the ratio of the masses of nitrogen in the two compounds should be a ratio of small whole numbers. Let's take the ratio

$$\frac{2.62}{0.655}$$

Dividing the numerator and denominator by 0.655, we get

$$\frac{4.00}{1.00}$$

which is indeed a ratio of small whole numbers.

EXAMPLE 1.12	**USING THE FACTOR–LABEL METHOD IN A CHEMICAL CALCULATION**
PROBLEM	Sulfur forms two compounds with fluorine. In one of them it is observed that 0.447 g of sulfur is combined with 1.06 g of fluorine and in the other, 0.438 g of sulfur is combined with 1.56 g of fluorine. Show that these data illustrate the law of multiple proportions.
SOLUTION	We can't examine the ratio of sulfur masses immediately because the masses of fluorine with which they are combined are different. We therefore must calculate how much sulfur will combine with the *same mass* of fluorine in the two compounds. The data given in the problem provide us with relationships between the masses of sulfur and fluorine that combine with each other to form each compound. In the first compound, we could say that 1.06 g of fluorine are *equivalent* to 0.447 g of sulfur—equivalent in the sense that whenever we find 1.06 g of fluorine in a sample of this compound, we know that we will also find 0.447 g of sulfur. We will find it convenient to express this kind of an equivalence—a *chemical equivalence*—in the following way:

Some people prefer to write this as 1.06 g fluorine = 0.447 g sulfur. The equals sign in this case should be read "is equivalent to."

1.06 g fluorine ~ 0.447 g sulfur

where the symbol ~ means "is equivalent to."

For the second compound, we can write the following equivalence:

1.56 g fluorine ~ 0.438 g sulfur

The relationships expressed by these equivalencies allow us to construct conversion factors that we can use in calculations. For example, the masses of sulfur and fluorine that combine in the second compound give us these *two* conversion factors

$$\frac{1.56 \text{ g fluorine}}{0.438 \text{ g sulfur}} \quad \text{and} \quad \frac{0.438 \text{ g sulfur}}{1.56 \text{ g fluorine}}$$

Now let's see how we can use these conversion factors to solve the problem at hand. For the first compound, we know that 0.447 g of sulfur combines with 1.06 g of fluorine. Let us calculate the mass of sulfur that would be combined with 1.06 g of fluorine in a sample of the second compound. We can state this as follows:

$$1.06 \text{ g fluorine} \sim ? \text{ g sulfur}$$

To perform the calculation, we multiply 1.06 g of fluorine by the conversion factor that allows us to eliminate the units "g fluorine."

$$1.06 \text{ g fluorine} \times \left(\frac{0.438 \text{ g sulfur}}{1.56 \text{ g fluorine}} \right) \sim 0.298 \text{ g sulfur}$$

We now know the masses of sulfur that combine with the *same mass* of fluorine (1.06 g) in these two compounds. The next step is to look at the ratio of these masses, that is,

$$\frac{0.447}{0.298} = \frac{0.447/0.298}{0.298/0.298} = \frac{1.50}{1.00}$$

which is the same as 3/2, a ratio of small whole numbers.

The use of chemical equivalencies, and conversion factors derived from them, will be found extensively in later chapters. You should make a special effort now to understand the concept of chemical equivalence by working the problems on the laws of definite proportions and multiple proportions at the end of the chapter.

1.9 ATOMIC WEIGHTS

The key to the success of Dalton's atomic theory was the concept that each element has a characteristic atomic mass, and the ability of the theory to explain so well the laws of chemistry prompted chemists to measure them. At this point, however, a serious problem arose.

Because of their small size, there certainly was no way to determine the masses of individual atoms. All that scientists could hope to do was to arrive at a set of relative atomic weights. (Note that the terms atomic mass and atomic weight are used interchangeably.) We might determine, for instance, that in one compound formed between carbon and oxygen, 3.0 g of carbon were combined with 4.0 g of oxygen. In other words, oxygen contributed $\frac{4}{3}$ (or $1\frac{1}{3}$) times as much mass toward the compound as the carbon did. If this substance is composed of molecules, each containing one atom of carbon and one atom of oxygen, then it follows that each oxygen atom must weigh $1\frac{1}{3}$ times as much as one carbon atom; thus, it appears that we have established the relative masses of these two elements.

The arguments just presented rest on a very critical assumption, that is, that the compound we were discussing is composed of one atom of carbon for each atom of oxygen. If this assumption about the number of carbon and oxygen atoms in a molecule is false, we have arrived at the wrong relative weights. Because of this kind of difficulty, a self-consistent table of atomic weights could not be developed until a method was devised for determining chemical formu-

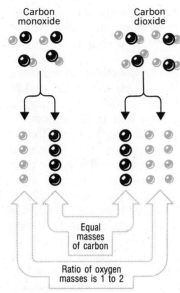

Carbon monoxide   **Carbon dioxide**

Equal masses of carbon

Ratio of oxygen masses is 1 to 2

Figure 1.16

The law of multiple proportions. Here are four molecules each of carbon monoxide and carbon dioxide. Each contains the same number of carbon atoms, ◕, and thus the same mass of carbon. The masses of oxygen that are combined with the carbon are in a whole number (1 to 2) ratio.

Scientists who work with substances composed of very large molecules, such as proteins, often refer to an atomic mass unit as a *dalton* (1 amu = 1 dalton).

las. This didn't happen until about 60 years after Dalton had introduced his theory.

A complete table of atomic weights appears on the inside front cover of this book. This is a table of relative atomic weights with the masses of atoms expressed in units called **unified atomic mass units** (u). The size of this unit is chosen in a rather arbitrary way. To see this, let us return to our discussion of the compound between carbon and oxygen.

Let us assume, for the moment, that molecules of the substance are in fact composed of one atom of carbon and one of oxygen; then, as mentioned, one oxygen atom weighs 1.33 times as much as one carbon atom. If we were to define the unified atomic mass unit as equal to the mass of one carbon atom, we would then assign carbon an atomic weight of 1.00 u and oxygen an atomic weight of 1.33 u. Since we might find it more convenient to have these numbers as whole numbers (or as near to whole numbers as possible), we might define 1 u as equal to one-third of the mass of a carbon atom. If we did this, we would assign carbon an atomic weight of 3.00 u. Oxygen, which is $\frac{4}{3}$ times heavier, would have an atomic weight equal to 4.00 u. Thus, the size of the unified atomic mass unit, and hence the values that appear in a table of atomic weights, are really quite arbitrary.

As originally defined by chemists, 1 u was taken to be $\frac{1}{16}$ of the mass of an atom of naturally occurring oxygen. In this way, the masses of nearly all the elements come out to be approximately whole numbers. Oxygen was chosen as a standard because it forms compounds with nearly all the elements. However, as we will discuss in Chapter 3, not all atoms of an element have precisely the same mass, and naturally occurring oxygen was later found to be composed of a mixture of atoms of slightly different masses. Since the relative proportions of these different atoms (called **isotopes**) could conceivably change over a period of time, or be different at different locations, the entire atomic weight table could also change because the size of 1 u was based on this mixture. To avoid this problem, the unified atomic mass unit is currently defined as $\frac{1}{12}$ of the mass of one atom of a particular isotope of carbon.

Because atoms are so very tiny, the actual mass of 1 u is extremely small. For example, one atom of the reference isotope of carbon (called carbon-12) has a mass of 1.992×10^{-23} g. One unified atomic mass unit is one-twelfth of this, or 1.660×10^{-24} g. You might write this out in decimal form to get a feel for how really small the masses of atoms are.

1.10 SYMBOLS, FORMULAS, AND EQUATIONS

In a certain sense, learning chemistry is like learning a strange new language such as Greek (or Russian, if you happen to be Greek). We might compare the chemical symbols for the elements with an alphabet, the chemical formulas that we construct from the symbols with words, and the chemical equations that we write with sentences. In learning any new language, we must start at the beginning with the alphabet.

At the present time, a total of 108 different elements are known. Each element is identified by its name and can also be represented by its **chemical symbol.** Usually, the symbol bears a resemblance to the English name for the element. For instance, carbon = C, chlorine = Cl, nitrogen = N, and zinc = Zn. Some elements, however, have symbols that do not seem to correspond at all with their English names. In nearly all these cases, the elements have been known since the early history of chemistry when Latin was used as the universal language among scientists. For this reason, symbols for these elements are derived from their Latin names—for example, potassium (L. *kalium*) = K, sodium (*natrium*) = Na, silver (*argentum*) = Ag, mercury (*hydrargyrum*) = Hg, and copper (*cuprum*) = Cu. Regardless of the origin of the symbol, the first letter is always capitalized. A complete list of the elements with their chemical symbols appears on the inside front cover of the book.

A chemical compound is represented symbolically by its **chemical formula.** Water is represented by H_2O, carbon dioxide by CO_2, methane (natural gas) by CH_4, and aspirin by $C_9H_8O_4$. Chemical formulas also show the composition of substances. The subscripts in a formula give the relative numbers of atoms of each element that appear in the compound (when no subscript is written, a subscript of 1 is implied). The formula H_2O, for example, describes a substance containing two hydrogen atoms for every one oxygen atom. Similarly, the compound CH_4 contains one atom of carbon for every four atoms of hydrogen. Some compounds are more complex, and their formulas are written containing parentheses. For example, ammonium sulfate (a fertilizer) has the formula $(NH_4)_2SO_4$, which specifies the presence of two NH_4 units—for a total of two nitrogen atoms and eight hydrogen atoms—plus one sulfur atom and four oxygen atoms. As we will see later, there are good reasons for writing this formula as $(NH_4)_2SO_4$ instead of $N_2H_8SO_4$, although both represent the same number of atoms.

There are certain substances that form crystals that contain water molecules when their aqueous solutions are evaporated. These crystals are called **hydrates.** For example, copper(II) sulfate—an agricultural fungicide—forms blue crystals having five molecules of water for each copper(II) sulfate ($CuSO_4$). Its formula is written $CuSO_4 \cdot 5H_2O$. If the blue crystals are heated, water can be driven off to leave pure $CuSO_4$, which appears almost white. (See Figure 1.17.)

Chemical equations are written to show the chemical changes that occur during chemical reactions. For example, the equation

$$Zn + S \longrightarrow ZnS$$

describes a reaction (shown in Figure 1.18) in which zinc (Zn) reacts with sulfur (S) to produce zinc sulfide (ZnS),[8] a substance used on the inner surfaces of TV screens. The substances on the left side of the arrow are present before the reaction begins and are known as the **reactants.** The substances on the right of the arrow are formed by the reaction and are called the **products.** (In the example above, there is only one product.) The arrow is read as "react to yield" or simply "yield." This equation can be read as "zinc plus sulfur react to yield zinc

[8] The naming of chemical compounds is discussed in Chapter 5. For now, we use these names simply as labels.

Figure 1.17

(a) *Crystals of copper(II) sulfate,* $CuSO_4 \cdot 5H_2O$. (b) *When these crystals are heated, they lose water. The nearly white powder that remains is pure* $CuSO_4$.

(a)   (b)

Figure 1.18
Once ignited, a mixture of zinc and sulfur reacts violently to form zinc sulfide, ZnS.

A chemical equation gives a "before and after" view of a chemical reaction.

sulfide," or "zinc plus sulfur yield zinc sulfide," or "zinc reacts with sulfur to yield zinc sulfide." Sometimes, it is also desirable (or necessary) to indicate whether the reactants and products in a chemical reaction are solids, liquids, gases, or are dissolved in a solvent such as water. This is done by placing the letters s = solid, l = liquid, g = gas, aq = aqueous (water) solution in parentheses following the formulas of the substances in the equation. For instance, the equation

$$CaCO_3 \ (s) + H_2O \ (l) + CO_2 \ (g) \longrightarrow Ca(HCO_3)_2 \ (aq)$$

describes a reaction between solid calcium carbonate (limestone), liquid water, and gaseous carbon dioxide to give an aqueous solution of calcium bicarbonate. This is the reaction responsible for the dissolving of limestone by ground water containing carbon dioxide. It is one of the causes of "hard" water and the formation of limestone caves.

Many of the equations that we write will contain numbers called **coefficients** preceding the chemical formulas. An example is the reaction of hydrogen (H_2) with oxygen (O_2) to form water.

$$2H_2 + O_2 \longrightarrow 2H_2O$$

This equation is interpreted to mean that two hydrogen molecules plus one oxygen molecule (a coefficient of 1 is assumed when none is written) react to yield two molecules of H_2O. Such an equation is said to be **balanced** because it contains the same number of atoms of each element on both sides of the arrow. You will begin to learn how to balance equations in the next chapter.

EXAMPLE 1.13 COUNTING ATOMS IN CHEMICAL EQUATIONS

PROBLEM   The combustion of butane, the liquid fuel that can be seen sloshing around inside a disposable cigarette lighter, follows the chemical equation

$$2C_4H_{10} + 13O_2 \longrightarrow 8CO_2 + 10H_2O$$

How many atoms of oxygen are included among the product molecules?

SOLUTION   Both CO_2 and H_2O contain oxygen. One molecule of CO_2 contains 2 atoms of O, so 8 molecules contain $8 \times 2 = 16$ atoms of O. Similarly, one molecule of H_2O contains 1 atom of O, so 10 molecules contain 10 atoms of O. On the right side of the equation, therefore, there are $16 + 10 = 26$ atoms of oxygen. (Notice that this is the same number as in 13 molecules of O_2, each of which contains two atoms of oxygen.)

REVIEW QUESTIONS AND PROBLEMS

(Problems whose numbers are in blue have their answers in Appendix C at the back of the book; the more difficult problems are marked with asterisks.)

Scientific Method

1.1 What is the difference between a theory and a law?

1.2 Describe how a doctor applies the scientific method in treating an ill patient.

1.3 What distinguishes a qualitative observation from a quantitative one? Which is generally more useful?

1.4 Define data.

Measurement and Significant Figures

1.5 Why is it important for scientists to be concerned about reporting the correct number of significant figures in their measurements?

1.6 Define, in your own words, the terms *precision* and *accuracy*.

1.7 How many significant figures are there in the following numbers, all of which were obtained from measurement: 1.037 0, 0.000 417, 0.003 09, 100.1, 9.001 0?

1.8 Speed is defined as a ratio: distance traveled divided by the length of time required to go that distance (speed = distance/time). Suppose a car travels 346.2 km in 6.27 h.
   (a) What is its average speed? (Write all the digits given by your calculator.)
   (b) What is the maximum uncertainty in the distance traveled?
   (c) What is the maximum uncertainty in the time?
   (d) Show, by applying these uncertainties to the measured values, that the calculated speed should contain only three significant figures.

1.9 Express each of the following in scientific notation. Assume that any digits to the right of the last nonzero digit are *not* significant figures.
   (a) 1 250
   (b) 13 000 000
   (c) 60 230 000 000 000 000 000 000
   (d) 214 570
   (e) 31.47

1.10 Express each of the following in scientific notation.
   (a) 0.000 40
   (b) 0.000 000 000 3
   (c) 0.002 146
   (d) 0.000 032 8
   (e) 0.000 000 000 000 91

1.11 Write the following numbers in decimal notation.
   (a) 3×10^{10} (d) 3.4×10^{-7}
   (b) 2.54×10^{-5} (e) 0.0325×10^{6}
   (c) 122×10^{-2}

1.12 The length of a piece of land was measured to be 3000 m. Using scientific notation, express the measurement
   (a) to two significant figures
   (b) to three significant figures
   (c) in cm, to two significant figures

1.13 Perform the following computations, rounding answers to the proper number of significant figures. Assume all values are measured quantities.
   (a) 2.41×3.2
   (b) 4.025×18.2
   (c) $81.8 \div 104.2$
   (d) $3.476 + 0.002$
   (e) $81.4 - 0.002$

1.14 Perform the following computations, expressing the answers in scientific notation rounded to the proper number of significant figures. All values are from measurements.
   (a) $(2.047 \times 10^{8}) + (14.33 \times 10^{8})$
   (b) $(12.4 \times 10^{8}) + (92.3 \times 10^{7})$
   (c) $(42.003 \times 10^{5}) - (3.25 \times 10^{3})$
   (d) $118.45 - (0.033 \times 10^{3})$
   (e) $1.00 + (3.75 \times 10^{-8})$

1.15 Perform the following computations, rounding the answers to the correct number of significant figures. All values are measured.
   (a) $(341.7 - 22) + (0.00224 \times 814{,}005)$
   (b) $(82.7 \times 143) + (274 - 0.00653)$
   (c) $(3.53 \div 0.084) - (14.8 \times 0.046)$
   (d) $(324 \times 0.0033) + (214.2 \times 0.0225)$
   (e) $(4.15 + 82.3) \times (0.024 + 3.000)$
   (f) $0.2510 \times (15.50 - 12.75)$

1.16 Perform the following computations, expressing the answers in scientific notation rounded to the proper number of significant figures. Assume all values are from measurements.
   (a) $(3.42 \times 10^{8}) \times (2.14 \times 10^{6})$
   (b) $(1.025 \times 10^{6}) \times (14.8 \times 10^{-3})$
   (c) $(143.7) \times (84.7 \times 10^{16})$
   (d) $(5274) \times (0.33 \times 10^{-7})$
   (e) $(8.42 \times 10^{-7}) \times (3.211 \times 10^{-19})$

1.17 Perform the following computations, expressing the answers in scientific notation rounded to the proper number of significant figures. All values are measured.
   (a) $(12.45 \times 10^{6}) \div (2.24 \times 10^{3})$
   (b) $822 \div 0.028$
   (c) $(635.4 \times 10^{-5}) \div (42.7 \times 10^{-14})$
   (d) $(31.3 \times 10^{-12}) \div (8.3 \times 10^{-6})$
   (e) $(0.74 \times 10^{-9}) \div (825.3 \times 10^{18})$

1.18 Perform the following computations, expressing the answers in scientific notation rounded to the proper number of significant figures. All values are from measurement.
   (a) $(8.3 \times 10^{-6}) \times (4.13 \times 10^{-7}) \div (5.411 \times 10^{-12})$
   (b) $[(3.125 \times 10^{-6}) + (5.127 \times 10^{-5})] \times (6.72 \times 10^{8})$
   (c) $[14.39 + (2.43 \times 10^{1})] \div 1275$
   (d) $[(1.583 \times 10^{-4}) - (0.00255)] \times [(142.3) + (0.257 \times 10^{2})]$
   (e) $(0.0000425) \div [0.0008137 + (2.65 \times 10^{-5})]$

1.19 One kilometre is 1000 m (1 km = 1000 m). How many significant figures are in each of the numbers in this relationship?

Units and Conversions

1.20 The SI units for mass and volume are the kilogram and cubic metre. However, in the laboratory we generally use the units gram and cubic centimetre. Why?

1.21 Perform the following conversions:
   (a) 1.40 m to cm
   (b) 2800 mm to m
   (c) 185 cm^3 to dm^3
   (d) 18 g to kg
   (e) 15 cm^2 to m^2
   (f) 322 km to cm
   (g) 45 m s^{-1} to km h^{-1}
   (h) 1 km^3 to m^3

(i) 72 km h^{-1} to mm s^{-1}

(j) 25.33 kJ to J

1.22 What power of ten is implied by the following SI prefixes? What symbol is used to stand for each of them? (Try to answer this question without referring to Table 1.2.)

(a) pico

(b) mega

(c) centi

(d) nano

(e) kilo

1.23 Why were the freezing point and boiling point of water chosen as reference temperatures for the original definition of the Celsius temperature scale?

1.24 Fill in the blank with the correct unit.

(a) 3.2_____ = 3.2 × 10^{-9} m

(b) 42_____ = 4.2 × 10^{-2} m

(c) 7.3_____ = 0.0073 g

(d) 12.5_____ = 125 mm

(e) 3.5_____ = 3.5 × 10^{-3} cm^3

(f) 0.84_____ = 840 cm^3

1.25 Force and weight are both represented in SI by the derived unit newton (N). Show that the expressions $F = ma$ and $W = mg$ both involve the same SI base units, and thus lead to the same derived unit.

1.26 After shopping for a sports car, a certain wealthy freshman decided that she would choose between a new two-passenger, six-cylinder, Smokebelcher, which gives 12 km/L, and an old 1974, eight-cylinder, 10-passenger Pferdburper (with automatic transmission), which the owner guaranteed would consume only 9.0 L/100 km. Which car would be more economical to operate in terms of fuel used?

1.27 A student has just returned from Germany with a car that he purchased while on vacation. The speedometer is calibrated in kilometres per hour (km h^{-1}). As he drives away from the pier, he notices a sign that posts the speed limit at 13 m s^{-1}. What is the maximum speed that he can reach, in km h^{-1}, without having to worry about receiving a speeding ticket?

1.28 After receiving a speeding ticket, the student in the preceding question is informed by the police officer that the courthouse is located "690 000 cm straight down the road. You can't miss it." The odometer in the student's car measures kilometres (km). How far must he travel, in kilometres, before arriving at the courthouse?

1.29 In the country of Ferdovia, the Ferds thrive on potatoes. The average Ferd earns 142 thrubs (the local currency of Ferdovia) per week and spends approximately $\frac{1}{14}$ of his yearly income on potatoes. If potatoes cost 2 thrubs per kilogram, how many kilograms of potatoes does the average Ferd consume each year?

1.30 Gallium metal has one of the largest liquid ranges of any element. It melts at 30°C and boils at 1983°C. What are its melting and boiling points in kelvins?

1.31 Tungsten, used as filaments in electric light bulbs, has a melting point of 5927°C. What is its melting point expressed in kelvins?

1.32 Solid carbon dioxide (dry ice) has a temperature of 195 K. What is its temperature on the Celsius scale?

*1.33 Naphthalene (used in mothballs) has a melting point of 80°C and a boiling point of 218°C. Suppose that this substance was used to define a new temperature scale on which the melting point of naphthalene was 0°N and its boiling point was 100°N. What would be the freezing point and boiling point of water in °N? What general equation could we use to relate temperatures in °C and °N?

1.34 Suppose you had a thermometer designed to measure large changes in temperature, and that the markings on the thermometer were at 10-degree intervals set 1 mm apart. Explain why this thermometer would be inappropriate for determining whether a person's body temperature is normal.

Properties of Matter

1.35 What is the difference between an extensive and an intensive property? Can you think of any examples not mentioned in the text?

1.36 Consider two samples of matter, one a copper penny and the other a piece of window glass. Give three properties that would be different for the two samples. Are there any properties in which they are alike?

1.37 How is mass different from weight? Why do we use mass instead of weight to specify the quantity of matter in a given sample?

1.38 One of the substances found in gasoline is called heptane. From what you know about gasoline,

(a) suggest one physical property of heptane.

(b) suggest one chemical property of heptane.

1.39 Two samples of matter have exactly the same physical and chemical properties. What conclusion can be drawn regarding the two samples?

Density and Specific Gravity

1.40 What is the difference between density and specific gravity? What are their common metric units?

1.41 A block of magnesium had a mass of 14.3 g and a volume of 8.46 cm^3. What is the density of magnesium?

1.42 Water was placed into a graduated cylinder until the volume read 25.0 cm^3. An irregularly shaped piece of metal weighing 50.8 g was placed in the cylinder and completely submerged. The water level rose to the 36.2-cm^3 mark. What is the density of the metal?

*1.43 Titanium is an important structural metal used in aircraft because of its strength and low density. A solid cylinder of titanium 2.48 cm in diameter and 4.75 cm long was found to have a mass of 104.2 g. Calculate the density of titanium.

1.44 Lead is a well-known "heavy" metal. It has a density of 11.35 g cm^{-3}.
(a) What is the mass of 12.0 cm^3 of lead?
(b) What is the volume occupied by 155 g of lead?

1.45 Chloroform, $CHCl_3$, a liquid once used as an anesthetic, has a density of 1.492 g cm^{-3}.
(a) What is the volume of 10.00 g of $CHCl_3$?
(b) What is the mass of 10.00 cm^3 of $CHCl_3$?

*1.46 A glass vessel that can be repeatedly filled with precisely the same volume of liquid is called a **pycnometer.** A certain pycnometer, when empty and dry, weighed 25.296 g. When filled with water at 25°C, the pycnometer and water weighed 39.914 g. When filled with a liquid of unknown composition, the pycnometer and its contents weighed 33.485 g. At 25°C the density of water is 0.9970 g cm^{-3}.
(a) What is the volume of the pycnometer?
(b) What is the density of the unknown liquid?

1.47 Suppose that you were going to prepare a table of densities for 100 substances in which the density of each substance is to be expressed in five sets of units. How many numerical values must appear in this table? How many numerical values would be needed to express this same information using specific gravities and the density of water in different units?

1.48 The density of water is 1.00×10^3 kg m^{-3}. Isopropyl alcohol, sold in drug stores as rubbing alcohol, has a density of 7.87×10^2 kg m^{-3}.
(a) What is the specific gravity of isopropyl alcohol?
(b) What is the density of isopropyl alcohol in the units g cm^{-3}?

1.49 Propylene glycol, a substance used in nontoxic antifreeze for freshwater systems of campers and boats, has a specific gravity of 1.04. Water has a density of 1.00×10^3 kg m^{-3}. What would be the mass of the contents of a 1.00×10^4 m^3 tank car filled with propylene glycol?

1.50 Chemical Bond (the scientifically inclined brother of James) was faced with a chilling choice of beverage. Before him were three beakers containing clear colorless liquids. One was water, but the other two were fatally poisonous. Unaware of Chemical's scientific background, his foes thought they would give him a sporting chance in choosing which liquid to drink. He was told that the first beaker contained 275 cm^3 of liquid, which weighed 0.275 kg. The second contained 245 cm^3 of liquid, which weighed 389 g. The third contained 265 cm^3 of liquid, which weighed 2.99×10^5 mg. Which liquid should our daring detective have chosen?

Energy and Specific Heat

1.51 How does potential energy differ from kinetic energy?

1.52 What is meant by endothermic and exothermic?

1.53 If the potential energy of an object decreases as it is moved away from another object, what kind of force (attractive or repulsive) must exist between the two?

1.54 What is the difference between heat and temperature?

1.55 What are the units of specific heat? What is the value of the specific heat of water? Why do oceans and large lakes affect the temperatures of surrounding land masses?

1.56 A truck having a mass of 4500 kg is moving at a speed of 1.79 m s^{-1}. Calculate its kinetic energy in joules.

*1.57 Calculate the kinetic energy, in joules, of a 65-kg athlete running at a speed of 24 km h^{-1}.

1.58 Calculate the kinetic energy, in joules and in kilojoules, of a 255-kg motorcycle traveling at 80.0 km h^{-1}.

*1.59 A 10.6-megagram automobile is traveling at 55 km h^{-1}. What is its kinetic energy in joules? If all this energy were converted to heat, how much water could have its temperature raised by 10°C?

1.60 A car, whose mass is 1500 kg, skids to a halt from a speed of 60 m s^{-1}. How many joules of heat are generated by friction between the car and the pavement?

*1.61 A packing crate full of machinery has a mass of 1400 kg. It is moved a distance of 40 m by a person exerting a force of 510 N (1 N = 1 kg m s^{-2}). How many joules of work have been accomplished?

1.62 How much will the temperature of 150 g of liquid water change if it loses 35.0 J? How much will its temperature change if it gains 40.0 J?

1.63 How many joules are required to raise the temperature of 500 g of liquid water by 24°C?

*1.64 A copper penny weighing 3.14 g is heated to 100°C in boiling water. It is quickly dried and placed into a Styrofoam cup containing 10.00 g of water at 25.0°C. Copper has a specific heat capacity of 0.385 J g^{-1} °C^{-1}. What will be the final temperature of the penny and water? (Assume that a negligible quantity of heat is absorbed by the cup.) *Hint:* The law of conservation of energy requires that the number of joules lost by the metal is equal to the number of joules gained by the water.

Elements, Compounds, and Mixtures

1.65 How does an element differ from a compound? How are elements and compounds different than mixtures?

1.66 How is a physical change different than a chemical change?

1.67 What is the definition of a solution? How many phases can be present in a solution?

1.68 Two samples of fine beach sand were analyzed. One was found to consist of 3.44 g of silicon and 3.91 g of oxygen. The other consisted of 6.42 g of silicon and 7.30 g of oxygen. Do these data suggest that sand is a compound, or is it a mixture?

1.69 There are many examples of homogeneous and heterogeneous mixtures in the world around us. How would you classify sea water, air (unpolluted), smog, smoke, club soda, black coffee, a ham sandwich?

1.70 Identify the phases that exist in a copper pan containing two iron nails, a quart of water, and four glass marbles.

1.71 Describe how chromatography works.

1.72 Suppose a mixture contained salt, copper powder, and iron filings. Describe how you could separate and isolate the components of this mixture.

Dalton's Atomic Theory and Atomic Weights

1.73 State the postulates of Dalton's atomic theory.

1.74 Which chemical laws did Dalton's theory explain? Which law did it predict?

1.75 What is the most important atomic property in Dalton's atomic theory?

1.76 Distinguish between the terms atom and molecule.

1.77 How is the currently accepted unified atomic mass unit defined?

1.78 If the unified atomic mass unit were defined in such a manner that a single fluorine atom weighed 1 u, what would be the atomic weights of carbon and hydrogen?

1.79 Suppose that elements A and B formed a compound, and that for every 3.0 g of A there are 16.0 g of B. Can you calculate the relative mass of atoms of A and B from these data? Explain your answer.

*1.80 In a certain compound, 6.92 g of X were found, combined with 0.584 g of carbon. If the atomic weight of carbon = 12.0 u and if four atoms of X are combined with one atom of carbon, calculate the atomic weight of X.

Laws of Definite Proportions and Multiple Proportions

1.81 State, in words, the following laws:
(a) the law of conservation of mass
(b) the law of definite proportions
(c) the law of multiple proportions

1.82 Cyclopropane, a very effective anesthetic, contains the elements carbon and hydrogen combined in a ratio of 1.0 g of hydrogen to 6.0 g of carbon. If a given sample of cyclopropane was found to contain 24 g of hydrogen, how many grams of carbon would it contain?

1.83 Three samples of a solid substance composed of elements X and Y were prepared. The first was found to contain 4.31 g X and 7.69 g Y; the second was composed of 35.9% X and 64.1% Y; it was observed that 0.718 g X reacted with Y to form 2.00 g of the third sample. Show how these data demonstrate the law of definite composition.

1.84 Two samples of Freon (a coolant used in refrigerators and air conditioners) were analyzed. In one sample, 1.00 g of carbon was found to be combined with 6.33 g of fluorine and 11.67 g of chlorine. In the second sample, 2.00 g of carbon were found to be combined with 12.66 g of fluorine and 23.24 g of chlorine. What are the ratios of the masses of carbon to fluorine, carbon to chlorine, and fluorine to chlorine in the two samples? Show that these data support the law of definite composition.

1.85 Copper forms two oxides. In one of them, there are 1.26 g of oxygen combined with 10.0 g of copper. In the other, there are 2.52 g of oxygen combined with 10.0 g of copper. Show that these data illustrate the law of multiple proportions.

1.86 Two compounds are formed between phosphorus and oxygen. 1.50 g of one compound was found to contain 0.845 g of phosphorus, while a 2.50-g sample of the other contained 1.09 g of phosphorus. Show that these data are consistent with the law of multiple proportions.

*1.87 In Example 1.11 the first compound (2.62 g of N, 1.50 g O) has the formula N_2O (the substance is nitrous oxide). Suggest a possible formula for the second compound.

*1.88 In a sample of the compound MnO, 4.00 g of oxygen are combined with 13.7 g of manganese. How many grams of oxygen would be combined with 7.85 g of manganese in the compound MnO_2?

Symbols, Formulas, and Equations

1.89 Write chemical symbols for the following:
(a) iron   (f) calcium
(b) sodium   (g) nitrogen
(c) potassium   (h) neon
(d) phosphorus   (i) manganese
(e) bromine   (j) magnesium

1.90 Give the name of each of these elements:
(a) Ag   (f) Au
(b) Cu   (g) Cr
(c) S   (h) W
(d) Cl   (i) Ni
(e) Al   (j) Hg

1.91 How many atoms of each kind are represented in the following formulas:

(a) K_2S

(b) Na_2CO_3

(c) $K_4Fe(CN)_6$

(d) $(NH_4)_3PO_4$

(e) $Na_3Ag(S_2O_3)_2$

1.92 Plaster is composed of calcium sulfate ($CaSO_4$), which exists as a hydrate containing two water molecules for each $CaSO_4$. Write the formula for the hydrate.

1.93 Potassium alum (often just called alum) is used medically as an astringent—a substance that shrinks and drives blood from tissue. Its formula is $KAl(SO_4)_2 \cdot 12H_2O$. How many atoms of each element are represented by this formula?

1.94 Which of the following do not represent balanced equations?

(a) $ZnCl_2 + NaOH \rightarrow Zn(OH)_2 + NaCl$

(b) $CuCO_3 + 2HCl \rightarrow CuCl_2 + CO_2 + H_2O$

(c) $Fe_2O_3 + 2CO \rightarrow 2Fe + 2CO_2$

(d) $NH_4NO_3 \rightarrow N_2O + 2H_2O$

CHAPTER 2

STOICHIOMETRY: CHEMICAL ARITHMETIC

The proper fuel-to-air mixture for a race car engine can be as important as the driver's skill in winning a race. Chemicals, like the fuel and oxygen of the air, always combine in definite fixed ratios. In this chapter we will learn how these ratios are determined and used.

In Chapter 1 you were introduced to many important basic concepts in chemistry: the ideas of atoms and atomic weight, elements and compounds, and several major chemical laws. Chemists, physicists, and biologists have developed these ideas so that today they routinely think of chemical, physical, and biological processes taking place between atoms and molecules on a submicroscopic, atomic scale. However, we cannot see individual atoms or molecules, and in the laboratory it is necessary to work with huge numbers of these small particles. In this chapter we will see that there are ways of carrying out chemical reactions so that observations on a large (macroscopic) scale can be readily translated into the language of the atomic world. That this can be done with relative ease is really quite amazing.

In order to study chemical compounds effectively in the laboratory, it is necessary to have a knowledge of the quantitative relationships that exist among the amounts of the substances that enter into chemical reactions. **Stoichiometry** (derived from the Greek *stoicheion* = element and *metron* = measure) is the term that we use to refer to all the quantitative aspects of chemical composition and reaction. We will now see how chemical formulas are determined and how chemical equations prove useful for predicting the proper amounts of reactants that must be mixed to obtain a complete reaction—one in which there is not an excess amount of any reactant.

The ability to perform chemical calculations with ease, and in a routine way, is an essential part of learning chemistry. Usually, when students have difficulty later in the course, their problems can be traced to weaknesses in their understanding of the concepts developed in this chapter. Therefore, as you study the material covered here, concentrate on *understanding* the concepts. From an understanding will follow the ability to solve chemistry problems, but if you concentrate on solving problems without understanding what you're doing, you will almost certainly find yourself in trouble later in the course.

2.1 CHEMICAL SYMBOLS AND FORMULAS—ANOTHER LOOK

In the last chapter, we introduced you to chemical symbols and formulas. We will be using them extensively from now on, so let's look once again at how we interpret them. As an example, consider the symbol Cl, which is the chemical symbol for the element chlorine. One way of interpreting the symbol for chlorine is simply as an abbreviated form of the element's name. Thus, we could use Cl in place of the word chlorine, and sometimes this is done.

The most common use of chemical symbols is in writing chemical formulas, where the symbol stands for one atom of the element. For instance, in nature the element chlorine comes to us in the form of molecules that each contain two chlorine atoms. We write the formula of a chlorine molecule as Cl_2 to show this.

The elements hydrogen, oxygen, nitrogen, fluorine, chlorine, bromine, and iodine occur naturally as molecules of H_2, O_2, N_2, F_2, Cl_2, Br_2, and I_2, respectively.

In chemical problems, we frequently have to specify quantities of substances, and in doing so we often use chemical formulas. For example, we might write "25.0 g Cl_2." In this case, we really mean "25.0 g of Cl_2 *molecules*." On the other hand, if we write "25.0 g Cl," we mean "25.0 g of Cl *atoms*." Omitting the words molecules or atoms in such expressions is common practice, and as we proceed in our discussions we will begin to follow this practice, too. If you begin to become confused by this, recall what you've read here. It may help you considerably.

2.2 THE MOLE

We know that atoms react to form molecules in simple whole-number ratios. Hydrogen and oxygen atoms, for instance, combine in a 2-to-1 ratio to form water, H_2O; carbon and oxygen atoms combine in a 1-to-1 ratio to form carbon monoxide, CO. However, it is impossible to work with individual atoms because they are so tiny. Therefore, in any real-life laboratory situation, we must increase the size of these quantities to the point where we can see them and weigh them.

One way of enlarging the reaction is to work with dozens, instead of individual atoms.

$$1 \text{ atom C} \quad + \quad 1 \text{ atom O} \quad \longrightarrow \quad 1 \text{ molecule CO}$$
$$1 \text{ dozen C atoms} + 1 \text{ dozen O atoms} \longrightarrow 1 \text{ dozen CO molecules}$$
$$(12 \text{ atoms C}) \quad\quad (12 \text{ atoms O}) \quad\quad (12 \text{ molecules CO})$$

Notice that a 1-to-1 ratio of *dozens* of atoms is exactly the same as a 1-to-1 ratio of the atoms themselves. If we were to take 2 dozen carbon atoms and 2 dozen oxygen atoms (another 1-to-1 ratio of *dozens*), we again could be sure that there would be equal numbers of atoms of carbon and oxygen (a 1-to-1 ratio of *atoms*). In fact, it doesn't matter how many dozens of each kind of atom we take; all we have to be careful about is that we have equal numbers of dozens, so that a 1-to-1 dozen ratio is maintained.

This is such an important concept that it's worth exploring for another case. Let's consider the substance water, H_2O. If we were dealing with individual atoms, we could write the equation

$$2 \text{ atoms of H} + 1 \text{ atom of O} \longrightarrow 1 \text{ molecule of } H_2O$$

We could then scale up the size of the reaction by working with dozens of hydrogen and oxygen atoms

$$2 \text{ dozen H atoms} + 1 \text{ dozen O atoms} \longrightarrow 1 \text{ dozen } H_2O \text{ molecules}$$

or

$$4 \text{ dozen H atoms} + 2 \text{ dozen O atoms} \longrightarrow 2 \text{ dozen } H_2O \text{ molecules}$$

or

$$6 \text{ dozen H atoms} + 3 \text{ dozen O atoms} \longrightarrow 3 \text{ dozen } H_2O \text{ molecules}$$

In each case, we maintain a 2-to-1 ratio of H to O atoms by maintaining a 2-to-1 ratio of dozens of these atoms.

It should be obvious now that if we had some way of counting atoms by the dozen, we could take dozens of them in a ratio that is exactly equal to the desired atom ratio, and in so doing we would be assured of having the proper atom ratio. Unfortunately, a dozen atoms or molecules is still much too small to work with, so we must find a still larger unit. The "chemist's dozen" is called the **mole** (symbolized **mol**). It is composed of 6.022×10^{23} objects (we will say more about the origin of this number, called **Avogadro's number,** later).

$$1 \text{ dozen} = 12 \text{ objects}$$

$$1 \text{ mol} = 6.022 \times 10^{23} \text{ objects}$$

The same reasoning that we can use with the dozen applies equally to the mole. The mole is simply a larger collection.

$$1 \text{ mol of C atoms} + 1 \text{ mol of O atoms} \longrightarrow 1 \text{ mol of CO molecules}$$

or

$$1 \text{ mol C} \quad + \quad 1 \text{ mol O} \quad \longrightarrow \quad 1 \text{ mol CO}$$
$$\mathbf{(6.022 \times 10^{23}} \quad \mathbf{(6.022 \times 10^{23}} \quad \mathbf{(6.022 \times 10^{23}}$$
$$\mathbf{atoms\ C)} \quad\quad \mathbf{atoms\ O)} \quad\quad \mathbf{molecules\ CO)}$$

We see that when we take 1 mol of carbon atoms and 1 mol of oxygen atoms, we have equal numbers of carbon and oxygen atoms and can construct enough CO molecules to fill our mole "box" exactly.

A very important thing to notice here is that the same whole-number ratios that apply to the individual atoms and molecules also apply exactly to the numbers of *moles* of atoms and molecules. Everything is simply increased by the same factor. For example, to form carbon tetrachloride, CCl_4, we know that

$$1 \text{ atom C} + 4 \text{ atoms Cl} \longrightarrow 1 \text{ molecule } CCl_4$$

We can immediately enlarge this to moles

$$1 \text{ mol C} + 4 \text{ mol Cl} \longrightarrow 1 \text{ mol } CCl_4$$

By taking moles of carbon and chlorine atoms in a 1-to-4 ratio, we can be sure of having one carbon atom for every four chlorine atoms. And if our goal is simply keeping the carbon-to-chlorine atom ratio equal to 1-to-4, we can work with *any* number of moles of carbon atoms, just as long as the number of moles of chlorine atoms is 4 times larger. Thus, if we started with 2 mol C atoms, we would need 8 mol Cl atoms

$$2 \text{ mol C} + 8 \text{ mol Cl} \longrightarrow 2 \text{ mol } CCl_4$$

or, if we started with 5 mol C atoms, we would need 20 mol Cl atoms

$$5 \text{ mol C} + 20 \text{ mol Cl} \longrightarrow 5 \text{ mol } CCl_4$$

In each case, there is a 1-to-4 ratio of carbon to chlorine atoms.

To summarize, then, the *ratio* by which moles of substances react is the same as the *ratio* by which their atoms and molecules react. This simple idea forms the basis for *all* quantitative chemical reasoning.

Remember, 1 mol C means 1 mol of C atoms. What does 1 mol CCl_4 mean?

Students who have difficulty learning chemistry often have not learned to think problems through in terms of moles.

EXAMPLE 2.1 DETERMINING MOLE RATIOS FROM CHEMICAL FORMULAS

PROBLEM What mole ratio of carbon to chlorine must be chosen to prepare the substance C_2Cl_6 (hexachloroethane), which is a solvent used to manufacture explosives.

SOLUTION The atom ratio in C_2Cl_6 is

$$\frac{2 \text{ atoms C}}{6 \text{ atoms Cl}} = \frac{1}{3}$$

To prepare C_2Cl_6, the carbon to chlorine *atom* ratio must be maintained at 1:3. *The mole ratio must also be 1:3.* Carbon and chlorine must be combined in a ratio of 1 mol of C to 3 mol of Cl.

The atom ratio in a chemical formula such as C_2Cl_6 establishes a number of mole ratios that are useful for constructing conversion factors that can be employed in solving problems. We've already seen one of these, which can be written as either

$$\frac{2 \text{ mol C}}{6 \text{ mol Cl}} \quad \text{or} \quad \frac{6 \text{ mol Cl}}{2 \text{ mol C}}$$

There are others as well. The formula tells us that 1 molecule of C_2Cl_6 contains 2

Of course, we could also simplify these ratios to give

$$\frac{1 \text{ mol C}}{3 \text{ mol Cl}} \quad \text{and} \quad \frac{3 \text{ mol Cl}}{1 \text{ mol C}}$$

In fact, we often do.

atoms of C and that it also contains 6 atoms of Cl. We can scale this up immediately to moles: 1 mol C_2Cl_6 contains 2 mol C and 6 mol Cl. This gives two equivalencies

$$1 \text{ mol } C_2Cl_6 \sim 2 \text{ mol C}$$

$$1 \text{ mol } C_2Cl_6 \sim 6 \text{ mol Cl}$$

In other words, any time we have a mole of C_2Cl_6 molecules, there will be two moles of carbon atoms in it, and any time we have a mole of C_2Cl_6 molecules, there will be six moles of chlorine atoms in it. That's what we mean by an equivalence ($\sim$). As in Chapter 1, we can use these equivalencies to form conversion factors

$$\frac{1 \text{ mol } C_2Cl_6}{2 \text{ mol C}} \quad \text{or} \quad \frac{2 \text{ mol C}}{1 \text{ mol } C_2Cl_6}$$

and

$$\frac{1 \text{ mol } C_2Cl_6}{6 \text{ mol Cl}} \quad \text{or} \quad \frac{6 \text{ mol Cl}}{1 \text{ mol } C_2Cl_6}$$

The following examples show how we use them.

EXAMPLE 2.2 USING MOLE RATIOS AS CONVERSION FACTORS

PROBLEM   How many moles of carbon atoms are needed to combine with 4.87 mol chlorine atoms to form the substance C_2Cl_6?

SOLUTION   The problem can be stated as follows

$$4.87 \text{ mol Cl} \sim ? \text{ mol C}$$

To solve the problem, we must multiply the 4.87 mol Cl by one of the conversion factors relating moles of carbon to moles of chlorine in this compound. As we saw above, we have two of them to choose from

$$\frac{2 \text{ mol C}}{6 \text{ mol Cl}} \quad \frac{6 \text{ mol Cl}}{2 \text{ mol C}}$$

We choose the one that allows us to cancel the units "mol Cl."

$$4.87 \text{ mol Cl} \times \left(\frac{2 \text{ mol C}}{6 \text{ mol Cl}} \right) = 1.62 \text{ mol C}$$

We need 1.62 mol C. Notice that we are entitled to report three significant figures. The numbers in the mole ratio are exact numbers because the atom ratio is exact—in general, atoms combine in exact whole number ratios when they form compounds.

EXAMPLE 2.3 USING MOLE RATIOS AS CONVERSION FACTORS

PROBLEM   How many moles of carbon are in 2.65 mol C_2Cl_6?

SOLUTION   This time we use the equivalence

$$1 \text{ mol C}_2\text{Cl}_6 \sim 2 \text{ mol C}$$

which can be used to construct a conversion factor to set up the arithmetic.

$$2.65 \; \cancel{\text{mol C}_2\text{Cl}_6} \times \left(\frac{2 \text{ mol C}}{1 \; \cancel{\text{mol C}_2\text{Cl}_6}} \right) \sim 5.30 \text{ mol C}$$

Thus, 2.65 mol C_2Cl_6 contains 5.30 mol C.

2.3   MEASURING MOLES OF ATOMS

In a chemical reaction, atoms or molecules combine in whole number ratios, and we also have seen that moles of these substances react in the same whole number ratios. In this sense, then, the mole might be called a *chemical unit*. Its size is large enough so that a mole of atoms or molecules represents an amount that is easily worked with in the laboratory, but unfortunately there is no device that allows us directly to count out atoms in multiples of Avogadro's number. We must, therefore, have a way of translating these chemical units to *laboratory units*—something that can be measured in the lab.

Earlier it was stated that a mole consists of 6.022×10^{23} objects. This rather odd number was not selected deliberately—instead, it just happens to be the number of atoms in a sample of any element that has a mass in grams that is numerically equal to that element's atomic weight.[1] For example, the atomic weight of carbon is 12.011, so 1 mol of carbon atoms has a mass of 12.011 g. Similarly, the atomic weight of oxygen is 15.9994, so a mole of oxygen atoms has a mass of 15.9994 g.

A table of atomic weights gives us the number of grams in 1 mol of the atoms of any element.

$$1 \text{ mol C} = 12.011 \text{ g C}$$
$$1 \text{ mol O} = 15.9994 \text{ g O}$$

Thus, it is the balance that becomes our tool for measuring moles. To obtain one mole of any element, all we need to do is look up the element's atomic weight. The number we find is the number of grams of the element that we must take to have 1 mole of it.

Converting between grams and moles is a routine calculation that you must learn to do quickly. Some examples of this kind of calculation, along with the use of the mole in chemical calculations, are shown in the following sample problems.

---

[1] Historically, the mole was first defined as the amount of an element having a mass in grams numerically equal to its atomic weight. Later, it was discovered experimentally that there are 6.022×10^{23} atoms in 1 mole of an element. We will see how Avogadro's number can be measured in later chapters.

The mole, one of the seven base units of SI, is formally defined as the amount of substance of a system which contains as many elementary entities as there are atoms in exactly 0.012 kg of carbon-12 (the most common isotope of carbon). The full meaning of this definition will become clear when atomic structure is discussed in greater detail (Chapter 3). For now, notice that SI does not directly define the number of particles in a mole, but simply specifies a *method* by which the number can be experimentally determined.

One mole of four different elements: sulfur, iron, mercury, and copper. Each sample contains the same number of atoms.

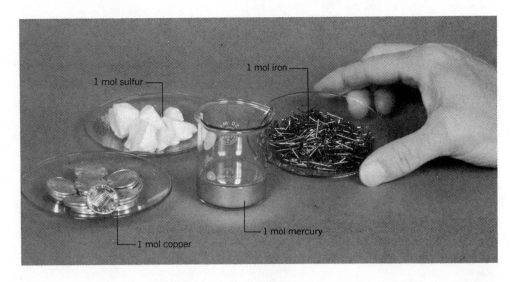

EXAMPLE 2.4 CONVERTING FROM GRAMS TO MOLES

PROBLEM How many moles of silicon, Si, are in 30.5 g Si? Silicon is an element used to make transistors.

SOLUTION Our problem is one of converting the units of grams of Si to moles of Si, that is, 30.5 g Si = (?) mol Si. We know from the table of atomic weights that

$$1 \text{ mol Si} = 28.1 \text{ g Si}$$

To convert from g Si to moles, we must multiply 30.5 g Si by a factor that contains the units g Si in the denominator, that is,

$$\frac{1 \text{ mol Si}}{28.1 \text{ g Si}}$$

When we set up this problem.

$$30.5 \text{ g Si} \times \left(\frac{1 \text{ mol Si}}{28.1 \text{ g Si}}\right) = 1.09 \text{ mol Si}$$

Thus, 30.5 g Si = 1.09 mol Si.

EXAMPLE 2.5 CONVERTING FROM MOLES TO GRAMS

PROBLEM How many grams of Cu are there in 2.55 mol of Cu?

SOLUTION From the table of atomic weights,

$$1 \text{ mol Cu} = 63.5 \text{ g Cu}$$

Cu is the symbol for copper.

The conversion factor must have the units mol Cu in the denominator, that is,

$$\frac{63.5 \text{ g Cu}}{1 \text{ mol Cu}}$$

Setting up the problem, we have

$$2.55 \text{ mol Cu} \times \left(\frac{63.5 \text{ g Cu}}{1 \text{ mol Cu}}\right) = 162 \text{ g Cu}$$

EXAMPLE 2.6   USING MOLE RELATIONSHIPS

PROBLEM   How many moles of Ca are required to react with 2.50 mol of Cl to produce the compound $CaCl_2$ (calcium chloride), used to melt ice on roads in the winter.

SOLUTION   Our problem: 2.50 mol Cl ~ (?) mol Ca.

We know from the formula that 1 atom of Ca combines with 2 atoms of Cl. Thus,

$$1 \text{ atom Ca} \sim 2 \text{ atoms Cl}$$

Because moles combine in the same ratio as atoms,

$$1 \text{ mol Ca} \sim 2 \text{ mol Cl}$$

We obtain our answer as follows:

$$2.50 \text{ mol Cl} \times \left(\frac{1 \text{ mol Ca}}{2 \text{ mol Cl}} \right) \sim 1.25 \text{ mol Ca}$$

EXAMPLE 2.7   USING THE MOLE/MASS RELATIONSHIP

PROBLEM   How many grams of Ca must react with 41.5 g Cl to produce $CaCl_2$?

SOLUTION   Our problem: 41.5 g Cl ~ (?) g Ca.
What quantities do we need and what do we know?
We know that

$$1 \text{ mol Ca} \sim 2 \text{ mol Cl} \qquad \text{(Why?)}$$

Also,

$$\left. \begin{array}{l} 1 \text{ mol Cl} = 35.5 \text{ g Cl} \\ 1 \text{ mol Ca} = 40.1 \text{ g Ca} \end{array} \right\} \quad \text{(atomic weight table)}$$

Here is the solution to the problem:

First we calculate how many moles of Cl are available. Then we calculate how many moles of Ca are needed. Finally, we calculate the mass of this many moles of Ca.

$$41.5 \text{ g Cl} \times \left(\frac{1 \text{ mol Cl}}{35.5 \text{ g Cl}} \right) = 1.17 \text{ mol Cl}$$

$$1.17 \text{ mol Cl} \times \left(\frac{1 \text{ mol Ca}}{2 \text{ mol Cl}} \right) \sim 0.585 \text{ mol Ca}$$

$$0.585 \text{ mol Ca} \times \left(\frac{40.1 \text{ g Ca}}{1 \text{ mol Ca}} \right) = 23.5 \text{ g Ca}$$

And here are all these steps written together.

Stringing together conversion factors makes the arithmetic simple to perform using a calculator.

$$41.5 \text{ g Cl} \times \left(\frac{1 \text{ mol Cl}}{35.5 \text{ g Cl}} \right) \times \left(\frac{1 \text{ mol Ca}}{2 \text{ mol Cl}} \right) \times \left(\frac{40.1 \text{ g Ca}}{1 \text{ mol Ca}} \right) \sim 23.5 \text{ g Ca}$$

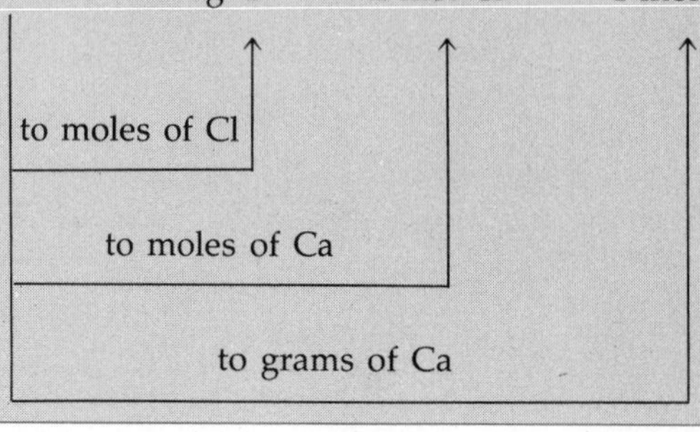

to moles of Cl

to moles of Ca

to grams of Ca

In this last example we have strung together a series of conversion factors. When we set up the problem, we link these together by considering what units must be eliminated by cancellation. Thus, the first factor had to have g Cl in the denominator and took us as far as moles of Cl. The second factor had to have moles of Cl in the denominator and would give us moles of Ca if we stopped at that point. The third factor was required to have moles Ca in the denominator. At this point we stop since our units are now those of the answer, and we have only to perform the arithmetic to obtain the correct result.

Avogadro's number and the mole

We have noted that the number of particles in one mole, which is called Avogadro's number, is equal to 6.022×10^{23}. This is a tremendous number; written in ordinary decimal notation, it is 602 200 000 000 000 000 000 000. Because there are so many atoms in one mole of an element, the mass of an individual atom is extremely small. The next example illustrates how Avogadro's number can be used to calculate properties of individual atom-sized particles.

EXAMPLE 2.8 USING AVOGADRO'S NUMBER IN CALCULATIONS

PROBLEM   What is the mass of one atom of calcium?

SOLUTION   When a calculation deals with some property of individual atoms or molecules and relates that property to things that we can measure in the laboratory, Avogadro's number is usually needed. Here we can restate the problem as

$$1 \text{ atom Ca} \sim (?) \text{ g Ca}$$

Notice that we are being asked to express the property of one atom (its mass) in terms of a quantity (grams) that we can measure in the laboratory. We need Avogadro's number to relate the tiny submicroscopic world of atoms to the large world of the laboratory. We have the relationships

$$6.02 \times 10^{23} \text{ atoms Ca} = 1 \text{ mol Ca (to 3 significant figures)}$$

$$1 \text{ mol Ca} = 40.1 \text{ g Ca}$$

The solution to the problem can then be set up as

$$1 \text{ atom Ca} \times \left(\frac{1 \text{ mol Ca}}{6.02 \times 10^{23} \text{ atoms Ca}} \right) \times \left(\frac{40.1 \text{ g Ca}}{1 \text{ mol Ca}} \right) \sim 6.66 \times 10^{-23} \text{ g Ca}$$

2.4 MEASURING MOLES OF COMPOUNDS: MOLECULAR WEIGHTS AND FORMULA WEIGHTS

As with elements, the balance can also be used indirectly to measure moles of compounds. The simplest way of obtaining the mass of one mole of a substance is merely to add up the atomic weights of all the elements present in the compound. If the substance is composed of molecules (for example, CO_2, H_2O, or NH_3), the sum of the atomic weights is called the **molecular weight**. Thus, the molecular weight of CO_2 is obtained as

$$
\begin{array}{lll}
\text{C} & 1 \times 12.0 \text{ u} = & 12.0 \text{ u} \\
2 \text{ O} & 2 \times 16.0 \text{ u} = & \underline{32.0 \text{ u}} \\
CO_2 & \text{total} & 44.0 \text{ u}
\end{array}
$$

Similarly, the molecular weight of $H_2O = 18.0$ u and that of $NH_3 = 17.0$ u. The mass of 1 mol of a substance is obtained simply by writing its molecular weight followed by the units, grams. Thus,

$$1 \text{ mol } CO_2 = 44.0 \text{ g}$$

$$1 \text{ mol } H_2O = 18.0 \text{ g}$$

$$1 \text{ mol } NH_3 = 17.0 \text{ g}$$

In later chapters we will encounter many compounds that do not contain separate, distinct molecules. We will find that when certain atoms react, they often gain or lose negatively charged particles called **electrons**. Sodium and chlorine happen to react this way. When sodium chloride, NaCl, is formed from the elements, each Na atom loses one electron and each Cl atom gains one. Since Na and Cl are electrically neutral to start, these atoms acquire a charge when NaCl is formed. These are written as Na^+ (positive because Na has lost a negatively charged electron) and Cl^- (negative because Cl has gained an electron). *Atoms or groups of atoms that have acquired an electrical charge are called* **ions**. Since solid NaCl is composed of Na^+ and Cl^- ions, it is said to be an **ionic compound**.

This entire topic is explored further in Chapters 4 and 5. For now, it is only necessary for you to know that compounds that are ionic do not contain molecules. Their formulas simply state the ratio of the different atoms in the substance. In NaCl, the atoms are in a one-to-one ratio. In the ionic compound $CaCl_2$, the ratio of Ca to Cl atoms is 1 to 2 (relax—at this point you were not expected to know that $CaCl_2$ is ionic). Rather than refer to molecules of NaCl or $CaCl_2$, we use the term **formula unit** to specify the pair of ions in NaCl (Na^+ and Cl^-) or the set of three ions in $CaCl_2$.

For ionic compounds, the sum of the atomic weights of the elements present in one formula unit is known as the **formula weight**. For NaCl, this is $22.99 + 35.45 = 58.44$. One mole of NaCl (6.022×10^{23} formula units of NaCl) would contain 58.44 g NaCl. Use of the term formula weight, of course, is not restricted to ionic compounds. It can also be applied to molecular substances, in which case the terms formula weight and molecular weight mean the same thing.

Na	22.99 u
Cl	35.45 u
NaCl	58.44 u

1 mol sodium chromate Na_2CrO_4

1 mol water (H_2O)

1 mol sodium chloride NaCl

1 mol copper sulfate $CuSO_4 \cdot 5H_2O$

One mole of four different compounds. Each contains the same number of formula units, although the number of atoms is different from sample to sample. This is because the number of atoms per formula unit is different for each compound.

The term **molar mass** applies equally to both ionic compounds and molecular substances; in either case the term refers to the mass (in grams) of one mole of the substance specified. The composition of the substance as either molecules or ions does not affect its molar mass value.

EXAMPLE 2.9 CONVERSIONS BETWEEN MOLES AND GRAMS FOR COMPOUNDS

PROBLEM Sodium carbonate, Na_2CO_3, is a very important industrial chemical used in making glass.

(a) What is the mass (in grams) of 0.250 mol Na_2CO_3?

(b) How many moles of Na_2CO_3 are in 132 g Na_2CO_3?

SOLUTION To answer these questions we need the formula weight of Na_2CO_3. We calculate this from the atomic weights of its elements

In this example, calculating the formula weight to the nearest 0.1 u gives enough significant figures to perform the calculations properly.

2 Na	$2 \times 23.0 =$ 46.0
1 C	$1 \times 12.0 =$ 12.0
3 O	$3 \times 16.0 =$ 48.0
Total	106.0

The formula weight is 106.0; therefore

$$1 \text{ mol } Na_2CO_3 = 106.0 \text{ g } Na_2CO_3$$

This can be used to make conversion factors relating grams and moles of Na_2CO_3, which we need to answer the questions.

(a) To convert 0.250 mol Na_2CO_3 to grams, we set up the units to cancel.

$$0.250 \text{ mol } Na_2CO_3 \times \left(\frac{106.0 \text{ g } Na_2CO_3}{1 \text{ mol } Na_2CO_3} \right) = 26.5 \text{ g } Na_2CO_3$$

(b) Again, we set the units to cancel.

$$132 \text{ g } Na_2CO_3 \times \left(\frac{1 \text{ mol } Na_2CO_3}{106.0 \text{ g } Na_2CO_3} \right) = 1.25 \text{ mol } Na_2CO_3$$

2.5 PERCENTAGE COMPOSITION

A very simple and also often very useful computation is the calculation of the **percentage composition** of a compound—that is, the percentage of the total mass (also called the percent by mass) contributed by each element. The procedure to determine the percentage composition is illustrated in Example 2.10.

EXAMPLE 2.10 CALCULATING PERCENTAGE COMPOSITION FROM A CHEMICAL FORMULA

PROBLEM What is the percentage composition of chloroform, $CHCl_3$, a substance once used as an anesthetic?

SOLUTION The total mass of 1 mol of $CHCl_3$ is obtained from the molecular weight

In general

% by mass =

$\dfrac{\text{mass of part}}{\text{mass of whole}} \times 100$

or

$$(12.01 + 1.008 + 3 \times 35.45) \text{ u} = 119.37 \text{ u}$$

$$(12.01 \text{ g} + 1.008 \text{ g} + 3 \times 35.45 \text{ g}) = 119.37 \text{ g}$$

$$\% \text{ C} = \frac{\text{mass carbon}}{\text{mass } CHCl_3} \times 100$$

$$\% \ C = \frac{12.01 \text{ g C}}{119.37 \text{ g CHCl}_3} \times 100 = 10.06\% \text{ C}$$

$$\% \ H = \frac{1.008 \text{ g H}}{119.37 \text{ g CHCl}_3} \times 100 = 0.844\% \text{ H}$$

$$\% \ Cl = \frac{106.35 \text{ g Cl}}{119.37 \text{ g CHCl}_3} \times 100 = 89.09\% \text{ Cl}$$

total % = 100.00

A similar calculation can be used to determine the mass of an element in a given sample of a compound. This is illustrated in Example 2.11.

EXAMPLE 2.11 CALCULATING THE MASS OF AN ELEMENT IN A SAMPLE OF A COMPOUND

PROBLEM Calculate the mass of iron (Fe) in a 10.0-g sample of rust, Fe_2O_3.

SOLUTION To solve this problem, we can first calculate the fraction of Fe_2O_3 that is Fe. From the formula, we see that one formula unit of Fe_2O_3 contains 2 atoms of Fe. Therefore, we know that 1 mol of Fe_2O_3 contains 2 mol of Fe. Also,

$$1 \text{ mol Fe} = 55.85 \text{ g Fe}$$

$$1 \text{ mol Fe}_2O_3 = 159.7 \text{ g Fe}_2O_3$$

We can always calculate the mass of 1 mol of a substance if we know its chemical formula.

Therefore, 1 mol Fe_2O_3 (159.7 g Fe_2O_3) contains 2 mol Fe (111.7 g Fe). The fraction of Fe_2O_3 that is Fe is thus

$$\frac{\text{mass of Fe}}{\text{mass of Fe}_2O_3} = \frac{111.7 \text{ g Fe}}{159.7 \text{ g Fe}_2O_3}$$

Finally, the mass of Fe in the 10.0-g sample of Fe_2O_3 is equal to the sample mass multiplied by the fraction of the sample that is Fe.

$$10.0 \text{ g Fe}_2O_3 \times \left(\frac{111.7 \text{ g Fe}}{159.7 \text{ g Fe}_2O_3} \right) \sim 6.99 \text{ g Fe}$$

2.6 CHEMICAL FORMULAS

There are different kinds of chemical formulas, and each conveys certain kinds of information. These can include the elemental composition, the relative numbers of each kind of atom present, the actual numbers of atoms of each kind in a molecule of the substance, or the structure of a molecule of the substance. For convenience, we can classify formulas according to the information that they provide.

A formula that simply gives the relative number of atoms of each element present in a formula unit is called a **simplest formula**. It is also called an **empirical formula** because it is normally derived from the results of some experimental analysis. The formulas NaCl, H_2O, and CH_2 are empirical formulas.

A formula that states the actual number of each kind of atom found in a molecule is called a **molecular formula**. H_2O is a molecular formula (as well as an empirical formula) since a molecule of water contains 2 atoms of H and 1 atom of O. The formula C_2H_4 is a molecular formula for a substance (ethylene) containing 2 atoms of carbon and 4 atoms of hydrogen. Note that the simplest formula of this compound is CH_2 because the carbon-to-hydrogen ratio is $1:2$. The simplest formula CH_2 is not unique to C_2H_4. however. A substance whose empirical formula is CH_2 could have as a molecular formula CH_2, C_2H_4, C_3H_6, and so on.

It is possible for many different compounds to have the same empirical formula.

A third type of formula is a **structural formula,** for example,

$$\begin{array}{c} H \\ | \\ H-C-C \overset{O}{\diagup} O-H \\ | \\ H \end{array}$$

acetic acid (present in vinegar)

In a structural formula the dashes between the different atomic symbols represent the "chemical bonds" that hold the atoms together in the molecule. We will see more of them in Chapter 5. A structural formula gives us information about the way in which the atoms in a molecule are linked together and provides information that allows us to write the molecular and empirical formulas. Thus, for acetic acid shown above, we can also write its molecular formula ($C_2H_4O_2$) and its empirical formula (CH_2O).

The most desirable kind of formula to have, of course, is the structural formula, because it also contains all the information provided by the other two types. However, in chemistry, as in the rest of life, there's no "free lunch"—we never get something for nothing! The more information a formula conveys, the more difficult it is to arrive at experimentally. We will see how empirical and molecular formulas are derived, but most of the procedures involved in the determination of structural formulas are beyond the scope of this book.

2.7 EMPIRICAL FORMULAS

Since the simplest formula gives the relative numbers of atoms present in a compound, it must also give the relative number of moles of each element. Therefore, to obtain the empirical formula for a compound, we must determine the number of moles of each of its elements that are present in a particular sample. Then we can calculate the simplest whole-number ratio of the moles to find the subscripts. The following examples illustrate how this is done.

EXAMPLE 2.12 CALCULATING AN EMPIRICAL FORMULA

PROBLEM A sample of a brown-colored gas that is a major air pollutant is found to contain 2.34 g of N and 5.34 g of O. What is the simplest formula of the compound?

SOLUTION We proceed by calculating the number of moles of each element present. We know that

$$1 \text{ mol N} = 14.0 \text{ g N} \quad \text{(Why?)}$$

$$1 \text{ mol O} = 16.0 \text{ g O} \quad \text{(Why?)}$$

Therefore,

$$2.34 \text{ g N} \times \left(\frac{1 \text{ mol N}}{14.0 \text{ g N}} \right) = 0.167 \text{ mol N}$$

$$5.34 \text{ g O} \times \left(\frac{1 \text{ mol O}}{16.0 \text{ g O}} \right) = 0.334 \text{ mol O}$$

We might write our formula $N_{0.167}O_{0.334}$. It does indeed tell us the relative number of moles of N and O; however, since the formula should have meaning on a molecular level where whole numbers of atoms are combined, the subscripts must be integers. If we divide each subscript by the smallest one, we obtain

$$N_{\frac{0.167}{0.167}}O_{\frac{0.334}{0.167}} = NO_2$$

EXAMPLE 2.13   CALCULATING AN EMPIRICAL FORMULA
FROM PERCENTAGE COMPOSITION

PROBLEM   What is the empirical formula of a compound composed of 43.7% P and 56.3% O by mass?

SOLUTION   It is quite common to have the results of a chemical analysis in the form of percentage composition by mass. The simplest way to proceed in such a case is to imagine having a 100-g sample of the compound. From the analysis, this sample would contain 43.7 g P and 56.3 g O (notice that the percents become grams of compound). Now that we know the masses of phosphorus and oxygen in the same sample, we convert the masses to moles and proceed as before.

$$43.7 \text{ g P} \times \left(\frac{1 \text{ mol P}}{31.0 \text{ g P}}\right) = 1.41 \text{ mol P}$$

$$56.3 \text{ g O} \times \left(\frac{1 \text{ mol O}}{16.0 \text{ g O}}\right) = 3.52 \text{ mol O}$$

Our formula is

$$P_{1.41}O_{3.52} = P_{\frac{1.41}{1.41}}O_{\frac{3.52}{1.41}} = PO_{2.50}$$

Whole numbers may be obtained by doubling each of these values. Thus, the empirical formula is P_2O_5.

Usually, it is not possible to decompose a compound into its elements and then weigh them, so a different strategy must be used. As a rule, this involves causing the compound to undergo some chemical reaction that divides the elements from each other and isolates them in compounds having known chemical formulas. Such analyses require very careful work to ensure that all of a given element in the original sample is isolated in the compound of known formula. The next two examples illustrates this for typical analyses of compounds composed of the elements carbon, hydrogen, and oxygen.

EXAMPLE 2.14   DETERMINING THE COMPOSITION OF A
COMPOUND BY A CHEMICAL ANALYSIS

PROBLEM   A 1.025-g sample of a compound that contains only carbon and hydrogen was burned in oxygen to give carbon dioxide and water vapor as products. These products were trapped separately and weighed. It was found that 3.01 g of CO_2 and 1.85 g of H_2O were formed in the reaction. What is the empirical formula of the compound?

SOLUTION   At first glance this problem may seem very difficult, but let's examine what we know and what we can calculate.

We already know how to calculate the number of moles of C in the 3.01 g of CO_2, and we also can calculate the number of moles of H in 1.85 g of H_2O. Since all the carbon in the CO_2 and all the hydrogen in the water came from the original sample, these calculations will give the number of moles of each of these elements in the sample. This is precisely the information that we need to calculate the empirical formula.

Now that we know what has to be done, let's perform the calculations. The number of moles of carbon atoms in 3.01 g of CO_2 is

$$3.02 \text{ g } CO_2 \times \left(\frac{1 \text{ mol } CO_2}{44.0 \text{ g } CO_2} \right) \times \left(\frac{1 \text{ mol C}}{1 \text{ mol } CO_2} \right) = 0.0686 \text{ mol C}$$

and the number of moles of hydrogen atoms in 1.85 g of H_2O is

$$1.85 \text{ g } H_2O \times \left(\frac{1 \text{ mol } H_2O}{18.0 \text{ g } H_2O} \right) \times \left(\frac{2 \text{ mol H}}{1 \text{ mol } H_2O} \right) = 0.206 \text{ mol H}$$

The empirical formula is therefore $C_{\frac{0.0686}{0.0686}} H_{\frac{0.206}{0.0686}}$ which is the same as CH_3.

EXAMPLE 2.15 DETERMINING THE COMPOSITION OF A COMPOUND BY AN INDIRECT METHOD

PROBLEM A 0.1000-g sample of ethyl alcohol (grain alcohol), known to contain only carbon, hydrogen, and oxygen, was allowed to react completely with oxygen to form the products CO_2 and H_2O. These products were trapped separately and weighed. 0.1910 g of CO_2 and 0.1172 g of H_2O were found. What is the empirical formula of the compound?

SOLUTION The solution to this problem requires a slightly different approach than we used in Example 2.14. This is because the oxygen in the original sample is distributed between the CO_2 and the H_2O along with some *additional* oxygen that was required to form the products. In other words, not all the oxygen in the products came from the original sample. Now that we realize the problem, let's see how we can solve it.

Following the same procedure used in Example 2.11, we can compute the mass of carbon and the mass of hydrogen in the CO_2 and H_2O that were formed from the compound when it reacted. Since the only source of C and H was the original compound, the difference between the mass of the compound taken (0.1000 g) and the total mass of carbon and hydrogen must be the mass of oxygen in the original 0.1000 g. (We must obtain the mass of oxygen in this way, rather than from the total mass of oxygen in the H_2O and CO_2, since only a portion of the oxygen in the products came from the original compound.) Once we know the masses of carbon, hydrogen, and oxygen in the 0.1000 g-sample, we can calculate how many moles there are of each in the 0.1000 g and, hence, the empirical formula of the unknown compound. Thus, having planned our course of action, we now proceed with the computations. The molar masses of CO_2 and H_2O are 44.0 and 18.0 g mol^{-1}, respectively. The fraction of the mass of CO_2 that is carbon is

$$\frac{12.0 \text{ g C}}{44.0 \text{ g } CO_2}$$

Likewise, the fraction of H_2O that is hydrogen is equal to

$$\frac{2.01 \text{ g H}}{18.0 \text{ g } H_2O}$$

The mass of carbon in the original compound is equal to the mass of CO_2 multiplied by the fraction of the mass that is due to carbon.

$$0.1910 \text{ g } CO_2 \times \left(\frac{12.0 \text{ g C}}{44.0 \text{ g } CO_2} \right) \sim 0.0521 \text{ g C}$$

Similarly,

$$0.1172 \text{ g H}_2\text{O} \times \left(\frac{2.01 \text{ g H}}{18.0 \text{ g H}_2\text{O}} \right) \sim 0.0131 \text{ g H}$$

In order to calculate the mole ratios, we must know the amounts of C, H, and O, in the *same sized sample.*

The total mass contributed by the carbon and hydrogen in the sample is

$$0.0521 \text{ g C} + 0.0131 \text{ g H} = 0.0652 \text{ g}$$

The mass of oxygen = 0.1000 g − 0.0652 g = 0.0348 g O. Next, we calculate the number of moles of C, H, and O.

$$0.0521 \text{ g C} \times \left(\frac{1 \text{ mol C}}{12.0 \text{ g C}} \right) = 4.34 \times 10^{-3} \text{ mol C}$$

Similar calculations for hydrogen and oxygen give 1.31×10^{-2} mol H and 2.17×10^{-3} mol O. The empirical formula is therefore

$$C_{0.00434}H_{0.0131}O_{0.00217} = C_{\frac{0.00434}{0.00217}} H_{\frac{0.0131}{0.00217}} O_{\frac{0.00217}{0.00217}}$$

or

$$C_2H_6O$$

2.8 MOLECULAR FORMULAS

Ionic compounds don't have molecular formulas because they don't contain molecules.

Not only does the molecular formula provide the information contained in the empirical formula, but it also tells us how many atoms of each element are present in a molecule of a substance. Remember that an empirical formula of CH_2 is found for any molecule that possesses twice as many hydrogen atoms as carbon atoms. To distinguish among all the possible choices, we require the molecular weight of the compound. This is because *the molecular weight is always an integral multiple of the empirical formula weight* (see Table 2.1). To find the number of times that the empirical formula is repeated in the molecular formula, we simply divide the experimentally determined molecular weight[2] by the empirical formula weight.

Table 2.1
Molecular weights as multiples of the empirical formula weight

Formula	Molecular Weight
CH_2	$14.0 = 1 \times 14.0$
C_2H_4	$28.0 = 2 \times 14.0$
C_3H_6	$42.0 = 3 \times 14.0$
C_4H_8	$56.0 = 4 \times 14.0$
C_nH_{2n}	$n \times 14.0$

EXAMPLE 2.16 DETERMINING THE MOLECULAR FORMULA OF A COMPOUND

PROBLEM   A colorless liquid used in rocket engines, whose empirical formula is NO_2, has a molecular weight of 92.0. What is its molecular formula?

---

[2] We will see how molecular weights can be measured in Chapters 9 and 11.

SOLUTION The formula weight of NO_2 is 46.0. The number of times the empirical formula, NO_2, occurs in the compound is

$$\frac{92.0}{46.0} = 2$$

The molecular formula is then $(NO_2)_2 = N_2O_4$ (dinitrogen tetroxide).

N_2O_4 is the preferred answer because $(NO_2)_2$ implies a knowledge of the structure of the molecule (i.e., that two NO_2 units are somehow joined together).

2.9 BALANCING CHEMICAL EQUATIONS

Recall that a chemical equation is a shorthand description of the changes that occur during a chemical reaction—a sort of before-and-after picture of what happens. One of the most useful properties of a chemical equation is that it allows us to determine the quantitative relationships that exist among reactants and products. To be helpful in this way, however, the equation must be balanced; that is, it must obey the law of conservation of mass by having the same number of atoms of each kind on both sides of the arrow.

In order to minimize errors, writing a balanced chemical equation should always be considered a two-step process.

STEP 1. First write an unbalanced equation with correct formulas for all the reactants and products. (At this time you're not expected to know the formulas for compounds, so they will be given to you. We will discuss how to write formulas later in the book.)

STEP 2. Balance the equation by adjusting the coefficients that precede the formulas. *During this step, you are not permitted to change the subscripts in any of the formulas.* To do so would change the nature of the substances.

There is never any excuse for having an improperly balanced equation, since it is always possible, by counting atoms on each side of the equation, to determine whether the equation is, in fact, balanced.

Most simple chemical equations can be easily balanced by inspection. This involves examining the equation and adjusting the coefficients until equal numbers of each element are present among both the reactants and products. For example, consider the reaction of sodium carbonate (Na_2CO_3) with hydrochloric acid (HCl) to produce sodium chloride (NaCl), carbon dioxide (CO_2), and water.

To obtain a properly balanced chemical equation, we proceed as follows:

1. We write the unbalanced equation

$$Na_2CO_3 + HCl \longrightarrow NaCl + H_2O + CO_2$$

2. Coefficients are introduced to balance the equation. This is where you will probably need some practice so that you can learn to balance an equation quickly. Although there are no set rules to tell you where to start, it is often best to seek out the most complex formula in the equation and begin there by giving it a coefficient of 1. In this case, we start with the Na_2CO_3. Since there are two Na atoms in this formula on the left, there must also be two Na atoms on the right, so place a 2 in front of the formula NaCl. This gives us

$$Na_2CO_3 + HCl \longrightarrow 2NaCl + H_2O + CO_2$$

Hydrochloric acid is added to a solution of sodium carbonate. Among the products is the gas CO_2, which is seen bubbling from the reaction mixture.

Balancing an equation by inspection proceeds partly by trial and error and becomes easier with practice.

Now there are two Cl atoms on the right but only one on the left, so we place a 2 in front of the HCl.

$$Na_2CO_3 + 2HCl \longrightarrow 2NaCl + H_2O + CO_2$$

A fast inspection reveals that this question is now balanced.

It was stated above that a balanced equation obeys the law of conservation of mass. For the equation that we have just balanced, and in fact for any equation, there is an infinite number of sets of coefficients that fulfill this requirement. Thus, the equations

$$2Na_2CO_3 + 4HCl \longrightarrow 4NaCl + 2CO_2 + 2H_2O$$

$$5Na_2CO_3 + 10HCl \longrightarrow 10NaCl + 5CO_2 + 5H_2O$$

are also balanced. The usual practice, however, is to use the smallest possible set of whole-number coefficients, although there are occasions, as we will see, when other choices are advantageous.

EXAMPLE 2.17   BALANCING AN EQUATION BY INSPECTION

PROBLEM   Balance the following equation for the combustion of octane, C_8H_{18}, a component of gasoline:

$$C_8H_{18} + O_2 \longrightarrow CO_2 + H_2O$$

SOLUTION   Inspection of the equation quickly suggests that we must adjust the coefficients preceding CO_2 and H_2O to balance C and H. Carbon can be balanced by placing an 8 before the CO_2; hydrogen can be balanced by placing a 9 before the H_2O ($9H_2O$ contains 18 H atoms because each H_2O contains 2 H atoms). This gives

$$C_8H_{18} + O_2 \longrightarrow 8CO_2 + 9H_2O$$

Now we can work on the oxygen. On the right there are 25 O atoms ($2 \times 8 + 9 = 25$). On the left the O atoms come in pairs. This means that we must have $12\frac{1}{2}$ pairs (O_2 molecules) to have 25 O atoms on the left. This gives us

$$C_8H_{18} + 12\tfrac{1}{2}O_2 \longrightarrow 8CO_2 + 9H_2O$$

Finally, we can eliminate the fractional coefficient by doubling each coefficient.

$$2C_8H_{18} + 25O_2 \longrightarrow 16CO_2 + 18H_2O$$

2.10   CALCULATIONS BASED ON CHEMICAL EQUATIONS

A chemical equation can be interpreted in several ways. Consider, for example, the balanced equation for the combustion of ethanol, C_2H_5OH, the alcohol used in gasohol

$$C_2H_5OH + 3O_2 \longrightarrow 2CO_2 + 3H_2O$$

On a molecular, submicroscopic level we can view this as the reaction between individual molecules.

1 molecule C_2H_5OH + 3 molecules $O_2 \longrightarrow$ 2 molecules CO_2 + 3 molecules H_2O

But we can just as easily scale this up to lab-sized amounts because mole quantities react in the same ratio as the individual molecules.

1 mol C_2H_5OH + 3 mol $O_2 \longrightarrow$ 2 mol CO_2 + 3 mol H_2O

The key point is that *the coefficients in a chemical equation provide the ratios in which moles of one substance react with or form moles of another.* The coefficients provide us with a whole host of different chemical equivalencies that can be used to form conversion factors. In this example, we can write six such equivalencies:

$$1 \text{ mol } C_2H_5OH \sim 3 \text{ mol } O_2$$

$$1 \text{ mol } C_2H_5OH \sim 2 \text{ mol } CO_2$$

$$1 \text{ mol } C_2H_5OH \sim 3 \text{ mol } H_2O$$

$$3 \text{ mol } O_2 \sim 2 \text{ mol } CO_2$$

$$3 \text{ mol } O_2 \sim 3 \text{ mol } H_2O$$

$$2 \text{ mol } CO_2 \sim 3 \text{ mol } H_2O$$

Thus, the balanced equation provides quantitative relationships among all the reactants and products. Let's look at a few sample calculations based on the reaction for the combustion of ethanol.

EXAMPLE 2.18   USING A CHEMICAL EQUATION IN A CALCULATION INVOLVING MOLES

PROBLEM   How many moles of oxygen are needed to burn 1.80 mol C_2H_5OH according to the balanced equation

$$C_2H_5OH + 3O_2 \longrightarrow 2CO_2 + 3H_2O$$

SOLUTION   The coefficients of the equation give us the relationship

$$1 \text{ mol } C_2H_5OH \sim 3 \text{ mol } O_2$$

which we can use for a conversion factor. We set up the arithmetic so the units mol C_2H_5OH cancel.

$$1.80 \text{ mol } C_2H_5OH \times \left(\frac{3 \text{ mol } O_2}{1 \text{ mol } C_2H_5OH} \right) = 5.40 \text{ mol } O_2$$

We need 5.40 mol O_2.

EXAMPLE 2.19   USING A CHEMICAL EQUATION IN A CALCULATION INVOLVING MOLES

PROBLEM   How many moles of CO_2 will be formed when 0.274 mol C_2H_5OH burns?

SOLUTION   Now we look at the coefficients for C_2H_5OH and CO_2, which give us

$$1 \text{ mol } C_2H_5OH \sim 2 \text{ mol } CO_2$$

Then we set up the arithmetic to get the proper units in the answer.

$$0.274 \text{ mol } C_2H_5OH \times \left(\frac{2 \text{ mol } CO_2}{1 \text{ mol } C_2H_5OH} \right) = 0.548 \text{ mol } CO_2$$

EXAMPLE 2.20   USING A CHEMICAL EQUATION IN A CALCULATION INVOLVING MOLES

PROBLEM   How many moles of water will form when 3.66 mol CO_2 are produced during the combustion of C_2H_5OH?

SOLUTION The coefficients in the equation for the combustion of C_2H_5OH tell us that

$$2 \text{ mol } CO_2 \sim 3 \text{ mol } H_2O$$

Therefore,

$$3.66 \text{ mol } CO_2 \times \left(\frac{3 \text{ mol } H_2O}{2 \text{ mol } CO_2} \right) = 5.49 \text{ mol } H_2O$$

When we actually carry out a chemical reaction in the lab, we usually are interested in *laboratory units*—grams. But we can always translate from grams to moles or moles to grams if we have the chemical formulas of the substances involved. You learned to do that in Sections 2.3 and 2.4.

EXAMPLE 2.21 USING A CHEMICAL EQUATION IN A CALCULATION INVOLVING GRAMS

PROBLEM Freshly exposed aluminum surfaces react with oxygen to form a tough aluminum oxide coating that protects the metal from further corrosion. The reaction is

$$4Al + 3O_2 \longrightarrow 2Al_2O_3$$
$$\textbf{aluminum oxide}$$

How many grams of O_2 are required to react with 0.300 mol of Al?

SOLUTION Our problem can be stated simply as

$$0.300 \text{ mol Al} \sim (?) \text{ g } O_2$$

To solve it, we first convert moles of Al to moles of O_2. Then we can change moles of O_2 to grams of O_2. According to the balanced equation

$$4 \text{ mol Al} \sim 3 \text{ mol } O_2$$

Therefore, the number of moles of O_2 required is

$$0.300 \text{ mol Al} \times \left(\frac{3 \text{ mol } O_2}{4 \text{ mol Al}} \right) \sim 0.225 \text{ mol } O_2$$

To convert moles of O_2 to grams of O_2, we use the relationship

$$1 \text{ mol } O_2 = 32.0 \text{ g } O_2$$

A common error is to write the formula for oxygen as O_2 and then use 16.0 as the molecular weight. Be careful about this.

$$0.225 \text{ mol } O_2 \times \left(\frac{32.0 \text{ g } O_2}{1 \text{ mol } O_2} \right) = 7.20 \text{ g } O_2$$

We could have combined these two steps.

$$0.300 \text{ mol Al} \times \left(\frac{3 \text{ mol } O_2}{4 \text{ mol Al}} \right) \times \left(\frac{32.0 \text{ g } O_2}{1 \text{ mol } O_2} \right) \sim 7.20 \text{ g } O_2$$

EXAMPLE 2.22 USING A CHEMICAL EQUATION IN A CALCULATION INVOLVING GRAMS

PROBLEM From the chemical equation in the last example, calculate the number of grams of Al_2O_3 that could be produced if 12.5 g of O_2 completely react with aluminum.

SOLUTION Our problem:

$$12.5 \text{ g } O_2 \sim (?) \text{ g } Al_2O_3$$

The solution to this problem requires that we establish the chemical equivalence between O_2 and Al_2O_3. This kind of equivalence is *always* found from the coefficients in the balanced equation and is expressed in our chemical units, moles. From the equation

$$3 \text{ mol } O_2 \sim 2 \text{ mol } Al_2O_3$$

We now need the relationships permitting us to translate between laboratory units, grams, and chemical units, moles. These are

$$1 \text{ mol } O_2 = 32.0 \text{ g } O_2$$

$$1 \text{ mol } Al_2O_3 = 102 \text{ g } Al_2O_3$$

These three relationships can now be applied as conversion factors. We arrange them so that the proper units cancel.

$$12.5 \text{ g } O_2 \times \left(\frac{1 \text{ mol } O_2}{32.0 \text{ g } O_2} \right) \times \left(\frac{2 \text{ mol } Al_2O_3}{3 \text{ mol } O_2} \right) \times \left(\frac{102 \text{ g } Al_2O_3}{1 \text{ mol } Al_2O_3} \right) \sim 26.6 \text{ g } Al_2O_3$$

Notice that in solving this problem we made use of the mole ratio specified by the coefficients of the balanced equation. In general, there is no way to go from grams of one substance to grams of another without going through moles. The diagram below summarizes the steps involved.

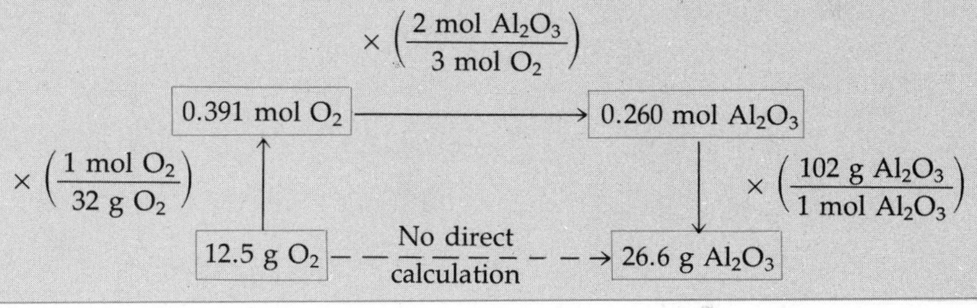

2.11 LIMITING-REACTANT CALCULATIONS

If arbitrary quantities of reactants are chosen when a chemical reaction is carried out, it is quite likely that one of the reactants will be completely consumed before the others are used up. For example, if 5 mol of H_2 and 1 mol of O_2 are mixed and allowed to react following the equation

$$2H_2 + O_2 \longrightarrow 2H_2O$$

After the reaction is finished, there will be 3 mol of H_2 left over.

only 2 mol of H_2 will react, completely consuming the 1 mol of O_2. After the O_2 is gone, no further reaction can occur and no further product can be formed. The amount of product is therefore limited by the reactant that disappears first. The reactant that disappears first (the reactant that is used up completely) is called the **limiting reactant.** Any other reactants are present in excess, and some fraction of the amounts originally available will be left over after the reaction is finished. The following two examples illustrate how we can determine which is the limiting reactant and then calculate how much product is formed.

EXAMPLE 2.23	SOLVING LIMITING REACTANT PROBLEMS
PROBLEM	Zinc and sulfur react to form zinc sulfide, a substance used in phosphors that coat the inner surfaces of TV picture tubes. The equation for the reaction is

$$Zn + S \longrightarrow ZnS$$

In a particular experiment 12.0 g of Zn are mixed with 6.50 g of S and allowed to react.

(a) Which is the limiting reactant?

(b) How many grams of ZnS can be formed from this particular reaction mixture?

(c) How many grams of which element will remain unreacted in this experiment?

SOLUTION

(a) The equation gives us the mole relationship

$$1 \text{ mol Zn} \sim 1 \text{ mol S}$$

To determine the limiting reactant, we first convert our reactant quantities to moles.

$$12.0 \text{ g Zn} \times \left(\frac{1 \text{ mol Zn}}{65.4 \text{ g Zn}} \right) = 0.183 \text{ mol Zn}$$

$$6.50 \text{ g S} \times \left(\frac{1 \text{ mol S}}{32.1 \text{ g S}} \right) = 0.202 \text{ mol S}$$

We can also determine the limiting reactant based on the amount of sulfur available. 0.202 mol S requires 0.202 mol Zn. Since we only have 0.183 mol Zn, all the Zn will be used up, so Zn is the limiting reactant.

Because these elements combine in a 1-to-1 ratio, 0.183 mol Zn requires 0.183 mol S. We see that there is more S than is required and that all of the Zn is able to react. *Zn is therefore the limiting reactant.*

(b) The amount of product formed depends only on the amount of limiting reactant. After the Zn has been used up, no more ZnS can form, so we use the amount of Zn (0.183 mol) to figure out the amount of ZnS that is produced. From the equation, we can write

$$1 \text{ mol Zn} \sim 1 \text{ mol ZnS}$$

Therefore, 0.183 mol Zn will form 0.183 mol ZnS. The mass of product is

$$0.183 \text{ mol ZnS} \times \left(\frac{97.5 \text{ g ZnS}}{1 \text{ mol ZnS}} \right) = 17.8 \text{ g ZnS}$$

(c) To calculate the mass of unreacted sulfur, we first subtract the number of moles of sulfur that reacted from the number of moles of sulfur initially available.

$$\text{moles S unreacted} = (0.202 \text{ mol} - 0.183 \text{ mol}) = 0.019 \text{ mol S}$$

Now we convert to grams.

$$0.019 \text{ mol S} \times \left(\frac{32.1 \text{ g S}}{1 \text{ mol S}} \right) = 0.61 \text{ g S} \qquad \text{left over}$$

EXAMPLE 2.24   SOLVING LIMITING REACTANT PROBLEMS

PROBLEM

Ethylene, C_2H_4, burns in air to form CO_2 and H_2O according to the equation

$$C_2H_4 + 3O_2 \longrightarrow 2CO_2 + 2H_2O$$

How many grams of CO_2 will be formed when a mixture containing 1.93 g C_2H_4 and 5.92 g O_2 is ignited?

SOLUTION

The relationship between C_2H_4 and O_2, as specified by the chemical equation, is in units of moles

$$1 \text{ mol } C_2H_4 \sim 3 \text{ mol } O_2$$

To solve our problem, we first convert our given quantities to moles.

$$1.93 \text{ g } C_2H_4 \times \left(\frac{1 \text{ mol } C_2H_4}{28.0 \text{ g } C_2H_4} \right) = 0.0689 \text{ mol } C_2H_4 \text{ available}$$

$$5.92 \text{ g O}_2 \times \left(\frac{1 \text{ mol O}_2}{32.0 \text{ g O}_2} \right) = 0.185 \text{ mol O}_2 \text{ available}$$

Let's see now whether there is sufficient O_2 to react with all the C_2H_4 (alternatively, we could see if there is enough C_2H_4 to react with all the O_2).

$$0.0689 \text{ mol C}_2\text{H}_4 \times \left(\frac{3 \text{ mol O}_2}{1 \text{ mol C}_2\text{H}_4} \right) \sim 0.207 \text{ mol O}_2 \text{ needed to consume all the C}_2\text{H}_4$$

0.185 mol O_2 requires 0.0617 mol C_2H_4. Since there is excess C_2H_4, the limiting reactant is O_2.

The calculation tells us that 0.207 mol of O_2 is required, but only 0.185 mol of O_2 is available. Therefore, O_2 is the limiting reactant.

We now use the limiting reactant to calculate the amount of product formed.

$$0.185 \text{ mol O}_2 \times \left(\frac{2 \text{ mol CO}_2}{3 \text{ mol O}_2} \right) \times \left(\frac{44.0 \text{ g CO}_2}{1 \text{ mol CO}_2} \right) \sim 5.43 \text{ g CO}_2$$

There is an interesting postscript to the last problem. Although not stated explicitly, our assumption was that even without sufficient oxygen to consume all the C_2H_4, any C_2H_4 that did react was converted completely to CO_2 and H_2O. If this were the case, one aspect of automotive air pollution would be removed. What actually happens when the hydrocarbon (in this case C_2H_4) is present in excess is that some of it is converted to CO. In the internal-combustion engine, gasoline (which is composed of a mixture of hydrocarbons) is burned in a limited supply of oxygen and the incomplete combustion therefore produces a mixture of CO, CO_2, and H_2O.

2.12 THEORETICAL YIELD AND PERCENTAGE YIELD

Sometimes a given set of reactants is able to produce more than one set of products, depending on reaction conditions. In the last paragraph, for instance, it was pointed out that the combustion of hydrocarbons in a limited supply of oxygen produces a mixture of products. Usually, the formation of side products (products other than those being sought) is undesirable, and three quantities that chemists are concerned with under these circumstances are the theoretical yield, the actual yield, and the percentage yield.

The **theoretical yield** *of a given product is the maximum yield that could be obtained if the reactants gave only that product.* In Example 2.24 we calculated the theoretical yield of CO_2, assuming that all the C_2H_4 that burned was converted entirely to CO_2 and H_2O.

The **percentage yield** is a measure of the efficiency of the reaction. It is defined as

The theoretical yield is a computed quantity; the actual yield is obtained by measuring the quantity of product actually formed in an experiment.

$$\text{percentage yield} = \frac{\text{actual yield}}{\text{theoretical yield}} \times 100$$

where the **actual yield** is the quantity of product that is actually produced in a given experiment. For example, suppose that in the reaction in Example 2.24 we had obtained only 3.48 g CO_2, with the remainder of the carbon as either CO or elemental carbon. The actual yield, then, was 3.48 g CO_2. The theoretical yield was 5.43 g CO_2, so the percentage yield of CO_2 would then be

$$\text{percentage yield CO}_2 = \frac{3.48 \text{ g CO}_2}{5.43 \text{ g CO}_2} \times 100$$

$$\text{percentage yield CO}_2 = 64.1\%$$

2.13 MOLAR CONCENTRATION

In many chemical reactions, both in the laboratory and in the world around us, one or more of the reactants are present in a solution—that is, they are dissolved in some fluid such as water. In our bodies, for instance, the blood dissolves nutrients and transports them to our cells where they undergo the complex chain of reactions called metabolism.

In Chapter 1 you learned that a solution is a homogeneous mixture and may be of variable composition. An example is salt dissolved in water, where the amount of salt dissolved in a given amount of water can vary from one solution to another. In discussions of solutions we frequently use the terms solvent and solute. Normally, we refer to the solvent as the component whose physical state doesn't change when the solution is formed. All the other components, which are dissolved in the solvent, are called **solutes**. In an aqueous salt solution, therefore, water is the solvent and salt is the solute. The term **concentration** is used to describe the relative amounts of solute and solvent in a solution. A solution in which a large amount of solute is dissolved in the solvent is said to have a high concentration of the solute.

Two words that are often used in comparing the concentrations of solutions are *concentrated* and *dilute*. A **concentrated solution** has a relatively high concentration of solute; a **dilute solution** has a relatively low concentration. An important thing to remember is that concentrated and dilute are relative terms. A solution that contains 0.01 g of NaCl per cubic decimetre is dilute compared to one that contains 0.10 g of NaCl per dm^3, but the solution containing 0.10 g of NaCl per dm^3 is dilute compared to one having 10 g of NaCl per dm^3.

As we will see later in the book, there are various ways of expressing the concentration of a solution quantitatively. The one that is the most useful chemically for dispensing specific amounts of a dissolved solute is called **molar concentration** or **molarity**. This is defined as the ratio of the number of moles of solute in the solution divided by the volume of the solution expressed in cubic decimetres (litres).

Molar concentration is an intensive quantity formed as a ratio of two extensive quantities: moles of solute and volume of solution.

$$\text{molar concentration} = \frac{\text{mol solute}}{dm^3 \text{ solution}}$$

A solution that contains 1.00 mol of NaCl in 1.00 dm^3 of solution is a 1.00 **molar**, or 1.00 M solution. If it contained 2.00 mol NaCl per cubic decimetre, it would be a 2.00 M solution. Let's look at a simple calculation showing how the molarity of a solution is computed.

EXAMPLE 2.25	CALCULATING THE MOLAR CONCENTRATION OF A SOLUTION
PROBLEM	A 2.00-g sample of sodium hydroxide, NaOH (the major ingredient in the drain cleaner Drano®), was dissolved in water to produce a total volume of exactly 200 cm^3 of solution. What is the molar concentration of this sodium hydroxide solution?
SOLUTION	To calculate the molar concentration we must take the ratio of moles of solute to cubic decimetres of solution. Therefore, we first convert the 2.00 g NaOH (the solute) into moles. Since the molar mass of NaOH is 40.0 g mol^{-1},

$$2.00 \text{ g NaOH} \times \left(\frac{1 \text{ mol NaOH}}{40.0 \text{ g NaOH}} \right) = 0.0500 \text{ mol NaOH}$$

The volume of the solution is 200 cm^3, which is the same as 0.200 dm^3. Therefore, the molar concentration of the solution is

$$\text{molar concentration} = \frac{0.0500 \text{ mol NaOH}}{0.200 \text{ dm}^3}$$

$$= 0.250 \text{ mol dm}^{-3}$$

$$= 0.250 \, M$$

Molar concentration is such a useful concentration unit because if we know the molar concentration of a particular solution, we can dispense a desired number of moles of the solute just by measuring out the proper volume. For instance, suppose we had a large container filled with a 0.250 M solution of NaOH. Let's also suppose that in a particular reaction we needed exactly 0.250 mol of NaOH. The label on the bottle tells us that each dm^3 of the solution contains 0.250 mol NaOH, so all we have to do is measure 1.00 dm^3 of the solution and we have the necessary 0.250 mol of NaOH. If we only needed 0.125 mol NaOH, we would only need 0.500 dm^3 of the solution. On the other hand, if we needed 0.500 mol NaOH, we would need 2.00 dm^3 of the solution. The following examples show how we use molar concentration in calculations.

EXAMPLE 2.26

CALCULATING THE VOLUME OF A SOLUTION THAT CONTAINS A SPECIFIC AMOUNT OF SOLUTE

PROBLEM

How many cubic centimetres of 0.250 M NaOH solution are needed to provide 0.0200 mol NaOH?

SOLUTION

In solving problems of this type we use molar concentration as a conversion factor. A label that reads 0.250 M NaOH can be translated into either of two ratios

The first step is to translate the label.

$$\frac{0.250 \text{ mol NaOH}}{1.00 \text{ dm}^3 \text{ solution}} \quad \text{or} \quad \frac{1.00 \text{ dm}^3 \text{ solution}}{0.250 \text{ mol NaOH}}$$

If we want the volume in cubic centimetres, these can be expressed as

$$\frac{0.250 \text{ mol NaOH}}{1000 \text{ cm}^3 \text{ solution}} \quad \text{or} \quad \frac{1000 \text{ cm}^3 \text{ solution}}{0.250 \text{ mol NaOH}}$$

Now, let's restate our problem as

$$0.0200 \text{ mol NaOH} \sim (?) \text{ cm}^3 \text{ NaOH}$$

Conversion between dm^3 and cm^3 is something you should be able to do effortlessly. If you can't, then practice!

The units tell us which conversion factor to use.

$$0.0200 \cancel{\text{ mol NaOH}} \times \left(\frac{1000 \text{ cm}^3 \text{ solution}}{0.250 \cancel{\text{ mol NaOH}}} \right) = 80.0 \text{ cm}^3 \text{ solution}$$

Thus, if we measure out 80.0 cm^3 of this solution, it will contain the desired 0.0200 mol NaOH.

EXAMPLE 2.27

CALCULATING THE AMOUNT OF SOLUTE IN A SOLUTION OF KNOWN MOLAR CONCENTRATION

PROBLEM

How many grams of NaOH are in 50.0 cm^3 of 0.400 M NaOH solution?

SOLUTION

This time we want to find the mass of solute in a portion of the solution. We begin by finding the number of moles of NaOH in the solution; then we convert that to grams. First we have to translate the label.

$$0.400 \ M \ \text{NaOH means} \ \frac{0.400 \ \text{mol NaOH}}{1000 \ \text{cm}^3 \ \text{solution}}$$

Then

$$50.0 \ \text{cm}^3 \ \text{solution} \times \left(\frac{0.400 \ \text{mol NaOH}}{1000 \ \text{cm}^3 \ \text{solution}} \right) = 0.0200 \ \text{mol NaOH}$$

Finally

$$0.0200 \ \text{mol NaOH} \times \left(\frac{40.0 \ \text{g NaOH}}{1 \ \text{mol NaOH}} \right) = 0.800 \ \text{g NaOH}$$

Thus, 50.0 cm^3 of this solution contains 0.800 g NaOH

Sometimes it is necessary to prepare a solution having a specific concentration while you're working in the lab. This is really not very difficult, as the following example shows.

EXAMPLE 2.28 PREPARING A SOLUTION THAT HAS A SPECIFIED MOLAR CONCENTRATION

PROBLEM How many grams of silver nitrate, AgNO$_3$, are needed to prepare 500 cm^3 of a 0.300 M AgNO$_3$ solution?

SOLUTION This kind of problem really asks, "How many grams of AgNO$_3$ are in 500 cm^3 of a 0.300 M AgNO$_3$ solution?" If you know the answer, you would know how much solute to use to make the solution. We therefore proceed as before. First, we translate the label.

$$0.300 \ M \ \text{AgNO}_3 \ \text{means} \ \frac{0.300 \ \text{mol AgNO}_3}{1000 \ \text{cm}^3 \ \text{solution}}$$

Then

$$500 \ \text{cm}^3 \ \text{solution} \times \left(\frac{0.300 \ \text{mol AgNO}_3}{1000 \ \text{cm}^3 \ \text{solution}} \right) = 0.150 \ \text{mol AgNO}_3$$

The molar mass of AgNO$_3$ is 170 g mol^{-1}. Therefore

$$0.150 \ \text{mol AgNO}_3 \times \left(\frac{170 \ \text{g AgNO}_3}{1 \ \text{mol AgNO}_3} \right) = 22.5 \ \text{g AgNO}_3$$

To actually prepare the solution described in Example 2.28, we would have to dissolve the 25.5 g AgNO$_3$ in enough water to give a *final volume* of exactly 500 cm^3. Measuring volumes with this accuracy is accomplished using a volumetric flask (Figure 2.1). The flask contains the specified volume when it is filled to the mark etched around the neck. Figure 2.2 illustrates the sequence of steps involved in preparing the solution.

As a final point, in preparing the solution in the preceding example, the volume of the solution is adjusted to a *final volume* of 500 cm^3. We didn't simply add 500 cm^3 of water to the silver nitrate, because this would give a final volume just a bit larger than 500 cm^3 (both the solute and solvent take up space in the

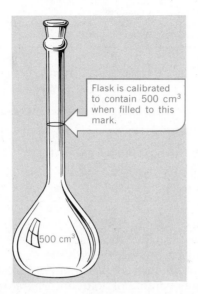

Figure 2.1
A 500-cm^3 volumetric flask.

Flask is calibrated to contain 500 cm^3 when filled to this mark.

500 cm^3

(a)   (b)   (c)   (d)   (e)

Figure 2.2
Preparation of a solution having a particular molar concentration. (a) The solute is accurately weighed into a volumetric flask. (b) Distilled water is added. (c) The flask is stoppered and the solution is swirled to dissolve the solute. (d) Distilled water is added carefully to bring the volume up to the mark etched around the neck of the flask. (e) The flask is stoppered again and then shaken to make the solution uniform.

flask). If we had actually added 500 cm^3 of water, the concentration would have been just a bit less than the desired 0.300 *M* because the solute would be spread over a slightly larger volume than expected.

2.14 PREPARING SOLUTIONS BY DILUTION

In the course of performing routine laboratory operations, it is not uncommon to have to dilute solutions—to make them less concentrated by the addition of solvent. For example, many laboratory chemicals are purchased in the form of

Table 2.2
Concentrated laboratory reagents

Reagent	Density (g cm^{-3})	Mass Percent	Molar Concentration
Sulfuric acid (H_2SO_4)	1.84	96	18
Hydrochloric acid (HCl)	1.18	36	12
Phosphoric acid (H_3PO_4)	1.7	85	15
Nitric acid (HNO_3)	1.42	70	16
Acetic acid ($HC_2H_3O_2$)	1.05	100	17.5
Aqueous ammonia (NH_3)	0.90	28	15

Chemicals stocked in the laboratory are frequently referred to as **reagents.**

concentrated aqueous solutions (Table 2.2) because that is the most economical way to buy them. Usually, these solutions are too concentrated to use, and therefore they must be made more dilute.

The process of dilution simply involves spreading a given amount of solute throughout a larger volume of solution. Therefore, when a concentrated solution is diluted by the addition of solvent, the number of moles of the solute in the mixture remains constant. This is the key to working problems dealing with dilution.

If we multiply a solution's molar concentration, M, by its volume, V, we obtain the number of moles of solute.

$$M \cdot V = \frac{mol}{\cancel{dm^3}} \times \cancel{dm^3} = mol$$

In using the relationship in Equation 2.1, the volumes can be expressed in either cm^3 or dm^3. If cm^3 is used, the product $M \cdot V$ gives *millimoles*, but during dilution the number of millimoles of solute doesn't change either.

Since the number of moles of solute stays the same while the solution is diluted, the product of the initial molar concentration and volume (M_iV_i) must equal the product of the final molar concentration and volume (M_fV_f). This gives us the useful equation

$$M_iV_i = M_fV_f \qquad [2.1]$$

The next three sample problems illustrate how this equation is used.

EXAMPLE 2.29 **WORKING DILUTION PROBLEMS**

PROBLEM How much water must be added to 25.0 cm^3 of 0.500 M KOH solution to produce a solution whose concentration is 0.350 M?

SOLUTION Our equation is

$$M_iV_i = M_fV_f$$

$$M_i = 0.500 \; M \quad M_f = 0.350 \; M$$

$$V_i = 25.0 \; cm^3 \quad V_f = ?$$

Solving for V_f and substituting, we get

$$V_f = \frac{(0.500 \; M)(25.0 \; cm^3)}{0.350 \; M}$$

$$V_f = 35.7 \; cm^3$$

Since the initial volume was 25.0 cm^3, we must *add* 10.7 cm^3. (We are assuming that volumes are additive. When working with dilute solutions, this assumption is generally quite valid.)

EXAMPLE 2.30 WORKING DILUTION PROBLEMS

PROBLEM What volume of concentrated H_2SO_4 (18.0 M) is required to prepare 750 cm^3 of 3.00 M H_2SO_4 solution?

SOLUTION Again

$$M_iV_i = M_fV_f$$

$$M_i = 18.0\ M \quad M_f = 3.00\ M$$

$$V_i = ? \qquad V_f = 750\ cm^3$$

Solving for V_i gives

$$V_i = \frac{M_fV_f}{M_i}$$

$$V_i = \frac{(3.00\ M)(750\ cm^3)}{18.0\ M}$$

$$V_i = 125\ cm^3$$

To prepare the solution, 125 cm^3 of the concentrated H_2SO_4 is diluted to a total final volume of 750 cm^3.

EXAMPLE 2.31 WORKING DILUTION PROBLEMS

PROBLEM Suppose that 200 cm^3 of water were added to 450 cm^3 of a solution labeled 0.600 M HNO_3. What will be the new concentration of the solute in the final solution?

SOLUTION Once again

$$M_iV_i = M_fV_f$$

$$M_i = 0.600\ M \quad M_f = ?$$

$$V_i = 450\ cm^3 \quad V_f = 200\ cm^3 + 450\ cm^3 = 650\ cm^3$$

In using Equation 2.1, volumes can be in any convenient unit. The only requirement is that V_i and V_f have the *same* units.

Solving for M_f gives

$$M_f = \frac{M_iV_i}{V_f}$$

$$= \frac{(0.600\ M)(450\ cm^3)}{(650\ cm^3)}$$

$$= 0.422\ M$$

The concentration of HNO_3 in the final solution is 0.422 M.

A very important safety note applies to Example 2.30. When concentrated chemicals are diluted, a large quantity of heat is sometimes liberated. This is especially true for sulfuric acid. **To absorb this heat safely, you must *always add the concentrated acid to the water,* never the reverse.** If water is added to the concentrated acid, so much heat is liberated that it can cause the water to boil suddenly, spattering the acid. If you happen to be standing in the way, this could definitely ruin your whole day!

2.15   THE STOICHIOMETRY OF REACTIONS IN SOLUTIONS

As you might expect, in order for two chemicals to react with each other, their particles must come in contact. This requirement is the reason that chemists routinely carry out reactions in solution. In a solution, the solute particles (either molecules or ions) are spread uniformly throughout the mixture. When solutions of reactants are combined, this uniformity of mixing allows the particles to intermingle freely, and because so many reactant particles come in contact with each other all at one time, reactions in solution are often quite rapid. For example, if crystals of lead nitrate, $Pb(NO_3)_2$, and sodium chromate, Na_2CrO_4, are mixed, no noticeable chemical changes are observed. This is because the reactants contact each other only where the surfaces of their crystals touch. However, if these two chemicals are first dissolved in water, and then their aqueous solutions are mixed, a yellow solid having the formula $PbCrO_4$ is produced immediately. This reaction is shown in Figure 2.3. The insoluble product of the reaction, $PbCrO_4$, is called lead chromate and is used as a pigment in oil and water colors.

A solid substance that is formed in a solution is called a **precipitate.** Such substances are said to be insoluble.

In Chapter 8, we will explore in detail exactly what takes place in a solution when substances such as $Pb(NO_3)_2$ and Na_2CrO_4 react. For now, our purpose is to focus on some typical problems that deal with the stoichiometry of such reactions.

The quantitative relationships that apply to reactions in solution are the same as those for reactions that occur anywhere else. The coefficients in the balanced equation provide the mole ratios needed to solve stoichiometry problems. The differences, when they exist, are in the *laboratory units* that are used to measure amounts of reactants.

Consider the equation for the reaction pictured in Figure 2.3.

$$Pb(NO_3)_2(aq) + Na_2CrO_4(aq) \longrightarrow PbCrO_4(s) + 2NaNO_3(aq)$$

In Section 2.10, you learned to calculate the quantities of reactants needed for a given reaction by converting moles to grams. We can take the same approach to reactions in solution, too. In this case, the stoichiometry problems are no different than those we have done earlier. For example, if we wish to allow 0.100 mol of $Pb(NO_3)_2$ to react with 0.100 mol of Na_2CrO_4, the first step is to convert these amounts to grams using the appropriate molar masses. The molar mass of $Pb(NO_3)_2$ is 331 g mol^{-1} and the molar mass of Na_2CrO_4 is 162 g mol^{-1}, so 0.100 mol of $Pb(NO_3)_2$ weighs 33.1 g and 0.100 mol of Na_2CrO_4 weighs 16.2 g. To carry out the reaction, we would dissolve these quantities of the reactants in water and then combine the solutions to obtain the desired products.

Figure 2.3
(a) *Beakers containing solutions of* $Pb(NO_3)_2$ *and* Na_2CrO_4.
(b) *The solution of* Na_2CrO_4 *being poured into the solution of* $Pb(NO_3)_2$. *The reaction produces a precipitate of* $PbCrO_4$.

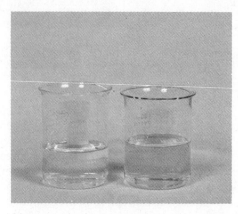

(a)

(b)

When reactions take place in solution, however, we have another way of measuring amounts of reactants—by measuring volumes of solutions of known concentration. The following examples illustrate how useful molar concentration is in solving problems that deal with reactions in solution.

EXAMPLE 2.32 USING MOLAR CONCENTRATION IN STOICHIOMETRY PROBLEMS

PROBLEM Aluminum hydroxide, $Al(OH)_3$, one of the antacid ingredients in Maalox, can be prepared by the reaction of aluminum sulfate, $Al_2(SO_4)_3$, and sodium hydroxide, NaOH. The balanced chemical equation for the reaction is

$$Al_2(SO_4)_3(aq) + 6NaOH(aq) \longrightarrow 2Al(OH)_3(s) + 3Na_2SO_4(aq)$$

What volume of 0.200 M NaOH solution is needed to completely react with 3.50 g $Al_2(SO_4)_3$?

SOLUTION As in any stoichiometry problem, the amounts of the reactants (in moles) are related to each other by the coefficients in the balanced equation. The first step, therefore, is to calculate the number of moles of $Al_2(SO_4)_3$ (molar mass = 342.2 g mol^{-1}).

$$3.50 \text{ g } Al_2(SO_4)_3 \times \left(\frac{1 \text{ mol } Al_2(SO_4)_3}{342.2 \text{ g } Al_2(SO_4)_3} \right) = 1.02 \times 10^{-2} \text{ mol } Al_2(SO_4)_3$$

Next, we use the coefficients to establish the necessary mole ratio.

$$1.02 \times 10^{-2} \text{ mol } Al_2(SO_4)_3 \times \left(\frac{6 \text{ mol NaOH}}{1 \text{ mol } Al_2(SO_4)_3} \right) = 6.12 \times 10^{-2} \text{ mol NaOH}$$

The final step is to calculate the volume of the NaOH solution that contains this number of moles. Translating the molar concentration yields the two conversion factors

$$\frac{0.200 \text{ mol NaOH}}{1000 \text{ cm}^3 \text{ solution}} \quad \text{and} \quad \frac{1000 \text{ cm}^3 \text{ solution}}{0.200 \text{ mol NaOH}}$$

It is the second one that we need to make the units cancel correctly.

$$6.12 \times 10^{-2} \text{ mol NaOH} \times \left(\frac{1000 \text{ cm}^3 \text{ solution}}{0.200 \text{ mol NaOH}} \right) = 306 \text{ cm}^3 \text{ solution}$$

To carry out the reaction, we would dissolve the 3.50 g of $Al_2(SO_4)_3$ in water and then add 306 cm^3 of the NaOH solution to give complete reaction.

EXAMPLE 2.33 USING MOLAR CONCENTRATION IN STOICHIOMETRY PROBLEMS

PROBLEM Chalk is composed of calcium carbonate, $CaCO_3$. This water-insoluble compound is formed when a solution of calcium chloride, $CaCl_2$, is added to a solution of sodium carbonate, Na_2CO_3. The reaction is

$$CaCl_2(aq) + Na_2CO_3(aq) \longrightarrow CaCO_3(s) + 2NaCl(aq)$$

What volume of 0.250 M $CaCl_2$ is needed to react completely with 50.0 cm^3 of 0.150 M Na_2CO_3 solution?

SOLUTION From the volume and molar concentration of the Na_2CO_3 solution, we can calculate the number of moles of Na_2CO_3 that are available. A concentration of 0.150 M Na_2CO_3 means

$$\frac{0.150 \text{ mol } Na_2CO_3}{1000 \text{ cm}^3 \text{ solution}}$$

Therefore

$$50.0 \text{ cm}^3 \text{ solution} \times \left(\frac{0.150 \text{ mol Na}_2\text{CO}_3}{1000 \text{ cm}^3 \text{ solution}} \right) = 7.50 \times 10^{-3} \text{ mol Na}_2\text{CO}_3 \ .$$

Since the coefficients of $CaCl_2$ and Na_2CO_3 are the same in the equation, the amount of $CaCl_2$ needed will also be 7.50×10^{-3} mol. The final step, therefore, is to calculate the volume of the $CaCl_2$ solution that contains this number of moles of solute. The concentration gives us these conversion factors

$$\frac{0.250 \text{ mol CaCl}_2}{1000 \text{ cm}^3 \text{ solution}} \quad \text{and} \quad \frac{1000 \text{ cm}^3 \text{ solution}}{0.250 \text{ mol CaCl}_2}$$

The second one gives us

$$7.50 \times 10^{-3} \text{ mol CaCl}_2 \times \left(\frac{1000 \text{ cm}^3 \text{ solution}}{0.250 \text{ mol CaCl}_2} \right) = 30.0 \text{ cm}^3 \text{ solution}$$

The required volume of this $CaCl_2$ solution is 30.0 cm^3.

EXAMPLE 2.34 LIMITING REACTANTS IN SOLUTION REACTIONS

PROBLEM Silver bromide, AgBr, is the light sensitive chemical in photographic film. This insoluble compound forms when an aqueous solution of silver nitrate, $AgNO_3$, is mixed with an aqueous solution of calcium bromide, $CaBr_2$.

$$2AgNO_3(aq) + CaBr_2(aq) \longrightarrow 2AgBr(s) + Ca(NO_3)_2(aq)$$

How many grams of solid AgBr will be formed if 50.0 cm^3 of $0.180 \ M$ $AgNO_3$ are mixed with 60.0 cm^3 of $0.0850 \ M$ $CaBr_2$?

SOLUTION When you're given both the volume and the molar concentration of a solution, you're actually given the number of moles of solute. That is because the product of molar concentration and volume (in dm^3) yields moles. Since we are given the molar concentration and volume for *both* reactants, we effectively have the number of moles of each. In Section 2.11 you learned that in such cases it is necessary to determine the limiting reactant, and then calculate the amount of product based on the amount of the limiting reactant that is available.

molar concentration ×
volume (dm^3) = moles

$\left(\dfrac{\text{moles}}{dm^3} \right) \times dm^3 = \text{moles}$

Now that we know the strategy we have to use, let's solve the problem. First, we calculate the number of moles of each reactant in their respective solutions.

Recognizing the problem as a limiting reactant type is the first step in solving it correctly.

For $AgNO_3$

$$50.0 \text{ cm}^3 \text{ solution} \times \left(\frac{0.180 \text{ mol AgNO}_3}{1000 \text{ cm}^3 \text{ solution}} \right) = 9.00 \times 10^{-3} \text{ mol AgNO}_3$$

For $CaBr_2$

$$60.0 \text{ cm}^3 \text{ solution} \times \left(\frac{0.0850 \text{ mol CaBr}_2}{1000 \text{ cm}^3 \text{ solution}} \right) = 5.10 \times 10^{-3} \text{ mol CaBr}_2$$

Next, we find out which is the limiting reactant. Let's determine how many moles of $CaBr_2$ are needed to react with all the $AgNO_3$.

$$9.00 \times 10^{-3} \text{ mol AgNO}_3 \times \left(\frac{1 \text{ mol CaBr}_2}{2 \text{ mol AgNO}_3} \right) = 4.50 \times 10^{-3} \text{ mol CaBr}_2 \text{ needed}$$

Notice that the amount of $CaBr_2$ available (5.10×10^{-3} mol) is larger than the amount needed. This means $CaBr_2$ will be left over, so $AgNO_3$ is the limiting reactant.

Finally, we calculate the amount of AgBr formed from the available $AgNO_3$.

$$9.00 \times 10^{-3} \text{ mol AgNO}_3 \times \left(\frac{2 \text{ mol AgBr}}{2 \text{ mol AgNO}_3} \right) = 9.00 \times 10^{-3} \text{ mol AgBr}$$

The molar mass of AgBr is 187.8 g mol^{-1}, so

$$9.00 \times 10^{-3} \text{ mol AgBr} \times \left(\frac{187.8 \text{ g AgBr}}{1 \text{ mol AgBr}} \right) = 1.69 \text{ g AgBr}$$

The mass of AgBr formed by mixing these two solutions is 1.69 g.

REVIEW QUESTIONS AND PROBLEMS

(Problems whose numbers are in blue have their answers in Appendix C at the back of the book; the more difficult problems are marked with asterisks.)

The Mole and Simple Mole Relationships

2.1 In what sense are the mole, the dozen, and the gross related to each other? How many things are in a mole?

2.2 What does *1 mol of carbon* mean?

2.3 What is the mole ratio of sulfur to oxygen in the air pollutant SO_2 (sulfur dioxide)?

2.4 In baking soda, $NaHCO_3$, what is the mole ratio of
(a) sodium to hydrogen
(b) sodium to carbon
(c) sodium to oxygen
(d) hydrogen to carbon
(e) hydrogen to oxygen
(f) carbon to oxygen

2.5 Butane, the volatile liquid fuel in disposable cigarette lighters, has the formula C_4H_{10}.
(a) What is the mole ratio of C to H in this compound?
(b) How many moles of C are in 3.00 mol C_4H_{10}?
(c) How many moles of H are in 0.250 mol C_4H_{10}?
(d) How many moles of C must combine with 0.600 mol H to make this compound?
(e) Hydrogen occurs in nature as *diatomic* molecules, H_2. How many moles of H_2 have enough H atoms to react with 0.600 mol C to give C_4H_{10}?

2.6 Iron pyrite, FeS_2, forms beautiful golden crystals that are known as "fool's gold."
(a) How many moles of sulfur would be needed to combine with 1.00 mol Fe to form FeS_2?
(b) How many moles of iron are needed to combine with 1.44 mol S to form FeS_2?
(c) How many moles of sulfur are in 3.00 mol FeS_2?
(d) How many moles of FeS_2 are needed to give 3.00 mol Fe?

2.7 Ordinary sand is composed chiefly of silica, a compound in which there are two oxygen atoms for each silicon atom.
(a) What is the formula for silica?

(b) How many atoms of oxygen would be needed to combine with 25 atoms of silicon to form silica?
(c) How many moles of oxygen atoms would be needed to combine with 25 mol of silicon atoms to form silica?
(d) If you had 4.50 mol silica, how many moles of silicon and oxygen atoms would there be in it?

2.8 How many moles of S are in 1.00 mol of As_2S_3?

2.9 How many moles of O are in 1.50 mol of Cr_2O_3?

2.10 Based only on the amount of carbon available, how many moles of CO_2 could be liberated from 1.00 mol of limestone, $CaCO_3$?

2.11 How many moles of $BaSO_4$ could be made from 1.25 mol of $Al_2(SO_4)_3$?

Measuring Moles of Atoms and Compounds

2.12 Give the mass of 1.00 mol of atoms of each of the following elements:
(a) magnesium
(b) carbon
(c) iron
(d) chlorine
(e) sulfur
(f) strontium

2.13 How many moles of atoms are in 50.0 g of each of the following elements?
(a) sodium
(b) arsenic
(c) chromium
(d) aluminum
(e) potassium
(f) silver

2.14 Why is the term formula weight preferred for some substances, rather than molecular weight?

2.15 Give the molar mass of the following:
(a) MgO
(b) $CaCl_2$
(c) PCl_5
(d) S_2Cl_2
(e) Na_3PO_4

2.16 Give the molar mass of the following:
(a) SiO_2 (quartz)
(b) $Mg(OH)_2$ (in milk of magnesia)
(c) $MgSO_4 \cdot 7H_2O$ (epsom salts)
(d) $Ca_2Mg_5(Si_4O_{11})_2(OH)_2$ (asbestos)
(e) $C_6H_8O_6$ (vitamin C)
(f) $C_{12}H_{22}O_{11}$ (sucrose—cane sugar)

2.17 What is the mass of 1.35 mol of caffeine, $C_8H_{10}N_4O_2$?

2.18 What is the mass of 2.33 mol of penicillin, $C_{16}H_{18}O_4N_2S$?

2.19 What is the mass of 6.30 mol of lead sulfate, $PbSO_4$?

2.20 What is the mass of 0.144 mol of TiO_2, a pigment used in white paint?

2.21 How many moles of sodium bicarbonate, $NaHCO_3$ (baking soda), are in a 242-g sample?

2.22 How many moles of butane, C_4H_{10}, are in 1.40×10^3 g of butane?

2.23 How many moles of sulfuric acid, H_2SO_4, are in 85.3 g of H_2SO_4?

2.24 A substance with the formula $PbHAsO_4$ has been used in insecticides, and on the label it appears under the name lead arsenate. How many moles of this poison are in 25.0 g?

2.25 How many moles of potassium are in 125 g of KCl?

2.26 How many moles of S are in 632 g of iron pyrite, FeS_2?

2.27 When coal containing iron pyrite, FeS_2, is burned, all the sulfur is converted to the air pollutant, sulfur dioxide, SO_2. How many moles of FeS_2 would have to react to produce 1.00 kg of SO_2?

Using Avogadro's Number

2.28 What is the mass, in grams, of one atom of Fe? What is the mass of one molecule of SO_2?

2.29 If a sample of ethylene, C_2H_4, contains 3.50×10^{17} atoms of carbon, how many atoms of hydrogen does it contain? How many grams of carbon and hydrogen are in the sample?

2.30 Ordinary table sugar is sucrose, $C_{12}H_{22}O_{11}$. What is the mass of one molecule of sucrose? How many times more massive is a molecule of sucrose than one atom of carbon? How many molecules of sucrose are in 25.0 g of sucrose? What is the total number of atoms in 25.0 g of sucrose?

2.31 How many atoms of carbon are in 4.00×10^{-8} g of propane, C_3H_8?

2.32 Carbon atoms have a diameter of approximately 1.5×10^{-8} cm. If carbon atoms were laid in a row 3.0 cm long, what would be the total mass of carbon?

2.33 Calculate the mass of Cu required to react with 5.00×10^{20} molecules of S_8 to form Cu_2S.

Percentage Composition

2.34 Calculate the percentage composition of each of the following:
(a) $FeCl_3$ (d) $(NH_4)_2HPO_4$
(b) Na_3PO_4 (e) Hg_2Cl_2
(c) $KHSO_4$

2.35 Calculate the percentage composition of each of the following:
(a) (benzene) C_6H_6
(b) (ethyl alcohol) C_2H_5OH
(c) (potassium dichromate) $K_2Cr_2O_7$
(d) (xenon tetrafluoride) XeF_4
(e) (calcium carbonate) $CaCO_3$

2.36 Calculate the mass of nitrogen in 30.0 g of the amino acid, glycine, CH_2NH_2COOH.

2.37 Calculate the mass of hydrogen in 12.0 g of NH_3.

2.38 A 12.5-g sample of a compound containing only phosphorus and sulfur was analyzed and found to contain 7.04 g of phosphorus and 5.46 g of sulfur. What is the percentage composition of this compound?

2.39 A 4.25-g sample of a compound that contains only carbon, hydrogen, and oxygen was burned in an atmosphere of pure O_2. This produced 9.34 g of CO_2 and 5.09 g of H_2O. (a) Calculate the masses of C and H that were in the original 4.25-g sample. (b) What was the mass of O in the original sample? (c) What is the percentage composition of this compound?

Chemical Formulas

2.40 State the difference between a structural formula, a molecular formula, and an empirical formula?

2.41 What is the empirical formula of each of the following?
(a) $(NH_4)_2S_2O_8$ (e) $C_3H_8O_3$
(b) Fe_2O_3 (f) $C_6H_{12}O_6$
(c) Al_2Cl_6 (g) Hg_2SO_4
(d) C_6H_6

2.42 Ethylene glycol, used as permanent antifreeze, has the structural formula

$$
\begin{array}{ccccc}
 & H & H & & \\
 & | & | & & \\
H-O-&C&-C&-O-H \\
 & | & | & & \\
 & H & H & &
\end{array}
$$

What is its molecular formula and empirical formula?

2.43 Why is the simplest formula for a substance called an empirical formula? What is the dictionary definition of empirical?

2.44 A sample of an air pollutant composed of sulfur and oxygen was found to contain 1.40 g sulfur and 2.10 g oxygen. What is the empirical formula of the compound?

2.45 The freon propellant from an aerosol can was analyzed. A sample of it contained 0.423 g C, 2.50 g Cl, and 1.34 g F. What is the empirical formula of this substance?

2.46 What is the empirical formula of the compound described in Problem 2.38?

2.47 What is the empirical formula of the compound described in Problem 2.39?

2.48 A dry-cleaning fluid composed of carbon and chlorine was found to have the composition: 14.5% C, 85.5% Cl (by mass). What is the empirical formula of this compound?

2.49 Arsenic reacts with oxygen to form a compound that is 75.7% arsenic and 24.3% oxygen, by mass. What is the empirical formula of this compound?

2.50 A 1.31-g sample of sulfur was allowed to react with an excess of chlorine to produce 4.22 g of a product that contains only sulfur and chlorine. What is the empirical formula of the compound?

2.51 A substance was found to be composed of 60.8% sodium, 28.5% boron, and 10.5% hydrogen. What is the empirical formula of the compound?

2.52 Vanillin is composed of carbon, 63.2%; hydrogen, 5.26%; and oxygen, 31.6%. What is the empirical formula of vanillin?

2.53 A 0.537-g sample of an organic compound containing only carbon, hydrogen, and oxygen was burned in air to produce 1.030 g of CO_2 and 0.632 g H_2O. What is the empirical formula of the compound?

*2.54 A 1.35-g sample of a substance containing carbon, hydrogen, nitrogen, and oxygen was burned to produce 0.810 g H_2O and 1.32 g CO_2. In a separate reaction, all the nitrogen in 0.735 g of the substance was converted to ammonia. This gave 0.284 g of NH_3. Determine the empirical formula of the substance.

*2.55 An organic compound was synthesized and a sample of it was analyzed and found to contain only C, H, N, O, and Cl. It was observed that when a 0.150-g sample of the compound was burned, it produced 0.138 g of CO_2 and 0.0566 g of H_2O. All the nitrogen in a different 0.200-g sample of the compound was converted to NH_3, which had a mass of 0.0238 g. Finally, the chlorine in a 0.125-g sample of the compound was converted to Cl^- and by reacting it with $AgNO_3$, all the chlorine was recovered as AgCl. The AgCl, when dried, had a mass of 0.251 g.

(a) Calculate the mass percent of each element in the compound.
(b) Determine the empirical formula for the compound.

2.56 The following are empirical formulas and molar masses for five compounds. What are their molecular formulas?
(a) NaS_2O_3,   270.4 g mol^{-1}
(b) C_3H_2Cl,   147.0 g mol^{-1}
(c) C_2HCl,   181.4 g mol^{-1}
(d) Na_2SiO_3,   732.6 g mol^{-1}
(e) $NaPO_3$,   305.9 g mol^{-1}

*2.57 Citric acid, the substance that makes lemon juice sour, is composed of only carbon, hydrogen, and oxygen. When a 0.5000-g sample of citric acid was burned, it produced 0.6871 g of CO_2 and 0.1874 g of H_2O. The molecular weight of the compound is 192. What are the empirical and the molecular formulas for citric acid?

2.58 Polystyrene, a common plastic, is composed of many styrene units linked together as shown below.

The basic styrene unit is shown within brackets. The subscript n means that this unit is repeated many times. A particular sample of this plastic was found to have an average molecular weight of 1 million. What is the average number of styrene units in a chain?

Balancing Equations

2.59 What important chemical law is obeyed by a balanced chemical equation?

2.60 Balance the following equations by inspection:
(a) $ZnS + HCl \rightarrow ZnCl_2 + H_2S$
(b) $HCl + Cr \rightarrow CrCl_2 + H_2$
(c) $Al + Fe_3O_4 \rightarrow Al_2O_3 + Fe$
(d) $H_2 + Br_2 \rightarrow HBr$
(e) $Na_2S_2O_3 + I_2 \rightarrow NaI + Na_2S_4O_6$

2.61 Balance the following equations:
(a) $LaCl_3 + Na_2CO_3 \rightarrow La_2(CO_3)_3 + NaCl$
(b) $NH_4Cl + Ba(OH)_2 \rightarrow BaCl_2 + NH_3 + H_2O$
(c) $Ca(OH)_2 + H_3PO_4 \rightarrow Ca_3(PO_4)_2 + H_2O$
(d) $La_2(CO_3)_3 + H_2SO_4 \rightarrow La_2(SO_4)_3 + H_2O + CO_2$
(e) $Na_2O + (NH_4)_2SO_4 \rightarrow Na_2SO_4 + H_2O + NH_3$

2.62 Balance the following equations:
(a) $C_4H_{10} + O_2 \rightarrow CO_2 + H_2O$
(b) $C_7H_6O_2 + O_2 \rightarrow CO_2 + H_2O$
(c) $P_4O_{10} + H_2O \rightarrow H_3PO_4$
(d) $FeS_2 + O_2 \rightarrow Fe_2O_3 + SO_2$
(e) $NH_3 + O_2 \rightarrow NO + H_2O$

2.63 Balance the following equations:
(a) $Fe + HCl \rightarrow H_2 + FeCl_2$
(b) $PbO_2 + HCl \rightarrow H_2O + PbCl_2 + Cl_2$
(c) $Fe_2O_3 + H_2SO_4 \rightarrow Fe_2(SO_4)_3 + H_2O$
(d) $NO_2 + H_2O \rightarrow NO + HNO_3$
(e) $C_2H_6S + O_2 \rightarrow CO_2 + H_2O + SO_2$

Calculations Based on Chemical Equations

2.64 Acetylene, which is used as a fuel in welding torches, is produced in a reaction between calcium carbide and water

$$CaC_2 + 2H_2O \longrightarrow Ca(OH)_2 + C_2H_2 \ (g)$$
calcium   **acetylene**
carbide

(a) How many moles of C_2H_2 would be produced from 2.50 mol of CaC_2?
(b) How many grams of C_2H_2 would be formed from 0.500 mol of CaC_2?
(c) How many moles of water would be consumed when 3.20 mol of C_2H_2 are formed?
(d) How many grams of $Ca(OH)_2$ are produced when 28.0 g of C_2H_2 are formed?

2.65 Consider the following balanced equation:

$$6ClO_2 + 3H_2O \longrightarrow 5HClO_3 + HCl$$

(a) How many moles of $HClO_3$ are produced from 14.3 g of ClO_2?
(b) How many grams of H_2O are needed to produce 5.74 g of HCl?
(c) How many grams of $HClO_3$ are produced when 4.25 g of ClO_2 are added to 0.853 g of H_2O?

2.66 White phosphorus, composed of P_4 molecules, is used in military incendiary devices because it ignites spontaneously when exposed to air. The product of reaction with oxygen is P_4O_{10}.
(a) Write a balanced chemical equation for the reaction of P_4 with air. (Remember that oxygen in the air is present as O_2 molecules.)

(b) How many moles of P_4O_{10} can be produced using 0.500 mol O_2?
(c) How many grams of P_4 are needed to produce 50.0 g P_4O_{10}?
(d) How many grams of P_4 will react with 25.0 g O_2?

2.67 Hydrazine, N_2H_4, and hydrogen peroxide, H_2O_2, have been used as rocket propellants. They react according to the equation

$$7H_2O_2 + N_2H_4 \longrightarrow 2HNO_3 + 8H_2O$$

(a) How many moles of HNO_3 are formed from 0.0250 mol N_2H_4?
(b) How many moles of H_2O_2 are required if 1.35 mol H_2O are to be produced?
(c) How many moles of H_2O are formed if 1.87 mol HNO_3 are produced?
(d) How many moles of H_2O_2 are required to react with 22.0 g N_2H_4?
(e) How many grams of H_2O_2 are needed to produce 45.8 g HNO_3?

2.68 When iron is produced from its ore, Fe_2O_3, the reaction is

$$Fe_2O_3 + 3CO \longrightarrow 2Fe + 3CO_2$$

(a) How many moles of CO are needed to produce 35.0 mol Fe?
(b) How many moles of Fe_2O_3 react if 4.50 mol CO_2 are formed?
(c) How many grams of Fe_2O_3 must react to give 0.570 mol Fe?
(d) How many moles of CO are needed to react with 48.5 g Fe_2O_3?
(e) How many grams of Fe are formed when 18.6 g CO react?

2.69 During the naval battles of the South Pacific in World War II, the U.S. Navy produced smokescreens by spraying titanium tetrachloride into the moist air where it reacted according to the equation

$$TiCl_4 + 2H_2O \longrightarrow TiO_2 + 4HCl$$

The dense smoke was caused by the TiO_2.
(a) How many moles of H_2O are needed to react with 6.50 mol $TiCl_4$?
(b) How many moles of HCl are formed when 8.44 mol $TiCl_4$ react?
(c) How many grams of TiO_2 are formed from 14.4 mol $TiCl_4$?
(d) How many grams of HCl are formed when 85.0 g of $TiCl_4$ react?

2.70 The insecticide DDT (which ecologists now recognize as a serious environmental pollutant) is manufactured in a reaction between chlorobenzene and chloral

$$2C_6H_5Cl + C_2HCl_3O \longrightarrow C_{14}H_9Cl_5 + H_2O$$
chlorobenzene   **chloral**   **DDT**

How many kilograms of DDT can be produced from 1000 kg of chlorobenzene?

2.71 Aspirin (which many students take after working on chemistry problems) is prepared by the reaction of salicylic acid ($C_7H_6O_3$) with acetic anhydride ($C_4H_6O_3$) according to the reaction

$$C_7H_6O_3 + C_4H_6O_3 \longrightarrow C_9H_8O_4 + C_2H_4O_2$$
(aspirin)

How many grams of salicylic acid must be used to prepare two aspirin tablets, each containing 325 mg aspirin?

2.72 Dimethylhydrazine, $(CH_3)_2NNH_2$, has been used as a fuel in the Apollo lunar descent module, with liquid N_2O_4 as the oxidizer. The products of the reaction between these two in the rocket engine are H_2O, CO_2, and N_2.
(a) Write a balanced chemical equation for the reaction.
(b) Calculate the number of kilograms of N_2O_4 required to burn 50.0 kg of dimethylhydrazine.

2.73 The fermentation of sugar to produce ethyl alcohol follows the equation

$$C_6H_{12}O_6 \xrightarrow{\text{yeasts}} 2C_2H_5OH + 2CO_2$$

What is the maximum mass of alcohol that can be obtained from 500 g of sugar?

*2.74 White lead, a pigment used in lead-based paints, is manufactured by the reactions

$$2Pb + 2HC_2H_3O_2 + O_2 \longrightarrow 2Pb(OH)C_2H_3O_2$$
$$6Pb(OH)C_2H_3O_2 + 2CO_2 \longrightarrow$$
$$Pb_3(OH)_2(CO_3)_2 + 2H_2O + 3Pb(C_2H_3O_2)_2$$
white lead

(a) Starting with 20.0 g of Pb, how many grams of white lead can be prepared?
(b) How many grams of CO_2 will be required if 14.0 g of O_2 are consumed in the first reaction? Assume that all the Pb in the first reaction is completely converted to the products of the second reaction.

2.75 Phosphate rock, $Ca_3(PO_4)_2$, is treated with sulfuric acid to produce phosphate fertilizer

$$Ca_3(PO_4)_2 + 2H_2SO_4 + 4H_2O \longrightarrow$$
$$Ca(H_2PO_4)_2 + 2CaSO_4 \cdot 2H_2O$$
phosphate fertilizer

How many kilograms of sulfuric acid are required to react with 25.0 kg of phosphate rock?

Limiting Reactant Calculations

2.76 What is meant by the term *limiting reactant*? In words, describe how the limiting reactant is identified.

2.77 Consider the reaction

$$Fe(s) + 2HCl(aq) \longrightarrow FeCl_2(aq) + H_2(g)$$

In an experiment, 0.40 mol Fe and 0.75 mol HCl are combined.
(a) Which is the limiting reactant?
(b) How many moles of H_2 will be formed?
(c) How many moles of the reactant in excess will remain after the reaction has stopped?

2.78 Aluminum (Al) reacts with sulfuric acid (H_2SO_4), which is the acid in automobile batteries, according to the equation

$$2Al + 3H_2SO_4 \longrightarrow Al_2(SO_4)_3 + 3H_2$$

If 20.0 g Al are put into a solution containing 115 g of H_2SO_4,
(a) which is the limiting reactant?
(b) how many moles of H_2 will be formed?
(c) how many grams of $Al_2(SO_4)_3$ will be formed?
(d) how many grams of the reactant in excess will be left over?

2.79 Under appropriate conditions, acetylene, C_2H_2, and HCl react to form vinyl chloride, C_2H_3Cl. This substance is used to manufacture polyvinyl chloride (PVC) plastics and has been shown to be carcinogenic. The equation for the reaction is

$$C_2H_2 + HCl \longrightarrow C_2H_3Cl$$

In a given instance, 35.0 g of C_2H_2 are mixed with 51.0 g of HCl.
(a) Which is the limiting reactant?
(b) How many grams of C_2H_3Cl are formed?
(c) How many grams of the reactant in excess remain after the reaction is completed?

*2.80 Freon-12, a gas used as a refrigerant, is prepared by the reaction

$$3CCl_4 + 2SbF_3 \longrightarrow 3CCl_2F_2 + 2SbCl_3$$
Freon-12

If 150 g of CCl_4 are mixed with 100 g of SbF_3,
(a) how many grams of CCl_2F_2 can be formed?
(b) how many grams of which reactant will remain after reaction has ceased?

*2.81 Silver tarnishes in the presence of hydrogen sulfide (rotten egg odor) and oxygen because of the reaction

$$4Ag + \underset{\substack{\text{hydrogen} \\ \text{sulfide}}}{2H_2S} + O_2 \longrightarrow \underset{\substack{\text{(silver} \\ \text{sulfide,} \\ \text{black)}}}{2Ag_2S} + 2H_2O$$

How much Ag_2S could be obtained from a mixture of 0.950 g Ag, 0.140 g H_2S, and 0.0800 g O_2?

2.82 Phosgene, $COCl_2$, was once used as a war gas. It is poisonous because when it is inhaled, it reacts with water in the lungs to produce hydrochloric acid, HCl, which causes severe lung damage, leading ultimately to death. The chemical reaction is

$$COCl_2 + H_2O \longrightarrow CO_2 + 2HCl$$

(a) How many moles of HCl are produced by the reaction of 0.430 mol of $COCl_2$?
(b) How many grams of HCl are produced when 11.0 g of CO_2 are formed?
(c) How many moles of HCl will be formed if 0.200 mol of $COCl_2$ are mixed with 0.400 mol of H_2O?

*2.83 Acetylene, C_2H_2, can react with two molecules of Br_2 to form $C_2H_2Br_4$ by the series of reactions

$$C_2H_2 + Br_2 \longrightarrow C_2H_2Br_2$$

$$C_2H_2Br_2 + Br_2 \longrightarrow C_2H_2Br_4$$

If 5.00 g of C_2H_2 are mixed with 40.0 g of Br_2, what masses of $C_2H_2Br_2$ and $C_2H_2Br_4$ will be formed? Assume that all the C_2H_2 has reacted.

Theoretical Yield and Percentage Yield

2.84 What is meant by the *theoretical yield* for a given reaction mixture? What is the meaning of *percentage yield*? When you carry out a reaction, how do you determine the *actual yield*?

2.85 In an experiment, a student allowed benzene, C_6H_6, to react with excess bromine, Br_2, in an attempt to prepare bromobenzene, C_6H_5Br. This reaction also produced, as a by-product, dibromobenzene, $C_6H_4Br_2$. On the basis of the equation

$$C_6H_6 + Br_2 \longrightarrow C_6H_5Br + HBr$$

(a) What is the maximum quantity of C_6H_5Br that the student could have hoped to obtain from 15.0 g of benzene? (This is the theoretical yield.)
(b) In this experiment, the student obtained 2.50 g of $C_6H_4Br_2$. How much C_6H_6 was *not* converted to C_6H_5Br?
(c) What was the student's actual yield of C_6H_5Br?
(d) Calculate the percentage yield for this reaction.

*2.86 In a reaction between methane, CH_4, and chlorine, Cl_2, four products can be formed: CH_3Cl, CH_2Cl_2, $CHCl_3$, and CCl_4. In a particular instance, 20.8 g of CH_4 were allowed to react with excess Cl_2 and gave 5.0 g CH_3Cl, 25.5 g CH_2Cl_2, and 59.0 g $CHCl_3$. All the CH_4 reacted.
(a) How many grams of CCl_4 were formed?
(b) On the basis of available CH_4, what is the theoretical yield of CCl_4?
(c) What is the percentage yield of CCl_4?

(d) How many grams of Cl_2 reacted with the CH_4? (*Note:* The hydrogen that is displaced from the carbon also combines with Cl_2 to form HCl.)

*2.87 A chemist wishes to synthesize a certain compound that has a molecular weight of 100. The synthesis requires six consecutive steps, each giving a 50% yield (computed on a *mole* basis). If the chemist begins with 30.0 g of starting material having a molecular weight of 80.0, how many grams of final product will be obtained? How many grams of starting material will be required to produce 10.0 g of final product?

Solutions and Molar Concentration

2.88 Define the following terms:
(a) solution
(b) solvent
(c) solute
(d) concentration
(e) dilute
(f) concentrated

2.89 Define molar concentration.

2.90 To make 1.00 dm^3 of a 1.00 M solution of the sugar glucose, $C_6H_{12}O_6$, requires 180 g $C_6H_{12}O_6$. Describe how you would actually go about preparing this solution.

2.91 Trisodium phosphate, Na_3PO_4, is a very powerful, but caustic, cleaning agent. If a bottle containing a solution of Na_3PO_4 were labeled 0.20 M Na_3PO_4, how could you use that information in the form of a conversion factor?

2.92 Calculate the molar concentration of the solute in the following solutions:
(a) 0.250 mol NaCl in 0.400 dm^3 of solution
(b) 1.45 mol sucrose in 345 cm^3 of solution
(c) 195 g H_2SO_4 in 875 cm^3 of solution
(d) 80.0 g KOH in 0.200 dm^3 of solution

2.93 Calculate the molar concentration (molarity) of the following solutions:
(a) 1.35 mol NH_4Cl in a total volume of 2.45 dm^3
(b) 0.422 mol $AgNO_3$ in a total volume of 742 cm^3
(c) 3.00×10^{-3} mol KCl in 10.0 cm^3 of solution
(d) 4.80×10^{-2} g $NaHCO_3$ in 25.0 cm^3 of solution

2.94 A millimole (mmol) is 10^{-3} mol. Suppose that each cubic centimetre of a solution contained 0.250 mmol of NaCl, that is, its concentration is 0.250 mmol cm^{-3}.
(a) Calculate the concentration of this solution in units of mol dm^{-3}.

(b) What is the molar concentration of this solution?

(c) In what way are the numerical values of concentration related when the concentration is reported in mol dm^{-3}, and mmol cm^{-3}?

2.95 Suppose a solution of lithium carbonate, Li_2CO_3, a drug used to treat manic depressives, is labeled 0.150 M.

(a) How many moles of Li_2CO_3 are present in 250 cm^3 of this solution?

(b) How many grams of Li_2CO_3 are in 630 cm^3 of the solution?

(c) How many cubic centimetres of the solution would be needed to supply 0.0100 mol Li_2CO_3?

(d) How many cubic centimetres of the solution would be needed to provide 0.0800 g of Li_2CO_3?

2.96 A solution was labeled 0.375 M KOH.

(a) How many cubic centimetres of this solution are needed to give 0.100 mole KOH?

(b) How many moles of KOH are in 45.0 cm^3 of this solution?

(c) How many cubic centimetres of this solution are needed to give 10.0 g of KOH?

(d) How many grams of KOH are in each cubic centimetre of the solution?

2.97 Calcium acetate, $Ca(C_2H_3O_2)_2$, is the substance used along with methyl alcohol to make "canned heat." How many grams of calcium acetate are needed to prepare 2.00 L of 0.250 M solution?

2.98 How many grams of potassium nitrate, KNO_3, are needed to prepare 250.0 cm^3 of a 3.00×10^{-2} M solution?

*2.99 The formula for epsom salts is $MgSO_4 \cdot 7H_2O$. How many grams of epsom salts are needed to prepare 500 cm^3 of a solution that can be labeled 0.150 M $MgSO_4$?

Dilution of Solutions

2.100 How does the number of moles of solute in a solution change as the solution is diluted?

2.101 What is the proper procedure for diluting a solution of a concentrated reagent such as H_2SO_4 (sulfuric acid)?

2.102 What volume of 18.0 M H_2SO_4 must be added to 100 cm^3 of H_2O to give a solution of 5.0 M H_2SO_4?

2.103 What volume of concentrated NH_3 must be used to prepare 250 cm^3 of 0.500 M NH_3?

2.104 What volume of concentrated H_2SO_4 must be used to produce 400 cm^3 of 3.0 M H_2SO_4 solution?

2.105 To what volume must 100 cm^3 of 0.500 M H_2SO_4 be diluted to give 0.200 M H_2SO_4?

2.106 How much water must be added to 85.0 cm^3 of 1.00 M H_3PO_4 to produce 0.650 M H_3PO_4?

*2.107 To what volume must 250 cm^3 of 1.40 M H_2SO_4 be diluted to give a solution, 25.0 cm^3 of which is able to completely react with 15.0 cm^3 of 0.750 M NaOH? The equation for the reaction is

$$2NaOH + H_2SO_4 \longrightarrow Na_2SO_4 + 2H_2O$$

*2.108 How many cubic centimetres of 1.00 M HCl must be added to 50.0 cm^3 of 0.500 M HCl to give a solution whose concentration is 0.600 M?

Stoichiometry of Reactions in Solution

2.109 In Problem 2.81, it was mentioned that aluminum reacts with sulfuric acid.

$$2Al(s) + 3H_2SO_4(aq) \longrightarrow Al_2(SO_4)_3(aq) + 3H_2(g)$$

(a) How many cubic centimetres of 0.200 M H_2SO_4 are needed to react completely with 3.50 g Al?

(b) If a large piece of aluminum (an excess amount of Al) was placed in 400 cm^3 of 0.200 M H_2SO_4 solution, how many moles of H_2 would be produced by this reaction?

2.110 The light sensitive compound in photographic film is usually silver bromide, AgBr. It is made commercially by the reaction of solutions of sodium bromide, NaBr, and silver nitrate, $AgNO_3$.

$$NaBr(aq) + AgNO_3(aq) \longrightarrow AgBr(s) + NaNO_3(aq)$$

If 300 cm^3 of 0.250 M NaBr are mixed with 200 cm^3 of 0.400 M $AgNO_3$,

(a) which is the limiting reactant?

(b) how many grams of AgBr are formed?

2.111 Lye, which is sodium hydroxide (NaOH), can be neutralized by sulfuric acid. The reaction is

$$2NaOH(aq) + H_2SO_4(aq) \longrightarrow Na_2SO_4(aq) + 2H_2O$$

(a) How many cubic centimetres of 0.200 M H_2SO_4 are needed to react completely with 25.0 cm^3 of 0.400 M NaOH?

(b) How many cubic centimetres of 0.100 M NaOH are needed to react completely with 50.0 cm^3 of 0.270 M H_2SO_4?

(c) If 40.0 cm^3 of 0.300 M NaOH are added to 15.0 cm^3 of 0.350 M H_2SO_4, how many moles of Na_2SO_4 will be formed?

2.112 Magnesium hydroxide, $Mg(OH)_2$ is the white milky substance in milk of magnesia. When NaOH is added to a solution of $MgCl_2$, $Mg(OH)_2$ is formed.

$$2NaOH(aq) + MgCl_2(aq) \longrightarrow Mg(OH)_2(s) + 2NaCl(aq)$$

(a) How many cubic centimetres of 0.300 M NaOH

are needed to react with 75.0 cm^3 of 0.200 M MgCl$_2$?

(b) How many grams of Mg(OH)$_2$ can be formed if an excess of NaOH is added to 50.0 cm^3 of 0.600 M MgCl$_2$?

(c) How many grams of Mg(OH)$_2$ will be formed if 30.0 cm^3 of 0.200 M MgCl$_2$ are added to 100 cm^3 of 0.140 M NaOH solution?

*2.113 The silver nitrate (AgNO$_3$) in 20.00 cm^3 of a certain solution was allowed to react with sodium chloride according to the equation AgNO$_3$(aq) + NaCl(aq) → AgCl(s) + NaNO$_3$(aq). The AgCl was collected and dried, producing 0.2867 g AgCl. What was the molar concentration of the original AgNO$_3$ solution?

CHAPTER 3

THE PERIODIC TABLE AND THE MAKEUP OF ATOMS

Neon signs such as these in the Smithsonian Institution Museum of History and Technology are glass tubes filled with various gases. High voltage electricity passing through them causes the gases to give off various colors of light. The study of the flow of electricity through similar tubes, which is described in this chapter, was one of the first steps in unlocking the secrets of the internal structures of atoms.

Dalton's atomic theory was a milestone in the development of chemistry. It successfully explained the laws of chemistry that were known then, and as you saw in Chapter 1, it even predicted a new law—the law of multiple proportions. In fact, all the calculations that you learned to do in the last chapter are based on one of the central points of Dalton's theory, that atoms of a given element have a characteristic atomic mass.

Although Dalton's theory was marvelously successful in explaining the quantitative laws of chemistry, it could not explain why substances react as they do. To find the answers to these kinds of questions, scientists had to search for relationships among the properties of the elements and study atoms more closely.

We begin this chapter by introducing you to some of the elements and the kinds of properties that can be used to characterize them. Properties such as these are the facts of chemistry—facts that ultimately were among the clues that led to the development of the modern periodic table, which is located on the inside front cover of the book. These facts also form the basis of modern theories of atomic structure, because theory has to explain them. Beyond serving as a foundation for theory, however, these facts are also a part of the chemical knowledge that you should take with you from this course.

3.1   SOME PROPERTIES OF THE ELEMENTS

The range of properties shown by the elements is tremendous. At room temperature, some are gases, two are liquids, and the rest are solids. Some are metallic and some are not, and some have properties in between. Some are hard and some are soft; some are very dense and others have very low densities. With such variety, we have to look for ways to classify these properties so that it is possible to make some sense out of them.

One of the simplest methods of classification is to divide the elements into three categories: **metals, nonmetals,** and **metalloids.** The elements in each of these categories have certain distinctive characteristics.

Metals

Everyone has seen metals of one kind or another—an iron nail, aluminum foil, copper wire, or a "chrome-plated" bumper on a car, for example. And you are no doubt familiar with some of the properties that characterize metals, even if you haven't thought about them very much. One of these, for instance, is the distinctive appearance that metals have. They shine with a luster that is so characteristic that it's called a *metallic luster*.

Metals are also similar in their abilities to deform without breaking when hit with a hammer and to stretch when pulled. All metals have both these properties to some degree. The ability to deform when hammered is called **malleability,** and some metals, such as gold, can be hammered or squeezed into extremely thin sheets. *Gold leaf* (Figure 3.1), for example, consists of gold with a small amount of silver and copper that has been beaten into sheets that are so thin (about 90 nm) that they are translucent; some light can actually be seen passing through them. Malleability is also a property that a blacksmith relies on when forging a horseshoe, and a silversmith uses the malleability of silver in hammering a design into a fine silver tray.

The ability of a metal to stretch when pulled from opposite directions is called **ductility.** This property is used in the manufacture of wire, which is illustrated in Figure 3.2. The metal to be made into wire, which might be steel, copper, or brass, is first formed into a rod. One end is tapered, fed through a

Figure 3.1

A very thin film of pure gold leaf is transferred from a paper backing to a sculpture in Rockefeller Center in New York City. The gold provides a nontarnishing decorative finish to the work.

Figure 3.2

Copper wire is pulled through smaller and smaller dies and becomes thinner. The ductility of copper makes this possible.

die, and attached to a pulling device on the other side. The metal is then drawn through the die where it undergoes a reduction in size and an increase in length.

Everyone knows that metals are good conductors of electricity. They are also good conductors of heat. If you have ever touched a metal object that has been been lying in the sun for a while, you know how very hot it feels. In fact, it feels much hotter than other objects alongside that are not metallic. The reason is that as your hand absorbs heat from the metal, heat travels quickly from the neighboring parts of the object to replace it. Nonmetallic objects don't feel as hot because when your hand removes heat, it can't be replaced rapidly, and the part of the object in contact with your hand becomes cooler.

More than 70% of the elements are metals, and although there are some similarities among them, there are many differences, too. Some metals are quite common and we encounter them nearly every day in their elemental forms, that is, uncombined with other elements. The metals mentioned previously (iron, aluminum, copper, and chromium) are only a few examples. There are other metals that are so reactive only chemists, or chemistry students, ever have an opportunity to see them. For example, in Chapter 1 we mentioned the reaction between sodium and chlorine to form sodium chloride, common table salt. Sodium is a very reactive metal that combines not only with chlorine, but avidly with both oxygen and moisture. In Figure 3.3, we see a bar of sodium, heavily tarnished on the outside, but the element's bright metallic luster is revealed by the freshly cut slice of the metal.

The range of chemical reactivity of metals is very broad. Sodium is typical of one extreme, while gold is typical of the other. Jewelry is made from gold for several reasons, one of which is the fact that it doesn't tarnish when exposed to air and moisture. This very low reactivity, combined with gold's high electrical conductivity, accounts for one of this metal's most important commercial uses: the plating of electrical contacts in computers and other electronic devices.

Besides chemical reactivity, metals differ in certain physical properties such as hardness and melting point. Some metals are very hard, and some are very soft. Chromium and iron are examples of hard metals; gold and lead are examples of soft ones. Sodium is also a soft metal; in Figure 3.3, we see it being cut with a knife. The extremes of melting point are even more impressive. Tungsten has the highest melting point of any element, 3400°C, which accounts for its use as the filament in electric light bulbs. Mercury has the lowest melting point of any metal, −38.9°C, which means that it is a liquid at room temperature. As you know, mercury is the fluid commonly used in thermometers.

Figure 3.3

The typical metallic luster of this freshly exposed surface of sodium will quickly fade as the metal tarnishes as a result of the reaction of the sodium with oxygen and moisture in the air.

Figure 3.4

Two forms of the element carbon. (a) Graphite, a black powder. (b) A brilliant diamond of gem quality.

(a)

(b)

Nonmetals

Most of the nonmetallic elements (the nonmetals) are rarely encountered in our daily activities in their elemental forms; instead, they are usually found in compounds. One nonmetal that most people have seen is carbon, which occurs in nature in two different forms (Figure 3.4). The more common variety is called graphite. This is the form that we find in charcoal briquets and the lead in lead pencils. The less common and more valuable form of carbon is diamond. Graphite and diamond have properties that are quite different from those that we associate with metals. Neither has the luster of a metal, and neither is malleable or ductile.

Other nonmetals that you have encountered are oxygen and nitrogen, which are the principal components of the atmosphere. Usually, you are not aware of their existence, because they are colorless gases and you can't see them. Oxygen and nitrogen occur as **diatomic molecules,** molecules containing two atoms each. Other nonmetallic elements form similar molecules, and most are gases, also. These are hydrogen (H_2), fluorine (F_2), and chlorine (Cl_2). Bromine (Br_2) and iodine (I_2) are also diatomic, but bromine is a liquid and iodine is a solid at room temperature.

Just as the properties of the metals cover a broad range, so do the properties of the nonmetals. As we've seen, some are gases and one (bromine) is a liquid. There are others that are solids; carbon is just one example. Besides differing in these physical properties, nonmetals differ from each other in their chemical properties. Fluorine, for example, is extremely reactive while helium is *inert* (totally unreactive). These differences, which are very important, will be explored in more detail later in the book.

Metalloids

Metalloids are elements that have properties that are intermediate between those of metals and those of nonmetals. The best known example is the element silicon. Two others are arsenic (As) and antimony (Sb), which are shown in Figure 3.5. In terms of outward appearances, these elements have something of a metallic look about them, but their dark color gives them away. They certainly differ in appearance from typical metals such as iron or silver.

Metalloids are typically semiconductors—they conduct electricity, but not nearly as well as metals. These semiconductor properties are especially valuable in the electronics industry, because they make possible all the microelectronic

Figure 3.5
Elemental antimony, and arsenic.

devices found in hand-held calculators and microcomputers. Except for their electrical properties, however, the metalloids are much more like nonmetals than metals.

3.2   THE FIRST PERIODIC TABLE

Chemical and physical properties like those described in the last section were discovered early on in the history of chemistry. Scientists, even as early as 1800, had accumulated significant quantities of information about the elements known to them. This knowledge, however, existed for the most part as isolated and unrelated facts that needed to be correlated in some fashion before their total significance could be grasped. Early attempts at a classification of the elements met with only limited success, and it wasn't until 1869 that the forerunner of our modern periodic table was devised. This resulted from the work of two chemists, a Russian named Dmitri Mendeleev and a German named Julius Lothar Meyer. Both worked independently and produced similar tables at about the same time. Mendeleev presented the results of his work to the Russian Chemical Society in the early part of 1869, but Meyer's table didn't appear until December of that same year. Because Mendeleev had the good fortune of publishing first, he is usually given credit for the periodic table.

Mendeleev was a chemistry teacher, and while he was preparing a textbook for his students, he discovered that if he arranged the elements in order of increasing atomic weight, elements with similar properties occurred at periodic intervals. For example, he could pick out the elements lithium (Li), sodium (Na), potassium (K), and rubidium (Rb). Each of these elements forms a water-soluble compound with chlorine that has the general formula, MCl, where M stands for Li, Na, K, and so on. Although this is an interesting fact by itself, what is especially significant is that if we examine the elements that immediately follow Li, Na, K, and Rb in the list, they form another group of similar elements. Thus, Be follows Li, Mg follows Na, Ca follows K, and Sr follows Rb. These elements form the compounds $BeCl_2$, $MgCl_2$, $CaCl_2$, and $SrCl_2$. Recognizing this, Mendeleev divided the list into a series of rows and stacked them so that those elements having similar properties are arranged in vertical columns. The result was the first periodic table, which is illustrated in Figure 3.6.

Dmitri Mendeleev.

	Group I	Group II	Group III	Group IV	Group V	Group VI	Group VII	Group VIII
1	H 1							
2	Li 7	Be 9.4	B 11	C 12	N 14	O 16	F 19	
3	Na 23	Mg 24	Al 27.3	Si 28	P 31	S 32	Cl 35.5	
4	K 39	Ca 40	— 44	Ti 48	V 51	Cr 52	Mn 55	Fe 56, Co 59 Ni 59, Cu 63
5	(Cu 63)	Zn 65	— 68	— 72	As 75	Se 78	Br 80	
6	Rb 85	Sr 87	?Yt 88	Zr 90	Nb 94	Mo 96	— 100	Ru 104, Rh 104 Pd 105, Ag 100
7	(Ag 108)	Cd 112	In 113	Sn 118	Sb 122	Te 128	I 127	
8	Cs 133	Ba 137	?Di 138	?Ce 140	—	—	—	— — — —
9	—	—	—	—	—	—	—	
10	—	—	?Er 178	?La 180	Ta 182	W 184	—	Os 195, Ir 517 Pt 198, Au 199
11	(Au 199)	Hg 200	Tl 204	Pb 207	Bi 208	—		
12	—	—	—	Th 231	—	U 240	—	— — — —

Figure 3.6

Mendeleev's Periodic Table (1871). The numbers appearing with the symbols are atomic weights.

Mendeleev was convinced that the periodic recurrence of properties was related to atomic weight. He thought that the atomic weight of Te had been measured incorrectly, so he placed Te where it belonged based on its chemical properties.

When Mendeleev constructed his table, not all of the elements had yet been discovered. He realized this, because in order to always have similar elements in the same column, or **group,** he was forced to leave occasional blanks in his table. It was also necessary for him to reverse the atomic-weight order of tellurium (Te) and iodine (I), whose atomic weights in 1869 were thought to be 128 and 127 amu, respectively. Mendeleev placed them in the table in reverse order (according to atomic weights) because their properties dictated that tellurium belongs in Group VI and iodine in Group VII. (The groups are numbered with Roman numerals for ease of identification.)

One of the benefits of Mendeleev's table was that it became possible to predict some of the properties of the missing elements. This was because elements in any particular column had to have similar properties. For example, germanium, which lies below silicon and above tin in Group IV, had not been discovered when Mendeleev assembled his table. Therefore, a space appears at this spot in the chart. On the basis of its position, Mendeleev predicted that the properties of this element, which he called "eka-silicon," should lie intermediate between those of silicon and tin. Table 3.1 shows how closely his predicted properties were to those found for germanium when it was discovered in 1886.

3.3 ATOMIC NUMBERS AND THE MODERN PERIODIC TABLE

If you look at the modern periodic table in Figure 3.7, you will see a column of elements that did not appear in Mendeleev's table. This is the rightmost column entitled "noble gases." These are elements with virtually no chemical reactivity,

Table 3.1
Predicted properties of eka-silicon and observed properties of germanium

Property	Silicon	Tin	Predicted for Eka-silicon (Es)	Currently Accepted for Germanium
Atomic weight (u)	28	118	72	72.59
Density (g cm^{-3})	2.33	7.28	5.5	5.3
Melting point (°C)	1410	232	High	947
Physical form at room temperature	Gray nonmetal	White metal	Gray metal	Gray metal
Reaction with acids and alkalies	Acid—no reaction Alkalies—slow reaction	Slow attack by acids and by alkalies	Very slow attack by acids and alkalies	Will react with concentrated acids and with concentrated alkalies
Number of chemical bonds usually formed	4	4	4	4
Formula of chloride	SiCl$_4$	SnCl$_4$	EsCl$_4$	GeCl$_4$
Density of chloride (g cm^{-3})	1.50	2.23	1.9	1.88
Boiling point of chloride (°C)	57.6	114	100	84

and because they formed no known compounds, scientists in Mendeleev's time were totally unaware of their existence. When these elements were finally discovered, it was found that the atomic weight of argon (Ar) is slightly larger than the atomic weight of potassium. Nevertheless, potassium clearly belonged with the other elements in Group I, and argon clearly belonged in the group with the other noble gases. Once again, it was necessary to place a pair of elements in the table in reverse atomic-weight order.

When it was discovered that argon and potassium had to be placed in the periodic table in reverse order, scientists realized that the case of tellurium and iodine was no accident. It was clear that the quantity determining the position of an element in the table is *not* its atomic weight. The true basis of the periodic table lay elsewhere and finally became apparent with the work of a young British scientist, Henry Moseley.

Moseley investigated the X rays that different elements could be caused to emit and found that he could assign a number to each element based on the properties of the X rays that it gives off. He called this number the element's **atomic number.** (We will see later in this chapter that the atomic number is an important quantity in describing the detailed structure of an atom.) What was especially significant about Moseley's work was that an element's atomic num-

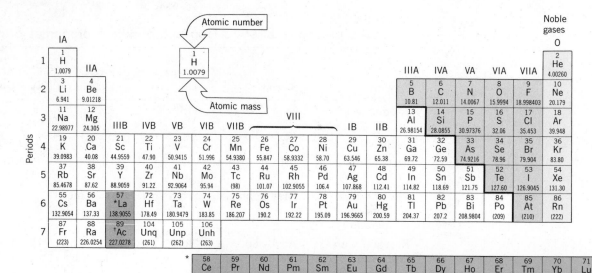

Figure 3.7

The modern periodic table of the elements.

Members of a group in the periodic table bear resemblance to each other as do members of a family—hence the term, *family of elements.*

ber was found to be the same as its position-number in the periodic table. This leads to the modern statement of the **periodic law:** *When the elements are arranged in order of increasing atomic number, a periodic repetition of physical and chemical properties occurs.*

Over a period of time, the periodic table has undergone some refinements. The one in use today (sometimes called the "long" form of the periodic table) is shown in Figure 3.7. The numbers written above the chemical symbols are atomic numbers, and those below are atomic weights. Like Mendeleev's table, it is constructed of a number of vertical columns called **groups,** each containing a *family of elements.* These groups are identified by a Roman numeral and a letter, either A or B[1]. The horizontal rows are called **periods** and are designated by Arabic numerals. Thus, the elements hydrogen and helium are members of the first period; lithium through neon are members of the second period; and so on.

The groups labeled with the letter A (Groups IA through VIIA) and Group 0 are referred to collectively as the **representative elements.** Those labeled with the letter B (Groups IB through VIIB) plus Group VIII (actually composed of the three short columns in the center of the table) constitute the **transition elements.** Similarities between properties of the A- and B-group elements exist, although the similarities are often very weak.

Finally, we see that there are two long rows of elements lying just below the main part of the table. These elements, called the **inner transition elements,** actually belong in the body of the table, as shown in Figure 3.8. They are usually placed below simply to conserve space. Notice in Figure 3.8 that the first row of inner transition elements (58 through 71) follows lanthanum (La) and is part of the sixth period. Because they follow lanthanum, these elements are collectively called the **lanthanides.** They are also known as the **rare earth elements** because of their very low relative abundance in the earth's crust. The second row of inner

---

[1] There is some controversy over the numbering and lettering of the groups in the periodic table. The system used in this book closely follows Mendeleev's and is preferred by many educators. An alternative designation of A and B groups has been used in Europe, which led the I.U.P.A.C. to adopt still a third system in which the groups are numbered sequentially from 1 to 18 (from left to right).

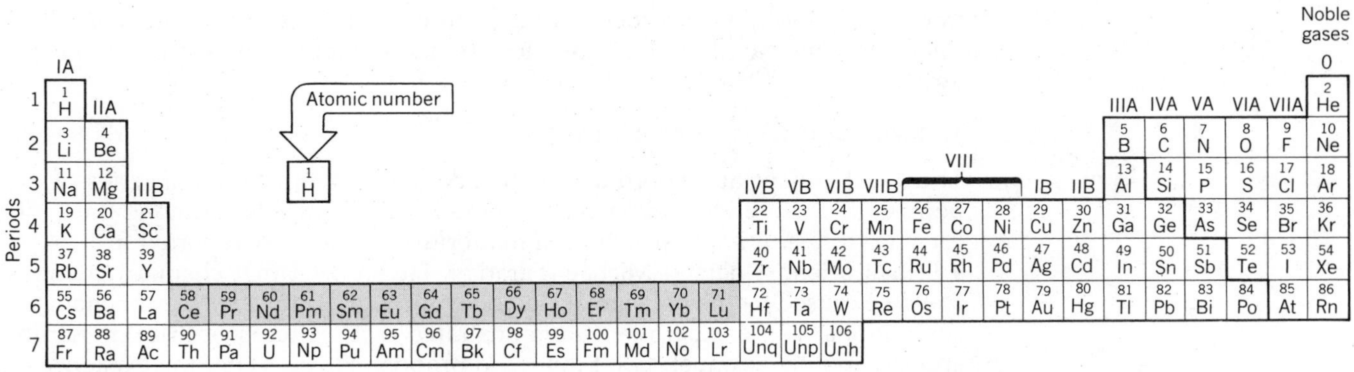

	IA																					Noble gases 0
1	1 H	IIA															IIIA	IVA	VA	VIA	VIIA	2 He
2	3 Li	4 Be															5 B	6 C	7 N	8 O	9 F	10 Ne
3	11 Na	12 Mg	IIIB						VIII				IB	IIB			13 Al	14 Si	15 P	16 S	17 Cl	18 Ar
4	19 K	20 Ca	21 Sc		IVB	VB	VIB	VIIB	22 Ti	23 V	24 Cr	25 Mn	26 Fe	27 Co	28 Ni	29 Cu	30 Zn	31 Ga	32 Ge	33 As	34 Se	35 Br	36 Kr
5	37 Rb	38 Sr	39 Y						40 Zr	41 Nb	42 Mo	43 Tc	44 Ru	45 Rh	46 Pd	47 Ag	48 Cd	49 In	50 Sn	51 Sb	52 Te	53 I	54 Xe
6	55 Cs	56 Ba	57 La	58 Ce ... 71 Lu					72 Hf	73 Ta	74 W	75 Re	76 Os	77 Ir	78 Pt	79 Au	80 Hg	81 Tl	82 Pb	83 Bi	84 Po	85 At	86 Rn
7	87 Fr	88 Ra	89 Ac	90 Th ... 103 Lr					104 Unq	105 Unp	106 Unh												

Atomic number → 1 H

Inner transition rows:
Period 6: 58 Ce, 59 Pr, 60 Nd, 61 Pm, 62 Sm, 63 Eu, 64 Gd, 65 Tb, 66 Dy, 67 Ho, 68 Er, 69 Tm, 70 Yb, 71 Lu
Period 7: 90 Th, 91 Pa, 92 U, 93 Np, 94 Pu, 95 Am, 96 Cm, 97 Bk, 98 Cf, 99 Es, 100 Fm, 101 Md, 102 No, 103 Lr

Figure 3.8

The modern periodic table with the two rows of inner transition elements in their proper locations.

Alkali comes from the Arabic *alqili*, meaning *ashes of saltwort*. Saltwort is a shrub that grows in coastal areas. Its ashes were once a major source of sodium carbonate (soda ash).

Sodium hydroxide (NaOH) is called *caustic* soda.

Most of the elements are metals.

transition elements (90 through 103) follows actinium (Ac) and is part of the seventh period. These elements are collectively known as the **actinides.**

Certain families of elements are characterized by names as well as by their group numbers. For example, the Group IA elements are frequently spoken of as the **alkali metals** because certain of their compounds are caustic or "alkaline." The Group IIA elements are called the **alkaline earth metals;** these elements are found in minerals and certain of their compounds are caustic, too. The Group VIIA elements are called the **halogens,** a name derived from the Greek meaning "salt former." Finally, the Group 0 elements are the **noble gases** (they are also sometimes called the *inert gases*) because of their extremely limited ability to react chemically.

In Section 3.1, we saw that we could classify an element as a metal, nonmetal, or metalloid. In the periodic table, elements on the left are metals and those on the right are nonmetals. The heavy step-shaped line drawn from boron (B) to astatine (At) approximately represents the boundary between metallic and nonmetallic behavior, and those elements adjacent to this line are the metalloids. Thus, there is a gradual transition from metallic to nonmetallic behavior in this region. Going from left to right across a period, the elements change from metallic to nonmetallic. The same kind of transition is observed when going from top to bottom in some groups—for example, Group IVA. At the top is carbon, a nonmetal, and at the bottom is lead, a metal.

The periodic table is probably the most useful aid that chemists have at their disposal. It can be used to correlate all sorts of useful information, and it helps us see important trends. The variation in the metallic character of the elements across a period and down a group, discussed in the preceding paragraph, is just one example. From the standpoint of the development of a theory concerning the structure of the atom, the periodic table represents a compilation of experimental data that must be explained. A successful theory must somehow account for the way the table is structured. For example, *why* are there only two elements in period 1, *why* are there eight each in periods 2 and 3, and so on. It must also explain *why* it is that elements in any given group exhibit similar properties.

3.4 THE STRUCTURES OF ATOMS

When Dalton proposed his atomic theory, he imagined atoms to be the simplest forms of matter. According to him, they were tiny, indestructible particles. Now, of course, we know better. Atoms are not indivisible; they can and have been split into a whole host of different subatomic particles. Fortunately for us, in chemistry we are concerned with only three of them: electrons, protons, and neutrons. The way these tiny particles are distributed in atoms is called **atomic structure.**

The electrical nature of matter

The development of modern atomic theory came about through a series of experiments, each of which added a bit more to our knowledge. Among the earliest clues to the ultimate structure of matter was a discovery reported in 1834 by a British scientist named Michael Faraday. He showed that chemical changes could be caused by the passage of electricity through water solutions of chemical compounds. These experiments demonstrated that matter is electrical in nature and led another scientist, G.J. Stoney, to propose 40 years later the existence of particles of electricity that he called **electrons.**

Gas discharge-tube experiments

Toward the end of the nineteenth century, physicists began to investigate the flow of electricity in devices called **gas discharge tubes.** These are glass tubes with a metal **electrode** sealed into each end, as illustrated in Figure 3.9. It was discovered that when a high electrical potential is applied across the electrodes and air is partially removed from the tube with a vacuum pump, a flow of electricity—an *electrical discharge*—takes place through the gas. While this discharge is occurring, the gas in the tube glows. Neon signs, in fact, are modern versions of gas discharge tubes in which the gas in the tube is neon or another gas, instead of air.

Ordinary fluorescent lamps are gas discharge tubes containing mercury vapor. Light given off by the mercury when the discharge occurs causes a phosphor on the inner surface of the lamp to give off white light.

Physicists were very curious about what was taking place inside these tubes during the electrical discharge, and they performed various experiments to investigate the phenomenon. In one of these, they placed a screen coated with a zinc sulfide phosphor (similar to the material used to coat the inside of a TV picture tube) between the electrodes and found that it glowed on the side facing the negative electrode. This suggested that the discharge originates at the negative electrode (which they called the **cathode**) and moves in the direction of the positive electrode (the **anode**). Because these "rays" came from the cathode, they were named **cathode rays.**

In further experiments, it was found that cathode rays:

1. Normally travel in straight lines.
2. Cast shadows.
3. Can spin a pinwheel placed in their path, which suggests that cathode rays are composed of particles.
4. Heat a metal foil placed between the electrodes.

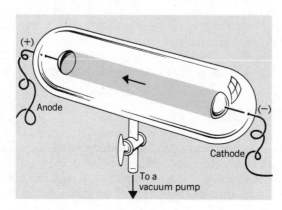

Figure 3.9

A gas discharge tube.

Figure 3.10

Some properties of cathode rays. (a) Cathode rays are deflected by an electric field and (b) by a magnetic field.

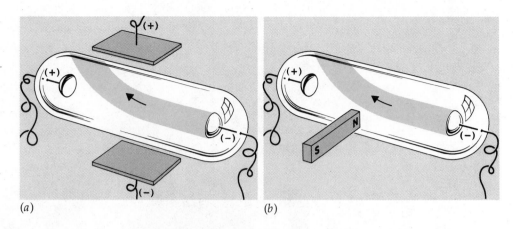

(a)   (b)

5. Can be deflected by electric or magnetic fields. The direction in which they are deflected indicates that the particles are electrically charged and that the charge is negative. This is illustrated in Figure 3.10.

A modern television picture tube.

Quantitative information came in 1897 when J.J. Thomson, another British scientist, used a cathode-ray tube quite similar to present-day TV picture tubes to measure the ratio of the charge to the mass of the cathode ray particles. This device is shown schematically in Figure 3.11. Particles generated at the cathode are accelerated toward the anode, which has a hole in it. Some pass through the hole and continue on, striking the phosphor-coated face of the tube at B where they produce a bright spot. If oppositely charged plates are placed above and below the tube, the beam is deflected toward the positive plate and strikes the face at A. The extent of deflection that the particle undergoes will be *directly proportional* to its charge—a particle with a large negative charge will be attracted to the positive plate more than one with a small charge. The extent of deflection will also be *inversely proportional* to the mass of the particle, because a massive particle will be less affected by the electrostatic attraction as it passes between the plates than a particle of smaller mass. This is similar to the differences between the way a breeze affects the path of a golf ball and the way it affects the path of a much lighter Ping-Pong ball. The influences of charge and mass on the extent of deflection can be combined by saying that the observed deflection depends on the *ratio* of the particles charge, e, to its mass, m. This is the **charge-to-mass ratio,** symbolized as e/m.

The path of a moving charged particle is affected by both electric and magnetic fields.

If a magnetic field is generated at right angles to the electric field, as shown in Figure 3.11, the cathode rays are deflected in a direction exactly opposite to

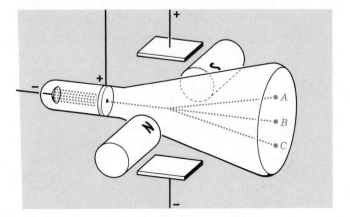

Figure 3.11

A cathode-ray tube used to measure the charge-to-mass ratio of the electron.

The monitor screen used to view the output of a microcomputer is called a CRT, meaning "Cathode Ray Tube."

that caused by the electrically charged plates. In the absences of the electric field, the beam is bent by the magnetic field so that it collides with the surface of the tube at *C*.

In practice, Thomson applied a magnetic field of known strength across the tube and noted the deflection of the electron beam. Charge was then applied to the plates until the beam was brought back to its original point of impact, *B*. From the magnitude of the electric and magnetic fields, Thomson calculated the charge-to-mass ratio, e/m, for the particles to be -1.76×10^8 coulombs/gram. The **coulomb (C)** is the SI unit of electric charge. It is equal to the quantity of charge that moves past a given point in a wire when an electric current of 1 **ampere** (1 **A**) flows for 1 second. In more familiar terms, if you have an ordinary 100-watt light bulb burning, it is using a current of about 0.83 A. It takes 1.2 seconds for one coulomb of charge to pass through the bulb. A coulomb, therefore, represents a fairly large quantity of electric charge, and an especially significant result of Thomson's experiments was that the same value of the charge-to-mass ratio was always obtained for the cathode rays regardless of the composition of the electrodes or the nature of the residual gas in the apparatus. These findings suggested that the negatively charged particles in the cathode rays are fundamental building blocks of all matter. Such particles are called **fundamental particles,** and the particles in cathode rays are, in fact, the electrons described by Stoney.

3.5   THE CHARGE ON THE ELECTRON

The value of e/m for the electron, -1.76×10^8 C g^{-1}, suggests that the electron either has an enormous electrical charge or its mass is very small. The results of Thomson's experiments, however, could not reveal which alternative is correct. An independent measurement of either the mass or the charge was necessary.

The problem was resolved when the charge on the electron was determined by means of a rather clever experiment performed in 1908 by R.A. Millikan at the University of Chicago. In his apparatus, illustrated in Figure 3.12, a fine mist of

Figure 3.12

Millikan's oil drop experiment.

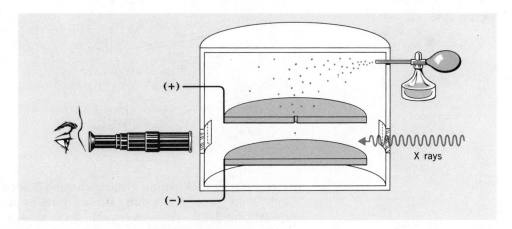

X rays are a high-energy form of light and can eject electrons from particles that absorb them. That's what makes large doses of X rays dangerous.

oil droplets was sprayed above a pair of parallel metal plates. As droplets settled through a hole in the upper plate, the air between the plates was briefly irradiated with X rays. Electrons, knocked from gas atoms by the X rays, were picked up by the oil droplets, thereby giving them a negative charge. Millikan found that by placing an electric charge on the plates (upper plate positive, lower plate negative), the downward motion of negatively charged drops could be slowed or even stopped. A knowledge of the mass of a drop—measured by observing its rate of fall in the absence of the electric field—and the quantity of charge on the plates required to keep that drop suspended permitted Millikan to calculate the quantity of charge on the drop.

After he performed this experiment many times, Millikan observed that the quantity of charge on the oil drops was always a multiple of -1.60×10^{-19} C. He reasoned that since the oil drops could only pick up whole numbers of electrons, the total charge on any drop must be a multiple of the charge on a single electron. This suggested that the charge on the electron is -1.60×10^{-19} C. Once its charge was known, the electron's mass could then be calculated from the already known charge-to-mass ratio

$$\frac{e}{m} = -1.76 \times 10^8 \text{ C g}^{-1}$$

Solving for m, the mass of the electron, we get

$$m = \frac{e}{-1.76 \times 10^8 \text{ C g}^{-1}}$$

Substituting $e = -1.60 \times 10^{-19}$ C gives

$$m = \frac{-1.60 \times 10^{-19} \text{ C}}{-1.76 \times 10^8 \text{ C g}^{-1}}$$

$$= 9.11 \times 10^{-28} \text{ g}$$

About 5.2 billion billion (5.2 $\times$ 10^{18}) electrons pass through a 100-watt electric light bulb each second it is lit.

If you write this number in standard decimal form, you might be able to appreciate how incredibly tiny the electron is.

3.6 POSITIVE PARTICLES AND THE MASS SPECTROMETER

Ordinary things that we encounter every day are electrically neutral. Therefore, since negatively charged electrons are a part of everything, positively charged particles must also exist in all matter. The search for these positive particles began with experiments using specially designed gas discharge tubes with per-

Figure 3.13

Canal rays. Positive ions passing through holes in the cathode appear as canal rays at the rear of the electrode. Bursts of light are seen when they collide with a phosphor on the end of the tube.

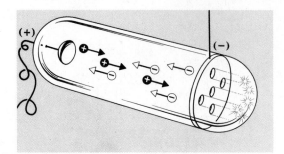

forated cathodes. When an electric current flowed through the tube, streamers of light were observed coming from the holes at the rear of the negative electrode (Figure 3.13). These were called canal rays.

During an electric discharge, electrons emitted from the cathode collide with atoms of the residual gas in the tube, knocking electrons off them. When they lose electrons the atoms become positively charged ions (an ion, you recall, is an electrically charged particle that is formed when electrons are added to or removed from a neutral atom or molecule). These positive ions are attracted toward the cathode. Although most of them collide with the cathode, some travel through the perforations and emerge at the rear where they are observed as the canal rays. If the rear wall of the discharge tube is coated with a phosphor, flashes of light can also be seen where they hit the wall.

The instrument used to investigate the properties of positive ions is called the **mass spectrometer,** which is illustrated schematically in Figure 3.14. The sample to be studied is introduced as a gas at A, where an electrical discharge across electrodes B and C causes the gas to be ionized (i.e., converted to ions). The positive ions produced are then accelerated through the wire grid E and formed into a narrow beam by the slits F and G. This beam of ions flows between the poles of a powerful magnet that acts to deflect the particles along a circular path.

The degree to which the paths of the positive ions are affected by the magnetic field is determined by the charge-to-mass ratio of the ions. Heavy ions are deflected less than light ones, so the extent of deflection is *inversely proportional* to an ion's mass (the larger the mass, the smaller the deflection). Ions that have a large charge are deflected more than those with a small charge, so the extent of deflection is *directly proportional* to an ion's charge. By adjusting the strength of

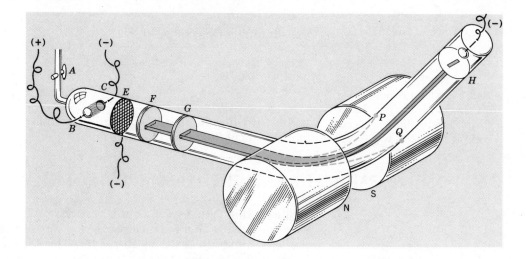

Figure 3.14

The mass spectrometer.

the magnetic field, ions with any desired e/m ratio can be focused on the detector at H. Ions with higher e/m ratios are deflected more (for example, to P), while those with smaller e/m ratios are deflected less (for example, to Q).

If the ion beam entering the magnetic field consists of a mixture of ions of different masses, but with the same charge, it is divided into a number of beams, each of which contains ions of the same mass. This separation of ions of different masses produces a **mass spectrum,** just as the separation of white light into its different colors by a prism gives a visible light spectrum. Modern mass spectrometers are capable of very accurately measuring the masses of the ions in a mass spectrum, and many of the atomic masses in the table inside the front cover of the book are derived from mass spectral data.

When scientists first studied the data obtained from mass spectrometers, they found the following:

1. Positive ions always have e/m ratios that are *much smaller* than that of the electron. This means that they are either much more massive than the electron (that is, m is very large) or they carry very small positive charges (that is, e is small). Since they are formed from neutral atoms by the loss of electrons, the charge that they carry is either equal in magnitude to the electron's charge or is some whole-number multiple of it. Therefore, in order to have a much smaller e/m than the electron, their masses must be much larger.

2. The value of the e/m ratio depends on the nature of the gas introduced into the mass spectrometer, which shows that not all positive ions have the same e/m.

Discovery of the proton

The lightest positive particle that is ever found in a mass spectrometer is the hydrogen ion, which is produced by stripping an electron from a hydrogen atom. It is also found that the masses of other positive ions are very nearly whole-number multiples of the mass of the hydrogen ion. Thus, it was concluded that the hydrogen ion is a fundamental particle of positive charge that is present in all atoms, and it was given the name **proton,** which is derived from the Greek word *protos* meaning *first*. A hydrogen atom, therefore, consists of one proton and one electron, and if we compare the mass of the proton to that of the electron, we find the proton to be 1836 times more massive than an electron. Since atoms are ordinarily electrically neutral, they must have equal numbers of electrons and protons, so the electrons constitute only a small fraction of an atom's total mass.

Atoms that are more massive than hydrogen contain more than one proton, and each atom of a given element has the same number of protons. The number of protons in an atom of an element is called the element's **atomic number.** We will see later how atomic numbers were determined.

Expressing electrical charges

When ions are formed from neutral atoms, the atoms lose or gain whole numbers of electrons. Therefore, the charge carried by an ion is a multiple of the charge on one electron, and is either positive or negative depending on whether electrons are lost or gained. Rather than write the charge on an ion as a multiple of 1.60×10^{-19} C, it is simpler to define a unit of electrical charge as equal in magnitude to 1.60×10^{-19} C. Then we can express the charge on the particle as plus or minus the appropriate number of these units. On this scale, for example, the electron itself has a charge of $1-$ (meaning one unit of negative charge). A

We will discuss the production of a visible light spectrum in more detail in Chapter 4.

proton, which has an actual charge of $+1.60 \times 10^{-19}$ C, has one unit of positive charge, or a charge of 1+.

We commonly indicate the charge on an ion by writing the number of units of positive or negative charge as a superscript on the right side of the chemical symbol. Thus, He^{2+} is an ion formed from helium by removing two electrons from the neutral atom. The ion O^{2-} is formed from an oxygen atom by the addition of two electrons. Normally, if the charge is either 1+ or 1−, we omit the number 1 and just write the charge as + or −. For example, when a sodium atom loses one electron, it forms the ion Na^+, and when a chlorine atom gains one electron, it forms the ion Cl^-.

3.7   RADIOACTIVITY

We discuss radioactivity in more detail in Chapter 22.

Some atoms of some elements are not stable. They spontaneously emit radiations of various types. This phenomenon, called **radioactivity,** was discovered by Henri Becquerel in 1896. Radioactive substances emit three important types of radiation.

1. Alpha radiation, composed of He^{2+} ions called alpha particles (α particles).
2. Beta radiation, consisting of electrons. When emitted by a radioactive substance, they are called beta particles (β particles).
3. Gamma radiation (γ rays), consisting of highly energetic, very penetrating light waves. (They are similar to X rays).

The phenomenon of radioactivity was yet another bit of evidence that atoms were not indestructible particles and that they contained still simpler parts.

3.8   THE NUCLEAR ATOM

Rutherford was so surprised by the results of these scattering experiments that he said it was as if a cannonball had been fired at a piece of tissue paper and the projectile had bounced back to hit the gunner.

One of the most significant developments in our understanding of the structure of the atom was provided by Ernest Rutherford in 1911. Prior to this time, it was thought that the atom had a nearly uniform density throughout, with the electrons being buried in a glob of positive charge, much like raisins in a pudding. With this picture of a rather mushy atom in mind, Rutherford assigned one of his students the task of measuring the scattering of α particles that were aimed at a thin gold foil. From his earlier experiments, Rutherford expected the α particles to pass through the foil virtually undisturbed, which would be consistent with a more or less uniform distribution of positive and negative charges. Nevertheless, he suggested that the student check to see if any α particles were scattered at large angles, and he was astonished to learn that some were. In fact, it was found that some α particles came almost straight back toward the source, which meant that they had encountered something positively charged and extremely massive (Figure 3.15).

There was only one way that Rutherford could explain why most of the α particles passed easily through the foil, but a few were deflected at extremely large angles. He concluded that the atom contained a very small, extremely dense positively charged **nucleus** that held all the atom's protons and nearly all the atom's mass. Since the nucleus holds all the atoms positive charge, it also follows that the electrons in the atom are distributed somehow outside the nucleus in the remaining volume of the atom.

It is very difficult to imagine how really small a nucleus is. Its diameter is approximately 10^{-13} cm, compared to the atom itself whose diameter is of the order of 10^{-8} cm. To appreciate these relative sizes, imagine that the atom could

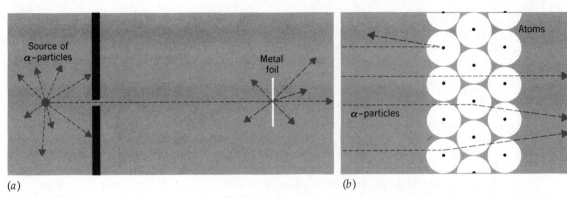

Figure 3.15

(a) *Scattering of α particles by a metal foil.* (b) *Deflection of* *α particles due to repulsions between positively charged* *α particles and positively charged nuclei.*

be enlarged so the nucleus was about the size of an orange (about 8 cm in diameter). If you were standing by this nucleus, the outer parts of the atom would be approximately *7 km* away. That is an "atomic" diameter of 14 km, compared to a nuclear diameter of 8 cm!

Because nearly all the mass of an atom is packed into the tiny nucleus, the density of nuclear material is enormous—about 10^{14} g cm^{-3}. To give you some idea of how dense this is, if all of the nuclei in the crude oil cargo of one of the world's largest supertankers could be crammed together until they were touching, they would only occupy approximately 0.004 cm^3—about the volume of a tenth of a drop of water—but they would have a mass over 180 gigagrams $(1.8 \times 10^8$ kg)!

3.9 THE NEUTRON

By making careful measurements of the scattering of α particles by various metal foils, Rutherford was able to estimate the number of protons in the nuclei of the atoms. When he did this, he found that only about half of the nuclear mass could be accounted for by protons. He therefore suggested that the nucleus also contains another kind of particle—one that is electrically neutral and has a mass very nearly equal to that of a proton. The existence of this particle was confirmed in 1932 by the English scientist J. Chadwick, who bombarded beryllium with α particles and found that highly energetic, uncharged particles were emitted. These particles, which are called **neutrons,** have a mass only slightly larger than that of the proton.

The properties of the three major particles found in an atom are described in Table 3.2. In summary, an atom is composed of a dense nucleus containing

Table 3.2
Some properties of subatomic particles

	Mass		Charge	
	Grams	Unified Atomic Mass Units (u)	Coulombs	Electronic Charge Units
Proton	1.67×10^{-24}	1.007276	$+1.602 \times 10^{-19}$	1+
Neutron	1.67×10^{-24}	1.008665	0	0
Electron	9.11×10^{-28}	0.0005486	-1.602×10^{-19}	1−

protons and neutrons. These particles provide nearly all the atom's mass. The nucleus is surrounded in some fashion by the atom's electrons, which are somehow distributed throughout the remaining volume of the atom.

3.10 ISOTOPES

In Chapter 1 we noted that not all atoms of the same element have identical masses, which is contrary to Dalton's original hypothesis. We referred to these different kinds of atoms as **isotopes.** The existence of isotopes is a common phenomenon, and most of the elements occur naturally as mixtures of isotopes.

Each atom of a given element has the same atomic number, but the number of neutrons can vary.

As we will see, the properties of an element are determined almost entirely by the number and distribution of it electrons. *Therefore, it is the atomic number (or number of protons) that serves, indirectly, to distinguish an atom of one element from an atom of another, because the number of electrons must equal the number of protons in an electrically neutral atom.* In other words, an atom's atomic number identifies which element it is. Any mass differences that exist between atoms of the same element arise from different numbers of neutrons.

A particular isotope of an element is identified by specifying its atomic number, Z, and its **mass number,** A. The mass number is the *sum* of the number of protons and neutrons in the atom. The number of neutrons present can therefore be obtained from the difference, $A - Z$.

We indicate an atom symbolically by writing its mass number as a superscript and its atomic number as a subscript. Both precede the atomic symbol

$$^{A}_{Z}X$$

The atomic number is often omitted in specifying an isotope because the atomic symbol, in effect, supplies the same information. Thus, $^{12}_{6}C$ and ^{12}C both represent carbon-12.

For example, a carbon atom (atomic number = 6) that has six neutrons would have a symbol of $^{12}_{6}C$. It is this isotope of carbon, incidentally, that serves as the basis of the current scale of atomic weights; that is, the mass of one atom of $^{12}_{6}C$ is defined as exactly 12 unified atomic mass units (u).

It should be noted that, except for carbon-12, there is a difference between an isotope's mass number and its actual mass. For example, the isotope ^{16}O has a mass number of 16, which means that the sum of the number of protons and neutrons is 16. However, the actual mass of an atom of ^{16}O is 15.994 91 u. The reasons for the discrepancy are somewhat complex and will be discussed in Chapter 22. The main point now is that you realize the mass and mass number have slightly different values.

The difference between an isotope's mass (in u) and its mass number is usually very small.

The fact that atoms of a given element can differ in the number of neutrons they contain has had practical applications. One example is in archeological dating. Radioactive isotopes, such as ^{14}C, undergo nuclear changes that cause them to be transformed into other elements. Atoms of carbon-14, for example, are unstable, and over a period of time each of them emits a β-particle and becomes a stable atom of $^{14}_{7}N$. This gradual transformation of the atoms of $^{14}_{6}C$ in a sample by the emission of β-particles is called the **radioactive decay** of this isotope. From the known rate of this decay and the relative abundances of ^{14}C and ^{12}C (a nonradioactive isotope of carbon) in both living and fossil material, the age of the fossil can be estimated. The details of radioactive decay and how it is applied to archeological dating will be discussed in greater detail in Chapter 22.

Another application of radioactive isotopes is in chemotherapy. Treatment of thyroid cancer, for instance, is accomplished by administering carefully controlled doses of radioactive ^{131}I, which tends to concentrate in the thyroid gland where the radiation produced by the ^{131}I causes destruction of cancerous cells. In this case, use is made of the body's natural tendency to concentrate iodine (either radioactive or nonradioactive) in the thyroid gland.

Very accurate measurements of the masses of isotopes and their relative abundances are made using a mass spectrometer.

As we noted above, nearly all elements as they are found in nature occur as mixtures of isotopes. For example, the element copper is found to contain the naturally occurring isotopes $^{63}_{29}Cu$ and $^{65}_{29}Cu$ whose masses have been accurately determined to be 62.9298 and 64.9278 u, respectively. Their relative abundances are 69.09% and 30.91%. The observed atomic weight of copper, 63.55, is obtained as an average of the isotopic masses, weighted according to the relative abundances of each isotope, as shown in Example 3.1.

EXAMPLE 3.1 CALCULATING THE AVERAGE ATOMIC WEIGHT FROM ISOTOPIC MASSES AND RELATIVE ABUNDANCES

PROBLEM Using the data supplied in the paragraph above, calculate the average atomic mass of copper.

SOLUTION The arithmetic involved in solving this problem is very simple, once you understand what has to be done. Suppose that we had one mole of naturally occurring copper. The data above tell us that it is composed of two isotopes, ^{63}Cu and ^{65}Cu. In addition, their relative abundances tell us that 69.09% of the atoms would be ^{63}Cu and 30.91% of the atoms would be ^{65}Cu. That's the same as saying that we would have 69.09% of a mole of ^{63}Cu and 30.91% of a mole of ^{65}Cu. We can easily calculate these masses, because we have the atomic masses of each of the isotopes.

The mass of a mole of ^{63}Cu is 62.9298 g, so the mass of 69.09% of a mole is

$$62.9298 \text{ g} \times 0.6909 = 43.48 \text{ g}$$

The mass of a mole of ^{65}Cu is 64.9278 g, so the mass of 30.91% of a mole is

$$64.9278 \text{ g} \times 0.3091 = 20.07 \text{ g}$$

Therefore, in one mole of naturally occurring copper there is 43.48 g of ^{63}Cu and 20.07 g of ^{65}Cu, so the total mass of the mole of copper atoms is

$$43.48 \text{ g} + 20.07 \text{ g} = 63.55 \text{ g}$$

If a mole of naturally occurring copper has a mass of 63.55 g, then the average atomic weight of copper must be 63.55 u.

REVIEW QUESTIONS AND PROBLEMS

(Problems whose numbers are in blue have their answers in Appendix C at the back of the book; the more difficult problems are marked with asterisks.)

Metals, Nonmetals, and Metalloids

3.1 Define *malleability*. Why is this an important property to a blacksmith?

3.2 Define *ductility*. What commercial process makes use of this property?

3.3 Besides malleability and ductility, give three other properties of metals.

3.4 Compare gold, iron, and sodium with respect to their reactivity toward air and moisture. Why isn't iron used to make jewelry?

3.5 The reaction of sodium with water produces gaseous hydrogen and sodium hydroxide, NaOH. Write a balanced chemical equation for this reaction.

3.6 Only two metals are colored. Which are they? (You should be able to answer this based on firsthand experience.)

3.7 Why do metals feel so hot compared to nonmetallic objects, when left in the summer sun?

3.8 Give an industrial application of gold that relies on this metal's very low degree of chemical reactivity.

3.9 Which metal has the highest melting point? What is one of its uses? Which metal has the lowest melting point? What is one of its uses?

3.10 Which nonmetals occur as diatomic molecules? Write their formulas. Which of these are gases, which is a liquid, and which is a solid?

3.11 What are the names of the two elemental forms of carbon? How are they alike, and how are they different?

3.12 Referring to Figure 1.12 on page 26, compare those properties of copper and sulfur that are evident from the photograph. Is it evident which is a metal and which is a nonmetal?

3.13 Which two nonmetals are the principal constituents of the atmosphere?

3.14 Which is the most chemically reactive nonmetal?

3.15 In what ways are the metalloids similar to metals, and how do they differ from metals?

The First Periodic Table

3.16 What was the basis on which Mendeleev constructed his periodic table?

3.17 Why were there blanks left in Mendeleev's periodic table?

3.18 Why was Mendeleev able to predict the properties of the elements that belonged in the blanks in his table?

3.19 The element gallium was not known when Mendeleev constructed his table. Here are some properties of aluminum (just above gallium) and indium (just below gallium) along with those actually observed for gallium.

	Aluminum	Indium	Gallium
Atomic weight	27	115	70
Melting point	660°C	157°C	30°C
Boiling point	2467°C	2000°C	2403°C
Formula of chloride	$AlCl_3$	$InCl_3$	$GaCl_3$
Formula of oxide	Al_2O_3	In_2O_3	Ga_2O_3
Melting point of chloride	190°C	586°C	78°C

What would Mendeleev have predicted for these properties of gallium? How do they compare with those actually observed?

3.20 Besides Te and I, Ar and K, find *three* other places in the modern periodic table where elements occur in reverse atomic-weight order.

3.21 Which group of elements appears in the modern periodic table, but did not appear in Mendeleev's table? Why?

The Modern Periodic Table

3.22 Make a sketch of the modern periodic table in which the lanthanides and actinides are placed in the body of the table in their proper locations.

3.23 What do we call a column of elements in the periodic table? What do we call a row of elements?

3.24 State the modern version of the periodic law.

3.25 Which of the following are representative elements: Mg, Ti, Fe, Se, Ni, Br?

3.26 Which of the following are transition elements: Sr, Ru, As, W, Ag, Al?

3.27 Which elements constitute the inner transition elements?

3.28 Which of these elements is a halogen: Na, Ca, Fe, F, As?

3.29 Write the formulas of the molecules formed by the halogens in their elemental state.

3.30 Which of these elements is an alkali metal: Br, K, O, S, N?

3.31 Which of these elements is an alkaline earth metal: Cu, B, Ba, Ne, Se?

3.32 Which of the following are metals: Ta, Nd, Se, F, Cs?

3.33 Write the symbols for the metalloids.

3.34 How does the metallic character of the elements vary from left to right across a period? How does it vary from top to bottom to Group VA?

3.35 Identify the metals, nonmetals, and metalloids in period 4. Identify them in Group VA.

3.36 From what you've learned earlier in this chapter, which of the following metals would be especially reactive toward oxygen and moisture: Fe, Pt, K, Mn, or Sn? On what information do you base your answer?

3.37 Radium, a highly radioactive element, forms a water-soluble compound with chlorine. From what you've learned in this chapter, what would you expect its formula to be?

3.38 You encounter certain metals quite often on a day-to-day basis. Identify those in the following list that are transition metals:
(a) aluminum   (e) tin
(b) iron   (f) silver
(c) chromium   (g) gold
(d) copper   (h) lead

Properties of the Electron

3.39 What discovery was made in 1834 by Michael Faraday? What did it suggest about the nature of matter?

3.40 Sketch and label a typical gas discharge tube.

3.41 What properties are observed for cathode rays?

3.42 What is meant by the term *fundamental particle?* What suggests that the particles making up cathode rays are fundamental particles?

3.43 As a charged particle passes between a pair of electrically charged plates, the extent of deflection that it experiences is directly proportional to its charge. Explain why this is so. Explain why the extent of deflection is inversely proportional to the particle's mass.

3.44 What is the definition of a coulomb? From the measured electrical charge on an electron, calculate the number of coulombs of charge carried by 1 mol of electrons.

3.45 What is the mass in grams of 1 mol of electrons? Hydrogen atoms each contain one electron. From the atomic mass of hydrogen, calculate the percentage of the mass of an "average" hydrogen atom that is contributed by an electron.

3.46 In Millikan's experiment, the charge on the oil droplets was always found to be a multiple of -1.60×10^{-19} C. Suppose that this experiment were repeated and the following values were obtained.

-3.20×10^{-19} C -2.40×10^{-19} C
-5.60×10^{-19} C -7.20×10^{-19} C
-6.40×10^{-19} C

On the basis of these data, what would be the charge on the electron?

Positive Particles

3.47 What is the origin of the positive particles that exist in a gas discharge tube?

3.48 What are canal rays? What are they composed of?

3.49 Why was the hydrogen ion assumed to be a fundamental particle (i.e., a particle that cannot be broken into something simpler)? What name is given to this particle?

3.50 How are particles of different mass separated in a mass spectrometer?

3.51 From the data in Table 3.2, calculate the charge-to-mass ratio of the proton. How does this value compare to the charge-to-mass ratio of an electron?

3.52 In terms of atomic structure, what is an atom's atomic number?

3.53 What distinguishes an atom of nitrogen from an atom of oxygen?

3.54 Aluminum atoms form ions having a charge equal to $+4.8 \times 10^{-19}$ C. How should we write the symbol for this ion?

3.55 The element selenium forms an ion with a charge of -3.2×10^{-19} C. How should we write the symbol for this ion?

The Nuclear Atom and Nuclear Particles

3.56 Why did the results of Rutherford's scattering experiments lead him to conclude that all of an atom's positive charge is concentrated in a very small, very dense nucleus?

3.57 Describe the three important kinds of radiation that are observed to be emitted by radioactive substances.

3.58 An alpha particle is the nucleus of a helium atom. In a mass spectrometer, would its path be deflected more or less than that of a proton? Explain your answer.

3.59 Give the charge, in coulombs, carried by each of the following:
(a) an alpha particle
(b) a proton
(c) an electron
(d) a beta particle
(e) a gamma ray
(f) a neutron

3.60 The proton (the nucleus of a hydrogen atom) has a mass of 1.67×10^{-24} g. Suppose that its diameter is 1.00×10^{-13} cm. Calculate the density of the nucleus, assuming it is spherical in shape. (*Note:* A sphere's volume, V, can be calculated from its radius, r, by the equation $V = \frac{4}{3}\pi r^3$.)

3.61 If the diameter of a nucleus is approximately 10^{-13} cm and the diameter of an atom is approximately 2×10^{-8} cm, calculate approximately the percentage of the volume of an atom that is occupied by its nucleus.

*3.62** The earth has a mass of 5.98×10^{24} kg and a diameter of approximately 13 000 km. What would the diameter of the earth (in km) be if it had the same mass but was composed entirely of nuclear material? (Use the density of nuclear material calculated in Problem 3.60.)

3.63 Why did Rutherford propose the existence of neutrons?

3.64 For each of the following, give the approximate mass in u and the charge in units of 1.60×10^{-19} C.
(a) proton   (c) electron
(b) neutron   (d) α particle

Isotopes

3.65 Explain why the atomic masses of some elements (such as Cl or Cu, for instance) are so far from whole numbers.

3.66 What is the difference between an atom's mass number and its actual atomic mass?

3.67 What are the numbers of protons, neutrons, and electrons in each of the following: $^{132}_{55}Cs$, $^{115}_{48}Cd^{2+}$, $^{194}_{81}Tl$, $^{105}_{47}Ag^+$, $^{78}_{34}Se^{2-}$?

3.68 What are the numbers of protons, neutrons, and electrons in each of the following: $^{131}_{56}Ba$, $^{109}_{48}Cd^{2+}$, $^{36}_{17}Cl^-$, $^{63}_{28}Ni$, $^{170}_{69}Tm$?

3.69 Write the appropriate symbol for each of the following isotopes:

(a) $Z = 26$, $A = 55$ (d) $Z = 71$, $A = 170$

(b) $Z = 37$, $A = 86$ (e) $Z = 70$, $A = 169$

(c) $Z = 81$, $A = 204$

3.70 How many neutrons are in each of the atoms in Question 3.69?

3.71 What isotope serves as the current standard for the atomic mass scale?

3.72 The element Eu occurs naturally as a mixture of 47.82% $^{151}_{63}Eu$, whose mass is 150.9 u, and 52.81% $^{153}_{63}Eu$, whose mass is 152.9 u. Calculate the average atomic mass of Eu.

3.73 Naturally occurring boron consists of two isotopes, ^{10}B with a mass of 10.012 94 and ^{11}B with a mass of 11.009 31. The abundance of ^{10}B is 19.6% and that of ^{11}B is 80.4%. Calculate the average atomic mass of B.

3.74 Naturally occurring lead is composed of four isotopes. Their abundances and masses are given below. Calculate the average atomic mass of lead.

Isotope	Mass (u)	Abundance (%)
^{204}Pb	203.973	1.48
^{206}Pb	205.9745	23.6
^{207}Pb	206.9759	22.6
^{208}Pb	207.9766	52.3

*3.75 Naturally occurring chlorine is composed of ^{35}Cl with atomic mass of 34.96885 and ^{37}Cl with atomic mass of 36.96590. The average atomic mass of Cl is 35.453. What are the percentages of each isotope in naturally occurring chlorine?

*3.76 Ordinary silver is a mixture of two isotopes, ^{107}Ag with a mass of 106.9041 u and ^{109}Ag with a mass of 108.9047 u. The average relative atomic weight of silver is 107.868 u. What are the relative abundances, expressed as percents, of these two silver isotopes?

CHAPTER 4

ELECTRONIC STRUCTURE AND THE PERIODIC TABLE

Much of our knowledge about the internal structures of atoms has as its origin the fact that atoms are only able to lose energy in certain specific amounts, and that this energy can appear as light of particular colors. In a laser, one of the most intriguing new tools of science and technology, many atoms lose energy all at once, which produces a very intense light beam. One of its uses is in the recording (shown above) and reconstruction of three-dimensional images called holographs, a technique that may someday bring us 3-D television and movies.

In Chapter 3 you learned that atoms are not the simplest forms of matter. They are composed of still simpler things: protons and neutrons that are located in a tiny nucleus in the center of the atom, surrounded by enough electrons to make the atom electrically neutral. This picture of atomic structure is sufficient to explain some properties of the elements—the existence of isotopes, for example— but it still doesn't explain their chemical and physical properties. Why, for example, is sodium so much more like the element below it, potassium, than the element alongside, magnesium? In terms of atomic structure, what is it that causes us to place sodium and potassium in the same group?

When atoms react with each other, it is only their outer parts that ever come into contact. Their nuclei are so tiny and buried so deeply inside the atoms that they never come close to each other. Therefore, since it is the electrons that occupy the outer regions of the atom, chemical similarities and differences must somehow be related to the way that the electrons are arranged. This arrangement of electrons is called the atom's **electronic structure.**

We begin this chapter with a discussion of the experimental facts that helped solve the mystery of electronic structure and led ultimately to our currently accepted theory. Our goal is to be able to describe the electronic structure of an atom and relate it to the atom's position in the periodic table. From this knowledge we can begin to understand some of the trends in the chemical and physical properties of the elements.

4.1   ELECTROMAGNETIC RADIATION AND ATOMIC SPECTRA

The key that has permitted the deduction of the electronic structures of the elements is an analysis of the light that atoms emit when they are energized, either by being heated in a flame or by passing an electric discharge through them. However, before we can discuss this, we must first learn what light is.

Light in all its varied forms—X rays, visible light, infrared and ultraviolet radiation, and radio and TV waves—is called **electromagnetic radiation.** It is a form of energy that travels through space as a wave at a constant speed called the speed of light, 3.00×10^8 m s^{-1}. Light waves, like waves in general, are characterized by certain properties. One is the intensity of the wave, which is measured by its **amplitude,** the maximum height of its peaks, as shown in Figure 4.1. Another characteristic of the wave is its **wavelength,** λ, the distance between consecutive peaks or troughs in the wave. This is also illustrated in Figure 4.1. A third quantity that characterizes a wave is its **frequency,** ν, which is the number of peaks that pass a given point in one second as the wave moves by. For any wave, a relationship between its wavelength and its frequency exists.

$$(\text{wavelength}) \times (\text{frequency}) = \text{speed of the wave}$$

For light waves, the speed is represented by the symbol c. This gives the useful equation

$$\lambda \cdot \nu = c \qquad\qquad\qquad\qquad [4.1]$$

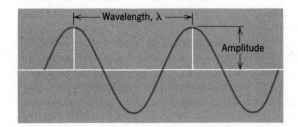

Figure 4.1

Properties of a wave.

Figure 4.2

The electromagnetic spectrum.

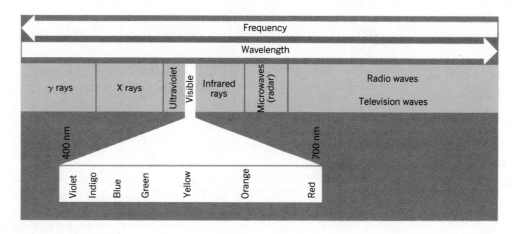

The visible spectrum ranges from approximately 400 to 700 nm.

Light waves come in all sorts of wavelengths, giving rise to the **electromagnetic spectrum** that is illustrated in Figure 4.2. The units used to express these wavelengths depend on the region of the spectrum in which the radiation occurs. CB radios, for example, broadcast waves having wavelengths of about 11 metres, so metres are convenient units to express these wavelengths. Visible light, which encompasses only the very narrow portion of the electromagnetic spectrum to which our eyes are sensitive, has wavelengths of the order of several hundred nanometres, so this unit is convenient for reporting the lengths of visible light waves.

Frequency, in general, describes the number of events *per unit of time*. If we are discussing family gatherings, for instance, they might take place *three times per year*. If we are speaking of an alternating electric current, we find that the current reverses direction *120 times per second*. In each case, the notion of frequency is in the *per unit of time*, and the unit that we associate with frequency has the dimension of 1/time or time^{-1}. In SI, the base unit of time is the second, s. Therefore, the SI unit of frequency is 1/s or s^{-1}. This unit is given the name **hertz,** and its symbol is **Hz.**

$$1 \text{ Hz} = 1 \text{ s}^{-1}$$

EXAMPLE 4.1   CONVERTING FREQUENCY TO WAVELENGTH

PROBLEM   A ham radio operator broadcasts at a frequency of 14.2 MHz (megahertz). What is the wavelength of the radio waves put out by the transmitter?

SOLUTION   Since we wish to calculate the wavelength, let's solve Equation 4.1 for λ.

$$\lambda = \frac{c}{\nu}$$

The speed of light can be expressed as 3.00×10^8 m s^{-1}, and the frequency can be written $\nu = 14.2 \times 10^6$ Hz or 14.2×10^6 s^{-1} (remember, *mega* means $\times 10^6$). Substituting these values gives

$$\lambda = \frac{3.00 \times 10^8 \text{ m s}^{-1}}{14.2 \times 10^6 \text{ s}^{-1}} = 21.1 \text{ m}$$

EXAMPLE 4.2   CONVERTING FREQUENCY TO WAVELENGTH

PROBLEM   What is the wavelength, in nanometres, of green light having a frequency of 6.67×10^{14} Hz?

SOLUTION | Again we solve Equation 4.1 for λ ($\lambda = c/\nu$). Using $c = 3.00 \times 10^8$ m s^{-1} and $\nu = 6.67 \times 10^{14}$ s^{-1}, we get

$$\lambda = \frac{3.00 \times 10^8 \text{ m s}^{-1}}{6.67 \times 10^{14} \text{ s}^{-1}} = 4.50 \times 10^{-7} \text{ m}$$

The SI prefix *nano* means $\times 10^{-9}$.

Since 1 nm = 10^{-9} m,

$$\lambda = 4.50 \times 10^{-7} \text{ m} \times \left(\frac{1 \text{ nm}}{10^{-9} \text{ m}}\right)$$

$$= 450 \text{ nm}$$

EXAMPLE 4.3 | CONVERTING WAVELENGTH TO FREQUENCY

PROBLEM | What is the frequency of infrared radiation that has a wavelength of 1.25×10^3 nm?

SOLUTION | First we solve Equation 4.1 for ν ($\nu = c/\lambda$). If we use $c = 3.00 \times 10^8$ m s^{-1}, we must express the wavelength in metres.

$$\lambda = 1.25 \times 10^3 \text{ nm} \times \left(\frac{10^{-9} \text{ m}}{1 \text{ nm}}\right)$$

$$= 1.25 \times 10^{-6} \text{ m}$$

Now we can solve for the frequency.

$$\nu = \frac{3.00 \times 10^8 \text{ m s}^{-1}}{1.25 \times 10^{-6} \text{ m}}$$

$$= 2.40 \times 10^{14} \text{ s}^{-1}$$

The frequency is 2.40×10^{14} Hz.

Atomic spectra

Gas discharge tubes were discussed on page 90.

If sunlight or the light from an ordinary electric light bulb is formed into a narrow beam and then passed through a prism onto a screen, a rainbow of colors is produced (Figure 4.3). By refraction, the prism splits the white light into a spectrum that is composed of light of all colors. This "rainbow" is called a **continuous spectrum,** because all the wavelengths of light are present. However, if the light is emitted by a gas discharge tube containing a gas such as hydrogen or sodium vapor, or if the light comes from a compound heated in a flame, the spectrum produced is quite different in appearance, as shown in Figure 4.4 on p. 108. Instead of a rainbow of colors, only a few colored lines are seen. These lines are the image of the slit used to form the narrow beam of light, and because of the appearance of the spectrum, it is often referred to as a **line spectrum.** Because the light is emitted by an element that has been *excited* (energized), it is also referred to as an **atomic emission spectrum,** or simply an **emission spectrum,** or **atomic spectrum.**

It is quite obvious that the light given off by an element when it produces an emission spectrum does not consist of radiation of all wavelengths. The dark regions of the spectrum correspond to wavelengths for which light is *not* emitted. Each element has its own characteristic emission spectrum that can be used to identify it. For example, you have probably seen crime shows on TV in which the police take paint chips from the clothing of a hit-and-run victim for analysis.

Figure 4.3

Refraction of white light by a prism separates the light into a continuous spectrum, which contains light of all wavelengths, or colors.

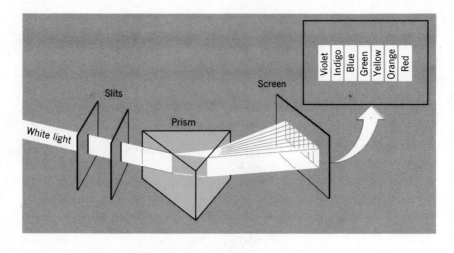

Mercury vapor lamps, which are nearly as efficient as sodium vapor lamps, give a bluish-white light.

Johann Balmer was a Swiss mathematician and physicist who taught secondary school in Basel, Switzerland. He also worked part time at the University of Basel teaching geometry. He died in 1898.

If the elements contained in the paint, along with their relative amounts, match a similar analysis of paint scrapings taken from a suspect's car, there is fairly strong evidence that the suspect's car was involved in the crime. Such an analysis can be done easily by using atomic spectra, because each element in the paint produces its own characteristic set of emission lines.

The atomic spectra of certain elements are also used in modern street lighting. In older incandescent lighting, in which a tungsten filament is heated white-hot by an electric current, much of the emitted energy is in the form of infrared light. Unfortunately, our eyes are not sensitive to this form of radiation, so this energy is wasted—only a relatively small fraction of the electrical energy actually appears as visible light. Modern street lights consist of high intensity mercury or sodium vapor lamps in which the light is produced by the atomic emission spectra of these elements. In these lamps, most of the electrical energy appears as light in the visible portion of the spectrum, and they are therefore much more efficient. For example, sodium emits intense light at a wavelength of 589 nm, which is yellow. The golden glow of some street lights is from this emission line of sodium, which is produced in high-pressure sodium vapor lamps. There is another, much more yellow light emitted by low-pressure sodium vapor lamps that you may also have seen. Both kinds of sodium lamps are very energy efficient, and many communities have switched to this kind of lighting to save money on the costs of electricity.

The occurrence of line spectra baffled physicists for many years. In 1884 Balmer found that there was a relatively simple equation that could be used to calculate the wavelengths of all the lines in the visible spectrum of hydrogen:

$$\frac{1}{\lambda} = 109\ 678\ \text{cm}^{-1} \left(\frac{1}{2^2} - \frac{1}{n^2} \right) \tag{4.2}$$

where λ is the wavelength in cm and n is an integer that can have values of 3, 4, 5, 6, . . . , ∞. By choosing a particular value of n, the wavelength of a line in the spectrum can be calculated. Thus, when $n = 3$,

$$\frac{1}{\lambda} = 109\ 678\ \text{cm}^{-1} \left(\frac{1}{4} - \frac{1}{9} \right)$$

Figure 4.4

Formation of the emission spectrum of hydrogen. The four lines whose wavelengths are listed are those that can be seen in the visible portion of the hydrogen spectrum.

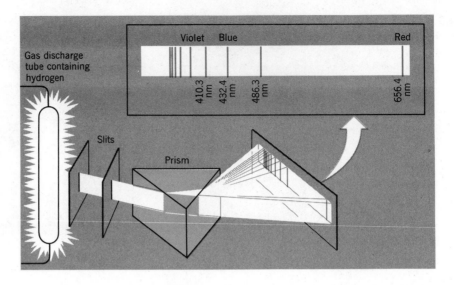

$$\frac{1}{\lambda} = 15\ 233\ \text{cm}^{-1}$$

or

$$\lambda = 6.565 \times 10^{-5}\ \text{cm} = 656.5\ \text{nm}$$

Similarly, when $n = 4$, 5, and 6, we compute λ to be 486.3, 432.4, and 410.3 nm. These values, as we can see from Figure 4.4, are equal to the wavelengths of the lines in the visible portion of the hydrogen spectrum. All the lines related by Equation 4.2 constitute what is called the *Balmer series*.

The hydrogen spectrum in Figure 4.4 lists only the lines that appear in the visible region of the spectrum. Light is also emitted by hydrogen in the infrared and ultraviolet regions. The wavelengths of these other series of lines can be fitted to the general equation (called the Rydberg equation)

$$\frac{1}{\lambda} = 109\ 678\ \text{cm}^{-1} \left(\frac{1}{n_1^2} - \frac{1}{n_2^2} \right) \qquad [4.3]$$

where n_1 and n_2 are integers that may assume values of 1, 2, 3, . . . , ∞ with the requirement that n_2 is always greater than n_1. Thus, when $n_1 = 1$, the values of n_2 can be 2, 3, 4, . . . , ∞ and the lines in the **Lyman series** are obtained. When $n_1 = 2$ and $n_2 = 3$, 4, 5, . . . , ∞, we get the Balmer series. These and other series are summarized in Figure 4.5 and Table 4.1.

Table 4.1

Series of lines in the hydrogen spectrum

Series	n_1	n_2
Lyman	1	2, 3, 4, . . . , ∞
Balmer	2	3, 4, 5, . . . , ∞
Paschen	3	4, 5, 6, . . . , ∞
Brackett	4	5, 6, 7, . . . , ∞
Pfund	5	6, 7, 8, . . . , ∞

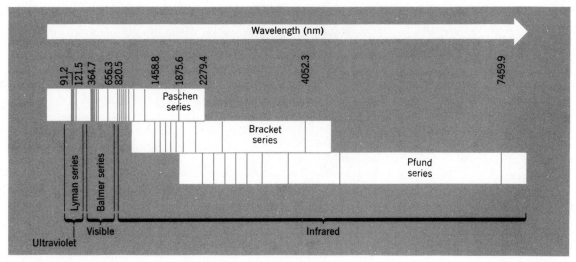

Figure 4.5

Series of lines in the hydrogen spectrum.

EXAMPLE 4.4 USING THE RYDBERG EQUATION TO CALCULATE THE WAVELENGTHS OF LINES IN THE HYDROGEN SPECTRUM

PROBLEM Calculate the wavelength of the third line in the Brackett series for hydrogen.

SOLUTION The Rydberg equation gives the reciprocal of wavelength

$$\frac{1}{\lambda} = 109\ 678\ \text{cm}^{-1} \left(\frac{1}{n_1^2} - \frac{1}{n_2^2} \right)$$

For the Brackett series (Table 4.1), $n_1 = 4$. The third line in the series would correspond to $n_2 = 7$. Substituting gives

$$\frac{1}{\lambda} = 109\ 678\ \text{cm}^{-1} \left(\frac{1}{4^2} - \frac{1}{7^2} \right)$$

$$= 109\ 678\ \text{cm}^{-1} (0.042\ 091\ 8)$$

$$= 4616.55\ \text{cm}^{-1}$$

Taking the reciprocal gives

$$\lambda = 2.166\ 12 \times 10^{-4}\ \text{cm}$$

Expressed in nanometres (1 nm = 10^{-9} m)

$$\lambda = 2166.12\ \text{nm}$$

For many years X rays were called roentgen rays.

Henry Moseley's death during the invasion of Gallipoli in WWI led Britain to assign noncombat duties to its scientists during WWII.

A postscript to this story of atomic spectra is the discovery of atomic numbers by Henry Moseley. In 1895, Wilhelm Roentgen (1845–1923) discovered that when high energy electrons in a discharge tube collide with the anode, a very penetrating kind of radiation is produced. Roentgen called this radiation X rays. An X-ray tube is illustrated in Figure 4.6. Moseley discovered that the frequencies of the X rays produced by the tube depended on the material used for the anode. Thus, each element produces its own characteristic X-ray spectrum. In analyzing the frequencies of these X rays, Moseley found that they could be related to the location of the elements in the periodic table. As noted earlier

Figure 4.6
Production of X rays. A high-voltage electric discharge across the electrodes causes the anode to emit X rays.

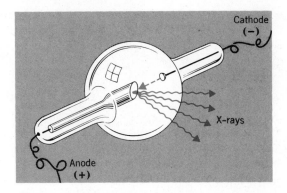

(page 87), he was able to assign an integer—the atomic number—that was the same as an element's position-number in the table. Experiments by Rutherford and his students enabled Moseley to conclude that this atomic number represented the number of protons in the nucleus.

4.2 THE BOHR THEORY OF THE HYDROGEN ATOM

Early attempts to account for the existence of line spectra on the basis of the motion of electrons in the atom met with complete failure. If an electron is moving around a nucleus, it must be following a curved path; otherwise it would simply leave the atom. However, a particle moving at a constant speed along a curved path undergoes an acceleration[1] and, according to the accepted laws of physics at that time, a charged particle (such as the electron) that undergoes an acceleration should continuously lose energy by emitting electromagnetic radiation. In fact, that's how radio and TV signals are broadcast. Electricity is pumped up and down an antenna at the proper frequency. The starting and stopping (acceleration and deceleration) of the charge causes the antenna to radiate the electromagnetic waves. In terms of the atom, the known laws of physics implied that the electron should gradually lose energy and spiral in toward the nucleus, resulting in the collapse of the atom. Since atoms do not collapse, physicists were faced with a problem that challenged their most fundamental theories.

The way out of this problem found its origin in the work of Max Planck (1900) and Albert Einstein (1905). They had demonstrated that, in addition to

[1] For example, consider what happens to an object, such as a coin, when it is placed on the edge of a rapidly spinning phonograph record. Anyone who has attempted this has found that the object is thrown outward, away from the center of the record as shown in the figure below. The coin obviously experiences a force (called centrifugal force) that causes it to fly off the edge. Since the coin possesses mass, and there is the relationship that *force = mass × acceleration,* the coin must also be experiencing an acceleration as it moves in its *circular* path at a constant speed at the edge of the record.

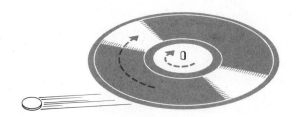

A coin is thrown outward from the center of a spinning record.

possessing wave properties, light also has particle properties. Thus there are instances when light behaves as if it were composed of tiny packets, or **quanta,** of energy (later called **photons**). The energy, E, of the photon emitted or absorbed by a substance is proportional to the frequency of the light, ν. These two quantities are related by the equation

$$E_{\text{photon}} = h\nu \tag{4.4}$$

where h is a proportionality constant called **Planck's constant,** which has a value of 6.63×10^{-34} joule seconds (the units are a product of energy × time).

In 1913, a Danish physicist Neils Bohr incorporated the ideas of Planck and Einstein into a theory that enabled him to calculate the energy of the electron in the hydrogen atom, and from this he was able to account for the atomic spectrum of hydrogen. Unfortunately, the theory failed to calculate energies correctly for any atoms more complex than hydrogen and it has since been replaced by a more successful one. Bohr's theory is instructive to look at briefly, though, because it illustrates how theories about the submicroscopic world of atoms develop and how they are tested.

Bohr's approach to the structure of the atom was simply to postulate that because atoms do not collapse and *because light is emitted by an atom only at certain frequencies* (which means that only certain specific energy changes occur), the electron in an atom can possess only certain, restricted quantities of energy. This is often phrased in a somewhat more esoteric way by saying that the energy of the electron is *quantized*. This means that the electron can have only certain discrete quantities of energy and none in between. We express this by saying that the electron is restricted to specific **energy levels** in the atom.

Bohr's theoretical model of the atom imagined that the electron travels around the nucleus in orbits of fixed size and energy (Figure 4.7a). From this model he mathematically derived an equation for the energy of the electron that had the form

$$E = -A\frac{1}{n^2} \tag{4.5}$$

If the electron can't have energies between its quantized values, it cannot lose energy a little at a time and spiral into the nucleus.

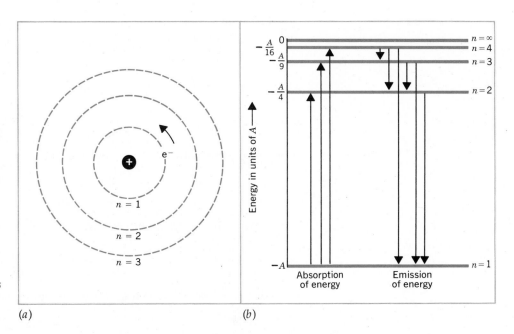

Figure 4.7

Bohr's model of the atom. (a) The electron is able to travel along certain specific orbits of fixed energy. This part of Bohr's theory was wrong. (b) The energy of the electron changes by specific quantities when the electron goes from one energy level to another. This part of Bohr's theory was correct and is incorporated in our currently accepted theory.

in which the constant A could be evaluated from a knowledge of the mass and charge of the electron and Planck's constant. The value of A is 2.18×10^{-18} joule. The quantity n is an integer, called a **quantum number,** that can have only whole-number values of 1, 2, 3, and so on, up to infinity. The quantum number serves to identify the orbit of the electron, and the energy of an electron in a particular orbit depends on the value of n (Figure 4.7b). The lowest energy level occurs when $n = 1$, since this yields the largest value for the fraction $1/n^2$ and thus the most negative (and therefore lowest) E. The idea of a negative energy seems rather odd at first glance. Actually, the minus sign occurs because of a rather arbitrary choice of the zero point on the energy scale. We will learn later that we can only measure differences in energy, so the choice of where the zero point is placed is really unimportant.

With his theory Bohr had created a model describing how the electron behaves in an atom. His theory, just like any other theory, must be able to be checked experimentally; otherwise we cannot know if it's wrong. Of course, there is no way actually to observe the electron. Therefore, indirect evidence must be used to check the validity of the model. To do this, Bohr mathematically derived an equation for the wavelengths of the light emitted by hydrogen when it produces its atomic spectrum. According to Bohr, when energy is absorbed by an atom, for example in an electric discharge, the electron is raised in energy from one level to another, and when the electron returns to a lower energy level, a photon is emitted whose energy is equal to the difference between the two levels (see Figure 4.7). If we take n_2 to be the quantum number of the upper level and n_1 to be that of the lower (so that $n_2 > n_1$), the difference in energy, ΔE, between the two is

$$\Delta E = E_{n_2} - E_{n_1} \qquad [4.6]$$

$$\Delta E = \left(-A\frac{1}{n_2{}^2}\right) - \left(-A\frac{1}{n_1{}^2}\right)$$

which can be written as

$$\Delta E = A\left(\frac{1}{n_1{}^2} - \frac{1}{n_2{}^2}\right) \qquad [4.7]$$

If this energy difference appears as a photon, it would have a frequency, ν, that could be calculated from Equation 4.4,

$$\Delta E = h\nu$$

which, by incorporating Equation 4.1, can be expressed as

$$\Delta E = h\frac{c}{\lambda} = hc\frac{1}{\lambda}$$

Substituting this into Equation 4.7, we get

$$hc\frac{1}{\lambda} = A\left(\frac{1}{n_1{}^2} - \frac{1}{n_2{}^2}\right) \qquad [4.8]$$

which, upon rearrangement, yields

$$\frac{1}{\lambda} = \frac{A}{hc}\left(\frac{1}{n_1{}^2} - \frac{1}{n_2{}^2}\right) \qquad [4.9]$$

The quantity A/hc has a value of 109 730 cm^{-1}, so our final equation is

$$\frac{1}{\lambda} = 109\ 730\ \text{cm}^{-1} \left(\frac{1}{n_1^2} - \frac{1}{n_2^2} \right) \qquad [4.10]$$

Using a theoretical model to derive an equation that allows a computed value to be compared with a measured value is an approach frequently followed by theoreticians.

Comparing Equations 4.3 and 4.10, we see that they are virtually identical. The Rydberg equation (4.3) is obtained from experimental observation while Equation 4.10 is derived from theory. This match between theory and experiment would suggest that Bohr was on the right track. Unfortunately, this approach was not at all successful with atoms more complex than hydrogen; however, his introduction of the notion of quantum numbers and quantized energy levels played a significant role in the development of our understanding of atomic structure.

EXAMPLE 4.5 CALCULATING ENERGY CHANGES IN THE HYDROGEN ATOM

PROBLEM Calculate the energy required to remove an electron from the lowest energy level of the hydrogen atom to produce the H^+ ion.

SOLUTION The lowest energy level has $n = 1$. The electron becomes free of the atom if it is raised to the level with $n = \infty$. The energy required to raise the electron from $n = 1$ to $n = \infty$ is given by Equation 4.6,

$$\Delta E = E_\infty - E_1$$

or, as shown in Equation 4.7,

$$\Delta E = A \left(\frac{1}{1^2} - \frac{1}{\infty^2} \right)$$

Substituting the value of A, 2.18×10^{-18} J, gives

$$\Delta E = 2.18 \times 10^{-18}\ \text{J} \left(\frac{1}{1^2} - \frac{1}{\infty^2} \right)$$

The value of $1/\infty^2$ is zero and $1^2 = 1$. Therefore,

$$\Delta E = 2.18 \times 10^{-18}\ \text{J}$$

EXAMPLE 4.6 CALCULATING ENERGY CHANGES IN THE HYDROGEN ATOM

PROBLEM Calculate the energy liberated when an electron drops from the fifth to the second energy level in hydrogen.

SOLUTION We have $n_2 = 5$, $n_1 = 2$.

$$\Delta E = 2.18 \times 10^{-18}\ \text{J} \left(\frac{1}{2^2} - \frac{1}{5^2} \right)$$

$$\Delta E = 2.18 \times 10^{-18}\ \text{J} \left(\frac{1}{4} - \frac{1}{25} \right)$$

$$\Delta E = 4.58 \times 10^{-19}\ \text{J}$$

The energy liberated is 4.58×10^{-19} J.

4.3 THE WAVE NATURE OF MATTER: WAVE MECHANICS

The currently accepted theory explaining the behavior of electrons in atoms is called **wave mechanics,** which has its roots in a hypothesis put forward by Louis de Broglie in 1924. De Broglie suggested that if light can behave in some instances as if it were composed of particles, perhaps particles, at times, exhibit properties that we normally associate with waves.

De Broglie's argument proceeded as follows. Einstein had shown that the energy equivalent, E, of a particle of mass, m, is equal to

$$E = mc^2 \qquad [4.11]$$

where c is the speed of light. A photon whose energy is E could thus be said to have an effective mass equal to m. Max Planck had shown that the energy of a photon is given by Equation 4.4,

$$E = h\nu = \frac{hc}{\lambda}$$

Equating these two gives

$$\frac{hc}{\lambda} = mc^2$$

When we solve for λ, the wavelength, we obtain

$$\lambda = \frac{h}{mc}$$

If this equation also applies to particles, such as the electron, the equation can be written as

$$\lambda = \frac{h}{m\nu} \qquad [4.12]$$

where we have replaced c, the speed of light, by v, the speed of the particle.

Experimental evidence for this dual wave-particle nature of matter exists in the form of a phenomenon called diffraction, a property that can only be explained by wave motion. If light is allowed to pass through a very thin slit whose width is about the same as the wavelength of the light, the slit behaves as if it were a tiny light source, scattering light in all directions. This phenomenon is called **diffraction.** If two of these slits are placed alongside each other, each behaves as a separate source of light. When a screen is placed so that this light falls on it, we observe a pattern, called a **diffraction pattern,** that consists of light and dark areas as shown in Figure 4.8. In the bright areas, light waves that arrive

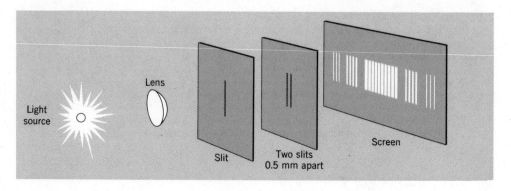

Figure 4.8

Production of a diffraction pattern.

Figure 4.9

Constructive and destructive interference.

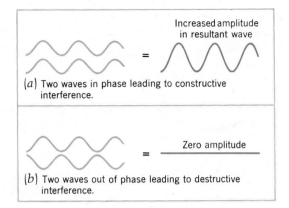

(a) Two waves in phase leading to constructive interference.

Increased amplitude in resultant wave

(b) Two waves out of phase leading to destructive interference.

Zero amplitude

from each hole are *in phase;* that is, the peaks and troughs of the two waves are lined up so that the amplitudes of the waves add together to produce a resultant wave of greater intensity. This is illustrated in Figure 4.9a. In the darkened areas, waves that arrive from the two slits are *out of phase* with each other, which means that the peaks of one wave coincide with the troughs of the other. When this happens, the amplitudes of the waves cancel (Figure 4.9b), so that zero intensity (darkness) is observed. Similar diffraction patterns can be produced with certain particles, including electrons, protons, and neutrons. Since diffraction can only be explained as a property of waves, this confirms the wave nature of matter. The reason that the wave nature of matter was not discovered earlier is that objects large enough to see, either with the naked eye or with the aid of a microscope, possess so much mass that their wavelengths are much too short to be observed.

EXAMPLE 4.7 APPLYING THE DE BROGLIE RELATIONSHIP

PROBLEM What is the wavelength of a grain of sand that has a mass of 0.000010 g and is moving at a speed of 0.010 m s^{-1} (36 m h^{-1})?

SOLUTION We must use Equation 4.12,

$$\lambda = \frac{h}{mv}$$

Planck's constant has a value of

$$h = 6.63 \times 10^{-34} \text{ J s} \quad (\text{J s} = \text{joule} \times \text{second})$$

Since 1 J = 1 kg m^2 s^{-2}, we can write h as

$$h = 6.63 \times 10^{-34} \text{ (kg m}^2\text{/s}^2\text{) s} = 6.63 \times 10^{-34} \text{ kg m}^2\text{/s}$$

We are given that

$$m = 1.0 \times 10^{-5} \text{ g} = 1.0 \times 10^{-8} \text{ kg}$$
$$v = 0.010 \text{ m s}^{-1}$$

Substituting these quantities into our equation, we have

$$\lambda = \frac{6.63 \times 10^{-34} \text{ kg m}^2\text{/s}}{(1.0 \times 10^{-8} \text{ kg})(0.010 \text{ m/s})} = 6.6 \times 10^{-24} \text{ m}$$

This wavelength is far too small to be detected by any device existing at this time. Larger objects have even larger masses and therefore still smaller wavelengths.

Figure 4.10

Vibrations on a guitar string.
(a) An open string is plucked.
(b) A harmonic produced by
briefly touching the string at its
midpoint as it is plucked. (c) A
still higher harmonic.

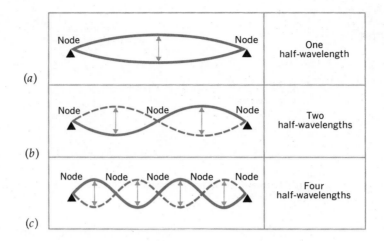

We saw in Section 4.2 that one of the significant results of the Bohr theory of the atom was the introduction of integer quantum numbers. If we consider the electron to be a wave that surrounds the nucleus, we find that the appearance of integers occurs in a very natural way. To see how this might happen, let's first consider some simpler waves—these that occur on a guitar string. When a string is plucked it moves up and down in the center, but its ends remain motionless as shown in Figure 4.10a. At the ends, the height of the wave—its amplitude—is zero. These points of zero amplitude are called **nodes,** and any waves that have stationary nodes, whether on a guitar string or not, are called **standing waves.**

If you've ever played the guitar, you know that even without using your finger to press the string at the frets along the neck of the instrument, it is possible to play a variety of notes called harmonics. For example, if the string is briefly touched at its midpoint as it is plucked, a note an octave higher is produced. Figure 4.10b shows that when this harmonic is played there are three nodes along the string. Still other harmonics produce even more nodes.

In examining the waves in Figure 4.10, we can see that there are certain restrictions on their allowed wavelengths along the string because each end of the string must always be a node. This means that there must be a *whole number* of *half-wavelengths* repeated along the length of string. Waves for which this is not true simply cannot exist on the string. What we see, therefore, is that a whole number—sort of a "musical quantum number"—arises automatically when we consider the standing waves possible on a guitar string.

A similar kind of situation exists in atoms because the electron's matter waves are standing waves, too. The shapes of the matter waves are much different than those on a guitar string, of course, because the atom is three-dimensional and the conditions restricting the locations of the nodes are also much different. Nevertheless, the restrictions that determine what electron waves can exist give rise quite naturally to integer quantum numbers. In three dimensions, however, there are three quantum numbers instead of only one.

In 1926, Erwin Schrödinger (1887–1961) applied mathematics to the investigation of the standing waves in the hydrogen atom and began a field of study called wave mechanics or **quantum mechanics.** The mathematics here is quite advanced, so we will avoid it completely and only look at the results of the theory.

Schrödinger solved a mathematical equation called a wave equation.[2] He

If the length of the string is L, then $n(\lambda/2) = L$ where n is a whole number. The allowed wavelengths along the string, therefore, are $\lambda = 2L/n$.

Schrödinger shared the Nobel Prize for Physics in 1933 with Paul Dirac, another physicist, for their pioneering work in quantum mechanics.

---

[2] The wave equation can really be solved only for a very few species. Fortunately, one of them is the hydrogen atom, and the results obtained for hydrogen, as it turns out, can be extended quite successfully to the other elements in the periodic table.

obtained a set of mathematical functions called **wave functions** (usually represented by the Greek letter psi, ψ) that described the shapes and energies of the electron waves. Each of these different possible waves is called an **orbital** (to distinguish it from Bohr's orbits). *Each orbital in an atom has a characteristic energy and is viewed as describing a region around the nucleus where the electron can be expected to be found.* The wave functions describing the orbitals are themselves characterized by the values of three quantum numbers (as we hinted earlier).

According to wave mechanics, each of the energy levels in an atom is associated with one or more orbitals. In atoms that contain more than one electron, the distribution of the electrons about the nucleus is determined by the number and kind of orbitals occupied. Therefore, in order to investigate the way the electrons are arranged in space, we must first examine the energy levels in the atom. This is best accomplished through a discussion of the quantum numbers.

1. The **principal quantum number, *n*.** The energy levels in an atom are arranged roughly into main levels, or **shells,** as determined by the principal quantum number, n. The larger the value of n, the greater the average energy of the levels belonging to the shell. We will also see that n determines the size of the orbitals. As in the Bohr theory, n may have values of 1, 2, 3, . . . , and so on up to infinity. Letters are also associated with these shells as shown below.

> The larger the value of *n*, the greater the average distance of the electron from the nucleus.

Principal quantum number	1	2	3	4 . . .
Letter designation		K	L	M N . . .

For example, sometimes the shell with $n = 1$ is referred to as the K shell.

2. The **azimuthal quantum number, *l*.** Wave mechanics predicts that each main shell is composed of one or more subshells, or sublevels, each of which is specified by a secondary quantum number, l, called the azimuthal quantum number. As we will see, this quantum number determines the shape of an orbital and, to a certain degree, its energy. For any given shell, l may have values of 0, 1, 2, and so on, up to a maximum of $n - 1$ for that shell. Thus, when $n = 1$, the largest (and only) value of l that is allowed is $l = 0$. Therefore, the K shell consists of only one subshell. When $n = 2$, two values of l occur, $l = 0$ and $l = 1$; hence the L shell is composed of two subshells. The values of l that occur for each value of n are summarized in the table at the left.

Notice that *the number of subshells in any given shell is simply equal to the value of n for that shell.*

For the purposes of discussing the distribution of electrons in an atom, it is common practice to associate letters with the various values of l:

n	l
1	0
2	0, 1
3	0, 1, 2
4	0, 1, 2, 3
.	.
.	.
.	.
n	0, 1, 2, . . . , $n - 1$

Value of l	0	1	2	3	4	5	6 . . .
Subshell designation	s	p	d	f	g	h	i . . .

The first four letters find their historical origin in the study of the atomic spectra of the alkali metals (lithium through cesium). In these spectra four series of lines are observed and they are termed the *s*harp, *p*rincipal, *d*iffuse, and *f*undamental series, hence the letters s, p, d, and f. For $l = 4$, 5, 6, and so on, we just continue with the alphabet. For our purpose, however, we will be interested only in s, p, d, and f subshells, because they are the only ones populated by electrons in atoms in their **ground state** (state of lowest energy).

To specify a subshell within a given shell, we write the value of n for the shell followed by the letter designation of the subshell. For example, the s subshell of the second shell ($n = 2$, $l = 0$) would be called the 2s subshell. Similarly, the p subshell of the second shell ($n = 2$, $l = 1$) would be the 2p subshell.

3. The **magnetic quantum number, m_l.** Each subshell is composed of one or more orbitals. An orbital within a particular subshell is distinguished by its value of m_l, which serves to determine its orientation in space relative to the other orbitals. The magnetic quantum number derives its name from the fact that it can be used to explain the appearance of additional lines in atomic spectra produced when atoms are caused to emit light while in a magnetic field. It has integer values that range between $-l$ and $+l$. When $l = 0$, only one value of m_l is permitted, $m_l = 0$; therefore an s subshell consists of only one orbital (we call it an s orbital). A p subshell ($l = 1$) contains three orbitals corresponding to m_l equal to -1, 0, and $+1$. In a similar fashion, we find that a d subshell ($l = 2$) is composed of five orbitals and an f subshell ($l = 3$), seven. This is summarized in Table 4.2. Notice the simple progression in the numbers of orbitals per subshell: 1, 3, 5, 7, and so on.

In describing the orbitals (electron waves) in an atom, we can assign each one a set of values for n, l, and m_l. For example, one wave will have $n = 1$, $l = 0$, and $m_l = 0$. This lets us identify it as a 1s orbital, and we think of the 1s orbital as being *occupied* by that electron. In a certain sense, it is as if we viewed the atom as a sort of parking garage for electrons, where each orbital is a possible parking space corresponding to a particular wave shape and energy. When an electron is in one of these parking spaces, it has the wave shape and energy of that particular orbital.

The energies of these shells, subshells, and orbitals in atoms having more than one electron are perhaps best illustrated by means of Figure 4.11. There are several points about this diagram worth noting. First, we see that the average energy of the shells increases with increasing value of the principal quantum number, n. Thus the shell with $n = 1$ lies lowest in energy; above that there is the shell with $n = 2$ (composed of the 2s and 2p subshells); higher still we find the shell with $n = 3$, and so on.

Also note that as n becomes larger, the spacing between successive shells

Hydrogen is the only element in which all orbitals with the same value of n have the same energy. That's why Bohr only needed one quantum number in his theory and why the Bohr theory didn't work for other atoms.

Table 4.2
Summary of quantum numbers

Principal Quantum Number, n (Shell)	Azimuthal Quantum Number, l (Subshell)	Subshell Designation	Magnetic Quantum Number, m_l (Orbital)	Number of Orbitals in Subshell
1	0	1s	0	1
2	0	2s	0	1
	1	2p	-1 0 $+1$	3
3	0	3s	0	1
	1	3p	-1 0 $+1$	3
	2	3d	-2 -1 0 $+1$ $+2$	5
4	0	4s	0	1
	1	4p	-1 0 $+1$	3
	2	4d	-2 -1 0 $+1$ $+2$	5
	3	4f	-3 -2 -1 0 $+1$ $+2$ $+3$	7

Figure 4.11

Approximate energy-level diagram for atomic orbitals in multielectron atoms. We use this diagram to determine which orbitals in an atom are populated by electrons.

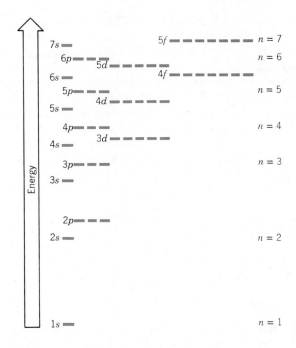

becomes less, as illustrated on the right side of Figure 4.11. Because of this narrowing energy separation, we begin to observe overlap among the subshells of the third and higher shells. The $4s$ subshell, for example, lies lower in energy than the $3d$ subshell. This overlap is even more pronounced in higher shells, where the $5s$ subshell lies below the $4d$, the $6s$ and $4f$ below the $5d$, and the $7s$ and $5f$ below the $6d$.

In Figure 4.11 we have indicated each orbital by means of a dash. Each s subshell is shown as a single dash to stress that it is composed of only one orbital. Likewise, p subshells are shown as three dashes, d subshells as five dashes, and f subshells as seven dashes. Observe that all the orbitals of a given subshell are shown to have the same energy.

The sequence of energy levels described by Figure 4.11 turns out to be of critical importance in determining the arrangement of electrons in an atom. Before discussing this, however, we must look at yet another quantum number.

4.4 ELECTRON SPIN AND THE PAULI EXCLUSION PRINCIPLE

In addition to the three quantum numbers, n, l, and m_l, which come directly from the solution of the wave equation, there is yet another, called the **spin quantum number, m_s.** This quantum number arises because the electron behaves as if it were spinning (in much the same manner as the earth spins about its axis). The circular motion of electric charge that results causes the electron to act as a tiny electromagnetic, just as passing an electric current through a wire wrapped about a nail causes the nail to become magnetic (see Figure 4.12). Since the electron can only spin in either of two directions, m_s may have only two values. These turn out to be $+\frac{1}{2}$ and $-\frac{1}{2}$, although the actual values are not really important to us.

We find, therefore, that each electron in an atom can be assigned a set of values for its four quantum numbers, n, l, m_l, and m_s, which determine the orbital in which the electron will be found and the direction in which the electron will be spinning. There is a restriction, however, on the values that these

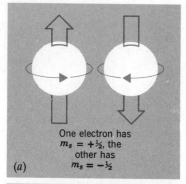

One electron has $m_s = +\frac{1}{2}$, the other has $m_s = -\frac{1}{2}$

(a)

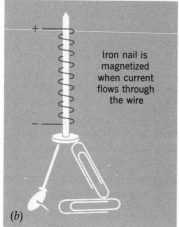

Iron nail is magnetized when current flows through the wire

(b)

Figure 4.12

The spin of the electron. (a) The electron behaves as if it is spinning about an axis through its center. (b) A spinning charge produces a magnetic field just as the circulation of charge through a wire wrapped about a nail causes the nail to become magnetic.

quantum numbers can have. This is expressed as the **Pauli exclusion principle,** which states that *no two electrons in any one atom may have all four quantum numbers the same.* This means that if we choose a particular set of values for n, l, and m_l corresponding to a particular orbital (for example, $n = 1$, $l = 0$, $m_l = 0$; the 1s orbital), we are able to have only two electrons with different values of the spin quantum number, m_s (that is, either $m_s = +\frac{1}{2}$ or $m_s = -\frac{1}{2}$). In effect, *this limits the number of electrons in any given orbital to two, and it also requires that the spins of these two electrons be in opposite directions.*

Because the Pauli exclusion principle leads to a restriction of a maximum of two electrons in any orbital, the maximum number of electrons that can be accommodated in s, p, d, and f subshells can be summarized as follows.

Subshell	Number of Orbitals	Maximum Number of Electrons
s	1	2
p	3	6
d	5	10
f	7	14

The maximum number of electrons permitted in any shell is equal to $2n^2$. For example, the K shell ($n = 1$) can hold up to two electrons and the L shell ($n = 2$) can hold a maximum of eight.

The spin of the electron is also responsible for most of the magnetic properties that we find associated with atoms and molecules. Materials that are **diamagnetic** experience no attraction for another magnet.[3] In these substances there are the same number of electrons of each spin, so their magnetic effects cancel. **Paramagnetic** substances, on the other hand, are weakly attracted to a magnetic field. In these materials there are more electrons of one spin than the other (as will always be true when an atom or molecule has an odd number of electrons) and total cancellation does not occur. The extra electrons of one spin cause the atom or molecule, as a whole, to behave as if it were itself a tiny magnet. **Ferromagnetic** substances, of which iron is the most common example, owe their very strong magnetic behavior to interactions among paramagnetic atoms in the solid state. Ferromagnetism is about 1 million times stronger than paramagnetism. This phenomenon is discussed in more detail in Chapter 21.

4.5 THE ELECTRON CONFIGURATIONS OF THE ELEMENTS

The lowest energy distribution of electrons is an atom's ground state.

The way the electrons are distributed among the orbitals of an atom is its **electronic structure** or **electron configuration.** As we've suggested earlier, this is determined by the order in which the subshells occur on the scale of increasing energy. The reason is that in an atom in its ground state, the electrons will be found in the lowest energy levels available. In hydrogen, for instance, the single electron will be located in the 1s subshell because it is this level that has the lowest energy. To indicate that the 1s subshell is populated by one electron, we use a superscript (in this case, 1), on the subshell designation. Thus we would denote the electron configuration of hydrogen as $1s^1$. As we proceed in this discussion, it will also be necessary to keep tabs on the electron spins. One method that is often employed is to symbolize an electron with its spin in one

---

[3] They are, in fact, repelled slightly by a magnetic field. This is a result of the motion of the electrons in the atom and is not associated with the electron's spin.

direction by an arrow pointing up, ↑, and an electron with opposite spin as an arrow pointing down, ↓. To indicate the distribution of electrons among the orbitals of the atom, we will place the arrows over bars that symbolize orbitals. Hydrogen, for example, is represented as

$$\text{H} \quad \frac{\uparrow}{1s}$$

This kind of representation of the electron configuration is usually called an **orbital diagram.**

To obtain the electron configurations of the other elements in the periodic table, let us imagine that we are able to proceed from one atom to the next by adding a proton and any necessary neutrons to the nucleus, followed by an electron that we place into the lowest available energy level. As you follow this discussion, refer both to the periodic table and the energy-level diagram in Figure 4.11. A complete table of the electron configurations of the elements is contained in Table 4.3.

Hydrogen is the simplest element, consisting of just a single proton and one electron. The next element, atomic number, 2, is helium. Here there are two electrons to consider and, because the $1s$ orbital can accommodate them both, the electronic structure of helium is $1s^2$ and its orbital diagram is

$$\text{He} \quad \frac{\uparrow\downarrow}{1s}$$

When two electrons have opposite spins, they are said to be paired.

Note that in placing the electrons in the same orbital, we have indicated that their spins are in opposite directions as required by the Pauli exclusion principle. We refer to this by saying that their spins are *paired*, or simply that the electrons are paired.

The next two elements following He are Li and Be, which have three and four electrons, respectively. In each element, the first two electrons will enter the $1s$ subshell and, since no more than two electrons can occupy an s subshell, the remaining electron(s) must occupy the $2s$ subshell. The electron configurations of Li and Be, then, are Li, $1s^2 2s^1$ and Be, $1s^2 2s^2$. We could also show this as

$$\text{Li} \quad \frac{\uparrow\downarrow}{} \quad \frac{\uparrow}{}$$
$$\text{Be} \quad \frac{\uparrow\downarrow}{1s} \quad \frac{\uparrow\downarrow}{2s}$$

We can also write these configurations as

Li [He] $2s^1$
Be [He] $2s^2$

Since both Li and Be have a completed $1s$ subshell, which corresponds to the electron configuration of He, they can also be written as

$$\text{Li} \quad [\text{He}] \quad \frac{\uparrow}{}$$
$$\text{Be} \quad [\text{He}] \quad \frac{\uparrow\downarrow}{2s}$$

Here we focus our attention on the electronic structure of the outermost shell—the shell with highest n—that, in chemical reactions, is responsible for chemical changes. Electrons in shells below the outer shell are said to be **core electrons.** In this example, the inner filled $1s$ subshell is called the helium core. We will frequently find it useful to consider only those electrons that occur outside a core of electrons that corresponds to the electron configuration of one of the noble gases.

Table 4.3
The electron configurations of the elements

Atomic Number				Atomic Number				Atomic Number			
1	H	$1s^1$		36	Kr	[Ar]	$4s^23d^{10}4p^6$	71	Lu	[Xe]	$6s^24f^{14}5d^1$
2	He	$1s^2$		37	Rb	[Kr]	$5s^1$	72	Hf	[Xe]	$6s^24f^{14}5d^2$
3	Li	[He]	$2s^1$	38	Sr	[Kr]	$5s^2$	73	Ta	[Xe]	$6s^24f^{14}5d^3$
4	Be	[He]	$2s^2$	39	Y	[Kr]	$5s^24d^1$	74	W	[Xe]	$6s^24f^{14}5d^4$
5	B	[He]	$2s^22p^1$	40	Zr	[Kr]	$5s^24d^2$	75	Re	[Xe]	$6s^24f^{14}5d^5$
6	C	[He]	$2s^22p^2$	41	Nb	[Kr]	$5s^14d^4$	76	Os	[Xe]	$6s^24f^{14}5d^6$
7	N	[He]	$2s^22p^3$	42	Mo	[Kr]	$5s^14d^5$	77	Ir	[Xe]	$6s^24f^{14}5d^7$
8	O	[He]	$2s^22p^4$	43	Tc	[Kr]	$5s^24d^5$	78	Pt	[Xe]	$6s^14f^{14}5d^9$
9	F	[He]	$2s^22p^5$	44	Ru	[Kr]	$5s^14d^7$	79	Au	[Xe]	$6s^14f^{14}5d^{10}$
10	Ne	[He]	$2s^22p^6$	45	Rh	[Kr]	$5s^14d^8$	80	Hg	[Xe]	$6s^24f^{14}5d^{10}$
11	Na	[Ne]	$3s^1$	46	Pd	[Kr]	$4d^{10}$	81	Tl	[Xe]	$6s^24f^{14}5d^{10}6p^1$
12	Mg	[Ne]	$3s^2$	47	Ag	[Kr]	$5s^14d^{10}$	82	Pb	[Xe]	$6s^24f^{14}5d^{10}6p^2$
13	Al	[Ne]	$3s^23p^1$	48	Cd	[Kr]	$5s^24d^{10}$	83	Bi	[Xe]	$6s^24f^{14}5d^{10}6p^3$
14	Si	[Ne]	$3s^23p^2$	49	In	[Kr]	$5s^24d^{10}5p^1$	84	Po	[Xe]	$6s^24f^{14}5d^{10}6p^4$
15	P	[Ne]	$3s^23p^3$	50	Sn	[Kr]	$5s^24d^{10}5p^2$	85	At	[Xe]	$6s^24f^{14}5d^{10}6p^5$
16	S	[Ne]	$3s^23p^4$	51	Sb	[Kr]	$5s^24d^{10}5p^3$	86	Rn	[Xe]	$6s^44f^{14}5d^{10}6p^6$
17	Cl	[Ne]	$3s^23p^5$	52	Te	[Kr]	$5s^24d^{10}5p^4$	87	Fr	[Rn]	$7s^1$
18	Ar	[Ne]	$3s^23p^6$	53	I	[Kr]	$5s^24d^{10}5p^5$	88	Ra	[Rn]	$7s^2$
19	K	[Ar]	$4s^1$	54	Xe	[Kr]	$5s^24d^{10}5p^6$	89	Ac	[Rn]	$7s^26d^1$
20	Ca	[Ar]	$4s^2$	55	Cs	[Xe]	$6s^1$	90	Th	[Rn]	$7s^26d^2$
21	Sc	[Ar]	$4s^23d^1$	56	Ba	[Xe]	$6s^2$	91	Pa	[Rn]	$7s^25f^26d^1$
22	Ti	[Ar]	$4s^23d^2$	57	La	[Xe]	$6s^25d^1$	92	U	[Rn]	$7s^25f^36d^1$
23	V	[Ar]	$4s^23d^3$	58	Ce	[Xe]	$6s^24f^15d^1$	93	Np	[Rn]	$7s^25f^46d^1$
24	Cr	[Ar]	$4s^13d^5$	59	Pr	[Xe]	$6s^24f^3$	94	Pu	[Rn]	$7s^25f^6$
25	Mn	[Ar]	$4s^23d^5$	60	Nd	[Xe]	$6s^24f^4$	95	Am	[Rn]	$7s^25f^7$
26	Fe	[Ar]	$4s^23d^6$	61	Pm	[Xe]	$6s^24f^5$	96	Cm	[Rn]	$7s^25f^76d^1$
27	Co	[Ar]	$4s^23d^7$	62	Sm	[Xe]	$6s^24f^6$	97	Bk	[Rn]	$7s^25f^9$
28	Ni	[Ar]	$4s^23d^8$	63	Eu	[Xe]	$6s^24f^7$	98	Cf	[Rn]	$7s^25f^{10}$
29	Cu	[Ar]	$4s^13d^{10}$	64	Gd	[Xe]	$6s^24f^75d^1$	99	Es	[Rn]	$7s^25f^{11}$
30	Zn	[Ar]	$4s^23d^{10}$	65	Tb	[Xe]	$6s^24f^9$	100	Fm	[Rn]	$7s^25f^{12}$
31	Ga	[Ar]	$4s^23d^{10}4p^1$	66	Dy	[Xe]	$6s^24f^{10}$	101	Md	[Rn]	$7s^25f^{13}$
32	Ge	[Ar]	$4s^23d^{10}4p^2$	67	Ho	[Xe]	$6s^24f^{11}$	102	No	[Rn]	$7s^25f^{14}$
33	As	[Ar]	$4s^23d^{10}4p^3$	68	Er	[Xe]	$6s^24f^{12}$	103	Lr	[Rn]	$7s^25f^{14}6d^1$
34	Se	[Ar]	$4s^23d^{10}4p^4$	69	Tm	[Xe]	$6s^24f^{13}$				
35	Br	[Ar]	$4s^23d^{10}4p^5$	70	Yb	[Xe]	$6s^24f^{14}$				

At beryllium, which has four electrons, the $2s$ subshell is completed. The fifth electron of boron ($Z = 5$) must then enter the next lowest available subshell, which is the $2p$. This gives boron the configuration $1s^22s^22p^1$. Similarly, the fifth and sixth electrons of carbon must enter the $2p$ subshell; thus, we represent carbon as $1s^22s^22p^2$. However, if we examine the distribution of the electrons

Notice that all the orbitals of the *p* subshell are shown, even though not all of them are occupied by electrons.

over the various orbitals, we face a choice; the electrons could be arranged in the following three ways[4]:

C   [He]   ↿⇂   ↿⇂   __ __

or

C   [He]   ↿⇂   ↑   ↓   __

or

C   [He]   ↿⇂   ↑   ↑   __
           2s   2p

The last two electrons can be paired in the same orbital, paired in different orbitals, or arranged so that their spins are in the same direction (unpaired).

As it turns out, experiments show that the last diagram gives the electron configuration that is lowest in energy. **Hund's rule** summarizes this experimental evidence: *Electrons entering a subshell containing more than one orbital will be spread out over the available orbitals with their spins in the same direction.* For nitrogen (Z = 7), therefore, the electron configuration would be written as $1s^2 2s^2 2p^3$, and its ground state would have the orbital diagram

N   [He]   ↿⇂   ↑   ↑   ↑
           2s   2p

N   [He] $2s^2 2p^3$

Finally, the elements oxygen, fluorine, and neon (Z = 8, 9, and 10, respectively) lead to the completion of the 2*p* subshell.

O   [He]   ↿⇂   ↿⇂   ↑   ↑

F   [He]   ↿⇂   ↿⇂   ↿⇂   ↑

Ne   [He]   ↿⇂   ↿⇂   ↿⇂   ↿⇂
            2s   2p

O   [He] $2s^2 2p^4$

F   [He] $2s^2 2p^5$

Ne   [He] $2s^2 2p^6$

After the 2*p* subshell is filled at Ne, the next lowest available energy level is the 3*s*. This becomes populated with Na and Mg (Z = 11 and 12). After this the 3*p* subshell is gradually filled by the next six electrons as we complete the configurations of the atoms Al through Ar (Z = 13 through 18). Then, since the 4*s* subshell lies at lower energy than the 3*d*, it is occupied next by the nineteenth and twentieth electrons of K and Ca (Z = 19 and 20).

Examination of Figure 4.11 reveals that after the 4*s* subshell is completed, additional electrons begin to populate the 3*d* subshell. Scandium, therefore, will have the electron configuration $1s^2 2s^2 2p^6 3s^2 3p^6 4s^2 3d^1$. Normally, we order the subshells to place all those with the same value of the principal quantum number together. For scandium, this gives

Sc $1s^2 2s^2 2p^6 3s^2 3p^6 3d^1 4s^2$

and the orbital diagram is

Sc   [Ar]   ↑   __ __ __ __   ↿⇂
                    3d   4s

---

[4] These are the only three possibilities that we have to consider because in an isolated atom, each of the *p* orbitals is equivalent in energy. Thus the arrangements

↿⇂   __ __   __   ↿⇂   __   __ __   ↿⇂
     2p   2p   2p

are indistinguishable from one another experimentally.

As we proceed through Ti and V (Z = 22 and 23), two more electrons are added to the 3d subshell; however, when we get to Cr (Z = 24), we find the structure

Cr   [Ar]   ↑ ↑ ↑ ↑ ↑   ↑
                    3d   4s

instead of

Cr   [Ar]   ↑ ↑ ↑ ↑ __   ↑↓
                    3d   4s

This unexpected result occurs because a half-filled or completely filled subshell possesses an extra, added stability. The origin of this extra stability is very complex; so we cannot discuss it here. Nevertheless, the phenomenon is quite important and should be kept in mind. We see it again in period 4, for example, when we get to copper. On the basis of our energy-level diagram in Figure 4.11, we would predict copper to have the electron configuration

Cu   [Ar]   ↑↓ ↑↓ ↑↓ ↑↓ ↑   ↑↓
                    3d   4s

The actual structure of the ground state is given by

Cu   [Ar]   ↑↓ ↑↓ ↑↓ ↑↓ ↑↓   ↑
                    3d   4s

Similar exceptions occur elsewhere. For example, Ag and Au have filled *d* subshells just as copper does.

By transferring an electron from the 4s to the 3d subshell of copper, one filled and one half-filled subshell are produced, instead of the filled 4s and the neither filled nor half-filled 3d subshell that we initially would predict. Because the electron configurations of Cr and Cu are not predictable by our rules, they must be remembered as exceptions.

After the 3d subshell is completed at atomic number 30 (zinc), the 4p subshell is gradually filled as we proceed from Ga to Kr (Z = 31 to 36). This is followed by the completion of the 5s subshell from Rb to Sr (Z = 37, 38); the 4d subshell as we progress across the second row of transition elements (Z = 39 to 48); the 5p from In to Xe (Z = 49 to 54); and the 6s with Cs and Ba (Z = 55, 56).

Based on the energy-level sequence in Figure 4.11, we would expect that after the 6s subshell had been filled the 4f subshell should be populated next. Actually, at La (Z = 57) the last electron enters the 5d subshell instead. The 4f subshell is filled afterward, with a few minor irregularities. As we go to higher and higher shells, these irregularities become more frequent because the spacing between subshells becomes smaller and smaller. As we proceed from atom to atom, the energies of the various subshells shift about somewhat as the nuclear charge increases and as the electron populations of the subshells change. The result is that it is difficult to predict accurately the electron configuration of elements of very high atomic number. Nevertheless, we can account for the occurrence of the lanthanide elements by the filling of the 4f subshell (an f subshell can accommodate 14 electrons and the lanthanide series consists of 14 elements). Likewise, we can account for the actinide elements as the result of the filling of the 5f subshell.

4.6 THE PERIODIC TABLE AND ELECTRON CONFIGURATIONS

In the last section we saw that the results of wave mechanics could be used to predict the electron configurations of the elements. These electron configurations are based on theory and, to be considered useful and valid, they must somehow manifest themselves in obvious ways. One of the strongest supports for the assignment of electron configurations is the periodic table itself. Recall that, in constructing the current periodic table, elements were arranged under each other in groups because of their similar chemical properties. For example, all the elements in Group IA are metals that form ions with a charge of 1+ when they react. If we examine the electron configurations of these elements, we see that the outer shell (shell of highest n) for each has only one electron in an s subshell.

Li	[He] $2s^1$	Rb	[Kr] $5s^1$
Na	[Ne] $3s^1$	Cs	[Xe] $6s^1$
K	[Ar] $4s^1$		

Be	[He] $2s^2$	Sr	[Kr] $5s^2$
Mg	[Ne] $3s^2$	Ba	[Xe] $6s^2$
Ca	[Ar] $4s^2$		

Similarly, all the elements in Group IIA have an outer-shell electron configuration that we might generalize as ns^2. In fact, by examining any group within the periodic table, we see that all the elements in the group possess essentially identical outer-shell electronic structures—only their values of n differ. It is not really surprising that similar electronic structures lead to similar chemical and physical properties.

Because the properties of the elements depend on their electron configurations, it is important for you to develop the ability to write them down. There are several ways to remember the sequence in which the various levels are filled; however, the best aid is the periodic table itself. As we have just seen, the order of filling the energy levels can be used to account for the structure of the periodic table. We can also work in the other direction and use the periodic table to deduce electronic structures.

If we look back over the procedure for determining electronic structures, we find that for any element in Groups IA and IIA the final electron is added to an s subshell, and that the principal quantum number of that subshell is the same as the period number. Sodium, for instance, is a period 3 element and has its outer electron in the $3s$ subshell. For elements in Groups IIIA to Group 0 the last electron is added to a p subshell whose value of n is also the same as the period number. For example, as we complete the electron configurations of the period 2 elements boron through neon, the final electron is placed in a $2p$ subshell. In the case of the transition elements, the final electron that we add is placed into a d subshell with n equal to *one less* than the period number. For example, with iron (a fourth-period element), the last electron enters a $3d$ subshell. Finally, notice that the electronic structure of an inner transition element (i.e., one from the lanthanide or actinide series) is completed by placing an electron in an f subshell whose principal quantum number is *two* less than the period number.

We can use these observations by making the periodic table tell us which subshells are filled as we build up the electron configuration of an atom. As before, we begin with hydrogen and we proceed through the elements in the periodic table in order of increasing atomic number until we arrive at the element in which we are interested. As we move across a given period, we add electrons to an s subshell when we pass through Groups IA and IIA, and to a p subshell when we pass through Groups IIIA through Group 0. The value of n for these subshells is the same as the period number. As we pass through a row of transition elements we fill a d subshell with n equal to the period number minus 1, and as we move across a row of inner transition elements we fill an f subshell with n equal to the period number minus two. This is summarized in Figure 4.13, and the following examples illustrate how this method works.

Figure 4.13

The use of the periodic table to predict electron configurations. Shaded areas depict the subshells that are filled to obtain the configuration of lead (see Example 4.9).

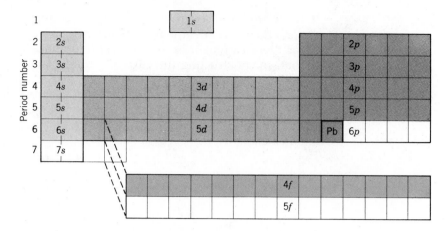

EXAMPLE 4.8 PREDICTING ELECTRON CONFIGURATIONS

PROBLEM What is the electron configuration of antimony (Sb)?

SOLUTION Antimony has atomic number 51. To reach this element we have to pass completely through periods 1, 2, 3, and 4, and part of the way across period 5. As we do this, here is what we get:

Period 1—fill the $1s$ subshell, which gives $\qquad 1s^2$

Period 2—fill the $2s$ and $2p$ subshells $\qquad 2s^2 2p^6$

Period 3—fill the $3s$ and $3p$ subshells $\qquad 3s^2 3p^6$

Period 4—fill the $4s$, $3d$, and $4p$ subshells (in that order) $\qquad 4s^2 3d^{10} 4p^6$

Period 5—to get as far as cadmium ($Z = 48$), we have to fill the $5s$ and
   $4d$ subshells. Then we have to move three spaces into the p
   region to get to Sb. This gives $\qquad 5s^2 4d^{10} 5p^3$

Putting this all together, we have

$$1s^2 2s^2 2p^6 3s^2 3p^6 4s^2 3d^{10} 4p^6 5s^2 4d^{10} 5p^3$$

If we collect all the subshells of a given shell together we obtain

$$1s^2 2s^2 2p^6 3s^2 3p^6 3d^{10} 4s^2 4p^6 4d^{10} 5s^2 5p^3$$

We can also write the configuration by showing the noble gas core plus the electrons outside of it

$$[Kr]4d^{10} 5s^2 5p^3$$

EXAMPLE 4.9 PREDICTING ELECTRON CONFIGURATIONS

PROBLEM What is the electron configuration of lead?

SOLUTION Lead has atomic number 82. This means that in building up the lead atom we cross through periods 1 to 5 and part of 6. Proceeding from left to right across one period after another, we fill, in order, the subshells $1s$, $2s$, $2p$, $3s$, $3p$, $4s$, $3d$, $4p$, $5s$, $4d$, $5p$, $6s$, $4f$, $5d$, and finally end by placing two electrons into the $6p$ subshell. Taking into account the maximum population of each subshell, we obtain as the electron configuration of lead,

$$1s^2 2s^2 2p^6 3s^2 3p^6 4s^2 3d^{10} 4p^6 5s^2 4d^{10} 5p^6 6s^2 4f^{14} 5d^{10} 6p^2$$

As before, we might prefer to write all subshells of a given shell together. Thus for lead we would have

$$1s^2 2s^2 2p^6 3s^2 3p^6 3d^{10} 4s^2 4p^6 4d^{10} 4f^{14} 5s^2 5p^6 5d^{10} 6s^2 6p^2$$

When we begin to discuss the chemical bonds formed between atoms, we often will be interested only in the electron population of the outer shell, or **valence shell,** of an atom. (Valence is a word that relates to the number of chemical bonds an atom is able to form.) This is because it is only the outer parts of atoms that come into contact with each other when the atoms combine to form compounds. If all we are interested in is the configuration of an atom's valence shell, we do not have to work through the entire electron configuration. Instead, we can locate the element in the periodic table and immediately find the information we need. For example, suppose we wished to know only the outer-shell electron configuration for tin, Sn. We find this element in Group IVA and period 5. The period number tells us that the outer shell has $n = 5$, so we are only interested in populated orbitals that have 5 as their principal quantum number. Crossing period 5 as far as Sn, we would place two electrons in the $5s$, ten in the $4d$, and finally two in the $5p$.

$$\text{Sn} \quad [\text{Kr}] \quad 4d^{10}5s^25p^2$$

We ignore the electrons in the $4d$ because this subshell is not part of the valence shell, so the valence shell configuration of tin is

$$5s^25p^2$$

EXAMPLE 4.10 PREDICTING OUTER-SHELL ELECTRON CONFIGURATIONS

PROBLEM
What is the outer-shell configuration of silicon?

SOLUTION
Silicon (Si) is in period 3; therefore, the outer shell is the third shell. To get to Si in period 3 we fill the $3s$ subshell and place two electrons in the $3p$ subshell. This gives us

$$\text{Si} \quad 3s^23p^2$$

EXAMPLE 4.11 CONSTRUCTING ORBITAL DIAGRAMS BY USE OF THE PERIODIC TABLE

PROBLEM
Use the periodic table to construct the orbital diagram of the valence shell of tellurium, Te.

SOLUTION
First, let's determine the electron configuration of the valence shell. Then, we can translate that into the appropriate orbital diagram. We begin by noting that Te is in period 5, which means that its valence shell is the 5th shell—we are only interested in orbitals that are part of the 5th shell. As we cross the 5th row, we first fill the $5s$ orbital (which gives $5s^2$); then, we fill the $4d$ orbital (which gives $4d^{10}$); and then we place 4 electrons into the $5p$ subshell (which gives $5p^4$). The electron configuration of Te is therefore

$$\text{Te} \quad [\text{Kr}] \ 4d^{10}5s^25p^4$$

and the valence shell configuration is

$$\text{Te} \quad 5s^25p^4$$

Notice that the $4d$ subshell is *not* part of the valence shell. Also, note that the valence shell includes both the $5s$ and $5p$ subshells.

The next step is to translate this into an orbital diagram. We begin by drawing one dash for the $5s$ orbital and *three* dashes for the $5p$ orbital. Don't forget to show all the orbitals of a given subshell, even if not all of them have electrons. The best way to start is to draw the dashes for the orbitals and then fill in the electrons. For Te, we get

Te __   __ __ __
    5s   5p

Now we place 2 electrons in the 5s orbital, being sure to indicate that they are paired. Next, we place electrons in the 5p orbitals, one orbital at a time until they are each half-filled (which accounts for 3 of the 4 electrons); then we add the 4th electron to one of the half-filled 5p orbitals with its spin paired. This gives the completed orbital diagram:

$$\text{Te} \quad \underline{\uparrow\downarrow} \qquad \underline{\uparrow\downarrow} \;\; \underline{\uparrow} \;\; \underline{\uparrow}$$
$$\phantom{\text{Te} \quad} 5s \qquad\qquad 5p$$

4.7   THE SPATIAL DISTRIBUTION OF ELECTRONS

In wave mechanics, our view of where the electron is likely to be found around the nucleus is far different from the idea of circular orbits imagined by Bohr. This is a consequence of the **uncertainty principle** of Heisenberg, which states that if we attempt to measure, at the same time, both the position and momentum of a particle, our measurements will be subject to errors that are related to one another by the equation

$$\Delta x \cdot \Delta(mv) \geq \frac{h}{4\pi} \qquad\qquad [4.13]$$

This equation states that the product of the uncertainty in the position of the particle, Δx, times the uncertainty in its momentum, $\Delta(mv)$, must be greater than or equal to Planck's constant divided by 4π. What this really means is that we are limited in our ability to know simultaneously where the electron is and where it is going. It leads us, instead, to refer to the probability of finding the electron in some small element of volume at various places around the nucleus. More specifically, it is the square of the wave function, ψ^2, that gives this probability.

On the basis of this concept let's look at the **probability distribution**—the way the probability varies from one place to another—for the single electron in the 1s orbital of the hydrogen atom. A graph of ψ^2 as a function of the distance from the nucleus, r, is shown in Figure 4.14. Notice that those regions in which the probability of observing the electron is greatest lie close to the nucleus and that the probability decreases as we move away from the nucleus, gradually approaching zero as r approaches infinity. In other words, we might expect to find the electron in the hydrogen atom almost anywhere, but most of the time it stays fairly close to the nucleus, effectively surrounding it in a cloud of electronic charge (in fact, the terms *electron cloud* and *charge cloud* are frequently used when describing an electron distribution). The electron spends most of its time in regions where the probability of finding it is high, and there the concentration of charge, which we call **electron density,** is large. In other regions the charge is thinly spread and the electron density small.

There are several ways to indicate the distribution of charge in an orbital. One way is to plot ψ^2 as we have already done. Another, in two dimensions, is

The density described in Chapter 1 is *mass density,* the quantity of mass packed into a given volume. The *electron density* is the quantity of electrical charge packed into a given volume.

Figure 4.14
Probability of finding the electron (ψ^2) as a function of distance from the nucleus for the 1s orbital of hydrogen.

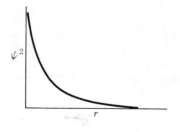

to illustrate the charge cloud as shown in Figure 4.15, where the darker shaded areas represent regions of higher electron density.

Higher-energy s orbitals differ in some respects from the 1s orbital. Figure 4.16 compares the way the electron density varies for 1s, 2s, and 3s orbitals. For the 2s orbital, notice that as we move out from the nucleus the electron density drops to zero, then increases again before gradually decreasing once more. We learned earlier that those places where the amplitude or intensity of a wave drops to zero are called nodes. Thus the electron wave for an electron in a 2s orbital has a node just like waves on a guitar string. In Figure 4.16 we see that the 2s orbital has one node and the 3s orbital has two nodes.

The shapes of s orbitals

Despite their differences, all s orbitals have an important property in common. If we draw a surface on which the probability of observing the electron is constant, the surface has the shape of a sphere. *All s orbitals have a spherical "shape."* The main difference is that as we go to higher values of n, the sphere within which we find most (for example, 90%) of the electron density becomes larger. In other words, the size of the charge cloud gets larger with increasing principal quantum number, not only for s orbitals, but also for p, d, and f orbitals. This means that electrons in orbitals of higher n will be at a greater average distance from the nucleus and that the atom gets larger as its higher-energy subshells become populated.

The shapes of p orbitals

The "shape" of the electron cloud characteristic of a 2p orbital is illustrated in Figure 4.17. We see that for a 2p orbital the electron density is not distributed in a spherically symmetrical manner about the nucleus as it is in an s orbital. Instead, it is concentrated in particular regions along a straight line passing through the nucleus. Electron density occurs on both sides of the nucleus so that an electron in a 2p orbital spends part of its time on each side of the atom.

Higher-energy p orbitals contain nodes just as the higher-energy s orbitals. A cross section of a 3p orbital is illustrated in Figure 4.18. Despite these differences there is a concentration of electron density along certain specific directions in all p orbitals. As a result, all p orbitals have definite directional properties. As we will see, these allow us to understand why molecules have the shapes they do. Because we are interested only in the directional properties of the orbitals,

Nodes are a natural part of standing waves.

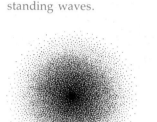

Figure 4.15

A representation of the charge cloud in hydrogen.

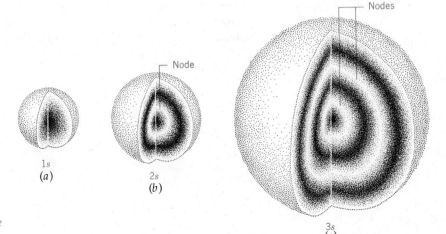

Figure 4.16

Electron density distribution in the 1s, 2s, and 3s orbitals of an atom.

Figure 4.17

The shape of a 2p orbital. (a) A dot distribution drawing shows that the electron density is concentrated in two regions that lie on opposite sides of the nucleus. (b) The kind of drawing that we will use in this book to represent a p orbital.

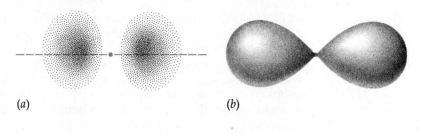

(a)

(b)

Figure 4.18

Cross section of a 3p orbital.

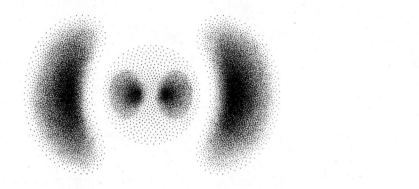

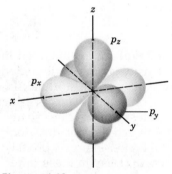

Figure 4.19

The three p orbitals of a given p subshell point in directions that are mutually perpendicular. By imagining that they point along a set of xyz axes, we can label them p_x, p_y, and p_z.

we will represent all *s* orbitals as spheres and all *p* orbitals as a pair of dumbbell-shaped lobes pointing in opposite directions from the nucleus.

A *p* subshell is composed of three *p* orbitals, each having the same shape. They differ from one another only in the directions in which their electron density is concentrated. These directions lie at right angles to one another as shown in Figure 4.19. Because the *p* orbitals can be drawn on a set of *xyz* axes, we identify the orbitals by the notation p_x, p_y, p_z.

The shapes of *d* orbitals

The *d* orbitals are a bit more complicated than *p* orbitals. Because of this, it is difficult to draw all of them together on the same set of axes, as we have done for the *p* orbitals. In Figure 4.20, they are shown separately, but in an atom, they are really superimposed on the same nucleus.

The first thing to notice about the *d* orbitals is that they all do not look alike. Four of them, labeled d_{xy}, d_{xz}, d_{yz}, and $d_{x^2-y^2}$, have the same shape, and each is composed of four lobes of electron density. The difference is that they point in different directions. The fifth, labeled d_{z^2}, consists of two large lobes of electron density pointing in opposite directions along the *z* axis, plus a donut of electron

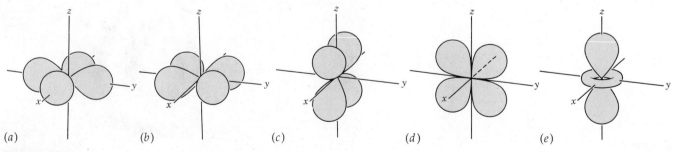

(a)   (b)   (c)   (d)   (e)

Figure 4.20

Directional properties of the d orbitals. (a) $d_{x^2-y^2}$, (b) d_{xy}, (c) d_{xz}, (d) d_{yz}, and (e) d_{z^2}.

density in the xy plane. The labels for the d orbitals are not as descriptive as those for the p orbitals and actually have their origin in the mathematics of wave mechanics. Although we will use these labels when we wish to refer to the various d orbitals, you need not worry about where they come from.

The shapes of the f orbitals are even more complex than the d orbitals. Fortunately, they are only needed in discussions of the chemistry of the inner transition elements, which is not included in this book. Therefore, we will not burden you further with the shapes of the f orbitals.

4.8 THE VARIATION OF PROPERTIES WITH ATOMIC STRUCTURE

Within the periodic table, many of the properties of the elements vary in a more or less regular fashion as we proceed from left to right across a period or from top to bottom in a group. We saw this in Chapter 3 where we noted a change in the metallic character of the elements within a period and within a group. Most of these variations can be accounted for directly in terms of variations in the electronic structures of the elements. In this section, we will look briefly at three important properties whose variations are easily seen and which are related to each other.

Atomic size

As we've mentioned previously, atoms are very tiny—they have diameters of about 10^{-10} m. A unit that has been used for many years to report the sizes of atoms, ions, or molecules is the **angstrom (Å).**

$$1 \text{ Å} = 10^{-8} \text{ cm} = 10^{-10} \text{ m}$$

This is not a recognized SI unit and in current scientific literature these dimensions are usually expressed either in **nanometres** (nm) or **picometres** (pm).

$$1 \text{ nm} = 10^{-9} \text{ m}$$
$$1 \text{ pm} = 10^{-12} \text{ m}$$

We will express atomic and molecular dimensions in picometers.

In any discussion of atomic size, we immediately face a basic problem of definition: Exactly what do we mean by the size or radius of an atom? We have seen that the electron density in an atom does not end abruptly at some particular distance from the nucleus. Instead it trails off gradually, approaching zero at very large distances from the center of the atom. Because of this it is difficult to define precisely what we mean by the size of an atom. One attempt to solve this problem might be to take the radius of an atom as half the distance between neighboring atoms when the element is present in its most dense form (i.e., most highly compacted form, which is usually the solid). Even this definition, however, is complicated because when atoms are attached to each other by chemical bonds, as they are in molecules like H_2 or Cl_2, they approach each other more closely than nonbonded atoms do (e.g., the noble gases when they are frozen). Also, the atomic radius that we measure for atoms of a pure element will not necessarily be the same in compounds. For example, carbon atoms in diamond (pure carbon) are separated by a distance of 154 pm[5] and we would thus assign carbon a radius of 77 pm. In the ethane molecule, C_2H_6, the carbon–carbon distance is also 154 pm; however, in ethylene, C_2H_4, and acetylene, C_2H_2, we find carbon–carbon distances equal to 137 and 120 pm, respectively.

[5] We will see how interatomic distances are obtained in Chapter 10.

These lead to atomic radii for carbon of 69 and 60 pm, both considerably smaller than 77 pm.

Despite this difficulty of definition, we can compare the atomic radii of the elements if they are measured under circumstances that lead to essentially similar kinds of bonds between their atoms. In Figure 4.21 we illustrate the variation of atomic radius with atomic number. We see that as we proceed down within a group, the size of atoms generally increases, and that as we proceed from left to right across a period, a gradual decrease in size is observed.

In order to interpret these trends within the periodic table in terms of electronic structure, we must look at the factors determining the size of the outer shells of atoms, that is, the average distance at which electrons in the outer shell occur. As we have discussed in the preceding section, one factor is the principal quantum number of the outer shell (recall that the electron occurs at increasingly larger distances from the nucleus with increasing value of n).

The size of the outer shell also depends on the quantity of positive charge that the outer electrons feel. The higher the positive nuclear charge, the more the outer electrons are pulled toward the center of the atom. In multielectron atoms, the positive charge felt by the outer electrons is always less than the full nuclear charge. This is because electrons in inner shells partially shield those in the outer shell. Stated another way, the negative charge of the electrons in the core, whose electron density lies partially between the nucleus and the valence shell, partly offsets the positive charge of the nucleus. The residual net charge felt by the outer valence electrons is called the **effective nuclear charge.** In a sodium atom, for example, there are 10 electrons in the neon core. These partially shield the outer $3s$ electron of Na from the positive charge of the 11 protons in the nucleus. As a result, the $3s$ electron of sodium feels an effective nuclear charge that is much less than $+11$. (Actually, it is only about $+2.8$.)

Figure 4.21

A graph of atomic radius versus atomic number.

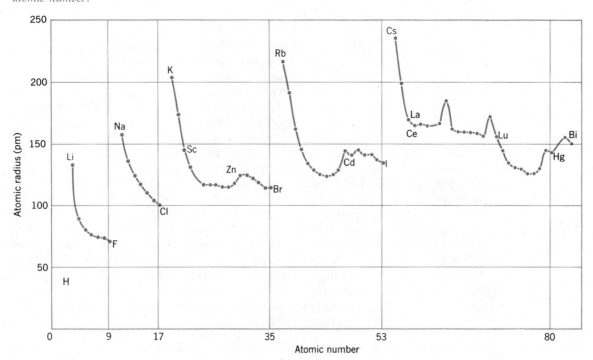

As we proceed downward from one atom to another within a group, each successive element has its outer electrons in a shell with a larger value of n. The effective nuclear charge felt by these electrons stays nearly the same, so the dominant effect is an increase in size that accompanies an increase in the value of the principal quantum number of the outer-shell orbitals.

For the representative elements, as we move from left to right across a period, we add electrons to the same shell and simultaneously increase the nuclear charge. Because they are in the *same* shell, the outer-shell electrons do not shield each other from the nucleus very well, so the effective nuclear charge experienced by any one electron in the outer shell increases. This increase in effective nuclear charge leads to a greater attraction for the outer-shell electrons, so they are pulled in closer to the nucleus, and the sizes of the atoms decrease.

The variation in atomic size across a period is not always a smooth one, as the irregularities seen for period 6 in Figure 4.22 demonstrate.

The variation in size as we pass through a row of transition elements is much less than among the representative elements. This is because electrons are being added to an inner shell as the nuclear charge gets larger. In the first row of the transition elements, for instance, the outer electrons occur in a $4s$ subshell, but each successive electron is added to the inner $3d$ subshell as we proceed across the table. The inner-shell electrons are almost completely effective at shielding the outer shell from the nuclear charge, so the outer $4s$ electrons experience only a very gradual increase in effective nuclear charge across this region of the periodic table. Therefore, only small changes in size occur.

The lanthanide contraction

As we move from left to right across a row of transition elements, there is a gradual decrease in size because the inner electrons that are being added are not completely effective at shielding the outer electrons. A similar phenomenon also occurs among the inner transition elements—for example, the lanthanides. For these elements, the electrons being added go into the $4f$ subshell. Although they shield the outer $6s$ electrons quite well, the shielding isn't perfect, so the outer electrons experience a very gradual increase in effective nuclear charge, and the sizes of the atoms decrease on going from Ce to Lu.

If you look back at Figure 3.8 on page 89, you see that the lanthanide elements fall between La (in Group IIIB) and Hf (in Group IVB). In the fifth period, however, there is no similar row of elements. On moving from Y to Zr, there is an increase of one proton to the nucleus and only one electron to the atom, so only a small decrease in size occurs. However, in the sixth period, a much larger size decrease occurs between La and Hf because of the intervening lanthanide elements. This additional decrease in size, which is known as the **lanthanide contraction,** causes Hf to be the same size as Zr, even though Hf is lower down in the group.

The effects of the lanthanide contraction extend beyond Hf in the sixth period. All the rest of the transition elements in the sixth period are nearly the same size as the elements above them in the fifth period. One result of this is that the sixth-period transition metals are all extremely dense. Their atoms are the same size as the atoms above them, but they pack a lot more mass. Even beyond the transition elements, we see the influence of the lanthanide contraction with unusually dense elements such as lead and bismuth.

The sizes of ions

In the next chapter we will see that many elements react to produce compounds by the formation of ions. In general, positive ions are smaller than the neutral atoms from which they are formed, while negative ions are larger than neutral

Table 4.4
Atomic and ionic radii (in picometres)

		Positive Ions						Negative Ions		
		Atomic Radius	Ionic Radius	Charge				Atomic Radius	Ionic Radius	Charge
Group IA	Li	135	60	(+1)		Group VIIA	F	64	136	(−1)
	Na	154	95	(+1)			Cl	99	181	(−1)
	K	196	133	(+1)			Br	114	195	(−1)
	Rb	211	148	(+1)			I	133	216	(−1)
	Cs	225	169	(+1)		Group VIA	O	66	140	(−2)
Group IIA	Be	90	31	(+2)			S	104	184	(−2)
	Mg	130	65	(+2)			Se	117	198	(−2)
	Ca	174	99	(+2)			Te	137	221	(−2)
	Sr	192	113	(+2)		Group VA	N	70	171	(−3)
	Ba	198	135	(+2)			P	110	212	(−3)
Group IIIA	Al	143	50	(+3)						
	Ga	122	62	(+3)						
	In	162	81	(+3)						

Elements That Form More Than One Ion					
Fe	126	Fe^{2+}	76	Fe^{3+}	64
Co	125	Co^{2+}	78	Co^{3+}	63
Cu	128	Cu^{+}	96	Cu^{2+}	69

atoms (Table 4.4). The decrease in size that accompanies the creation of a positive ion is often a result of the removal of all the electrons from the outer shell of the atom. This gives the ion an electron configuration that is the same as a noble gas. For example, when sodium atoms react, each loses its single 3s electron to produce an Na^+ ion whose electronic structure consists of the neon core. The outer shell at this point has its principal quantum number equal to 2 and therefore the outer-shell electrons in an Na^+ ion are at a smaller average distance from the nucleus than the 3s electron in an Na atom.

When negative ions are produced from neutral atoms, electrons are added to the outer shell without any change in the nuclear charge. Each additional electron provides some degree of shielding for the other electrons, so the effective nuclear charge felt by any one electron in the outer shell decreases. At the same time, the presence of additional electrons in the outer shell increases the repulsions between electrons. Both of these factors tend to cause the outer shell to expand in size, causing the negative ion to be larger than the neutral atom.

Particles become smaller when electrons are removed from them, and larger when electrons are added.

Ionization energy

*The **ionization energy** is defined as the energy required to remove an electron from an isolated gaseous atom in its ground state.* We can represent this process by an equation such as

$$Na(g) \longrightarrow Na^+(g) + e^-$$

This is an endothermic process because the electron is attracted to the positive nucleus; therefore, energy must be supplied to remove it. Since all atoms other than hydrogen possess more than one electron, they also have more than one ionization energy. The quantity of energy required to remove the first electron

Work must be done to pull the electron from the neutral atom.

is called the first ionization energy, that required to remove the second electron is called the second ionization energy, and so forth. As we might expect, successive ionization energies increase in magnitude because the species from which the electron is removed becomes progressively more positively charged. For example, the first ionization energy involves the removal of an electron from a neutral atom, but the second ionization energy involves the removal of an electron from an ion whose charge is 1+.

Table 4.5 contains successive ionization energies for the first 20 elements in the periodic table. They are given in units of kilojoules per mole (kJ mol^{-1}) and represent energies needed to remove electrons from 1 mol of gaseous atoms. An examination of the data in this table points out the great stability associated with an electron core having a noble gas electron configuration. We see, for example, that for a Group IA element the first ionization energy is relatively low and that the second ionization energy is very much greater. For the Group IIA elements a large increase in ionization energy occurs after two electrons have been removed, while for Group IIIA elements the break occurs after the third electron has been lost. In fact, we see that in general a very large jump in ionization energy always occurs after an atom has lost a number of electrons that is numerically equal to its group number. Since a Group IA element contains one electron outside a noble gas electron configuration, a Group IIA element, two, and so on, these larger increases in ionization energy must reflect the extreme difficulty encountered in trying to break into the noble gas structure that lies below the outer shell.

Table 4.5

Ionization energies of the first 20 elements (kJ mol^{-1})

	First	Second	Third	Fourth	Fifth	Sixth	Seventh	Eighth
H	1 312							
He	2 371	5 247						
Li	520	7 297	11 810					
Be	900	1 757	14 840	21 000				
B	800	2 430	3 659	25 020	32 810			
C	1 086	2 352	4 619	6 221	37 800	47,300		
N	1 402	2 857	4 577	7 473	9 443	53 250	64 340	
O	1 314	3 391	5 301	7 468	10 980	13 320	71 300	84 050
F	1 681	3 375	6 045	8 418	11 020	15 160	17 860	92 000
Ne	2 080	3 963	6 276	9 376	12 190	15 230	—	—
Na	495.8	4 565	6 912	9 540	13 360	16 610	20 110	25 490
Mg	737.6	1 450	7 732	10 550	13 620	18 000	21 700	25 660
Al	577.4	1 816	2 744	11 580	15 030	18 370	23 290	27 460
Si	786.2	1 577	3 229	4 356	16 080	19 790	23 780	29 250
P	1 012	1 896	2 910	4 954	6 272	21 270	25 410	29 840
S	999.6	2 260	3 380	4 565	6 996	8 490	28 080	31 720
Cl	1 255	2 297	3 850	5 146	6 544	9 330	11 020	33 600
Ar	1 520	2 665	3 947	5 770	7 240	8 810	11 970	13 840
K	418.8	3 069	4 600	5 879	7 971	9 619	11 380	14 950
Ca	589.5	1 146	4 941	6 485	8 142	10 520	12 350	13 830

Figure 4.22

The variation of first ionization energy with atomic number.

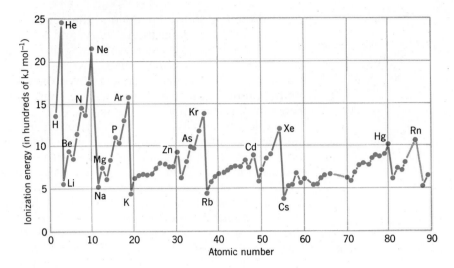

The variation of the first ionization energy across periods and down groups, illustrated in Figure 4.22, quite closely parallels the trends in atomic size. This really should not be too surprising, since the energy required to remove an electron from an atom completely should depend in part on how far away it is from the nucleus. In addition, the same factors responsible for causing an outer shell to contract in size as we proceed across a period will also lead to the electron being held more tightly. Thus, as we proceed down within a group (e.g., the alkali metals), the increase in size that occurs is accompanied by a decrease in ionization energy. As we move across a period, from left to right, the increased effective nuclear charge experienced by the outer-shell electrons causes the shell to shrink in size and also makes it more difficult to remove an electron.

If we examine more closely the trend in ionization energy across a period, we note some irregularities. In period 2, for example, we might expect a uniform increase in ionization energy as we go from Li to Ne. What we actually find, however, is that the ionization energy of beryllium is larger than that of boron and the ionization energy of nitrogen is larger than that of oxygen. These reversals can also be explained by the electronic structures of the elements.

In the case of beryllium, the first electron to be removed comes from the filled 2s subshell, whereas the electron removed first from boron is in the singly occupied 2p subshell. The 2p subshell is higher in energy than the 2s, so the 2p electron of boron is more easily removed than a 2s electron of beryllium.

When we get to nitrogen, we find that we have a half-filled 2p subshell (electronic structure of nitrogen is $1s^2 2s^2 2p^3$), while in oxygen the 2p subshell is occupied by four electrons. The fourth electron in this 2p subshell is in an orbital already occupied by another electron, so it experiences considerable electron-electron repulsion. As a result this electron is more easily removed than one of the electrons in a singly occupied orbital in the nitrogen atom. Note that the same inverted order of values for the ionization energy also occurs in periods 3 and 4, where the ionization energy of phosphorus is greater than for sulfur and that for arsenic is greater than for selenium.

Electron affinity

*The **electron affinity** is the energy that is released or absorbed when an electron is added to a neutral gaseous atom in its ground state.* Such a process occurs, for example,

Table 4.6
Electron affinities^a for the representative elements (kJ mol^{-1})

IA	IIA	IIIA	IVA	VA	VIA	VIIA
H						
−73						
Li	Be	B	C	N	O	F
−60	≈+100	−27	−122	≈+9	−141	−328
Na	Mg	Al	Si	P	S	Cl
−53	≈+30	−44	−134	−72	−200	−348
K	Ca	Ga	Ge	As	Se	Br
−48	—	−30	−120	−77	−195	−325
Rb	Sr	In	Sn	Sb	Te	I
−47	—	−30	−121	−101	−190	−295
Cs	Ba	Tl	Pb	Bi	Po	At
−45	—	−30	−110	−110	−183	−270

^{a} Negative values mean that the process $M + e^- \rightarrow M^-$, is exothermic.

when a chlorine atom picks up an electron to become a negative ion

$$Cl(g) + e^- \longrightarrow Cl^-(g)$$

The electron affinity (like the ionization energy) applies to isolated atoms and usually represents an exothermic process. This is so because we are placing the electron into an environment where it experiences the attraction of the nucleus. We can see how the addition of an electron to an atom would release energy by considering the reverse process, pulling the electron away from the attractive force of the nucleus. If removing the electron requires work (i.e., is endothermic), the opposite process would release energy.

> The potential energy of an electron drops as it becomes attached to an atom to which it is attracted. This difference in potential energy is the energy that's released.

There are instances when more than one electron is added to the outer shell of the atom. For example, oxygen reacts to form the ion, O^{2-}, which requires that an oxygen atom pick up two electrons. The first electron enters a neutral atom, but the second electron, which gives the ion a charge of 2−, must be forced onto an already negative ion. This requires work; therefore we find that the second electron affinity of an atom is an endothermic quantity. Table 4.6 contains electron affinities for some representative elements. The table is not complete because electron affinities are difficult to measure and, for many elements, have not been accurately determined.

> Adding more than one electron to an isolated atom is always endothermic overall.

As with the ionization energy, the variations in electron affinity generally parallel the variations in atomic size. This is because we are considering the placement of an electron into the outer shell of the atom. The closer the electron can get to the nucleus, the greater the effect of the nuclear charge. Therefore, atoms that are very small and have outer shells that experience a high effective nuclear charge (i.e., elements in the upper right of the periodic table) have very large electron affinities. On the other hand, atoms that are large and whose outer shells feel the effect of a small effective nuclear charge (such as the elements in Groups IA and IIA) have small electron affinities.

If we examine Table 4.6, we find that the nonmetals of the second period have electron affinities that are less exothermic than the elements of the same group just below them in period 3. This is a bit of a surprise because in a given group, size increases going down. Apparently, the crowding of electrons in the outer shell of a period 2 element makes the mutual repulsions of the electrons substantially greater than in the outer shell of a period 3 element. Therefore, even though the electron that is added to a period 2 element gets closer to the nucleus than one added to a period 3 element, the greater repulsions in the smaller outer shell lead to a lesser *net* amount of energy being evolved.

Finally, carbon has a rather substantial exothermic electron affinity, while that of nitrogen is actually endothermic. With carbon, the entering electron can occupy a vacant 2*p* orbital and therefore experiences only minimal electron-electron repulsions. With nitrogen, however, an additional electron must be placed into an orbital that is already occupied by an electron. The greater electron-electron repulsions that result, as well as the loss of the extra stability that a nitrogen atom has because of its half-filled *p* subshell, cause the electron affinity to be an endothermic quantity for nitrogen.

REVIEW QUESTIONS AND PROBLEMS

(Problems whose numbers are in blue have their answers in Appendix C at the back of the book; the more difficult problems are marked with asterisks.)

Electromagnetic Radiation

4.1 Sketch a wave and label its wavelength and its amplitude. What relationship exists between the wavelength and the frequency of a light wave?

4.2 What is the value of the speed of light in metres per second?

4.3 What is the SI derived unit of frequency? Why are the wavelengths of radio waves given in metres and those of visible light given in nanometres? What wavelength limits encompass the visible portion of the electromagnetic spectrum?

4.4 How do the wavelengths of infrared light and ultraviolet light compare to the wavelengths of visible light?

4.5 Arrange the following in order of increasing wavelength: visible light, microwaves, ultraviolet light, X rays, infrared light, TV waves, gamma rays.

4.6 Which type of radio broadcasts have a higher frequency, FM or AM? (If necessary, refer to the local radio listings in a newspaper.)

4.7 Calculate the wavelength of light whose frequency is 8.0×10^{15} Hz. Calculate the frequency of light whose wavelength is 200.0 nm.

4.8 A radar transmitter broadcasts electromagnetic radiation in the microwave region of the spectrum. A typical radar wave has a frequency of 9.40×10^9 Hz. What is its wavelength (a) in metres and (b) in centimetres?

4.9 Radio station WCBS in New York broadcasts their FM signal at a frequency of 101.1 megahertz (MHz). Their AM signal is broadcast at 880 kilohertz (kHz). What are the wavelengths of these signals expressed in metres?

4.10 One of the lines in the spectrum of mercury (produced, for example in a mercury vapor lamp) is green and has a wavelength of 546 nm. What is the frequency of this line?

Atomic Spectra

4.11 What is the difference between a continuous spectrum and a line spectrum? Describe how you could produce each of them.

4.12 From the point of view of atomic structure, what is the significance of line spectra?

4.13 What other terms are used to describe line spectra?

4.14 Mercury vapor lamps, used for highway and street lights, produce light corresponding to the atomic spectrum of mercury. One of the lines in the spectrum is green and has a wavelength of 546 nm. What is the frequency of this line?

4.15 Use the data in Figure 4.4 to calculate the frequencies of the visible lines in the atomic spectrum of hydrogen.

4.16 What is the wavelength of the light that gives sodium vapor lamps their characteristic yellow color? What is the frequency of this light?

4.17 Why is it possible to use atomic emission spectra to determine the elemental composition of a substance?

4.18 Use the Rydberg equation (Equation 4.3) to calculate the wavelengths of the first two lines in the Pfund series of the hydrogen spectrum.

4.19 Use the Rydberg equation (Equation 4.3) to calculate the ionization energy of hydrogen and compare your answer to the results of Example 4.5.

Bohr Theory

4.20 Why does Planck's constant have units of *energy × time?*

4.21 What is a photon?

4.22 Calculate the energy in joules of a photon having a frequency of 3×10^{15} Hz. If a photon has an energy of 2×10^{-20} J, what is its wavelength?

4.23 Calculate the energy contained in one photon of yellow light emitted by sodium with a wavelength of 589 nm. What is the energy of a mole of these photons, expressed in kilojoules? How much could the temperature of 10.0 kg of water be raised by the energy in 1 mol of these photons?

4.24 Why does Planck's relationship (Equation 4.4), along with atomic emission spectra, suggest the existence of energy levels in atoms?

4.25 Describe Bohr's model of the atom. What initial evidence was there that Bohr's theory might be correct? Why was his theory eventually discarded?

4.26 How does Bohr's theory explain the emission and absorption of light by a hydrogen atom?

4.27 Use the Rydberg equation to calculate the wavelength (in nanometres) of the spectral line of hydrogen that would result when an electron drops from the fourth Bohr orbit to the second, and from the sixth Bohr orbit to the third.

4.28 How many joules must be supplied to raise an electron from the first Bohr orbit to the third?

Wave Mechanics

4.29 State the de Broglie relationship.

4.30 What is the effective mass of a mole of photons that have a wavelength of 589 nm?

*4.31 How many kilograms of water could be heated from 0°C to 100°C by converting 1.0 g of matter entirely into energy? Recall that it takes 4.184 J of energy to raise the temperature of 1.0 g of water by 1°C. (See also Section 1.5.)

*4.32 Calculate the kinetic energy in joules of an electron with a wavelength of 0.10 nm.

*4.33 How long would it take a 2.0 g bullet to travel the length of a 10-cm gun barrel if it had a wavelength of 0.10 nm?

4.34 What is a diffraction pattern? How is it produced?

4.35 Why don't we observe the wave properties of large objects such as baseballs or airplanes?

4.36 What direct evidence is there for the wave properties of the electron?

4.37 What is a standing wave? What are nodes?

Quantum Numbers

4.38 What is the symbol for the principal quantum number? What are its allowed values? What physical characteristic of the electron wave is associated with the value of its principal quantum number?

4.39 What is the symbol for the azimuthal quantum number? What are its allowed values? What characteristic of an electron wave is associated with the value of its azimuthal quantum number?

4.40 What is the symbol for the magnetic quantum number? What are its allowed values? What characteristic of an electron wave is associated with the value of its magnetic quantum number?

4.41 What is an orbital?

4.42 How many subshells are in the fourth shell?

4.43 If the value of n for an electron is 5, what subshells are possible for this electron?

4.44 If the largest value of m_l for an electron is 3, what kind of subshell must it be in?

4.45 How many orbitals are there in the shell with $n = 5$?

4.46 What does the term *ground state of an atom* mean?

4.47 How many electrons can be accommodated in each of the following types of subshells: s, p, d, f, g, h? What is the lowest value of n for a shell that has an h subshell? What are the allowed values of m for an h subshell?

4.48 Within a given shell, how do the energies of the s, p, d, and f subshells compare?

4.49 Why did Bohr's theory work for hydrogen, but fail for atoms with two or more electrons?

Electron Spin

4.50 What property of an electron is associated with its "spin"?

4.51 What are the only values that m_s can have?

4.52 What is the Pauli exclusion principle? What significance does it have in determining the electronic structure of an atom?

4.53 Give the values of n, l, m_l, and m_s for each electron in a filled L shell.

4.54 How many electrons can be placed into the M shell of an atom?

4.55 What do we mean when we say that two electrons are paired?

4.56 What magnetic property is associated with unpaired electrons in an atom, molecule, or ion? What is the magnetic property associated with the complete pairing of all the electrons in an atom, molecule, or ion?

Electron Configurations

4.57 What is Hund's rule?

4.58 Use the periodic table as a guide in writing the complete electron configurations of these elements: P, Ni, As, Ba, Rh, Ho, Sn.

4.59 Write the complete electron configuration for Rb, Sn, Br, Cr, and Cu.

4.60 Give the outer-shell electron configuration for K, Al, F, S, Tl, and Bi.

4.61 Use the periodic table to arrive at the electronic structure of the outer shells of the atoms of Si, Se, Sr, Cl, O, S, As, and Ga.

4.62 Why are there no elements in period 3 between Mg and Al?

4.63 How many electrons are in p-orbitals in (a) As, (b) Si, (c) Ru?

4.64 Draw the orbital diagram for (a) phosphorus and (b) calcium.

4.65 Draw orbital diagrams for each element in the first row of transition elements (Z = 21 through 30). Indicate which are paramagnetic and which are diamagnetic.

4.66 Draw the orbital diagram for the valence shell of (a) Sn, (b) Br, (c) Ba.

4.67 Which of the following atoms are diamagnetic: (a) Cd, (b) Ge, (c) Pt, (d) Sr, (e) Kr?

4.68 How many orbitals are there in the valence shell of phosphorus?

Shapes of Orbitals

4.69 What is the major difference between a Bohr orbit and an orbital? In what way is the Heisenberg uncertainty principle involved in this comparison?

4.70 On a single set of Cartesian coordinate axes, sketch the shapes of the three p orbitals. Label them p_x, p_y, and p_z.

4.71 How do the $1s$ and $2s$ orbitals differ? How are they alike?

4.72 How does the shape of an s orbital differ from that of a p orbital? Sketch the shapes of these kinds of orbitals.

4.73 On separate sets of axes, sketch the shapes of the following orbitals: (a) d_{xy}, (b) $d_{x^2-y^2}$, (c) d_{z^2}

4.74 On the basis of the mutual repulsion of electrons and the spatial orientations of the p orbitals in a given subshell, the correct orbital diagram for the ground state of nitrogen seems quite reasonable. Why?

Periodic Trends in Properties

4.75 What difficulties are there in defining the size of an atom or an ion? What units are used for expressing atomic size?

4.76 A potassium atom has a diameter of about 406 pm. Express this in metres and nanometres.

4.77 Choose the largest atom: Ge, Sb, Sn, As.

4.78 Choose the larger species in each pair:
(a) S or Se
(b) C or N
(c) Fe^{2+} or Fe^{3+}
(d) O^+ or O^-
(e) S or S^{2-}

4.79 Explain the variation in ionic size observed for the series, N^{3-}, O^{2-}, and F^- (Table 4.4) in terms of the effective nuclear charge and electron-electron repulsions experienced by their outer-shell electrons.

4.80 What is the lanthanide contraction? How might this be used to explain why the elements in the sixth period following the lanthanides have higher ionization energies than the elements directly above them in the fifth period (e.g., the ionizaton energy of Pt = 870 kJ mol^{-1}, while that of Pd = 805 kJ mol^{-1})?

4.81 In Table 4.4, we find some elements that form more than one positive ion. In each case the ion with the greater positive charge is smaller. Why is this so?

4.82 Define *ionization energy* and *electron affinity*.

4.83 How can we explain the variation in ionization energy across a period in the periodic table?

4.84 Choose the species with the larger ionization energy.
(a) Li or Be
(b) Be or B
(c) C or N
(d) N or O
(e) Ne or Na
(f) S or S^+
(g) Na^+ or Mg^+

Check your answers by referring to Table 4.5.

4.85 Draw a graph, on a set of axes like that below, of the ionization energy versus the number of electrons removed from the atom for each of the elements Li, C, O, S, and Ne.

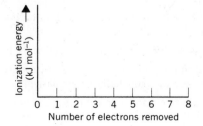

4.86 Choose the species with the more exothermic electron affinity:
   (a) S or Cl
   (b) S or S^-
   (c) P or As
   (d) O or S

4.87 Why is the second electron affinity for an element always an endothermic quantity?

4.88 The Group VIIA elements have electron affinities that are considerably larger than those of the Group VIA elements. What does this suggest about the stability of the noble gas electron configuration?

*4.89 If all the energy required to remove the electrons from 1 mol of H atoms was used instead to heat water, how many grams of water could have their temperature increased by 25°C?

CHAPTER 5

CHEMICAL BONDING: GENERAL CONCEPTS

Candle wax and rock salt (sodium chloride) have quite different properties, which reflect the kinds of attractive forces—the chemical bonds—that hold the atoms together. In this chapter we will examine some of the properties of these substances and why atoms form different kinds of chemical bonds.

In Chapter 4 we spent considerable time discussing the electronic structures of atoms and their relationships to some atomic properties. A property possessed by almost all atoms is the ability to combine with others to produce more complex species. Hydrogen atoms, for example, combine with each other to form molecules of hydrogen, H_2, and they combine with oxygen atoms to form water, H_2O. Sodium and chlorine combine with each other to produce sodium chloride, NaCl. In all these compounds, the atoms are held to each other by relatively strong forces of attraction that we call **chemical bonds.** The way atoms form these chemical bonds is related to their electronic structures, and the kinds of bonds that exist within compounds is the principal factor determining their chemical properties. Ultimately, they also are responsible for the physical properties of substances.

The theories and language used to describe chemical bonds have evolved over the years. In this chapter we begin with the simplest concepts. Although these have been followed by more elaborate theories based on wave mechanics, they still serve many uses in discussing the structures and shapes of molecules. This is due directly to their simplicity.

5.1 PROPERTIES OF IONIC AND MOLECULAR COMPOUNDS

Chemical bonds are something that we can't actually see. We know they exist because atoms do combine with each other, and they do form compounds. Of necessity, therefore, explanations of chemical bonds are theories, and as you know from Chapter 1, theories explain experimental facts. Let's choose two compounds and look at the kinds of facts that ultimately must be explained.

Two compounds that you are familiar with and that are remarkably different in their chemical and physical properties are sodium chloride, NaCl, and eicosane, $C_{20}H_{42}$. Sodium chloride, of course, is the chemical name for common table salt. Eicosane is the chemical name for one of the compounds in a mixture of substances commonly called parafin wax—the material from which candles are made, and which is used for sealing jars of homemade jams and jellies. What are some of the properties of these substances that allow us to distinguish between them?

If you sprinkle some salt crystals on a piece of paper and study them under a magnifying glass, you will see that sodium chloride is a white crystalline material. The crystals are brittle and shatter to a powder when you crush them. You also know that sodium chloride is soluble in water, and you are probably aware that it melts at a high temperature—in fact, sodium chloride melts at about 800°C, and it boils at over 1400°C.

Eicosane has all the properties that we associate with wax. If you study a piece of wax, you will see that it lacks the crystalline appearance of salt. If you attempt to crush it, it just flattens out; wax is soft and easily deformed. Eicosane is insoluble in water, but it does dissolve in substances such as gasoline or paint thinner—solvents in which salt is very insoluble. In contrast to salt, eicosane melts at a low temperature, 37°C (which just happens to be normal body temperature), and it boils at only 343°C. Eicosane also burns in air, which is quite different from the behavior of sodium chloride.

Another property of salt that distinguishes it from compounds like eicosane and provides a clue to the kinds of particles found within it is its ability to conduct electricity when melted. This property can be easily tested with the apparatus shown in Figure 5.1, which consists of two electrodes connected in series with the light bulb. When the apparatus is plugged into an electrical outlet and contact is made across the electrodes, the light bulb lights.

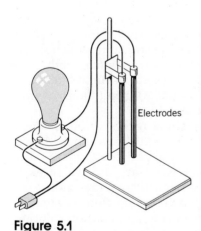

Electrodes

Figure 5.1

An apparatus to test electrical conductivity. The electrodes are connected in series with the light bulb, and when electrical contact is made across them, the bulb can light.

Figure 5.2

Neither solid NaCl (a) *nor solid eicosane* (b) *conducts electricity.* (c) *Molten eicosane doesn't conduct either, but molten NaCl* (d) *does.*

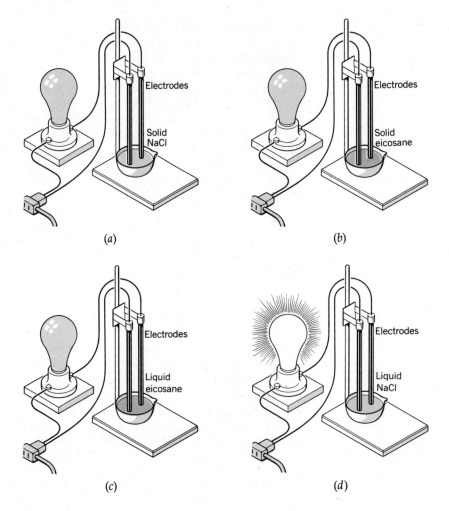

As illustrated in Figure 5.2, when the electrodes are dipped into crystals of solid salt or shavings of solid eicosane, the light bulb doesn't light, which shows that neither of these solids conducts electricity. However, when both compounds are melted and the electrodes are dipped into them, the light bulb lights for the molten salt, but it remains unlit for the eicosane. Molten (liquid) sodium chloride conducts electricity, but liquid eicosane does not.

Sodium chloride and eicosane are compounds that represent opposite extremes in the way that atoms are able to form bonds with each other. At one extreme, the atoms combine to form electrically neutral particles that are called **molecules.** This is the way the atoms in eicosane combine, and we speak of eicosane as a **molecular compound.** The other principal way that atoms can combine is by the transfer of one or more electrons from one atom to another. This is what happens when sodium and chlorine react to give sodium chloride. When atoms gain or lose electrons, they form ions, and in the solid compound that forms, there are both positive and negative ions in a ratio that gives an electrically neutral substance. We mentioned these kinds of compounds in Chapter 2 when we described the reason that we use the term formula weight instead of molecular weight for some substances. A compound such as sodium chloride, which is composed of ions, is called an **ionic compound.**

The electrical behavior of salt and of eicosane can be understood rather

The physical properties of ionic and molecular compounds will be described and explained in further detail in later chapters.

easily in terms of the kinds of particles within them. In the solid state, neither conducts because there are no electrically charged particles that are able to move. In solid eicosane, there are only neutral molecules and in sodium chloride, the ions are frozen in place. When melted, the eicosane still is unable to conduct because the liquid consists of neutral molecules that cannot transport electrical charge. But when sodium chloride is melted, the ions are set free and are able to move about. These charged particles can transport charge through the liquid, so molten sodium chloride does conduct.

The properties of sodium chloride and eicosane are typical of many ionic and molecular compounds. Ionic compounds tend to be brittle, have high melting points, and are able to conduct electricity in the liquid state. Molecular compounds, on the other hand, tend to be soft, have low melting points, and do not conduct electricity, either when melted or in the solid state. Now that we know what some of these properties are, we can turn our attention to explaining why and how certain combinations of atoms produce molecular substances, and why other atoms form ionic compounds.

5.2 LEWIS SYMBOLS

G. N. Lewis received his Ph.D. from Harvard and later served as professor of chemistry and dean at the University of California.

When atoms interact to form a bond, only their outer portions come into contact; therefore, only their outer electron configurations are usually important. To keep tabs on the outer-shell (also called valence-shell[1]) electrons, we use a special type of notation called **Lewis symbols,** named after the American chemist, Gilbert N. Lewis (1875–1946).

To construct the Lewis symbol for an element, we write its atomic symbol surrounded by a number of dots (or X's or circles, etc.), each of which represents one electron in the atom's valence shell. For example, the element hydrogen, which has one electron in its valence shell, is given the Lewis symbol, H·. Any atom, in fact, with one electron in its outer shell has a similar Lewis symbol. This includes any element in Group IA of the periodic table, so that each of the elements Li, Na, K, Rb, Cs, and Fr has a Lewis symbol that we might generalize as X· (where X = Li, Na, etc.). Generalized Lewis symbols for the representative elements are given in Table 5.1.

In general, the number of valence electrons that an atom of a representative element has is equal to its group number. Therefore, we see that the group number is also equal to the number of dots in the atom's Lewis symbol. This is useful to remember because it makes writing the Lewis symbol for an element very simple. Notice that in Table 5.1 the number of unpaired electrons for atoms in Groups IIA, IIIA, and IVA doesn't agree with the predictions you would make by writing their electron configurations. The Lewis symbols are written this way only because when the atoms form bonds, they *behave* as though they have the number of unpaired electrons shown by their Lewis symbols. We will see that Lewis symbols are useful in discussing bonds between atoms. The formulas we draw with them are called either *Lewis structures* or *electron-dot formulas.*

Lewis symbols usually are *not* used for the transition elements or inner transition elements.

EXAMPLE 5.1	WRITING THE LEWIS SYMBOL FOR AN ATOM
PROBLEM	What is the Lewis symbol for germanium (Z = 32)?

---

[1]As noted in the previous chapter, *valence* is a term sometimes associated with chemical bonding. It normally describes an atom's bond-forming capacity. Unfortunately, the term has been used with more than one meaning, so its definition has become ambiguous. For this reason, the term is rarely used today.

Table 5.1
Lewis symbols for A-Group elements

Group	IA	IIA	IIIA	IVA	VA	VIA	VIIA
Symbol	$X\cdot$	$\cdot X\cdot$	$\cdot \overset{\cdot}{X}$	$\cdot \overset{\cdot}{X}\cdot$	$\cdot \overset{\cdot}{X}:$	$\cdot \overset{\cdot\cdot}{X}:$	$\cdot \overset{\cdot\cdot}{X}:$

SOLUTION   Germanium is in Group IVA and therefore has four valence electrons. Its Lewis symbol has four dots that we arrange symmetrically around the chemical symbol.

$\cdot \overset{\displaystyle \cdot}{\underset{\displaystyle \cdot}{Ge}} \cdot$

5.3 THE IONIC BOND

Chemical bonds can be divided into two general categories: **ionic** (or **electrovalent**) **bonds** and **covalent bonds.**

Ions are formed when one or more electrons are transferred from the valence shell of one atom to the valence shell of another. The atom that loses electrons becomes a positive ion (a **cation**) while the atom that acquires electrons becomes negatively charged (an **anion**). *The ionic bond* results from the attraction between the oppositely charged ions.

An example of the formation of an ionic substance is the reaction between atoms of lithium and fluorine. The electronic structures of these are

$$\text{Li} \quad 1s^2 2s^1$$

and

$$\text{F} \quad 1s^2 2s^2 2p^5$$

When they react the lithium atom loses the electron from its $2s$ subshell and becomes Li^+. Notice that the lithium ion has an electron configuration that is the same as the noble gas helium.

$$\text{Li } (1s^2 2s^1) \longrightarrow \text{Li}^+ (1s^2) + e^-$$

The electron lost by the lithium atom is picked up by the fluorine atom, which becomes a fluoride ion, F^-. This ion has an electron configuration identical to that of the noble gas neon.

$$\text{F } (1s^2 2s^2 2p^5) + e^- \longrightarrow \text{F}^- (1s^2 2s^2 2p^6)$$

Once formed, the Li^+ and F^- ions attract one another because of their opposite charges. It is this attraction between the ions that constitutes the ionic bond.

When a reaction between lithium and fluorine actually takes place in real life, huge numbers of atoms are involved and equally huge numbers of ions are formed. These ions pack themselves together to form the ionic solid, LiF. It is important to remember that an ionic solid such as this does not contain discrete molecules, but instead contains ions packed so that the attractive forces between ions of opposite charge are maximized while repulsive forces between ions of the same charge are minimized. In LiF, for example, each cation (Li^+) is sur-

Cation has three syllables: cat'i·on. Anion also has three syllables: an'i·on.

The configuration of helium is $1s^2$.

The configuration of neon is $1s^2 2s^2 2p^6$.

Attractions and repulsions between electrically charged particles are called *electrostatic* attractions and repulsions.

Figure 5.3

In a crystal of lithium fluoride, LiF, each Li$^+$ ion is surrounded by six F$^-$ ions and each F$^-$ ion is surrounded by six Li$^+$ ions. The large spheres are F$^-$ ions and the small spheres are Li$^+$ ions.

Li$^+$ [$:\ddot{\text{F}}:$]$^-$ is the Lewis structure for LiF.

rounded by, and attracted equally by six anions (F$^-$), as shown in Figure 5.3.

The Lewis symbols introduced in the last section can be used to illustrate the transfer of electrons that occurs during the formation of an ionic compound. For instance, we can show the reaction of Li and F as

$$\text{Li} \cdot + \cdot \ddot{\text{F}} : \longrightarrow \text{Li}^+ \left[: \ddot{\text{F}} : \right]^-$$

The brackets appearing around the fluorine on the right indicate that all eight electrons are the exclusive property of the fluoride ion, F$^-$.

When Li and F react, electrons are lost or gained until both atoms have a noble gas electron configuration. Except for helium, this corresponds to an outer-shell configuration of ns^2np^6 (a total of eight electrons in the outer shell). In the last chapter we saw that the electronic structures of the noble gases possess a great deal of stability. The tendency for atoms to achieve this stable electron arrangement forms the basis of the so-called **octet rule,** which simply states that *an atom tends to gain or lose electrons until there are eight electrons in its outer shell.*

In the case of ionic bonding, the octet rule really works well only for the metals in Groups IA and IIA (except Li and Be), and for the nonmetals. Here we find that the metals lose electrons and the nonmetals gain them. The charges on the ions that these elements form, which are given in Table 5.2, are determined by the numbers of electrons that the atoms must lose or gain to reach an octet.

Failure of the octet rule

Except for the metals in Groups IA and IIA and aluminum, the octet rule doesn't work well for cations. When a transition element or **post-transition element** (an element to the right of a row of transition elements, such as tin or lead) forms a positive ion, the outer-shell electron configuration is generally not the same as that of a noble gas.

When a positive ion is formed from an atom, electrons are *always* lost first from the shell with the largest value of n. This means that a transition element always loses electrons from its outer s subshell before any electrons are lost from

Table 5.2

Some common cations and anions

Ions with noble gas electron configurations

1+	2+	3+	3−	2−	1−
Li^+		Al^{3+}	N^{3-}	O^{2-}	F^-
Na^+	Mg^{2+}		P^{3-}	S^{2-}	Cl^-
K^+	Ca^{2+}			Se^{2-}	Br^-
Rb^+	Sr^{2+}			Te^{2-}	I^-
Cs^+	Ba^{2+}				

its outermost d subshell. For example, the zinc ion, Zn^{2+}, is formed when a zinc atom loses its outer $4s$ electrons.

$$Zn~([Ar]~3d^{10}4s^2) \longrightarrow Zn^{2+}~([Ar]~3d^{10}) + 2e^-$$

The electron configuration of the Zn^{2+} ion also can be rewritten as

$$Zn^{2+}~[Ne]~3s^23p^63d^{10}$$

and we see that its outer shell, $3s^23p^63d^{10}$, does not have the usual noble gas configuration, ns^2np^6. It does have one thing in common with a noble gas configuration, however—all the subshells in the outer shell are complete. Because of this similarity, the $ns^2np^6nd^{10}$ configuration is called a **pseudonoble gas configuration.**

Because the octet rule fails for so many metals, the ions that they form are difficult to predict. They just have to be learned through practice. Table 5.3 contains some of the more common cations of the transition and post-transition metals. If you work the exercises at the end of the chapter, they will help you learn the formulas of these ions.

Writing formulas for ionic compounds

The ratio in which the atoms of two elements combine to form an ionic compound is determined by the charges acquired by the cation and anion. This is because the compound formed must be electrically neutral. If we diagram the

Table 5.3

Cations formed by some transition and post-transition elements

Chromium	Cr^{2+} Cr^{3+}	Gold	Au^+ Au^{3+}
Manganese	Mn^{2+} Mn^{3+}	Zinc	Zn^{2+}
		Cadmium	Cd^{2+}
Iron	Fe^{2+} Fe^{3+}	Mercury	Hg_2^{2+} Hg^{2+}
Cobalt	Co^{2+} Co^{3+}	Tin	Sn^{2+} Sn^{4+}
Nickel	Ni^{2+}	Lead	Pb^{2+} Pb^{4+}
Copper	Cu^+ Cu^{2+}	Bismuth	Bi^{3+}
Silver	Ag^+		

reaction with Lewis symbols, we can see how this happens. Consider, for example, the reaction of calcium with chlorine. To arrive at a noble gas configuration, a calcium atom (Group IIA) must lose two electrons and each chlorine atom (Group VIIA) must pick up one electron.

$$Ca\!: + \quad \begin{array}{c} \cdot\ddot{Cl}\!: \\[4pt] \cdot\ddot{Cl}\!: \end{array} \longrightarrow Ca^{2+},\ 2\left[:\ddot{Cl}\!:\right]^{-}$$

$$\text{or}$$
$$CaCl_2$$

We see that two chlorine atoms must react with each calcium atom to produce one Ca^{2+} ion and two Cl^- ions. The neutral compound, calcium chloride, has the formula $CaCl_2$. Similar reasoning leads us to expect a compound between Li and O to have the formula Li_2O (the compound is called lithium oxide).

$$\begin{array}{c} Li\cdot \\[4pt] + \ \cdot\ddot{O}\!: \longrightarrow 2Li^{+},\ \left[:\ddot{O}\!:\right]^{2-} \\[4pt] Li\cdot \end{array}$$

In writing the formulas of ionic compounds such as lithium fluoride, calcium chloride, or lithium oxide, three general rules are followed:

1. The positive ion is always written first in the formula. This is just a custom that chemists always follow.

2. The ratio of positive to negative ions is always chosen so that the total positive and negative charges are equal—the compound must be electrically neutral.

3. The smallest set of subscripts that gives electrical neutrality is always chosen. We always write empirical formulas for ionic compounds. This is because ionic compounds do not contain discrete molecules.

Applying the second rule sometimes requires a little juggling. Consider, for example, the compound formed when oxygen in the air reacts with an aluminum surface to form a thin film of aluminum oxide. Aluminum, in Group IIIA, loses three electrons to achieve a noble gas structure and produces the ion Al^{3+}. Oxygen, on the other hand, forms the ion O^{2-}. To produce a neutral compound, two Al^{3+} ions must be combined with three O^{2-} ions. In that way the total positive charge [$2 \times (3+) = 6+$] is the same as the total negative charge [$3 \times (2-) = 6-$]. Therefore, aluminum oxide has the formula Al_2O_3.

EXAMPLE 5.2	WRITING THE FORMULA FOR AN IONIC COMPOUND
PROBLEM	What is the formula of an ionic compound formed between Mg and P?
SOLUTION	Magnesium is in Group IIA and therefore will lose two electrons to achieve a noble gas configuration. The ion formed is Mg^{2+}. Phosphorus is found in Group VA and must acquire three electrons to attain a noble gas structure. This produces the ion P^{3-}. To

make the compound neutral, the total positive or negative charge must be a multiple of both the 2+ and 3−. The smallest number that is divisible by both 2 and 3 is 6, so

$$(2+) \times 3 = 6+$$
$$(3-) \times 2 = \underline{6-}$$
$$\text{net} \quad 0$$

The formula has to be Mg_3P_2.

There is a very simple shortcut that you can use to obtain the formula of an ionic compound. It involves just exchanging superscripts for subscripts.

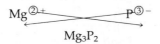

The total positive charge is 6+; the total negative charge is 6−. If you use this method to obtain the formula, remember that ionic compounds are always represented by empirical formulas. Be sure that the subscripts are the simplest set of whole numbers. For example, using this method with Mg^{2+} and O^{2-} gives Mg_2O_2 initially. This must be reduced to MgO.

Compounds that contain polyatomic ions

There are many substances that contain **polyatomic ions**—ions composed of more than one atom. The formulas of these compounds are also determined by

> You must know the names of the polyatomic ions so that you can name compounds containing them, and you need to know their charges to write correct formulas for their compounds.

Table 5.4
Some common polyatomic ions

Cations	
NH_4^+	Ammonium
H_3O^+	Hydronium

Anions (alternate names in parentheses)

CO_3^{2-}	Carbonate	ClO_2^-	Chlorite
HCO_3^-	Hydrogen carbonate (bicarbonate)	ClO^- (or OCl^-)	Hypochlorite
$C_2O_4^{2-}$	Oxalate	PO_4^{3-}	Phosphate (orthophosphate)
CN^-	Cyanide	HPO_4^{2-}	Hydrogen phosphate (hydrogen orthophosphate)
NO_3^-	Nitrate		
NO_2^-	Nitrite	$H_2PO_4^-$	Dihydrogen phosphate (dihydrogen orthophosphate)
OH^-	Hydroxide		
SO_4^{2-}	Sulfate	CrO_4^{2-}	Chromate
HSO_4^-	Hydrogen sulfate (bisulfate)	$Cr_2O_7^{2-}$	Dichromate
SO_3^{2-}	Sulfite	MnO_4^-	Permanganate
HSO_3^-	Hydrogen sulfite (bisulfite)	$C_2H_3O_2^-$	Acetate
ClO_4^-	Perchlorate		
ClO_3^-	Chlorate		

the relative numbers of cations and anions that must be present in order to achieve a neutral solid. Table 5.4 lists some common polyatomic ions. You should be sure to learn their formulas, charges, and names.

The atoms within a polyatomic ion are held to each other by covalent bonds, which we will discuss later. A few of these ions are highly colored and impart their characteristic colors to compounds (and aqueous solutions) containing them. Let us now look at some examples of how the formulas of ionic compounds containing this type of ion are obtained.

EXAMPLE 5.3 WRITING THE FORMULA FOR AN IONIC COMPOUND THAT CONTAINS A POLYATOMIC ION

PROBLEM What is the formula for the ionic substance containing the ions Na^+ and CO_3^{2-}?

SOLUTION In order for the compound to be neutral the number of positive charges must equal the number of negative charges. This requires two Na^+ ions per CO_3^{2-} ion. The compound (called sodium carbonate) therefore has the formula Na_2CO_3. One of its uses is in making glass.

EXAMPLE 5.4 WRITING THE FORMULA FOR AN IONIC COMPOUND THAT CONTAINS A POLYATOMIC ION

PROBLEM Calcium phosphate, an ionic compound that is an important component of many fertilizers, contains the ions Ca^{2+} and PO_4^{3-}. What is its formula?

SOLUTION The total number of positive or negative charges represented in the formula must be divisible by both 2 and 3. The smallest number that meets this requirement is $2 \times 3 = 6$. Thus, there must be six positive and six negative charges in the formula. This is achieved by taking three Ca^{2+} ions and two PO_4^{3-} ions, so the formula of calcium phosphate is $Ca_3(PO_4)_2$.

We get this same answer exchanging superscripts for subscripts.

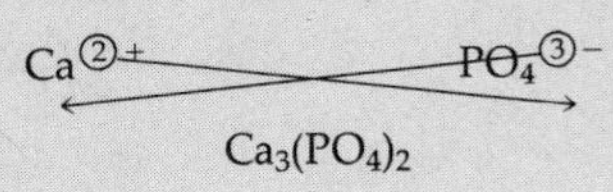

$$Ca^{\textcircled{2}+} \diagdown \diagup PO_4^{\textcircled{3}-}$$
$$Ca_3(PO_4)_2$$

5.4 FACTORS THAT INFLUENCE THE FORMATION OF IONIC COMPOUNDS

In general, low energy implies stability.

We normally associate a lowering of the potential energy of a system with an increase in its stability. For example, a ruler standing on its end will fall on its side, and in the process its potential energy decreases as it achieves a more stable configuration. When atoms combine to form an ionic compound, they also undergo changes that bring them to a more stable condition. In fact, it is a drive toward lowering the potential energy of the particles that causes the electron transfer to take place.

The actual reaction between the elements lithium and fluorine is not as simple as we pictured it in the last section. Lithium does not exist as simple atoms, but rather as a solid; fluorine occurs as a gas composed of F_2 molecules. Nevertheless, an important feature of the reaction is the transfer of electrons from lithium to fluorine. Let's look briefly at the energy changes to see why it occurs.

Interpreting the coefficients as moles, the equation for the formation of 1 mol of lithium fluoride is

$$Li(s) + \tfrac{1}{2}F_2(g) \longrightarrow LiF(s)$$

The coefficient of $\tfrac{1}{2}$ is permitted here because we are dealing in mole-sized quantities, that is

$$1 \text{ mol } Li(s) + \tfrac{1}{2} \text{ mol } F_2(g) \longrightarrow 1 \text{ mol } LiF(s)$$

To analyze the factors that contribute to the energy change in this reaction we will *imagine* that the change occurs in a sequence of steps, whose net result is simply the reaction we have written here. The law of conservation of energy allows us to do this. No matter how many steps it takes to get from one place to another along different paths, the net change in energy must be the same; otherwise we could create or destroy energy by running back and forth along different paths. This kind of application of the law of conservation of energy will appear frequently in the coming chapters.

Let's suppose that instead of combining $Li(s)$ with $\tfrac{1}{2}F_2(g)$ directly, we could follow the lower path in Figure 5.4. This path, called a Born-Haber cycle, consists of five different steps, each having an energy change associated with it. Let's look at each step.

Step 1. $\qquad\qquad\qquad\qquad$ $Li(s) \longrightarrow Li(g)$

Vaporization of 1 mol of $Li(s)$ to produce lithium atoms in a vapor where they are so far apart that they are effectively isolated atoms. The energy required for this endothermic process has been measured to be 155 kJ.

Step 2. $\qquad\qquad\qquad\qquad$ $\tfrac{1}{2}F_2(g) \longrightarrow F(g)$

Decomposition of $\tfrac{1}{2}$ mol of $F_2(g)$ to give 1 mol of fluorine atoms. The energy required for this endothermic process is 79 kJ.

Step 3. $\qquad\qquad\qquad\qquad$ $Li(g) \longrightarrow Li^+(g) + e^-$

Removal of the outer electrons from the Li atoms. This is the first ionization energy of Li and has a value of 520 kJ. This produces 1 mol of $Li^+(g)$. This step is also endothermic.

Figure 5.4

Born–Haber cycle for LiF. Numbers correspond to processes described in the text.

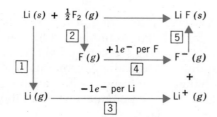

Step 4. $F(g) + e^- \longrightarrow F^-(g)$

Addition of an electron to each F atom. This is the electron affinity of F and is equal to 328 kJ. This exothermic step produces 1 mol of $F^-(g)$.

Step 5. $Li^+(g) + F^-(g) \longrightarrow LiF(s)$

Bringing together the mole of Li^+ and F^- to give 1 mol of LiF(s). Energy is also released in this step[2] and is called the **lattice energy**. It is equal to 1016 kJ.

Just as the overall reaction is the net result of Steps 1 through 5, the overall energy change is the net result of the energy changes in Steps 1 through 5. Steps 1, 2, and 3 are endothermic—energy must be put into the system to vaporize Li(s), decompose $F_2(g)$, and remove electrons from Li(g). The total input is 754 kJ. The last two steps (4 and 5) are exothermic—energy is released from the system when electrons are added to F(g) and when Li^+ and F^- come together to form LiF(s). The total energy released is 1344 kJ.

The net energy change is the difference between the energy put into the system and the energy released. Here we see that 590 kJ more are released than must be added. Therefore the net reaction is exothermic. This means that the system's energy is lowered and LiF(s) is considerably more stable than Li(s) and $F_2(g)$. In fact, when we examine the numbers, we see that it is the large energy lowering effect of the lattice energy that is primarily responsible for the stability of this compound. This is the case with other ionic solids as well.

By a similar analysis, we can also understand why electron transfer between lithium and fluorine ceases once Li^+ and F^- have been formed—that is, once these ions have acquired a noble gas configuration. Removing an additional electron from lithium requires breaking into the noble gas core, which demands the input of an enormous quantity of energy. Placing an additional electron on the fluoride ion is also very expensive in terms of energy primarily because the electron must be placed into the next higher energy level (the 3s orbital), and also because the electron must be forced onto an already negative ion. These very endothermic processes—breaking into the noble gas core of Li^+ and exceeding the octet of F^-—require much more energy than can be gotten back in the form of lattice energy, so an ionic solid composed of Li^{2+} and F^{2-} is unstable and never even forms. Similar analyses for other ionic compounds of the representative elements show that the lattice energy is only able to offset the endothermic contributions to the net energy change up to the point where the ions have achieved a noble gas configuration.

At this point we can ask: What conditions most favor the formation of an ionic substance? From our analysis we see that the most stable ionic compounds will result when elements of low ionization energy combine with elements of high electron affinity, or when the lattice energy of the resulting compound is very large, or both. Under these conditions more energy is given off by the exothermic processes than is absorbed by the endothermic ones, with the net result that the total potential energy contained within the reacting species decreases.

Since metals generally have rather low ionization energies and electron affinities, they tend to lose electrons to form cations; nonmetals, on the other hand, have large ionization energies and electron affinities, so they usually acquire electrons to produce anions. For this reason, most compounds formed

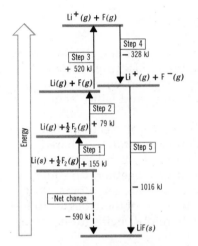

An energy diagram for the formation of LiF(s) from Li(s) and $\frac{1}{2}F_2(g)$. Going up corresponds to endothermic changes; going down corresponds to exothermic ones. Since more energy is released, overall, than is absorbed, the net reaction is exothermic (by 590 kJ).

Ionic compounds tend to be formed between metals and nonmetals.

---

[2] Since energy is required to separate positive and negative ions, energy must obviously be released when they are brought together. The reaction $Li^+ + F^- \rightarrow LiF$, therefore, is exothermic.

Figure 5.5

The electron distribution in the H₂ molecule.

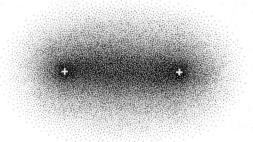

between metals and nonmetals are ionic, particularly the substances formed when an element from Group IA or IIA reacts with an element in the upper right corner of the periodic table (excluding Group 0).

5.5 THE COVALENT BOND

In many instances the formation of an ionic substance is not energetically favorable. For example, the creation of a cation may require too large an energy input (ionization energy) to be recovered by the energy released when the anion is formed and the ionic solid is produced (electron affinity and lattice energy). In these situations a **covalent bond** is formed.

A covalent bond results from the *sharing* of a *pair* of electrons between atoms. The binding force results from the attraction between these shared electrons and the positive nuclei of the atoms entering into the bond. In this sense, the electrons serve as a sort of glue cementing the atoms together. Consider, for example, the formation of the H₂ molecule from two hydrogen atoms. As the atoms approach each other, the single 1s electron on each of them begins to feel the attraction of both nuclei. The electron density therefore begins to shift to the region between the nuclei, as shown in Figure 5.5.

If we examine the energy changes accompanying the formation of the bond, we find that as the atoms come together the energy begins to decrease. This is because the electrons are coming closer to another positive nucleus to which they are also attracted (remember how potential energy changes between particles that attract each other). The energy curve for the molecule is shown in Figure 5.6. Notice that at small internuclear distances the energy rises steeply.

At small distances, the repulsions between nuclei outweigh the attractions that the nuclei have for the electron density between them.

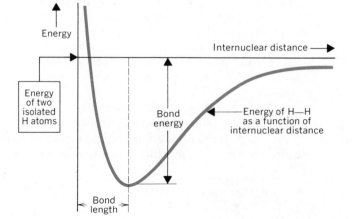

Figure 5.6

Energy diagram for the formation of H₂ from two hydrogen atoms. The bond energy is 435 kJ mol⁻¹ of H₂ and the bond length is 75 pm.

This is caused by the repulsions between the nuclei. The most stable (lowest energy) distance of separation between the two nuclei occurs when the energy is a minimum. At this point the attractions and repulsions are balanced. The depth of this minimum is the quantity of energy that must be supplied to separate the atoms and is called the **bond energy.** The distance between the nuclei when the energy is a minimum is called the **bond length** or **bond distance.**

When two atoms such as hydrogen share a pair of electrons, the spins of the electrons become paired. This is an important aspect of the creation of a covalent bond. Each H atom completes its valence shell by acquiring a share of an electron from another atom. We can indicate the formation of H_2, using Lewis symbols, as

$$H\cdot \; + \; H\cdot \longrightarrow H:H$$

in which the pair of electrons in the bond is shown as a pair of dots between the two H atoms. Often a dash is used instead of the pair of dots, so the H_2 molecule may be represented as H—H.

The number of covalent bonds that an atom forms can often be explained simply by counting the number of electrons required to achieve a noble gas configuration. For example, a carbon atom has four electrons in its valence shell. To attain a noble gas configuration, it usually acquires, through sharing, four additional electrons. Therefore, a carbon atom usually forms four covalent bonds, and with hydrogen it forms the molecule CH_4, which is called methane (the chief component of natural gas).

$$\cdot\dot{C}\cdot \; + \; 4H\times \longrightarrow \begin{matrix} & \overset{\times\cdot}{H} & \\ H & \overset{}{\underset{\cdot\times}{\times C \times}} & H \\ & H & \end{matrix}$$

Nitrogen, which has five valence electrons, has to gain only three electrons through sharing to complete an octet; therefore, nitrogen forms three covalent bonds with hydrogen to form the ammonia molecule, NH_3. In a similar fashion it is easy to see why the formula for water is H_2O and that for hydrogen fluoride is HF.

$$H \overset{\times}{\underset{\underset{H}{\times\cdot}}{N}} H \qquad H \overset{\times}{\underset{\underset{H}{\times\cdot}}{\ddot{O}}} : \qquad H \overset{\times}{\underset{\cdot\cdot}{\ddot{F}}} :$$

In each of the molecules that we've discussed up to this point, the atoms have been joined by covalent bonds that each consist of a single pair of electrons. These are called **single bonds.** It is also possible for a pair of atoms to share two or even three pairs of electrons. For example, in carbon dioxide the carbon shares two pairs of electrons with each oxygen atom. These are **double bonds.**

$$\ddot{O}::C::\ddot{O}$$

The electrons in the bonds are counted as belonging to each atom, so we can count eight electrons around carbon and around each oxygen.

8 electrons

---

*A covalent bond is sometimes called an **electron pair bond**.*

One dash stands for two electrons with their spins paired.

$$\begin{matrix} & H & \\ & | & \\ H & -C- & H \\ & | & \\ & H & \end{matrix}$$

$$\begin{matrix} H & -\overset{}{N}- & H \\ & | & \\ & H & \end{matrix}$$

$$\begin{matrix} H & -\ddot{O}: \\ & | \\ & H \end{matrix}$$

$$H-\ddot{F}:$$

Using dashes to represent a pair of electrons, we can also draw this as

$$\ddot{O}=C=\ddot{O}$$

Nitrogen is an example of a molecule with a **triple bond.**

$$:N:::N: \quad \text{or} \quad :N\equiv N:$$

If we count all six electrons between the atoms as belonging to both, each atom has an octet.

Exceptions to the octet rule

There are many examples of covalent compounds that fail to obey the octet rule. For instance, the molecule $BeCl_2$, which exists in the gas phase at high temperatures, is formed by the pairing of the two Be valence electrons with electrons on two chlorine atoms[3]

$$:\ddot{C}l\cdot + {}^{\times}Be^{\times} + \cdot\ddot{C}l: \longrightarrow :\ddot{C}l \underset{\times}{\times} Be \underset{\times}{\times} \ddot{C}l:$$

In this molecule the Be atom has only four electrons in its valence shell. Another example is BCl_3

$$
\begin{array}{c}
:\ddot{C}l: \\
\overset{\times}{} \\
:\ddot{C}l \underset{\times}{\times} B \underset{\times}{\times} \ddot{C}l:
\end{array}
$$

Molecules having atoms that are surrounded by less than an octet of electrons are rare and must simply be learned as exceptions. The method we will learn later for drawing Lewis structures doesn't work for them.

There are many molecules in which the central atom has more than eight electrons in its valence shell. Two typical examples are PCl_5 and SF_6. To form covalent bonds between the central atom and each of the surrounding atoms, more than four pairs of electrons are needed. In PCl_5, for example, there are five covalent bonds; in SF_6 there are six. The central atom in each of these molecules uses all its valence electrons to form covalent bonds.

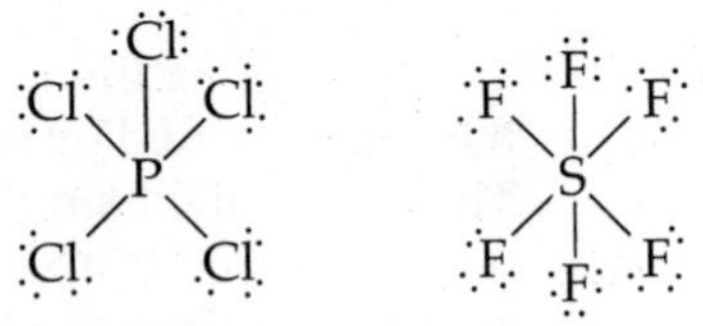

In these compounds, both phosphorus and sulfur have exceeded the number of electrons required for a noble gas electron configuration. This can occur with these elements because, in each case, the valence shell can accommodate more than eight electrons (both P and S are in the third period and the third shell can contain up to 18 electrons because of the availability of the relatively low-energy $3d$ subshell). Elements in the second period (Li to Ne) never form compounds with more than eight electrons in their valence shell because the second shell cannot have more than an octet.

[3] Even though $BeCl_2$ is formed from elements in Groups IIA and VIIA, it is covalent instead of ionic. The reason is discussed in Chapter 18.

5.6 DRAWING LEWIS STRUCTURES

Lewis structures for covalently bonded molecules and polyatomic ions are very useful. One reason, as we will see in the next chapter, is that they allow us to predict the shapes of molecules or polyatomic ions. Therefore, you should learn how to draw them.

The first step in drawing a Lewis structure is deciding which atoms are attached to each other. For example, in CO_2, we must know that there are two O atoms bonded to the C atom, and that the molecule does not have a structure such as O—O—C. The only way of being sure, of course, is to obtain the structural information experimentally, and this is what a chemist would do if a new compound has been discovered. (There are methods for determining molecular structures, but they are beyond the scope of this book and we won't attempt to discuss them.)

Even in the absence of concrete structural data, the arrangement of atoms often can be inferred from the formula of the species. In simple binary molecules and polyatomic ions, such as CO_2 or CO_3^{2-}, the central atom usually appears first in the formula. For example, carbon is the central atom in both CO_2 and CO_3^{2-}. This generalization also holds for NH_3, NO_2, NO_3^-, SO_3, and SO_4^{2-}. Unfortunately, it is not true for H_2O and H_2S (in which H atoms are bound to O and S, respectively). Nor is it true for molecules such as HClO (in which the O is the central atom) or ions like SCN^- (in which C is central). The structure of the molecule is therefore not always obvious. If you must guess, the most symmetrical arrangement of atoms has the greatest chance of being correct. Once we know the arrangement of atoms in the molecule, however, we can then go about distributing the valence electrons. The procedure for doing this can be summarized in the following steps, which we will illustrate by some examples.

1. *Count all the valence electrons of the atoms.* If the species is an ion, add an additional electron for each negative charge or subtract an electron for each positive charge.

2. *Place one pair of electrons in each bond.*

3. *Complete the octets of the atoms bonded to the central atom.* (Remember, however, that the valence shell of any hydrogen atom is complete with only two electrons.)

4. *Place any additional electrons on the central atom in pairs.*

5. *If the central atom still has less than an octet, you must form multiple bonds so that each atom has an octet.*

Now let's look at some examples that show how this procedure works.

It's useful to remember that hydrogen can never be a central atom because it forms only one covalent bond.

EXAMPLE 5.5 DRAWING LEWIS STRUCTURES

PROBLEM What is the Lewis structure for CCl_4 (carbon tetrachloride, a substance once used as a dry cleaning solvent until it was found to be very toxic)?

SOLUTION First we need the arrangement of atoms. The formula suggests that it is

$$\begin{array}{ccc} & Cl & \\ Cl & C & Cl \\ & Cl & \end{array}$$

This arrangement of atoms, showing which atoms are bonded to each other, is sometimes called the **skeletal structure.**

Now we count the valence electrons:

$$\text{carbon (Group IVA) contributes } 4e^- \qquad 4e^-$$
$$\text{chlorine (Group VIIA) contributes } 7e^- \text{ each } \underline{28e^-}$$
$$\text{Total} \qquad 32e^-$$

We begin distributing the electrons by placing a pair in each bond. This gives

$$\begin{array}{c} \text{Cl} \\ \text{Cl} : \overset{..}{\underset{..}{\text{C}}} : \text{Cl} \\ \text{Cl} \end{array}$$

This has used $8e^-$, so there are $32e^- - 8e^- = 24e^-$ left. We now complete the valence shells of the Cl atoms.

$$\begin{array}{ccc}
 & : \overset{..}{\text{Cl}} : & \\
: \overset{..}{\underset{..}{\text{Cl}}} : \overset{..}{\underset{..}{\text{C}}} : \overset{..}{\underset{..}{\text{Cl}}} : & \text{or} & : \overset{..}{\underset{..}{\text{Cl}}} \text{---} \text{C} \text{---} \overset{..}{\underset{..}{\text{Cl}}} : \\
 & : \overset{..}{\underset{..}{\text{Cl}}} : & : \overset{..}{\underset{..}{\text{Cl}}} :
\end{array}$$

This has used all $24e^-$, so none are left. Each atom has an octet and, therefore, we can stop. This is the Lewis structure for CCl_4.

EXAMPLE 5.6   DRAWING LEWIS STRUCTURES

PROBLEM   What is the Lewis structure for SF_4?

SOLUTION   We begin by choosing an arrangement of atoms. Once again, the formula provides a clue to the skeletal structure

$$\begin{array}{c} \text{F} \\ \text{F} \quad \text{S} \quad \text{F} \\ \text{F} \end{array}$$

Next, we count valence electrons: 6 from sulfur and 7 from each fluorine gives a total of $6 + 28 = 34$ electrons. As before, we start by placing a pair in each bond.

$$\begin{array}{c} \text{F} \\ \text{F} : \overset{..}{\underset{..}{\text{S}}} : \text{F} \\ \text{F} \end{array}$$

This leaves us with $34 - 8 = 26$ electrons. Next we complete the valence shells of fluorine.

$$\begin{array}{c} : \overset{..}{\underset{..}{\text{F}}} : \\ : \overset{..}{\underset{..}{\text{F}}} : \overset{..}{\underset{..}{\text{S}}} : \overset{..}{\underset{..}{\text{F}}} : \\ : \overset{..}{\underset{..}{\text{F}}} : \end{array}$$

This has used $24e^-$, so there are still $2e^-$ left. Step 4 tells us to place them on the sulfur (the central atom). Rearranging the fluorines to make room for the extra dots on sulfur gives us the dot structure for SF_4.

EXAMPLE 5.7 DRAWING LEWIS STRUCTURES

PROBLEM   What is the Lewis structure for the carbonate ion, CO_3^{2-}?

SOLUTION   This time we have an ion whose formula suggests the atom arrangement

$$O$$
$$O \quad C \quad O$$

Next we count valence electrons; carbon supplies $4e^-$, each oxygen supplies $6e^-$, and the negative charge adds another $2e^-$. The total is $4 + 18 + 2 = 24$ electrons. Placing a pair in each bond gives

$$O$$
$$O : \ddot{C} : O$$

This has used $6e^-$, leaving us with $18e^-$. These are used to complete the octets around oxygen.

$$:\ddot{O}:$$
$$:\ddot{O} : \ddot{C} : \ddot{O}:$$

This time we've run out of electrons, but carbon doesn't have an octet. What we must do now is move one of the unshared pairs on an oxygen into one of the bonds (it doesn't matter which one).[4]

$$:\ddot{O}: \qquad\qquad :\ddot{O}:$$
$$:\ddot{O} : C : \ddot{O}: \quad \text{gives} \quad :\ddot{O}::C: \ddot{O}:$$

By creating the double bond, we increase the number of electrons around carbon without actually taking them away from the oxygen.

Notice that the oxygen hasn't lost the pair, but carbon completes its octet by sharing it. Finally, we should be sure to indicate the charge on the ion by enclosing the Lewis structure in brackets with the charge outside.

$$\left[\begin{array}{c} :\ddot{O}: \\ | \\ :\ddot{O}=C-\ddot{O}: \end{array} \right]^{2-}$$

EXAMPLE 5.8 DRAWING LEWIS STRUCTURES

PROBLEM   Draw the electron-dot formula for the poisonous gas, hydrogen cyanide, HCN (C is the central atom).

SOLUTION   There are 10 valence electrons to distribute (1 from H, 4 from C, 5 from N). First we place a pair in each bond.

$$H : C : N$$

[4] Applying this approach to drawing the Lewis structures of BCl_3 or $BeCl_2$ would give

$$Cl-B=Cl \quad \text{and} \quad Cl=Be=Cl$$
$$|$$
$$Cl$$

However, in many of their chemical properties, these molecules *behave* as though they only contain single bonds, so the structures with double bonds are not acceptable, even though they satisfy the octet rule. As mentioned earlier, BCl_3 and $BeCl_2$ are exceptions to the octet rule and must be learned as special cases.

This accounts for 4 electrons. As we add the remaining $6e^-$, we must keep in mind that the valence shell of H can hold only $2e^-$. No more electrons can be placed around H because it already has 2. Completing the octet of nitrogen gives

$$H : C : \overset{\cdot\cdot}{\underset{\cdot}{N}} :$$

Once again the central atom doesn't have an octet. This can be corrected as follows:

$$H : C : \overset{\cdot\cdot}{\underset{\cdot\cdot}{N}} : \longrightarrow H : C ::: N :$$

The HCN molecule contains a triple bond.

5.7 BOND ORDER AND SOME BOND PROPERTIES

At the beginning of Section 5.5, we described two features that characterize a covalent bond: bond length and bond energy. The bond length, you recall, is the distance between the nuclei of the two atoms joined by the bond. The bond energy is the energy needed to pull the atoms apart and produce neutral fragments. For a diatomic molecule such as H_2, this represents the process

$$H : H \ (g) \longrightarrow H \cdot (g) + H \cdot (g)$$

while in a molecule such as C_2H_6 the carbon–carbon bond energy represents the energy needed to cause the reaction

It should not be surprising to learn that the magnitudes of the bond length and the bond energy differ for bonds between different atoms. Some bonds are strong and some are weak; some have long bond lengths and some bonds are short.

One factor that affects the bond length and the bond energy is the amount of electron density between the nuclei. A convenient way of expressing this is by giving the **bond order**—*the number of covalent bonds that exist between a pair of atoms.* Consider, for example, the following molecules:

ethane   ethylene   acetylene

The carbon–carbon bond order in ethane is 1, in ethylene it is 2, and in acetylene it is 3.

As long as we are dealing with bonds between the same elements, we can relate bond length and bond energy to the bond order. As the bond order between a pair of atoms increases, additional electron density is placed between the two nuclei, which causes them to be pulled together. Therefore, *the bond length decreases as the bond order increases.* Increasing the bond order also makes it

There are ways of experimentally measuring bond lengths and bond energies.

Table 5.5
Variation of bond properties with bond order

Bond	Bond Order	Average Bond Length (pm)	Average Bond Energy (kJ mol^{-1})	Average Vibrational Frequency (Hz)
C—C	1	154	370	3.0×10^{13}
C=C	2	137	699	4.9×10^{13}
C≡C	3	120	960	6.6×10^{13}
C—O	1	143	350	3.2×10^{13}
C=O	2	123	750	5.2×10^{13}
C—N	1	147	300	3.7×10^{13}
C≡N	3	116	730	6.8×10^{13}

more difficult to pull the bonded atoms apart. Therefore, *the bond energy increases as the bond order increases.* Data that illustrate this are shown in Table 5.5.

Another bond property related to the bond order is the **vibrational frequency** of the atoms joined by the bond. The atoms within a molecule are not stationary; they are in constant motion. This motion can be resolved into two basic types: vibration in which a pair of atoms move toward and away from each other along a line joining their centers, much as two balls connected by a spring (Figure 5.7a); and bending in which the angle between the three atoms alternately increases and decreases (Figure 5.7b). For simplicity we will restrict our discussion to vibrational motion.

There are two factors that affect the frequency of vibration (i.e., the number of vibrations per second). One is the masses of the atoms bonded together and

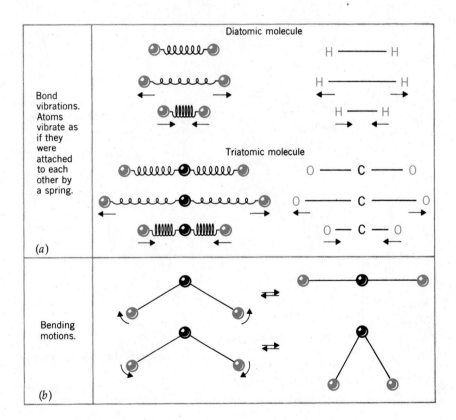

Figure 5.7

The motion of atoms within molecules.

the other is the bond order. *For a given pair of atoms, as the bond order increases, so does the vibrational frequency.* This is because increasing the bond order increases the attractive forces holding the nuclei together, in effect, stiffening the "spring" between the two atoms.

Today the measurement of the vibrational frequencies of bonds is really quite simple. It happens that these vibrational frequencies are about the same as the frequency of infrared radiation, and when infrared light shines on a substance, radiation is absorbed if it has the same frequency as the vibrational frequency of a bond. By observing which frequencies are selectively removed from the infrared spectrum, we can deduce these vibrational frequencies. The data recorded in the right column of Table 5.5 were obtained in this way.

Virtually every modern chemistry laboratory has one or more instruments to measure and record infrared absorption spectra.

In complex molecules there are many different vibrational modes available to the atoms and many different frequencies are absorbed from the infrared "rainbow." The infrared absorption spectra of most molecules are therefore quite complicated. Nevertheless, an experienced chemist often finds such absorption spectra extremely valuable as an aid in deducing molecular structure. In addition, each molecule, because of its unique structure, gives rise to its own characteristic absorption spectrum, which can be used to identify the compound and thus serves as a sort of fingerprint. Examples of infrared absorption spectra of some drugs are shown in Figure 5.8.

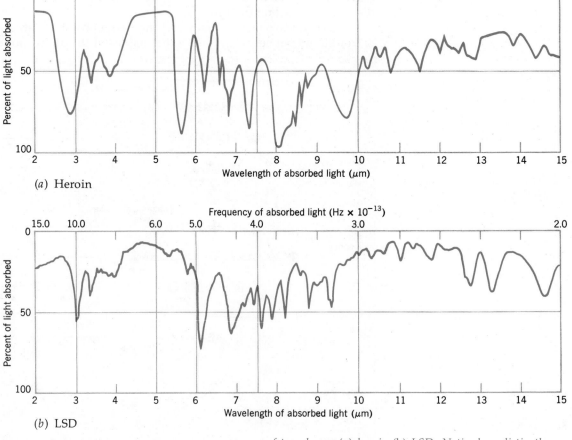

Figure 5.8
Infrared absorption spectra can be easily measured. The spectra are of two drugs: (a) heroin (b) LSD. Notice how distinctly different the spectra are.

EXAMPLE 5.9   RELATING BOND PROPERTIES TO BOND
ORDER

PROBLEM   Compare the carbon–oxygen bond lengths, bond energies, and bond vibrational frequencies in the molecules

$$\begin{array}{ccc} & H & \\ & | & \\ H-C-\ddot{O}-H & \end{array} \qquad \begin{array}{c} H \;\; :\ddot{O}: \;\; H \\ | \quad \| \quad | \\ H-C-C-C-H \\ | \qquad\quad | \\ H \qquad\quad H \end{array}$$

methyl alcohol   acetone

SOLUTION   The C—O bond order in methyl alcohol is 1, and in acetone it is 2. This means that compared to methyl alcohol, the C—O bond in acetone will be shorter, have a larger bond energy, and have a higher vibrational frequency.

5.8   RESONANCE

In an SO_2 molecule, there is nothing physically that makes any one S—O bond special. Nature, therefore, treats them equally by distributing the double bond equally between the two bonds.

When we draw the Lewis structure for the SO_2 molecule following the rules given in Section 5.6, we obtain

$$:\ddot{O}-\ddot{S}=\ddot{O}:$$

Based on our discussion in the last section, we would expect the two sulfur–oxygen bonds to be different. The Lewis structure suggests that one should be shorter, stronger, and have a larger vibrational frequency than the other. However, all experimental evidence indicates that the two sulfur–oxygen bonds are identical. In fact, it appears that they have properties that are about midway between what would be expected for a S—O single bond and a S—O double bond. Quite clearly, there is something wrong with the Lewis structure that we've drawn. We might attempt to remedy the situation by dividing one of the pairs of electrons in the double bond, placing one in each bond, but this isn't satisfactory either because it suggests that there are unpaired electrons in the SO_2 molecule. Experimentally it can be demonstrated that SO_2 is diamagnetic, which means that all its electrons are paired.

The way that we get around this problem is as follows. When we are drawing the Lewis structure for SO_2, we reach this stage

$$:\ddot{O}-\ddot{S}-\ddot{O}:$$

Next, we realize that the sulfur has less than an octet, so it is necessary to create a double bond in the structure. However, we have two choices—we can place it in the bond at the left, or in the bond at the right. Let's draw both structures.

$$:\ddot{O}=\ddot{S}-\ddot{O}: \qquad :\ddot{O}-\ddot{S}=\ddot{O}:$$

Obviously, neither structure is satisfactory, but scientists have found that they can use both of these structures together to represent the actual structure of the molecule. The solution to the problem is called **resonance.** We say that the actual structure of the molecule does not correspond to either of the two structures that we have drawn, but instead, to a structure somewhere in between that has properties of both. This true structure, which we can't actually draw with Lewis symbols, is said to be a **resonance hybrid,** and the two structures that we can draw are said to be **resonance structures** or **contributing structures.**

Because the S—O bonds are the same, we say they are *equivalent.*

The bond properties of the resonance hybrid are also intermediate between the properties of the bonds of the contributing structures. In SO_2, for example, the actual bond lengths and bond energies lie between those for a single bond and a double bond. A rough estimate of where they lie can be gotten by calculating the average S—O bond order. This is done by obtaining the total number of bonds between the sulfur and one of the oxygens in all the resonance structures, and then dividing this total by the number of resonance structures. For example, if we choose the bond to the left of the sulfur in our two resonance structures, we see that there are two bonds in the first structure and one bond there in the second. The total, therefore, is three. Since there are two contributing structures, the average bond order is $3/2 = 1.5$. As we noted earlier, this is what is suggested by experimental evidence.

Fractional bond orders are not only possible, but actually occur in many different compounds.

It is really quite unfortunate that the word resonance was used to describe this phenomenon, because the impression often is that the actual structure of the molecule fluctuates back and forth between the two contributing structures. This definitely is not so! The actual structure of SO_2 never corresponds to the one on the left and it never corresponds to the one on the right. Sulfur dioxide has only one, single electronic structure that is intermediate between the two contributing structures. Our difficulty is that we just can't draw it adequately by using Lewis symbols. The problem is somewhat like trying to describe the beast you would get if you were able to cross a dog and a cat. When you try to picture this offspring, you visualize it with characteristics of both parents. But you don't think of it as a dog one instant and a cat the next.

Sometimes, we need more than one resonance structure to explain the structure of a molecule or ion. When we draw the Lewis structure of SO_3, for instance, we get to this point.

Sulfur doesn't have an octet, so we need to create a double bond. However, this time we have three choices, so we must draw three resonance structures.

A double-headed arrow is used here to indicate resonance.

The electronic structure of some ions must also be represented by resonance. For example, the nitrate ion, NO_3^-, and the carbonate ion, CO_3^{2-}, have the same number of valence electrons as SO_3 and therefore have similar resonance structures.

Resonance is certainly not restricted to inorganic compounds. In proteins, for example, amino acids are linked together in long chains by "peptide bonds."

glycine
(an amino acid)

peptide bond
(a peptide)

There is evidence that the C—N bond in the peptide linkage actually lies somewhere between a single bond and a double bond. To explain this it is suggested that the peptide bond is a resonance hybrid of structures such as

Structure 2 is obtained by rearranging the electrons in Structure 1 this way

5.9   COORDINATE COVALENT BONDS

When a nitrogen atom combines with three hydrogen atoms to form the molecule NH_3, the nitrogen atom has completed its octet. We might expect, therefore, that the maximum number of covalent bonds that we would observe a nitrogen atom to form is three. There are instances, however, when nitrogen may have more than three covalent bonds. In the ammonium ion, NH_4^+, which is formed in the reaction

the nitrogen is covalently bound to four hydrogen atoms. When the additional bond between the H^+ and the N atom is created, both of the electrons in the bond come from the nitrogen. *This type of bond, in which a pair of electrons from one atom is shared by two atoms, is called a* **coordinate covalent bond.** It is important that you remember that the coordinate covalent bond is really no different, once formed, than any other covalent bond and that our distinction is primarily aimed at keeping track of electrons; that is, it is "bookkeeping."

When Lewis structures are written using dashes to represent electron pairs, the coordinate covalent bond is sometimes indicated by means of an arrow pointing away from the atom supplying the electron pair. For example, the product of the reaction of boron trichloride, BCl_3, and ammonia, NH_3, is a substance known as an **addition compound** because it is formed by simply adding one molecule to another.

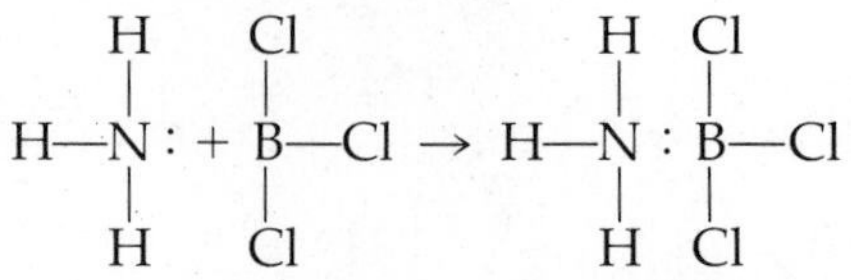

$$
\begin{array}{ccccccc}
& H & Cl & & & H & Cl \\
& | & | & & & | & | \\
H\!-\!N: & + & B\!-\!Cl & \rightarrow & H\!-\!N: & B\!-\!Cl \\
& | & | & & & | & | \\
& H & Cl & & & H & Cl
\end{array}
$$

To show that the electron pair shared between the B and N originates on the nitrogen, the Lewis structure of this addition compound can be written

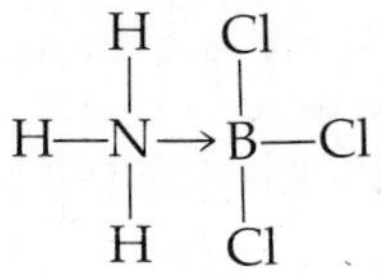

$$
\begin{array}{ccc}
H & & Cl \\
| & & | \\
H\!-\!N & \rightarrow & B\!-\!Cl \\
| & & | \\
H & & Cl
\end{array}
$$

Using this type of notation we are tempted to write the structure of the NH_4^+ ion as

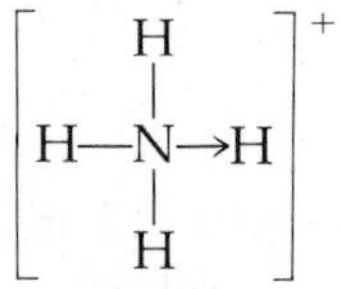

$$
\left[
\begin{array}{c}
H \\
| \\
H\!-\!N\!\rightarrow\!H \\
| \\
H
\end{array}
\right]^{+}
$$

Remember, once the bond is formed, it doesn't matter where the electrons in it came from.

This gives the impression that one of the N—H bonds is different from the other three. It has been shown experimentally, however, that all four N—H bonds are identical. Therefore, to avoid conveying false impressions, the NH_4^+ ion is written simply as

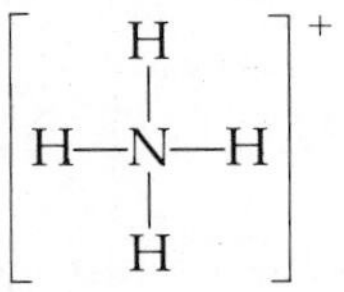

$$
\left[
\begin{array}{c}
H \\
| \\
H\!-\!N\!-\!H \\
| \\
H
\end{array}
\right]^{+}
$$

5.10 POLAR MOLECULES AND ELECTRONEGATIVITY

When we realize that each element has a different nuclear charge and electron configuration, it is not unreasonable to expect that atoms of different elements have different abilities to attract electrons when they form chemical bonds. **Electronegativity** *is the term used to describe an atom's attraction for the electrons in a bond.* It is important not to confuse this term with electron affinity, which is an energy term and refers to an isolated atom.

When two identical atoms combine, as for instance in H_2, both have the same electronegativity. Since each atom is equally capable of attracting the electron pair in the bond, the pair will be shared equally and will spend, on the average, 50% of its time in the vicinity of each nucleus. Therefore, each H atom has around it two electrons 50% of the time. When averaged out, this is the same as one electron all of the time. The averaged one electron will completely neutralize the positive charge on each nucleus, and each atom in H_2 carries a net charge of zero.

If the electronegativities of the two atoms joined by a bond are different, the electron pair will spend more of its time around the more electronegative element. For example, consider the HCl molecule. It happens that chlorine is more electronegative than hydrogen, so the pair of electrons in the H—Cl bond spends more time around chlorine than around hydrogen. This means that the Cl atom acquires a slight negative charge and the H atom, a slight positive charge. We indicate this as

$$\overset{\delta+}{H}\overset{\delta-}{—\ddot{\underset{\cdot\cdot}{Cl}}}:$$

δ is the lowercase Greek letter delta.

where $\delta+$ and $\delta-$ are meant to indicate partial positive and negative charges.

Within a molecule, equal positive and negative charges separated by a distance constitute a **dipole.** Therefore, the HCl molecule, with its centers of positive and negative charge, is a dipole and is said to be **polar.** In fact, any diatomic molecule (a molecule formed from two atoms) formed from two elements of different electronegativity will be polar.

A dipole is defined quantitatively by its **dipole moment,** the product of the charge on either end of the dipole times the distance between the charges. A very polar molecule is one with a large dipole moment, while a nonpolar molecule will have no dipole moment at all.

When three or more atoms are bonded together, it is possible to have a nonpolar molecule even though there are polar bonds present. Carbon dioxide is an example. The CO_2 molecule is linear and may be represented as

$$\underset{\delta-\quad\delta+}{\overset{\cdot\cdot\quad\delta+\quad\delta-}{:\ddot{O}=C=\ddot{O}:}}$$

showing that oxygen is more electronegative than carbon.

The overall dipole moment of a molecule arises as a sum of the individual bond dipoles within the molecule, which add together like vectors. In CO_2 these bond dipoles are oriented in opposite directions, and they exactly cancel each other.

$$:\ddot{O}=C=\ddot{O}:$$
$$\overset{\longleftarrow+\quad+\longrightarrow}{}$$

(An arrow with a plus sign on one end is used to represent the bond dipole.)

In the water molecule, which happens to have a bent or angular shape, the two bond dipoles do not cancel each other entirely, but rather are partially additive. As a result, the H_2O molecule does have a net dipole moment (indicated by the heavy arrow) and is polar.

We will have more to say about the effect of molecular structure on the polarity of molecules in the next chapter.

We would like to have some quantitative measure of electronegativity so that we can make predictions concerning the polarity of bonds. One approach toward this, taken by R. S. Mulliken in 1934, uses the average of the ionization energy and electron affinity. A very electronegative element has a very high ionization energy, so it is difficult to remove its electrons. It also has a very high electron affinity, so a very stable species results when electrons are added. On the other hand, an element of low electronegativity will have a low ionization energy and low electron affinity so that it loses electrons readily and has little tendency to pick them up. Unfortunately, it is very difficult to measure the

Elements with low electronegativity are often said to be **electropositive** because they tend to give their electrons away rather easily.

electron affinity of an element. Therefore, this method of assigning electronegativities is not universally applicable.

The most widely used scale of electronegativities was developed by Linus Pauling (a winner of two Nobel prizes and an advocate of using vitamin C to ward off the common cold). He observed that when atoms of different electronegativities are combined, their bonds are stronger than expected. Presumably two factors contribute to the bond strength. One of them is the covalent bonding between the atoms. The other is an additional binding produced by an attraction between the oppositely charged ends of the bond dipole. The extra bond strength, then, was attributed to this additional binding and Pauling used this concept to develop his table of electronegativities (Table 5.6).

The actual numerical values in Table 5.6 are not very important. What is important is the *difference* in the electronegativities of the atoms joined by a bond. When the difference is small, the bond is relatively nonpolar; when the difference is large, a very polar bond is formed. And when the difference between the electronegativities of two combining atoms is very large, the electron pair will spend virtually 100% of its time about the more electronegative element. This is the same as saying that an electron is transferred from the atom of low electronegativity to that of high electronegativity. The result, of course, is an ionic bond. It is clear that the degree of ionic character in a bond can vary between zero (for example, H_2) to essentially 100% depending on the electronegativity difference between the bonded atoms. There is no sharp dividing line between ionic and covalent bonding. When the electronegativity difference is 1.7, the bond is about 50% ionic.

Finally, electronegativity trends within the periodic table are worth noting. We see that the most electronegative elements are located in the upper right portion of the table; the least electronegative are found in the lower left. This is consistent with the trends in ionization energy (IE) and electron affinity (EA) discussed in Chapter 4, where we saw that elements with the largest IE and EA are in the upper right region of the periodic table and those with the smallest IE and EA are in the lower left. It is also consistent with our observations that atoms from opposite ends of the periodic table—lithium and fluorine, for example—form bonds that are essentially 100% ionic, and that atoms such as carbon and oxygen form covalent bonds that are only somewhat polar.

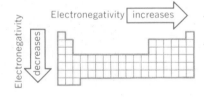

Table 5.6
The complete electronegativity scale[a]

Li	Be											B	C	N	O	F
1.0	1.5											2.0	2.5	3.0	3.5	4.0
Na	Mg											Al	Si	P	S	Cl
0.9	1.2											1.5	1.8	2.1	2.5	3.0
K	Ca	Sc	Ti	V	Cr	Mn	Fe	Co	Ni	Cu	Zn	Ga	Ge	As	Se	Br
0.8	1.0	1.3	1.5	1.6	1.6	1.5	1.8	1.8	1.8	1.9	1.6	1.6	1.8	2.0	2.4	2.8
Rb	Sr	Y	Zr	Nb	Mo	Tc	Ru	Rh	Pd	Ag	Cd	In	Sn	Sb	Te	I
0.8	1.0	1.2	1.4	1.6	1.8	1.9	2.2	2.2	2.2	1.9	1.7	1.7	1.8	1.9	2.1	2.5
Cs	Ba	La–Lu	Hf	Ta	W	Re	Os	Ir	Pt	Au	Hg	Tl	Pb	Bi	Po	At
0.7	0.9	1.1–1.2	1.3	1.5	1.7	1.9	2.2	2.2	2.2	2.4	1.9	1.8	1.8	1.9	2.0	2.2
Fr	Ra	Ac	Th	Pa	U	Np–No										
0.7	0.9	1.1	1.3	1.5	1.7	1.3										

Also shown: H 2.1 (centered, boxed)

[a]From Linus Pauling: *The Nature of the Chemical Bond, Third Edition.* Copyright © 1960 by Cornell University. Used by permission of the publisher, Cornell University Press.

5.11 THE NAMING OF CHEMICAL COMPOUNDS

If you leaf through a chemical handbook, such as the *Handbook of Chemistry and Physics*, you will find an enormous number of compounds listed there. Yet these represent only a fraction of all the compounds that have been discovered, and each year the list grows longer. Naming these compounds represents a real challenge because it is important that each unique substance have its own unique name. In addition, the names can't be chosen in a haphazard way; otherwise there would be no possible way for anyone to remember them all. For this reason, a systematic procedure for naming chemical compounds has been developed.

A system of names used in a particular branch of knowledge, like chemistry, is called *nomenclature.*

In this section we will describe, in a somewhat abbreviated way, how **inorganic compounds** are named. These are compounds whose structures are not primarily determined by the linking together of carbon atoms. Such carbon compounds are called **organic compounds** and are discussed in Chapter 23.

You will discover during your study of chemistry that not every compound is named according to the system. Some very common substances, such as water (H_2O) and ammonia (NH_3), were known long before the systematic nomenclature was developed and are best recognized by their common (or *trivial*) names. Trivial names are also used for extremely complex compounds where the names derived on a systematic basis are very long, complex, and cumbersome.

Binary compounds

A **binary compound** is composed of atoms of only two different elements. In naming these substances the less electronegative (more metallic) element is specified first by giving its ordinary English name. The name of the second element (almost always a nonmetal) is obtained by adding the suffix *ide* to its stem, as shown in Table 5.7. Some typical examples are

NaCl	sodium chloride
SrO	strontium oxide
Al_2S_3	aluminum sulfide
Mg_3P_2	magnesium phosphide
HBr	hydrogen bromide

There are many metals that commonly form more than one positive ion. Chromium and iron are two examples. Chromium forms the ions Cr^{2+} and Cr^{3+}, and iron forms the ions Fe^{2+} and Fe^{3+}. When compounds of these elements are named, it is necessary to specify which of the positive ions are present, and there are two methods that are used for that purpose. In the older method, the

Table 5.7
Names of anions derived from nonmetals

Group IVA	Group VA	Group VIA	Group VIIA
C^{4-}; *carbide*[a]	N^{3-} *nitride*	O^{2-} *oxide*	F^- *fluoride*
Si^{4-}; *silicide*	P^{3-} *phosphide*	S^{2-} *sulfide*	Cl^- *chloride*
	As^{3-} *arsenide*	Se^{2-} *selenide*	Br^- *bromide*
		Te^{2-} *telluride*	I^- *iodide*

[a]Carbon also forms a number of complex carbides, for example, C_2^{2-} in CaC_2.

suffixes *-ic* and *-ous* are used to specify the higher and lower charged ions. Thus, the ions of chromium would be specified as

$$Cr^{3+} \quad \text{chromic} \qquad CrCl_3 \quad \text{chromic chloride}$$
$$Cr^{2+} \quad \text{chromous} \qquad CrCl_2 \quad \text{chromous chloride}$$

If the symbol for the metal is derived from its original Latin name, we generally use its Latin stem when naming the ion. For example, the ions of iron are named as follows:

$$Fe^{2+} \quad \text{ferrous} \qquad FeCl_2 \quad \text{ferrous chloride}$$
$$Fe^{3+} \quad \text{ferric} \qquad FeCl_3 \quad \text{ferric chloride}$$

Other examples of the use of the older system are shown in Table 5.8. Notice that mercury is an exception to the rule of using the Latin stem in the name. Also, notice that the older system of nomenclature simply distinguishes between the higher and lower charges: It does not specify what the charge is on the metal ion. This is a serious flaw that has been overcome in the newer method of naming these ions.

The preferred method of naming ions like those in Table 5.8 is called the **Stock system,** named after the German chemist Alfred Stock (1876–1946) who developed it. In the Stock system, a Roman numeral equal to the charge on the metal ion is written in parentheses after the common English name for the element. Thus, Fe^{2+} would be named iron(II) and Fe^{3+} would be named iron(III). The alternative names for the compounds $FeCl_2$ and $FeCl_3$ are therefore

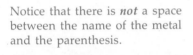

Notice that there is *not* a space between the name of the metal and the parenthesis.

$$FeCl_2 \quad \text{ferrous chloride} \quad \text{or} \quad \text{iron(II) chloride}$$
$$FeCl_3 \quad \text{ferric chloride} \quad \text{or} \quad \text{iron(III) chloride}$$

In using the Stock system, it is important to remember that the Roman numeral is the *charge on the ion,* not necessarily the subscript for the metal ion. For example,

$$Cu_2O \quad \text{is copper(I) oxide}$$
$$CuO \quad \text{is copper(II) oxide}$$

Even though the Stock system is preferred today, it is still necessary to know the older system as well. For example, if an experiment calls for iron(III) chloride, it is likely that you will only find it in a bottle labeled ferric chloride.

In naming binary covalent compounds—compounds generally formed between two nonmetals—a third system of nomenclature is used to specify the numbers of atoms in the formula. This system uses Greek prefixes to indicate the

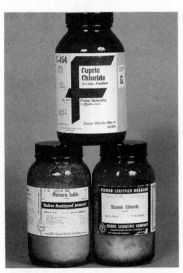

Bottles of chemicals labeled using the older system of nomenclature.

Table 5.8
Metals that commonly form two ions

Chromium	Manganese	Iron	Cobalt	Lead
Cr^{2+} chromous	Mn^{2+} manganous	Fe^{2+} ferrous	Co^{2+} cobaltous	Pb^{2+} plumbous
Cr^{3+} chromic	Mn^{3+} manganic	Fe^{3+} ferric	Co^{3+} cobaltic	Pb^{4+} plumbic

Copper	Tin	Mercury		
Cu^+ cuprous	Sn^{2+} stannous	Hg_2^{2+} mercurous (note that there are two Hg atoms)		
Cu^{2+} cupric	Sn^{4+} stannic	Hg^{2+} mercuric		

numbers of atoms of each kind in the formula of the molecule. These prefixes, along with their meanings, are

di-	two	penta-	five	octa-	eight
tri-	three	hexa-	six	nona-	nine
tetra-	four	hepta-	seven	deca-	ten

In naming these compounds, the less electronegative element is also specified first, followed by the more electronegative element. The rule for forming the names and an illustration of how it is applied to the compound P_4O_{10} is as follows:

$$\begin{bmatrix} \text{Prefix indicating} \\ \text{number of atoms} \\ \text{of first element} \end{bmatrix} - \begin{bmatrix} \text{Name of} \\ \text{first} \\ \text{element} \end{bmatrix} - \begin{bmatrix} \text{Prefix indicating} \\ \text{number of atoms} \\ \text{of second element} \end{bmatrix} - \begin{bmatrix} \text{Root name} \\ \text{of second} \\ \text{element} \end{bmatrix} - \text{ide}$$

tetra · · · phosphorus   deca · · · ox · · · ide

Thus, P_4O_{10} is tetraphosphorus decaoxide. Some other examples are

NO_2 nitrogen dioxide
N_2O_4 dinitrogen tetroxide
        (the *a* in *tetra* is dropped for ease of pronunciation)
N_2O_5 dinitrogen pentoxide
PCl_3 phosphorus trichloride
PCl_5 phosphorus pentachloride
S_2Cl_2 disulfur dichloride

In some instances, the prefix *mono-*, meaning *one*, is also used when it is desired to avoid ambiguity.

CO_2 carbon dioxide
CO carbon monoxide

Compounds containing polyatomic ions

In Section 5.3 we saw that many ions contain more than one atom and are therefore referred to in general as polyatomic ions. These species enter into ionic compounds as discrete units and generally stay intact in most chemical reactions. A list of these is given in Table 5.4. As with binary compounds, substances that contain these ions are always named with the positive ion first. Some examples are

Usually the *-ide* ending specifies a monatomic anion such as Cl^- (chlor*ide*). Hydroxide ion, OH^-, and cyanide ion, CN^-, are two exceptions.

| Na_2CO_3 | sodium carbonate | $Ba(OH)_2$ | barium hydroxide |
| $Ca(C_2H_3O_2)_2$ | calcium acetate | $(NH_4)_2SO_4$ | ammonium sulfate |

The Stock system is also preferred when the metal can form more than one positive ion.

	Stock System	Old System
$MnSO_4$	manganese(II) sulfate	manganous sulfate
$Fe_2(C_2O_4)_3$	iron(III) oxalate	ferric oxalate

Binary acids

Among the important classes of compounds that we will discuss later in the book are substances called acids. Perhaps you have already encountered some common ones in the laboratory—hydrochloric acid, for example—and you surely have experienced the sour taste of citric acid in lemon juice and acetic acid in vinegar. As we will see, acids are substances that release H^+ ions when they are dissolved in water.

An important kind of acid is formed when a binary compound of hydrogen and a nonmetal (for example, hydrogen chloride, HCl, or hydrogen sulfide, H_2S) is dissolved in water. Water solutions of these compounds are called binary acids (or sometimes *hydro acids*). They are named as *hydro . . . ic acid*, where the stem of the name of the nonmetal is inserted in place of the dotted line. Examples are

The gas HF is *hydrogen fluoride;* its aqueous solution is called *hydrofluoric acid.* Similar statements apply to the other nonmetal–hydrogen compounds in this list.

HF	hydro*fluor*ic acid
HCl	hydro*chlor*ic acid
HBr	hydro*brom*ic acid
HI	hydr*iod*ic acid
H_2S	hydro*sulfur*ic acid

When an acid is allowed to react with hydroxide ion (a reaction called **neutralization**), an ionic compound is formed. For example,

Anions such as Cl^- can be formed directly from the element, as when Cl_2 reacts with Na to form NaCl, or they can be formed by neutralization of an acid with a base.

$$NaOH + HCl \longrightarrow H_2O + NaCl$$

$$2KOH + H_2S \longrightarrow 2H_2O + K_2S$$

Ionic compounds such as NaCl and K_2S are called salts. The word salt is not reserved just for sodium chloride, although this is the common household name for NaCl. In chemistry, the term **salt** is used for *any* ionic compound not containing oxide ion or hydroxide ion. Notice that the salts formed from *hydro . . . ic acids* contain monoatomic anions that end in the suffix *-ide*.

Oxoacids

Oxoacids are acids that contain hydrogen, oxygen, and at least one other element (usually a nonmetal). Sulfuric acid, H_2SO_4 is an example. Notice that the hydrogen is specified first in the formula (we will have more to say about this in Chapter 7). Often, an element is able to form more than one oxoacid. Sulfur, for example, forms the two acids H_2SO_3 and H_2SO_4, which differ in the number of oxygen atoms that they contain. In naming these acids, the one with the greater number of oxygens is given the ending *-ic*, and the one with the lesser number of oxygens is given the ending *-ous*. Thus, we have

H_2SO_4	sulfuric acid
H_2SO_3	sulfurous acid

Notice that the prefix *hydro-* is not used in the name of an oxoacid.

Although polyatomic ions are formed by neutralization of oxoacids, they can also be made by other reactions.

Compounds produced by the neutralization of oxoacids contain the polyatomic ions given in Table 5.4. Notice that the anion derived from the "ic" acid has a name that ends in *-ate*, whereas the anion that comes from the "ous" acid ends in *-ite*.

H_2SO_4	sulfuric acid	SO_4^{2-}	sulfate
H_2SO_3	sulfurous acid	SO_3^{2-}	sulfite
HNO_3	nitric acid	NO_3^-	nitrate
HNO_2	nitrous acid	NO_2^-	nitrite
$HClO_3$	chloric acid	ClO_3^-	chlorate
$HClO_2$	chlorous acid	ClO_2^-	chlorite

Some nonmetals form more than two oxoacids. This is the case with chlorine, bromine, and iodine. For these acids, we use the prefix *hypo-* for the acid that has fewer oxygens than the "ous" acid, and we use the prefix *per-* for the acid that has more oxygens than the "ic" acid. For example,

hypochlorous acid	$HClO$	ClO^-	hypochlorite
chlorous acid	$HClO_2$	ClO_2^-	chlorite
chloric acid	$HClO_3$	ClO_3^-	chlorate
perchloric acid	$HClO_4$	ClO_4^-	perchlorate

Parent Acid	Typical Acid Salts
H_2SO_4	$NaHSO_4$
H_2CO_3	$NaHCO_3$
H_3PO_4	NaH_2PO_4
	Na_2HPO_4

Partial neutralization of an acid that is capable of furnishing more than one H^+ per acid molecule gives salts that are called **acid salts.** Some examples are shown in the table at the left. When only one acid salt is formed (as with H_2SO_4 or H_2CO_3), the salt can be named by adding the prefix *bi-* to the name of the anion of the acid.

$NaHSO_4$	sodium *bi*sulfate
$NaHCO_3$	sodium *bi*carbonate

The salt can also be named by specifying the presence of the H by writing "hydrogen."

$NaHSO_4$	sodium hydrogen sulfate
NaH_2PO_4	sodium dihydrogen phosphate
Na_2HPO_4	sodium hydrogen phosphate (disodium hydrogen phosphate)

Note the use of the prefix *di-* to indicate the number of hydrogen atoms (as well as to remove ambiguity as to the number of Na in the last formula).

REVIEW QUESTIONS AND PROBLEMS

(Problems whose numbers are in blue have their answers in Appendix C at the back of the book; the more difficult problems are marked with asterisks.)

Properties of Ionic and Molecular Compounds

5.1 What are chemical bonds?

5.2 In general, how do hardness and melting point compare for molecular and ionic compounds?

5.3 Compare the electrical conductivity of molecular and ionic compounds (a) in the solid state and (b) in the liquid state.

5.4 What exists in ionic compounds that explains their electrical conductivity when melted?

Lewis Symbols

5.5 What is the purpose of Lewis symbols?

5.6 Write Lewis symbols for (a) selenium, (b) bromine, (c) aluminum, (d) barium, (e) germanium, and (f) phosphorus.

5.7 Why are all the Lewis symbols of the elements in a given group the same?

5.8 (a) How many unpaired electrons does a carbon atom have in its ground state?
(b) Draw the Lewis symbol for carbon. How many unpaired electrons does it suggest?
(c) Why is the discrepancy between the answers to (a) and (b) allowed?

5.9 What is the valence shell of an atom?

5.10 How many valence electrons are there in an atom of: (a) arsenic, (b) iodine, (c) calcium, (d) tin, and (e) sulfur?

Ionic Bonding

5.11 What is an *ionic bond?*

5.12 Define *cation* and *anion.*

5.13 Write the electron configuration of the following ions: (a) Ba^{2+}, (b) Se^{2-}, (c) Al^{3+}, (d) Na^+, and (e) Br^-.

5.14 Which noble gases have the same electron configurations as the ions in the preceding question?

5.15 What is the origin of the *octet rule?* What is the octet rule?

5.16 Draw Lewis structures for the following ionic compounds: (a) BaO, (b) Na$_2$O, (c) KF, (d) CaS, and (e) Mg$_2$C.

5.17 Without referring to Table 5.2, write the symbols of the ions formed by the following elements: (a) strontium, (b) nitrogen, (c) oxygen, (d) chlorine, (e) potassium, and (f) magnesium.

5.18 What is a post-transition element?

5.19 Give the electron configurations of these ions: (a) Zn^{2+}, (b) Sn^{2+}, (c) Bi^{3+}, (d) Cr^{2+}, (e) Fe^{3+}, and (f) Ag$^+$.

5.20 What is the *pseudonoble gas* configuration? Give *three* ions that have such a configuration in their outer shells.

5.21 Which of the following pairs of ions have *exactly* the same electron configurations: (a) K$^+$ and Cl$^-$, (b) Na$^+$ and Br$^-$, (c) Li$^+$ and F$^-$, (d) Sr^{2+} and Br$^-$, (e) Mg^{2+} and C^{4-}?

5.22 Indicate how the electron configuration changes for each atom when the following ionic compounds are formed from the elements: K$_2$O, Mg$_3$N$_2$, Na$_2$S, and BaBr$_2$.

Formulas of Ionic Compounds

5.23 Write the formula for the ionic compound formed from (a) calcium and bromine, (b) potassium and sulfur, (c) aluminum and oxygen, (d) barium and phosphorus, and (e) magnesium and nitrogen.

5.24 What is wrong with the following formulas? (a) RbO, (b) SK$_2$, (c) Ca$_2$O$_2$, (d) Br$_2$K.

5.25 Use Lewis symbols and the octet rule to explain why the formula for the ionic compound formed from Mg and C is Mg$_2$C.

5.26 Write the formulas of all the ionic compounds formed by the following metals with chloride ion, Cl$^-$, and with sulfide ion, S^{2-}:
(a) chromium   (e) nickel
(b) manganese   (f) copper
(c) iron   (g) silver
(d) cobalt

5.27 Write the formulas of all the ionic compounds formed by the following metals with bromide ion, Br$^-$, and oxide ion, O^{2-}:
(a) gold   (e) tin
(b) zinc   (f) lead
(c) cadmium   (g) bismuth
(d) silver

5.28 Give the correct name for each of the following polyatomic ions: (a) NH$_4^+$, (b) CO$_3^{2-}$, (c) CrO$_4^{2-}$, (d) SO$_3^{2-}$, (e) C$_2$H$_3$O$_2^-$.

5.29 Write the correct formula, including the charge, for each of the following polyatomic ions:
(a) cyanide ion   (g) oxalate ion
(b) perchlorate ion   (h) dichromate ion
(c) permanganate ion   (i) sulfate ion
(d) nitrate ion   (j) bicarbonate ion
(e) phosphate ion   (k) sulfite ion
(f) hydroxide ion   (l) nitrite ion

5.30 Write formulas for compounds composed of the following pairs of ions: (a) Na$^+$, CO$_3^{2-}$; (b) Ca^{2+}, ClO$_3^-$; (c) Sr^{2+}, S^{2-}; (d) Cr^{3+}, Cl$^-$; (e) Ti^{4+}, ClO$_4^-$.

5.31 Write the formulas for the ionic compounds formed by each of the metals in Table 5.3 with the anions specified in Question 5.28. (This is a lot of work, but well worth the effort.)

5.32 Write formulas for compounds composed of the following pairs of ions:
(a) Fe^{3+}, HPO$_4^{2-}$ (d) Cu^{2+}, C$_2$H$_3$O$_2^-$
(b) K$^+$, N^{3-} (e) Ba^{2+}, SO$_3^{2-}$
(c) Ni^{2+}, NO$_3^-$

5.33 On the basis of their electron configurations, why do many of the transition elements form ions with a charge of 2+?

5.34 Why do elements from period 2 never exceed an octet in their valence shells?

Factors That Control the Formation of Ionic Compounds

5.35 What is the name of the energy term associated with the reaction, Na$^+(g)$ + Cl$^-(g)$ → NaCl(s)? Is the process indicated in this chemical equation endothermic or exothermic?

5.36 Construct a Born–Haber cycle for the formation of KBr(s) from K(s) and Br$_2$(l). Indicate which steps are endothermic and which are exothermic.

5.37 Why is KF(s) more stable than K(s) and F$_2$(g)?

5.38 Use a Born–Haber cycle to show that the reaction

$$K(s) + \tfrac{1}{2}Cl_2(g) \longrightarrow KCl(s)$$

is exothermic. The following energies are known (a negative sign indicates the process is exothermic): For K(s) → K(g), 90.0 kJ; for $\tfrac{1}{2}$Cl$_2$(g) → Cl(g), 119 kJ; for K(g) → K$^+$(g), 419 kJ; for Cl(g) → Cl$^-$(g), −348 kJ; for K$^+$(g) + Cl$^-$(g) → KCl(s), −704.2 kJ.

*5.39 Given the following data, calculate the lattice energy of CaCl$_2$ in kilojoules per mole. Energy needed to vaporize 1 mole of Ca(s) = 192 kJ; first ionization energy of Ca = 589.5 kJ mol^{-1}; second ionization energy of Ca = 1146 kJ mol^{-1}; electron affinity of Cl = −348 kJ mol^{-1}; bond energy of Cl$_2$ = 238 kJ mol^{-1} of Cl—Cl bonds; energy change for the reaction, Ca(s) + Cl$_2$(g) → CaCl$_2$(s), −795 kJ mol^{-1} of CaCl$_2$(s) formed.

*5.40 Given the following data, calculate the electron affinity of Br. The energy change for the reaction, $Na(s) + \frac{1}{2}Br_2(l) \rightarrow NaBr(s)$, is -360 kJ. The energy needed to vaporize 1 mol of $Na(s)$ is 109 kJ. The energy needed to vaporize 1 mol of $Br_2(l)$ is 31 kJ. The ionization energy of $Na(g)$ is 495.8 kJ/mol^{-1}. The bond energy of Br_2 is 192 kJ/mol^{-1} of Br—Br bonds. The lattice energy of NaBr is -734.3 kJ mol^{-1}.

Drawing Lewis Structures

5.41 Which atoms are bonded to which other atoms in the following molecules or ions? (Sketch the *skeletal structure* of each of them.) (a) NH_3, (b) PCl_3, (c) SCl_2, (d) NO_2^-, (e) BrF_5, (f) PCl_4^+.

5.42 How many valence electrons are in each of the species in the preceding question?

5.43 Why, in general, do we not choose hydrogen as the central atom in a molecule?

5.44 Draw Lewis structures for the molecules PCl_3, SiH_4, BCl_3, H_2S, C_3H_8, and CO.

5.45 Draw Lewis structures for the molecules Cl_2, SO_2, OF_2, SnH_4, C_2H_4, and SCl_2.

5.46 Draw Lewis structures for the ions Cl^-, S^{2-}, ClO^-, ClO_4^-, SO_3^{2-}, and PO_4^{3-}.

5.47 Draw Lewis structures for the ions NO_3^-, NO^+, NO_2^-, and CO_3^{2-}.

5.48 Draw Lewis structures for SeF_6, SeF_4, ICl_3, $AsCl_5$, ICl_2^-, ICl_4^-, and XeF_4.

5.49 Which of the following compounds do not obey the octet rule: ClF_3, OF_2, SF_4, SO_2, IF_7, NO_2, BCl_3?

Covalent Bonding

5.50 What is the difference between an ionic bond and a covalent bond?

5.51 Fluorine and chlorine atoms can combine to form the covalently bonded molecule ClF. Why, energetically, is ionic bonding *not* favored for ClF?

5.52 How many hydrogen atoms would be expected to bond covalently to each of the following atoms: (a) Ge, (b) S, (c) Br, (d) Si, (e) P?

5.53 Use Lewis symbols to diagram the reaction for the formation of the covalently bonded molecules NH_3, H_2O, and HF from their atoms.

5.54 What is a double bond? What is a triple bond?

5.55 How many unpaired electrons are there in a molecule of (a) NH_3, (b) H_2O, (c) CH_4, (d) HCl?

5.56 Define *bond energy* and *bond length*.

5.57 Write the Lewis structures of $BeCl_2$ and BCl_3. In each molecule, how many electrons are in the valence shell of the central atom?

5.58 How many orbitals are in the valence shell of (a) carbon, (b) nitrogen, (c) phosphorus, (d) chlorine?

5.59 What is the maximum number of covalent bonds that can be formed by an atom of nitrogen? Could a phosphorus atom form more covalent bonds than a nitrogen atom? Explain your answer.

Bond Order and Bond Properties

5.60 Define *bond order*.

5.61 What are the various bond orders in the following molecules?
(a) CCl_4
(b) HCN (See Example 5.8.)
(c) CO_2
(d) NO^+
(e)

(methyl isocyanate, the chemical that caused the tragedy in Bhopal, India, in December 1984.)

5.62 How does bond energy vary with bond order? Why?

5.63 How does bond length vary with bond order? Why?

5.64 Describe the bending and vibrational motions with the SO_2 molecule.

5.65 How does the bond vibrational frequency vary with bond order? Why? How are bond vibrational frequencies measured?

5.66 The C—C bond length in a series of compounds was found to be as follows: compound 1, 154 pm; compound 2, 137 pm; compound 3, 146 pm; and compound 4, 140 pm. Arrange these in order of increasing C—C bond order. How would you expect the C—C bond energies to vary?

Resonance

5.67 What is a resonance hybrid? Why is the concept of resonance used?

5.68 Draw the resonance structures of SO_3, NO_3^- and CO_3^{2-}; SO_2 and NO_2^-.

5.69 Draw the resonance structures of HNO_3, SeO_2, SeO_3, and N_2O_4.

Structures:

5.70 Draw resonance structures for $C_2O_4^{2-}$, CH_3COO^-, and N_3^-.

Structures:

5.71 In Section 5.8 we noted that there is evidence that the C—N bond in the peptide linkage possesses some double-bond character. What kind of experimental evidence would be expected to confirm this?

5.72 For the molecules SO_2 and SO_3, compare the SO bond energies, bond lengths, and S—O vibrational frequencies.

5.73 How would you expect the N—O bond distance in NO_2^- to compare with that in NO_3^-?

5.74 What would you expect the value of the bond order in SO_3 to be?

5.75 Compare the C—O bond properties—bond order, bond energy, bond length, and vibrational frequencies—in the following. (*Hint:* Draw resonance structures when necessary.)

Coordinate Covalent Bonds

5.76 What is a coordinate covalent bond? How does it differ from other covalent bonds?

5.77 Use Lewis symbols to show the formation of a coordinate covalent bond in the reaction

$$AlCl_3 + Cl^- \longrightarrow AlCl_4^-$$

5.78 The molecule CH_4 would not be expected to participate in the formation of a coordinate covalent bond. Why?

5.79 The BF_3 molecule reacts with F^- to form BF_4^-. Use Lewis symbols to explain this reaction in terms of coordinate covalent bonding.

Electronegativity

5.80 Define electronegativity. What is the difference between electronegativity and electron affinity?

5.81 Define polar, dipole, and dipole moment.

5.82 What trends in electronegativity occur in the periodic table? What correlation, if any, exists between ionization energy and electronegativity?

5.83 Use the periodic table to decide which of the following bonds should be more polar:
(a) P—F or P—O
(b) Al—S or Al—Cl
(c) Se—Cl or Se—Br

5.84 Which of the following contain bonds that are more than 50% ionic: $AlCl_3$, MgO, Al_2O_3, NF_3, CsF, $FeCl_2$, SO_2, Ca_3P_2, Mg_2Si? (Use the data in Table 5.6.)

5.85 Which of the following have bonds that are less than 50% ionic: NH_3, MnF_2, BCl_3, $MgCl_2$, BeI_2, NaH? (Use the data in Table 5.6.)

5.86 Arrange the following compounds in order of increasing ionic character of their bonds: SO_2, H_2S, SF_2, OF_2, ClF_3, H_2Se, F_2. (Use the data in Table 5.6.)

5.87 Which one of the following is most electropositive: Al, Rb, Mg, N?

5.88 The O—S—O bond angle in SO_2 is less than 180°. Explain why this is a polar molecule. Which end of the molecule carries the positive charge?

5.89 The XeF_4 molecule has a square planar structure. The fluorine atoms are located at the corners of a square, with the Xe located in the center of the square. Are the bonds polar? Is the molecule polar?

*5.90 Below are calculated and experimental bond energies for the hydrogen halides. Use these data to show that the electronegativities of the halogens decrease from F to I.

Calculated and experimental bond energies

	Calculated (kJ mol^{-1})	Experimental (kJ mol^{-1})
HF	295	565
HCl	337	431
HBr	310	360
HI	290	300

*5.91 Hydrogen is more electronegative than any of the Group IA elements. Based on this statement and the data below, show that the electronegativities of the alkali metals decrease from Li to Rb.

Calculated and experimental bond energies

	Calculated Bond Energy (kJ mol^{-1})	Experimental Bond Energy (kJ mol^{-1})
Li—H	272	238
Na—H	256	200
K—H	244	180
Rb—H	242	160

Naming Compounds

5.92 Name the following (when appropriate, use *both* the Stock system and the old system):
(a) NaBr
(b) CaO
(c) $FeCl_3$
(d) $CuCO_3$
(e) CBr_4
(f) P_4O_6
(g) $AsCl_5$
(h) $Mn(HCO_3)_2$
(i) $NaMnO_4$
(j) O_2F_2

5.93 Write chemical formulas for the following compounds:
(a) aluminum nitrate
(b) iron(II) sulfate
(c) ammonium dihydrogen phosphate
(d) iodine pentafluoride
(e) phosphorus trichloride
(f) dinitrogen tetroxide
(g) potassium permanganate
(h) magnesium hydroxide
(i) hydrogen selenide
(j) sodium hydride

5.94 Name the following:
(a) Cr_2O_3
(b) $Mg(H_2PO_4)_2$
(f) $AlPO_4$
(g) Mg_3N_2

(c) $Cu(NO_3)_2$
(d) $CaSO_4$
(e) $Ba(OH)_2$
(h) PbC_2O_4
(i) $(NH_4)_2CO_3$
(j) $K_2Cr_2O_7$

5.95 Write formulas for the following:
(a) titanium(IV) oxide
(b) silicon tetrachloride
(c) calcium selenide
(d) potassium nitrate
(e) aluminum sulfate
(f) nickel(II) bicarbonate
(g) sodium bisulfate
(h) ammonium dichromate
(i) calcium acetate
(j) strontium hydroxide

5.96 Write formulas for the following:
(a) ferric sulfate
(b) ferrous chloride
(c) mercurous nitrate
(d) cuprous chloride
(e) stannic chloride
(f) cobaltous hydroxide
(g) auric chloride
(h) chromic acetate

5.97 Write the names for each of the compounds in Exercise 5.31. When appropriate, follow the Stock system of nomenclature.

CHAPTER 6

COVALENT BONDING AND MOLECULAR STRUCTURE

The unique structure of elemental carbon in its graphite form permits specially grown graphite fibers to have properties of strength and flexibility that are especially useful in tennis rackets and fishing rods. Here we see a graphite fiber-reinforced fishing rod bending to the weight of a large fish in the Florida Keys. The special properties of graphite are a result of the way carbon atoms are bonded to each other in its structure, which is one of the topics explored in this chapter.

In Chapter 5 we found that chemical bonds can be broadly classified into two main categories: ionic bonds and covalent bonds. The ionic bond arises as a purely electrostatic attraction between oppositely charged particles and is therefore nondirectional. This means that the arrangement of ions in a cluster is determined simply by the balancing of attractive and repulsive forces between the ions, not by their electronic structures. The covalent bond, on the other hand, has very definite directional properties. Covalently bound substances, such as molecules or polyatomic ions, have characteristic shapes that are usually retained when these substances undergo physical changes such as melting or vaporization.

The *shapes* of molecules, which means the way their atoms are arranged in space, affect many of their physical and chemical properties. In Chapter 5, for example, we learned that molecular shape can determine whether or not a molecule is polar—a phenomenon that will be explored further in this chapter. Later you will learn that molecular polarity has a very strong influence on such physical properties as melting point and boiling point. Molecular shape can also affect chemical properties. In biological systems, such as our own bodies, the chemical reactions that keep us alive (and even allow us to study chemistry) depend on the very precise fitting together of molecules. If this fit is destroyed, which generally is what happens in cases of poisoning, the organism dies. Thus, an understanding of molecular geometry and the factors that affect it are critical to our understanding of chemistry.

So far, the simple picture of a covalent bond as a pair of dots shared between two atoms has given us no information about molecular structure. In this chapter we will first see how Lewis structures can be used to predict molecular shapes with a surprisingly high degree of accuracy. Then we will examine modern theories of bonding that attempt to answer the "why" and "how" questions—*why* molecules have the shapes that they do, and *how* can atoms actually share electrons with each other. You should keep in mind throughout these discussions that each theory represents an attempt to describe the *same* physical phenomenon. None of the theories is perfect—otherwise we would only have to consider one of them. Each has its usefulness, and each has its weaknesses. The theory you apply in a particular circumstance depends largely on what aspect of the covalent bond you are attempting to explain and, to some extent, your own feelings about the validity of the various theories.

6.1 MOLECULAR SHAPES

Although there is an enormous number of different molecules, the number of different ways that atoms arrange themselves around each other is rather limited. This makes understanding the shapes of molecules much easier, because we are able to describe these shapes with a relatively small number of terms.

Virtually all molecules have shapes that can be considered to be derived from a basic set of five different geometries. Therefore, before we discuss theories that predict them or explain them, it is important for you to fully understand these structures and have a feel for what they present in three dimensions. Let's examine them closely before we proceed further.

The names for some of these structures may be unfamiliar to you. Nevertheless, you should learn to associate the name with the three-dimensional figures they represent.

1. *Linear.* A linear arrangement of atoms occurs when they are all in a straight line. The angle formed between two bonds that go to the same central atom, which we call the **bond angle,** is 180°.

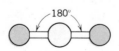

2. *Planar triangular.* A planar triangular arrangement of four atoms has them all in the same plane. The central atom is surrounded by three others located at the corners of a triangle. The bond angles are all 120°.

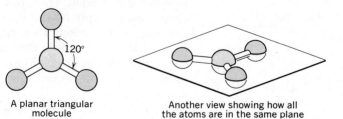

A planar triangular molecule

Another view showing how all the atoms are in the same plane

3. *Tetrahedral.* A tetrahedron is a four-sided pyramid having equilateral triangles as faces. In a tetrahedral molecule, the central atom is located in the center of this tetrahedron and four other atoms are located at the corners. The bond angles are all equal and have values of 109.5°.

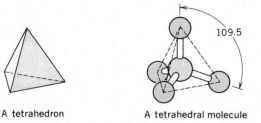

A tetrahedron

A tetrahedral molecule

4. *Trigonal bipyramidal.* A trigonal bipyramid consists of two trigonal pyramids (pyramids with triangular bases similar to tetrahedrons) that share a common face.

A trigonal bipyramid

In a trigonal bipyramidal molecule, a central atom is surrounded by five others. The central atom is located in the center of the triangular plane shared by the upper and lower pyramids. The five atoms attached to the central atom are located at the five corners. In this kind of molecule, the bond angles are not all the same. Between any two bonds that lie in the central triangular plane, the bond angle is 120°. The angle is only 90° between a bond in the central triangular plane and a bond that points to the top or bottom of the trigonal bipyramid.

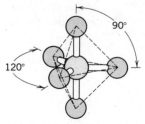

A trigonal bipyramidal molecule

When we draw a trigonal bipyramidal molecule we normally sketch a triangle as it would look tilted backward so that we would be looking at it from its edge. Then we draw lines to the top and bottom of the trigonal bipyramid.

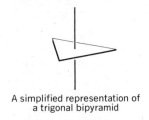

A simplified representation of
a trigonal bipyramid

5. *Octahedral.* An octahedron is a geometrical figure that has eight faces. We can think of it as two square pyramids sharing a common square base. Notice that the figure has only six corners even though it has eight faces.

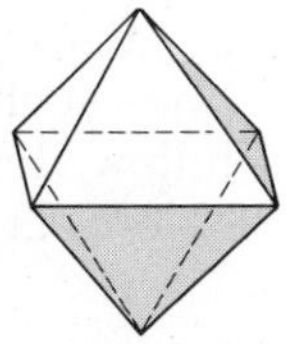

An octahedron

Remember, an octahedral molecule has only six atoms joined to a central atom.

In an octahedral molecule, the central atom is surrounded by six others. The central atom is located in the center of the square plane that passes through the middle of the octahedron. The six atoms bonded to it are at the six corners of the octahedron. The angle between any pair of adjacent bonds is the same and has a value of 90°.

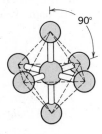

The square plane, tilted back and viewed in perspective, resembles a parallelogram. Draw this first. Then draw a line from the center upward to the top apex of the octahedron. Finally, draw a line from the center to the bottom apex, but don't show the part hidden by the square plane in the center.

A simplified drawing of an octahedron generally shows the square plane in the center, as it would look tilted back, with lines running to the top and bottom of the octahedron.

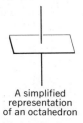

A simplified
representation
of an octahedron

Before going on to the next section, you should practice drawing each of the five structures described here. If you understand them well, it will make the rest of this chapter easier to understand, too.

6.2 VALENCE SHELL ELECTRON-PAIR REPULSION THEORY

One of the primary goals of chemical bonding theory is to explain and (we hope) to predict molecular structure. A theory that is exceedingly simple in its concept and remarkably successful in its ability to predict molecular geometry accurately is called the **valence shell electron-pair repulsion theory (VSEPR theory)**. In applying it, it is not necessary to employ the notion of atomic orbitals at all. We will see, instead, that if an electron-dot structure can be drawn for a molecule, its general shape can be predicted.

When we seek to predict the shape of a molecule, we look for a way to decide how atoms or groups of atoms (for which we will use the general term **ligand**) are arranged around a central atom. For example, in a molecule such as SO_2, how are the oxygen atoms (the ligands) arranged around the sulfur atom? Are the three atoms in a straight line (i.e., is the molecule linear), or is the bond angle something less than 180°? To decide such questions, the VSEPR theory proposes that the geometric arrangement of ligands around the central atom is determined *solely* by the repulsions among the electron pairs in the valence shell of the central atom. According to the theory, these electron pairs assume positions such that the repulsions between them are a minimum, and the ligands follow along. To see how this works, let's begin by considering the simple molecule $BeCl_2$. Its electron-dot structure is given as

$$:\overset{..}{Cl} \times Be \times \overset{..}{Cl}:$$

where the crosses are beryllium electrons and the dots are chlorine electrons. This particular molecule, you recall, violates the octet rule, and there are only two pairs of electrons in the valence shell of Be. According to the VSEPR theory, these electron pairs will arrange themselves to be as far apart as possible, so that the repulsion between them is at a minimum. When there are two electron pairs in the valence shell, this minimal repulsion occurs when they are located on opposite sides of the nucleus, so that we have

$$\overset{\times}{} —Be— \overset{\times}{}$$

In the $BeCl_2$ molecule, the ligands (i.e., the chlorine atoms) are attached to the Be by the sharing of these electron pairs. This means that the chlorines must be placed where the electron pairs are, and the molecule should therefore have the *linear* structure

$$180°$$
$$Cl—Be—Cl$$

This is, in fact, the shape of the $BeCl_2$ molecule in the gas phase.

We can also extend this reasoning to situations involving double or triple bonds. For instance, the CO_2 molecule has the dot structure

$$:\overset{..}{O}::C::\overset{..}{O}:$$

where we see that there are double bonds between C and O. Both pairs of electrons in a double bond must be confined to the same general region in the valence shell of an atom—otherwise, it wouldn't be a *double* bond. Therefore, in terms of their effect on determining molecular geometry, a group of four electrons in a double bond behaves much like a group of two electrons in a single bond. In the valence shell of carbon, therefore, we have *two* groups of four, and

If necessary, review the procedures for writing Lewis structures on page 157.

these groups will locate themselves on opposite sides of the carbon nucleus so that the repulsions between them are a minimum

$$:\!\!-\!\!C\!\!-\!\!:$$

As before, the ligands (in this case, oxygen) are attached to the central atom through these electron pairs and we again have a *linear* structure.

$$:\ddot{O}\!\!=\!\!C\!\!=\!\!\ddot{O}:$$

More than two pairs (or groups of pairs) of electrons

When there are more than two pairs (or groups of pairs) of electrons in the valence shell we find other geometric arrangements as shown in Figure 6.1. Electron pairs arranged in the valence shell in this manner lead to minimum repulsions. Let us see how we can use these electron-pair arrangements to predict molecular structure.

Notice that these geometries are the same five that we discussed in Section 6.1.

Three groups of electrons in the valence shell

In Chapter 5 we saw that the molecule BCl_3 has the dot structure

$$:\overset{..}{\underset{..}{Cl}}:$$
$$:\overset{..}{\underset{..}{Cl}}:B:\overset{..}{\underset{..}{Cl}}:$$

Thus there are three electron pairs around boron. According to Figure 6.1, we therefore expect the three chlorine atoms to be arranged around the boron atom at the corners of a planar equilateral triangle. Experimentally, this is the structure that is found for BCl_3.

BCl_3 is said to be a planar triangular molecule.

Now let's consider the molecule SO_2. The electron-dot structure for one of its two resonance structures is

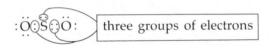

Around the sulfur there are three groups of electrons—two groups each with one pair, and one group with two pairs (the double bond). To have minimum repulsions these groups of electrons are situated at the corners of a planar triangle with the sulfur in the center.

$$\overset{..}{\underset{\diagup\diagdown}{S}}$$

Attaching the oxygen atoms, one to a single pair and one to the double pair, we have

$$\overset{..}{\underset{:\ddot{O}.\ddot{O}.}{S}}$$

We use the terms *molecular shape, molecular structure, molecular geometry,* and *arrangement of atoms* interchangeably. They all refer to how the atoms in a molecule are arranged in space relative to each other.

Thus the theory predicts that the two oxygen atoms and the sulfur do not lie in a straight line. How then should we describe the structure?

When we give the shape of the SO_2 molecule, or of any other molecule, we always describe how the *atoms* in the molecule are arranged around the central

Figure 6.1

Arrangements of electron pairs that lead to minimum repulsions.

Number of Electron Pairs	Geometric Arrangement of Electron Pairs	
2	Linear	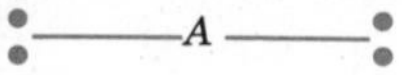
3	Planar triangular	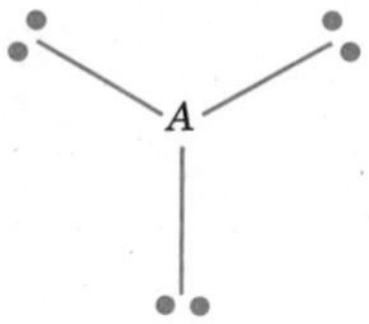
4	Tetrahedral	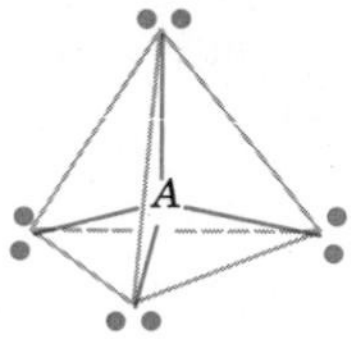
5	Trigonal bipyramidal	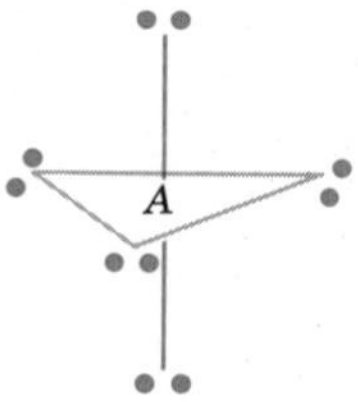
6	Octahedral	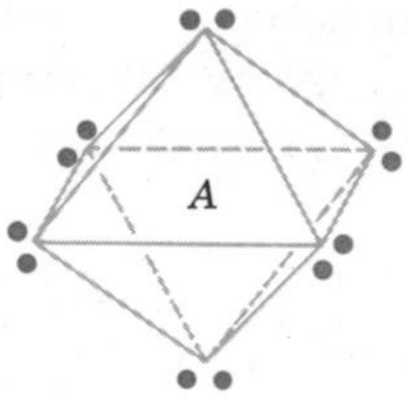 or 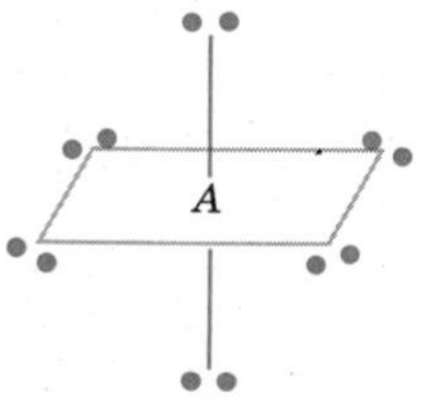

We can determine the arrangement of atoms in a molecule experimentally, but not the geometrical arrangement of their electrons.

atom, *not* how the electrons are arranged. Therefore, even though the electron groups in the valence shell of sulfur are presumed to be in a triangle, we *do not* describe the SO_2 molecule as triangular. Instead, we say that it is nonlinear, or bent, or angular, or V-shaped, or use some other description that simply says the three atoms are not arranged in a straight line.

An important aspect of the structure of the SO_2 molecule is the unshared pair of electrons in the valence shell of the sulfur. Such an unshared pair of electrons is called a **lone pair,** and lone pairs on the central atom have a strong influence on the shapes of molecules. In the case of SO_2, we see that it forces the bond pairs away from a linear arrangement and squeezes them toward each other. We will see similar effects in other molecules.

In summary, when there are three groups of electrons around an atom, they are arranged at the corners of a triangle. If they are all bonded to ligands, we have a molecule that we might generalize as AX_3, in which A stands for the central atom and X stands for a ligand. The structure of an AX_3 molecule is planar triangular. If only two groups are bonded, leaving one lone pair, we have a species AX_2E, where we use E to represent the lone pair. In an AX_2E molecule, the atomic nuclei are situated so as to give a nonlinear structure. Figure 6.2

Figure 6.2

Geometries of molecules or ions in which the central atom has three electron pairs (or groups of pairs) in its valence shell.

TYPE	EXAMPLE	STRUCTURE	DESCRIPTION
AX_3	BCl_3	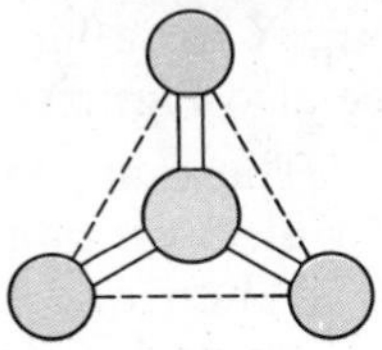	Planer triangular
AX_2E	SO_2	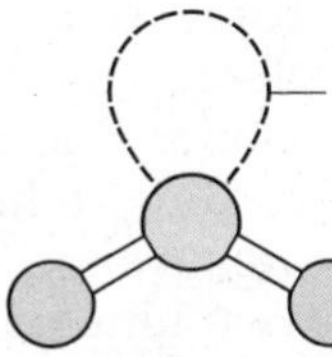 Lone pair	Nonlinear angular bent

illustrates the structures that we find for these kinds of molecules or polyatomic ions. In the figure, the lone pair on the AX_2E species is shown as an electron cloud. Notice that we have omitted from this figure, as well as from our discussion, molecules with the formula AXE_2. This would be a diatomic molecule, and when only two atoms are bonded to each other there's only one way for them to be connected. Only when there are three or more atoms in a molecule or ion do we have a choice of geometries.

Four groups of electrons in the valence shell

If an atom has four pairs of electrons in its valence shell, minimum repulsions occur if they are arranged tetrahedrally. We have just seen that when there are three electron pairs (or groups of pairs) in the valence shell of a central atom, two possible molecular shapes can occur depending on whether or not one of the pairs is a lone pair. For molecules in which the central atom has four pairs in its

TYPE	EXAMPLE	STRUCTURE	DESCRIPTION
AX_4	CH_4	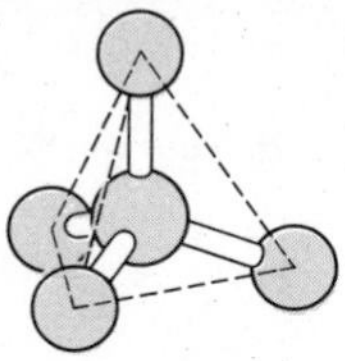	Tetrahedral
AX_3E	NH_3	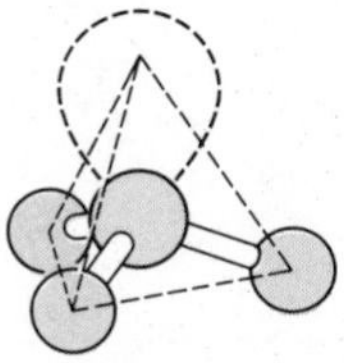	Pyramidal
AX_2E_2	H_2O	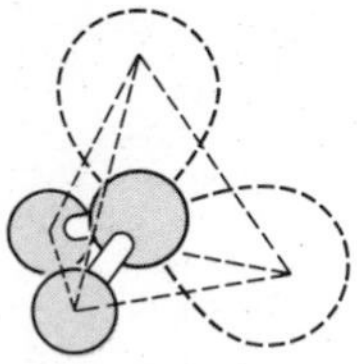	Nonlinear

Figure 6.3

Geometries of molecules in which the central atom has four pairs of electrons in its valence shell.

valence shell there are three possible molecular shapes—all of which are derived from the tetrahedral arrangement of the electrons. Once again using *A* for the central atom, *X* for a ligand, and *E* for a lone pair, we can represent these as follows (see also Figure 6.3).

- AX_4 These are tetrahedral molecules with ligands bonded by all four electron pairs. An example is methane, CH_4.
- AX_3E When one lone pair is present, a *trigonal pyramidal* molecule is formed. This is a molecule shaped like a pyramid with a triangular base. An example is ammonia, NH_3, that has the nitrogen atom at the top of the pyramid and the three hydrogens around the base. Notice that we describe the shape of the ammonia molecule according to how the atoms are arranged, not by how the electrons are arranged.
- AX_2E_2 Two lone pairs gives a nonlinear or angular structure—for example, water.

Five electron pairs

Five electron pairs will have minimum repulsions if they are arranged at the corners of a trigonal bipyramid, as shown in Figure 6.1. This gives four possible structures, depending on the number of lone pairs, as illustrated in Figure 6.4.

TYPE	EXAMPLE	STRUCTURE	DESCRIPTION
AX_5	PCl_5 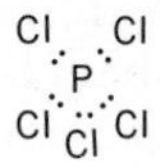	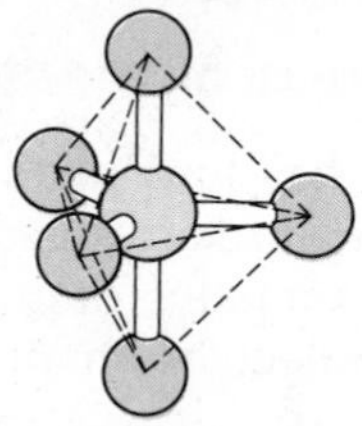	Trigonal bipyramidal
AX_4E	$TeCl_4$ 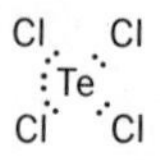	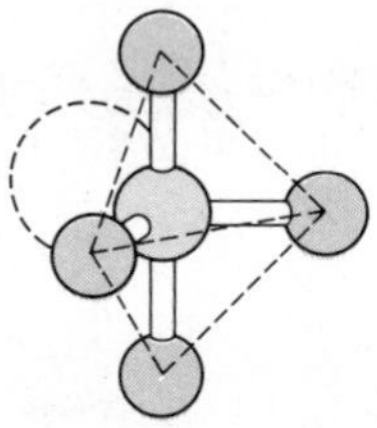	Distorted tetrahedral
AX_3E_2	ClF_3 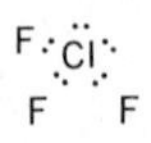	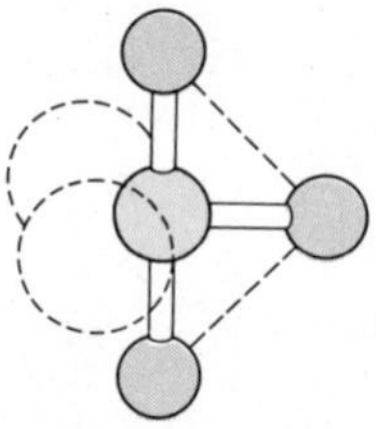	T-shaped
AX_2E_3	I_3^- 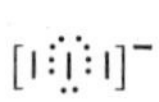	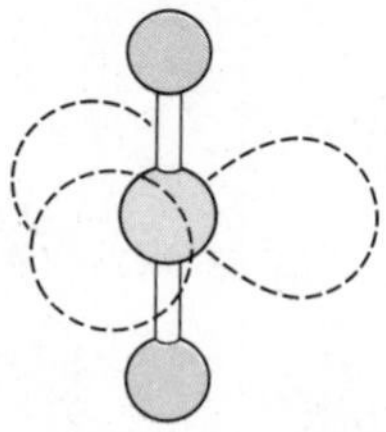	Linear

Figure 6.4
Molecular structures that result when the central atom has five electron-pair groups in its valence shell.

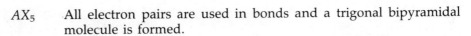
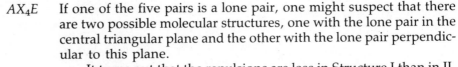

AX_5 All electron pairs are used in bonds and a trigonal bipyramidal molecule is formed.

AX_4E If one of the five pairs is a lone pair, one might suspect that there are two possible molecular structures, one with the lone pair in the central triangular plane and the other with the lone pair perpendicular to this plane.

It turns out that the repulsions are less in Structure I than in II. To understand this, we have to first realize that a lone pair is larger in volume than a pair of electrons in a bond. In a bond, the electrons are under the influence of two positive nuclei, but in a lone pair the electrons are attracted to only one nucleus. The greater total nuclear charge felt by the bonding pair causes it to be pulled into a smaller volume, so effectively it takes up less space than a lone pair. Because the lone pair is larger than the bonding pair, it exerts a greater repulsion toward other pairs in the valence shell.

Now that we realize that a lone pair creates a greater repulsion than does a bonding pair, let's examine Structures I and II to see which will give the least repulsions. We do this by examining how the bonds in the structures are oriented relative to the lone pair. In Structure I, the lone pair is alongside two bonds that are at angles of 90° (those that point up and down) and two bonds that are at angles of 120° (those that are in the triangular plane of the trigonal bipyramid along with the lone pair). In Structure II, the lone pair is alongside *three* bonds at angles of 90°, and the fourth bond is at an angle of 180°. In terms of their influence on structure, it is only the repulsions due to the nearest neighbors that are important. In Structure I, the lone pair has only two nearest-neighbor bonds, but in Structure II it has three. Structure I, therefore, has the lesser total repulsion, so it is preferred. In fact, it is *always* found that the lone pairs prefer to be in the triangular plane of the trigonal bipyramid, even when there are two or three lone pairs.

The shape of the AX_4E molecule is difficult to describe—the term we will use is *distorted tetrahedral*.

AX_3E_2 This structure has two lone pairs in the central triangular plane and the atoms are arranged in the form of the letter T drawn on its side (⊣). The molecule is described as being *T-shaped*.

AX_2E_3 Now there are three lone pairs in the triangular plane and the atoms are in a straight line—the structure is described as *linear*.

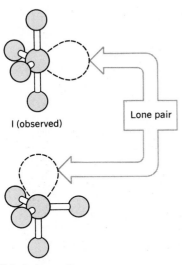

I (observed)

Lone pair

II (not observed)

Six electron pairs

These have minimum repulsions when arranged octahedrally (Figure 6.1). This gives *five* possibilities: AX_6, AX_5E, AX_4E_2, AX_3E_3, and AX_2E_4. Only the first three are actually observed to occur, however. These are shown in Figure 6.5.

AX_6 When all electron pairs are used for bonding, an *octahedral* structure is formed.

AX_5E The atoms in this structure are at the corners of a pyramid having a square base, so the structure is described as being *square pyramidal*.

AX_4E_2 With two lone pairs, minimum repulsions occur if they are as far apart as possible. This gives an arrangement of atoms that we describe as *square planar*.

Figure 6.5
Molecular structures that result when the central atom has six electron-pair groups in its valence shell.

TYPE	EXAMPLE	STRUCTURE	DESCRIPTION
AX_6	SF_6 F F F:S:F F F	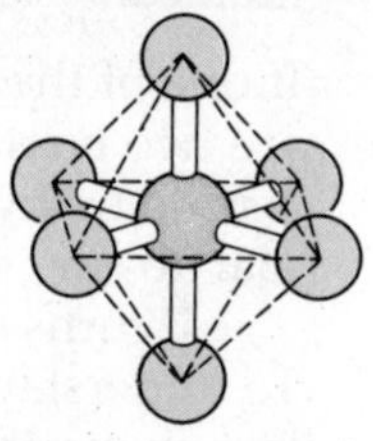	Octahedral
AX_5E	IF_5 F F :I:F F F	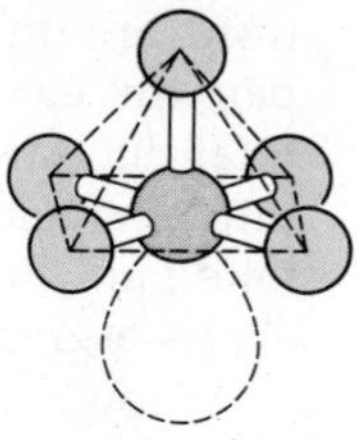	Square pyramidal
AX_4E_2	ICl_4^- [Cl Cl Cl:I:Cl]$^-$	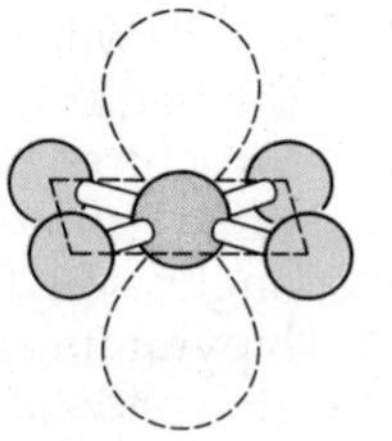	Square planar

Practice drawing the various structures. It will help you to visualize them.

Let's look now at some examples that illustrate how we use the VSEPR theory to predict the shapes of molecules and ions. In doing this, it is best if you are able to visualize the structure and identify it by its name. If this is an overwhelming problem for you, the structure can be found by identifying the number of ligands and lone pairs around the central atom, and then referring to Table 6.1.

Table 6.1
Summary of molecular shapes

Type of Molecule or Ion	Shape
AX_2	Linear
AX_3	Planar triangular
AX_2E	Nonlinear (angular, bent)
AX_4	Tetrahedral
AX_3E	Trigonal pyramidal
AX_2E_2	Nonlinear (angular, bent)
AX_5	Trigonal bipyramidal
AX_4E	Distorted tetrahedral
AX_3E_2	T-shaped
AX_2E_3	Linear
AX_6	Octahedral
AX_5E	Square pyramidal
AX_4E_2	Square planar

EXAMPLE 6.1 PREDICTING THE SHAPE OF AN ION BY THE
VSEPR THEORY

PROBLEM   What is the shape of the sulfate ion, SO_4^{2-}?

SOLUTION   We begin by drawing the Lewis structure for the ion following the procedure given in
Chapter 5. This gives

The sulfur has four electron pairs in its valence shell, which must be arranged tetrahedrally.

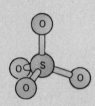

Attaching the oxygen atoms gives a tetrahedral ion.

We can get the same answer, of course, by recognizing (from the Lewis structure) that
SO_4^{2-} is an AX_4 species, which is tetrahedral.

EXAMPLE 6.2 PREDICTING THE SHAPE OF AN ION BY THE
VSEPR THEORY

PROBLEM   Formate ion, HCO_2^-, comes from formic acid, the substance that produces the stinging
sensation in bites from fire ants. The dot structure of formate ion is

What is its shape?

SOLUTION   The double bond behaves just like a single bond for purposes of predicting molecular
shape. This ion has three groups of electrons around the carbon and they are arranged
in a planar triangular fashion. Since all groups are used in bonding, the ion (an AX_3
type) has a planar triangular shape.

EXAMPLE 6.3 PREDICTING THE SHAPE OF A MOLECULE BY
THE VSEPR THEORY

PROBLEM   Arsenic, a well-known poison, can be detected by converting its compounds to the
unstable substance AsH_3 (arsine), which decomposes easily on a clean hot glass surface

where it deposits a mirrorlike coating of pure arsenic. What is the shape of an AsH_3 molecule?

SOLUTION   The first step is always to draw the Lewis structure. This is

$$H{-}\overset{\cdot\cdot}{\underset{|}{As}}{-}H$$
$$H$$

The four electron pairs are arranged tetrahedrally, but only three of them are used in bonds; one is a lone pair. The molecular structure that we get is

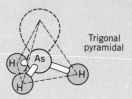

Trigonal pyramidal

We get the same answer recognizing AsH_3 as an AX_3E molecule.

6.3 POLARITY OF MOLECULES AND MOLECULAR STRUCTURE

In Section 5.10, we discussed very briefly how shape can influence whether or not a molecule with polar bonds is a polar molecule. Molecular polarity is an important property because many of the physical properties substances depend on it. This is because polar molecules attract each other. When a large number of polar molecules are together, they tend to orient themselves so that the partially positive end of one is near the partially negative end of another. The electrostatic attraction of opposite charges leads to a net attraction between the molecules. In Chapter 10 we will see how these attractions influence such properties as boiling point and ease of evaporation. But first, let's look more closely at the way molecular polarity is related to molecular structure. We'll begin with a quick review of some of the principles developed in Section 5.10.

A covalent bond will be polar when the two atoms joined by the bond differ in electronegativity, and you should recall that the atom with the larger electronegativity is the one that carries the partial negative charge. The other carries an equal partial positive charge.

In order for a molecule to be polar, it must be a dipole. This means that opposite ends of the molecule must carry *opposite* electrical charges. If there are no charges on opposite ends, or if the charges are of the same sign, the molecule is not a dipole, and it is not considered polar.

When a molecule consists of only two atoms, it must be polar if the bond is polar. This has to be true because the two atoms in the molecule have equal and opposite partial charges. However, if there are three or more atoms in a molecule, it is possible for it to be nonpolar, even when the bonds are polar. In Section 5.10 we discussed two simple cases illustrating this, carbon dioxide and water.

Carbon dioxide has the Lewis structure

$$:\!\overset{\cdot\cdot}{O}{=}C{=}\overset{\cdot\cdot}{O}\!:$$

A collection of polar molecules orient themselves so that the attractions are greater than the repulsions.

Repulsive forces

Attractive forces

and a linear shape. The bonds in CO_2 are polar, with the more electronegative oxygen carrying the negative charge in each case. Because the molecule is linear, these partially negative oxygens are found at opposite ends of the molecule, so the opposite ends carry the same electrical charge. This means, of course, that the molecule is nonpolar.

As we noted in Section 5.10, the overall polarity of the molecule also can be considered to result from the interaction of the various polar bonds within the molecule. To analyze these effects, we view the individual **bond dipoles**—the dipoles *within* the molecule produced by the unequal sharing of electrons in the bonds—as vectors that can either reinforce or cancel each other. As you saw on page 167, the bond dipoles in CO_2 point in opposite directions and cancel completely, which tells us that CO_2 is nonpolar.

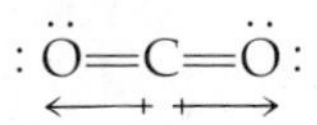

$:\ddot{O}=C=\ddot{O}:$ (Bond dipoles, ↦, cancel because they point in opposite directions.)

On page 167, you saw a similar analysis for water. In this case, the individual bond dipoles do not cancel entirely, because the molecule is nonlinear. Instead, they partially add in one direction to give a net dipole for the molecule.

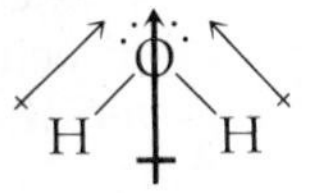

(Heavy arrow is the net dipole for the molecule.)

The differences between water and carbon dioxide illustrate how important molecular structure is in determining molecular polarity. This is one of the reasons why the VSEPR theory is so valuable: Its ability to predict molecular structure enables us to also make predictions about the polarity of molecules. To do this, let's begin by analyzing how bond dipoles are expected to interact in some simple arrangements, and then we can extend these simple cases to the study of more complex structures. In each of these discussions, we assume that all the atoms bonded to the central atom are the *same*.

A linear arrangement of bond dipoles

We have already examined this arrangement of bond dipoles in discussing the CO_2 molecule. As we will see, other molecular structures have this arrangement of dipoles within them, so it is useful to remember that a pair of equivalent bond dipoles pointing in opposite directions cancel each other.

← + →   (Cancellation of the bond dipoles is obvious here.)

A planar triangular arrangement of bond dipoles

Here we have bond dipoles arranged as shown below.

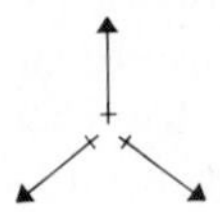

Although not as obvious as with the linear arrangement, complete cancellation of the bond dipoles occurs here, too. The two dipoles pointing downward at an

If the atoms bonded to the central atom are not the same, the individual bonds will differ in their polarities and cancellation of the bond dipoles can't occur.

angle cancel the one pointing up, and so forth. Because of this, we expect that planar triangular molecules such as BCl_3 and SO_3 should be nonpolar, and indeed, they are. In fact, any planar triangular molecule will be nonpolar, if the central atom is bonded to three other identical atoms. (Of course, if the three peripheral atoms are not the same, then the bonds will differ in their polarities and complete cancellation cannot occur. A molecule AX_2Y, in which atoms X and Y are different, will therefore be polar.)

A tetrahedral arrangement of bond dipoles

In this structure the bond dipoles point toward the corners of a tetrahedron.

This arrangement also leads to complete cancellation of the dipoles. Each individual bond dipole is cancelled by the effects of the other three that point partially in the opposite direction. As a result, tetrahedral molecules such as CCl_4 or CF_4 are nonpolar, even though they have polar bonds.

Trigonal bipyramidal and octahedral arrangements of bond dipoles

To analyze these structures, we can make use of what we've learned so far about simpler ones. As you can see in Figure 6.6, the trigonal bipyramidal structure actually consists of a triangular arrangement of bonds in a plane through the center, plus a pair of bonds pointing in opposite directions perpendicular to this plane. We have seen that a planar triangular distribution of bond dipoles leads to cancellation, so the three bonds in the triangular plane in the center of the trigonal bipyramid contribute nothing to the polarity of the molecule. We have also seen that a pair of bond dipoles pointing in opposite directions cancel each other, so the two bonds that are perpendicular to the triangular plane in the trigonal bipyramid also cancel. Overall, then, we get complete cancellation of all the bond dipoles, which means that trigonal bipyramidal molecules such as PCl_5 will be nonpolar, regardless of how polar their bonds happen to be.

Symmetrical octahedral molecules such as SF_6 are nonpolar, too. In the octahedral structure (Figure 6.7), there are three pairs of bonds pointing in opposite directions. Total cancellation occurs for each pair, so all the individual bond dipoles in the octahedral structure are cancelled and a nonpolar molecule is the result.

Molecules that have lone pairs on the central atom

In most cases, when there are lone pairs in the valence shell of the central atom, complete cancellation of the bond dipoles does not occur. Consider, for example, the case of SO_2, which has a V-shaped structure and one lone pair on the central sulfur atom.

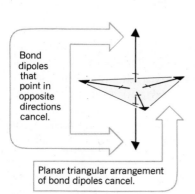

Figure 6.6

The bond dipoles in a trigonal bipyramidal AX_5 molecule cancel to give a nonpolar molecule.

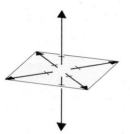

Figure 6.7

In an octahedral AX_6 molecule, there are three sets of two bonds. Within each set, the linearly arranged bond dipoles cancel, so the molecule is nonpolar.

Figure 6.8

In a linear AX_2E_3 molecule and a square planar AX_4E_2 molecule, bond dipoles cancel to give nonpolar molecules.

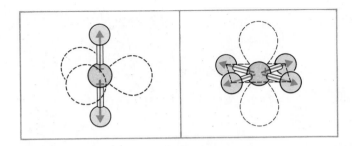

For SO_2, the two bond dipoles that point partially in one direction are not offset by a third one in the opposite direction. As with the water molecule, the bond dipoles are partially additive and the molecule as a whole is polar.

A similar imbalance is found in most of the other structures that have lone pairs on the central atom. There are two exceptions, however. One is the linear AX_2E_3 structure and the other is the square planar AX_4E_2 structure, which are shown in Figure 6.8. For these two, the bond dipoles are symmetrically arranged and their effects cancel completely to give nonpolar molecules.

EXAMPLE 6.4 PREDICTING WHETHER OR NOT A MOLECULE IS POLAR

PROBLEM   Is the SiF_4 molecule polar?

SOLUTION   The first step is to write the Lewis structure and then determine the shape of the molecule by using the VSEPR theory. By following the rules we developed in Chapter 5, the Lewis structure of the molecule is

$$
\begin{array}{c}
:\ddot{F}: \\
| \\
:\ddot{F}\!-\!Si\!-\!\ddot{F}: \\
| \\
:\ddot{F}:
\end{array}
$$

The VSEPR theory tells us that the molecule is tetrahedral. We've seen that in this structure all the bond dipoles cancel, so we can say with confidence that SiF_4 is a nonpolar molecule.

EXAMPLE 6.5 PREDICTING WHETHER OR NOT A MOLECULE IS POLAR

PROBLEM   Is the chloroform molecule, $CHCl_3$, polar?

SOLUTION   The Lewis structure for chloroform is

$$
\begin{array}{c}
:\ddot{Cl}: \\
| \\
:\ddot{Cl}\!-\!C\!-\!H \\
| \\
:\ddot{Cl}:
\end{array}
$$

and it is a tetrahedral molecule. However, not all the bonds in the molecule are the same. Chlorine is a more electronegative atom than hydrogen, so the C—Cl bonds are more polar than the C—H bond. Because the bond dipoles are not all equal, complete cancellation cannot take place, so the molecule will be polar.

EXAMPLE 6.6 PREDICTING WHETHER OR NOT A MOLECULE
IS POLAR

PROBLEM   Is the PCl_3 molecule polar?

SOLUTION   As before, we begin with the Lewis structure.

$$:\ddot{C}l—\overset{\displaystyle P}{\underset{\displaystyle \underset{\cdot\cdot}{:\ddot{C}l:}}{|}}—\ddot{C}l:$$

VSEPR theory tells us that the molecule has a trigonal pyramidal shape, and we note that there is a lone pair of electrons on the phosphorus atom. In this structure, the bond dipoles do not cancel, and we can expect that the molecule will be polar.

EXAMPLE 6.7 PREDICTING WHETHER OR NOT A MOLECULE
IS POLAR

PROBLEM   Is the XeF_4 molecule polar?

SOLUTION   First we draw the Lewis structure.

$$\underset{\displaystyle :\ddot{F}:}{\overset{\displaystyle :\ddot{F}:}{:\ddot{F}—Xe—\ddot{F}:}}$$

The VSEPR theory tells us that it is a square planar molecule, and we have seen that in this structure, all the bond dipoles cancel, even though there are lone pairs in the valence shell of the xenon. As a result, we conclude that XeF_4 is nonpolar.

6.4 VALENCE BOND THEORY

The electron-pair repulsion theory is a useful device for predicting molecular geometry, but it still doesn't answer the basic questions: How do atoms share electrons between their valence shells and how are these electrons able to avoid each other? To find the answers we must look to the results of quantum mechanics to see how the orbitals of atoms interact with each other when bonds are formed.

There are two important approaches to chemical bonding based on the results of quantum mechanics. One of these, called the **valence bond theory,** permits us to retain our picture of individual atoms coming together to form a covalent bond. The other, called **molecular orbital theory,** views a molecule as a set of positive nuclei with orbitals that extend over the entire molecule. The electrons that populate these molecular orbitals do not belong to any individual atoms but, instead, to the molecule as a whole. We will look at the molecular orbital theory in more detail in Section 6.9.

When the valence bond and molecular orbital theories are extended and refined, both give essentially the same results.

The basic postulate of the valence bond theory is that when two atoms come together to form a covalent bond, an atomic orbital of one atom overlaps with an atomic orbital of the other. By **overlap,** we mean that the two orbitals share some common region in space. The pair of electrons that we have come to associate with a covalent bond is shared between the two atoms in this region of overlap. Another important aspect of the theory is that the strength of the covalent bond,

Figure 6.9

Formation of H_2 by overlap of 1s orbitals.

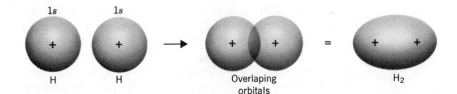

as measured by the amount of energy needed to break it, is proportional to the *amount* of overlap of the atomic orbitals—the greater the overlap, the stronger the bond, and the more the potential energy of the atoms is lowered as the bond is formed. As a result, the atoms in a molecule tend to position themselves so that maximum orbital overlap exists.

Let's see how this theory applies to some familiar compounds. The simplest of these is the hydrogen molecule, which is formed from two hydrogen atoms each of which has a single electron in a 1s orbital. According to valence bond theory, we would view the H—H bond as resulting from the overlap of the two 1s orbitals, as shown in Figure 6.9.[1] The electron density that this gives in the molecule is the same as that described in Chapter 5 (Figure 5.5).

In the HF molecule we have a somewhat different state of affairs. Fluorine has the valence-shell electron configuration

$$\text{F} \quad \underline{\underset{2s}{\uparrow\downarrow}} \quad \underline{\underset{}{\uparrow\downarrow}} \, \underline{\underset{2p}{\uparrow\downarrow}} \, \underline{\uparrow}$$

where we find one of the 2p orbitals occupied by a single electron. It is with this partially occupied 2p orbital that the hydrogen 1s orbital overlaps, as illustrated in Figure 6.10. In this case the hydrogen electron and the fluorine electron can pair up and be shared between the two nuclei. Note that the 1s orbital of the hydrogen atom does not overlap with an already filled atomic orbital on fluorine because then there would be three electrons in the bond (two from the fluorine 2p orbital and one from the hydrogen 1s orbital). This situation is not permitted. In the valence bond theory, *only two electrons with their spins paired may be shared in one set of overlapping orbitals.*

Suppose that we now consider the molecule H_2O. Here we have two hydrogen atoms bound to a single oxygen atom. The outer-shell electron configuration of oxygen is

$$\text{O} \quad \underline{\underset{2s}{\uparrow\downarrow}} \quad \underline{\underset{}{\uparrow\downarrow}} \, \underline{\uparrow} \, \underline{\underset{2p}{\uparrow}}$$

and it shows that there are two unpaired electrons in p orbitals. This allows the two hydrogen atoms, with their electrons in 1s orbitals, to bond to the oxygen by means of the overlap of their 1s orbitals with these partially filled oxygen p orbitals (Figure 6.11). We can represent this using the following orbital diagram:

$$\text{O (in } H_2O) \quad \underline{\underset{2s}{\uparrow\downarrow}} \quad \underline{\underset{}{\uparrow\downarrow}} \, \underline{\underset{}{\uparrow\downarrow}} \, \underline{\underset{2p}{\uparrow\downarrow}}$$

where the colored arrows represent the electrons from the hydrogen atoms. Since the p orbitals are oriented at 90° to one another, we expect the H—O—H bond angle in water to also be 90°. Actually, this angle is 104.5°. One explanation for this discrepancy (we will see another one later) is that since the O—H bonds

Figure 6.10

Formation of HF by the overlap of the partially filled fluorine 2p orbital with the 1s orbital of hydrogen.

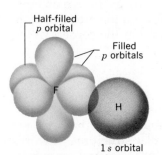

Figure 6.11

Bonding in H_2O. Overlap of two half-filled oxygen 2p orbitals with the hydrogen 1s orbitals.

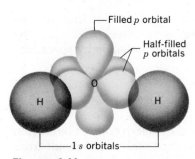

[1] In illustrations throughout this chapter, we will be looking at schematic representations of various kinds of orbitals, which may differ considerably from the actual shapes of the orbitals. This is being done to keep the illustrations simple. For the most part, we are interested primarily in the directional properties of the orbitals rather than their actual shapes.

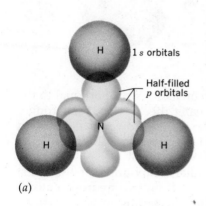

(b)

Figure 6.12
Bonding in NH_3 gives a trigonal pyramidal molecule. (a) Overlap of the 2p orbitals of nitrogen with 1s orbitals of hydrogen. (b) Trigonal pyramidal shape of the NH_3 molecule.

are highly polar, the H atoms carry a substantial positive charge and therefore repel one another. This factor tends to increase the H—O—H angle. However, since the best overlap between the hydrogen 1s orbitals and the oxygen 2p orbitals occurs at an angle of 90°, the H—O—H angle cannot increase too much without a considerable decrease in overlap, which would produce a substantial loss in bond strength. There are thus two factors working in opposition to each other, one tending to increase the bond angle and one tending to reduce it to 90°. It appears that a balance is obtained when the angle is 104.5°.

Qualitatively, the valence bond theory can account for the geometry of the water molecule. We can also apply the theory to the ammonia molecule with reasonable success. Nitrogen, being in Group VA, has three unpaired electrons in its *p* subshell.

$$\text{N} \quad \underset{2s}{\underline{\uparrow\downarrow}} \quad \underset{2p}{\underline{\uparrow}\;\underline{\uparrow}\;\underline{\uparrow}}$$

Three hydrogen atoms can form bonds to nitrogen by overlapping their 1s orbitals with the half-filled *p* orbitals as shown in Figure 6.12*a*. The orbital diagram shows how the nitrogen completes its valence shell by this process.

$$\text{N (in } NH_3) \quad \underset{2s}{\underline{\uparrow\downarrow}} \quad \underset{2p}{\underline{\uparrow\downarrow}\;\underline{\uparrow\downarrow}\;\underline{\uparrow\downarrow}} \qquad \text{(Colored arrows = H electrons.)}$$

As in the water molecule, the H—N—H angles are larger than the expected 90°, having values in this case of 107°; as with H_2O, we might attempt to explain this angle in terms of repulsions between the hydrogens. In any event, we obtain a picture of the NH_3 molecule that has a pyramidal shape, with the nitrogen atom at the apex of the pyramid and the three hydrogen atoms at the corners of the base (Figure 6.12*b*).

6.5 HYBRID ORBITALS

The very simple picture of the overlap of half-filled atomic orbitals that we have just developed cannot be used to account for all molecular structures. It works well for H_2 and HF, but is only marginally acceptable for water and ammonia. When we get to methane, it breaks down completely. With carbon, we would initially expect only two bonds to be formed with hydrogen since the valence shell of carbon contains only two unpaired electrons.

$$\text{C} \underset{2s}{\underline{\uparrow\downarrow}} \quad \underset{2p}{\underline{\uparrow}\;\underline{\uparrow}\;\underline{}}$$

The species CH_2, however, does not exist as a stable molecule. Instead, the simplest compound between carbon and hydrogen is methane, which has the formula CH_4. Attempting to explain the structure of this molecule by spreading the electrons out to give

$$\underset{2s}{\underline{\uparrow}} \quad \underset{2p}{\underline{\uparrow}\;\underline{\uparrow}\;\underline{\uparrow}}$$

suggests that three of the C—H bonds will be formed by the overlap of hydrogen 1s orbitals with carbon 2p orbitals, while the remaining bond would result from the overlap of the carbon 2s orbital with a hydrogen 1s orbital. This fourth C—H bond should certainly be different from the other three bonds, because it is formed from different kinds of orbitals. It has been found experimentally, however, that *all* four C—H bonds are identical and the molecule has a structure in

There are no simple atomic orbitals arranged in a tetrahedral fashion.

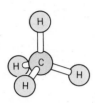

Figure 6.13
The structure of methane, CH₄.

An atom forms hybrid orbitals by mixing appropriate combinations of its atomic orbitals.

It takes energy to form hybrid orbitals, but this energy is more than paid back by the much stronger bonds formed by them.

The various kinds of hybrid orbitals are identified by indicating with superscripts the numbers of each kind of atomic orbitals used to form the hybrid.

which the carbon atom lies at the center of a tetrahedron with the hydrogen atoms located at the four corners (Figure 6.13). Apparently, the orbitals that carbon uses to form bonds in molecules like CH_4, and that other atoms use to form bonds in the more complex trigonal bipyramidal and octahedral structures, are not just simple atomic orbitals. The question is: What kinds of orbitals are they?

The solution to this apparent dilemma is found in the mathematics of wave mechanics. In that theory, the solution of Schrödinger's wave equation provides us with a series of wave functions, ψ, each of which describes a different atomic orbital. It is the property of these mathematical functions that when they are squared, they enable us to calculate the probability of locating the electron at some point in space around the nucleus and, in fact, the spheres and figure-eights that we have been drawing roughly correspond to pictorial representations of the probability distributions predicted by the wave functions for s and p orbitals.

What is important to us here is that it is possible to combine these wave functions by appropriately adding or subtracting them to give new functions referred to as **hybrid orbitals.** In this way, two or more atomic orbitals are mixed together to produce a new set of orbitals. In general, these hybrid orbitals possess different directional properties than the pure atomic orbitals from which they are created. For example, Figure 6.14 illustrates the result of combining a $2s$ orbital with a $2p$ orbital to give a new set of two sp hybrid orbitals. In this drawing, notice we have indicated that the wave function for a p orbital has positive numerical values in some regions about the nucleus and negative values in others.[2] The s orbital, on the other hand, has the same algebraic sign everywhere. Therefore, when these wave functions are alternatively added and subtracted, the new orbitals that result become larger in those regions where both functions have the same sign and smaller in regions where they are of opposite sign. In effect, by constructive and destructive interference of the electron waves corresponding to the s and p orbitals, two new orbitals are formed.

Hybrid orbitals such as these possess some very interesting properties. We see for each orbital that one lobe is much larger than the other, and because of this, a hybrid orbital can overlap well in only one direction—the direction in which the orbital protrudes the most. A hybrid orbital is therefore very strongly directional in its ability to enter into covalent bond formation. Furthermore, because the large lobe of a hybrid orbital extends out farther from the nucleus than an unhybridized orbital does, it is able to overlap much more effectively with an orbital of another atom. Therefore, bonds formed from hybrid orbitals tend to be stronger than those formed from ordinary atomic orbitals.

Thus far we have examined what occurs when one s and one p orbital are mixed together. Other combinations of orbitals are also possible, with the number of orbitals in a hybrid set, as well as their orientations, determined by which atomic orbitals are combined. Table 6.2 contains a listing of the sets of hybrid orbitals that can be used to explain most of the molecular structures we encounter in this book. Their directional properties, which you no doubt recognize by now, are illustrated in Figure 6.15 on page 199. Notice that the number of each kind of atomic orbital included in a combination is specified by an appropriate superscript on the atomic orbital type. Thus the sp^3d^2 hybrids are formed from one s orbital, three p orbitals, and two d orbitals.

---

[2] Although difficult to visualize, the amplitude of the electron wave in certain orbitals can be positive in some regions and negative in others. This is similar to the displacement of the string on a guitar being positive (upward) in some places and negative (downward) in others for the different harmonics that we looked at in Figure 4.10 on Page 116.

Figure 6.14

*Formation of two sp hybrid orbitals from an s and a p orbital.
(a) s and p orbitals drawn separately. (b) s and p orbitals before hybridization. (c) Two sp hybrid orbitals are formed (drawn separately). (d) The two sp hybrid orbitals drawn together to show their directional properties. Note that one orbital points to the left, the other to the right.*

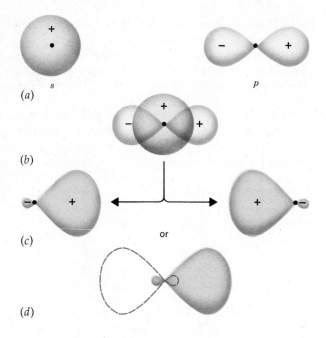

Let us see now how we can use the information contained in Table 6.2 and Figure 6.15 to account for the structures of some typical molecules. We might begin with a molecule such as BeH_2. It would have the dot structure

$$H{:}Be{:}H$$

The electronic structure of beryllium's valence shell is

$$Be \quad \underset{2s}{\underline{\uparrow\downarrow}} \quad \underset{2p}{\underline{\quad}\ \underline{\quad}\ \underline{\quad}}$$

In order to form two covalent bonds with H atoms, the Be atom must provide two half-filled (i.e., singly occupied) orbitals. This can be accomplished by creating a pair of *sp* hybrids and placing one electron in each of them.

$$Be \quad \underbrace{\underline{\uparrow}\ \underline{\uparrow}}_{sp} \quad \underbrace{\underline{\quad}\ \underline{\quad}}_{\text{Unhybridized } 2p \text{ orbitals}}$$

VSEPR theory would predict that BeH_2 should be linear.

The atom's valence electrons are distributed over the hybrid orbitals by following Hund's rule (page 123).

Table 6.2

Hybrid orbitals

Hybrid Orbitals	Number of Orbitals	Orientation
sp	2	Linear
sp^2	3	Planar triangle
sp^3	4	Tetrahedral
sp^3d	5	Trigonal bipyramidal
sp^3d^2	6	Octahedral

Figure 6.15

Directional properties of hybrid orbitals. The minor lobes have been omitted for the sake of clarity.

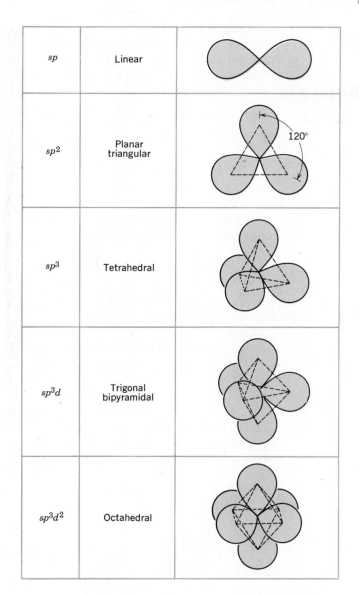

sp	Linear	
sp^2	Planar triangular	
sp^3	Tetrahedral	
sp^3d	Trigonal bipyramidal	
sp^3d^2	Octahedral	

The two H atoms can then bond to the beryllium atom by the overlap of their respective singly occupied s orbitals with the singly occupied Be sp hybrids as shown in Figure 6.16. The orbital diagram for the molecule is

Be (in BeH$_2$)   ⇅ ⇅   __ __   (Colored arrows = H electrons.)
sp p

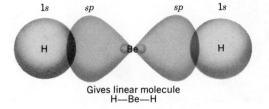

Gives linear molecule
H—Be—H

Figure 6.16

The bonding in BeH$_2$.

Because of the orientation of the sp hybrid orbitals, the H atoms are forced to lie on opposite sides of the Be, and a linear H—Be—H molecule results.

Let us now return to our problem of the structure of CH_4. If we use hybrid orbitals on the carbon atom, we find that in order to provide four orbitals with which hydrogen $1s$ orbitals can overlap, we must use a set of sp^3 hybrids.

$(s + p + p + p)$ gives a set of four equivalent sp^3 hybrid orbitals.

$$\text{C} \quad \underset{2s}{\underline{\uparrow\downarrow}} \quad \underset{2p}{\underline{\uparrow}\ \underline{\uparrow}\ \underline{\quad}} \quad \text{(Unhybridized)}$$

$$\underset{sp^3}{\underline{\uparrow}\ \underline{\uparrow}\ \underline{\uparrow}\ \underline{\uparrow}} \quad \text{(Hybridized)}$$

In Figure 6.15 we see that these orbitals point toward the vertices of a tetrahedron. Therefore, when the four hydrogen atoms are attached to the carbon by orbital overlap with these sp^3 hybrids

$$\text{C (in } CH_4) \quad \underset{sp^3}{\underline{\uparrow\downarrow}\ \underline{\uparrow\downarrow}\ \underline{\uparrow\downarrow}\ \underline{\uparrow\downarrow}}$$

a tetrahedral molecule results, as shown in Figure 6.17. This is in agreement with the structure that is found by experiment.

As another example, consider the molecule SF_6. Sulfur, being in Group VIA, has six valence electrons distributed over the $3s$ and $3p$ subshells

$$\text{S} \quad \underset{3s}{\underline{\uparrow\downarrow}} \quad \underset{3p}{\underline{\uparrow\downarrow}\ \underline{\uparrow}\ \underline{\uparrow}} \quad \underset{3d}{\underline{\quad}\ \underline{\quad}\ \underline{\quad}\ \underline{\quad}\ \underline{\quad}}$$

Until now we've omitted the $3d$ subshell when writing the electron configuration of a period-3 element.

Here we have shown the empty $3d$ subshell as well as the $3s$ and $3p$ subshells that contain electrons. In order for sulfur to form six covalent bonds to fluorine, six half-filled orbitals must be created. This can be accomplished by using two of the unoccupied $3d$ orbitals, and an sp^3d^2 hybrid set is formed.

$$\text{S} \quad \underset{3s}{\underline{\uparrow\downarrow}} \quad \underset{3p}{\underline{\uparrow\downarrow}\ \underline{\uparrow}\ \underline{\uparrow}} \quad \underset{3d}{\underline{\quad}\ \underline{\quad}\ \underline{\quad}\ \underline{\quad}\ \underline{\quad}}$$

$(s + p + p + p + d + d)$ gives a set of six equivalent sp^3d^2 hybrid orbitals.

$$\text{S} \quad \underset{sp^3d^2}{\underline{\uparrow}\ \underline{\uparrow}\ \underline{\uparrow}\ \underline{\uparrow}\ \underline{\uparrow}\ \underline{\uparrow}} \quad \underset{3d}{\underline{\quad}\ \underline{\quad}\ \underline{\quad}} \quad \begin{array}{l}\text{(Hybridized)}\\ \text{(Not used in hybrids)}\end{array}$$

$$\text{S (in } SF_6) \quad \underset{sp^3d^2}{\underline{\uparrow\downarrow}\ \underline{\uparrow\downarrow}\ \underline{\uparrow\downarrow}\ \underline{\uparrow\downarrow}\ \underline{\uparrow\downarrow}\ \underline{\uparrow\downarrow}} \quad \underset{3d}{\underline{\quad}\ \underline{\quad}\ \underline{\quad}} \quad \begin{array}{l}\text{(Colored arrows =}\\ \text{F electrons.)}\end{array}$$

(Not used in hybrids)

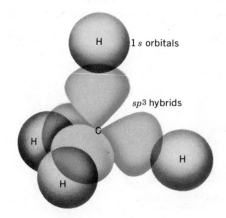

Figure 6.17

The formation of methane by overlap of hydrogen 1s orbitals with carbon sp^3 hybrids.

The sp^3d^2 orbitals point toward the corners of an octahedron, which explains the octahedral geometry of SF_6.

Molecules in which there are lone pairs on the central atom

In our earlier discussion we viewed the structure of H_2O and NH_3 as resulting from the use of half-filled p atomic orbitals on the oxygen and nitrogen atoms, respectively. An alternative view of the bonding in these molecules employs sp^3 hybrid orbitals on the central atom. In the tetrahedral set of hybrids, the orbitals are oriented at angles of 109.5° to each other. The bond angles in water (104.5°) and ammonia (107°) are not too different from the tetrahedral angle and, using water as an example, we might consider the molecule to result from the overlap of hydrogen 1s orbitals with two partially occupied sp^3 orbitals on the oxygen atom.

O   ⇅   ⇅ ↑ ↑   (Unhybridized)
   ‾‾ ‾‾‾‾‾‾‾
   2s   2p

   ⇅ ⇅ ↑ ↑   (Hybridized)
   ‾‾‾‾‾‾‾‾‾‾
      sp^3

O (in H_2O)   ⇅ ⇅ ⇅ ⇅   (Colored arrows = H electrons.)
              ‾‾‾‾‾‾‾‾‾‾‾‾
                  sp^3

Notice that only two of the hybrid orbitals are involved in bond formation while the other two contain nonbonded lone pairs of electrons. In the case of ammonia, three of the sp^3 orbitals are used in bonding while the fourth orbital contains a lone pair of electrons (Figure 6.18). There is, in fact, rather strong experimental evidence to indicate that this lone pair does indeed project out from the nitrogen atom as implied in this picture of the NH_3 molecule. It is worth noting that in our previous description of NH_3 we found this lone pair of electrons in an s orbital that would have spread the electron pair symmetrically about the nucleus.

In the case of H_2O and NH_3, the H—X—H bond angles (104.5° and 107°, respectively) are less than the tetrahedral angle of 109° observed in the molecule, CH_4. One way to account for this is to consider the influence of the lone-pair electrons present in hybrid orbitals of the central atom. A pair of electrons in a bond is attracted to two nuclei and, therefore, might be expected to occupy a smaller effective volume than a pair of electrons in a nonbonded orbital, which experience the attraction of only one nucleus. The lone-pair electrons, then, because of their greater space requirement, tend to crowd together the electron

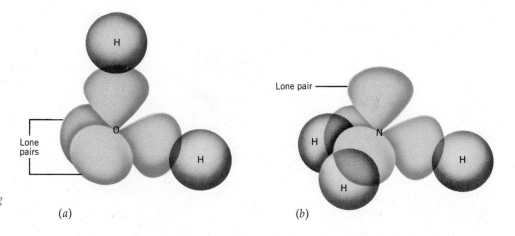

Figure 6.18

The use of sp³ *hybrids for bonding in (a) H₂O and (b) NH₃.*

(a)

(b)

pairs located in the bonds and hence reduce the bond angle to something less than 109°. On this basis we anticipate a greater reduction in bond angle for water than for ammonia, since water has two lone pairs while ammonia has only one.

Valence bond theory and the VSEPR theory

By now you have probably noticed that the orientations of the hybrid orbitals in Figure 6.15 are the same as the orientations that give minimum repulsions between electron pairs described in the valence shell electron-pair repulsion theory, and that the two theories give identical results. For example, the shapes of methane, water, and ammonia predicted by the VSEPR theory are the same as those accounted for by using sp^3 hybrids in valence bond theory. Both theories use a tetrahedral arrangement of electron pairs in these molecules. This useful correlation gives us a simple way of anticipating the kinds of hybrid orbitals that an atom will use in a particular molecule. For instance, let's look again at SF_6. If we draw a dot structure for the molecule, we get

The VSEPR theory predicts that the six electron pairs around sulfur should be arranged octahedrally. Now we can ask ourselves, "What kinds of hybrid orbitals have an octahedral geometry?" The answer, of course, is sp^3d^2, and that is exactly what we used in our explanation of the structure of SF_6 by valence bond theory. We see, therefore, that the VSEPR theory can be used to help us choose the right kinds of hybrid orbitals to use in the valence bond theory. The two theories complement one another very nicely in explaining the bonding in molecules.

EXAMPLE 6.8   **USING THE VALENCE BOND THEORY TO EXPLAIN BONDING**

PROBLEM   How can Valence Bond theory explain the bonding in the PCl_3 molecule?

SOLUTION   Since we aren't told what the structure of the molecule is, let's begin by using the VSEPR theory. This requires that we generate the Lewis structure of the molecule, which is

The VSEPR theory predicts that the electron pairs should be arranged tetrahedrally around the phosphorus atom. This would require that the phosphorus atom use sp^3 hybrid orbitals. We next examine the electronic structure of phosphorus. The orbital diagram for the atom is

Creating sp^3 hybrids would give the following before bond formation:

Each chlorine has the following electronic structure:

$$\text{Cl}\quad \underline{\underset{3s}{\uparrow\downarrow}}\quad \underline{\underset{}{\uparrow\downarrow}}\,\underline{\underset{3p}{\uparrow\downarrow}}\,\underline{\uparrow}$$

The P—Cl bonds would therefore be formed by the overlap of an unpaired electron in a chlorine $3p$ orbital with an unpaired electron in an sp^3 orbital on the phosphorus atom. We also see that the lone pair on the phosphorus atom occupies one of the sp^3 orbitals.

$$\text{P (in PCl}_3)\quad \underline{\uparrow\downarrow}\;\underline{\uparrow\downarrow}\;\underline{\uparrow\downarrow}\;\underline{\uparrow\downarrow}\qquad \text{(Colored arrows are chlorine electrons.)}$$
$$sp^3$$

EXAMPLE 6.9 **USING THE VALENCE BOND THEORY TO EXPLAIN BONDING**

PROBLEM   Determine the kind of hybrid orbitals used by sulfur in SF_4 and explain the bonding in this molecule according to valence bond theory.

SOLUTION   Let's use the VSEPR theory to help us choose hybrid orbitals. This means that we first need the Lewis structure for SF_4. Following our usual procedure, we get

$$\overset{\displaystyle :\ddot{F}:\quad:\ddot{F}:}{\underset{\displaystyle :\ddot{F}:}{:\ddot{F}-\ddot{S}:}}$$

Notice that there are five electron pairs around the sulfur. The VSEPR theory tells us that they should be located at the corners of a trigonal bipyramid, and the hybrid orbital set that is trigonal bipyramidal is sp^3d.

Now we examine the electronic structure of sulfur.

We have included the vacant $3d$ subshell because we know we will need a d orbital to form the sp^3d hybrids.

$$\text{S}\quad \underline{\underset{3s}{\uparrow\downarrow}}\quad \underline{\underset{}{\uparrow\downarrow}}\,\underline{\uparrow}\,\underline{\underset{3p}{\uparrow}}\quad \underline{\ }\,\underline{\ }\,\underline{\underset{3d}{\ }}\,\underline{\ }\,\underline{\ }$$

Forming the sp^3d hybrid gives

$$\text{S}\quad \underline{\uparrow\downarrow}\;\underline{\uparrow}\;\underline{\uparrow}\;\underline{\uparrow}\;\underline{\underset{sp^3d}{\uparrow}}\quad \underline{\ }\,\underline{\ }\,\underline{\underset{3d}{\ }}\,\underline{\ }\,\underline{\ }$$

Notice that we have enough half-filled orbitals to form the four bonds to fluorine.

$$\text{S (in F}_4)\quad \underline{\uparrow\downarrow}\;\underline{\uparrow\downarrow}\;\underline{\uparrow\downarrow}\;\underline{\uparrow\downarrow}\;\underline{\underset{sp^3d}{\uparrow\downarrow}}\quad \underline{\ }\,\underline{\ }\,\underline{\underset{3d}{\ }}\,\underline{\ }\,\underline{\ }\qquad \text{(Colored arrows = F electrons.)}$$

This gives us our bonding picture for SF_4, in which a lone pair occupies one of the hybrid orbitals. The structure of SF_4 is shown in Figure 6.19

Figure 6.19

The structure of SF_4. Note the lone pair of electrons in the sp^3d hybrid orbital. The VSEPR theory, by predicting the structure, even tells us which of the hybrid orbitals houses the lone pair.

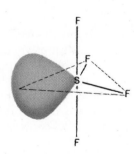

EXAMPLE 6.10 | USING THE VALENCE BOND THEORY TO
EXPLAIN BONDING

PROBLEM | The H_2S molecule is nonlinear, with an H—S—H bond angle of approximately 92°. Explain the structure of this molecule by using the valence bond theory.

SOLUTION | In this example, we are given actual data about the molecule. Whatever explanation we devise, it must adequately explain the facts. With this in mind, let's begin as before. The Lewis structure for H_2S is

$$H—\ddot{\underset{..}{S}}—H$$

The VSEPR theory would predict that the electron pairs are arranged tetrahedrally, and that in turn would suggest sp^3 hybrids. However, there is a problem with this. The angle between bonds in a tetrahedral structure is 109.5°, but the bond angle in H_2S is only 92°. Is there a better explanation for the kinds of orbitals used by the sulfur? Let's look once again at the electronic structure of sulfur.

This time we haven't bothered to include the vacant 3d subshell because it isn't needed.

$$S \quad \underset{3s}{\underline{\uparrow\downarrow}} \quad \underset{3p}{\underline{\uparrow\downarrow} \ \underline{\uparrow} \ \underline{\uparrow}}$$

Notice that the two unpaired electrons sulfur needs for bonding are in p orbitals. By now you should know that p orbitals are at 90° angles to each other, which is very close to the bond angle that we are trying to explain. Therefore, it appears that in this molecule, the sulfur atom *does not* use hybrid orbitals. The bonding picture for the molecule is therefore simply

$$S \quad \underset{3s}{\underline{\uparrow\downarrow}} \quad \underset{3p}{\underline{\uparrow\downarrow} \ \underline{\uparrow\downarrow} \ \underline{\uparrow\downarrow}} \quad \text{(Colored arrows are hydrogen electrons.)}$$

This example illustrates how we must be careful in applying the various bonding theories. In the absence of actual data about the structure of the molecule, we can use the VSEPR theory to anticipate the structure and then derive a valence bond explanation of the bonding. Usually this explanation will be correct. The case of H_2S, however, illustrates that the VSEPR theory can be wrong about bond angles. When there are actual data about the structure of the molecule, the data must be explained properly, regardless of what the VSEPR theory may predict.

Coordinate covalent bonds

Before moving on, a word should be said about the coordinate covalent bond. An example of this, you remember, is provided by the ammonium ion.

$$\left[\begin{array}{c} H \\ \cdot\times \\ H \overset{\times}{\cdot} N \overset{}{\cdot} H \\ \cdot\times \\ H \end{array} \right]^{+} \quad \text{or} \quad \left[\begin{array}{c} H \\ | \\ H—N{\rightarrow}H \\ | \\ H \end{array} \right]^{+}$$

According to valence bond theory, two electrons are shared between two overlapping orbitals. It doesn't matter, however, where the electrons come from. If one comes from each of the overlapping orbitals, an ordinary covalent bond is formed. If one orbital is empty and the other is filled, both electrons can come from the filled orbital and a coordinate covalent bond is formed. Thus we can imagine the coordinate covalent bond in the ammonium ion to be formed by the overlap of an *empty* 1s orbital centered on a proton (a hydrogen ion, H^+) with the

Figure 6.20

Formation of a coordinate covalent bond by the overlap of an empty orbital on the hydrogen with a filled orbital on the nitrogen.

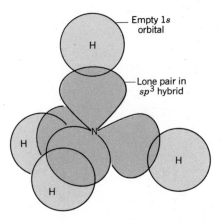

completely filled lone-pair orbital on the nitrogen of an ammonia molecule as shown in Figure 6.20. The electron pair is then shared in the region of orbital overlap. Once the bond is formed, of course, it is a full-fledged covalent bond whose properties do not depend on its origin. Therefore, the four N—H bonds in NH_4^+ are identical and the ion is usually represented as simply

$$\begin{bmatrix} & H & \\ & | & \\ H & -N- & H \\ & | & \\ & H & \end{bmatrix}^+$$

This same argument can be extended to other coordinate covalent bonds as well.

6.6 MULTIPLE BONDS

Double and triple bonds occur when two and three pairs of electrons, respectively, are shared between two atoms. As examples, we have seen the molecules ethylene, C_2H_4, and acetylene, C_2H_2.

$$\begin{array}{cc} H & H \\ \diagdown & \diagup \\ C=C & \qquad H—C\equiv C—H \\ \diagup & \diagdown \\ H & H \end{array}$$

ethylene　　　　　　**acetylene**

The bonding in ethylene is usually interpreted in the following way. In order to form bonds to three other atoms (two hydrogens and one carbon), each carbon atom employs a set of sp^2 hybrids.

$$C \quad \underline{\uparrow\downarrow} \quad \underline{\uparrow} \ \underline{\uparrow} \ \underline{}$$
$$\quad\quad 2s \qquad\quad 2p$$

gives

$$C \quad \underline{\uparrow} \ \underline{\uparrow} \ \underline{\uparrow} \quad \underline{\uparrow}$$
$$\qquad\quad sp^2 \qquad\quad p \text{ (Unhybridized)}$$

Two of these hybrid orbitals are used for overlap with hydrogen $1s$ orbitals while the third sp^2 orbital overlaps with a similar sp^2 orbital on the other carbon atom, as shown in Figure 6.21a. This, then, accounts for all of the C—H bonds in C_2H_4 as well as *one* of the electron pairs shared between the two carbons.

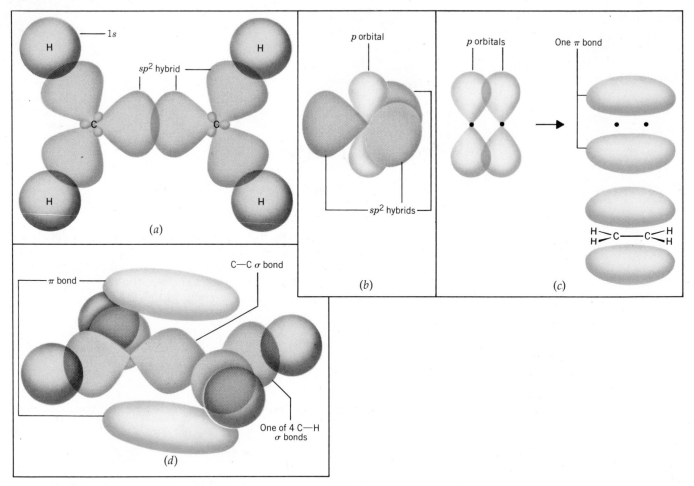

Figure 6.21

The bonding in ethylene, C_2H_4. (a) Overlap of hydrogen 1s orbitals with sp^2 hybrid orbitals on carbon. The carbon atoms are also bound to each other by overlap of sp^2 hybrid orbitals. (b) Unhybridized p orbital is perpendicular to the plane of the sp^2 hybrid orbitals. (c) Formation of the π bond by sideways overlap of p orbitals. (d) A drawing showing the entire ethylene molecule.

The σ bond is sandwiched between the two halves of the π bond like a frankfurter between the two halves of a bun.

Because of the way the sp^2 orbitals are created, each carbon atom also has an unhybridized p atomic orbital that is perpendicular to the plane of the sp^2 orbitals and that projects above and below the plane of these hybrids (Figure 6.21*b*). When the two carbon atoms are joined together, these p orbitals approach each other sideways and, in addition to the bond formed from the overlap of sp^2 orbitals, a second bond is formed in which the electron cloud is concentrated above and below the carbon–carbon axis. This we see illustrated in Figure 6.21*c*.

In terms of this interpretation, the double bond in ethylene consists of two distinctly different kinds of bonds, and to differentiate between them a specific notation is employed. A bond that concentrates electron density along the line joining the bound nuclei is called a **σ bond (sigma bond)**. The overlap of the sp^2 orbitals of adjacent carbons therefore gives rise to a σ bond. The bond that is formed by the sideways overlap of two p orbitals, and that provides electron density above and below a plane containing both of the bound nuclei, is called a **π bond (pi bond)**. Thus in ethylene we find the double bond consists of one σ

Figure 6.22

Formation of the two π bonds in acetylene. Each carbon has two unhybridized p orbitals that overlap with the p orbitals on its neighbor.

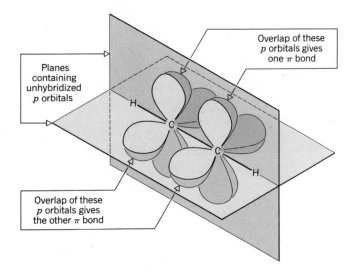

bond and one π bond. Notice that in this double bond the two electron pairs manage to avoid one another by occupying different regions in space.

Another point to note is that bonds formed by the overlap of the hydrogen 1s orbitals with carbon sp^2 hybrid orbitals (Figure 6.21a) also concentrate electron density along a line joining bound atoms. Therefore, these C—H bonds would also be termed σ bonds.

In acetylene each carbon is bound to only two other atoms, a hydrogen and a carbon atom. Two orbitals are needed for this purpose, and a pair of sp hybrid orbitals are used.

$$C \quad \underset{}{\uparrow\downarrow} \quad \uparrow \quad \uparrow \quad \underline{}$$

$$\underbrace{\uparrow \quad \uparrow}_{sp} \quad \underbrace{\uparrow \quad \uparrow}_{p}$$

This leaves two singly occupied unhybridized p orbitals on each carbon that are mutually perpendicular as well as perpendicular to the sp hybrids. When the carbon atoms join by way of σ bond formation between an sp hybrid orbital on each carbon, the p orbitals can also overlap to yield two π bonds that surround the axis joining the carbon nuclei (Figures 6.22 and 6.23). A triple bond therefore consists of one σ and two π bonds. The two π bonds in acetylene (or in any triple

Figure 6.23

The triple bond in acetylene consists of one σ bond and two π bonds. (a) Two π bonds. (b) Cylindrical electron distribution about the bond axis.

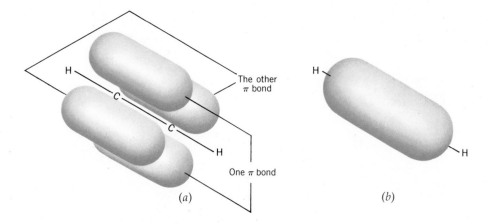

bond) give a total electron distribution that is cylindrical about the bond axis. This is shown in Figure 6.23*b*.

Summary

In arriving at the structure of a molecule, such as ethylene or acetylene, *the shape of the molecular framework is determined by the σ bonds that arise from the overlap of the hybrid orbitals.* Double and triple bonds in a structure result from additional π bonds. In summary, we find the following:

single bond—one σ bond
double bond—one σ + one π bond
triple bond—one σ + two π bonds

EXAMPLE 6.11   **IDENTIFYING SIGMA AND PI BONDING, AND THE KINDS OF HYBRID ORBITALS USED BY ATOMS IN A MOLECULE**

PROBLEM   Identify the kinds of hybrid orbitals used by the atoms in acetic acid whose structure is

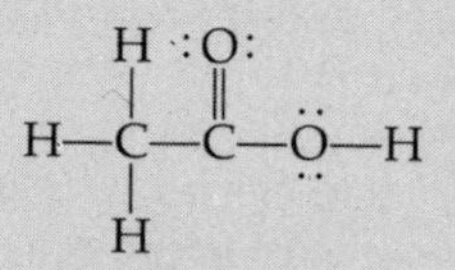

H :O:
| ||
H—C—C—Ö—H
| ..
H

What kinds of bonds (σ, π) exist between the atoms?

SOLUTION   To identify the kinds of hybrid orbitals that an atom uses, we simply count the number of groups of electrons around the atom. Then we choose a hybrid set that has that number of orbitals. For example, the carbon on the left has four bonds (four pairs) and uses sp^3 orbitals. The carbon next to it has three groups of electrons, which means it uses sp^2 orbitals. The doubly bonded oxygen has three groups of electrons, so its uses sp^2 orbitals, too. Finally, the singly bonded oxygen has four pairs around it and must use sp^3 hybrids. (No doubt you've noticed that we haven't said anything about hydrogen—hydrogen always uses only its 1*s* orbital for bonding.)

Hydrogen only forms σ bonds.

Now we can identify the kinds of bonds in the molecule.

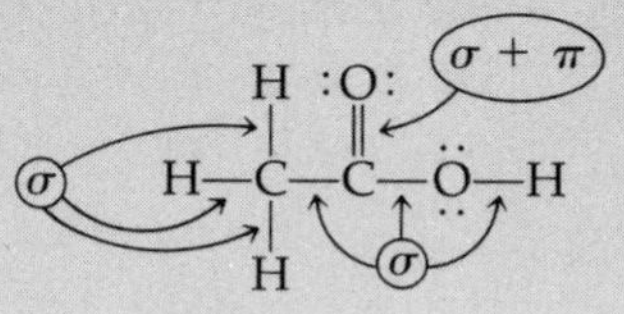

6.7   RESONANCE

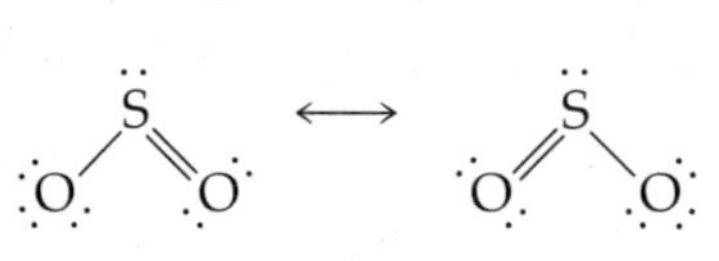

In Chapter 5 we saw that there are instances in which we cannot draw a single satisfactory Lewis dot formula for a molecule or ion. Some examples, you might remember, are SO_2, SO_3, and NO_2^-. Sulfur dioxide, for instance, was drawn as

and it was stated that the actual electronic structure of this molecule corresponds to a resonance hybrid of these two structures.

Electron dot formulas, as we have drawn them, closely correspond to the valence bond pictures developed in the preceding sections. Each pair of dots drawn between two atoms represents a pair of electrons shared in a region where atomic orbitals of the bonded atoms overlap. When we draw one of the resonance structures of SO_2, we are therefore referring to a bonding picture in which the S—O bond consists of a single σ bond while the other is composed of one σ and one π bond.

When the valence bond theory was developed, it was recognized that there were numerous instances in which a single valence bond structure was inadequate in accounting for the molecular structure, so the concept of resonance evolved. The inability in these cases to draw a single picture to describe the electron density in the molecule is one drawback of valence bond theory. Nevertheless, the correspondence between the valence bond structures that are based on orbital overlap and the simple electron-dot formulas make the valence bond concept a very useful one.

6.8 SINGLE BONDS VERSUS MULTIPLE BONDS: THE MOLECULAR STRUCTURES OF THE ELEMENTAL NONMETALS

We have seen that atoms are able to share electrons in two basic ways. One is by the formation of σ bonds, which frequently involves the overlap of hybrid orbitals. The second is by the formation of π bonds, which normally requires the sideways overlap of unhybridized p orbitals. (Pi bonds can also be formed by d orbitals, but we do not discuss these in this book.)

Not all atoms have the same tendencies to form π bonds. The ability of an atom to form π bonds determines its ability to form multiple bonds, and this in turn greatly affects the kinds of molecular structures that the element produces. One of the most striking illustrations of this is the molecular structures of the elemental nonmetals and metalloids.

If we take a brief look at the electronic structures of the nonmetals and metalloids, we see that their atoms generally contain valence shells that are only partially filled. Exceptions to this, of course, are the noble gases, which have electronically complete valence shells. We have also seen that atoms tend to complete their valence shells by bond formation. Therefore, in their elemental states, most of the nonmetallic elements do not exist as single atoms, but instead form bonds to each other and produce more complex molecular species. (The only nonmetals that actually occur in nature as collections of single atoms are the noble gases.)

One of the controlling factors in determining the complexity of the molecular structures of the nonmetals and metalloids is their abilities to form multiple bonds. Small atoms, such as those in period 2, are able to approach each other closely. As a result, effective sideways overlap of their p orbitals can occur, and these atoms form strong π bonds. Therefore, carbon, nitrogen, and oxygen are able to form multiple bonds about as easily as they are able to form single bonds. On the other hand, when the atoms are large—which is the case for atoms from periods 3, 4, and so on—π-type overlap is relatively ineffective, and these elements seem to have very little tendency to form π bonds at all. Rather than form a double bond consisting of one σ bond and one π bond, these elements would rather form two separate σ bonds. This leads to a useful generalization: *Elements in period 2 are able to form multiple bonds fairly readily, while elements below them in periods 3, 4, 5, and 6 have a tendency to prefer single bonds.* Let's look at some of the consequences of this.

Oxygen and nitrogen have 6 and 5 electrons, respectively, in their valence shells. This means that an oxygen atom needs two electrons to complete its valence shell, and a nitrogen atom needs three. Although a perfectly satisfactory Lewis structure for O_2 can't be drawn, experimental evidence suggests that the oxygen molecule does possess a double bond. The nitrogen molecule, which we discussed earlier, contains a triple bond. Oxygen and nitrogen, because of their small size, are capable of multiple bonding because they are able to form strong π bonds. This allows them to form a sufficient number of bonds with just a *single* neighbor to complete their valence shells, so they are able to form diatomic molecules.

Oxygen, in addition to forming the stable species O_2, also can exist in another exceedingly reactive molecular form called **ozone,** which has the formula O_3. The structure of ozone can be represented as a resonance hybrid

*In ozone, oxygen also partici-
pates in π bonding.*

This unstable molecule can be generated by the passage of an electric discharge through ordinary O_2, and the pungent odor of ozone can often be detected in the vicinity of high-voltage electrical equipment. It is also formed in limited amounts in the upper atmosphere by the action of ultraviolet radiation from the sun on O_2. The presence of ozone in the upper atmosphere shields the earth and its creatures from exposure to intense and harmful ultraviolet light.

The existence of an element in more than one form, either as the result of differences in molecular structure as with O_2 and O_3, or as the result of differences in the packing of molecules in the solid, is a phenomenon called **allotropy.** The different forms of the element are called allotropes—O_2 is one allotrope of oxygen, and O_3 is another.

Let's turn our attention now to another period 2 nonmetal, carbon. This element has four electrons in its valence shell, so it must share four electrons to complete its octet. There is no way for carbon to form a quadruple bond, so a simple C_2 species is not stable under ordinary conditions. Instead, carbon completes its octet in either of two different ways, and two allotropic forms of elemental carbon are found. One of these is **diamond.** In diamond, each carbon atom uses sp^3 hybrid orbitals to form covalent bonds to four other carbon atoms at the corners of a tetrahedron. Each of those atoms, in turn, are bonded through sp^3 hybrid orbitals to three more, and so on, as illustrated in Figure 6.24*a*. This produces a gigantic three-dimensional network of interconnected covalent bonds. A diamond crystal of gem quality, such as the one shown in Figure 6.24*b*, therefore consists of a huge number of carbon atoms covalently bonded together

Figure 6.24
(a) *The structure of diamond.*
(b) *Gem-quality diamond.*

(a)   (b)

in one enormous molecule. Since breaking a diamond crystal involves rupturing a very large number of covalent bonds, diamond is extremely hard.

The second allotrope of carbon is called **graphite.** This is the form that probably is more familiar to you. In graphite, the carbon atoms are arranged in the form of hexagonal rings connected together in large planar sheets, perhaps somewhat reminiscent of chicken wire (Figure 6.25). Each carbon is surrounded by three nearest neighbors at angles of 120° from one another. This means that the molecular framework is based on σ bonds produced by the overlap of sp^2 hybrid orbitals on the carbon atoms. On each of the carbons throughout the entire structure, there remains an unhybridized p orbital containing an unpaired electron. These p orbitals are ideally situated to form π bonds with their neighbors, and one of the many resonance forms of a graphite sheet is illustrated in Figure 6.26.

In the total graphite structure, shown in Figure 6.27, the planes of carbon atoms are stacked in layers so that each carbon atom lies above another in every second layer. Within any given layer, adjacent carbon atoms are fairly close together (141 pm), while the spacing between successive planes is much greater (335 pm). The different planes of carbon atoms are not held together by covalent bond, but instead by much weaker forces. As a result, the layers are able to slide over one another with relative ease, and as you may know, graphite has many applications as a dry lubricant. As every school child is taught, the lead in a lead pencil is actually graphite, and when one writes with a pencil, the layers of graphite slide off one another and onto the paper.

In graphite, carbon exhibits multiple bonding, as do nitrogen and oxygen in their molecular forms. As we noted earlier, their abilities to do this reflects their ability to form strong π bonds—a requirement for the formation of a double or triple bond. When we move to the third and successive periods, a different state of affairs exists. Here, we have much larger atoms that are not able to form strong π bonds. These elements prefer to form only single bonds, and the molecular structures of the free elements reflect this.

Elements of Group VIIA

Each of the elements in Group VIIA are diatomic in the free state. This is so because π bonding is not necessary in any of their molecular structures. Chlorine, for example, requires just one electron to complete its octet, so it only needs to form one covalent bond with another atom. Therefore, it forms one σ bond to another chlorine atom and is able to exist as diatomic Cl_2. Bromine and iodine form diatomic Br_2 and I_2 molecules for the same reason. The structures of the remaining nonmetals and metalloids are considerably more complex, however.

Elements of Group VIA

Below oxygen in Group VIA is sulfur, which has the Lewis symbol

$$\cdot \ddot{S} :$$

It requires two electrons to obtain an octet, so it must form two covalent bonds. But sulfur doesn't form π bonds well, so it prefers to form two single bonds to different sulfur atoms. Each of these also prefers to bond to two different sulfur atoms and this gives rise to a

$$-\ddot{S}-\ddot{S}-\ddot{S}-\ddot{S}-$$

sequence. Actually, in sulfur's most stable form, the sulfur atoms are arranged in an 8-member ring to give a molecule with the formula S_8. The S_8 ring has a

The carbon in soot, coal, and charcoal briquettes is the graphite allotrope.

C   ⇅   ↑ ↑
    —   — —
    2s   2p

hybridization

↑ ↑ ↑   ↑
— — —   —
  sp^2 p

The unhybridized p orbitals are perpendicular to the planar sheet of carbon atoms.

In naturally occurring graphite, gas molecules such as O_2 become trapped easily between the layers of carbon atoms. These small molecules serve as submicroscopic ball bearings that allow the layers to slide over each other.

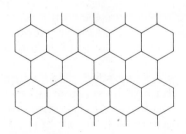

Figure 6.25
Sigma-bond framework of a graphite sheet.

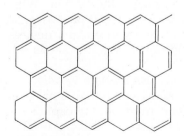

Figure 6.26
One of the resonance structures for graphite.

Figure 6.27
Total graphite structure.

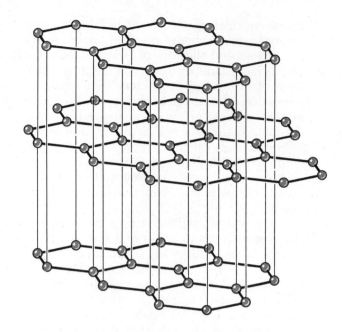

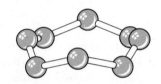

Figure 6.28
The structure of the S_8 ring.

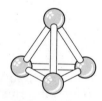

Figure 6.29
The structure of white phosphorus, P_4.

By contrast, the strong triple bond in N_2 makes elemental nitrogen very *unreactive* except at high temperatures.

puckered crownlike shape, which is illustrated in Figure 6.28. Selenium, below sulfur in Group VIA, also forms Se_8 rings in one of its allotropic forms. And selenium and tellurium also exist in a gray form in which there are long Se_x and Te_x chains (the subscript x is a large number).

Elements of Group VA

The elements below nitrogen in Group VA all have five valence electrons. Phosphorus is an example

$$\cdot \overset{\displaystyle \cdot}{\underset{\displaystyle \cdot}{P}} \cdot$$

To achieve a noble gas structure, it must acquire three more electrons. Since there is little tendency for phosphorus to form multiple bonds, as nitrogen does when it forms N_2, the octet is completed by the formation of three single bonds to three different phosphorus atoms.

The simplest elemental form of phosphorus is a waxy solid called **white phosphorus.** It consits of P_4 molecules in which each phosphorus atom lies at a corner of a tetrahedron, as illustrated in Figure 6.29. Notice that in this structure each phosphorus is bound to three others. This particular allotrope of phosphorus is very reactive because of the very small P—P—P bond angle of 60°. At this small angle, the orbitals of the phosphorus atoms don't overlap very well, so the bonds are weak. This means that breaking a P—P bond occurs easily, and this is the first step in causing this molecule to react with something. As a result, P_4 molecules are readily attacked by other chemicals, especially oxygen. White phosphorus is so reactive toward oxygen that it ignites and burns spontaneously in air. For this reason, white phosphorus is used in military incendiary devices, and you've probably seen movies in which exploding phosphorus shells produce arching showers of smoking particles.

A second allotrope of phosphorus that is much less reactive is called **red phosphorus.** At the present time, its structure is unknown, although it has been suggested that it contains P_4 tetrahedra joined at the corners. Red phosphorus is also used in explosives and fireworks, and it is mixed with fine sand and used on the striking surface of matchbooks. As a match is drawn across the surface, friction ignites the phosphorus, which then ignites the ingredients in the tip of the match.

Streamers of white smoke arc through the air as a phosphorus bomb explodes.

A third allotrope of phosphorus is called **black phosphorus,** which is formed by heating white phosphorus at very high pressures. This variety has a layer structure in which each phosphorus atom in a layer is covalently bonded to three others in the same layer. As in graphite, these layers are stacked one atop another, with only weak forces between the layers. As you might expect, black phosphorus has many similarities to graphite.

The elements arsenic and antimony, which are just below phosphorus in the periodic table, are also able to form somewhat unstable yellow allotropic forms containing As_4 and Sb_4 molecules, but their most stable forms have a metallic appearance with structures similar to black phosphorus.

Elements of Group IVA

Finally, we look at the heavier elements in Group IVA: silicon and germanium. To complete their octets, each must form four covalent bonds. Unlike carbon, however, they have no apparent tendency to form multiple bonds, so they don't form allotropes that have a graphite structure. Instead, each of them forms a solid with a structure similar to diamond.

Summary

In this section we have examined the structures of the elemental forms of the nonmetallic elements. Some of these are simply diatomic molecules, but others are quite complex. This complexity can be traced to two factors. One is the number of electrons that the atom of the element needs to complete its octet. The second is the preference of the heavier (and larger) nonmetals and metalloids for single bonds. This preference prevents them from forming diatomic molecules and causes them to form more complex structures.

6.9 MOLECULAR ORBITAL THEORY

In our discussion of the electronic structures of atoms in Chapter 4 we saw that around an atomic nucleus there exists a set of atomic orbitals. The ground state electronic structure of a particular atom was derived by feeding the appropriate number of electrons into this set of atomic orbitals such that (1) no more than two electrons populated a single orbital, (2) each electron was placed into the lowest energy orbital available, and (3) electrons were spread out as much as possible, with unpaired spins, over orbitals of the same energy.

Figure 6.30

The combination of atomic 1s orbitals to give bonding and antibonding molecular orbitals.

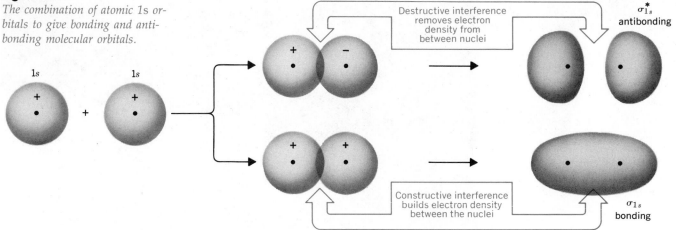

An atom is one positive center (a nucleus) surrounded by a set of atomic orbitals; a molecule is a cluster of positive centers surrounded by a set of molecular orbitals.

Molecular orbital theory proceeds in much this same way. According to this theory, a molecule contains a certain arrangement of atomic nuclei, and spread out over these nuclei is a set of **molecular orbitals.** The electronic structure of the molecule is obtained by feeding the appropriate number of electrons into these molecular orbitals following the same rules that apply to the filling of atomic orbitals.

No one is quite sure what shapes molecular orbitals have in any particular molecule or ion. What *appears* to be an approximately correct picture is obtained by combining the atomic orbitals that reside on the nuclei composing the molecule. These combinations are achieved by considering the constructive interference and destructive interference of the electron waves of the atoms in the molecule. This is shown in Figure 6.30 for the 1s orbitals on two identical nuclei. Notice that when the amplitudes of the two waves are *added,* the resulting molecular orbital has a shape that concentrates electron density between the two nuclei. Electrons placed in such a molecular orbital tend to hold the nuclei together and stabilize a molecule. For that reason this orbital is called a **bonding molecular orbital.** Because the electron density in the orbital is centered along the line joining the atomic nuclei, it is also a σ-type of orbital. Since it is derived in this case from two 1s atomic orbitals, we refer to it as the σ_{1s} **molecular orbital.**

You will also observe in Figure 6.30 that a second molecular orbital is obtained by the destructive interference of the electron waves. In this instance a molecular orbital is produced that places the maximum electron density outside the region between the two nuclei. If the electrons of a molecule are placed into this molecular orbital, they do not help to cement the nuclei together. In fact, the attractions between the electrons and the nuclei are less than the mutual repulsions of the electrons in this kind of molecular orbital and this produces a net increase in energy rather than a decrease. Consequently, electrons placed into this molecular orbital lead to a destabilization of the molecule and, as a result, the orbital is said to be **antibonding.** This antibonding orbital also has its greatest electron density along the line that passes through the two nuclei and is thus a σ-type orbital. Its antibonding character is denoted by an asterisk superscript; thus it is called the σ_{1s}^* **molecular orbital.** As you might expect, we can also draw similar pictures for the combination of any pair of s orbitals; therefore, in a diatomic molecule we have σ_{2s}, σ_{2s}^*, σ_{3s}, σ_{3s}^*, . . . , molecular orbitals.

In a molecule, the *p* orbitals are also capable of interacting to produce bonding and antibonding molecular orbitals, as illustrated in Figure 6.31. Here we

Figure 6.31
Formation of molecular orbitals from atomic p orbitals.

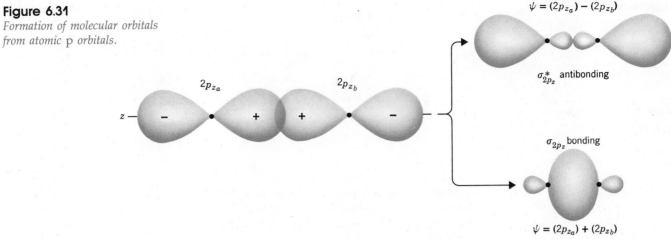

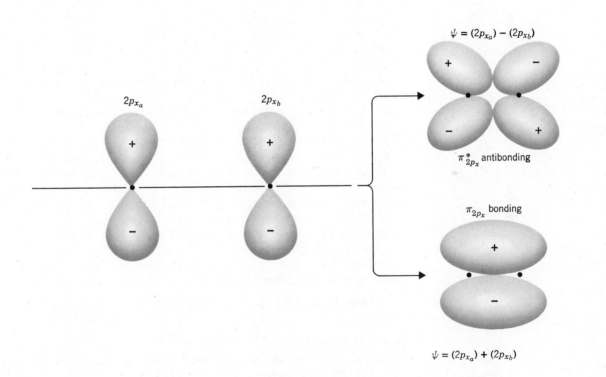

have arbitrarily chosen to denote the internuclear axis as the z axis of our coordinate system, so the p orbitals that point toward one another correspond to p_z orbitals. Again we find that one combination of orbitals gives a bonding molecular orbital, with electron density placed between the two nuclei, while the second combination places most of the electron density outside the region between the nuclei. The p_z orbitals, like s orbitals, form σ-type molecular orbitals, and for $2p_z$ orbitals they would be labeled σ_{2p_z} and $\sigma^*_{2p_z}$.

Having chosen the z axis at the internuclear axis, we find that the p_x and p_y orbital on the two nuclei of our molecule are forced to overlap in a sideways fashion to produce π and π^* molecular orbitals (Figure 6.31). Also keep in mind that the π_{p_x} and $\pi^*_{p_x}$ orbitals are the same as the π_{p_y} and $\pi^*_{p_y}$ orbitals, except that they are situated at 90° to each other when viewed down the molecular axis.

Figure 6.32

(a) *The energies of the bonding and antibonding σ_{1s} molecular orbitals.* (b) *Bonding in H_2.*

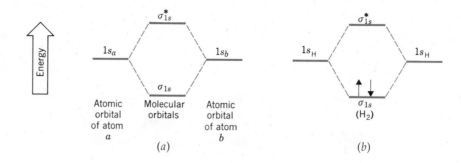

For a diatomic molecule, we have now examined the shapes of the molecular orbitals that can be considered to arise as a consequence of the overlap of atomic orbitals. To discuss the electronic structure of a diatomic molecule, however, we must know the relative energies of these orbitals. Once this has been established, we can then proceed with filling the orbitals with electrons, following the rules mentioned earlier.

Let us first consider the σ_{1s} and σ_{1s}^* orbitals. Electrons placed into the bonding orbital lead to stable bond formation and, therefore, an energy lower than that of two separate atoms. On the other hand, electrons placed into the antibonding orbital lead to a destabilization of the molecule and thus to a state higher in energy than the atoms from which the molecule is formed. We can represent this schematically as shown in Figure 6.32a, where the energies of the atomic orbitals of the separate atoms appear on either side of the energy-level diagram while the energies of the molecular orbitals appear in the center.

Using this simple diagram we can examine the bonding in the H_2 molecule. There are two electrons in H_2 that we place in the lowest-energy molecular orbital, the σ_{1s}, (Figure 6.32b). The electron distribution in H_2 is therefore that described by the shape of the σ_{1s} orbital. Notice that this picture is the same as that developed in the valence bond view of H_2. This should not be too surprising since both theories are attempting to describe the same molecular species.

Before moving on, let us also see why the molecule He_2 does *not* exist. The species He_2 would have four electrons, two of which would be placed into the σ_{1s} orbital. The other two would be forced to occupy the σ_{1s}^* orbital. The pair of electrons in the antibonding orbital would cancel out the stabilizing influence of the bonding pair. As a result, the **net bond order,** which we can define as

$$\binom{\text{Net bond}}{\text{order}} = \frac{(\text{no. of } e^- \text{ in bonding MOs}) - (\text{no. of } e^- \text{ in antibonding MOs})}{2}$$

has a value of zero for He_2. Since the bond order in He_2 is zero, He_2 is not a stable molecule and is not observed to exist under normal conditions.

For diatomic molecules of second period elements we really only need to consider molecular orbitals derived from the interaction of the 2s and 2p orbitals. The 1s orbitals are essentially buried beneath the valence-shell orbitals and are therefore not involved to any appreciable extent in the bonding in these species. The energy-level diagram for the molecular orbitals created from the 2s and 2p orbitals is shown in Figure 6.33a.[3] Let's see how this energy-level diagram can be used to account for the bonding in the molecules N_2, O_2, and F_2.

Nitrogen is in Group VA and, therefore, each nitrogen atom contributes five electrons to the N_2 molecule from its valence shell. This means that we must

---

[3] The relative energies of the σ_{2p_z} and the π_{2p_x}, π_{2p_y} set actually shift about somewhat as we cross the second period. This diagram will give qualitatively correct results for any species you will encounter here.

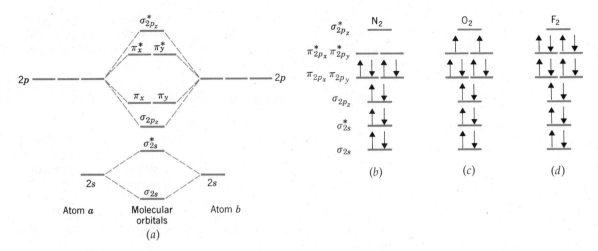

Figure 6.33
(a) *Energies of molecular orbitals formed from atomic orbitals having* n = 2 *in diatomic molecules.* (b–d) *Molecular orbital electron configurations of* N_2, O_2, *and* F_2.

place ten electrons into our set of molecular orbitals. As shown in Figure 6.33*b*, two electrons enter the σ_{2s}, two go into the σ_{2s}^*, two more into the σ_{2p_z} and, finally, two into each of the bonding π orbitals, π_{2p_x} and π_{2p_y}. As before, the two σ_{2s}^* antibonding electrons cancel the effect of the σ_{2s} bonding electrons, leaving us with a net total of six bonding electrons (two each in the σ_{2p_z}, π_{2p_x}, and π_{2p_y} orbitals). If, as usual, we take two electrons to represent a "bond," we find that N_2 is held together by a triple bond composed of one σ and two π bonds. As with H_2, we arrive at the same resultant description of the bonding in N_2 with both the valence bond and molecular orbital theories.

The real mark of success for molecular orbital theory is seen in its description of the O_2 molecule. This species is found experimentally to be paramagnetic with two unpaired electrons. In addition, its bond length and bond energy suggest that there is a double bond between the two oxygen atoms. An attempt to derive a valence bond picture for O_2, however, gives us

A dot structure such as

·Ö:Ö·

has the correct number of unpaired electrons, but it is unsatisfactory because there is only an O—O single bond.

$$\ddot{\text{O}} :: \ddot{\text{O}}$$

where, to satisfy the octet rule and give a double bond, all of the electrons appear in pairs.

The molecular orbital description of O_2 is seen in Figure 6.33*c*. The first 10 of the 12 valence electrons populate all the same molecular orbitals as in N_2. The final two electrons must then be placed in the $\pi_{2p_x}^*$ and $\pi_{2p_y}^*$ antibonding orbitals. However, because these two orbitals are of the same energy, the electrons spread themselves out with their spins in the same direction. These two antibonding π electrons cancel the effects of two of the π-bonding electrons, so in the final analysis we see that O_2 is held together by a *net* double bond (one σ and one net π bond). Also notice that the molecule is predicted to have two unpaired electrons, in precise agreement with experimental evidence.

Finally, with F_2 (which contains two more electrons than O_2), we find that the two π^* antibonding orbitals are filled (Figure 6.33*d*). This leaves one net single bond, and once again the valence bond and molecular orbital theories give the same result.

The success of the molecular orbital theory is not restricted merely to diatomic molecules. In more complex molecules, however, the energy level diagrams are much more difficult to predict, so we will not attempt to extend the theory any further in this manner.

Delocalized molecular orbitals

A useful concept in molecular orbital theory is the idea that molecular orbitals (and therefore, chemical bonds) can extend over more than two nuclei. It is this aspect in particular that sets molecular orbital theory apart from the valence bond theory, in which only two different atoms are permitted to share a pair of electrons. It also permits the molecular orbital theory to avoid the concept of resonance.

One of the most important examples of this is the benzene molecule, C_6H_6, which consists of six carbon atoms arranged in a ring as shown in the skeletal structure below.

Valence bond theory treats this molecule as a resonance hybrid of the two structures

which are usually drawn as

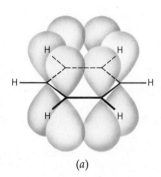

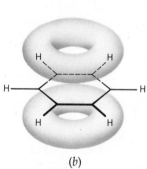

(a)

(b)

Figure 6.34

(a) *At each carbon in the benzene ring there is a p orbital that is perpendicular to the plane of the ring.* (b) *Overlap of the unhybridized p orbitals produces a donut-shaped molecular orbital. The ring is sandwiched between the top and bottom halves of the molecular orbital.*

It is understood that at each apex of the hexagon, there is a carbon that is also bonded to a hydrogen.

If we examine any one carbon atom in the benzene ring, we see that it is bonded to three atoms: two carbons and a hydrogen. This means that it uses sp^2 hybrid orbitals, and it also means that at each carbon atom in the ring there is an unhybridized p orbital perpendicular to the plane of the ring. This is illustrated in Figure 6.34a. According to molecular orbital theory, each of these p orbitals overlaps with *two* other p orbitals—one on each of the two neighboring carbon atoms. This gives one large circular π-type bond that extends around the six-member ring, with part of the bond below the plane of the ring and the other part above the plane of the ring as illustrated in Figure 6.34*b*. In effect, the ring is sandwiched between two donut-shaped electron clouds.

A molecular orbital that extends over three or more nuclei, such as the large π electron cloud in the benzene molecule, is said to be a **delocalized** molecular orbital, meaning that the electron pair is not *localized* between two atoms. By permitting delocalized bonds, molecular orbital theory is able to provide a single electronic picture of the bonding, even in molecules like benzene or SO_2. As a result, there is no need in molecular orbital theory to resort to the somewhat clumsy concept of resonance.

While we are on the topic of benzene, you might notice how similar its structure is to the basic repeating unit in the graphite structure, shown on page 211. In fact, in graphite there is a large delocalized π electron cloud that extends across the top and bottom of each layer of carbon atoms. Because these electrons are not localized between individual atoms, they can be caused to flow fairly easily across the entire layer. This ability of the electrons to move through the delocalized electron cloud is what is responsible for the relatively high electrical conductivity of graphite. An electron can be pumped into one end of the electron cloud as another is pulled out of the other end.

REVIEW QUESTIONS AND PROBLEMS

(Problems whose numbers are in blue have their answers in Appendix C at the back of the book.)

Molecular Shapes

6.1 Sketch the shapes of the five common molecular shapes. (If necessary, refer to the drawings in Section 6.1.)

6.2 What is the bond angle in a linear molecule?

6.3 What are the bond angles in planar triangular, tetrahedral, and octahedral molecules?

6.4 What are the bond angles in a trigonal bipyramidal molecule?

6.5 How many atoms are bonded to the central atom in a molecule that has
(a) a tetrahedral shape?
(b) an octahedral shape?
(c) a trigonal bipyramidal shape?

VSEPR Theory

6.6 What is the basic postulate of the valence shell electron-pair repulsion theory?

6.7 Use the valence shell electron-pair repulsion theory to predict the geometry of each of the following (in each case, the central atom is written first).
(a) NF_3
(b) PH_4^+
(c) CO_3^{2-}
(d) NO_2^-
(e) $SeCl_4$
(f) ICl_2^-
(g) BrF_5
(h) CCl_4
(i) $AlCl_6^{3-}$
(j) $SbCl_5$
(k) $SnCl_2$

6.8 For each of the following, use the valence shell electron-pair repulsion theory to predict (1) the arrangement of electron-pair groups around the central atom (written first in the formula) and (2) the molecular shape.
(a) CH_2O
(b) ClO_3^-
(c) SiF_6^{2-}
(d) PCl_3
(e) $SOCl_2$
(f) $SnCl_4$
(g) SO_3
(h) BF_4^-
(i) OF_2
(j) XeF_4

6.9 Use the valence shell electron-pair repulsion theory to predict for each of the following: (1) the geometric arrangement of electron pairs around the central atom (written first in the formulas), (2) the molecular shape.
(a) ClO_2^-
(b) SCl_2
(c) $SbCl_6^-$
(d) PCl_4^+
(e) IF_4^-
(f) PO_4^{3-}
(g) CH_3^+
(h) ICl_3
(i) NO_3^-
(j) AsH_3
(k) $POCl_3$

6.10 Describe the changes in molecular geometry that take place during the following reactions:
(a) $BF_3 + F^- \rightarrow BF_4^-$
(b) $PCl_5 + Cl^- \rightarrow PCl_6^-$
(c) $ICl_3 + Cl^- \rightarrow ICl_4^-$
(d) $SF_2 + F_2 \rightarrow SF_4$
(e) $C_2H_2 + H_2 \rightarrow C_2H_4$
(f) $Cl-\overset{O}{\underset{\|}{C}}-Cl \longrightarrow COCl^+ + Cl^-$

6.11 On p. 187, it was suggested that lone electron pairs tend to repel rather strongly electron pairs between bonded atoms; in other words, lone-pair/bond-pair repulsions are greater than bond-pair/bond-pair repulsions. This will lead to structural distortions of some of the idealized molecular geometries pictured in Figures 6.2, 6.3, 6.4, and 6.5. Predict the nature of these distortions and sketch the shapes of the resulting molecules.

Polarity of Molecules

6.12 Why are some covalent bonds polar? What is a dipole?

6.13 Why can a molecule that has polar bonds be a non-polar molecule?

6.14 SO_2 is polar, but SO_3 is nonpolar. Why?

6.15 Which of the following are polar molecules? (The central atom is written first in each of them.)
(a) SCl_2 (d) $POCl_3$ (g) SF_4 (j) BrF_5
(b) BF_3 (e) PCl_5 (h) XeO_3
(c) ICl_3 (f) XeF_2 (i) $SnCl_4$

6.16 A certain molecule with the formula AX_3 is nonpolar, even though A and X differ substantially in electronegativity. What is the likely geometry of the AX_3 molecule?

Valence Bond Theory and Hybrid Orbitals

6.17 What are the basic postulates of the valence bond theory?

6.18 What is meant by orbital overlap? What is its importance in covalent bond formation?

6.19 Use valence bond theory to explain the bonding in the Cl_2 molecule. Make a sketch that shows the orbital overlap involved in the formation of the Cl—Cl bond. What happens to the spins of the electrons as the Cl—Cl bond is formed?

6.20 Use valence bond theory to explain the bonding in the HCl molecule. Use a sketch to show the overlap of orbitals that leads to the formation of the bond.

6.21 Based on previous discussions in Chapter 5, describe what happens to the potential energy of the H and Cl atoms as they come together to form the H—Cl bond. Sketch a graph showing how the potential energy varies with internuclear distance. In HCl, the average bond length is 127 pm, and the bond energy is 431 kJ mol^{-1}. Indicate these values on your sketch.

6.22 What is a hybrid orbital? How is the shape of a hybrid orbital different than the shape of a "pure" atomic orbital? Why do atoms usually prefer to use hybrid orbitals in bond formation?

6.23 What are the geometries of the following kinds of hybrid orbitals?
(a) sp (b) sp^2 (c) sp^3 (d) sp^3d (e) sp^3d^2

6.24 What angles exist between the orbitals in
(a) sp^3 hybrids
(b) sp^2 hybrids
(c) sp hybrids
(d) sp^3d^2 hybrids

6.25 Why is it necessary to suggest carbon's use of hybrid orbitals in attempting to explain the structure of CH_4?

6.26 What evidence suggests that oxygen and nitrogen use sp^3 hybrid orbitals for bonding in H_2O and NH_3?

6.27 In arsine, AsH_3, the H—As—H bond angles are very close to 90°. Based on this information, explain the bonding in AsH_3 according to the valence bond theory.

6.28 The H—Se—H bond angle in H_2Se is 91°. According to valence bond theory, what kinds of orbitals does Se use for bonding in this molecule?

6.29 In PF_3, the F—P—F bond angle is approximately 104°. Why is it more reasonable to suggest that phosphorus uses sp^3 hybrids for bonding, rather than pure p orbitals? Diagram the outer-shell electron configuration of P and F. Show how hybridization takes place and how the bonds are formed.

6.30 In PF_3, is there any way to tell *for sure* whether fluorine uses a pure p orbital for bonding or some kind of hybrid orbital? Explain your answer.

6.31 Describe the bonding in the molecule $SnCl_2$. (Begin by using the VSEPR theory to predict its shape.)

6.32 Use the VSEPR theory to predict the shapes of each of the following. Then, use the orbital diagrams for the central atom to explain the bonding in each species.
(a) BCl_3 (d) $AlCl_6^{3-}$ (g) PCl_3
(b) NH_4^+ (e) $BeCl_2$ (h) TeF_4
(c) PCl_5 (f) $SbCl_6^-$ (i) ClO_4^-

6.33 Use your answers to Question 6.7 to suggest the types of hybrid orbitals that would be used in valence bond theory to account for these geometries.

6.34 What kind of hybrid orbitals would be used by the central atom in each species in Question 6.8?

6.35 Diagram the bonding in $SbCl_5$. What kind of hybrid orbitals are involved in the bonding?

6.36 It is possible for SiF_4 to react with F^- to give SiF_6^{2-} but it is not possible for CF_4 to form CF_6^{2-}. Why?

6.37 We have discussed the reaction, $BCl_3 + NH_3 \rightarrow Cl_3BNH_3$ earlier (Chapter 5). What kind of hybrid orbitals are used by B and N before and after reaction? How does the geometry change about B and N as the reaction occurs?

6.38 Which of the species in Question 6.32 have one or more bonds that would be considered to have been formed by way of coordinate covalent bonding? How, if at all, does a coordinate covalent bond differ from a normal covalent bond once it has been formed?

6.39 $SnCl_4$ is a volatile liquid composed of individual $SnCl_4$ molecules. Describe the bonding that is expected in this molecule.

6.40 Hybrid orbitals are not symmetrical about the nucleus. They concentrate electron density on the side of the nucleus where the orbital is large. Lone electron pairs in hybrid orbitals therefore are expected to contribute to the dipole moment of the molecule. It is observed experimentally that NF_3 is nearly a nonpolar molecule; NH_3 is very polar. The electronegativity difference between N and F is nearly the same as that between N and H. How does this support the view that in both NF_3 and NH_3 the nitrogen uses sp^3 hybrid orbitals?

Multiple Bonds and Resonance

6.41 Describe a σ bond; a π bond. What constitutes a double bond? A triple bond?

6.42 What kind of hybrid orbitals are used by each atom in the molecule below? What kinds of bonds (σ, π) occur between the atoms?

$$H-C\equiv C-C=C-C-\ddot{O}-C-H$$

6.43 Describe the bonding in the N_2 molecule according to the valence bond theory.

6.44 Use valence bond theory to explain the bonding in the cyanide ion, CN^-.

6.45 Make a sketch showing the overlap of orbitals to form the σ and π bonds in the formaldehyde molecule

$$\begin{array}{c}H\\ \diagdown\\ C=\ddot{O}:\\ \diagup\\ H\end{array}$$

6.46 Draw resonance structures for the nitrite ion, NO_2^-. What kind of hybrid orbitals are used by the nitrogen atom in this ion?

Molecular Structures of the Nonmetals

6.47 Why are the period-2 nonmetals able to form much stronger π bonds than the nonmetals of period 3? Why does a period-3 nonmetal prefer to form several σ bonds instead of one σ bond and several π bonds?

6.48 Even though the nonmetals of periods 3, 4, and 5 do not tend to form π bonds, each of the halogens is able to exist as a diatomic molecule (Cl_2, Br_2, I_2). Why?

6.49 What are *allotropes*?

6.50 What are the two allotropes of oxygen?

6.51 Describe the molecular structure of elemental sulfur.

6.52 What is the molecular structure of white phosphorus?

6.53 Why is white phosphorus so chemically reactive?

6.54 Compare the structures of graphite and diamond. What kinds of hybrid orbitals are used by carbon in each of these?

6.55 Silicon only forms a diamondlike molecular structure. Why doesn't it form a structure similar to graphite?

Molecular Orbital Theory

6.56 From the point of view of molecular orbital theory, in what way is the electronic structure of an atom similar to the electronic structure of a molecule? In what way do they differ?

6.57 Sketch the shapes of the σ_{1s} and σ^*_{1s} molecular orbitals. How do their energies compare? Why is the σ^*_{1s} molecular orbital said to be antibonding?

6.58 Sketch the shapes of the σ_{2p_z} and $\sigma^*_{2p_z}$ molecular orbitals. Sketch the shapes of the π_{2p_x} and π_{2p_y} molecular orbitals.

6.59 Describe the bonding in the N_2 molecule according to molecular orbital theory.

6.60 Predict the relative stabilities of the species N_2^+, N_2, and N_2^-. What is the net bond order in each of them? From our earlier discussion in Chapter 5, how would you expect the bond lengths in these species to compare?

6.61 Predict the relative stabilities of the species O_2^+, O_2, and O_2^-. Based on the relationship between bond order and bond properties, which we discussed in Chapter 5, predict the variations of bond lengths and bond vibrational frequencies for these species.

6.62 How would you expect the bond order, bond energy, bond length, and bond vibrational frequency of the peroxide ion, O_2^{2-}, to compare to that of O_2?

6.63 Use Figure 6.33a to draw molecular orbital energy-level diagrams for Li_2, Be_2, B_2, and C_2. Which of these should not exist, which should be paramagnetic?

6.64 What can you predict about the stabilities of the species in Question 6.63 when one electron is (a) removed from each, (b) added to each?

6.65 How does molecular orbital theory avoid the concept of resonance?

6.66 The species H_2^+ and He_2^+ have been observed. Use molecular orbital theory to account for their existence.

6.67 Give the valence bond and molecular orbital descriptions for the following species that can be drawn as two or more resonance structures: (a) SO_2, (b) NO_3^-, and (c)

$$H-C\begin{array}{c}\diagup O\\ \diagdown O^-\end{array}$$

CHAPTER 7

CHEMICAL REACTIONS AND THE PERIODIC TABLE

Rain water made acidic by dissolved carbon dioxide seeps through cracks in limestone deposits to create huge underground caverns such as the Carlsbad Caverns in New Mexico, shown here. The chemical reactions involved in the hollowing-out of the cavern are among the typical acid-base reactions discussed in this chapter.

The real "meat" of chemistry is chemical reactions. It is here that we get to see amazing changes in properties as various substances combine and interact with each other to form new materials. We mentioned this in Chapter 1 when we described the reaction of sodium with chlorine to form sodium chloride. The reactants include a soft, typically electrically conductive metal (sodium) and a pale yellow-green gas with a choking odor (chlorine). Either of these substances taken internally will cause severe medical problems, or even death. Yet when they are allowed to combine, the product of the reaction is a substance our bodies cannot do without, sodium chloride.

You've been introduced to a number of chemical reactions during the last several chapters, but they have been used primarily to illustrate some particular principle—for example, calculations dealing with the stoichiometry of reactions. However, principles such as stoichiometry, the electronic structures of atoms, or chemical bonding are developed to help us *understand* chemical reactions and the properties of substances produced in them. In this chapter, therefore, we will view chemical reactions from a different perspective than before. We will examine chemical reactions for their own sake.

When you consider that there are many millions of chemical compounds, it is clear that there also must be millions of different chemical reactions. Learning about them individually would be hopelessly complex. Therefore, chemists categorize them based on certain similarities, and in this chapter you will encounter several such classes of reactions. You will also see that for a given type of reaction, there are certain trends that are most easily followed by using the periodic table as a guide, which illustrates once again how useful the periodic table can be in the study of chemistry.

7.1 REACTIONS IN SOLUTION

In Chapter 2 we noted that many reactions are carried out in solution simply for practical reasons. When the reactants are dissolved in a solvent, their particles become uniformly distributed throughout the solution and are able to intermingle freely. This permits the reaction to occur as rapidly as possible.

One of the most important solvents for chemical reactions is water. It certainly is a common substance, and it is also a good solvent for many different kinds of chemicals, both molecular and ionic. Much attention has been paid to reactions in aqueous solutions, partly because of the availability of water as a solvent and also the recognition of water as a medium in which biochemical reactions take place. As a result, most of our discussions of reactions in solution will focus on aqueous systems.

Solution terminology

Many of the terms that we use in discussing solutions were introduced to you in Chapter 2. These include the terms *solvent* and *solute*. Recall that the solvent is generally taken to be the substance present in largest proportion, and all the other substances are considered solutes. In solutions that contain water, however, water is nearly always taken to be the solvent, even when it is present in relatively small amounts. For example, a mixture of 96% H_2SO_4 and 4% H_2O by mass is called "concentrated sulfuric acid," which implies that a large quantity of sulfuric acid is *dissolved in* a small amount of water—water is taken to be the solvent and H_2SO_4 is taken to be the solute.

Another set of terms mentioned in Chapter 2 is *concentrated* and *dilute*. A concentrated solution has a *relatively* large proportion of solute to solvent, and a dilute solution has a *relatively* smaller proportion of solute to solvent. The emphasis here is that these are relative terms: A particular solution is considered concentrated when compared to some other solution that has a lower proportion of solute to solvent.

In most cases, there is a limit to the amount of a solute that can dissolve in a given amount of solvent at a particular temperature. For example, if we add sodium chloride to 100 cm^3 of water at 0°C, only 35.7 g of the salt will dissolve, regardless of how much we place into the water. Any excess NaCl will simply settle to the bottom of the container. A solution that contains as much dissolved solute as it can hold while in contact with excess solute is said to be a **saturated solution,** and the quantity of solute needed to give a saturated solution with a given quantity of solvent is called the **solubility** of that particular solute. Thus, the solubility of sodium chloride in water at 0°C is 35.7 g of NaCl per 100 cm^3 of water. Usually a solute's solubility changes with temperature. For example, at 100°C the solubility of NaCl is 39.1 g/100 cm^3 of H$_2$O. This means that we must always specify the temperature when stating the solubility.

If a particular solution contains less solute than is needed for saturation, it is said to be an **unsaturated solution.** An example would be a solution of 20 g of NaCl in 100 cm^3 of H$_2$O at 0°C. An unsaturated solution is capable of dissolving more solute—in this case, an additional 15.7 g of NaCl could be dissolved in each 100 cm^3.

It is important to remember that the terms saturated and unsaturated are not related in any direct way to the terms concentrated and dilute. For example, a saturated solution of silver chloride at room temperature contains only 0.000 089 g AgCl/100 cm^3 of water and would certainly be considered dilute. On the other hand, it would take about 500 g of lithium chlorate, LiClO$_3$, to form a saturated solution in 100 cm^3 of water at this same temperature. A solution containing 400 g of LiClO$_3$ generally would be considered concentrated, even though it is unsaturated.

Finally, there are some substances that frequently form **supersaturated solutions,** solutions that contain more solute than ordinarily required for saturation. Sodium acetate, NaC$_2$H$_3$O$_2$, is an example. At 0°C, this compound is soluble in water to the extent of 119 g/100 cm^3, but its solubility increases greatly with increasing temperature. If a hot unsaturated solution that contains more than 119 g of NaC$_2$H$_3$O$_2$ per 100 cm^3 is cooled to 0°C, the excess solute should separate from the solution and settle to the bottom, but usually it does not. The excess solute remains in the solution, and the solution is supersaturated. Supersaturated solutions are unstable; if a tiny crystal of NaC$_2$H$_3$O$_2$ is added, additional solute crystallizes on this "seed" crystal until the concentration of the solution drops to the point of saturation (see Figure 7.1).

If a particular supersaturated solution contained 150 g NaC$_2$H$_3$O$_2$ per 100 cm^3, 31 g of the solute would crystallize from each 100 cm^3 when a tiny seed crystal is added.

Figure 7.1

Supersaturation. (a) Supersaturated solution. (b) Introduction of a seed crystal. (c) Excess solute crystallizes on the seed.

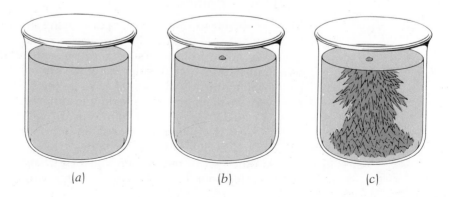

(a)   (b)   (c)

Electrolytes

Earlier, we noted that water is generally a good solvent for ionic compounds. Solutions of these substances possess some unusual properties, one of which is that they conduct electricity. This can be demonstrated with a conductivity apparatus like that shown in Figure 7.2. If the electrodes are immersed in pure water, the bulb doesn't light because water is a very poor conductor of electricity. However, if a typical ionic solid such as sodium chloride is added to the water, the bulb will burn brightly as soon as the solid begins to dissolve. Compounds such as NaCl, which give electrically conducting solutions, are called **electrolytes.**

How do we explain the electrical conductivity of solutions of ionic compounds? In the solid state, these substances are composed of positive and negative ions that are held in place in a rigid framework by electrostatic forces. When they are dissolved in water, however, these solids break apart, or **dissociate,** to yield ions that are more or less free to roam about in the solution. It is the presence of these freely moving ions that imparts to the solution the ability to conduct electricity.

Arrhenius received the 1903 Nobel Prize in chemistry for his theory of electrolytes.

This explanation of electrolytes was first proposed by Svante Arrhenius, one of Sweden's most famous chemists. It's interesting to note that he was nearly denied his doctoral degree in 1884 at the university in Upsala, Sweden for making such a revolutionary suggestion. Nevertheless, his theory has withstood the test of time because it explains so successfully the unusual properties of salt solutions.

When an ionic compound dissolves in water, the ions become surrounded by water molecules and are said to be **hydrated,** which we indicate by writing (*aq*) after their formulas. For example, the dissociation of sodium chloride that occurs when the solid is dissolved in water can be represented by the equation

$$NaCl(s) \longrightarrow Na^+(aq) + Cl^-(aq)$$

Often, for simplicity, we will leave off the labels (*s*) and (*aq*) if doing so creates no confusion.

The production of ions in solutions is not limited to ionic compounds. There are many molecular substances that react with water to produce ions and there-

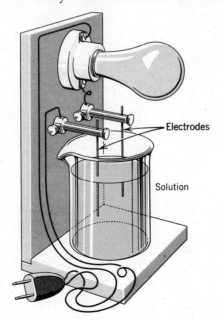

Figure 7.2

The conductivity apparatus described in Chapter 5 also can be used to test the electrical conductivity of solutions, as illustrated here.

fore qualify as electrolytes. Hydrogen chloride is a typical example. When HCl gas is dissolved in water, the following reaction takes place:

$$HCl(g) + H_2O \longrightarrow H_3O^+(aq) + Cl^-(aq)$$

This reaction is properly called an **ionization reaction** because it produces ions where there were none before. (However, it is often referred to as dissociation, simply to avoid having to use different terms for ionic and molecular electrolytes.) The reaction occurs by the transfer of a proton, or hydrogen ion (H^+), from the HCl molecule to the water molecule to produce the **hydronium ion,** H_3O^+, and a chloride ion, Cl^-.

The name proton is frequently used to mean hydrogen ion. The removal of hydrogen's single electron gives just a bare proton.

$$H\!:\!\ddot{C}\!l\!: + :\!\ddot{O}\!:\!H \longrightarrow \left[H\!:\!\ddot{O}\!:\!H\right]^+ + \left[:\!\ddot{C}\!l\!:\right]^-$$
$$H H$$

Thus, even though hydrogen chloride exists by itself as discrete molecules of HCl (liquid HCl doesn't conduct electricity), when it is dissolved in water, it produces ions and becomes an electrolyte.

As we will see, the hydronium ion is one of the most important species to consider in discussions of chemical reactions in aqueous solutions. It is useful to think of it as a proton that has associated itself with a water molecule. We can do this because when the hydronium ion reacts, it does so by giving up the proton, which leaves the water molecule as one of the products. In a sense, the H_2O of the hydronium ion just serves as a carrier for the H^+ ion. For this reason, the hydronium ion is often written simply as H^+, and we often speak of H_3O^+ ion as hydrogen ion. By leaving out the H_2O of the H_3O^+ ion, we can write the dissociation of HCl as

H^+ is the active ingredient in H_3O^+.

$$HCl(aq) \longrightarrow H^+(aq) + Cl^-(aq)$$

Even though we write H^+, *always* remember that there is at least one, and probably several more, H_2O molecules associated with the proton in aqueous solutions.[1]

EXAMPLE 7.1 **WRITING EQUATIONS FOR IONIZATION REACTIONS**

PROBLEM Hydrogen bromide, HBr, is an electrolyte in water and gives hydrogen ion and bromide ion in solution. Write an equation for the ionization reaction.

SOLUTION The reaction is between HBr molecules and water molecules.

$$HBr(aq) + H_2O \longrightarrow H_3O^+(aq) + Br^-(aq)$$

If we write the hydronium ion in its abbreviated form as H^+, the equation can be written

$$HBr(aq) \longrightarrow H^+(aq) + Br^-(aq)$$

Strong and weak electrolytes—chemical equilibrium

The two examples of electrolytes discussed above, NaCl and HCl, are essentially completely dissociated in aqueous solution; 1 mol of NaCl gives 1 mol of Na^+ ions and 1 mol of Cl^- ions, and 1 mol of HCl gives 1 mol of H_3O^+ ions and

---

[1] There is, in fact, evidence that suggests that the H^+ ion may exist as $H_9O_4^+$, that is, $H_3O(H_2O)_3^+$, in aqueous solution.

In general, ionic compounds are strong electrolytes: They are essentially 100% dissociated in water.

1 mol of Cl^- ions. Substances such as NaCl and HCl, which are completely dissociated in solution, are called **strong electrolytes.**

There are also many molecular substances that have no tendency at all to undergo ionization when dissolved in water. Sugar and ethyl alcohol are two relatively common examples. When these compounds are dissolved in water, their molecules just mix with the water molecules to give a uniform solution, but the solution does not contain any ions as a result of the reaction of the solute with H_2O. Since these solutes do not produce ions in the solution, the solution doesn't conduct electricity and these solutes are called **nonelectrolytes.**

Between the extremes of strong electrolytes and nonelectrolytes exists a large collection of compounds called **weak electrolytes.** These compounds produce aqueous solutions that conduct electricity, but only very weakly. An example is acetic acid, $HC_2H_3O_2$, which is the substance that gives vinegar its sour taste. If the electrodes of the conductivity apparatus are dipped into a solution of this solute, the bulb glows dimly.

In a solution of acetic acid, only a small fraction of all the acetic acid molecules are present as ions, produced by the reaction

$$HC_2H_3O_2(aq) + H_2O \longrightarrow H_3O^+(aq) + C_2H_3O_2^-(aq)$$

For example, in a 1.0 M solution of $HC_2H_3O_2$, only about 0.42% of the solute has undergone this reaction. The rest of the acetic acid exists as uncharged molecules.

Some additional examples of weak electrolytes are found in Table 7.1. Notice that water itself is included in this table because it is actually a very weak electrolyte by virtue of the reaction

$$H_2O + H_2O \rightleftharpoons H_3O^+ + OH^-$$

This slight ionization of water plays a very important role in many chemical reactions in which water is a solvent. Special attention is given to this topic in Chapter 15.

The reason for the limited degree of dissociation of weak electrolytes is worth discussing at this point because it illustrates one of the most important concepts in chemistry, one to which we will devote three chapters at a later time (Chapters 14, 15, and 16).

In a solution of acetic acid, molecules of $HC_2H_3O_2$ are constantly colliding with molecules of water and, in each encounter, there is a certain probability that a proton will be transferred from an $HC_2H_3O_2$ molecule to a water molecule to yield H_3O^+ and $C_2H_3O_2^-$ ions. There are also encounters, in this solution, between acetate ions and hydronium ions. When these ions meet, there is a high probability that an H_3O^+ ion will lose a proton to a $C_2H_3O_2^-$ ion to reform $HC_2H_3O_2$ and H_2O molecules. Thus, in this solution we have two reactions occurring simultaneously:

$$HC_2H_3O_2 + H_2O \longrightarrow H_3O^+ + C_2H_3O_2^- \tag{1}$$

and

$$H_3O^+ + C_2H_3O_2^- \longrightarrow HC_2H_3O_2 + H_2O \tag{2}$$

Dynamic implies activity. The opposing reactions don't cease at equilibrium; they occur at equal rates.

When the rate at which the ions are formed by reaction 1 is equal to the rate at which they disappear by reaction 2, their concentrations in the solution will no longer change with time. In fact, the concentrations of all the species will remain constant from this point on, even though if we followed any particular $C_2H_3O_2$ unit in the solution, it would sometimes exist as a $C_2H_3O_2^-$ ion and at other times as a $HC_2H_3O_2$ molecule. Such a state of affairs is called **equilibrium.** It is said to be a **dynamic equilibrium** because things are continually happening in

Table 7.1
Some weak electrolytes

Substance	Dissociation Reaction	Percent Dissociation of the Solute in a 1.00 M Solution
Water	$H_2O + H_2O \rightarrow H_3O^+ + OH^-$	1.8×10^{-7} (55.5 mol H_2O per dm^3)
Acetic acid	$HC_2H_3O_2 + H_2O \rightarrow H_3O^+ + C_2H_3O_2^-$	0.42
Ammonia	$NH_3 + H_2O \rightarrow NH_4^+ + OH^-$	0.42
Hydrogen cyanide	$HCN + H_2O \rightarrow H_3O^+ + CN^-$	2.0×10^{-3}

the solution—two reactions are taking place: ions reacting to yield molecules and molecules reacting to produce ions.

To indicate chemical equilibrium in a reacting system, we use a set of double arrows, $\rightleftharpoons$, in the chemical equation. Thus the equilibrium that we have been discussing is expressed as

$$HC_2H_3O_2 + H_2O \rightleftharpoons H_3O^+ + C_2H_3O_2^-$$

The use of this notation implies that the *forward reaction* (the reaction going from left to right) is occurring at the same rate as the *reverse reaction* (the reaction going from right to left). In a solution of acetic acid, these rates just happen to become equal when only a small fraction of all the acetic acid molecules has undergone ionization. We say that the extent of ionization is small and that the **position of equilibrium**—the relative proportions of the reactants and products—*lies to the left*, in the direction of the molecular form of the solute. In other words, almost everything is present in nonionized form.

For strong electrolytes, such as HCl, the reaction of the ions to form neutral molecules has nearly no tendency to occur. When an H_3O^+ ion encounters a Cl^- ion in the solution, nothing happens to them. Therefore, when HCl is dissolved in water, only the forward reaction takes place, and in just an instant all the HCl is converted to ions. The solute has become 100% ionized. When we write the equation for the reaction of a strong electrolyte with water, we omit the reverse arrow because the reverse reaction doesn't occur. Therefore, for HCl we write

$$HCl(aq) + H_2O \longrightarrow H_3O^+(aq) + Cl^-(aq)$$

The concept of a dynamic equilibrium is very important. All processes, both chemical and physical, tend to move toward a state of equilibrium, and we will use this concept many times in later chapters to analyze physical changes as well as chemical reactions.

The specific chemical properties of a substance determine the relative tendencies to occur of the forward and reverse reactions. Some factors that influence these properties are discussed later in the chapter.

EXAMPLE 7.2 | **WRITING EQUATIONS FOR THE IONIZATION OF WEAK ELECTROLYTES**

PROBLEM | Formic acid, $HCHO_2$, is a weak electrolyte. Write an equation for its ionization in water.

SOLUTION | This compound ionizes just like acetic acid, and because it is a weak electrolyte, the ionization is an equilibrium. The equation is therefore

$$HCHO_2(aq) + H_2O \rightleftharpoons H_3O^+(aq) + CHO_2^-(aq)$$

We can also write the equation as follows if we use the abbreviated form of the hydronium ion:

$$HCHO_2(aq) \rightleftharpoons H^+(aq) + CHO_2^-(aq)$$

Ionic reactions

Many of the chemical reactions that you carry out in the laboratory portion of your chemistry course involve reactions between ions in solution. In fact, anyone who uses water as a solvent eventually encounters such reactions.

A typical example of an ionic reaction is the one that occurs when solutions of sodium chloride and silver nitrate are mixed, which is shown in Figure 7.3. As one solution is added to the other, a white solid, silver chloride, is formed. If the sodium chloride solution contains 1 mol of NaCl and the silver nitrate solution contains 1 mol of $AgNO_3$, then 1 mol of AgCl is formed and the solution contains 1 mol of dissolved $NaNO_3$. If we wished, we could separate the silver chloride from the solution by filtering the mixture as shown in Figure 7.4. When the clear solution passing through the filter—the **filtrate**—is evaporated, crystalline sodium nitrate is left behind.

The chemical equation that represents the change that has occurred during this reaction is

$$NaCl(aq) + AgNO_3(aq) \longrightarrow AgCl(s) + NaNO_3(aq)$$

This kind of reaction, in which cations and anions have changed partners, is known as **metathesis,** or **double replacement** (Cl^- has replaced NO_3^- or Na^+ has replaced Ag^+). The equation we have written to describe the reaction is called a **molecular equation,** because all the reactants and products are written

Figure 7.3

(a) *Two solutions, one containing silver nitrate and the other* *sodium chloride.* (b) *As one solution is added to the other, a white solid composed of silver chloride, AgCl, is formed.*

Figure 7.4
Separating a precipitate from a solution by filtration. (a) The mixture is guided into the filter by pouring it down a glass rod. The beaker below catches the filtrate that passes through the filter. (b) The last traces of the precipitate are rinsed from the beaker by a stream of water from a wash bottle.

NaCl exists as ions in both the solid and its aqueous solutions.

as if they were molecules. A more accurate representation of the reaction as it actually occurs in solution is given by the **ionic equation.** We have seen that a solution of NaCl does not contain molecules but, instead, consists of Na^+ and Cl^- ions dispersed throughout the solvent. Similarly, a silver nitrate solution contains Ag^+ and NO_3^- ions. When these two solutions are mixed, solid AgCl is formed by a combination of the Ag^+ and Cl^- ions. We refer to such a solid formed in a solution as the result of a chemical reaction as a **precipitate.** The solution of sodium nitrate that remains contains Na^+ and NO_3^- ions and we can write the ionic equation as

$$Na^+(aq) + Cl^-(aq) + Ag^+(aq) + NO_3^-(aq) \longrightarrow AgCl(s) + Na^+(aq) + NO_3^-(aq)$$

In this equation we have shown all the soluble ionic substances as being dissociated in solution. The formula for silver chloride is written in molecular form because its ions are no longer separate—they are together in the solid.

If we examine the ionic equation we have just written, we see that Na^+ and NO_3^- do not actually undergo any change during the course of the reaction. The same Na^+ and NO_3^- ions are present after the chemical reaction as before and they have, in a sense, just "gone along for the ride." For this reason, ions that do not change during reaction are frequently called **spectator ions.** Since they do not take part in the reaction, we may eliminate them from the equation to arrive at the **net ionic equation,** that is, the equation for the net change that has taken place:

$$Ag^+(aq) + Cl^-(aq) \longrightarrow AgCl(s)$$

This net ionic equation is useful in more than one way. First, it focuses our attention on the species that participate in the important changes occurring in the solution. Second, it tells us any substance that produces Ag^+ ions in solution will react with any other substance that gives Cl^- ions in solution to yield a precipitate of AgCl. For example, we would predict that a precipitate of AgCl would also form if we mixed solutions of potassium chloride and silver fluoride. (AgF is soluble in water even though AgCl is not.[2]) This is, in fact, precisely what takes place. The molecular, ionic, and net ionic equations for this reaction are as follows:

$$AgF(aq) + KCl(aq) \longrightarrow AgCl(s) + KF(aq) \qquad \text{(Molecular)}$$

$$Ag^+(aq) + F^-(aq) + K^+(aq) + Cl^-(aq) \longrightarrow AgCl(s) + K^+(aq) + F^-(aq) \qquad \text{(Ionic)}$$

$$Ag^+(aq) + Cl^-(aq) \longrightarrow AgCl(s) \qquad \text{(Net Ionic)}$$

Each of these kinds of equations is useful in its own way; none is the best way of representing the reaction. The form that we use in any particular instance depends on what aspect of the reaction we wish to focus our attention. If we want to think about the net chemical change that occurs, the net ionic equation is best. However, if we want to work with these chemicals in carefully measured amounts, it is the stoichiometry of the molecular equation that is important.

EXAMPLE 7.3 WRITING IONIC AND NET IONIC EQUATIONS FOR METATHESIS REACTIONS

PROBLEM   If a solution of sodium hydroxide is added to a solution of iron(III) chloride, a precipitate of iron(III) hydroxide is formed. After the reaction, the solution contains dissolved sodium chloride. Write a molecular equation for the metathesis reaction, and then write the appropriate ionic and net ionic equations for the reaction.

SOLUTION   The first step is to write the correct formulas of the reactants and products. (You should be able to do this based on your study of nomenclature.) Sodium hydroxide is NaOH; iron(III) chloride is $FeCl_3$; iron(III) hydroxide is $Fe(OH)_3$; and sodium chloride is NaCl. The balanced molecular equation is

$$3NaOH(aq) + FeCl_3(aq) \longrightarrow Fe(OH)_3(s) + 3NaCl(aq)$$

To write the ionic equation, we "dissociate" all the soluble compounds and write the formula of the insoluble $Fe(OH)_3$ in molecular form. This gives

$$3Na^+ + 3OH^- + Fe^{3+} + 3Cl^- \longrightarrow Fe(OH)_3(s) + 3Na^+ + 3Cl^-$$

(For simplicity, we've omitted the (aq) after the formulas of the ions in solution. However, we've kept (s) after the formula for the iron(III) hydroxide to call attention to the fact that it is a precipitate.)

To obtain the net ionic equation, we drop any spectator ions from the ionic equation.

[2] The solubility of AgCl is very low, 0.000 089 g/100 cm³, and AgCl can be considered, for most purposes, to be insoluble; that is, the amount of AgCl in solution can generally be considered negligible. For comparison, the solubility of AgF in water is approximately 185 g/100 cm³ at room temperature.

Notice that there are $3Na^+$ on both sides and also $3Cl^-$ on both sides. If we omit these, we get the net ionic equation.

$$3OH^- + Fe^{3+} \longrightarrow Fe(OH)_3(s)$$

Usually, we prefer to write the cation first. This equation would normally be written

$$Fe^{3+} + 3OH^- \longrightarrow Fe(OH)_3(s)$$

7.2 ACIDS AND BASES IN AQUEOUS SOLUTIONS

Among the various substances that Arrhenius recognized as electrolytes are compounds that we call acids and bases. These include some of our most common and important chemicals. For example, the sour taste of vinegar and lemon juice results from acids. They contain acetic acid and citric acid, respectively. We all need ascorbic acid in our diets; it's also called vitamin C. Sulfuric acid, the acid used as the electrolyte in the fluid in automobile batteries, ranks well above any other industrial chemical in production volume each year.

Among important bases we find household ammonia. Sodium hydroxide, which is also a base, goes by the common name lye. It is available on supermarket shelves and is found in oven cleaners and drain cleaner Drano. Even sodium bicarbonate and milk of magnesia, which we take to soothe an upset (acid) stomach, are bases.

Acids and bases have certain properties that help us identify them. For example, solutions of acid have a characteristic sour taste. That's why lemon juice and vinegar are sour. On the other hand, bases such as the magnesium hydroxide in milk of magnesia have a bitter taste. (*Caution:* Even though these are general properties of acids and bases, you should *never* test for acids and bases by tasting a chemical in the laboratory. It might not be healthy!) Another property of acids and bases is their effect on **indicators,** chemicals whose colors depend on the acidity or basicity of their solutions. A typical example is the dye called litmus. Litmus is a chemical that has a blue color in a basic solution and a pink color in an acidic solution. It is common to have vials of litmus paper—thin strips of paper impregnated with litmus—available in the lab to test the acidity or basicity of solutions. If pink litmus paper is dipped into a solution that is basic, it turns blue, and if blue litmus paper is dipped into a solution that is acidic, it turns pink.

Stated in modern terms, the Arrhenius definitions of acids and bases are as follows: *An **acid** is a substance that increases the concentration of hydronium ion, H_3O^+, in an aqueous solution, and a **base** is a substance that increases the concentration of hydroxide ion, OH^-.* Let's examine more closely some examples of each of these kinds of chemicals.

Acids

In general, acids are molecular substances that produce hydronium ion by reaction with water. For example, hydrogen chloride is an acid, because when it is dissolved in water it reacts with the solvent to produce H_3O^+

$$HCl(aq) + H_2O \longrightarrow H_3O^+(aq) + Cl^-(aq)$$

If we use H^+ as an abbreviation for the hydronium ion and leave out the molecule of water that carries the H^+, we can also write this reaction as

$$HCl(aq) \longrightarrow H^+(aq) + Cl^-(aq)$$

Aqueous solutions of hydrogen chloride are called *hydrochloric acid*. As you learned in the last section, HCl is a strong electrolyte; its reaction with water is complete. Because a mole of HCl releases a mole of H_3O^+, hydrochloric acid is a **strong acid.** In other words, any reasonably concentrated solution of HCl contains a large concentration of hydronium ion.

There are also many acids that are weak electrolytes. Acetic acid, $HC_2H_3O_2$, is an example. Recall that this acid reacts with water according to the equation

$$HC_2H_3O_2(aq) + H_2O \rightleftharpoons H_3O^+(aq) + C_2H_3O_2^-(aq)$$

or more simply

$$HC_2H_3O_2(aq) \rightleftharpoons H^+(aq) + C_2H_3O_2^-(aq)$$

This is an equilibrium, and in a solution of $HC_2H_3O_2$ only a small fraction of the solute is dissociated into ions. This means that the concentration of H_3O^+ in the solution is low. As a result, acetic acid and other acids that are weak electrolytes are called **weak acids.**

Notice that in the formula for acetic acid, $HC_2H_3O_2$, hydrogen appears twice—once at the beginning of the formula and once in the middle. In writing the formula for an acid, it is common practice to write the *acidic hydrogens*—those that are able to transfer to water molecules to give H_3O^+ ions—first in the formula. Other hydrogens that are not "acidic" are written later. Even though the acetic acid molecule contains four hydrogen atoms, only one of them is able to react with a water molecule to form H_3O^+.

Both HCl and $HC_2H_3O_2$ are able to furnish only one hydrogen ion (or proton) per molecule of the acid. Such acids are said to be **monoprotic acids.** There are also many acids that are able to furnish more than one proton per molecule of the acid. As a class, they are referred to as **polyprotic acids.** Two examples are sulfuric acid, H_2SO_4 and phosphoric acid, H_3PO_4.

Sulfuric acid is also called a **diprotic acid** because each molecule of it is able to give up two protons. This happens in two distinct steps

$$H_2SO_4 \longrightarrow H^+ + HSO_4^-$$

$$HSO_4^- \rightleftharpoons H^+ + SO_4^{2-}$$

Similarly, phosphoric acid, which is an example of a **triprotic acid,** dissociates in three separate steps.

$$H_3PO_4 \rightleftharpoons H^+ + H_2PO_4^-$$

$$H_2PO_4^- \rightleftharpoons H^+ + HPO_4^{2-}$$

$$HPO_4^{2-} \rightleftharpoons H^+ + PO_4^{3-}$$

Notice that the second step in the dissociation of H_2SO_4 is an equilibrium (only about 10% of the HSO_4^- is actually dissociated). Despite this, sulfuric acid is considered a strong acid because the first dissociation step is complete. Phosphoric acid, which is used in manufacturing fertilizers and detergents and various soft drinks such as colas, is a weak acid because all three of its dissociation steps are equilibria that do not proceed very far toward completion.

All the acids that we have examined so far contain hydrogen in their formulas. However, there are other substances that do not contain hydrogen, yet still produce acidic solutions when dissolved in water. A common example is carbon dioxide. When dissolved in water (in a carbonated beverage, for example), it reacts as follows:

$$CO_2 + H_2O \rightleftharpoons H_2CO_3$$

The compound H_2CO_3 is called carbonic acid; it is a weak diprotic acid that dissociated in two steps, the first of which is

$$H_2CO_3 \rightleftharpoons H^+ + HCO_3^-$$

This particular acid, incidently, is responsible for the large limestone caverns that exist in various parts of the world. Groundwater, made acidic by CO_2 dissolved in it from the atmosphere, trickles through the limestone ($CaCO_3$) with which it reacts.

$$CaCO_3(s) + H_2CO_3(aq) \longrightarrow Ca^{2+}(aq) + 2HCO_3^-(aq)$$

The product of the reaction is *soluble* calcium bicarbonate that flows away in the groundwater, leaving huge caverns behind.

The reaction of CO_2 with water is typical of many of the nonmetal oxides; they react with water to form oxoacids. Another example is sulfur dioxide, an air pollutant released by burning high-sulfur fuels. It reacts with water as follows:

$$SO_2 + H_2O \longrightarrow H_2SO_3$$

This is one of the reactions that is responsible for *acid rain,* an environmental problem that is particularly severe in the northeastern U.S., southern Canada, and heavily industrialized areas of Europe. Rain water, passing through the atmosphere polluted by sulfur dioxide and other oxides, such as SO_3 and NO_2, becomes acidic by reactions of this type.

A nonmetal oxide that is able to react with water to form an oxoacid is called an **acidic anhydride,** when means "acid without water." Table 7.2 contains a list of common acids and acidic anhydrides, and it also shows their reactions with water. Those that are marked with an asterisk are strong acids; you should learn these. It happens that there are very few strong acids, so the list is short. Then, if you come across an acid that isn't on the list of strong ones, you can be fairly sure it is weak.

These reactions were responsible for the limestone cavern shown in the photo at the beginning of this chapter.

Anhydride comes from the Greek anydros meaning water-less.

Bases

There are two principal kinds of bases: ionic hydroxides and molecular substances that react with water to produce OH^-. Sodium hydroxide and calcium hydroxide are typical ionic hydroxides. In the solid state they consist of a metal ion and hydroxide ion, and when they dissolve in water they dissociate.

$$NaOH(s) \longrightarrow Na^+(aq) + OH^-(aq)$$

$$Ca(OH)_2(s) \longrightarrow Ca^{2+}(aq) + 2OH^-(aq)$$

As is typical of ionic compounds when they dissolve in water, this dissociation is complete, so ionic metal hydroxides are **strong bases.**

The most common molecular substance that is a base is ammonia, NH_3. It reacts with water in an equilibrium.

$$NH_3(aq) + H_2O \rightleftharpoons NH_4^+(aq) + OH^-(aq)$$

In this case, a proton is transferred from a water molecule to an ammonia molecule.

Aqueous solutions of ammonia are sometimes called ammonium hydroxide, which suggests the formula NH_4OH. The species NH_4OH does not exist, however.

$$\text{H}\!:\!\overset{\cdot\cdot}{\text{N}}\!:\ +\ \text{(H)}\!:\!\overset{\cdot\cdot}{\underset{\cdot\cdot}{\text{O}}}\!:\ \rightleftharpoons\ \left[\text{H}\!:\!\overset{\text{H}}{\underset{\text{H}}{\text{N}}}\!:\!\text{H}\right]^+ +\ \left[:\!\overset{\cdot\cdot}{\underset{\cdot\cdot}{\text{O}}}\!:\!\text{H}\right]^-$$

This reaction is indicated as an equilibrium because only a small fraction of the NH_3 in the solution is present as NH_4^+ and OH^- at any given instant.

Table 7.2

Some acids and bases

Acids (Those followed by an asterisk are strong electrolytes and are completely ionized in aqueous solution.)

Monoprotic acids:	HF	Hydrofluoric acid	$HClO_3$	Chloric acid*
$HX \rightarrow H^+ + X^-$	HCl	Hydrochloric acid*	$HClO_4$	Perchloric acid*
	HBr	Hydrobromic acid*	HIO_4	Periodic acid*
	HI	Hydriodic acid*	HNO_3	Nitric acid*
	HOCl	Hypochlorous acid	HNO_2	Nitrous acid
	$HClO_2$	Chlorous acid	$HC_2H_3O_2$	Acetic acid

Diprotic acids:	H_2SO_4	[a]Sulfuric acid*	H_2S	Hydrosulfuric acid
$H_2X \rightarrow H^+ + HX^-$	H_2SO_3	Sulfurous acid	H_3PO_3	Phosphorous acid (only
$HX^- \rightarrow H^+ + X^{2-}$	H_2CO_3	Carbonic acid		two hydrogens can be
	$H_2C_2O_4$	Oxalic acid		removed as protons)

Triprotic acids:	H_3PO_4	Phosphoric acid (orthophosphoric acid)
$H_3X \rightarrow H^+ + H_2X^-$		
$H_2X^- \rightarrow H^+ + HX^{2-}$		
$HX^{2-} \rightarrow H^+ + X^{3-}$		

Typical acidic oxides	SO_2	$SO_2 + H_2O \rightarrow H_2SO_3$
(nonmetal oxides)	SO_3	$SO_3 + H_2O \rightarrow H_2SO_4$
	N_2O_3	$N_2O_3 + H_2O \rightarrow 2HNO_2$
	N_2O_5	$N_2O_5 + H_2O \rightarrow 2HNO_3$
	P_4O_6	$P_4O_6 + 6H_2O \rightarrow 4H_3PO_3$
	P_4O_{10}	$P_4O_{10} + 6H_2O \rightarrow 4H_3PO_4$

Bases

Molecular bases	NH_3	[b]Ammonia	($NH_3 + H_2O \rightleftharpoons NH_4^+ + OH^-$)
(Weak bases)	N_2H_4	Hydrazine	($N_2H_4 + H_2O \rightleftharpoons N_2H_5^+ + OH^-$)
	NH_2OH	Hydroxylamine	($NH_2OH + H_2O \rightleftharpoons NH_3OH^+ + OH^-$)

Ionic bases	Metal hydroxides	$M(OH)_n \rightarrow M^{n+} + nOH^-$
(Strong bases)	NaOH	
	$Ca(OH)_2$	

Typical basic oxides	Na_2O	
(metal oxides)	K_2O }	$M_2O + H_2O \rightarrow 2MOH$
	CaO	
	SrO }	$MO + H_2O \rightarrow M(OH)_2$
	BaO	

[a] The second dissociation of H_2SO_4 is that of a weak acid.

[b] Aqueous solution of ammonia are sometimes referred to as ammonium hydroxide. Commercially prepared solutions of NH_3, for example, are labeled in this way.

Ammonia is therefore a weak electrolyte. It is also said to be a **weak base,** because its solutions contain only relatively small concentrations of OH^-. In general, molecular bases are weak bases.

Earlier in this section you learned that nonmetal oxides are acidic anhydrides. Metal oxides have just the opposite property—they are **basic anhydrides.** An example is sodium oxide, Na_2O. This is actually an ionic compound containing Na^+ and O^{2-}. When this compound dissolves in water, the oxide ion reacts to form hydroxide ion.

$$O^{2-} + H_2O \longrightarrow 2OH^-$$

This reaction is complete, so any soluble metal oxide immediately reacts with water to give a solution of its hydroxide.

$$Na_2O(s) + H_2O \longrightarrow 2Na^+(aq) + 2OH^-(aq)$$

Table 7.2 also contains a list of bases and basic anhydrides, and it shows their reactions with water.

Neutralization

The most important reaction that acids and bases undergo is their reaction with each other, a reaction called **neutralization.** In aqueous solutions, the neutralization reaction between a strong acid and a strong base takes the form of the net ionic equation

$$H_3O^+ + OH^- \longrightarrow 2H_2O$$

or using H^+ as an abbreviation for the hydronium ion,

$$H^+ + OH^- \longrightarrow H_2O$$

If we examine the molecular equation for a typical acid-base reaction, the reaction between sodium hydroxide and hydrochloric acid,

$$NaOH + HCl \longrightarrow NaCl + H_2O$$

we can arrive at another useful generalization. *The products of a neutralization reaction in aqueous solution are a salt and water.*

The formation of water in a neutralization reaction has such a strong tendency to occur that acids will react with both insoluble metal hydroxides and oxides. For example, the neutralization of stomach acid, which is HCl, by insoluble magnesium hydroxide in milk of magnesia follows the net ionic equation

The driving force behind this reaction is strong enough to cause insoluble bases to dissolve in acids and insoluble acids to dissolve in bases.

$$Mg(OH)_2(s) + 2H^+(aq) \longrightarrow Mg^{2+}(aq) + H_2O$$

In a somewhat similar reaction, acid is used to remove rust (Fe_2O_3) from the surface of steel prior to its being given a protective coating of tin or zinc. Even though the iron(III) oxide is insoluble, it reacts with the acid.

$$Fe_2O_3(s) + 6H^+(aq) \longrightarrow 2Fe^{3+}(aq) + 3H_2O$$

Acid salts

In the reaction of NaOH with HCl, there is only one possible salt that can be formed. When a base reacts with a polyprotic acid, however, two or more salts may be possible, corresponding to different extents of neutralization of the acid. Consider, for example, the reaction of NaOH with H_2SO_4. If 2 mol NaOH are added to a solution that contains 1 mol H_2SO_4, complete neutralization of the acid takes place and the salt Na_2SO_4 is formed.

$$2NaOH + H_2SO_4 \longrightarrow Na_2SO_4 + 2H_2O$$

On the other hand, if only 1 mol NaOH is added to 1 mol H_2SO_4, the products are 1 mol water and 1 mol $NaHSO_4$.

$$NaOH + H_2SO_4 \longrightarrow NaHSO_4 + H_2O$$

Similar reactions can occur with tripotic acids such as H_3PO_4, and salts such as the following can be obtained:

Na_3PO_4 trisodium phosphate (sodium phosphate)
Na_2HPO_4 disodium hydrogen phosphate
NaH_2PO_4 sodium dihydrogen phosphate

Recall that salts such as these, which are the result of partial neutralization of a polyprotic acid, are called **acid salts.** Acid salts such as $NaHSO_4$ are acidic and can undergo further neutralization. Thus, $NaHSO_4$ is able to react with NaOH.

$$NaHSO_4 + NaOH \longrightarrow Na_2SO_4 + H_2O$$

7.3 BRØNSTED–LOWRY ACIDS AND BASES

Figure 7.5 is a photograph of open bottles of concentrated ammonia and concentrated hydrochloric acid beside each other. Above the bottles is a smoke consisting of microscopic crystals of ammonium chloride, NH_4Cl, formed by the reaction of gaseous HCl with gaseous NH_3. (The HCl escapes from the concentrated hydrochloric acid, and the NH_3 escapes from the concentrated aqueous ammonia solution.) The equation for the reaction is

$$NH_3(g) + HCl(g) \longrightarrow NH_4Cl(s)$$

The product of this reaction, NH_4Cl, is the same salt that is formed if the solutions in the bottles are mixed.

Figure 7.5

The reaction of gaseous hydrogen chloride with gaseous ammonia, each coming from its concentrated aqueous solution, produces a white cloud of ammonium chloride.

Since the reaction of HCl with NH_3 gives the same product regardless of whether or not a solvent is present, it seems reasonable to consider both to be acid-base reactions. However, the Arrhenius definition is inadequate for the job. In the gas phase, there are no hydronium ions or hydroxide ions, so we can't interpret the gaseous reaction as an acid-base reaction in the same way that we analyzed others in the previous section. We obviously need a more general definition of acids and bases.

A somewhat more general approach to acids and bases was proposed independently in 1923 by the Danish chemist J. N. Brønsted and the British chemist T. M. Lowry. They gave the following definitions: *An **acid** is a substance that is able to donate a proton (a hydrogen ion, H^+) to some other substance, and a **base** is a substance that is able to accept a proton from an acid.* Stated more briefly, an acid is a proton donor and a base is a proton acceptor.

Based on this definition, we can now analyze the reaction between gaseous HCl and gaseous NH_3. This reaction can be diagrammed using Lewis structures as follows:

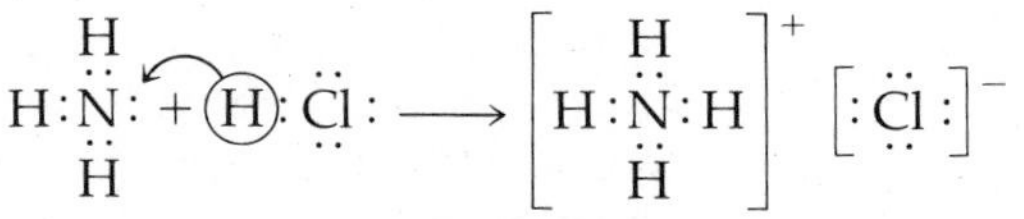

$$H\!:\!\overset{\displaystyle H}{\underset{\displaystyle H}{N}}\!: + \,\overset{\frown}{\text{(H)}}\,:\!\overset{..}{\underset{..}{Cl}}\!: \longrightarrow \left[H\!:\!\overset{\displaystyle H}{\underset{\displaystyle H}{N}}\!:\!H \right]^{+} \; \left[:\!\overset{..}{\underset{..}{Cl}}\!: \right]^{-}$$

Since a proton is transferred from the HCl to the NH_3, this is clearly an acid-base reaction in the Brønsted–Lowry sense. In the reaction, HCl is the acid because it is the proton donor, and NH_3 is the base because it is the proton acceptor. Thus, under both the Arrhenius and Brønsted–Lowry definitions, HCl is an acid and NH_3 is a base.

The reaction between HCl and NH_3 illustrates the more general nature of the Brønsted–Lowry definition of acids and bases. Any reaction in which a proton is transferred from one particle to another is an acid-base reaction; hydronium ion and hydroxide ion do not have to be involved. In fact, acid-base reactions do not even require a solvent.

Conjugate acids and bases

Let us consider a reaction similar to others that we discussed earlier, but this time we will view it in terms of the Brønsted–Lowry definition. Hydrogen fluoride, HF, is a weak acid in water and undergoes ionization that proceeds by the following reaction:

$$HF + H_2O \longrightarrow H_3O^+ + F^-$$

In this reaction, the HF is functioning as an acid because it is donating a proton to the water molecule. Water, on the other hand, is functioning as a *base* because it is accepting a proton from the HF.

Unlike HCl, which is a strong acid, HF is a weak acid and is incompletely ionized in water. Because of this, there is an equilibrium involved in the ionization, which properly should be written

$$HF + H_2O \rightleftharpoons H_3O^+ + F^-$$

Let's examine the reverse reaction—the reaction of the hydronium ion with fluoride ion—more closely.

$$H_3O^+ + F^- \longrightarrow HF + H_2O$$

This reverse reaction is also a Brønsted–Lowry acid-base reaction, with the hydronium ion serving as the acid and the fluoride ion serving as the base. There-

The reason why HCl is a strong acid, while HF is weak, will be discussed in the next section.

fore, in the equilibrium there are two acids and two bases, one of each on either side of the arrow.

$$HF + H_2O \rightleftharpoons H_3O^+ + F^-$$
acid   base   acid   base

When the acid HF reacts, it forms the base F^-. These two substances are related to each other by the loss or gain of a *single* proton and constitute a **conjugate acid-base pair.** We say that F^- is the **conjugate base** of HF, and HF is the **conjugate acid** of F^-. In this equilibrium, we also can identify H_2O and H_3O^+ as a conjugate pair. Water is the conjugate base of H_3O^+, and H_3O^+ is the conjugate acid of H_2O.

An acid differs from its conjugate base by *only one* proton. All the other atoms are the same.

Another example of a Brønsted–Lowry acid-base reaction occurs in aqueous solutions of ammonia.

$$NH_3 + H_2O \rightleftharpoons NH_4^+ + OH^-$$

In this case water serves as an acid by giving up a proton to a molecule of NH_3, which thereby acts as a base. In the reverse reaction, on the other hand, NH_4^+ is the acid and OH^- is the base. Again we have two acid-base conjugate pairs: NH_3 and NH_4^+ is one of them, and H_2O and OH^- is the other.

In general, we can represent any Brønsted–Lowry acid-base reaction as

$$\text{acid } (X) + \text{base } (Y) \rightleftharpoons \text{base } (X) + \text{acid } (Y)$$

where acid (X) and base (X) represent one conjugate pair and acid (Y) and base (Y) the other. Notice that the members of a conjugate pair differ *only by one proton*. They are otherwise the same. Also, within a conjugate pair, the acid has one more hydrogen (actually, H^+) than the base.

EXAMPLE 7.4	WRITING THE FORMULAS FOR CONJUGATE ACIDS AND BASES
PROBLEM	(a) What is the formula for the conjugate base of $N_2H_5^+$?
	(b) What is the formula for the conjugate acid of NH_2^-?
SOLUTION	(a) The conjugate base has *one* less proton than the acid, so the conjugate base of $N_2H_5^+$ must have only 4 hydrogens and it must have *one* less positive charge. Therefore, the conjugate base is N_2H_4.
	(b) We have to add a proton to NH_2^- to get its conjugate acid. The acid will therefore have *one* more hydrogen and *one* more positive charge. Since the charge on the base is -1, adding a positive charge gives a net charge of zero. The conjugate acid is NH_3.

EXAMPLE 7.5	IDENTIFYING CONJUGATE ACIDS AND BASES
PROBLEM	Identify the acid-base conjugate pairs in the reaction
	$$H_2S + (CH_3)_2NH \rightleftharpoons (CH_3)_2NH_2^+ + HS^-$$
	For each pair, which is the acid and which is the base?
SOLUTION	Our first task is to identify the conjugate pairs. We know that the *only* difference between them is the number of hydrogens and the charge; everything else must be the same. Scanning the equation, we see that H_2S and HS^- both have sulfur in them, and the other substances in the equation do not have sulfur. Therefore, H_2S and HS^- must represent one of the conjugate pairs. We also see that the HS^- ion can be formed from H_2S by the loss of a proton, so H_2S is the conjugate acid and HS^- is the conjugate base.

Comparing the other two substances, $(CH_3)_2NH$ and $(CH_3)_2NH_2^+$, we see that they are identical except for the additional H^+ possessed by $(CH_3)_2NH_2^+$. Therefore, $(CH_3)_2NH_2^+$ is the conjugate acid and $(CH_3)_2NH$ is the conjugate base.

$$H_2S + (CH_3)_2NH \rightleftharpoons (CH_3)_2NH_2^+ + HS^-$$

| acid | base | acid | base |

Perhaps you noticed that in the reactions of HF and NH_3 with water, water functions as a base in one instance and as an acid in the other. A substance, such as water, that can serve as either an acid or a base, depending on conditions, is said to be **amphiprotic** or **amphoteric**. Water is not the only substance that behaves this way. For example, water, acetic acid, and ammonia all undergo reactions with themselves to produce ions.

$$H_2O + H_2O \rightleftharpoons H_3O^+ + OH^-$$

$$HC_2H_3O_2 + HC_2H_3O_2 \longrightarrow H_2C_2H_3O_2^+ + C_2H_3O_2^-$$

$$NH_3 + NH_3 \rightleftharpoons NH_4^+ + NH_2^-$$

$$\textbf{(acid) + (base)} \rightleftharpoons \textbf{(acid) + (base)}$$

When a substance reacts with itself to form ions, it is called an **autoionization reaction**. These particular reactions can be diagrammed with Lewis structures that illustrate how one molecule functions as an acid while the other behaves as a base.[3]

In each case, the substance is playing the role of both an acid and a base.

Solutions of metal ions in water

An interesting example of a Brønsted-Lowry acid-base reaction occurs in aqueous solutions that contain metal ions with high positive charges. For example,

---

[3] In general, an autoionization reaction involves the creation of a cation-anion pair from two neutral molecules of the same substance. This occurs by the transfer of an atom and some charge from one particle to another, but the atom that is transferred doesn't have to be a proton. For example, the following is also an autoionization reaction.

$$2PCl_5 \longrightarrow [PCl_4^+][PCl_6^-]$$

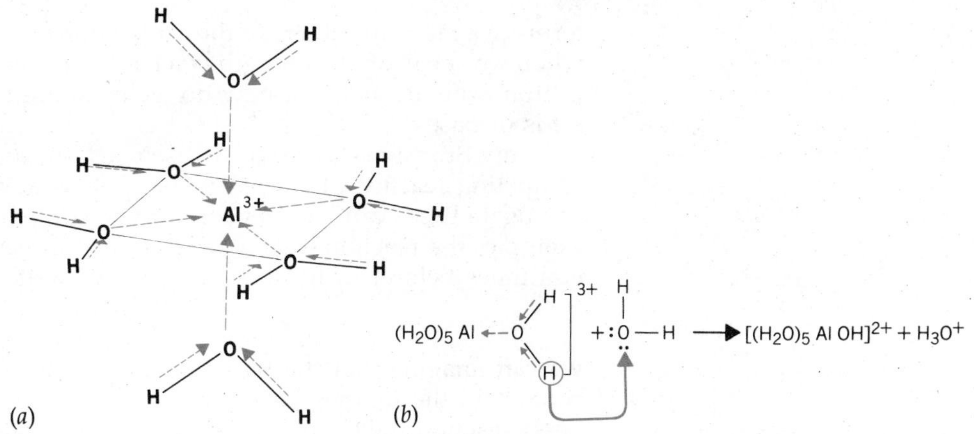

Figure 7.6
(a) *Dotted lines in color indicate how electron density is drawn toward the highly charged Al^{3+} ion and away from the O—H bonds. This increases the polarity of the O—H bonds and makes it easier to remove the hydrogens as H$^+$. (b) A proton is transferred from one of the water molecules surrounding the Al^{3+} ion to a water molecule in the solvent. In this reaction, the Al(H$_2$O)$_6$$^{3+}$ ion is serving as a Brønsted acid.*

solutions of the salt AlCl$_3$ are acidic, as are solutions of salts that contain Cr^{3+} and Fe^{3+}.

When a salt is dissolved in water, it dissociates completely and the positive and negative ions become surrounded by water molecules—they become hydrated. Each negative ion has a cluster of water molecules around it, with the positive ends of the water dipoles pointing at the negative ion. Similarly, each positive ion is surrounded by a number of water molecules arranged so that the negative ends of their dipoles point at the positive ion. If the salt contains a metal ion with a large positive charge, such as the aluminum ion, Al^{3+}, the water molecules that surround the cation are held especially tightly, and the positive ion with its layer of water molecules moves around in the solution as a single unit. In fact, when salts of Al^{3+} are cystallized from aqueous solutions, their crystals contain the octahedral Al(H$_2$O)$_6$$^{3+}$ ion, and it is believed that this ion exists in aqueous solutions of aluminum salts as well.

The acidity of solutions that contain ions such as Al^{3+}, Cr^{3+}, and Fe^{3+} is explained as follows. The high charge on the metal ion draws electron density toward itself and away from the oxygen atoms of the water molecules that surround it. These oxygen atoms, in turn, draw electron density from the O—H bonds (Figure 7.6). This makes the O—H bonds even more polar than they are in an ordinary water molecule. In other words, the hydrogen atoms of the water molecules surrounding the metal ion carry an even greater partial positive charge than the hydrogen atoms in an ordinary water molecule. As a result, a hydrogen is rather easily removed as H$^+$ from an ion such as Al(H$_2$O)$_6$$^{3+}$ and the ion is acidic. Its reaction with water can be viewed as a Brønsted-Lowry acid-base reaction.

$$\underset{\text{acid}}{Al(H_2O)_6^{3+}} + \underset{\text{base}}{H_2O} \rightleftharpoons \underset{\text{base}}{Al(H_2O)_5OH^{2+}} + \underset{\text{acid}}{H_3O^+}$$

7.4   THE STRENGTHS OF ACIDS AND BASES: PERIODIC TRENDS

At various times in this chapter we've mentioned that some acids are strong and others are weak. When we talk of the strength of an acid in Brønsted–Lowry terms, we mean the ability of the acid to donate a proton to another species. And when we speak of the strength of a base, we mean its relative ability to capture a proton from an acid. But how do we compare these abilities among a series of acids or bases?

Any Brønsted–Lowry acid-base reaction can be viewed as two opposing or competing reactions between pairs of acids and bases. In one sense, the two conjugate bases can be considered to be competing for a proton. Consider, for example, the reaction of HCl with water—a reaction that we've discussed several times before. If, for the moment, we write the reaction as an equilibrium

$$HCl + H_2O \rightleftharpoons H_3O^+ + Cl^-$$

we can imagine that the two bases, H_2O and Cl^-, are in competition for the proton. In the forward direction, the H_2O captures the proton, and in the reverse reaction the Cl^- ion captures it. You've already learned that HCl is completely ionized in water, which means that the forward reaction goes essentially all the way toward completion, while the reverse reaction hardly proceeds at all. In equilibrium terms, we say that the position of equilibrium lies far to the right, in favor of the products. This must mean that the H_2O has a far greater affinity for protons than the Cl^- ion, because in the competition the H_2O wins—the water is capable of capturing essentially all the available H^+. We express this relative ability to pick up a proton by stating that water is a stronger base than the chloride ion.

We can also compare the relative strengths of the two acids in this reaction, HCl and H_3O^+. Because the position of equilibrium lies so far to the right, HCl must be far more effective at donating protons than H_3O^+—after all, there are no HCl molecules left in the solution. This means that hydrogen chloride is a stronger acid than hydronium ion.

It is often desirable to compare the strengths of two or more acids relative to the same base. This is what we do when we compare the strengths of HCl and HF in water. In water, HCl is essentially 100% ionized, but in a 1 M solution of HF only about 3% of the HF is ionized.

$$HCl + H_2O \longrightarrow H_3O^+ + Cl^- \quad (100\%)$$

$$HF + H_2O \rightleftharpoons H_3O^+ + F^- \quad \text{(about 3\% in 1 } M \text{ HF)}$$

This information suggests that in water, HCl is a stronger acid than HF, because the HCl is more effective at giving its protons to water molecules. You can also see that the conclusion arrived at here is in agreement with our previous definitions of strong and weak acids: Strong acids are completely ionized in water, and weak acids are incompletely ionized.

Trends in the strengths of acids

The factors that determine acid strengths are relatively simple to understand, and many of them can be related to the compositions and structures of the acids, as well as to the positions of key elements within the periodic table.

Oxoacids An oxoacid such as H_2SO_4 is acidic in water because a hydrogen from one of its O—H bonds is released as a hydrogen ion, H^+. This becomes attached to a water molecule to produce a hydronium ion.

In a molecule of H_2SO_4, or in any oxoacid, the O—H bonds are polar because oxygen is more electronegative than hydrogen. This means that the hydrogen carries a partial positive charge. Anything that can affect how polar this bond is will also affect the acidity. Making the O—H bond more polar, for instance, gives the hydrogen a greater partial positive charge and makes it easier for it to come off as H^+.

There are two major factors that affect the polarity of the O—H bonds in an oxoacid. One is the number of lone oxygen atoms bonded to the central nonmetal atom. For example, we can compare the two oxoacids of sulfur, H_2SO_4 and H_2SO_3, whose structures are

lone oxygens ⟶ and ⟵ **lone oxygen**

Each undergoes ionization in water by transferring its protons, in two successive steps, to water molecules. Thus, for H_2SO_4 we have in the first step,

H_2SO_4 is a strong acid.

With H_2SO_4 this step proceeds essentially to completion, while for H_2SO_3, on the other hand, the first step,

H_2SO_3 is a weak acid.

takes place only to a very limited degree (about 11% for a 1 M solution). In each of these substances the sulfur is bonded to two O—H groups. In H_2SO_4, however, the sulfur is also attached to two other lone oxygen atoms, while the sulfur in H_2SO_3 is bonded to only one such O atom. We know that oxygen is a very electronegative element, and we expect these S—O bonds to be polar, with the negative charge concentrated about the oxygen end of the dipole. In other words, the oxygen atoms attached to the sulfur draw electron density away so that the sulfur acquires a partial positive charge, the magnitude of which will be greater in H_2SO_4 than H_2SO_3,

This positive charge on the sulfur tends to draw electron density from the S—OH bonds. Since the electronegative oxygen does not wish to lose electrons, the overall net effect is that some electronic charge will be withdrawn from the O—H bonds. This electron-withdrawing effect will be greater in H_2SO_4 than H_2SO_3 because in H_2SO_4 the sulfur bears a greater positive charge and is therefore better able to draw electrons to itself. This means that in H_2SO_4 the hydrogen atoms carry a higher positive charge than they do in H_2SO_3 and are thus more easily removed as H^+. As a result, H_2SO_4 is a stronger acid than H_2SO_3.

This phenomenon is illustrated even more graphically if we consider the oxoacids of chlorine

Hypochlorous acid	HOCl	H—Ö—Cl:
Chlorous acid	$HClO_2$	H—Ö—Cl—Ö:
Chloric acid	$HClO_3$	H—Ö—Cl—Ö: :O:
Perchloric acid	$HClO_4$	:O: H—Ö—Cl—Ö: :O:

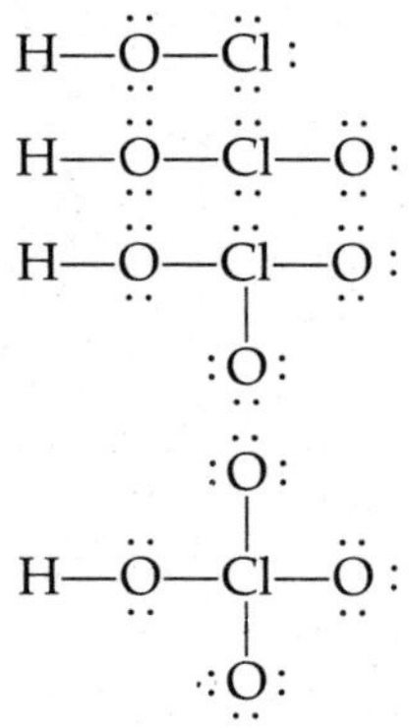

As the number of oxygen atoms surrounding the chlorine increases, so does the acidity. Thus HOCl is a relatively weak acid, $HClO_2$ is stronger, $HClO_3$ is essentially fully dissociated in water, and $HClO_4$ is just about the strongest acid there is.[4]

EXAMPLE 7.6   COMPARING THE STRENGTHS OF ACIDS

PROBLEM   Nitrous acid, HNO_2, is formed in the stomach from nitrites that are used as preservatives in meats. There is considerable concern that it may be able to react with certain biological molecules to produce carcinogenic (cancer-causing) agents. How does the acid strength of HNO_2 compare with that of nitric acid, HNO_3?

SOLUTION   First let's draw the Lewis structures for HNO_2 and HNO_3. Each would have one O—H bond (the hydrogen is bonded to an oxygen in an oxoacid) and any other oxygens are lone oxygens.

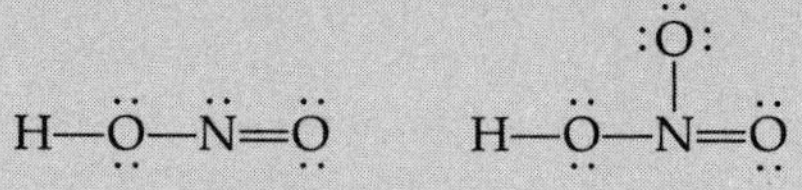

        H—Ö—N=Ö   H—Ö—N=Ö
        nitrous acid   **nitric acid**

---

[4] This reasoning can also be extended to other compounds. For example, acetic acid has the formula

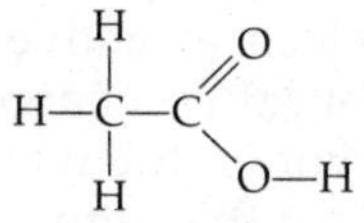

acetic acid

and is a weak acid. At any instant in a 1 M solution, only about 0.4% of the acetic acid molecules have reacted to form H_3O^+ and $CH_3CO_2^-$ ions. However, if a hydrogen in the CH_3 group is replaced with a very electronegative element such as chlorine,

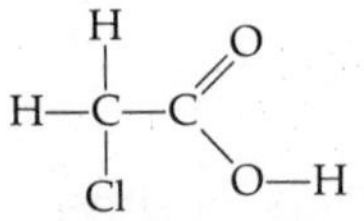

chloroacetic acid

the electron-attracting power of the chlorine increases the polarity of the O—H bond and makes the acid stronger. In a 1 M solution of chloroacetic acid, about 3% of the molecules have reacted to form H_3O^+ and $CH_2ClCO_2^-$ ions.

Since HNO_3 has more lone oxygens than HNO_2, the O—H bond in HNO_3 should be more polar than the one in HNO_2. Therefore HNO_3 should be the stronger acid (which it is).

The second major factor that affects the acidity of an oxoacid is the electronegativity of the central atom, and here we can find some trends in the periodic table. As the central atom, X, in a molecule such as

$$\ddot{\text{O}}\text{—}X\text{—}\ddot{\text{O}}\text{—H}$$

becomes more electronegative, electron density is funneled toward X and away from the O—H bond. Therefore, as the electronegativity of X increases, so does the acidity of the molecules.

Within a group, we know that the electronegativity increases from bottom to top. This means that the acidity of oxoacids having the same general formula should also increase from bottom to top in a group, which they do. For instance, phosphoric acid (H_3PO_4) is a stronger acid than arsenic acid (H_3AsO_4). Similarly, sulfuric acid (H_2SO_4) is stronger than selenic acid (H_2SeO_4).

We also learned that the electronegativity increases from left to right in a period, so the strengths of the oxoacids should increase in that direction, too. In fact, there is a very rapid increase in acid strength from left to right, not only because of electronegativity changes, but also because the formulas of the acids change. In period 3, for example, we find the following oxoacids in which the central atoms are each surrounded by four oxygen atoms:

$$H_3PO_4 \qquad\qquad H_2SO_4 \qquad\qquad HClO_4$$

$$\text{H—}\ddot{\text{O}}\text{—P—}\ddot{\text{O}}\text{—H} \qquad \text{H—}\ddot{\text{O}}\text{—S—}\ddot{\text{O}}\text{—H} \qquad \text{H—}\ddot{\text{O}}\text{—Cl—}\ddot{\text{O}}$$

The electronegativity increases from phosphorus to chlorine and the number of lone oxygens also increase—both are factors favoring an increase in acid strength. Thus, sulfuric acid is stronger than phosphoric acid, and perchloric acid is stronger than sulfuric acid. Figure 7.7 summarizes the trends in the strengths of the oxoacids.

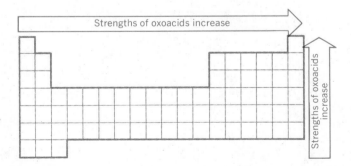

Figure 7.7

Variations within the periodic table of the strengths of oxoacids having similar formulas.

Binary acids

We can also find periodic trends in the strengths of the binary acids of the nonmetals. Recall that these acids are water solutions of nonmetal hydrides—for example, HF, HCl, and H_2S. They produce acidic solutions by reacting with water. For instance,

$$H-\ddot{O}: + H:\ddot{Cl}: \longrightarrow \left[H-\overset{|}{\underset{H}{\ddot{O}}}-H\right]^+ + \left[:\ddot{Cl}:\right]^-$$

When we consider binary acids derived from elements belonging to the same group, an increase in acidity down the group is observed experimentally. Thus the hydrohalides increase in acidity in the order HF < HCl < HBr < HI.

At first glance this order of acidity seems opposite to what we might predict. We know, for instance, that fluorine is more electronegative than chlorine; therefore the HF bond is more polar than the HCl bond. Consequently, the H in HF is more positively charged than the H in HCl. Thus we are tempted to predict that HF should lose a proton more readily than HCl. This, however, is precisely the *reverse* of their acid strengths in water, where HF is weak and HCl is essentially 100% ionized.

The solution to this puzzle can be understood by realizing that there are actually two opposing factors contributing to the acidity of these compounds. One is, indeed, the ionic character of the H—X bond; the other is the H—X bond strength. As we proceed down within a group, the nonmetal becomes progressively larger and there is an accompanying rapid decrease in the strength of the H—X bond. This weakening of the H—X bond turns out to be more than sufficient to compensate for the decrease in the polarity of the bonds, and a net increase in acid strength is observed.

When we look at the acidity of the binary hydrogen compounds of elements in the same period, for example, NH_3, H_2O, and HF, the dominant factor becomes the polarity of the H—X bond. As we go from left to right within the period there is little change in the size of the nonmetal and relatively little change in the H—X bond energy. There is, however, a very dramatic increase in the ionic character of the H—X bond, which is reflected in a rapid increase in acidity from ammonia to hydrogen fluoride. This is summarized in Figure 7.8.

> In general, bond strengths tend to decrease as the bonded atoms become larger.

7.5 LEWIS ACIDS AND BASES

The Brønsted–Lowry definition of acids and bases is more general than the Arrhenius definition because it removes the restriction of having to deal with reactions in aqueous solution. However, even the Brønsted–Lowry concept is

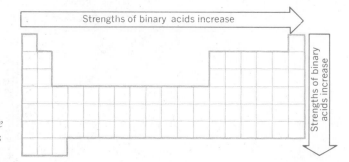

Figure 7.8
Variations within the periodic table of the strengths of the binary acids of the nonmetals.

restricted in scope, since it limits discussion of acid-base phenomena to proton-transfer reactions. There are many reactions that have all the earmarks of acid-base reactions but that do not fit the Brønsted–Lowry mold. The approach taken by the U.S. chemist Gilbert N. Lewis further extends the acid-base concept to cover these cases.

In the Lewis definition of acids and bases: *A* **base** *is defined as a substance that can donate a* **pair** *of electrons to the formation of a covalent bond. An* **acid** *is a substance that can accept a pair of electrons to form the bond.*

A simple example of a Lewis acid-base reaction is the reaction of a proton with a hydroxide ion,

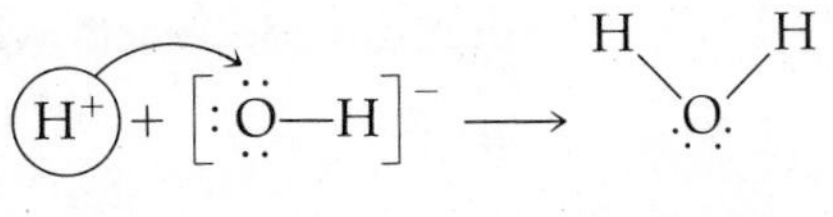

$$(H^+) + \left[\ddot{:}O{-}H\right]^- \longrightarrow \ \overset{H \qquad H}{\underset{\ddot{\ddot{O}}}{\diagdown \diagup}}$$

The hydroxide ion is the Lewis base because it furnishes the pair of electrons that become shared with the hydrogen. The hydrogen ion, on the other hand, is the Lewis acid because it accepts a share of the pair of electrons when the O—H bond is created.

Another example is the reaction between BF_3 and ammonia,

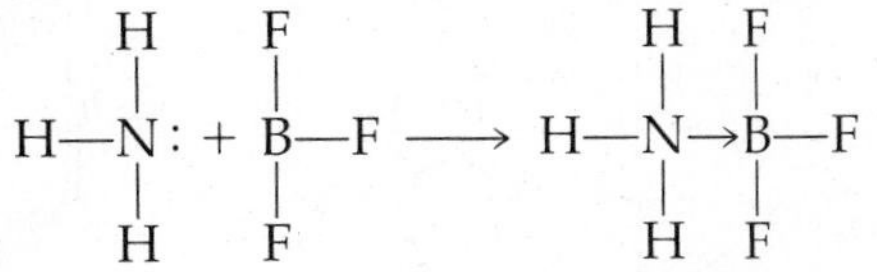

$$\underset{\overset{|}{H}}{\overset{\overset{H}{|}}{H{-}N}}\!: + \underset{\overset{|}{F}}{\overset{\overset{F}{|}}{B{-}F}} \longrightarrow \underset{\overset{|}{H} \quad \overset{|}{F}}{\overset{\overset{H}{|} \quad \overset{F}{|}}{H{-}N{\rightarrow}B{-}F}}$$

In this case the NH_3 functions as the base and BF_3 serves as the acid. Compounds containing elements with incomplete valence shells, such as BF_3 or $AlCl_3$, tend to be Lewis acids, while compounds or ions that have lone pairs of electrons can behave as Lewis bases. When the Lewis acid-base reaction occurs, a coordinate covalent bond is formed.

Still other examples of Lewis acid-base reactions are provided by the reactions of metal oxides with nonmetal oxides. Recall that metal oxides, in water, produce hydroxides. For instance,

$$Na_2O + H_2O \longrightarrow 2NaOH$$

Nonmetal oxides react to form acids as illustrated by the reaction

$$SO_3 + H_2O \longrightarrow H_2SO_4$$

When these two solutions are mixed, neutralization occurs, with the production of the solvent plus a salt,

$$2NaOH + H_2SO_4 \longrightarrow 2H_2O + Na_2SO_4$$

The production of Na_2SO_4 from Na_2O and SO_3 can take place directly without the introduction of any water whatsoever, as shown by the equation,

$$Na_2O(s) + SO_3(g) \longrightarrow Na_2SO_4(s)$$

According to the Lewis definition, this also is a neutralization reaction between a Lewis base (oxide ion) and a Lewis acid (sulfur trioxide):

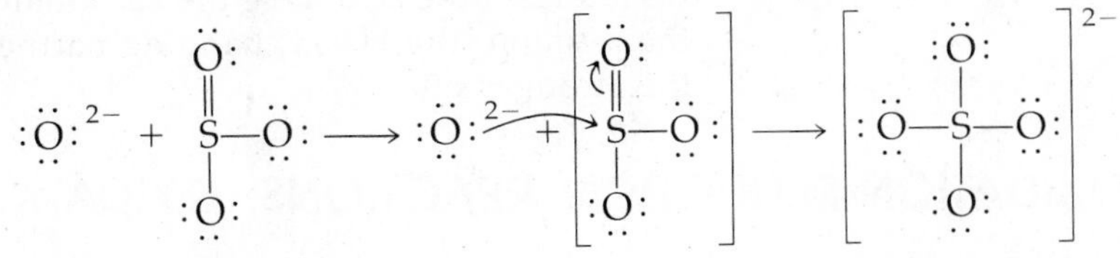

base　　**acid**　　　　　　　　　　　　　　　　　　　　　**sulfate ion**

Margin notes:

This is the same G. N. Lewis who originated the Lewis symbols and structures you learned to draw in Chapter 5.

Lewis base—electron pair donor.
Lewis acid—electron pair acceptor.

Recall that NH_3BF_3 is an example of an addition compound.

A pair of electrons in the double bond moves to the oxygen on top, allowing the sulfur to accept the share of a pair of electrons from the oxide ion.

In this case we find that some electronic rearrangement must take place as the oxygen becomes attached to the sulfur. Nevertheless, the overall change can be viewed as a neutralization reaction.

Reactions of this type, between an oxide such as CaO and SO_2 or SO_3, are important for the removal of sulfur oxides from the gases produced by combustion of high-sulfur fuels. For example,

$$CaO(s) + SO_2(g) \longrightarrow CaSO_3(s)$$

A reaction quite analogous to that just described has been used on spacecraft to remove carbon dioxide from the air breathed by the astronauts. In this case carbon dioxide reacts with LiOH,

$$CO_2(g) + LiOH(s) \longrightarrow LiHCO_3(s)$$
lithium bicarbonate

(Lithium hydroxide is used because of the very low atomic weight of Li. This results in many moles of LiOH per kilogram.) This reaction also can be viewed as a Lewis acid-base reaction

bicarbonate ion

The reaction between Na_2O and SO_3 illustrates the limitations of the Brønsted–Lowry concept. Since no protons are involved in the reaction, it would never be classified as an acid-base reaction under the Brønsted–Lowry definition.

Thus far we have looked only at simple acid-base neutralizations. The acid-base reactions discussed in Section 7.3 can also be treated from the Lewis point of view. In the Brønsted–Lowry theory these reactions were regarded as competitions in which the strongest acid prevails by losing its proton. Under the Lewis definition these reactions are looked on as constituting the displacement of one base (the weaker one) by another. Referring to our reaction of HCl with H_2O, for example,

$$HCl + H_2O \longrightarrow H_3O^+ + Cl^-$$

the Lewis theory interprets the change as the result of replacing the Cl^- ion in HCl by the stronger base, H_2O.

In other words, the stronger base, H_2O, pushes out the weaker one, Cl^-. Here we interpret the acid to be the H^+ ion instead of the entire HCl molecule, and in the reaction, the H^+ is changing partners as it moves from the weaker base to the stronger one.

7.6 OXIDATION-REDUCTION REACTIONS; OXIDATION NUMBERS

In the formation of an ionic bond between atoms of Li and F, we saw that an electron is transferred from Li to F to produce Li^+ and F^- ions. In the formation of HCl from H_2 and Cl_2, we saw that a polar covalent bond is formed in which an electron was partially transferred from hydrogen to chlorine. Very many chemical reactions are of this type—they involve some transfer of electronic charge from one atom to another. Some other examples include the rusting of iron, the metabolism of foods by our bodies, and the chemical changes that take place inside batteries. Because electron transfer is such a common and important process, chemists have defined terms that apply specifically to these changes. These are

>**Oxidation**—a loss of electrons
>**Reduction**—a gain of electrons

Thus in the formation of LiF, Li undergoes oxidation by losing an electron and F undergoes reduction by acquiring an electron. In a similar fashion, when the HCl molecule is formed by the reaction of hydrogen and chlorine, the H atom loses some electronic charge to the Cl atom and is oxidized while the Cl atom becomes reduced.

Lithium, in its reaction with fluorine, is said to be a **reducing agent** because it supplies the electron that the fluorine requires in order to be reduced—that is, it is the agent that has allowed reduction to occur. The fluorine, on the other hand, by accepting the electron from Li, permits oxidation to take place and thus is said to be an **oxidizing agent.** In a similar manner, we would consider hydrogen the reducing agent and chlorine the oxidizing agent when these two elements react to produce HCl. In general, *oxidizing agents acquire electrons and become reduced, while reducing agents lose electrons and become oxidized.*

Oxidation and reduction are always discussed together because, in any reaction, whenever one substance loses electrons, another substance picks them up. We know that this is true because we never observe electrons as a product of a chemical reaction, nor are electrons ever consumed when a chemical change occurs. Thus, oxidation is always accompanied by reduction. As a result, the simple abbreviated term, **redox,** is often used in discussing these reactions.

Chemists have devised a bookkeeping system using what are called **oxidation numbers** to keep track of electrons during chemical reactions. An oxidation number can be defined as *the charge that an atom would have if both of the electrons in each bond were assigned to the more electronegative element.* The term **oxidation state** is also used, interchangeably, with the term oxidation number.

In the substance LiF, because an electron has, in fact, been transferred to the F atom, the oxidation number assigned to Li^+ is +1. The oxidation number of fluorine in the F^- ion is −1. (For oxidation numbers, we will write the sign before the number. This is to distinguish them from actual electrical charges, for which we write the number before the algebraic sign.)

In the HCl molecule, because Cl is more electronegative than hydrogen, we assign an oxidation number of +1 to H and −1 to Cl, as if the electron pair were in the sole possession of the Cl atom.

In a nonpolar molecule such as H_2, where both atoms are the same and therefore have the same electronegativity, it is senseless to assign the electron pair to either atom since no electron transfer has occurred. In this case each H atom is assigned an oxidation number of zero.

The following set of rules has been developed to aid us in assigning oxidation numbers to the various atoms in a compound.

To be able to assign oxidation numbers properly, you must learn these rules and how to apply them.

Rules for assigning oxidation numbers

1. The oxidation number of any element in its elemental form is zero, regardless of the complexity of the molecule in which it occurs. Thus the atoms in Ne, F_2, P_4, and S_8 all have oxidation numbers of zero.

2. The oxidation number of any simple ion (one atom) is equal to the charge on the ion. The ions Na^+, Al^{3+}, and S^{2-} have oxidation numbers of $+1$, $+3$, and -2, respectively.

3. The sum of all the oxidation numbers of all the atoms in a neutral compound is zero. For a polyatomic ion, the algebraic sum of the oxidation numbers must be equal to the ion's charge.

In addition to these basic rules, we can use the fact that certain elements have the same oxidation numbers in all (or nearly all) their compounds.

4. Fluorine always has an oxidation number of -1 in its compounds.

5. The elements in Group IA (except hydrogen) always have an oxidation number of $+1$ in compounds.

6. The elements in Group IIA always have an oxidation number of $+2$ in compounds.

7. A Group VIIA element has an oxidation number of -1 in binary compounds (compounds that contain only two different elements) with metals. For example, Cl has an oxidation number of -1 in $FeCl_2$, $CrCl_3$, and NaCl.

8. Oxygen almost always has an oxidation number of -2 in its compounds.

9. Hydrogen almost always has an oxidation number of $+1$ in its compounds.

10. For familiar polyatomic ions, such as SO_4^{2-} or NO_3^-, the charge on the ion can be taken as the ion's *net* oxidation number.

In rules 8 and 9 the words "almost always" are used because there are some exceptions. For example, in binary compounds with fluorine, oxygen must have a positive oxidation number because fluorine is *always* -1. Somewhat more common are peroxides, such as the O_2^{2-} ion and H_2O_2 that contain an O—O bond, in which oxygen is assigned an oxidation number of -1. Oxygen also forms compounds called superoxides that contain the O_2^- ion in which the oxidation number of oxygen is $-\frac{1}{2}$. Finally, hydrogen forms binary compounds with some metals—for example, NaH—in which hydrogen is given an oxidation number of -1. Although you should know of these exceptions, they are rare, and we normally assign oxidation numbers of -2 to oxygen and $+1$ to hydrogen when we see them in compounds.

EXAMPLE 7.7   DETERMINING OXIDATION NUMBERS

PROBLEM   What are the oxidation numbers of all of the atoms in KNO_3 (potassium nitrate, a substance used to make gun powder)?

SOLUTION   We know that the sum of the oxidation numbers of all of the atoms must be equal to zero (the charge on KNO_3—rule 3).

$$
\begin{array}{lll}
K & 1 \times (+1) = +1 & \text{(rule 5)} \\
N & 1 \times (x) = \ \ x & \\
O & 3 \times (-2) = \underline{-6} & \text{(rule 8)} \\
\textit{sum of oxidation numbers} = \ \ 0 & \text{(rule 3)}
\end{array}
$$

x must equal $+5$ in order for the sum to be zero.

EXAMPLE 7.8 DETERMINING OXIDATION NUMBERS

PROBLEM What is the oxidation number of sulfur in $Na_2S_4O_6$ (sodium tetrathionate)?

SOLUTION Again, the sum of the oxidation numbers must be zero.

$$
\begin{array}{lrl}
Na & 2 \times (+1) = & +2 \\
S & 4 \times (x) = & 4x \\
O & 6 \times (-2) = & \underline{-12} \\
& \text{sum} = & 0
\end{array}
$$

$$
4x = +10 \quad \text{or} \quad x = \frac{+10}{4} = +\frac{5}{2}
$$

Notice that the oxidation number of an atom need not be an integer.

EXAMPLE 7.9 DETERMINING OXIDATION NUMBERS

PROBLEM What is the oxidation number of Cr in the $Cr_2O_7{}^{2-}$ ion?

SOLUTION This time the sum of the oxidation numbers must equal -2 (rule 3).

$$
\begin{array}{lrl}
Cr & 2 \times (x) = & 2x \\
O & 7 \times (-2) = & \underline{-14} \\
& \text{sum} = & -2
\end{array}
$$

Therefore,

$$
2x = +12
$$

$$
x = +6
$$

It is very important to keep in mind that oxidation numbers were developed simply for bookkeeping. Except for simple ions like Na^+ or F^-, the charges are fictitious. For example, the Cl in HCl does not carry a 1− charge as its oxidation number would imply; it carries only a partial negative charge (the actual charge is only about 0.17−).

7.7 BALANCING REDOX EQUATIONS USING OXIDATION NUMBERS

Oxidation-reduction processes form a very important class of chemical reactions. They take place between many inorganic and organic compounds, and they are extremely important in biochemical systems where they provide the mechanism for energy transfer in living organisms.

As a rule, the stoichiometry of redox reactions tends to be more complicated than for reactions that do not involve electron transfer. As a result, chemical equations for oxidation-reduction reactions are often complex and are difficult to balance by inspection. Fortunately, there are methods that can be applied to aid in balancing these equations.

In the previous section it was noted that during redox reactions electrons are never produced as a product, nor are they necessary as a reactant. *When a chemical reaction involves oxidation-reduction, the total number of electrons lost in the oxidation process must equal the total number gained during reduction.* We can use this fact to help us balance equations for this type of reaction. One procedure we can use is called the **oxidation number change method.** To illustrate the steps involved, let's balance the equation

$$
HCl + K_2Cr_2O_7 \longrightarrow KCl + CrCl_3 + Cl_2 + H_2O
$$

Step 1 *Assign oxidation numbers to all the atoms in the equation.* We will write them below the chemical symbols to avoid confusing them with actual charges.

$$HCl + K_2Cr_2O_7 \longrightarrow KCl + CrCl_3 + Cl_2 + H_2O$$
$$\substack{+1-1} \qquad \substack{+1+6-2} \qquad \substack{+1-1} \quad \substack{+3-1} \quad \substack{0} \quad \substack{+1-2}$$

Step 2 *Identify which atoms change oxidation number and insert temporary coefficients, if necessary, so that we have the same number of these atoms on both sides of the equation.*

We see that chromium changes from +6 to +3. We also see that some chlorine changes from −1 to 0 and some doesn't change at all; for now, we will only look at the chlorine that does. To make the number of chromium atoms and chlorine atoms that change the same on both sides, we put a temporary coefficient of 2 in front of HCl and a 2 in front of $CrCl_3$. This gives

> These temporary coefficients help us figure the number of electrons transferred in Step 3.

$$2HCl + K_2Cr_2O_7 \longrightarrow KCl + 2CrCl_3 + Cl_2 + H_2O$$
$$\substack{+1-1} \qquad \substack{+1+6-2} \qquad \substack{+1-1} \quad \substack{+3-1} \quad \substack{0} \quad \substack{+1-2}$$

Step 3 *Compute the total change in oxidation number for both the oxidation and reduction.*

We have *two* chlorines changing from −1 to 0; the total change corresponds to a loss of $2e^-$. Next we see that two chromiums change from +6 to +3; this total change corresponds to a gain of $6e^-$ $(2 \times 3e^-)$.

$$2e^- \text{ lost (total)}$$

$$6e^- \text{ gained (total)}$$

$$2HCl + K_2Cr_2O_7 \longrightarrow KCl + 2CrCl_3 + Cl_2 + H_2O$$
$$\substack{-1} \qquad \substack{+6} \qquad\qquad \substack{+3} \qquad \substack{0}$$

Step 4 *Make the total loss and gain of electrons the same by multiplying coefficients by appropriate factors.*

If we multiply the number of electrons lost by 3, there will be a total of $6e^-$ lost as well as gained. We accomplish this by also multiplying the coefficients of HCl and Cl_2 by 3. This gives

$$3 \times 2e^- = 6e^- \text{ lost}$$

$$6e^- \text{ gained}$$

$$3 \times 2 \downarrow$$
$$6HCl + K_2Cr_2O_7 \longrightarrow KCl + 2CrCl_3 + 3Cl_2 + H_2O$$

Step 5 *Finally, balance the rest of the equation by inspection.* There are 2 potassiums on the left, which require a coefficient of 2 for KCl. Then we add up all the chlorines on the right (there are fourteen), so we change the coefficient of HCl to 14. Finally, we place a 7 in front of H_2O to balance the H and O. The final equation is

$$14HCl + K_2Cr_2O_7 \longrightarrow 2KCl + 2CrCl_3 + 3Cl_2 + 7H_2O$$

The procedure outlined above will work on most oxidation-reduction equations, although simple equations (for example, $H_2 + Cl_2 \rightarrow 2HCl$) are more easily balanced by inspection.

7.8 BALANCING REDOX EQUATIONS BY THE ION-ELECTRON METHOD

Many redox reactions that we encounter in both the laboratory and around the home involve ions. Laundry bleach, for example, contains hypochlorite ion, OCl^-, as its active ingredient. The bleaching action occurs by an ionic reaction—the oxidation of colored materials by the OCl^- ion in the wash water. Earlier in this chapter we saw that net ionic equations are useful for focusing our attention on the key changes that occur in ionic reactions involving metathesis. As you might expect, they are also helpful in analyzing redox reactions between ions.

A procedure particularly well suited for balancing net ionic equations for redox reactions is called the **ion-electron method.** It is based on the principle of divide and conquer: The equation is split into two simpler parts, called **half-reactions,** that are balanced separately and then recombined to give the final balanced net ionic equation.

As an example, consider the reaction between solutions of $SnCl_2$ and $HgCl_2$, which gives insoluble Hg_2Cl_2 as one product and Sn^{4+} in solution as the other. In applying the ion-electron method, we begin by writing a *skeleton equation* that shows only those substances that are actually involved in the reaction; we leave out any spectator ions. For the reaction at hand, the reactants are Sn^{2+}, Hg^{2+}, and Cl^-. The products are Hg_2Cl_2 and Sn^{4+}. The skeleton equation is therefore

$$Sn^{2+} + Hg^{2+} + Cl^- \longrightarrow Sn^{4+} + Hg_2Cl_2$$

Next, we divide the equation into two half-reactions. These are

$$Sn^{2+} \longrightarrow Sn^{4+}$$
$$Hg^{2+} + Cl^- \longrightarrow Hg_2Cl_2$$

Dividing the skeleton equation into the half-reactions is the most difficult part of using the ion-electron method, so let's review how it is accomplished. Notice that in the skeleton equation there are *two* products. We choose one of them as a product in one of the half-reactions, and the other as the product in the second-half reaction. Now look at the first half-reaction. Since there is tin on the right, we almost must have tin on the left—the same elements must appear on opposite sides of the arrow; otherwise it would be impossible to balance the half-reaction. For the second half-reaction, the product contains both mercury and chlorine, so we require both mercury (Hg^{2+}) and chlorine (Cl^-) on the left.

The next step is to balance the half-reactions, and there are two requirements to be met here. *In order for any equation involving ions to be balanced,*

1. There must be the same number of atoms of each kind on both sides of the equation.
2. The net charge must be the same on both sides of the equation.

The first requirement is met by placing coefficients into the half-reactions to balance the atoms. The first equation already contains one tin on each side, so nothing needs be done to it. The second equation is balanced according to atoms by placing coefficients of 2 in front of both Hg^{2+} and Cl^-. This gives

$$Sn^{2+} \longrightarrow Sn^{4+}$$
$$2Hg^{2+} + 2Cl^- \longrightarrow Hg_2Cl_2$$

The second requirement is met by balancing the half-reactions according to charge. This is accomplished by adding as many electrons as necessary to the more positive side (or sometimes the less negative side). For the first half-reaction, we add two electrons to the right to give a net charge of 2+ on both sides.

$$Sn^{2+} \longrightarrow Sn^{4+} + 2e^-$$

Charge: 2+   (4+) + (2−) = 2+
                                     (net)

For the second half-reaction, we must add two electrons to the left to give a net charge of zero on both sides.

$$2e^- + 2Hg^{2+} + 2Cl^- \longrightarrow Hg_2Cl_2$$

Charge: (2−) + (4+)   + (2−)   0

Net = 0

This gives the two balanced half-reactions

$$Sn^{2+} \longrightarrow Sn^{4+} + 2e^-$$

$$2e^- + 2Hg^{2+} + 2Cl^- \longrightarrow Hg_2Cl_2$$

The final step is to make the number of electrons gained in the second half-reaction equal the number lost in the first. In this particular example, we don't actually have to do anything because the necessary condition is already met. Now the half-reactions are recombined by adding them together.

$$Sn^{2+} \longrightarrow Sn^{4+} + 2e^-$$

$$\underline{2e^- + 2Hg^{2+} + 2Cl^- \longrightarrow Hg_2Cl_2}$$

$$2e^- + Sn^{2+} + 2Hg^{2+} + 2Cl^- \longrightarrow Sn^{4+} + Hg_2Cl_2 + 2e^-$$

Finally, we cancel anything that is the same on both sides of the equation. In this and any other equation balanced by this method, the electrons *must* cancel. If they don't, you've made a mistake. The finished equation is

$$Sn^{2+} + 2Hg^{2+} + 2Cl^- \longrightarrow Sn^{4+} + Hg_2Cl_2$$

Notice that both the atoms and the charge are in balance.

Reactions that involve H^+ or OH^-

In many redox reactions in aqueous solution, H^+ or OH^- are either consumed or produced. These reactions also usually involve water as either a product or a reactant. For example, if concentrated hydrochloric acid is poured over potassium permanganate ($KMnO_4$), the chloride ion from the acid is oxidized to elemental chlorine by the permanganate ion, which is reduced to give Mn^{2+} in the solution. During this reaction, hydrogen ion is consumed as it combines with the oxygen atoms of the permanganate to give molecules of water. During the reaction, therefore, the amount of hydrogen ion in the solution decreases along with the amounts of chloride and permanganate ions.

Not only are H^+ and OH^- reactants or products in many reactions, their presence or absence can also affect the other products of the reactions. For example, if permanganate ion is used as an oxidizing agent in an acidic solution, the reduction product is generally Mn^{2+}. But if the solution is basic, the permanganate ion is reduced to insoluble MnO_2. Therefore, when we carry out a redox reaction, it is important to know something about the acidity or basicity of the solution.

When we use the ion-electron method for balancing equations, it is not necessary to know if H^+ or OH^- is a reactant or a product, or if water is con-

Of course, the hydrogen ion is more accurately represented as H_3O^+, but for simplicity we use the hydronium ion's abbreviation, H^+.

sumed or produced. All we need to know is whether the reaction is taking place in an acidic or a basic solution. The act of balancing the equation then tells us which of these species, if any, are involved.

Reactions that occur in an acidic solution

In any acidic aqueous solution, two of the major species are H_2O and H^+. These can be used in the ion-electron method to help us balance hydrogen and oxygen atoms in half-reactions. The general approach is essentially the same as that used on balancing the equation for the reaction between Sn^{2+}, Hg^{2+}, and Cl^-. It is very systemmatic and can be broken down into the following steps.

Steps in the ion-electron method for acidic solutions

Step 1. Divide the skeleton equation into half-reactions.

Step 2. Balance atoms other than oxygen and hydrogen.

Step 3. Next, balance the oxygens in each half-reaction by adding water molecules to the side that needs oxygen atoms. Add one H_2O for each oxygen needed.

Step 4. Balance the hydrogen atoms in each half-reaction by adding H^+ to the side that needs hydrogen. Add one H^+ for each hydrogen needed.

Step 5. Balance the charge in each half-reaction by adding electrons to the appropriate side.

Step 6. Multiply each half-reaction by appropriate factors to make the number of electrons gained equal the number lost.

Step 7. Add the two half-reactions together.

Step 8. Cancel anything that is the same on both sides of the equation.

If you follow these rules you will always obtain a properly balanced equation.

It is important to follow these steps in the sequence given here.

To see how these rules apply, let's balance the equation for the reaction of HCl with $KMnO_4$, which we mentioned above. In this reaction Cl^- is oxidized to Cl_2 and MnO_4^- is reduced to Mn^{2+}. The skeleton equation, which does not contain either H^+ or H_2O, is

$$Cl^- + MnO_4^- \longrightarrow Cl_2 + Mn^{2+} \text{ (acidic solution)}$$

You will always be told whether the solution is acidic or basic.

Step 1 *Divide the equation into half-reactions.*

$$Cl^- \longrightarrow Cl_2$$
$$MnO_4^- \longrightarrow Mn^{2+}$$

Except for hydrogen and oxygen, each side of a given half-reaction must begin with the same elements; otherwise it can't be balanced.

Step 2 *Balance atoms other than H and O.* In the first half-reaction we place a coefficient of 2 in front of Cl^-. We don't have to do anything to the second half-reaction.

$$2 Cl^- \longrightarrow Cl_2$$
$$MnO_4^- \longrightarrow Mn^{2+}$$

Step 3 *Balance oxygens by adding H_2O to the side that needs O.* In the second half-reaction, there are 4 oxygens on the left and none on the right. We therefore have to add $4H_2O$ to the right side. Then the oxygens will balance.

$$2Cl^- \longrightarrow Cl_2$$
$$MnO_4^- \longrightarrow Mn^{2+} + 4H_2O$$

Step 4 *Balance hydrogens by adding H^+ to the side that needs H.* The right side of the second half-reaction has a total of 8 hydrogens; there are none on the left. Therefore, we add $8H^+$ to the left side. (Be sure you remember to write the + sign on the hydrogen ion.)

$$2Cl^- \longrightarrow Cl_2$$

$$8H^+ + MnO_4^- \longrightarrow Mn^{2+} + 4H_2O$$

Note that all the atoms now balance.

Step 5 *Balance the charge by adding electrons.* In this step we make the net charge the same on both sides by following the same procedure as in the reaction between Sn^{2+}, Hg^{2+}, and Cl^-. In the first half-reaction we have to add 2 electrons to the right side. In the second half-reaction we must add 5 electrons to the left side. (Before we add electrons, the net charge on the left is 7+ and the net charge on the right is 2+. To make them the same, we add $5e^-$ to the left.)

$$2Cl^- \longrightarrow Cl_2 + 2e^-$$

$$5e^- + 8H^+ + MnO_4^- \longrightarrow Mn^{2+} + 4H_2O$$

Step 6 *Make the number of electrons gained equal the number lost.* This can be accomplished by multiplying the first half-reaction through by 5 and the second through by 2. There will then be $10e^-$ lost and gained.

$$5(2Cl^- \longrightarrow Cl_2 + 2e^-)$$

$$2(5e^- + 8H^+ + MnO_4^- \longrightarrow Mn^{2+} + 4H_2O)$$

$5 \times (2e^-) = 10e^-$
$2 \times (5e^-) = 10e^-$

Step 7 *Add the two half-reactions.* When we do this, we will not bother to bring down the electrons, because we know they will cancel. They have to—we went to the trouble to make them the same in each half-reaction.

$$5(2Cl^- \longrightarrow Cl_2 + 2e^-)$$

$$\underline{2(5e^- + 8H^+ + MnO_4^- \longrightarrow Mn^{2+} + 4H_2O)}$$

$$10Cl^- + 16H^+ + 2MnO_4^- \longrightarrow 5Cl_2 + 2Mn^{2+} + 8H_2O$$

Step 8 *Cancel anything that is the same on both sides.* In this case there is nothing to be cancelled, so we are finished. The final balanced equation is

$$10Cl^- + 16H^+ + 2MnO_4^- \longrightarrow 5Cl_2 + 2Mn^{2+} + 8H_2O$$

The following is another example.

EXAMPLE 7.10	BALANCING AN EQUATION FOR A REACTION THAT TAKES PLACE IN AN ACIDIC SOLUTION
PROBLEM	Balance the following equation by the ion-electron method $$Cr_2O_7^{2-} + H_2S \longrightarrow Cr^{3+} + S \text{ (acidic solution)}$$
SOLUTION	We simply follow the steps outlined above.
Step 1.	Divide the equation into half-reactions. $$Cr_2O_7^{2-} \longrightarrow Cr^{3+}$$ $$H_2S \longrightarrow S$$

Step 2.   Balance the Cr in the first equation.

$$Cr_2O_7^{2-} \longrightarrow 2Cr^{3+}$$

$$H_2S \longrightarrow S$$

Step 3.   Balance oxygens with H_2O.

$$Cr_2O_7^{2-} \longrightarrow 2Cr^{3+} + 7H_2O$$

$$H_2S \longrightarrow S$$

Step 4.   Balance hydrogens with H^+.

$$14H^+ + Cr_2O_7^{2-} \longrightarrow 2Cr^{3+} + 7H_2O$$

$$H_2S \longrightarrow S + 2H^+$$

Step 5.   Balance the charge with electrons.

$$6e^- + 14H^+ + Cr_2O_7^{2-} \longrightarrow 2Cr^{3+} + 7H_2O$$

$$H_2S \longrightarrow S + 2H^+ + 2e^-$$

Steps 6 and 7.   Multiply the second equation through by 3 to make the electrons gained equal to the electrons lost. Then add the half-reactions together.

$$6e^- + 14H^+ + Cr_2O_7^{2-} \longrightarrow 2Cr^{3+} + 7H_2O$$

$$\underline{3 \times (H_2S \longrightarrow S + 2H^+ + 2e^-)}$$

$$14H^+ + Cr_2O_7^{2-} + 3H_2S \longrightarrow 2Cr^{3+} + 7H_2O + 3S + 6H^+$$

Step 8.   Cancel anything that's the same on both sides. Six H^+ can be canceled (removed) from each side, which leaves $8H^+$ on the left. The final equation is therefore

$$8H^+ + Cr_2O_7^{2-} + 3H_2S \longrightarrow 2Cr^{3+} + 7H_2O + 3S$$

Redox reactions in basic solutions

We have just seen that H_2O and H^+ are used to balance half-reactions that occur in acidic solution. In a basic solution the dominant species are H_2O and OH^-, so these are the species that should be used to achieve material balance. Although you can use H_2O and OH^- directly,[5] the simplest technique is to first balance the reaction as if it occurred in acidic solution, and then perform the conversion described below to adjust it to conform to conditions in basic solution.

Suppose we wished to balance the following half-reaction taking place in a basic solution.

$$Pb \longrightarrow PbO$$

First we balance it as if it occurred in an acidic solution.

$$H_2O + Pb \longrightarrow PbO + 2H + 2e^-$$

The conversion to basic solution follows these three steps.

[5] To balance half-reactions in basic solution, the following rules can be used:
1. *To balance a hydrogen atom we add* **one** *H_2O molecule to the side of the half-reaction deficient in hydrogen, and to the other side we add* **one** *hydroxide ion.*
2. *To balance one oxygen atom we add* **two** *hydroxide ions to the side deficient in oxygen and* **one** *H_2O molecule to the other side.*

Step 1 For each H^+ that must be eliminated from the equation, add an OH^- to both sides of the equation. In this example, we have to eliminate $2H^+$, so we add $2OH^-$ to each side.

$$H_2O + Pb + 2OH^- \longrightarrow PbO + 2H^+ + 2OH^- + 2e^-$$

Step 2 Combine H^+ and OH^- to form H_2O. We have $2H^+$ and $2OH^-$ on the right, which give $2H_2O$.

$$H_2O + Pb + 2OH^- \longrightarrow PbO + \underbrace{2H_2O}_{2H^+ + 2OH^-} + 2e^-$$

Step 3 Cancel any H_2O that are the same on both sides. We can cancel one H_2O from each side. The final balanced half-reaction in basic solution is

$$Pb + 2OH^- \longrightarrow PbO + H_2O + 2e^-$$

EXAMPLE 7.11 BALANCING AN EQUATION FOR BASIC SOLUTION

PROBLEM Balance the following reaction for the oxidation of plumbite ion, $Pb(OH)_3^-$, to lead dioxide by hypochlorite ion in basic solution.

$$Pb(OH)_3^- + OCl^- \longrightarrow PbO_2 + Cl^-$$

SOLUTION First, the equation is balanced as though it occurred in acidic solution. We begin by dividing it into half-reactions.

$$Pb(OH)_3^- \longrightarrow PbO_2$$
$$OCl^- \longrightarrow Cl^-$$

Now we balance them according to atoms.

$$Pb(OH)_3^- \longrightarrow PbO_2 + H_2O + H^+$$
$$2H^+ + OCl^- \longrightarrow Cl^- + H_2O$$

Then we balance the charge by adding electrons.

$$Pb(OH)_3^- \longrightarrow PbO_2 + H_2O + H^+ + 2e^-$$
$$2e^- + 2H^+ + OCl^- \longrightarrow Cl^- + H_2O$$

Since the number of electrons lost in the first half-reaction is already equal to the number gained in the second, we can add them.

$$2e^- + 2H^+ + OCl^- + Pb(OH)_3^- \longrightarrow Cl^- + 2H_2O + PbO_2 + H^+ + 2e^-$$

Next, we cancel electrons and any other substances that are the same on both sides. The equation is now balanced for acidic solution.

$$H^+ + OCl^- + Pb(OH)_3^- \longrightarrow Cl^- + 2H_2O + PbO_2$$

Next, we perform the three-step conversion to basic solution. First we add to *each side* the same number of OH^- as there are H^+ in the equation.

$$\underbrace{H^+ + OH^-}_{\text{forms } H_2O} + OCl^- + Pb(OH)_3^- \longrightarrow Cl^- + 2H_2O + PbO_2 + OH^-$$

On the left, H^+ and OH^- become H_2O.

$$H_2O + OCl^- + Pb(OH)_3^- \longrightarrow Cl^- + 2H_2O + PbO_2 + OH^-$$

Then we cancel one H_2O from each side to get the final balanced equation.

$$OCl^- + Pb(OH)_3^- \longrightarrow Cl^- + H_2O + PbO_2 + OH^-$$

7.9 METALS AS REDUCING AGENTS

The elements and their compounds undergo many oxidation-reduction reactions. In the remainder of this chapter we will take a brief look at some that are typical of certain classes of elements or compounds. As you will see, these reactions can often be related to an element's position within the periodic table, which helps us organize and remember chemical facts. We begin with some properties of the metals.

Metals are elements that have low ionization energies and small electron affinities—they lose electrons relatively easily and have little tendency to gain them. When they react with other elements, therefore, metals tend to form positive ions by the loss of electrons; in other words, they become oxidized. This means, of course, that they serve as reducing agents.

The reaction of metals with acids

One of the characteristic ways that metals behave as reducing agents is in their reactions with acids. A typical example is the reaction of zinc with hydrochloric acid or sulfuric acid.

$$Zn(s) + 2HCl(aq) \longrightarrow ZnCl_2(aq) + H_2(g)$$

$$Zn(s) + H_2SO_4(aq) \longrightarrow ZnSO_4(aq) + H_2(g)$$

The net reaction for both of these is the same.

$$Zn(s) + 2H^+(aq) \longrightarrow Zn^{2+}(aq) + H_2(g)$$

In the reaction, zinc is oxidized and hydrogen ion is reduced; therefore, zinc is the reducing agent and hydrogen ion is the oxidizing agent.

In aqueous solutions of HCl or H_2SO_4 the only oxidizing agent is H^+; under ordinary conditions neither Cl^- nor SO_4^{2-} is reduced. Acids such as HCl and H_2SO_4, in which the only effective oxidizing agent is H^+, are called **nonoxidizing acids.** (This may seem strange, since the acid does attack metals by oxidation, but the term is used to differentiate these acids from others in which the anion of the acid is also an oxidizing agent.)

Some other common metals that also react with nonoxidizing acids are iron, magnesium, and aluminum. In each case, the reaction gives hydrogen and the metal ion in solution.

$$Fe(s) + 2H^+(aq) \longrightarrow Fe^{2+}(aq) + H_2(g)$$

$$Mg(s) + 2H^+(aq) \longrightarrow Mg^{2+}(aq) + H_2(g)$$

$$2Al(s) + 6H^+(aq) \longrightarrow 2Al^{3+}(aq) + 3H_2(g)$$

In general, then, for metals that react with nonoxidizing acids

$$metal + H^+ \longrightarrow metal\ ion + H_2(g)$$

As implied in the preceding paragraph, not all metals are able to be oxidized by hydrogen ion. Two common metals that fit into this category are copper and

An ordinary iron nail dissolving in hydrochloric acid. The bubbles are H_2.

If you spill battery acid, H_2SO_4, on the metal parts of your car, the acid will attack the metal unless it is washed off with water.

silver—if either of these metals are placed in a solution of hydrochloric acid, nothing happens. This simply reflects the fact that some metals, such as copper and silver, are more difficult to oxidize than others, and that H^+ simply is not up to the job. A stronger oxidizing agent than H^+ is needed to oxidize these metals.

One acid capable of dissolving copper and silver is nitric acid, HNO_3. This is an example of an **oxidizing acid;** in addition to H^+, a solution of this acid also contains the nitrate ion, which is a stronger oxidizing agent than H^+. The violent reaction of copper with HNO_3 is shown in Figure 7.9. A reddish-brown gas is produced that is nitrogen dioxide, NO_2, formed in the reaction

$$Cu(s) + 2NO_3^-(aq) + 4H^+(aq) \longrightarrow Cu^{2+}(aq) + 2NO_2(g) + 2H_2O$$

When HNO_3 serves as an oxidizing agent, the products depend, to a degree, on how concentrated the acid is. For example, with copper there are the following reactions.

With dilute HNO_3

$$3Cu(s) + 2NO_3^-(aq) + 8H^+(aq) \longrightarrow 3Cu^{2+}(aq) + 2NO(g) + 4H_2O$$

With concentrated HNO_3

$$Cu(s) + 2NO_3^-(aq) + 4H^+(aq) \longrightarrow Cu^{2+}(aq) + 2NO_2(g) + 2H_2O$$

A similar reaction occurs with silver. Notice that H_2 is not among the products of these reactions. Instead, nitrate ion is reduced to either NO or NO_2.

Earlier we mentioned sulfuric acid as an example of a nonoxidizing acid, and this is indeed how sulfuric acid behaves in dilute aqueous solutions. However, when concentrated and hot, sulfuric acid is also an oxidizing agent. For example, hot concentrated sulfuric acid reacts with copper as follows.

Hot concentrated sulfuric acid is a dangerous solution.

$$Cu + 2H_2SO_4 \xrightarrow{\text{hot}} CuSO_4 + SO_2 + 2H_2O$$

Periodic trends in the ease of oxidation of metals

The ease of oxidation is an important property of metals. Many of their practical applications depend on it (or rather, the *lack* of it). This is because the air-oxidation of metals, which we often call *corrosion*, produces compounds that lack

Figure 7.9

A copper penny reacts violently with concentrated nitric acid, as this sequence of photographs show. The dark vapors are ni- *trogen dioxide, the same gas that gives smog its characteristic color.*

The tendency of steel to react with air and moisture to form rust is a problem often solved by applying a protective coat of paint. Here the historic Williamsburg bridge, which connects Manhattan Island with Brooklyn in New York City, is receiving its periodic paint job.

metallic properties. Corrosion therefore wipes out the desirable properties for which metals are often chosen. For this reason, extremely reactive elements such as those of Group IA have very few practical uses, and none that would require exposure to the atmosphere.

We are able to tolerate a moderate ease of oxidation in some metals—iron, for example—because they have such desirable physical properties, but if severely corrosive conditions are to be encountered, iron must somehow be protected. Huge sums of money are spent annually, for instance, to paint steel structures such as bridges.

In Chapter 4 you learned that there are periodic trends in the ionization energies of the elements. The ionization energy is one of the properties that determines how easily metals lose electrons when they form compounds, so it should be no surprise that there are also periodic trends in the ease of oxidation of metals. These are illustrated in Figure 7.10. The ease of oxidation only roughly

Figure 7.10

General trends in the ease with which metals are oxidized. Elements with generally low ionization energies are easily oxidized.

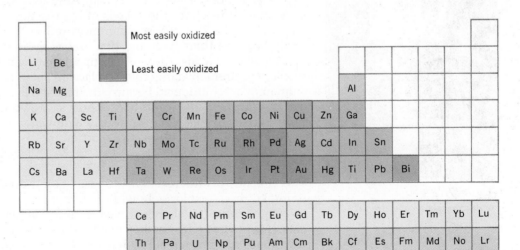

(a)

(b)

(c)

Figure 7.11
Relative rates of reaction of HCl with (a) iron, (b) zinc, and *(c) magnesium. Zinc reacts more rapidly than iron, and magnesium reacts more rapidly than zinc.*

Figure 7.12
Sodium reacts violently with water. The heat of the reaction ignites the sodium metal and the hydrogen gas produced by the reaction.

parallels the ionization energy, however, because the ionization energy applies to isolated atoms. In chemical reactions, metals normally occur as solids, so the ionization energy is only one factor involved. In studying Figure 7.10, you should learn where the very reactive metals are, as well as the general location of very unreactive ones.

For metals that are attacked by nonoxidizing acids, the ease of oxidation parallels the vigor with which they react with protons. In general, the more easily oxidized the metal, the faster it evolves H_2 (temperature and acid concentration being constant). For example, iron, zinc, and magnesium all react with HCl to liberate H_2. The general reaction is the same

$$M(s) + 2HCl(aq) \longrightarrow MCl_2(aq) + H_2(g)$$

where *M* stands for Fe, Zn, and Mg. Although the products are essentially the same in each reaction, the speeds of the reactions differ by quite a lot, as illustrated in Figure 7.11. These differences reflect the fact that magnesium is more easily oxidized than zinc, and that zinc is more easily oxidized than iron.

Of all the metals, those of Group IA are the most easily oxidized. In fact, it would be extremely dangerous to place an alkali metal such as sodium or potassium in hydrochloric acid because the reaction is explosively violent. These metals, due to their low ionization energies, are so easily oxidized by sources of protons that they all react vigorously with water to liberate hydrogen, as shown in Figure 7.12. For sodium, the reaction is

$$2Na(s) + 2H_2O \longrightarrow 2Na^+(aq) + 2OH^-(aq) + H_2(g)$$

The violence of this reaction increases going down the group. Rubidium, for example, reacts explosively with water.

The Group IIA elements have slightly larger ionization energies than the elements in Group IA, so they generally are not as easily oxidized. Calcium and the elements below it react with water, but less vigorously than the alkali metals.

$$Ca(s) + 2H_2O \longrightarrow Ca^{2+}(aq) + 2OH^-(aq) + H_2(g)$$

Mg does react slowly with boiling water to release H_2.

At room temperature, magnesium doesn't react with water because of an insoluble oxide coating that forms on its surface when the metal is exposed to air. Beryllium doesn't react with water at all; it is even less reactive than magnesium.

Most of the transition metals of period 4 react with dilute nonoxidizing acids such as HCl and H_2SO_4. Iron and zinc are examples that were discussed earlier. An exception to this general rule is copper, which requires an oxidizing acid such as HNO_3. The post–transition metals (the metals in Groups IIIA, IVA, and VA) are also reasonably reactive and react with nonoxidizing acids (although the reaction of lead is slow).

Electrical contacts in low-voltage circuits are gold-plated because even a little corrosion can stop the flow of electricity.

In Figure 7.10, we see that the least reactive metals are those in the lower center portion of the block of transition metals. Here we find platinum (Pt), iridium (Ir), and gold (Au)—metals so unreactive that they are not even attacked by concentrated nitric acid. Their superior resistance to chemical attack led them to be referred to as **noble metals.** They do dissolve slowly, however, in a mixture of concentrated HCl and concentrated HNO_3 called **aqua regia.** (Alchemists, using Latin, named this mixture *royal water* because it alone attacked gold, a *noble metal.*)

The activity series for metals

The reaction of an acid with a metal is characteristic of a much broader class of chemical reactions in which one element displaces another from a compound. Some people call them **single displacement reactions.** Another example of this kind of reaction is the change that occurs when a strip of metallic zinc is dipped into a solution containing copper(II) sulfate (Figure 7.13). After a while, the zinc strip has acquired a heavy deposit of reddish-brown copper, and the blue color of the copper(II) ion in the solution has faded. If we were to analyze the solution, it would be found to contain zinc ion. The reaction that has occurred is

$$Zn(s) + Cu^{2+}(aq) \longrightarrow Cu(s) + Zn^{2+}(aq)$$

Notice the similarity to the reaction of zinc with hydrogen ion.

$$Zn(s) + 2H^+(aq) \longrightarrow H_2(g) + Zn^{2+}(aq)$$

Reactions like that of zinc with copper(II) ion also allow us to rank metals according to their ease of oxidation. For example, we have just seen that zinc is able to reduce copper(II) ion in a solution, but if we dip a copper strip into a solution that contains Zn^{2+}, nothing happens (as we see in Figure 7.14).

$$Cu(s) + Zn^{2+}(aq) \longrightarrow \text{No reaction}$$

Figure 7.13

The reaction of zinc with copper(II) ion. (Left) A piece of shiny zinc next to a beaker containing a copper(II) sulfate solution. (Center) When the zinc is placed in the solution, copper(II) *ions are reduced to the free metal while the zinc dissolves. (Right) After a while the zinc becomes coated with a layer of copper.*

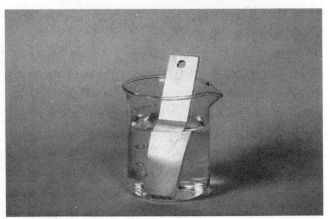

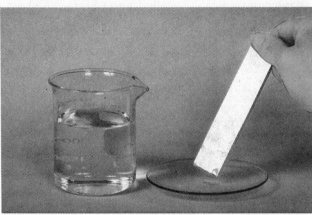

Figure 7.14

Although metallic zinc will displace copper from a solution containing Cu^{2+} ion, metallic copper will not displace Zn^{2+} from its solutions. Here we see that the copper bar is unaffected by being dipped into a solution of zinc sulfate.

Thus, zinc can displace copper from copper compounds, but copper can't displace zinc from zinc compounds. Stated another way, zinc willingly gives up electrons to copper(II) ion, but copper will not give up electrons to zinc ion. What this means is that zinc is more easily oxidized than copper (a fact that we've already established by comparing the effects of H^+ on zinc and copper metals.)

By comparing the abilities of metals to displace others from compounds, it is possible to place the metals in order of decreasing ease of oxidation. For example, consider the following reactions that are observed to take place experimentally:

$$Fe(s) + Pb^{2+}(aq) \longrightarrow Fe^{2+}(aq) + Pb(s)$$

$$Mg(s) + Fe^{2+}(aq) \longrightarrow Mg^{2+}(aq) + Fe(s)$$

$$Pb(s) + Cu^{2+}(aq) \longrightarrow Pb^{2+}(aq) + Cu(s)$$

The first reaction tells us that Fe is more easily oxidized than Pb; the second that Mg is more easily oxidized than Fe; and the third that Pb is more easily oxidized than Cu. Of these four metals, therefore, magnesium is most easily oxidized and copper is least easily oxidized. Place in order of decreasing ease of oxidation, they are

$$Mg > Fe > Pb > Cu$$

A list of metals arranged this way is called an **activity series,** and a more extensive version is given in Table 7.3. Those metals at the top of the table are most easily oxidized, and those at the bottom are least easily oxidized. Notice that the alkali and alkaline earth metals are high on the list, indicating their extreme ease of oxidation. Also notice that the noble metals are at the bottom of the list, which is indicative of their high resistance to oxidation.

One of the benefits of the activity series is that we can use it to determine the outcome of single displacement reactions. Any metal on the list is able to displace a metal below it from its compounds. For example, magnesium is above iron in the series. This means that magnesium is more easily oxidized than iron, and it is a better reducing agent than iron. If magnesium metal is placed into a solution of an iron compound, the magnesium will be oxidized and the iron ion will be reduced.

EXAMPLE 7.12 USING THE ACTIVITY SERIES TO PREDICT CHEMICAL REACTIONS

PROBLEM What will happen if a piece of chromium is dipped into a solution of silver nitrate?

SOLUTION If we refer to the activity series in Table 7.3, we see that chromium is above silver in the table. Therefore, chromium is able to reduce silver ion to metallic silver. The net ionic equation for the reaction is

$$Cr(s) + 3Ag^+(aq) \longrightarrow Cr^{3+}(aq) + 3Ag(s)$$

Notice three silver ions react for each chromium atom that reacts. This is necessary to make the net charge on both sides of the equation the same—a requirement for a balanced equation.

EXAMPLE 7.13 USING THE ACTIVITY SERIES TO PREDICT CHEMICAL REACTIONS

PROBLEM What will happen if a piece of lead is dipped into a solution that contains aluminum sulfate?

SOLUTION As before, we examine the activity series. This time we find that lead is *below* aluminum in the series, which means that lead is unable to reduce aluminum ion. As a result, there is no reaction in this mixture.

By this time you have probably noticed that hydrogen, H_2, is also listed in the activity series. Its location sets off those metals that are able to be oxidized by hydrogen ion. Any metal above hydrogen in the series is able to reduce H^+ to give H_2, so any metal above hydrogen will react with nonoxidizing acids such as HCl. Furthermore, any metal from sodium up to the top of the list will react with water to release H_2.

7.10 NONMETALS AS OXIDIZING AGENTS

In the last section you learned that metals serve as reducing agents when they react to form compounds. Nonmetals often behave in just the opposite way by functioning as oxidizing agents. In these instances, the nonmetal atom gains one or more electrons and acquires a negative oxidation number. For example, chlorine reacts with metals to form ionic chlorides that contain the Cl^- ion, and

Table 7.3
The activity series of metals

Metal	Oxidation Reaction	
Lithium	$Li \longrightarrow Li^+ + e^-$	
Cesium	$Cs \longrightarrow Cs^+ + e^-$	
Rubidium	$Rb \longrightarrow Rb^+ + e^-$	
Potassium	$K \longrightarrow K^+ + e^-$	
Barium	$Ba \longrightarrow Ba^{2+} + 2e^-$	
Strontium	$Sr \longrightarrow Sr^{2+} + 2e^-$	
Calcium	$Ca \longrightarrow Ca^{2+} + 2e^-$	
Sodium	$Na \longrightarrow Na^+ + e^-$	
Magnesium	$Mg \longrightarrow Mg^{2+} + 2e^-$	Ease of Oxidation Decreases
Aluminum	$Al \longrightarrow Al^{3+} + 3e^-$	
Manganese	$Mn \longrightarrow Mn^{2+} + 2e^-$	
Zinc	$Zn \longrightarrow Zn^{2+} + 2e^-$	
Chromium	$Cr \longrightarrow Cr^{3+} + 3e^-$	
Iron	$Fe \longrightarrow Fe^{2+} + 2e^-$	
Cadmium	$Cd \longrightarrow Cd^{2+} + 2e^-$	
Cobalt	$Co \longrightarrow Co^{2+} + 2e^-$	
Nickel	$Ni \longrightarrow Ni^{2+} + 2e^-$	
Tin	$Sn \longrightarrow Sn^{2+} + 2e^-$	
Lead	$Pb \longrightarrow Pb^{2+} + 2e^-$	
Hydrogen	$H_2 \longrightarrow 2H^+ + 2e^-$	
Copper	$Cu \longrightarrow Cu^{2+} + 2e^-$	
Silver	$Ag \longrightarrow Ag^+ + e^-$	
Mercury	$Hg \longrightarrow Hg^{2+} + 2e^-$	
Platinum	$Pt \longrightarrow Pt^{2+} + 2e^-$	
Gold	$Au \longrightarrow Au^{3+} + 3e^-$	

when oxygen reacts with carbon to form CO_2, the oxygen acquires an -2 oxidation number.

There are variations in the tendency of the nonmetals to act as oxidizing agents—some tend to acquire electrons more readily than others. As with the metals, there are periodic trends that also parallel the overall variations in ionization energy and electron affinity. Those elements with the larger ionization energies (and electron affinities) tend to be more easily reduced, and therefore they are the better oxidizing agents. Since ionization energy and electron affinity increase from left to right across a period and from bottom to top in a group, the best oxidizing agents are located in the upper right of the periodic table.

It is useful to know the specific trends in oxidizing ability across a period and down a group. Within a given period, the strongest oxidizing agents are located at the right. For example, in period 2 the strongest oxidizing agent is fluorine, and the second best is oxygen. Similarly, in period 3, the strongest oxidizing agent is chlorine. Within a group, the elements that are the best oxidizing agents are at the tops of their respective groups. Among the halogens, for example, the best oxidizing agent is fluorine, followed by chlorine, and so forth. In Group VIA, the best oxidizing agent is oxygen, followed by sulfur, and so on.

In the previous section you saw that a metal that is easily oxidized is able to displace one that is less easily oxidized from compounds. In other words, a stronger reducing agent can displace a weaker one. A similar phenomenon is observed among the nonmetals. An element that is a better oxidizing agent is able to displace a weaker oxidizing agent from compounds. For example, if fluorine is passed over sodium chloride, the more powerful oxidizing agent, fluorine, displaces the weaker one, chlorine, by oxidizing Cl^- to Cl_2.

$$F_2(g) + 2NaCl(s) \longrightarrow Cl_2(g) + 2NaF(s)$$

In a similar way, chlorine is able to oxidize bromide ion and iodide ion to molecular bromine and iodine, respectively.

$$Cl_2 + 2Br^- \longrightarrow 2Cl^- + Br_2$$
$$Cl_2 + 2I^- \longrightarrow 2Cl^- + I_2$$

7.11 MOLECULAR OXYGEN AS AN OXIDIZING AGENT

One of the most common and powerful oxidizing agents—in the laboratory, commerce, and our everyday existence—is molecular oxygen, O_2. Scientists realized very early on that oxygen combines with many different substances, and they called such reactions "oxidation." The removal of oxygen from a compound was called "reduction." It wasn't until some time later that the general nature of redox reactions was appreciated, and the term oxidation was then extended to cover any change in which an element, compound, or ion loses electrons to an oxidizing agent.

Although the reactions of oxygen are varied, there are some that are typical of specific classes of reactants.

Reactions of metals with oxygen

The formation of metal oxides by the direct reaction of metals with oxygen is a common event. In fact, in the form of corrosion, it represents both a nuisance and a source of economic waste in our modern society. Iron reacts with oxygen in the presence of moisture to form rust, an iron oxide whose crystals contain water molecules in variable amounts (as indicated by the variable coefficient x in the following equation).

Rust is a *hydrated* iron(III) oxide.

$$2Fe(s) + \tfrac{3}{2}O_2(g) + xH_2O(l) \longrightarrow Fe_2O_3{\cdot}xH_2O(s)$$

Aluminum, another common structural metal, also forms an oxide by direct reaction with oxygen in the air.

$$2Al(s) + \tfrac{3}{2}O_2(g) \longrightarrow Al_2O_3(s)$$

As you learned earlier, aluminum is even more easily oxidized than iron, so a fresh aluminum surface reacts rapidly with O_2 and quickly becomes covered by a thin oxide layer. Unlike rust on iron, however, the Al_2O_3 adheres tightly to the aluminum surface, effectively protecting it from further attack.

The corrosion of iron and aluminum in the air is a slow reaction. Some reactions of metals with oxygen are much more rapid. They evolve lots of heat and light and are generally referred to as **combustion**. An example is the reaction of magnesium with oxygen.

$$2Mg(s) + O_2(g) \longrightarrow 2MgO(s)$$

This chemical change produces not only a great deal of thermal energy, but also intense light—a fact that accounts for one of the important uses of magnesium—

Figure 7.15

Combustion of the fine magnesium wire inside the flashbulb produces a flash of light and leaves the interior of the bulb coated with magnesium oxide.

Steel is cut by a stream of pure oxygen whose reaction with red-hot iron produces enough heat to melt the metal and send a shower of burning steel sparks flying.

in flares and flashbulbs. A flashbulb (Figure 7.15) contains a fine magnesium wire in an atmosphere of pure oxygen. The flashbulb is fired by passing a small electric current through the wire, which heats it and sets off the combustion reaction. Afterwards, the interior of the bulb is coated with a thin deposit of the white powder MgO.

Although the corrosion of iron is a slow reaction, iron can be made to react much more rapidly by raising the temperature. For example, the cutting of steel by an acetylene torch is accomplished by first heating the steel to a high temperature with an oxygen-acetylene flame. Once the metal is very hot, the acetylene is turned off and the steel is bathed in a stream of pure oxygen. At this high temperature, the iron burns rapidly in the pure oxygen, and in the process a large quantity of heat is evolved that melts the steel and sends showers of sparks flying.

Reactions of nonmetals with oxygen

Oxygen also combines directly with most of the nonmetals to form covalent oxides. An example, with which you are surely familiar, is the reaction of O_2 with carbon (in the form of charcoal briquets or coal). In the presence of an excess amount of O_2, the product is carbon dioxide

$$C(s) + O_2(g) \longrightarrow CO_2(g)$$

but if only a limited supply of oxygen is available, significant amounts of carbon monoxide are formed

$$2C(s) + O_2(g) \longrightarrow 2CO(g)$$

Because of this second reaction, bags of charcoal contain the warning shown in the photograph in Figure 7.16.

Carbon monoxide itself is able to react with oxygen and form CO_2.

$$2CO(g) + O_2(g) \longrightarrow 2CO_2(g)$$

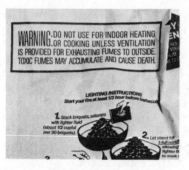

Figure 7.16

The reaction of elemental carbon with a limited supply of oxygen can produce dangerous carbon monoxide. This is why bags of charcoal bricquettes contain a warning about the hazards of using them indoors.

The reaction is quite exothermic, and CO is used industrially as a fuel because it can be made cheaply from coal and because it is easily transported through pipelines.

Two other nonmetals that react readily with oxygen are sulfur and phosphorus. Sulfur burns in the air with a blue flame and produces sulfur dioxide, a gas with an irritating choking odor.

$$S(s) + O_2(g) \longrightarrow SO_2(g)$$

This reaction is the first step in the synthesis of sulfuric acid, which is discussed in detail later in this book.

In Chapter 6 we discussed the allotropes of phosphorus—red phosphorus and white phosphorus. Both burn in oxygen to give the oxide P_4O_{10}, although the reaction of white phosphorus is spontaneous; it ignites spontaneously when the P_4 is exposed to air.

$$P_4(s) + 5O_2(g) \longrightarrow P_4O_{10}(s)$$
white
phosphorus

The large $:N\equiv N:$ bond energy prevents N_2 from reacting directly with O_2.

Not all the nonmetals react directly with oxygen to form oxides. Nitrogen is an example of one that doesn't, which explains why our atmosphere—a mixture of nitrogen and oxygen—is stable. Attempts to ignite mixtures of N_2 and O_2 are unsuccessful because reactions between them are endothermic. However, if air is heated to very high temperatures, such as in an automobile engine, a small amount of the oxide NO is produced. When released to the atmosphere through the auto exhaust, this compound begins a chain of reactions eventually causing smog.

Reactions of organic compounds with oxygen

Organic compounds are generally considered to be the compounds of carbon. Their number is enormous, and it is certainly not our intention to discuss them at length here. Our goal is merely to look at the typical kinds of reactions that they undergo when they are burned in O_2.

The simplest organic compounds are called *hydrocarbons*—compounds composed only of carbon and hydrogen. The simplest hydrocarbon is methane, CH_4, which is the chief component of natural gas. Hydrocarbons are also the main constituents of gasoline, kerosene, diesel oil, and candle wax.

Methane and other hydrocarbons burn readily in air. When sufficient oxygen is available, the products of combustion are carbon dioxide and water.

$$CH_4 + 2O_2 \longrightarrow CO_2 + 2H_2O$$
methane

$$2C_4H_{10} + 13O_2 \longrightarrow 8CO_2 + 10H_2O$$
butane
(in disposable
cigarette lighters)

$$2C_8H_{18} + 25O_2 \longrightarrow 16CO_2 + 18H_2O$$
octane
(in gasoline)

However, in a more limited supply of oxygen, the products can include carbon monoxide. For example

$$2CH_4 + 3O_2 \longrightarrow 2CO + 4H_2O$$

With an extremely limited supply of oxygen, only the hydrogen combines to form water, and a sooty flame containing elemental carbon is produced.

$$CH_4 + O_2 \longrightarrow C + 2H_2O$$

Carbon formed by this reaction is called lampblack. Its production is an important use of methane.

Organic compounds often contain elements in addition to carbon and hydrogen. If they contain oxygen, this is incorporated in the CO_2 and H_2O formed in the combustion reaction. For example, the combustion of methyl alcohol,

CH_3OH, the fuel in Sterno, follows the equation

$$2CH_3OH + 3O_2 \longrightarrow 2CO_2 + 4H_2O$$

Similarly, ethyl alcohol, C_2H_5OH, which is also known as grain alcohol and is the alcohol used to prepare the automotive fuel gasohol, burns according to the equation

$$C_2H_5OH + 3O_2 \longrightarrow 2CO_2 + 3H_2O$$

An important source of the sulfur dioxide found in polluted air is the presence of sulfur-containing organic compounds in coal and fuels made from high-sulfur crude oils. When these organic compounds burn, the sulfur in them is oxidized to SO_2. An example is the combustion of a compound called propyl mercaptan, C_3H_7SH, which is a strong–smelling compound that evolves from freshly chopped onions.

$$C_3H_7SH + 6O_2 \longrightarrow 3CO_2 + 4H_2O + SO_2$$

Other sulfur-containing compounds react is a similar way to yield the same products.

REVIEW QUESTIONS AND PROBLEMS

(Problems whose numbers are in blue have their answers in Appendix C at the back of the book; the more difficult problems are marked with asterisks.)

Reactions in Solution—General

7.1 Why are solutions usually employed for carrying out chemical reactions?

7.2 Given the meaning of the terms: solvent, solute, concentrated, dilute, saturated, supersaturated, unsaturated.

7.3 What is the meaning of the term solubility?

7.4 Why do we specify the temperature when giving the solubility of a solute in a particular solvent?

7.5 Can a saturated solution be dilute? Explain your answer.

7.6 How can a supersaturated solution be changed to a saturated solution?

7.7 The solubility of sugar in water increases with increasing temperature. Describe how you would prepare a supersaturated solution of sugar in water.

7.8 At 25°C, the solubility of $PbSO_4$ in water is 0.043 g dm^{-3}. How would you describe (concentrated, dilute, saturated, unsaturated, supersaturated) a solution that contains 0.050 g of $PbSO_4$ per dm^3?

Electrolytes

7.9 What is an electrolyte? What is a nonelectrolyte?

7.10 How can one distinguish between a strong and a weak electrolyte? What are present in solutions of electrolytes that are absent in solutions of nonelectrolytes?

7.11 Write chemical equations for the dissociation of the following strong electrolytes: KCl, $(NH_4)_2SO_4$, Na_3PO_4, $NaOH$, HCl.

7.12 Why is the hydronium ion often abbreviated H^+?

7.13 What is a dynamic equilibrium?

7.14 Cadmium sulfate, $CdSO_4$, is an exception to the rule that salts are strong electrolytes. In a 1.00 M solution, $CdSO_4$ is only 7% dissociated. Write an equation to represent the equilibrium dissociation of this compound.

7.15 Write an equation to represent the equilibrium dissociation of water.

7.16 What is the meaning of the term *position of equilibrium?*

7.17 What can be said about the position of equilibrium in (a) the dissociation of water; (b) the dissociation of HCl?

7.18 Use Lewis symbols to diagram the reaction of the strong electrolyte HBr with water.

*7.19 In the equilibrium ionization of acetic acid in water, how would you expect the rate of the reverse reaction (i.e., $C_2H_3O_2^- + H_3O^+ \rightarrow HC_2H_3O_2 + H_2O$) to change as the solution is diluted by adding more solvent? How should the rate of the forward reaction be affected? What should happen to the percentage ionization of $HC_2H_3O_2$ as the solution is made more dilute?

Ionic Reactions

7.20 What is the term used to describe a solid that is formed in a solution during a chemical reaction?

7.21 What is a metathesis reaction? What other name is often applied to this kind of reaction?

7.22 What is meant by the term *spectator ion?*

7.23 How can a precipitate be separated from a reaction mixture?

7.24 What are the differences between a molecular equation, an ionic equation, and a net ionic equation?

7.25 Of what value is a net ionic equation?

7.26 The compounds K_2CO_3, Na_2CO_3, KCl, and $NaCl$ are soluble in water, but $CaCO_3$ is not. Given the following molecular equation:

$$CaCl_2(aq) + Na_2CO_3(aq) \longrightarrow CaCO_3(s) + 2NaCl(aq)$$

write the ionic and net ionic equations for the reaction between solutions of $CaCl_2$ and K_2CO_3.

7.27 Write ionic and net ionic equations for the following:
(a) $CuCl_2(aq) + Pb(NO_3)_2(aq) \rightarrow$
$$Cu(NO_3)_2(aq) + PbCl_2(s)$$
(b) $FeSO_4(aq) + 2NaOH(aq) \rightarrow$
$$Fe(OH)_2(s) + Na_2SO_4(aq)$$
(c) $ZnSO_4(aq) + BaCl_2(aq) \rightarrow ZnCl_2(aq) + BaSO_4(s)$
(d) $2AgNO_3(aq) + K_2SO_4(aq) \rightarrow$
$$Ag_2SO_4(s) + 2KNO_3(aq)$$
(e) $(NH_4)_2CO_3(aq) + CaCl_2(aq) \rightarrow$
$$2NH_4Cl(aq) + CaCO_3(s)$$

7.28 Write *balanced* ionic and net ionic equations for the following:
(a) $Cu(NO_3)_2(aq) + NaOH(aq) \rightarrow$
$$Cu(OH)_2(s) + NaNO_3(aq)$$
(b) $BaCl_2(aq) + Al_2(SO_4)_3(aq) \rightarrow$
$$BaSO_4(s) + AlCl_3(aq)$$
(c) $Hg_2(NO_3)_2(aq) + HCl(aq) \rightarrow$
$$Hg_2Cl_2(s) + HNO_3(aq)$$
(d) $Bi(NO_3)_3(aq) + Na_2S(aq) \rightarrow$
$$Bi_2S_3(s) + NaNO_3(aq)$$
(e) $CaCl_2(aq) + Na_2SO_4(aq) \rightarrow$
$$CaSO_4(s) + NaCl(aq)$$

Acids and Bases in Aqueous Solutions

7.29 What are some properties of acids and bases?

7.30 For aqueous solutions, how are acids and bases defined?

7.31 How can you test a solution in the laboratory to find out whether it is basic?

7.32 What is the difference between a strong acid and a weak acid?

7.33 Make a list of the names and chemical formulas of the strong acids. (Refer to Table 7.2, if necessary.) Write equations for their ionization in water.

7.34 Write the equations for the stepwise ionization of the following acids in water: (a) H_2SO_3 and (b) H_3AsO_4.

7.35 Define these terms: (a) monoprotic acid, (b) diprotic acid, (c) triprotic acid, and (d) polyprotic acid.

7.36 What is an acidic anhydride? What is a basic anhydride?

7.37 For each of the following, state whether their aqueous solutions would be expected to be acidic or basic: (a) N_2O_3 (b) CaO (c) P_4O_{10} (d) SeO_2 (e) Cs_2O

7.38 The anhydride of nitric acid is N_2O_5. Write a balanced equation showing the reaction of N_2O_5 with water to give HNO_3.

7.39 What acid is formed by the reaction of P_4O_{10} with H_2O?

7.40 A newly discovered element is found to react with oxygen to form an oxide that, when dissolved in water, causes litmus to turn blue. Would the element be classed as a metal or a nonmetal? Why?

7.41 What are the two classes of substances that are bases in an aqueous solution?

7.42 Write an equation for the reaction of the oxide ion with H_2O.

7.43 What is the name of the base formed by the reaction of K_2O with water?

7.44 Write an equation for the ionization of ammonia in water? Is ammonia a strong or a weak base?

7.45 Hydrazine, H_2H_4, is a weak base in water. Write an equation that shows how this molecule reacts to produce a basic solution.

7.46 The Lewis formula of hydrazine is shown below. Use Lewis symbols to show why solutions of N_2H_4 in water are basic.

$$H-\overset{\displaystyle ..}{N}-\overset{\displaystyle ..}{N}-H$$
$$\quad\ \ |\quad\ \ |$$
$$\quad\ \ H\quad\ H$$

7.47 What is the net ionic equation for the reaction of a strong acid with a strong base?

7.48 Complete and balance the following equations:
(a) $KOH + HCl \rightarrow$
(b) $NaOH + HC_2H_3O_2 \rightarrow$
(c) $NH_3(aq) + HCl \rightarrow$
(d) $CuO + HBr \rightarrow$
(e) $Fe_2O_3 + H_2SO_4 \rightarrow$

7.49 What is an acid salt? Write the formulas of all the salts that can be formed by the reaction of KOH with H_3PO_4. Name each of these salts.

Brønsted–Lowry Acids and Bases

7.50 What are the Brønsted–Lowry definitions of an acid and a base?

7.51 What is the conjugate base of (a) NH_3, (b) NH_4^+, (c) $HC_2H_3O_2$, (d) H_3PO_4, (e) HNO_3?

7.52 What is the conjugate acid of (a) HSO_4^-, (b) SO_4^{2-}, (c) H_2O, (d) Cl^-, (e) CHO_2^-?

7.53 Why is the Brønsted–Lowry definition of acids and bases less restrictive than the Arrhenius definition?

7.54 Identify the two acid-base conjugate pairs in each of the following reactions:
(a) $C_2H_3O_2^- + H_2O \rightleftharpoons OH^- + HC_2H_3O_2$
(b) $HF + NH_3 \rightleftharpoons NH_4^+ + F^-$
(c) $Zn(OH)_2 + 2OH^- \rightleftharpoons ZnO_2^{2-} + 2H_2O$
(d) $Al(H_2O)_6^{3+} + OH^- \rightleftharpoons Al(H_2O)_5OH^{2+} + H_2O$
(e) $N_2H_4 + H_2O \rightleftharpoons N_2H_5^+ + OH^-$
(f) $NH_2OH + HCl \rightleftharpoons NH_3OH^+ + Cl^-$
(g) $O^{2-} + H_2O \rightleftharpoons 2OH^-$
(h) $H^- + H_2O \rightleftharpoons H_2 + OH^-$
(i) $NH_2^- + N_2H_4 \rightleftharpoons NH_3 + N_2H_3^-$
(j) $HNO_3 + H_2SO_4 \rightleftharpoons H_3SO_4^+ + NO_3^-$

7.55 Identify the acid-base conjugate pairs in each of the following reactions:
(a) $HClO_4 + N_2H_4 \rightleftharpoons N_2H_5^+ + ClO_4^-$
(b) $HSO_3^- + H_3PO_3 \rightleftharpoons H_2SO_3 + H_2PO_3^-$
(c) $C_5H_5NH^+ + (CH_3)_3N \rightleftharpoons C_5H_5N + (CH_3)_3NH^+$
(d) $CO_3^{2-} + H_2O \rightleftharpoons HCO_3^- + OH^-$
(e) $HCHO_2 + C_7H_5O_2^- \rightleftharpoons C_7H_5O_2H + CHO_2^-$
(f) $H_2C_2O_4 + CH_3NH_2 \rightleftharpoons HC_2O_4^- + CH_3NH_3^+$
(g) $H_2CO_3 + H_2O \rightleftharpoons HCO_3^- + H_3O^+$
(h) $C_2H_5OH + NH_2^- \rightarrow C_2H_5O^- + NH_3$
(i) $NO_2^- + N_2H_5^+ \rightleftharpoons HNO_2 + N_2H_4$
(j) $HCN + H_2SO_4 \rightleftharpoons H_2CN^+ + HSO_4^-$

7.56 Write autoionization reactions for the following solvents:
(a) $H_2O(l)$ (b) $NH_3(l)$ (c) $HCN(l)$

7.57 What would be the formula of the conjugate acid of dimethylamine, $(CH_3)_2NH$? What would be the formula of its conjugate base?

7.58 In water, the HCO_3^- ion is amphiprotic. Write equilibria to illustrate this.

7.59 Explain why a solution that contains the ion $Cr(H_2O)_6^{3+}$ is acidic.

Strengths of Acids

7.60 Which is the stronger acid?
(a) H_2SO_3 or $HClO_3$
(b) HNO_3 or H_2CO_3
(c) H_3AsO_4 or H_3PO_4

(d) 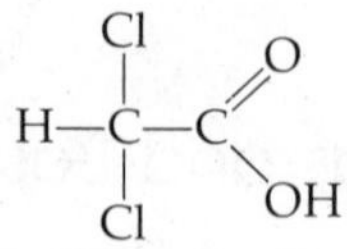 or

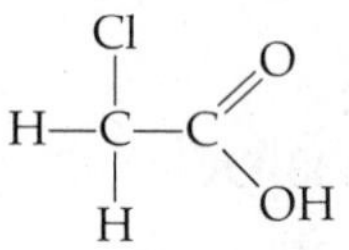

(e) 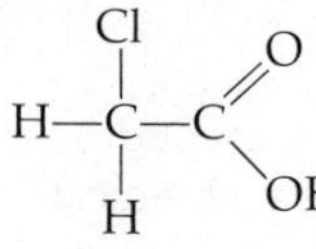 or

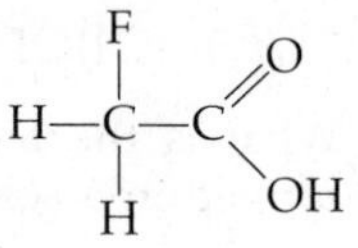

(f) 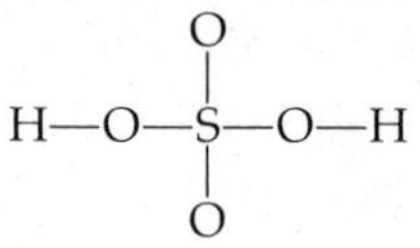 or

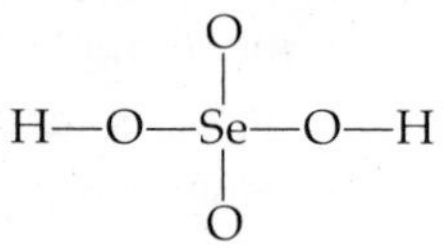

(g) $HClO_3$ or $HBrO_3$
(h) $HOBr$ or $HBrO_3$

7.61 Which is the stronger acid?
(a) H_2S or H_2Se
(b) H_2Se or HBr
(c) PH_3 or NH_3

7.62 Which compound would you expect to be more acidic, CH_3OH or CH_3SH?

7.63 Why is HNO_3 a stronger acid than HNO_2?

7.64 Why is HCl a stronger acid than HF?

7.65 Why is HCl a stronger acid than H_2S?

Lewis Acids and Bases

7.66 What is the Lewis definition of a base? What is a Lewis acid?

7.67 Explain why the reaction $H^+ + NH_3 \rightarrow NH_4^+$ is a Lewis acid-base reaction. Which is the Lewis acid and which is the Lewis base?

7.68 Boron trichloride, BCl_3, reacts with diethyl ether, $(C_2H_5)_2O$, to form an *addition compound*, which we can write as $Cl_3B\leftarrow O(C_2H_5)_2$. Use electron-dot formulas to interpret this reaction as a Lewis acid-base neutralization.

7.69 Explain why the reaction of CO_2 with H_2O to produce H_2CO_3, which we can also write as $CO(OH)_2$, can be viewed as a Lewis acid-base neutralization.

7.70 Silver ion reacts with ammonia to form an ion having the formula $Ag(NH_3)_2^+$ in which two ammonia molecules are bound to the Ag^+ ion. Explain how this can be viewed as a Lewis acid-base reaction.

Oxidation-Reduction: Oxidation Numbers

7.71 Define oxidation, reduction, oxidation state, oxidizing agent, and reducing agent.

7.72 Assign oxidation numbers to each atom in $KClO_2$, $BaMnO_4$, Fe_3O_4, O_2F_2, IF_5, $HOCl$, $CaSO_4$, $Cr_2(SO_4)_3$, O_3, and Hg_2Cl_2.

7.73 Assign oxidation numbers to each atom in H_2SO_4, CBr_4, OF_2, H_2O_2 (hydrogen peroxide), $CrCl_3$, Mn_2O_7, $KMnO_4$, $H_2C_2O_4$, $KClO_3$, and $LiNO_3$.

7.74 Identify the following changes as either oxidation or reduction:

(a) MnO_2 to MnO_4^- (d) OCl^- to ClO_3^-

(b) BiO_3^- to Bi^{3+} (e) N_2O_4 to N_2O

(c) SO_2 to SO_3

7.75 Many biological processes involve oxidation and reduction. For example, ethyl alcohol (grain alcohol) is metabolized in a series of oxidation steps that involve the following carbon-containing compounds:

7.76 Assign an average oxidation number to carbon in each of these compounds.

Oxidation-Number-Change Method

Balance the following molecular equations by the oxidation-number-change method:

(a) $HNO_3 + Zn \rightarrow Zn(NO_3)_2 + H_2O + NH_4NO_3$

(b) $K + KNO_3 \rightarrow N_2 + K_2O$

(c) $C_3H_7OH + Na_2Cr_2O_7 + H_2SO_4 \rightarrow$
$Cr_2(SO_4)_3 + Na_2SO_4 + H_2O + HC_3H_5O_2$

(d) $H_2S + HNO_3 \rightarrow S + NO + H_2O$

(e) $Fe(OH)_2 + O_2 + H_2O \rightarrow Fe(OH)_3$

7.77 Identify the oxidizing agent and the reducing agent in each of the reactions in the preceding question.

7.78 Balance the following molecular equations by the oxidation-number-change method:

(a) $Cu + HNO_3 \rightarrow Cu(NO_3)_2 + NO + H_2O$

(b) $MnO_2 + HBr \rightarrow Br_2 + MnBr_2 + H_2O$

(c) $(CH_3)_2CHOH + CrO_3 + H_2SO_4 \rightarrow$
$(CH_3)_2CO + Cr_2(SO_4)_3 + H_2O$

(d) $PbO_2 + Sb + NaOH \rightarrow PbO + NaSbO_2 + H_2O$

(e) $NO_2 + H_2O \rightarrow HNO_3 + NO$ (NO_2 is both oxidized *and* reduced)

7.79 Identify the substance oxidized and the substance reduced in each of the reactions in the preceding question.

Ion-Electron Method

7.80 Balance the following equations by the ion-electron method. All reactions occur in an acidic solution.

(a) $Cu + NO_3^- \rightarrow Cu^{2+} + NO$

(b) $Zn + NO_3^- \rightarrow Zn^{2+} + NH_4^+$

(c) $Cr + H^+ \rightarrow Cr^{3+} + H_2$

(d) $Cr_2O_7^{2-} + H_3AsO_3 \rightarrow Cr^{3+} + H_3AsO_4$

(e) $I^- + SO_4^{2-} \rightarrow I_2 + H_2S$

(f) $Ag^+ + AsH_3 \rightarrow H_3AsO_4 + Ag$

(g) $S_2O_8^{2-} + HNO_2 \rightarrow NO_3^- + SO_4^{2-}$

(h) $MnO_2 + Br^- \rightarrow Br_2 + Mn^{2+}$

(i) $S_2O_3^{2-} + I_2 \rightarrow I^- + S_4O_6^{2-}$

(j) $IO_3^- + HSO_3^- \rightarrow I^- + SO_4^{2-}$

7.81 Balance the following equations by the ion-electron method. All the reactions take place in an acidic solution.

(a) $Cr_2O_7^{2-} + CH_3CH_2OH \rightarrow Cr^{3+} + CH_3CHO$

(b) $PbO_2 + Cl^- \rightarrow Pb^{2+} + Cl_2$

(c) $Mn^{2+} + BiO_3^- \rightarrow MnO_4^- + Bi^{3+}$

(d) $ClO_3^- + HAsO_2 \rightarrow H_3AsO_4 + Cl^-$

(e) $PH_3 + I_2 \rightarrow H_3PO_2 + I^-$

(f) $MnO_4^- + S_2O_3^{2-} \rightarrow S_4O_6^{2-} + Mn^{2+}$

(g) $Mn^{2+} + PbO_2 \rightarrow MnO_4^- + Pb^{2+}$

(h) $As_2O_3 + NO_3^- \rightarrow H_3AsO_4 + N_2O_3$

(i) $P + Cu^{2+} \rightarrow Cu + H_2PO_4^-$

(j) $MnO_4^- + H_2S \rightarrow Mn^{2+} + S$

7.82 Balance the following equations by the ion-electron method. All the reactions take place in a basic solution.

(a) $CN^- + AsO_4^{3-} \rightarrow AsO_2^- + CNO^-$

(b) $CrO_2^- + HO_2^- \rightarrow CrO_4^{2-} + OH^-$

(c) $Zn + NO_3^- \rightarrow Zn(OH)_4^{2-} + NH_3$

(d) $Cu(NH_3)_4^{2+} + S_2O_4^{2-} \rightarrow SO_3^{2-} + Cu + NH_3$

(e) $N_2H_4 + Mn(OH)_3 \rightarrow Mn(OH)_2 + NH_2OH$

(f) $MnO_4^- + C_2O_4^{2-} \rightarrow MnO_2 + CO_3^{2-}$

(g) $ClO_3^- + N_2H_4 \rightarrow NO_3^- + Cl^-$

*7.83 Balance the following equations by the ion-electron method:

(a) $P_4 \rightarrow PH_3 + H_2PO_2^-$ (basic solution)

(b) $Cu + Cl^- + As_4O_6 \rightarrow CuCl + As$ (acidic solution)

(c) $IPO_4 \rightarrow I_2 + IO_3^- + H_2PO_4^-$ (acidic solution)

(d) $NO_2 \rightarrow NO_3^- + NO$ (acidic solution)

(e) $Br_2 \rightarrow Br^- + BrO_3^-$ (basic solution)

(f) $HSO_2NH_2 + NO_3^- \rightarrow N_2O + SO_4^{2-}$ (acidic solution)

(g) $ClO_3^- + Cl^- \rightarrow Cl_2 + ClO_2$ (acidic solution)

(h) $ClO_2 \rightarrow ClO_2^- + ClO_3^-$ (basic solution)

(i) $Se \rightarrow Se^{2-} + SeO_3^{2-}$ (basic solution)

(j) $ICl \rightarrow IO_3^- + I_2 + Cl^-$ (acidic solution)

(k) $FNO_3 \rightarrow O_2 + F^- + NO_3^-$ (basic solution)

(l) $Fe(OH)_2 + O_2 \rightarrow Fe(OH)_3$ (basic solution)

7.84 Balance the following equations. Both reactions occur in a basic solution.

(a) $Zn \rightarrow Zn(OH)_4^{2-} + H_2$

(b) $CrO_2^- + HO_2^- \rightarrow CrO_4^{2-}$

Metals as Reducing Agents

7.85 What kinds of oxidation states do metals have in virtually all their compounds?

7.86 Why do metals nearly always function as reducing agents when they react?

7.87 Why is HCl called a nonoxidizing acid?

7.88 Manganese reacts with HCl to give H_2 gas and Mn^{2+} in solution. Write a balanced equation for the reaction.

7.89 Write a balanced equation for the reaction of water with (a) sodium, (b) rubidium, (c) strontium.

7.90 Write an equation for the reaction of HBr(aq) with aluminum.

7.91 Why is HNO_3 referred to as an oxidizing acid?

7.92 Write balanced net ionic equations for the reaction of silver with (a) concentrated HNO_3 and (b) dilute HNO_3.

7.93 Why is nitric acid used to dissolve silver instead of hydrochloric acid?

7.94 If zinc is allowed to react with very dilute nitric acid, zinc is oxidized to Zn^{2+}, while the NO_3^- ion can be reduced to NH_4^+. Write a balanced net ionic equation for the reaction.

7.95 When hot and concentrated, H_2SO_4 is a potent oxidizing agent. It oxidizes zinc to Zn^{2+} and the sulfate can be reduced to H_2S. Write a balanced net ionic equation for the reaction.

7.96 Where, in general, are the easily oxidized metals located in the periodic table? Where do we find the metals that are least easily oxidized?

7.97 Based on trends in ionization energy, which element in each of the following pairs should be more easily oxidized?
(a) Rb or Sr   (c) Na or Al
(b) Na or Rb   (d) Al or Ca

7.98 Why do the metals below magnesium in Group IIA have very few practical applications?

7.99 Why are gold and platinum known as noble metals?

7.100 Write a balanced equation for the reaction of hydrochloric acid with (a) magnesium and (b) aluminum.

7.101 Write balanced equations for these reactions:
(a) chromium + hydrochloric acid →
chromium(III) chloride + hydrogen
(b) nickel + sulfuric acid →
nickel(II) sulfate + hydrogen

7.102 Write net ionic equations for the reactions in Questions 7.100 and 7.101. What is the oxidizing agent in each of these reactions?

7.103 Would you expect calcium to react more or less rapidly than magnesium with HCl? What is the basis for your answer?

7.104 What is *aqua regia*?

7.105 What is a *single displacement reaction*?

7.106 Based on the activity series in Table 7.3, predict the outcome of the following reactions:

(a) $Al(s) + Zn^{2+}(aq) \rightarrow$
(b) $Sn(s) + Cu^{2+}(aq) \rightarrow$
(c) $Ag(s) + Co^{2+}(aq) \rightarrow$
(d) $Mn(s) + Pb^{2+}(aq) \rightarrow$
(e) $Cu(s) + Mg^{2+}(aq) \rightarrow$
(f) $Hg(l) + H^+(aq) \rightarrow$
(g) $Ni(s) + H^+(aq) \rightarrow$
(h) $Cd(s) + H_2O \rightarrow$
(i) $Ba(s) + H_2O \rightarrow$
(j) $H_2(g) + Pt^{2+}(aq) \rightarrow$

Nonmetals as Oxidizing Agents

7.107 Arrange the following lists of nonmetals in order of increasing strength of the nonmetals as oxidizing agents:
(a) O, F, N, C
(b) I, Cl, Br, F

7.108 Complete the following equations. If no reaction occurs, Write N.R.
(a) $F_2 + Cl^- \rightarrow$
(b) $Br_2 + Cl^- \rightarrow$
(c) $I_2 + Cl^- \rightarrow$
(d) $Br_2 + I^- \rightarrow$

7.109 In general, how does the oxidizing ability of the nonmetals vary (a) across a period and (b) down a group?

Oxygen as an Oxidizing Agent

7.110 What is combustion?

7.111 What is rust? Write a chemical equation for its formation.

7.112 What is the product of corrosion of aluminum? Why does this cause less of a problem than the rusting of iron?

7.113 What chemical reaction takes place inside a flashbulb?

7.114 Write a chemical equation for the reaction of oxygen with (a) iron, (b) lithium, (c) calcium, (d) magnesium, (e) aluminum.

7.115 Write a chemical equation for the reaction of oxygen (present in plentiful supply) with (a) carbon, (b) sulfur, (c) phosphorus.

7.116 Write a chemical equation for the reaction of carbon in a limited supply of oxygen.

7.117 Write chemical equations for the complete combustion of (a) C_9H_{20}, (b) $C_2H_4(OH)_2$, (c) $(CH_3)_2S$.

7.118 If CH_4 is burned in a limited supply of O_2, what are the products of combustion?

7.119 One of the ingredients in candle wax is eicosane, $C_{20}H_{42}$. When candles burn, they usually produce a yellow sooty flame. Write the various chemical equations that illustrate the combustion of eicosane under these conditions.

CHAPTER 8

IONIC REACTIONS IN SOLUTION—A CLOSER LOOK

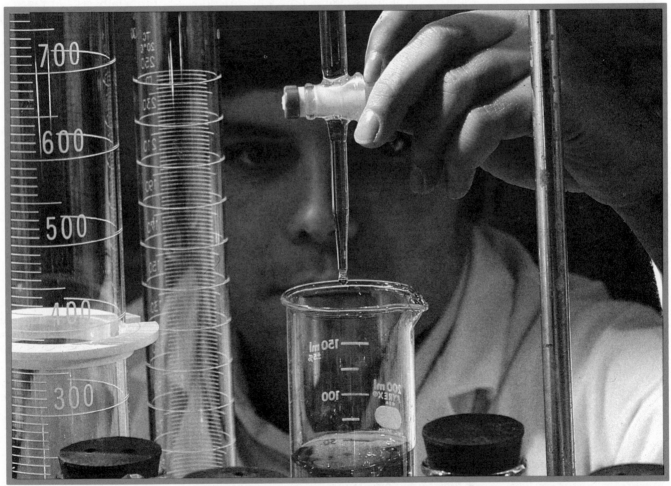

Ionic reactions have many practical uses, especially in laboratories that specialize in chemical analysis. The analytical technique seen here is called titration and is one of the topics discussed in this chapter.

Chemical reactions between ions are commonplace, not only in the chemistry lab, but in the home as well. For example, if you come from a locale that has "hard water," you know it is difficult to form a lather with soap. Instead, the soap tends to form an insoluble scum with the water. This action is the result of an ionic reaction. Hard water is formed when ground water seeps through deposits of limestone (calcium carbonate) or gypsum (calcium sulfate). Although these compounds are nearly insoluble, small quantities of them do dissolve. Since they are ionic compounds, they dissociate fully in solution, so the tap water contains calcium ion, Ca^{2+}, and either sulfate or carbonate ion. An example of soap is the ionic compound called sodium stearate, $NaC_{18}H_{35}O_2$. When this soap is dissolved in water, it also dissociates just like any other ionic compound. If the water contains calcium ions, the stearate ions react with them to form insoluble calcium stearate, which is the scum.

In this chapter, we will examine metathesis reactions like the one between calcium ions and stearate ions to learn why they occur and even how to predict them. We will also examine how ionic reactions—both metathesis and redox—can be of use in the lab.

8.1 METATHESIS REACTIONS: WHY THEY OCCUR

Recall that a metathesis reaction (also called a double replacement reaction) involves two compounds in solution and an exchange of cations between the two anions. An example is the reaction between silver nitrate and sodium chloride, for which we can write the following molecular, ionic, and net ionic equations.

Molecular equation

$$AgNO_3(aq) + NaCl(aq) \longrightarrow AgCl(s) + NaNO_3(aq)$$

Ionic equation

$$Ag^+(aq) + NO_3^-(aq) + Na^+(aq) + Cl^-(aq) \longrightarrow AgCl(s) + Na^+(aq) + NO_3^-(aq)$$

Net ionic equation

$$Ag^+(aq) + Cl^-(aq) \longrightarrow AgCl(s)$$

The reason this reaction occurs is because a precipitate of silver chloride is formed. In a sense, it is the formation of the precipitate that is the driving force for the reaction, and we can tell that a net reaction takes place because when we drop the spectator ions from the ionic equation, a net ionic equations exists. We can contrast this to the situation that exists when solutions of KCl and $NaNO_3$ are mixed. First, let's attempt to write a molecular equation for a metathesis reaction. We do this by exchanging cations between the chloride ion and the nitrate ion.

The spectator ions are Na^+ and NO_3^-.

$$KCl + NaNO_3 \longrightarrow KNO_3 + NaCl$$

Both of the reactants in this equation are soluble in water, as are both of the products. Let's make a habit of indicating this by writing (*aq*) after the formulas of the soluble substances.

$$KCl(aq) + NaNO_3(aq) \longrightarrow KNO_3(aq) + NaCl(aq)$$

To write metathesis equations correctly, you must be able to recognize polyatomic ions and know their electrical charges. If necessary, review Table 5.4 on page 150.

Now, let's construct the ionic equation. To do this, we write all soluble strong electrolytes in dissociated form. Because all the compounds in the equation are salts, we know they are all strong electrolytes, so the ionic equation is

$$K^+(aq) + Cl^-(aq) + Na^+(aq) + NO_3^-(aq) \longrightarrow$$
$$K^+(aq) + NO_3^-(aq) + Na^+(aq) + Cl^-(aq)$$

When we compare the left and right sides of this equation, we see that they are the same except for the sequence in which we have written the ions. If we cross out the spectator ions, there is nothing left, which tells us that there is no net chemical change in this mixture. Therefore, we would say that KCl does not react with $NaNO_3$ when their solutions are mixed. All we obtain is a mixture of the ions.

The formation of a precipitate is just one of *three* general driving forces in metathesis reactions. A net ionic equation also will exist if one of the products of the reaction is a weak electrolyte, or if one of the products is a gas. Let's examine all three of these in a bit more detail.

An ionic reaction will occur only if there is a net ionic equation.

Precipitation reactions

In the examples given above, we see that the formation of a precipitate in the reaction between $AgNO_3$ and NaCl prevents us from cancelling all the ions as spectator ions. As a result, a net ionic equation exists. On the other hand, if all the reactants and products are soluble, all the ions cancel and there is no reaction. In principle, therefore, if we knew the solubilities of all the compounds that could be formed between pairs of cations and anions, we could *predict*, based on the formation of a precipitate, when chemical reactions would occur.

Predicting metathesis reactions based on solubilities is not as simple as it may seem at first glance. This is because there is no sharp distinction between soluble and insoluble compounds. Certainly, substances such as sodium chloride would be considered soluble, and substances such as silver chloride insoluble. But between these extremes there are compounds like $PbCl_2$ and $AgC_2H_3O_2$, which are of intermediate solubility, and therefore are referred to as *partially soluble*, or *slightly soluble*.

Whether or not a precipitate of any particular salt will form when solutions of reactants are mixed depends on the concentrations of the ions that make up the salt. If these ion concentrations are large enough to make the reaction mixture supersaturated with respect to the salt, then a precipitate will form. If the reaction mixture isn't supersaturated, then no precipitate will form. Generally, a salt is said to be insoluble if a precipitate will form even if the concentrations of its ions are very small. However, if the potential product of the reaction is of moderate solubility, the concentrations of the ions may not be large enough to cause precipitation, and no reaction would be observed. For a partially soluble salt, a reaction will only be observed if the solutions of the reactants are fairly concentrated.

This entire discussion has been on a qualitative level. A quantitative treatment of solubilities of ionic solids is reserved until Chapter 16. For now, we will use the following **solubility rules** as a rough guide in predicting the course of metathesis reactions. In constructing these rules, we have divided all salts into only two categories: soluble and insoluble. In doing so, we've included the slightly soluble substances among the insoluble compounds, the assumption being that our predictions will apply to reaction mixtures in which the ion concentrations are approximately 0.1 M or larger.

If the concentrations of the ions are 0.1 M or larger, even the slightly soluble salts will precipitate, and a net reaction will be observed.

Solubility Rules

Salts that are soluble

1. All salts of the alkali metals are soluble.
2. All salts of the ammonium ion, NH_4^+, are soluble.
3. All salts of the following anions are soluble: nitrate ion (NO_3^-), chlorate ion (ClO_3^-), perchlorate ion (ClO_4^-), and acetate ion ($C_2H_3O_2^-$).

Salts that are generally soluble, with some exceptions

4. All chlorides, bromides, and iodides are soluble except those of Ag^+, Pb^{2+}, and Hg_2^{2+}. (Note that mercury in the +1 oxidation state exists as the diatomic ion Hg_2^{2+}.)
5. All sulfates (SO_4^{2-}) are soluble except those of Ca^{2+}, Sr^{2+}, Ba^{2+}, and Pb^{2+}.

Salts that are generally insoluble, with some exceptions

6. All metal oxides are insoluble, except for those of the alkali metals, Ca^{2+}, Sr^{2+}, and Ba^{2+}. Recall that soluble metal oxides are basic anhydrides and react with water to give hydroxide ion.

$$O^{2-} + H_2O \longrightarrow 2OH^-$$

Therefore, soluble metal oxides react with water to give their hydroxides in solution. For example

$$CaO(s) + H_2O \longrightarrow Ca^{2+}(aq) + 2OH^-(aq)$$

7. All hydroxides are insoluble, except for those of the alkali metals, Ca^{2+}, Sr^{2+}, and Ba^{2+}.
8. All carbonates (CO_3^{2-}), phosphates (PO_4^{3-}), sulfides (S^{2-}), and sulfites (SO_3^{2-}) are insoluble, except for those of NH_4^+ and the alkali metals. (Of course, since you know that all ammonium salts and all salts of the alkali metals are soluble, you already know these exceptions.)

A solution of a soluble metal oxide is really a solution of that metal's hydroxide. For example, a solution of Na_2O is really a solution of $NaOH$.

EXAMPLE 8.1 **USING THE SOLUBILITY RULES**

PROBLEM
Are the following salts soluble or insoluble, based on the solubility rules given above: (a) Na_2S, (b) $AgBr$, (c) $Cu(C_2H_3O_2)_2$?

SOLUTION
(a) Na_2S contains the ions Na^+ and S^{2-}. The cation is an alkali metal, so we need go no further. All salts of the alkali metals are soluble. Therefore, we conclude that Na_2S is soluble.

(b) $AgBr$ contains the ions Ag^+ and Br^-. This time we recognize the anion as one listed among the "soluble with exceptions." All salts of Br^- are soluble except those of Ag^+, Pb^{2+}, and Hg_2^{2+}. Since the cation is among the exceptions, this salt is insoluble.

(c) $Cu(C_2H_3O_2)_2$ contains the ions Cu^{2+} and $C_2H_3O_2^-$. Salts of the acetate ion, $C_2H_3O_2^-$, are included in the soluble category. Therefore, we know that $Cu(C_2H_3O_2)_2$ is soluble.

You should practice applying the solubility rules by working the exercises at the end of the chapter. Once you know them, you are ready to tackle problems like those illustrated in the next two examples.

EXAMPLE 8.2 **APPLYING THE SOLUBILITY RULES IN PREDICTING METATHESIS REACTIONS**

PROBLEM
Would a chemical reaction occur if solutions of $FeCl_3$ and KOH are mixed? If so, give the net ionic equation.

SOLUTION
The first step is to write an equation for the metathesis reaction that might occur. In doing this, be very careful to write the correct formulas of the products. Some students,

To write the correct formulas of the products, you must first decide what ions exist among the reactants. Then, combine them to form electrically neutral formula units.

for example, might be tempted to write the products as FeOH and KCl$_3$; *these are wrong.* You must take into account the charges on the ions to be sure that the formulas represent electrically neutral compounds. Since FeCl$_3$ is composed of Fe^{3+} and Cl$^-$ ions and KOH is composed of K$^+$ and OH$^-$ ions, the correct formulas of the products are Fe(OH)$_3$ and KCl. Assembling this into a balanced molecular equation gives

$$FeCl_3 + 3KOH \longrightarrow 3KCl + Fe(OH)_3$$

On the basis of our solubility rules, KCl is soluble but Fe(OH)$_3$ is not. Therefore, we would expect a precipitate of Fe(OH)$_3$ to form in the mixture and the KCl would remain dissociated in solution. The ionic equation for the reaction is

$$Fe^{3+}(aq) + 3Cl^-(aq) + 3K^+(aq) + 3OH^-(aq) \longrightarrow Fe(OH)_3(s) + 3K^+(aq) + 3Cl^-(aq)$$

Cancelling spectator ions (K$^+$ and Cl$^-$), we have the net ionic equation

$$Fe^{3+}(aq) + 3OH^-(aq) \longrightarrow Fe(OH)_3(s)$$

EXAMPLE 8.3

APPLYING THE SOLUBILITY RULES IN PREDICTING METATHESIS REACTIONS

PROBLEM

Would a chemical reaction be expected to occur if solutions containing NH$_4$NO$_3$ and Pb(C$_2$H$_3$O$_2$)$_2$ are mixed?

SOLUTION

To answer the question we must attempt to write a net ionic equation. First, we start with the molecular equation, exchanging NO$_3^-$ for C$_2$H$_3$O$_2^-$ and vice versa when we write the products.

$$2NH_4NO_3 + Pb(C_2H_3O_2)_2 \longrightarrow 2NH_4C_2H_3O_2 + Pb(NO_3)_2$$

Rules 2 and 3 tell us that all the reactants *and* products are soluble. The ionic equation can then be written

$$2NH_4^+(aq) + 2NO_3^-(aq) + Pb^{2+}(aq) + 2C_2H_3O_2^-(aq) \longrightarrow 2NH_4^+(aq) + 2C_2H_3O_2^-(aq) + Pb^{2+}(aq) + 2NO_3^-(aq)$$

If we cancel ions that are the same on both sides of the equation, everything disappears. Consequently, the answer to the question is that there is no net chemical reaction.

Reactions in which a weak electrolyte is formed

In any solution of a weak electrolyte, it is observed that only a small fraction of the solute is dissociated (ionized). Most of it exists in the form of molecules, rather than as ions. In a 1 M solution of acetic acid, for example, only about 0.42% of the acid exists as H$^+$ and C$_2$H$_3$O$_2^-$ ions, which means that the other 99.58% is present in the solution as molecules of HC$_2$H$_3$O$_2$. The reason for this, as you may recall, is that there is little tendency for the molecules of HC$_2$H$_3$O$_2$ to react with the solvent to form ions, but there is a strong tendency for the ions to react with each other to form molecules. As a result, the rates of the opposing reactions in the equilibrium

$$HC_2H_3O_2(aq) \rightleftharpoons H^+(aq) + C_2H_3O_2^-(aq)$$

can only become equal if most of the C$_2$H$_3$O$_2$ units are present in molecules.

If H$^+$ and C$_2$H$_3$O$_2^-$ ions are brought together in large numbers in a solution, an unstable situation exists. There is essentially no HC$_2$H$_3$O$_2$, so the ions will be combining at a much faster rate than they are being produced. Therefore, in just a brief instant nearly all the ions will disappear as they are replaced in the solution by molecules of the acid.

This is precisely what happens when solutions of the strong electrolytes HCl and $NaC_2H_3O_2$ are mixed. The ionic reaction is

$$H^+(aq) + Cl^-(aq) + Na^+(aq) + C_2H_3O_2^-(aq) \longrightarrow Na^+(aq) + Cl^-(aq) + HC_2H_3O_2(aq)$$

Notice that we've written the formula for acetic acid in nonionized (molecular) form. This is because most of the $HC_2H_3O_2$ exists this way. If we now cancel spectator ions, the net reaction is

$$H^+(aq) + C_2H_3O_2^-(aq) \longrightarrow HC_2H_3O_2(aq)$$

The driving force in this reaction—the reason a reaction occurs—is the decrease in the number of ions that takes place when the two fully dissociated reactants form the partially dissociated product.

One of the most important classes of reactions in aqueous solution that takes place by the formation of a weak electrolyte is acid-base neutralization. This reaction, you recall, involves the formation of water, a *very* weak electrolyte. If solutions of HCl and NaOH are mixed, the net ionic equation for the neutralization is

$$H^+(aq) + OH^-(aq) \longrightarrow H_2O$$

Water is such a weak electrolyte that its formation can cause insoluble oxides to dissolve in acids and weak acids to react with bases. For example, iron(III) oxide dissolves in a strong acid such as HCl as follows.

Molecular equation

$$Fe_2O_3(s) + 6HCl(aq) \longrightarrow 2FeCl_3(aq) + 3H_2O$$

Ionic equation

$$Fe_2O_3(s) + 6H^+(aq) + 6Cl^-(aq) \longrightarrow 2Fe^{3+}(aq) + 6Cl^-(aq) + 3H_2O$$

Net ionic equation

$$Fe_2O_3(s) + 6H^+(aq) \longrightarrow 2Fe^{3+}(aq) + 3H_2O$$

The reaction of acetic acid with sodium hydroxide follows these equations.

Molecular equation

$$HC_2H_3O_2(aq) + NaOH(aq) \longrightarrow NaC_2H_3O_2(aq) + H_2O$$

Ionic equation

$$HC_2H_3O_2(aq) + Na^+(aq) + OH^-(aq) \longrightarrow Na^+(aq) + C_2H_3O_2^-(aq) + H_2O$$

Net ionic equation

$$HC_2H_3O_2(aq) + OH^-(aq) \longrightarrow C_2H_3O_2^-(aq) + H_2O$$

EXAMPLE 8.4	**WRITING EQUATIONS FOR REACTIONS THAT INVOLVE THE FORMATION OF A WEAK ELECTROLYTE**
PROBLEM	Copper(II) oxide dissolves slowly in acetic acid to give a solution of copper(II) acetate. Write molecular, ionic, and net ionic equations for the reaction.
SOLUTION	The reactants are CuO and $HC_2H_3O_2$. We also know that an acid reacts with a metal oxide to give a salt and water. Therefore, the molecular equation is

$$CuO + 2HC_2H_3O_2 \longrightarrow Cu(C_2H_3O_2)_2 + H_2O$$

Now let's work on the ionic equation. To do this, we have to write all soluble strong electrolytes in dissociated form, and all weak electrolytes and insoluble compounds in molecular form. From the solubility rules, we know that CuO is insoluble in water (rule 6), and that $Cu(C_2H_3O_2)_2$ is soluble (rule 3). The ionic equation therefore becomes

$$CuO(s) + 2HC_2H_3O_2(aq) \longrightarrow Cu^{2+}(aq) + 2C_2H_3O_2^-(aq) + H_2O$$

This is also the net ionic equation because there are no spectator ions to cancel.

Figure 8.1
A solution of HCl being poured into a solution of Na_2CO_3.

Reactions in which a gas is formed

In some cases, the molecular species formed in a metathesis reaction is either insoluble in water and bubbles out as a gas, or it decomposes into a substance that evolves as a gas. For example, when HCl is added to a solution of Na_2S, one of the products is the weak electrolyte H_2S. H_2S, however, is a gas that has a low solubility in water, so it bubbles out of the reaction mixture. The molecular equation for the reaction is

$$2HCl(aq) + Na_2S(aq) \longrightarrow H_2S(g) + 2NaCl(aq)$$

for which the net ionic equation is

$$2H^+(aq) + S^{2-}(aq) \longrightarrow H_2S(g)$$

Another example is the reaction between HCl and Na_2CO_3, which produces the weak electrolyte H_2CO_3.

$$2HCl(aq) + Na_2CO_3(aq) \longrightarrow H_2CO_3(aq) + 2NaCl(aq)$$

The ionic equation for the reaction is

$$2H^+(aq) + 2Cl^-(aq) + 2Na^+(aq) + CO_3^{2-}(aq) \longrightarrow H_2CO_3(aq) + 2Na^+(aq) + 2Cl^-(aq)$$

and the net ionic equation is

$$2H^+(aq) + CO_3^{2-}(aq) \longrightarrow H_2CO_3(aq)$$

Carbonic acid, H_2CO_3, is unstable in large concentrations and decomposes readily to give CO_2 and H_2O. Carbon dioxide is not very soluble in water, and it escapes as a gas as shown in Figure 8.1. The equation for the decomposition of H_2CO_3 is

$$H_2CO_3(aq) \longrightarrow H_2O + CO_2(g)$$

Therefore, the net overall ionic equation for the reaction between hydrogen ion and carbonate ion is

$$2H^+(aq) + CO_3^{2-}(aq) \longrightarrow CO_2(g) + H_2O$$

A similar reaction occurs when bicarbonates are treated with acids.

$$H^+(aq) + HCO_3^-(aq) \longrightarrow H_2CO_3(aq) \longrightarrow CO_2(g) + H_2O$$

Both soluble and insoluble carbonates react with acids to liberate carbon dioxide.

The formation of CO_2 in the reaction of acids with carbonates is a sufficiently strong driving force for reaction to cause insoluble carbonates to react. For example, insoluble calcium carbonate, found in limestone, reacts with acids such as HCl to release CO_2.

$$CaCO_3(s) + 2HCl(aq) \longrightarrow CaCl_2(aq) + CO_2(g) + H_2O$$

The ionic and net ionic equations are

$$CaCO_3(s) + 2H^+(aq) + 2Cl^-(aq) \longrightarrow Ca^{2+}(aq) + 2Cl^-(aq) + CO_2(g) + H_2O$$

and

$$CaCO_3(s) + 2H^+(aq) \longrightarrow Ca^{2+}(aq) + CO_2(g) + H_2O$$

The reactions of acids with carbonates and bicarbonates are quite common. For instance, some people take sodium bicarbonate (baking soda) to soothe an upset stomach. The bicarbonate ion reacts with stomach acid (HCl) to give carbon dioxide, which makes the person burp. Another example is the product Alka Seltzer, which contains (among other things) sodium bicarbonate and a weak acid, citric acid. When the tablet is placed in water and begins to dissolve, the citric acid reacts with the sodium bicarbonate to give the fizz.

Acid spills in the lab can be safely neutralized by sprinkling solid $NaHCO_3$ on them, because the hydrogen ion combines with the bicarbonate ion to form harmless CO_2 and H_2O.

A list of some other gases that are evolved in metathesis reactions is given in Table 8.1. Among these, ammonia deserves special mention. Ammonia has quite a large solubility in water, so if it is formed in only small amounts, very little actually leaves the solution. Its presence can still be detected, however, because of the odor of ammonia that escapes from the solution.

EXAMPLE 8.5 WRITING EQUATIONS FOR REACTIONS THAT INVOLVE THE FORMATION OF A GAS

PROBLEM   Write a net ionic equation for the reaction between NH_4NO_3 and $Ba(OH)_2$

SOLUTION   We begin as usual by writing a double replacement reaction.

$$2NH_4NO_3 + Ba(OH)_2 \longrightarrow 2NH_4OH + Ba(NO_3)_2$$

The molecular species NH_4OH (ammonium hydroxide) does not really exist; it is fictitious. It actually is $NH_3 + H_2O$ (note that there are one N, five H, and one O in both NH_4OH and $NH_3 + H_2O$). We should therefore rewrite the equation as

$$2NH_4NO_3 + Ba(OH)_2 \longrightarrow 2NH_3 + 2H_2O + Ba(NO_3)_2$$

The ionic equation is obtained by applying the solubility rules and also noting that NH_3 is a gas (Table 8.1).

$$2NH_4^+(aq) + 2NO_3^-(aq) + Ba^{2+}(aq) + 2OH^-(aq) \longrightarrow 2NH_3(g) + 2H_2O + Ba^{2+}(aq) + 2NO_3^-(aq)$$

Finally, cancelling spectator ions gives

$$2NH_4^+(aq) + 2OH^-(aq) \longrightarrow 2NH_3(g) + 2H_2O$$

which becomes

$$NH_4^+(aq) + OH^-(aq) \longrightarrow NH_3(g) + H_2O$$

when the coefficients are reduced to the simplest set of whole numbers.

In this section we have developed ways for you to predict the outcome of a large number of ionic reactions. *To apply these methods, you must know the solubility rules on page 277, you must be able to recognize weak electrolytes as taught in Chapter 7, and you must know the contents of Table 8.1.* Learning these things will require a lot of work and practice on your part, but it is worthwhile because the knowledge gained is a valuable part of what you should take with you from this course.

Table 8.1
Gases that are formed in metathesis reactions

Gas	Typical Reaction in Which It Is Produced
CO_2	$Na_2CO_3 + 2HCl \longrightarrow H_2CO_3 + 2NaCl$ $H_2CO_3 \longrightarrow H_2O + CO_2(g)$ *Net equation:* $CO_3^{2-} + 2H^+ \longrightarrow CO_2(g) + H_2O$
SO_2	$Na_2SO_3 + 2HCl \longrightarrow H_2SO_3 + 2NaCl$ $H_2SO_3 \longrightarrow H_2O + SO_2(g)$ *Net equation:* $SO_3^{2-} + 2H^+ \longrightarrow H_2O + SO_2(g)$
NH_3	$NH_4Cl + NaOH \longrightarrow NH_3(g) + H_2O + NaCl$ *Net equation:* $NH_4^+ + OH^- \longrightarrow NH_3(g) + H_2O$
H_2S	$Na_2S + 2HCl \longrightarrow H_2S(g) + 2NaCl$ *Net equation:* $S^{2-} + 2H^+ \longrightarrow H_2S(g)$
NO NO_2	$NaNO_2 + HCl \longrightarrow HNO_2 + NaCl$ $2HNO_2 \longrightarrow H_2O + NO_2(g) + NO(g)$ *Net equation:* $2NO_2^- + 2H^+ \longrightarrow H_2O + NO_2(g) + NO(g)$

8.2 THE PREPARATION OF INORGANIC SALTS BY METATHESIS REACTIONS

An important aspect of chemistry is **synthesis,** the *planned* preparation of compounds from other materials. Chemical companies carry out such processes commercially to make all sorts of chemicals for industrial applications. Similarly, pharmaceutical companies conduct synthesis programs to manufacture drugs of various kinds. And both industrial and academic research laboratories specialize in the development of methods for preparing new and interesting chemicals—chemicals that might later be synthesized commercially on a much larger scale.

Inorganic salts—compounds like NaCl and KNO_3—are often used in the laboratory for a variety of purposes. Sometimes a particular salt that is needed isn't available from the stockroom, so either it must be purchased or synthesized from other available chemicals. For someone familiar with the topics covered in the previous section, synthesis can be quicker, easier, and less expensive than a purchase. But how do we proceed?

Suppose we needed to make sodium bromide, NaBr, for some particular purpose. How could we do it? If we planned to use a metathesis reaction, we would need a source of sodium ion and a source of bromide ion. Could we, therefore, simply mix a solution of NaCl with a solution of KBr? The answer is no. Although the resulting mixture would contain the constituents of NaBr, all the ions Na^+, K^+, Cl^-, and Br^- would be in the solution together, and it would be nearly impossible to isolate pure NaBr from the mixture. *We are looking for reactions in which one product is easily separated from the other, a condition that is met if one of the products is either a precipitate, a gas, or the solvent, water.*

Precipitation reactions

Precipitation reactions are particularly useful when the desired product is insoluble in water. For instance, suppose that we wished to prepare silver bromide, AgBr, the light sensitive substance used in most photographic film. This substance is insoluble and could be formed by the reaction

$$Ag^+(aq) + Br^-(aq) \longrightarrow AgBr(s)$$

Now that we know this, the trick is to transform the net ionic equation into a molecular equation so that we can choose the right chemicals with which to work. In doing this, let's look at the requirements we wish the chemicals to have.

1. Both reactants must be soluble. This means we must choose a soluble silver salt and a soluble bromide salt.

2. The other product in the metathesis reaction must be soluble so that the AgBr can be separated from the reaction mixture without being contaminated by another precipitate. This means that the salt formed from the anion of the silver compound and the cation of the bromide compound must also be soluble.

To be able to choose chemicals that fit these requirements, we must know the solubility rules well. One likely choice of reactants is $AgNO_3$ (all nitrates are soluble, so both conditions 1 and 2 are sure to be fulfilled) and NaBr (all salts of Na^+ are soluble). The molecular equation would then be

$$AgNO_3(aq) + NaBr(aq) \longrightarrow AgBr(s) + NaNO_3(aq)$$

The last step in preparing the AgBr would be to separate it from the mixture by filtration. In this preparation, it isn't necessary to worry too much about stoichiometry because if we want to prepare 1 mol of AgBr, we can add 1 mol of $AgNO_3$ to a solution containing *at least* 1 mol of NaBr. Any excess NaBr beyond that required to react completely with the $AgNO_3$ remains dissolved and does not contaminate the AgBr precipitate.

When the desired product remains in solution, greater attention must be paid to the stoichiometry of the reaction. For example, to prepare NaBr (which is water-soluble) by a precipitation reaction, we might mix solutions of Na_2SO_4 and $BaBr_2$, since $BaSO_4$ is insoluble. The net reaction is

$$Ba^{2+}(aq) + SO_4^{2-}(aq) \longrightarrow BaSO_4(s)$$

and the molecular equation is

$$Na_2SO_4(aq) + BaBr_2(aq) \longrightarrow BaSO_4(s) + 2NaBr(aq)$$

To prepare pure NaBr, we must react equal numbers of moles of Na_2SO_4 and $BaBr_2$. The solid $BaSO_4$ could then be filtered from the mixture and the volume of the resulting solution reduced by evaporation to the point at which nearly all the desired product crystallizes.

There are some instances when we can use relative degrees of insolubility in synthetic procedures. For instance, the silver halides decrease in solubility in this order: AgCl, AgBr, and AgI. If a solution containing KBr is stirred with solid AgCl, the solid is converted to the less soluble AgBr, and Cl^- replaces Br^- in solution. The net effect is that AgCl is converted to AgBr.

$$AgCl(s) + Br^-(aq) \longrightarrow AgBr(s) + Cl^-(aq)$$

To apply this technique, however, we must have a more detailed knowledge of solubilities than is provided by the solubility rules on page 277.

EXAMPLE 8.6	PREPARING A SALT BY A PRECIPITATION REACTION
PROBLEM	How could lead sulfate, $PbSO_4$ be prepared by a precipitation reaction?
SOLUTION	*Answering questions of this kind requires a thorough knowledge of the solubility rules.* Lead sulfate is insoluble and can be formed by the net ionic reaction

$$Pb^{2+}(aq) + SO_4^{2-}(aq) \longrightarrow PbSO_4(s)$$

To prepare it, we need a soluble lead salt and a soluble sulfate. Furthermore, the other product in the metathesis must be soluble. There are undoubtedly several alternatives. One of them is

$$Na_2SO_4(aq) + Pb(NO_3)_2(aq) \longrightarrow PbSO_4(s) + 2NaNO_3(aq)$$

$Pb(NO_3)_2$ was chosen because all nitrates are soluble. This ensures a soluble lead salt as a reactant *and* a soluble second product. After choosing $Pb(NO_3)_2$, we can pick any soluble sulfate as the other reactant.

Neutralization reactions

Another useful method of preparing salts is by the neutralization of an acid with a base. Suppose we wanted to prepare NaBr. In this case, we choose an acid that gives us the anion (HBr) and a base that gives us the cation (NaOH). When they are combined, the following reaction occurs:

$$NaOH + HBr \longrightarrow H_2O + NaBr$$

If we are careful to allow precisely 1 mol of NaOH to react with 1 mol of HBr, the only product in the aqueous solution is NaBr.

Metal oxides can be used in place of hydroxides in similar reactions. For example, the reaction

$$ZnO(s) + 2HClO_4(aq) \longrightarrow Zn(ClO_4)_2(aq) + H_2O$$

The reaction, $O^{2-} + 2H^+ \rightarrow H_2O$, has such a strong tendency to occur that insoluble oxides dissolve in acids.

could be used to prepare zinc perchlorate.

There is an advantage to using insoluble oxides or hydroxides as starting materials in these reactions. If an excess of the oxide or hydroxide is used, the limiting reactant will be the acid, and none of it will remain unreacted in the solution. This means that the only solute in the solution after the reaction is complete will be the desired salt.

EXAMPLE 8.7	PREPARING A SALT BY A NEUTRALIZATION REACTION
PROBLEM	How could we prepare calcium acetate, $Ca(C_2H_3O_2)_2$, one of the substances used to make Sterno (a brand of "canned heat"), by a neutralization reaction?
SOLUTION	We could use either the hydroxide or oxide of calcium, plus acetic acid. If we used the oxide, it would react with water to form the hydroxide, which could then be neutralized by the addition of acetic acid.

$$CaO(s) + H_2O \longrightarrow Ca(OH)_2(aq)$$
$$Ca(OH)_2(aq) + 2HC_2H_3O_2(aq) \longrightarrow Ca(C_2H_3O_2)_2(aq) + 2H_2O$$

EXAMPLE 8.8	PREPARING A SALT BY THE REACTION OF AN ACID WITH AN INSOLUBLE OXIDE
PROBLEM	How could we prepare 1 mol of copper(II) nitrate from copper(II) oxide and nitric acid?
SOLUTION	Copper(II) oxide is CuO, and it is insoluble in water. The reaction is therefore

$$CuO(s) + 2HNO_3(aq) \longrightarrow Cu(NO_3)_2(aq) + H_2O$$

If we make HNO_3 the limiting reactant, all the acid will be consumed and some of the

CuO will remain after all the nitric acid has reacted. This means that the only soluble solute in the reaction mixture will be $Cu(NO_3)_2$. When the reaction mixture is filtered, the filtrate will be essentially a solution of pure $Cu(NO_3)_2$, and evaporation of the solution will yield crystals of the pure salt. The proper procedure, therefore, is to use slightly more than 1 mol of CuO and exactly 2 mol of HNO_3. This amount of acid will produce exactly 1 mol of $Cu(NO_3)_2$ in the solution and leave some unreacted CuO.

Reactions in which one product is a gas

A very convenient way to prepare inorganic salts is provided by the reaction between an acid and a metal carbonate. Recall that the carbonate ion reacts with hydrogen ion to produce carbon dioxide and water,

$$CO_3{}^{2-} + 2H^+ \longrightarrow H_2O + CO_2(g)$$

This reaction proceeds readily regardless of whether the source of carbonate ion is a soluble or insoluble salt.

The approach here is very similar to the use of an insoluble oxide as a starting material.

The fact that most metal carbonates are insoluble leads to a very simple method for the preparation of pure salts. The synthesis of copper(II) bromide, for example, can be accomplished by the addition of HBr to a suspension of the insoluble $CuCO_3$ until nearly all the starting material reacts according to the equation

$$CuCO_3(s) + 2HBr(aq) \longrightarrow CuBr_2(aq) + H_2O + CO_2(g)$$

When the mixture is filtered to remove the undissolved $CuCO_3$, the solution that remains contains $CuBr_2$, which is virtually free of contaminants. The product can then be recovered by evaporation of the water.

The preparation of salts from soluble carbonates is also feasible; once again, however, attention must be paid to the stoichiometry of the reaction to obtain pure products. For example, NaBr can be made from Na_2CO_3 and HBr.

$$Na_2CO_3(aq) + 2HBr(aq) \longrightarrow 2NaBr(aq) + H_2O + CO_2(g)$$

The quantity of *pure* NaBr that can be crystallized from the reaction mixture depends on how closely a 1:2 mole ratio is maintained between the reactants Na_2CO_3 and HBr.

EXAMPLE 8.9	PREPARING A SALT BY THE REACTION OF A CARBONATE WITH AN ACID
PROBLEM	How could we prepare $Ca(ClO_4)_2$ by a reaction in which one product is a gas?
SOLUTION	We can allow $CaCO_3$ to react with $HClO_4$.

$$CaCO_3(s) + 2HClO_4(aq) \longrightarrow Ca(ClO_4)_2(aq) + CO_2(g) + H_2O$$

To obtain a pure product we use an excess of $CaCO_3$, which is filtered from the mixture after reaction has ceased.

EXAMPLE 8.10	USING A COMBINATION OF REACTIONS TO PREPARE A SALT BY METATHESIS
PROBLEM	How could we prepare $Cu(ClO_4)_2$ from $CuSO_4$?

SOLUTION
Often we can combine a series of reactions to give a final desired product. In this case, we know that we can prepare $Cu(ClO_4)_2$ from $Cu(OH)_2$ and $HClO_4$. Also, $Cu(OH)_2$ is insoluble and can be prepared from $CuSO_4$ by a precipitation reaction. One set of reactions is

$$CuSO_4(aq) + 2NaOH(aq) \longrightarrow Cu(OH)_2(s) + Na_2SO_4(aq)$$

$$Cu(OH)_2(s) + 2HClO_4(aq) \longrightarrow Cu(ClO_4)_2(aq) + 2H_2O$$

In practice, we would add the NaOH to the $CuSO_4$ and filter the insoluble $Cu(OH)_2(s)$. This solid would then be allowed to react with less than a stoichiometric amount of $HClO_4$ so that a little excess $Cu(OH)_2$ remains unreacted. This would be filtered to give a solution containing nearly pure $Cu(ClO_4)_2$.

8.3 STOICHIOMETRY OF IONIC REACTIONS

Toward the end of Chapter 2 we discussed some of the applications of stoichiometry to chemical reactions in solution, and if you look back at page 70, you will see that these included metathesis reactions. At that time, however, we could not treat them as ionic reactions because you had not yet been introduced to the concepts of ions and electrolytes. Now that we've overcome that restriction, we can examine how stoichiometric principles can be applied to ionic substances and reactions in solution.

Concentrations

In working quantitatively with solutes in a solution, it is necessary to know the solute's concentration. There are many ways of expressing concentration, and each has its advantages for specific applications. One way of expressing concentration, for example, is percent composition by mass, or *mass/mass percent*. Concentrated sulfuric acid, as we noted some time ago, is composed of 96% H_2SO_4 by mass. This gives the composition of the solution in parts per hundred by mass. In other words, it tells us how many grams of the solute are present in 100 g of solution.

EXAMPLE 8.11
WORKING WITH MASS PERCENT AS A CONCENTRATION UNIT

PROBLEM
How would you prepare a solution that is 5.00% (by mass) NaCl in water?

SOLUTION
The label tells us that there should be 5.00 g NaCl in 100 g of the solution. To prepare the solution, we add 95.0 g of water to 5.00 g of NaCl. Since the density of water is very close to 1.00 g cm^{-3}, we can use 95.0 cm^3 of water and save the trouble of weighing the water.

Analytical methods are becoming so sensitive that some substances can be detected at concentration levels of parts per billion (ppb).

A somewhat similar unit that is frequently used to express very small concentrations (e.g., the concentrations of impurities or pollutants in air or water) is **parts per million, ppm,** by volume or by mass. For instance, a typical carbon monoxide level in heavy smog is approximately 40 ppm by volume, whereas a typical nitrogen oxide concentration is about 0.2 ppm by volume.

EXAMPLE 8.12 CALCULATING A CONCENTRATION IN PARTS PER MILLION

PROBLEM Atmospheric SO_2, produced by the combustion of high-sulfur fuels, is an important air pollution problem. The amount of SO_2 is the air may be determined by bubbling the air through an acidic solution of $KMnO_4$, which oxidizes the SO_2 to SO_4^{2-}. In a typical analysis, 500 dm^3 of air with a density of 1.20 g dm^{-3} are passed through the $KMnO_4$ solution and 1.50×10^{-5} mol of $KMnO_4$ are reduced to Mn^{2+}. What is the concentration of SO_2 in parts per million (by mass)?

SOLUTION We first must have a balanced equation. The ion-electron method gives us

$$2MnO_4^- + 5SO_2 + 2H_2O \longrightarrow 2Mn^{2+} + 5SO_4^{2-} + 4H^+$$

If 1.50×10^{-5} mol of MnO_4^- is reduced, this requires (based on the stoichiometry of the reaction)

$$1.50 \times 10^{-5} \text{ mol MnO}_4^- \times \left(\frac{5 \text{ mol SO}_2}{2 \text{ mol MnO}_4^-} \right) \sim 3.75 \times 10^{-5} \text{ mol SO}_2$$

The mass of SO_2 in the air is

$$3.75 \times 10^{-5} \text{ mol SO}_2 \times \left(\frac{64.1 \text{ g SO}_2}{1 \text{ mol SO}_2} \right) \sim 2.40 \times 10^{-3} \text{ g SO}_2$$

The 500-dm^3 sample of air contained 2.40×10^{-3} g SO_2.

The concentration in parts per million is obtained as

$$\frac{\text{mass of SO}_2}{\text{total mass of air sample}} \times 10^6$$

The mass of the air sample can be obtained from its density

$$500 \text{ dm}^3 \times \left(\frac{1.20 \text{ g}}{\text{dm}^3} \right) \sim 600 \text{ g air}$$

Therefore, the concentration of the SO_2 pollutant is

$$\frac{2.40 \times 10^{-3} \text{ g SO}_2}{600 \text{ g air}} \times 10^6 = 4.00 \text{ ppm}$$

The most useful concentration unit for working problems in stoichiometry is molar concentration, or molarity. This was treated rather extensively in Section 2.13, and if necessary, you should review this section to freshen your memory. Molar concentration is particularly well suited for dealing with the stoichiometry of reactions in solution, and some of the aspects of this were examined in Chapter 2.

When working with ionic compounds and reactions in solution, one of the kinds of calculations that you should be able to perform routinely is determining the molar concentration of a particular ion in a solution of a strong electrolyte. For example, suppose you were given a solution labeled "1.00 M $CaCl_2$." What are the concentrations of Ca^{2+} ion and Cl^- ion in this solution? To answer this question, you have to realize that the given concentration refers to the number of moles of the salt per dm^3, and you also have to remember that salts are completely dissociated in an aqueous solution. In this case, there is 1.00 mol of $CaCl_2$ per dm^3, and $CaCl_2$ dissociates in solution as follows:

$$CaCl_2 \longrightarrow Ca^{2+} + 2Cl^-$$

One mole of $CaCl_2$ dissociates to give *one* mole of Ca^{2+} and *two* moles of Cl^-, so

Molar concentration is useful because it expresses the amount of solute in chemical units: moles.

there is one mole of Ca^{2+} per dm^3 and two moles of Cl^- per dm^3. Therefore, we also could label this solution "$1.00\ M\ Ca^{2+}$" and "$2.00\ M\ Cl^-$."

EXAMPLE 8.13

CALCULATING THE CONCENTRATION OF AN ION IN A SOLUTION

PROBLEM

What are the aluminum ion and sulfate ion concentrations in a $0.240\ M$ solution of $Al_2(SO_4)_3$?

SOLUTION

Each formula unit of $Al_2(SO_4)_3$ dissociates into two Al^{3+} ions and three SO_4^{2-} ions. Therefore, the number of moles of Al^{3+} is two times the number of moles of $Al_2(SO_4)_3$ specified. Similarly, the number of moles of SO_4^{2-} is three times the number of moles of $Al_2(SO_4)_3$ that are given. This means that

$$Al^{3+} \text{ concentration} = 2 \times (0.240\ M) = 0.480\ M$$
$$SO_4^{2-} \text{ concentration} = 3 \times (0.240\ M) = 0.720\ M$$

Stoichiometry

Once you have learned to calculate the concentrations of ions in a solution of a salt, you can easily use a net ionic equation to solve stoichiometry problems. This is illustrated in Example 8.14.

EXAMPLE 8.14

USING A NET IONIC EQUATION IN STOICHIOMETRY PROBLEMS

PROBLEM

Consider the net ionic equation for the reaction of chloride ion with permanganate ion in an acidic solution.

$$2MnO_4^-(aq) + 10Cl^-(aq) + 16H^+(aq) \longrightarrow 2Mn^{2+}(aq) + 5Cl_2(g) + H_2O$$

What volume (in cm^3) of $0.350\ M\ CaCl_2$ solution would be needed to generate 1.25 g of Cl_2 gas?

SOLUTION

First we calculate the number of moles of Cl_2. From the atomic weight table, the molar mass of Cl_2 is 70.9 g mol^{-1}. Therefore

$$1.25\ \cancel{g\ Cl_2} \times \left(\frac{1\ mol\ Cl_2}{70.9\ \cancel{g\ Cl_2}}\right) = 1.76 \times 10^{-2}\ mol\ Cl_2$$

$1.76 \times 10^{-2}\ \cancel{mol\ Cl_2} \times$
$\left(\dfrac{10\ mol\ Cl^-}{5\ \cancel{mol\ Cl_2}}\right) =$
$3.52 \times 10^{-2}\ mol\ Cl^-$

From the coefficients in the equation, it is clear that the formation of this amount of Cl_2 would require the oxidation of $2(1.76 \times 10^{-2}) = 3.52 \times 10^{-2}\ mol\ Cl^-$.

Now we have to calculate the volume of the $CaCl_2$ solution needed. A simple way of doing this is first to calculate the chloride ion concentration in the $CaCl_2$ solution. Since one $CaCl_2$ gives two Cl^- ions, the chloride ion concentration is two times the given concentration of the salt.

$$Cl^- \text{ concentration} = 2 \times (0.350\ M) = 0.700\ M\ Cl^-$$

Recall that molarity gives us a conversion factor that can be used in either of two ways. In this case, we can write

$$\frac{0.700\ mol\ Cl^-}{1000\ cm^3\ soln} \quad \text{or} \quad \frac{1000\ cm^3\ soln}{0.700\ mol\ Cl^-}$$

We use the second factor to obtain the answer.

$$3.52 \times 10^{-2}\ \cancel{mol\ Cl^-} \times \left(\frac{1000\ cm^3\ soln}{0.700\ \cancel{mol\ Cl^-}}\right) = 50.3\ cm^3\ soln$$

The volume of $CaCl_2$ solution needed is $50.3\ cm^3$.

The same principles that we applied in solving the problem in Example 8.14 can be used in working limiting reactant problems involving ionic reactions. This is illustrated by the following example.

EXAMPLE 8.15 WORKING LIMITING REACTANT PROBLEMS FOR IONIC REACTIONS

PROBLEM

Suppose that 20.0 cm³ of 0.150 M $Al_2(SO_4)_3$ are added to 30.0 cm³ of 0.200 M $BaCl_2$ solution, yielding a precipitate of $BaSO_4$.

$$Ba^{2+}(aq) + SO_4^{2-}(aq) \longrightarrow BaSO_4(s)$$

(a) How many grams of $BaSO_4$ will be formed in the reaction?

(b) What will be the concentrations of any ions remaining in the reaction mixture after the reaction is complete?

SOLUTION

In a problem of this type, it is best to begin by calculating the number of moles of each of the ions in the solution. The number of moles of aluminum sulfate in its solution is

$$20.0 \text{ cm}^3 \text{ soln} \times \left(\frac{0.150 \text{ mol } Al_2(SO_4)_3}{1000 \text{ cm}^3 \text{ soln}} \right) = 3.00 \times 10^{-3} \text{ mol } Al_2(SO_4)_3$$

and the number of moles of barium chloride in its solution is

$$30.0 \text{ cm}^3 \text{ soln} \times \left(\frac{0.200 \text{ mol } BaCl_2}{1000 \text{ cm}^3 \text{ soln}} \right) = 6.00 \times 10^{-3} \text{ mol } BaCl_2$$

Therefore, based on the formulas of the salts, we have

Al^{3+}	6.00×10^{-3} mol
SO_4^{2-}	9.00×10^{-3} mol
Ba^{2+}	6.00×10^{-3} mol
Cl^-	1.20×10^{-2} mol

Now we are ready to solve both parts of the problem.

6.00×10^{-3} mol Ba^{2+} requires only 6.00×10^{-3} mol SO_4^{2-}.

(a) The reaction involves Ba^{2+} and SO_4^{2-}, which react in a 1:1 mole ratio. Examining the numbers of moles of each, we can see that there is more sulfate ion than needed to react with all the barium ion. Therefore, Ba^{2+} is the limiting reactant, and we calculate the mass of product formed based on it.

$$6.00 \times 10^{-3} \text{ mol } Ba^{2+} \times \left(\frac{1 \text{ mol } BaSO_4}{1 \text{ mol } Ba^{2+}} \right) \times \left(\frac{233.4 \text{ g } BaSO_4}{1 \text{ mol } BaSO_4} \right) = 1.40 \text{ g } BaSO_4$$

(b) To calculate the concentrations of the ions in the final solution, we have to take into account the fact that one solution dilutes the other when the two of them are mixed. Therefore, we need the final volume. This is the sum of the volumes of the two solutions that were mixed, and it equals 50.0 cm³, or 0.0500 dm³.

Molar concentration =

$\dfrac{\text{moles of ion}}{\text{total volume of soln}}$

The concentrations of the spectator ions, Al^{3+} and Cl^-, are

$$\frac{6.00 \times 10^{-3} \text{ mol } Al^{3+}}{0.0500 \text{ dm}^3 \text{ soln}} = 0.120 \text{ M } Al^{3+}$$

$$\frac{1.20 \times 10^{-2} \text{ mol } Cl^-}{0.0500 \text{ dm}^3 \text{ soln}} = 0.240 \text{ M } Cl^-$$

In the final solution there is also some left-over sulfate ion that wasn't able to react with Ba^{2+}. The number of moles of this ion in the final solution is equal to the initial number of moles minus the number of moles that reacted.

$$\text{moles } SO_4^{2-} \text{ remaining} = (9.00 \times 10^{-3} \text{ mol}) - (6.00 \times 10^{-3} \text{ mol})$$
$$= 3.00 \times 10^{-3} \text{ mol } SO_4^{2-}$$

The sulfate ion concentration is therefore

$$\frac{3.00 \times 10^{-3} \text{ mol } SO_4^{2-}}{0.0500 \text{ dm}^3 \text{ soln}} = 0.0600 \text{ M } SO_4^{2-}$$

Because Ba^{2+} was the limiting reactant, it was used up completely, so its concentration is practically zero. (Actually, a very small amount of $BaSO_4$ still remains in solution. In Chapter 16, this will be treated quantitatively. For now, however, you can take the concentration of the limiting reactant to be essentially zero after the reaction is over.)

8.4 CHEMICAL ANALYSIS AND TITRATIONS

One of the problems frequently faced in chemistry is the determination of the composition of a substance, or of a mixture of substances. For example, in Chapter 2 you learned that the calculation of an empirical formula requires information about the amounts of each of the elements combined in a sample of the compound. When a substance appears as the product of a chemical reaction, it doesn't simply stand up and declare its composition—this information must be obtained experimentally. The name that we give to the experimental determination of chemical composition is **chemical analysis.**

Chemical analysis is not just something of interest to a research chemist who makes new compounds. Chemical, pharmaceutical, and cosmetic companies employ chemists whose sole job it is to analyze samples of the products being manufactured, so that quality can be maintained. Similarly, mining companies rely on analytical chemists to determine the compositions of the raw materials taken from the earth, so that they can be properly processed to obtain the desired end products. And environmental chemists use chemical analysis to determine whether pollutants are present in the air and water, and if so, what their concentrations are. In fact, one very important facet of chemistry is the investigation of new and better ways to achieve chemical analyses, particularly of compounds present in mixtures where difficulties in separating the components present a major obstacle to success.

Ionic reactions in solution can frequently be used to advantage in performing chemical analyses, where the kind of reaction used depends on the specific nature of the material being analyzed. Sometimes a precipitation reaction is convenient, as in Example 8.16 below. In this type of analysis, a reaction is chosen that allows us to separate the target of the analysis from the rest of the sample being studied. In Example 8.16, for instance, the goal is to analyze for sodium sulfate in the presence of sodium chloride. The reaction that is chosen is one that allows the separation of the sulfate from the chloride, so the amount of sulfate can be determined. That is why Ba^{2+} is added to a solution of the sample—$BaSO_4$ is insoluble but $BaCl_2$ is soluble.

EXAMPLE 8.16	USING A PRECIPITATION REACTION IN A CHEMICAL ANALYSIS
PROBLEM	A white powder was known to be a mixture of NaCl and Na_2SO_4. A sample of the powder weighing 1.244 g was dissolved in water and a solution of $Ba(NO_3)_2$ was added

until the precipitation of $BaSO_4$ was complete. The reaction mixture was filtered carefully to be sure that none of the precipitate was lost, and the $BaSO_4$ was then dried and found to have a mass of 0.851 g. What was the percent (by mass) of Na_2SO_4 in the original sample?

SOLUTION

This is really a very simple problem. Let's examine how we will solve it. First, we can calculate the number of moles of $BaSO_4$ that were obtained. From this, we can calculate the number of moles of Na_2SO_4 in the sample. We can do this because all the sulfate ion in the Na_2SO_4 was recovered in the $BaSO_4$ precipitate. Once we know the number of moles of Na_2SO_4 in the sample, we can calculate its mass in grams and then the percent by mass.

Now that we know how to proceed, the calculations are simple. The number of moles of $BaSO_4$ (molar mass 233.4 g mol^{-1}) is

$$0.851 \text{ g } \cancel{BaSO_4} \times \left(\frac{1 \text{ mol } BaSO_4}{233.4 \text{ g } \cancel{BaSO_4}} \right) = 3.65 \times 10^{-3} \text{ mol } BaSO_4$$

$BaSO_4$ is formed in the reaction

$$Ba^{2+}(aq) + SO_4^{2-}(aq) \longrightarrow BaSO_4(s)$$

$3.65 \times 10^{-3} \cancel{\text{mol } BaSO_4} \times$
$\left(\dfrac{1 \cancel{\text{mol } SO_4^{2-}}}{1 \cancel{\text{mol } BaSO_4}} \right) \times$
$\left(\dfrac{1 \text{ mol } Na_2SO_4}{1 \cancel{\text{mol } SO_4^{2-}}} \right) =$
$3.65 \times 10^{-3} \text{ mol } Na_2SO_4$

so each mole of $BaSO_4$ is formed from one mole of sulfate ion. The number of moles of SO_4^{2-} in the sample, therefore, is also 3.65×10^{-3} mol. From the formula Na_2SO_4, we see that each mole of this compound yields one mole of sulfate, so the number of moles of Na_2SO_4 is 3.65×10^{-3}, too. The molar mass of Na_2SO_4 is 142.0 g mol^{-1}, so the number of grams of Na_2SO_4 in the original sample was

$$3.65 \times 10^{-3} \cancel{\text{mol } Na_2SO_4} \times \left(\frac{142.0 \text{ g } Na_2SO_4}{1 \cancel{\text{mol } Na_2SO_4}} \right) = 0.518 \text{ g } Na_2SO_4$$

The percent by mass of Na_2SO_4 in the sample is obtained as

$$\text{percent } Na_2SO_4 = \frac{\text{mass of } Na_2SO_4}{\text{mass of sample}} \times 100$$

Substituting values gives

$$\text{percent } Na_2SO_4 = \frac{0.518 \text{ g } Na_2SO_4}{1.244 \text{ g sample}} \times 100$$
$$= 41.6\% \text{ } Na_2SO_4$$

The sample is 41.6% Na_2SO_4.

Titrations

Titration is an analytical procedure that allows us to measure the amount of one solution needed to react exactly with the contents of another solution. Such analyses, which involve the measurements of volumes of solutions of reactants, are called **volumetric analyses.** In a titration, one of the solutions that contains a reactant is placed in a **buret**—a long tube fitted at one end with a valve (called a **stopcock**) and precisely graduated in volume units (see Figure 8.2). The solution in the buret is called the **titrant,** and during a titration this solution is delivered slowly through the stopcock into another vessel that contains a solution of the other reactant. The titrant is added until complete reaction is signaled by an **indicator,** a substance that generally is added to the solution in the receiving vessel and which undergoes some sort of color change when the reaction is over. The color change of the indicator marks the **endpoint** in the titration, so named because it is at this point that the delivery of the titrant is stopped and the volume of the titrant used in the reaction recorded.

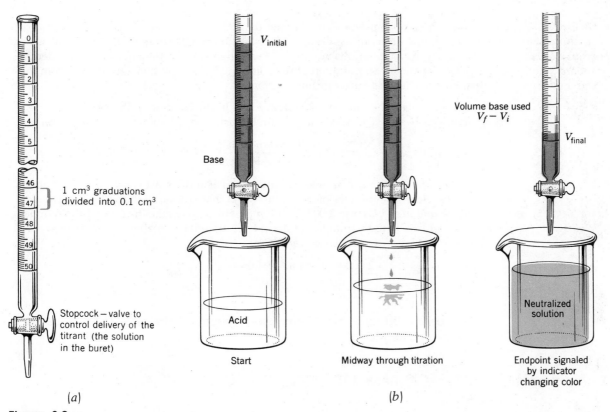

Figure 8.2

(a) *A buret.* (b) *The titration of an acid with a base.*

One kind of reaction that is frequently used in titrations is acid-base neutralization. Normally, the solution of the base is the titrant and the solution of the acid is placed in the receiving flask or beaker. The indicator is a substance that has one color in an acidic solution and another color in a basic solution. Litmus is such a substance—you may recall that in the presence of an acid, litmus is pink, and in the presence of a base, litmus is blue. Another indicator you may use in the laboratory is phenolphthalein (the second "ph" is silent when you pronounce it). Phenolphthalein is colorless in an acidic solution and pink in a basic solution.

EXAMPLE 8.17   CALCULATIONS INVOLVING TITRATIONS

PROBLEM   A solution of sodium hydroxide was prepared having a concentration of approximately 0.1 M. It was desired to measure the concentration of this solution accurately, so a 20.00 cm^3 portion of a 0.1000 M solution of HCl was carefully measured into a beaker and a few drops of phenolphthalein was added to it. A buret was filled with the NaOH solution and used to titrate the HCl solution. The titration required 18.47 cm^3 of the base in order to reach the endpoint. What is the concentration of the NaOH solution?

SOLUTION   Since the problem deals with a chemical reaction, let's begin with a balanced chemical equation.

$$NaOH(aq) + HCl(aq) \longrightarrow NaCl(aq) + H_2O$$

Determining the exact concentration of a solute in a solution is called *standardizing the solution.* A standard solution is one with a precisely known concentration.

To solve the problem, we need to know the exact ratio of moles of NaOH to dm³ of solution. In the titration, we used 18.47 cm³ of the NaOH solution. If we can calculate the amount of NaOH in this volume, we will then have all the information needed to calculate the molar concentration of the solution. We can get this information from the quantities of chemicals used in the reaction.

First, we calculate the number of moles of HCl in the solution in the beaker *before* the titration was begun. This is obtained from the acid's initial volume and concentration.

$$20.00 \; \cancel{cm^3 \; HCl \; soln} \times \left(\frac{0.1000 \; mol \; HCl}{1000 \; \cancel{cm^3 \; HCl \; soln}} \right) = 2.000 \times 10^{-3} \; mol \; HCl$$

2.000×10^{-3} mol NaOH was in the 18.47 cm³ of titrant.

The coefficients of the equation tell us that the HCl and NaOH combine in a 1:1 mole ratio. Therefore, the number of moles of NaOH that reacted with the HCl in order to reach the endpoint was also 2.000×10^{-3} mol. Finally, we take the ratio of the number of moles of NaOH to the number of dm³ of solution that contained it. Since 18.47 cm³ equals 0.01847 dm³, the molar concentration is

$$\frac{2.000 \times 10^{-3} \; mol \; NaOH}{0.01847 \; dm^3 \; soln} = 0.1083 \; M \; NaOH$$

Acid-base titrations find many uses in chemical analyses. This is illustrated in the next example.

EXAMPLE 8.18 **USING ACID-BASE TITRATIONS IN A CHEMICAL ANALYSIS**

PROBLEM   A sample of an analgesic drug was analyzed for aspirin, a monoprotic acid, $HC_9H_7O_4$, by titration with a base. In a titration, a 0.500-g sample of the drug required 21.50 cm³ of 0.100 *M* NaOH for complete neutralization. What percentage of the drug was aspirin?

SOLUTION   Let's begin with a chemical equation for the reaction.

$$HC_9H_7O_4 + NaOH \longrightarrow NaC_9H_7O_4 + H_2O$$

From the volume of the base and its concentration we can calculate the number of moles of base consumed.

$$21.50 \; \cancel{cm^3 \; soln} \times \left(\frac{0.1000 \; mol \; NaOH}{1000 \; \cancel{cm^3 \; soln}} \right) = 0.002\;15 \; mol \; NaOH$$

$0.002\;15 \; \cancel{mol \; NaOH} \times$
$\left(\dfrac{1 \; mol \; HC_9H_7O_4}{1 \; \cancel{mol \; NaOH}} \right) =$
$0.002\;15 \; mol \; HC_9H_7O_4$

The coefficients of $HC_9H_7O_4$ and NaOH are the same in the chemical equation, so the number of moles of each that react are the same. Therefore, the number of moles of $HC_9H_7O_4$ in the sample was $0.002\;15$ mol $HC_9H_7O_4$. The molar mass of aspirin is $180.2 \; g \; mol^{-1}$, so the number of grams of aspirin in the sample was

$$0.002\;15 \; \cancel{mol \; HC_9H_7O_4} \times \left(\frac{180.2 \; g \; HC_9H_7O_4}{1 \; \cancel{mol \; HC_9H_7O_4}} \right) = 0.387 \; g \; HC_9H_7O_4$$

The percent aspirin is found as

$$percent \; aspirin = \frac{mass \; aspirin}{mass \; sample} \times 100$$

Substituting values gives

$$percent \; aspirin = \frac{0.387 \; g \; HC_9H_7O_4}{0.500 \; g \; sample} \times 100$$
$$= 77.4\%$$

8.5 OXIDIZING AND REDUCING AGENTS IN THE LABORATORY

In the last section we discussed one of the laboratory applications of metathesis and neutralization reactions—chemical analyses. There also are many occasions when redox reactions are useful in the lab. These include both synthesis and analysis. In these applications, there are certain oxidizing and reducing agents that are used more often than others, either because of their general availability or because of their special properties.

Oxidizing agents

There are many relatively strong oxidizing agents that could be used in laboratory applications. In choosing an oxidizing agent, however, one of the criteria is always ease of use, and this tends to exclude certain substances most of the time. For example, chlorine is quite a powerful oxidizing agent, but it can't be used in an open laboratory because it is a poisonous gas—special precautions must be taken to use chlorine safely. Therefore, in most applications we would avoid chlorine as an oxidizing agent of choice.

The three most common oxidizing agents in the laboratory are the permanganate ion, MnO_4^-, chromate ion, CrO_4^{2-}, and the dichromate ion, $Cr_2O_7^{2-}$. These are all very powerful oxidizing agents and should be handled with care. They are quite effective in supplying oxygen and should not be allowed to contact organic materials because of the potential for fire.

Permanganate Ion. The permanganate ion itself is purple, and that is the color exhibited by solutions containing this ion (Figure 8.3). Generally, the permanganate ion is available as the purplish-black potassium salt, $KMnO_4$.

When permanganate ion functions as an oxidizing agent, the manganese is reduced from the +7 oxidation state. However, the oxidation state of Mn in the product depends on the acidity of the solution. If the reduction takes place in a strongly acidic solution, the manganese is reduced to the nearly colorless Mn^{2+} ion according to the half-reaction

$$8H^+(aq) + MnO_4^-(aq) + 5e^- \longrightarrow Mn^{2+}(aq) + 4H_2O$$

This can lead to a very dramatic color change if the other reactants and products in the redox reaction have little or no color themselves. In Figure 8.3 we see a

Figure 8.3

A solution of $KMnO_4$ being poured into a stirred acidic solution containing Fe^{2+}.

solution of $KMnO_4$ being poured in a solution that contains Fe^{2+} and a large concentration of sulfuric acid. The purple of the MnO_4^- ion disappears almost immediately as the solutions mix and the following reaction occurs

$$5Fe^{2+}(aq) + 8H^+(aq) + MnO_4^-(aq) \longrightarrow Mn^{2+}(aq) + 4H_2O + 5Fe^{3+}(aq)$$

If the permanganate ion is reduced in a neutral or slightly basic solution, the product of the reduction is generally insoluble manganese dioxide, MnO_2. The half-reaction for this reduction is

$$2H_2O + MnO_4^-(aq) + 3e^- \longrightarrow MnO_2(s) + 4OH^-(aq)$$

Chromate Ion and Dichromate Ion The sodium and potassium salts of these ions are commonly found on laboratory shelves. Both chromate and dichromate ions contain chromium in the +6 oxidation state, and they can be converted from one to the other by adjusting the acidity of the solution. If a solution containing the yellow chromate ion is made acidic, the CrO_4^{2-} is converted to the red-orange $Cr_2O_7^{2-}$.

$$2CrO_4^{2-}(aq) + 2H^+(aq) \longrightarrow Cr_2O_7^{2-}(aq) + H_2O$$
chromate ion   dichromate ion

On the other hand, if the solution contains dichromate ion and is made basic, the $Cr_2O_7^{2-}$ is converted to CrO_4^{2-}.

$$Cr_2O_7^{2-}(aq) + 2OH^-(aq) \longrightarrow 2CrO_4^{2-}(aq) + H_2O$$

Because of these reactions, when used in an acidic solution, the active oxidizing agent is $Cr_2O_7^{2-}$. If the solution is basic, however, the oxidizing agent is CrO_4^{2-}.

Regardless of the acidity of the solution, when these ions act as oxidizing agents, the chromium is reduced to the +3 oxidation state. However, the formula of the product that contains the chromium does depend on how acidic or basic the solution is. In acidic solutions, the chromium is reduced to Cr^{3+} ion. In slightly basic solutions the reduction product is insoluble $Cr(OH)_3$. And in a very basic solution, the chromate ion is reduced to the ion CrO_2^-, which is called the chromite ion.

Acidic solution

$$6e^- + 14H^+(aq) + Cr_2O_7^{2-}(aq) \longrightarrow 2Cr^{3+}(aq) + 7H_2O$$

Slightly basic solution

$$3e^- + 4H_2O + CrO_4^{2-}(aq) \longrightarrow Cr(OH)_3(s) + 5OH^-(aq)$$

Very basic solution

$$3e^- + 2H_2O + CrO_4^{2-}(aq) \longrightarrow CrO_2^-(aq) + 4OH^-(aq)$$

Redox titrations and chemical analysis

Redox reactions are often used in chemical analyses, and one of the most useful oxidizing agents for titrations is potassium permanganate. As we mentioned above, solutions of this salt are purple because they contain the purple MnO_4^- ion. We also noted that when permanganate ion is reduced in an acidic solution, the manganese is changed to Mn^{2+}, which is nearly colorless. (It is actually a very pale pink, but in dilute solutions the color is extremely difficult to detect.) This sharp change in color allows MnO_4^- to serve as its own indicator in a titration, provided the other substances involved in the reaction are not deeply colored themselves. Here is how it works: As the $KMnO_4$ solution is added to the reaction mixture from the buret, the MnO_4^- is reduced and the purple color immediately fades. This continues as long as any reducing agent is present in the reaction mixture. But as soon as the last traces of the reducing agent have been oxidized, the next drop of $KMnO_4$ solution added has nothing with which to react, so the solution takes on a noticeable pink color because of the presence of the small amount of unreacted MnO_4^-. The use of MnO_4^- in a typical analysis is illustrated in the next example.

EXAMPLE 8.19 CHEMICAL ANALYSIS USING A REDOX TITRATION

PROBLEM A 1.000-g sample of an iron ore containing Fe_2O_3 was dissolved in acid and all the iron converted to Fe^{2+}. The solution was titrated with 90.40 cm^3 of 0.020 00 M $KMnO_4$ to give Fe^{3+} and Mn^{2+} among the products. What percent of the ore was Fe_2O_3?

SOLUTION As usual, we will need a balanced chemical equation for the reaction. Actually, the reaction described in this problem is the one shown in Figure 8.3, and the chemical equation for it is

$$5Fe^{2+}(aq) + 8H^+(aq) + MnO_4^-(aq) \longrightarrow Mn^{2+}(aq) + 4H_2O + 5Fe^{3+}(aq)$$

To solve the problem, we begin by calculating the number of moles of MnO_4^- that reacted.

$$90.40 \ \text{cm}^3 \ \text{MnO}_4^- \ \text{soln} \times \left(\frac{0.020 \ 00 \ \text{mol MnO}_4^-}{1000 \ \text{cm}^3 \ \text{MnO}_4^- \ \text{soln}} \right) = 1.808 \times 10^{-3} \ \text{mol MnO}_4^-$$

By now you should realize that the next step is to use the coefficients in the equation to calculate the amount of iron that reacted, and from that, the mass of Fe_2O_3 in the sample.

$$1.808 \times 10^{-3} \ \text{mol MnO}_4^- \times \left(\frac{5 \ \text{mol Fe}^{2+}}{1 \ \text{mol MnO}_4^-} \right) \times \left(\frac{1 \ \text{mol Fe}_2O_3}{2 \ \text{mol Fe}^{2+}} \right) \times \left(\frac{159.7 \ \text{g Fe}_2O_3}{1 \ \text{mol Fe}_2O_3} \right) = 0.7218 \ \text{g Fe}_2O_3$$

Since the sample weighed 1.000 g

$$\text{percent Fe}_2O_3 = \frac{\text{mass Fe}_2O_3}{\text{mass sample}} \times 100$$

$$= \frac{0.7218 \ \text{g Fe}_2O_3}{1.000 \ \text{g sample}} \times 100$$

$$= 72.18\% \ \text{Fe}_2O_3$$

Reducing agents

Laboratory reducing agents can be chosen from a rather wide variety of substances. If a strong reducing agent is required for some particular purpose, one of the more active metals can be used—magnesium or zinc, for example. (As you learned in the last chapter, metals react by becoming oxidized, which makes them reducing agents.) There are some drawbacks, however; the reaction must take place on the metal's surface and it is somewhat difficult to control both the speed of the reaction and the amount of reducing agent consumed. Usually, therefore, when redox reactions are carried out in solution, reducing agents are chosen that are soluble in water. In this way they can react with an oxidizing agent in a homogeneous environment.

Tin(II) A mild reducing agent, which is used in various analytical procedures to reduce a metal ion from a higher to a lower oxidation state, is Sn^{2+} ion. For example, in the analysis of an iron ore, such as the one described in Example 8.19, the iron is generally present in the +3 oxidation state in the ore. To be titrated by permanganate ion, however, it must be reduced to Fe^{2+}, and this can be accomplished by treating the Fe^{3+} solution with $SnCl_2$. The half-reaction for the oxidation of Sn^{2+} is very simply

$$Sn^{2+}(aq) \longrightarrow Sn^{4+}(aq) + 2e^-$$

Sulfites and Bisulfites Salts that contain the sulfite ion, SO_3^{2-}, or the bisulfite ion, HSO_3^-, are often used as convenient reducing agents. These anions come from the neutralization (complete or partial) of sulfurous acid, H_2SO_3.

When sulfite or bisulfite ions are oxidized, the product is sulfate ion. If the solution is basic, the reactant is SO_3^{2-}, regardless of whether the original solute contained sulfite or bisulfite ion. This is because the bisulfite ion is itself slightly acidic, and in the presence of base it is neutralized to SO_3^{2-}. On the other hand, if the solution is acidic, the form of the reactant is HSO_3^- or even H_2SO_3. This is because protons are forced onto the SO_3^{2-} ion in the presence of acid, with the number of H^+ added depending on how acidic the reaction mixture is.

The oxidation of bisulfite ion in an acidic solution follows the half-reaction

$$HSO_3^-(aq) + H_2O \longrightarrow SO_4^{2-}(aq) + 3H^+(aq) + 2e^-$$

The oxidation of sulfite ion in a basic solution occurs more easily than the oxidation of bisulfite in an acidic solution, so in basic solution it is a better reducing agent. The reduction takes place according to this half-reaction.

$$SO_3^{2-}(aq) + 2OH^-(aq) \longrightarrow SO_4^{2-}(aq) + H_2O + 2e^-$$

Thiosulfate Ion Another sulfur-containing reducing agent that has laboratory applications is the thiosulfate ion, $S_2O_3^{2-}$. If attacked by a strong oxidizing agent, the $S_2O_3^{2-}$ ion is oxidized to sulfate ion. This is what happens, for example, if chlorine gas is bubbled into a solution of $Na_2S_2O_3$. The reaction is

$$4Cl_2 + S_2O_3^{2-} + 5H_2O \longrightarrow 8Cl^- + 2SO_4^{2-} + 10H^+$$

This reaction makes $S_2O_3^{2-}$ especially useful for trapping Cl_2 gas that might otherwise be released into the atmosphere.

The reaction of $S_2O_3^{2-}$ with I_2 leads to one of the most useful analytical procedures involving a redox titration. Iodine is a less powerful oxidizing agent than Cl_2, and when it reacts with thiosulfate ion, it gives the $S_4O_6^{2-}$ ion (tetrathionate ion).

$$I_2 + 2S_2O_3^{2-} \longrightarrow 2I^- + S_4O_6^{2-}$$

Actually, this reaction is almost always carried out in the presence of an excess amount of iodide ion, which reacts with the I_2 to form the I_3^- ion.

$$I_2 + I^- \longrightarrow I_3^-$$

I_3^- is the **triiodide ion.**

This is done because I_2 itself is insoluble in water, and by forming the I_3^- ion, the iodine is held in solution where it can react smoothly and quickly. Thus, the net reaction in the presence of excess iodide ion is

$$I_3^- + 2S_2O_3^{2-} \longrightarrow 3I^- + S_4O_6^{2-}$$

What makes this reaction so useful in a titration is that its endpoint is very easy to detect. Iodine has the peculiar property of being absorbed onto the surface of starch molecules, forming a deep blue-black colored species with the starch. The color of this *iodine-starch complex* is so intense that it permits us to detect the presence of I_2, even if the iodine concentration is very low. In a typical titration, the starch is added to the solution containing the iodine in the receiving vessel, which gives the solution a blue color. Then the thiosulfate solution is added gradually from a buret until the blue color just disappears, signaling the endpoint.

The reaction of $S_2O_3^{2-}$ with I_2 not only allows us to determine the amount of iodine in a sample, but also the amount of any substance that can produce iodine by a chemical reaction. For example, the amount of hypochlorite ion, OCl^-, in bleach can be determined by allowing the bleach to react with an excess amount of iodide ion. The OCl^- ion oxides the I^- to I_2, and by measuring the amount of I_2 produced we can calculate the amount of OCl^- that must have been in the bleach sample. This is described in Example 8.20.

EXAMPLE 8.20   A REDOX TITRATION USING THE IODINE-THIOSULFATE REACTION

PROBLEM   Household laundry bleach is a dilute solution of NaOCl (sodium hypochlorite) in water. A 1.500-g sample of bleach was treated with excess I^-. This caused the I^- to be oxidized to I_2, which formed soluble I_3^- with the excess I^- in the solution. The net reaction was

$$3I^- + OCl^- + 2H^+ \longrightarrow I_3^- + Cl^- + H_2O$$

The I_3^- formed in this reaction was titrated with 0.05000 M $S_2O_3^{2-}$ using starch as an indicator. The titration required 42.32 cm^3 of the $S_2O_3^{2-}$ solution. What was the percent NaOCl by mass in the bleach?

SOLUTION   First, let's write the equation for the reaction of the I_3^- with the $S_2O_3^{2-}$.

$$I_3^- + 2S_2O_3^{2-} \longrightarrow 3I^- + S_4O_6^{2-}$$

If we look at the reaction above and the reaction given in the statement of the problem, we see that from the amount of $S_2O_3^{2-}$ used in the titration, we can calculate the amount of I_3^- that reacted, and from that the amount of OCl^- in the bleach sample.

The amount of $S_2O_3^{2-}$ in the titrant used was

$$42.32 \; \cancel{cm^3 \; soln} \times \left(\frac{0.05000 \; mol \; S_2O_3^{2-}}{1000 \; \cancel{cm^3 \; soln}} \right) = 2.116 \times 10^{-3} \; mol \; S_2O_3^{2-}$$

The coefficients of the equation tell us that the number of moles of I_3^- that reacted is half the number of moles of $S_2O_3^{2-}$ that reacted.

$$2.116 \times 10^{-3} \; \cancel{mol \; S_2O_3^{2-}} \times \left(\frac{1 \; mol \; I_3^-}{2 \; \cancel{mol \; S_2O_3^{2-}}} \right) = 1.058 \times 10^{-3} \; mol \; I_3^-$$

From the equation for the reaction of I_3^- with OCl^-, we can calculate the number of moles of OCl^- that were responsible for the formation of the I_3^-.

$$1.058 \times 10^{-3} \; \cancel{mol \; I_3^-} \times \left(\frac{1 \; mol \; OCl^-}{1 \; \cancel{mol \; I_3^-}} \right) = 1.058 \times 10^{-3} \; mol \; OCl^-$$

The formula NaOCl tells us that the 1.058×10^{-3} mol OCl^- came from $1.058 \times$

10^{-3} mol NaOCl, which has a molar mass of 74.44 g mol^{-1}. The mass of NaOCl in the bleach sample was therefore

$$1.058 \times 10^{-3} \text{ mol NaOCl} \times \left(\frac{74.44 \text{ g NaOCl}}{1 \text{ mol NaOCl}} \right) = 7.876 \times 10^{-2} \text{ g NaOCl}$$

The percent NaOCl in the sample is

$$\text{percent NaOCl} = \frac{\text{mass NaOCl}}{\text{mass sample}} \times 100$$

$$= \frac{7.876 \times 10^{-2} \text{ g NaOCl}}{1.500 \text{ g sample}} \times 100$$

$$= 5.251\% \text{ NaOCl}$$

REVIEW QUESTIONS AND PROBLEMS

(Problems whose numbers are in blue have their answers in Appendix C at the back of the book; the more difficult problems are marked with asterisks.)

Metathesis Reactions

8.1 What are the three driving forces for metathesis reactions?

8.2 Without looking at the solubility rules, indicate whether the following are soluble or insoluble: KCl, $(NH_4)_2SO_4$, $AgNO_3$, $PbSO_4$, $Mn(OH)_2$, $FePO_4$, $CaCO_3$, $Zn(ClO_4)_2$, $Ba(C_2H_3O_2)_2$, NiO.

8.3 Without looking at the solubility rules, indicate whether the following are soluble or insoluble: KNO_3, $FeCl_2$, $NiCO_3$, $(NH_4)_2HPO_4$, Hg_2Cl_2, $Al(OH)_3$, PbI_2, CuI_2, $SrBr_2$, CoS.

8.4 Write ionic and net ionic equations for the following:
(a) $Al(OH)_3(s) + 3HCl(aq) \rightarrow AlCl_3(aq) + 3H_2O$
(b) $CuCO_3(s) + H_2SO_4(aq) \rightarrow$
$$CuSO_4(aq) + H_2O + CO_2(g)$$
(c) $Cr_2(CO_3)_3(s) + 6HNO_3(aq) \rightarrow$
$$2Cr(NO_3)_3(aq) + 3H_2O + 3CO_2(g)$$

8.5 Write net ionic equations for each of the following:
(a) $NaBr + AgNO_3 \rightarrow AgBr + NaNO_3$
(b) $CoCO_3 + 2HNO_3 \rightarrow Co(NO_3)_2 + CO_2 + H_2O$
(c) $NaC_2H_3O_2 + HNO_3 \rightarrow NaNO_3 + HC_2H_3O_2$
(d) $Pb(NO_3)_2 + (NH_4)_2SO_4 \rightarrow PbSO_4 + 2NH_4NO_3$
(e) $H_2S + Cu(NO_3)_2 \rightarrow 2HNO_3 + CuS$
(f) $NaOH + NH_4Cl \rightarrow NH_3 + H_2O + NaCl$

8.6 Write balanced net ionic equations for each of the following:
(a) $CoS + HCl \rightarrow H_2S + CoCl_2$

(b) $PbCO_3 + HNO_3 \rightarrow H_2O + CO_2 + Pb(NO_3)_2$
(c) $PbCO_3 + H_2SO_4 \rightarrow PbSO_4 + H_2O + CO_2$
(d) $SnCl_2 + NaOH \rightarrow Sn(OH)_2 + NaCl$
(e) $Ag_2O + HCl \rightarrow AgCl + H_2O$
(f) $MgSO_4 + NiCl_2 \rightarrow MgCl_2 + NiSO_4$

8.7 Write molecular, ionic, and net ionic equations for the reaction (if any) between
(a) Na_2SO_4 and $BaCl_2$
(b) $Ca(NO_3)_2$ and $(NH_4)_2CO_3$
(c) $NaC_2H_3O_2$ and HNO_3
(d) NaOH and $CuCl_2$
(e) $(NH_4)_2CO_3$ and HNO_3

8.8 Write molecular, ionic, and net ionic equations for the reaction (if any) between
(a) H_2SO_4 and $BaSO_4$ (d) $MgSO_4$ and LiOH
(b) NH_4Br and $MnSO_4$ (e) $AgC_2H_3O_2$ and KCl
(c) K_2S and $Ni(C_2H_3O_2)_2$

8.9 Write molecular, ionic, and net ionic equations for any reaction that would occur between
(a) AgBr and KI   (d) K_2SO_3 and HCl
(b) $BaCl_2$, SO_2, and H_2O (e) $BaCO_3$ and H_2SO_4
(c) $Na_2C_2O_4$ and HCl

*8.10 Based on what you've learned in Chapters 7 and 8, predict what will occur if CO_2 is bubbled into a solution of NaOH. Use chemical equations to describe any reactions that will take place.

Preparation of Salts by Metathesis Reactions

8.11 Describe how you would prepare the following by a metathesis reaction in which one product is a precipitate: (a) $NH_4C_2H_3O_2$, (b) $Fe_3(PO_4)_2$, (c) $CuCO_3$, (d) Na_2SO_4, (e) $PbSO_4$.

8.12 Describe how you would prepare the following by a metathesis reaction in which one product is a precipitate: (a) $Mg(OH)_2$, (b) $BaCl_2$, (c) $Fe(C_2H_3O_2)_2$, (d) $Ni(ClO_4)_2$, (e) $BaSO_3$.

8.13 Describe how you would prepare the following by a metathesis reaction in which one product is a gas: (a) $CaCl_2$, (b) $Mn(ClO_4)_2$, (c) $BaSO_4$, (d) $NaNO_3$, (e) $NH_4C_2H_3O_2$.

8.14 Describe how you would prepare the following by a neutralization reaction: (a) $Ca(NO_3)_2$, (b) $Na_2C_2O_4$, (c) $Fe(HSO_4)_2$, (d) $Al(ClO_4)_3$, (e) $NiBr_2$.

8.15 How could you prepare
(a) $CuCl_2$ from $Cu(NO_3)_2$
(b) $BaCl_2$ from $BaBr_2$
(c) $NaClO_4$ from Na_2SO_4
(d) $Mg(C_2H_3O_2)_2$ from $MgCl_2$
(e) Na_2CO_3 from Na_2SO_3

Stoichiometry of Ionic Reactions

8.16 What is the meaning of the concentration unit *percent by mass?*

8.17 What is the meaning of the concentration unit *parts per million?*

8.18 Modern analytical techniques have achieved such sensitivities that pollutants and potential cancer-causing agents can be detected at the levels of *parts per billion, ppb.* What does this concentration term mean?

8.19 The concentration of fluoride ion in sea water is approximately 0.001 g per 1000 g of sea water. Express this concentration in
(a) percent by mass
(b) parts per million
(c) parts per billion

8.20 Mercury is an extremely toxic substance that deactivates enzyme molecules that promote biochemical reactions. A 25.0-g sample of tuna fish taken from a large shipment was analyzed for this substance and found to contain 2.1×10^{-5} mol of Hg. By law, foods having a mercury content above 0.50 ppm cannot be sold (they cannot even be given away). Determine whether this shipment of tuna must be confiscated.

8.21 What is the molar concentration of each of the following:
(a) 1.50 mol of NaCl in 2.00 dm^3 of solution
(b) 0.248 mol of KCN in 250 cm^3 of solution

(c) 0.750 mol of H_2SO_4 in 1.35 dm^3 of solution
(d) 85.5 g of HNO_3 in 1.00 dm^3 of solution
(e) 44.5 g of $NH_4C_2H_3O_2$ in 600 cm^3 of solution

8.22 How many moles of solute are in: (a) 250 cm^3 of 0.100 M KCl; (b) 1.65 dm^3 of 1.40 M $HClO_4$; (c) 0.0250 dm^3 of 0.0100 M $HC_2H_3O_2$.

8.23 How many grams of Na_2CO_3 are required to prepare 300 cm^3 of a 0.150 M solution?

8.24 How many grams of $Ba(OH)_2$ are required to prepare 250 cm^3 of a solution that has a hydroxide ion concentration of 0.300 M?

8.25 Pure nitric acid has a density of 1.513 g cm^{-3}. What is its molar concentration?

8.26 A solution of $MgSO_4$ contains 22.0% $MgSO_4$ by mass and contains 273.8 g of the salt per dm^3. What is the density of the solution? What is the molar concentration of the solution?

8.27 A particular pollutant was present in water at a concentration of 825 ppm. What was the concentration in percent by mass? If the pollutant was benzene, C_6H_6, what was its molar concentration (assume that 1000 cm^3 of solution has a mass of essentially 1000 g)?

8.28 What are the molar concentrations of the ions in each of the following salt solutions:
(a) 0.100 M LiCl
(b) 0.250 M $CaCl_2$
(c) 1.20 M $(NH_4)_2SO_4$
(d) 0.600 M $NaHSO_4$
(e) 0.400 M $Fe_2(SO_4)_3$

8.29 What are the molar concentrations of the ions in each of these salt solutions?
(a) 0.0250 M $Ba(OH)_2$
(b) 0.300 M $Cd(NO_3)_2$
(c) 0.400 M Na_2HPO_4
(d) 0.100 M $Cr_2(SO_4)_3$
(e) 0.0450 M $Hg_2(NO_3)_2$

8.30 The concentration of SO_4^{2-} is 0.100 M in a solution of Na_2SO_4. What is the molar concentration of Na_2SO_4?

8.31 The concentration of Cl^- is 0.160 M in a solution of $FeCl_3$. What is the molar concentration of $FeCl_3$?

8.32 The concentration of Al^{3+} is 0.140 M in a solution of $Al_2(SO_4)_3$. What is the molar concentration of $Al_2(SO_4)_3$?

8.33 How many moles of each kind of ions are present in
(a) 50.0 cm^3 of 0.200 M NaCl
(b) 30.0 cm^3 of 0.160 M $CaCl_2$
(c) 27.0 cm^3 of 0.650 M Na_2SO_4
(d) 135 cm^3 of 0.820 M $(NH_4)_2SO_4$
(e) 75.0 cm^3 of 0.250 M $Al_2(SO_4)_3$

8.34 What are the molar concentrations of the ions in a solution prepared by dissolving 10.45 g $CuSO_4 \cdot 5H_2O$ in a volume of 150.0 cm^3 of solution?

8.35 What volume (in cm^3) of 0.100 M NaOH is needed to react with 5.00×10^{-3} mol of H_2SO_4 to give Na_2SO_4?

8.36 What volume (in cm^3) of 1.250 M HNO_3 contains enough nitric acid to dissolve a 3.22-g copper penny? The net ionic equation is

$$3Cu(s) + 8H^+(aq) + 2NO_3^-(aq) \longrightarrow$$
$$3Cu^{2+}(aq) + 2NO(g) + 4H_2O$$

8.37 Copper(II) carbonate dissolves in perchloric acid ($HClO_4$).
(a) Write a net ionic equation for the reaction.
(b) How many milliliters of 1.35 M $HClO_4$ are needed to prepare 5.25 g of $Cu(ClO_4)_2$?
(c) How many grams of $CuCO_3$ are needed to prepare 5.25 g $Cu(ClO_4)_2$?

8.38 What volume (in cm^3) of 0.300 M NaOH is required to react with 500 cm^3 of 0.170 M H_3PO_4 to yield (a) Na_3PO_4, (b) Na_2HPO_4, (c) NaH_2PO_4?

8.39 How many cubic centimetres of 0.100 M $BaCl_2$ are required to react with 25.0 cm^3 of 0.200 M H_2SO_4?

8.40 How many milliliters of 0.1000 M $BaCl_2$ are required to react completely with 25.0 cm^3 of 0.200 M $Fe_2(SO_4)_3$?

8.41 A 0.244-g sample of benzoic acid (a monoprotic acid) requires 20.0 cm^3 of 0.100 M NaOH for complete neutralization. Calculate the molar mass of the acid.

8.42 20.0 cm^3 of 0.200 M $AgNO_3$ were added to 30.0 cm^3 of 0.200 M NaCl.
(a) What is the net ionic equation for the reaction?
(b) How many moles of precipitate are formed?
(c) What is the mass of the precipitate?
(d) What are the concentrations of each of the remaining ions in the final solution?

8.43 What mass of AgCl will be formed if 25.0 cm^3 of 0.050 M HCl are added to 100 cm^3 of 0.050 M $AgNO_3$?

8.44 50.0 cm^3 of 0.240 M $BaCl_2$ were added to 45.0 cm^3 of 0.180 M $Fe_2(SO_4)_3$.
(a) What mass of $BaSO_4$ was formed?
(b) What are the concentrations of the remaining ions in the final solution?

8.45 A 0.500-g sample of $Cr_2(SO_4)_3$ was dissolved in water and an excess of dilute NaOH was added, precipitating $Cr(OH)_3$. This precipitate was collected by filtration. What volume (in cm^3) of 0.400 M HNO_3 is needed to dissolve the precipitate?

8.46 Caproic acid, a foul-smelling substance found in certain excretions of male goats during breeding season, has an empirical formula of C_3H_6O. A 0.100-g sample of the acid required 17.2 cm^3 of 0.0500 M NaOH for complete reaction. Assuming that the acid is monoprotic, calculate (a) its molar mass; (b) its molecular formula.

8.47 If 380 cm^3 of 0.273 M $Ba(OH)_2$ are added to 500 cm^3 of 0.520 M HCl, will the mixture be acidic or basic? Calculate the concentration of H^+ (or OH^- if the solution is basic) in the final mixture. Assume that volumes are additive.

8.48 40.0 cm^3 of 0.270 M $Ba(OH)_2$ are added to 25.0 cm^3 of 0.330 M $Al_2(SO_4)_3$.
(a) Write the chemical equation for the reaction that occurs.
(b) What total mass of precipitate is formed?
(c) What is the concentration of each of the ions remaining in solution?

Chemical Analysis and Titrations

8.49 What is the purpose of a *chemical analysis*?

8.50 A 0.249-g sample of a compound containing titanium and chlorine was dissolved in water and treated with silver nitrate solution. The silver chloride that formed had a mass of 0.694 g after being filtered, washed, and dried. What is the empirical formula of the original compound?

8.51 A certain lead ore contains the compound $PbCO_3$. A sample of this ore weighing 1.526 g was treated with nitric acid, which dissolved the $PbCO_3$, and the resulting solution was then treated with Na_2SO_4. This gave a precipitate of $PbSO_4$ that was dried and found to weigh 1.081 g. What is the percent by mass of $PbCO_3$ in the ore?

*8.52 A 1.850-g sample of a mixture of $CuCl_2$ and $CuBr_2$ was dissolved in water and mixed thoroughly with a 1.800-g portion of AgCl. After the reaction, the solid, which now consisted of a mixture of AgCl and AgBr, was filtered, washed, and dried. Its mass was found to be 2.052 g. What percent by mass of the original mixture was $CuBr_2$?

*8.53 Silver iodide is less soluble than silver bromide, which in turn is less soluble than silver chloride. A 0.2000-g sample that was known to be a mixture of NaCl, NaBr, and NaI was dissolved in water and an excess of $AgNO_3$ was added. The precipitate containing AgCl, AgBr, and AgI was filtered, dried, and had a mass of 0.4120 g. This solid was then placed in water and treated with NaBr solution, which converted any AgCl to AgBr (but it did not affect the AgI). The precipitate that now consisted of AgBr and AgI was filtered, dried, and found to weigh 0.4881 g. It was then placed in water and treated with NaI, which converted the AgBr to AgI. When the solid (now consisting entirely of AgI) was filtered and dried, its mass was 0.5868 g. What were the mass percentages of NaCl, NaBr, and NaI in the original sample?

8.54 Describe the following: (a) buret, (b) titration, (c) titrant, and (d) endpoint.

8.55 What is the function of an indicator? What color is phenolphthalein in (a) an acidic solution and (b) a basic solution?

8.56 A 15.00 cm^3 portion of a solution of H_2SO_4 of unknown concentration was titrated with 0.150 M NaOH. The titration required 21.30 cm^3 of the base. Assuming complete neutralization of the H_2SO_4, what was the acid's molar concentration?

8.57 A 1.030-g portion of a mixture containing $CaCO_3$, $CaSO_4 \cdot 2H_2O$, and $BaSO_4$ was heated, which caused the $CaCO_3$ to decompose into CaO and CO_2. The resulting solid was treated with water, which caused the CaO to dissolve, forming $Ca(OH)_2$. This solution required 37.25 cm^3 of 0.120 M HCl for complete neutralization in a titration. What was the percent by mass of $CaCO_3$ in the original sample?

*8.58 A sample of rock containing limestone ($CaCO_3$) was heated, converting the $CaCO_3$ to CaO. This was treated with H_2O to give $Ca(OH)_2$, which was then titrated with HCl. In one analysis, a 0.2000-g sample, taken *after* converting the $CaCO_3$ to CaO as described above, required 30.3 cm^3 of 0.1000 M HCl for complete neutralization. What was the mass percent $CaCO_3$ in the original rock?

8.59 Ascorbic acid (vitamin C) is a diprotic acid having the formula, $H_2C_6H_6O_6$. A sample of a vitamin supplement was analyzed by titrating a 0.1000-g sample dissolved in water with 0.0200 M NaOH. A volume of 15.2 cm^3 of the base was required to completely neutralize the ascorbic acid. What was the mass percent ascorbic acid in the sample?

*8.60 A mixture of the monoprotic acids, lactic acid, $HC_3H_5O_3$ (found in sour milk) and caproic acid, $HC_6H_{11}O_2$, (found in excretions from the goat) was titrated with 0.0500 M NaOH. A 0.1000-g sample of the mixture required 20.4 cm^3 of the base. What is the mass of each acid in the sample?

*8.61 A mixture of $MgCO_3$ and $CaCO_3$ (a dolomitic limestone used in agriculture) was heated to produce MgO and CaO. A 2.000-g sample of this oxide mixture was allowed to react with 100 cm^3 of 1.00 M HCl. The excess HCl required 19.6 cm^3 of 1.00 M NaOH for complete neutralization. What were the percentages of $CaCO_3$ and $MgCO_3$ in the original limestone sample?

*8.62 An amino acid isolated from a piece of animal tissue was believed to be glycine, $NH_2CH_2CO_2H$. A 0.0500-g sample was treated in such a way that all the nitrogen in it was converted to ammonia. This NH_3 was added to 50.0 cm^3 of 0.05000 M HCl, neutralizing part of the acid.

$$NH_3 + HCl \longrightarrow NH_4Cl$$

The acid remaining in the solution was titrated with 0.0600 M NaOH, which required 30.57 cm^3 of the base for neutralization.

(a) How many moles of HCl were neutralized by the NH_3?

(b) How many grams of nitrogen were in the 0.0500-g sample?

(c) What was the mass percent nitrogen in the sample? How does it compare with the percentage nitrogen calculated for glycine?

Oxidizing and Reducing Agents in the Laboratory

8.63 Chlorine, Cl_2, is a powerful oxidizing agent, but it is seldom used in that capacity in the lab. Why?

8.64 What color is characteristic of solutions that contain (a) CrO_4^{2-}, (b) $Cr_2O_7^{2-}$, (c) MnO_4^-.

8.65 Complete and balance net ionic equations for the following reactions:
(a) $SO_3^{2-} + CrO_4^{2-}$ (slightly basic solution)
(b) $Sn^{2+} + MnO_4^-$ (acidic solution)
(c) $S_2O_3^{2-} + Cl_2$ (acidic solution)

8.66 What would be the correct balanced net ionic equation for the reaction between sodium sulfite and sodium chromate if the reaction takes place in an acidic solution?

8.67 What is the net ionic equation for the reaction of Na_2SO_3 and $KMnO_4$ in a basic solution?

8.68 What is the net ionic equation for the reaction of I_3^- with $Na_2S_2O_3$?

8.69 What reaction would occur if SO_3^{2-} is oxidized by CrO_4^{2-} in a very basic solution?

Titrations Involving Oxidation-Reduction

8.70 Why is starch used as an indicator in titrations involving I_3^- and $S_2O_3^{2-}$?

8.71 Why is $KMnO_4$ a convenient titrant in redox reactions?

8.72 A 1.362-g sample of an iron ore that contains Fe_3O_4 was dissolved in acid and all the iron was reduced to Fe^{2+}. The solution was then acidified with H_2SO_4 and titrated with 39.42 cm^3 of 0.0281 M $KMnO_4$ solution.
(a) What is the net ionic equation for the reaction between MnO_4^- and Fe^{2+}?
(b) What is the percentage Fe_3O_4 in the sample?

8.73 A 2.385-g mixture of $CaCl_2$ and NaCl was dissolved in water and treated with a solution of sodium oxalate, $Na_2C_2O_4$, which produced a precipitate of calcium oxalate, CaC_2O_4. This precipitate was filtered from the mixture and then dissolved in HCl to give $H_2C_2O_4$.

$$CaC_2O_4(s) + 2H^+(aq) \longrightarrow Ca^{2+}(aq) + H_2C_2O_4(aq)$$

The $H_2C_2O_4$ was titrated with $KMnO_4$, giving CO_2

and Mn^{2+} as products. The titration required 19.64 cm^3 of 0.2000 M KMnO$_4$.

(a) How many moles of CaC$_2$O$_4$ had precipitated?

(b) What was the percent by mass of CaCl$_2$ in the original sample?

8.74 Brass is an alloy of copper and zinc. A particular sample of brass weighing 0.244 g was dissolved in nitric acid. Afterwards, the solution containing the Cu^{2+} and Zn^{2+} was treated with an excess of a solution containing KI. The copper reacted as follows:

$$2Cu^{2+}(aq) + 5I^-(aq) \longrightarrow 2CuI(s) + I_3^-(aq)$$

The I_3^- formed in this reaction was titrated (using starch as an indicator) with 25.34 cm^3 of 0.1000 M Na$_2$S$_2$O$_3$.

(a) How many moles of I_3^- had formed in the reaction above?

(b) What was the percent by mass of copper in this sample of brass?

CHAPTER 9
PROPERTIES OF GASES

Hot air balloons are a beautiful sight as they reach for the sky. Gases such as air expand when they are heated, and as they expand they become less dense. It is the buoyant effect of the surrounding cooler air that lifts the balloons skyward. The behavior of gasses when they are heated is just one of the properties of gases discussed in this chapter.

Matter is capable of existing in three different physical forms or **states:** solid, liquid, and gas. In this chapter and the next, we will examine the physical and chemical characteristics of these states and the transformations that occur among them. This chapter deals with the gaseous state, in which the **intermolecular forces of attraction**—the attractions one molecule experiences toward others around it—are weak. These weak forces allow the rapid, independent movement of the molecules and cause the physical behavior of a gas to be nearly independent of its chemical composition. Instead, the behavior of a gas is controlled by its volume, pressure, temperature, and number of moles. Since these variables are of paramount importance, we will begin our discussion by taking a close look at them as they apply to the gaseous state.

9.1 VOLUME AND PRESSURE

When a gas is introduced into a container, the molecules move freely within it and occupy the container's entire volume. As a result, the volume of a gas is given simply by specifying the volume of the vessel in which it is held. Because gases mix freely with one another, when there are several gases in a mixture, the volume of each component is the same as the volume occupied by the entire mixture.

Pressure is defined as force per unit area; it is an intensive quantity formed as a ratio of two extensive quantities: force and area. The derived SI unit of force is the newton (N). (For perspective, a 102-g mass at Earth's surface is attracted downward by gravity with a one-newton force.) If a 100-N force is exerted on a piston whose total area is 100 m^2, the pressure acting on each square metre is 100 N/100 m^2 = 1 N m^{-2}. A pressure of 1 N m^{-2} is termed a pascal (Pa). This is a relatively small pressure unit; it's roughly equivalent to the pressure exerted on a slice of bread by a thin layer of butter, for example.

*The **unit area** in this discussion is 1 m^2.*

If the same 100-N force is exerted on a smaller area, for example, 1 m^2, the pressure is greatly magnified (Figure 9.1). Now the pressure is 100 N m^{-2}, or 100 Pa. The dependence of pressure on both force and the area over which it is spread has been experienced firsthand by anyone who has ever stepped on a nail. A 50.0-kg person (who experiences a 490-N force downward at Earth's surface) stepping on even a dull nail having a point area of 1.0 mm^2 will experience a pressure of *490 000 000 N m^{-2} (Pa) or 490 megapascals (MPa)!* This is more than enough to cause the nail to puncture the skin.

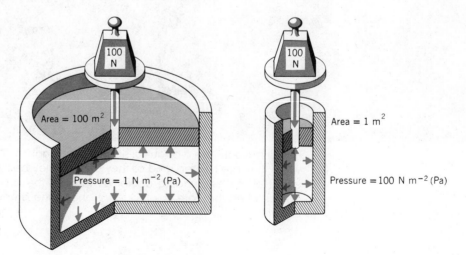

Figure 9.1

Pressure. The fluid in the cylinder exerts the same pressure on all the walls of the container.

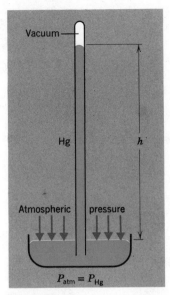

Figure 9.2

A barometer. The atmospheric pressure supports a column of mercury of height h.

Normal barometric pressure at sea level is approximately 760 mmHg.

If the pressure generated by a piston is applied to a fluid (a gas or liquid), as illustrated in Figure 9.1, it is transmitted uniformly in all directions so that all the walls of the container experience the same pressure. If the piston is supported by the fluid, then the fluid also exerts an equal pressure on the piston as well as the other walls of the container.

The ability of trapped gases to exert a pressure is demonstrated when you inflate an automobile tire. In most cases the four tires that support the car are inflated to a pressure of approximately 200 kPa above the pressure of the surrounding atmosphere. The reason a tire becomes flat when it "springs a leak" is because gases flow from a region of high pressure to a region of lower pressure; in our example, this flow is from inside the tire to the atmosphere.

The pressure of the atmosphere

The atmosphere of the earth is a mixture of gases that exerts a pressure called the atmospheric pressure. We measure this pressure using a device called a **barometer.** Figure 9.2 shows a barometer constructed by filling a glass tube about one meter long with mercury and inverting it (without spilling any) into a dish of mercury so that the open end is submerged. In this figure we see that the mercury in the tube does not completely pour out when it is inverted; instead it maintains a particular height (h) above the reservoir. What keeps the mercury in the tube is the pressure of the atmosphere pushing on the surface of the mercury in the dish. The height of the mercury column is found to be independent of the diameter and length of the glass tube, as long as a space appears over the mercury. This space, for all practical purposes, is a vacuum ($P \approx 0$). The height of the column does change, however, when the atmospheric pressure changes. For example, when a storm approaches, the atmospheric pressure drops and the column becomes shorter. In fact, it is the height of such a mercury column that is reported as the barometric pressure in weather forecasts on radio and TV.

To measure atmospheric pressure we compare the various pressures acting along a *reference level*, which we choose as the surface of the reservoir. At this level outside the inverted tube, the pressure is caused by the downward force of the gases in the atmosphere (P_{atm}). Inside the tube the pressure at the reference level is caused by the downward pull of gravity on the mercury in the column (P_{Hg}). When these two opposing pressures are exactly equal ($P_{Hg} = P_{atm}$), the mercury in the column remains stationary. Atmospheric pressure is directly related to the length (h) of the column of mercury in a barometer and can therefore be expressed in units of millimetres of mercury (mmHg). A **standard atmosphere (atm),** defined as a pressure of exactly 101 325 Pa (101.325 kPa), is equal to the pressure that would support a 760-mm column of mercury measured at 0°C.[1] Thus 1 atm = 760 mmHg = 101.325 kPa. Even though laboratory barometers may remain calibrated in mmHg units, the SI pressure unit (pascal) retains the advantage and convenience of being directly related to "force per unit area" (N m^{-2}) dimensions. The standard atmosphere (atm) has long served as a reference pressure in chemical thermodynamics (Chapter 12).

A barometer similar to that described above could also be constructed using water as a liquid. The length of the column of water would be consid-

---

[1] The length of the column of mercury, which is supported by atmospheric pressure, varies with both the density of the mercury and the pull of gravity on the mercury in the column. Since density varies with temperature and the pull of gravity varies with altitude, in this definition of the standard atmosphere, it is necessary to specify a reference temperature (0°C) as well as a reference altitude (sea level).

erably greater than that of a column of mercury, because the atmospheric pressure would be supporting a less dense liquid ($d_{water} = 1.00$ g cm^{-3}, $d_{Hg} = 13.6$ g cm^{-3}).

EXAMPLE 9.1 | **CALCULATING THE HEIGHT OF A BAROMETER FILLED WITH WATER**

PROBLEM | If water was the liquid in a barometer, what would be the length (h) of the water column at 1 atm of pressure?

SOLUTION | The density of Hg is 13.6 times that of H_2O. This means that to have equal masses of mercury and water, the volume of water must be 13.6 times as great as the volume of mercury. Since we are comparing columns of the same diameter, the water column would have to be 13.6 times as long as the Hg column to contain 13.6 times as much volume. Thus

$$1.00 \text{ mm Hg} = 13.6 \text{ mm } H_2O$$

Then

$$1 \text{ atm} = 760 \text{ mm Hg} \times \left(\frac{13.6 \text{ mm } H_2O}{1.00 \text{ mm Hg}} \right) = 1.03 \times 10^4 \text{ mm } H_2O$$

In general, as the density of the liquid being supported in a column by some external pressure increases, the length of the column of liquid decreases.

Measuring pressures of trapped gases

It is often desirable to know the pressure of a gas present in a closed system (for example, the pressure of gases produced during a chemical reaction). The instrument normally used for these pressure measurements is called a **manometer.** An open-end manometer (Figure 9.3) is simply a U-shaped tube containing some liquid, such as mercury. One arm of the tube is connected to a system whose pressure is to be measured while the other arm remains open to the atmosphere. When the pressure of the gas inside the system (P_{gas}) is equal to P_{atm}, the level of the liquids in both arms will be the same, as shown in Figure 9.3a. If the pressure of the gas is greater than P_{atm}, the mercury in the left arm will be forced downward, causing the mercury in the right arm to rise (Figure 9.3b). We obtain the pressure of the gas in this system by comparing the pressures exerted in both arms at a reference level, h_0, which is chosen to be the height of the shortest column. The pressure exerted on the left column when $P_{gas} > P_{atm}$ is simply P_{gas}, while at the same level in the right arm the pressure is P_{atm} plus the pressure exerted by the column of mercury that rises above the reference level, P_{Hg}. When the levels are stationary, the pressures at the reference level on both sides are equal and

$$P_{gas} = P_{atm} + P_{Hg}$$

The atmospheric pressure (P_{atm}) is found with a barometer, and P_{Hg} is simply the difference in the heights of the two mercury columns. Similarly, when $P_{gas} < P_{atm}$, shown in Figure 9.3c, the pressure in the left arm at the reference level is $P_{gas} + P_{Hg}$, while in the right column the pressure is P_{atm}. In this case, when the columns are stationary

$$P_{gas} + P_{Hg} = P_{atm}$$

In problems dealing with open-end manometers, it is usually best to draw a picture and then decide whether to add or subtract the pressure difference from the atmospheric pressure.

Figure 9.3
An open-end manometer. (a) The pressure of the trapped gas is equal to the atmosphere pressure. (b) The gas pressure is greater than atmospheric pressure. (c) The gas pressure is less than atmospheric pressure.

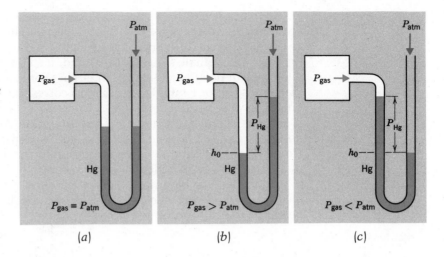

(a)   (b)   (c)

so that

$$P_{gas} = P_{atm} - P_{Hg}$$

Therefore, when $P_{gas} < P_{atm}$, the pressure of the gas in the system is found by subtracting the difference in the heights of the columns from atmospheric pressure.

A closed-end manometer is generally used for the measurement of low pressures (usually much smaller than atmospheric pressure). This manometer consists of a U-shaped tube with one arm closed and the other connected to the system as shown in Figure 9.4. When the pressure of the gas in the system is equal to P_{atm}, the right arm is completely filled while the left arm is only partially filled. If the pressure of the gas in the system is reduced, the level in the left arm will increase, which will cause the level in the right arm to decrease, as shown in Figure 9.4*b*. At the reference level the pressure exerted on the left arm is P_{gas}, while on the right arm the pressure is P_{Hg} (the space above the mercury is a vacuum). When the columns are stationary, $P_{gas} = P_{Hg}$, and the pressure exerted by the gas in the system is simply found as the difference in the heights of liquid in the two arms of the manometer.

A closed-end manometer is convenient because you don't have to measure the atmospheric pressure.

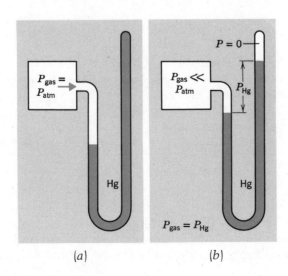

(a)   (b)

Figure 9.4
A closed-end manometer. (a) When the gas pressure equals atmospheric pressure, the mercury is forced to the top of the right arm of the manometer. (b) The pressure of the gas can be measured directly if it is considerably smaller than atmospheric pressure.

Mercury is not the only fluid ever used in manometers, especially if low pressures are to be measured. This is because mercury, with its large density, gives a very small difference in the heights of the columns in a manometer when the pressure is small. On the other hand, if a liquid of lower density is used, the difference in heights can be much larger. For example, in Example 9.1 we saw that a pressure of 1 mmHg is equal to a corresponding water height of 13.6 mm. This is because water is only 1/13.6 as dense as mercury, so the column must be 13.6 times as high to exert the same pressure. A similar effect would be produced if we used a fluid that was half as dense as mercury—a pressure of 1 mmHg, would produce a height of 2 mm for the other fluid.

There is a very simple relationship between the heights of columns of fluids in manometers (or barometers) and their densities. Consider two liquids A and B. If, for a given pressure, the height of the column of liquid A is h_A and the height of the column of liquid B is h_B, then

d_A and d_B are the densities of A and B.

$$h_B = h_A \times \frac{d_A}{d_B}$$

[9.1]

EXAMPLE 9.2 USING FLUIDS OTHER THAN MERCURY IN A MANOMETER

PROBLEM   A liquid having a density of 1.15 g cm^{-3} was used in an open-end manometer. In a particular experiment, a difference of heights was measured to be 14.7 mm, with the arm connected to the container filled with gas lower than the arm open to the atmosphere. The atmospheric pressure was 756.00 mmHg. What was the pressure of the gas?

SOLUTION   This condition corresponds to that shown in Figure 9.3a, so we must add the difference in pressure to the atmospheric pressure to obtain the pressure of the trapped gas.

$$P_{gas} = P_{atm} + P_{fluid}$$

But first, we must convert the difference in heights to torr. This means that we have to calculate the height of a mercury column that would exert the same pressure as 14.7 mm of the fluid in the manometer. We can do this by applying Equation 9.1. The density of mercury is 13.6 g cm^{-3}. Therefore

$$h_{Hg} = h_{fluid} \times \frac{d_{fluid}}{d_{Hg}}$$

$$h_{Hg} = 14.7 \text{ mm} \times \frac{1.15 \text{ g cm}^{-3}}{13.6 \text{ g cm}^{-3}}$$

$$= 1.24 \text{ mm}$$

A height of 1.24 mm Hg corresponds to a pressure difference of 1.24 mmHg. The pressure of the gas is therefore

$$P_{gas} = 756.00 \text{ mmHg} + 1.24 \text{ mmHg}$$

$$= 757.24 \text{ mmHg} (100.96 \text{ kPa})$$

9.2 BOYLE'S LAW

The way in which the volume of a gas is related to its pressure was first described in detail in 1662 by an Irish chemist and physicist, Robert Boyle (1627–1691). Using an apparatus similar to that in Figure 9.5, Boyle found that at

Figure 9.5

Boyle's law apparatus. The pressure on the trapped gas, as measured by the height h *of the mercury column, is increased by adding mercury to the U-tube. The volume of the trapped gas decreases as the pressure is raised.*

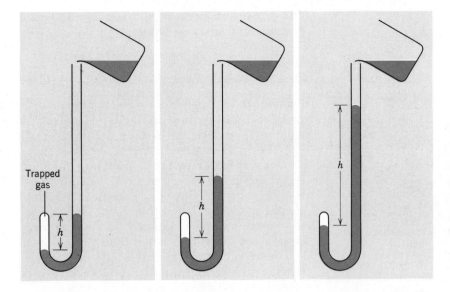

constant temperature the volume of a fixed quantity of trapped gas decreases as the pressure on the gas is increased. Anyone who has ever used a bicycle pump is aware of this inverse relationship between the pressure and volume of a gas. As the piston of the pump is forced downward, the gas is compressed into a smaller volume while its pressure is raised (Figure 9.6). If we allow the compressed gas to escape, we can, for instance, use it to inflate a tire.

Boyle repeated his experiments many times with several different gases and found his initial observation to be a universal property of all of them. The results of his experiments can be formulated into **Boyle's law,** which states that *at a constant temperature, the volume occupied by a fixed quantity of gas is inversely proportional to the applied pressure.* This can be expressed mathematically as

$$V \propto \frac{1}{P}$$

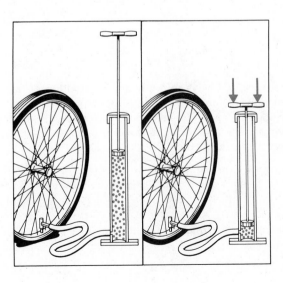

Figure 9.6

Compression of a gas by decreasing its volume using a bicycle pump is an example of Boyle's law.

The proportionality can be made into an equality by the introduction of a proportionality constant. Thus

$$V = \text{constant} \cdot \frac{1}{P}$$

or
$$PV = \text{constant} \qquad [9.2]$$

Equation 9.2 states that for a given quantity of a gas at constant temperature, the product of its pressure and volume is a constant. If the pressure increases, the volume must decrease to keep the product of $P \times V$ constant.

The inverse relationship between P and V predicted from Boyle's law is shown graphically by the colored line in Figure 9.7. Real gases, such as hydrogen or oxygen or nitrogen, do not follow this predicted behavior exactly, however, as shown by the white line. At very high pressures the measured volume is always somewhat larger than that calculated from Boyle's law. At low pressures the measured and calculated volumes approach each other—real gases obey Boyle's law quite well when their pressures are low.

A hypothetical gas that would obey Boyle's law perfectly under all conditions is called an **ideal gas.** Real gases approach ideal gas behavior at low pressures but they are said to exhibit nonideal behavior when their properties deviate from Boyle's law. The degree of nonideality differs for different real gases.

Under conditions usually encountered in the laboratory, and to the degree of precision of most of our calculations, gases generally behave ideally; that is, they obey Boyle's law. Equation 9.2, therefore, is useful in calculating the effect of a pressure change on the volume of a gas at constant temperature.

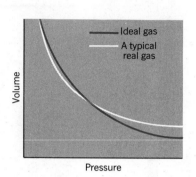

Figure 9.7

Boyle's law. The colored line illustrates how the pressure and volume of an ideal gas are related. The white line shows how a typical real gas behaves. At low pressure the behavior of the real gas approaches that of the ideal gas.

EXAMPLE 9.3 APPLYING THE PRESSURE-VOLUME LAW (BOYLE'S LAW)

PROBLEM   If 100 cm³ of a gas, originally at 100 kPa, is compressed to a pressure of 125 kPa at constant temperature, what will its final volume be?

SOLUTION   In problems of this type, which deal with two sets of conditions, it is best to begin by setting up a table containing all the data

	Initial (i)	Final (f)
Pressure (P)	100 kPa	125 kPa
Volume (V)	100 cm³	?

There are two ways to approach the solution of this problem. One method is to use Equation 9.2 and simply substitute numerical values. The other relies on an *understanding* of how P and V are related to each other to set up the arithmetic. Although they both give the same answers, you can improve your understanding of gas behavior by practicing the second method.

METHOD 1   According to Boyle's law, the PV products in both initial and final states are equal to the same constant; therefore they must be equal to each other.

$$P_i V_i = P_f V_f$$

Solving for the unknown final volume, we get

$$V_f = V_i \left(\frac{P_i}{P_f} \right) \qquad [9.3]$$

We now substitute values from our table.

$$V_f = 100 \text{ cm}^3 \times \left(\frac{100 \text{ kPa}}{125 \text{ kPa}} \right)$$

and

$$V_f = 80.0 \text{ cm}^3$$

METHOD 2   We have just seen (Equation 9.3) that V_f is related to V_i by a ratio of pressures,

$$V_f = V_i \times (\text{ratio of pressures})$$

If you use this reasoning approach, you will gain a better understanding of the gas laws.

We can use an understanding of Boyle's law to determine what this pressure ratio must be. In this problem, the pressure is increasing from 100 kPa to 125 kPa. We know that this will cause the volume to decrease, so V_f must be smaller than V_i. This means that we must multiply V_i by a factor smaller than one. Of the two possible pressures ratios, (125 kPa/100 kPa) and (100 kPa/125 kPa), only the second satisfies this condition. Therefore we multiply the initial volume, 100 cm³, by 100 kPa/125 kPa.

$$V_f = 100 \text{ cm}^3 \times \left(\frac{100 \text{ kPa}}{125 \text{ kPa}} \right)$$

$$V_f = 80.0 \text{ cm}^3$$

9.3   CHARLES' LAW

In 1787, a French chemist named Jacques Alexander Charles studied the effects of temperature changes on the volume of a confined gas held at a constant pressure. He discovered that as the temperature of a gas is increased, the gas expands to fill a larger volume. In fact, this is what lifts hot air balloons like those in the photograph at the beginning of this chapter. When the air is heated, it expands and becomes less dense. Just as a piece of wood, which is less dense than water, floats on water, so the low density hot air trapped in the balloon floats in the higher density cool air surrounding it.

When the data obtained in experiments similar to those performed by Charles are plotted, we obtain a graph like that shown in Figure 9.8. Here, the

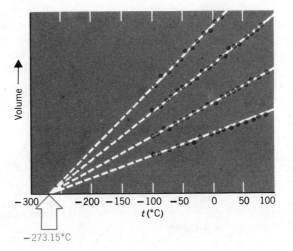

Figure 9.8
Charles' law plot of V *versus* t(°C).

volume of the gas is plotted against the temperature in degrees Celsius. The colored points correspond to typical data, and the white lines are drawn to most closely fit the data. Each line represents data collected on a different sized sample of gas. Since all gases eventually condense to a liquid if cooled sufficiently, only the solid portions of the lines correspond to temperatures at which measurements are possible—at lower temperatures the gas condenses. However, if these lines are extrapolated (extended) to a volume of zero, they all meet at the same temperature, $-273.15°C$. What is especially significant is that this same behavior is exhibited by all gases; when plots of volume versus temperature are extrapolated to zero volume ($V = 0$), the temperature axis is always crossed at $-273.15°C$. This point represents the temperature at which all gases, if they did not condense, would have a volume of zero, and below which they would have a negative volume. Negative volumes are impossible, of course, so it was reasoned that this must be the lowest possible temperature. It is called **absolute zero.**

Absolute zero represents the zero point on the kelvin temperature scale, as we saw in Chapter 1. The kelvin scale is also called the **absolute temperature scale,** and to obtain the kelvin temperature we add 273.15 to the Celsius temperature,

$$T(\text{K}) = t(°\text{C}) + 273.15$$

When you encounter a capital T in an equation, it almost always stands for the absolute temperature.

For most purposes, we will use only three significant figures, so $0°C = 273$ K and we simply use the approximate relationship,

$$T(\text{K}) = t(°\text{C}) + 273$$

The absolute temperature scale is *always* used when the temperature enters numerically into a computation involving the pressures and volumes of gases.

The straight lines in Figure 9.8 suggests that at constant pressure the volume of a gas is directly proportional to its temperature, providing the temperature is expressed in the proper units. The relationship that exists between the volume of a gas and its temperature is summarized by **Charles' law:** *At constant pressure, the volume of a given quantity of a gas is directly proportional to its absolute temperature.* Writing Charles' law mathematically, we have

A straight-line graph results when there is a direct proportionality between the two quantities being plotted.

$$V \propto T \qquad\qquad [9.4]$$

Making the proportionality an equality and rearranging, we obtain

$$\frac{V}{T} = \text{constant} \qquad\qquad [9.5]$$

If Charles' law were strictly obeyed, gases would not condense when they are cooled. Therefore, condensation is considered nonideal behavior, and all real gases behave more and more nonideally as their condensation temperatures are approached. This means that *gases behave in an ideal fashion only at relatively high temperatures and low pressures.* An example of the application of Charles' law is shown below.

EXAMPLE 9.4 APPLYING THE TEMPERATURE-VOLUME LAW (CHARLES' LAW)

PROBLEM A sample of a gas occupies 250 cm³ at 27°C. What volume will it occupy at 35°C if there is no change in pressure?

SOLUTION   Once again we should set up our table of data.

	Initial (i)	Final (f)
Volume (V)	250 cm³	?
Temperature (T)	27 + 273 = 300 K	35 + 273 = 308 K

It is absolutely essential that you express temperature in kelvins in these calculations.

Note that we have converted to absolute temperatures.

Equation 9.5 implies that

$$\frac{V_i}{T_i} = \frac{V_f}{T_f}$$

If we solve for V_f, we find that the initial and final volumes are related as

You can also use

$$V_f = V_i\left(\frac{T_f}{T_i}\right)$$

$$V_f = V_i \times \text{(ratio of absolute temperatures)}$$

Without actually solving the equation, we can obtain the correct temperature ratio through reasoning. Gases expand when heated. Since the temperature is increasing from 300 K to 308 K, the final volume must be larger than the initial volume. The temperature ratio must be a fraction with a value greater than one. This requires the larger temperature in the numerator. Therefore,

$$V_f = 250 \text{ cm}^3 \times \left(\frac{308 \text{ K}}{300 \text{ K}}\right)$$

$$V_f = 257 \text{ cm}^3$$

9.4   GAY-LUSSAC'S LAW

If you read the label on an aerosol can, you will find a warning such as "Do not incinerate" or "Do not store above 50°C." These warnings are necessary because the pressure of a confined gas rises when the gas is heated. If the pressure becomes too large, the container can burst and cause serious injury. Joseph Gay-Lussac, a contemporary of Jacques Charles, was the first to discover that the pressure of a fixed amount of gas is directly proportional to its absolute temperature if the volume of the gas is held constant,

$$P \propto T$$

or

$$\frac{P}{T} = \text{constant} \qquad [9.6]$$

This is a mathematical statement of the **law of Gay-Lussac.**

EXAMPLE 9.5   APPLYING THE TEMPERATURE-PRESSURE LAW (GAY-LUSSAC'S LAW)

PROBLEM   What would be the pressure of a gas, originally at 115 kPa, if the temperature is lowered from 35°C to 25°C at a constant volume?

SOLUTION   Equation 9.6 implies that

$$\frac{P_i}{T_i} = \frac{P_f}{T_f}$$

We should now be able to write that

$$P_f = P_i \times \text{(ratio of temperatures)}$$

Tabulating the data, we obtain

In these calculations, be sure to express temperature in kelvins.

	(i)	(f)
P	115 kPa	?
T	308 K	298 K

Since the temperature is decreasing, the pressure will decrease. This requires the temperature ratio to be smaller than one.

You can also use

$$P_f = P_i\left(\frac{T_f}{T_i}\right)$$

$$P_f = 115 \text{ kPa} \times \left(\frac{298 \text{ K}}{308 \text{ K}}\right)$$

$$P_f = 111 \text{ kPa}$$

9.5   THE COMBINED GAS LAW

The equations corresponding to Boyle's law, Charles' law, and Gay-Lussac's law can be incorporated into one single equation that is useful for many computations. That is

$$\frac{P_iV_i}{T_i} = \frac{P_fV_f}{T_f} \qquad [9.7]$$

Notice that if $T_i = T_f$, the temperature may be dropped and Equation 9.7 reduces to a statement of Boyle's law (that is, $P_iV_i = P_fV_f$ at constant temperature). Similarly, if $P_i = P_f$, the equation reduces to Charles' law and if $V_i = V_f$, the equation reduces to Gay-Lussac's law. As in each of the separate laws, the combined gas law holds only if the amount of gas is not changed.

When working with gases it is useful to define a reference set of conditions of temperature and pressure. These conditions, known as **standard temperature and pressure,** or simply **STP,** are 0°C (273 K) and 1 atm (760 torr).

EXAMPLE 9.6   USING THE COMBINED GAS LAW

PROBLEM   What would be the volume of a gas at STP if it was found to occupy a volume of 255 cm^3 at 25°C and 85.0 kPa?

SOLUTION   In this problem both temperature and pressure are changing. To compute the final volume we must combine Boyle's law and Charles' law. Equation 9.7 does this,

$$\frac{P_iV_i}{T_i} = \frac{P_fV_f}{T_f}$$

If we solve this equation for V_f, we find

Of course, you could also use

$$V_f = V_i\left(\frac{P_i}{P_f}\right)\left(\frac{T_f}{T_i}\right)$$

$$V_f = V_i \times \text{(pressure ratio)} \times \text{(temperature ratio)}$$

Once again we can use reasoning to set up these ratios. First let's tabulate the data.

	Initial (i)	Final (f)	
V	255 cm^3	?	
P	85.0 kPa	101.3 kPa	STP
T	298 K	273 K	

There is a pressure increase, which should tend to cause the volume of the gas to decrease. The pressure ratio should therefore be smaller than one, which requires the larger pressure in the denominator.

$$V_f = 255 \text{ cm}^3 \times \left(\frac{85.0 \text{ kPa}}{101.3 \text{ kPa}}\right) \times \text{(temperature ratio)}$$

The temperature is decreasing. According to Charles' law this should cause a further decrease in the volume. The temperature ratio must also be smaller than one and the larger temperature must be in the denominator.

$$V_f = 255 \text{ cm}^3 \times \left(\frac{85.0 \text{ kPa}}{101.3 \text{ kPa}}\right) \times \left(\frac{273 \text{ K}}{298 \text{ K}}\right)$$

$$V_f = 196 \text{ cm}^3$$

The volume at STP is 196 cm^3.

EXAMPLE 9.7   USING THE COMBINED GAS LAW

PROBLEM   A sample of a gas exerts a pressure of 82.5 kPa in a 300-cm^3 container at 25°C. What pressure would the same gas sample exert in a 500-cm^3 container at 50°C.

SOLUTION   As usual, begin by tabulating the data.

	Initial (i)	Final (f)
P	82.5 kPa	?
V	300 cm^3	500 cm^3
T	298 K	323 K

This time the relationship is

You could also use

$$P_f = P_i\left(\frac{V_i}{V_f}\right)\left(\frac{T_f}{T_i}\right)$$

$$P_f = P_i \times \text{(volume ratio)} \times \text{(temperature ratio)}$$

The volume is increasing from 300 cm^3 to 500 cm^3, which should favor a pressure decrease. Therefore the volume ratio must be smaller than one.

$$P_f = 82.5 \text{ kPa} \times \left(\frac{300 \text{ cm}^3}{500 \text{ cm}^3}\right) \times \text{(temperature ratio)}$$

The temperature is increasing from 298 K to 323 K. This favors a pressure increase, so the temperature ratio must be larger than one.

$$P_f = 82.5 \text{ kPa} \times \left(\frac{300 \text{ cm}^3}{500 \text{ cm}^3}\right) \times \left(\frac{323 \text{ K}}{298 \text{ K}}\right)$$

$$P_f = 49.5 \text{ kPa}$$

9.6 DALTON'S LAW OF PARTIAL PRESSURES

When two or more gases that do not react chemically are placed in the same container, the pressure exerted by each gas in the mixture is the same as it would be if it were the only gas in the container. The pressure exerted by each gas in a mixture is called its **partial pressure** and, as observed by John Dalton, the total pressure is equal to the sum of the partial pressures of each gas in the mixture. This statement, known as **Dalton's law of partial pressures,** can be expressed as

$$P_T = p_a + p_b + p_c + \cdots$$

We can't actually measure the individual partial pressures; we can only measure the total pressure.

where P_T is the total pressure of the mixture (which could be measured with a manometer) and p_a, p_b, and p_c are the partial pressures of gases a, b, and c, respectively. For example, if nitrogen, oxygen, and carbon dioxide were placed in the same vessel, the total pressure of the mixture would be

$$P_T = p_{N_2} + p_{O_2} + p_{CO_2}$$

Thus, if the partial pressure of nitrogen was 25 kPa, that of oxygen 35 kPa, and that of carbon dioxide 45 kPa, the total pressure of the mixture would be

$$P_T = 25 \text{ kPa} + 35 \text{ kPa} + 45 \text{ kPa}$$

$$P_T = 105 \text{ kPa}$$

Dalton's law can be useful in determining the pressure resulting from the mixing of two gases that were originally in separate containers, as shown by the following example.

| EXAMPLE 9.8 | USING DALTON'S LAW OF PARTIAL PRESSURES |

PROBLEM   If 200 cm^3 of N_2 at 25°C and a pressure of 35 kPa are mixed with 350 cm^3 of O_2 at 25°C and a pressure of 45 kPa, so that the resulting volume is 300 cm^3, what would be the final pressure of the mixture at 25°C?

SOLUTION   From Dalton's law we know that we can treat each gas in the mixture as if it were the only gas present. Therefore, we can calculate *independently* the new pressures of N_2 and O_2 when they are placed in the 300-cm^3 container. Since there is no temperature change, we have simply a Boyle's law calculation for each gas.

Following the methods used in Examples 9.3 through 9.7, we first set up our tables of data.

For N$_2$	(i)	(f)		For O$_2$	(i)	(f)
p	35 kPa	?		p	45 kPa	?
V	200 cm^3	300 cm^3		V	350 cm^3	300 cm^3

For each calculation we can write

$$p_f = p_i \times (\text{ratio of volumes})$$

Since the volume of the N_2 is increasing, its pressure must decrease; p_f must be less than p_i. This requires a volume ratio smaller than one, which means that the larger volume must be in the denominator. Thus

$$p_{N_2} = 35 \text{ kPa} \times \left(\frac{200 \text{ cm}^3}{300 \text{ cm}^3}\right)$$

$$p_{N_2} = 23 \text{ kPa}$$

For O_2 the volume is decreasing; p_f must be greater than p_i. This requires a volume ratio larger than one.

$$p_{O_2} = 45 \text{ kPa} \times \left(\frac{350 \text{ cm}^3}{300 \text{ cm}^3} \right)$$

$$p_{O_2} = 52 \text{ kPa}$$

The total pressure of the mixture is the sum of the partial pressures.

$$P_T = p_{N_2} + p_{O_2} = 23 \text{ kPa} + 52 \text{ kPa}$$

$$P_T = 75 \text{ kPa}$$

This can only be done if the gas has a low solubility in water.

Gases prepared in the laboratory are quite often collected by the displacement of water, as shown in Figure 9.9. A gas collected in this manner becomes "contaminated" with water molecules that evaporate into the gas. These water molecules also exert a pressure called the **vapor pressure.** For reasons that we will discuss in Chapter 10, the vapor pressure of water depends *only* on the temperature of the liquid water (Table 9.1). The pressure of the water vapor contributes to the total pressure of the "wet" gas, so we can write

$$P_T = p_{\text{gas}} + p_{H_2O}$$

If the level of the water is the same inside the collection flask as outside, as shown in Figure 9.9, then the pressure inside must also be the same as outside, namely, atmospheric pressure. The atmospheric pressure can be determined with a barometer, and the vapor pressure of water can be obtained from Table 9.1 if the temperature of the liquid is known. The partial pressure of the pure gas is therefore

$$p_{\text{gas}} = P_T - p_{H_2O}$$

EXAMPLE 9.9 COLLECTING A GAS OVER WATER

PROBLEM A student generates oxygen gas in the laboratory and collects it in a manner similar to that shown in Figure 9.9. She collects the gas at 25°C until the levels of the water inside and outside the flask are equal. If the volume of the gas is 245 cm³ and the atmospheric pressure is 98.5 kPa:

(a) What is the pressure of O_2 gas in the "wet" gas mixture at 25°C?
(b) What would be the volume of dry oxygen at STP?

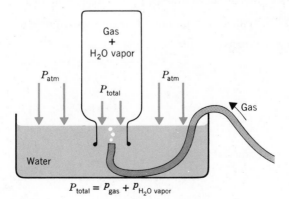

Figure 9.9

Collection of a gas by displacement of water.

Table 9.1
Vapor pressure of water
as a function of temperature

Temp. (°C)	Press. (kPa)	Temp. (°C)	Press. (kPa)	Temp. (°C)	Press. (kPa)
0	0.61	18	2.07	40	7.37
1	0.65	19	2.20	45	9.59
2	0.71	20	2.33	50	12.33
3	0.76	21	2.49	55	15.73
4	0.81	22	2.64	60	19.92
5	0.87	23	2.81	65	25.00
6	0.93	24	2.97	70	31.16
7	1.00	25	3.17	75	38.54
8	1.07	26	3.36	80	47.34
9	1.15	27	3.56	85	57.81
10	1.23	28	3.77	90	70.10
11	1.31	29	4.00	95	84.54
12	1.40	30	4.24	96	87.67
13	1.49	31	4.49	97	90.94
14	1.60	32	4.76	98	94.30
15	1.71	33	5.03	99	97.75
16	1.81	34	5.32	100	101.32
17	1.93	35	5.63	101	105.00

SOLUTION

(a) For gases collected in this manner we know that

$$P_T = p_{gas} + p_{H_2O}$$

Substituting p_{O_2} for p_{gas} and rearranging, we have

$$p_{O_2} = P_T - p_{H_2O}$$

According to Table 9.1, the partial pressure of water at 25°C is 3.17 kPa. The atmospheric pressure was given as 98.5 kPa. Therefore, the partial pressure of O_2 is

$$p_{O_2} = 98.5 \text{ kPa} - 3.17 \text{ kPa}$$

$$p_{O_2} = 95.3 \text{ kPa} \qquad \text{(rounded to the correct number of significant figures)}$$

This is the pressure exerted by the oxygen alone.

(b) This part of the problem is a combined Boyle's law–Charles' law calculation.

First, tabulate the data as follows:

	(i)	(f)	
V	245 cm³	?	
P	95.3 kPa	101.3 kPa	STP
T	298 K	273 K	

Next, recall that

$$V_f = V_i \times (\text{ratio of pressures}) \times (\text{ratio of temperatures})$$

The pressure change should tend to decrease the volume; therefore, the pressure ratio should be less than one. The temperature decrease should also decrease the volume; therefore, the temperature ratio should be less than one. Thus

$$V_f = 245 \text{ cm}^3 \times \left(\frac{95.3 \text{ kPa}}{101.3 \text{ kPa}} \right) \times \left(\frac{273 \text{ K}}{298 \text{ K}} \right)$$

$$V_f = 211 \text{ cm}^3 \text{ at STP}$$

9.7 CHEMICAL REACTIONS BETWEEN GASES

Many gases are able to undergo chemical reactions with each other. Hydrogen and oxygen can combine to form water. Nitrogen and hydrogen, under appropriate conditions, combine to form ammonia. If these reactions are carried out so that the volumes of the gaseous reactants and products are measured at the same temperatures and pressures, the volumes are related in a simple fashion. For example, when hydrogen and oxygen are placed in a vessel and allowed to react with each other to form gaseous water, two volumes of hydrogen always react with one volume of oxygen to form two volumes of gaseous water. This can be expressed in the form of an equation as

2 volumes hydrogen + 1 volume oxygen $\longrightarrow$ 2 volumes gaseous water

Similarly, when one volume of hydrogen reacts with one volume of chlorine, two volumes of hydrogen chloride gas are produced; that is,

1 volume hydrogen + 1 volume chlorine $\longrightarrow$ 2 volumes hydrogen chloride

For the reaction between hydrogen and nitrogen to form ammonia, we find this relationship.

1 volume nitrogen + 3 volumes hydrogen $\longrightarrow$ 2 volumes ammonia

Experimental observations such as these form the basis of **Gay-Lussac's law of combining volumes,** which states that *the volumes of gaseous substances that are consumed and produced in a chemical reaction are in ratios of small whole numbers, provided the volumes are measured under the same conditions of temperature and pressure.*

The significance of Gay-Lussac's observation was later recognized by Amadeo Avogadro. He proposed what is now known as **Avogadro's principle:** *Under conditions of constant temperature and pressure, equal volumes of gas contain equal numbers of molecules.* Since equal numbers of molecules mean equal numbers of moles, the number of moles of any gas is related directly to its volume:

$$V \propto n \qquad\qquad [9.8]$$

where n is the number of moles of gas. On this basis Gay-Lussac's law is easily understood because the volumes of gaseous reactants and products occur in the same ratios as the coefficients in the balanced equation.[2] For example,

$$2H_2(g) + O_2(g) \longrightarrow 2H_2O(g)$$

2 volumes + 1 volume $\longrightarrow$ 2 volumes

[2] In fact, these observations were used to show that gases such as H_2, O_2, and Cl_2 must be *at least* diatomic and that their subscripts must be even. The only way *two* volumes of hydrogen chloride could be formed from *one* volume of hydrogen and *one* volume of chlorine is if each molecule of hydrogen and chlorine contained two atoms of H and two atoms of Cl, respectively, and hydrogen chloride was HCl. If hydrogen chloride was H_2Cl_2, then each molecule of hydrogen and chlorine would have to be composed of four atoms, and so forth.

Table 9.2
Molar volumes of several real gases at STP

Substance	Molar Volume (dm^3)
Oxygen, O_2	22.397
Nitrogen, N_2	22.402
Hydrogen, H_2	22.433
Helium, He	22.434
Argon, Ar	22.397
Carbon dioxide, CO_2	22.260
Ammonia, NH_3	22.079

From Avogadro's principle we expect 1 mol of any gas to occupy the same volume at a given temperature and pressure. It has been found by experiment that the average volume occupied by 1 mol of a gas at STP is 22.4 dm^3. We will assume this is the **molar volume** of an ideal gas at STP. For real gases the molar volume actually fluctuates about this average, as shown in Table 9.2.

EXAMPLE 9.10 **STOICHIOMETRY OF GASEOUS REACTIONS**

PROBLEM What volume of O_2, at STP, is required for the complete combustion of 4.50 dm^3 of butane, C_4H_{10}, at STP? Butane is the fuel in disposable cigarette lighters.

SOLUTION As with any problem in stoichiometry, we should first write a balanced equation for the reaction. This is

$$2C_4H_{10} + 13O_2 \longrightarrow 8CO_2 + 10H_2O$$

To solve the problem we can compute the number of moles of C_4H_{10} using the molar volume at STP.

$$4.50 \; \text{dm}^3 \; C_4H_{10} \times \left(\frac{1 \; \text{mol} \; C_4H_{10}}{22.4 \; \text{dm}^3 \; C_4H_{10}} \right) \sim 0.201 \; \text{mol} \; C_4H_{10}$$

Next we calculate the number of moles of O_2 required, using the coefficients in the equation,

$$0.201 \; \text{mol} \; C_4H_{10} \times \left(\frac{13 \; \text{mol} \; O_2}{2 \; \text{mol} \; C_4H_{10}} \right) \sim 1.31 \; \text{mol} \; O_2$$

Finally we can calculate the volume of O_2, again using the molar volume of a gas at STP.

$$1.31 \; \text{mol} \; O_2 \times \left(\frac{22.4 \; \text{dm}^3 \; O_2}{1 \; \text{mol} \; O_2} \right) \sim 29.3 \; \text{dm}^3 \; O_2$$

The volume of O_2 required is 29.3 dm^3.

In problems involving gaseous reactants or products at the *same temperature and pressure*, we can take a shortcut to the answer. If we set up the calculation above with all the conversion factors strung together, we have

$$4.50 \; \text{dm}^3 \; C_4H_{10} \times \left(\frac{1 \; \text{mol} \; C_4H_{10}}{22.4 \; \text{dm}^3 \; C_4H_{10}} \right) \times \left(\frac{13 \; \text{mol} \; O_2}{2 \; \text{mol} \; C_4H_{10}} \right) \times \left(\frac{22.4 \; \text{dm}^3 \; O_2}{1 \; \text{mol} \; O_2} \right) \sim 29.3 \; \text{dm}^3 \; O_2$$

The number 22.4 appears in both numerator and denominator and therefore cancels. The volumes of reactants (or products) are simply related by the coefficients in the equation. Thus in this problem we could state that

$$2 \text{ dm}^3 \text{ C}_4\text{H}_{10} \sim 13 \text{ dm}^3 \text{ O}_2$$

Remember, this only works if the volumes are compared at the same T and P.

This is a consequence of Gay-Lussac's law. Realizing this, we see that the solution to the problem could have been obtained as

$$4.50 \text{ dm}^3 \text{ C}_4\text{H}_{10} \times \left(\frac{13 \text{ dm}^3 \text{ O}_2}{2 \text{ dm}^3 \text{ C}_4\text{H}_{10}} \right) \sim 29.3 \text{ dm}^3 \text{ O}_2$$

EXAMPLE 9.11 STOICHIOMETRY OF GASEOUS REACTIONS

PROBLEM   The drain cleaner, Drano, contains small bits of aluminum, which react with NaOH (the main ingredient in this product) to produce bubbles of hydrogen. These bubbles presumably are designed to stir the mixture and hasten its action. What volume (in cm^3) of H_2, measured at STP, will be released when 0.150 g of Al are dissolved? The chemical equation is

$$2\text{Al} + 2\text{OH}^- + 2\text{H}_2\text{O} \longrightarrow 3\text{H}_2 + 2\text{AlO}_2^-$$

SOLUTION   First we calculate the number of moles of Al that react.

$$0.150 \text{ g Al} \times \left(\frac{1 \text{ mol Al}}{27.0 \text{ g Al}} \right) = 0.005\ 56 \text{ mol Al}$$

Next we calculate the number of moles of H_2 produced.

$$0.005\ 56 \text{ mol Al} \times \left(\frac{3 \text{ mol H}_2}{2 \text{ mol Al}} \right) \sim 0.008\ 34 \text{ mol H}_2$$

Since 1 mol H_4 = 22.4 dm^3 H_2 at STP,

$$0.008\ 34 \text{ mol H}_2 \times \left(\frac{22.4 \text{ dm}^3 \text{ H}_2}{1 \text{ mol H}_2} \right) = 0.187 \text{ dm}^3 \text{ H}_2$$

Expressed in cubic centimetres, the answer is 187 cm^3 H_2.

9.8   THE IDEAL GAS LAW

We have thus far discussed three volume relationships that an ideal gas obeys. These are

$$\text{Boyle's Law} \qquad V \propto \frac{1}{P}$$

$$\text{Charles' law} \qquad V \propto T$$

$$\text{Avogadro's law} \qquad V \propto n$$

We can combine these to obtain

$$V \propto n \left(\frac{1}{P} \right) (T)$$

or

$$V \propto \frac{nT}{P} \qquad\qquad [9.9]$$

Equation 9.9 reduces to an expression of Boyle's law during a process in which the volume changes as a result of a pressure change only. Since n and T remain constant, volume is proportional only to pressure; that is,

$$V \propto \frac{1}{P} \qquad \text{at constant } n \text{ and } T$$

Similarly, we can see that Equation 9.9 becomes Charles' law for a volume change at constant n and P.

$$V \propto T \qquad \text{at constant } n \text{ and } P$$

Avogadro's principle is seen as

$$V \propto n \qquad \text{at constant } T \text{ and } P$$

The proportionality in Equation 9.9 is made into an equality by the introduction of a proportionality constant, R, called the **universal gas constant.** Equation 9.9 then becomes

$$V = \frac{nRT}{P}$$

or, as it is usually written,

$$PV = nRT \qquad\qquad\qquad [9.10]$$

Equation 9.10 is obeyed exactly only by the hypothetical ideal gas and is a mathematical statement of the **ideal gas law.** It is also called the **equation of state for an ideal gas** because it relates those variables (P, V, n, T) that specify the physical properties of the gas. If any three of these are given, the fourth variable can have only one value as determined by Equation 9.10.

When a real gas comes very close to obeying the ideal gas law its behavior is said to be ideal. Fortunately, most real gases are nearly ideal under conditions of temperature and pressure normally encountered in the laboratory, so the ideal gas law can be used quite accurately to describe their behavior. The only time Equation 9.10 should not be used is when it is necessary to perform very accurate computations.

To use the ideal gas law we must have a value for the gas constant, R. This can be computed by inserting appropriate values for P, V, n, and T into Equation 9.10 and solving for R. For 1 mol of an ideal gas at STP, $P = 101.325$ kPa ($101\ 325$ N m^{-2}), $V = 22.4$ dm^3 (0.0224 m^3), $n = 1$ mol, and $T = 273$ K. Solving for R, we get

$$R = \frac{PV}{nT}$$

$$R = \frac{(101\ 325 \text{ N m}^{-2})(0.0224 \text{ m}^3)}{(1 \text{ mol})(273 \text{ K})}$$

$$R = 8.31 \text{ N m mol}^{-1} \text{ K}^{-1}$$

Using a more precise value for the molar volume and standard temperature, we'd find that $R = 8.314$ N m mol^{-1} K^{-1}. Since 1 N m is defined as one joule (J) of energy, R can be expressed (perhaps somewhat surprisingly) as 8.314 J mol^{-1} K^{-1}. This expression for R will be useful in later thermodynamic work (Chapter 12), even though it does not seem to relate directly to pressure-volume gas measurements. If the calculation of R is repeated with $P = 101.325$ kPa and $V = 22.4$ dm^3, we find $R = 8.314$ kPa dm^3 mol^{-1} K^{-1}. Other values of R are possible, as Example 9.12 suggests.

Usual laboratory conditions are approximately 100 kPa and 25°C.

More precise measurements give $R = 8.314$ N m mol^{-1} K^{-1}.

EXAMPLE 9.12   CHANGING THE UNITS OF THE GAS CONSTANT

PROBLEM   What is the value of R when pressure is expressed in standard atmospheres (atm) and volume is expressed in cubic decimetres?

SOLUTION   This is simply a unit-conversion problem making use of the relationship

$$1 \text{ atm} = 101.325 \text{ kPa}$$

$$R = \left(\frac{8.314 \text{ kPa dm}^3}{\text{mol K}} \right) \times \left(\frac{1 \text{ atm}}{101.325 \text{ kPa}} \right)$$

$$= 0.082\,06 \text{ atm dm}^3 \text{ mol}^{-1} \text{ K}^{-1}$$

When you use R in a computation, be sure the units cancel correctly.

The choice of the value of R to be used in a given computation is governed by the units of P and V. You will find it best, perhaps, to learn one value of R ($8.314 \text{ kPa dm}^3 \text{ mol}^{-1} \text{ K}^{-1}$) and to convert P and V to units that can be used with that value of R.

The following are some examples of the application of the ideal gas law.

EXAMPLE 9.13   USING THE IDEAL GAS LAW

PROBLEM   What volume will 25.0 g of O_2 occupy at 20°C and a pressure of 89.0 kPa?

SOLUTION   From the ideal gas law,

$$V = \frac{nRT}{P}$$

We will use $R = 8.314 \text{ kPa dm}^3 \text{ mol}^{-1} \text{ K}^{-1}$. Tabulating our data, we get

P	89.0 kPa
V	?
n	25.0 g of $O_2 \times \dfrac{1 \text{ mol of } O_2}{32.0 \text{ g of } O_2} = 0.781 \text{ mol } O_2$
T	20 + 273 = 293 K

Substituting, we obtain

$$V = \frac{(0.781 \text{ mol}) \times 8.314 \text{ kPa dm}^3 \text{ mol}^{-1} \text{ K}^{-1} \times (293 \text{ K})}{(89.0 \text{ kPa})}$$

$$V = 21.4 \text{ dm}^3$$

EXAMPLE 9.14   DETERMINING THE MOLAR MASS OF A GAS

PROBLEM   A student collected natural gas from a laboratory gas jet at 25°C in a 250-cm³ flask until the pressure of the gas was 73.5 kPa. The gas sample mass was 0.118 g at a temperature of 25°C. From these data, calculate the molar mass of the gas.

SOLUTION Determining molar mass is one of the most useful applications of the ideal gas law. To do this, we need to know the number of moles of gas in the sample. We can determine the number of moles of gas present using the ideal gas law,

$$n = \frac{PV}{RT}$$

Again we use $R = 8.314 \text{ kPa dm}^3 \text{ mol}^{-1} \text{ K}^{-1}$. Our data are

P	73.5 kPa
V	$250 \text{ cm}^3 \times \dfrac{1 \text{ dm}^3}{1000 \text{ cm}^3} = 0.250 \text{ dm}^3$
n	?
T	$25 + 273 = 298 \text{ K}$

Substituting, we obtain

$$n = \frac{(73.5 \text{ kPa}) \times (0.250 \text{ dm}^3)}{(8.314 \text{ kPa dm}^3 \text{ mol}^{-1} \text{ K}^{-1}) \times (298 \text{ K})}$$

$$n = 0.007\ 42 \text{ mol}$$

To calculate the molar mass we must determine the mass of 1 mol of the substance. We now know that

$$0.118 \text{ g} = 0.007\ 42 \text{ mol}$$

The mass of 1 mol, therefore, is

A molar mass can be calculated if you know the number of moles in a given mass of the compound. Simply take the ratio of grams to moles.

$$1 \text{ mol} \times \left(\frac{0.118 \text{ g}}{0.007\ 42 \text{ mol}} \right) = 15.9 \text{ g}$$

Thus the molar mass is 15.9 g mol^{-1}. This natural gas sample is really methane, CH_4, having a molar mass of 16.0 g mol^{-1}.

EXAMPLE 9.15 DETERMINING THE MOLAR MASS OF A GAS FROM ITS DENSITY

PROBLEM A student measured the density of a gaseous compound to be 1.34 g dm^{-3} at 25°C and 1.000 atm, and was told that the compound was composed of 79.8% carbon and 20.2% hydrogen by mass.

(a) What is the empirical formula of the compound?
(b) What is its molar mass?
(c) What is the molecular formula of the compound?

SOLUTION (a) Following the procedure outlined in Section 2.7, we find the empirical formula of the carbon–hydrogen compound. If we assume a 100-g sample,

$$79.8 \text{ g C} \times \left(\frac{1 \text{ mol C}}{12.0 \text{ g C}} \right) = 6.65 \text{ mol C}$$

$$20.2 \text{ g H} \times \left(\frac{1 \text{ mol H}}{1.01 \text{ g H}} \right) = 20.0 \text{ mol H}$$

The empirical formula is $C_{\frac{6.65}{6.65}} H_{\frac{20.0}{6.65}}$ or CH_3, which would give an empirical formula weight of $CH_3 = 15.0$.

(b) The density gives the mass of 1 dm^3 of the gas. 1.00 dm^3 = 1.34 g

To calculate molar mass we need a relationship between mass and moles. Following the procedure in Example 9.14, we get

P	1.000 atm $\times \dfrac{101.3 \text{ kPa}}{1 \text{ atm}}$ = 101.3 kPa
V	1.00 dm^3
n	?
T	25 + 273 = 298 K

$$n = \frac{PV}{RT} = \frac{(101.325 \text{ kPa}) \times (1.00 \text{ dm}^3)}{(8.314 \text{ kPa dm}^3 \text{ mol}^{-1} \text{ K}^{-1}) \times (298 \text{ K})}$$

$$n = 0.0409 \text{ mol}$$

Thus

$$0.0409 \text{ mol} = 1.34 \text{ g}$$

The mass of 1 mol is

$$1 \text{ mol} \times \left(\frac{1.34 \text{ g}}{0.0409 \text{ mol}} \right) = 32.8 \text{ g}$$

The molar mass is 32.8 g mol^{-1}.

(c) We see that this is approximately twice the empirical formula weight, which means that the molecular formula must be

$$(CH_3)_2 \quad \text{or} \quad C_2H_6$$

This is a substance called ethane.

An equation relating the molar mass of a gas to its density directly can easily be derived from the ideal gas law and used to solve problems such as part (b) of this last example. We know, for example, that the number of moles of a substance is obtained by dividing its mass, in grams, by the molar mass.

$$\text{number of moles } (n) = \frac{\text{number of grams } (g)}{\text{molecular weight } (M)}$$

or simply,

$$n = \frac{g}{M}$$

Substituting for n in the ideal gas law gives

$$PV = \frac{g}{M} RT$$

which can be rearranged to solve for M.

Some people find it easy to remember equations like 9.11 and 9.12. Others find it easier to approach problems as shown in Examples 9.14 and 9.15. The choice is up to you.

$$M = \frac{g}{V} \frac{RT}{P} \qquad [9.11]$$

Density (d) is a ratio of mass to volume, g/V. This allows us to write Equation 9.11 as

$$M = d \frac{RT}{P} \qquad [9.12]$$

To solve part (b) of Example 9.15 we can substitute the given values for d, R, T, and P into this equation.

$$M = \frac{1.34 \text{ g}}{\text{dm}^3} \frac{(8.314 \text{ kPa dm}^3 \text{ mol}^{-1} \text{ K}^{-1})(298 \text{ K})}{(101.3 \text{ kPa})}$$

$$M = 32.8 \text{ g mol}^{-1}$$

9.9 GRAHAM'S LAW OF EFFUSION

If two gases are placed in the same container, they spread and mix spontaneously—a process that is called **diffusion.** Anyone who has driven past a skunk that didn't quite make it across the road has experienced this phenomenon personally. It isn't long before the occupants of the car are in need of a little fresh air.

A process somewhat similar to diffusion is called **effusion.** This is a process by which a gas, under pressure, escapes from a vessel by passing through a very small opening, as illustrated in Figure 9.10. Effusion is also responsible for the fate of helium-filled rubber balloons. Perhaps as a child you brought home such a balloon and found that by the next morning much of the helium had escaped. Actually, the helium had effused through tiny pores in the rubber.

Thomas Graham (1805–1869), a British chemist, studied the rates of effusion of various gases through porous plugs of plaster of paris. He found that if he measured these rates under the same conditions of temperature and pressure, the rates were inversely proportional to the *square roots* of the densities of the gases. This statement, known as **Graham's law,** can be expressed mathematically as

$$\text{rate of effusion} \propto \sqrt{\frac{1}{d}}$$

By taking the ratio, we see that the proportionality constant (whatever it may be) cancels from numerator and denominator.

The rate of effusion of two gases (labeled simply A and B) can be compared by dividing the rate of one by the other; that is,

$$\frac{\text{rate of effusion } (A)}{\text{rate of effusion } (B)} = \sqrt{\frac{d_B}{d_A}}$$

Looking back at Equation 9.12, we see that the density of a gas is directly proportional to its molecular weight. This allows us to rewrite the ratio of effusion rates as

Graham's law works for diffusion, too.

$$\frac{\text{rate of effusion } (A)}{\text{rate of effusion } (B)} = \sqrt{\frac{d_B}{d_A}} = \sqrt{\frac{M_B}{M_A}} \qquad [9.13]$$

where M_A and M_B are the molar mass of gases A and B, respectively.

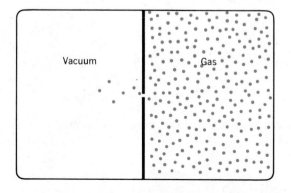

Figure 9.10
Effusion of a gas into a vacuum.

EXAMPLE 9.16 CALCULATING RELATIVE RATES OF EFFUSION

PROBLEM
Which gas will effuse faster, ammonia or carbon dioxide? What are their relative rates of effusion?

SOLUTION
The molar mass of CO_2 is 44 g mol^{-1} and that of NH_3 is 17 g mol^{-1}. Therefore NH_3 will effuse faster. We can calculate how much faster with Equation 9.13.

$$\frac{\text{rate of effusion (NH}_3)}{\text{rate of effusion (CO}_2)} = \sqrt{\frac{M_{CO_2}}{M_{NH_3}}} = \sqrt{\frac{44}{17}} = 1.6$$

Therefore the rate of effusion of NH_3 is 1.6 times faster than the rate of CO_2.

9.10 KINETIC MOLECULAR THEORY AND THE GAS LAWS

We have seen that the study of gases produced a variety of different gas laws. The people who discovered these laws could not help wondering what gases must be composed of to behave as they do. What is the origin of the pressure of a gas? Why are gases so compressible? Why do they expand when heated? And why do light gases effuse more rapidly than heavy ones? These questions, among others, illustrate the need for a theoretical model of a gas.

The **kinetic molecular theory** evolved in an attempt to explain the physical behavior of substances. According to this theory, any given sample of matter is composed of a huge number of small particles (molecules or individual atoms) that are in constant, random motion. In addition, the theory proposes that the average kinetic energy that these particles possess is proportional to the absolute temperature of the sample. As we will see, these two basic postulates apply to all three states of matter: solids, liquids, and gases. What we want to do now is to see how they explain the properties of a gas. Although the postulates can be used to actually derive the ideal gas law mathematically, we will only use them qualitatively.

The pressure-volume relationship: Boyle's law

The most striking quality of a gas is its compressibility. The molecules envisioned in the kinetic molecular theory must therefore be very tiny and very far apart in a gas so that there is plenty of empty space between them. Only in this way could they be so easily crowded together. As these tiny particles fly about, they collide with each other and the walls of the container. Each impact with the wall exerts a tiny push, and the cumulative effects of enormous numbers of such impacts each second on each square centimeter of the wall gives rise to the pressure of the gas.

With this model of a gas in mind, we can now explain Boyle's law. If we halve the volume of a gas, we pack twice as many molecules into each cubic centimetre. Over each square centimetre of wall there are now twice as many molecules as before, and therefore there must be twice as many molecule-wall collisions each second. This means that the pressure has doubled, as illustrated in Figure 9.11. If halving the volume doubles the pressure, then pressure and volume are inversely proportional to each other, which is a statement of Boyle's law.

An ideal gas, you recall, would obey Boyle's law exactly under all conditions. This means that no matter how tightly packed the molecules were, it would always be possible to halve their volume by doubling the pressure. The

At 0°C, the average speed of an oxygen molecule is about 1600 km h^{-1}.

Figure 9.11

When the volume of a gas is halved on going from (a) to (b), twice as many molecules are crowded into each tiny unit of volume. This produces twice as many molecule-wall collisions each second, and that doubles the pressure.

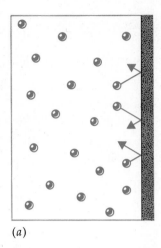
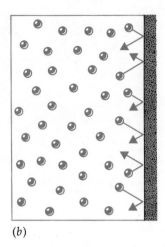

(*a*)   (*b*)

only way this could happen over and over again, of course, is if the gas was composed of particles having no volume themselves, so that the entire volume would be empty space. Real molecules do have finite volumes, however, so no real gas could obey Boyle's law perfectly, especially at high pressure. We will discuss the consequences of this further in the next section.

Distribution of molecular speeds

The second postulate of the kinetic molecular theory is that the average kinetic energy of a collection of molecules is proportional to their absolute temperature. Before we can see how this applies to the gas laws, we have to take a close look at the distribution of molecular energies implied by the use of the term *average kinetic energy*.

In a gas (or a liquid or a solid), molecules are in constant motion. Because they continually collide with each other, there is a wide range of different molecular speeds. For example, occasionally a collision might leave a particular molecule almost motionless, until it suffers another collision and is sent on its way again an instant later. Another molecule might receive several "rear-end" collisions in a row that give it a very high speed. In this way the speeds of molecules are forever changing through collisions with others. At any instant, some molecules will be moving slowly and some will be going very fast, although most will have speeds in between.

Because each molecule has a kinetic energy equal to $\frac{1}{2}mv^2$, where m is its mass and v is its speed, there is also a distribution of kinetic energies associated with the distribution of molecular speeds. Figure 9.12 illustrates the shape of this kinetic energy distribution at three different temperatures. Plotted vertically are the fractions of all the molecules that have the particular kinetic energies given along the horizontal axis. For example, at zero K.E., which corresponds to molecules standing still, this fraction is essentially zero since very few, if any, molecules are motionless at any instant. The fraction having a particular K.E. increases as we move to higher energies (higher speeds) and eventually becomes a maximum. At still higher kinetic energies the fraction decreases and gradually approaches zero again at kinetic energies corresponding to very fast-moving molecules. The curve does not go all the way to zero, however, because there is virtually no upper limit to molecular speeds, other than the speed of light.

The maximum on this curve represents the kinetic energy possessed by the largest fraction of molecules. This kinetic energy would be found most frequently (that is, with the greatest probability) if we were able to examine mole-

This is called a Maxwell–Boltzmann distribution.

Figure 9.12

The distribution of kinetic energies in a collection of molecules at three different temperatures.

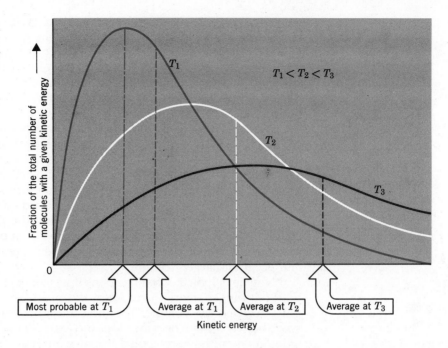

cules at random; hence it is called the *most probable kinetic energy*. The average kinetic energy occurs at a higher value than the most probable kinetic energy because the curve is not symmetrical. Just as a few "curve breakers" in a chemistry class tend to raise the class average on exams, the high-velocity molecules shift the average K.E. above the most probable K.E.

When the temperature of a substance is raised, the curve changes so that the average K.E. increases, as shown in Figure 9.12. More molecules have high speeds and fewer have low ones and, on the average, the molecules move faster. Thus when heat is added to a substance to raise its temperature, the energy goes into increasing the average kinetic energy and increases the speeds of the particles.

The relationship between kinetic energy and temperature also leads quite naturally to the concept of an absolute zero. As kinetic energy is removed from a substance, its molecules move more and more slowly. If all the molecules were to cease moving, their average kinetic energy would be zero and, since negative kinetic energies are impossible (a molecule cannot be going slower than when it is standing still), the temperature of the substance would also be at its lowest point. It is this temperature that we refer to as absolute zero—the temperature when all molecular motion has ceased.[3] It should be understood, however, that electronic motion would still continue at absolute zero. Even though the molecules would be motionless, the electrons would still be "whizzing" about their respective nuclei.

Temperature and the gas laws

You've probably begun to wonder what all of this has to do with the gas laws. Let's look at Gay-Lussac's pressure-temperature law first. This law states that if

---

[3] In fact, even at absolute zero there is still a residual molecular motion that is required by the Heisenberg uncertainty principle (Chapter 4). This principle states that we cannot simultaneously know exactly the position and momentum of a particle. We will see in the next chapter that the average positions of particles in the solid state can be determined, and if the particles were motionless we would also know that their momentum was zero, thereby violating the uncertainty principle.

we keep the volume constant, the pressure is directly proportional to the absolute temperature. In other words, when the temperature goes up, so does the pressure. How can we explain this? According to kinetic theory, raising the temperature increases the average kinetic energy of the molecules, so the molecules move faster. This means that they will strike the wall more frequently, and that when they hit the wall the average force of impact will be greater. These factors make the pressure increase.

We can also explain Charles' law, which says that the volume increases when we raise the temperature, provided we keep the pressure constant. We have just seen that raising the temperature causes more molecules to hit the wall each second and also causes the force of impact of the molecules with the walls to increase. The only way to keep the pressure constant is to simultaneously decrease the number of collisions per second with each square centimetre of wall. This can be accomplished by allowing the gas to expand so fewer molecules are over each square centimetre of wall. In other words, to keep the pressure of a gas constant as we raise its temperature, we have to allow it to expand and occupy a larger volume.

Now let's look at what happens if we cool a gas. Following the reasoning in the preceding paragraph, to keep the pressure constant as the gas cools, we must decrease the volume. The molecules move more and more slowly and the space between them becomes less and less. All real gases eventually condense to a liquid when they are cooled because attractive forces between the molecules eventually cause "sticky" collisions. An ideal gas, however, would not condense regardless of how much we cooled it, so another property of the molecules of an ideal gas is that they have no intermolecular attractions. Consequently, *an ideal gas is a hypothetical substance whose molecules have no volume and no intermolecular attractive forces.*

When the molecules stick together they settle to the bottom of the container as a liquid.

Graham's law

The postulate of the kinetic theory relating average kinetic energy to temperature can be used very simply to derive Graham's law. Suppose we have two different gases, A and B. If they are both at the same temperature, the average kinetic energies of their molecules must be the same. This means that

$$\overline{K.E._A} = \overline{K.E._B}$$

or

m_A and m_B are the masses of molecules A and B.

$$\tfrac{1}{2}m_A\overline{v_A^2} = \tfrac{1}{2}m_B\overline{v_B^2} \qquad [9.14]$$

where $\overline{v^2}$ is called the **mean square speed** of the molecules, and is the average of the speeds-squared of all the molecules; that is,

$$\overline{v^2} = \frac{v_1^2 + v_2^2 + v_3^2 + \cdots}{n_T}$$

where v_1, v_2, v_3, and so on, represent the speeds of molecules 1, 2, 3, and so on, and n_T is the total number of molecules present. Equation 9.14 can be rearranged to give

$$\frac{\overline{v_A^2}}{\overline{v_B^2}} = \frac{m_B}{m_A}$$

Taking the square root of both sides, we have that

$$\frac{\overline{v_A}}{\overline{v_B}} = \sqrt{\frac{m_B}{m_A}} \qquad [9.15]$$

where $\bar{v}$ is called the root-mean-square speed. For any molecule, its actual mass is proportional to its molecular weight—that's the principle upon which the atomic weight table is based. This means that

$$M \propto m$$

Therefore, we can substitute M_A and M_B for m_A and m_B into Equation 9.15 and arrive at

$$\frac{\overline{v_A}}{\overline{v_B}} = \sqrt{\frac{M_B}{M_A}}$$

The rate at which gases effuse should be directly proportional to the velocity of their molecules, with faster molecules effusing at a higher rate. Thus we are led to conclude that

$$\frac{\text{rate of effusion }(A)}{\text{rate of effusion }(B)} = \frac{\overline{v_A}}{\overline{v_B}} = \sqrt{\frac{M_B}{M_A}}$$

or simply

$$\frac{\text{rate of effusion }(A)}{\text{rate of effusion }(B)} = \sqrt{\frac{M_B}{M_A}}$$

which is Graham's law.

Avogadro's principle

Equal volumes of gas at the same temperature and pressure have equal numbers of molecules. That is how we previously stated Avogadro's principle. But we could have also stated this principle as follows: Equal numbers of gas molecules occupying the same volume at the same temperature exert the same pressure. We can explain this by noting that the average force of impact of the molecules colliding with a given area of the wall depends on their average kinetic energy, and therefore on their temperature. If the temperature of two gas samples is the same, then the average kinetic energy of their molecules must be equal, and if the number of molecules per unit volume is the same, then it follows that their pressures must also be the same.

Dalton's law of partial pressures

Molecules of an ideal gas are unaware of each other's existence, except when they collide, because they have no attractions for each other. In a mixture of gases, each behaves independently and exerts a pressure that is the same as it would exert if it were alone. The cumulative effect of the individual partial pressures is the total pressure.

Avogadro's principle can also be applied to gas mixtures. For example, consider the earth's atmosphere, where about 1 of every 5 molecules are O_2 and 4 of every 5 are N_2. Since one-fifth of the molecules are oxygen, only one-fifth of the pressure is contributed by O_2; the other four-fifths of the pressure is contributed by N_2.

The partial pressure of a gas is related quantitatively to the total pressure by its **mole fraction** (usually given by the symbol x), which is the number of moles of the gas in question divided by the total number of moles of gas in the mixture. For some gas A,

$$x_A = \frac{\text{number of moles of } A}{\text{total number of moles of gas in the mixture}} \qquad [9.16]$$

If p_A is the partial pressure of A and P_T is the total pressure, then

$$p_A = x_A P_T \qquad [9.17]$$

In the atmosphere, for example, in each 5 mol of air there are 1 mol O_2 and 4 mol N_2. Therefore,

$$x_{O_2} = \frac{1\ \text{mol}}{5\ \text{mol}} = 0.2$$

$$x_{N_2} = \frac{4\ \text{mol}}{5\ \text{mol}} = 0.8$$

If the total pressure of a sample of air were 50 kPa, then

$$p_{O_2} = 0.2\ (50\ \text{kPa})$$
$$= 10\ \text{kPa}$$
$$p_{N_2} = 0.8\ (50\ \text{kPa})$$
$$= 40\ \text{kPa}$$

9.11 REAL GASES

The differences between a real gas and a hypothetical ideal gas became apparent in the last section. Molecules of an ideal gas are abstract points in space and have no volume, whereas a real gas is composed of actual molecules whose atoms occupy some space. The effects of this were seen in Figure 9.7, where we found that the volume occupied by a real gas at high pressure is larger than the volume an ideal gas would occupy under the same conditions.

We also learned in the last section that molecules of an ideal gas would have no attractive forces between them and they could be cooled to absolute zero without condensing to a liquid. Molecules of a real gas do attract each other, however. Their behavior is similar to that shown in Figure 9.13. As the gas is cooled, its volume begins to fall below the Charles' law value. Then, suddenly, the substance condenses to a liquid with a much smaller volume. At still lower temperatures it freezes to a solid. Another manifestation of the attractive forces between gas molecules is the cooling that occurs when a compressed gas is allowed to expand freely into a vacuum. As the gas expands, the average distance of separation between the molecules increases. Since there are forces of

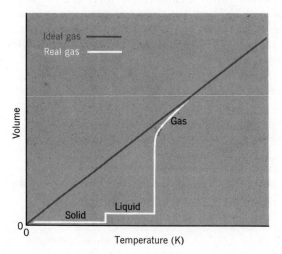

Figure 9.13

Variation of volume with temperature for real and ideal gases.

attraction between them, moving the molecules apart requires work (energy). The source of this energy is the kinetic energy of the gas—in the process of expansion kinetic energy is converted to potential energy. This removal of kinetic energy leads, of course, to a decrease in the average kinetic energy of the gas and, since the average kinetic energy is directly related to temperature, the gas becomes cooler.

Because real gases deviate from ideal behavior, especially at high pressure and low temperature, the ideal gas law can't be used to make highly accurate calculations. One way to improve the accuracy is to modify the ideal gas law to take into account the factors that cause a real gas to differ from an ideal gas.

Suppose that the gas molecules in a container could be stopped and allowed to settle to the bottom. We would see that part of the volume of the container is occupied by the gas molecules. The remaining free space is somewhat less than the volume of the container. If, in this hypothetical case, another gas molecule was added, it could move in the free space, but not the entire volume of the container. This same situation exists when the molecules are moving.

In an ideal gas the molecules would have no volume themselves, so the ideal gas would be entirely empty space into which other molecules could be squeezed when it is compressed. If we associate the empty space available in a real gas with this *ideal volume*, V_{ideal}, then the measured volume of the real gas, V_{meas}, is actually slightly larger than V_{ideal} by a quantity that is related to the size of the real molecules. According to J.D. van der Waals (1837–1923), a Dutch physicist, the measured volume is

> The volume within which the molecules cannot move is called the *excluded volume*.

$$V_{meas} = V_{ideal} + nb$$

where b is the correction due to the excluded volume per mole and n is the number of moles of gas. Solving for the ideal gas volume we have

$$V_{ideal} = V_{meas} - nb \qquad [9.18]$$

Van der Waals also included a correction to the pressure that takes into account the attractive forces between the molecules of a real gas. In an ideal gas, the molecules would travel in straight lines because there are no attractive forces to cause their paths to curve. But in a real gas, the attractive forces do cause molecules to curve and change directions as they pass by each other, as shown in Figure 9.14. As a result, the molecules of a real gas travel longer distances to reach the walls, and this takes more time. This means that the real gas molecules don't strike the walls as frequently as the molecules of an ideal gas. A lower collision frequency means, of course, that the real gas will exert a smaller pressure than the ideal gas. Van der Waals corrected for this with the term n^2a/V^2,

> In a real gas, the attractive forces cause the molecules to linger somewhat within the body of the gas, so the molecules don't strike the walls as often.

Figure 9.14

(a) In an ideal gas the molecules travel in straight lines. (b) In a real gas the paths curve as one molecule passes close to another, because the molecules attract each other. Asterisks indicate the points at which molecules pass close to each other.

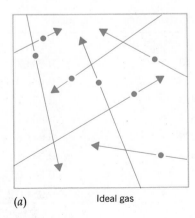

(a)   Ideal gas

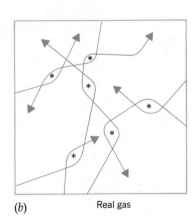

(b)   Real gas

which he added to the measured pressure to bring it up to the pressure that an ideal gas would exert.

$$P_{ideal} = P_{meas} + \frac{n^2a}{V^2} \qquad [9.19]$$

The quantity a is proportional to the strengths of the attractive forces, and we see that the stronger the attractions, the more must be added to the measured pressure to make it as large as the ideal gas pressure. This is reasonable if we consider that when the attractive forces are very large, the molecules will undergo large deflections in their paths. This means that they will tarry longer within the body of the gas, thereby striking the walls less frequently.

Substituting these corrected pressures and volumes (from Equations 9.19 and 9.18, respectively) into the ideal gas equation gives us

$$\left(P + \frac{n^2a}{V^2}\right)(V - nb) = nRT \qquad [9.20]$$

Van der Waals received the 1910 Nobel Prize in physics for this equation.

in which all symbols stand for measured quantities. This equation is called the **van der Waals equation of state for a real gas.** It is more complex than the ideal equation, but it does work well for many gases over fairly wide ranges of temperature and pressure.

The values of the constants a and b depend on the nature of the gas because the molecular volumes and the molecular attractions vary from gas to gas. Some typical values of a and b are found in Table 9.3. We see that molecules containing many atoms, such as C_2H_5OH, have large values of b. This is not surprising, since such molecules would be expected to be larger than molecules containing only a few atoms.

The variation among the values of a reflects variations in the strengths of the intermolecular attractions. It is easy to understand why polar molecules such as NH_3, H_2O, CH_3OH, and C_2H_5OH attract each other. These molecules are dipoles that tend to align themselves so that the partial positive charge on one attracts the partial negative charge on another (Figure 9.15). The attractions between nonpolar molecules, such as O_2, CH_4, and C_2H_6, or between isolated atoms such as helium and the other noble gases, are more difficult to explain. We will study all these intermolecular attractions in detail in the next chapter.

From the discussion above, we see that the values of a and b enable us to expand our knowledge about the molecules of which a real gas is composed. The van der Waals constants for a gas are obtained by making careful measurements of P, V, and T and then choosing values of a and b that make the van der Waals

Table 9.3
van der Waals constants for real gases

	a (dm^6 kPa mol^{-2})	b (dm^3 mol^{-1})
He	3.4	0.023 7
O_2	138	0.031 8
NH_3	423	0.037 1
H_2O	553	0.030 5
CH_4	228	0.042 8
C_2H_6	556.2	0.063 80
CH_3OH	964.9	0.067 02
C_2H_5OH	1218	0.084 07

Figure 9.15

Electrostatic interactions between dipoles. The attractions outweigh the repulsions, so the molecules feel a net attraction toward each other.

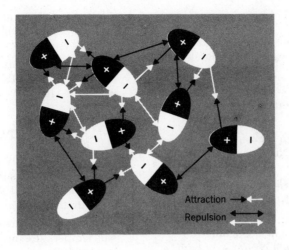

equation produce the best match with the experimental data. In this sense, *a* and *b* are experimentally determined quantities that enable us to check our theories on molecular size and attractions.

REVIEW QUESTIONS AND PROBLEMS

(Problems whose numbers are in blue have their answers in Appendix C at the back of the book; the more difficult problems are marked with asterisks.)

Pressures of Gases; Barometers and Manometers

9.1 Define pressure. Explain why the height of mercury in a barometer is independent of the cross-sectional area of the tube.

9.2 What is the SI unit of pressure and how is it related to the standard atmosphere?

9.3 An open-end manometer containing mercury was connected to a vessel containing a gas at a pressure of 99 kPa. The atmospheric pressure was 102 kPa. Sketch a diagram showing the relative heights of the mercury in each arm of the manometer.

9.4 What is the advantage of using a closed-end manometer for pressure measurements?

9.5 Why is mercury a useful substance for use in barometers and manometers?

9.6 Suppose that in the closed-end manometer shown in Figure 9.4 the mercury in the closed arm was 15.8 cm higher than in the arm connected to the container full of gas. What is the pressure of the gas in the container expressed in kPa?

9.7 An open-end manometer was connected to a flask containing a gas at an unknown pressure. The mercury in the arm open to the atmosphere was 65 mm higher than in the closed end. The atmospheric pressure was 97.7 kPa (733 mmHg). What was the pressure of the gas?

9.8 An open-end manometer was connected to a vessel containing a gas at a pressure of 71.3 kPa. The atmospheric pressure was 103.2 kPa. What was the difference in height (in millimetres) between the mercury levels in the manometer?

9.9 A flask containing a gas was connected to both a closed-end and an open-end manometer. In the closed-end manometer the mercury in the sealed arm was 755 mm above the level in the arm connected to the flask. In the open-end manometer, the arm connected to the gas was 17 mm higher than the side open to the air. What was the atmospheric pressure in kPa?

9.10 A manometer connecting two flasks (labeled *A* and *B*) contains an oil having a density of 0.847 g cm^{-3}. The oil in the arm connected to flask *A* is 74.0 cm higher than the oil in the arm connected to flask *B*. The gas in flask *A* has a pressure of 97.7 kPa (836 mmHg). What is the pressure of the gas in Flask *B*?

9.11 A person digs a well and finds water 11 m below the ground. If the average atmospheric pressure at the well is 1 atm, will the person be able to draw water from the well using a pump, mounted at ground level, that works by suction? Explain your answer.

Boyle's Law

9.12 State Boyle's law in words. Is the law obeyed exactly by all gases? What is a gas that exactly obeys Boyle's law called?

9.13 A gas has a volume of 350 mL at 99 kPa. What will its volume be at 120 kPa if the temperature remains constant?

9.14 A sample of SO_2 occupies 1.45 dm^3 at 2.75 atm. If we assume no temperature change, what volume will this gas occupy at 107 kPa?

9.15 A gas is compressed at constant temperature from a volume of 540 cm^3 to 320 cm^3. If the initial pressure was 63.3 kPa, what is the final pressure?

9.16 A gas exerts a pressure of 175 kPa in a container having a volume of 35.0 m^3. What will its pressure be if the gas is transferred to a 40.0 m^3 container at the same temperature?

9.17 A bicycle pump has a barrel that is 75.0 cm long. If we assume that on the upstroke air is drawn in at a pressure of 101 kPa, how long must the downstroke be (in centimetres) to raise the pressure of the air to 555 kPa, if the temperature of the air remains constant? (That's about the pressure in the tire of a 10-speed bike.)

Charles' Law

9.18 State Charles' law in words. In terms of Charles' law, why is −273.15°C the lowest possible temperature?

9.19 If you want to be sure that the tires of your car are properly inflated, you should check them before beginning a trip rather than after having driven on them for a long period. Why?

9.20 Turbocharging is a way of increasing the power output from a gasoline or diesel engine. The exhaust gases are used to turn a turbine connected to a compressor. The compressor forces more air into the cylinders so that more fuel can be burned and more power can be produced. Compressing the air, however, heats it, and even more power can be produced if the air from the compressor is cooled before it is let into the cylinders. Why?

9.21 What makes a hot-air balloon rise?

9.22 At 25°C and 1 atm a gas occupies a volume of 1.50 dm^3. How many liters will it occupy at 100°C and 1 atm?

9.23 A balloon has a volume of 2.0 dm^3 indoors at a temperature of 25°C. If it is taken outdoors on a very cold winter day when the temperature is −28.9°C, what will its volume be? Assume constant air pressure within the balloon.

9.24 What would be the final volume of a 2.00-dm^3 sample of a gas that is heated from 26°C to 100°C at constant pressure?

9.25 A sample of O_2 occupies 285 cm^3 at 25°C. At what Celsius temperature will it occupy 350 cm^3 if the pressure remains constant?

9.26 At what temperature will a gas sample occupy 0.850 dm^3 at 112 kPa pressure if it occupies 400 cm^3 at 32°C and 112 kPa?

Gay-Lussac's Law

9.27 Why is it dangerous to incinerate an aerosol can?

9.28 A gas exerts a pressure of 47.0 kPa at 20°C. What pressure will it exert if its temperature is raised to 40°C without a change in volume?

9.29 A gas has a pressure of 87.0 kPa at 25°C. To what Celsius temperature must it be heated to raise its pressure to 110 kPa?

*9.30 An automobile tire is inflated to a *gauge pressure* of 205 kPa at 18°C. (The actual pressure is 205 kPa *above* atmospheric pressure.) After a trip the temperature of the tire has risen to 54°C. What will be the gauge pressure of the air in the tire (in kPa), if we assume that no air has leaked out and that the volume of the tire hasn't changed? Assume the atmospheric pressure is 101 kPa

Combined Gas Law

9.31 What is STP?

9.32 If a gas, originally in a 50.0-cm^3 container at a pressure of 86.0 kPa, is transferred to another container whose volume is 65.0 cm^3, what would be its new pressure if
(a) there were no temperature change?
(b) the temperature of the first container were 25°C and that of the second were 35°C?

9.33 A 300-cm^3 sample of a gas exerts a pressure of 60.0 kPa at 27°C. What pressure would it exert in a 200-cm^3 container at 20°C?

9.34 A 2.00-dm^3 sample of a gas originally at 25°C, and a pressure of 93.3 kPa, is allowed to expand to a volume of 5.00 dm^3. If the final pressure of the gas is 78.0 kPa, what is its final temperature?

9.35 The density of CO_2 is 1.96 g cm^{-3} at 0°C and 1 atm. Determine its density at 86.7 kPa and 25°C.

9.36 If a 50.0-cm^3 sample of gas exerts a pressure of 60.0 kPa at 35°C, what volume will it occupy at STP?

Dalton's Law

9.37 What is Dalton's law of partial pressures? Can partial pressures actually be measured directly?

9.38 A 1.00-dm^3 mixture of gases is produced from 1.00 dm^3 of N_2 at 27 kPa, 1.00 dm^3 of O_2 at 67 kPa, and 1.00 dm^3 of Ar at 21 kPa. What is the pressure of the mixture?

9.39 A 1.00-dm^3 flask is filled by placing in it the contents of a 2.00-dm^3 flask of N_2 at 40.0 kPa and a 2.00-dm^3 flask of H_2 at 11.0 kPa. What is the pressure of the mixture in the 1.00-dm^3 flask?

9.40 What would be the total pressure of a mixture prepared by adding 20.0 cm^3 of N_2 at 0°C and 98.7 kPa plus 30.0 cm^3 of O_2 at 0°C and 85.3 kPa to a 50.0-cm^3 container at 0°C?

9.41 A mixture of N_2 and O_2 has a volume of 100 cm^3 at a temperature of 50°C and a pressure of 107.0 kPa. It was prepared by adding 50.0 cm^3 of O_2 at 60°C and 53.5 kPa with X cm^3 of N_2 at 40°C and 53.5 kPa. What is the value of X?

9.42 A gas is collected by the displacement of water until the total pressure inside a 100-cm^3 flask is 93.3 kPa at 25°C. Calculate the volume of dry gas at STP.

9.43 A mixture of N_2 and O_2 in a 200-cm^3 vessel exerts a pressure of 96.0 kPa at 35°C. If there are 0.0020 mol of N_2 present:
(a) What is the mole fraction of N_2 (see Equation 9.17)?
(b) What is the partial pressure of N_2?
(c) What is the partial pressure of O_2?
(d) How many moles of O_2 are present?

9.44 The air exhaled by an average human being might have the following typical composition, expressed in terms of partial pressures: N_2, 75.9 kPa; O_2, 15.5 kPa; CO_2, 3.7 kPa; water vapor, 6.2 kPa. What are the mole fractions of each gas?

9.45 What volume (cm^3) of CO_2 at 30°C and 90 kPa must be added to a 500-cm^3 container of N_2 at 20°C and 100 kPa to give a mixture having a pressure of 110 kPa at 20°C?

9.46 Calculate the volume occupied by 0.0244 g of O_2 if it were collected over water at 23°C and at a total pressure of 98.7 kPa.

*9.47 Three gases were added to the same 10-dm^3 container to give a total pressure of 107 kPa at 30°C. If the mixture contained 8.0 g of CO_2, 6.0 g of O_2, and an unknown mass of N_2, calculate the following:
(a) the total number of moles of gas in the container
(b) the mole fraction of each gas
(c) the partial pressure of each gas
(d) the mass of N_2 in the container

*9.48 A gas, at a total pressure of 107 kPa and a volume of 500 cm^3 over water at 35°C, is compressed to a volume of 250 cm^3, also over water at 35°C. Calculate the final pressure of the wet gas.

*9.49 During a rainstorm in July in New York City the humidity was found to be 100%, which means that the air was saturated with water vapor. The atmospheric pressure was 98.7 kPa and the temperature was 31°C. Dry air has an average molar mass of 28.8 g mol^{-1}. Calculate the mass of water in 1.00 dm^3 of the air during the storm.

9.50 280 cm^3 of gas are collected over water at 20°C. The water level inside the collection bottle is 28.4 mm higher than the water level outside. The atmospheric pressure is 102 kPa. What would be the volume of dry gas at STP?

Molar Volume

9.51 What is Gay-Lussac's law of combining volumes? What is Avogadro's principle?

9.52 Calculate the volume occupied, *at STP*, by
(a) 0.200 mol O_2
(b) 12.4 g Cl_2
(c) a mixture of 0.100 mol N_2 and 0.050 mol O_2

9.53 Calculate the mass of 245 cm^3 of SO_2 at STP.

9.54 What is the density of butane, C_4H_{10}, at STP?

9.55 The density of a gas was found to be 1.96 g dm^{-3} at STP. What is its molar mass?

Ideal Gas Law

9.56 Suppose that the ideal gas law were $PV^2 = nR/T^2$. What would be the units of R if P is in kPa and V in dm^3?

9.57 What determines the choice of the value of R that you should use in an ideal gas law calculation?

9.58 What is the value of the gas constant in the units Pa dm^3 mol^{-1} K^{-1}?

9.59 In the laboratory a student filled a 250-cm^3 container with an unknown gas until a pressure of 1.00 atm was obtained. He then found that the sample of gas weighed 0.164 g. Calculate the molar mass of the gas if the temperature in the laboratory was 25°C.

9.60 Calculate the pressure, in kPa, that would be exerted by 25 kg of steam (H_2O) in a 1000 dm^3 boiler at 200°C if we assume ideal gas behavior.

9.61 The density of a gas was found to be 1.81 g dm^{-3} at 30°C and 1.00 atm. What is its molar mass?

9.62 Calculate the volume occupied by 0.234 g of NH_3 at 30°C and a pressure of 86.0 kPa.

9.63 A chemist observed a gas being evolved in a chemical reaction and collected some of it for analyses. It was found to contain 80.0% carbon and 20.0% hydrogen by mass. It was also observed that 500 cm^3 of the gas at 1 atm and 0°C weighed 0.6695 g.
(a) What is the empirical formula of the gaseous compound?
(b) What is its molar mass?
(c) What is its molecular formula?

*9.64 A 0.2000-g sample of a fishy smelling liquid known to contain only carbon, hydrogen, and nitrogen was burned and produced 0.482 g CO_2 and 0.271 g H_2O. A second sample weighing 0.2500 g was treated in such a way that all the nitrogen in the substance was converted to N_2. This gas was collected and found to occupy 42.3 cm^3 at 26.5°C and 100.6 kPa.

(a) What are the percentages of carbon, hydrogen, and nitrogen in the compound?

(b) What is the empirical formula of the compound?

9.65 A sample of N_2 in a flask at 25°C exerts a pressure of 70.0 kPa. When 0.100 g of O_2 are added to this flask, the pressure rises to 101 kPa. The temperature remains constant and there is no reaction between the N_2 and O_2. How many grams of N_2 are in the flask?

*9.66 The product PV has the dimensions of energy. Given the data, $1\,J = 1\,N\,m$, $1\,atm = 101\,325\,Pa$, and $1\,Pa = 1\,N\,m^{-2}$, calculate the number of joules equal to 1 dm^3 atm. What is the value of the gas constant, R, in $J\ mol^{-1}\ K^{-1}$ and $cal\ mol^{-1}\ K^{-1}$?

Gas Stoichiometry

9.67 In the reaction, $N_2(g) + 3H_2(g) \rightarrow 2NH_3(g)$, what volume of N_2, measured at STP, is required to produce 400 cm^3 of NH_3, measured at STP? What volume of H_2 at STP is required?

9.68 In the reaction,

$$2NO(g) + 2H_2(g) \longrightarrow 2H_2O(g) + N_2(g)$$

how many cm^3 of N_2, measured at STP, would be produced from (a) 0.001 40 mol NO, (b) 1.3×10^{-3} g H_2?

9.69 Oxygen gas, generated in the reaction, $KClO_3 \rightarrow KCl + O_2$ (unbalanced), was collected over water at 30°C in a 150-cm^3 vessel until the total pressure was 80.0 kPa.

(a) How many grams of dry O_2 were produced?

(b) How many grams of $KClO_3$ were consumed in the reaction?

9.70 Nitric acid is produced by dissolving NO_2 in water according to the equation,

$$3NO_2(g) + H_2O(l) \longrightarrow 2HNO_3(l) + NO(g)$$

How many cm^3 of NO_2 at 25°C and 103 kPa are required to produce 10.0 g of HNO_3?

*9.71 120 cm^3 of NH_3 at 25°C and 100.0 kPa were mixed with 165 cm^3 of O_2 at 50°C and 85.0 kPa and transferred to a 300-cm^3 reaction vessel where they were allowed to react according to the equation

$$4NH_3(g) + 5O_2(g) \longrightarrow 4NO(g) + 6H_2O(g)$$

What will be the total pressure in the reaction vessel at 150°C after the reaction is over? Assume the reaction goes to completion.

9.72 Calculate the maximum volume of CO_2, at 100.0 kPa and 28°C, that could be produced by allowing 500 cm^3 of CO, at 105.0 kPa and 15°C, to react with 500 cm^3 of O_2 at 110.0 kPa and 0°C.

*9.73 Ozone, O_3, is an important species in the chain of reactions that lead to the production of smog. In an ozone analysis, 2.0×10^5 dm^3 of air at STP were drawn through a solution of NaI where the O_3 undergoes the reaction,

$$O_3 + 2I^- + H_2O \longrightarrow O_2 + I_2 + 2OH^-$$

The I_2 formed was titrated with 0.0100 M $Na_2S_2O_3$ with which it reacts.

$$I_2 + 2S_2O_3^{2-} \longrightarrow 2I^- + S_4O_6^{2-}$$

In the analysis, 0.420 cm^3 of the $Na_2S_2O_3$ solution was required to completely react with all of the I_2.

(a) Calculate the number of moles of I_2 that reacted with the $S_2O_3^{2-}$ solution.

(b) How many moles of I_2 were produced in the first reaction?

(c) How many moles of O_3 were contained in the 200 000 dm^3 of air?

(d) What volume would the O_3 occupy at STP?

(e) What is the concentration of O_3, in parts per million by volume, in the air sample?

9.74 An important reaction in the production of nitrogen fertilizers is the oxidation of ammonia,

$$4NH_3(g) + 5O_2(g) \xrightarrow{500°C} 4NO(g) + 6H_2O(g)$$

What volume of O_2, measured at 25°C and 90.7 kPa, must be used to produce 100 dm^3 of NO at 500°C and 100.0 kPa?

9.75 A student collected 35.0 cm^3 of O_2 over water at 25°C and a total pressure of 99.3 kPa from the decomposition of a 0.2500-g sample known to contain a mixture of KCl and $KClO_3$. The reaction that produced the oxygen was

$$2KClO_3 \longrightarrow 2KCl + 3O_2$$

(a) How many moles of O_2 were collected?

(b) How many grams of $KClO_3$ were decomposed?

(c) What percent of the sample (by mass) was $KClO_3$?

Graham's Law

9.76 What is the difference between *diffusion* and *effusion?*

9.77 What is Graham's law of effusion?

9.78 Compare the rates of effusion of He and Ne. Which gas effuses faster, and how much faster?

9.79 If, at a particular temperature, the average speed of CH_4 molecules is 1600 km h^{-1}, what would be the average speed of CO_2 molecules at the same temperature?

9.80 The rate of effusion of an unknown gas was determined to be 2.92 times faster than that of NH_3. What is the approximate molar mass of the unknown gas?

Kinetic Theory

9.81 What are the postulates of the kinetic molecular theory?

9.82 In terms of kinetic theory, what is the origin of the pressure of a gas?

9.83 In qualitative terms, why do molecules of gases with low molecular weights diffuse faster than gases with high molecular weights, provided they are at the same temperature?

9.84 In terms of the kinetic theory, how should the rate of diffusion of a gas be affected by an increase in temperature? Explain.

9.85 Why does the pressure of a gas increase when the volume is decreased (at constant temperature)?

9.86 Why does the pressure of a gas increase when the temperature is increased, provided the volume is constant?

9.87 If a warm object is placed in contact with a cool object, heat transfer occurs until they both come to the same temperature. How can the kinetic molecular theory account for this heat transfer and these temperature changes that occur?

9.88 Sketch a graph showing the distribution of kinetic energies for a gas at two different temperatures. For each temperature indicate the most probable K.E. and the average K.E. Why is the curve not symmetrical?

Real Gases

9.89 What is meant by nonideal behavior of a gas? Under what conditions is this behavior most evident?

9.90 Explain why most gases cool on expansion into a vacuum.

9.91 What physical significance do the constants a and b have in the van der Waals equation of state for real gases?

9.92 Why did van der Waals subtract a correction from the measured volume? Why did he add a correction to the measured pressure?

9.93 Use the van der Waals equation to calculate the pressure exerted by 1.000 mol of He at 0°C in a volume of 22.400 dm^3. Compare this to the pressure an ideal gas would exert under these same conditions.

9.94 Use the van der Waals equation to calculate the pressure exerted by 1.000 mol of C_2H_6 at 0°C in a volume of 22.400 dm^3. Compare this to the pressure of an ideal gas under these same conditions.

*9.95 Calculate the molar volume (in dm^3) of O_2 at STP from the van der Waals equation and compare it to the value in Table 9.2. (*Hint:* The volume can be obtained by successive approximations if you solve for the V in the term, $V - nb$.)

CHAPTER 10

STATES OF MATTER AND INTERMOLECULAR FORCES

The transitions of water among its solid, liquid, and gaseous states are common events that we often take for granted until they produce a scene as striking as this. Water vapor condensing directly to the solid state on an ice-cold window pane produced this pattern of frost crystals. Why crystals such as these have such form and regularity is one of the topics that we will explore in the pages ahead as we study the properties of the states of matter and the transitions among them.

The properties of gases, which we studied in the last chapter, are vastly different than those of the other two states of matter, liquids and solids. For one thing, we can't even see a gas, except if it's colored. We would be totally unaware that gases such as those in our atmosphere exist if we couldn't feel a breeze and see it rustle the leaves on a tree, or if we hadn't discovered that gases exert a pressure. Yet who would fail to recognize a lake full of water or a gigantic icy glacier?

The water vapor in the air (which we recognize as humidity), the water of a lake, and the water frozen in a glacier are all forms of the same chemical substance. They are each composed of molecules of water and have the same set of chemical properties. It is their physical properties that make them seem so different. In this chapter we will study why gases, liquids, and solids are so different, and what important properties liquids and solids have. We will also study how the three states of matter change from one to another and what this tells us about them.

10.1 COMPARING THE PROPERTIES OF GASES, LIQUIDS, AND SOLIDS

According to the kinetic molecular theory, all forms of matter are composed of small, rapidly moving particles. In Chapter 9 we saw how this postulate could be used to explain certain properties of gases. These same particles also exist in liquids and solids, and there are two main reasons why gases, liquids, and solids differ so much from each other. One is the tightness of the packing of the particles, and the other is the strengths of the attractive forces between them. Although both factors are interrelated and affect each of the physical properties, certain properties are influenced more by one than by the other. Two properties particularly affected by how tightly the particles are packed are the compressibility and the rates of diffusion. Properties particularly influenced by the strengths of the intermolecular attractive forces are shapes, volumes, and the ability to flow; surface tension; and rate of evaporation.

Compressibility

In a gas, the molecules are widely separated so there is a lot of empty space into which they may be crowded. As a result, gases are very compressible. The molecules in a liquid or a solid, however, are tightly packed and there is very little empty space between them. For this reason, increasing the pressure has hardly any effect on their volume, and they are virtually incompressible.

The incompressibility of liquids is an important and useful property. Many types of hydraulic machinery depend on it to transmit enormous forces that lift

Compressed gases are used in many practical applications, too.

The incompressibility of liquids is used in hydraulic machinery to exert enormous forces. Here we see one of the largest hydraulic forging presses in the United States, located at the Alcoa plant in Cleveland, Ohio. It can exert forces up to 400 meganewtons.

Figure 10.1

Diffusion can occur more rapidly in a gas than in a liquid because a molecule moves farther between collisions in a gas.

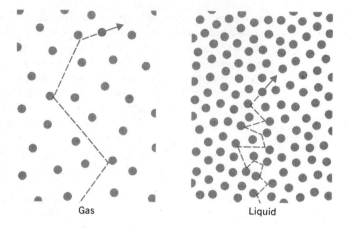

Gas   Liquid

and move heavy things. You depend on it yourself when you step on the brakes of a car. The force exerted by your foot is first magnified and then transmitted by an oil through the brake lines to the wheels where it causes the brake shoes to rub against a surface and stop the car. If air gets into the brake lines, the force you exert with your foot simply compresses the air, and the car won't stop very quickly at all. (Disaster!)

Diffusion

Molecules diffuse rapidly in a gas, compared to a liquid or a solid, because they move relatively long distances between collisions. They get where they're going quickly because there are relatively few interruptions along their paths, as shown in Figure 10.1. When two liquids mix, however, the molecules of one diffuse throughout the molecules of the other at a much slower rate than when two gases are mixed. We can observe the diffusion of two liquids by dropping a small quantity of ink into some water. As shown in Figures 10.2a and 10.2b, when the ink drop strikes the water we see it as a concentrated dot, which slowly spreads throughout the liquid. Diffusion takes place because the molecules in both liquids are able to move throughout the container. However, since the molecules in a liquid are so close together, the average distance that they

Rates of diffusion increase with increasing temperature because molecules move faster at higher temperatures.

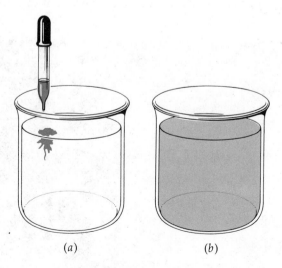

Figure 10.2

Diffusion in liquids (a) An ink drop is placed into water. (b) Ink has spread throughout the liquid.

(a)   (b)

travel between collisions—their *mean free path*—is very short. The molecules undergo billions of collisions before going very far and these constant interruptions in their paths keep them from spreading rapidly throughout the liquid.

Diffusion within solids is even much slower than in liquids. Not only are the molecules very tightly packed, but they are also held quite rigidly in place. They are not free to roam about, even though each may vibrate and bounce around inside the small cavity that it occupies within a crystal. Only molecules (or ions, if the solid is ionic) that have very large kinetic energies are able to muscle their way past their neighbors, so diffusion is extremely slow at ordinary temperatures.

High temperature solid-state diffusion is important in manufacturing electronic components such as transistors.

Volume and shape

The most obvious property of gases, liquids, and solids is the way they behave when transferred from one container to another. A gas, as we learned, expands to completely fill whatever container it is placed in. A liquid, however, retains a constant volume, but conforms to the shape of its container. Gases and liquids are both fluids; they flow and can be pumped from place to place. A solid, however, is not a fluid, and maintains both its shape and volume.

In a gas, the intermolecular attractive forces are so weak that the rapidly moving molecules can easily overcome them and expand to fill a container. The strengths of these forces are much larger in a liquid, however, and are responsible for holding the molecules close together. In a solid the attractive forces are still larger and hold the molecules more or less firmly in place so that they cannot move over and around each other. This prevents solids from flowing like gases and liquids.

Surface tension

Have you ever noticed how raindrops form beads of water on a freshly waxed car? Have you ever wondered why moist grains of sand stick together, but fall apart if either dry or completely submerged in water? These are phenomena that are caused by a property of liquids called surface tension.

In a liquid, each molecule moves about always under the influence of its neighbors. A molecule within the liquid is completely surrounded by others to which it is attracted (Figure 10.3a). However, a molecule at the surface is not completely surrounded, and feels attractions only to molecules below and beside it (Figure 10.3b). Molecules at the surface thus feel a net attraction in a direction toward the interior of the liquid. For a molecule to come to the surface, it must overcome this attraction. In other words, its potential energy must in-

The units of surface tension are *energy per unit area.* The surface tension is a measure of the energy needed to create a unit area of surface.

Figure 10.3
Intermolecular attractive forces in liquids. (a) A molecule near the center of the liquid. (b) A molecule at the surface of the liquid.

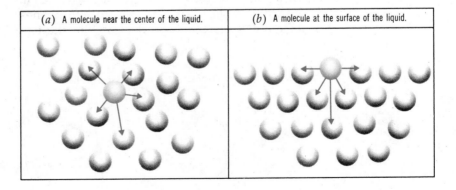

| (*a*) A molecule near the center of the liquid. | (*b*) A molecule at the surface of the liquid. |

Liquids tend to form spherical drops because a sphere has the smallest surface area for a given volume.

Manufacturing perfectly spherical ball bearings in the weightlessness of space is a potential dividend of the space program that relies on the surface tension of liquid metals.

Certain insects can walk on water because the surface of the water behaves like a skin that resists puncture.

crease—in a sense, work must be done to pull it to the surface. Making the surface of a liquid larger, therefore, requires an input of energy, and the quantity of energy needed is proportional to the liquid's **surface tension.**

The lowest energy (most stable) state for a given volume of liquid is when its surface area is a minimum. This gives the fewest high-energy surface molecules. The shape that satisfies this condition is a sphere, which is why raindrops are nearly spherical. All liquids strive to minimize their surface areas and tend toward spherical shapes as much as possible. This natural tendency is what allows you to fire polish glass tubing in the lab. As the glass softens, the sharp edges become rounded because the attractive forces within the glass tend to reduce the surface area. In Figure 10.4 we see why moist grains of sand are drawn together. In doing so, the film of water between them achieves a lower energy by reducing its surface area. Pulling the grains of sand apart requires that the surface tension be overcome, so the particles tend to stay together.

The magnitude of a liquid's surface tension depends on the strengths of the attractive forces between its molecules. When the attractive forces are large, the surface tension is large. Surface tension is also a function of the temperature of the liquid. Increasing the temperature (which increases the kinetic energy of the individual molecules) decreases the effectiveness of the intermolecular attractive forces, so surface tension decreases as the temperature is raised.

Figure 10.4

Grains of sand such as these are drawn together when wet because the film of water between them tends to reduce its surface area, thereby achieving a lower energy.

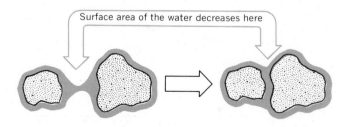

A property that we often think of when a liquid is mentioned is its ability to **wet** things—to spread across their surfaces as a thin film. Water, for example, spreads smoothly across a clean glass surface. This ability reflects a similarity among the strengths of the attractive forces between individual water molecules and between water molecules and the glass surface. Because these attractive forces are similar, little effort must be expended to spread the water molecules across the surface of the glass, so the surface tension is easily overcome and the water wets the glass. However, if the glass has a thin film of grease or oil on it, to which the water molecules are only weakly attracted, the weak attractions can't overcome the water's surface tension and beads of water form. The water can't wet the greasy surface—a fact that is especially annoying with an oily automobile windshield and greasy lab glassware.

Detergents contain substances called surfactants that lower the surface tension of water and allow the detergent solution to spread more easily over greasy surfaces.

Evaporation

In a liquid or a solid, just as in a gas, the molecules are constantly undergoing collisions, giving rise to a distribution of individual molecular velocities and, of course, kinetic energies. Even at room temperature, a small percentage of the molecules are moving with relatively high kinetic energies. If some of these faster-moving molecules possess enough kinetic energy to overcome the attractive forces within the liquid or solid, they can escape through the surface into the gaseous state—they **evaporate.**

The evaporation of a liquid is something everyone has seen. A small amount of spilled gasoline or nail polish remover quickly disappears, and rain puddles evaporate after a summer shower. Solids can also evaporate, although most of them that we encounter under normal conditions don't seem to disappear rapidly, if at all. But have you ever seen dry ice? It is composed of solid carbon dioxide. It is called *dry* ice because the solid doesn't melt; it simply evaporates. Have you ever wondered what happens to the crystals of naphthalene (moth flakes) that you sprinkle in a drawer or garment bag? They evaporate, too. Direct conversion of a solid to a gas, without melting, is called **sublimation.** Solid carbon dioxide and naphthalene are two substances that readily sublime at atmospheric pressure.

Figure 10.5 represents a typical distribution of the kinetic energies of the molecules in a liquid. The shaded area corresponds to the fraction of the total number of molecules that possess sufficient kinetic energy to evaporate. Just as removing the smart students from a chemistry class will lower the class average on exams, the loss of the higher-energy fraction because of evaporation leads to a lowering of the average kinetic energy of the remaining molecules. Since the temperature is directly proportional to the average kinetic energy, this results in a decrease in the temperature of a liquid as it evaporates. For example, we have all felt cool after a bath, because the evaporation of water from the body has

Figure 10.5

Kinetic energy distribution in a liquid.

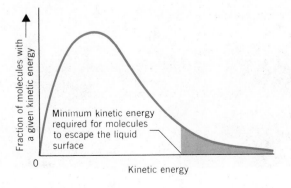

Fraction of molecules with a given kinetic energy

Minimum kinetic energy required for molecules to escape the liquid surface

0

Kinetic energy

Figure 10.6

Kinetic energy distribution at low temperature (a) and high temperature (b). The same minimum kinetic energy is needed to escape at both temperatures, but the total fraction of molecules having at least this much energy (the shaded area) is larger at the higher temperature.

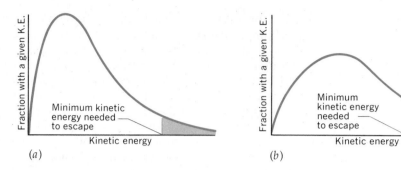

drawn heat from us. In fact, evaporation of perspiration provides the body with a mechanism for controlling body temperature.

Several things affect the rates at which liquids evaporate. One, of course, is the temperature. We know that warm water evaporates faster than cold water. The reason can be seen in Figure 10.6. Notice that the shaded area under the curve, which is the *total* fraction of the molecules that have enough kinetic energy to escape from the liquid, is larger at the higher temperature. In a given period of time, more molecules evaporate simply because more have enough energy to get away.

Another factor affecting the rate of evaporation is surface area. Increasing the surface area brings more high-speed molecules close to the surface so they can escape before losing kinetic energy through collisions.

A third factor is the strengths of the intermolecular attractions. Figure 10.7 compares two liquids at the same temperature. In one (liquid *A*) the attractive forces are very strong, so only very high-speed molecules have enough kinetic energy to overcome the attractions and escape. There are relatively few of these molecules, so the rate of evaporation is slow. In the other liquid the attractive forces are weak, and a much larger fraction have the energy needed to escape. Therefore, this liquid evaporates quickly. The attractive forces in most solids are much greater than in liquids, which explains why solids usually do not evaporate easily.

Steam issues from cooling towers at the Three Mile Island nuclear reactor prior to the accident there. The temperature lowering effect produced by the evaporation of water is often used commercially for cooling purposes.

Figure 10.7

Two liquids, A and B, at the same temperature. The total fraction of molecules (shaded area) having enough energy to escape from the liquid is greater for liquid B than for A.

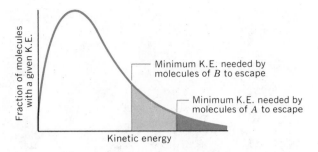

10.2 INTERMOLECULAR ATTRACTIVE FORCES

Attractions within molecules (chemical bonds) determine chemical properties; attractions between molecules determine physical properties.

The nonideal behavior of real gases described in Chapter 9, and the properties of liquids and solids, such as those discussed in the last section, illustrate the importance of understanding the origins and relative strengths of intermolecular attractions. These ultimately determine nearly all the physical properties of a substance. Knowing how the strengths of intermolecular attractions relate to chemical composition and structure therefore allows us to understand the behavior of substances and even anticipate the kinds of materials that have particular desirable properties.

Dipole-dipole forces

Polar molecules have ends that are oppositely charged. In a collection of these molecules, the individual dipoles tend to orient themselves so that the partial positive charge on one is near the partial negative charge on others. Because the molecules are constantly moving and colliding with each other, this alignment is far from perfect, particularly in liquids and gases. Nevertheless, the attractions between the oppositely charged ends of the dipoles outweigh the repulsions between like-charged ends, and a net overall attraction exists between them (Figure 10.8).

The energy required to separate a pair of dipoles is proportional to $1/d^3$, where d is the distance between dipoles.

Dipole-dipole attractions are normally considerably weaker than ionic or covalent bonds—they are only about 1% as strong. Their strength also decreases very rapidly as the distance between the dipoles increases, so their effects between the widely spaced molecules in a gas are very much less than between tightly packed molecules in a liquid or a solid. This is why the molecules of a gas behave almost as though there were no attractive forces at all.

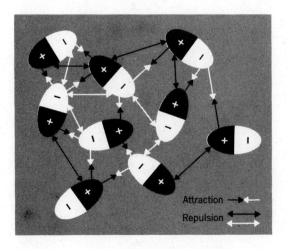

Figure 10.8

Electrostatic interactions between dipoles.

Figure 10.9

Hydrogen bonding in water.
(a) The water molecule is very
polar. (b) Hydrogen bonding pro-
duces strong attractions between
the water molecules.

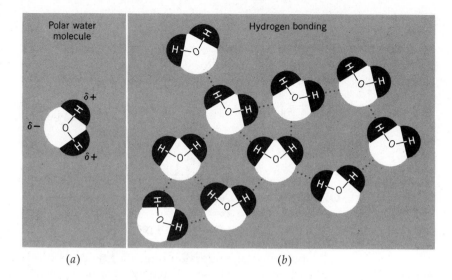

(a)   (b)

Hydrogen bonding

A particularly strong dipole-dipole attraction occurs when hydrogen is cova-lently bonded to a very small highly electronegative element such as fluorine, oxygen, or nitrogen. In these instances extremely polar molecules are formed in which the small hydrogen atom carries a substantial partial positive charge. Because the positive end of this dipole can closely approach the negative end of a neighboring dipole, the force of attraction between the two is quite large. This special kind of dipole interaction is called a **hydrogen bond,** and is about 5 to 10% as strong as an ordinary covalent bond.

Hydrogen bonding is a very important type of attractive force. In water, for example, the molecules interact strongly with each other by hydrogen bonding (Figure 10.9). This produces attractive forces that are much stronger than those between other molecules of similar size and mass, and is what makes water a liquid at room temperature. Hydrogen bonding is also responsible for control-ling the orientation of water molecules in ice (Figure 10.10), where we see that each water molecule is surrounded tetrahedrally by four others to which it is held by hydrogen bonds. This causes ice to have a very "open" structure, and makes ice less dense than liquid water. That's why ice cubes and icebergs float (much to the distress of the captain of the *Titanic*).

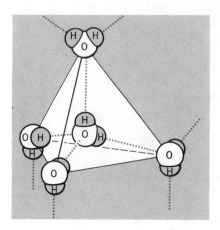

Figure 10.10

Hydrogen bonding (dotted lines)
between water molecules in ice.

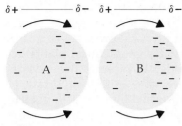

Figure 10.11
*London forces. As the instantane-
ous dipole in atom A forms, it
induces a dipole in atom B.*

The strength of London forces,
as measured by the energy
needed to separate the particles, is proportional to $1/d^6$.

London forces

Even uncombined atoms and nonpolar molecules experience weak attractions.
We know this because substances like helium and hydrogen can be condensed
to a liquid if they are cooled to a sufficiently low temperature. Attractive forces
must exist to hold them together in the liquid state. These attractions are called
London forces, after the German physicist Fritz London who provided an explanation of them.

When electrons move about in an atom or molecule, their motion is somewhat random, so that at any given instant there is a chance that more electrons
will be on one side of the particle than the other. At that particular instant the
particle will be a dipole. We call it an *instantaneous dipole* because its existence is
only momentary. In a collection of atoms or molecules the electron motions in
nearby neighboring particles are not entirely independent. As the negative end
of an instantaneous dipole begins to form in one of them, it pushes electrons
away in the particle alongside, as shown in Figure 10.11. We say that the instantaneous dipole *induces* a dipole in its neighbor. As you can see, because of the
way the dipoles are formed, they attract each other, and that attraction produces
a momentary tug that helps hold them together.

Normally, London forces are rather weak because of their fleeting existence.
They are present between all particles—ions as well as polar and nonpolar molecules—but play a very minor role in the attractions between ions. This is because
the forces of attraction between ions are so strong. London forces do play a
significant role in the attractions between all kinds of molecules, especially nonpolar ones.

10.3 HEAT OF VAPORIZATION

Earlier we noted that liquids become cool when they evaporate because the
departure of high-energy molecules lowers the average kinetic energy of those
that remain behind. If we wished to keep the temperature of the liquid constant
during evaporation, we would have to add heat. This is to replace the energy
taken away by the molecules that evaporate, and we can think of it as the energy
needed to cause the evaporation, at a constant temperature, of a given amount
of the liquid. Normally, the amount of liquid is taken to be a mole, and *the
quantity of energy needed to cause the evaporation of one mole of the liquid is called its*
molar heat of vaporization, or simply its **heat of vaporization** (the word *molar* is
understood).

The molar heat of vaporization is represented symbolically as $\Delta H_{\text{vaporization}}$
(or simply ΔH_{vap}). The Greek letter Δ is usually used to symbolize a change. In
this case it is a change in the total quantity of energy (represented by the letter
H) contained in the substance that evaporates. In a formal sense, this change in
energy is defined as the difference between the quantity of energy contained in
the substance in its final state (vapor) and the quantity of energy that it contained in its initial state (liquid). Thus

$$\Delta H_{\text{vap}} = H_{\text{vapor}} - H_{\text{liquid}}$$

H_{vapor} is the energy of the
vapor and H_{liquid} is the energy
of the liquid.

In actual practice, neither H_{vapor} nor H_{liquid} can actually be measured. This is no
tragedy, however, because we are really only interested in their difference,
ΔH_{vap}, which can be measured. The formal definition given above serves instead to define the algebraic sign of ΔH_{vap}. Converting a liquid to a vapor involves pulling the molecules away from each other, and that increases the potential energy of the molecules. This means that the potential energy of the
vapor is larger than the potential energy of the liquid, so H_{vapor} is larger than

H_{liquid}. The algebraic sign of ΔH_{vap} is therefore positive. (In general, ΔH is a positive quantity for endothermic changes and negative for exothermic ones.)

The molar heat of vaporization is an important physical property of a substance. Engineers have to know about it when they design chemical plants because they must know how much energy will be needed to vaporize the solvents used in various processes. The heat of vaporization of water is important to meteorologists because much of the solar energy absorbed by the earth goes to evaporating water from the oceans. In a sense, this energy becomes stored in the water vapor and when the water condenses it is released. This is the origin of the enormous quantities of energy contained in violent hurricanes and other rain storms.

The term heat of vaporization *usually means* molar heat of vaporization.

EXAMPLE 10.1 USING THE HEAT OF VAPORIZATION

PROBLEM The heat of vaporization of water is 40.6 kJ mol^{-1}. How much heat energy is required to convert 1.00 dm^3 of water to steam?

SOLUTION Water has a density of 1.00 g cm^{-3}. Therefore, 1.00 dm^3 of water has a mass of 1.00×10^3 g. The problem, as is generally the case, is one of unit conversion.

$$1.00 \times 10^3 \text{ g H}_2\text{O} \sim (?) \text{ kJ}$$

This can be set up as

$$1.00 \times 10^3 \text{ g H}_2\text{O} \times \left(\frac{1 \text{ mol H}_2\text{O}}{18.0 \text{ g H}_2\text{O}}\right) \times \left(\frac{40.6 \text{ kJ}}{1 \text{ mol H}_2\text{O}}\right) \sim 2260 \text{ kJ (to three significant figures)}$$

Thus, 2260 kJ of heat are required.

A useful feature of the heat of vaporization is that its magnitude provides a good measure of the strengths of the attractive forces in a liquid. This is easy to understand. When the intermolecular attractions are strong, a lot of energy must be supplied to pull the molecules apart as the liquid is changed to a gas.

Table 10.1 lists some values of ΔH_{vap} for various substances. We can use these data to gain some insight into the way chemical composition and structure influence intermolecular attractions. For example, if we look at the series of hydrocarbons, CH$_4$ through C$_{10}$H$_{22}$, we observe a steady increase in ΔH_{vap} with an increase in molecular weight. These compounds are nonpolar; therefore the only attractive forces that exist between their molecules are London forces, so these must also increase from CH$_4$ to C$_{10}$H$_{22}$. The reason can be seen by examining the structures of the molecules. These hydrocarbons are chainlike molecules, as shown in Figure 10.12, and as we proceed from CH$_4$ to C$_{10}$H$_{22}$ the length of the carbon chain increases. This has the effect of increasing the number of loca-

Hydrocarbons are compounds that consist of only carbon and hydrogen.

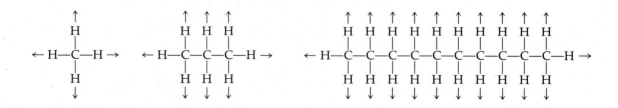

Figure 10.12

Attractive forces increase with increasing chain length. There are more points along the molecule that can be attracted to other molecules nearby.

Table 10.1
Heats of vaporization and boiling points
of various substances

Compound	ΔH_{vap} (kJ mol^{-1})	Boiling Point (°C)
CH_4	9.20	−161
C_2H_6	14	−89
C_3H_8	18.1	−30
C_4H_{10}	22.3	0
C_6H_{14}	28.6	68
C_8H_{18}	33.9	125
$C_{10}H_{22}$	35.8	160
F_2	6.52	−188
Cl_2	20.4	−34.6
Br_2	30.7	59
HF	30.2	17
HCl	15.1	−84
HBr	16.3	−70
HI	18.2	−37
H_2O	40.6	100
H_2S	18.8	−61
NH_3	23.6	−33
PH_3	14.6	−88
SiH_4	12.3	−112

X = halogen atom in HX

tions along the molecule where London forces may occur with other molecules. A long chainlike molecule is therefore held in more places than a short molecule, and more energy must be supplied to remove such long-chain molecules from the liquid. The result is that as the chain length increases, ΔH_{vap} increases.

Another factor that influences the strengths of London forces is molecular size. If we examine molecules of the same general formula, such as the halogens (F_2, Cl_2, Br_2), we find that large molecules have a greater ΔH_{vap} than small molecules. As we proceed from F_2 to Br_2, the atoms that make up the molecules become larger, so the molecules also become larger. As the size increases, the outer electrons become farther from the nuclei and are not held as tightly. Because of this the electron cloud of a large molecule is more easily distorted, or *polarized*, and it is easier to create the instantaneous dipoles that are responsible for the London forces. (The ease of distortion of the electron cloud is referred to as **polarizability**.) The result is that the London forces are stronger between molecules composed of large, easily polarized atoms such as bromine than between molecules composed of small atoms such as fluorine. That is why ΔH_{vap} increases from F_2 to Br_2.

When we look at the hydrogen halides, HF through HI, we find that the expected variation of ΔH_{vap} with molecular size is reversed between HF and HCl. In fact, HF has a considerably higher heat of vaporization than any of the other HX compounds. This anomalous behavior is attributed to the presence of hydrogen bonding in HF. We see the same inverted order of ΔH_{vap} for H_2O and H_2S and for NH_3 and PH_3, where, again, hydrogen bonding is significant for

H_2O and NH_3 but not for H_2S and PH_3. Oxygen, fluorine, and nitrogen are all very small and are the most electronegative elements in the periodic table, while the elements below them are much larger and much less electronegative. It is not surprising, therefore, that hydrogen bonding is important only for H_2O, HF, and NH_3. Normal behavior is reached in Group IVA hydrides, where ΔH_{vap} for CH_4 is less than ΔH_{vap} for SiH_4. Here, neither CH_4 nor SiH_4 have any tendency to hydrogen bond because they are nonpolar.

10.4 VAPOR PRESSURES OF LIQUIDS

When a liquid evaporates from an open container, all the liquid will eventually disappear because the molecules that enter the vapor simply diffuse away into the atmosphere. However, if the liquid is placed into a closed container, the molecules that evaporate cannot escape and accumulate in the vapor space above the liquid. There they exert a pressure, just like any other gas molecules, and we call this pressure the **vapor pressure.**

If we study how the vapor pressure changes when a liquid is placed into an evacuated container, we find that it initially rises and then gradually becomes constant. In addition, for a given liquid at a given temperature, the final value of the vapor pressure is always the same, regardless of the size of the container or the amount of liquid in it, just as long as some liquid is still present when the limiting pressure is reached.

The behavior of the vapor pressure just described is explained in the following way. When the liquid is introduced into the container, it begins to evaporate and molecules of the substance start to accumulate above the liquid in the vapor space. These molecules collide with the walls of the container and produce the pressure. As time passes, more and more molecules of the substance enter the vapor phase, and the pressure rises. But during this time something else is happening. One of the walls surrounding the gas space is the liquid itself, and when molecules from the gas strike the liquid's surface, there is a high probability that they will become trapped there. This is because the kinetic energy of an incoming molecule tends to become scattered among the liquid molecules at the surface, and there is little likelihood that the incoming molecule will retain enough kinetic energy to escape again. As a result, gaseous molecules of the substance that strike the liquid's surface return to the liquid state. Thus, we not only have evaporation taking place, but condensation as well.

The rate at which molecules evaporate from the liquid is determined by the liquid's temperature, which we will assume is being kept constant. Therefore,

Figure 10.13

(a) *A liquid begins to evaporate into a closed container.* (b) *Equilibrium is reached when the rate of evaporation equals the rate of condensation.*

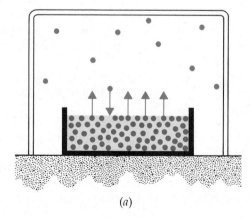

(a)

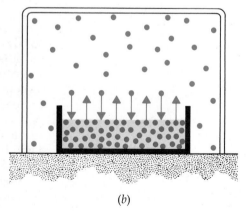

(b)

Figure 10.14

Measurement of vapor pressure. When a small amount of liquid is introduced above the mercury in a barometer, the vapor pressure of the liquid forces the mercury down. (a) No liquid above the mercury. (b) H_2O. (c) Ethyl alcohol, C_2H_5OH. (d) Diethyl ether, $(C_2H_5)_2O$.

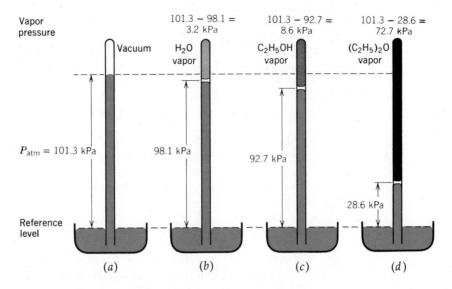

The concept of a dynamic equilibrium was introduced earlier in our discussion of weak electrolytes.

the rate of evaporation is constant. This means that molecules are moving into the vapor at a constant rate. The rate of condensation, however, depends on the concentration of molecules in the vapor—the more molecules per unit volume, the more collisions they will make with the surface of the liquid. When the liquid is first placed in the container, none of its molecules are in the vapor, so there is no condensation taking place. As a result, initially there will be a net influx of molecules into the gas phase as the liquid evaporates. However, as the number of molecules in the vapor increases, the rate of condensation will also increase, and this will continue until the liquid is evaporating as fast as its vapor is condensing (Figure 10.13). Once this happens, there will no longer be any change with time in the number of molecules in the gas. (For each hundred molecules that leave the liquid, one hundred return.) Since the number of molecules in the vapor has become constant, so has the pressure. At this point, the liquid is in *dynamic equilibrium* with its vapor, and the pressure exerted by the vapor once equilibrium has been established is called the **equilibrium vapor pressure** of the liquid.

One way that we can measure the vapor pressure of a liquid (usually, the term *vapor pressure* is used to mean *equilibrium vapor pressure*—the word "equilibrium" is understood) is by the use of a barometer, as shown in Figure 10.14. First the height of the mercury in the barometer is measured accurately. Next, the liquid whose vapor pressure is to be determined is carefully added to the barometer by means of an eye dropper, and allowed to rise to the top of the mercury in the column, as shown in Figures 10.14*b*, 10.14*c*, and 10.14*d* (most liquids are less dense than mercury and will therefore float on the mercury surface). The space above the mercury column in Figure 10.14*a* is, for all practical purposes a vacuum[1], so there is no pressure exerted on the top of the mercury column. The space above the mercury in Figures 10.14*b*, 10.14*c*, and 10.14*d*, however, contains a very small amount of liquid plus its vapor. As the liquid evaporates, the pressure of the trapped vapor pushes mercury out of the barometer and causes the level of the mercury in the column to fall, and when the liquid and vapor are finally in equilibrium, the height of the mercury column becomes stationary.

The total pressure exerted at the reference level outside each barometer will be the atmospheric pressure, P_{atm}. The total pressure exerted within the barom-

---

[1] Mercury itself does have a finite vapor pressure (about 10^{-4} kPa at room temperature) and, therefore, should never be left in an open container because of its high toxicity.

eter is P_{Hg}, the pressure resulting from the pull of gravity on the mercury in the column, plus P_{vapor}, the pressure exerted by the vapor in equilibrium with its liquid. The additional pressure exerted by the weight of the small amount of liquid on top of the column is negligibly small. Therefore, at equilibrium in each barometer

$$P_{atm} = P_{Hg} + P_{vapor}$$

In Figure 10.14*a*, $P_{vapor} = 0$; therefore, $P_{atm} = P_{Hg} = 101.3$ kPa. In Figure 10.14*b*, 10.14*c*, and 10.14*d*, $P_{Hg} = 98.1$ kPa, 92.7 kPa, and 28.6 kPa, respectively. Therefore, at 25°C the vapor pressure of water is 3.2 kPa, that of ethyl alcohol is 8.6 kPa, and that of diethyl ether is 72.7 kPa.

Factors that affect the vapor pressure

There are two principal factors that determine the magnitude of the vapor pressure. One is the nature of the attractive forces in the liquid itself, and the other is the temperature. Both of these influence the rate at which molecules evaporate. In liquids where the intermolecular attractions are strong, only molecules that have very large kinetic energies will be able to escape, and the rate of evaporation will be low. On the other hand, if the intermolecular attractions are weak, a large fraction of the molecules will have sufficient kinetic energy to leave and the rate of evaporation will be large. When the rate of evaporation is small, only a small concentration of vapor molecules will be needed in order to have the rate of condensation equal to the rate of evaporation, but if the rate of evaporation is large, a large concentration of vapor molecules will have to exist to reach equilibrium. This means that liquids that have large intermolecular attractions will have low vapor pressures, and vice versa.

If we examine the three liquids whose vapor pressure measurements were described above, we see that water has the lowest vapor pressure and diethyl ether has the largest. This means that the attractive forces are largest in water and smallest in the ether. This is reasonable, because in water there are strong hydrogen bonds, and in the ether, which is nearly nonpolar, there are only relatively weak London forces.

The effect of temperature on the vapor pressure can be seen in Figure 10.15, which contains a graph of vapor pressure versus temperature for ether, alcohol, and water. We see that as the temperature is increased, the vapor pressure rises. The reason for this is easily understood by recalling that raising the temperature increases the rate of evaporation. As you know by now, this leads to an increase in the vapor pressure.

Vapor pressure-temperature curves like those in Figure 10.15 do not continue indefinitely. Consider, for example, what happens as a sealed container half full of liquid is heated. At first there is a distinct boundary or surface between the more dense liquid and the less dense vapor. As the temperature is raised, more liquid evaporates and more molecules accumulate in the vapor. Since the number of molecules per cubic centimetre is increasing, the density of the vapor is also increasing. At the same time, the liquid becomes less dense—it expands just as the liquid mercury in a thermometer does. As the temperature rises, therefore, the increasing density of the vapor gradually approaches the decreasing density of the liquid. Eventually a temperature is reached at which they become identical: Both have the same number of molecules per cubic centimetre and there is really no difference between them. If we were watching the liquid-vapor boundary in the sealed container, we would see that it suddenly disappears when the two phases become the same.

Figure 10.15

Vapor pressure curves. The vapor pressure of a liquid increases with increasing temperature.

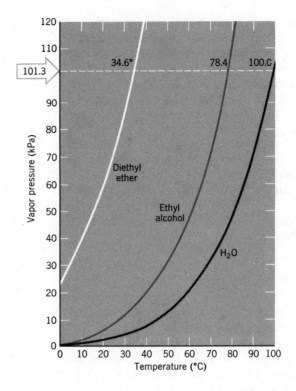

Above its critical temperature, a substance behaves as a gas.

The highest temperature at which a distinct liquid phase exists is called the **critical temperature,** symbolized T_c. The vapor pressure at the critical temperature is called the **critical pressure,** P_c. Vapor pressure-termperature curves terminate at T_c and P_c and the point at the end of the curve is the **critical point.** Critical temperatures and pressures of some substances are given in Table 10.2.

Each substance has its own characteristic values for T_c and P_c that are controlled by the strengths of the intermolecular attractions. When these attractions are very weak, as with helium, the molecules must be slowed down a great deal before they will stick together when they collide. This means that the gas must be cooled to a low temperature for it to condense.

Factors that do not affect the vapor pressure

We have seen that the vapor pressure of a liquid depends on the nature of the liquid and its temperature. What happens to the vapor pressure if the volume of

Table 10.2

Some critical temperatures and pressures

Compound	T_c (°C)	P_c (MPa)
Methane (CH_4)	−82.1	4.64
Ethane (C_2H_6)	32.2	4.88
Benzene (C_6H_6)	288.9	4.92
Ammonia	132.5	11.40
Carbon dioxide	31	7.39
Water	374.1	22.06
Helium	−267.8	0.23

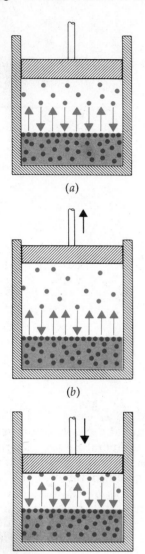

Figure 10.16

Effect of volume changes on vapor pressure. (a) Equilibrium between liquid and vapor. (b) No equilibrium. Rate of evaporation is greater than rate of condensation. (c) No equilibrium. Rate of condensation is greater than rate of evaporation.

the vapor is changed? Suppose that a liquid-vapor equilibrium has been established in the apparatus in Figure 10.16a. The arrows indicate that evaporation and condensation are occurring at equal rates. Now suppose that the piston is pulled up, as shown in Figure 10.16b. This expansion will cause a sudden drop in the pressure of the vapor, which means that fewer collisions will be occurring each second with each square centimetre of the walls. Each square centimetre of the liquid's surface will now be receiving fewer returning vapor molecules, but we've done nothing to affect the rate of evaporation. As a result, evaporation will be occurring faster than condensation. More liquid will therefore evaporate until the concentration of molecules in the vapor is sufficient to make the rate of condensation equal to the rate of evaporation once again. At that point, equilibrium will be reestablished and, provided the temperature hasn't changed, the vapor pressure will have returned to its previous value. At this newly established equilibrium the larger volume of gas is now occupied by more molecules. The vapor pressure will be the same as before the volume change occurred, but the volume of the liquid will be slightly smaller.

Decreasing the volume of the vapor by lowering the piston (Figure 10.16c) will also disturb the equilibrium. Increasing the pressure of the vapor will cause an increase in the number of collisions each second with each square centimetre of the walls. This, in turn, will lead to an increase in the rate of condensation, but will have essentially no effect on the rate of evaporation. The rate at which the molecules leave the vapor will consequently be greater than the rate at which molecules leave the liquid. This imbalance in rates causes the pressure exerted by the vapor to decrease and the volume of the liquid to increase. Eventually, the rate of condensation will decrease to a point where it exactly equals the rate of evaporation, reestablishing equilibrium. At this new equilibrium the smaller vapor volume, caused by the movement of the piston, will be occupied by fewer gaseous molecules. The vapor pressure will have returned to its initial value, and the volume of the liquid will have increased slightly.

From this discussion we see that the *vapor pressure of a liquid is independent of the volume of the container, provided that there is some liquid present so that an equilibrium can be established.*

Because the vapor pressure of a liquid is independent of the volume of the vapor it is possible to condense a gas simply by decreasing its volume, provided the temperature of the gas is below its critical temperature. For example, suppose a sample of water vapor had a pressure of 70 kPa at 100°C. If this gas were compressed, it would obey Boyle's law and its pressure would rise until it reached 101.3 kPa—the equilibrium vapor pressure of water at 100°C. Further compression would then cause some of the water vapor to condense to a liquid, because only in this way could the pressure be prevented from rising above 101.3 kPa.

If the temperature of the gas is above its critical temperature, then the gas must be cooled before it can be condensed. For example, methane cannot be liquefied at *room temperature* by increasing its pressure because room temperature (approximately 25°C) is above methane's critical temperature. At room temperature, methane can exist in only one phase—as a gas. However, if methane is cooled first to −82.1°C or lower, compression of the gas will eventually cause some to condense. At −82.1°C, for instance, some liquid will begin to form when the pressure on the methane reaches 4.64 MPa. As the volume continues to be decreased, more and more liquid will be formed and the pressure will remain constant at 4.64 MPa, because that is the vapor pressure of methane at this temperature.

Liquefied natural gas (mostly methane), which is transported in large tanker vessels such as the one shown here, must be kept below −82.1°C, which is methane's critical temperature.

Le Châtelier's principle

The dynamic equilibrium between a liquid and its vapor can be represented by the equation

$$\text{liquid} \rightleftharpoons \text{vapor} \qquad [10.1]$$

Here the double arrows mean that the rates of evaporation and condensation are equal. If we in any way disturb this system so that the equilibrium is upset, a change occurs that tends to bring the system back to equilibrium. For example, we've seen that if we suddenly increase the volume of the vapor, the pressure drops and equilibrium is upset. In response, more of the liquid evaporates until the pressure is restored to its equilibrium value. In Equation 10.1, this corresponds to a change from left to right (i.e., liquid → vapor) and results in a new position of equilibrium in which there is less liquid and more vapor. In a sense, the position of equilibrium has shifted to the right in response to the disturbance imposed on it.

The way an equilibrium system behaves when it is disturbed can be predicted in a very simple fashion by applying a principle proposed in 1888 by the French chemist, Henry Le Châtelier (1850–1936). **Le Châtelier's principle** states that *when a system in a state of dynamic equilibrium is disturbed by some outside influence that upsets the equilibrium, the system responds by undergoing a change in a direction that reduces the disturbance and, if possible, brings the system back to equilibrium.*

To see how this works, let's apply Le Châtelier's principle to the effect that a volume increase has on a liquid-vapor equilibrium. When the piston of the apparatus in Figure 10.16 is pulled upward, the volume of the vapor is increased and the pressure drops. Since the pressure is no longer equal to the equilibrium vapor pressure, the system is no longer at equilibrium. How can this system counteract the disturbance and bring the pressure back to its equilibrium value? The answer, of course, is to produce more vapor, which will exert more pres-

This analysis by Le Châtelier's principle is certainly much less tedious than our previous analysis of the same phenomenon.

sure.[2] In other words, some additional liquid will evaporate and produce more vapor. We can also say that this volume increase shifts the position of equilibrium in Equation 10.1 to the right.

The effect of a temperature change on the liquid-vapor equilibrium can also be predicted with Le Châtelier's principle. This is simple if we include the energy change as part of the reaction. Since the conversion of liquid to vapor is endothermic, heat is absorbed during the change and can be included as a reactant in the equation.

$$\text{heat} + \text{liquid} \rightleftharpoons \text{vapor}$$

Now, suppose the temperature of a liquid-vapor equilibrium system is increased. This is accomplished by adding heat (with a Bunsen burner, for example). The system can counteract the buildup of heat energy by changing some liquid to vapor—a change that absorbs heat. The increased amount of vapor will exert more pressure, so our conclusion is that the vapor pressure will increase with increasing temperature (which, of course, it does). In summary, Le Châtelier's principle predicts that a temperature increase will shift the position of equilibrium in the direction of an endothermic change. It also is not difficult to see that a temperature decrease will shift an equilibrium in the direction of an exothermic change.

10.5 BOILING POINT

A glance at the photograph in Figure 10.17 tells you that the liquid in the beaker is boiling. Large bubbles are forming within the liquid and rising to the surface. When a bubble is formed, the liquid that originally occupied this space is pushed aside and the level of the liquid in the container is forced to rise against the downward pressure exerted by the atmosphere. In other words, it is the pressure exerted by the vapor inside the bubble that pushes the surface of the liquid up against the opposing atmospheric pressure. This can occur only when the vapor pressure of the liquid becomes equal to the prevailing atmospheric pressure. If it were less, the atmospheric pressure would cause the bubble to collapse. The temperature at which the liquid boils—its **boiling point**—is therefore the temperature at which its vapor pressure equals the atmospheric pressure.

As long as bubbles are forming within the liquid—that is, as long as the liquid is boiling—the vapor pressure of the liquid is equal to the atmospheric pressure. Since the vapor pressure remains constant, the temperature of the boiling liquid also stays the same. An increase in the rate at which heat is supplied to the boiling liquid simply causes bubbles to form more rapidly. The liquid boils away more quickly, but the temperature does not increase.

The normal boiling points of water, ethyl alcohol, and diethyl ether can be read from the graph in Figure 10.15.

It should be obvious from this discussion that the boiling point of a liquid depends on the prevailing atmospheric pressure—the larger the atmospheric pressure is, the higher the temperature must be to give a matching vapor pressure. The boiling point of a liquid at 1 atm (101.325 kPa) is referred to as its **normal boiling point.** For water, the normal boiling point is 100°C. At higher pressures its boiling point is greater; at lower pressures (for example, on a mountaintop) its boiling point is less. Boiling points given in reference tables are always normal boiling points, unless otherwise stated.

The constant temperature maintained by a boiling liquid is relied on when we use water to cook foods. While water boils, its temperature remains constant. As long as water surrounds the food, we know that the food won't burn. A pressure cooker also takes advantage of the fact that the boiling point changes

---

[2] If the volume of the vapor is increased sufficiently, all the liquid will evaporate and equilibrium cannot be reestablished. This will occur, for example, if the piston in Figure 10.16 is removed entirely so that the liquid is open to the atmosphere.

Figure 10.17
Just a casual glance at the water in this beaker would tell you that it is boiling. Why?

with pressure. These cookers are time-savers because they allow foods to be prepared at a much faster rate than possible in an open pot. The lid on a pressure cooker forms a tight seal on the pot and is equipped with a pressure-relief valve to prevent the pot from exploding. The heat supplied by the stove causes more and more liquid water to evaporate and the pressure inside the kettle increases until steam begins to exit from the relief valve. Since the pressure inside the cooker at this point is higher than 1 atm, the water boils at a temperature above 100°C, so foods cook faster.

The boiling point is another property that gives a good estimate of how strong the attractive forces are between the molecules in a liquid. Liquids whose molecules attract each other strongly tend to have high boiling points, and vice versa. In Table 10.1, for example, we see that among the hydrocarbons the boiling points rise along with the heats of vaporization.

The dependence of boiling point on intermolecular attractive forces is also seen in Figure 10.18, in which the boiling points of some hydrogen compounds

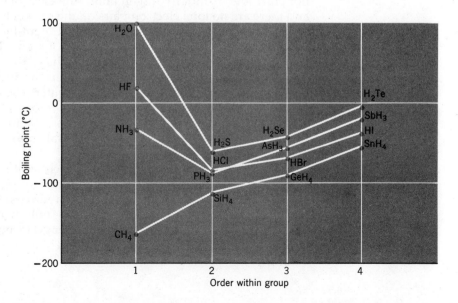

Figure 10.18
Boiling points of hydrogen compounds of Group IVA, VA, VIA, and VIIA elements.

Figure 10.19

Hydrogen bonding in HF and H_2O. (a) Each HF molecule has only one hydrogen atom that can hydrogen bond to something else. (b) Each H_2O molecule has two hydrogen atoms that can hydrogen bond to other H_2O molecules.

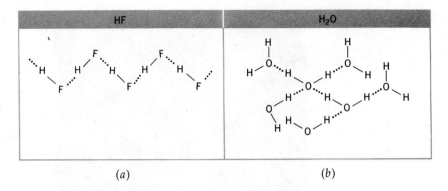

(a) (b)

of the elements in Groups IVA, VA, VIA, and VIIA are compared. Let's look at the compounds of Group IVA first because they form a nearly ideal pattern. We see from the figure that as the atomic weights of the elements in Group IVA increase, so do the boiling points of their hydrogen compounds. Using reasoning similar to that in Section 10.3, we know that as the sizes of the molecules increase from CH_4 to SnH_4, the London forces also increase. We expect, therefore, that the boiling points of these compounds should increase, as they actually do, in the direction of increasing molecular weights.

Except for the first members of the hydrogen compounds of Groups VA, VIA, and VIIA, the same trend is also observed—that is, increasing boiling point with increasing atomic weight of the element in the group. The first member of each of these groups, however, has a relatively high boiling point, much higher than expected from molecular weights alone. They, therefore, must possess attractive forces in addition to those of the London type. The position of each of the first members in relation to the rest of the group can be attributed to the presence of hydrogen bonding. We saw in Section 10.3 that the strongest hydrogen bonds form when hydrogen is bonded to a very small electronegative element. Therefore, the contribution of hydrogen bonding is expected to be the strongest for the first members of Groups VA, VIA, and VIIA, and becomes relatively unimportant for the remaining members. Methane, CH_4, which is nonpolar and cannot hydrogen bond (the electronegativity of carbon is too low and there are no lone electron pairs on the carbon to which hydrogen bonds can be formed) follows the normal, nearly straight-line pattern, within its group.

You may have noticed that water has a higher boiling point than hydrogen fluoride, even though HF is more polar than H_2O. The reason seems to be the number of hydrogen bonds that they can form. In HF, each molecule is hydrogen bonded to two others, whereas in water, each molecule can be hydrogen bonded to four others (Figure 10.19). Therefore, even though HF forms somewhat stronger hydrogen bonds than water, the total strength of four hydrogen bonds to a water molecule exceeds the total strength of two hydrogen bonds to an HF molecule.

The hydrogen bonding in NH_3 is much weaker than in H_2O or HF because of the considerably lower electronegativity of nitrogen. In addition, the nitrogen in NH_3 has only one lone pair to which a hydrogen bond can be formed, so each molecule, on the average, can be held by only two hydrogen bonds: one *from* another NH_3 molecule and one *to* another NH_3 molecule. Altogether, the total strength of the hydrogen bonding in NH_3 is so small that NH_3 has a lower boiling point than either HF or H_2O.

10.6   FREEZING POINT

At the freezing point (melting point), there is an equilibrium between solid and liquid.

$$\text{liquid} \rightleftharpoons \text{solid} + \text{heat}$$

Anyone who has ever made a tray of ice cubes in a refrigerator realizes that liquids freeze if heat is removed from them. You also know that ice cubes melt when they absorb heat. For any substance at a given pressure there is a characteristic temperature at which the liquid and solid can coexist in equilibrium. This is called either the **freezing point** or **melting point,** depending on whether you imagine approaching it from a high or low temperature. At the freezing point (melting point) the rate at which particles leave the solid and enter the liquid is the same as the rate at which particles leave the liquid and join the solid. If heat is added, some solid melts and more liquid is formed, but the temperature stays the same as long as both phases are present. Similarly, if some heat is removed, some liquid freezes and more solid forms—again without a temperature change.

As with evaporation and condensation, there are energy changes associated with freezing and melting. The **molar heat of crystallization,** ΔH_{cryst} *is the quantity of energy that must be removed from 1 mol of a liquid to convert it to a solid at the same temperature.* For melting, which is also called *fusion,* there is the **molar heat of fusion,** ΔH_{fus}—*the energy needed to melt 1 mol of a solid.* Following the definition given for the molar heat of vaporization, we can express these as

The actual values of H_{solid} and H_{liquid} can't be measured. Only their difference can be obtained experimentally.

$$\Delta H_{\text{cryst}} = H_{\text{solid}} - H_{\text{liquid}}$$

$$\Delta H_{\text{fus}} = H_{\text{liquid}} - H_{\text{solid}}$$

We see that the numerical values of ΔH_{cryst} and ΔH_{fus} must be identical; one is simply the negative of the other.

The size of the molar heat of fusion (or crystallization) is a measure of the difference in the strengths of the attractive forces between the liquid and solid, and it is always much smaller than the molar heat of vaporization, as shown in Table 10.3. When a solid melts there are relatively small changes in the distances between the molecules. Therefore, only small changes in potential energy are involved. When a liquid is converted to a gas, however, the intermolecular distances increase tremendously and large energy changes occur. This means that the quantity of energy (ΔH_{fus}) required to cause the molecules of a solid to overcome their attractive forces and form a liquid is small compared to the energy (ΔH_{vap}) required for liquid molecules to move apart, forming a gas.

Fusion means melting. The thin metal band in an electrical fuse protects a circuit by melting if too much current is passed through it. On the right is a fuse that has done its job.

Table 10.3

Heats of fusion and vaporization

Substance	ΔH_{fus} (kJ mol^{-1})	ΔH_{vap} (kJ mol^{-1})
Water	5.98	40.6
Benzene	9.92	30.7
Chloroform	12.4	31.9
Diethyl ether	6.86	26.0
Ethanol	7.61	38.6

EXAMPLE 10.2 USING THE HEAT OF FUSION

PROBLEM Calculate the energy, in kilojoules, necessary to melt 1.00 g of ice.

SOLUTION As before, we have a unit conversion requiring ΔH_{fus}. The problem reduces to

$$1.00 \text{ g } H_2O \sim (?) \text{ kJ}$$

From Table 10.3, $\Delta H_{fus} = 5.98$ kJ mol^{-1} for H_2O. Therefore,

$$1.00 \text{ g } H_2O \times \left(\frac{1 \text{ mol } H_2O}{18.0 \text{ g } H_2O} \right) \times \left(\frac{5.98 \text{ kJ}}{1 \text{ mol } H_2O} \right) \sim 0.332 \text{ kJ}$$

The energy needed is 0.332 kJ.

10.7 CRYSTALLINE SOLIDS

When most substances freeze, or when they are created in a precipitation reaction, they form crystals that have highly regular, symmetrical shapes. You have probably seen photographs of snowflakes similar to those in Figure 10.20. Notice how symmetrical each ice crystal is and how each has a characteristic hexagonal form, although the individual fine details are different.

The highly regular surface features of a crystal are a reflection of an orderly repeating pattern of atoms, molecules, or ions that exist within it. This order has made possible detailed analyses of the structures of solids and has led to much of our knowledge of the shapes of molecules and the sizes of atoms and ions.

Figure 10.20

Snowflakes are crystals of ice. Like other crystals, they have regular features that reflect the ordered arrangement of the particles within them.

X-ray diffraction

In 1912, a German physicist named Max von Laue pointed out that a crystal could serve as a three-dimensional diffraction grating if the wavelength of the incident radiation were of the same order of magnitude as the distance between particles in the solid. This condition is fulfilled by X rays, which have wavelengths of approximately 100 pm (0.1 nm).

When a crystal is bathed in X rays, each atom of the crystal within the path of an X ray absorbs some of its energy and then reemits it in all directions. Thus each atom is a source of secondary wavelets, and the X rays are said to be scattered by the atoms. These secondary wavelets from the different sources interfere with each other, either reinforcing or cancelling each other. In certain directions the waves emanating from nearly all the atoms in any orderly array are in phase—that is, the peaks and troughs of the waves coincide as shown in Figure 10.21*a*—and intense beams of X rays are observed in these directions. In all other directions the waves from various atoms are out of phase (Figure 10.21*b*) and cancel each other; thus no intensity is detected.

Two English scientists, William Bragg and his son Lawrence, treated the diffraction of X rays as if the process were reflection. In Bragg's treatment, the X rays that penetrate a crystal are thought of as being reflected by successive layers of particles within the substance (Figure 10.22). We can see from this diagram that beams reflected from deeper layers must travel farther to reach the detector. For there to be any intensity at the detector these waves have to be in phase with those reflected from the upper layers, which must mean that the extra distance traveled by the more penetrating beam has to be some integral multiple of the wavelength of the X rays.

Bragg showed that in order to observe any intensity in the emerging X rays, a relatively simple relationship had to be fulfilled. This relationship, known as the **Bragg equation,** is

$$2d \sin \theta = n\lambda \qquad [10.2]$$

where *d* is the spacing between the successive layers that are reflecting the X rays, θ is the angle at which the X rays enter and leave the particular set of layers, λ is the wavelength of the X rays, and *n* is an integer (that is, *n* = 1, or 2,

In the older scientific literature, X-ray wavelengths and atomic dimensions are given in angstroms. Today, the SI units nanometres or picometres are preferred.

$$1 \text{ Å} = 0.1 \text{ nm}$$

$$1 \text{ Å} = 100 \text{ pm}$$

The Braggs received the 1915 Nobel Prize for physics for their work.

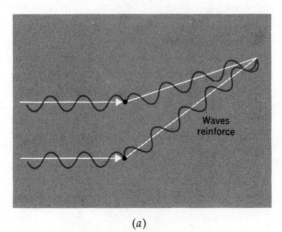

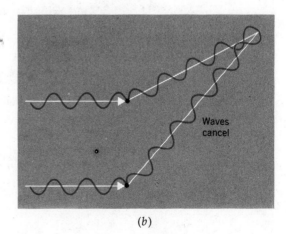

(a)   (b)

Figure 10.21

Diffraction. Scattered waves reinforce each other only at certain angles. (a) In phase. (b) Out of phase.

Figure 10.22
Bragg's law. Bragg showed that the waves from different planes of atoms are in phase only when $2d \sin \theta = n\lambda$.

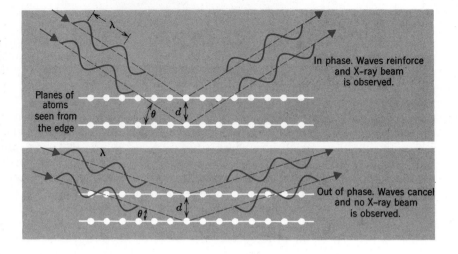

Planes of atoms seen from the edge

In phase. Waves reinforce and X-ray beam is observed.

Out of phase. Waves cancel and no X-ray beam is observed.

or 3, etc.). The Bragg equation serves as the basis for the study of crystalline structure by X-ray diffraction.

In practice, X rays of known wavelength are directed at a crystal and the angles at which they are reflected are recorded—for example, on a piece of photographic film (Figure 10.23). By measuring the angles at which the X rays are reflected, it is a simple matter to calculate the distances between planes of atoms within a crystal, as illustrated in Example 10.3. If, in addition, the intensities of the reflected X rays are measured, a crystallographer may be able to deduce, through a rather complex procedure, the actual positions of atoms within the solid. In this way the molecular structures of many substances have been found. In recent years, X-ray diffraction has become a powerful tool in biochemistry by which the structures of even very complex molecules have been investigated. For example, Rosalind Franklin's X-ray data led James Watson, Francis Crick, and Maurice Wilkins to their deduction of the double-helix structure of DNA—a feat that won Watson, Crick, and Wilkins the Nobel Prize in 1962.

Dorothy Hodgkin won the Nobel Prize in 1964 for her X-ray structure determination of vitamin B_{12}.

EXAMPLE 10.3 WORKING WITH BRAGG'S LAW

PROBLEM X rays of wavelength 154 pm (0.154 nm) strike a crystal and are observed to be reflected at an angle of 22.5°. Assuming that $n = 1$, calculate the spacing between the planes of atoms that are responsible for this reflection.

SOLUTION We wish to calculate d. Solving Equation 10.2 for d, we have

$$d = \frac{n\lambda}{2 \sin \theta}$$

From the data, $n = 1$, $\lambda = 154$ pm, $\theta = 22.5°$. Substituting, we get

$$d = \frac{(1)(154 \text{ pm})}{2 \sin(22.5)}$$

$$d = \frac{154 \text{ pm}}{2(0.383)}$$

$$= 201 \text{ pm}$$

Figure 10.23
Production of an X-ray diffraction pattern.

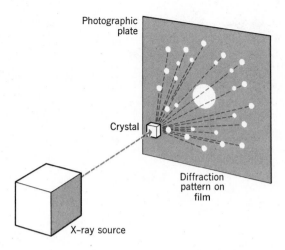

10.8 LATTICES

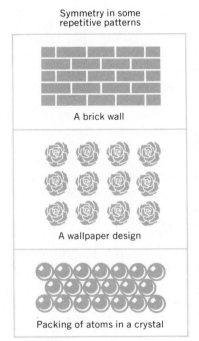

Symmetry in some repetitive patterns

A brick wall

A wallpaper design

Packing of atoms in a crystal

Any repetitive pattern has a symmetrical aspect about it, whether it be the stacking of bricks in a wall, a wallpaper design, or the orderly packing of particles in a crystal. For example, certain repeating distances between the elements of the pattern can be easily recognized, and the lines along which the elements of the pattern are repeated are at certain angles to each other.

To avoid having to deal with the details of a repeating structure, and concentrate on its symmetrical features, it is convenient to describe the structure simply in terms of a set of points that have the same repeat distances as the structure, arranged along lines oriented at the same angles. This kind of pattern of points is called a **lattice,** and when applied to the description of the packing of particles in a solid, we often use the term **crystal lattice.**

In a crystal the number of particles is enormous. If you could imagine being in the center of even the tiniest crystal, you would find that the particles go on as far as you could see in every direction. Describing the positions of all these particles or their lattice points is impossible and, fortunately, unnecessary. All we need to do is to describe the basic repeating unit of the lattice, which we call a unit cell. To see this, let's begin in two dimensions.

Figure 10.24*a* illustrates a two-dimensional lattice. It is a *square lattice* because the lattice points lie at the corners of squares. The repeating unit of this lattice—its **unit cell**—is the square drawn in white. If we began with this unit cell, we could produce the entire lattice simply by moving it repeatedly left and right, and up and down by distances equal to its edge length. In this sense, all the properties of a lattice are contained in the properties of its unit cell.

Figure 10.24
(a) *A portion of a two-dimensional square lattice with its unit cell drawn in white.* (b) *A design based on a square lattice.*

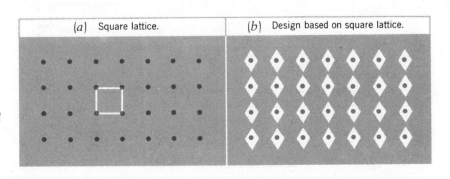

(*a*) Square lattice. | (*b*) Design based on square lattice.

Figure 10.25

A simple cubic lattice.

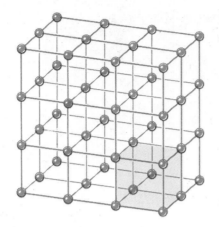

An important fact about lattices is that the same *kind* of lattice can be used to describe many different designs. For example, if we decided to place a diamond at each lattice point, we could create a wallpaper design like that shown in Figure 10.25*b*. A different wallpaper pattern could be created by placing a rose at each lattice point, or by changing the lengths of the edges of the unit cell.

The extension of the lattice concept to crystals in three dimensions is quite straightforward. In Figure 10.25 we see an example of a simple cubic lattice in which a unit cell is shaded in color. By associating a particular chemical environment with each lattice point we can arrive at a chemical structure, and by varying the chemical environment about each point we can create an infinite number of chemical structures all based on the same lattice.

The unit cells of all three-dimensional lattices are similar in that they all have eight corners, as shown in Figure 10.26. Unit cells differ by the lengths of their edges (*a*, *b*, and *c*) and the angles opposite them (α, β, and γ). In 1848, Auguste Bravais showed that *only* 14 different kinds of lattices are possible. These can be divided among seven basic crystal systems, whose properties are described in Table 10.4. What this means is that the crystals of all the millions of different chemical compounds ever discovered can be described by just this small set of lattices. It is this fact that has made a detailed understanding of the solid state possible.

Three common kinds of lattices are characterized by the cubic unit cells shown in Figure 10.27. The simplest is called a **simple cubic unit cell** or a **primitive cubic unit cell.** It has lattice points only at the corners. When oxygen is frozen it has a primitive cubic lattice with this kind of unit cell.

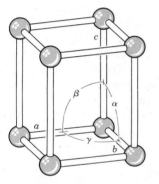

Figure 10.26

A three-dimensional unit cell. The edges a, b, *and* c *intersect at angles* α, β, *and* γ.

Table 10.4
**Properties of the unit cells
of the seven crystal systems**

System	Edge Lengths	Angles
Cubic	$a = b = c$	$\alpha = \beta = \gamma = 90°$
Tetragonal	$a = b \neq c$	$\alpha = \beta = \gamma = 90°$
Orthorhombic	$a \neq b \neq c$	$\alpha = \beta = \gamma = 90°$
Monoclinic	$a \neq b \neq c$	$\alpha = \beta = 90° \neq \gamma$
Triclinic	$a \neq b \neq c$	$\alpha \neq \beta \neq \gamma$
Rhombohedral	$a = b = c$	$\alpha = \beta = \gamma \neq 90°$
Hexagonal	$a = b \neq c$	$\alpha = \beta = 90°; \gamma = 120°$

Figure 10.27
The three unit cells belonging to the cubic system. (a) Simple cubic. (b) Body-centered cubic. (c) Face-centered cubic.

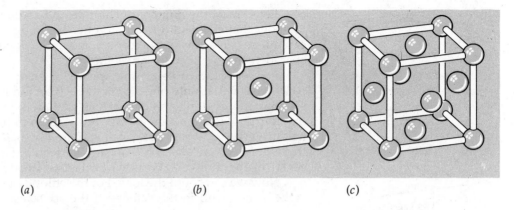

(a)   (b)   (c)

A **body-centered cubic unit cell** has lattice points at the corners, plus a lattice point in the center of the cell. Some common metals—chromium, iron, and tungsten, for example—form crystals with a body-centered cubic lattice.

A **face-centered cubic unit cell** has lattice points at the eight corners and one in the center of each of its six faces. This produces a very common kind of lattice that is found in crystals of such metals as nickel, copper, silver, gold, and aluminum.

Rock salt is NaCl.

Cubic lattices are not only characteristic of many elements. Some important and familiar compounds also have cubic lattices. For example, one of our most common compounds, sodium chloride, forms crystals having a face-centered cubic lattice. A portion of a NaCl crystal is shown in Figure 10.28 along with its unit cell. Notice how the chloride ions are located at positions corresponding to the lattice points, with the sodium ions squeezed in between. This kind of packing of cations and anions is called the *rock salt structure*. Similar structures are found for many of the other alkali halides as well—for instance, KCl and LiCl.

The unit cell could also be drawn with Na^+ at the lattice points and Cl^- in between.

Among the factors that determine the kind of lattice and structure an ionic compound can form are the relative sizes of the ions and the ratio of the numbers of anions to cations in the crystal. The question of size is rather complex, so we won't discuss it further, but the importance of the anion/cation ratio is not very difficult to see.

Since a crystal is composed of a large number of unit cells, whatever the anion/cation ratio is in the crystal as a whole, it must also be the same in the unit

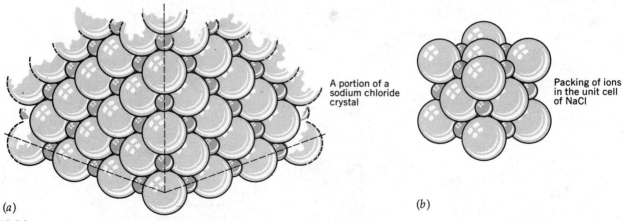

A portion of a sodium chloride crystal

Packing of ions in the unit cell of NaCl

(a)   (b)

Figure 10.28
The crystal structure of sodium chloride. (a) A portion of a NaCl crystal showing the packing of chloride ions (large spheres) and sodium ions (small spheres). (b) The arrangement of ions in the face-centered cubic unit cell of NaCl.

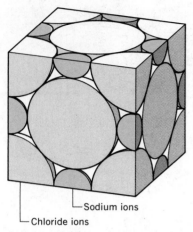

Sodium ions

Chloride ions

Figure 10.29

The faces of the unit cell slice through the ions at the corners, the edges, and the centers of the faces. Only parts of these ions are within this unit cell.

cell. Let's count the number of chloride and sodium ions in the unit cell of NaCl to show that their ratio is one-to-one. In doing this, however, we have to be careful, because ions at the corners, along the edges, and in the face-centers are shared with one or more other unit cells.

An ion at the corner of the unit cell is shared with seven others. In Figure 10.29 we see that only one-eighth of such an ion is in a given unit cell. We also see that an ion along an edge, which is shared among four unit cells, has only one-fourth of itself in a given unit cell. An ion in the center of a face contributes half to a given unit cell, because it is shared between two of them. In addition to these, there is one Na^+ ion that can't be seen in Figure 10.29. It is located in the center of the unit cell, as shown in Figure 10.30.

Now we can count the ions. For the chloride ions, we see that they are at the eight corners and in the centers of the six faces.

$$8 \text{ corners} \times \tfrac{1}{8} Cl^- \text{ per corner} = 1 Cl^-$$
$$6 \text{ faces} \times \tfrac{1}{2} Cl^- \text{ per face} = 3 Cl^-$$
$$\text{Total} \quad 4 Cl^-$$

Thus altogether there are four chloride ions contained within the unit cell. For the sodium ions, we have one along each of the 12 edges of the cube, which each contribute one-quarter to the unit cell, plus one in the center that is entirely within the unit cell.

$$12 \text{ edges} \times \tfrac{1}{4} Na^+ \text{ per edge} = 3 Na^+$$
$$1 Na^+ \text{ in center} = 1 Na^+$$
$$\text{Total} \quad 4 Na^+$$

The number of sodium ions in the unit cell is also four. This means, therefore, that the Na^+ and Cl^- ions are in a one-to-one ratio, which is necessary for the crystal to be electrically neutral.

Any substance that crystallizes with the rock salt structure *must* have a one-to-one anion/cation ratio. Sodium chloride and the other alkali halides have formulas that satisfy this condition, and many of them form crystals having this

This does not mean that all substances with a 1-to-1 anion/cation ratio must crystallize with the rock salt structure.

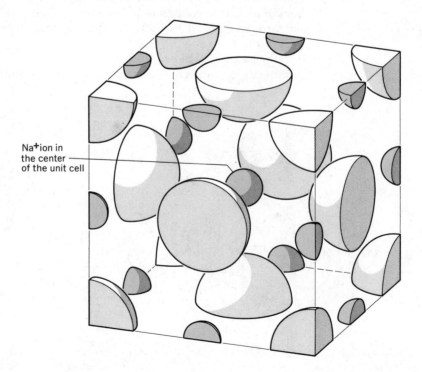

Na^+ ion in the center of the unit cell

Figure 10.30

Exploded view of the NaCl unit cell showing the Na^+ ion in the center.

structure. Calcium oxide, CaO, also has the rock salt structure, which is allowed because of its formula. However, $CaCl_2$ or Al_2O_3 could not possibly form crystals with the rock salt structure because their anion/cation ratios forbid it. Thus, we see that the formula of a compound places certain restrictions on the kinds of crystal structures it can and cannot have.

EXAMPLE 10.4 **CALCULATING THE NUMBER OF IONS PER UNIT CELL**

PROBLEM   Sodium forms crystals having a body-centered cubic lattice. How many sodium atoms are contained within one unit cell?

SOLUTION   The unit cell has sodium atoms at the eight corners, each of which is shared among a total of eight cells, plus one sodium atom in the center.

$$8 \text{ corners} \times \tfrac{1}{8}\text{Na per corner} = 1\text{Na}$$
$$1 \text{ Na in the center} \qquad = \underline{1\text{Na}}$$
$$\text{Total} = 2\text{Na}$$

Each unit cell has two sodium atoms within it.

Atomic and ionic radii

The study of the structures of crystals has yielded much useful information. The applications to the determination of molecular structures—even those as complex as DNA—have already been mentioned, although they are far too complex to discuss in any detail here. We can see an illustration of the usefulness of studying crystals, however, by examining how information about unit cells can be used to calculate the radii of atoms and ions.

It was noted in the last section that copper is a metal that crystallizes with a face-centered cubic lattice. X-ray diffraction measurements show that the unit cell has an edge length of 362 pm (this is length AC in Figure 10.31). Copper atoms are in contact along the line joining points A and B (the face diagonal). This distance corresponds to four times the radius, r, of a copper atom. From geometry we know that

$$\overline{AB} = \sqrt{2}(\overline{AC}) = \sqrt{2}(362 \text{ pm})$$
$$\overline{AB} = 512 \text{ pm}$$

Therefore,

$$4r = 512 \text{ pm}$$
$$r = 128 \text{ pm}$$

The radius of a copper atom in metallic copper is therefore 128 pm. In a similar fashion we can also use the results of X-ray diffraction to determine the radii of ions in ionic crystals.

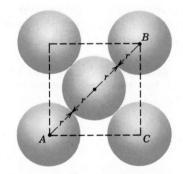

Figure 10.31
Copper atoms in the face of a unit cell.

10.9 TYPES OF CRYSTALS

We have seen that there is only a limited number of ways of arranging particles in a crystalline solid. The particular arrangements, as well as the physical properties of the solid, are determined by the types of particles present at the lattice points and the nature of the attractive forces between them. We can divide crystals into types: molecular, ionic, covalent, and metallic.

Molecular crystals

In molecular crystals, either molecules or individual atoms occupy lattice sites. The attractive forces between them are of the type described in Section 10.2 and are much weaker than the covalent bonds that exist within individual molecules. London forces are present in crystals of nonpolar substances such as Ar, O_2, naphthalene (moth crystals), and CO_2 (dry ice). In crystals of polar molecules such as SO_2, the dominant forces are a result of dipole-dipole attractions, and in solids such as ice (H_2O), NH_3, and HF, the molecules are held in place by hydrogen bonding. Since these are relatively weak forces (compared to covalent or ionic attractions), molecular crystals tend to have small lattice energies and are easily deformed; we say that they are soft. Also, relatively little thermal energy is required to overcome these attractions, and molecular solids generally tend to have low melting points.

Molecular crystals are poor conductors of electricity because the electrons are bound to individual molecules and are not able to move freely through the solid.

Ionic crystals

In an ionic crystal, such as NaCl, there are ions located at lattice sites and the binding between them is mainly electrostatic (which is essentially nondirectional). As a result, the kind of lattice that is formed is determined mostly by the relative sizes of the ions and their charges. When the crystal forms, the ions arrange themselves to maximize attractions and minimize repulsions.

Because electrostatic forces are strong, ionic crystals have large lattice energies. They are often hard and are characterized by relatively high melting points. They are also very brittle. When struck they tend to shatter, because as planes of ions slip by one another they pass from a condition of mutual attraction to one of mutual repulsion. This is illustrated in Figure 10.32.

In the solid state, ionic compounds are poor conductors of electricity because the ions are held rigidly in place. When melted, however, the ions are free to move about and ionic substances become good conductors.

Figure 10.32

An ionic crystal breaks when struck. (a) Attraction between ions opposite each other. (b) When struck, part of the crystal slips past the rest. Ions of the same sign face each other. (c) The repulsive forces push the crystal apart.

Attraction, ions of opposite sign face each other

(a)

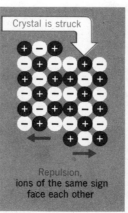

Crystal is struck

Repulsion, ions of the same sign face each other

(b)

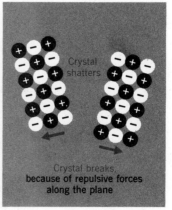

Crystal shatters

Crystal breaks, because of repulsive forces along the plane

(c)

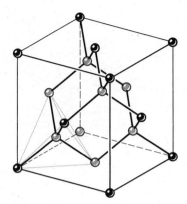

Figure 10.33

The unit cell of diamond. Note how the carbon atoms inside the unit cell are each covalently bonded to four others. The same applies to each of the carbon atoms in the crystal.

Covalent crystals

In a covalent crystal there is a network of covalent bonds between the atoms that extends throughout the entire solid. An example of such a substance is diamond, whose structure is shown in Figure 10.33. Diamond is a form of elemental carbon in which each atom is covalently bonded to four nearest neighbors. Other common examples are carborundum (silicon carbide, SiC) and quartz (silicon dioxide, SiO_2, commonly recognized as a major constituent of many sands).

Because of the interlocking framework of covalent bonds, covalent crystals have very high melting points and are usually extremely hard. Diamond, of course, is the hardest substance known and is used in grinding and cutting tools. Silicon carbide is like diamond, except that half of the carbon atoms in the structure have been replaced by silicon atoms. It too is very hard and is used as an abrasive in sandpaper, as well as in other grinding and cutting applications.

Covalent crystals are poor conductors of electricity because the electrons in the solid are localized in the covalent bonds and are not free to move through the crystal.

Metallic crystals

The simplest picture of a metallic crystal has positive ions (nuclei plus core electrons) situated at lattice points, with the valence electrons belonging to the crystal as a whole instead of to any single atom. The solid is held together by the electrostatic attraction between the lattice of positive ions and this sort of "sea of electrons." These electrons may move freely, so we find metals to be good conductors of electricity. Since the melting points and hardness of metals vary over wide ranges, there must also be, at least in some cases, some degree of covalent bonding between atoms in the solid.

Table 10.5 summarizes the important properties of these different kinds of solids.

Table 10.5

Types of solids

	Molecular	Ionic	Covalent	Metallic
Chemical units at lattice sites	Molecules or atoms	Positive and negative ions	Atoms	Positive ions
Forces holding the solid together	London forces, dipole-dipole, hydrogen bonds	Electrostatic attraction between + and − ions	Covalent bonds	Electrostatic attraction between + ions and electron "sea"
Some properties	Soft, generally low melting, nonconductors	Hard, brittle, high melting, non-conductors (but conduct when melted)	Very hard, high melting, non-conductors	Hard to soft, low to high melting, high luster, good conductors
Some examples	CO_2 (dry ice) H_2O (ice) $C_{12}H_{22}O_{11}$ (sugar) I_2	NaCl (salt) $CaCO_3$ (limestone; chalk) $MgSO_4$ (in Epsom salt)	SiC (carborundum) C (diamond) WC (tungsten carbide, used in cutting tools)	Na, Fe, Cu, Hg

Figure 10.34
Liquid crystals. (a) Packing in a nematic liquid crystal. (b) Packing in a smectic liquid crystal. (c) Packing in a cholesteric liquid crystal.

(a)   *(b)*   *(c)*

10.10 LIQUID CRYSTALS

In a typical liquid, the molecules have no orderly pattern. They are able to move past each other easily, so liquids are able to flow. In a typical solid, the molecules are highly ordered, but they are also held rigidly in place, so solids do not flow. Certain substances, in a range of temperatures just above their melting points, exhibit properties characteristic of both of these states, and are therefore called **liquid crystals.** They are fluid, but their molecules are arranged in a highly ordered way. At temperatures above this range this order is lost and they become like any other liquid.

There are three kinds of liquid crystals: nematic, smectic, and cholesteric. They are all composed of rodlike molecules, but they differ in the kind of order that exists within them. In a **nematic liquid crystal** the rodlike molecules are arranged like loosely packed soda straws (Figure 10.34a). The order within a **smectic liquid crystal** is even greater. Here there are parallel rodlike molecules arranged in layers, as shown in Figure 10.34b. In a **cholesteric liquid crystal** the molecules are in layers too, but they are aligned parallel to the layers in a nematic fashion, with slightly different orientations from one layer to the next, as illustrated in Figure 10.34c.

What makes liquid crystals so interesting and potentially useful is the way they affect light. If you've ever seen an oil slick on water, you've probably noticed a rainbow of colors that seems to be reflected from it. This is actually caused by diffraction of light from the surfaces of the oil and water. When the oil layer is very thin, all but a narrow band of wavelengths is lost as reflected waves by destructive interference. We see only those waves that are reflected in phase from the two layers—the top of the oil layer and the top of the water layer beneath the oil. This same phenomenon occurs when light reflects from the layers in a cholesteric liquid crystal, but what is especially interesting is that the reflected color changes with temperature. The reason is that the distance between the layers changes with temperature. This allows these substances to be used for a sort of "color mapping" of temperature regions on an object. One medical application is locating a vein by the warmer skin temperature that it produces.

The substances used in the liquid crystal displays found in pocket calculators and wristwatches are nematic liquid crystals. Their optical characteristics are affected by an electric field. In this case a thin film of the liquid crystal is sandwiched between transparent electrodes arranged on glass in special patterns. When a particular electrode segment is energized, the orientations of the molecules in the liquid crystal are changed and the segment becomes opaque. By activating appropriate segments in this way, various numbers or letters can be formed. An important advantage of these displays is that they use very little energy, so the batteries used to power them last a long while.

10.11   HEATING AND COOLING CURVES; CHANGES OF STATE

When heat is added to a solid, initially at some temperature below its melting point, the temperature begins to rise. Once the melting point is reached, the temperature levels off—it stays constant until all the solid has melted. When more heat is added, the liquid becomes warmer until it reaches its boiling point. As heat is supplied to the boiling liquid, the temperature stays constant again until the liquid has all been converted to a gas. Then the temperature can rise further as still more heat is added. Figure 10.35, which illustrates this graphically for 1 mol of a typical substance, is called a **heating curve.**

In those portions of the heating curve where the temperature is rising, heat that is added is increasing the average kinetic energy of the molecules. But what is happening during melting and boiling—those portions of the curve where the temperature stays the same? Because the temperature stays constant, the average kinetic energy also remains the same. This means that the heat energy being added must be increasing the potential energies of the molecules.

If the K.E. doesn't change, then the P.E. must change if energy is added.

When most solids melt, there is a slight expansion in volume. This means that the particles must be getting slightly further apart. Since there are attractive forces between the particles, energy must be supplied to separate them.[3] This energy is the heat of fusion. Similarly, when a liquid vaporizes, the molecules go from the closely packed liquid state to their widely spaced distribution in a vapor. This also requires an energy input to overcome the attractive forces—the heat of vaporization. In both melting and vaporization it is the distance between molecules that attract each other that is changing, and as we learned in Chapter 1, this involves changes in potential energy. Thus, ΔH_{fus} and and ΔH_{vap} are energy changes that affect the potential energies of particles.

A cooling curve, like that in Figure 10.36, is essentially the opposite of a heating curve. During condensation and freezing, the temperature stays constant while the potential energies of the particles decrease as they come together.

Some liquids do not follow a smooth transition into the solid state, but instead give rise to a cooling curve such as that shown in Figure 10.37. As the temperature of the liquid drops, it eventually reaches point *A*, the expected

[3] A most important exception to this, of course, is ice, in which the solid is less dense (more expanded) than the liquid. Here energy supplied to the solid disrupts some of the hydrogen bonding that exists in the solid ice and the open structure of ice collapses to give a more dense liquid.

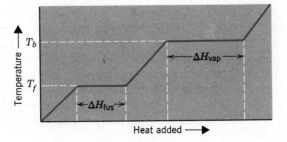

Figure 10.35

A typical heating curve for 1 mol of a substance. T_f is the freezing point and T_b is the boiling point.

Figure 10.36

A typical cooling curve for 1 mol of a substance.

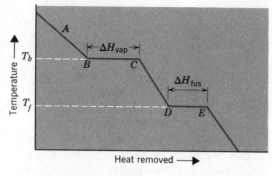

Figure 10.37

Supercooling. As the liquid is cooled, its temperature drops below the freezing point. After a short time, freezing begins and the temperature rises to the freezing point.

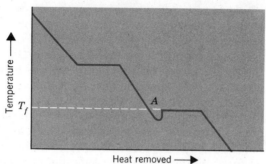

freezing point of the substance. The molecules, however, may not be oriented properly to fit into the crystalline lattice, and random motion continues as heat is further withdrawn from the liquid. As a result, the temperature of the liquid drops below its expected freezing point and the liquid is said to be **supercooled.** Once a small number of molecules have achieved the correct pattern, a tiny crystal is formed that serves as a seed on which additional molecules may rapidly accumulate. Potential energy is suddenly released as this crystal quickly grows, and the energy that is evolved increases the average kinetic energy of the molecules in the liquid and solid. Therefore, the temperature of the system rises again until it returns to the freezing point, after which the substance behaves normally. Further removal of heat eventually leads to the complete conversion of the liquid to a solid.

Some substances, such as glass, rubber, and many plastics, never do achieve a crystalline state when their liquids solidify on cooling. These compounds consist of long chainlike molecules that intertwine in the liquid. As they are cooled their molecules move so slowly that they never do find the proper orientation to form a crystalline solid, and an **amorphous solid** results instead. The term amorphous comes from the Greek word meaning "without shape," and if you've ever examined pieces of broken glass, you will recall that the surfaces are not uniform and flat like those of a crystal. Instead, they curve in unpredictable ways. This is because amorphous "solids" are actually supercooled liquids and lack the internal order found within crystals.

A property of supercooled liquids is that they continue to flow, although very slowly, even at room temperature. For example, very old glass exhibits greater crystallinity, when examined by X-ray diffraction, than freshly formed glass does, showing that molecules are slowly finding their way into a crystalline lattice. A supercooled liquid familiar to many children is Silly Putty. In many ways it behaves like a solid, particularly when forced to flow rapidly (for exam-

(a) *Silly Putty flows slowly like a very viscous liquid.* (b) *If it is stretched quickly, it fractures like a solid.*

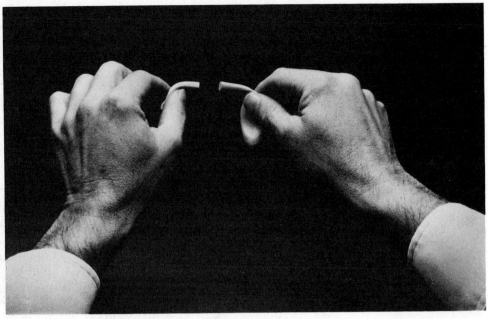

(a)

(b)

ple, it breaks when pulled apart suddenly). In other ways it flows like a liquid. Supercooled liquids such as glass, Silly Putty, and plastics in general, do not have sharp, well-defined melting points, but instead gradually soften when heated.

10.12 VAPOR PRESSURES OF SOLIDS

In Section 10.1 you learned that solids are able to evaporate just like liquids; the process is called sublimation. In a crystalline solid the molecules vibrate about their lattice positions and continually undergo collisions with their nearest neighbors. This gives rise to a distribution of kinetic energies in the solid, just as it does in a liquid or a gas. A small fraction of the molecules at the surface of a solid possess large kinetic energies, large enough to be able to overcome the attractive forces within the solid and break away from the surface, entering the gaseous phase above. That's how sublimation occurs.

When sublimation takes place in a closed container, more and more molecules enter the gaseous state and the pressure exerted by the vapor increases. Meanwhile, the slower-moving gaseous molecules that are colliding with the surface of the solid become trapped and return to the solid state. In time, the rate at which the molecules leave will exactly equal the rate at which they return, and a dynamic equilibrium will be established. The pressure exerted by a vapor in equilibrium with its solid is known as the **equilibrium vapor pressure of the solid.** Just as in liquids, the vapor pressure of a solid depends on the ease with which the molecules are able to enter the gaseous state. For example, the attractive forces are stronger in ionic solids than molecular solids and, as expected, we find that the vapor pressures of ionic solids are generally very much lower than those of molecular solids.

A practical application of sublimation that you've probably heard of is freeze drying. Freeze-dried instant coffee, for example, is manufactured by first freezing a batch of brewed coffee and then removing the ice component by vacuum. The vacuum creates an atmosphere of diminished pressure in which ice readily sublimes. Removing the water in this way preserves the delicate heat-sensitive molecules that give coffee its taste, so the quality of the product is improved. Solid foods—even entire meals—are also freeze-dried, both to protect their flavor as well as prevent spoilage, since bacteria that might cause harm cannot grow and multiply in the absence of moisture. This allows freeze-dried foods to be stored without refrigeration, and they are easily reconstituted by adding water to them. Campers often carry these products because of their convenience.

10.13 PHASE DIAGRAMS

The vapor pressure of a solid, like that of a liquid, is a function of its temperature. Increasing the temperature of a solid-vapor equilibrium, according to Le Châtelier's principle, leads to a shift in the position of the equilibrium that will occur with the absorption of heat. The production of vapor from the solid is an endothermic process; therefore as the temperature rises, more of the solid will evaporate and more of the vapor will be produced until equilibrium is once again attained. As a result, the equilibrium vapor pressure of a solid increases with increasing temperature until eventually a temperature is reached at which the solid melts. Further increases in temperature, beyond this point, will then give rise to a liquid-vapor equilibrium curve that terminates at the critical temperature of the substance. If, using water as an example, we plot the vapor pressure versus temperature for the solid-vapor equilibrium and the liquid-vapor equilibrium on the same graph, we produce Figure 10.38. Each point along the solid curve represents specific combinations of temperatures and pressures that must be achieved for the solid to be in equilibrium with its vapor. Likewise, points along the liquid curve represent combinations of temperatures and pressures required for the liquid to be in equilibrium with its vapor. The point of intersection of these two curves, called the **triple point,** corresponds to a unique temperature and pressure at which all *three* states of matter (solid, liquid, and gas) coexist in equilibrium with each other simultaneously. The triple point occurs at

The temperature at the triple point of water is the SI reference temperature 273.16 K.

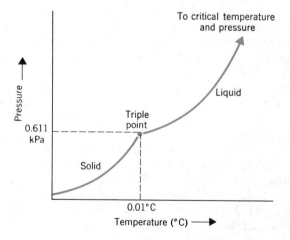

Figure 10.38

Solid and liquid vapor pressure curves for water.

Figure 10.39

Phase diagram for water (somewhat distorted). T_f is the normal freezing point and T_b is the normal boiling point. The slope of the solid-liquid line is exaggerated. The actual slope is much less to the left (3.3 MPa is required to lower the melting point of ice by only 1°C).

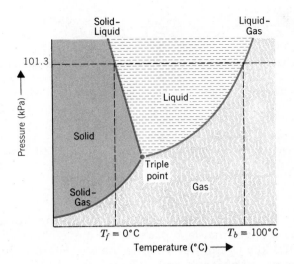

a temperature and pressure that depend on the nature of the substance in question. For example, the triple point of water occurs at a temperature of 0.01°C and a pressure of 0.611 kPa (0.006 03 atm), while the triple-point temperature of carbon dioxide is −57°C and the triple-point pressure is 530 kPa (5.2 atm).

There is still another equilibrium that can be represented on the same graph. This line corresponds to the combinations of temperatures and pressures that must be maintained to achieve a solid-liquid equilibrium. At a pressure of 101.3 kPa (1 atm), the melting point of water is 0°C; therefore, the solid-liquid equilibrium line passes through both the triple point and the normal melting point as shown in Figure 10.39. This graph is called a **phase diagram** because it allows us to pinpoint temperatures and pressures at which the various phases exist, as well as those conditions under which equilibrium can occur. For instance, at a pressure of 1 atm, water exists as a solid at all temperatures below 0°C, and in fact the region bounded by the solid-liquid equilibrium line and the solid-vapor equilibrium line corresponds to all the temperatures and pressures at which water exists as a solid. Similarly, in the region bounded by the solid-liquid and liquid-vapor equilibrium lines, the substance can exist only as a liquid, while to the right of both the solid-vapor and liquid-vapor lines the substance must be a gas.

Table 10.6 lists some randomly chosen temperatures and pressures and the physical states of water that we can predict from its phase diagram. You might verify these predictions to illustrate how to use a phase diagram.

Table 10.6
Physical states of water at
random temperatures and pressures

Temperature (°C)	Pressure (kPa)	State
25	100	Liquid
0	200	Liquid
0	50	Solid
100	50	Gas

EXAMPLE 10.5 INTERPRETING A PHASE DIAGRAM

PROBLEM What phase of water would exist at a temperature of 5°C and a pressure of 0.5 kPa?

SOLUTION To answer this question, we have to locate 5°C and 0.5 kPa on the phase diagram. Here it is best to examine Figure 10.38, which gives a more detailed view of the region around the triple point. Notice that 0.5 kPa is below the pressure corresponding to the triple point, and that 5°C lies to the right of the triple point. This places this point in the "gas" region of the phase diagram, so at this temperature and pressure, water would exist as a gas.

To gain a further insight into the meaning of a phase diagram, let us follow the changes that take place as we move along a line of constant pressure, say 101.3 kPa (1 atm), by varying the temperature. In Figure 10.40, point A lies in the region of the diagram where a sample of the substance would exist entirely as a solid, as shown in Figure 10.41a. When the temperature rises to point B in Figure 10.40, the solid begins to melt and an equilibrium between the solid and liquid can occur (Figure 10.41b). At a still higher temperature, point C, all the solid will have been converted to a liquid (Figure 10.41c); and, when the liquid-vapor line is encountered at point D in Figure 10.40, vapor may at last begin to form and an equilibrium can exist (Figure 10.41d). Finally at a sufficiently high temperature, such as point E, all the water will exist in the vapor state (Figure 10.41e).

We could also proceed with a similar analysis in which the temperature is held constant and the pressure is permitted to change. For example, at point F in Figure 10.40, the water would exist entirely as a gas (Figure 10.42a). At a higher pressure, point G in Figure 10.40, a solid-vapor equilibrium would exist (Figure 10.42b), and above that pressure, at point H, all the water would be converted to a solid (Figure 10.42c). As the pressure is increased further, we encounter the solid-liquid line at point B in Figure 10.40, where we again have an equilibrium as represented by Figure 10.42d. At still higher pressures, the water will melt so that at point I all the water is present in the liquid state (Figure 10.42e).

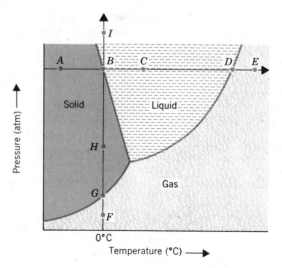

Figure 10.40

Phase diagram for water (not drawn to scale).

Figure 10.41

Raising the temperature at a constant pressure of 1 atm. Temperatures correspond to points A to E in Figure 10.40.

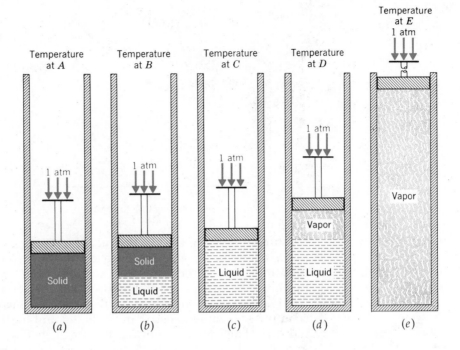

In the phase diagram for water we see that the solid-liquid equilibrium line slants to the left. This is a direct consequence of the fact that liquid water at 0°C has a higher density than the solid. Le Châtelier's principle requires that an increase in pressure on a system at equilibrium will lead to the production of the more dense phase; that is, a rise in pressure favors the packing together of molecules—quite a reasonable expectation. This means that if we have solid and liquid water at equilibrium and we increase the pressure while holding the tem-

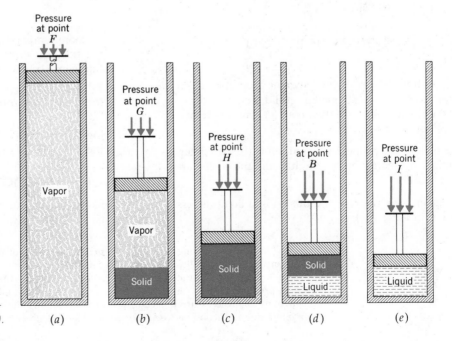

Figure 10.42

Raising the pressure at a constant temperature of 0°C. Pressures correspond to points on Figure 10.40.

Figure 10.43

Phase diagram for carbon dioxide.

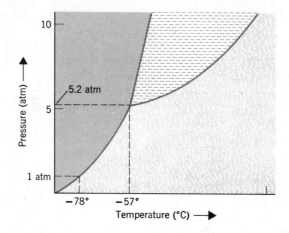

perature at 0°C, we should produce the higher-density, liquid phase.[4] On the phase diagram, a rise in pressure at constant temperature amounts to moving upward along a vertical line. We can move from the solid-liquid equilibrium line upward into a region of all liquid only if the solid-liquid line leans to the left.

Water is quite an unusual substance. For nearly all other compounds the solid phase is more dense than the liquid and for these substances the solid-liquid line slants to the right, as is shown in the phase diagram for CO_2 that appears in Figure 10.43. An interesting feature of this phase diagram is that the entire liquid range lies above a pressure of 1 atm; therefore, it is impossible to form liquid CO_2 at atmospheric pressure. Instead, as the gas is cooled, the solid-vapor equilibrium line is encountered at −78°C and the vapor is converted directly to the solid. This also explains why dry ice sublimes rather than melts at ordinary pressures.

At room temperature, CO_2 exists as a liquid at the high pressure inside a CO_2 fire extinguisher.

REVIEW QUESTIONS AND PROBLEMS

(Problems whose numbers are in blue have their answers in Appendix C at the back of the book; the more difficult problems are marked with asterisks.)

General Properties of the States of Matter

10.1 State how the following physical properties differ for the three states of matter—solid, liquid, and gas: (a) density, (b) rate of diffusion, (c) compressibility, (d) ability to flow.

10.2 Which of the properties in Question 10.1 are determined primarily by how tightly the molecules are packed?

10.3 Explain why the rate of diffusion in a liquid is less than in a gas.

10.4 At room temperature, diffusion in solids is virtually nonexistent. Why is this so? Why does solid-state diffusion occur at very high temperatures?

10.5 What is surface tension? Why do liquids tend to form *spherical* droplets?

10.6 As a child, you probably had the experience of filling a glass above its rim with water. Why doesn't the water simply overflow?

Evaporation and Sublimation

10.7 Sketch the kinetic energy distribution for a liquid at two temperatures. Indicate the minimum K.E. re-

[4] In fact, it had long been thought that this melting of water that occurs at high pressure is responsible for our ability to skate on ice. It was believed that the high pressure produced by the skater's weight concentrated on the sharp edge of a blade causes the ice just beneath the blade to melt, producing a thin film of liquid water that serves as a lubricant and allows the skate to slide smoothly on the ice. Current feeling, however, is that this film of water is most probably the result of melting due to friction between the moving skate blade and the ice.

quired for molecules to escape the liquid. On the basis of this diagram, explain why liquids evaporate faster at higher temperatures.

10.8 Why does evaporation lead to a lowering of temperature?

10.9 Clothes dry more rapidly on a dry day than on a humid day. They also dry more rapidly if there is a breeze blowing than if the air is still. Why?

10.10 At high altitudes, ice and snow gradually disappear without melting. Why?

10.11 Why does warm water evaporate more quickly than cold water?

10.12 Solid iodine vaporizes without melting if warmed gently. What is this process called?

10.13 Salt, NaCl, is harvested from seawater by evaporation of water from large shallow ponds. What is the purpose of spreading the water over such large areas?

10.14 At a given temperature, methyl alcohol—the fuel in canned heat—evaporates much more rapidly than propylene glycol—a food additive. Which substance has the weaker intermolecular attractions?

10.15 Can you think of any common household products that sublime?

Intermolecular Attractions

10.16 What are dipole-dipole attractions? Why are they weaker in a gas than in a liquid or solid?

10.17 What are hydrogen bonds? Why are they most important in compounds containing N—H, O—H, and F—H bonds?

10.18 What are London forces?

10.19 What type (or types) of intermolecular attractive forces are found in the following:
(a) HCl   (e) NO
(b) Ar   (f) CO_2
(c) CH_4 (g) H_2S
(d) HF   (h) SO_2

10.20 How is the structure of ice affected by hydrogen bonding?

10.21 If you were asked to compare the strengths of the intermolecular attractive forces in liquid A with those in liquid B, what types of data would you collect?

10.22 How should the strengths of London forces vary from helium to argon in Group 0?

10.23 What is polarizability? How is it related to the strengths of intermolecular attractions?

10.24 What evidence is there for hydrogen bonding in H_2O, HF, and NH_3?

Heats of Vaporization

10.25 What is the formal definition of the molar heat of vaporization? Why is it unimportant that we can't measure the total energy of either a liquid or its vapor?

10.26 Suppose that two substances, X and Y, have heats of vaporization equal to 38 and 28 kJ mol^{-1}, respectively. Which compound would you expect to have the higher boiling point? Which compound would be less likely to exhibit hydrogen bonding?

10.27 Among the hydrocarbons, CH_4 through $C_{10}H_{22}$, both ΔH_{vap} and boiling points increase with increasing molecular weight, even though the molecules are composed of the same kinds of atoms. How can this be explained?

10.28 How would you expect the heats of vaporization to vary among the compounds, PH_3, AsH_3, SbH_3?

10.29 How would you expect the heats of vaporization to vary among the compounds, H_2S, H_2Se, H_2Te?

10.30 What is a reasonable explanation for the fact that ΔH_{vap} is larger for H_2O than HF?

10.31 The molar heat of vaporization of water is greater than that of any of the components of gasoline. Yet, if gasoline is spilled on your hand, it gives a greater cooling effect than if water is spilled. Why?

10.32 What is the source of energy in a thunderstorm?

10.33 Steam at 100°C produces a much more severe burn than an equivalent amount of liquid water at 100°C. Why?

10.34 *Trouton's rule* states that the ratio of the heat of vaporization to the boiling point (in kelvins) is approximately a constant. Verify this for the hydrocarbons CH_4 through $C_{10}H_{22}$ in Table 10.1. What conclusions can you draw concerning the relationship between ΔH_{vap} and boiling point?

10.35 Calculate the heat necessary to convert 55.0 g of ethanol (ethyl alcohol, C_2H_5OH) from liquid to vapor. ΔH_{vap} = 38.6 kJ mol^{-1}.

10.36 How much energy is necessary to melt 35.0 g of benzene (C_6H_6)? ΔH_{fus} = 9.92 kJ mol^{-1}.

10.37 A 14.5-g sample of liquid mercury required 4.29 kJ to completely convert it to vapor at the same temperature. What is ΔH_{vap} of Hg in kJ mol^{-1}?

*10.38 A 68.2-kg skater skids to a halt from a speed of 16 km h^{-1}. If we assume that all of his energy appears as frictional heat transferred to ice at 0°C, how many grams of ice will be melted?

*10.39 A student (with very slow reflexes) holds his hand in a stream of steam at 100°C until exactly 1.00 g of water has condensed. If this water then cools to 40°C, how much energy has been absorbed by the student's hand?

*10.40 A 10.0-g cube of solid benzene (C_6H_6) at its melting point is introduced into 50.0 g of H_2O at 30.0°C. Given that ΔH_{fus} for C_6H_6 is 9.92 kJ mol^{-1}, to what temperature will the water have cooled by the time all of the benzene has melted?

*10.41 A 50.0-g ice cube at 0.00°C is added to 10.0 g of steam at 100.0°C. What will be the final temperature of the 60.0 g of water?

Vapor Pressure

10.42 What is meant by the term *equilibrium vapor pressure?*

10.43 What are the two principal factors that affect the observed vapor pressure of a liquid?

10.44 At room temperature, the vapor pressure of acetone (in nail polish remover) is approximately 30 kPa, whereas the vapor pressure of ethyl alcohol is approximately 8 kPa at this same temperature. Which of these substances possesses the stronger intermolecular attractions?

10.45 Why does the vapor pressure of a liquid increase with increasing temperature?

10.46 Why is the vapor pressure a measure of the strengths of intermolecular attractive forces in a liquid?

10.47 Explain why decreasing the volume of a container does not alter the vapor pressure of a liquid.

10.48 The vapor pressure of a liquid does not depend on the liquid's surface area. Explain why this is so.

10.49 A glass of cola, or other beverage, with ice in it often becomes wet on the outside. Explain why.

10.50 At −20°C there is less water in 1 m^3 of air at 100% humidity than at 20°C when the humidity is 100%. Why is this so?

10.51 When warm moist air is forced to rise over mountains, clouds form and rain frequently falls. On the basis of the concepts in this chapter, explain why this happens.

10.52 Define the terms: critical temperature, critical pressure.

10.53 What happens to the boundary between liquid and vapor at temperatures above the critical temperature?

10.54 To what temperature must helium be cooled before it can be condensed to a liquid?

10.55 If you shake a CO_2 fire extinguisher on a cool day, you can feel a liquid sloshing around inside. On a hot summer day, when the temperature is around 30°C, there is no sloshing when the same extinguisher is shaken. Why?

10.56 Why does blowing gently across the surface of a hot cup of coffee help to cool it?

Le Châtelier's Principle

10.57 State Le Châtelier's principle.

10.58 Using Le Châtelier's principle, predict the effect of a change in temperature on the equilibria:
(a) solid + heat $\rightleftharpoons$ liquid
(b) liquid + heat $\rightleftharpoons$ vapor

10.59 How will an increase in pressure effect the equilibrium: solid $\rightleftharpoons$ vapor?

Boiling Point

10.60 Define: boiling point, normal boiling point.

10.61 From Figure 10.15, estimate the boiling point of water at a pressure of 70 kPa.

10.62 At the top of Mount Everest in the Himalayas, which is 8.9 km above sea level, the atmospheric pressure is approximately 36 kPa. At what temperature would water boil at that altitude?

10.63 Explain why compounds with strong intermolecular attractive forces have higher boiling points than compounds with weak intermolecular attractive forces.

Crystals and X-Ray Diffraction

10.64 What external features do crystals exhibit?

10.65 What is an amorphous solid? How does it differ from a crystalline solid?

10.66 What is a lattice? What is a unit cell? Why can one kind of lattice be used to describe many different chemical structures?

10.67 What is the Bragg equation? What do the symbols in the equation stand for?

10.68 What quantities determine the kind of lattice to which a particular unit cell belongs?

10.69 Sketch and name the three types of cubic lattices.

10.70 Describe the rock salt structure. What kind of lattice does it belong to? How many formula units are there per unit cell? Could a salt like K_2S crystallize in the rock salt structure? Explain your answer.

10.71 Calculate the angles at which X rays of wavelength 229 pm will be observed to be reflected from crystal planes spaced (a) 1000 pm apart, (b) 200 pm apart. Assume that $n = 1$.

10.72 Calculate the interplanar spacings (in picometers) that correspond to reflections at $\theta = 20.0°$, 27.4°, and 35.8° by X rays of wavelength 0.141 nm. Assume that $n = 1$.

10.73 From the following list of angles, determine the angles at which X rays of wavelength 141 pm, diffracted from planes of atoms 200 pm apart, are in phase: $\theta = 17.3°$, 20.5°, 44.8°, and 55.3°.

10.74 Chromium, used to protect and beautify other metals, crystallizes in a body-centered cubic structure in which the Cr atoms are in contact along the body diagonal of the unit cell. The edge of the unit cell is 288.4 pm. Calculate the atomic radius of a Cr atom in picometres.

*10.75 Chromium crystallizes with a body-centered lattice. Its density is 7.19 g cm^{-3} and the unit cell edge is 288.4 pm. Use these data to compute Avogadro's number.

10.76 Gold crystallizes with a face-centered cubic lattice. The length of the unit cell edge is 407.86 pm. What is the atomic radius of a gold atom in picometres?

10.77 Aluminum crystallizes in a face-centered cubic structure. If the Al atom has an atomic radius of 143 pm, what is the length of the unit cell edge in Al in picometres?

10.78 CsCl forms a simple cubic lattice in which there are Cs$^+$ ions at the corners of the unit cell and a Cl$^-$ ion in the center of the cell. The cation/anion contact occurs along the body diagonal of the unit cell. The length of the unit cell edge is 412.3 pm. The Cl$^-$ ion has a radius of 181 pm. What is the radius of the Cs$^+$ ion in picometers?

10.79 RbCl has the rock salt structure shown in Figure 10.28. The unit-cell edge length is 658 pm. Cations and anions are in contact along the edges. The ionic radius of the chloride ion is 181 pm. Calculate the ionic radius of the Rb$^+$ ion in picometres.

10.80 Silver has an atomic radius of 144 pm. What would be the density of Ag in g cm^{-3} if it were to crystallize in the following structures: (a) simple cubic, (b) body-centered cubic, (c) face-centered cubic? The actual density of Ag is 10.6 g cm^{-3}. Which of these corresponds to the correct structure for Ag?

*10.81 Refer to the diagram below to derive the Bragg equation. Remember that the extra distance traveled by the more penetrating beam must be an integral multiple of the wavelength in order to have constructive interference.

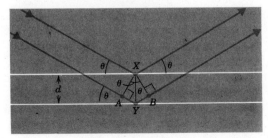

$$\sphericalangle XAY = \sphericalangle XBY = 90°$$

*10.82 Calculate the quantity of vacant (unoccupied) space (in pm^3) in a primitive cubic, a body-centered cubic,

and a face-centered cubic packing of identical spheres of diameter 100 pm.

10.83 LiBr has the rock salt structure in which Br$^-$ ions, centered at lattice points, are in contact. Calculate the ionic radii of Br$^-$ and Li$^+$ in picometres if the unit-cell edge is 550 pm. Why is the accepted value for the ionic radius of Li$^+$ (60 pm) smaller than the value that you just computed?

10.84 CsCl crystallizes with a cubic unit cell of edge length 412.3 pm. The density of CsCl is 3.99 g cm^{-3}. Show that the unit cell cannot be face-centered or body-centered.

10.85 Metallic sodium crystallizes with a body-centered lattice. The element has a density of 0.97 g cm^{-3}. What is the length of the edge of the unit cell in Na expressed in nanometres?

10.86 Calcium fluoride crystallizes with a cubic lattice. The unit cell has an edge length of 546.26 pm. The density of CaF$_2$ is 3.180 g cm^{-3}. How many formula units of CaF$_2$ must there be per unit cell?

10.87 NaCl (which has the rock salt structure) has a density of 2.165 g cm^{-3}. The ionic radius of Cl$^-$ is 181 pm. What is the ionic radius of Na$^+$ in picometres?

Crystal Types

10.88 Identify the kinds of chemical units associated with each of the following kinds of crystals: molecular, metallic, covalent, ionic. Describe the properties of each of them. What kinds of attractive forces exist between the chemical units in these crystals?

10.89 Indicate which type of crystal (ionic, covalent, etc.) each of the following would form upon solidification: (a) O$_2$ (b) H$_2$S (c) Pt (d) KCl (e) Ge (f) Al$_2$(SO$_4$)$_3$ (g) Ne.

10.90 Indicate which type of crystal (ionic, covalent, etc.) each of the following would form upon solidification: (a) Br$_2$ (b) LiF (c) MgO (d) Cr (e) SiO$_2$ (f) PH$_3$ (g) NaOH.

10.91 SnCl$_4$ is a colorless liquid having a boiling point of 114°C and a melting point of −33°C. SnCl$_2$, on the other hand, is a white solid that melts at 246°C. What type of solid (ionic, covalent, etc.) is most likely formed when SnCl$_4$ solidifies?

10.92 Elemental boron is extremely hard (nearly as hard as diamond) and has a melting point of 2300°C. It is a poor conductor of electricity at room temperature. What kind of solid would you expect for boron based on these properties?

10.93 Paraffin (wax) is low-melting, soft, and is a nonconductor in both the solid and liquid states. What kind of solid is expected for paraffin?

10.94 OsO_4 has a melting point of 39.5°C and is a nonconductor of electricity in the molten state. It boils at 130°C. What kind of solid is expected for OsO_4?

10.95 $CaCO_3$ (calcite) is brittle. It decomposes, before it melts, at a temperature of about 900°C. What kind of solid is likely for calcite?

Liquid Crystals

10.96 What is a liquid crystal? How is it similar to a liquid? How is it similar to a crystal?

10.97 What are the three kinds of liquid crystals? How do they differ?

10.98 What is the origin of the color of cholesteric liquid crystals? Why does the color change with temperature?

10.99 Which type of liquid crystal is used in the LCD displays of calculators and watches?

Changes of State: Heating and Cooling Curves

10.100 What equilibrium exists at the melting point of a substance?

10.101 What is the difference between a substance's melting point and freezing point?

10.102 What does *fusion* mean?

10.103 What is the difference between the molar heat of fusion and the molar heat of crystallization?

10.104 Explain why for any given substance, ΔH_{fus} is smaller than ΔH_{vap}.

10.105 Aluminum has a melting point of 660°C and a boiling point of 1800°C. Its $\Delta H_{fus} = 10.7$ kJ mol^{-1} and $\Delta H_{vap} = 225$ kJ mol^{-1}. Sketch and label a heating curve for aluminum.

10.106 When a supercooled liquid begins to freeze, its temperature rises. Why doesn't the temperature ever rise above the melting point of the substance?

10.107 Why doesn't glass have a sharp melting point?

10.108 Is it possible to have only *liquid* water in a container at 0°C?

10.109 Explain, on a molecular level, why the temperature remains constant as heat is added to vaporize a liquid at its boiling point.

Phase Diagrams

10.110 At a pressure of 101 kPa a new compound was found to melt at 25°C and boil at 95°C. The triple point of the substance was determined to occur at a pressure of 20 kPa and a temperature of 20°C. Sketch the phase diagram for this substance. Label, on your drawing, the solid, liquid, and vapor regions as well as the solid-liquid, liquid-vapor, and solid-vapor equilibrium lines.

10.111 On the basis of the phase diagram in Question 10.110, describe the changes you would observe if, at a constant temperature of 22°C, the pressure on a sample of the compound is gradually increased from 2 to 200 kPa. What would be observed if the same process were to occur at a constant temperature of 10°C?

10.112 Sketch the heating curve you would expect to find when 1 mol of the compound described in Question 10.110 is heated at a constant rate under a constant pressure of 101 kPa. On your drawing, indicate the melting point and boiling point of the substance. Also, label the intervals that correspond to ΔH_{fus} and ΔH_{vap}.

10.113 What can we conclude about the relative densities of the liquid and solid phases of the compound in Question 10.110.

10.114 With the aid of the phase diagram in Figure 10.43, predict the physical state of carbon dioxide under each of the following conditions of temperature and pressure:

Temperature (°C)	Pressure (atm)
−80	1.0
−60	1.0
−56	10.0
−56	2.0
−65	5.0
−40	10.0

10.115 Iodine, I_2, sublimes without melting when heated in an open container at atmospheric pressure. What can be said about the triple point of I_2?

10.116 Use Le Châtelier's principle to predict how variations in pressure will affect the melting point of: (a) water, (b) carbon dioxide.

CHAPTER 11

PHYSICAL PROPERTIES OF SOLUTIONS AND COLLOIDS

This is a common winter sight in northern climates, as a snow plow scatters salt behind it to melt the remaining ice and snow. This is just one of many practical applications of the effects that solutes have on the physical properties of solutions.

In the previous two chapters we discussed the physical properties associated with the three states of matter. For the most part, however, these discussions applied to pure substances, and it is only rarely that we encounter pure materials, either in our daily activities or in the laboratory. Usually, the chemicals we meet are found in mixtures of various kinds.

The physical properties of mixtures are often quite different than those of their isolated, pure components. It is not uncommon to take advantage of this for a variety of practical purposes. For example, steel is a mixture of iron and other elements, such as carbon and other metals. By combining these ingredients in controlled proportions, the finished product can have properties such as hardness and corrosion resistance that differ substantially from those of iron itself.

In this chapter we study the physical properties of mixtures. We begin by examining the various kind of mixtures that can occur and what effect the sizes of the particles have on the properties of a mixture. One common and special kind of mixture is a solution, and we have already discussed the importance of solutions as a medium for carrying out chemical reactions. In this chapter we will spend much of our time studying another aspect of solutions—the effect that a solute has on a solution's physical properties. Many of these phenomena have very useful laboratory applications, such as the determination of molecular weights. They also have been applied to many practical problems, such as the refining of crude oil and the desalination of seawater.

11.1 KINDS OF MIXTURES: SUSPENSIONS, COLLOIDS, AND SOLUTIONS

In Chapter 1 you learned that any particular sample of matter can be classified as either a pure substance or a mixture. Pure substances include the elements and all the compounds that are formed from them. They are characterized by their constant composition. All samples of pure water, for example, are composed of the two elements hydrogen and oxygen in a ratio of 1 g of hydrogen for each 8 g of oxygen. What is special about mixtures is their variable composition. There seems to be no end to the variety and complexity of mixtures because they can be composed of any number of components in any proportions by mass.

One of the principal features that differentiates one kind of mixture from another is the sizes of the particles. Accordingly, mixtures are divided into three general categories: suspensions, colloids, and solutions.

Suspensions

In a suspension, relatively large particles of at least one component are distributed throughout another. Examples are fine sand suspended in water, or snow being blown about through the air, or a precipitate in a reaction mixture. In all these cases, the sizes of the suspended particles are large enough to be seen, either with the naked eye or a microscope. Furthermore, if not continually agitated, the particles of a suspension will settle under the influence of gravity, although the rate at which they settle depends on their size. Coarse sand will settle rapidly in water, but fine mud will settle at a considerably slower rate.

In the laboratory we often find it necessary to separate suspended precipitates from a reaction mixture. One method is filtration. The mixture containing the suspended material is passed through filter paper as described on page 230. Sometimes we also take advantage of the tendency of a suspension to settle under the influence of gravity, but we help the process along by using a centrifuge (Figure 11.1). In a centrifuge, a mixture is spun rapidly, and the centrifugal force thus produced behaves as a very powerful artificial gravity that drives the precipitate to the bottom of the container.

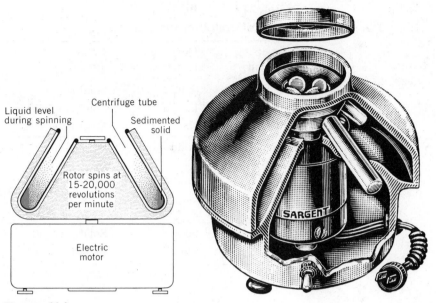

Figure 11.1

Cutaway views of typical laboratory centrifuges. A suspended solid is driven toward the bottom of the centrifuge tubes by the centrifugal force created when the rotor spins at high speed.

The physical properties of suspensions, such as the freezing point or vapor pressure of a suspension of a solid in a liquid, are little affected by the suspended particles. Thus, muddy water freezes at 0°C, just as pure water does. The suspended particles are too large, and their number is too small compared to the number of molecules of water in the mixture to have any measurable effect.

Solutions

Compared to suspensions, solutions stand at the opposite end of the particle size "spectrum." In a solution, all the particles—those of both the solute and the solvent—are of the dimensions of individual molecules or ions. These particles are distributed uniformly throughout each other to yield a single homogeneous phase.

Because of the intimate way that the particles of a solute distribute themselves among those of a solvent, the physical properties of a solution often differ quite a bit from those of the solvent alone. In fact, most of this chapter will focus on the effects that a solute has on the physical properties of a solution. Before we turn to that, however, there is a third kind of mixture to discuss, which has some unique properties.

Colloids

Colloid is derived from the Greek *kolla*, meaning glue. Old-time glues were colloidal dispersions in water.

Colloids, also called **colloidal dispersions** or **colloidal suspensions,** are mixtures that are intermediate between true solutions and suspensions. An example is homogenized milk, which consists of very tiny drops of butterfat dispersed in an aqueous phase that also contains casein (a protein) and a few other ingredients. In a colloid such as milk, the solutelike particles are larger than those in a solution, but smaller than the floating particles in a suspension. Because of the way the sizes of colloidal particles compare to the dimensions of the particles of

Table 11.1
Types of colloidal dispersions

Dispersing Medium	Dispersed Phase	Colloid Type	Examples
Solid	Solid	Solid sol	Pearls, opals
Solid	Liquid	Solid emulsion	Cheese, butter
Solid	Gas	Solid foam	Pumice, marshmallow
Liquid	Solid	Sol, gel	Starch in water, jello, paint
Liquid	Liquid	Emulsion	Milk, mayonnaise
Liquid	Gas	Foam	Whipped cream, shaving cream
Gas	Solid	Solid aerosols	Smoke, dust
Gas	Liquid	Liquid aerosols	Clouds, mist, fog

The particles in a colloid are generally too small to be removed by filtration through filter paper. The particles pass through the pores of the filter.

Milk freezes at very nearly 0°C, just as pure H_2O does.

the medium in which they are distributed, we don't use the terms solute and solvent. Instead we refer to the **dispersed phase** and the **dispersing medium.**

Typically, colloidal particles range in size from about 1 nm to 1000 nm. Usually, they consist of collections of many molecules or ions, although many of the large molecules in living systems, such as proteins, fall into this size range as well. Even though the particles are larger than those in a true solution, they are still small enough so that constant collisions with the surrounding medium keep them suspended for long periods. One of the general properties of colloids, therefore, is their stability toward separating under the influence of gravity. In fact, some colloids appear to be stable indefinitely. As with suspensions, the relative number of colloid particles in a mixture is small compared to the number of particles of the dispersing medium. Because of this, most of the physical properties of colloids differ very little from those of the dispersing medium.

Table 11.1 describes the various combinations of phases that can be combined to give colloidal dispersions. As you can see from the examples given, we

A spectacular scene is produced by the Tyndall effect as a haze of colloidal-sized water droplets scatters sunlight that pierces a cloud cover.

(a)   *(b)*   *(c)*

Figure 11.2

In this sequence of photographs, we see (a) *a beam of light passing through pure water,* (b) *through a clear solution, and* (c) *through a colloidal dispersion of starch in water.*

come across colloids on a daily basis. You might also notice that all the combinations are possible, except for a colloid formed by a gas dispersed in a gas. Since all gases mix uniformly at the molecular level, gases only form solutions with each other.

The particles in a colloid are too small to be visible with either the naked eye or an ordinary microscope. Nevertheless, they do have an influence on visible light; the particle sizes are just right to cause light to be scattered at large angles. When the concentration of colloid particles is large, this scattering can make the mixture opaque: Light is not allowed to pass through. Milk is an example. Incoming light is scattered by the particles and absorbed, so it never has a chance to exist. When less concentrated, the colloidal dispersion can appear cloudy, and if dilute enough, it can even appear transparent. A dilute colloidal dispersion of starch in water, for example, appears to be as transparent as a solution.

> The larger the particles, the larger the angle at which they scatter a beam of light.

The difference between a colloid (even a dilute one) and a true solution can be seen if we view, from the side, a focused beam of light as it passes through them both, as shown in Figure 11.2. The beam of light is invisible as it passes through the water and the solution, but its path is clearly seen as it goes through the colloidal dispersion. This is because the colloidal particles are able to scatter the light away from the direction of the beam, and it is this scattered light that makes the path of the light beam visible. In a solution, the particles of the solute are too small to cause this scattering, so the light beam can't be detected when viewed from the side. The scattering of light that is so typical of colloidal particles is a phenomenon called the **Tyndall effect.**

Stability of colloidal dispersions

In order for a colloid to be stable, its particles must be prevented from sticking to each other when they collide. If they do stick, the particles will grow in size and eventually separate from the mixture. For emulsions (liquids dispersed in liquids), stability is achieved by the action of an **emulsifying agent.** Two common examples of emulsions are milk and mayonnaise. Both consist of an oil dispersed in an aqueous phase. As you know, oil and water "don't mix," and if you shake a mixture of them, afterward they tend to separate quickly into two distinct phases. In mayonnaise, this is prevented by the addition of egg yolk, which forms a protective layer around the tiny drops of vegetable oil when the mixture is whipped. Casein in milk serves a similar purpose by preventing the coalescing of the tiny droplets of butterfat.

> Many consumer products, including salad dressings, contain emulsifiers to keep them homogeneous.

Colloids of solids in liquids (sols) are often stabilized by the adsorption of ions onto the surfaces of the colloid particles. (**Adsorption** is a process whereby something sticks to the surface of something else.) For example, a beautiful red

Stabilization of an $Fe_2O_3 \cdot xH_2O$ sol by adsorption of Fe^{3+} ions on the surfaces of the colloidal particles. Because the particles carry charges of the same sign, they repel each other and do not collide.

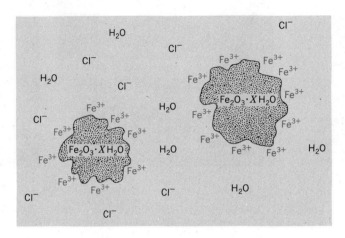

sol is formed if $FeCl_3$ is added slowly to boiling water. This occurs because of a chemical reaction in which the hydrated iron(III) ions lose water and hydrogen ions to form a *hydrated oxide*, $Fe_2O_3 \cdot xH_2O$, that contains a variable amount, x, of water of hydration. The equation for the change can be written

$$2Fe^{3+}(aq) + (x + 3)H_2O \longrightarrow Fe_2O_3 \cdot xH_2O(sol) + 6H^+(aq)$$

As the particles of the sol begin to grow, they adsorb Fe^{3+} ions on their surfaces, which makes them positively charged. Because each of the oxide particles acquires the *same* electrical charge, they repel each other. As a result, they no longer collide, so they stop growing. By the time this has happened, they have reached colloidal size.

In aerosols such as smoke, the colloidal particles also pick up electrical charges, but these tend to be from static electricity. Nevertheless, the effect is the same. Because their electrical charges are of the same sign, they repel each other and don't stick together when they collide.

Destabilizing colloids

Colloids can be made unstable by countering those things that stabilize them. When this happens, the particles can come together and grow, and this causes them to separate, or coagulate. Sometimes this coagulation happens by accident, and other times we deliberately destabilize a colloid. For example, the first step in making cheese is curdling the milk.

Sols such as the one formed by the hydrated iron(III) oxide can be coagulated by adding an electrolyte capable of neutralizing the charges on the surfaces of their particles. The addition of a solution containing phosphate ion, for example, will coagulate the sol just mentioned. The negatively charged PO_4^{3-} ions gather around the positively charged Fe^{3+} ions on the surface of the colloidal particles. This effectively neutralizes the charges on the particles and allows them to collide and grow, which ultimately leads to their precipitation. Colloidal clays carried by rivers are precipitated by this same kind of action when they meet the salt water of the sea. River deltas like that at the mouth of the Mississippi have been formed, in part, in this way.

Aerosols consisting of solids dispersed in the air are also separated by neutralizing their electrical charges. Smoke and dust can be removed from the air by passing the mixture over an electrically charged wire grid that carries a charge opposite in sign to that carried by the colloidal particles. The particles are attracted to the grid where their charges are neutralized, which allows them to precipitate.

The first step in making cheese is the addition of rennet to milk, which destabilizes the colloidal dispersion and causes the milk to curdle.

11.2 TYPES OF SOLUTIONS

The most common type of solution we encounter consists of a solute dissolved in a liquid, so most of our attention will be directed toward solutions of this kind. Liquid solutions can be prepared by dissolving a solid in a liquid (for example, NaCl in water), a liquid in a liquid (for example, ethylene glycol in water—antifreeze solution), or a gas in a liquid (for example, carbonated beverages, which contain dissolved carbon dioxide).

In addition to liquid solutions it is possible to have solutions of gases, such as the atmosphere that surrounds the earth, as well as solid solutions that are formed when a substance is dissolved in a solid. The properties of gaseous solutions were discussed in Section 9.6 under the heading "Dalton's Law of Partial Pressures," and nothing more need be said about them here. Solid solutions, of which many **alloys** (mixtures of metals) are examples, are of two types. **Substitutional solid solutions** exist in which atoms, molecules, or ions of one substance take the place of particles of another substance in a crystalline lattice, as shown in Figure 11.3*a*. Zinc sulfide and cadmium sulfide form such mixtures

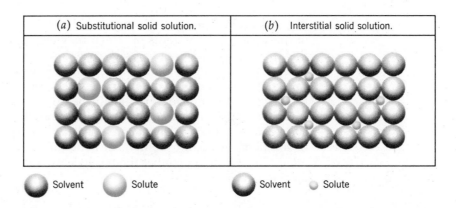

Figure 11.3
Two types of solid solutions:
(a) substitutional solid solution;
(b) interstitial solid solution.

Cutting tools tipped with tungsten carbide are often used by machinists like the one shown here operating a metal lathe.

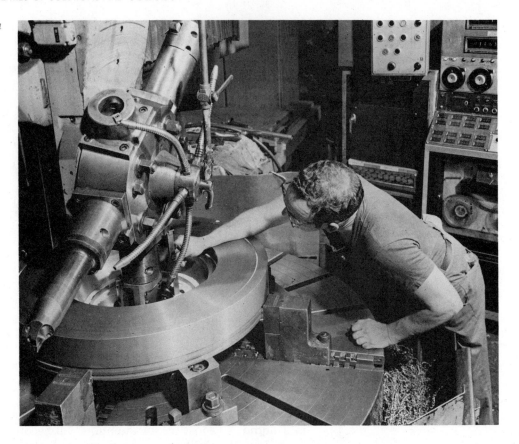

in which cadmium ions randomly replace zinc ions in the ZnS lattice. Another example is brass—a substitutional solid solution of copper and zinc.

Interstitial solid solutions constitute the other type and are formed by placing atoms of one kind into voids, or interstices, that exist between atoms in the host lattice. This is illustrated in Figure 11.3*b*. Tungsten carbide, WC, an extremely hard substance that has found many uses in cutting tools designed for machining steels, is an example of an interstitial solid solution in which the tungsten atoms are arranged in a face-centered cubic pattern with carbon atoms in *octahedral holes*—spaces within the crystal where the carbon atoms are surrounded by six tungsten atoms at the vertices of an octahedron.

11.3 CONCENTRATION UNITS

The physical properties of solutions are determined by the relative proportions of the various components of which they are composed. We have already seen that there are a variety of ways of expressing the concentration of one substance in another. For example, in Chapter 8 we discussed molar concentration and we saw that it is useful for the purpose of dealing with stoichiometry in solution. Molar concentration, defined again below for the sake of completeness, was created to satisfy a need. In a similar fashion, it has been found that certain other concentration units are most convenient for interpreting the physical properties of solutions. A point to remember about all concentration units is that they are *ratios*. The way to remember them is to learn the units associated with numerator and denominator.

Mole fraction

The mole fraction unit of concentration appeared in our discussion of Dalton's law of partial pressures in Section 9.10. It is defined as the number of moles of a particular component of the solution divided by the total number of moles of all the substances present in the mixture,

$$x_A = \frac{n_A}{n_A + n_B + n_C + \cdots}$$

Recall that the symbol x stands for mole fraction.

For example, a solution composed of 2.0 mol of water and 3.0 mol of ethanol (C_2H_5OH) has a mole fraction of water given by

$$x_{H_2O} = \frac{2.0 \text{ mol } H_2O}{2.0 \text{ mol } H_2O + 3.0 \text{ mol } C_2H_5OH} = \frac{2.0 \text{ mol}}{5.0 \text{ mol}}$$

$$x_{H_2O} = 0.40$$

Similarly, the mole fraction of ethanol in the mixture is

$$x_{C_2H_5OH} = \frac{3.0 \text{ mol}}{5.0 \text{ mol}} = 0.60$$

We see that the sum of all the mole fractions is equal to one, as of course it must be.

Another frequently used term is **mole percent** (abbreviated **mol %**), which is simply equal to 100 × mole fraction. Thus, the mixture above is composed of 40 mol % water and 60 mol % ethanol.

Mass fraction

The **mass fraction** specifies the fraction of the total mass of a solution that is contributed by a particular component. A mixture composed of 12.5 g of water and 37.5 g of ethanol has a mass fraction of water, w_{H_2O}, given by

$$w_{H_2O} = \frac{12.5 \text{ g } H_2O}{12.5 \text{ g } H_2O + 37.5 \text{ g } C_2H_5OH} = \frac{12.5 \text{ g}}{50.0 \text{ g}}$$

$$w_{H_2O} = 0.250$$

In a similar fashion, we find that the mass fraction of ethanol in the mixture is 0.750.

Parts per million equals mass fraction $\times 10^6$.

This mass percent can be indicated as follows:
25% w/w water
75% w/w ethanol

Mass percent, which is perhaps best thought of as *parts per hundred by mass*, is equal to the mass fraction multiplied by 100. It is more frequently used than mass fraction. The solution above is described as composed of 25.0% water and 75.0% ethanol, by mass. This means that in 100.0 g of a solution with this concentration there are 25.0 g of water and 75.0 g of ethanol. If there were 100.0 kg of this solution, it would contain 25.0 kg of water and 75.0 kg of ethanol.

Molar Concentration

Molar concentration is the ratio of the number of moles of solute to the *total volume* of the solution. It is expressed in moles per cubic decimetre (or litre). If 0.500 mol of HCl is dissolved in 250 cm^3 of solution, the molar concentration is

$$\frac{0.500 \text{ mol solute}}{0.250 \text{ dm}^3 \text{ solution}} = 2.00 \text{ mol dm}^{-3} = 2.00 \ M$$

To prepare a 1.000 M solution of sucrose ($C_{12}H_{22}O_{11}$) we would place 1.000 mol of sucrose (342.3 g) into a volumetric flask calibrated to contain 1.0000 dm^3 when filled to the line etched around its neck (Figure 11.4). Some water is added, the mixture is stirred to dissolve the solute, and then more water is carefully added to bring the total solution volume up to the etched mark. At this point we have 1.000 mol of sugar dissolved in a total volume of 1.0000 dm^3. The concentration is 1.00 mol dm^{-3}, or 1.000 molar (1.000 M).

Conversions between concentration units

The concentration units of mole fraction and mass fraction (or mass percent) can be easily converted from one to another. All that we require is the molar masses of the solvent and solute. In performing these conversions, we begin by imagining a certain quantity of the solution and then, from the way the given concentration unit is defined, we "divide" the solution into its components: solute and solvent. Once the quantities of solute and solvent are known, we can use them to calculate the concentration in the new units desired.

EXAMPLE 11.1 CONVERTING FROM MASS PERCENT TO MOLE FRACTION

PROBLEM An aqueous solution of Epsom salts is composed of 20.0% magnesium sulfate by mass. What is the mole fraction of $MgSO_4$ and H_2O?

SOLUTION When given either mass percent or mass fraction, we begin by imagining that we have a certain total *mass* of solution. This allows us to use the mass percent or mass fraction to divide the solution into the masses of solute and solvent. To make the arithmetic easy, let's imagine that we have 100.0 g of the magnesium sulfate solution. The mass percent value—20.0% $MgSO_4$—tells us immediately that this 100.0 g of solution must contain 20.0 g $MgSO_4$ and 80.0 g H_2O. Now that we have masses of $MgSO_4$ and H_2O, we can solve the problem.

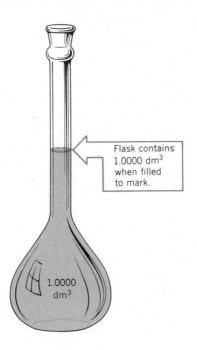

Flask contains 1.0000 dm^3 when filled to mark.

1.0000 dm^3

Figure 11.4

A volumetric flask.

We know that 100 g of the solution contains 20.0 g of $MgSO_4$ and 80.0 g of water. Since the mole fraction of $MgSO_4$ represents a ratio of moles of solute to moles of solute plus solvent, we first convert 20.0 g of $MgSO_4$ to moles of $MgSO_4$.

$$20.0 \text{ g } MgSO_4 \times \left(\frac{1 \text{ mol } MgSO_4}{120.4 \text{ g } MgSO_4} \right) = 0.166 \text{ mol } MgSO_4$$

The number of moles of water in the solution is

$$80.0 \text{ g } H_2O \times \left(\frac{1 \text{ mol } H_2O}{18.0 \text{ g } H_2O} \right) = 4.44 \text{ mol } H_2O$$

The mole fraction of $MgSO_4$ is found by dividing the number of moles of $MgSO_4$ by the total number of moles of both the solute and solvent.

$$x_{MgSO_4} = \frac{0.166 \text{ mol}}{4.44 \text{ mol} + 0.166 \text{ mol}} = \frac{0.166 \text{ mol}}{4.61 \text{ mol}}$$

$$x_{MgSO_4} = 0.0360$$

It follows that the mole fraction of water must be

$$x_{H_2O} = 1.0000 - 0.0360 = 0.9640$$

EXAMPLE 11.2 CONVERTING FROM MOLE FRACTION TO MASS PERCENT

PROBLEM Benzene (C_6H_6) and chloroform ($CHCl_3$) are solvents that have proven to be highly toxic. They are mutually soluble in each other. In a certain solution of benzene and chloroform, the mole fraction of C_6H_6 is 0.450. What is the mass percent of C_6H_6 in this mixture?

SOLUTION When we are given the mole fraction, we begin by imagining that we have enough of the mixture so that there is exactly 1 mol of particles. In this 1 mol of mixture, therefore, we must have 0.450 mol of C_6H_6.

$$1.000 \text{ mol mixture} \times \left(\frac{0.450 \text{ mol } C_6H_6}{1 \text{ mol mixture}} \right) \sim 0.450 \text{ mol } C_6H_6$$

The amount of $CHCl_3$ must be the difference between the total number of moles and the number of moles that are C_6H_6.

$$\text{moles } CHCl_3 = \text{total moles} - \text{moles } C_6H_6$$

$$= 1.000 \text{ mol} - 0.450 \text{ mol}$$

$$= 0.550 \text{ mol}$$

We have now divided the solution into its components. To compute mass percent C_6H_6, we need the mass of C_6H_6 and the total mass of the mixture.

$$0.450 \text{ mol } C_6H_6 \times \left(\frac{78.1 \text{ g } C_6H_6}{1 \text{ mol } C_6H_6} \right) = 35.1 \text{ g } C_6H_6$$

$$0.550 \text{ mol } CHCl_3 \times \left(\frac{119.4 \text{ g } CHCl_3}{1 \text{ mol } CHCl_3} \right) = 65.7 \text{ g } CHCl_3$$

The total mass of the mixture is

$$35.1 \text{ g} + 65.7 \text{ g} = 100.8 \text{ g}$$

The percent of C_6H_6 can be found as

$$\%\,C_6H_6 = \frac{\text{mass } C_6H_6}{\text{mass of mixture}} \times 100$$

$$\%\,C_6H_6 = \frac{35.1 \text{ g}}{100.8 \text{ g}} \times 100 = 34.8\%$$

To perform conversions between mole fraction and mass percent, molar masses are the only data required. In order to convert these concentration units to molar concentration, we need to know the density of the solution.

EXAMPLE 11.3   CONVERTING FROM MOLAR CONCENTRATION TO MASS PERCENT

PROBLEM   The painful sting of ant bites is caused by formic acid injected under the skin by the ant. Calculate the mass percent of formic acid ($HCHO_2$) in a solution that is 1.099 M $HCHO_2$. The density of the solution is 1.0115 g cm^{-3}.

SOLUTION   To calculate mass percent, we need the mass of solute and the total mass of the solution. Let's begin by examining the meaning of molar concentration. For this solution

$$1.099\ M\ HCHO_2 \quad \text{means} \quad \frac{1.099 \text{ mol } HCHO_2}{1000 \text{ cm}^3 \text{ soln}}$$

From the molar mass of $HCHO_2$, we can calculate the number of grams of solute.

$$1.099 \text{ mol } HCHO_2 \times \left(\frac{46.03 \text{ g } HCHO_2}{1 \text{ mol } HCHO_2}\right) = 50.59 \text{ g } HCHO_2$$

From the density, which relates the mass of *solution* to its volume, we can calculate the total mass of the 1000.0 cm^3 of solution.

$$1000.0 \text{ cm}^3 \text{ soln} \times \left(\frac{1.0115 \text{ g soln}}{1 \text{ cm}^3 \text{ soln}}\right) = 1011.5 \text{ g soln}$$

Now we can calculate the percent solute in the solution.

$$\%\,HCHO_2 = \frac{50.59 \text{ g}}{1011.5 \text{ g}} \times 100$$

$$= 5.001\%$$

You've probably come to realize that in order to solve problems like those in the last three examples it is absolutely essential that you know how the various concentration units are defined. If you've acquired this knowledge and begin the problem correctly, these conversions are not difficult.

11.4   THE SOLUTION PROCESS IN LIQUID SOLUTIONS

Experience has taught us that substances differ widely in their solubilities in various solvents. For instance, we all know that oil and water don't mix, and that to remove an oil stain from clothing a solvent such as naphtha must be used. It is also generally known that sodium chloride (table salt) will dissolve in water but not in gasoline. What accounts for these differences in behavior? The answer lies in a close examination of the solution process.

Solutions of liquids in liquids

When one substance dissolves in another, particles of the solute—either molecules or ions, depending on the nature of the solute—must be distributed throughout the solvent and, in a sense, the solute particles in the solution occupy positions that are normally taken by solvent molecules. In a liquid, molecules are packed together very closely and interact strongly with their neighbors. The ease with which a solute particle may replace a solvent molecule depends on the relative forces of attraction of solvent molecules for each other, solute particles for each other, and the strength of the solute-solvent interactions. For example, in a solution formed between benzene (C_6H_6) and carbon tetrachloride (CCl_4), both species are nonpolar and experience only relatively weak London forces. As it happens, the strengths of the attractive forces between pairs of benzene molecules and pairs of carbon tetrachloride molecules are of nearly the same magnitude as between molecules of benzene and carbon tetrachloride. For this reason, molecules of benzene can replace CCl_4 molecules in solution with ease. As a result, these two substances are completely soluble in all proportions. We say they are **miscible.**

This explains why oil and grease are soluble in CCl_4 and other dry-cleaning agents.

Now let's look at what happens when we attempt to dissolve water in CCl_4. Water is a very polar substance that interacts strongly with other water molecules through the formation of hydrogen bonds. By comparison, the strength of the attractive forces between water and the nonpolar CCl_4 molecules is much weaker. If we attempt to disperse water molecules throughout CCl_4, we find that when the water molecules encounter one another they tend to stick together simply because they attract each other much more strongly than they do molecules of the solvent. This clumping together continues until the two substances have formed two distinct phases: one consisting of water with a very small amount of CCl_4 in it, and the other, CCl_4 containing a small quantity of H_2O.

Water and ethanol, C_2H_5OH, interact with each other by hydrogen bonding, too.

When two polar substances, such as ethanol and water, are mixed, we again have a situation in which the solute-solute forces of interaction are of compara-

Water and carbon tetrachloride form a two-phase mixture. The layer on the bottom is the more dense CCl_4. An orange dye has been added to the water and a little iodine dissolved in the CCl_4 to make the two layers easier to see.

ble strength to the solvent-solvent attractive forces, and in which the solute and solvent molecules interact strongly with each other. Once again a condition exists where the solute particles can readily replace those of solvent and, hence, water and ethanol are miscible.

Between the two extremes of complete miscibility (benzene-carbon tetra-chloride; water-ethanol) and virtually total immiscibility (CCl_4-H_2O), we have many substances that are only partially soluble in one another. For instance, in Table 11.2 the solubilities of a series of different alcohols are listed in moles of solute (alcohol) per 100 g of water. As we proceed to higher molecular weights, the polar OH group represents an ever smaller portion of the molecules; as a result, alcohol molecules become less like water as they become larger. Parallel-ing the increasing size of the nonpolar hydrocarbon portions of these alcohols, we observe a corresponding decrease in their solubilities in water.

Solutions of solids in liquids

When we analyze what occurs when a solid dissolves in a liquid, somewhat different factors must be considered. In a solid, the molecules or ions are ar-ranged in a very regular pattern and the attractive forces are at a maximum. In order for the solute particles to enter into a solution, the solute-solvent forces of attraction must be sufficient to overcome the attractive forces that hold the solid together.

In molecular crystals, the attractive forces are relatively weak, being of di-pole-dipole or London type, and are rather easily overcome. Substances whose crystals are held together by London forces (the weakest of all) are therefore able to dissolve to appreciable extents in nonpolar solvents. However, they are not soluble to any great degree in polar solvents for the same reasons that nonpolar liquids are insoluble in polar solvents—that is, the polar solvent molecules at-tract each other too strongly to be pushed aside and replaced by molecules to which they are only weakly attracted. For example, solid iodine, which is com-posed of nonpolar I_2 molecules, is fairly soluble in CCl_4 (giving rise to a beautiful violet solution) but only very slightly soluble in water (where it forms a pale yellow-brown solution).

When the solid is composed of polar molecules or ions, we find that they are insoluble in nonpolar solvents. The weak solute-solvent attractions, compared to the strong attractions within the crystal, are not sufficient to tear apart the lattice. This is why sugar, which consists of molecules tightly bound to each other by hydrogen bonding, is insoluble in solvents like gasoline. However,

Table 11.2
Solubilities of some alcohols in water

Substance	Formula	Solubility (mol solute/100 g H_2O)
Methanol	CH_3OH	∞
Ethanol	C_2H_5OH	∞ } Completely miscible
Propanol	C_3H_7OH	∞
Butanol	C_4H_9OH	0.12
Pentanol	$C_5H_{11}OH$	0.031
Hexanol	$C_6H_{13}OH$	0.0059
Heptanol	$C_7H_{15}OH$	0.0015

Figure 11.5

Hydration of ions in solution.

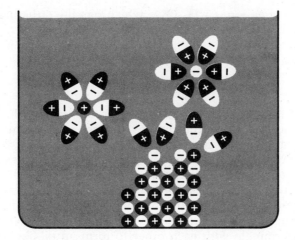

As you know, sodium chloride is also insoluble in nonpolar solvents such as gasoline.

sugar is soluble in water because this solvent attracts sugar molecules in much the same way that sugar molecules attract each other.

In ionic solids the attractive forces are especially strong, so it takes an extremely polar solvent such as water to cause them to dissolve. Even moderately polar solvents such as methyl alcohol or ethyl alcohol are not up to the job, and salts such as NaCl are insoluble in them, but soluble in water.

When an ionic substance dissolves in water, the ions adjacent to one another in the solid become separated and are surrounded by water molecules. In Chapter 7 we represented this dissociation of the solute by an equation such as

$$NaCl(s) \longrightarrow Na^+(aq) + Cl^-(aq)$$

The hydration of ions was mentioned in Chapter 7 when we discussed electrolytes.

This orientation of water molecules may actually extend through several layers.

In Figure 11.5 we take a closer look at what occurs during this process. In the immediate vicinity of a positive ion, the surrounding water molecules are oriented so that the negative ends of their dipoles point in the direction of the positive charge. The water molecules surrounding a negative ion have their positive ends directed at the ion. An ion enclosed within this ''cage'' of water molecules is said to be **hydrated** and, in general, when a solute particle becomes surrounded by molecules of a solvent we say that it is **solvated;** hydration is a special case of the more general phenomenon of solvation.

The layer of oriented water molecules that surrounds an ion helps to neutralize the ion's charge and serves to keep ions of opposite charge from attracting each other strongly over large distances within the solution. In a sense, the solvent insulates the ions from each other. Nonpolar solvents do not dissolve ionic compounds because they can neither tear an ionic lattice apart nor do they offer any shielding for the ions. In a nonpolar solvent, ions quickly congregate and separate from the solution as the solid.

In summary, substances that exhibit *similar* intermolecular attractive forces tend to be soluble in one another. This observation is often stated very simply as ''like dissolves like.'' Nonpolar substances are soluble in nonpolar solvents, while polar or ionic compounds dissolve in polar solvents.

An octahedral hydrated aluminum ion, $Al(H_2O)_6^{3+}$.

Soaps and detergents

One of the more practical applications of the solubility relationships described above is the use of soaps and detergents to remove dirt and oil from fabrics. Long before anyone knew any modern chemistry, people had learned to make soap by reacting animal fats with aqueous basic solutions. The reaction of a base with fats liberates anions of substances called fatty acids. An example is the

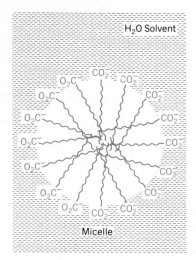

Figure 11.6

The formation of micelles in soap solutions. The hydrophobic tails of the anions of fatty acids intermingle while the hydrophilic heads are turned outward to face the solvent water.

strearate ion, $C_{18}H_{35}O_2^-$, which has the structure

$$CH_3-CH_2-CH_2-CH_2-CH_2-CH_2-CH_2-CH_2-CH_2-CH_2-CH_2-CH_2-CH_2-CH_2-CH_2-CH_2-CH_2-\overset{\overset{\displaystyle O}{\|}}{C}-O^-$$

hydrocarbon tail   **anionic head**

Notice that one end of this particle consists of a long hydrocarbon chain and the other consists of the charge-carrying $-CO_2^-$ unit. The hydrocarbon end of the stearate ion resembles other hydrocarbons that we have seen, such as the molecules in gasoline or motor oil. As you know, these do not tend to be soluble in water, but they do dissolve in each other. We describe this behavior by saying that the molecules are **hydrophobic**, which strictly translated means "water fearing." Hydrophobic substances avoid water, and this is exactly what happens in a solution that contains soap. The hydrophobic tails of many fatty acid anions intermingle with each other to form an oil-like globule, with their electrically charged **hydrophilic** ("water loving") heads pointing outward toward the solvent, as illustrated in Figure 11.6. These collections of fatty acid anions, which are colloidal in size, are called **micelles.**

The ability of soaps to remove oil and grease from fabrics is also traced to the principle "like dissolves like." When soap contacts an oil stain on a cloth fiber, the hydrophobic tails of the anions dissolve in the oil. The oil is gradually separated from the fiber and encapsulated in micelles that trap bits of oil inside. This emulsifies the oil and holds it in suspension so it can be carried away with the wash water (Figure 11.7).

Synthetic detergents are very similar to the salts of fatty acids found in soap, except they are manufactured chemically from materials other than animal fats. Examples include salts called sodium alkylbenzenesulfonates, which have the general structure

$$CH_3(CH_2)_x-\!\!\!\bigotimes\!\!\!-SO_3^- \; Na^+$$

Their advantage over natural soaps is that they work in hard water. At the beginning of Chapter 8 you learned that hard water contains ions such as Ca^{2+} that form a precipitate with the anions found in soap. This removes the anions and thereby reduces the effectiveness of the soap solution. The anions of syn-

Figure 11.7

The action of soap on grease that is attached to a surface such as a cloth fiber. (a) The hydrocarbon tails of the soap anions dissolve in the grease. (b) The grease spot gradually breaks up and becomes pincushioned by the soap anions. (c) Small bits of grease are held in colloidal suspension by the soap. The anionic heads keep the grease from coalescing because the particles carry the same electrical charge. (From J. R. Holum, Fundamentals of General, Organic, and Biological Chemistry, 2nd ed., p. 459. Copyright © 1982, John Wiley & Sons, used with permission.)

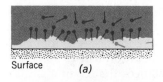

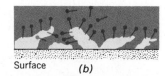

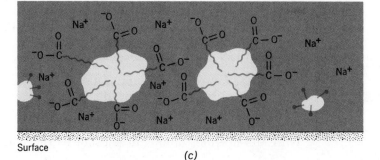

thetic detergents don't precipitate in the presence of Ca^{2+}, however, so their cleansing action is not affected by hard water.

11.5 HEATS OF SOLUTION

The solution process nearly always occurs with either an absorption or release of energy. For example, when potassium iodide is dissolved in water the mixture becomes cool, indicating that for potassium iodide the solution process is endothermic. On the other hand, when lithium chloride is added to water the mixture becomes warm, signifying that the solution process in this case evolves heat and is therefore exothermic. *The quantity of energy that is absorbed or released when a substance enters solution is called the heat of solution and is given the symbol, ΔH_{soln}.* As in our definitions of the heat of vaporization and heat of fusion, ΔH_{soln} represents a difference. It is the difference between the energy possessed by the solution after it has been formed and the energy that the components of the solution had before they were mixed; that is,

$$\Delta H_{soln} = H_{soln} - H_{components}$$

We will learn why in Chapter 12.

Neither H_{soln}, nor $H_{components}$ can actually be measured, but their difference, ΔH_{soln}, can be. When energy is evolved during the solution process, the resulting solution possesses less energy than the components from which it was prepared, so the difference represented by ΔH_{soln} is a negative number. Conversely, an endothermic solution process would have a positive ΔH_{soln}. Heats of solution for some typical ionic solids in water are shown in Table 11.3.

The magnitude of the heat of solution can provide us with information about the relative forces of attraction between the various particles that make up a solution. To analyze the factors that contribute to the absorption or evolution of energy, let us imagine that we could create the solution in a stepwise fashion.

Solutions of liquids in liquids

When one liquid dissolves in another, we can imagine that the molecules of the solvent are caused to move apart so as to allow room for the solute molecules.

Table 11.3
Heats of solution

Substance	Heat of Solution[a] (kJ mol^{-1} solute)
KCl	17.2
KBr	19.9
KI	20.3
LiCl	−37.0
LiI	−59.0
LiNO$_3$	−1.3
AlCl$_3$	−321
Al$_2$(SO$_4$)$_3$·6H$_2$O	−230
NH$_4$Cl	−16
NH$_4$NO$_3$	26

[a] At "infinite" dilution. The heat of solution depends, to an extent, on the concentration of the solution produced. A negative sign signifies an exothermic process.

Although this three-step process is purely hypothetical, it allows us to analyze the factors that contribute to ΔH_{soln}.

$\Delta H_{soln} = 0$ for an ideal solution.

Similarly, for the solute to enter solution, its molecules must also become separated so that they can take their places in the mixture. Since there are attractive forces between molecules in both the solvent and solute, the process of separating their molecules requires an input energy—that is, work must be done on both the solute and solvent to separate their molecules from one another. Finally, as the solute and solvent, in their expanded states, are brought together, energy is released because of the attractions that exist between the solute and solvent molecules.[1] This sequence of steps we have just described is illustrated in Figure 11.8.

In some substances, such as benzene and carbon tetrachloride, the intermolecular attractive forces are of very nearly the same magnitude; therefore, these compounds form solutions with virtually no evolution or absorption of heat. Solutions in which the solute-solute, solute-solvent, and solvent-solvent interactions are all the same are called **ideal solutions.** The energy changes that occur along the series of steps that we have devised to arrive at the solution are shown graphically in Figure 11.9. We see that for an ideal solution the energy released in the final step is the same as that absorbed in the first two; thus the net change is zero.

When the molecules of the solute and solvent attract each other more strongly than they do molecules of their own kind, more energy can be released as the expanded solute and solvent are brought together than was required to separate them in the first place (Figure 11.10). Under these circumstances the overall solution process can result in the evolution of heat and be exothermic. This occurs when acetone, an important solvent (nail polish remover), and water are mixed.

When the solute-solvent attractive forces are weaker than those between pure solute or pure solvent, the formation of a solution requires a net input of energy. This is because more energy is absorbed in separating the molecules in the first two steps than is recovered when the solute and solvent are brought

[1] Recall from Chapter 1 that when particles that attract one another are pulled apart, work must be supplied to increase the potential energy. When these particles are brought together again, the same quantity of energy is released.

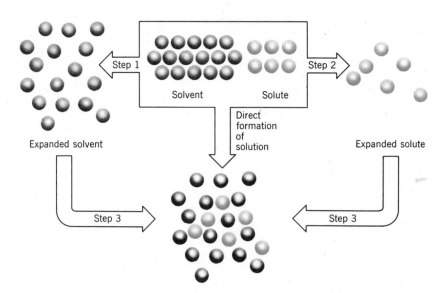

Figure 11.8

Formation of a liquid-liquid solution.

Figure 11.9

Energy changes to produce an ideal solution. The actual solution process follows the direct path, although the net result is the same both ways.

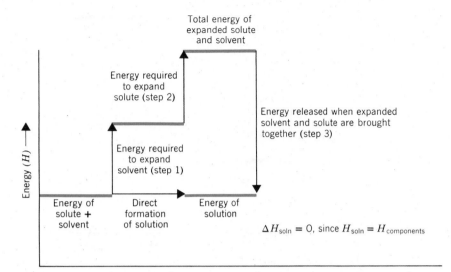

together in the third (Figure 11.11). In this case the solution becomes cool as it is formed, signifying that an endothermic change has occurred. This happens, for example, when ethanol and a hydrocarbon solvent such as hexane are mixed. The nonpolar hexane molecules come between the hydrogen bonded ethanol molecules, effectively destroying the hydrogen bonds. This absorbs energy and the solution becomes cool.

Solutions of solids in liquids

We can approach the energetics of the solution of a solid in a liquid in much the same way as we did for a liquid in a liquid. We recognize that the act of dissolving a compound such as KI in water involves removing the K^+ and I^- ions from the solid and placing them in an environment where they are surrounded by water molecules. Suppose, now, that we could separate these two steps. The first step involves separating the ions so that they are infinitely far apart, and the second amounts to taking these now isolated ions and placing them into water where they become surrounded by molecules of the solvent. To accomplish the first step, it is necessary to break apart the lattice by pulling the ions away from one another, so the process

$$KI(s) \longrightarrow K^+(g) + I^-(g)$$

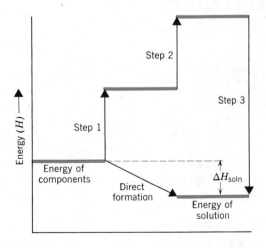

Figure 11.10

An exothermic solution process.

Figure 11.11

An endothermic solution process.

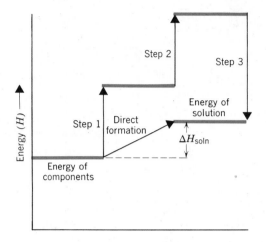

This instant cold pack contains capsules of ammonium nitrate, NH$_4$NO$_3$, and water. When they are broken and their contents mixed, the salt dissolves and produces an amazing cooling effect.

requires an input of energy; that is, it is endothermic. Recall that the quantity of energy required to tear apart the solid to obtain isolated particles is called the lattice energy. Ionic substances, because of the very strong electrostatic attractions between oppositely charged ions, have quite large lattice energies. Molecular solids, however, have small lattice energies because of their relatively weak intermolecular attractive forces.

In the second step of our solution process we imagine that the K$^+$ and I$^-$ ions are brought into water where they become hydrated. As described in the last section, an ion in water is surrounded by a "cage" of oriented water molecules; therefore, a hydrated ion experiences a net attraction for the solvent dipoles. Because of this attraction, a quantity of energy, known as the **hydration energy,** is released when an ion is placed into water.[2] We can indicate the change that occurs in this hydration step as

$$K^+(g) + I^-(g) + xH_2O \longrightarrow K^+(aq) + I^-(aq)$$

which tells us that some number, x, of water molecules becomes bound to the potassium and iodide ions as they become hydrated in aqueous solution.

The overall change that takes place when KI dissolves in water can be represented as the sum of the two steps we have just considered.

$$KI(s) \longrightarrow K^+(g) + I^-(g)$$
$$\underline{K^+(g) + I^-(g) + xH_2O \longrightarrow K^+(aq) + I^-(aq)}$$
$$KI(s) + K^+(g) + I^-(g) + xH_2O \longrightarrow K^+(g) + I^-(g) + K^+(aq) + I^-(aq)$$

or, after canceling species that appear on both sides,

$$KI(s) + xH_2O \longrightarrow K^+(aq) + I^-(aq)$$

The heat of solution, ΔH_{soln}, corresponds to the net energy change that occurs and is equal to the difference between the quantity of energy supplied during the first step and the quantity of energy evolved during the second (Figure 11.12). If the lattice energy is greater than the hydration energy, a net input of energy is required when the substance dissolves and the process is endothermic. This occurs with KI. Conversely, when the hydration energy exceeds the lattice energy, a situation that occurs with LiCl, more energy is released when the ions become hydrated than is required to break up the ionic lattice, and an exothermic change is observed when the solid dissolves.

Remember that the solution is actually formed by the direct path in Figure 11.12. The alternative path is just hypothetical, but it helps us understand the energy change that occurs along the direct path.

---

[2] In general, for a solvent other than water, this energy is called the **solvation energy.**

Figure 11.12

Energy changes that occur when KI dissolves in H₂O.

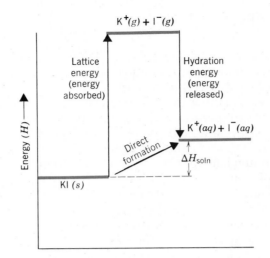

The relationship between the lattice energy and hydration energy can be seen in Table 11.4. Note that the agreement between the calculated and experimentally determined heats of solution is far from perfect. This is a result of inaccuracies in the theoretical models used in computing these energy quantities. Nevertheless, we still can see that when theory predicts a large exothermic change the experimental quantity corresponds to a large release of energy. Similarly, when the theory predicts a trend in ΔH_{soln}, as with KCl, KBr, and KI, the measured values follow the same trend. The agreement is therefore sufficient to support the arguments made above.

The preceding explanations allow us to understand the energy changes that take place during the solution process. Unfortunately, however, it is very difficult to predict ahead of time, in any particular case, whether the formation of a solution will be exothermic or endothermic. This is because the same factors that lead to a high hydration energy also tend to produce a high lattice energy. The extent to which an ion is attracted to a solvent dipole increases as the ion becomes smaller, because a smaller ion can get closer to a solvent molecule than a larger one. The interaction of the solvent with an ion also increases as the charge on the ion becomes greater. However, the degree to which ions attract each other in the solid also grows with decreasing size and increasing charge, so that as the hydration energy becomes larger so does the lattice energy. Thus we have two factors that are affected in the same way by changes in size and charge, and it is virtually impossible to predict in advance which effect will predominate.

Table 11.4
Lattice energies, hydration energies, and heats of solution for some alkali halides

Compound	Lattice Energy (kJ mol^{-1})	Hydration Energy (kJ mol^{-1})	Calculated ΔH_{soln} (kJ mol^{-1})	Measured ΔH_{soln} (kJ mol^{-1})
LiCl	+833	−883	−50	−37.0
LiBr	+787	−854	−67	−49.0
NaCl	+766	−770	−4	+3.9
NaBr	+728	−741	−13	−0.602
KCl	+690	−686	+4	+17.2
KBr	+665	−657	+8	+19.9
KI	+632	−619	+13	+20.3

11.6 SOLUBILITY AND TEMPERATURE

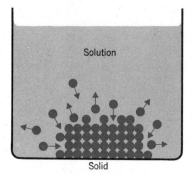

Figure 11.13

Solubility as an equilibrium state. The solute dissolves at the same rate as it crystallizes.

In Chapter 7 we defined the solubility of a substance as the amount of solute required to produce a saturated solution in some particular quantity of solvent. At a given temperature a saturated solution in contact with undissolved solute represents another example of a state of dynamic equilibrium. As illustrated in Figure 11.13, particles of the solute are constantly passing into the solution and, at the same time, the solute particles already in the solution are continually colliding with and sticking to the undissolved solute. Although we show this equilibrium here for a solid dissolved in a liquid, the same concept applies to any type of solution (except gases—all gases are completely miscible).

Since a saturated solution in contact with excess solute constitutes a state of dynamic equilibrium, when the system is disturbed the effect of the disturbance can be predicted using Le Châtelier's principle. A change in temperature corresponds to such a disturbance, and in Chapter 10 we saw that a rise in temperature favors a shift in the position of equilibrium in a direction that will absorb heat. Therefore, if the dissolving of additional solute into an already saturated solution absorbs energy, the solubility of that substance increases as the temperature is raised. Conversely, if placing additional solute into the saturated solution is an exothermic process, the solute will become less soluble as the temperature is increased.

In general, the solubility of most solid and liquid substances in a liquid solvent increases with increasing temperature. For gases in liquids, the opposite behavior is observed. The solution process for a gas in a liquid is nearly always exothermic, because the solute particles are already separated from each other and the dominant heat effect arises from the solvation that occurs when the gas dissolves. Le Châtelier's principle predicts that a rise in temperature will favor an endothermic change. For a gas, this takes place when it leaves a solution. Therefore, we expect a gas to become less soluble as the temperature of the liquid in which it is dissolved becomes higher. For example, in bringing water to a boil, tiny bubbles appear on the surface of the pot before boiling begins. These bubbles contain air that is driven out of solution as the water becomes hot. We also use this general solubility behavior of gases when we store opened bottles of carbonated beverages in the refrigerator. These liquids retain their dissolved CO_2 longer when they are kept cold because CO_2 is more soluble in them at low temperatures. Analysis of the quantity of dissolved gases in streams, lakes, and rivers reveals still another example of this phenomenon. The concentration of dissolved oxygen, which is imperative to marine life, decreases in the summer months, compared to when similar analyses are performed during the winter months—all other conditions being equal, of course.

Certain species of fish die if the water becomes too warm because the amount of dissolved oxygen becomes too small.

Rock candy is formed by the crystallization of sugar from a saturated solution that is slowly cooled.

Fractional crystallization

Figure 11.14 illustrates graphically the way in which solubility changes with temperature for a variety of typical solids in water. From these solubility curves it is evident that the variation of solubility with temperature is quite different for different substances. For some substances, such as KNO_3, the solubility changes very rapidly with temperature, while for others the change is more gradual. These differences in behavior provide the basis for a useful laboratory technique called **fractional crystallization,** which is often used to separate impurities from the products of a chemical reaction.

In this technique the impure product is first dissolved in a small amount of hot solvent—generally one in which the desired product is less soluble than the impurities. As the hot solution is allowed to cool, the pure product separates from the mixture, leaving the impurities behind. Finally, the crystals of the

Figure 11.14

Solubility curves for typical solids in H₂O.

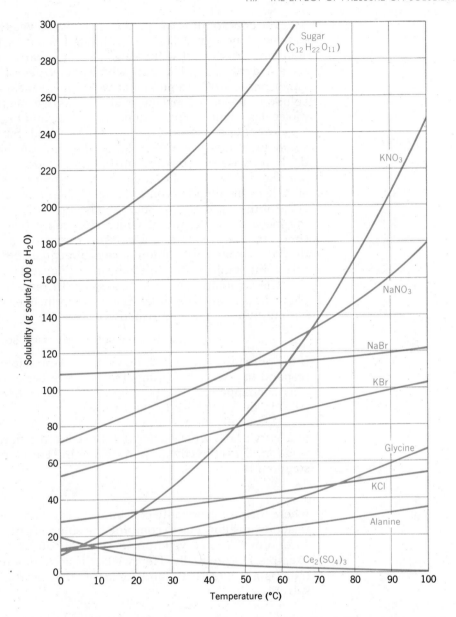

product are filtered from the cool solution and dried. The amount of pure product that can be recovered in this fashion depends on the concentrations of the impurities and their solubilities relative to that of the desired material.

11.7 THE EFFECT OF PRESSURE ON SOLUBILITY

In general, pressure has very little effect on the solubility of liquids or solids in liquid solvents. The solubility of gases, however, always increases with increasing pressure. Carbonated beverages, for example, are bottled under pressure to ensure a high concentration of CO_2, and once the bottle has been opened, the beverage quickly loses its carbonation unless it is recapped. The same phenomenon is responsible for decompression sickness, also known as the bends. When a deep-sea diver or a tunnel worker comes to the surface too quickly, nitrogen and oxygen that has dissolved in the blood at high pressure is suddenly released in the form of bubbles in the blood vessels. This is very painful and in extreme cases can even cause death.

Astronauts have to worry about their space suits tearing for the same reason.

The effect of pressure on the solubility of a gas is not difficult to understand. Let's imagine that a liquid is saturated with a gaseous solute, and that this solution is in contact with the gas at some particular pressure. Once again we have a dynamic equilibrium where molecules of the solute are entering the vapor phase at the same rate at which molecules from the gas are entering the solution, as shown in Figure 11.15a. As we might expect, the rate at which molecules go into solution depends on the number of collisions per second that they experience with the surface of the liquid. Similarly, the rate at which the solute molecules leave the solution depends on their concentration. If we suddenly increase the pressure of the gas, we pack the molecules closer together and the number of collisions per second that the gas molecules make with the surface of the liquid becomes larger. When this occurs, the rate at which molecules of the gas enter the solution also grows larger, but without a corresponding increase in the rate at which they leave (Figure 11.15b). As a result, the concentration of solute molecules in the solution rises until the rate at which they are leaving the solution once again equals the rate at which they enter. At this point equilibrium is reestablished (Figure 11.15c).

The solubility behavior of gases with respect to pressure also can be explained easily in terms of Le Châtelier's principle. We might represent the equilibrium by the following equation:

$$\text{solute}(g) + \text{solvent}(l) \rightleftharpoons \text{solution}(l)$$

If the pressure is increased, Le Châtelier's principle tells us that the system will respond in a way that brings the pressure back down toward its initial value. This can happen if more gas dissolves, because then there would be fewer molecules in the gas phase to exert pressure.

Quantitatively, the influence of pressure on the solubility of a gas is given by **Henry's law,** which states that the concentration of the gaseous solute in the solution, C_g, is directly proportional to the partial pressure of the gas above the solution; that is,

$$C_g = k_g p_g \qquad\qquad [11.1]$$

where the proportionality constant, k_g, is called the **Henry's law constant.** This relationship allows us to compute the solubility of a gas at some particular pressure, provided that we know its solubility at some other pressure, as shown in Example 11.4. Actually, Henry's law is accurate only for relatively low concentrations and pressures, and for gases that do not react significantly with the solvent.

Figure 11.15

The effect of pressure on the solubility of a gas. (a) An equilibrium exists between a gas and its solution. (b) The equilibrium is upset. (c) Equilibrium is restored when more gas dissolves.

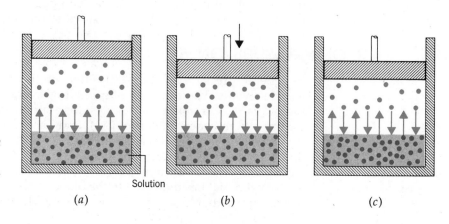

Solution

(a)　　　　(b)　　　　(c)

EXAMPLE 11.4   USING HENRY'S LAW TO CALCULATE THE
SOLUBILITY OF A GAS

PROBLEM   At 25°C, oxygen gas collected over water at a *total* pressure of 101 kPa is soluble to the
extent of 0.0393 g dm^{-3}. What would its solubility be if its partial pressure over water
were 107 kPa?

SOLUTION   The data given us permit the calculation of the Henry's law constant if we know the
partial pressure of oxygen above the solution. The total pressure is the sum of the partial
pressures of the H_2O vapor and oxygen,

$$P_{total} = p_{H_2O} + p_{O_2}$$

From Table 9.1 we find the vapor pressure of water to be 3.17 kPa at 25°C; therefore, the
partial pressure of oxygen is

$$p_{O_2} = P_{total} - p_{H_2O}$$

$$p_{O_2} = 101 \text{ kPa} - 3 \text{ kPa} = 98 \text{ kPa}$$

The Henry's law constant is obtained as the ratio

$$k_{O_2} = \frac{C_{O_2}}{p_{O_2}}$$

$$k_{O_2} = \frac{0.0393 \text{ g dm}^{-3}}{98 \text{ kPa}} = 4.0 \times 10^{-4} \frac{g}{dm^3 \text{ kPa}}$$

Now we can use Henry's law to determine that at a partial pressure of 107 kPa the
solubility of oxygen is

$$C_{O_2} = \left(4.0 \times 10^{-4} \frac{g}{dm^3 \text{ kPa}}\right)(107 \text{ kPa})$$

$$C_{O_2} = 0.043 \text{ g dm}^{-3}$$

11.8   VAPOR PRESSURES OF SOLUTIONS

The formation of a solution has very little effect on the *chemical* properties of its
components. Sodium, for instance, reacts with the water in an aqueous solution
to yield exactly the same products as when it reacts with distilled water. The
physical properties of substances, however, are often dramatically altered when
they become part of a solution. The same water that can freeze and crack the
block of an automobile engine at −20°C will remain liquid if it is mixed with
ethylene glycol—antifreeze.

Among the many physical properties of liquids that are affected by the
formation of a solution is the vapor pressure. When a nonvolatile solute—that
is, one that has no tendency to escape from a solution—is dissolved in a liquid
solvent, the solvent's vapor pressure is lowered. If, in addition, the solute is a
nonelectrolyte and doesn't dissociate, the extent to which the vapor pressure is
lowered depends on the mole fraction of the solute. The vapor pressure of the
solvent above the solution, which in this case we can call the vapor pressure of
the solution, $P_{solution}$, is given by **Raoult's law**

Raoult's law is easy to use if
you remember the definitions
of the various terms.

$$P_{solution} = x_{solvent} P^0_{solvent} \qquad [11.2]$$

where $x_{solvent}$ is the mole fraction of the solvent in the solution and $P^0_{solvent}$ is the
vapor pressure of the pure solvent. For example, a solution that contains
95 mol % water and 5 mol % of a nonvolatile solute such as sugar has $x_{H_2O} =$

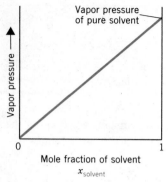

If a solution obeyed Raoult's law for all concentrations, its vapor pressure would vary linearly from zero up to the vapor pressure of the pure solvent.

0.95. At a temperature where the vapor pressure of pure water is 40 kPa, the vapor pressure of the solution will be

$$P_{solution} = 0.95 \ (40 \ kPa)$$

$$= 38 \ kPa$$

It is not difficult to see why Raoult's law holds. Figure 11.16*a* illustrates the condition in which a pure solvent is in equilibrium with its vapor. This vapor, as described in Section 10.4, exerts a pressure that is ultimately determined by the fraction of the total number of molecules at the surface of the liquid that have enough kinetic energy to escape. If we now look at a solution containing a nonvolatile solute (Figure 11.16*b*), we find that a portion of the solvent molecules at the surface has been replaced by molecules of the solute. Since the entire system—solvent plus solute—is at a single temperature, all the molecules in the solution belong to a single distribution of kinetic energies. In both the solution and the pure solvent, the same fraction of surface molecules has more than the minimum kinetic energy that *solvent* molecules need to break away from the liquid, but in the solution only a *portion* of that fraction is actually composed of molecules of the solvent. The others are solute molecules. The result is that there are fewer molecules at the surface of the solution capable of leaving than at the surface of the pure solvent. Therefore, the rate of evaporation of solvent molecules from a solution is less than from the pure solvent. This means that for the solution, fewer molecules are needed in the vapor to give an equal rate of condensation, so the vapor pressure is lower.

Because only the solvent can evaporate, the fraction of molecules at the surface of the solution that can escape into the vapor depends on that *fraction* of all the molecules at the surface that are solvent molecules—that is, the ratio of the number of moles of solvent particles to the total number of moles of particles that compose the surface. This ratio, of course, is the mole fraction of the solvent. If the solution were composed of 95 mol % solvent, then we expect to find only 95% of the molecules at the surface to belong to the solvent. This means that the rate of evaporation from the solution is expected to be only 95% of that for the solvent alone. The equilibrium vapor pressure should therefore be reduced to 95% of that for the pure solvent, which is the same result we obtain by the application of Raoult's law.

Figure 11.16
Molecular view of Raoult's law.
(a) *Pure solvent;* (b) *solution.*

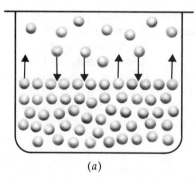

(*a*)

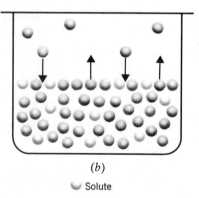

(*b*)

Solute

EXAMPLE 11.5   APPLYING RAOULT'S LAW

PROBLEM   10.0 g of the paraffin $C_{20}H_{42}$ were dissolved in 50.0 g of benzene, C_6H_6. At 53°C, the vapor pressure of pure benzene is 40 kPa. What is the vapor pressure of the solution at this temperature?

SOLUTION   To calculate the vapor pressure of the solution, we need the mole fraction of the *solvent*. The first thing to do therefore is to calculate the number of moles of each component. The molar mass of $C_{20}H_{42}$ is 282 g mol^{-1}, and the molar mass of C_6H_6 is 78.1 g mol^{-1}.

$$10.0 \text{ g C}_{20}\text{H}_{42} \times \left(\frac{1 \text{ mol C}_{20}\text{H}_{42}}{282 \text{ g C}_{20}\text{H}_{42}} \right) = 0.035 \text{ mol C}_{20}\text{H}_{42}$$

$$50.0 \text{ g C}_6\text{H}_6 \times \left(\frac{1 \text{ mol C}_6\text{H}_6}{78.1 \text{ g C}_6\text{H}_6} \right) = 0.640 \text{ mol C}_6\text{H}_6$$

The mole fraction of benzene in the mixture is therefore

$$x_{C_6H_6} = \frac{0.640 \text{ mol}}{0.676 \text{ mol}} = 0.947$$

Raoult's law states that

$$P_{\text{solution}} = x_{\text{solvent}} P^0_{\text{solvent}}$$

Substituting values gives

$$P_{\text{solution}} = 0.947 \ (40 \text{ kPa})$$

$$= 38 \text{ kPa}$$

Solutions containing more than one volatile component

In many solutions, such as benzene and carbon tetrachloride, for example, both solute and solvent have appreciable tendencies to undergo evaporation. In this case, the vapor will contain both solute and solvent molecules, and the vapor pressure of the solution will be the sum of the partial pressures exerted by each component. If we follow the same line of reasoning as above, we conclude that the partial pressure of any component above such a mixture is also given by Raoult's law. Thus the partial pressure of component A, p_A, is given by

$$p_A = x_A P_A^0 \qquad [11.3]$$

where P_A^0 is the vapor pressure of pure A and X_A is its mole fraction in the solution. Similarly, the partial pressure of a second component, p_B is given as

$$p_B = x_B P_B^0 \qquad [11.4]$$

Finally, the total vapor pressure of a mixture of A and B is given by Dalton's law,

$$P_T = p_A + p_B \qquad [11.5]$$

Substituting Equations 11.3 and 11.4 into Equation 11.5 gives

$$P_T = x_A P_A^0 + x_B P_B^0$$

Figure 11.17a is a plot of the partial pressures of A and B, and the total vapor pressure as a function of solution composition for such a two-component mixture.

EXAMPLE 11.6 CALCULATING THE VAPOR PRESSURE OF A SOLUTION OF TWO VOLATILE COMPONENTS

PROBLEM A mixture was prepared that contained 50.0 g of carbon tetrachloride, CCl_4, and 50.0 g of chloroform, $CHCl_3$. At 50°C, the vapor pressure of pure CCl_4 is 42.3 kPa and that of pure $CHCl_3$ is 70.1 kPa. What is the vapor pressure of the mixture at 50°C?

SOLUTION To solve this problem, we must use Raoult's law to calculate the partial pressures of each component above the solution. Then the total vapor pressure is simply the sum of the partial pressures. We begin, therefore, by calculating the mole fractions of each of the components.

$$50.0 \text{ g } CCl_4 \times \left(\frac{1 \text{ mol } CCl_4}{153.8 \text{ g } CCl_4} \right) = 0.325 \text{ mol } CCl_4$$

$$50.0 \text{ g } CHCl_3 \times \left(\frac{1 \text{ mol } CHCl_3}{119.4 \text{ g } CHCl_3} \right) = 0.419 \text{ mol } CHCl_3$$

The mole fractions are therefore

$$x_{CCl_4} = \frac{0.325 \text{ mol}}{0.744 \text{ mol}} = 0.437$$

$$x_{CHCl_3} = \frac{0.419 \text{ mol}}{0.744 \text{ mol}} = 0.563$$

Now we can calculate the partial pressures of each component above the solution. For the CCl_4 we have

$$p_{CCl_4} = x_{CCl_4} P^0_{CCl_4}$$

$$= (0.437) \times (42.3 \text{ kPa})$$

$$= 18.5 \text{ kPa}$$

and for $CHCl_3$

$$p_{CHCl_3} = x_{CHCl_3} P^0_{CHCl_3}$$

$$= (0.563) \times (70.1 \text{ kPa})$$

$$= 39.5 \text{ kPa}$$

The total vapor pressure is the sum of these.

$$P_{soln} = p_{CCl_4} + p_{CHCl_3}$$

$$= 18.5 \text{ kPa} + 39.5 \text{ kPa}$$

$$= 58.0 \text{ kPa}$$

The vapor pressure of the solution is 58.0 kPa.

Ideal and nonideal solutions

Actually, very few mixtures really obey Raoult's law very closely over wide ranges of composition. Benzene and carbon tetrachloride, a pair of substances that do form such mixtures, are said to yield **ideal solutions**. Mixtures that deviate from Raoult's law are called **nonideal**. When the vapor pressure of a mixture is greater than that predicted, it is said to exhibit a **positive deviation** from Raoult's law (Figure 11.17b); conversely, when a solution gives a lower vapor pressure than we would expect from Raoult's law, it is said to show a **negative deviation** (Figure 11.17c).

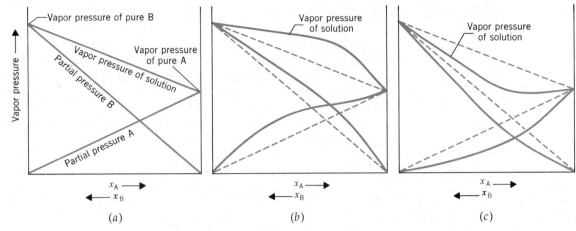

Figure 11.17

Vapor pressure of a two-component system. (a) *Ideal solution;* (b) *positive deviations from Raoult's law;* (c) *negative deviations from Raoult's law.*

As we saw in our discussion of heats of solution, the origin of nonideal behavior lies in the relative strengths of the interactions between molecules of the solute and solvent. When the attractive forces between the solute and solvent molecules are weaker than those between solute molecules or between solvent molecules, neither the solute nor solvent particles are held as tightly in the solution as they are in the pure substances. The escaping tendency of each is therefore greater in the solution than in the solute or solvent alone. As a result, the partial pressures of both of them over the solution are greater than predicted by Raoult's law, and the solution exhibits a larger vapor pressure than expected.

Just the opposite effect is produced when the solute-solvent interactions are stronger than the solute-solute or solvent-solvent interactions. Each substance, in the presence of the other, is held more tightly than in the pure materials, and their partial pressures over a solution are therefore less than Raoult's law would predict. The result is that such a solution exhibits a negative deviation from ideality.

Since, in a solution that shows positive deviations from ideal behavior, the forces of attraction between solute and solvent are weaker than those between both solute molecules and solvent molecules, the formation of these solutions occurs with the absorption of energy (Section 11.5). Conversely, of course, mixtures that exhibit negative deviations from Raoult's law are formed with the evolution of heat. This is summarized in Table 11.5.

Table 11.5
Summary of solution properties

Relative Attractive Forces	ΔH_{soln}	Temperature Change When Solution Is Formed	Deviations from Raoult's Law	Example
A–A, B–B = A–B	Zero	None	None (ideal solution)	Benzene–chloroform
A–A, B–B < A–B	Negative (exothermic)	Increase	Negative	Acetone–water
A–A, B–B > A–B	Positive (endothermic)	Decrease	Positive	Ethanol–hexane

11.9 FRACTIONAL DISTILLATION

In a simple distillation process—one that could be used to separate sodium chloride and water, for example—a volatile solvent is vaporized from a solution and subsequently condensed to provide a pure liquid (see Figure 1.14 on page 28). If the process is continued, eventually all the solvent will be removed and only the solid solute will remain.

The separation of mixtures of volatile liquids into their components presents more of a problem. A technique that can frequently be used successfully to accomplish this task is called **fractional distillation.**

Let's suppose that we had a mixture of two volatile liquids, A and B, that form an ideal solution. This mixture will boil when the sum of the partial pressures of A and B equals the prevailing atmospheric pressure; that is, when

$$P_{atm} = p_A + p_B$$

The boiling points of various mixtures of A and B will increase gradually from that of the more volatile component (let us say, A) to that of the less volatile one, B, as shown in Figure 11.18.

Suppose, now, that when 1.00 mol of A is mixed with 2.00 mol of B, the resulting mixture boils (at 101.4 kPa) at a temperature at which the vapor pressure of *pure A* is 152 kPa and that of *pure B* is 76.0 kPa. Under these conditions the partial pressure of A is

$$p_A = x_A P_A{}^0$$

$$p_A = \left(\frac{1.00 \text{ mol } A}{1.00 \text{ mol } A + 2.00 \text{ mol } B}\right) \times (152 \text{ kPa})$$

$$p_A = \left(\frac{1.00 \text{ mol}}{3.00 \text{ mol}}\right) \times (152 \text{ kPa})$$

$$p_A = (0.333)(152 \text{ kPa}) = 50.7 \text{ kPa}$$

Similarly, the partial pressure of B would be

$$p_B = \left(\frac{2.00 \text{ mol}}{3.00 \text{ mol}}\right) \times (76.0 \text{ kPa})$$

$$p_B = 50.7 \text{ kPa}$$

The sum of p_A and p_B is 101.4 kPa as, of course, it must be if the solution is to boil.

What can we say about the composition of the vapor? In Section 9.10, under our discussion of Dalton's law of partial pressures, it was stated that the partial pressure of a gas in a mixture is equal to its mole fraction multiplied by the total pressure *exerted by the gas*. We can write this as

$$p_A = x_{A(vapor)} P_{T(vapor)}$$

where $x_{A(vapor)}$ is the mole fraction of A in the vapor and $P_{T(vapor)}$ is the total vapor pressure. In the vapor over our solution, the partial pressure of each gas is 50.7 kPa and the total pressure is 101.4 kPa. This means that the mole fraction of both A and B *in the vapor* must be 0.500. In the liquid the mole fraction of A was only 0.333, so the vapor contains a greater proportion of the more volatile component (A) than the solution. In fact, any time we boil a mixture of these two substances, the vapor will be richer than the solution in the more volatile compound.

On our boiling-point diagram we can indicate the composition of the vapor by the upper curve drawn in Figure 11.19. Here, points corresponding to the

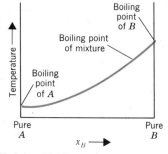

Figure 11.18

Boiling-point curve for a mixture of A *and* B.

Note that this equation applies to the *vapor.*

When a liquid solution made from two volatile liquids is boiled, the vapor is always richer in the more volatile component.

Figure 11.19

Boiling-point diagram for a two-component mixture. A liquid of composition x_1 boils at temperature T_1 and gives a vapor having the composition x_2. When this vapor is condensed and reheated, it boils at temperature T_2 and gives a vapor with composition x_3.

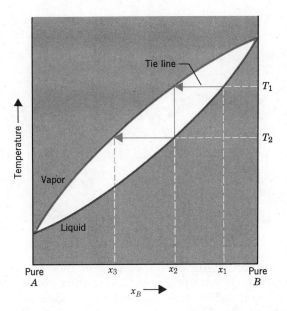

compositions of liquid and vapor in equilibrium can be obtained by drawing a horizontal line, called a **tie line,** between the curve for the liquid and that for the vapor. When the composition of the mixture is x_1, it boils at a temperature T_1 to provide a vapor that has a composition x_2. If this vapor is condensed and then reheated, it will boil at a temperature T_2 and give a vapor whose composition is x_3. Repetition of this process will produce fractions ever richer in *A*. This procedure is called **fractional distillation,** and is useful not only in the laboratory, where it is employed for purifying the products of chemical reactions, but also industrially. For instance, the petroleum industry uses fractional distillation to separate crude oil into its various components, which include gasoline, kerosene, oils, and paraffin.

Solutions that exhibit large deviations from Raoult's law

There are some solutions that exhibit very large deviations from ideality; as a result they cannot be totally separated into their components even by fractional distillation. Ethanol (grain alcohol) and water form such a mixture. Solutions of these two substances have such large positive deviations from Raoult's law that there is a maximum in the vapor pressure curve and hence a minimum in the boiling-point diagram as shown in Figure 11.20. A solution with such a minimum boiling point is called a **minimum-boiling azeotrope.** Fractional distillation

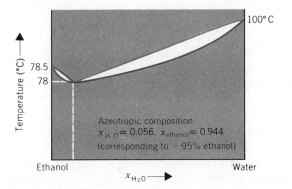

Figure 11.20

Boiling-point diagram for water-ethanol mixtures (not drawn to scale).

The alcohol obtained by distilling a water-alcohol mixture cannot be used for blending with gasoline to make gasohol because the alcohol must be water free. Other methods must be used to make dry "absolute alcohol."

of solutions lying on either side of this azeotropic composition is capable of separating them into, at best, one pure component plus a solution having the minimum boiling point. As any moonshiner will agree, ethyl alcohol-water mixtures (obtained by fermentation of sugars, for example) are rich in water. Fractional distillation is able to concentrate the alcohol to, at best, the azeotropic composition of approximately 95% by volume of ethyl alcohol.[3] Once this composition has been achieved, the liquid and vapor have the same composition, and no additional fractionation takes place.

There are also solutions that show large negative deviations from ideality and therefore have a minimum in their vapor pressure curves. This leads to a maximum on the boiling point diagram and hence to a **maximum-boiling azeotrope.** Hydrochloric acid, for instance, forms a maximum-boiling azeotrope having the approximate composition, 20% HCl and 80% H_2O by mass, with a boiling point of 109°C.

11.10 COLLIGATIVE PROPERTIES OF SOLUTIONS

Properties that depend on the number of particles of solute in a solution, instead of on their specific chemical nature, are called **colligative properties.**[4] Vapor pressure is one of these. We have seen that, according to Raoult's law, the addition of a nonvolatile solute to a substance causes its vapor pressure to be lowered. In our explanation of how this happened, nothing was said about the specific nature of the solute other than it was incapable of escaping from the solution and that it was undissociated.

What other effects do the solute particles have on the physical properties of a solution? To answer this question we have to examine how the vapor pressure lowering affects the phase diagram of the solvent. This is illustrated in Figure 11.21 for an aqueous solution of a nonvolatile solute. Notice that at every temperature the vapor pressure of the solution lies below the vapor pressure of the pure solvent, as indicated by the dashed line. Because of this, the solution must be heated to a temperature above the normal boiling point of the solvent in order

---

[3] This is too strong to consume without dilution. A 95% solution of ethyl alcohol is 190 proof. Aged whiskey that is 86 proof is only 43% alcohol, by volume.

[4] Derived from the latin, *colligare*, to collect. These properties are determined by the number of particles in the entire "collection," not by what the particles are composed of.

Figure 11.21

Effect of a nonvolatile solute on the phase diagram of H_2O.

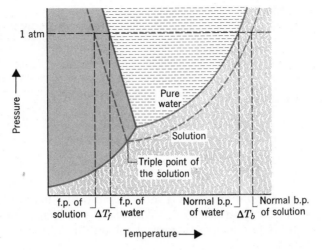

1 atm

Pressure

Pure water

Solution

Triple point of the solution

f.p. of solution ΔT_f | f.p. of water   Normal b.p. of water ΔT_b | Normal b.p. of solution

Temperature

The solid vapor-pressure curve for the solution is the same as the solid vapor-pressure curve for the solvent.

for its vapor pressure to be equal to 1 atm. In other words, the normal boiling point of the solution is higher than that of the water itself. The extent that the boiling point is raised is indicated on the diagram by ΔT_b. We call this increase in the boiling point the *boiling point elevation*.

Examining Figure 11.21 we also find that the solution has a new triple point that occurs at the intersection of the vapor pressure curve for the solution and the solid vapor pressure curve for the pure solvent. Generally, the solute particles don't fit into the lattice formed by the solvent when it freezes, so the solid that forms is the pure solvent. As a result, there is no separate solid vapor pressure curve for the solution. As you learned in Chapter 10, the solid-liquid equilibrium line (which defines the freezing point as a function of pressure) rises from the triple point. Since the new triple point for the solution lies to the left of that for the pure solvent, the freezing point of the solution is lower than the freezing point of the solvent. The extent that the freezing point is lowered (the *freezing point depression*) is shown on the diagram as ΔT_f.

In summary, the presence of a solute increases the liquid range of the solution by both raising the boiling point and lowering the freezing point. One of the most common practical applications of this phenomenon is the use of an antifreeze solution in the radiator of an automobile. The solute is usually ethylene glycol, $C_2H_4(OH)_2$, which is completely miscible with water and has a very low vapor pressure itself—it is a nonvolatile solute. When dissolved in water, it both lowers the freezing point and raises the boiling point. In the winter it protects a car by preventing the liquid in the radiator from freezing, as water would if it were used instead. In hot summer weather, the antifreeze solution also protects the radiator from boiling over as easily as it would if it were filled with pure water.

For dilute solutions it has been found that the extent to which the boiling point is raised, and the freezing point is lowered, depends on the *mole fraction* of the solute (x_A) in the solution,

$$\Delta T_b = \left(\frac{RT_b^2}{\Delta H_{vap}} \right) x_A \qquad [11.6]$$

$$\Delta T_f = \left(\frac{RT_f^2}{\Delta H_{fus}} \right) x_A \qquad [11.7]$$

where R (perhaps surprisingly) is the universal gas constant (8.314 J mol^{-1} K^{-1}), T_b and T_f are the normal boiling point and freezing point values for the pure solvent (in kelvins), and ΔH_{vap} and ΔH_{fus} are, respectively, the molar

Table 11.6
Boiling and freezing constants for some solvents

Solvent	Boiling Point (°C)	ΔH_{vap} (kJ mol^{-1})	Melting Point (°C)	ΔH_{fus} (kJ mol^{-1})
Water	100.0	40.7	0.0	6.01
Benzene	80.1	30.8	5.5	9.83
Camphor	—	—	179	6.84
Acetic acid	118.2	24.3	17	11.5

heat of vaporization (J mol^{-1}) and molar heat of fusion (J mol^{-1}) of the solvent. The magnitudes of T_b, T_f, ΔH_{vap}, and ΔH_{fus} are characteristic of each solvent. Table 11.6 lists some typical solvents and their respective values for these quantities.

Although the derivation of these two equations is well beyond the scope of this text, notice that for a given solvent all the terms within the parentheses of equations 11.6 and 11.7 are constant. That is,

$$\Delta T_b = k_b x_A \qquad [11.6a]$$

$$\Delta T_f = k_f x_A \qquad [11.7a]$$

where k_b and k_f are characteristic constants for the solvent used. The extent of boiling temperature elevation or freezing temperature lowering (in either K or °C units, since temperature differences are involved) is thus directly proportional to the mole fraction of nonvolatile solute added. Doubling the mole fraction of solute will, for example, double the change in the boiling and freezing temperatures. Example 11.7 shows how these equations can be applied to a specific solution system.

EXAMPLE 11.7 CALCULATING THE FREEZING AND BOILING POINTS OF A SOLUTION

PROBLEM   What would be the freezing point and boiling point of a solution containing 6.50 g of ethylene glycol ($C_2H_6O_2$), commonly used as an automotive antifreeze, in 200.0 g of water?

SOLUTION   To determine ΔT_f and ΔT_b we must find the mole fraction of solute—the ratio of moles of $C_2H_6O_2$ to total moles of $C_2H_6O_2$ and H_2O in the solution. The number of moles of $C_2H_6O_2$ is

$$6.50 \text{ g C}_2\text{H}_6\text{O}_2 \times \left(\frac{1 \text{ mol C}_2\text{H}_6\text{O}_2}{62.1 \text{ g C}_2\text{H}_6\text{O}_2}\right) = 0.105 \text{ mol C}_2\text{H}_6\text{O}_2$$

The number of moles of solvent is

$$200.0 \text{ g H}_2\text{O} \times \left(\frac{1 \text{ mol H}_2\text{O}}{18.02 \text{ g H}_2\text{O}}\right) = 11.10 \text{ mol H}_2\text{O}$$

The mole fraction of solute is therefore

$$x_A = \frac{\text{mol C}_2\text{H}_6\text{O}_2}{\text{mol C}_2\text{H}_5\text{O}_2 + \text{mol H}_2\text{O}} = \frac{0.105 \text{ mol}}{0.105 \text{ mol} + 11.10 \text{ mol}} = 9.38 \times 10^{-3}$$

From Table 11.6 we find for H_2O that $\Delta H_{fus} = 6.01$ kJ mol^{-1} and $\Delta H_{vap} = 40.7$ kJ mol^{-1}, $T_f = 0.00$°C (273.15 K) and $T_b = 100.00$°C (373.15 K). Since $R = 8.314$ J mol^{-1} K^{-1}, the

ΔH and T values must be expressed in compatible J mol^{-1} and K units. Thus

$$\Delta T_f = \frac{RT_f^2}{\Delta H_{fus}}x_A = \frac{(8.314 \text{ J mol}^{-1}\text{ K}^{-1})(273.15 \text{ K})^2}{6.01 \times 10^3 \text{ J mol}^{-1}} \times 9.38 \times 10^{-3}$$

$$\Delta T_b = \frac{RT_b^2}{\Delta H_{vap}}x_A = \frac{(8.314 \text{ J mol}^{-1}\text{ K}^{-1})(373.15 \text{ K})^2}{40.7 \times 10^3 \text{ J mol}^{-1}} \times 9.38 \times 10^{-3}$$

$$\Delta T_f = 0.968 \text{ K} = 0.968°C$$

$$\Delta T_b = 0.267 \text{ K} = 0.267°C$$

The freezing and boiling points of the solution are then $-0.978°C$ and $100.267°C$. We see that solutions considerably more concentrated than this (approximately 3%) are necessary to protect an automobile's cooling system in frigid weather. See Problem 11.92.

If a knowledge of the mole fraction of solute permits us to determine the extent to which the boiling point and freezing point differ from those of the pure solvent, then it should also be possible to calculate the composition of a solution from measured values of ΔT_b and ΔT_f. This aspect of these colligative properties proves particularly useful, because with it we can measure molar masses experimentally. This, combined with analytical data on percent composition, enables us to find molecular formulas. Example 11.8 illustrates how to apply this concept.

EXAMPLE 11.8   DETERMINING A MOLAR MASS FROM COLLIGATIVE PROPERTIES

PROBLEM   A 5.50-g sample of a compound whose empirical formula is C_3H_3O was dissolved in exactly 250.0 g of benzene. It was found that the freezing point of the solution was 1.20°C below that of pure benzene. Determine (a) the molar mass and (b) the molecular formula of this compound.

SOLUTION   (a) From Table 11.6, T_f for benzene is 5.5°C (278.6 K), and ΔH_{fus} is 9.83 kJ mol^{-1} (9.83 × 10^3 J mol^{-1}). If we solve Equation 11.7 for the mole fraction of solute, x_A, we obtain

$$x_A = \left(\frac{\Delta H_{fus}}{RT_f^2}\right)\Delta T_f$$

Upon substituting the values for the molar heat of fusion, universal gas constant, freezing point of solvent, and the freezing-point depression, we have

$$x_A = \frac{(9.83 \times 10^3 \text{ J mol}^{-1})(1.02 \text{ K})}{(8.314 \text{ J mol}^{-1}\text{K}^{-1})(278.6 \text{ K})^2} = 1.55 \times 10^{-2}$$

This mole fraction value represents the ratio of moles of unknown solute (n_{ukn}) to total moles of solute plus moles of benzene. Total moles of benzene is

$$250.0 \text{ g } C_6H_6 \times \left(\frac{1 \text{ mol } C_6H_6}{78.11 \text{ g } C_6H_6}\right) = 3.201 \text{ mol } C_6H_6$$

Thus we have

$$x_A = \frac{n_{ukn}}{n_{ukn} + n_{benzene}} = \frac{n_{ukn}}{n_{ukn} + 3.201} = 1.55 \times 10^{-2}$$

Solving for moles of solute

$$n_{ukn} = 0.0155(n_{ukn}) + 0.0496$$

$$0.9845(n_{ukn}) = 0.0496$$

$$n_{ukn} = \frac{0.0496}{0.9845} = 0.0504 \text{ mol solute}$$

Finally, since molar mass is expressed as grams per mole, the molar mass of solute must be

$$\frac{5.50 \text{ g}}{0.0504 \text{ mol}} = 109 \text{ g mol}^{-1}$$

(b) Now that we know the molar mass, we can determine the molecular formula as in Chapter 2. The molecular formula must contain the empirical formula repeated an integral number of times. The molar mass is therefore an integral multiple of the empirical formula weight, which for C_3H_3O is 55.0 u. Since the molar mass that we have found is twice this value, the molecular formula must be $(C_3H_3O)_2$ or $C_6H_6O_2$.

In practice, molar masses cannot be determined by this method as accurately as we have implied. However, an error even as large as 10% (that is, measured molar masses ranging, in this case, from about 100 to 120 g mol^{-1}) certainly still permits us to choose among the possibilities, C_3H_3O, $C_6H_6O_2$, and $C_9H_9O_3$, with their corresponding molar masses, 55.0, 110, and 165 g mol^{-1}. For this reason, this technique has proven to be a valuable tool in chemistry.

11.11 OSMOTIC PRESSURE

Osmosis is a process whereby a solvent passes from a dilute solution into a more concentrated one by moving through a thin film that selectively permits the pasage of the solvent, but restricts the passage of the solute. Such films are called **semipermeable membranes,** and typical examples include certain types of parchment paper and some gelatinlike inorganic substances. A similar phenomenon called **dialysis,** which occurs at cell walls in plants and animals, allows the passage of water, small ions, and small molecules but restricts passage of large molecules such as proteins. Osmosis is the limiting case of dialysis.

In the process of osmosis there is a drive toward equalization of concentrations between the two solutions in contact with one another across the mem-

A patient uses a portable artificial kidney machine. It cleanses the blood of impurities by dialysis.

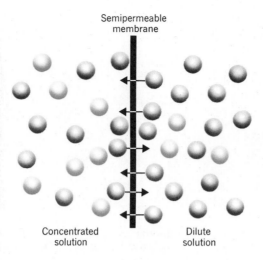

Semipermeable
membrane

Concentrated
solution

Dilute
solution

brane. The rate of passage of solvent molecules through the membrane into the more concentrated solution is greater than their rate of passage in the opposite direction, presumably because at the surface of the membrane the solvent concentration is greater in the more dilute solution (Figure 11.22). We observe a similar effect if two solutions, with unequal concentrations of a nonvolatile solute, are placed in a sealed enclosure, as shown in Figure 11.23. The rate of evaporation from the dilute solution is greater than that from the concentrated solution, but the rate of return to each is the same (both solutions are in contact with the same gas phase). As a result, neither solution is in equilibrium with the vapor. In the dilute solution molecules are evaporating faster than they are condensing, while in the concentrated solution the reverse occurs. Consequently, there is a gradual net transfer of solvent from the dilute solution into the more concentrated one until they both achieve the same concentration.

If we perform an osmosis experiment using the apparatus in Figure 11.24, in which we have a solution in compartment A and pure water in B, the passage of solvent from B to A will slowly increase the volume of A and decrease the volume of B. As this occurs, the height of the liquid in the capillary of compartment A will rise while the height of the liquid in the other capillary will drop, and there will be a pressure difference between the two solutions that depends on the difference in these heights, Δh. Now, the ease with which a solvent molecule can be transferred from B to A, and from A to B, also depends on this pressure difference. As the pressure on side A increases, it becomes increasingly more difficult to squeeze another solvent molecule into this solution. It also becomes

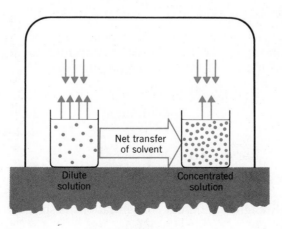

Net transfer
of solvent

Dilute
solution

Concentrated
solution

Figure 11.24

Measurement of osmotic pressure.

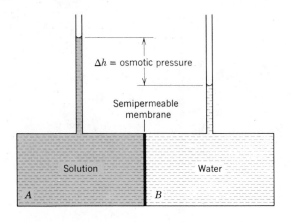

II is the capital Greek letter pi.

$R = 8.314$ kPa dm^3 mol^{-1} K^{-1}

easier for a water molecule to "pop out" of the solution into the compartment with pure water. Therefore, as Δh becomes larger, the net rate of transfer of water from B to A decreases, and at some point a value of Δh is reached at which water molecules are moving across the membrane at equal rates—equilibrium is achieved. The pressure difference between the two compartments at equilibrium is called the **osmotic pressure** of the *solution* and is symbolized by the Greek letter, Π.

For solutions that are dilute, it can be shown that the osmotic pressure is proportional to the molar concentration (M) of the solute, and that the proportionality constant is RT, where R is the gas constant and T is the absolute temperature.

$$\Pi = MRT \qquad [11.8]$$

Another form of this equation can be obtained if we realize that the molar concentration of the solute is obtained as the ratio of the number of moles of solute (n) to the volume of the solution (V).

$$M = \frac{n}{V} \qquad [11.9]$$

Substituting this into Equation 11.8 gives

$$\Pi = \frac{n}{V}RT$$

or

$$\Pi V = nRT \qquad [11.10]$$

Jacobus Henricus van't Hoff, a Dutch chemist, received the first Nobel Prize for Chemistry in 1901.

It is interesting to note the similarity between Equation 11.10 (called the van't Hoff equation) and the ideal gas law,

$$PV = nRT$$

The magnitude of the osmotic pressure, even in very dilute solutions, is quite large. For instance, with a concentration of 0.010 mol of solute particles per dm^3 (0.010 M) at room temperature (298 K), the osmotic pressure would be

$$\Pi = MRT$$

$$\Pi = \left(\frac{0.010 \text{ mol}}{1 \text{ dm}^3}\right) \times \left(\frac{8.314 \text{ kPa dm}^3}{\text{mol K}}\right) \times 298 \text{ K}$$

$$\Pi = 25 \text{ kPa}$$

This pressure is sufficient to support a column of water nearly 2.5 m high! Because the osmotic pressure that can be developed between solutions of only slightly different concentrations is so great, it is very important that fluids added to the body intravenously not alter significantly the osmotic pressure of the blood. If the blood fluids become too dilute, the osmotic pressure that develops within the blood cells can cause them to rupture. On the other hand, if the fluids are too concentrated, water will diffuse out of the cells and they will no longer function properly. For this reason care is taken to use solutions with the same osmotic pressure as the solution within the cells. Solutions that have the same osmotic pressure are called **isotonic** solutions.

The large differences in pressure developed between solutions of very similar concentrations provide us with a method of measuring the very large molar masses of polymers (both of synthetic and biological origin). Freezing-point lowering and boiling-point elevation just won't work in these cases. For instance, a solution containing even as much as 150 g of a solute whose molar mass is 30 000 g mol^{-1} in 1000 g of water produces a solution with a mole fraction of solute of only 0.000 09. The freezing-point depression of this solution is approximately 0.009 degrees. Moreover, in most cases it is not even possible to dissolve this much solute, so that in practice the freezing-point changes are virtually undetectable.

A solution containing only 15 g of this solute per 1000 g of water at 25°C, however, would have an easily measured osmotic pressure. Because the solution is so dilute, we can assume that 1000 g of the solvent water will produce approximately 1000 cm^3 of solution (1.00 dm^3), so we can express the concentration as 5.0×10^{-4} M with very little error. Now we can calculate the osmotic pressure.

$$\Pi = \left(\frac{5.0 \times 10^{-4}\ \text{mol}}{\text{dm}^3} \right) \times \left(\frac{8.314\ \text{kPa dm}^3}{\text{mol K}} \right) \times 298\ \text{K}$$

$$\Pi = 1.2\ \text{kPa}$$

This corresponds to a pressure of 9.1 mmHg. If we assume that the solution has a density of 1 g cm^{-3}, this pressure will support a column of the liquid 12.3 cm high. A height such as 12.3 cm is very easily measured with accuracy, so osmotic pressure measurements are a useful tool for the determination of molar masses of large molecules.

The osmosis process can be stopped and even reversed by the application of pressure equal to or greater than the osmotic pressure of the solution. This reversal of osmosis is used to desalinate seawater. An apparatus for this process is shown schematically in Figure 11.25. If pressure were not applied to this system, osmosis would occur from left to right—that is, molecules would be transferred from freshwater to saltwater (Figure 11.25a). However, when a pressure exceeding Π is applied, the osmosis is driven in the reverse direction (Figure 11.25b), which forces water molecules out of the saline solution, leaving the impurities behind. One type of membrane that is strong enough to withstand these pressures is a film of cellulose acetate placed over a suitable support. Cellulose acetate is permeable to water but impermeable to the ions and impurities in seawater. Desalination plants have been constructed thus far that can produce as much as 170 m^3 of fresh water daily.

Many large ships use reverse osmosis to produce freshwater from the sea.

11.12 SOLUTIONS OF ELECTROLYTES

For simplicity, we have limited our discussion thus far to solutions that do not contain an electrolyte. The reason for this is that the vapor pressure lowering, freezing-point depression, boiling-point elevation, and osmotic pressure depend on the *number* of particles present in the solution. One mole of a nonelectrolyte, such as sugar, when placed in 100 mol of water yields 1 mol of particles,

Figure 11.25

Desalination by reverse osmosis.
(a) Osmosis with no pressure applied to the saline solution.
(b) Reverse osmosis occurs when a pressure larger than Π is applied to the solution.

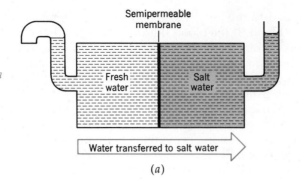

(a)

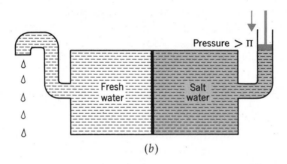

(b)

and a 0.01-mole fraction sucrose solution would have a certain freezing point that is below that of pure water. However, a solution containing 1 mol of an electrolyte such as NaCl contains 2 mol of particles—1 mol of Na^+ ions and 1 mol of Cl^- ions. As a result, a solution labeled "0.01 mole fraction NaCl" actually contains twice as many solute particles as a 0.01-mole fraction solution of a nonelectrolyte, and theoretically should have a freezing-point depression twice that of the 0.01-mole fraction sucrose solution. In a similar fashion, 1 mol of $CaCl_2$ would produce 3 mol of ions in solution, so that a 0.01-mole fraction $CaCl_2$ solution should have a freezing-point depression three times as great as a 0.01-mole fraction solution of sucrose. (Actually, these predictions for NaCl and $CaCl_2$ are not entirely accurate, as we will see later, but they are close.) Similar predictions also hold fairly well for the other colligative properties.

EXAMPLE 11.9 **CALCULATING THE VAPOR PRESSURE OF A SOLUTION OF AN ELECTROLYTE**

PROBLEM A solution of calcium chloride was prepared by dissolving 25.0 g of $CaCl_2$ in exactly 500 g of H_2O. What is the expected vapor pressure of this solution at 80°C? At 80°C, water has a vapor pressure of 47.3 kPa. What would the vapor pressure of the solution be if $CaCl_2$ were not an electrolyte?

SOLUTION To solve the problem we must use Raoult's law.

$$P_{solution} = x_{solvent}P^0_{solvent}$$

so we need to calculate the mole fraction of water. For the $CaCl_2$,

$$25.0 \text{ g } CaCl_2 \times \left(\frac{1 \text{ mol } CaCl_2}{111 \text{ g } CaCl_2}\right) = 0.225 \text{ mol } CaCl_2$$

and for water

$$500 \text{ g } H_2O \times \left(\frac{1 \text{ mol } H_2O}{18.0 \text{ g } H_2O}\right) = 27.8 \text{ mol } H_2O$$

Before we compute x_{H_2O}, we have to realize that the 0.255 mol $CaCl_2$ will produce three times as many moles of ions, that is, 0.675 mol ions. Therefore,

$$x_{H_2O} = \frac{27.8 \text{ mol } H_2O}{27.8 \text{ mol } H_2O + 0.675 \text{ mol ions}} = 0.975$$

The vapor pressure of the solution should therefore be

$$P_{solution} = 0.975 \times 47.3 \text{ kPa}$$

$$= 46.1 \text{ kPa}$$

If $CaCl_2$ were not an electrolyte, the mole fraction of water would have been

$$x_{H_2O} = \frac{27.8 \text{ mol } H_2O}{27.8 \text{ mol } H_2O + 0.225 \text{ mol } CaCl_2} = 0.993$$

and

$$P_{solution} = 0.993 \times 47.3 \text{ kPa}$$

$$= 47.0 \text{ kPa}$$

As you might expect, weak electrolytes produce effects on the colligative properties that lie between those of nonelectrolytes and strong electrolytes. For example, acetic acid undergoes reaction with water to establish the equilibrium

$$HC_2H_3O_2 + H_2O \rightleftharpoons H_3O^+ + C_2H_3O_2^-$$

If we ignore the H_2O molecule that serves as the carrier for the H^+, this can be written in a simpler way as

$$HC_2H_3O_2 \rightleftharpoons H^+ + C_2H_3O_2^-$$

Because there is this equilibrium, only a portion of the acetic acid is present as its ions while the rest exists as molecules, so at equilibrium, 1 mol of this solute exists as more than 1 mol of particles but less than 2. We can use the boiling-point elevation and freezing-point depression to determine the extent to which such a weak electrolyte is dissociated.

EXAMPLE 11.10

USING FREEZING-POINT DEPRESSION TO DETERMINE THE PERCENT DISSOCIATION OF A WEAK ELECTROLYTE

PROBLEM

Very careful measurement reveals that a solution of 1.00 mol HF in 1.00 kg H_2O has a freezing point of $-1.88°C$. What percent of the HF is dissociated into H^+ and F^- ions in this solution?

SOLUTION

The dissociation of hydrogen fluoride can be represented by

$$HF \rightleftharpoons H^+ + F^-$$

We will call the amount of HF that has undergone dissociation in 1.00 kg of water, x.

number of moles of HF dissociated = x

The number of moles of HF remaining must be the difference between the amount of HF that we put into the solution and the amount dissociated; that is, $1.00 - x$.

number of moles of HF remaining at equilibrium = $1.00 - x$

We see from the chemical equation above that for every mole of HF that dissociates, we produce 1 mol of H^+ and 1 mol of F^-. When x moles of HF dissociate, we must therefore form x mol of H^+ and x mole of F^-, so that at equilibrium we also have number of moles of $H^+ = x$

number of moles of $F^- = x$

The *total* number of moles of particles in 1.00 kg of solvent is the sum of the moles contributed by the HF, H^+, and F^-. This total is

total number of moles of particles $= (1.00 - x) + x + x$

$$= 1.00 + x$$

Now, from the freezing-point depression of this solution, 1.88°C, we can calculate the mole fraction of total solute *particles* (HF, H^+, and F^-) in the water solution.

$$x_A = \left(\frac{\Delta H_{fus}}{RT_f^2}\right) \Delta T_f = \frac{(6.01 \times 10^3 \text{ J mol}^{-1})(1.88 \text{ K})}{(8.314 \text{ J mol}^{-1}\text{K}^{-1})(273.15 \text{ K})^2} = 0.0182$$

The total moles of water present can be found

$$1.00 \text{ kg H}_2\text{O}\left(\frac{10^3 \text{ g H}_2\text{O}}{1 \text{ kg H}_2\text{O}}\right)\left(\frac{1 \text{ mol H}_2\text{O}}{18.0 \text{ g H}_2\text{O}}\right) = 55.5 \text{ mol H}_2\text{O}$$

The total moles of solute particles are now calculated, using the definition of mole fraction

$$x_A = \frac{n_{solute}}{n_{solute} + n_{solvent}} = \frac{n_{solute}}{n_{solute} + 55.5} = 0.0182$$

$$n_{solute} = 0.0182(n_{solute}) + 1.01$$

$$0.9818(n_{solute}) = 1.01$$

$$n_{solute} = \frac{1.01}{0.9818} = 1.03 \text{ mol solute particles}$$

Therefore,

1.03 mol of particles $= (1.00 + x)$ mol of particles

and thus

$$x = 0.03 \text{ mol}$$

The fraction of HF dissociated is equal to the number of moles that have broken apart (0.03 mol) divided by the total number of moles of HF placed in the solution (1.00 mol).

$$\text{fraction dissociated} = \frac{0.03}{1.00} = 0.03$$

The percent of the HF dissociated is 3%.

Solutions of electrolytes have larger values of ΔT_f and ΔT_b than we might initially have expected because of dissociation. There are also instances in which the freezing-point depression and boiling-point elevation are smaller than we would at first predict. This occurs when **association** (the opposite of dissociation) takes place between solute particles in the solution. For instance, a solution of 1 mol (122 g) of benzoic acid in 1.00 kg of benzene produces a freezing-point depression only slightly more than half the expected depression, implying that there are only about half as many particles in the solution as we anticipated. Since this approximately $\frac{1}{2}$ mol of particles has a mass of 122 g, the apparent molar mass is about 240 g mol^{-1}. Therefore, when association takes place, the measured molar masses are actually higher than we would predict.

In this particular example, the anomalous behavior of benzoic acid in benzene is attributed to hydrogen bonding between benzoic acid molecules to form a **dimer** (a particle created from two identical simpler units).

benzoic acid   benzoic acid dimer—hydrogen bonds shown as dotted lines

Interionic attractions

Earlier in this section we mentioned that the actual freezing points of solutions of electrolytes such as NaCl and CaCl$_2$ are not exactly the same as those calculated if we assume complete dissociation. This happens because the ions in a solution are not really completely independent particles. Although the solvent shields them from each other's charge, this shielding is not perfect, and it becomes worse as the solution becomes more concentrated—that is, as the average distance between the ions becomes smaller. As a result, a concentrated solution behaves as if it has fewer ions than a dilute solution, and the ions' effectiveness at altering the properties of the solution (boiling point, freezing point, osmotic pressure) diminishes as the concentration of the solute becomes larger. Consequently, ionic compounds behave as if they are less fully dissociated in concentrated solutions than when they are dilute.

Quantitatively, the degree to which an electrolyte behaves as if it were dissociated can be expressed by the **van't Hoff factor**, i. This quantity can be defined as the ratio of the observed freezing-point depression produced by a solution to the freezing point that the solution would exhibit if the solute were a nonelectrolyte.

$$i = \frac{(\Delta T_f) \text{ measured}}{(\Delta T_f) \text{ calculated as nonelectrolyte}}$$

Table 11.7 has values of the i factor for several strong electrolytes. For NaCl, KCl, and MgSO$_4$, i approaches 2 as the solution becomes more dilute. For K$_2$SO$_4$, i approaches 3, as we expect.

Table 11.7
Values of the van't Hoff factor at various concentrations

Salt	(mol solute/kg H$_2$O)			i Factor if Completely Dissociated
	0.1	0.01	0.001	
NaCl	1.87	1.94	1.97	2.00
KCl	1.85	1.94	1.98	2.00
K$_2$SO$_4$	2.32	2.70	2.84	3.00
MgSO$_4$	1.21	1.53	1.82	2.00

It is interesting to compare the effects of the changes on the ions on the interionic attractions. For NaCl, the value of i changes by about 5% going from 0.1 mol solute/kg H_2O to 0.001 mol solute/kg H_2O. For K_2SO_4, which has a doubly charged SO_4^{2-} ion, i changes by about 22% for the same dilution. When there are two doubly charged ions in $MgSO_4$, the i factor changes by about 50% for the same dilution. These observations are not surprising, because as the charges on the cations and anions become larger, so should their attractions for each other. This leads to a lower degree of independence of the ions as their charges increase.

REVIEW QUESTIONS AND PROBLEMS

(Problems whose numbers are in blue have their answers in Appendix C at the back of the book; the more difficult problems are marked with asterisks.)

Mixtures; Suspensions and Colloids

11.1 What differentiates a mixture from a pure substance?

11.2 What range of particle sizes are found in suspensions? Give two examples of suspensions.

11.3 What range of particle sizes are found in colloids? What are these dimensions expressed in millimetres?

11.4 How can a suspension of a solid in a liquid be separated? (Give two methods.)

11.5 How does a centrifuge work?

11.6 Describe the Tyndall effect. What kind of mixture displays the Tyndall effect? Why?

11.7 For the following colloids, identify the nature of (1) the dispersing phase, (2) the dispersed phase, and (3) the kind of colloid:
(a) styrofoam
(b) cream
(c) lard
(d) jelly
(e) liquid rubber cement

11.8 What is an emulsifying agent?

11.9 Suppose you wished to determine whether a clear, colorless mixture was a solution or a colloid. What simple test could you perform?

11.10 What effect does electrical charge have on stabilizing a colloidal dispersion? How can a colloid that is stabilized in this way be coagulated?

11.11 When $AgNO_3$ is first added to a solution of NaCl, the mixture has a milky appearance that persists until enough Ag^+ has been added to react with all the Cl^-. At that point the precipitate coagulates and settles to the bottom of the container. Based on what you've learned in Section 11.1, explain these observations.

11.12 What is the one combination of phases for the dispersing agent and the dispersed phase that is incapable of yielding a colloidal dispersion? Why?

Types of Solutions

11.13 What kinds of solutions are possible?

11.14 How do the sizes of the particles in a solution compare to those in a colloidal dispersion or a suspension?

11.15 What is a substitutional solid solution? Give an example.

11.16 What is an interstitial solid solution? Give an example.

11.17 Suppose a solid solution is formed between two substances, one whose particles are very large and the other whose particles are very small. Which type of solid solution is this likely to be?

Concentration Units

11.18 State the definitions of the following units: mole fraction, mole percent, mass fraction, mass percent, molar concentration.

11.19 What feature do all concentration units have in common?

11.20 Calculate the mole fraction, mass fraction, and mass percent of glycerin in a solution prepared by dissolving 45.0 g of glycerin, $C_3H_5(OH)_3$, in 100.0 g of H_2O.

11.21 A mixture is prepared from 45.0 g of benzene (C_6H_6) and 80.0 g of toluene (C_7H_8). Calculate (a) the mass percent of each component, (b) the mole fraction of each component.

11.22 A solution containing 121.8 g of $Zn(NO_3)_2$ per dm^3

has a density of 1.107 g cm^{-3}. Calculate (a) the mass percent of $Zn(NO_3)_2$ in the solution, (b) the mole fraction of $Zn(NO_3)_2$, (c) the molar concentration of the solution.

11.23 What are the mole fraction and mass percent of a solution prepared by dissolving 0.30 mol of $CuCl_2$ in 40.0 mol of H_2O?

11.24 A sample of drinking water was found to be severely contaminated with chloroform, $CHCl_3$, a known carcinogen. The level of contamination was 12.4 ppm (by mass).
(a) Express this in percent by mass.
(b) What is the molar concentration of the $CHCl_3$ in the water?

11.25 A solution of isopropyl alcohol (rubbing alcohol), C_3H_7OH, in water has a mole fraction of alcohol equal to 0.250. What is the mass percent alcohol in the solution?

11.26 The solubility of baking soda, $NaHCO_3$, in water at 20°C is 9.6 g/100 g of H_2O. What is the mole fraction of $NaHCO_3$ in a saturated solution?

11.27 A saturated solution of NaCl at 30°C has a mole fraction of NaCl of 0.101. What is the mass fraction of NaCl in the solution?

11.28 A solution of sodium carbonate was prepared containing 14.0% Na_2CO_3 by mass. What is the mole fraction of Na_2CO_3 in this solution?

11.29 The World Health Organization's drinking water standards specify that the maximum permissible concentration of magnesium ion in drinking water is 150 mg dm^{-3}. What does this correspond to in terms of molar concentration?

11.30 Concentrated sulfuric acid is 96.0% H_2SO_4 by mass. What are the mole fractions of H_2SO_4 and water?

11.31 An aqueous solution of ammonium nitrate contained 2.25 mol NH_4NO_3 per kilogram of pure water. What was the mass percent of ammonium nitrate in the solution and what were the mole fractions of ammonium nitrate and water?

11.32 A solution of benzene, C_6H_6, dissolved in chloroform, $CHCl_3$, had a mole fraction of C_6H_6 equal to 0.240.
(a) What was the mole percent $CHCl_3$ in the solution?
(b) What were the mass percents of $CHCl_3$ and C_6H_6 in the solution?

11.33 An antifreeze solution is prepared from 222.6 g of ethylene glycol, $C_2H_4(OH)_2$, and 200.0 g of water. Its density is 1.072 g cm^{-3}. Calculate the molar concentration of the solution.

11.34 A 4.03 M solution of ethylene glycol, $C_2H_4(OH)_2$, has a density of 1.045 g cm^{-3}. Calculate the mass percent $C_2H_4(OH)_2$ and mole fraction of $C_2H_4(OH)_2$.

11.35 Below is a list of the most abundant ions in seawater.

Ion	mol/kg pure H_2O
Chloride	0.566
Sodium	0.486
Magnesium	0.055
Sulfate	0.029
Calcium	0.011
Potassium	0.011
Bicarbonate	0.002

Calculate the mass, in grams, of each component contained in 4.00 dm^3 of seawater with a density of 1.024 g cm^{-3}. What is the total mass of ions in this sample?

*11.36 Suppose that you wish to prepare a solution containing 10.0% Na_2CO_3 by mass. The bottle of chemical that you have lists the contents as $Na_2CO_3 \cdot 10H_2O$. How many grams of the hydrate would be needed to prepare 50.0 g of the 10.0% Na_2CO_3 solution?

The Solution Process

11.37 Why do nonpolar solutes tend to be insoluble in polar solvents?

11.38 Molecular H_2 and O_2 are relatively insoluble in water, but ammonia is quite soluble. Why?

11.39 Which gas, NH_3 or PH_3, would you expect to be more soluble in water? Why?

11.40 As applied to solubility, what is the significance, on a molecular level, of the phrase "like dissolves like"?

11.41 Frequently a liquid that is soluble in water can be made less soluble (and can therefore be made to separate as a distinct phase) by addition of salt to the solution. How can this be explained?

11.42 Small amounts of water in the fuel tank of an automobile can cause severe difficulties in engine performance. The problem can be overcome by adding "dry gas" to the fuel. The "dry gas" consists mostly of methyl alcohol, CH_3OH, which allows the water to dissolve in the gasoline. Can you explain how the dry gas accomplishes this?

11.43 What is meant when we say an ion is hydrated? What is the meaning of the term solvation?

11.44 What role does water play in dissolving an ionic compound?

11.45 What is a micelle? Why does soap form micelles? How does a soap or detergent remove a grease spot from clothing?

11.46 What advantages do synthetic detergents have over soap?

Heats of Solution

11.47 How is the heat of solution defined? What happens to the temperature of a mixture as a solution is formed if the ΔH_{soln} is negative?

11.48 The combined attractions between ethanol molecules in pure ethanol and CCl_4 molecules in pure CCl_4 are larger than the attractions between ethanol and CCl_4 molecules. When ethanol dissolves in CCl_4, is ΔH_{soln} positive or negative?

11.49 Compare the definitions of an ideal gas and an ideal solution.

11.50 Discuss the relationship between lattice energy and hydration energy in determining the magnitude and sign of the heat of solution of a solid in a liquid.

11.51 On the basis of size and charge, choose the ion in each of the following pairs with the larger hydration energy:
(a) Na^+ or K^+ (d) Fe^{2+} or Fe^{3+}
(b) F^- or Cl^- (e) S^{2-} or Cl^-
(c) K^+ or Ca^{2+}

11.52 Why is ΔH_{soln} for gases nearly always negative?

11.53 The heat of solution for $AlCl_3$ is -321 kJ mol^{-1}. What is the probable reason for this highly exothermic ΔH_{soln}?

11.54 For NaCl, $\Delta H_{soln} = +4.94$ kJ mol^{-1}. For this salt, which is larger: the lattice energy or the hydration energies of the ions?

11.55 Why is it so difficult to predict whether the heat of solution of a solid in a liquid will be endothermic or exothermic?

11.56 Use the data in Table 11.3 to calculate the quantity of heat liberated by dissolving 10.0 g of $AlCl_3$ in 1.00 dm^3 water.

11.57 How many joules are absorbed by dissolving 115 g of NH_4NO_3 in 100 cm^3 of H_2O?

The Effect of Temperature on Solubility

11.58 The solution process for KI in water is endothermic (ΔH_{soln} is positive). Would you expect KI to become more or less soluble as the temperature is increased? Explain.

11.59 Describe, qualitatively, the procedure called fractional crystallization.

11.60 On the basis of the information provided in Figure 11.14, predict which solid will separate first from a solution containing equal masses of KNO_3 and KBr when the solution is gradually evaporated at a temperature of 70°C. What will occur if the solution is gradually evaporated at 20°C?

11.61 How many grams of $NaNO_3$ will precipitate if a saturated solution of $NaNO_3$ in 200 g of H_2O at 70°C is cooled to 25°C?

11.62 How many grams of water at 80°C are required to dissolve 35.0 g of NaBr?

11.63 Why do gases always become less soluble in liquids as the temperature is raised?

The Effect of Pressure on Solubility

11.64 Why do gases become more soluble in liquids as the pressure is increased?

11.65 Why do pressure changes have very little effect on the solubility of a solid in a liquid?

11.66 The partial pressure of ethane over a saturated solution containing 6.56×10^{-2} g of ethane is 100.1 kPa. What is its partial pressure when the saturated solution contains 5.00×10^{-2} g of ethane?

11.67 The Henry's law constant for a gas dissolved in water was found to be 4.88×10^{-4} g dm^{-3} kPa^{-1} at 25°C. In an experiment the gas was collected over water and its concentration was found to be 0.0478 g dm^{-3}. What was the *total* pressure of gas above the solution?

11.68 Methane (natural gas) has a solubility in water at 25°C of 2.09×10^{-4} g dm^{-3} at a pressure of 98.1 kPa. If we assume Henry's law is valid, what would be the concentration of methane (in g dm^{-3}) in ground water deep below the earth's surface at a pressure of 100 MPa and a temperature of 25°C?

*11.69 Air contains approximately 20% O_2 by volume. The Henry's law constant for O_2 at 25°C is 4.01×10^{-4} g dm^{-3} kPa^{-1}. Calculate the mass of O_2 per dm^3 of water in a stream that has a temperature of 25°C if the atmospheric pressure is 101 kPa. (Assume equilibrium with the atmosphere.)

Vapor Pressures of Solutions: Raoult's Law

11.70 Explain, on a molecular level, why the vapor pressure of the solvent is expected to be directly proportional to its mole fraction in the solution (Raoult's law).

11.71 What are meant by positive and negative deviations from Raoult's law?

11.72 How is the sign of ΔH_{soln} related to positive and negative deviations from Raoult's law?

11.73 The vapor pressure of benzene (C_6H_6) at 25°C is 12.5 kPa. What will be the vapor pressure at 25°C, of a solution prepared by dissolving 56.4 g of the nonvolatile solute, $C_{20}H_{42}$, in 1.000 kg of benzene?

11.74 The vapor pressure of pure methyl alcohol at 30°C is 21.3 kPa. What mole fraction of glycerol (a nonvolatile nondissociating solute) would be required to lower the vapor pressure to 17.3 kPa?

11.75 Heptane (C_7H_{16}) has a vapor pressure of 105 kPa at 100°C. At this same temperature, octane (C_8H_{18}) has a vapor pressure of 47 kPa. What will be the vapor pressure of a mixture of 25.0 g of heptane and 35.0 g of octane? Assume ideal solution behavior.

11.76 At 25°C the vapor pressures of benzene (C_6H_6) and toluene (C_7H_8) are 12.5 and 3.59 kPa, respectively. At what applied pressure will a solution prepared from 60.0 g of benzene and 40.0 g of toluene boil at 25°C?

11.77 The vapor pressure of a mixture containing 400 g of carbon tetrachloride and 43.3 g of an unknown substance is 18.3 kPa at 30°C. The vapor pressure of pure carbon tetrachloride at 30°C is 19.1 kPa, while that of the pure unknown is 11.3 kPa. What is the approximate molar mass of the unknown?

11.78 A solution containing 8.3 g of a nonvolatile nondissociating substance dissolved in 1 mol of chloroform, $CHCl_3$, has a vapor pressure of 68.1 kPa. The vapor pressure of pure $CHCl_3$ at the same temperature is 70.1 kPa. Calculate (a) the mole fraction of the solute, (b) the number of moles of solute, (c) the molar mass of the solute.

11.79 To what temperature (in °C) would a solution containing 150 g of glycerol, $C_3H_5(OH)_3$, in 100 g of H_2O have to be heated to have a vapor pressure of 12.1 kPa? Assume ideal solution behavior, $C_3H_5(OH)_3$ to be a nonvolatile solute, and refer to Table 9.1.

Boiling-Point Diagram and Fractional Distillation

11.80 Describe, qualitatively, the procedure called fractional distillation.

11.81 If we refer to Figure 11.19, approximately how many times must boiling, followed by condensation of the resulting vapor, be repeated in order to obtain a portion of liquid having a mole fraction of A of at least 0.80 if the original mole fraction of A was 0.20?

11.82 Benzene has a boiling point of 80.1°C; carbon tetrachloride boils at 76.8°C. Sketch a boiling-point diagram for benzene-carbon tetrachloride mixtures. Assume ideal solution behavior.

11.83 Water and butyl alcohol form an azeotrope that boils at 92.4°C at 1 atm. At this same pressure butyl alcohol boils at 117.8°C. The composition of the azeotrope is 28.4 mol % butyl alcohol, 71.6 mol % water. Sketch the boiling-point diagram for butyl alcohol-water mixtures. Do these substances show positive or negative deviations from ideality?

Boiling-Point Elevation and Freezing-Point Depression

11.84 What is a *colligative property?*

11.85 We've described how the addition of a nonvolatile solute to a solvent reduces the escaping tendency of the solvent from the solution, and in Section 11.10 we saw that this leads to a boiling-point elevation. On a molecular level, account for the fact that the presence of a solute also reduces the tendency of the solvent to escape from the liquid onto the solid. Explain why a lower temperature must be achieved to establish equilibrium between the solid solvent and the solution than between the pure solid and liquid solvent.

11.86 What will be the freezing point and boiling point of an aqueous solution containing 55.0 g of glycerol, $C_3H_5(OH)_3$, dissolved in 250 g of water? Glycerol is a nonvolatile, undissociated solute.

11.87 What is the molar mass and molecular formula of a nondissociating compound whose empirical formula is C_4H_2N if 3.84 g of the compound in 500 g of benzene give a freezing-point depression of 0.307°C?

11.88 A solution containing 16.9 g of a nondissociating substance in 250 g of water has a freezing point of −0.744°C. The substance is composed of 57.2% C, 4.77% H, and 38.1% O. What is the molecular formula of the compound?

11.89 How many grams of glucose, $C_6H_{12}O_6$ (a nondissociating solute), are required to lower the freezing point of 150 g of H_2O by 0.750°C? What will be the boiling point of this solution?

11.90 An aqueous solution freezes at −2.47°C. What is its boiling point?

11.91 Calculate the freezing point in °C of an aqueous solution of a weak electrolyte ($x_A = 1.80 \times 10^{-3}$) that is 7.5% dissociated.

*11.92 The cooling system of an automobile usually contains a solution of antifreeze prepared by mixing equal volumes of ethylene glycol, $C_2H_4(OH)_2$, and water. The density of ethylene glycol is 1.113 g cm^{-3}. Calculate the freezing point of this mixture.

On the label of the antifreeze container it is said that this mixture will protect your engine to a temperature of $-36°C$. How does your computed freezing point compare with this?

*11.93 What is the percent dissociation of a weak electrolyte, HX, in water if a solution with 4.48×10^{-3} mole fraction of the substance has a freezing point of $-0.500°C$?

Osmotic Pressure

11.94 What is the difference between osmosis and dialysis?

11.95 What is a *semipermeable membrane*?

11.96 What are *isotonic solutions*? Why must the solute concentration be carefully controlled in intravenous feeding?

11.97 Calculate the osmotic pressure, in kPa, of an aqueous solution containing 5.0 g of sucrose, $C_{12}H_{22}O_{11}$, per dm^3 at 25°C.

11.98 A solution of 0.400 g of a polypeptide in 1.00 dm^3 of an aqueous solution has an osmotic pressure at 27°C of 0.499 kPa. What is the approximate molar mass of this polymer?

Solutions of Electrolytes

11.99 If we assume complete dissociation, what would the expected freezing point be of a solution containing 0.0018 mole fraction MgSO$_4$?

11.100 What would be the expected freezing point of 0.0018-mole fraction CaCl$_2$ solution?

11.101 If the measured freezing-point depression of a solution is *less* than that calculated for a solute, what does this suggest?

11.102 What would be the osmotic pressure, in kPa, of a 0.010 *M* aqueous solution of the electrolyte, NaCl, at 25°C? (Assume 100% dissociation of NaCl in water).

11.103 Calculate the *i* factor for the weak electrolyte, HF, in Example 11.10. What conclusions would you draw about the *i* factors for weak electrolytes?

11.104 What is the osmotic pressure of sea water at 25°C? What is the minimum pressure, in MPa, needed to desalinate seawater by reverse osmosis? (See the data in Problem 11.35.)

*11.105 A solution containing 0.1000 mol sugar per 1.000 kg H$_2$O was placed in the left compartment of the apparatus described in Figure 11.24, and a solution containing 0.1000 mol of acetic acid, HC$_2$H$_3$O$_2$, per 1.000 kg H$_2$O was placed in the right compartment. At 25°C, the osmotic pressure of the HC$_2$H$_3$O$_2$ solution was then measured to be 3.32 kPa larger than the sugar solution. What fraction of the acetic acid is dissociated in this solution?

Interionic Attractions

11.106 What would the actual freezing point be of a 0.0018-mole fraction MgSO$_4$ solution, if we take into account interionic attractions?

11.107 Based on the data in Table 11.7, which 1:1 electrolyte (one positive ion to one negative ion) appears to be *least* fully dissociated in concentrated solutions? How does this agree (or disagree) with what might be predicted based on the charges on the ions involved?

11.108 On the basis of what you have learned in this chapter, how would you interpret an *i* factor having a value less than 1.00?

11.109 If we assume equal concentrations in water, which of the following salts would you expect to appear to be least fully dissociated: KCl, NiCl$_2$, or Al$_2$(SO$_4$)$_3$?

11.110 If we assume complete dissociation, what *i* factor would you expect for each of the salts in Question 11.109?

CHAPTER 12

CHEMICAL THERMODYNAMICS

The Hotel Madison in Boston is demolished in seconds by explosives placed strategically on its ground floor. Once begun, this spontaneous collapse was sure to happen because it was accompanied by both a decrease in potential energy and an increase in disorder. In this chapter we will see that these two factors ultimately control the fate of all chemical and physical changes.

In a study of chemistry it's natural to question why certain chemical reactions take place and others do not. Certainly, it would be nice if we could predict ahead of time what will happen when several chemicals are mixed. Then we could sit at home and do our chemistry, instead of going to the lab. Unfortunately (or fortunately, depending on your love for lab work), chemistry hasn't evolved to that point yet, but we do know what controls the outcome of a reaction.

There are two factors that determine whether or not we will observe a particular reaction, either in the laboratory or elsewhere. Thermodynamics, studied in this chapter, tells us whether a change is possible—it answers the question, "Will the reaction occur by itself, without outside help?" It also tells us what the position of equilibrium will be when the composition of the reaction mixture ceases to change. Chemical kinetics, the subject of Chapter 13, is concerned with the speeds at which chemical changes take place. Both of these factors, spontaneity and speed, must be in our favor if we hope to observe the formation of the products of a chemical change. For example, thermodynamics predicts that at room temperature hydrogen gas and oxygen gas should react to produce water. However, a mixture of H_2 and O_2 is stable virtually indefinitely (provided no one strikes a match). This is so because, at room temperature, hydrogen and oxygen react at such an extremely slow rate that even though their reaction to produce water is spontaneous, it takes nearly forever to proceed to completion.

Thermo implies heat; *dynamics* implies movement or change.

Thermodynamics is basically concerned with the energy changes that accompany chemical and physical processes. Historically, it evolved without a detailed knowledge of the structure of matter; in fact, this is one of its strongest points. In this chapter we will take a rather informal approach to the subject in an effort to avoid mathematical formalism, and we will develop many of the concepts of thermodynamics by considering changes that take place on a molecular level.

12.1 SOME COMMONLY USED TERMS

Before we proceed, we must first establish the meaning of some frequently used terms. A word that has been used rather loosely in previous sections is **system.** By system we mean *that particular portion of the universe on which we wish to focus our attention.* Everything else we call the **surroundings.** For example, if we wished to consider the changes taking place in a solution of sodium chloride and silver nitrate, our system is the solution, while the beaker and everything else around the solution are considered the surroundings.

A system and its surroundings are separated from each other by a boundary, or interface, and it is across this boundary that we consider the exchange of energy. Usually the boundary is real and apparent, as the surface of a solution and the walls of a beaker. Sometimes, however, the boundary is imaginary and chosen simply for convenience. For example, suppose we were to consider the Earth as a system and the rest of the universe as the surroundings. We know that the atmosphere doesn't just end at some particular altitude, so there isn't a real boundary between system and surroundings. In this case, we would choose arbitrarily some particular altitude, where the atmosphere has become very thin, as the place where our system ends and the surroundings begin.

If a change occurs so that heat cannot be transferred across the interface between the system and its surroundings, we speak of it as an **adiabatic** process. An example is a reaction carried out in an insulated container, such as a Thermos bottle. An explosive reaction is another example of an adiabatic process. Such a reaction occurs so rapidly that the heat energy produced cannot be readily dissipated. The heat build-up raises the products to very high temperatures, and these products fly apart rapidly, pushing walls, ceilings, and so on (the surroundings) before them.

When thermal contact is maintained between a system and its surroundings, heat can flow between them and it is frequently possible to keep the system at a constant temperature while a change takes place. In this case the change is said to be **isothermal.** The human body possesses an elaborate temperature-control system that maintains a constant body temperature, so biochemical reactions taking place within us are therefore essentially isothermal.

To discuss the changes that occur in a system it is necessary to define its properties very precisely before and after the change occurs. We do this by specifying the **state** of the system—that is, some particular set of conditions of pressure, temperature, number of moles of each component, and their physical form (for example, gas, liquid, solid, or crystalline form). Once these variables are specified, all the properties of the system are fixed. Thus a knowledge of these quantities permits us to define unambiguously the properties of a system. For instance, if we have two samples of pure liquid water, each consisting of 1 mol and each at the same temperature and pressure, we know that all the properties of each sample will be identical (volume, density, surface tension, vapor pressure, etc.).

The quantities P, V, and T are called **state functions** or **state variables.** This is because they serve to determine the physical state of a given system, and in a particular state their values do not depend on the prior history of the sample. For example, the volume of 1 mol of water at 25°C and 1 atm (101.325 kPa) does not depend on what its temperature or pressure might have been at some time in the past. Furthermore, on going from one state to another the changes in P, V, and T do not depend on how the sample is treated. If the temperature of a sample of water is changed from 25°C to 35°C, it doesn't matter if the sample is first cooled to 0°C and then warmed to 35°C, or whether the temperature is increased directly from 25°C to 35°C. In the final state the temperature is the same regardless of the path taken between the initial and final conditions, and the change in temperature, ΔT, therefore depends *only* on the temperatures of those initial and final states.

There are some instances in which the interrelationships between the state functions can be expressed in equation form to give an **equation of state.** The equation of state for an ideal gas, $PV = nRT$, is an example. We have also seen the van der Waals equation of state, which can be applied with reasonable success to real gases.

Another quantity we will use is called the **heat capacity**—*the quantity of heat energy required to raise the temperature of a given quantity of a substance by one degree Celsius.* The units of heat capacity are $J\ °C^{-1}$. The **specific heat** represents the heat capacity per gram; that is, it is *the quantity of heat necessary to raise the temperature of 1 g of a substance by 1°C.* The specific heat of water is $4.184\ J\ g^{-1}\ °C^{-1}$. We also speak of the **molar heat capacity**—*the heat necessary to raise the temperature of 1 mol of a substance by 1 degree.* Since the molar mass of water is $18.0\ g\ mol^{-1}$, the molar heat capacity of water is $75.3\ J\ mol^{-1}\ °C^{-1}$.

The magnitude of the change in a state function is independent of how a system goes from one state to another.

EXAMPLE 12.1 CALCULATING THE HEAT CAPACITY OF A WATER BATH

PROBLEM   What would be the heat capacity expressed in $kJ\ °C^{-1}$ of a water bath containing $4.00\ dm^3$ of water? The specific heat of water is $4.184\ J\ g^{-1}\ °C^{-1}$.

SOLUTION

For simplicity, we've ignored the heat capacity of the container that holds the water.

We wish to calculate the number of joules needed to raise the temperature of the water in the bath by 1.00°C. If we assume the density of water is 1.00 g cm^{-3}, the mass of water is 4000 g. Therefore,

$$\text{heat capacity of bath} = 4000 \text{ g} \times \frac{4.184 \text{ J}}{\text{g °C}} \times \frac{1 \text{ kJ}}{1000 \text{ J}}$$

$$= 16.7 \text{ kJ °C}^{-1}$$

12.2 THE FIRST LAW OF THERMODYNAMICS

In thermodynamics we study the energy changes that occur when systems pass from one state to another. Repeated observations by many scientists over many years have led to the conclusion that in any process, energy is neither created nor destroyed. This conclusion is the basis of the *law of conservation of energy* that we discussed in Chapter 1.

An important consequence of this law is that if we bring a system from one state to another, the net energy change must be the same regardless of the path that we follow. Stated another way, the difference in energy between two states is a constant, and we obtain the same energy difference regardless of how we cause the system to change from one state to another. If this were not true, we could create energy in violation of the law of conservation of energy. For example, suppose we could liberate 800 J of energy by condensing a certain amount of water vapor at 1 atm and 100°C. Let's also suppose that we could find a way of using only 400 J of energy to vaporize this same amount of water at the same temperature and pressure. If this were possible, we could continually vaporize and condense the water, and on each cycle we would obtain, free, 400 J. We would have invented a *perpetual motion machine*, a device that creates energy. But because the law of conservation of energy is a *fact*, perpetual motion schemes such as this are impossible. (In fact, the U.S. Patent office refuses to issue patents on them unless the inventor is able to supply a working model.) The **first law of thermodynamics** recognizes the impossibility of perpetual motion machines by stating that *if a system undergoes some series of changes that ultimately brings it back to its original state, the net energy change is zero.*

After this rule was adopted, the Patent Office stopped receiving patent applications for perpetual motion machines.

Normally we are not very interested in processes that bring us back to where we started. Instead we are concerned with how the energy changes when a system is transformed from one state to another. Thermodynamics defines a quantity called the **internal energy,** E, that is used in describing these changes when they take place in either chemical or physical systems. The internal energy is the total energy of the system—the total of all the energies that it possesses as a consequence of the kinetic energy of its atoms, ions, or molecules, plus all the potential energy that arises from the binding forces between the particles making up the system. A change in the internal energy, ΔE, is defined as

$$\Delta E = E_{\text{final}} - E_{\text{initial}}$$

Notice that we have used the same convention here (final minus initial) as in our previous discussions of energy changes (heats of vaporization, heats of solution, etc.). Here, too, we can't actually determine E_{final} or E_{initial}. The reason, which we promised you in our earlier discussions, is that we don't know how fast a system or its particles are moving, and we have no way of knowing the effects of all the attractive forces on the system. For instance, any system that we study moves as the earth revolves around its axis. The earth, in turn, revolves around the sun, which moves through the galaxy, which moves through space. How fast, nobody knows, so we can't calculate the system's kinetic energy. Nor can we figure the total potential energy caused by all the attractive forces between the system and the rest of the universe. But as we've noted before, this is no terrible crisis because we are only interested in how the energy *changes*, and this we can measure.

When a system changes from one state to another, there are two ways for it to exchange energy with its surroundings. One is for it to gain or lose heat energy. If the system absorbs heat, its energy rises, and if it loses heat, its energy drops. The second way for the system to exchange energy with its surroundings is to do work or have work done on it. If the system does work, its energy drops (just as you have less energy after doing work). On the other hand, if work is done on the system, its energy rises. For example, when you (the surroundings) do work on a watch (the system) by winding its spring, the energy of the watch increases. The energy bookkeeping for both heat and work is taken care of by the equation

$$\Delta E = q - w \qquad\qquad [12.1]$$

where q is defined as the *heat absorbed from the surroundings* by a system when it undergoes a change, and w is defined as the *work done by the system on its surroundings*. This is also a form of the first law of thermodynamics and simply states that the change in the internal energy is equal to the difference between the quantity of energy *gained* by the system in the form of heat and the quantity of energy *removed* in the form of work that is performed on the surroundings.[1]

Since Equation 12.1 deals with the transfer of quantities of energy, it is necessary to establish sign conventions to avoid confusion in our bookkeeping. Heat *added to* a system and work *done by* a system are considered positive quantities. Thus, if a certain change is accompanied by the absorption of 50 J of heat and the expenditure of 30 J of work, $q = +50$ J and $w = +30$ J. The change in internal energy of the system is

$$\Delta E_{system} = (+50 \text{ J}) - (+30 \text{ J})$$

or

$$\Delta E_{system} = +20 \text{ J}$$

Thus the system has undergone a net increase in energy amounting to +20 J. How about the surroundings?

When the system gains 50 J, the surroundings lose 50 J; therefore $q = -50$ J for the surroundings. When the system performs work, it does so on the surroundings. We say that the surroundings have done negative work, and $w = -30$ J for the surroundings. The change in the internal energy of the surroundings is thus

$$\Delta E_{surroundings} = (-50 \text{ J}) - (-30 \text{ J})$$

$$\Delta E_{surroundings} = -20 \text{ J}$$

The change in the internal energy of the system is thus equal, but opposite in sign, to ΔE for the surroundings. This has to be so in order to satisfy the law of conservation of energy.

Margin notes:

ΔE is equal to the energy input minus the energy output.

One statement of the first law of thermodynamics is that $\Delta E = 0$ for a change in which a system begins and ends in the same state.

---

[1] You may find that some thermodynamicists prefer to write Equation 12.1 as $\Delta E = q + w$. In doing so, they define w as the quantity of energy gained if the system has work done on it. In other words, this equation states that an increase in the internal energy is the sum of the energy gained by the system in the form of heat, plus the energy that the system gains by having work done on it. If you think about this for a while, you will realize that both equations are equivalent. It is just the meaning (and algebraic sign) of w that differs. Now that you are aware of this, we shall continue to use q and w as defined in the text and by Equation 12.1.

In summary,

q positive ($q > 0$); heat is added to the system.
q negative ($q < 0$); heat is evolved by (removed from) the system.
w positive ($w > 0$); the system performs work—energy is removed.
w negative ($w < 0$); work is done on the system—energy is added.

The internal energy is a state function, and the magnitude of ΔE therefore depends only on the initial and final states of the system and not the path taken between them. This is very much the same as the change in your bank balance that occurs between the beginning and the end of a month. During any given month the change in the balance is brought about as the combined results of some number of deposits and withdrawals. If the total number of dollars provided by the deposits exceeds those removed by the withdrawals, your balance increases. However, the net change in your balance at the end of the month depends only on the initial and final amounts of money in the bank, not on the individual transactions during the month. There is an infinite number of combinations of deposits and withdrawals that could lead to the same change in your balance. The same sort of relationship exists among ΔE, q, and w. The sign and magnitude of ΔE are controlled only by the values of E in the initial and final states. For any given change, ΔE, there are many different paths that can be followed with their own characteristic values of q and w. However, for the same initial and final states, the difference between q and w is always the same. Therefore, even though ΔE is a state function, q and w are not.

q and w are like the deposits and withdrawals. E is like the bank balance.

Work

Before we can study the chemical or physical implications of the first law of thermodynamics, we have to examine how a system can do work. One kind of work a system can perform is electrical work. That's what happens when energy is withdrawn from a battery to power a wristwatch or start a car. Chemical reactions in the battery produce energy that pushes electrons through a wire, and this electrical energy can do work for us.

Another kind of work that is especially important in chemical changes is accomplished when a system expands its volume against an opposing pressure. Work is always accomplished when an opposing force is pushed through some distance. Expressed mathematically,

$$\text{work} = \text{force} \times \text{distance}$$

Pressure is defined as force per unit area, and in Figure 12.1 we see an external pressure P exerted by a force F spread over the area A of a piston.

$$P = \frac{F}{A}$$

The volume of the gas in the cylinder is equal to its cross-sectional area, A, multiplied by the height of the column of gas, h.

$$V = Ah$$

When the gas expands and pushes back the piston, A remains the same but h changes. The volume change is therefore

$$\Delta V = V_f - V_i$$

$$\Delta V = Ah_f - Ah_i$$

$$\Delta V = A(h_f - h_i) = A(\Delta h)$$

The product of pressure times volume change is

Figure 12.1

Pressure-volume work.

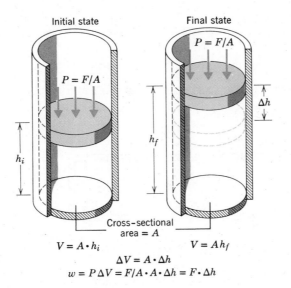

$$P\,\Delta V = \frac{F}{A}A(\Delta h) = F\,\Delta h$$

We see that $P\,\Delta V$ is equivalent to force (F) times distance (Δh) and, therefore, equals work.

$$w = P\,\Delta V \qquad\qquad [12.2]$$

This expansion work is performed by any system—it doesn't have to be a gas—when it expands against an external pressure imposed by the surroundings. Conversely, when a system contracts under the influence of an external pressure, work is performed on the system. With SI units, pressure is expressed in pascals (newtons per square metre)

$$1\ \text{Pa} = 1\ \text{N/m}^2$$

and volume is expressed in cubic metres, m^3. Therefore, the units of $P\,\Delta V$ are

$$\text{Pa}\ \text{m}^3 = \frac{\text{N}}{\text{m}^2}\cdot\text{m}^3 = \text{N}\cdot\text{m}$$

but $1\ \text{N}\cdot\text{m} = 1\ \text{J}$. Therefore,

$$1\ \text{Pa}\ \text{m}^3 = 1\ \text{J}$$

Let's look now at a couple of examples that illustrate what the first law of thermodynamics tells us. First, any particular change, whether it be physical or chemical, has a certain net energy change associated with it that is determined solely by the energies of the initial and final states. Once we have picked what these states are, the value of ΔE is fixed. But there are many different sets of q and w that produce the same ΔE. In other words, the quantity of work that we can obtain from a given change depends on how we carry it out.

A physical example that illustrates this is the expansion of a gas. For simplicity, let's imagine an ideal gas expanding in such a way that its temperature remains constant. This can be accomplished, for example, by placing a cylinder of the gas in a large vat of water held at a constant temperature by a thermostat.

If the temperature is constant during the expansion, then the average kinetic energy of the molecules stays the same. Furthermore, since there are no attractive forces between the particles of an ideal gas, there is no potential en-

Figure 12.2

Expansion of a gas against an opposing pressure equal to zero.
(a) *Before expansion.* P = 1000 kPa; V = 1.00 dm³.
(b) *After expansion, final state.* P = 100 kPa; V = 10.0 dm³.

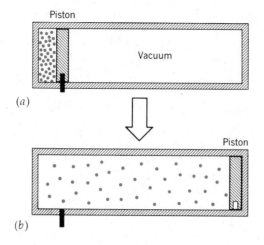

Since the change in K.E. and P.E. are both zero, $\Delta E = 0$.

ergy. This means that when an ideal gas expands or contracts at constant temperature there is no net change in its total energy, so $\Delta E = 0$. Therefore

$$\Delta E = 0 = q - w$$

$$q = w$$

This tells us that if the gas does work by expanding, it absorbs an equivalent quantity of heat energy from the surroundings.

This really only holds for an ideal gas. For a real gas ΔE is small, but not equal to zero.

If you've ever lifted a heavy box, you know that the quantity of work you have to do depends on how heavy the box is—that is, on how hard the box pushes back. So it is with the expansion of a gas. In Figure 12.2a we see a compressed gas pushing against a piston held by a pin through the cylinder wall. On the other side of the piston is a vacuum. When the pin is pulled out of the piston, you know what will happen. The compressed gas will push the piston to the opposite end of the cylinder (Figure 12.2b). But the gas won't do any work, because the piston doesn't push back; there is no opposing pressure during the expansion. Thus $w = 0$ and $q = 0$.

Now let's look at the expansion pictured in Figure 12.3. Here we begin with the same gas, at the same temperature as before, and at the same initial pressure and volume (1000 kPa and 1.00 dm³). We also finish at the same set of conditions as before, so ΔE must again equal zero and $q = w$. This time, however, we have a constant opposing pressure of 100 kPa. When the pin is pulled back and the gas expands, it has to push against this opposing pressure and it therefore does work. The quantity of work is equal to $P \, \Delta V$. Since the final volume is 10.0 dm³, ΔV equals 9.0 dm³. Therefore,

$$P_iV_i = P_fV_f$$
$$(100 \text{ kPa})(1 \text{ dm}^3) = (100 \text{ kPa})(10 \text{ dm}^3)$$

$$w = P \, \Delta V = (100 \text{ kPa})(9.0 \text{ dm}^3) \qquad w = (100 \times 10^3 \text{ N m}^{-2}) \times (9 \times 10^{-3} \text{ m}^3)$$

opposing pressure

$$w = 900 \text{ N m} = 900 \text{ J}$$

This 900 J of work done by the gas is equal to 900 J of heat energy that the gas absorbed from the surroundings.

Heat and work in chemical systems

The two expansions described by Figures 12.2 and 12.3 have the same ΔE and show that the values of q and w depend on how a change is carried out. But what has this got to do with chemistry? How does it apply to a chemical system? To answer this question, let's consider an ordinary automobile battery in which a

Figure 12.3

Expansion of an ideal gas against a constant opposing pressure of 100 kPa. (a) Initial state: P_{gas} = 1000 kPa, V_{gas} = 1.00 dm³. (b) Final state: P_{gas} = 100 kPa, V_{gas} = 10.0 dm³.

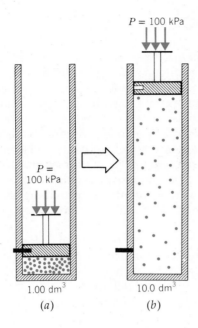

1.00 dm³   10.0 dm³
(a)   (b)

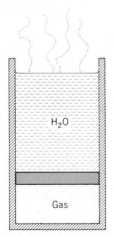

Figure 12.4

A reversible expansion of a gas. As water molecules gradually evaporate, the pressure decreases and the gas expands. The process will be reversed if water molecules begin to condense into the liquid instead of evaporate.

redox reaction provides the source of electricity. What we discover here is that the energy that we get in the form of work depends on how we discharge the battery. If we simply take a steel bar and place it across the terminals, sparks fly and the battery gets very hot—all the energy produced by the chemical reaction inside the battery appears as heat and we've done no work. On the other hand, if we connect the battery to an electric motor, like the starter of a car, we can use some of the energy of the reaction in the battery to crank the engine. This rotates parts and moves pistons. Work is accomplished because the motor offers some resistance to the flow of electricity. It is as if the motor supplied an opposing pressure against which the electrical "pressure" could work.

What is particularly interesting is that if very careful measurements are made, we find that the quantity of work that can be done by a battery depends on how fast it is discharged. If we discharge it quickly by offering little resistance to the flow of electricity, heat is generated that can't be used as work. As we increase the resistance, we slow down the rate at which energy is removed, but more of it can appear as work. The limit of this is where the energy is withdrawn infinitely slowly by having the battery just barely able to overcome the opposing resistance—in this case the work done would be a maximum. Any change such as this in which the driving force is virtually balanced by an opposing force is called a **reversible process.** This is because increasing the opposing force just slightly would reverse the direction of the change. An example in a physical system is the piston-cylinder apparatus in Figure 12.4. As the water evaporates one molecule at a time, the pressure opposing the expansion is very gradually decreased to match the internal pressure of the gas. The internal pressure, which is the driving force for the expansion, is continually balanced by the opposing pressure, and if a molecule of water vapor were to condense, the pressure would increase and a slight compression of the gas would occur.

A very important concept, which we will return to later, is that *the maximum work available from any change is obtained if the change occurs by a reversible process.* Unfortunately, reversible processes like the infinitely slow discharge of a battery take forever to occur. All real, spontaneous changes do not take place by reversible processes and the work that can be derived from them is always less than the theoretical maximum.

12.3 HEATS OF REACTION: THERMOCHEMISTRY

Virtually every chemical reaction occurs with either the absorption or evolution of energy. These changes reflect the differences between the potential energies associated with the bonds in the reactants and the products. For instance, when two hydrogen atoms come together to form an H_2 molecule, energy is evolved because the total potential energy of the nuclei and electrons in the H_2 molecule is lower than the total potential energy of these particles in two isolated H atoms. This same energy, when added to an H_2 molecule, can break it apart, and thus represents the bond energy discussed in Chapter 5. We see, therefore, that the measurement of the quantity of energy evolved or absorbed when a chemical reaction occurs has the potential of providing us with very fundamental information concerning the stability of molecules and the strengths of chemical bonds.

If we conduct a chemical reaction in a closed container of fixed volume, the system undergoing reaction cannot perform pressure-volume work on the surroundings because $\Delta V = 0$; hence $P\,\Delta V = 0$. Any heat absorbed or evolved under these circumstances (let's call it q_v, meaning heat absorbed at constant volume) is precisely equal to the change in the internal energy of the system,

$$\Delta E = q_v \qquad [12.3]$$

Stated another way, ΔE is equal to the heat absorbed or evolved by the system under conditions of constant volume (i.e., the *heat of reaction at constant volume*). When the reaction is endothermic, both q and ΔE are positive; for an exothermic process q and ΔE are negative.

Experimentally, ΔE can be measured by using a device called a **bomb calorimeter** (Figure 12.5). The apparatus consists of a strong steel bomb into which the reactants (for example, H_2 and O_2) are placed. The bomb is then immersed in an insulated bath containing a precisely known quantity of water. There the reaction is set off by a small heater wire within the bomb, and heat is evolved. The entire system is permitted to come to thermal equilibrium, at which point the calorimeter (bomb and water) will be at a higher temperature than before reaction. By carefully measuring the temperature of the water before

A bomb calorimeter is normally used to measured ΔE for exothermic reactions.

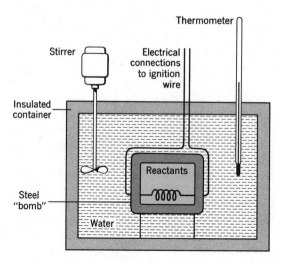

Figure 12.5
Bomb calorimeter.

and after reaction, and by knowing the heat capacity of the calorimeter (including the bomb and the water), the quantity of heat evolved by the chemical reaction can be computed.

EXAMPLE 12.2   CALCULATING THE HEAT OF REACTION DETERMINED IN A BOMB CALORIMETER

PROBLEM Hydrogen (0.100 g) and oxygen (0.800 g) are compressed into a 1.00-dm^3 bomb, which is then placed into the water in a calorimeter. Before the reaction is set off, the temperature of the water is 25.000°C; after reaction the temperature rises to 25.155°C. The heat capacity of the calorimeter (bomb, water, etc.) is 90.8 kJ °C^{-1}. What is ΔE for this reaction?

SOLUTION The change in temperature that occurs is 0.155°C. From the heat capacity we can find the number of joules evolved.

$$\text{Heat evolved, } q = \left(90.8 \frac{kJ}{°C}\right)(0.155°C)$$

Therefore

$$q_v = -14.1 \text{ kJ}$$

Hence,

$$\Delta E = -14.1 \text{ kJ}$$

Notice that q_v and ΔE are both negative because the system—the H_2 and O_2—evolved energy when they reacted.

An extensive property depends on the size of the sample being studied. An intensive property is independent of sample size.

For the reaction in Example 12.2, the magnitude of ΔE depends on the amount of H_2 and O_2 reacted; that is, ΔE is an extensive quantity. We can convert this to an intensive quantity—one that is characteristic of the reaction between any amounts of H_2 and O_2—by calculating the heat evolved *per mole* of product formed. In Example 12.2 we produced 0.0500 mol of H_2O. Therefore, we say that $\Delta E = -14.1$ kJ/0.0500 mol of H_2O, or $\Delta E = -282$ kJ mol^{-1}.

Reactions at constant pressure; the enthalpy change

The heat evolved at constant volume permits us to compute ΔE. However, most changes that are of practical interest to us take place in open containers at essentially constant atmospheric pressure. Under these conditions rather sizable volume changes can occur. For example, when 2 mol of gaseous H_2 react with 1 mol of gaseous O_2 to produce 2 mol of liquid water, at a constant pressure of 100 kPa, the volume changes from about 67 dm^3 to 0.036 dm^3. If we imagine this change to take place in a cylinder with a piston exerting a constant pressure of 100 kPa, then as the reaction proceeds to completion, the surroundings perform work on the system, the magnitude of which is the $P \Delta V$ product of very nearly 6700 J (6.7 kJ). This extra energy absorbed when the system is compressed can be released as additional heat when the reaction is carried out at constant pressure, so the heat of reaction at constant pressure is more exothermic than ΔE by about 6.7 kJ.

To avoid the necessity of considering PV work when heats of reaction are measured at constant pressure, we define a new thermodynamic function called the **heat content,** or **enthalpy,** *H.* This is given as

Enthalpy is from the Greek word *enthalpein,* meaning *to heat in.*

$$H = E + PV \tag{12.4}$$

페이지

For a change at constant pressure,

$$\Delta H = \Delta E + P\,\Delta V \qquad\qquad [12.5]$$

If only PV work is involved in the change, we know that

$$\Delta E = q - P\,\Delta V$$

Substituting this into Equation 12.5, we have

$$\Delta H = (q - P\,\Delta V) + P\,\Delta V$$

$$\Delta H = q_p \qquad\qquad [12.6]$$

If heat is evolved, the sign of q_p is negative, and so is the sign of ΔH.

Thus we see that ΔH is equal to the heat, q_p, absorbed or evolved at constant pressure.

The enthalpy of a system is a state function, just like the internal energy, so the magnitude of ΔH depends only on the enthalpies of the initial and final states. Thus we can write

$$\Delta H = H_{\text{final}} - H_{\text{initial}}$$

Here we use the same symbolism as in our earlier discussions of ΔH_{vap}, ΔH_{fus}, and so on. Those quantities, in fact, correspond to the enthalpy changes associated with vaporization, fusion, and so forth.

Differences between ΔE and ΔH

In many instances the differences between ΔH and ΔE are small, particularly for chemical reactions. When a reaction occurs in which all the reactants and products are liquids or solids, only very small changes in volume take place. As a result, $P\,\Delta V$ is very small and ΔH has very nearly the same magnitude as ΔE. When chemical reactions occur in which gases are either consumed or produced, much larger volume changes occur, and the $P\,\Delta V$ product is also much greater. Even in these cases, however, ΔE is usually so large compared to the $P\,\Delta V$ term that ΔE and ΔH are still nearly the same. We see this in Example 12.3.

EXAMPLE 12.3 CONVERTING BETWEEN ΔH AND ΔE

PROBLEM When 2.00 mol of H_2 and 1.00 mol of O_2 at 100°C and 1 atm, react to produce 2.00 mol of gaseous water at 100°C and 1 atm, a total of 484.5 kJ are evolved. What are (a) ΔH, and (b) ΔE for the production of a single mole of $H_2O(g)$?

SOLUTION (a) Since the reaction

$$2H_2(g) + O_2(g) \longrightarrow 2H_2O(g)$$

is occurring at constant pressure,

$$q_p = \Delta H = \frac{-484.5 \text{ kJ}}{2 \text{ mol } H_2O}$$

The minus sign, remember, signifies that the reaction is exothermic. For the production of 1 mol of water,

$$\Delta H = -242.2 \text{ kJ mol}^{-1}$$

(b) To calculate ΔE from ΔH, we have to use Equation 12.5. Solving for ΔE we have

$$\Delta E = \Delta H - P\,\Delta V$$

Thus we have to subtract the $P\,\Delta V$ product from ΔH. For simplicity, we will assume the gases to be ideal (this causes little error). Then, taking V_i and n_i to be the initial volume and initial number of moles of reactants at a given P and T, we obtain

$$PV_i = n_iRT$$

Similarly, for the final state (the products)

$$PV_f = n_fRT$$

The pressure-volume work in the process is given by

$$PV_f - PV_i = P(V_f - V_i) = P\,\Delta V$$

This is equal to

$$P\,\Delta V = n_fRT - n_iRT$$

$$P\,\Delta V = (n_f - n_i)RT = (\Delta n)RT$$

We are interpreting the equation on a mole basis.

The quantity Δn = (number of moles of gaseous products) − (number of moles of gaseous reactants). Using the coefficients in the chemical equation in part (a), we see that there are two moles of gaseous products and three moles of gaseous reactants. Therefore,

$$\Delta n = 2.00 \text{ mol} - 3.00 \text{ mol} = -1.00 \text{ mol}$$

Next, taking $R = 8.314 \text{ J mol}^{-1} \text{ K}^{-1}$ to obtain the desired energy units, we get

$$P\,\Delta V = (-1.00 \text{ mol})(8.314 \text{ J mol}^{-1} \text{ K}^{-1})(373 \text{ K})$$

$$P\,\Delta V = -3100 \text{ J} = -3.10 \text{ kJ (rounded)}$$

What we have calculated, based on the chemical equation, is the $P\,\Delta V$ work for the formation of 2 mol of $H_2O(g)$. For one mole, $P\,\Delta V = -1.55$ kJ. Now we can calculate ΔE in kJ mol^{-1}.

$$\Delta E = \Delta H - P\,\Delta V$$

$$\Delta E = -242.2 \text{ kJ mol}^{-1} - (-1.55 \text{ kJ mol}^{-1})$$

$$\Delta E = -240.6 \text{ kJ mol}^{-1}$$

Notice that ΔH and ΔE are really not very different, even in this reaction involving a change in the number of moles of gas. They differ by only about 0.6%.

In this last example we see that the PV work involved in a chemical reaction in which gases are either consumed or produced can be calculated by the simple expression

$$\text{pressure-volume work} = (\Delta n)RT \qquad [12.7]$$

Remember that Δn is *the change in the number of moles of **gas** on going from reactants to products*. Equation 12.7 does not apply to reactions in which only liquids or solids are involved. For such reactions the volume changes are extremely small and the $P\,\Delta V$ work is usually negligible compared to other energy changes that take place. For such reactions ΔE and ΔH are therefore essentially identical.

12.4 HESS'S LAW OF HEAT SUMMATION

Since enthalpy is a state function, the magnitude of ΔH for a chemical reaction does not depend on the path taken by the reactants as they proceed to form the products. Let's consider, for example, the conversion of 1 mol of liquid water at 100°C and 1 atm to 1 mol of vapor at 100°C and 1 atm. This process absorbs 41 kJ of heat for each mole of H_2O vaporized and, hence, $\Delta H = +41$ kJ. We can represent this reaction as

$$H_2O(l) \longrightarrow H_2O(g) \qquad \Delta H = +41 \text{ kJ}$$

An equation written in this manner, in which the energy change is also shown, is called a **thermochemical equation** and is nearly always interpreted on a mole basis. Here, for instance, we see that 1 mol of $H_2O(l)$ is converted to 1 mol of $H_2O(g)$ by the absorption of 41 kJ.

The value of ΔH for this change will always be +41 kJ, provided that we refer to the same pair of initial and final states. We could even go so far as to first decompose the 1 mol of liquid into gaseous hydrogen and oxygen and then recombine the elements to produce $H_2O(g)$ at 100°C and 1 atm. The net change in enthalpy would still be the same, +41 kJ. Consequently, it is possible to look at some overall change as the net result of a sequence of chemical reactions. The net value of ΔH for the overall process is merely the sum of all the enthalpy changes that take place along the way. These last statements constitute **Hess's law of heat summation.**

Thermochemical equations serve as a useful tool for applying Hess's law. For example, the thermochemical equations that correspond to the indirect path just described for the vaporization of water are

$$H_2O(l) \longrightarrow H_2(g) + \tfrac{1}{2}O_2(g) \quad \Delta H = +283 \text{ kJ}$$
$$H_2(g) + \tfrac{1}{2}O_2(g) \longrightarrow H_2O(g) \qquad \Delta H = -242 \text{ kJ}$$

Notice that fractional coefficients are allowed in thermochemical equations. This is because a coefficient of 1/2 is taken to mean 1/2 mol. (In ordinary chemical equations, however, fractional coefficients are avoided because they are meaningless on a molecular level; one cannot have half an atom or molecule and still retain the chemical identity of the species.)

The two equations above tell us that 283 kJ are required to decompose 1 mol of $H_2O(l)$ into its elements, and that 242 kJ are evolved when they recombine to produce 1 mol of $H_2O(g)$. The net change (the vaporization of one mole of water) is obtained by adding the two chemical equations together and then canceling any quantities that appear on both sides of the arrow.

$$H_2O(l) + \cancel{H_2(g)} + \cancel{\tfrac{1}{2}O_2(g)} \longrightarrow H_2O(g) + \cancel{H_2(g)} + \cancel{\tfrac{1}{2}O_2(g)}$$

or

$$H_2O(l) \longrightarrow H_2O(g)$$

We also find that the heat of the overall reaction is equal to the algebraic sum of the heats of reaction for the two steps.

$$\Delta H = +283 \text{ kJ} + (-242 \text{ kJ})$$
$$\Delta H = +41 \text{ kJ}$$

Thus, *when we add thermochemical equations to obtain some net change, we also add their corresponding heats of reaction.*

To describe the nature of these thermochemical changes, we can also illustrate them graphically (Figure 12.6). This type of figure is frequently called an

Figure 12.6

Enthalpy diagram for the reaction
$H_2O(l) \rightarrow H_2O(g)$.

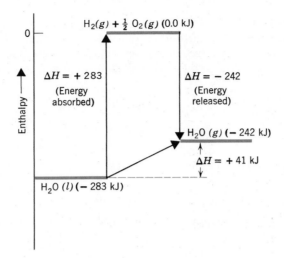

enthalpy diagram. Notice that we have chosen the enthalpy of the free elements as the zero point on the energy scale. This choice is entirely arbitrary because we are interested only in determining differences in H. In fact, we have no way at all of knowing absolute enthalpies, just as we have no way of knowing absolute internal energies. We can only measure ΔH. We used diagrams similar to Figure 12.6 in the last chapter in our discussion of heats of solution, so the concept of Hess's law should already be familiar.

EXAMPLE 12.4 APPLYING HESS'S LAW BY COMBINING THERMOCHEMICAL EQUATIONS

PROBLEM   The thermochemical equation for the combustion of acetylene, a fuel used in torches, is given by Equation (1).

We've numbered the ΔH's simply for ease of identification in the solution of the problem.

(1) $2C_2H_2(g) + 5O_2(g) \longrightarrow 4CO_2(g) + 2H_2O(l)$ $\Delta H_1 = -2602$ kJ

Ethane, another hydrocarbon fuel, reacts as follows:

(2) $2C_2H_6(g) + 7O_2(g) \longrightarrow 4CO_2(g) + 6H_2O(l)$ $\Delta H_2 = -3123$ kJ

and hydrogen and oxygen combine following this equation

(3) $H_2(g) + \frac{1}{2}O_2(g) \longrightarrow H_2O(l)$ $\Delta H_3 = -286$ kJ

All these data correspond to the same temperature and pressure: 25°C and 1 atm. Use these thermochemical equations to calculate the heat of hydrogenation of acetylene,

(4) $C_2H_2(g) + 2H_2(g) \longrightarrow C_2H_6(g)$ $\Delta H_4 = ?$

at 25°C and 1 atm.

SOLUTION   To solve this problem we must combine the given equations (1), (2), and (3) in such a way that when they are added, everything cancels except the formulas in the desired equation (4). This involves keeping an eye on the final equation as we rearrange the given equations so that they add up correctly. For example, we have $1C_2H_2(g)$ on the left of Equation (4), so we want to use Equation (1) with its coefficients divided by 2 (this gives Equation 5). We must also divide its ΔH by 2 because there are now only half as many moles involved. We also have $2H_2(g)$ on the left, so we have to take Equation (3) multiplied by 2, and we have to multiply its ΔH by 2 (this gives Equation 6). Finally, we have $1C_2H_6(g)$ on the right. To get it there, we must reverse Equation (2) and divide its coefficients by 2 (Equation 7). When we reverse an equation, we change the sign of its ΔH because it changes an exothermic reaction into an endothermic one. In this case, we also must divide it by 2. This gives us, therefore,

(5) $C_2H_2(g) + \frac{5}{2}O_2(g) \longrightarrow 2CO_2(g) + H_2O(l)$ $\Delta H_5 = -\dfrac{2602 \text{ kJ}}{2} = -1301 \text{ kJ}$

(6) $2H_2(g) + O_2(g) \longrightarrow 2H_2O(l)$ $\Delta H_6 = 2(-286 \text{ kJ}) = -572 \text{ kJ}$

(7) $2CO_2(g) + 3H_2O(l) \longrightarrow C_2H_6(g) + \frac{7}{2}O_2(g)$ $\Delta H_7 = +\dfrac{3123 \text{ kJ}}{2} = +1561 \text{ kJ}$

On the left, $\frac{5}{2}O_2 + O_2$ gives $\frac{7}{2}O_2$.

Adding Equations (5), (6), and (7) gives

$$C_2H_2(g) + 2H_2(g) + \tfrac{7}{2}O_2(g) + 2CO_2(g) + 3H_2O(l) \longrightarrow$$
$$2CO_2(g) + 3H_2O(l) + C_2H_6(g) + \tfrac{7}{2}O_2(g)$$

Canceling elements that are the same on both sides, we get

$$C_2H_2(g) + 2H_2(g) \longrightarrow C_2H_6(g)$$

which is the equation we want. Since it is obtained by adding Equations (5), (6) and (7), its ΔH (ΔH_4 in the statement of the problem) is obtained by adding the ΔH's of (5), (6) and (7).

$$\Delta H_4 = \Delta H_5 + \Delta H_6 + \Delta H_7$$

$$\Delta H_4 = (-1301 \text{ kJ}) + (-572 \text{ kJ}) + (1561 \text{ kJ})$$

$$= -312 \text{ kJ}$$

The heat of hydrogenation of acetylene is therefore -312 kJ.

Heats of formation

A particularly useful type of thermochemical equation corresponds to the formation of one mole of a substance from its elements. The enthalpy changes associated with these reactions are called **heats of formation** or **enthalpies of formation** and are denoted as ΔH_f. For example, thermochemical equations for the formation of liquid and gaseous water at 100°C and 1 atm are, respectively,

$$H_2(g) + \tfrac{1}{2}O_2(g) \longrightarrow H_2O(l) \qquad \Delta H_f = -283 \text{ kJ mol}^{-1}$$
$$H_2(g) + \tfrac{1}{2}O_2(g) \longrightarrow H_2O(g) \qquad \Delta H_f = -242 \text{ kJ mol}^{-1}$$

How can we use these equations to obtain the heat of vaporization of water? Clearly, we must reverse the first equation and then add it to the second. When we reverse this equation, we must also remember to change the sign of ΔH. If the formation of $H_2O(l)$ is exothermic, as indicated by a negative ΔH_f, the reverse process must be endothermic.

(Exothermic) $H_2(g) + \tfrac{1}{2}O_2(g) \longrightarrow H_2O(l)$ $\Delta H = \Delta H_f = -283 \text{ kJ}$

(Endothermic) $H_2O(l) \longrightarrow H_2(g) + \tfrac{1}{2}O_2(g)$ $\Delta H = -\Delta H_f = +283 \text{ kJ}$

When this last equation is added to that for the formation of $H_2O(g)$, we obtain

$$H_2O(l) \longrightarrow H_2O(g)$$

and the heat of reaction is

$$\Delta H = \Delta H_{f\ H_2O(g)} - \Delta H_{f\ H_2O(l)}$$

$$\Delta H = -242 \text{ kJ} - (-283 \text{ kJ}) = +41 \text{ kJ}$$

Notice that the heat of reaction for the overall change is equal to the heat of formation of the product *minus* the heat of formation of the reactant. In general, we can write that for any overall reaction

We will see that Equation 12.8 is a particularly useful form of Hess's law.

$$\Delta H_{\text{reaction}} = (\text{sum of } \Delta H_f \text{ of products}) - (\text{sum of } \Delta H_f \text{ of reactants}) \quad [12.8]$$

12.5   STANDARD STATES

Notice that the standard temperature for thermodynamic data is different than the standard temperature of 0°C used in calculations involving gases in Chapter 9.

The magnitude of ΔH_f depends on the conditions of temperature, pressure, and the physical state (gas, liquid, solid, crystalline form) of the reactants and products. For instance, at 100°C and 1 atm, the heat of formation of liquid water is -283 kJ mol^{-1}, while at 25°C and 1 atm, ΔH_f for $H_2O(l)$ is -286 kJ mol^{-1}. To avoid the necessity of always having to specify the conditions for which ΔH_f is recorded, and to permit comparisons between ΔH_f for various compounds, a standard set of conditions is chosen, usually 25°C and a pressure of 1 atm (101.325 kPa).[2] Under these conditions a substance is said to be in its **standard state.** Heats of formation of substances in their standard states are indicated as ΔH_f^0. For example, the standard heat of formation of liquid water $\Delta H_{f\,H_2O(l)}^0 = -286$ kJ/mol, and represents the heat liberated when H_2 and O_2, each in their natural form at 25°C and 1 atm, react to produce $H_2O(l)$ at 25°C and 1 atm.

We know energy is liberated because ΔH^0 is negative.

Table 12.1 contains standard heats of formation for a variety of different substances. Such a table is very useful because it permits us to use Equation 12.8 to calculate the **standard heats of reaction, ΔH^0**, for a very large number of different chemical changes. In performing these calculations, we arbitrarily take the ΔH_f^0 for an element in its natural, most stable form at 25°C and 1 atm to be equal to zero. Thus, in our computations, the zero point on the energy scale is again chosen to be that of the free elements. As mentioned before, because we speak only of *changes* in energy, the actual location of this zero point is unimportant. The following examples illustrate how the principles developed in the preceding two sections can be applied.

[2] In SI the atmosphere is not a recognized unit of pressure. For some time there had been a debate over an appropriate choice for the thermodynamic reference pressure, the desire being to choose a pressure in units of pascals that would not alter in any major way previously accepted values of thermodynamic quantities. Finally, the **bar** has been adopted as the standard reference pressure: 1 bar = 10^5 Pa. One bar differs from one atmosphere by only 1.3%, and for the thermodynamic quantities that we will deal with, their values at 1 atm and 1 bar differ by an insignificant quantity. We shall continue to refer to the standard pressure as 1 atm.

A more complete table of standard heats of formation is located in Appendix D.

Table 12.1
Standard heats of formation of some substances at 25°C and 1 atm

Substance	ΔH_f^0 (kJ mol^{-1})	Substance	ΔH_f^0 (kJ mol^{-1})
$Al_2O_3(s)$	−1676	$HBr(g)$	−36
$Br_2(l)$	0.00	$HI(g)$	+26
$Br_2(g)$	+30.9	$KCl(s)$	−436.0
$C(s, diamond)$	+1.88	$LiCl(s)$	−408.8
$CO(g)$	−110	$MgCl_2(s)$	−641.8
$CO_2(g)$	−394	$MgCl_2 \cdot 2H_2O(s)$	−1280
$CH_4(g)$	−74.9	$Mg(OH)_2(s)$	−924.7
$C_2H_6(g)$	−84.5	$NH_3(g)$	−46.0
$C_2H_4(g)$	+51.9	$N_2O(g)$	+81.5
$C_2H_2(g)$	+227	$NO(g)$	+90.4
$C_3H_8(g)$	−104	$NO_2(g)$	+34
$C_6H_6(l)$	+49.0	$NaF(s)$	−571
$CH_3OH(l)$	−238	$NaCl(s)$	−413
$HCHO_2(g)$	−363	$NaBr(s)$	−360
$CS_2(l)$	+89.5	$NaI(s)$	−288
$CS_2(g)$	+117	$Na_2O_2(s)$	−504.6
$CCl_4(l)$	−134	$NaOH(s)$	−426.8
$C_2H_5OH(l)$	−278	$NaHCO_3(s)$	−947.7
$CH_3CHO(g)$	−167	$Na_2CO_3(s)$	−1131
$HC_2H_3O_2(l)$	−487.0	$O_3(g)$	+143
$CaO(s)$	−635.5	$PbO_2(s)$	−277
$Ca(OH)_2(s)$	−986.6	$PbSO_4(s)$	−920.1
$CaSO_4(s)$	−1433	$SO_2(g)$	−297
$CuO(s)$	−155	$SO_3(g)$	−396
$Fe_2O_3(s)$	−822.2	$H_2SO_4(l)$	−813.8
$H_2O(l)$	−286	$SiO_2(s)$	−910.0
$H_2O(g)$	−242	$SiH_4(g)$	+33
$HF(g)$	−271	$ZnO(s)$	−348
$HCl(g)$	−92.5	$Zn(OH)_2(s)$	−642.2

EXAMPLE 12.5 CALCULATING ΔH^0 FOR A REACTION FROM STANDARD HEATS OF FORMATION

PROBLEM Many careful cooks keep sodium bicarbonate (baking soda) handy because it is a good extinguisher of oil or grease fires. Its decomposition products help smother the flames. The reaction is

$$2NaHCO_3(s) \longrightarrow Na_2CO_3(s) + H_2O(g) + CO_2(g)$$

Calculate ΔH^0 for the reaction.

SOLUTION   Equation 12.8 implies that

$$\Delta H^0 = (\text{sum } \Delta H_f^0 \text{ products}) - (\text{sum } \Delta H_f^0 \text{ reactants})$$

This means that we must add up all the heat evolved during the formation of the products from their elements and then subtract the heat evolved by the formation of the reactants from their elements.

Using data from Table 12.1, we see that the total enthalpy of formation of the products is

$$1 \text{ mol Na}_2\text{CO}_3(s) \times \left(\frac{-1131 \text{ kJ}}{1 \text{ mol Na}_2\text{CO}_3(s)}\right) = -1131 \text{ kJ}$$

$$1 \text{ mol H}_2\text{O}(g) \times \left(\frac{-242 \text{ kJ}}{1 \text{ mol H}_2\text{O}(g)}\right) = -242 \text{ kJ}$$

$$1 \text{ mol CO}_2(g) \times \left(\frac{-394 \text{ kJ}}{1 \text{ mol CO}_2(g)}\right) = -394 \text{ kJ}$$

$$\text{total } \Delta H_f^0 \text{ of products} = -1767 \text{ kJ}$$

For the reactant,

$$2 \text{ mol NaHCO}_3(s) \times \left(\frac{-947.7 \text{ kJ}}{1 \text{ mol NaHCO}_3(s)}\right) = -1895 \text{ kJ}$$

We have said that

$$\Delta H^0 = (\text{sum } \Delta H_f^0 \text{ products}) - (\text{sum } \Delta H_f^0 \text{ reactants})$$

Therefore

$$\Delta H^0 = -1767 \text{ kJ} - (-1895 \text{ kJ})$$

or

$$\Delta H^0 = +128 \text{ kJ}$$

Notice that in computing ΔH^0 for the overall reaction we have multiplied each ΔH_f^0 by the appropriate coefficient from the equation. This gives the total heat of reaction for the numbers of moles specified by the chemical equation.

EXAMPLE 12.6   **CALCULATING ΔH^0 FOR A REACTION FROM STANDARD HEATS OF FORMATION**

PROBLEM   Calculate ΔH^0 for the reaction

$$2\text{Na}_2\text{O}_2(s) + 2\text{H}_2\text{O}(l) \longrightarrow 4\text{NaOH}(s) + \text{O}_2(g)$$

How many kilojoules are liberated when 25.0 g Na_2O_2 react according to this equation?

SOLUTION   We use Equation 12.8.

$$\Delta H^0 = [4\Delta H_{f\,\text{NaOH}(s)}^0 + \Delta H_{f\,\text{O}_2(g)}^0] - [2\Delta H_{f\,\text{Na}_2\text{O}_2(s)}^0 + 2\Delta H_{f\,\text{H}_2\text{O}(l)}^0]$$

All the data are available in Table 12.1 except $\Delta H_{f\,\text{O}_2(g)}^0$, but we know (or *should* know) that the heat of formation of a pure element is zero. Therefore,

$$\Delta H^0 = \left[4 \text{ mol} \times \left(\frac{-426.8 \text{ kJ}}{1 \text{ mol NaOH}}\right) + 0.00\right]$$

$$- \left[2 \text{ mol} \times \left(\frac{-504.6 \text{ kJ}}{1 \text{ mol Na}_2\text{O}_2}\right) + 2 \text{ mol} \times \left(\frac{-286 \text{ kJ}}{1 \text{ mol H}_2\text{O}(l)}\right)\right]$$

$$\Delta H^0 = [-1707 \text{ kJ}] - [-1581 \text{ kJ}]$$

$$= -126 \text{ kJ}$$

To calculate the number of kilojoules liberated by 25.0 g Na_2O_2, we have to realize that the ΔH^0 that we calculated is the energy given off when 2 mol Na_2O_2 react. Therefore,

$$2 \text{ mol } Na_2O_2 \sim -126 \text{ kJ}$$

The molar mass of Na_2O_2 is 78.0 g mol^{-1}, so

$$25.0 \text{ g } Na_2O_2 \times \left(\frac{1 \text{ mol } Na_2O_2}{78.0 \text{ g } Na_2O_2}\right) \times \left(\frac{-126 \text{ kJ}}{2 \text{ mol } Na_2O_2}\right) \sim -20.2 \text{ kJ}$$

The reaction of 25.0 g Na_2O_2 liberates 20.2 kJ.

It is often impossible to measure directly the heat of formation of a compound. For example, no one has yet found a way to cause gaseous hydrogen and graphite (the most stable crystalline form of carbon) to react to form methane, CH_4, in a way that allows us to measure the energy change for the reaction. Therefore, in order to determine ΔH_f^0 for compounds such as methane, indirect methods must be used. One technique that works well with many carbon-containing compounds is to burn the compound in a bomb calorimeter and measure the heat of combustion. In these reactions, the heats of formation of the products are generally known, so the unknown value of ΔH_f^0 for the reactant can be calculated as illustrated in Example 12.7.

EXAMPLE 12.7 **USING THE HEAT OF COMBUSTION TO CALCULATE ΔH_f^0**

PROBLEM The combustion of 1 mol of benzene, $C_6H_6(l)$, to produce $CO_2(g)$ and $H_2O(l)$ liberates 3271 kJ when the products are returned to 25°C and 1 atm. What is the standard heat of formation of $C_6H_6(l)$ expressed in kilojoules per mole?

SOLUTION The equation for the combustion of 1 mol of C_6H_6 is

$$C_6H_6(l) + 7\tfrac{1}{2}O_2(g) \longrightarrow 6CO_2(g) + 3H_2O(l)$$

The standard heat of reaction, $\Delta H^0 = 3271$ kJ. From Equation 12.8, we know that

$$\Delta H^0 = [6\Delta H_f^0 {}_{CO_2(g)} + 3\Delta H_f^0 {}_{H_2O(l)}] - [\Delta H_f^0 {}_{C_6H_6(l)}]$$

Solving for the heat of formation of benzene, we obtain

$$\Delta H_f^0 {}_{C_6H_6(l)} = 6\Delta H_f^0 {}_{CO_2(g)} + 3\Delta H_f^0 {}_{H_2O(l)} - \Delta H^0$$

From Table 12.1 we can obtain the heats of formation of CO_2 and H_2O. Therefore,

$$\Delta H_f^0 {}_{C_6H_6(l)} = 6(-394) \text{ kJ} + 3(-286) \text{ kJ} - (-3271 \text{ kJ})$$

$$\Delta H_f^0 {}_{C_6H_6(l)} = +49 \text{ kJ}$$

Since 1 mol of C_6H_6 is involved,

$$\Delta H_f^0 = +49 \text{ kJ/mol}$$

12.6 BOND ENERGIES

We stated earlier that it should be possible to relate heats of reaction to changes in the potential energy associated with breaking and forming chemical bonds.

Strictly speaking, we should use ΔE for this purpose; however, since $P \Delta V$ contributions to ΔH are relatively small for chemical reactions, we can use ΔH in place of ΔE and still expect to obtain quite reasonable results. Therefore, we will use the terms bond energy and bond enthalpy interchangeably.

In Chapter 5 the bond energy was defined as the energy required to break a bond to produce neutral fragments. For a complex molecule, the energy needed to break all the bonds and reduce the gaseous molecule entirely to neutral gaseous atoms is called the **atomization energy.** Its value is the sum of all the bond energies in the molecule. Simple diatomic molecules, such as H_2, O_2, Cl_2, or HCl, possess only one bond; so the atomization energy is the same as the bond energy. For these simple cases the atomization energy can be obtained by studying the spectra produced when these molecules absorb or emit light. For more complex molecules, however, we employ an indirect method that makes use of measured heats of formation.

As an example, let's consider the molecule CH_4. If we use the same technique that was followed in Example 12.7, the standard heat of formation of $CH_4(g)$ can be determined experimentally to be -74.9 kJ mol^{-1}. This corresponds to the enthalpy change, ΔH_f^0, for the reaction,

$$C(s, \text{graphite}) + 2H_2(g) \longrightarrow CH_4(g)$$

We can envision an alternative path to take us from the free elements to the compound, methane, that follows a series of successive reactions,

(1) $\qquad C(s, \text{graphite}) \longrightarrow C(g) \qquad \Delta H_1$

(2) $\qquad 2H_2(g) \longrightarrow 4H(g) \qquad \Delta H_2$

(3) $\qquad C(g) + 4H(g) \longrightarrow CH_4(g) \qquad \Delta H_3$

The sum of these three reactions will give us our desired overall reaction, so the sum of their ΔH's must equal the ΔH of the overall reaction (that is, the ΔH_f^0 for CH_4).

$$\Delta H_f^0 = \Delta H_1 + \Delta H_2 + \Delta H_3 \qquad [12.9]$$

Steps 1 and 2 involve the formation of gaseous atoms from their elements, and their heats have been measured. They are found in Table 12.2 along with the heats of formation of gaseous atoms of some other typical elements. We've already noted that ΔH_f^0 for CH_4 has been determined, so the only unknown quantity is ΔH_3, which is the negative of the atomization energy (negative because Step 3 involves the *formation* of chemical bonds).

$$\Delta H_3 = -\Delta H_{\text{atom}} \text{ for } CH_4$$

Substituting and solving for the atomization energy in Equation 12.9 gives

$$\Delta H_{\text{atom}} = \Delta H_1 + \Delta H_2 - \Delta H_f^0$$

From the data in Table 12.2, we see that $\Delta H_1 = +715$ kJ (the heat of formation of gaseous carbon atoms). Similarly, $\Delta H_2 = 4(+218 \text{ kJ})$, which is four times the heat of formation of 1 mol of gaseous H atoms. Substituting these values and the heat of formation of CH_4 into the equation above yields

$$\Delta H_{\text{atom}} = (+715 \text{ kJ}) + (+872 \text{ kJ}) - (-74.9 \text{ kJ})$$

$$= +1662 \text{ kJ}$$

This quantity is the total energy needed to break all 4 C—H bonds in 1 mol of CH_4. Division by 4 provides us with an *average* bond energy of 415 kJ per mol

Table 12.2
Heats of formation of gaseous atoms from the elements in their standard states

Atom	ΔH_f^0 per mole of atoms kJ mol^{-1}
H	218
Li	161
Be	327
B	555
C	715
N	473
O	249
F	79.1
Na	108
Si	454
Cl	121
Br	112
I	107

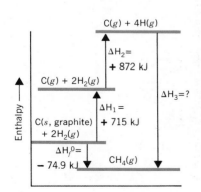

Table 12.3
Average bond energies

Bond	Bond Energy kJ mol^{-1}	Bond	Bond Energy kJ mol^{-1}
H—C	415	C=O	724
H—O	463	C—N	292
H—N	391	C=N	619
H—F	563	C≡N	879
H—Cl	432	C—C	348
H—Br	366	C=C	607
H—I	299	C≡C	833
C—O	356		

of C—H bonds. This value, along with some other average bond energies, appears in Table 12.3.

EXAMPLE 12.8 **CALCULATING BOND ENERGIES FROM THERMODYNAMIC DATA**

PROBLEM The heat of formation of ethylene, C_2H_4, is +51.9 kJ/mol. The structure of the molecule is

$$
\begin{array}{ccc}
\text{H} & & \text{H} \\
& \text{C}=\text{C} & \\
\text{H} & & \text{H}
\end{array}
$$

Assuming the C—H bond energy to be 415 kJ/mol, calculate the C=C bond energy.

SOLUTION Let's examine the two alternative paths shown in Figure 12.7 that take us from the reactants [C(s) and $H_2(g)$], to the product [$C_2H_4(g)$]. The sum of the energy changes in Steps 1, 2, and 3 must be the same as that for the direct path, ΔH_f^0.

$$\Delta H_1 + \Delta H_2 + \Delta H_3 = \Delta H_f^0$$

STEP 1 ΔH_1 is twice the heat of formation of a mole of carbon atoms.

$$\Delta H_1 = 2(715 \text{ kJ}) = 1430 \text{ kJ}$$

STEP 2 ΔH_2 is four times the heat of formation of a mole of hydrogen atoms.

$$\Delta H_2 = 4(218 \text{ kJ}) = 872 \text{ kJ}$$

STEP 3 ΔH_3 is the negative of the atomization energy.

$$\Delta H_3 = -\Delta H_{atom}$$

Therefore,

$$1430 \text{ kJ} + 872 \text{ kJ} + (-\Delta H_{atom}) = +51.9 \text{ kJ}$$

Solving for ΔH_{atom} yields

$$\Delta H_{atom} = 1430 \text{ kJ} + 872 \text{ kJ} - 51.9 \text{ kJ}$$

$$= 2250 \text{ kJ}$$

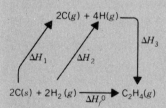

Figure 12.7

Alternative paths for the formation of ethylene from its elements in their standard states.

This is the energy required to break four moles of C—H bonds plus one mole of C=C bonds.

$$\Delta H_{\text{atom}} = 4\Delta H_{\text{C—H}} + \Delta H_{\text{C=C}}$$

Substituting and solving for $\Delta H_{\text{C=C}}$, we get

$$\Delta H_{\text{C=C}} = \Delta H_{\text{atom}} - 4\Delta H_{\text{C—H}}$$

$$= 2250 \text{ kJ} - 4(415 \text{ kJ})$$

$$= 590 \text{ kJ per mol of C=C bonds}$$

It's worth noting that this value is within 3% of the accepted average C=C bond energy of 607 kJ mol^{-1}.

A very important fact is that the average bond energies found in Table 12.3 can be used, in many cases, to compute heats of formation with a fair degree of accuracy, as illustrated in Example 12.9. It is very significant that a bond between two atoms has very nearly the same strength in one molecule as it does in another. This implies, for example, that nearly all C—H bonds are pretty much alike, whether in a small molecule such as CH_4 or a large complex one such as $C_{42}H_{86}$. The same applies to many other bonds as well. This phenomenon has greatly simplified the development of the modern theories about chemical bonding that we discussed in Chapter 6.

EXAMPLE 12.9 USING BOND ENERGIES TO ESTIMATE ΔH_f^0 FOR A COMPOUND

PROBLEM   Use the data in Tables 12.2 and 12.3 to compute the heat of formation of liquid ethyl alcohol in kilocalories per mole. This compound has a heat of vaporization, $\Delta H_{\text{vap}}^0 = 9.4$ kcal/mol and the structural formula

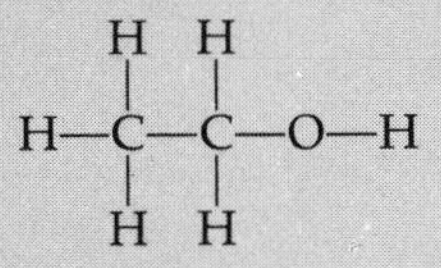

$$\begin{array}{ccc} & H & H \\ & | & | \\ H- & C-C & -O-H \\ & | & | \\ & H & H \end{array}$$

SOLUTION   We wish to determine ΔH^0 for the reaction,

$$2C(s, \text{graphite}) + 3H_2(g) + \tfrac{1}{2}O_2(g) \longrightarrow C_2H_5OH(l)$$

To compute ΔH_f, we follow the alternative path from reactants to products illustrated in Figure 12.8. Using the data in Table 12.2, we can compute the energy required to convert the reactants to gaseous atoms—that is, ΔH_A in Figure 12.8.

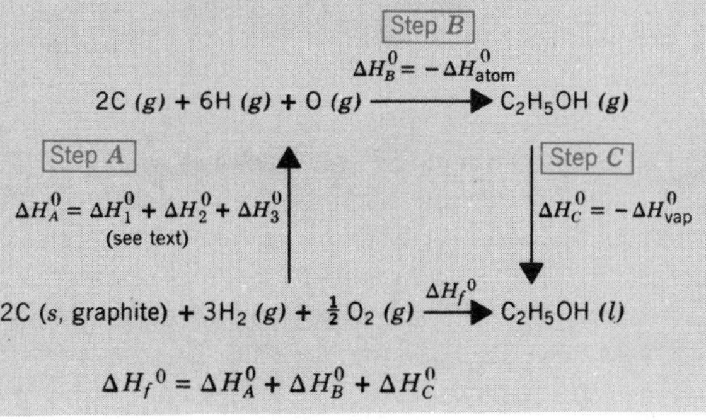

Figure 12.8

Alternative paths for the formation of liquid ethyl alcohol, $C_2H_5OH(l)$.

Step B
$$\Delta H_B^0 = -\Delta H_{\text{atom}}^0$$
$$2C(g) + 6H(g) + O(g) \longrightarrow C_2H_5OH(g)$$

Step A
$$\Delta H_A^0 = \Delta H_1^0 + \Delta H_2^0 + \Delta H_3^0$$
(see text)

Step C
$$\Delta H_C^0 = -\Delta H_{\text{vap}}^0$$

$$2C(s, \text{graphite}) + 3H_2(g) + \tfrac{1}{2}O_2(g) \xrightarrow{\Delta H_f^0} C_2H_5OH(l)$$

$$\Delta H_f^0 = \Delta H_A^0 + \Delta H_B^0 + \Delta H_C^0$$

$$2C(s, \text{graphite}) \longrightarrow 2C(g) \quad \Delta H_1^0 = 2\Delta H_{f\,C(g)}^0$$

$$= 2(+715 \text{ kJ}) = +1430 \text{ kJ}$$

$$3H_2(g) \longrightarrow 6H(g) \quad \Delta H_2^0 = 6\Delta H_{f\,H(g)}^0$$

$$= 6(+218 \text{ kJ}) = +1308 \text{ kJ}$$

$$\tfrac{1}{2}O_2(g) \longrightarrow O(g) \quad \Delta H_3^0 = \Delta H_{f\,O(g)}^0 = +249 \text{ kJ}$$

The total energy needed to give gaseous atoms, $\Delta H_A^0 = \Delta H_1^0 + \Delta H_2^0 + \Delta H_3^0 = +2987$ kJ.

Next, we compute the energy liberated when these atoms combined to form 1 mol of gaseous C_2H_5OH. This is the negative of the atomization energy of C_2H_5OH, which involves five C—H bonds, one C—C bond, one C—O bond, and one O—H bond. The energies are obtained from Table 12.3.

5(C—H)	5(415 kJ)
1(C—C)	348 kJ
1(C—O)	356 kJ
1(O—H)	463 kJ
ΔH_{atom}^0 for $C_2H_5OH(g)$ =	3242 kJ

Therefore, for Step B, $\Delta H_B^0 = -3242$ kJ

Finally, energy is liberated when $C_2H_5OH(g)$ is condensed to a liquid. The ΔH^0 for this process is the negative of ΔH_{vap}^0. Therefore $\Delta H_C^0 = -39$ kJ.

Now we can compute ΔH_f^0 for $C_2H_5OH(l)$ by adding the ΔH^0 values of each step in the alternative path.

$$\Delta H_f^0 = \Delta H_A^0 + \Delta H_B^0 + \Delta H_C^0$$

$$= +2987 \text{ kJ} + (-3242 \text{ kJ}) + (-39 \text{ kJ})$$

$$= -294 \text{ kJ}$$

Since the computation was performed for 1 mol, we can write

$$\Delta H_f^0 = -294 \text{ kJ mol}^{-1}$$

Comparing this to the value reported in Table 12.1,

$$\Delta H_f^0 = -278 \text{ kJ mol}^{-1}$$

we see that the agreement (within about 6%) is not really too bad, considering that we have assumed that each kind of bond has the same energy in *all* compounds.

12.7 ENERGY, ENTROPY, AND THE SPONTANEITY OF CHEMICAL AND PHYSICAL CHANGES

At the beginning of this chapter we said that thermodynamics is able to tell us when a reaction can occur spontaneously—that is, without continued outside assistance. To see how this can be accomplished, let's begin by finding out what factors are involved in a spontaneous change. One spontaneous process that we have all observed is a ball rolling down a hill. When it finally comes to rest at the bottom, its potential energy has decreased and it is in a more stable, lower energy state than before. We might conclude, therefore, that a process leading to

Exothermic changes like those occurring in this forest fire tend to be spontaneous.

a decrease in the energy of a system (here, a ball) should tend to be spontaneous. Indeed, many processes that are spontaneous do occur with the evolution of energy. For example, a mixture of hydrogen and oxygen, when ignited, reacts very rapidly to produce water. This chemical change is accompanied by the release of a large quantity of heat, so much, in fact, that the hot water vapor produced expands explosively.

Evolution of energy, however, is not the only criterion to be considered. There are many examples of changes that take place with the absorption of energy and yet are spontaneous. In the last chapter, we discussed heats of solution and saw that in many instances energy is absorbed when a salt dissolves in water. An example, you recall, is ammonium nitrate, the salt used in instant cold packs. The formation of the solution, although endothermic, is nevertheless spontaneous. What, then, is the driving force for this process that is capable of outweighing the endothermic energy effect that occurs?

When a solid such as NH_4NO_3 dissolves in water, the particles of the solute leave the well-ordered crystalline state and gradually diffuse throughout the liquid to produce a solution. In this final state the particles of the solute are in a more disordered, random condition than they were before they dissolved, as shown in Figure 12.9. Similarly, the solvent is in a more random state in the solution because the solvent molecules are, in a sense, dispersed throughout those of the solute as well.

Figure 12.9

An increase in disorder occurs when a crystalline solid dissolves to form a solution.

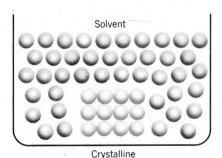
Solvent

Crystalline
solid

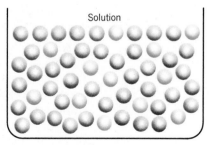
Solution

At this point we might pause and ask, "What exactly do randomness and disorder have to do with what happens to a system?" To answer this question, let's compare Figures 12.9a and 12.9b. If energy effects were completely unimportant, which of these distributions of particles would you expect to obtain if you were to toss molecules of the solvent and molecules of the solute into a container. Does it make sense that, by chance, all the solute particles should land in a stack that possesses the highly ordered structure of a crystal? This would be equivalent to throwing bricks over your shoulder and having them land as a perfect brick wall! Purely on statistical grounds, a crystal is a highly improbable arrangement of particles. Certainly, a statistically more probable distribution is one like that illustrated in Figure 12.9b, in which the particles of solute and solvent are jumbled together. The reason this is more probable is that there are so many different jumbled configurations. The crystal, on the other hand, has a low probability because the number of particle configurations is extremely small.

In any system, there is a tendency for the particles to move spontaneously from a state of low probability to one of higher probability. That is why an ordered deck of cards becomes scrambled if we toss it in the air, and it is the driving force behind the formation of a solution. This drive toward increased statistical probability, in fact, is one of the major factors in determining whether a particular change is spontaneous.

Entropy

The thermodynamic quantity that is related to the degree of randomness, or statistical probability of a system is called the **entropy** and is given the symbol S. The more random a system is, the greater its entropy, and based on our previous discussion, this means that an increase in entropy is a factor that favors spontaneity.

Several things influence the quantity of entropy that a substance has in a particular state, and frequently simple logic is all we need to compare one state with another. For example, if we compare a liquid with a solid, we know that the particles in a solid are highly ordered, while those in a liquid are disordered. Therefore, for a given substance at a particular temperature, its liquid state has a higher entropy than the solid. Similarly, if we compare a liquid with a gas, we see that the molecules in a liquid are confined to one region of the container (the bottom), but a gas can spread its molecules randomly throughout the entire vessel. Thus the gas has a higher entropy than the liquid. So for a given substance at a given temperature

$$S_{solid} < S_{liquid} < S_{gas}$$

Entropy changes

Entropy is a state function just like E and H, which means that the magnitude of a change in entropy, ΔS, depends only on the entropies of the system in its initial and final states.

$$\Delta S = S_{final} - S_{initial}$$

Notice that by following the usual practice, ΔS is defined as "final minus initial."

One way that a change in entropy can be brought about is by the addition of heat to a system. For example, consider a perfect crystal of carbon monoxide at absolute zero in which all the C—O dipoles are aligned in the same direction (Figure 12.10a). Because of the perfect alignment of the dipoles there is essentially perfect order in the crystal, and the entropy of the system is at a minimum. When heat is added to this crystal, the temperature rises above 0 K and thermal

If you toss marbles into a level box, it is improbable that they will all find themselves at one end. Similarly, it is less probable to have the particle distribution of a liquid than to have the particle distribution of a gas.

Figure 12.10
(a) *Perfect crystal of CO at 0 K.*
(b) *Crystal of CO above 0 K.*

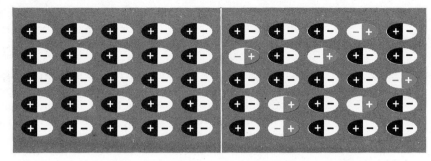

(a) *Perfect crystal of CO. Dipoles are all oriented in the same direction at 0 K.*

(b) *Crystal of CO above 0 K. Lattice vibrations cause some dipoles to rotate 180° and disorder within the lattice occurs. Disoriented dipoles are colored for emphasis.*

motion (vibrations) within the lattice cause some of the dipoles to become oriented in the opposite direction (Figure 12.10*b*). As a result, there is less order (more disorder), and the entropy of the crystal has obviously increased. Logically, the more heat added to the system, the greater the extent of disorder afterwards. It should not be surprising, therefore, to find that the entropy change, ΔS, is directly proportional to the quantity of heat added to the system. This heat is specified as q_{rev}—the quantity of heat that would be added to the system if the change follows a reversible path.

$$\Delta S \propto q_{rev}$$

The magnitude of ΔS is also *inversely* proportional to the temperature at which the heat is added. At low temperatures a given quantity of heat makes a large change in the relative degree of order. Near absolute zero, the addition of even a small quantity of heat causes the system to go from essentially perfect order to some degree of randomness—a very substantial change, so ΔS is large. If the same quantity of heat is added to the system at a higher temperature, however, the system goes from an already highly random state to one just slightly more random. This constitutes only a very small change in the *relative* degree of disorder and, hence, only a small entropy change. Thus, it can be shown that a change in entropy is finally given as

$$\Delta S = \frac{q_{rev}}{T} \qquad [12.10]$$

where T is the absolute temperature at which q_{rev} is transferred to the system. Note that entropy has units of *energy/temperature*, joules per kelvin ($J\ K^{-1}$).

Often it is useful to make qualitative predictions about the entropy change for a given process. Just knowing whether ΔS is positive or negative can give us a clue as to whether the change might be spontaneous or not. For example, the collapse of the building shown in the opening photo of this chapter is obviously accompanied by a lowering of the energy, so that favors spontaneity. Also, the transformation from the ordered structure of the building to the jumbled pile of rubble certainly is accompanied by a large increase in randomness, or entropy. Armed with this knowledge, we would predict that such a collapse surely takes place spontaneously, even if we had not seen the destruction, because both of the factors that control spontaneity (energy change and entropy change) are in its favor.

For chemical reactions it is sometimes (but not always) relatively easy to anticipate the sign of ΔS. For example, consider the reaction

$$2NaHCO_3(s) \longrightarrow Na_2CO_3(s) + CO_2(g) + H_2O(g)$$

which occurs when sodium bicarbonate (baking soda) is heated. Among the reactants there is only a solid, and as we know, these have low entropies. Among the products we have three different substances, so the particles (atoms) are not as ordered as among the reactants, and that should give a higher entropy. In addition, two of the products are gases, which we know to have large entropies. Both of these facts suggest that the products probably have a larger entropy than the reactants, so ΔS should be positive.

EXAMPLE 12.10 **DETERMINING THE SIGN OF THE ENTROPY CHANGE QUALITATIVELY**

PROBLEM Ammonia is made by the reaction of hydrogen with nitrogen in the reaction

$$3H_2(g) + N_2(g) \longrightarrow 2NH_3(g)$$

What should be the sign of ΔS for this reaction?

SOLUTION On the left side of the equation we can count 4 moles of gas (3 mol H_2 plus 1 mol N_2). On the right side of the equation there are only 2 moles of gas. This means that as the products are formed there are fewer molecules in the system, which must also mean that the atoms are combining to form more complex particles. More complexity means more order and therefore lower entropy. As a result, we can conclude that there is a decrease in entropy associated with the formation of NH_3 in this reaction, and ΔS is negative.

12.8 THE SECOND LAW OF THERMODYNAMICS

The **second law of thermodynamics** provides us with a way of comparing the effects of the two driving forces involved in a spontaneous process—changes in energy and changes in entropy. One statement of the second law is that *in any spontaneous process there is always an increase in the entropy of the universe* ($\Delta S_{total} >$ 0). This increase takes into account entropy changes in both the system and its surroundings,

$$\Delta S_{total} = \Delta S_{system} + \Delta S_{surroundings}$$

The entropy change that occurs in the surroundings is brought about by the heat added to the surroundings divided by the temperature at which it is transferred. For a process at constant P and T, the heat added to the surroundings is equal to the *negative* of the heat added to the system, which is given by ΔH_{system}. Thus

This is required by the law of conservation of energy.

$$q_{surroundings} = -\Delta H_{system}$$

The entropy change for the surroundings is therefore

$$\Delta S_{surroundings} = \frac{-\Delta H_{system}}{T}$$

and the total entropy change for the universe is

$$\Delta S_{total} = \Delta S_{system} - \frac{\Delta H_{system}}{T}$$

or

$$\Delta S_{total} = \frac{T \, \Delta S_{system} - \Delta H_{system}}{T}$$

This can be rearranged to give

$$T \, \Delta S_{total} = -(\Delta H_{system} - T \, \Delta S_{system})$$

Since ΔS_{total} must be a positive number for a spontaneous change, the product $T \, \Delta S_{total}$ must also be positive. This means that the quantity in parentheses on the right, $(\Delta H_{system} - T \, \Delta S_{system})$, must be negative so that $-(\Delta H_{system} - T \, \Delta S_{system})$ may be positive. Thus, in order for a spontaneous change to take place, the expression $(\Delta H_{system} - T \, \Delta S_{system})$ must be negative.

> *T is positive and ΔS_{total} is positive, so $T \, \Delta S_{total}$ is positive.*

At this point it is convenient to introduce another thermodynamic state function, G, called the **Gibbs free energy.** This is defined as

$$G = H - TS$$

For a change at constant T and P, we write

$$\Delta G = \Delta H - T \, \Delta S \qquad [12.11]$$

From the argument presented in the preceding paragraph, we see that ΔG must be less than zero for a spontaneous process; that is, ΔG must have a negative value at constant T and P.

The Gibbs free energy charge, ΔG, represents a composite of the two factors contributing to spontaneity, ΔH and ΔS. For systems in which ΔH is negative (exothermic) and ΔS is positive (increased disorder accompanying the change), both factors favor spontaneity and the process will occur spontaneously at all temperatures. Conversely, if ΔH is positive (endothermic) and ΔS is negative (increase in order), ΔG will always be positive and the change cannot occur spontaneously at any temperature.

> *It is thermodynamically impossible for a building to form all by itself from a pile of rubble. Why?*

In situations where ΔH and ΔS are both positive, or both negative, Equation 12.11 shows that temperature plays the determining role in controlling whether or not a change can take place. In the first case (ΔH and ΔS both are positive), ΔG will be negative *only* at high temperatures, where $T \, \Delta S$ is greater in magnitude than ΔH. Therefore, the reaction will be spontaneous only at elevated temperatures. An example is the melting of ice, which is nonspontaneous at low temperatures (below 0°C) and spontaneous at high temperatures (above 0°C).

> *The negative $T \, \Delta S$ term will only be numerically smaller than the negative ΔH term when T is small.*

When ΔH and ΔS are both negative, ΔG will be negative only at low temperatures. An example of this is the freezing of water. We know that heat must be removed from the liquid to produce ice, so the process is exothermic with a negative ΔH. Freezing is also accompanied by an ordering of the water molecules as they leave the random liquid state and become part of the crystal. As a result, ΔS is also negative. The sign of ΔG is determined both by ΔH, which in this case is negative, and $T \, \Delta S$, which is also negative. To compute ΔG we must subtract a negative $T \, \Delta S$ from a negative ΔH. The result will be negative only at low temperature. Therefore, at 1 atm we observe H_2O to freeze spontaneously only below 0°C. Above 0°C the magnitude of $T \, \Delta S$ is greater than ΔH, and ΔG becomes positive. As a result, freezing is no longer spontaneous. Instead, the reverse process (melting) occurs.

The effects of the signs of ΔH and ΔS and the effect of temperature on spontaneity, can be summarized as follows.

ΔH	ΔS	Outcome
(−)	(+)	Spontaneous at all temperatures
(+)	(−)	Nonspontaneous regardless of temperature
(+)	(+)	Spontaneous only at high temperature
(−)	(−)	Spontaneous only at low temperature

The significance of the Gibbs free energy change in determining spontaneity can be seen in the practical world around us. A classic example is the production of synthetic diamonds. People had been fascinated by this problem ever since 1797, when it was found that diamond was simply a form of carbon. Over the years many experiments were devised in an attempt to convert graphite, the common form of carbon, into its much more valuable counterpart. However, as of 1938 no one had yet been able to accomplish this feat. At that time a careful thermodynamic analysis of the problem was performed, the results of which are summarized in Figure 12.11.

In this figure, $\Delta G/T$ is plotted along the vertical axis and temperature, in kelvins, is plotted along the horizontal axis. Since ΔG for the reaction,

$$C(s, \text{graphite}) \longrightarrow C(s, \text{diamond})$$

must be negative in order for the process to be spontaneous, diamond can be produced only at temperatures and pressures that lie *below* the zero on the $\Delta G/T$ scale. For example, at 470 K the conversion of graphite into diamond can only take place at pressures greater than or equal to 20 000 atm. We can also conclude from the figure that at a constant pressure of 20 000 atm the reaction is not spontaneous above 470 K.

This analysis served to define the limits of temperature and pressure that would permit the conversion to take place. That was not the end of the problem, however, because suitable materials had to be found that would allow the reaction to proceed at a measurable rate. In fact, it was not until 1955 that success was finally achieved, and today synthetic diamonds are an important industrial abrasive used in grinding and cutting tools.

The conversion of graphite to diamond represents only one example of how thermodynamics can serve as a guide to answering practical questions. The principles of thermodynamics have been applied over the years to such diverse problems as the design of steam engines and the development of fuel cells that are now used as a source of electric power in spacecraft. Dr. Frederick Rossini, a noted thermodynamicist, has pointed out that the balance between the simultaneous opposing drives toward security (low energy) and freedom (high entropy) that control the fate of chemical systems also seem to determine equilibrium in a stable society, illustrating, perhaps, the truly wide scope of thermodynamics.

Figure 12.11

Thermodynamics of graphite to diamond conversion. Adapted from Chemical and Engineering News, *April 5, 1971, p. 51. Used by permission.*

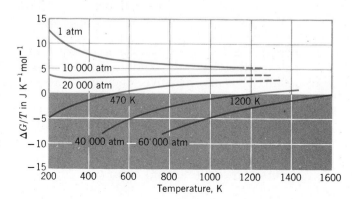

Another illustration of the impact of thermodynamics on our lives is the problem of *thermal pollution*, particularly in the neighborhood of power plants. It is a consequence of the second law of thermodynamics that any heat engine designed to convert heat into useful work must absorb heat from a high-temperature source. Some of this heat can be converted to work, but some of it must also be deposited in a low-temperature reservoir. Therefore, any device that tries to convert heat into useful work also discharges some heat into the environment. An automobile engine burns gasoline or diesel fuel, converts part of the resultant heat energy into work that propels the car, while the remainder is discharged to the radiator. Large power stations burn fuels (either fossil fuels such as coal or oil, or nuclear fuels such as uranium) to generate electrical power. In the process they too deposit relatively large quantities of heat into the environment. Cooling water from such power plants is heated to such an extent that it has been known to raise the temperature of surrounding rivers or lakes by as much as 13°C or more. These higher temperatures decrease the solubility of oxygen and kill some species of fish. It has also been found, however, that clams and oysters thrive in these warmer waters.

Several years ago a nuclear power station in New Jersey was forced to shut down for repairs during the winter. In this case, the halting of thermal pollution caused a large fish kill because many fish that normally live in warm water were suddenly "put out in the cold."

> In general, the spread of pollutants in the environment is virtually unstoppable once they are released because of the enormous entropy increase that occurs. The only practical way to limit pollution is to not let it begin.

12.9   FREE ENERGY AND USEFUL WORK

One of the most important applications of chemical reactions is in the production of energy in the form of useful work. This can, for example, take the form of combustion, in which the heat generated is used to create steam for the production of mechanical work, or perhaps electrical work drawn from a dry cell or storage battery. The quantity G is called the *free energy* because ΔG represents the *maximum* quantity of energy released in a process occurring at constant temperature and pressure that is free—or available—to perform useful work. We have already associated ΔG with the factors that lead to a drive for spontaneity. What

> The "free" in free energy does not mean *without cost*. This is the energy that is free in the sense of not being tied up for some other purpose.

The interior of a modern steam powered electric generating plant. Although efficient by engineering standards, the thermodynamic efficiency is only about 35 to 40%.

we see now is that this driving force in a chemical change can be harnessed to perform work for us.

The actual quantity of work obtained from any real spontaneous process is always less than the maximum predicted by ΔG. This is because real processes are always irreversible, and we saw earlier that the maximum work can be extracted only from a truly reversible change. The free-energy change gives us a goal at which to aim. The closer a given process is to reversibility, the greater the quantity of available work that can be used. However, even relatively efficient systems are able to harness only a small fraction of the available free energy. Living systems, for example, are able to convert only about 40% of the free energy available in the oxidation of glucose to other forms of stored chemical energy (for example, ATP).

Mechanical systems, such as electrical generators and gasoline engines, are even less efficient than living systems at converting energy into work.

12.10 STANDARD ENTROPIES AND FREE ENERGIES

The **third law of thermodynamics** states that the entropy of any pure crystalline substance at absolute zero is equal to zero. This makes sense because in a perfect crystal at absolute zero there is perfect order. Because of this, it is possible to actually measure the absolute quantity of entropy that a substance has in its standard state by summing q_{rev}/T increments from 0 K to 298 K (25°C). Table 12.4 contains a number of these standard entropies. Standard entropies can be used to calculate standard entropy changes by a Hess's law-type of calculation.

$$\Delta S^0_{reaction} = (\text{sum of } S^0 \text{ products}) - (\text{sum of } S^0 \text{ of reactants}) \quad [12.12]$$

EXAMPLE 12.11 **CALCULATING THE STANDARD ENTROPY CHANGE FROM TABULATED VALUES OF S^0**

PROBLEM   Calculate the standard entropy change (in units of $J\,K^{-1}$) for the reaction

$$2NaHCO_3(s) \longrightarrow Na_2CO_3(s) + CO_2(g) + H_2O(g)$$

SOLUTION   We use Equation 12.12 and the data in Table 12.4.

$$\Delta S^0_{reaction} = [S^0_{Na_2CO_3(s)} + S^0_{CO_2(g)} + S^0_{H_2O(g)}] - [2S^0_{NaHCO_3(s)}]$$

$$= [1\ mol(136\ J\ mol^{-1}\ K^{-1}) + 1\ mol(213.6\ J\ mol^{-1}\ K^{-1})$$
$$+ 1\ mol(188.7\ J\ mol^{-1}\ K^{-1})] - [2\ mol(155\ J\ mol^{-1}\ K^{-1})]$$

$$= (538\ J\ K^{-1}) - (310\ J\ K^{-1})$$

$$= +228\ J\ K^{-1}$$

It's interesting to note that this reaction also has a positive ΔH^0 (Example 12.5), so it falls into that category in which both ΔH and ΔS are positive. The decomposition of $NaHCO_3$ is nonspontaneous at low temperature but becomes spontaneous at high temperature, and that is why it can be used as a fire extinguisher.

From standard heats of formation and standard entropies we can also calculate **standard free energies of formation, ΔG^0_f.** For example, consider the formation of CO_2 from the elements, with all reactants and products in their standard states,

$$C(s, \text{graphite}) + O_2(g) \longrightarrow CO_2(g)$$

Table 12.4
Absolute entropies at 25°C and 1 atm

Substance	S^0 J mol^{-1} K^{-1}	Substance	S^0 J mol^{-1} K^{-1}
Al(s)	28.3	Mg(OH)$_2$(s)	63.1
Al$_2$O$_3$(s)	51.0	N$_2$(g)	191.5
Br$_2$(l)	152.2	NH$_3$(g)	192.5
Br$_2$(g)	245.4	N$_2$O(g)	220.0
C(s, graphite)	5.69	NO(g)	210.6
C(s, diamond)	2.4	NO$_2$(g)	240.5
CO(g)	197.9	Na(s)	51.0
CO$_2$(g)	213.6	NaF(s)	51.5
CH$_4$(g)	186.2	NaCl(s)	72.8
C$_2$H$_6$(g)	230	NaBr(s)	83.7
C$_2$H$_4$(g)	220	NaI(s)	91.2
C$_2$H$_2$(g)	201	NaHCO$_3$(s)	155
C$_3$H$_8$(g)	269.9	Na$_2$CO$_3$(s)	136
CCl$_4$(l)	214.4	O$_2$(g)	205.0
Cl$_2$(g)	223.0	Pb(s)	64.9
F$_2$(g)	202.7	PbO$_2$(s)	68.6
H$_2$(g)	130.6	PbSO$_4$(s)	149
H$_2$O(l)	70.0	S(s, rhombic)	31.8
H$_2$O(g)	188.7	SO$_2$(g)	248
HF(g)	173.5	SO$_3$(g)	256
HCl(g)	186.7	H$_2$SO$_4$(l)	157
HBr(g)	198.5	Si(s)	19
HI(g)	206	SiO$_2$(s)	41.8
I$_2$(s)	116.1	Zn(s)	41.8
Mg(s)	32.5	ZnO(s)	43.5

Table 12.1 gives us the standard enthalpy of formation of $CO_2(g)$, ΔH_f^0, as -394 kJ mol^{-1}. From the data in Table 12.4 we can calculate ΔS_f^0.

Notice that we have to calculate ΔS_f^0, they're not tabulated.

$$\Delta S_f^0 = S_{CO_2}^0 - (S_C^0 + S_{O_2}^0)$$

$$\Delta S_f^0 = 213.6 - (5.7 + 205.0) \text{ J mol}^{-1} \text{ K}^{-1} = +2.9 \text{ J mol}^{-1} \text{ K}^{-1}$$

We can then obtain ΔG_f^0 as

$$\Delta G_f^0 = \Delta H_f^0 - T\Delta S_f^0$$

At 25°C (298 K), then,

$$\Delta G_f^0 = -394 \text{ kJ mol}^{-1} - (298 \text{ K})(2.9 \text{ J mol}^{-1} \text{ K}^{-1})$$

$$\Delta G_f^0 = -394 \text{ kJ mol}^{-1} - 860 \text{ J mol}^{-1}$$

Converting entirely to kilojoules per mole gives

$$\Delta G_f^0 = (-394 - 0.9) \text{ kJ mol}^{-1}$$

$$\Delta G_f^0 = -395 \text{ kJ mol}^{-1}$$

Table 12.5
Standard free energies of formation of 25°C and 1 atm

Substance	ΔG_f^0 kJ mol^{-1}	Substance	ΔG_f^0 kJ mol^{-1}
$Al_2O_3(s)$	−1577	$HBr(g)$	−53.1
$AgNO_3(s)$	−32	$HI(g)$	+1.30
$C(s, diamond)$	+2.9	$H_2O(l)$	−237
$CO(g)$	−137	$H_2O(g)$	−228
$CO_2(g)$	−395	$MgCl_2(s)$	−592.5
$CH_4(g)$	−50.6	$Mg(OH)_2(s)$	−833.9
$C_2H_6(g)$	−33	$NH_3(g)$	−17
$C_2H_4(g)$	+68.2	$N_2O(g)$	+104
$C_2H_2(g)$	+209	$NO(g)$	+86.8
$C_3H_8(g)$	−23	$NO_2(g)$	+51.9
$CCl_4(l)$	−65.3	$HNO_3(l)$	−79.9
$C_2H_5OH(l)$	−175	$PbO_2(s)$	−219
$HC_2H_3O_2(l)$	−392	$PbSO_4(s)$	−811.3
$CaO(s)$	−604.2	$SO_2(g)$	−300
$Ca(OH)_2(s)$	−896.6	$SO_3(g)$	−370
$CaSO_4(s)$	−1320	$H_2SO_4(l)$	−689.9
$CuO(s)$	−127	$SiO_2(s)$	−856
$Fe_2O_3(s)$	−741.0	$SiH_4(g)$	+52.3
$HF(g)$	−273	$ZnO(s)$	−318
$HCl(g)$	−95.4		

This and other standard free energies of formation are given at Table 12.5.

Earlier in this chapter we saw that ΔH^0 for a reaction can be computed from standard heats of formation. The same rules also apply for the calculation of ΔG^0 using standard free energies of formation; that is,

$$\Delta G^0 = (\text{sum of } \Delta G_f^0 \text{ of products}) - (\text{sum of } \Delta G_f^0 \text{ of reactants}) \quad [12.13]$$

EXAMPLE 12.12

CALCULATING ΔG^0 FOR A REACTION FROM TABULATED VALUES OF ΔG_f^0

PROBLEM

Silane, SiH_4, is the silicon analog of the main constituent in natural gas—methane, CH_4. Like methane, silane burns in air. The product is silica—a solid quite unlike carbon dioxide.

Silica, SiO_2, is the major component in ordinary sand.

$$SiH_4(g) + 2O_2(g) \longrightarrow SiO_2(s) + 2H_2O(g)$$
$$\text{silicon dioxide}$$
$$\text{(silica)}$$

Calculate ΔG^0 for this reaction in kilojoules.

SOLUTION We have to apply Equation 12.13, using the data in Table 12.5.

$$\Delta G^0 = [\Delta G^0_{f\,SiO_2(s)} + 2\Delta G^0_{f\,H_2O(g)}] - [\Delta G^0_{f\,SiH_4(g)} + 2\Delta G^0_{f\,O_2(g)}]$$

(As with ΔH^0_f, we take the standard free energy of formation of a free element to be zero.) Therefore,

$$\Delta G^0 = \left[1\ \text{mol}\left(\frac{-856\ \text{kJ}}{\text{mol}}\right) + 2\ \text{mol}\left(\frac{-228\ \text{kJ}}{\text{mol}}\right)\right] - \left[1\ \text{mol}\left(\frac{+52.3\ \text{kJ}}{\text{mol}}\right) + 0\right]$$

$$= -1364\ \text{kJ}$$

12.11 FREE ENERGY AND EQUILIBRIUM

Earlier we said that ΔG determines the maximum quantity of energy that is available to perform useful work as a system passes from one state to another. As a reaction proceeds, its capacity to perform work, as measured by G, diminishes until finally, at equilibrium, the system is no longer able to supply additional work. This means that both reactants and products possess the same free energy, and therefore $\Delta G = 0$. We see then that the value of ΔG for a particular change determines where the system stands with respect to equilibrium. When ΔG is negative—meaning that the free energy of the system is decreasing—the reaction is spontaneous and proceeds in the forward direction toward a state of equilibrium. When ΔG is zero, the system is in a state of dynamic equilibrium, and when ΔG is positive, the reaction is really spontaneous in the reverse direction.

At this point, it should be reemphasized that although ΔG may predict that a particular process is spontaneous, nothing is implied about how rapid the change will be.

Figure 12.12 illustrates graphically the free energy changes in a typical chemical reaction. Notice that the free energy curve has a minimum that lies below the free energies of *both* the reactants and the products. Therefore, if we

Figure 12.12

The variation of free energy in a homogeneous chemical system as the reaction proceeds from pure reactants on the left to pure products on the right. The minimum in the curve represents the extent of reaction required for the system to achieve equilibrium.

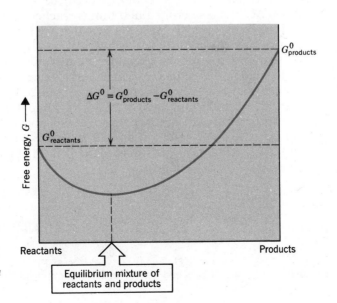

Figure 12.13

The position of equilibrium changes as the value of ΔG^0 changes. (a) Position of equilibrium in favor of the reactants. (b) Position of equilibrium intermediate between the reactants and products. (c) Position of equilibrium in favor of the products.

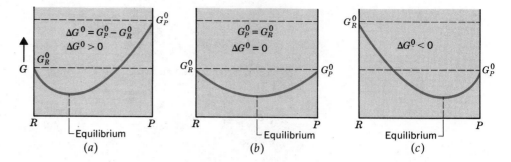

begin with pure reactants, we proceed in the direction of the products because this gives a lowering of the free energy—ΔG is negative. Also notice that if we begin with pure products, we proceed in the direction of the reactants—this lowers the free energy, too. Therefore, whether we begin with reactants or products, we *always* head in the direction of the minimum on the free energy curve. When the reaction gets there, equilibrium is reached ($\Delta G = 0$). No change in the composition can occur because going toward either the reactants or products involves going uphill on the free energy curve, and that is nonspontaneous.

In Figure 12.12 we also see ΔG^0 for the reaction. It is simply the difference between the free energies of the pure products and the pure reactants, but it doesn't tell us (directly) which way the reaction will proceed when the reaction mixture has some given composition. For instance, in this particular reaction ΔG^0 is positive because $G^0_{products}$ is larger than $G^0_{reactants}$. The sign and numerical value of ΔG^0 for the reaction is fixed—it is a constant for the reaction. It is the way that G changes with composition—that is, it's the slope of the free energy curve—that determines the direction the reaction will go.

At this point, you may ask, "Why bother with ΔG^0 at all?" The answer can be seen in Figure 12.13. In Figure 12.13*a* we see a reaction for which ΔG^0 is positive. In Figure 12.13*b*, the reaction has $\Delta G^0 = 0$, and in Figure 12.13*c*, the reaction has a negative ΔG^0. Notice how the location of the minimum on the free energy curve varies with ΔG^0. When ΔG^0 is positive, the minimum is near the reactants; hardly any reaction has occurred by the time equilibrium is reached. When ΔG^0 equals zero, equilibrium is reached about halfway between reactants and products, and when ΔG^0 is negative the reaction goes almost to completion by the time equilibrium is reached.

Now we can see the usefulness of ΔG^0 calculations. The value of ΔG^0 allows us to predict the ultimate outcome of a reaction if it is allowed to proceed to equilibrium. When ΔG^0 is positive by about 20 kJ or so, so little product will be present at equilibrium that we might conclude that nothing has happened in the reaction mixture. In other words, the reaction will appear to be nonspontaneous. On the other hand, when ΔG^0 is negative by about 20 kJ, the reaction will appear to have gone to completion. It is in this sense that ΔG^0 predicts the spontaneity of a reaction.

It is the algebraic sign of the change in G with a change in composition that corresponds to the ΔG that controls the spontaneity of a chemical reaction in one direction or the other.

EXAMPLE 12.13   USING ΔG^0 AS A GUIDE TO SPONTANEITY

PROBLEM   Is the reaction, $2SO_2(g) + O_2(g) \rightarrow 2SO_3(g)$ expected to be spontaneous at 25°C and 1 atm (i.e., will it proceed far toward completion by the time equilibrium is reached)?

SOLUTION   Let's calculate ΔG^0 for the reaction

$$\Delta G^0 = 2\Delta G^0_f \, SO_{3(g)} - [2\Delta G^0_f \, SO_{2(g)} + \Delta G^0_f \, O_{2(g)}]$$

$$= 2 \, \text{mol}\left(\frac{-370 \text{ kJ}}{\text{mol}}\right) - \left[2 \, \text{mol}\left(\frac{-300 \text{ kJ}}{\text{mol}}\right) + 0.0\right] = -740 \text{ kJ} - (-600 \text{ kJ})$$

$$= -140 \text{ kJ}$$

This very negative ΔG^0 indicates that the position of equilibrium is very largely in favor of SO_3, so the reaction should appear to be quite spontaneous. Actually, this is true but the reaction is very slow in the absence of agents called catalysts, which we will discuss in the next chapter.

REVIEW QUESTIONS AND PROBLEMS

(Problems whose numbers are in blue have their answers in Appendix C at the back of the book; the more difficult problems are marked with asterisks.)

Terminology of Thermodynamics

12.1   What does the word *thermodynamics* imply?

12.2   Define the following: system, surroundings, isothermal change, adiabatic change, state function, heat capacity, molar heat capacity, specific heat.

12.3   What is an *equation of state?*

12.4   What are some different kinds of work that a system can perform on its surroundings?

12.5   What is meant by the term spontaneous change?

12.6   What is a reversible process?

12.7   From an experimental point of view, how can a change be carried out isothermally? How can a change be carried out adiabatically?

First Law of Thermodynamics

12.8   State the first law of thermodynamics.

12.9   On a molecular basis, why is E a state function?

12.10   What is a perpetual motion machine? Why are such devices impossible?

12.11   Why is it not possible to measure or calculate E for a system?

12.12   A real gas usually cools slightly if it is allowed to expand adiabatically into an evacuated container.
(a) What should happen to the temperature of an ideal gas when it is treated in the same way?
(b) What conclusions can you draw about ΔE, q, and w for the *isothermal* expansion of a real gas?

12.13   Why is ΔE equal to zero for an isothermal expansion of an ideal gas?

12.14   Suppose that a steel spring is compressed, tied so it cannot expand, and then dissolved in hydrochloric acid. What happens to the potential energy stored in the spring?

12.15   What unit for work do the SI units of pressure and volume produce in calculations of pressure-volume work?

12.16   Why are we usually concerned more about changes in enthalpy than changes in internal energy?

12.17   A gas, initially under a pressure of 1500 kPa and having a volume of 10.0 dm^3, is permitted to expand isothermally in two steps. In the first step the external pressure is held constant at 750 kPa, and in the second step the external pressure is maintained at 100 kPa. What are q and w for each step? What is the net change in the internal energy for both the system and its surroundings? Assume ideal gas behavior.

12.18   A gas having an initial volume of 50.0 m^3 at an initial pressure of 200 kPa is allowed to expand against a constant opposing pressure of 100 kPa. Calculate the work done by the gas in kJ. If the gas is ideal and the expansion is isothermal, what is q for the gas in kilojoules?

12.19   If 500 cm^3 of a gas is compressed to 250 cm^3 under a constant external pressure of 300 kPa, and if the gas also absorbs 12.5 kJ, what are the values of q, w, and ΔE for the gas expressed in kJ? What is the value of ΔE for the surroundings?

12.20   In a particular change, a system absorbs 35 J of heat and has 40 J of work done on it.
(a) What are q, w, and ΔE for the system?
(b) What are q, w, and ΔE for the surroundings?

Calorimetry

12.21 What is a calorimeter?

12.22 What is a bomb calorimeter?

12.23 An electrical heater was used to determine the heat capacity of a calorimeter. It was found that when 1.347 kJ of electrical energy was converted to heat in the calorimeter, the temperature rose from 25.000°C to 26.135°C. What is the heat capacity in joules per degree Celsius?

12.24 A certain reaction in a bomb calorimeter liberated 14.3 kJ. If the initial temperature of the calorimeter was 25.000°C and the heat capacity of the calorimeter and its contents was 1.78×10^4 J °C^{-1}, what was the final temperature of the calorimeter?

12.25 Why is ΔE called the heat of reaction at constant volume? Why is ΔH called the heat of reaction at constant pressure?

12.26 A 0.100-mol sample of propane, a gas used for cooking in many rural areas, was placed into a bomb calorimeter with excess oxygen and ignited. The reaction was

$$C_3H_8(g) + 5O_2(g) \longrightarrow 3CO_2(g) + 4H_2O(l)$$

The initial temperature of the calorimeter was 25.000°C and its total heat capacity was 97.1 kJ °C^{-1}. The reaction raised the temperature of the calorimeter to 27.282°C.
(a) How many joules were liberated by the combustion of the propane?
(b) What is ΔE for the reaction expressed in kJ/mol^{-1} C$_3$H$_8$?

12.27 At 25°C and a constant pressure of 101 kPa, the reaction of 0.500 mol of OF$_2$ with water vapor, according to the equation

$$OF_2(g) + H_2O(g) \longrightarrow O_2(g) + 2HF(g)$$

liberates 162 kJ. Calculate ΔH in units of kJ mol^{-1} OF$_2$.

12.28 When 200 cm^3 of 1.00 M NaOH at 25.00°C were mixed with 150 cm^3 of 1.00 M HCl, also at 25.00°C, in a styrofoam "coffee-cup calorimeter," the temperature of the mixture rose to 30.00°C. Calculate ΔH in kilojoules for the neutralization of 1 mol of H$^+$ by 1 mol OH$^-$ (i.e., H$^+$ + OH$^-$ → H$_2$O(l)). Assume that the specific heat of each of the solutions is 4.184 J g^{-1} °C^{-1}.

12.29 Toluene, C$_7$H$_8$, is used in the manufacture of explosives such as TNT (trinitrotoluene). A 1.500-g sample of liquid toluene was placed in a bomb calorimeter along with excess oxygen. When the combustion of the C$_7$H$_8$ was initiated, the temperature of the calorimeter rose from 25.000°C to 26.413°C. The products of the combustion were CO$_2$(g) and H$_2$O(l), and the heat capacity of the calorimeter was 45.06 kJ °C^{-1}. Calculate ΔE for the reaction

$$C_7H_8(l) + 9O_2(g) \longrightarrow 7CO_2(g) + 4H_2O(l)$$

in kJ. (Assume that the coefficients in the equation stand for moles.)

Converting Between ΔE and ΔH

12.30 For what kinds of reactions is it safe to ignore differences between ΔE and ΔH? In general, do ΔE and ΔH differ by much?

12.31 Calculate ΔH for the reaction in Problem 12.26 in the units kJ mol^{-1} C$_3$H$_8$.

12.32 At 25°C and 1.00 atm, the reaction of 1.00 mol of CaO with water (shown below) evolves 62.3 kJ.

$$CaO(s) + H_2O(l) \longrightarrow Ca(OH)_2(s)$$

What are ΔH and ΔE, in kilojoules per mole of CaO, for this process given that the densities of CaO(s), H$_2$O(l), and Ca(OH)$_2$(s) at 25°C are 3.25 g cm^{-3}, 0.997 g cm^{-3}, and 2.24 g cm^{-3}, respectively? What does this tell you about the relative values of ΔH and ΔE when all substances are either liquids or solids?

***12.33** At 25°C, burning 0.200 mol of H$_2$ with 0.100 mol of O$_2$ to produce H$_2$O(l) in a bomb calorimeter raises the temperature of the apparatus 0.880°C. When 0.0100 mol of toluene, C$_7$H$_8$, is burned in this calorimeter, the temperature is raised by 0.615°C. The equation for the combustion reaction is

$$C_7H_8(l) + 9O_2(g) \longrightarrow 7CO_2(g) + 4H_2O(l)$$

Calculate ΔE for this reaction. Use ΔH_f^0 for H$_2$O(l) found in Table 12.1 to compute ΔE_f^0 for H$_2$O(l).

12.34 The reaction

$$Ca(s) + O_2(g) + H_2(g) \longrightarrow Ca(OH)_2(s)$$

has $\Delta H^0 = -897$ kJ. What is ΔE for this reaction?

12.35 The heat of vaporization ΔH_{vap} of H$_2$O at 25°C is 43.9 kJ mol^{-1}. Calculate q, w, and ΔE for the process.

***12.36** An important photochemical reaction in the production of smog is

$$NO_2(g) + h\nu \longrightarrow NO(g) + O(g)$$

If one quantum of energy is required to cause this reaction to occur, what must the wavelength of the light be? (Use the data in Tables 12.1 and 12.2, calculate ΔE for the process in kilojoules, then calculate ν in nanometres from the Planck relationship $\Delta E = h\nu$.)

Heats of Formation and Hess's Law

12.37 What is Hess's law of heat summation? What is meant by the term, standard state?

12.38 The following are two reactions showing the formation of 1 mol of SO_3:

$$SO_2(g) + \tfrac{1}{2}O_2(g) \longrightarrow SO_3(g)$$

$$S(s) + \tfrac{3}{2}O_2(g) \longrightarrow SO_3(g)$$

Should their enthalpy changes both be labeled ΔH_f^0 if they occur at 25°C and 1 atm? Explain.

12.39 Which of the following should have a more negative ΔH^0? Why?

$$2C(s) + 3H_2(g) + \tfrac{1}{2}O_2(g) \longrightarrow C_2H_5OH(g)$$

$$2C(s) + 3H_2(g) + \tfrac{1}{2}O_2(g) \longrightarrow C_2H_5OH(l)$$

*12.40 Based on the results of Problem 12.33, compute the standard heat of formation, ΔH_f^0 (in kJ mol^{-1}), of toluene. (Assume that the reaction was carried out at 25°C.)

12.41 The combustion of ethanol, $C_2H_5OH(l)$, to give $CO_2(g)$ and $H_2O(l)$ evolves 1.37×10^3 kJ per mol of $C_2H_5OH(l)$ when the products are returned to 25°C and 1 atm. Use this information and the data in Table 12.1 to calculate ΔH_f^0, in kJ mol^{-1}, for $C_2H_5OH(l)$.

12.42 Calculate ΔH_f^0 (in kJ mol^{-1}) for $C_3H_8(g)$ from the data in Problem 12.26.

12.43 Use the data in Table 12.1 and Hess's law to calculate ΔH^0, in kilojoules, for each of the following reactions:
(a) $2Al(s) + Fe_2O_3(s) \rightarrow Al_2O_3(s) + 2Fe(s)$
(b) $SiH_4(g) + 2O_2(g) \rightarrow SiO_2(s) + 2H_2O(g)$
(c) $CaO(s) + SO_3(g) \rightarrow CaSO_4(s)$
(d) $CuO(s) + H_2(g) \rightarrow Cu(s) + H_2O(g)$
(e) $C_2H_4(g) + H_2(g) \rightarrow C_2H_6(g)$

12.44 Use the data in Table 12.1 to calculate ΔH^0, in kilojoules, for each of the following reactions:
(a) $C_2H_2(g) + H_2(g) \rightarrow C_2H_4(g)$
(b) $SO_3(g) + H_2O(l) \rightarrow H_2SO_4(l)$
(c) $Mg(OH)_2(s) + 2HCl(g) \rightarrow MgCl_2 \cdot 2H_2O(s)$
(d) $CO_2(g) + H_2(g) \rightarrow CO(g) + H_2O(g)$
(e) $10N_2O(g) + C_3H_8(g) \rightarrow$
$\qquad\qquad 10N_2(g) + 3CO_2(g) + 4H_2O(g)$

12.45 Aerosol propellants are often chlorofluoromethanes (CFMs) such as Freon-11 ($CFCl_3$) and Freon-12 (CF_2Cl_2). It has been suggested that the continued use of these may ultimately deplete the ozone shield in the stratosphere, with catastrophic results to the inhabitants of our planet. In the stratosphere CFMs absorb high-energy radiation and produce Cl atoms that have a catalytic effect on removing ozone, O_3.

$O_3 + Cl \longrightarrow O_2 + ClO$		$\Delta H^0 = -126$ kJ
$ClO + O \longrightarrow Cl + O_2$		$\Delta H^0 = -268$ kJ
net	$O_3 + O \longrightarrow 2 O_2$	

The O atoms are present due to the dissociation of O_2 molecules by high-energy radiation. Calculate ΔH^0 for the net reaction for the removal of the ozone.

12.46 Given the following thermochemical equations:

$$2H_2(g) + O_2(g) \longrightarrow 2H_2O(l) \qquad \Delta H^0 = -571.5 \text{ kJ}$$

$$N_2O_5(g) + H_2O(l) \longrightarrow 2HNO_3(l) \qquad \Delta H^0 = -76.6 \text{ kJ}$$

$$\tfrac{1}{2}N_2(g) + \tfrac{3}{2}O_2(g) + \tfrac{1}{2}H_2(g) \longrightarrow HNO_3(l)$$
$$\Delta H^0 = -174.1 \text{ kJ}$$

calculate ΔH^0 for the reaction

$$2N_2(g) + 5O_2(g) \longrightarrow 2N_2O_5(g)$$

12.47 Given the following thermochemical equations:

$$Fe_2O_3(s) + 3CO(g) \longrightarrow 2Fe(s) + 3CO_2(g) \qquad \Delta H = -28 \text{ kJ}$$

$$3Fe_2O_3(s) + CO(g) \longrightarrow 2Fe_3O_4(s) + CO_2(g)$$
$$\Delta H = -59 \text{ kJ}$$

$$Fe_3O_4(s) + CO(g) \longrightarrow 3FeO(s) + CO_2(g)$$
$$\Delta H = +38 \text{ kJ}$$

calculate ΔH for the reaction,

$$FeO(s) + CO(g) \longrightarrow Fe(s) + CO_2(g)$$

without referring to the data in Table 12.1.

12.48 Use the results of Question 12.47 and the data in Table 12.1 to compute the standard heat of formation of FeO.

12.49 Acetylene, C_2H_2, a gas used in welding torches, is produced by the action of water on calcium carbide, CaC_2. Given the following thermochemical equations, calculate ΔH_f^0 for acetylene. Express your answer in both kJ/mol and kcal/mol.

$$CaO(s) + H_2O(l) \longrightarrow Ca(OH)_2(s) \qquad \Delta H^0 = -65 \text{ kJ}$$

$$CaO(s) + 3C(s) \longrightarrow CaC_2(s) + CO(g)$$
$$\Delta H^0 = +462 \text{ kJ}$$

$$CaC_2(s) + 2H_2O(l) \longrightarrow Ca(OH)_2(s) + C_2H_2(g)$$
$$\Delta H^0 = -126 \text{ kJ}$$

$$2C(s) + O_2(g) \longrightarrow 2CO(g) \qquad \Delta H^0 = -221 \text{ kJ}$$

$$2H_2O(l) \longrightarrow 2H_2(g) + O_2(g)$$
$$\Delta H^0 = +572 \text{ kJ}$$

12.50 Plaster of Paris, $CaSO_4 \cdot \tfrac{1}{2}H_2O$, is mixed with water with which it combines to produce gypsum, $CaSO_4 \cdot 2H_2O$. The reaction is exothermic, which explains why a plaster cast on a broken arm becomes warm as the cast hardens. Given that for $CaSO_4 \cdot \frac{1}{2}H_2O$, $\Delta H_f^0 = -1573$ kJ mol^{-1}, and for $CaSO_4 \cdot 2H_2O$, $\Delta H_f^0 = -2020$ kJ mol^{-1}, calculate ΔH^0 for the reaction,

$$CaSO_4 \cdot \tfrac{1}{2}H_2O(s) + \tfrac{3}{2}H_2O(l) \longrightarrow CaSO_4 \cdot 2H_2O(s)$$

12.51 Important reactions in the production of ozone in polluted air are

$$2NO(g) + O_2(g) \longrightarrow 2NO_2(g)$$

$$NO_2(g) \xrightarrow{h\nu} NO(g) + O(g)$$

$$O_2(g) + O(g) \longrightarrow O_3(g)$$

Calculate ΔH^0 for each of these processes using the data in Tables 12.1 and 12.2.

12.52 The body eliminates ethyl alcohol, C_2H_5OH, by oxidation to give water and the following series of carbon-containing products,

$$C_2H_5OH \xrightarrow{O_2} CH_3CHO \xrightarrow{O_2} HC_2H_3O_2 \xrightarrow{O_2} CO_2$$

Write balanced equations for each step in the oxidation and calculate its ΔH^0 in kilojoules. What is the overall ΔH^0 for complete oxidation to CO_2 and H_2O?

12.53 Use the data in Table 12.1 to determine how much heat is evolved in the combustion of 45.0 g of $C_2H_6(g)$ at 25°C to produce $CO_2(g)$ and $H_2O(g)$ under a constant pressure of 1.00 atm. Assume the products are returned to 25°C.

12.54 The average adult expends about 8000 kJ of energy per day for normal activity. If 1 g of carbohydrate provides 17 kJ of usable energy, how many grams of carbohydrates must be consumed to meet these caloric demands?

12.55 The evaporation of perspiration is one mechanism whereby the body disposes of excess thermal energy and manages to maintain a constant temperature. How many kilojoules are removed from the body by the evaporation of 10.0 g of H_2O at 25°C?

*12.56 Calculate the number of kilojoules liberated during the combustion of 1.00 dm^3 of octane (gasoline). The density of octane is 0.703 g cm^{-3}. What mass of hydrogen would have to be burned [giving $H_2O(l)$] to produce this same quantity of heat? If this H_2 were compressed to a pressure of 170 atm (17 MPa) at 25°C, what volume would it occupy? What does this suggest about the feasibility of a hydrogen fuel economy for the automobile? For octane, $C_8H_{18}(l)$, $\Delta H_f^0 = -208.4$ kJ mol^{-1}.

*12.57 It is estimated that the body generates up to 5900 kJ of thermal energy per hour during heavy physical exercise. If the only way that this excess energy

could be dissipated was through the evaporation of water, how many grams of water would have to evaporate per hour to keep the body temperature constant?

*12.58 How many grams of glucose, $C_6H_{12}O_6$, would have to be metabolized per hour (to give CO_2 and H_2O) in order to generate the excess thermal energy described in the preceding problem if it is assumed that 60% of the available energy appears as excess body heat? (The remaining 40% is used by the body to do mechanical work—moving of limbs, pumping of blood, etc.). For $C_6H_{12}O_6$, $\Delta H_{combustion} = -2820$ kJ mol^{-1}.

*12.59 How many dm^3 of natural gas (CH_4) at 25°C and 1 atm must be burned to provide sufficient energy to convert 250 cm^3 of H_2O at 20°C into steam at 100°C?

*12.60 The first ionization energy of gaseous sodium atoms is 494.1 kJ mol^{-1}, and adding electrons to gaseous chlorine atoms to form gaseous chloride ions liberates 351 kJ mol^{-1}. Use this information along with the heats of formation of gaseous Na and Cl atoms, and the heat of formation of NaCl(s), to calculate ΔH (in kilojoules) for the reaction

$$NaCl(s) \longrightarrow Na^+(g) + Cl^-(g)$$

The answer corresponds to the lattice energy of sodium chloride. (*Hint:* It will help if you write thermochemical equations for each process described in the problem.)

Bond Energies

12.61 What is meant by atomization energy? Write an equation representing the process for which ΔH is the atomization energy of H_2O. Indicate the physical state (gas, liquid, or solid) for each substance in the equation.

12.62 Why don't values of ΔH_f^0 computed from tabulated bond energies agree precisely with ΔH_f^0 measured experimentally?

12.63 Use the data in Tables 12.2 and 12.3 to calculate ΔH_f^0 for acetylene, $C_2H_2(g)$. The structure of the molecule is H—C≡C—H.

*12.64 Benzene is often written as a resonance hybrid of two equivalent structures,

The ΔH_f^0 for gaseous benzene has been determined from its heat of combustion to be $+82.8$ kJ mol^{-1}.

$$6C(s) + 3H_2(g) \longrightarrow C_6H_6(g) \qquad \Delta H_f^0 = +82.8 \text{ kJ mol}^{-1}$$

Use the data in Tables 12.2 and 12.3 to calculate ΔH_f^0 as kJ mol^{-1}. How does your calculated value compare with the experimental value? The difference between the calculated and experimental values is called the resonance energy. What might you conclude about the stability of species that exist as a composite of two or more resonance structures?

12.65 Use data in Tables 12.2 and 12.3 to calculate ΔH_f^0 (as kJ mol^{-1}) for gaseous propylene, the substance used to make the plastic polypropylene. The structure of propylene is

12.66 Use the average bond energies in Table 12.3 to compute the standard heat of formation of $C_3H_8(g)$ in kJ mol^{-1}. Its structure is

How well does your computed value compare with that reported in Table 12.1?

Entropy, Energy, and Spontaneity

12.67 How is the energy change in a reaction related to its likelihood of being spontaneous?

12.68 In what way is statistical probability related to spontaneity? How is entropy related to statistical probability?

12.69 If a spontaneous chemical reaction is endothermic, what can be said about ΔS for the reaction?

12.70 What are the units of ΔS?

12.71 Which of the following changes will produce an increase in the entropy of a system?
(a) an increase in temperature
(b) formation of gaseous products from solid reactants
(c) formation of a precipitate in a liquid solution
(d) an increase in the volume of the system
(e) the formation of a solid from a gas

12.72 What is the sign of the entropy change for each of the following processes?
(a) A solute crystallizes from a solution.
(b) Water evaporates.
(c) A deck of playing cards is shuffled.
(d) A card player is dealt 13 spades.
(e) Solid AgCl precipitates from a solution of $AgNO_3$ and NaCl.
(f) $^{235}_{92}U$ is extracted from a mixture of $^{235}_{92}U$ and $^{238}_{92}U$.

12.73 Predict the sign of the entropy change for the following reactions:
(a) $HCl(g) \rightarrow H(g) + Cl(g)$
(b) $2H_2(g) + O_2(g) \rightarrow 2H_2O(g)$
(c) $H_2(g) + I_2(s) \rightarrow 2HI(g)$
(d) $CaO(s) + 2NH_4Cl(s) \rightarrow 2NH_3(g) + CaCl_2(s)$
(e) $AgNO_3(aq) + NaCl(aq) \rightarrow AgCl(s) + NaNO_3(aq)$

12.74 Predict the sign of the entropy change for the following reactions:
(a) $Cl_2(g) + F_2(g) \rightarrow 2ClF(g)$
(b) $Na_2CO_3(aq) + 2HCl(aq) \rightarrow 2NaCl(aq) + H_2O(l) + CO_2(g)$
(c) $CO_2(g) + CaO(s) \rightarrow CaCO_3(s)$
(d) $4NH_3(g) + 3O_2(g) \rightarrow 2N_2(g) + 6H_2O(g)$
(e) $NH_3(g) + HCl(g) \rightarrow NH_4Cl(s)$

12.75 State the second law of thermodynamics.

12.76 What two criteria must be met in order for a process to be spontaneous, regardless of the temperature?

12.77 Why is it so difficult to remove pollutants from the environment?

12.78 ΔH and ΔS are nearly independent of temperature. Why is this not true for ΔG?

12.79 If we refer to Figure 12.11, what is the minimum pressure necessary for the conversion of graphite to diamond at a temperature of 200 K? Is it theoretically possible to change graphite to diamond at 1 atm?

Third Law of Thermodynamics

12.80 State the third law of thermodynamics.

12.81 Explain why the entropy of a pure substance is zero at 0 K. Would the entropy of an alloy such as brass be zero at 0 K? Explain your answer.

12.82 Which of the following reactions is accompanied by the greatest entropy change?
(a) $SO_2(g) + \frac{1}{2}O_2(g) \rightarrow SO_3(g)$
(b) $CO(g) + \frac{1}{2}O_2(g) \rightarrow CO_2(g)$

12.83 Calculate ΔS_f^0 for (a) $CCl_4(l)$; (b) $Mg(OH)_2(s)$; (c) $PbSO_4(s)$; (d) $NaHCO_3(s)$; and (e) $NH_3(g)$.

12.84 Use the results from Problems 12.43 and 12.86 to compute ΔS^0, in joules per kelvin, for the reactions in Problem 12.86.

12.85 Why is it possible to tabulate values of S^0, but not values of H^0?

Standard Free Energy Changes

12.86 Calculate ΔG^0 in kilojoules for each of the reactions in Exercise 12.43.

12.87 The standard free energy of formation of glucose is $\Delta G_f^0 = -910.2$ kJ mol^{-1}. Calculate ΔG^0 for the reaction,

$$C_6H_{12}O_6(s) + 6O_2(g) \longrightarrow 6CO_2(g) + 6H_2O(l)$$

12.88 Calculate ΔG^0 in kilojoules for the following reactions:
(a) $Pb(s) + PbO_2(s) + 2H_2SO_4(l) \rightarrow 2PbSO_4(s) + 2H_2O(l)$
(b) $CH_4(g) + 4Cl_2(g) \rightarrow CCl_4(l) + 4HCl(g)$
(c) $10N_2O(g) + C_3H_8(g) \rightarrow 10N_2(g) + 3CO_2(g) + 4H_2O(g)$

Free Energy and Useful Work

12.89 In terms of converting energy into useful work, what advantage and disadvantage does a reversible process offer?

12.90 What is the maximum quantity of useful work, expressed in kilojoules, that could be obtained at 25°C and 1 atm by the oxidation of propane, C_3H_8, according to the equation,

$$C_3H_8(g) + 5O_2(g) \longrightarrow 3CO_2(g) + 4H_2O(g)$$

Why is it that we always obtain less than this maximum quantity of work in any real process that uses propane as a fuel?

Free Energy and Equilibrium

12.91 Sketch a free energy diagram for a reaction that has a negative value of ΔG^0. In this reaction, does the position of equilibrium favor the reactants or products?

12.92 Explain why $\Delta G = 0$ for a system that is in a state of equilibrium.

12.93 The heat of fusion of water at 0°C is 6.02 kJ mol^{-1}; its heat of vaporization is 40.7 kJ mol^{-1} at 100°C. What are ΔS for the melting and boiling of 1 mol of water? Can you explain why ΔS_{vap} is greater than $\Delta S_{melting}$?

12.94 From the data in Tables 12.1 and 12.4, calculate the boiling point of liquid bromine in °C [i.e., the temperature at which $Br_2(l)$ and $Br_2(g)$ can coexist in equilibrium with each other when $p_{Br_2} = 1$ atm].

12.95 Why is it possible for a chemical reaction to occur spontaneously even though ΔG^0 for the reaction is positive?

12.96 Why can ΔG^0 be used to predict whether or not a reaction can be observed?

12.97 What relationship is there between ΔG^0 and the speed with which the products of a reaction are formed?

12.98 What two factors ultimately determine whether we will be able to observe the formation of products in a chemical reaction?

12.99 Describe the relationship between ΔG^0 and the position of equilibrium in a chemical reaction.

12.100 Which of the following reactions could *potentially* serve as a practical method for the preparation of NO_2? (*Note:* The equations are not balanced.)
(a) $N_2(g) + O_2(g) \rightarrow NO_2(g)$
(b) $HNO_3(l) + Ag(s) \rightarrow AgNO_3(s) + NO_2(g) + H_2O(l)$
(c) $NH_3(g) + O_2(g) \rightarrow NO_2(g) + NO(g) + H_2O(g)$
(d) $CuO(s) + NO(g) \rightarrow NO_2(g) + Cu(s)$
(e) $NO(g) + O_2(g) \rightarrow NO_2(g)$
(f) $H_2O(g) + N_2O(g) \rightarrow NH_3(g) + NO_2(g)$

CHAPTER 13

CHEMICAL KINETICS: THE STUDY OF THE RATES OF REACTIONS

The remains of a grain elevator in New Orleans, Louisiana, after it exploded in December, 1977, killing 35 people. Particle size is one of the factors that control the speeds of chemical reactions. Rapid combustion of very fine grain dust caused the explosive effects seen here. In this chapter we study the speeds of chemical reactions and the kinds of chemical information that comes from such studies.

It does not take long to find a reaction that thermodynamics predicts should proceed nearly to completion but yet is observed not to occur. We know from the last chapter that hydrogen and oxygen can be kept in contact with one another almost forever without forming noticeable amounts of water, even though their reaction to produce water is accompanied by a free-energy decrease. This is an example of a chemical change for which the speed of the reaction governs whether the formation of the products will or will not be observed.

Chemical kinetics, also referred to as chemical dynamics, is concerned with the speeds, or rates, of chemical reactions. In this area of chemistry we study the factors that control how rapidly chemical changes occur. These include the following:

1. *The nature of the reactants and products.* All other factors being equal, some reactions are just naturally fast and others are naturally slow, depending on the chemical makeup of the molecules or ions involved.
2. *The concentrations of the reacting species.* For two molecules to react with each other they must meet, and the probability that this will happen in a homogeneous mixture increases as their concentrations increase. For heterogeneous reactions—those in which the reactants are in separate phases—the rate also depends on the area of contact between the phases. Since many small particles have a much larger area than one large particle of the same total mass, decreasing the particle size increases the reaction rate. Sometimes this can have devastating results as we can see in the opening photograph of this chapter.
3. *The effect of temperature.* Nearly all chemical reactions take place faster when their temperatures are increased.
4. *The influence of outside agents called catalysts.* The rates of many reactions, including virtually all biochemical reactions, are affected by substances called catalysts that undergo no net chemical change during the course of the reaction.

Studying how these factors affect the rate of a reaction serves several purposes. For example, it allows us to adjust the conditions of a reaction system to obtain the products as quickly as possible. The importance of this in the commercial manufacture of chemicals is obvious. It also allows us to adjust conditions to make a reaction occur as slowly as possible. This is helpful, for instance, in controlling the growth of fungi and other microorganisms that spoil foods.

For chemists, one of the most significant benefits that comes from studying reaction rates is knowledge about the details of how chemical changes take place. We will see later in this chapter that a chemical reaction usually does not occur in one single step that involves the simultaneous collision of all the reactant molecules described in the balanced overall equation. Instead, the net change is the result of a sequence of simple reactions. This sequence is called the **mechanism** of the reaction, and studying reaction rates gives clues to what the mechanism is. In this way we gain insight into the fundamental reasons of why substances react the way they do.

Chopping a log into small pieces of kindling with a large total surface area makes a campfire easier to start.

13.1 REACTION RATES AND THEIR MEASUREMENT

Before we examine in more detail the factors that influence rates of reaction, let's be sure that we know what is meant by rate. In general, the rate (or speed) of any chemical reaction can be expressed as the ratio of the change in the concentration of a reactant (or product) to a change in time. This is exactly analogous to giving the speed of an automobile as the change in position (that is, the distance traveled) divided by its time of travel. Here the speed might be given as km h^{-1}.

With chemical reactions, the rate is usually expressed in moles per cubic decimetre (litre) per second.

$$\text{speed of auto} = \text{rate of travel} = \frac{\text{distance}}{\text{time}} = \frac{\text{kilometres}}{\text{hour}}$$

$$\text{rate of chemical reaction} = \frac{\text{change in concentration}}{\text{time}}$$

$$= \frac{\text{moles/dm}^3}{\text{second}} = \frac{\text{mol/dm}^3}{\text{s}}$$

$$= \text{mol dm}^{-3}\text{ s}^{-1}$$

To determine the rate of a given chemical reaction, we must measure how fast the concentration of a reactant or product changes. In practice, the species whose concentration is easiest to follow is determined at various time intervals. The simplest example is a reaction in which only one reactant undergoes a change to form a single product. An example is the conversion of cyclopropane to propylene.

Cyclopropane is used as a fast-acting anesthetic.

$$\underset{\textbf{cyclopropane}}{\underset{H_2C\!\!-\!\!-\!\!CH_2}{\overset{\overset{H_2}{C}}{\diagup\!\!\diagdown}}} \longrightarrow \underset{\textbf{propylene}}{H_3C\!\!-\!\!\overset{\overset{H}{|}}{C}\!\!=\!\!CH_2}$$

In general, the balanced equation for this type of reaction is

$$A \longrightarrow B \qquad\qquad [13.1]$$

When the reaction is carried out, no product (B) is present initially and, as time goes on, the concentration of B increases with a corresponding decrease in the concentration of A (Figure 13.1). An inspection of Figure 13.1 reveals that the rate of this chemical reaction changes with time. For instance, near the start of the reaction the concentration of A is decreasing rapidly and the concentration of B is rising rapidly. Much later during the reaction, however, only small changes in concentration occur with time, and the rate is therefore much less. In general, this type of behavior is observed with nearly every chemical reaction—as the reactants are consumed, the rate of reaction gradually decreases.

In more complex reactions, the rates of formation of the various products and the rates of disappearance of the various reactants are not all equal, but are related by the coefficients in the overall balanced equation. For example, consider the reaction

$$N_2(g) + 3H_2(g) \longrightarrow 2NH_3(g)$$

We see that for every molecule of N_2 that reacts, three molecules of H_2 react. This means that the hydrogen is disappearing three times as fast as the nitrogen. The coefficients also tell us that two molecules of NH_3 are formed from each N_2, so the rate at which NH_3 is formed must be twice as fast as the rate at which N_2 disappears.

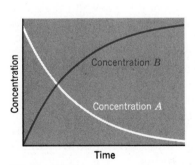

Figure 13.1

Change in the concentration of reactants and products with time for the reaction A → B.

Measurement of reaction rates

An accurate, quantitative estimate of the rate of reaction at any given moment during the reaction can be obtained from the slope of the tangent to the concen-

Figure 13.2

Estimation of the rate of reaction based on the change in concentration of B with time. The reaction is A → B.

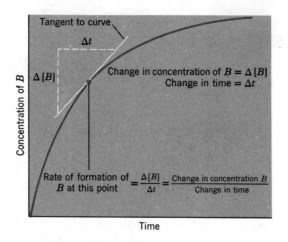

$$\frac{mol}{dm^3} = \text{molar concentration}$$

tration-time curve at that particular instant. This is shown in Figure 13.2. Square brackets, [ ], are used here to denote concentration in mol dm^{-3}. From the tangent to the curve we can write

$$\text{rate} = \frac{\Delta[B]}{\Delta t} \qquad [13.2]$$

We can also express the rate of the above reaction in terms of the concentration of the reactant A, because its concentration is also changing with time. The rate measured in terms of the concentration of A would be

$$\text{rate} = \frac{-\Delta[A]}{\Delta t} \qquad [13.3]$$

The minus sign indicates that the concentration of A is decreasing with time. A minus sign is always used whenever reactants are employed to express the rate.

When measuring the rate of any chemical reaction, the concentration that is monitored and the technique that is used to measure the change depend on the nature of the reaction. For example, for gaseous reactions the pressure can be followed, provided that there is a change in the number of moles of gas as the reaction proceeds. On the other hand, if a colored reactant or product is involved, the intensity of color can be monitored during the reaction. Whatever method of analysis is employed, it must be fast, accurate, and in no way interfere with the normal course of the reaction being studied.

EXAMPLE 13.1 CALCULATING RELATIVE RATES OF REACTION

PROBLEM   Ammonia can be made to burn according to the reaction

$$4NH_3(g) + 5O_2(g) \longrightarrow 4NO(g) + 6H_2O(g)$$

Suppose that at a particular moment during the reaction the ammonia is reacting at the rate of 0.24 mol dm^{-3} s^{-1}. (a) What is the rate at which oxygen is reacting? (b) What is the rate at which H$_2$O is being formed?

SOLUTION   The solution of this problem simply uses the coefficients of the equation to construct conversion factors relating the various numbers of moles of the reactants and products. We are given that

The negative sign, you recall, means that the concentration is *decreasing* with time.

$$\text{rate (for NH}_3) = \frac{-0.24 \text{ mol NH}_3}{dm^3 \text{ s}}$$

The coefficients in the equation allow us to construct the conversion factors

$$\frac{5 \text{ mol O}_2}{4 \text{ mol NH}_3} \quad \text{and} \quad \frac{6 \text{ mol H}_2\text{O}}{4 \text{ mol NH}_3}$$

We use these as follows:

(a) For the rate at which oxygen is consumed:

$$\text{rate (for O}_2) = \frac{-0.24 \text{ mol NH}_3}{\text{dm}^3 \text{ s}} \times \left(\frac{5 \text{ mol O}_2}{4 \text{ mol NH}_3} \right)$$

$$= \frac{-0.30 \text{ mol O}_2}{\text{dm}^3 \text{ s}}$$

(b) For the rate at which water is formed:

$$\text{rate (for H}_2\text{O}) = \frac{0.24 \text{ mol NH}_3}{\text{dm}^3 \text{ s}} \times \left(\frac{6 \text{ mol H}_2\text{O}}{4 \text{ mol NH}_3} \right)$$

$$= \frac{0.36 \text{ mol H}_2\text{O}}{\text{dm}^3 \text{ s}}$$

This rate is given with a positive sign because the concentration of H_2O is increasing.

13.2 RATE LAWS

In this section we begin to examine the factors that control the rate of reaction. Not all reactions take place at the same rate. Ionic reactions like those described in Chapter 7 are virtually instantaneous; the speed is determined by how rapidly we can mix the chemicals. Other reactions, such as the digestion of food, take place more slowly. These different rates exist primarily because of chemical differences among the reacting substances.

For any given reaction, one of the most important controlling influences is the concentrations of the reactants. Generally, if we follow a chemical reaction over a period of time, we find that its rate gradually decreases as the reactants are consumed. From this we conclude that the rate is related, in some way, to the concentrations of the reacting species. In fact, the rate is nearly always proportional to the concentrations of the reactants, each raised to some power. This means that for the general reaction

$$A \longrightarrow B$$

the rate can be written as

$$\text{rate} \propto [A]^x \qquad\qquad [13.4]$$

where the exponent, x, is called the **order of the reaction.** When $x = 1$ we have a first-order reaction. An example is the decomposition of cyclopropane mentioned earlier,

$$\text{rate} \propto [\text{cyclopropane}]^1$$

In some cases an exponent can even be negative, which means that increasing the concentration of that reactant *decreases* the reaction rate.

Second ($x = 2$), third ($x = 3$), and higher-order reactions are also possible, as are reactions in which x is a fraction. There are also examples of zero-order reactions, for which $x = 0$. For a zero-order reaction the rate is constant and does not depend on the concentrations of the reactants. An example is the decomposition of ammonia on a platinum or tungsten metal surface. The rate at which the ammonia decomposes is always the same, regardless of its concentration. Another example of a zero-order process is the elimination of ethyl alcohol by the body. Regardless of how much alcohol is present in the bloodstream, its rate of expulsion from the body is constant. Thus the rate is independent of concentration.

A very important fact is that there is not necessarily any direct relationship between the coefficients in the balanced chemical equation for a reaction and the order of the reaction. *The value of x can only be determined from experiment.*

If we consider a slightly more complex reaction, for example,

$$A + B \longrightarrow \text{products}$$

the rate usually depends on the concentrations of both A and B. Normally, increasing the concentration of either A or B will increase the reaction rate, and the rate is proportional to the product of the concentrations of A and B, each raised to some power.

$$\text{rate} \propto [A]^x[B]^y \qquad [13.5]$$

In this case, we say that the order of the reaction with respect to A is x and the order with respect to B is y. We can also describe the **overall order** of the reaction, which is the sum of the exponents on the concentration terms. In this example, the overall order is the sum $x + y$. Once again, x and y can have whole-number, fractional, negative or even zero values. When one of the exponents is zero, this simply means that the rate of the reaction does not depend on the concentration of that substance. For example, if the exponent y in Equation 13.5 were zero, the equation would become

$$\text{rate} \propto [A]^x[B]^0$$

Any quantity raised to the zero power is equal to 1, so the equation reduces to

$$\text{rate} \propto [A]^x(1)$$

or

$$\text{rate} \propto [A]^x$$

A real chemical example is the reaction

$$NO_2(g) + CO(g) \longrightarrow CO_2(g) + NO(g)$$

at temperatures below 225°C, the relationship between concentration and rate is

$$\text{rate} \propto [NO_2]^2$$

The rate is independent of the CO concentration but depends on the *square* of the NO_2 concentration. We say that the reaction is second-order with respect to NO_2 and zero-order with respect to CO. Notice that there is no relationship between the coefficients and the exponent. As mentioned previously, the order of a reaction can only be determined experimentally.

The proportionality represented by Equation 13.5 can be converted to an equality by introducing a proportionality constant, k, which we call the **rate constant**. The resulting equation, termed the **rate law** for the reaction, is

$$\text{rate} = k[A]^x[B]^y$$

For example, the rate law for the reaction between ICl and H_2,

$$2ICl(g) + H_2(g) \longrightarrow I_2(g) + 2HCl(g)$$

at 230°C has been found experimentally to be

$$\text{rate} = 0.163 \text{ dm}^3 \text{ mol}^{-1} \text{ s}^{-1} [ICl][H_2]$$

This reaction therefore, is *first order* with respect to both ICl and H_2 (hence *second order*, overall) and has as its rate constant $k = 0.163\ dm^3\ mol^{-1}\ s^{-1}$. It should be noted that this value of k applies only at 230°C for this particular reaction. Other reactions have other values of k, and, as we will see later, k varies with the temperature.

Determining the rate law

How can a rate law such as the one above be determined? One way is to perform a series of experiments in which the initial concentration of each reactant is systematically varied. Once again we can use as our example the simple reaction

$$A \longrightarrow B$$

The rate law for this reaction would take the form

$$rate = k[A]^x$$

If the reaction were first order, the value of x would be 1, and the rate expression would then be

$$rate = k[A]$$

For a first-order reaction, increasing the concentration by a factor of 2 increases the rate by a factor of 2^1.

This means that the rate of the reaction varies directly with the concentration of A raised to the first power. As a result, if we were to double the concentration of A from one experiment to another, we would also find that the rate increases by a factor of 2. We conclude, therefore, that *when the reaction rate is doubled by doubling the concentration of a reactant, the order with respect to that reactant is 1.*

Suppose, now, that the rate law were, instead,

$$rate = k[A]^2$$

In this instance, a twofold increase in the concentration would cause a *fourfold* increase in rate. To see this, let's imagine that the initial rate was measured with the concentration of A equal to, say a mol dm^{-3}. This rate would be given by

$$rate = k(a)^2$$

Now, if the reaction were repeated with $[A] = 2a$, the rate would be

$$rate = k(2a)^2$$

or

$$rate = 4ka^2$$

For a second-order reaction, increasing the concentration by a factor of 2 increases the rate by a factor of 2^2.

which is four times the previous rate. Thus, *if the rate is increased by a factor of 4 when the concentration of a reactant is doubled, the reaction is second-order with respect to that reactant. Similarly, we predict that the rate of a third-order reaction would undergo an eightfold increase when the concentration is doubled ($2^3 = 8$).*

The following examples illustrate how we can use these ideas to obtain the rate law for a reaction by varying the concentrations of reactants.

EXAMPLE 13.2 **DETERMINING THE RATE LAW FOR A REACTION**

PROBLEM   On the next page are some data collected in a series of experiments on the reaction of nitric oxide with bromine at 273°C:

$$2NO(g) + Br_2(g) \longrightarrow 2NOBr(g)$$

Experiment	Initial Concentration (mol dm^{-3})		Initial Rate of formation of NOBr (mol dm^{-3} s^{-1})
	NO	Br$_2$	
1	0.10	0.10	12
2	0.10	0.20	24
3	0.10	0.30	36
4	0.20	0.10	48
5	0.30	0.10	108

Determine the rate law for the reaction and compute the value of the rate constant.

SOLUTION   The rate law for the reaction will have the form

$$\text{rate} = k[NO]^x[Br_2]^y$$

To determine each exponent, we will study how the rate changes when the concentration of one reactant varies while that of the other reactant stays the same. For instance, when the NO concentration is held constant, we can see how changes in the Br$_2$ concentration affect the rate and thereby determine what y must be. The value of x is determined in a similar way. With this strategy in mind, let's study the data.

In experiments 1 to 3, the concentration of NO is constant and the concentration of Br$_2$ is varied. When the concentration of Br$_2$ is doubled (experiments 1 and 2), the rate is increased by a factor of 2; when it is tripled (experiments 1 and 3), the rate is increased by a factor of 3. The only way this could happen is if the concentration of Br$_2$ appears to the first power in the rate law.

Comparing experiments 1 and 4, we see that when the Br$_2$ concentration is held constant, the rate increases by a factor of 4 when the NO concentration is multiplied by 2. Similarly, raising the concentration of NO by a factor of 3 causes a ninefold increase in rate (experiments 1 and 5). This means that the exponent of the NO concentration in the rate law must be 2. Therefore,

$$\text{rate} = k[NO]^2[Br_2]$$

The rate constant can be evaluated using the data from any of these experiments. Working with experiment 1, we have

$$12 \text{ mol dm}^{-3} \text{ s}^{-1} = k(0.10 \text{ mol dm}^{-3})^2(0.10 \text{ mol dm}^{-3})$$

$$12 \text{ mol dm}^{-3} \text{ s}^{-1} = k(0.0010 \text{ mol}^3 \text{ dm}^{-9})$$

Solving for k, we get

$$k = \frac{12 \text{ mol dm}^{-3} \text{ s}^{-1}}{1.0 \times 10^{-3} \text{ mol}^3 \text{ dm}^{-9}} = 1.2 \times 10^4 \text{ dm}^6 \text{ mol}^{-2} \text{ s}^{-1}$$

You might wish to verify for yourself that the same rate constant is obtained from the other data. (That's why it's called the rate *constant*.)

EXAMPLE 13.3   DETERMINING THE RATE LAW FOR A REACTION

PROBLEM   The following data were collected for the reaction of *t*-butyl bromide, (CH$_3$)$_3$CBr, with hydroxide ion at 55°C.

$$(CH_3)_3CBr + OH^- \longrightarrow (CH_3)_3COH + Br^-$$

Experiment	Initial Concentration (M)		Initial Rate of Formation of $(CH_3)_3COH$ (mol dm^{-3} s^{-1})
	$(CH_3)_3CBr$	OH^-	
1	0.10	0.10	0.0010
2	0.20	0.10	0.0020
3	0.30	0.10	0.0030
4	0.10	0.20	0.0010
5	0.10	0.30	0.0010

What is the rate law and rate constant for this reaction?

SOLUTION   Based on the equation, we expect a rate law of the form

$$\text{rate} = k[(CH_3)_3CBr]^x[OH^-]^y$$

To obtain x and y, we follow the same approach as in the previous example.

Let's examine experiments 1, 2, and 3 first. In each of these the OH^- concentration is the same. Doubling the $(CH_3)_3CBr$ concentration doubles the rate; tripling it triples the rate. The order with respect to $(CH_3)_3CBr$ must therefore be 1.

In experiments 1, 4, and 5, the $(CH_3)_3CBr$ concentration is the same. Changing the OH^- concentration has no effect on the rate. This means that the reaction is zero order with respect to OH^-. Therefore,

$$\text{rate} = k[(CH_3)_3CBr]^1[OH^-]^0$$

Since anything raised to the zero power is 1,

$$\text{rate} = k[(CH_3)_3CBr]^1 \cdot 1$$

or

When no exponent is written, it's assumed to be 1.

$$\text{rate} = k[(CH_3)_3CBr]$$

The final rate law contains only the concentration of $(CH_3)_3CBr$, because this is the only concentration that affects the rate. To solve for the rate constant we can use the results of any of the experiments. Using experiment 1 and substituting the rate and concentration into the rate law give

$$0.0010 \text{ mol dm}^{-3} \text{ s}^{-1} = k(0.10 \text{ mol dm}^{-3})$$

$$k = \frac{0.0010 \text{ mol dm}^{-3} \text{ s}^{-1}}{0.10 \text{ mol dm}^{-3}}$$

$$= 0.010 \text{ s}^{-1}$$

In Example 13.1, the exponents in the rate law just happen to be the same as the coefficients in the balanced equation. This is not true in Example 13.2. Please keep in mind that the *only* way we can find the exponents in the rate law for a chemical reaction is by experimentally measuring the way that the concentrations of the reactants affect the rate. It is also important to remember that since temperature is another factor that influences the rate, a given value of k applies only at *one* temperature (the temperature at which it was measured).

13.3 CONCENTRATION AND TIME: HALF-LIVES

The rate law for a reaction tells us how the rate of a reaction is related to the concentrations of the reactants. By applying calculus to the rate law—which we won't attempt to go through here—an expression relating the concentration to time can be derived. For example, a first-order rate law such as

$$\text{rate} = k[A]$$

gives the expression

The expression $\ln x$ asks the question, to what power must the number e be raised to give x? The value of $e = 2.718\,28.\ldots$

$$\ln \frac{[A]_0}{[A]_t} = kt \qquad [13.6]$$

where $[A]_0$ is the initial concentration of A (at time t equal to zero), $[A]_t$ is the concentration at some time t after the beginning of the reaction, and the symbol "ln" tells us to take the natural logarithm of the ratio of $[A]_0$ divided by $[A]_t$. Natural logarithms occur frequently in the sciences, and if you haven't encountered them before, you should study Appendix A at the back of the book.

In terms of common, or base 10 logarithms, Equation 13.6 can be written

$$2.303 \log \frac{[A]_0}{[A]_t} = kt$$

You should learn the relationship $\ln x = 2.303 \log x$.

The factor 2.303 converts common logs to natural logs. If you have a scientific calculator, you will probably find it easier to work with natural logarithms.

Equation 13.6 or its counterpart in common logarithms is useful because it allows us to calculate, for a first-order reaction, the concentration of the reactant at any time during the course of the reaction, provided that we know the value of k. If we know $[A]_0$, $[A]_t$, and k, we can also compute t—the length of time that the reaction has progressed. We will see in Chapter 22 that this is useful in archaeological dating using radioactive isotopes.

EXAMPLE 13.4 **CALCULATING THE CONCENTRATION OF A REACTANT AT SOME TIME AFTER THE START OF THE REACTION**

PROBLEM At 400°C, the first-order conversion of cyclopropane into propylene has a rate constant of 1.16×10^{-6} s^{-1}. If the initial concentration of cyclopropane is 1.00×10^{-2} mol dm^{-3} at 400°C, what will its concentration be 24.0 hours after the reaction begins?

SOLUTION To solve this problem we use Equation 13.6. Our data are

$$[\text{cyclopropane}]_0 = 1.00 \times 10^{-2} \text{ mol dm}^{-3}$$

$$k = 1.16 \times 10^{-6} \text{ s}^{-1}$$

$$t = 24.0 \text{ h}$$

The first step is to solve for the ratio of concentrations.

$$\ln \frac{[\text{cyclopropane}]_0}{[\text{cyclopropane}]_t} = (1.16 \times 10^{-6} \text{ s}^{-1})(24.0 \text{ h})\left(\frac{3600 \text{ s}}{1 \text{ h}}\right)$$

$$= 0.100$$

Most scientific calculators handle natural logarithms and their antilogarithms easily. See Appendix A.

To obtain the concentration ratio we must take the antilog. If $\ln x = a$, then $x = e^a$. Therefore,

$$\frac{[\text{cyclopropane}]_0}{[\text{cyclopropane}]_t} = e^{0.100}$$

$$= 1.11$$

Now we solve for $[\text{cyclopropane}]_t$.

$$[\text{cyclopropane}]_t = \frac{[\text{cyclopropane}]_0}{1.11}$$

$$= \frac{1.00 \times 10^{-2} \text{ mol dm}^{-3}}{1.11}$$

$$= 9.01 \times 10^{-3} \text{ mol dm}^{-3}$$

After 24 hours the concentration of cyclopropane will have dropped to 9.01×10^{-3} M.

The equation relating concentration to time is different for different reaction orders. For instance, for a second-order reaction having the rate law

$$\text{rate} = k[B]^2$$

the relationship is

$$\frac{1}{[B]_t} - \frac{1}{[B]_0} = kt \qquad [13.7]$$

Even more complicated equations occur when the rate law is more complex, but we won't discuss them.

Half-lives

An important quantity, particularly for first-order reactions, is the **half-life, $t_{1/2}$**— *the length of time required for the concentration of the reactant to be decreased to half of its initial value.* At this point, $t = t_{1/2}$, and

$$[A]_t = \tfrac{1}{2}[A]_0$$

Substituting into Equation 13.6 gives

$$\ln \frac{[A]_0}{\tfrac{1}{2}[A]_0} = kt_{1/2}$$

$$\ln 2 = kt_{1/2}$$

$$0.693 = kt_{1/2}$$

Solving for $t_{1/2}$ gives

$$t_{1/2} = \frac{0.693}{k} \qquad [13.8]$$

Figure 13.3

A graph of a first-order reaction illustrating the concept of half-life.

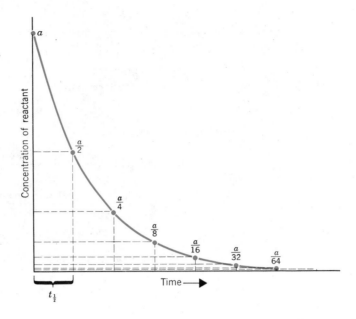

For a first-order reaction

Number of Half-lives	Fraction of Reactant Remaining
0	1
1	1/2
2	1/4
3	1/8
4	1/16
5	1/32
⋮	⋮
n	$1/2^n$

Equation 13.8 tells us that for a first-order reaction, $t_{1/2}$ depends only on k—it is constant throughout the reaction. If the half-life for a particular first-order reaction is one hour, then during the first hour the concentration drops to half of its initial value. During the second hour the concentration is again cut in half, so after a total of two hours the concentration is $\frac{1}{4}$ of its initial value (Figure 13.3).

The half-life of a second-order reaction differs from that of a first-order process by being concentration dependent. Following the same procedure as above, we find that a second-order reaction whose rate law is

$$\text{rate} = k[B]^2$$

has a half-life given by

$$t_{1/2} = \frac{1}{k[B]_0} \qquad [13.9]$$

This means that if we cut the concentration of B in half, the value of $t_{1/2}$ will double. Therefore, during the course of the reaction, each successive half-life is twice as large as the preceding one. If a second-order reaction such as this starts with a half-life of 20 min, the concentration at the beginning of the second half-life is half of the initial value, so the second half-life will be twice as long. Similarly, the third half-life will be twice as long as the second, and so forth.

EXAMPLE 13.5 CALCULATING THE HALF-LIFE FOR A REACTION

PROBLEM The decomposition of N_2O_5 dissolved in carbon tetrachloride is a first-order reaction. The chemical change is

$$2N_2O_5 \longrightarrow 4NO_2 + O_2$$

At 45°C the reaction was begun with an initial N_2O_5 concentration of 1.00 mol dm^{-3}. After 3.00 hours the N_2O_5 concentration had decreased to 1.21×10^{-3} mol dm^{-3}. What is the half-life of N_2O_5 expressed in minutes at 45°C?

SOLUTION To obtain the half-life, we must know the value of k. Since the reaction is first order we can use Equation 13.6

$$\ln\left(\frac{1.00 \text{ mol dm}^{-3}}{1.21 \times 10^{-3} \text{ mol dm}^{-3}}\right) = k(3.00 \text{ h})\left(\frac{3600 \text{ s}}{1 \text{ h}}\right)$$

$$\ln(826) = (1.08 \times 10^4 \text{ s})k$$

$$6.72 = (1.08 \times 10^4 \text{ s})k$$

Therefore,

$$k = \frac{6.72}{1.08 \times 10^4 \text{ s}}$$

$$= 6.22 \times 10^{-4} \text{ s}^{-1}$$

Now we can calculate the half-life

$$t_{1/2} = \frac{0.693}{6.22 \times 10^{-4} \text{ s}^{-1}}$$

$$= 1.11 \times 10^3 \text{ s}$$

This is 18.5 min.

13.4 COLLISION THEORY

For a chemical reaction to occur, the reacting molecules must collide with each other. This common-sense idea forms the basis of the **collision theory** of chemical kinetics. Basically, this theory states that the rate of a reaction is proportional to the number of collisions occurring each second between the reacting molecules:

$$\text{rate} \propto \frac{\text{number of collisions}}{\text{second}} \qquad [13.10]$$

As we will see shortly, this permits us to explain the dependence of the reaction rate on the concentrations of the reactants. In Section 13.6 we will also see that only a fraction of the total number of collisions each second are effective at producing a net chemical change, and that this fraction depends both on the nature of the reactants and the temperature.

At this point, let's see how collision theory accounts for the way the rate of a reaction depends on the concentrations of the reactants. Suppose that we have a reaction that occurs by the collision of two molecules, such as

$$A + B \longrightarrow \text{products}$$

In this case we are assuming that we know precisely what occurs between A and B—that is, we assume for the sake of this discussion that the products are formed in **bimolecular** (two-molecule) **collisions** between A and B.

According to the theory, the rate of the reaction is proportional to the number of collisions each second between molecules of A and B. If the concentration of A were doubled, then the number of A–B collisions per second would also be doubled because there would be twice as many A molecules that can collide with B. This would increase the rate by a factor of 2. Similarly, if the concentration of B were doubled, there would be a twofold increase in the number of A–B collisions per second and the rate would also increase by a factor of 2. From our previous discussion we conclude that the order with respect to each reactant is 1, so the rate law for this bimolecular collision process is

$$\text{rate} = k[A][B]$$

Now let's look at what would happen for a reaction of the type

$$2A \longrightarrow \text{products}$$

in which the reaction occurs by the collision of two A molecules. In this instance, if we double the concentration of A, we double the number of collisions per second that each *single A* molecule makes with its neighbors, because we have doubled the number of neighbors. However, we have also doubled the number of A molecules that are colliding. The number of A–A collisions per second has therefore doubly doubled—that is, increased by a factor of 2 squared. Consequently, the rate law for this bimolecular reaction between identical molecules is

> We are doubling both the number of molecules colliding and the number of collisions that each of them makes.

$$\text{rate} = k[A]^2$$

What we find, then, is that *if* we know what collision process is involved in the formation of products, we can predict, on the basis of collision theory, what the rate law for that process will be. *The exponents in the rate law are equal to the coefficients in the balanced equation for that collision process.*

At this point we might ask, why is it necessary to determine the rate law for a reaction experimentally? Why can't we simply use the coefficients of the balanced overall equation to deduce the rate law? The answer is that when we begin to study a reaction, we don't know what collision processes are involved. Are A–B collisions important or do A–A collisions determine the rate? We can't answer that question until we know the mechanism of the reaction, and that's our next topic.

13.5 REACTION MECHANISMS

The overall balanced equation for a reaction represents the net chemical change that occurs as the reaction proceeds to completion. This does not mean, however, that all the reactants must come together simultaneously to undergo a change that gives the products. In fact, the net change can (and usually does) actually represent the sum of a series of simpler reactions. These simpler reactions are referred to as **elementary processes.** The sequence of elementary processes that ultimately leads to the formation of the products is called the **reaction mechanism.** For example, it appears that the reaction

$$2NO + 2N_2 \longrightarrow 2H_2O + N_2$$

proceeds by the three-step mechanism

$$
\left.
\begin{array}{l}
2NO \longrightarrow N_2O_2 \\
N_2O_2 + H_2 \longrightarrow N_2O + H_2O \\
N_2O + H_2 \longrightarrow N_2 + H_2O
\end{array}
\right\} \quad \text{These are elementary processes.}
$$

The sum of these steps in the sequence does give us the overall balanced equation.

Reaction mechanisms, such as the one just described, are usually arrived at by bringing together both theory and experiment. For instance, suppose we wished to study the following hypothetical reaction in hopes of discovering the mechanism.

$$2A + B \longrightarrow C + D$$

We begin by first determining the rate law, perhaps by studying how the rate changes as we vary the concentrations of A and B as described earlier. Let us say it turned out to be

$$\text{rate} = k[A]^2[B]$$

Next, we attempt to propose a mechanism that, by the application of the principles of collision theory, gives us a predicted rate law that is the same as the one found by experiment.

Since we are beginners at proposing mechanisms, we might be tempted to propose a one-step mechanism in which two molecules of A and one of B come together simultaneously—that is, a three-body or *termolecular collision*. This process,

$$2A + B \longrightarrow C + D$$

indeed leads to the rate law

$$\text{rate} = k[A]^2[B]$$

when we take the coefficients to be equal to the exponents, and it is the same as the rate law found from experiment. But we must now ask ourselves, is this a realistic mechanism? A simultaneous three-body (termolecular) collision is statistically a very unlikely event, and it has been generally found that reactions that must proceed by such a path are very slow. As a result, a third-order reaction such as this, if it is fairly rapid, is usually interpreted as taking place by way of a series of simple bimolecular processes. (Back to the drawing board!)

One possible sequence of reactions is

$$2A \longrightarrow A_2$$
$$A_2 + B \longrightarrow C + D$$

Here we have two steps in which we propose that some relatively unstable intermediate, A_2, is first formed by the collision of two molecules of A. In a second step a reaction between A_2 and B produces the products C and D. Again the sum of these elementary processes gives us our net overall change.

Both of these reactions are unlikely to occur at the same rate, so let's suppose that the first reaction is slow, and that once the intermediate, A_2, is formed it reacts rapidly with B in the second step to produce the products. If this were true, the rate at which the final products appear is actually determined by how fast A_2 is produced. This first step serves as a bottleneck in the reaction path. We refer to this slowest step as the **rate-determining step** in the reaction because it governs how rapidly the overall reaction takes place. Because the rate-determining step is an elementary process (in this instance a bimolecular collision between two A molecules), we can predict with the aid of collision theory that its rate law should be

$$\text{rate} = k[A]^2$$

If this is the rate law for the rate-determining step, it will also be the rate law for the overall reaction. However, this rate law cannot be the correct one because it is not the same as the one determined from experiment. This does not necessarily mean that our mechanism is wrong, but it does mean that the first step cannot be the rate-determining step. Let us see what we would expect to observe if the second step were slow, instead of the first. In this case the rate law would be

$$\text{rate} = k[A_2][B] \qquad [13.11]$$

However, this rate law contains the concentration of the proposed intermediate

A mechanism must make chemical sense and also be capable of yielding the experimentally measured rate law. Proposing such mechanisms requires a lot of practice and chemical knowledge, plus a bit of "chemical intuition."

The reaction cannot proceed any faster than its slowest step.

(A_2) and the experimental rate law contains only the concentrations of reactants A and B. How can we express the concentration of A_2 in terms of A and B?

Once A_2 has been formed, it can react in either of two ways. Since we propose that A_2 is unstable (if it were stable we could isolate it and there would be no question at all about the path of the overall reaction) it can undergo decomposition to reform two molecules of A. The other possibility is that it undergoes a collision with B that leads to the formation of the products, C and D. Our mechanism therefore should include a reaction that allows A_2 to decompose.

$$A_2 \longrightarrow 2A$$

Our total mechanism is now

$$2A \longrightarrow A_2$$
$$A_2 \longrightarrow 2A$$
$$A_2 + B \longrightarrow C + D$$

Recall that in a dynamic equilibrium two opposing processes occur at equal rates.

If the rate at which the intermediate is formed from reactant A is equal to the rate at which A is formed from intermediate A_2, then these two reactions represent a state of dynamic equilibrium. We could therefore write our first two equations as an equilibrium, which would take the form

$$2A \rightleftharpoons A_2$$

We are now back to a two-step mechanism in which the first step is an equilibrium. Our mechanism is now

$$2A \rightleftharpoons A_2 \qquad \text{fast}$$
$$A_2 + B \longrightarrow C + D \qquad \text{slow}$$

Since, in an equilibrium situation, the rate of the forward reaction (rate_f) is equal to the rate of the reverse reaction (rate_r)

$$\text{rate}_f = k_f[A]^2 = \text{rate}_r = k_r[A_2]$$

or simply

$$k_f[A]^2 = k_r[A_2]$$

Solving this equation for $[A_2]$, we have

$$[A_2] = \frac{k_f[A]^2}{k_r}$$

We can now substitute this into Equation 13.11 and combine all the constants to give still another constant, let us say k'. This gives the rate expression

$$\text{rate} = k'[A]^2[B]$$

which does agree with the rate law found from experiment. Our proposed mechanism, therefore, *appears* to be a good one. However, a mechanism is, in essence, a theory. It is a sequence of steps that we dream up to explain the chemistry and to provide a rate law that agrees with experiment. It frequently happens, though, that more than one mechanism can be written to satisfy both criteria, so we can never be certain we have truly discovered the actual path of the reaction. Nevertheless, studying the kinetics of a reaction gives clues to what the mechanism may be and allows us to discard many alternatives. After that, we can only hope to gather further information that either supports (or proves wrong) our guess.

EXAMPLE 13.6   STUDYING THE MECHANISM OF A REACTION

PROBLEM   The decomposition of NO_2Cl is believed to involve the two-step mechanism

$$NO_2Cl \longrightarrow NO_2 + Cl$$
$$NO_2Cl + Cl \longrightarrow NO_2 + Cl_2$$

What would be the observed experimental rate law if the first step were slow and the second fast?

SOLUTION   If the first reaction is the slow step, it is also the rate-determining step. The rate law for the overall reaction should be the same as the rate law for the rate-determining step. Since only one molecule of NO_2Cl is involved, the rate law for the first reaction—as well as for the overall reaction—would be

$$rate = k[NO_2Cl]$$

13.6 EFFECTIVE COLLISIONS

If all the collisions that take place in a reaction vessel were effective in producing chemical change, all chemical reactions—including biochemical ones—would be over almost instantaneously. Since living creatures have finite life spans, it is clear that some factor (or factors) must intervene to decrease reaction rates to a reasonable level. Consider, for example, the decomposition of hydrogen iodide

$$2HI(g) \longrightarrow H_2(g) + I_2(g)$$

At a concentration of only 10^{-3} mol dm^{-3} of HI there are approximately 3.5×10^{28} collisions per dm^3 per second at 500°C. This is equivalent to 5.8×10^4 moles of collisions per dm^3 per second; if each of these collisions were effective, we would expect a rate of reaction of 5.8×10^4 mol dm^{-3} s^{-1}. Actually, the rate under these conditions is only about 1.2×10^{-8} mol dm^{-3} s^{-1}. This is smaller by a factor of approximately 5×10^{12} than we would observe if all collisions led to reaction, which means that only one out of every five thousand billion collisions actually leads to the formation of the products! Clearly, not all encounters between HI molecules result in the production of H_2 and I_2. In fact, only a very small fraction of the total number of collisions are effective. If we let Z be the total number of collisions that occur per second and f be the fraction of the total number of collisions that are effective, then the rate of a reaction would be

$$rate = fZ \qquad [13.12]$$

The fraction, f, is determined by the energies of the molecules that collide and, as we will see shortly, a certain minimum energy is required in order to cause a reaction to occur. In addition to this, in many instances the molecules must also collide with the proper orientation. The decomposition of a hypothetical AB molecule, whose collisions result in the formation of A_2 and B_2, can serve as an example.

$$2AB \longrightarrow A_2 + B_2$$

In order for the products A_2 and B_2 to be produced, the two atoms of A and two atoms of B must approach each other very closely so that $A-A$ and $B-B$ bonds can be formed. Suppose, now, that two $A-B$ molecules come together in a colli-

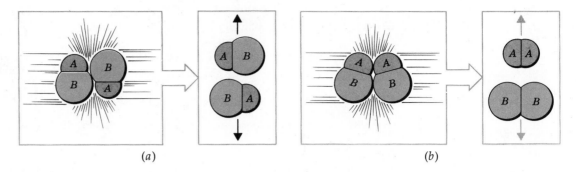

Figure 13.4

Collisions between A—B molecules. (a) *A collision that cannot produce a net chemical change.* (b) *A collision that can lead to a net reaction.*

sion oriented as shown in Figure 13.4a. We certainly do not expect this collision to be effective in forming the products. However, a collision in which the A–B molecules are aligned as shown in Figure 13.4b can lead to the creation of A–A and B–B bonds and, hence, a net chemical change. Thus the number of effective collisions—and, therefore, the rate of the reaction—is further decreased by a factor, p, that is a measure of the importance of the molecular orientations during collision:

$$\text{rate} = pfZ \qquad [13.13]$$

We have already seen that Z, the collision frequency, is proportional to the concentrations of the reacting molecules; therefore, in general,

$$Z = Z_0[A]^n[B]^m \ldots$$

where Z_0 is the collision frequency when all of the reactants are at unit concentration. Substituting this into Equation 13.13 gives us

$$\text{rate} = pfZ_0[A]^n[B]^m \ldots$$

or

$$\text{rate} = k[A]^n[B]^m \ldots$$

where $k = pfZ_0$. This then is the rate law derived from the principles of collision theory.

13.7 TRANSITION STATE THEORY

The collision theory, discussed in Section 13.4, focuses attention primarily on the relationship between reaction rate and the *number* of collisions per second between reactant molecules. In this way, the dependence of rate on concentration was explained. The **transition state theory** is concerned with what actually happens *during* a collision. It follows the energy and orientation of the reactant molecules as they collide, and in doing so seeks explanations of why so few collisions out of the many that occur are actually effective.

A collision between two molecules is quite unlike a collision between two billiard balls. The electron cloud of a molecule has no sharp boundary; its outer reaches are somewhat soft and "fuzzy." Therefore, when two molecules approach each other in a collision, the electron clouds experience a gradual increase in their mutual repulsions, and the molecules begin to slow down. As this happens, the kinetic energy of the molecules is gradually converted to potential

Figure 13.5
(a) When two slow-moving mole-cules collide, their electron clouds cannot interpenetrate much and they just bounce off each other, chemically unchanged. (b) When fast-moving molecules collide, atoms approach each other much more closely as their electron clouds interpenetrate. This can lead to bond making and bond breaking. The net change here is AB + C → A + BC.

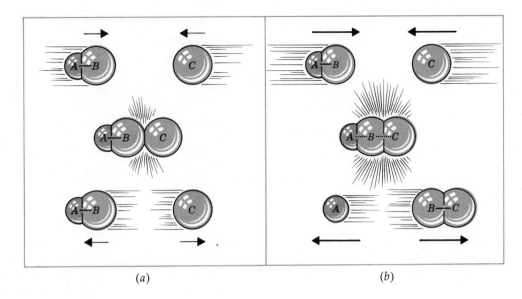

(a)   (b)

energy—somewhat like compressing a spring. If the pair of colliding molecules had little kinetic energy to begin with—that is, if they were not moving very fast—they come to a stop before their electron clouds have penetrated each other very much, and then they simply fly apart again, chemically unchanged (Figure 13.5a).

When two fast-moving molecules collide, they have a lot of kinetic energy that can be converted to potential energy. This means that they are able to overcome substantial forces of repulsion between their electron clouds, and they approach each other quite closely. As illustrated in Figure 13.5b, the interpenetration of the electron clouds that occurs permits a reshuffling of electrons. Old bonds are broken as new ones form and a net chemical change takes place.

From the preceding discussion we see that an effective collision—one that changes reactant molecules into product molecules—can only occur if the molecules collide with sufficient force. Expressed differently, there is a minimum kinetic energy that the molecules must possess jointly to overcome the repulsions between their electron clouds when they collide. This kinetic energy, which is changed to potential energy at the moment of impact, is called the **activation energy, E_a.**

The change in potential energy that takes place during the course of a reaction is shown in Figure 13.6. The horizontal axis is called the **reaction coordinate**—it follows the path of the reaction as molecules come together in a collision and product molecules emerge. In a sense, positions along the reaction coordinate represent the extent to which the reaction has progressed toward completion. On the left of this particular potential energy diagram we find two molecules of *AB*. As they approach each other, their potential energy increases to a maximum. As we continue toward the right along the reaction coordinate, the potential energy of the system decreases as the products, A_2 and B_2, move apart. When the A_2 and B_2 molecules are finally separated from one another, the total potential energy drops to essentially a constant value.

The activation energy for the decomposition of *AB* corresponds to the difference between the energy of the reactants and the maximum on the potential energy curve. Slow-moving molecules of *AB* do not possess sufficient energy to overcome this potential energy barrier, while fast-moving ones do.

Figure 13.6

Potential energy diagram for an exothermic reaction.

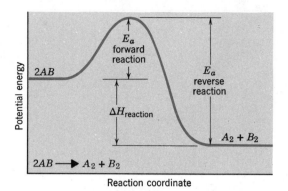

In a collision, the total energy (K.E. plus P.E.) is constant. If the P.E. becomes smaller after a collision, the K.E. must become larger.

In Figure 13.6 we have drawn the potential energy of the products so that it is lower than that of the reactants. The difference between them corresponds to the heat of reaction.[1] In this case, because the products are at a lower energy than the reactants, the reaction is exothermic. The energy released appears as an increase in the kinetic energy of the products; therefore, the temperature of the system rises as the reaction progresses.

In the reaction mixture there are also collisions between A_2 and B_2 molecules. Such collisions, if energetic enough, can reform AB molecules. In Figure 13.6 the activation energy for the reaction

$$A_2 + B_2 \longrightarrow 2AB$$

is indicated as the difference in energy between the products and the top of the potential energy hill. Since the forward reaction (the reaction from left to right) is exothermic, the reverse reaction is endothermic.

Figure 13.7 depicts the energy changes for a reaction that is endothermic in the forward direction. In this case the products are at a higher potential energy than the reactants. The net increase in potential energy that takes place as the products are formed occurs at the expense of kinetic energy. Consequently, there is a net overall decrease in the average kinetic energy as the reaction proceeds and the reaction mixture becomes cool.

An endothermic reaction is one in which K.E. is converted to P.E. In this sense a system absorbs or stores energy.

The species that exists at the top of the potential energy barrier during an effective collision corresponds to neither the reactants nor the products but, instead, to some highly unstable combination of atoms that we speak of as the

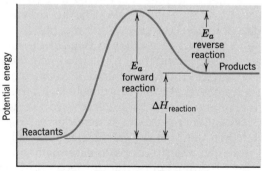

Figure 13.7

Potential energy diagram for an endothermic reaction.

[1] Strictly speaking, the difference in potential energy between the reactants and products corresponds to ΔE for the reaction, but we learned in Chapter 12 that ΔE and ΔH for a process are of very nearly the same magnitude.

Figure 13.8

Transition state theory and the potential energy diagram for a reaction.

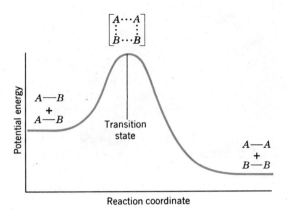

activated complex. This activated complex is said to exist in a **transition state** along the reaction coordinate (hence the name transition state theory).

Transition state theory views chemical kinetics in terms of the energy and geometry of the activated complex that, once it has formed, can come apart to yield the reactants again or go on to produce the products. For example, let us examine again the decomposition of hypothetical AB molecules to produce A_2 and B_2. The change that takes place along the reaction coordinate can be represented as

$$\begin{matrix} A \\ | \\ B \end{matrix} + \begin{matrix} A \\ | \\ B \end{matrix} \rightleftharpoons \begin{bmatrix} A \cdots A \\ \vdots \quad \vdots \\ B \cdots B \end{bmatrix} \longrightarrow \begin{matrix} A-A \\ + \\ B-B \end{matrix}$$

where we have used solid dashes to denote ordinary covalent bonds and dotted lines to symbolize the partially broken and partially formed bonds in the transition state, which is enclosed within brackets. Figure 13.8 illustrates this change as it occurs on the potential energy diagram for the reaction.

If the potential energy of the transition state is very high, then a great deal of energy must be available in a collision to form the activated complex. This results in a high activation energy and consequently a slow reaction. If it were possible somehow to produce an activated complex whose energy was closer to that of the reactants, the decreased activation energy would lead to a faster reaction rate.

13.8 EFFECT OF TEMPERATURE ON REACTION RATES

In nearly every instance an increase in temperature causes an increase in the rate of reaction and, as a very general rule of thumb, the rates of many reactions are about doubled by a 10°C rise in temperature. How can this behavior be explained?

According to kinetic theory, in any system there is a distribution of kinetic energies. In the last section we interpreted the activation energy to be the minimum kinetic energy required for a collision to be effective. All molecules having kinetic energies higher than this minimum are, therefore, capable of reacting. This can be illustrated for the kinetic energy distribution in a system as shown in Figure 13.9. The total fraction of all of the molecules having energies equal to or greater than E_a corresponds to the shaded portion of the area under the curve. If we compare this area for two different temperatures, we see that the total frac-

Figure 13.9

The effect of temperature on the number of molecules having kinetic energies greater than E_a.

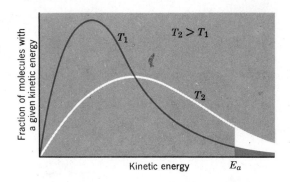

tion of molecules with sufficient kinetic energy to undergo effective collisions is greater at the higher temperature. As a result, the number of molecules that are capable of undergoing reaction increases with increasing temperature and, therefore, so does the reaction rate.

Measuring the activation energy

The magnitude of the rate constant, which is the rate of the reaction when all the concentrations have a value of one, depends on a number of factors. In Section 13.6 we saw that it depends in part on the collision frequency. Now we have learned that the rate of reaction also is affected by the orientations of the molecules during a collision and by the kinetic energy that they must have when they collide. These various factors are related quantitatively by the equation

$$k = Ae^{-E_a/RT}$$ [13.14]

You have seen the name Arrhenius before in our discussion of acids and bases in Chapter 7. In 1903, Arrhenius received the third Nobel Prize ever awarded in chemistry.

which is known as the **Arrhenius equation** after its discovered, the Swedish chemist, Svante Arrhenius. In the equation, e is the base of the natural logarithms (Appendix A), R is the gas constant, T is the absolute temperature and, of course, k is the rate constant and E_a is the activation energy. The factor A is a proportionality constant whose magnitude is related to the collision frequency and also to the importance of molecular orientations during a collision.

The Arrhenius equation provides us with a means of determining the value of the activation energy (as well as the factor A) from measurements of the rate constant at a minimum of two different temperatures. Taking the natural logarithm of Equation 13.14 gives

$$\ln k = \ln A - \frac{E_a}{RT}$$ [13.15]

We can compare this equation to the equation for a straight line.

$$\ln k = \ln A - \frac{E_a}{R}\left(\frac{1}{T}\right)$$
$$\updownarrow \qquad \updownarrow \qquad \updownarrow \quad \updownarrow$$
$$y \ = \ b \ + \ m \ \ x$$

Thus, a plot of $\ln k$ versus $1/T$ gives a straight line whose slope m is equal to $-E_a/R$ and whose intercept b with the ordinate (the vertical axis) is $\ln A$ (Figure 13.10).

Figure 13.10

Graphical determination of the activation energy, E$_a$. Points on the line represent the natural logarithms of experimentally measured rate constants at various temperatures. We determine the slope of the straight line that best fits the experimental data.

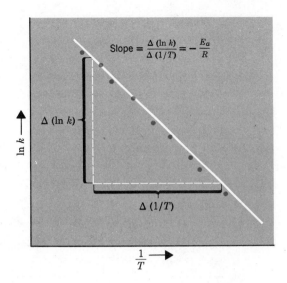

We can also obtain E_a from k at two temperatures by direct computation. For any temperature, T_1, Equation 13.14 becomes

$$k_1 = Ae^{-E_a/RT_1}$$

and for any other temperature, T_2, we can write

$$k_2 = Ae^{-E_a/RT_2}$$

Dividing k_1 by k_2, we have

$$\frac{k_1}{k_2} = \frac{Ae^{-E_a/RT_1}}{Ae^{-E_a/RT_2}}$$

or

$$\frac{k_1}{k_2} = e^{(E_a/R)[(1/T_2)-(1/T_1)]} \qquad [13.16]$$

Taking the natural logarithm of both sides, we get

$$\ln\left(\frac{k_1}{k_2}\right) = \frac{E_a}{R}\left(\frac{1}{T_2} - \frac{1}{T_1}\right) \qquad [13.17]$$

This can also be expressed in terms of common logarithms (base 10 logarithms) as

$$\log\left(\frac{k_1}{k_2}\right) = \frac{E_a}{2.303R}\left(\frac{1}{T_2} - \frac{1}{T_1}\right) \qquad [13.18]$$

Equations 13.17 and 13.18 can be used to compute E_a if rate constants at two different temperatures are known. They can also be used to calculate the rate constant of some specific temperature if E_a and k at some other temperature are known. The value of R that is used in either case depends on the units of the energy. When E_a is in joules, we use $R = 8.314$ J mol^{-1} K^{-1}.

EXAMPLE 13.7   CALCULATING THE ACTIVATION ENERGY FOR
A REACTION FROM RATE CONSTANTS AT TWO
TEMPERATURES

PROBLEM   At 300°C the rate constant for the reaction

is 2.41×10^{-10} s^{-1}. At 400°C, k equals 1.16×10^{-6} s^{-1}. What are the values of E_a (in kilojoules per mole) and A for this reaction?

SOLUTION   We can obtain E_a by substituting values of k_1, k_2, T_1, and T_2 into either Equation 13.17 or 13.18 and then solving for E_a. To avoid confusion, let's first tabulate our data.

	k	T
1	2.41×10^{-10} s^{-1}	300 + 273 = 573 K
2	1.16×10^{-6} s^{-1}	400 + 273 = 673 K

If you use a scientific calculator, you will find it best to use Equations 13.14 and 13.17 in calculations.

Notice that we have converted the temperatures to kelvins. Now let's use Equation 13.17 and work in natural logs. Substituting, we get

$$\ln \left(\frac{2.41 \times 10^{-10} \text{ s}^{-1}}{1.16 \times 10^{-6} \text{ s}^{-1}} \right) = \frac{E_a}{8.314 \text{ J mol}^{-1} \text{ K}^{-1}} \left(\frac{1}{673 \text{ K}} - \frac{1}{573 \text{ K}} \right)$$

$$\ln (2.08 \times 10^{-4}) = \frac{E_a}{8.314 \text{ J mol}^{-1} \text{ K}^{-1}} (0.00149 \text{ K}^{-1} - 0.00175 \text{ K}^{-1})$$

$$-8.48 = E_a (-3.1 \times 10^{-5} \text{ J}^{-1} \text{ mol})$$

$$E_a = \frac{-8.48}{-3.1 \times 10^{-5} \text{ J}^{-1} \text{ mol}}$$

$$= 2.7 \times 10^5 \text{ J mol}^{-1}$$

$$= 270 \text{ kJ mol}^{-1}$$

The easiest way to compute A is to use Equation 13.14[2] and solve for A.

$$A = \frac{k}{e^{-E_a/RT}} = k \, e^{E_a/RT}$$

---

[2] If you must use common logs and a log table, then use Equation 13.18 to calculate E_a. To obtain a value for A requires some manipulation of Equation 13.14. First we take the natural logarithm of both sides of the equation.

$$\ln k = \ln A - \frac{E_a}{RT}$$

or, in terms of common logarithms

$$2.303 \log k = 2.303 \log A - \frac{E_a}{RT}$$

Solving for $\log A$, we obtain

$$\log A = \log k + \frac{E_a}{2.303RT}$$

After substituting values and obtaining a value for $\log A$, take the antilogarithm to get A.

Using the first set of data and our calculated E_a, we get

$$A = (2.41 \times 10^{-10} \text{ s}^{-1})e^{(2.7 \times 10^5 \text{J mol}^{-1})/(8.314 \text{J mol}^{-1}\text{K}^{-1} \times 573\text{K})}$$

$$= (2.41 \times 10^{-10} \text{ s}^{-1}) \times (4.1 \times 10^{24})$$

$$= 9.9 \times 10^{15} \text{ s}^{-1}$$

Notice that A must have the same units as k.

13.9 CATALYSTS

A **catalyst** is a substance that increases the rate of a reaction without being consumed; after the reaction has ceased, it can be recovered from the reaction mixture chemically unchanged. The catalyst participates in the reaction by providing a lower-energy alternative mechanism for the production of the products. In Figure 13.11, note that the energy curve of the catalyzed reaction is drawn along a different reaction coordinate to emphasize that a different mechanism is involved. In addition, the energy barrier for the catalyzed path is lower than for the uncatalyzed reaction. This smaller activation energy means that in the reaction mixture there is a greater total fraction of molecules possessing sufficient kinetic energy to react (Figure 13.12). Therefore, in the presence of the catalyst there is an increased number of effective collisions. Of course, an increased number of effective collisions means a greater reaction rate.

Figure 13.11

Effect of a catalyst on the potential energy diagram. The catalyst changes the reaction mechanism by providing a different, low-energy mechanism for the formation of the products. ΔH is same for each path (it must be so because ΔH is a state function).

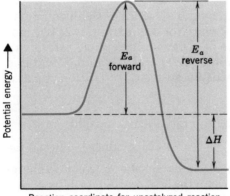

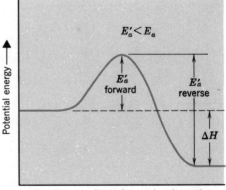

Figure 13.12

More molecules possess the minimum kinetic energy needed for an effective collision when the catalyst is present.

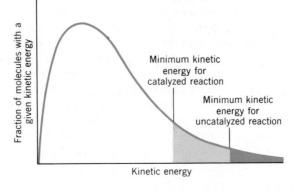

Since a catalyst emerges from a reaction chemically unchanged, it does not appear either as a reactant or a product in the overall balanced chemical equation, Instead, its presence is indicated by writing its name or formula over the arrow. For example, oxygen can be prepared by the thermal decomposition of potassium chlorate, $KClO_3$. In the absence of a catalyst the reaction is slow and the $KClO_3$ must be heated to a high temperature to cause it to decompose at a reasonable rate. However, if a small amount of manganese dioxide (MnO_2) is added to the $KClO_3$, the decomposition proceeds smoothly at a relatively low temperature. Analysis of the reaction mixture after the evolution of oxygen has ceased reveals that all of the MnO_2 added initially is still present, showing that MnO_2 has served as a catalyst. The equation for the catalyzed reaction is given as

$$2KClO_3 \xrightarrow{MnO_2} 2KCl + 3O_2$$

Even though a catalyst does not change the overall stoichiometry of a reaction, it does participate chemically by being consumed at one stage in the mechanism and produced again at a later stage. This regeneration of the catalyst permits the same catalyst to be used over and over again; therefore even a small amount of catalyst can have very profound effects on reaction rate. This phenomenon is particularly significant in biological systems, where practically every reaction is catalyzed by very small quantities of highly specific biochemical catalysts called enzymes.

Catalysts may be broadly classified into two categories: homogeneous and heterogeneous catalysts. A **homogeneous catalyst** is present in the same phase as the reactants and can serve to speed up the reaction by forming a reactive intermediate with one of the reactants. For example, the decomposition of *t*-butyl alcohol, $(CH_3)_3COH$, to produce water and isobutene, $(CH_3)_2C{=}CH_2$,

$$(CH_3)_3COH \longrightarrow (CH_3)_2C{=}CH_2 + H_2O$$

is catalyzed by the presence of small amounts of HBr. In the absence of HBr the activation energy for the reaction is 274 kJ mol^{-1}, and below 450°C the reaction takes place at a barely perceptible rate. In the presence of HBr an activation energy of only 127 kJ mol^{-1} is found, and it is possible that the catalyzed reaction proceeds by an attack of HBr on the alcohol

$$(CH_3)_3COH + HBr \longrightarrow (CH_3)_3CBr + H_2O$$

followed by the rapid decomposition of the *t*-butyl bromide, $(CH_3)_3CBr$,

$$(CH_3)_3CBr \longrightarrow (CH_3)_2C{=}CH_2 + HBr$$

Thus HBr provides an alternative low-energy path for the reaction and the mechanism for the reaction when HBr is present is different from when it is absent.

A **heterogeneous catalyst** is not in the same phase as the reactants, but provides a favorable surface on which the reaction can take place. An example of a reaction whose rate is increased by the presence of a heterogeneous catalyst is the reaction between hydrogen and oxygen to produce water. In the introduction to this chapter we pointed out that this reaction proceeds at a very slow rate when the two gases are mixed at room temperature. However, it has been found that the reaction proceeds at an appreciable rate when metals such as nickel, copper, or silver are present.

Adsorption means sticking to a surface.

Heterogeneous catalysts appear to function through a process whereby reactant molecules are adsorbed on a surface where the reaction then takes place. The high reactivity of hydrogen, in the presence of certain metals, for example, is thought to occur by the adsorption of H_2 molecules onto the catalytic surface. On the surface of the metal the bonds between hydrogen atoms are apparently

Figure 13.13

Production of hydrogen atoms on a metal surface. A hydrogen molecule collides with the surface where it is adsorbed and dissociates to produce H atoms.

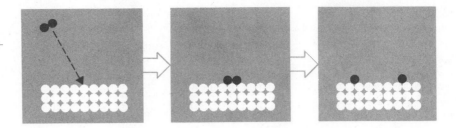

stretched or broken, as shown in Figure 13.13, so that the metal surface actually behaves as if it contained highly reactive hydrogen atoms.

Unless the reactant molecules can be adsorbed on the catalyst, no increase in reaction rate can occur. A substance whose presence during a reaction interferes with the adsorption process will therefore reduce the effectiveness of the catalysts and is thus called an **inhibitor.** These substances, by being strongly adsorbed on the catalytic surface, decrease the available space on which the reaction can occur. In some cases the catalyst eventually becomes useless and is said to be poisoned. The destruction of catalytic activity by poisoning is very important in biological systems, as we will see in Chapter 24.

There have been many commercial and industrial applications of heterogeneous catalysts. For example, small portable flameless heaters can be purchased (for heating camping tents and the like) in which the fuel and oxygen combine on a catalytic surface. The flameless combustion evolves the same amount of heat that would be generated if the fuel were burned directly, but because no flame is involved the heaters are much safer to operate. There are drawbacks, however. Unfortunately, the catalytic oxidation of the fuel is not totally efficient, and small amounts of carbon monoxide are produced. As a result, catalytic heaters must be used with care.

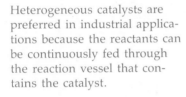

Heterogeneous catalysts are preferred in industrial applications because the reactants can be continuously fed through the reaction vessel that contains the catalyst.

Another very important application is in the control of auto exhaust emissions. Most gasoline-powered automobiles today are equipped with catalytic converters that employ a mixed metal oxide bed over which the exhaust gases pass after they are mixed with additional air (Figure 13.14). The catalyst quite effectively promotes the oxidation of CO and hydrocarbons to harmless CO_2 and H_2O. The catalysts in newer catalytic converters also remove nitrogen oxide pollutants by promoting their decomposition to nitrogen and oxygen. Catalytic mufflers suffer from the disadvantage of being poisoned by lead. As a result, lead-free fuels must be employed in autos fitted with this type of antipollution device. Another disadvantage is that they also catalyze the oxidation of SO_2 to SO_3, which then reacts with water vapor to produce a mist of sulfuric acid. Since SO_2 is produced from the combustion of high-sulfur fuels, this problem is a serious one and may ultimately lead to forced discontinuation of the use of catalytic mufflers.

13.10 CHAIN REACTIONS

The reactions that we have discussed so far have been rather simple and straightforward, with uncomplicated rate laws. There are some reactions, however, that have very complex rate laws and whose elementary processes involve extremely reactive intermediate substances that possess unpaired electrons. As a group these intermediates are called **free radicals** and might consist of single atoms or either neutral or electrically charged particles composed of a number of atoms. In chemical systems, molecules or ions can form free radicals either thermally (at high temperature) or by the absorption of photons of appropriate fre-

Figure 13.14
(a) *In a catalytic converter, air and exhaust gases are passed over a catalyst bed where CO is oxidized to CO_2 and nitrogen oxides are decomposed to N_2 and O_2. (b) A cutaway view of a catalytic converter used by General Motors Corporation on their passenger cars and light trucks.*

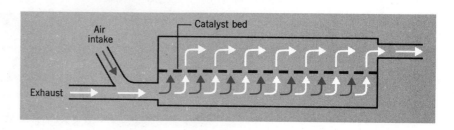

(a)

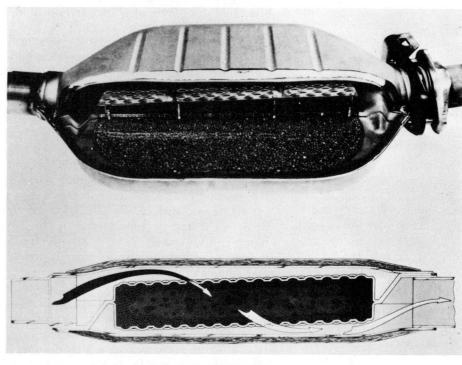

(b)

quencies. At high temperatures, the violent vibrational motions within the molecules can cause bonds to break, whereas in the absorption of radiation, the photons supply the energy to break chemical bonds. In either case, the fragments that are formed have unpaired electrons. (Normally, the energy required to break chemical bonds is such that the photons must have frequencies that place them in the ultraviolet or X-ray region of the spectrum. This is the reason that exposure to X rays or other forms of high-energy radiation is dangerous.)

As you learned in earlier chapters, electrons become paired when atoms form chemical bonds, and it is the drive toward the pairing of electrons that is responsible in part for the formation of bonds between atoms. This same drive is what makes free radicals so reactive; the particles seek out other atoms or molecules with which they are able to form bonds and thus pair their electrons. In these reactions it often happens that a product molecule is formed along with *another* free radical—a process that can be repeated over and over. Therefore once a free radical is formed through some initiation process, its reaction with other species is *self-propagating*, and the entire series of reactions that follows the production of the very reactive intermediate is called a **chain reaction.**

An example of a chain reaction is the mechanism that has been proposed to explain the rate law observed for the reaction between hydrogen and bromine. The overall reaction between these substances is

$$H_2 + Br_2 \longrightarrow 2HBr$$

In biological systems, free radicals have been blamed for a variety of ill effects. These range from the effects of aging to the formation of cancers.

If this reaction proceeded simply by a bimolecular collision between H_2 and Br_2, the expected rate law would be

$$\text{rate} = k[H_2][Br_2]$$

However, the actual rate law turns out to be

$$\text{rate} = k\frac{[H_2][Br_2]^{1/2}}{1 + [HBr]/k'[Br_2]}$$

which is very complex indeed. A mechanism that has been proposed to account for this rate law is the chain reaction shown below. A dot is used to represent the unpaired electron on a highly reactive atom or free radical.

1. $Br_2 \rightarrow 2Br\cdot$ — Initiation
2. $Br\cdot + H_2 \rightarrow HBr + H\cdot$ ⎫
3. $H\cdot + Br_2 \rightarrow HBr + Br\cdot$ ⎬ Propagation
4. $H\cdot + HBr \rightarrow H_2 + Br\cdot$ — Inhibition
5. $2Br\cdot \rightarrow Br_2$ — Termination

Reaction 1 is the thermal decomposition of diatomic bromine molecules to produce bromine atoms (the reactive intermediate). The overall reaction proceeds very slowly when the two reactants are mixed at room temperature. However, at high temperature Reaction 1 takes place to an appreciable extent and rapidly sets off the remaining reactions. Step 1 is therefore called the *initiation step* because it begins the chain. In Steps 2 and 3 the product HBr is formed as well as additional free atoms that serve to keep the reaction going. These steps are called *propagation steps* in the chain. Step 5, which leads only to the formation of a stable species, serves to end the chain and is called the *termination step*. Step 4 is referred to as an *inhibition step*, because its occurrence removes the product and thus decreases the overall rate of production of HBr. It is included in the mechanism because the presence of HBr decreases the reaction rate (note the appearance of [HBr] in the denominator of the rate law).

In general, chain reactions are very rapid; in fact, many explosive reactions appear to occur by chain mechanisms. The formation of a single reactive intermediate produces many product molecules before the chain is terminated. Consequently, the rate of formation of the products is many times greater than the rate of the initiation step alone.

REVIEW QUESTIONS AND PROBLEMS

(Problems whose numbers are in blue have their answers in Appendix C at the back of the book; the more difficult problems are marked with asterisks.)

General Facts About Reaction Rates

13.1 What are the four factors that control the rates of chemical reactions?

13.2 What effect does particle size have on the rate of a heterogeneous reaction?

13.3 What criteria must be met by methods used to study the rate of a reaction?

13.4 Make a list of five reactions that occur in the world around you and compare their rates. Try to think of some that are fast and some that are slow.

13.5 Why do we say that one of the factors that influences reaction rate is the nature of the reactants?

13.6 Which reaction, or set of reaction conditions, should give the larger rate of formation of $H_2(g)$:
(a) Magnesium or iron in 1.00 M HCl?
(b) Zinc in 1.00 M HCl, or in 0.10 M HCl?
(c) 1.0 g of powdered zinc in 1.00 M HCl or a 1.0-g zinc bar in 1.00 M HCl?
(d) An iron nail in 1.00 M HCl at 25°C, or in 1.00 M HCl at 40°C?

Reaction Rate and Its Measurement

13.7 Define *reaction rate*. What are the units?

13.8 For each of the following reactions, express the reaction rate in terms of the disappearance of the reactants and the appearance of the products. Predict the role of the coefficients in the balanced overall equation in determining the relative rates of disappearance of reactants and formation of products.
(a) $2H_2 + O_2 \rightarrow 2H_2O$
(b) $2NOCl \rightarrow 2NO + Cl_2$
(c) $NO + O_3 \rightarrow NO_2 + O_2$
(d) $H_2O_2 + H_2 \rightarrow 2H_2O$

13.9 Consider the reaction for the combustion of methane, CH_4,

$$CH_4(g) + 2O_2(g) \longrightarrow CO_2(g) + 2H_2O(g)$$

If the methane is burning at a rate of 0.16 mol dm^{-3} s^{-1}, at what rates are CO_2 and H_2O being formed?

13.10 For the reaction

$$4NH_3(g) + 3O_2(g) \longrightarrow 2N_2(g) + 6H_2O(g)$$

it was found that at a particular instant N_2 was being formed at a rate of 0.68 mol dm^{-3} s^{-1}.
(a) At what rate was water being formed?
(b) At what rate was NH_3 reacting?
(c) At what rate was O_2 being consumed?

13.11 The following data were collected for the reaction, $2A \rightarrow 4B + C$

Time (min)	Concentration of A (mol dm^{-3})	Concentration of B (mol dm^{-3})
0	1.000	0.000
10	0.800	0.400
20	0.667	0.667
30	0.571	0.858
40	0.500	1.000
50	0.444	1.112

Make a graph of the concentrations of A and B versus time (concentrations along the vertical axis, time along the horizontal axis). Estimate the rate of disappearance of A and the rate of formation of B at $t = 25$ min and $t = 40$ min. Compare the rates of disappearance of A and formation of B. What would you expect the rate of formation of C to be at $t = 25$ min and $t = 40$ min?

Rate Laws

13.12 What is a rate law? What factors affect the value of the rate constant for a given reaction?

13.13 What is meant by the *order of a reaction?*

13.14 What are the units of the rate constant for (a) a first-order reaction, (b) a second-order reaction, and (c) a third-order reaction?

13.15 The rate at which CO is removed from the earth's atmosphere by fungi in the soil is constant. What is the apparent order of this process?

13.16 What is the order with respect to each reactant and the overall order of the reactions described by the following rate laws:
(a) Rate $= k_1[A][B]$
(b) Rate $= k_2[E]^2$
(c) Rate $= k_3[G]^2[H]^2$

13.17 What would be the units of each of the rate constants in the preceding question if rate has the units *mol dm^{-3} s^{-1}?*

13.18 When the concentration of a reactant is doubled, by what factor would the rate of reaction be changed if the order with respect to that reactant were (a) 1, (b) 2, (c) 3, (d) 4, (e) $\frac{1}{2}$, (f) -2?

13.19 Suppose that when the concentration of a reactant was doubled the rate of reaction decreased by a factor of 2. What would be the exponent on the concentration term for that reactant in the rate law?

13.20 The rate constant for the reaction

$$2ICl + H_2 \longrightarrow I_2 + 2HCl$$

is 1.63×10^{-1} dm^3 mol^{-1} s^{-1}. The rate law is given by

$$\text{rate} = k[ICl][H_2]$$

What is the rate of the reaction for each of the sets of concentrations given below?

ICl Concentration (mol dm^{-3})	H$_2$ Concentration (mol dm^{-3})
0.25	0.25
0.25	0.50
0.50	0.50

13.21 The rate law for a reaction was found to be

$$\text{rate} = (2.35 \times 10^{-6} \text{ dm}^6 \text{ mol}^{-2} \text{ s}^{-1})[A]^2[B]$$

What would the rate of reaction be if:
(a) The concentrations of A and B were 1.00 mol dm^{-3}?
(b) $[A] = 0.250$ M, $[B] = 1.30$ M?

13.22 For the decomposition of dinitrogen pentoxide,

$$2N_2O_5 \longrightarrow 4NO_2 + O_2$$

the following data were collected:

N_2O_5 Concentration (mol dm^{-3})	Time (s)
5.00	0
3.52	500
2.48	1000
1.75	1500
1.23	2000
0.87	2500
0.61	3000

(a) Make a graph of the concentration of N_2O_5 versus time. Draw tangents to the curve at $t = 500$, 1000, and 1500 s. Determine the rate at these different reaction times.

(b) Determine the value of the rate constant at 500, 1000, and 1500 s, given the rate law, rate = $k[N_2O_5]$.

13.23 One of the reactions that can take place in polluted air is the reaction of nitrogen dioxide, NO_2, with ozone, O_3.

$$NO_2(g) + O_3(g) \longrightarrow NO_3(g) + O_2(g)$$

The following data were collected on this reaction at 25°C:

Initial NO_2 Concentration (mol dm^{-3})	Initial O_3 Concentration (mol dm^{-3})	Initial Rate of Formation of $O_2(g)$ (mol dm^{-3} s^{-1})
5.0×10^{-5}	1.0×10^{-5}	0.022
5.0×10^{-5}	2.0×10^{-5}	0.044
2.5×10^{-5}	2.0×10^{-5}	0.022

(a) What is the rate law for the reaction?

(b) What is the value of the rate constant (give the correct units)?

13.24 A certain reaction that follows the equation

$$2A(g) + B(g) \longrightarrow C(g) + 2D(g)$$

was studied at 25°C. The following data were obtained:

Initial Concentration of A (mol dm^{-3})	Initial Concentration of B (mol dm^{-3})	Initial Rate of Formation of C (mol dm^{-3} s^{-1})
0.010	0.010	1.20×10^{-3}
0.020	0.010	2.40×10^{-3}
0.030	0.010	3.60×10^{-3}
0.030	0.020	1.44×10^{-2}

(a) What is the rate law for the reaction?

(b) What is the rate constant at this temperature?

(c) If $[A] = 0.020$ M and $[B] = 0.060$ M, what will be the rate of formation of D?

13.25 At 27°C, the reaction, $2NOCl \rightarrow 2NO + Cl_2$, is observed to exhibit the following dependence of rate on concentration.

Initial NOCl Concentration (mol dm^{-3})	Initial Rate of Formation of NO (mol dm^{-3} s^{-1})
0.30	3.60×10^{-9}
0.60	1.44×10^{-8}
0.90	3.24×10^{-8}

(a) What is the rate law for the reaction?

(b) What is the rate constant?

(c) By what factor would the rate increase if the initial concentration of NOCl were increased from 0.30 to 0.45 M?

13.26 The reaction of NO with Cl_2 follows the equation

$$2NO + Cl_2 \longrightarrow 2NOCl$$

The following data were collected:

Initial NO Concentration (mol dm^{-3})	Initial Cl_2 Concentration (mol dm^{-3})	Initial Rate of Formation of NOCl (mol dm^{-3} s^{-1})
0.10	0.10	2.53×10^{-6}
0.10	0.20	5.06×10^{-6}
0.20	0.10	10.1×10^{-6}
0.30	0.10	22.8×10^{-6}

(a) What is the rate law for the reaction?

(b) What is the value of the rate constant (be sure to give the proper units)?

*13.27 For a first-order reaction, a graph of log[A] versus time (where A is a reactant) gives a straight line having a slope equal to $-k/2.303$. On the other hand, if the reaction is second-order with respect to A, a straight line is obtained when $1/[A]$ is plotted against time. In this case the slope of the line is equal to k. From this information, determine whether the reaction in Exercise 13.11 is first- or second-order. Calculate the rate constant for the reaction.

Concentration and Time; Half-Lives

13.28 Define *half-life of a reaction*.

13.29 How is the half-life of a first-order reaction affected by the concentrations of the reactants?

13.30 A reaction $2B(g) \rightarrow C(g)$ follows the rate law: rate = $k[B]^2$. Use what you have learned about half-lives for second-order reactions to construct a graph similar to Figure 13.3 for a second-order reaction. Take the initial concentration of B to be $1.00\ M$ and follow the reaction through four half-lives. You can use any arbitrary time scale for the horizontal axis.

13.31 The rate constant for the first-order decomposition of $N_2O_5(g)$ at 100°C is $1.46 \times 10^{-1}\ s^{-1}$.
  (a) If the initial concentration of N_2O_5 in a reaction vessel was $4.50 \times 10^{-3}\ mol\ dm^{-3}$, what will the concentration be 20.0 s after the decomposition begins?
  (b) What is the half-life (in seconds) of N_2O_5 at 100°C?
  (c) If the initial concentration of N_2O_5 was $4.50 \times 10^{-3}\ M$, what will the concentration be after three half-lives?

13.32 The decomposition of hydrogen iodide, HI, is second-order. At 500°C, the half-life of HI is 2.11 min when the initial HI concentration is $0.10\ M$. What will be the half-life (in minutes) when the initial HI concentration is $0.010\ M$?

13.33 If we refer to Exercise 13.32, what is the rate constant for the decomposition of HI at 500°C in the units, $dm^3\ mol^{-1}\ s^{-1}$?

13.34 The decomposition of C_2H_5Cl has the rate law: rate = $k[C_2H_5Cl]$. At 550°C, $k = 3.2 \times 10^{-2}\ s^{-1}$.
  (a) What is the half-life of this reaction at 550°C?
  (b) If the concentration of C_2H_5Cl was $0.010\ M$ after 1.00 min, what was the initial concentration of C_2H_5Cl?

Collision Theory and Reaction Mechanisms

13.35 How does collision theory account for the dependence of rate on the concentrations of the reactants?

13.36 Why can't we use the ideas of collision theory to predict in general the rate laws of chemical reactions? What must we know in order to predict the rate law?

13.37 What is a *reaction mechanism*?

13.38 Propane, C_3H_8, is a fuel used in many rural areas for cooking. It is known as LPG. It is unlikely that the combustion of propane occurs by a simple one-step mechanism,

$$C_3H_8(g) + 5O_2(g) \longrightarrow 3CO_2(g) + 4H_2O(g)$$

Explain why.

13.39 A mechanism for the reaction, $2NO + Br_2 \rightarrow 2NOBr$, has been suggested to be

Step 1 $NO + Br_2 \longrightarrow NOBr_2$

Step 2 $NOBr_2 + NO \longrightarrow 2NOBr$

  (a) What would be the rate law for the reaction if the first step in this mechanism were slow and the second fast?
  (b) What would be the rate law if the second step were slow, with the first reaction being a rapidly established dynamic equilibrium?
  (c) Experimentally, the rate law has been found to be

$$\text{rate} = k[NO]^2[Br_2]$$

What can we conclude about the relative rates of Steps 1 and 2?
  (d) Why do we not prefer a simple, one-step mechanism

$$NO + NO + Br_2 \longrightarrow 2NOBr$$

  (e) Can we, on the basis of the experimental rate law, definitely exclude the mechanism in part (d)?

13.40 The reaction, $NO_2(g) + CO(g) \rightarrow CO_2(g) + NO(g)$, appears to have the mechanism (at low temperature),

$$NO_2 + NO_2 \longrightarrow NO_3 + NO \qquad \text{slow}$$

$$NO_3 + CO \longrightarrow NO_2 + CO_2 \qquad \text{fast}$$

Explain why the reaction is zero-order with respect to CO.

13.41 The reaction of methyl bromide, CH_3Br, with OH^- appears to occur through a one-step mechanism involving collision of CH_3Br with OH^-,

$$CH_3Br + OH^- \longrightarrow CH_3OH + Br^-$$

The rate law for the reaction is found to be

$$\text{rate} = k[CH_3Br][OH^-]$$

In Example 13.3 we found that the rate law of the reaction of $(CH_3)_3CBr$ with OH^- has the rate law

$$\text{rate} = k[(CH_3)_3CBr]$$

Try to propose a mechanism that can account for the rate law for the reaction of $(CH_3)_3CBr$ with OH^-.

13.42 Suppose that the following sequence of reactions was proposed for a reaction:

Step 1 $2A \longrightarrow A_2$

Step 2 $A_2 + B \longrightarrow C + 2D$

  (a) What would be the overall net chemical reaction?
  (b) What would the rate law be if Step 1 were slow and Step 2 were fast?
  (c) What would the rate law be if Step 2 were slow and Step 1 were fast?

13.43 One of the reactions that occurs in polluted air in urban areas is $2NO_2(g) + O_3(g) \rightarrow N_2O_5(g) + O_2(g)$. It is believed that a species with the formula NO_3 is involved in the mechanism, and the observed rate law for the reaction is rate = $k[NO_2][O_3]$. Propose a mechanism for this reaction.

13.44 How do we know that not all collisions between reactant molecules lead to chemical change? What determines whether a particular collision will be effective?

13.45 How do the orientations of molecules influence whether a collision between them can be effective at producing chemical change?

13.46 One step in the mechanism for the decomposition of NO_2Cl into NO_2 and Cl_2 appears to be $NO_2Cl + Cl \rightarrow NO_2 + Cl_2$. Given the structure of NO_2Cl,

show how molecular orientation at the moment of a collision can be important in determining whether or not the products will be formed.

13.47 Nitric acid is one of the world's most important chemicals. One of its principal uses is in making fertilizers. In 1984, over 6.4×10^{12} g (6.4 Tg) of HNO_3 were produced, mostly by oxidation of ammonia. This reaction gives nitric oxide, NO, which then reacts with oxygen to form nitrogen dioxide, NO_2. Nitrogen dioxide finally reacts with water to form nitric acid. The oxidation of NO to NO_2 follows the reaction

$$2NO(g) + O_2(g) \longrightarrow 2NO_2(g)$$

and has as a rate law, rate = $k[NO]^2[O_2]$. Predict a possible mechanism for this reaction.

Transition State Theory

13.48 Draw a potential energy diagram for an endothermic reaction. Indicate on the drawing (a) the potential energy of the reactants, (b) the potential energy of the products, (c) the energies of activation for the forward and reverse reaction, (d) the heat of reaction.

13.49 What are meant by the terms, transition state and activated complex? Where on the potential energy diagram for a reaction will we find the transition state?

Effect of Temperature on Reaction Rate

13.50 Define the term, activation energy.

13.51 Explain qualitatively, in terms of the kinetic theory, why an increase in temperature leads to an increase in reaction rate.

13.52 Insects, which are cold-blooded animals whose changes in body temperature tend to follow changes in the temperature of their environment, become

quite sluggish in cool weather. On the basis of chemical kinetics, explain this phenomenon.

Calculations Involving the Activation Energy

13.53 The rate constants for the reaction between ICl and H_2 (Exercise 13.20) at 230 and 240°C have been found to be 0.163 and 0.348 dm^3 mol^{-1} s^{-1}, respectively. What are the values of E_a (in kilojoules per mole) and A for this reaction?

13.54 The rate constant for the reaction

$$CH_3I(g) + HI(g) \longrightarrow CH_4(g) + I_2(g)$$

at 200°C is 1.32×10^{-2} dm^3 mol^{-1} s^{-1}. At 275°C the rate constant is 1.64 dm^3 mol^{-1} s^{-1}. What is the activation energy (in kilojoules per mole) and the value of A?

13.55 The activation energy for the decomposition of HI

$$2HI(g) \longrightarrow H_2(g) + I_2(g)$$

is 182 kJ mol^{-1}. The rate constant for the reaction at 700°C is 1.57×10^{-3} dm^3 mol^{-1} s^{-1}. What is the value of the rate constant at 600°C?

13.56 The activation energy for the reaction

$$HI(g) + CH_3I(g) \longrightarrow CH_4(g) + I_2(g)$$

is 138 kJ mol^{-1}. At 200°C the rate constant has a value of 1.32×10^{-2} dm^3 mol^{-1} s^{-1}. What is the rate constant at 300°C?

*13.57 A chemist was able to determine that the rate of a particular reaction at 100°C was four times faster than at 30°C. The concentrations of the reactants were the same at both temperatures. Calculate the approximate energy of activation for the reaction in kJ mol^{-1}.

13.58 The decomposition of C_2H_5Cl is a first-order reaction having $k = 3.2 \times 10^{-2}$ s^{-1} at 550°C and $k = 9.3 \times 10^{-2}$ s^{-1} at 575°C. What is the activation energy, in kilojoules per mole, for this reaction?

13.59 The rate constant for the reaction

$$H_2(g) + I_2(g) \longrightarrow 2HI(g)$$

was measured at a series of temperatures. The data are given below

Temperature (°C)	k (dm^3 mol^{-1} s^{-1})
283	1.2×10^{-4}
302	3.5×10^{-4}
355	6.8×10^{-3}
393	3.8×10^{-2}
430	1.7×10^{-1}

Graphically determine the value of E_a in kJ mol^{-1} for this reaction.

*13.60 The development of a photographic image on film is a process controlled by the kinetics of the reduction of silver halide by a developer. The time required for development at a particular temperature is inversely proportional to the rate constant for the process. Below are published data on development times for Kodak's Tri-X film using Kodak D-76 developer. From these data, estimate the activation energy for the development process in kilojoules per mole.

Temperature (°C)	Time for Development (min)
18	10
20	9
21	8
22	7
24	6

Estimate the development time at 15°C.

*13.61 The cooking of an egg involves the denaturation of a protein called albumin, and the time required to achieve a particular degree of denaturation is inversely proportional to the rate constant for the process. This reaction has a high activation energy; $E_a = 418$ kJ mol^{-1}. Calculate how long it would take to cook a traditional three-minute egg on top of Mt. McKinley in Alaska on a day when the atmospheric pressure there is 47.3 kPa.

Catalysis

13.62 What is the difference between a homogeneous and a heterogeneous catalyst?

13.63 What is a heterogeneous catalyst? How does it function? What is an inhibitor?

13.64 How does a catalyst play a part in lowering the activation energy for a reaction?

13.65 What effect does a catalyst have on
(a) the heat of reaction?
(b) the potential energy of the reactants?
(c) the transition state?

Chain Reactions

13.66 What is a free radical? How can free radicals be formed?

13.67 The C—C bond energy in C_2H_6 is 348 kJ mol^{-1}. Calculate the frequency and wavelength (in nanometres) of light that would be able to promote the reaction

$$C_2H_6(g) \longrightarrow 2CH_3\cdot(g)$$

13.68 Why are free radicals dangerous in living organisms?

13.69 Why are chain reactions often so fast?

13.70 The decomposition of acetaldehyde, CH_3CHO, follows the overall reaction

$$CH_3CHO \longrightarrow CH_4 + CO$$

with small amounts of H_2 and C_2H_6 also being produced. The reaction is thought to proceed by a chain reaction involving free radicals (note again that a free radical is indicated by using a dot to represent its unpaired electron). A proposed mechanism is
(1) $CH_3CHO \rightarrow CH_3\cdot + CHO\cdot$
(2) $2CH_3\cdot \rightarrow C_2H_6$
(3) $CHO\cdot \rightarrow H\cdot + CO$
(4) $H\cdot + CH_3CHO \rightarrow H_2 + CH_3CO\cdot$
(5) $CH_3\cdot + CH_3CHO \rightarrow CH_4 + CH_3CO\cdot$
(6) $CH_3CO\cdot \rightarrow CH_3\cdot + CO$
Identify (a) the initiation step, (b) the propagation step(s), and (c) the termination step(s).

CHAPTER 14

CHEMICAL EQUILIBRIUM

A firefighter unknowingly applies Le Châtelier's principle as he breaths pure oxygen to combat a dose of carbon monoxide. The displacement of CO from hemoglobin in the blood by O_2 is a chemical equilibrium, and increasing the partial pressure of O_2 in the lungs shifts the position of equilibrium in favor of the O_2.

When a chemical reaction takes place spontaneously, the concentrations of the reactants and products change while the free energy of the system decreases. Eventually the free energy reaches a minimum, and as we learned in Chapter 12, the system comes to a state of equilibrium. If we follow the concentrations while this happens, we observe that they approach steady values, as shown in Figure 14.1. We find that the rate at which the reactants form the products approaches the rate at which the products form the reactants. When equilibrium is finally reached, both forward and reverse reactions are occurring at equal rates and the concentrations no longer change. It is the continuation of the forward and reverse reactions without a change in concentrations, you recall, that identifies this as a *dynamic equilibrium*.

All chemical systems tend toward equilibrium. In this chapter we will explore the quantitative relationships that can be used to describe the equilibrium state, and we will see how the principles of thermodynamics can be applied to a description of equilibrium.

14.1 THE EQUILIBRIUM LAW FOR A CHEMICAL REACTION

In the previous chapter you learned that for each chemical reaction there is a relationship between the rate of the reaction and the concentrations of the reactants. We called this the rate law for the reaction. For each chemical system, there is also a relationship among concentrations that determines the relative proportions of reactants and products at equilibrium. Unlike kinetics, however, this relationship can be obtained *directly* from the balanced chemical equation for the overall reaction. For a general reaction

a, *b*, *e*, and *f* are the coefficients of substances A, B, E, and F.

$$aA + bB \rightleftharpoons eE + fF$$

it is observed that at constant temperature the condition fulfilled at equilibrium is

$$\frac{[E]^e[F]^f}{[A]^a[B]^b} = K_c \qquad [14.1]$$

The equilibrium law is also referred to frequently as the *equilibrium expression* when it is discussed in mathematical terms.

The quantities written within square brackets denote equilibrium molar concentrations, and K_c is a constant called the **equilibrium constant.** The entire equation can be called the **equilibrium law** for the reaction, because when this system is at equilibrium, the value of the fraction on the left side of the equation must equal the value of K_c for the reaction at that particular temperature. If the value of the fraction doesn't equal K_c, then the system is *not* in a state of dynamic equilibrium.

From this discussion, you can see that a great deal of importance is placed on the fraction that appears on the left side of Equation 14.1. For historical

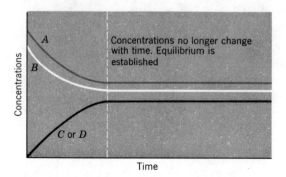

Figure 14.1

The approach to equilibrium for the reaction A + B → C + D.

reasons that we will not discuss further, this fraction is commonly called the **mass action expression.** The value of the fraction is called the **reaction quotient, Q,** and when the system is at equilibrium the reaction quotient has a value equal to K_c.

As we've noted above, the mass action expression is constructed using the coefficients in the balanced chemical equation as exponents on the appropriate concentrations. Let's consider a concrete example: the reaction of nitrogen with hydrogen to form ammonia. This is one of society's most important equilibria because it is used to capture nitrogen, taken from the air, in compounds that are used as nitrogen fertilizers. The equation is

$$N_2(g) + 3H_2(g) \rightleftharpoons 2NH_3(g)$$

The mass action expression for this reaction is written as

$$\frac{[NH_3]^2}{[N_2][H_2]^3}$$

This fraction, of course, always has some numerical value for this chemical system. For example, suppose some N_2 and H_2 were introduced into a container and permitted to react. Initially there would be no NH_3, so the value of the mass action expression would be zero. Then, as NH_3 is formed, the fraction would grow larger until, when equilibrium is finally reached, the value of the fraction would no longer change. It becomes equal to a value that we call the equilibrium constant, K_c.

It is always possible to write the equilibrium expression if you know the balanced chemical equation.

$$\frac{[NH_3]^2}{[N_2][H_2]^3} = K_c \text{ (at equilibrium)} \qquad [14.2]$$

The most important point about this equation is that at a given temperature the same numerical value of the mass action expression is always obtained for *any* system containing N_2, H_2, and NH_3 in equilibrium. *There are no restrictions on the individual concentrations of any reactant or product.* The only requirement for equilibrium is that when these concentrations are substituted into the mass action expression, the fraction is numerically equal to K_c. In effect, Equation 14.2 is a condition that must be fulfilled for N_2, H_2, and NH_3 to be in equilibrium with each other. This is why Equation 14.2 is spoken of as the equilibrium condition or equilibrium law for the reaction. The data in Table 14.1 illustrate this point.

For reactions involving gases, the partial pressures of the reactants and products are proportional to their molar concentrations. The equilibrium expression for these reactions can therefore be written using partial pressures instead

molar concentration $= \dfrac{\text{mol}}{\text{dm}^3} = \dfrac{n}{V}$

For an ideal gas,

$$\frac{n}{V} = \frac{P}{RT}$$

Table 14.1
Equilibrium concentrations (in mol dm^{-3}) at 500°C and the mass action expression for the reaction: $N_2(g) + 3H_2(g) \rightleftharpoons 2NH_3(g)$

$[H_2]$	$[N_2]$	$[NH_3]$	$\dfrac{[NH_3]^2}{[N_2][H_2]^3} = K_c$
0.150	0.750	1.23×10^{-2}	5.98×10^{-2}
0.500	1.00	8.66×10^{-2}	6.00×10^{-2}
1.35	1.15	4.12×10^{-1}	6.00×10^{-2}
2.43	1.85	1.27	6.08×10^{-2}
1.47	0.750	3.76×10^{-1}	5.93×10^{-2}
		Average	6.00×10^{-2}

of concentrations. For example, the equilibrium condition for the reaction between $N_2(g)$ and $H_2(g)$ can also be expressed as

$$\frac{p_{NH_3}^2}{p_{N_2}p_{H_2}^3} = K_P$$

We will use the symbol K_P to denote equilibrium constants derived from partial pressures and K_c to indicate equilibrium constants having molar concentrations in the mass action expression. In general, K_c and K_P are not numerically equal. We will discuss this further in Section 14.4.

We have written the mass action expression with the concentrations (or partial pressures) of the products in the numerator and those of the reactants in the denominator. Since this fraction is equal to a constant at equilibrium, its reciprocal must also be a constant. Thus

$$\frac{[NH_3]^2}{[N_2][H_2]^3} = K_c \qquad \frac{p_{NH_3}^2}{p_{N_2}p_{H_2}^3} = K_P$$

and

$$\frac{[N_2][H_2]^3}{[NH_3]^2} = \frac{1}{K_c} = K_c' \qquad \frac{p_{N_2}p_{H_2}^3}{p_{NH_3}^2} = \frac{1}{K_P} = K_P'$$

The mass action expression that is equal to K can always be constructed from the chemical equation.

Either form is a valid description of the equilibrium state. However, chemists have chosen, somewhat arbitrarily, always to write the equilibrium expression with the concentrations or partial pressures of the products appearing in the numerator. This allows us then to tabulate equilibrium constants without always having to state explicitly the form of the mass action expression. It is only necessary to specify the chemical equation and whether we are dealing with K_c or K_p.

14.2 THE EQUILIBRIUM CONSTANT

The equilibrium constant is a quantity that must be calculated from experimental data. One method involves the use of standard free energies of reaction to calculate a *thermodynamic equilibrium constant* and is outlined in Section 14.3. Another method involves the direct measurement of equilibrium concentrations that can then be substituted into the mass action expression to obtain a numerical value for K. We will look at a sample calculation of this type in Section 14.6.

Knowing the value of an equilibrium constant is useful because it allows us to perform computations that relate the concentrations of reactants and products in an equilibrium system. But even *without* doing calculations, the magnitude of K provides us with useful qualitative information about the extent to which a reaction proceeds toward completion. For example, consider the simple reaction

$$A \rightleftharpoons B$$

for which we write

$$\frac{[B]}{[A]} = K_c$$

Suppose that $K_c = 10$. This means that at equilibrium

$$\frac{[B]}{[A]} = 10 = \frac{10}{1}$$

which tells us that at equilibrium the concentration of B must be ten times larger than that of A. In other words, the position of equilibrium lies in favor of the product, B. On the other hand, if $K_c = 0.1$, then

$$\frac{[B]}{[A]} = 0.1 = \frac{1}{10}$$

In this case the equilibrium concentration of A would have to be ten times larger than the concentration of B at equilibrium, and the position of equilibrium would lie in favor of the reactant, A. It is a general rule that when K is large, the position of equilibrium lies far to the right. Conversely, when K is small, only relatively small amounts of the products are present in the system at equilibrium.

Let's look at two examples of real chemical reactions: first, the reaction of hydrogen with chlorine,

$$H_2(g) + Cl_2(g) \rightleftharpoons 2HCl(g)$$

for which $K_c = 4.4 \times 10^{32}$ at 25°C. This very large value of K tells us that at equilibrium the reaction will have proceeded far toward completion. If 1 mol each of H_2 and Cl_2 are combined, very little H_2 and Cl_2 will remain unreacted at equilibrium. By way of contrast, the decomposition of water vapor at room temperature (25°C),

$$2H_2O(g) \rightleftharpoons 2H_2(g) + O_2(g)$$

has $K_c = 1.1 \times 10^{-81}$. Examining this value of K_c, we can conclude that the decomposition takes place to only a very small degree, because in order to have such a very small value of K_c, the concentrations of the products (which appear in the numerator of the mass action expression) must be very small.

The same generalization applies for both K_c and K_P.

$$K_c = \frac{[HCl]^2}{[H_2][Cl_2]} = 4.4 \times 10^{32}$$

$$K_c = \frac{[H_2]^2[O_2]}{[H_2O]^2} = 1.1 \times 10^{-81}$$

14.3 THERMODYNAMICS AND CHEMICAL EQUILIBRIUM

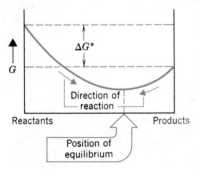

In Section 12.11 we saw, qualitatively, that there is a relationship between ΔG^0 for a reaction and the position of equilibrium. In addition, the direction in which a reaction proceeds toward equilibrium is determined by where the system lies with respect to the free-energy minimum. The reaction proceeds spontaneously only in a direction that gives rise to a decrease in free energy—that is, when ΔG is negative.

All of this is summed up quantitatively by the equation (which we will not attempt to justify)

$$\Delta G = \Delta G^0 + RT \ln Q \qquad [14.3]$$

or, in terms of common logs,

$$\Delta G = \Delta G^0 + 2.303\, RT \log Q$$

The symbol Q represents the reaction quotient for the reaction. For gases, Q is obtained from partial pressures; for reactions in solution, molar concentrations are used.[1]

Equation 14.3 tells us how ΔG varies with temperature and the relative proportions of reactants and products. For example, for the reaction,

$$2NO_2(g) \rightleftharpoons N_2O_4(g)$$

---

[1] Actually, to make Equation 14.3 fit exactly, "effective pressures" or "effective concentrations" must be used in Q. These are called **activities**. Fortunately, at low pressures in gaseous reactions, and low concentrations in solutions, the use of actual pressures and concentrations leads to only small errors.

Equation 14.3 would take the form

$$\Delta G = \Delta G^0 + RT \ln \left(\frac{p_{N_2O_4}}{p_{NO_2}^2} \right) \qquad [14.4]$$

At equilibrium the products and reactants have the same total free energy and $\Delta G = 0$ (Section 12.11), so Equation 14.4 becomes

$$0 = \Delta G^0 + RT \ln \left(\frac{p_{N_2O_4}}{p_{NO_2}^2} \right)$$

or

$$\Delta G^0 = -RT \ln \left(\frac{p_{N_2O_4}}{p_{NO_2}^2} \right)$$

At equilibrium for this reaction,

$$\frac{p_{N_2O_4}}{p_{NO_2}^2} = K_P$$

Therefore,

$$\Delta G^0 = -RT \ln K_P \qquad [14.5]$$

or

$$\Delta G^0 = -2.303 \, RT \log K_P$$

Equation 14.5, derived here for this specific example, applies to all reactions involving gases. For reactions in solution,

$$\Delta G^0 = -RT \ln K_c \qquad [14.6]$$

or

$$\Delta G^0 = -2.303 \, RT \log K_c$$

We now have a quantitative relationship between ΔG^0 and the equilibrium constant. The K computed using Equations 14.5 or 14.6 is sometimes called the **thermodynamic equilibrium constant.**

EXAMPLE 14.1   COMPUTING A THERMODYNAMIC
                   EQUILIBRIUM CONSTANT

PROBLEM   What is the thermodynamic equilibrium constant for the reaction,

$$2SO_2(g) + O_2(g) \rightleftharpoons 2SO_3(g)$$

at 25°C

SOLUTION   From the data in Table 12.5 we can obtain the standard free energies of formation of SO_3 and SO_2

$$\Delta G^0_{f \, SO_3} = -370 \frac{kJ}{mol}$$

$$\Delta G^0_{f \, SO_2} = -300 \frac{kJ}{mol}$$

By definition, $\Delta G^0_{f \, O_2} = 0.0$ kJ mol^{-1}.

Using these data, we can compute ΔG^0 for the reaction

$$\Delta G^0 = 2 \text{ mol} \times \left(-370 \frac{kJ}{\text{mol}}\right) - 2 \text{ mol} \times \left(-300 \frac{kJ}{\text{mol}}\right)$$

$$= -140 \text{ kJ}$$

Next, we solve Equation 14.5 for $\ln K_P$ (we are dealing with a gaseous reaction),

$$\ln K_P = \frac{-\Delta G^0}{RT}$$

Be sure ΔG^0 and R have exactly the same energy units.

We must express ΔG^0 in joules ($\Delta G^0 = -140\,000$ J), T in kelvins (298 K), and use $R = 8.314$ J mol^{-1} K^{-1}. Substituting numerical values gives

$$\ln K_p = \frac{-(-140\,000)}{(8.314)(298)}$$

$$= 56.5$$

Taking the antilogarithm gives

$K_P = e^{56.4} = 3 \times 10^{24}$

$$K_P = 3 \times 10^{24}$$

The magnitude of K for this reaction tells us that the position of equilibrium in the system should lie far in the direction of SO_3, and that at room temperature SO_2 should react almost completely with oxygen to form SO_3. This reaction is extremely slow at room temperature, however, but with a catalyst it becomes an important step in the industrial preparation of H_2SO_4. As mentioned in the last chapter, the same reaction takes place in the exhaust of an automobile equipped with a catalytic converter, but in this case the H_2SO_4 produced presents a health problem.

Atmospheric SO_2 produced by combustion of high-sulfur fuels is also slowly oxidized to SO_3.

Thermodynamic data can also be used to calculate equilibrium constants at temperatures other than 25°C. This is shown in Example 14.2.

EXAMPLE 14.2   CALCULATING K_P AT A TEMPERATURE OTHER THAN 25°C FROM THERMODYNAMIC DATA

PROBLEM   For the reaction, $2NO_2(g) \rightleftharpoons N_2O_4(g)$, $\Delta H^0_{298\text{ K}} = -56.9$ kJ and $\Delta S^0_{298\text{ K}} = -175$ J K^{-1}. Calculate K_P at 100°C

SOLUTION   To calculate K_P from Equation 14.5, we must have a numerical value for the equivalent of ΔG^0, but at 100°C instead of 25°C. Let's call this quantity $\Delta G'$. In Chapter 12, we learned that

$$\Delta G^0 = \Delta H^0 - (298 \text{ K}) \Delta S^0$$

And for some temperature other than 298 K, we can write

$$\Delta G' = \Delta H' - T \Delta S'$$

It happens that ΔH and ΔS vary only slightly with temperature, so for the purposes of many calculations we can assume temperature independence and write $\Delta H^0 = \Delta H'$ and $\Delta S^0 = \Delta S'$. Thus

$$\Delta G' = \Delta H^0 - T \Delta S^0$$

Therefore, at 100°C (373 K),

$$\Delta G'_{373} = -56\,900 \text{ J} - (373 \text{ K})(-175 \text{ J K}^{-1})$$

$$= +8380 \text{ J (rounded to three significant figures)}$$

Solving Equation 14.5 for ln K_P gives us

$$\ln K_P = \frac{-\Delta G'}{RT}$$

Once again we use $R = 8.314 \text{ J mol}^{-1} \text{ K}^{-1}$. Substituting numerical values, using $T = 373$ K, gives

$$\ln K_P = \frac{+8380}{(8.314)(373)}$$

$$= -2.70$$

$K_P = e^{-2.70} = 6.7 \times 10^{-2}$ Taking the antilogarithm yields

$$K_P = 6.7 \times 10^{-2}$$

The measurement of equilibrium constants also provides a very convenient method for obtaining thermodynamic data. This is illustrated in the next example.

EXAMPLE 14.3 **USING K_P TO OBTAIN THERMODYNAMIC DATA**

PROBLEM At 25°C it was found that $K_P = 7.13$ for the reaction

$$2NO_2(g) \rightleftharpoons N_2O_4(g)$$

What is ΔG^0 for this reaction in kilojoules?

SOLUTION We can calculate ΔG^0 by substituting appropriate values into Equation 14.5,

$$\Delta G^0 = -RT \ln K_P$$

Since we wish ΔG^0 in kilojoules, we must use $R = 8.314 \text{ J mol}^{-1} \text{ K}^{-1}$. As usual, T is the absolute temperature ($T = 298$ K in this example). Substituting numerical values, we get

$$\Delta G^0 = -(8.314)(298) \ln (7.13)$$

$$= -4870 \text{ J (to three significant figures)}$$

The value of ΔG^0 is in joules because R is in joules. To convert to kilojoules, simply divide by 1000:

$$\Delta G^0 = -4.87 \text{ kJ}$$

14.4 THE RELATIONSHIP BETWEEN K_P AND K_c

It was stated earlier that for reactions involving gases, K_P and K_c are not necessarily equal. For the general equation,

$$aA + bB \rightleftharpoons eE + fF$$

$$K_P = \frac{p_E{}^e p_F{}^f}{p_A{}^a p_B{}^b}$$

and

$$K_c = \frac{[E]^e[F]^f}{[A]^a[B]^b}$$

Concentration, you recall, has the units $mol\ dm^{-3}$, or n/V. Assuming ideal gas behavior, we can use the ideal gas law,

$$PV = nRT$$

to obtain the concentration of a gas, X, in a mixture as

$$[X] = \frac{n_X}{V} = \frac{p_X}{RT}$$

where p_X is its partial pressure. From this it follows that

$$p_X = [X]RT$$

Substituting this relationship into the expression for K_P, we have

$$K_P = \frac{p_E{}^e p_F{}^f}{p_A{}^a p_B{}^b} = \frac{[E]^e(RT)^e[F]^f(RT)^f}{[A]^a(RT)^a[B]^b(RT)^b}$$

This can be rearranged to give

$$K_P = \frac{[E]^e[F]^f}{[A]^a[B]^b}(RT)^{(e+f)-(a+b)}$$

or

When $\Delta n_g = 0$, $K_P = K_c$.

$$K_P = K_c(RT)^{\Delta n_g} \qquad [14.7]$$

where Δn_g *is the change in the number of moles of **gas** when going from reactants to products.*

Δn_g = (number of moles of gaseous products) − (number of moles of gaseous reactants)

Thus, K_P and K_c are related in a very simple fashion for reactions between ideal gases, a relationship that also holds adequately for many real gases.

EXAMPLE 14.4 CONVERTING FROM K_P TO K_c

PROBLEM   In Example 14.1 we determined the value of K_P for the reaction of SO_2 with O_2 to produce SO_3. What is K_c for this equilibrium at 25°C?

SOLUTION   Solving Equation 14.7 for K_c, we obtain

$$K_c = \frac{K_P}{(RT)^{\Delta n_g}} = K_P(RT)^{-\Delta n_g}$$

The chemical equation we are dealing with is

$$2SO_2(g) + O_2(g) \rightleftharpoons 2SO_3(g)$$

To calculate Δn_g, we interpret the coefficients as moles—there are two moles of gaseous products and three moles of gaseous reactants. Therefore,

$$\Delta n_g = (2 - 3)$$

$$= -1$$

In Example 14.1, we found $K_P = 3 \times 10^{24}$. From the equilibrium expression

$$\frac{p_{SO_3}^2}{p_{SO_2}^2 p_{O_2}} = K_P$$

We don't normally include units with K, but in this case it helps us choose the correct value for R.

Note that a K_p unit based on atm^{-1} is appropriate, since the ΔG^0 value from which it was calculated in Example 14.1 assumed that all gases were in their standard states ($P = 1$ atm). Since the units of R (8.314 dm^3 kPa mol^{-1} K^{-1}) require pressure units in kilopascals:

$$K_p = (3 \times 10^{24} \text{ atm}^{-1}) \times \left(\frac{1 \text{ atm}}{101.325 \text{ kPa}} \right) = 3 \times 10^{22} \text{ kPa}^{-1}$$

Now the value of K_c can be directly calculated

$$K_c = (3 \times 10^{22} \text{ kPa}^{-1})[(8.314 \text{ dm}^3 \text{ kPa mol}^{-1} \text{ K}^{-1})(298 \text{ K})]^{-(-1)}$$

Therefore,

$$K_c = 7 \times 10^{25} \text{ dm}^3 \text{ mol}^{-1}$$

14.5   HETEROGENEOUS EQUILIBRIA

Up to now our discussion has focused on homogeneous reactions—reactions in which all the reactants and products are in the same phase. Heterogeneous reactions, of which there are many examples, also eventually arrive at a state of equilibrium. A typical reaction that we might consider is the decomposition of solid $NaHCO_3$ to produce solid Na_2CO_3, gaseous CO_2, and gaseous H_2O.[2]

We learned that this reaction makes $NaHCO_3$ a good fire extinguisher.

$$2NaHCO_3(s) \rightleftharpoons Na_2CO_3(s) + CO_2(g) + H_2O(g)$$

We can write the equilibrium law for this reaction as

$$\frac{[Na_2CO_3(s)][CO_2(g)][H_2O(g)]}{[NaHCO_3(s)]^2} = K_c' \qquad [14.8]$$

For reasons that will be apparent shortly, we have temporarily indicated the equilibrium constant as K_c'.

In this reaction we have an equilibrium between the two gases, CO_2 and H_2O, and the two pure solid phases, $NaHCO_3$ and Na_2CO_3. We know by now that a pure solid substance such as $NaHCO_3$ is characterized by a density that is the same for all samples of $NaHCO_3$, regardless of their size. In addition, this density is unaffected by the nature of any chemical reaction that the substance is undergoing. This means that even during a chemical reaction the amount of $NaHCO_3$ per unit volume of the pure solid is always the same. In other words, the concentration of $NaHCO_3$ in pure solid $NaHCO_3$ is a constant. We cannot alter the number of moles per dm^3 of $NaHCO_3$ in the pure solid, nor can we change the concentration of Na_2CO_3 in pure solid Na_2CO_3. Therefore, the concentrations of these two substances in the equilibrium expression take on constant values that can be incorporated into the equilibrium constant. Rearranging Equation 14.8 to place all the constants on the same side gives

$$[CO_2(g)][H_2O(g)] = K_c' \underbrace{\frac{[NaHCO_3(s)]^2}{[Na_2CO_3(s)]}}_{\boxed{K_c}}$$

[2] Na_2CO_3 is produced commercially by this reaction. It is one of the most industrially important chemicals, ranking tenth in total production (about 8×10^9 kg produced annually), and is used in the manufacture of glass and many other important products.

or

$$[CO_2(g)][H_2O(g)] = K_c \qquad [14.9]$$

Thus we find that for heterogeneous reactions, *the equilibrium constant expression does not include the concentrations of pure solids*. Similarly, in reactions in which a reactant or product occurs as a pure liquid phase, the concentration of that substance in the pure liquid is also constant. As a result, *the concentrations of pure liquid phases also do not appear in an equilibrium constant expression*.[3] These simplifications apply *only* when we are dealing with *pure* condensed phases. When substances occur in liquid or solid solutions, their concentrations are variable and their concentration terms in the mass action expression therefore cannot be incorporated into K.

If we wish to work with K_P rather than K_c, we need to take into account only the substances present in the gas phase. For the decomposition of $NaHCO_3$, therefore, we have

$$K_P = p_{CO_2(g)} p_{H_2O(g)}$$

As noted in the last section, if we know K_c, we can evaluate K_P as

$$K_P = K_c(RT)^{\Delta n_g} \qquad [14.10]$$

where, for this reaction, $\Delta n_g = +2$.

EXAMPLE 14.5   WRITING EQUILIBRIUM EXPRESSIONS FOR HETEROGENEOUS REACTIONS

PROBLEM   What are the values of K_P and K_c for the reaction

$$H_2O(l) \rightleftharpoons H_2O(g)$$

at 25°C, given that the vapor pressure of water at 25°C equals 3.17 kPa?

SOLUTION   Since liquid water is a pure liquid phase, we can write

$$K_P = p_{H_2O(g)}$$

and

$$K_c = [H_2O(g)]$$

(a) From the relationships above,

$$K_P = p_{H_2O} = 3.17 \text{ kPa}$$

Note that this equilibrium expression states that the partial pressure of water must be a constant when the liquid and vapor are in equilibrium.

(b) We can evaluate K_c as

$$K_c = K_P(RT)^{-\Delta n_g}$$

---

[3] Thermodynamics handles this question in a slightly more elegant way by defining the activity of a pure solid or liquid as numerically equal to one. This simply makes terms involving pure solids or liquids disappear from the mass action expression. For instance, substituting values of 1 for $[NaHCO_3(s)]$ and $[Na_2CO_3(s)]$ in Equation 14.8 gives Equation 14.9 directly.

For this reaction, $\Delta n_g = 1$; therefore,

$$K_c = K_P(RT)^{-1} = \frac{K_P}{RT}$$

$$= \frac{3.17 \text{ kPa}}{(8.314 \text{ dm}^3 \text{ kPa mol}^{-1} \text{ K}^{-1})(298 \text{ K})}$$

or

$$K_c = 1.28 \times 10^{-3} \text{ mol dm}^{-3}$$

14.6 LE CHÂTELIER'S PRINCIPLE AND CHEMICAL EQUILIBRIA

The equilibrium expression, in the form of either K_P or K_c, can be used to perform numerical computations of various kinds dealing with equilibrium systems. This is discussed in the next section. Often, however, it is desirable simply to be able to predict how some disturbance imposed on a system from outside will influence the position of equilibrium. For instance, we may wish to predict, in a qualitative way, the conditions that favor the greatest production of products. Should we run our reaction at high or low temperature? Should the pressure on the system be high or low? These are questions we would like to answer quickly without having to perform tedious computations. We can do this by applying Le Châtelier's principle, which was introduced in Chapter 10.

Le Châtelier's principle: If a system in dynamic equilibrium is subjected to a disturbance that upsets the equilibrium, the system changes in such a way as to reduce the disturbance and, if possible, return to equilibrium.

Let's examine some of the ways that a chemical equilibrium can be upset and how the system is able to respond by changing the position of equilibrium (i.e., the relative proportions of reactants and products).

Changes in the concentration of a reactant or product

In a system such as

$$H_2(g) + I_2(g) \rightleftharpoons 2HI(g)$$

any change in the concentration of a reactant or product will cause the system to no longer be at equilibrium. As a result, a chemical change will occur that will return the system to equilibrium. From Le Châtelier's principle we know that if a system at equilibrium is disturbed, it will attempt to undergo some change to diminish the effect of the disturbance. For example, the addition of H_2 to an equilibrium mixture of H_2, I_2, and HI upsets the equilibrium, and the system responds by using up part of the additional H_2 by reaction with I_2 to produce more HI. When equilibrium has finally been reestablished, there will be a greater concentration of HI than before, and we say that for this reaction the position of equilibrium has been shifted to the right. This is illustrated in Figure 14.2. Notice that after equilibrium has been reestablished there continues to be more H_2 present than in the original reaction mixture. The system is never able to completely overcome the effect of a change in concentration. The final position of equilibrium differs from the original.

We can arrive at this same conclusion by considering the effect of added H_2

Figure 14.2
Addition of H$_2$ to the equilibrium, H$_2$ + I$_2$ ⇌ 2HI, increases the amount of HI and decreases the amount of I$_2$.

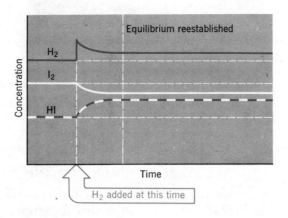

In a chemical equation, the position of equilibrium shifts away from a substance that's been added or toward a substance that's removed.

on the value of the mass action expression. For this reaction at equilibrium we have

$$\frac{[HI]^2}{[H_2][I_2]} = K_c$$

If H$_2$ is suddenly added to the system, the value of the denominator of the mass action expression becomes larger, and the entire fraction therefore becomes smaller than the equilibrium constant. The reaction that occurs to return the system to equilibrium must increase the value of the mass action expression until it once again is equal to K_c. For this to happen, the numerator must become larger and the denominator smaller. In other words, more HI will be formed at the expense of H$_2$ and I$_2$, and again we conclude that the addition of H$_2$ shifts the position of equilibrium in this reaction to the right.

By applying Le Châtelier's principle we can also predict the effect that removing a reactant or product will have on a system at equilibrium. For instance, if H$_2$ is somehow removed from the reaction vessel, the system will adjust by having some HI decompose in an effort to replenish the lost reactant. Thus, the position of equilibrium is shifted to the left when H$_2$ is removed.

Now we can use Le Châtelier's principle to predict what must be done to drive a reaction far toward completion—we can either add a large excess of one of the reactants or remove the products as they are formed. Recall that removing the products serves as the driving force for ionic reactions (Chapter 8) in which a product is either a precipitate, a gas, or a weak electrolyte. The creation of these products removes ions from solution and therefore forces the reaction to proceed toward completion.

Closer to home, we can see an application of Le Châtelier's principle in understanding the origin of tooth decay. Tooth enamel consists of an insoluble substance called hydroxyapatite, Ca$_5$(PO$_4$)$_3$OH. The dissolving of this substance from the teeth is called demineralization, and its formation is called remineralization. In the mouth there is an equilibrium

$$Ca_5(PO_4)_3OH(s) \underset{remineralization}{\overset{demineralization}{\rightleftharpoons}} 5Ca^{2+}(aq) + 3PO_4^{3-}(aq) + OH^-(aq)$$

which is established even with healthy teeth. However, when sugar is absorbed on teeth and ferments, H$^+$ is produced that upsets the equilibrium by combining with OH$^-$ to form water and with PO$_4^{3-}$ to form HPO$_4^{2-}$. Removing OH$^-$ and PO$_4^{3-}$ causes more of the Ca$_5$(PO$_4$)$_3$OH to dissolve, resulting in tooth decay. Fluoride helps prevent tooth decay by replacing OH$^-$ in hydroxyapatite. The resulting Ca$_5$(PO$_4$)$_3$F is very resistant to acid attack.

The effect of temperature on equilibrium

Up to now we have been careful to imply that the equilibrium constant for a reaction has a fixed numerical value only as long as the temperature remains constant. This is because temperature, as well as the concentrations of reactants and products, affects the position of equilibrium. However, the temperature, unlike the concentrations of reactants and products, affects the value of the equilibrium constant itself.

Let's look at the exothermic reaction of H_2 and N_2 to form NH_3. The equation for the formation of ammonia can be written as

$$3H_2(g) + N_2(g) \rightleftharpoons 2NH_3(g) + 92.0 \text{ kJ}$$

In a sense, we think of heat as a reactant or product in the equation.

where the heat of reaction is indicated as a product. If we have a system of these gases in equilibrium and wish to raise its temperature, we do so by adding heat to it from the surroundings. Le Châtelier's principle tells us that when we add this heat, the system will attempt to undergo a change that tends to use some of it up. Since the production of NH_3 is exothermic, its decomposition is endothermic. Therefore, raising the temperature will cause the position of equilibrium to shift to the left; it drives the reaction to a new position of equilibrium in which there is more N_2 and H_2 and less NH_3. *In general, an increase in temperature causes the position of equilibrium of an exothermic reaction to be shifted to the left, while that of an endothermic reaction is shifted to the right.*

We have just seen that an increase in temperature leads to a decrease in the concentration of NH_3 and an increase in the concentrations of both N_2 and H_2. This means that *at equilibrium* at the higher temperature the value of the mass action expression

$$\frac{[NH_3]^2}{[H_2]^3[N_2]}$$

will have decreased. Thus we find that for this exothermic reaction, K decreases with rising temperature. By the same token, for a reaction that is endothermic in the forward direction, K increases with increasing temperature.

Effect of pressure and volume changes on equilibrium

When the external pressure on a system is increased, it causes the volume of the system to decrease.

At constant temperature a change in the volume of a system also causes a change in pressure. We quite logically expect, therefore, that an increase in the external pressure on a system should favor any change that leads to a smaller volume (recall Boyle's law). We would not expect pressure changes to have any marked effect on the position of equilibrium in reactions where all the reactants and products are either solids or liquids, because these phases are virtually incompressible. However, pressure changes can have a very dramatic effect on equilibria that involve reactions in which gases are consumed or produced.

Let us again choose as an example the reaction for the formation of NH_3. If we have this system at equilibrium and suddenly decrease the volume of the container, we know that the pressure will go up. According to Le Châtelier's principle, we expect that a change in the system should occur that will reduce the pressure. How can this be brought about?

We know that the pressure of a gas is caused by collisions of the molecules with the walls of the container, and at a given temperature the greater the number of molecules per cubic centimetre, the greater the pressure. In the equilibrium

$$N_2(g) + 3H_2(g) \rightleftharpoons 2NH_3(g)$$

the number of molecules of gas decreases with the reaction proceeds from left to right—four reactant molecules produce two product molecules. This means that the pressure exerted by the gases in the system can be decreased if the position of equilibrium shifts to the right.[4] Thus *decreasing the volume of a mixture of gases that are in chemical equilibrium shifts the equilibrium in the direction of the fewest number of molecules of gas.*

Finally, note that when there are the same number of molecules of gaseous reactants and products on both sides of the equation, as in the reaction between H_2 and I_2,

$$H_2(g) + I_2(g) \rightleftharpoons 2HI(g)$$

pressure changes brought about by volume changes will not influence the amounts of the various substances present in the reaction mixture at equilibrium. This is because there is no way for the system to counteract pressure changes placed on it.

Changing the external pressure on a chemical system containing only liquids and solids has virtually no effect on the position of equilibrium.

Addition of an inert gas

If an inert (nonreacting) gas is introduced into a reaction vessel containing other gases at equilibrium, it will cause an increase in the total pressure within the container. This kind of pressure increase, however, will not affect the position of equilibrium because it will not alter the partial pressures or the concentrations of any of the substances already present.

Effect of a catalyst on the position of equilibrium

The catalyst affects the path of the reaction, but it doesn't alter the initial and final states of the system. This means that a catalyst can't change ΔG^0 for a reaction, and it can't change the equilibrium constant.

In Chapter 13 we saw that a catalyst affects a chemical reaction by lowering the activation energy barrier that must be overcome in order for the reaction to proceed. A catalyst affects the rate of a chemical change. It does not, however, affect the heat of reaction, and it is the heat of reaction, ΔH^0, along with the entropy change, ΔS^0, that determine ΔG^0, which in turn fixes the position of equilibrium at any given temperature. A catalyst merely speeds the approach to the position of equilibrium that is determined by ΔG^0.

14.7 EQUILIBRIUM CALCULATIONS

This section is intended to illustrate the types of computations that one might perform either to evaluate an equilibrium constant from measured concentrations or to use the equilibrium constant to calculate the concentrations of the reactants and products in a particular equilibrium mixture. First let's see how we might evaluate K in a typical experiment.

EXAMPLE 14.6	CALCULATING K_C FROM CONCENTRATION DATA AT EQUILIBRIUM
PROBLEM	The brown gas NO_2, an air pollutant, and the colorless gas N_2O_4 exist in equilibrium as indicated by the equation,

$$2NO_2 \rightleftharpoons N_2O_4$$

---

[4] By applying Le Châtelier's principle, we find that the production of ammonia from H_2 and N_2 is favored by high pressures and low temperatures. At low temperatures, however, the reaction is very slow; therefore in the industrial preparation of NH_3, pressures of 10 to 100 MPa and temperatures from 400 to 550°C are employed. Even though there is less NH_3 produced at equilibrium at these high temperatures, the speed of reaction is boosted to the point where the production of NH_3 is economically worthwhile.

In an experiment, 0.625 mol of N_2O_4 was introduced into a 5.00-dm^3 vessel and permitted to decompose until it reached equilibrium with NO_2. At equilibrium the concentration of N_2O_4 was 0.0750 M. What is K_c for this reaction?

SOLUTION The equilibrium constant expression for this reaction is

$$\frac{[N_2O_4]}{[NO_2]^2} = K_c$$

Remember, the equilibrium condition is only satisfied by equilibrium concentrations.

In order to calculate K_c, we must know the *equilibrium* concentrations of N_2O_4 and NO_2. In working out equilibrium problems, we will generally find it useful to set up a table like that below in order to establish quantities that correspond to equilibrium concentrations. The entries in the table are obtained by reasoning from the data provided in the problem. Remember that when using K_c, molar concentrations (that is, mol dm^{-3}) must be used. In this example the concentration of N_2O_4 initially was 0.625 mol/5.00 dm^3 = 0.125 M; the initial concentration of NO_2 was zero. These are the values in the first column of the concentration table.

In the statement of the problem we are told that the equilibrium concentration of N_2O_4 is 0.0750 M. The difference between the initial concentration (0.125 M) and the equilibrium concentration (0.0750 M) is the number of moles per dm^3 of N_2O_4 that decomposed.

$$0.125\ M - 0.0750\ M = 0.050\ M$$

This value is entered in the change column of the table with a minus sign to indicate that the N_2O_4 concentration has decreased.

The changes in the concentrations of N_2O_4 and NO_2—the values in the change column of the table—are related by the stoichiometry of the reaction. For each mole of N_2O_4 that decomposes, two moles of NO_2 are formed. Therefore, when the N_2O_4 concentration decreased by 0.050 M, the NO_2 concentration increased by $2 \times 0.050\ M = 0.10\ M$. To indicate the increase, the value is written with a plus sign.

When we construct the concentration table for an equilibrium problem in this way—using a minus sign to indicate a decrease in concentration and a plus sign to indicate an increase—the equilibrium concentration is obtained by adding the value in the change column algebraically to the initial concentration.

$$\text{For } N_2O_4 \qquad 0.125\ M - 0.050\ M = 0.075\ M$$
$$NO_2 \qquad 0.00\ M + 0.10\ M = 0.10\ M$$

	Initial Concentrations	Change	Equilibrium Concentrations
N_2O_4	0.125 M	−0.050 M	0.075 M
NO_2	0.00 M	+0.10 M	0.10 M

Now we can substitute the equilibrium concentrations into the mass action expression to calculate K_c.

$$\frac{(0.075)}{(0.10)^2} = K_c$$

and, finally,

$$K_c = 7.5$$

Knowledge of the equilibrium constant for a reaction allows us to calculate the concentrations or partial pressures of the substances present in a reaction mixture at equilibrium. The ease with which these computations can be carried out depends on the complexity of the mass action expression, the concentrations of the various species in the reaction mixture, and the magnitude of the equilibrium constant. We will look only at some of the more simple examples of problems of this type. The following sample problems, however, illustrate the type of reasoning employed in these computations, as well as some of the concepts that have been presented up to this point.

EXAMPLE 14.7 **USING K_P TO CALCULATE AN EQUILIBRIUM PRESSURE**

PROBLEM At 25°C, $K_P = 7.04 \times 10^{-2}$ kPa^{-1} for the reaction

$$2NO_2(g) \rightleftharpoons N_2O_4(g)$$

At equilibrium the partial pressure of NO$_2$ in a container is 15 kPa. What is the partial pressure of N$_2$O$_4$ in the mixture?

SOLUTION The first step in the solution of any equilibrium problem is to write down the equilibrium expression. For K_P, we have

$$K_P = \frac{p_{N_2O_4}}{p_{NO_2}^2} = 7.04 \times 10^{-2} \text{ kPa}^{-1}$$

We are given the equilibrium partial pressure of NO$_2$ (p_{NO_2} = 15 kPa). There is only one unknown quantity, $p_{N_2O_4}$. Substituting, we get

$$\frac{p_{N_2O_4}}{(15 \text{ kPa})^2} = 7.04 \times 10^{-2} \text{ kPa}^{-1}$$

$$p_{N_2O_4} = 7.04 \times 10^{-2} \text{ kPa}^{-1}(15 \text{ kPa})^2$$

$$= 16 \text{ kPa}$$

The partial pressure of N$_2$O$_4$ at equilibrium is 16 kPa.

EXAMPLE 14.8 **USING K_c TO CALCULATE AN EQUILIBRIUM CONCENTRATION**

PROBLEM At a temperature of 500°C, the equilibrium constant, K_c, for the nitrogen fixation reaction for the production of ammonia,

$$3H_2(g) + N_2(g) \rightleftharpoons 2NH_3(g)$$

has a value of 6.0×10^{-2}. If, in a particular reaction vessel at this temperature, there are 0.250 mol dm^{-3} of H$_2$ and 0.0500 mol dm^{-3} of NH$_3$ present at equilibrium, what is the concentration of N$_2$?

SOLUTION Let's first write the equilibrium constant expression. For this reaction we have

$$K_c = \frac{[NH_3]^2}{[H_2]^3[N_2]} = 6.0 \times 10^{-2}$$

We wish to calculate the concentration of N$_2$. This can be accomplished if we know the values of the equilibrium concentrations of both NH_3 and H_2 and, in this problem, these are given to us.

In this problem we don't need the concentration table because we're given K_c and all but one equilibrium concentration.

$$\left.\begin{array}{l} [NH_3] = 0.0500\ M \\ \\ [H_2] = 0.250\ M \end{array}\right\} \text{at equilibrium}$$

Substituting these numerical values into the mass action expression gives us

$$\frac{(0.0500)^2}{(0.250)^3[N_2]} = 6.0 \times 10^{-2}$$

If we solve for $[N_2]$, we get

$$[N_2] = \frac{(0.0500)^2}{(0.250)^3(6.0 \times 10^{-2})}$$

$$= 2.7\ M$$

The equilibrium concentration of N_2 is thus 2.7 mol dm^{-3}.

EXAMPLE 14.9 CALCULATING EQUILIBRIUM CONCENTRATIONS FROM K_c AND INITIAL CONCENTRATIONS

PROBLEM At 440°C the equilibrium constant for the reaction,

$$H_2(g) + I_2(g) \rightleftharpoons 2HI(g)$$

is 49.5. If 0.200 mol of H_2 and 0.200 mol of I_2 are placed into a 10.0-dm^3 vessel and permitted to react at this temperature, what will be the concentration of each substance at equilibrium?

SOLUTION Our equilibrium expression is

$$\frac{[HI]^2}{[H_2][I_2]} = 49.5$$

In this example we are given the *initial* concentrations of the reactants and products. These are

$$[H_2] = \frac{0.200\ \text{mol}}{10.0\ \text{dm}^3} = 0.0200\ M$$

$$[I_2] = \frac{0.200\ \text{mol}}{10.0\ \text{dm}^3} = 0.0200\ M$$

$$[HI] = 0.0\ M$$

If any HI had been present initially, its concentration would have been given in the statement of the problem.

Since no HI is present initially, we know that it will be formed from the violet-colored mixture of the H_2 and I_2 (the color is due to the I_2). Let's approach the problem by allowing x to be equal to the number of moles per dm^3 of H_2 that react. From the stoichiometry of the reaction we realize that this x mol dm^{-3} of H_2 will react with x mol dm^{-3} of I_2 to produce $2x$ mol dm^{-3} of HI. Thus the concentrations of H_2 and I_2 both decrease by x and the concentration of HI increases by $2x$. The equilibrium concentrations are obtained by applying these changes to the initial concentrations.

The coefficients of x *must* be in the same ratio as the coefficients in the balanced chemical equation.

	Initial Concentrations	Change	Equilibrium Concentrations
H_2	0.0200 M	$-x$	$(0.0200 - x)\ M$
I_2	0.0200 M	$-x$	$(0.0200 - x)\ M$
HI	0.0 M	$+2x$	$0.0 + 2x = 2x\ M$

Substituting the equilibrium quantities into the mass action expression gives

$$\frac{(2x)^2}{(0.0200 - x)(0.0200 - x)} = 49.5$$

or

$$\frac{(2x)^2}{(0.0200 - x)^2} = 49.5$$

In this case, we can take the square root of both sides of the equation to obtain

$$\frac{2x}{0.0200 - x} = 7.04$$

Solving for x, we obtain

$$2x = 7.04(0.0200 - x) = 0.141 - 7.04x$$

$$2x + 7.04x = 0.141$$

$$9.04x = 0.141$$

$$x = 0.0156$$

Finally, the equilibrium concentrations are

$$[H_2] = 0.0200 - 0.0156 = 0.0044\ M$$

$$[I_2] = 0.0200 - 0.0156 = 0.0044\ M$$

$$[HI] = 2(0.0156) = 0.0312\ M$$

In this last problem we employed some relatively simple algebra to help us arrive at the solution. Let us look at another example of this type.

EXAMPLE 14.10 CALCULATING EQUILIBRIUM CONCENTRATIONS FROM K_c AND INITIAL CONCENTRATIONS

PROBLEM   A 10.0-dm³ vessel is filled with 0.40 mol of HI at 440°C. What will be the concentration of H_2, I_2, and HI at equilibrium?

SOLUTION   In this problem we are concerned about the same equilibrium as in the previous example. The chemical equation is

$$H_2(g) + I_2(g) \rightleftharpoons 2HI(g)$$

and the equilibrium expression is

$$\frac{[HI]^2}{[H_2][I_2]} = 49.5$$

Initially, there is no H_2 or I_2 in the container, so their concentrations are given values of zero in the concentration table. The concentration of HI is 0.40 mol/10.0 dm³ = 0.040 M.

Next we need quantities in the change column. These are our unknowns, and we will use the symbol x once again. Since the magnitudes of the changes must be in the same ratio as the coefficients of the balanced equation, we can let the coefficients of x be equal to the coefficients in the chemical equation. Thus, the coefficients of x for the changes in the concentrations of H_2 and I_2 are each 1, and the coefficient of x for the change in the HI concentration is 2.

Next we have to decide whether the changes will be positive or negative. Since the initial concentrations of H_2 and I_2 are zero, they must increase. (They can't decrease if they are already zero.) Their changes are therefore positive. This means that the change for the HI must be negative, because if the H_2 and I_2 increase, they must be formed from the HI.

If the initial concentration of a reactant or product is zero, the change *must* be positive. The equilibrium concentration can't be negative.

	Initial Concentrations	Change	Equilibrium Concentrations
H_2	0.0 M	$+x$	$0.0 + x = x$ M
I_2	0.0 M	$+x$	$0.0 + x = x$ M
HI	0.040 M	$-2x$	$(0.040 - 2x)$ M

Substituting equilibrium quantities into the mass action expression gives us

$$\frac{(0.040 - 2x)^2}{(x)(x)} = 49.5$$

or

$$\frac{(0.040 - 2x)^2}{(x)^2} = 49.5$$

Taking the square root of both sides of the equation, we have

$$\frac{0.040 - 2x}{x} = 7.04$$

Solving for x, we get

$$0.040 - 2x = (x)(7.04)$$

$$x = 0.0044$$

We now calculate the equilibrium concentrations to be

$$[H_2] = x = 0.0044 \ M$$

$$[I_2] = x = 0.0044 \ M$$

$$[HI] = 0.040 - 2x = 0.031 \ M$$

Observe that we have obtained essentially the same answers in both Examples 14.9 and 14.10. If all the H_2 and I_2 in Example 14.9 had completely reacted, it would have produced 0.40 mol of HI—the same amount of HI that we began with in Example 14.10. We find, therefore, that the same equilibrium composition can be approached from either direction.

In the last two examples the solution of the algebra was simple because we were able to take the square root of both sides of the equation. You can't always expect the algebra to work out so easily, however, and in some cases it can really prove to be quite a challenge. Fortunately, though, when the equilibrium constant is either extremely large or extremely small, it is frequently possible to greatly reduce the difficulty of even rather complex algebra by making some simple approximations. Example 14.11 illustrates the kinds of simplifying approximations that we will find useful in the Chapter 15.

EXAMPLE 14.11 CALCULATING EQUILIBRIUM CONCENTRATIONS FROM K_c AND INITIAL CONCENTRATIONS WHEN K_c IS VERY SMALL

PROBLEM The equilibrium constant, K_c, for the decomposition of gaseous water at 500°C has a value of 6.0×10^{-28}. If 2.0 mol of H_2O is placed into a 5.0-dm^3 container, what will be the equilibrium concentrations of the three gases, H_2, O_2, and H_2O at 500°C?

SOLUTION The equation for the reaction is

$$2H_2O(g) \rightleftharpoons 2H_2(g) + O_2(g)$$

Therefore we can write

$$\frac{[H_2]^2[O_2]}{[H_2O]^2} = 6.0 \times 10^{-28}$$

The initial H_2O concentration is 2.0 mol/5.0 dm^3 = 0.40 M. As before, we let the coefficients of x be the same as the coefficients in the balanced chemical equation. Constructing our table, we get

	Initial Concentrations	Change	Equilibrium Concentrations
H_2O	0.40 M	$-2x$	$(0.40 - 2x)\ M$
H_2	0.0 M	$+2x$	$2x\ M$
O_2	0.0 M	$+x$	$x\ M$

Substituting equilibrium quantities into the mass action expression gives

$$\frac{(2x)^2(x)}{(0.40 - 2x)^2} = 6.0 \times 10^{-28}$$

Unless we can somehow simplify this equation, we have a real mess on our hands. Fortunately, in this case the problem can be made easy to solve.

Since K is very small, we know ahead of time that the reaction does not proceed very far toward completion. This means that very little H_2 and O_2 are formed, so x and $2x$ are going to be very small numbers. To simplify the algebra, we will make the assumption that $2x$ will be *much* smaller than 0.40, so that when $2x$ is subtracted from 0.40, the difference will still be very nearly 0.40 when rounded to the proper number of significant figures. If we neglect $2x$ when we compute the H_2O concentration, the equilibrium quantities become

We can only neglect a small x if it is *added to* or *subtracted from* some other value that's much larger.

$$[H_2O] = 0.40 - 2x \approx 0.40$$

$$[H_2] = 2x$$

$$[O_2] = x$$

Substituting these into the mass action expression gives

$$\frac{(2x)^2(x)}{(0.40)^2} = 6.0 \times 10^{-28}$$

from which we get

$$\frac{4x^3}{0.16} = 6.0 \times 10^{-28}$$

This is now simple to solve. First we solve for x^3,

$$x^3 = \frac{0.16}{4}(6.0 \times 10^{-28}) = 2.4 \times 10^{-29}$$

At this point x can be obtained by extracting the cube root. Many hand-held calculators can perform this operation directly or by raising 1.9×10^{-28} to the $\frac{1}{3}$ power.

$$\sqrt[3]{2.4 \times 10^{-29}} = (2.4 \times 10^{-29})^{1/3} = 2.9 \times 10^{-10}$$

0.40 − 0.000 000 000 58 = 0.399 999 999 42. When rounded, this gives 0.40.

We see that x is, in fact, very much smaller than 0.40, thus justifying our initial assumption. The final equilibrium concentrations are

$$[H_2] = 2x = 5.8 \times 10^{-10}\ M$$

$$[O_2] = x = 2.9 \times 10^{-10}\ M$$

$$[H_2O] = 0.40 - (5.8 \times 10^{-10}) = 0.40\ M$$

In working out a problem of this sort, look for any assumption that will make the algebra easier to handle. If the assumption you make is invalid, you will discover this when you check the assumption after obtaining a value for x. Sometimes no assumption of the kind we made above will be valid, and then

REVIEW QUESTIONS AND PROBLEMS

(Problems whose numbers are in blue have their answers in Appendix C at the back of the book; the more difficult problems are marked with asterisks.)

Equilibrium Expressions

14.1 What is meant by a *dynamic equilibrium?*

14.2 Write the mass action expression in terms of molar concentration for each of the following reactions:
(a) $N_2(g) + O_2(g) \rightleftharpoons 2NO(g)$
(b) $2NO(g) + O_2(g) \rightleftharpoons 2NO_2(g)$
(c) $2H_2(g) + S_2(g) \rightleftharpoons 2H_2S(g)$
(d) $2N_2O_5(g) \rightleftharpoons 4NO_2(g) + O_2(g)$
(e) $P_4O_{10}(g) + 6PCl_5(g) \rightleftharpoons 10POCl_3(g)$

14.3 Give the mass action expressions for the reactions in Question 14.2 in terms of partial pressures.

14.4 Write equilibrium expressions for K_P and K_c for each of the following reactions:
(a) $CO(g) + 2H_2(g) \rightleftharpoons CH_3OH(g)$
(b) $CO(g) + H_2O(g) \rightleftharpoons CO_2(g) + H_2(g)$
(c) $PCl_3(g) + Cl_2(g) \rightleftharpoons PCl_5(g)$
(d) $2NO_2(g) + 4H_2(g) \rightleftharpoons N_2(g) + 4H_2O(g)$
(e) $2H_2S(g) + 3O_2(g) \rightleftharpoons 2H_2O(g) + 2SO_2(g)$

14.5 Write equilibrium expressions for the reactions as written below:
(a) $H_2(g) + Cl_2(g) \rightleftharpoons 2HCl(g)$
(b) $\frac{1}{2}H_2(g) + \frac{1}{2}Cl_2(g) \rightleftharpoons HCl(g)$

How would the magnitude of the K for Reaction (a) compare with that for Reaction (b)?

14.6 Why do we always write the concentrations (or partial pressures) of the products in the numerator and those of the reactants in the denominator in the mass action expression?

14.7 Show that the following equilibrium data, obtained for the reaction

$$PCl_5(g) \rightleftharpoons PCl_3(g) + Cl_2(g)$$

demonstrate the constancy of the mass action expression for a system at equilibrium. What is K_c for this reaction?

Experiment	$[PCl_5]$	$[PCl_3]$	$[Cl_2]$
1	0.0023	0.23	0.055
2	0.010	0.15	0.37
3	0.085	0.99	0.47
4	1.00	3.66	1.50

Significance of K

14.8 What general information can be gathered by observing the magnitude of the equilibrium constant?

14.9 Arrange the following reactions in order of their increasing tendency to proceed toward completion:

(a) $4NH_3(g) + 3O_2(g) \rightleftharpoons 2N_2(g) + 6H_2O(g)$
$$K = 1 \times 10^{228}$$
(b) $N_2(g) + O_2(g) \rightleftharpoons 2NO(g)$ $K = 5 \times 10^{-31}$
(c) $2HF(g) \rightleftharpoons H_2(g) + F_2(g)$ $K = 1 \times 10^{-13}$
(d) $2NOCl(g) \rightleftharpoons 2NO(g) + Cl_2(g)$ $K = 4.7 \times 10^{-4}$

Thermodynamics and Equilibrium

14.10 What value would ΔG^0 have for a reaction if $K = 1$?

14.11 For reactions between gases, what kind of equilibrium constant is calculated from ΔG^0?

14.12 Using Equations 12.11 (p. 463) and 14.5, show that a straight line should be obtained if log K_P is plotted against $1/T$ (that is, log K_P along the vertical axis, $1/T$ along the horizontal axis). What does the slope of this line give? What does the value of log K_P at $1/T = 0$ (the y intercept of the line) give?

14.13 Use the data in Table 12.5 to calculate K_P at 25°C for the reaction $2HCl(g) + F_2(g) \rightleftharpoons 2HF(g) + Cl_2(g)$.

14.14 An air pollutant produced by burning high-sulfur fuels is sulfur dioxide. In smog, which contains appreciable amounts of NO_2, the sulfur dioxide can be oxidized to sulfur trioxide, which forms H_2SO_4 when it reacts with moisture. The reaction is

$$SO_2(g) + NO_2(g) \rightleftharpoons NO(g) + SO_3(g)$$

Use the data in Table 12.5 to calculate K_P for this reaction at 25°C.

14.15 The following thermodynamic data apply at 25°C:

Substance	ΔG_f^0 (kJ mol^{-1})
$NiSO_4 \cdot 6H_2O(s)$	−2222
$NiSO_4(s)$	−774
$H_2O(g)$	−228

(a) What is ΔG^0 for the reaction,

$$NiSO_4 \cdot 6H_2O(s) \rightleftharpoons NiSO_4(s) + 6H_2O(g)$$

(b) What is K_P for this reaction?

14.16 At 700 K, $\Delta G_{700\,K}' = -13.5$ kJ for the reaction, $CO(g) + 2H_2(g) \rightleftharpoons CH_3OH(g)$. Calculate the value of K_P for the reaction at 700 K.

14.17 The equilibrium constant, K_P, for the reaction, $COCl_2(g) \rightleftharpoons CO(g) + Cl_2(g)$, has a value of 4.62 kPa at 395°C. What is the value of $\Delta G_{668\,K}'$ (in kilojoules) for this reaction?

14.18 At 527°C the reaction

$$CO(g) + H_2O(g) \rightleftharpoons CO_2(g) + H_2(g)$$

has $K_P = 5.10$. What is $\Delta G_{800\,K}'$ for this reaction expressed in kilojoules?

14.19 Use the data in Tables 12.1 and 12.4 to compute $\Delta G_{773\,K}'$ (in kilojoules) and K_P at 500°C for the reaction

$$2HCl(g) \rightleftharpoons H_2(g) + Cl_2(g)$$

Assume that ΔH^0 and ΔS^0 are independent of temperature.

14.20 Use the data in Tables 12.1 and 12.4 to compute the temperature at which $K_P = 1$ for the reaction

$$C_2H_4(g) + H_2(g) \rightleftharpoons C_2H_6(g)$$

Assume that ΔH^0 and ΔS^0 are independent of temperature.

14.21 Methyl alcohol, CH_3OH, is a fuel that can be made from carbon monoxide (produced by burning coal) and steam. The equilibrium is

$$CO(g) + 2H_2(g) \rightleftharpoons CH_3OH(g)$$

At 427°C (700 K) a mixture of CO, H_2, and CH_3OH having the following partial pressures was prepared: $p_{CO} = 2 \times 10^{-3}$ atm, $p_{H_2} = 1 \times 10^{-2}$ atm, $p_{CH_3OH} = 3 \times 10^{-6}$ atm. For this reaction, $\Delta G_{700\,K}' = -13.5$ kJ. Use Equation 14.4 to determine whether this system is at equilibrium. If not, will the reaction proceed spontaneously to the left or the right?

Relationship Between K_P and K_c

14.22 For which of the reactions in Questions 14.2 and 14.4 would $K_P = K_c$?

14.23 The reaction

$$CO(g) + H_2O(g) \rightleftharpoons CO_2(g) + H_2(g)$$

is used industrially as a source of hydrogen. The value of K_c for this reaction at 500°C is 4.05. What is its value of K_P at this temperature?

14.24 The reaction

$$CH_4(g) + H_2O(g) \rightleftharpoons CO(g) + 3H_2(g)$$

is also a source of hydrogen. At 1500°C, its value of $K_c = 5.67$. What is its value of K_P at this temperature?

14.25 At 100°C, $K_P = 6.5 \times 10^{-2}$ for the reaction

$$2NO_2(g) \rightleftharpoons N_2O_4(g)$$

What is the value of K_c at this temperature?

Heterogeneous Equilibria

14.26 On the basis of the equilibrium

$$H_2O(l) \rightleftharpoons H_2O(g)$$

explain why the vapor pressure of water depends only on temperature and not the amount of liquid water in equilibrium with its vapor.

14.27 Why is it *not* necessary to include the concentrations of pure liquid or solid phases in the equilibrium constant expression?

14.28 Write equilibrium expressions for each of the following reactions:
(a) $CaCO_3(s) \rightleftharpoons CaO(s) + CO_2(g)$
(b) $Ni(s) + 4CO(g) \rightleftharpoons Ni(CO)_4(g)$
(c) $5CO(g) + I_2O_5(s) \rightleftharpoons I_2(g) + 5CO_2(g)$
(d) $Ca(HCO_3)_2(aq) \rightleftharpoons CaCO_3(s) + H_2O(l) + CO_2(g)$
(e) $AgCl(s) \rightleftharpoons Ag^+(aq) + Cl^-(aq)$

Le Châtelier's Principle

14.29 Consider the equilibrium $PCl_3(g) + Cl_2(g) \rightleftharpoons PCl_5(g)$. How would the following affect the position of equilibrium?
(a) addition of PCl_3
(b) removal of Cl_2
(c) removal of PCl_5
(d) decrease in the volume of the container
(e) addition of He without a change in volume

14.30 Which, if any, of the changes in Exercise 14.31 will change the value of the equilibrium constant for the reaction?

14.31 Indicate how each of the following changes affects the amount of H_2 in the system below, for which $\Delta H_{reaction} = +41$ kJ.

$$H_2(g) + CO_2(g) \rightleftharpoons H_2O(g) + CO(g)$$

(a) addition of CO_2
(b) addition of H_2O?
(c) addition of a catalyst
(d) increase in temperature
(e) decrease in the volume of the container

14.32 How will each of the changes in Exercise 14.31 affect the equilibrium constant?

14.33 Consider the equilibrium

$$2N_2O(g) + O_2(g) \rightleftharpoons 4NO(g)$$

How will the amount of NO at equilibrium be affected by
(a) adding N_2O?
(b) removing O_2?
(c) increasing the volume of the container?
(d) adding a catalyst?

14.34 For the reaction

$$4NH_3(g) + 3O_2(g) \rightleftharpoons 2N_2(g) + 6H_2O(l)$$

how will the amount of NH_3 at equilibrium be affected by
(a) adding O_2 to the system?
(b) adding N_2 to the system?
(c) removing H_2O from the system?
(d) decreasing the volume of the container?

14.35 Sketch a graph to show how the concentration of H_2, N_2, and NH_3 would change with time after N_2 had been added to a mixture of these gases initially at equilibrium.

14.36 In the equilibrium

$$CaCO_3(s) + heat \rightleftharpoons CaO(s) + CO_2(g)$$

how will the amount of $CaCO_3(s)$ change if
(a) CaO(s) is added?
(b) $CO_2(g)$ is added?
(c) the volume of the container is increased?
(d) the temperature is lowered?

Equilibrium Calculations

14.37 Referring to Exercise 14.7, calculate K_c for the reaction

$$PCl_3(g) + Cl_2(g) \rightleftharpoons PCl_5(g)$$

14.38 What is the value of K_P for the reaction

$$PCl_5(g) \rightleftharpoons PCl_3(g) + Cl_2(g)$$

Refer to the data in Exercise 14.7 (T = 298 K).

14.39 At 460°C, $K_c = 85.0$ for the reaction

$$SO_2(g) + NO_2(g) \rightleftharpoons NO(g) + SO_3(g)$$

A mixture of these gases has the following concentrations of the reactants and products: $[SO_2] = 0.040$ M, $[NO_2] = 0.50$ M, $[NO] = 0.30$ M, $[SO_3] = 0.020$ M. Is this system at equilibrium? If not, in which direction must the reaction proceed to reach equilibrium?

14.40 For the reaction $PCl_5(g) \rightleftharpoons PCl_3(g) + Cl_2(g)$, $K_c = 33.3$ at 760°C. In a container at equilibrium there are 1.29×10^{-3} mol dm^{-3} of PCl_5 and 1.87×10^{-1} mol dm^{-3} of Cl_2. Calculate the equilibrium concentration of PCl_3 in the vessel.

14.41 At 25°C, in a mixture of N_2O_4 and NO_2 in equilibrium at a total pressure of 85.5 kPa, the partial pressure of N_2O_4 is 57.0 kPa. Calculate for the reaction,

$$N_2O_4(g) \rightleftharpoons 2NO_2(g)$$

(a) K_P, (b) K_c, (c) $\Delta G^0_{298\ K}$ in kJ.

14.42 At a certain temperature the following equilibrium concentrations were found for the reactants and products in the reaction,

$$2HI(g) \rightleftharpoons H_2(g) + I_2(g)$$
$$[H_2] = 1.0 \times 10^{-3}\ M$$

$$[I_2] = 2.5 \times 10^{-2}\ M$$

$$[HI] = 2.2 \times 10^{-2}\ M$$

What is the value of K_c for this reaction?

14.43 In a particular experiment the following partial pressures were determined for the reaction at equilibrium

$$2NO(g) + Cl_2(g) \rightleftharpoons 2NOCl(g)$$

$$p_{NO} = 66 \text{ kPa} \qquad p_{Cl_2} = 18 \text{ kPa} \qquad p_{NOCl} = 15 \text{ kPa}$$

What is K_P for this reaction at the temperature at which the experiment was performed?

14.44 At 25°C, 0.0560 mol of O_2 and 0.020 mol N_2O were placed in a 1.00-dm^3 vessel and allowed to react according to the equation

$$2N_2O(g) + 3O_2(g) \rightleftharpoons 4NO_2(g)$$

When the system reached equilibrium, the concentration of the NO_2 was found to be 0.020 mol dm^{-3}.
(a) What were the equilibrium concentrations of N_2O and O_2?
(b) What is the value of K_c for this reaction at 25°C?

14.45 At 460°C, the reaction

$$SO_2(g) + NO_2(g) \rightleftharpoons NO(g) + SO_3(g)$$

has $K_c = 85.0$. What will be the equilibrium concentrations of the four gases if a mixture of SO_2 and NO_2 is prepared in which they each have an initial concentration of 0.0500 M?

14.46 For the reaction

$$H_2(g) + CO_2(g) \rightleftharpoons CO(g) + H_2O(g)$$

$K_c = 0.771$ at 750°C. If 1.00 mol of H_2 and 1.00 mol of CO_2 are placed into a 5.00-dm^3 container and permitted to react, what will be the equilibrium concentrations of all four gases?

14.47 Suppose a mixture of SO_2, NO_2, NO, and SO_3 was prepared at 460°C having the following initial concentrations: $[SO_2] = 0.0100 \ M$, $[NO_2] = 0.0200 \ M$, $[NO] = 0.0100 \ M$, and $[SO_3] = 0.0150 \ M$. At this temperature the reaction

$$SO_2(g) + NO_2(g) \rightleftharpoons NO(g) + SO_3(g)$$

has $K_c = 85.0$. What will be the equilibrium concentrations of the four gases?

14.48 The reaction

$$2CO_2 \rightleftharpoons 2CO + O_2$$

has $K_c = 6.4 \times 10^{-7}$ at 2000°C. If 1.0×10^{-3} mol of CO_2 is placed into a 1.0-dm^3 vessel at this temperature,
(a) What will be the equilibrium concentrations of CO and O_2?
(b) What fraction of the CO_2 will have decomposed?

*14.49 At 100°C the equilibrium constant, K_c, for the reaction

$$CO(g) + Cl_2(g) \rightleftharpoons COCl_2(g)$$

has a value of 4.6×10^9. If 0.20 mol of $COCl_2$ is placed into a 10.0-dm^3 flask at 100°C, what will be the concentrations of all species at equilibrium?

14.50 Sodium bicarbonate (baking soda) has many useful properties. Among them is the ability to serve as a fire extinguisher because of thermal decomposition to produce CO_2, which smothers the fire,

$$2NaHCO_3(s) \rightleftharpoons Na_2CO_3(s) + CO_2(g) + H_2O(g)$$

At 125°C the value of K_P for this reaction is $2.6 \times 10^3 \text{ kPa}^2$. What are the partial pressures of $CO_2(g)$ and $H_2O(g)$ in this system at equilibrium? Can you explain why $NaHCO_3$ is used in baking?

*14.51 In a 10.0-dm^3 mixture of H_2, I_2, and HI at equilibrium at 425°C there are 0.100 mol of H_2, 0.100 mol of I_2, and 0.740 mol of HI. If 0.50 mol of HI are now added to this system, what will be the concentrations of H_2, I_2, and HI once equilibrium has been reestablished?

*14.52 In Exercise 14.49 it was stated that at 100°C the value of K_c for the reaction, $CO(g) + Cl_2(g) \rightleftharpoons COCl_2(g)$, is 4.6×10^9. Suppose that 0.15 mol of CO and 0.30 mol of Cl_2 were placed into a 1.0-dm^3 vessel and allowed to react. What would the concentration be of each of the gases in the system at equilibrium? (*Hint:* First assume 100% reaction; then work backward toward equilibrium.)

*14.53 The production of NO by reaction of N_2 and O_2 in an automobile engine is an important source of nitrogen oxide pollution. At 1000°C the reaction, $N_2(g) + O_2(g) \rightleftharpoons 2NO(g)$, has $K_P = 4.8 \times 10^{-7}$. Suppose that the partial pressures of N_2 and O_2 in the cylinder of an engine after the gasoline vapor has been ignited are $p_{N_2} = 3.40$ MPa and $p_{O_2} = 0.41$ MPa. Assume that the temperature of the mixture is 1000°C. Calculate the partial pressure of NO in the mixture if the system has time to reach equilibrium.

*14.54 If it is assumed that the reactants and products in the preceding question are unable to react further when the exhaust gases are suddenly cooled as they exit the engine, calculate the partial pressure of the NO when the partial pressure of N_2 has dropped to 81 kPa and the temperature has dropped to 150°C.

*14.55 At a certain temperature $K_c = 7.5$ for the reaction

$$2NO_2(g) \rightleftharpoons N_2O_4(g)$$

If 2.0 mol of NO_2 are placed in a 2.0-dm^3 container and permitted to react, what will be the concentrations of NO_2 and N_2O_4 at equilibrium? What will be the equilibrium concentrations if the size of the container is doubled? Does this conform to what you would expect from Le Châtelier's principle?

CHAPTER 15

ACID-BASE EQUILIBRIA IN AQUEOUS SOLUTIONS

Controlling the acidity of a swimming pool with sodium bicarbonate is just one of the practical applications of the principles of acid-base equilibria discussed in this chapter.

In Chapter 7 we saw that many reactions of concern to us take place in aqueous solution. By now we also realize that chemical changes do not, in general, proceed entirely to completion, but instead approach a state of dynamic equilibrium. In this chapter and the next we will examine in greater detail, and on a quantitative basis, many of the ionic equilibria that can occur in aqueous solutions. We begin, in this chapter, with the equilibria and reactions of acids and bases. This is very important because of the amphiprotic nature of water itself, and because many compounds of biological interest, which occur in the aqueous environment of living systems, show acid-base properties.

15.1 THE IONIZATION OF WATER AND THE pH CONCEPT

In Chapter 7 we commented that water itself is a very weak electrolyte because of the reaction

$$H_2O + H_2O \rightleftharpoons H_3O^+(aq) + OH^-(aq)$$

This kind of reaction, in which two molecules of the solvent react with each other to form ions, is called **autoionization.** This is a very important equilibrium because it is present in any aqueous solution, regardless of what other reactions may also be taking place. Since the autoionization is an equilibrium, we can write an equilibrium expression for it. Following the approach of the last chapter this is

$$K = \frac{[H_3O^+][OH^-]}{[H_2O][H_2O]}$$

The molar concentration of water, which appears in the denominator of this expression, is very nearly constant ($\approx 55.6\ M$) in both pure water and dilute aqueous solutions. Therefore, $[H_2O]^2$ can be included with the equilibrium constant, K, on the left side of the equation. This gives

$$K \cdot [H_2O]^2 = [H_3O^+][OH^-]$$

The left side of this expression is the product of two constants which, of course, must also equal a constant. This combined constant is written as

$$K_w = K[H_2O]^2$$

Our equilibrium condition therefore becomes

$$K_w = [H_3O^+][OH^-]$$

Since $[H_3O^+][OH^-]$ is the product of ionic concentrations, K_w is called the **ion product constant** for water, or frequently simply the **ionization constant** or **dissociation constant** of water. At 25°C, $K_w = 1.0 \times 10^{-14}$, and this is one equilibrium constant that you should be sure to memorize.

K_w varies with temperature. At 37°C (body temperature)

$$K_w = 2.42 \times 10^{-14}$$

The equation for the autoionization of water is often simplified by omitting the water molecule that picks up the H^+ and abbreviating the hydronium ion as H^+. The chemical equation becomes

$$H_2O \rightleftharpoons H^+ + OH^- \tag{15.1}$$

(In this and most future chemical equations in this chapter we will omit (*aq*) after the formulas of the molecules and ions in the solution. These species are of course hydrated, but the equations are easier to work with if they are not cluttered by this notation.) When we use this simplified equation, the expression for the ionization constant for water is

$$K_w = [H^+][OH^-] \tag{15.2}$$

It is very important to remember that *in any aqueous solution the relationship expressed in Equation 15.2 must always be satisfied, regardless of any other equilibria that may also exist in the solution.*

Equation 15.2 can be used to calculate the molar concentrations of both the H^+ and OH^- ions in pure water. From the stoichiometry of the dissociation, we see that for each 1 mol of H^+ formed, 1 mol of OH^- is also produced. This means that at equilibrium, $[H^+] = [OH^-]$. If we let x equal the hydrogen ion concentration, then

$$x = [H^+] = [OH^-]$$

Substituting into Equation 15.2 gives

$$K_w = x \cdot x = x^2$$

or, because $K_w = 1.0 \times 10^{-14}$,

$$x^2 = 1.0 \times 10^{-14}$$

Taking the square root yields

$$x = 1.0 \times 10^{-7}$$

which means that the concentrations of hydrogen ion and hydroxide ion in pure water are

$$[H^+] = [OH^-] = 1.0 \times 10^{-7}\ M$$

Whenever the hydrogen ion concentration equals the hydroxide ion concentration, as it does in pure water, the solution is said to be *neutral.* An acid is a substance that makes the H^+ concentration greater than the OH^- concentration; conversely, a base makes the OH^- concentration greater than the H^+ concentration. However, remember that there is always *some* OH^- present in an acidic solution, just as there is always *some* H^+ present even if the solution is basic. At all times, Equation 15.2 is obeyed if the solution is at equilibrium.

In an aqueous solution of an acid we will often want to know what the H^+ concentration is. In these cases it is almost always safe to assume that essentially all the H^+ in the solution comes from the dissolved acid. In other words, it is usually safe to assume that the dissociation of water contributes a negligible amount of H^+ to the solution. This is because the presence of H^+ from an acid (for example, HCl) shifts the equilibrium

$$H_2O \rightleftharpoons H^+ + OH^-$$

to the left. Therefore, the amount of water that is dissociated in a solution of an acid is even less than in pure water, which means that the H^+ coming from the dissociation of water is less than $10^{-7}\ M$. Similarly, the OH^- concentration in a solution of a base can be calculated just from the concentration of the solute. The OH^- contributed by the dissociation of water is negligible. Example 15.1 illustrates this point.

EXAMPLE 15.1	**CALCULATING CONCENTRATIONS IN A SOLUTION OF A STRONG ACID**
PROBLEM	(a) What is the OH^- concentration in a 0.0010 M HCl solution? (b) What is the H^+ concentration derived from the dissociation of the solvent?
SOLUTION	(a) At equilibrium we must have

$$[H^+][OH^-] = 1.0 \times 10^{-14}$$

HCl is a strong acid and is essentially 100% dissociated.

$$HCl \longrightarrow H^+ + Cl^-$$

Therefore, 0.0010 mol of HCl per dm^3 gives 0.0010 mol of H^+ per dm^3. The total hydrogen ion concentration, then, is 0.0010 M *plus* the quantity contributed by the dissociation of water. Let's assume for the moment that this contribution is negligible, as suggested above, and that it can be ignored. This gives

$$[H^+] = 0.0010\ M + \text{(contribution from } H_2O) \approx 0.0010\ M$$

Solving for the hydroxide ion concentration, we get

$$[OH^-] = \frac{1.0 \times 10^{-14}}{[H^+]}$$

$$= \frac{1.0 \times 10^{-14}}{0.0010} = 1.0 \times 10^{-11}\ M$$

(b) The hydroxide ion in part (a) comes entirely from the dissociation of water. Therefore, the concentration of H^+ derived from H_2O must *also* be $1.0 \times 10^{-11}\ M$, as can be seen from the stoichiometry of Equation 15.1. Note that this value ($1.0 \times 10^{-11}\ M$) is indeed negligible compared to the H^+ concentration produced by the HCl ($1.0 \times 10^{-3}\ M$), so the assumption made in part (a) was valid.

The pH Concept

Hydrogen ion and hydroxide ion enter into many equilibria in addition to the dissociation of water, so it is frequently necessary to specify their concentrations in aqueous solutions. These concentrations may range from relatively high values to very small small ones (for example, 10 M to $10^{-14}\ M$), and a logarithmic notation[1] has been devised to simplify the expression of these quantities. In general, for some quantity X,

Notice that logs to the base 10 are used here, *not* natural logs.

$$pX = \log \frac{1}{X} = -\log X \qquad [15.3]$$

For example, if we wish to specify the hydrogen ion concentration in a solution, we speak of **pH.** This is defined as

$$pH = \log \frac{1}{[H^+]} = -\log [H^+]$$

In a solution where the hydrogen ion concentration is $10^{-3}\ M$, we therefore have

$$pH = -\log (10^{-3}) = -(-3)$$

or

$$pH = 3$$

Similarly, if the hydrogen ion concentration is $10^{-8}\ M$, the pH of the solution is 8.

Following the same approach for the hydroxide ion concentration, we can define the **pOH** of a solution as

$$pOH = -\log[OH^-]$$

Just as the H^+ and OH^- ion concentrations in a solution are related to each

[1] A discussion on the use of logarithms can be found in Appendix A.

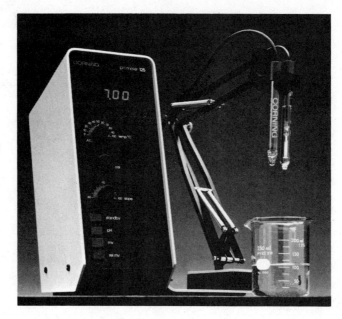

A pH meter. The instrument is first calibrated by dipping the pair of electrodes (one sensitive to the H^+ concentration and the other a reference electrode) into a solution of known pH, and then adjusting the scale appropriately. Afterwards, the pH of any other solution is obtained simply by dipping the electrodes into the solution and reading the pH from the scale.

other, so also are the pH and pOH. From the equilibrium expression for the dissociation of water,

$$\log K_w = \log[H^+] + \log[OH^-]$$

Multiplying through by -1 gives

$$(-\log K_w) = (-\log[H^+]) + (-\log[OH^-])$$

If we follow our definition, $-\log K_w = pK_w$. Therefore,

$$pK_w = pH + pOH$$

$$
\begin{aligned}
pK_w &= -\log K_w \\
&= -\log (1.0 \times 10^{-14}) \\
&= 14.00
\end{aligned}
$$

Since $K_w = 1.0 \times 10^{-14}$, $pK_w = 14.00$. This gives the useful relationship

$$pH + pOH = 14.00 \qquad [15.4]$$

In a neutral solution, $[H^+] = [OH^-] = 10^{-7}\ M$, and pH = pOH = 7.0, so that in a neutral solution we say that the pH = 7.0. In an acidic solution the hydrogen ion concentration is greater than $10^{-7}\ M$ (for example, $10^{-3}\ M$) and the pH is less than 7.0. By the same token, in basic solutions the $[H^+]$ is less than $10^{-7}\ M$ (for example, $10^{-10}\ M$) and the pH is greater than 7.0. This is summarized below.

	$[H^+]$	$[OH^-]$	pH	pOH
Acidic solution	$>10^{-7}$	$<10^{-7}$	<7	>7
Neutral solution	10^{-7}	10^{-7}	7	7
Basic solution	$<10^{-7}$	$>10^{-7}$	>7	<7

Many common substances are either acidic or basic, and their degree of acidity or basicity is conveniently expressed in terms of pH (Figure 15.1). Notice that substances having a pH less than 7—that is, those that are acidic—have characteristically sour tastes. Lemon juice contains citric acid, for example, and vinegar contains acetic acid. On the other hand, basic substances such as milk of magnesia—a suspension of $Mg(OH)_2$ in water—have a bitter taste. Although

Figure 15.1

The pH of some common materials.

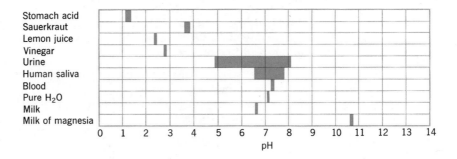

sour and bitter taste is the body's way of judging the acidity of foods, *never* taste chemicals in the laboratory; many of them are poisonous and will definitely ruin your day!

Most students today use scientific calculators for their calculations. These electronic marvels have keys that allow you to obtain both common logarithms and natural logarithms at the press of a button. We will assume that you have this kind of calculator to help you with your calculations, and in the next three example problems we will review briefly the procedures involved in calculating pH from [H$^+$] and [H$^+$] from pH. These are routine calculations that you should be able to perform almost without effort, and in future problems dealing with hydrogen ion concentrations we will assume you have mastered this topic. If you do not have a scientific calculator, don't be discouraged. A description of the use of log tables in obtaining pH from [H$^+$], and vice versa, is contained in Appendix A, where you will find the pH calculations in Examples 15.2 through 15.4 worked out using values taken from a table of common logarithms.

EXAMPLE 15.2 **CALCULATING THE pH OF A SOLUTION OF A STRONG ACID**

PROBLEM   What is the pH of a 0.0020 M solution of HCl?

SOLUTION   By now you should know that HCl is a strong acid, which means that it is 100% ionized. Therefore, in a 0.0020 M solution of HCl the hydrogen ion concentration is 0.0020 M and the Cl$^-$ concentration is 0.0020 M. To determine the pH, we need only know [H$^+$].

The pH is defined as

$$pH = -\log[H^+]$$

Substituting yields

$$pH = -\log(2.0 \times 10^{-3})$$

Performing this operation gives

$$pH = 2.70$$

Notice that the number of digits after the decimal is equal to the number of significant figures in the hydrogen ion concentration.

EXAMPLE 15.3 **CALCULATING THE pH OF A SOLUTION OF A STRONG BASE**

PROBLEM   What is the pH of a 5.0×10^{-4} M solution of NaOH?

SOLUTION   Metal hydroxides are strong bases, which means they are 100% dissociated. Therefore, in this solution the OH^- concentration is equal to 5.0×10^{-4} M. To calculate the pH, we can proceed in either of two ways:

1. We can calculate $[H^+]$ from K_w and the known value of $[OH^-]$, and then proceed as in the previous example.
2. We can use the $[OH^-]$ to calculate the pOH and then subtract this value from 14 to obtain the pH.

METHOD 1   We know that

$$K_w = [H^+][OH^-]$$

Therefore,

$$[H^+] = \frac{1.0 \times 10^{-14}}{5.0 \times 10^{-4}}$$

$$= 2.0 \times 10^{-11} \ M$$

From this we calculate the pH

$$pH = -\log(2.0 \times 10^{-11})$$

$$= 10.70$$

METHOD 2   By definition

$$pOH = -\log[OH^-]$$

In this problem

$$[OH^-] = 5.0 \times 10^{-4} \ M$$

Therefore,

$$pOH = -\log(5.0 \times 10^{-4})$$

$$= 3.30$$

The pH is therefore

$$pH = 14.00 - 3.30$$

$$= 10.70$$

EXAMPLE 15.4   CALCULATING $[H^+]$ AND $[OH^-]$ FROM pH

PROBLEM   A sample of orange juice was found to have a pH of 3.80. What were the H^+ and OH^- concentrations in the juice?

SOLUTION   The pH is defined as

$$pH = -\log[H^+]$$

The antilogarithm of this gives the following relationship:

$$[H^+] = 10^{-pH}$$

For the sample of orange juice, therefore,

$$[H^+] = 10^{-3.80}$$

$$= 1.6 \times 10^{-4}$$

(Check the direction booklet that came with your calculator to be sure you can take the antilogarithm of -3.80. Practice using your calculator until you can obtain the answer given above.)

Once we have obtained $[H^+]$, we can calculate $[OH^-]$ from K_w.

$$[OH^-] = \frac{1.0 \times 10^{-14}}{1.6 \times 10^{-4}}$$

$$= 6.3 \times 10^{-11} \, M$$

Alternatively, we could first calculate pOH from pH and then take the antilogarithm.

$$pOH = 14.00 - pH$$

$$= 14.00 - 3.80 = 10.20$$

$$[OH^-] = 10^{-pOH}$$

$$= 10^{-10.20}$$

$$= 6.3 \times 10^{-11} \, M$$

15.2 DISSOCIATION OF WEAK ACIDS AND BASES

Acetic acid, $HC_2H_3O_2$, is a typical example of a weak acid. In water it is only partially ionized and the molecules of the acid exist in dynamic equilibrium with the ions produced in the ionization reaction

$$HC_2H_3O_2 + H_2O \rightleftharpoons H_3O^+ + C_2H_3O_2^-$$

The equilibrium expression for this reaction is

$$K = \frac{[H_3O^+][C_2H_3O_2^-]}{[HC_2H_3O_2][H_2O]}$$

In all our calculations involving ionic equilibria, we will be dealing with relatively dilute solutions.

In dilute solutions, the concentration of H_2O is not appreciably different than it is in pure water, so we may safely take it to be a constant that can be included with K on the left side of the equal sign. That is,

$$K \times [H_2O] = K_a = \frac{[H_3O^+][C_2H_3O_2^-]}{[HC_2H_3O_2]}$$

where we have used K_a to represent the **acid dissociation constant** or **ionization constant.** This same equilibrium expression can be obtained if we simplify the dissociation by omitting the solvent. Thus, for the dissociation of acetic acid we would write

$$HC_2H_3O_2 \rightleftharpoons H^+ + C_2H_3O_2^-$$

and the equilibrium expression would then be

$$K_a = \frac{[H^+][C_2H_3O_2^-]}{[HC_2H_3O_2]}$$

In general, for any weak acid, HA, the simplified equation for the dissociation reaction can be written as

$$HA \rightleftharpoons H^+ + A^-$$

and its acid dissociation constant is given by

$$K_a = \frac{[H^+][A^-]}{[HA]}$$

This same approach can also be applied to weak bases. Normally, these are substances that react with water and pick up a hydrogen ion. An example is the weak base, ammonia.

$$NH_3 + H_2O \rightleftharpoons NH_4^+ + OH^-$$

If we omit the solvent, the **base ionization constant,** K_b, for this reaction is

$$K_b = \frac{[NH_4^+][OH^-]}{[NH_3]}$$

In general, for any weak base B the ionization equilibrium can be written as

$$B + H_2O \rightleftharpoons BH^+ + OH^-$$

and the expression for K_b is

$$K_b = \frac{[BH^+][OH^-]}{[B]}$$

The extent to which a weak acid or base undergoes ionization, as well as the value of the ionization constant, must be determined experimentally. (One way of doing this is to measure the pH of a solution prepared by dissolving a known quantity of the weak acid or base in a given volume of solution, as illustrated in Examples 15.5 and 15.6.) Ionization constants of a number of weak acids and bases are listed in Table 15.1. Notice that their values are quite small—ranging from 10^{-2} to 10^{-10}. Recalling that numbers this small can be simplified by applying the logarithmic notation of Equation 15.3, we can write for any K_a,

$$pK_a = -\log K_a$$

and for any K_b,

$$pK_b = -\log K_b$$

For example, the pK_a of acetic acid is

$$pK_a = -\log K_a = -\log(1.8 \times 10^{-5})$$
$$= 4.74$$

and for pyridine, a bad-smelling liquid,

$$pK_b = -\log (1.7 \times 10^{-9})$$
$$= 8.77$$

From our discussion in Section 14.2, we know that the smaller the value of an equilibrium constant, the smaller the extent of reaction. Therefore, the smaller the value of K_a or K_b, the smaller the extent of ionization and the weaker the acid or base. Relative strengths of acids and bases can also be indicated by their pK_a's and pK_b's. In this case the smaller the value of pK_a or pK_b, the *stronger* is the acid or base. Let's compare, for example, the pK_a's for acetic, chloroacetic, and dichloroacetic acid. Their pK_a's are

$HC_2H_3O_2$	$pK_a = 4.74$
$HC_2H_2ClO_2$	$pK_a = 2.85$
$HC_2HCl_2O_2$	$pK_a = 1.30$

The order of increasing acidity is, therefore,

acetic < chloroacetic < dichloroacetic acid

(A discussion of *why* the acidity of these substances increases in this fashion can be found on page 244).

Table 15.1
Ionization constants for some weak acids and bases

Weak Acid	Ionization	K_a	pK_a
Chloroacetic acid	$HC_2H_2O_2Cl \rightleftharpoons H^+ + C_2H_2O_2Cl^-$	1.4×10^{-3}	2.85
Hydrofluoric acid	$HF \rightleftharpoons H^+ + F^-$	6.5×10^{-4}	3.19
Nitrous acid	$HNO_2 \rightleftharpoons H^+ + NO_2^-$	4.5×10^{-4}	3.35
Formic acid	$HCHO_2 \rightleftharpoons H^+ + CHO_2^-$	1.8×10^{-4}	3.74
Lactic acid	$HC_3H_5O_3 \rightleftharpoons H^+ + C_3H_5O_3^-$	1.38×10^{-4}	3.86
Benzoic acid	$HC_7H_5O_2 \rightleftharpoons H^+ + C_7H_5O_2^-$	6.5×10^{-5}	4.19
Acetic acid	$HC_2H_3O_2 \rightleftharpoons H^+ + C_2H_3O_2^-$	1.8×10^{-5}	4.74
Butyric acid	$HC_4H_7O_2 \rightleftharpoons H^+ + C_4H_7O_2^-$	1.5×10^{-5}	4.82
Nicotinic acid	$HC_6H_4NO_2 \rightleftharpoons H^+ + C_6H_4NO_2^-$	1.4×10^{-5}	4.85
Propionic acid	$HC_3H_5O_2 \rightleftharpoons H^+ + C_3H_5O_2^-$	1.4×10^{-5}	4.85
Barbituric acid	$HC_4H_3N_2O_3 \rightleftharpoons H^+ + C_4H_3N_2O_3^-$	1.0×10^{-5}	5.00
Veronal (diethylbarbituric acid)	$HC_8H_{11}N_2O_3 \rightleftharpoons H^+ + C_8H_{11}N_2O_3^-$	3.7×10^{-8}	7.43
Hypochlorous acid	$HOCl \rightleftharpoons H^+ + OCl^-$	3.1×10^{-8}	7.51
Hydrocyanic acid	$HCN \rightleftharpoons H^+ + CN^-$	4.9×10^{-10}	9.31

Weak Base	Ionization	K_b	pK_b
Diethylamine	$(C_2H_5)_2NH + H_2O \rightleftharpoons (C_2H_5)_2NH_2^+ + OH^-$	9.6×10^{-4}	3.02
Methylamine	$CH_3NH_2 + H_2O \rightleftharpoons CH_3NH_3^+ + OH^-$	3.7×10^{-4}	3.43
Ammonia	$NH_3 + H_2O \rightleftharpoons NH_4^+ + OH^-$	1.8×10^{-5}	4.74
Hydrazine	$N_2H_4 + H_2O \rightleftharpoons N_2H_5^+ + OH^-$	1.7×10^{-6}	5.77
Hydroxylamine	$NH_2OH + H_2O \rightleftharpoons NH_3OH^+ + OH^-$	1.1×10^{-8}	7.97
Pyridine	$C_5H_5N + H_2O \rightleftharpoons C_5H_5NH^+ + OH^-$	1.7×10^{-9}	8.77
Aniline	$C_6H_5NH_2 + H_2O \rightleftharpoons C_6H_5NH_3^+ + OH^-$	3.8×10^{-10}	9.42

Let's look now at some examples showing how these equilibrium constants can be calculated.

EXAMPLE 15.5 USING THE pH OF A SOLUTION OF AN ACID TO CALCULATE K_a AND THE PERCENT DISSOCIATION

PROBLEM   A student prepared a 0.10 M acetic solution and measured its pH to be 2.88. Calculate K_a for acetic acid and determine its percent dissociation.

SOLUTION   The first step is to write the equation for the equilibrium.

$$HC_2H_3O_2 \rightleftharpoons H^+ + C_2H_3O_2^-$$

To evaluate K_a we must have equilibrium concentrations to substitute into the expression,

$$K_a = \frac{[H^+][C_2H_3O_2^-]}{[HC_2H_3O_2]}$$

From pH we can obtain the H^+ concentration.

$$pH = 2.88$$

$$[H^+] = 1.3 \times 10^{-3}\ M$$

$[H^+] = 10^{-2.88} = 1.3 \times 10^{-3}\ M$

The $[H^+]$ comes from the dissociation of $HC_2H_3O_2$ and, from the stoichiometry of the equation, we see that the concentrations of H^+ and $C_2H_3O_2^-$ must be equal because they are formed in a 1-to-1 ratio.

$$[H^+] = [C_2H_3O_2^-] = 1.3 \times 10^{-3}\ M$$

The concentration of undissociated $HC_2H_3O_2$ at equilibrium is equal to the original concentration, 0.10 M, *minus* the number of moles per dm^3 of acetic acid that have dissociated. At equilibrium, then, we have

Equilibrium Concentrations	
H^+	$1.3 \times 10^{-3}\ M$
$C_2H_3O_2^-$	$1.3 \times 10^{-3}\ M$
$HC_2H_3O_2$	$1.0 \times 10^{-1} - 0.013 \times 10^{-1} = 1.0 \times 10^{-1}\ M$

Note that when we compute the acetic acid concentration to the *proper number of significant figures*, the amount that has dissociated is negligible compared to the amount initially present. Thus,

$$0.10\ M - 0.0013\ M = (0.0987\ M) = 0.10\ M$$

Substituting the equilibrium concentrations into the expression for K_a, we have

$$K_a = \frac{(1.3 \times 10^{-3})(1.3 \times 10^{-3})}{(1.0 \times 10^{-1})}$$

$$= 1.7 \times 10^{-5}$$

This value of K_a differs from that given in Table 15.1 because of "round-off" error.

The **percent dissociation** of acetic acid in this solution is found by dividing the number of moles per dm^3 of $HC_2H_3O_2$ that have dissociated by the quantity of acetic acid that was available initially, all multiplied by 100.

$$\text{percent dissociation} = \frac{(\text{mol dm}^{-3}\ HC_2H_3O_2\ \text{dissociated})}{(\text{mol dm}^{-3}\ HC_2H_3O_2\ \text{available})} \times 100$$

$$= \frac{1.3 \times 10^{-3}\ M}{1.0 \times 10^{-1}\ M} \times 100 = 1.3\%$$

EXAMPLE 15.6 USING THE PERCENT DISSOCIATION TO CALCULATE AN EQUILIBRIUM CONSTANT

PROBLEM   A student prepared a 0.010 M NH_3 solution and, by a freezing-point-lowering experiment, determined that the NH_3 had undergone 4.2% ionization. Calculate the K_b for NH_3.

SOLUTION   Ammonia ionizes in water according to the reaction,

$$NH_3 + H_2O \rightleftharpoons NH_4^+ + OH^-$$

for which we write

$$K_b = \frac{[NH_4^+][OH^-]}{[NH_3]}$$

From the stoichiometry of the ionization we see that at equilibrium

$$[NH_4^+] = [OH^-]$$

Since the 0.010 M solution undergoes 4.2% ionization, the number of moles per dm^3 of these ions present at equilibrium is

The amount ionized per dm^3 is 4.2% of 0.010 mol dm^{-3}.

$$[NH_4^+] = [OH^-] = 0.042 \times 0.010\ M = 4.2 \times 10^{-4}\ M$$

The number of moles per dm^3 of NH$_3$ at equilibrium would be

$$[NH_3] = 1.0 \times 10^{-2} - 0.042 \times 10^{-2} = 0.958 \times 10^{-2}\ M$$

When this is rounded off to the appropriate number of significant figures, we have $[NH_3] = 1.0 \times 10^{-2}\ M$ (once again, the quantity lost by ionization is negligible). Our equilibrium concentrations are shown at the left.

	Equilibrium Concentrations
NH$_4^+$	$4.2 \times 10^{-4}\ M$
OH$^-$	$4.2 \times 10^{-4}\ M$
NH$_3$	$1.0 \times 10^{-2}\ M$

When these concentrations are substituted into the equation for K_b, we have

$$K_b = \frac{(4.2 \times 10^{-4})(4.2 \times 10^{-4})}{(1.0 \times 10^{-2})}$$

or

$$K_b = 1.8 \times 10^{-5}$$

In Examples 15.5 and 15.6 we computed K from a knowledge of equilibrium concentrations. We can also use our knowledge of K to calculate the concentrations in an equilibrium mixture. Let's look at some examples.

EXAMPLE 15.7   USING K_a TO CALCULATE THE CONCENTRATIONS OF SPECIES IN A SOLUTION OF A WEAK ACID

PROBLEM   What are the concentrations of all the species present in a 0.50 M HC$_2$H$_3$O$_2$ solution?

SOLUTION   First we write the chemical equation for the equilibrium,

$$HC_2H_3O_2 \rightleftharpoons H^+ + C_2H_3O_2^-$$

From Table 15.1 we find that $K_a = 1.8 \times 10^{-5}$ for HC$_2$H$_3$O$_2$. Therefore,

$$\frac{[H^+][C_2H_3O_2^-]}{[HC_2H_3O_2]} = 1.8 \times 10^{-5}$$

The quantities that we must substitute into this expression must represent equilibrium concentrations. In obtaining these we will construct a table as we did in Chapter 14. A solution labeled 0.50 M HC$_2$H$_3$O$_2$ would be prepared by dissolving 0.50 mol of HC$_2$H$_3$O$_2$ in 1.00 dm^3 of solution, so the label gives us the concentration of acetic acid that existed before any of it dissociated. The initial concentration of HC$_2$H$_3$O$_2$ is therefore taken to be 0.50 M. Initially the solution contained no C$_2$H$_3$O$_2^-$, and we will neglect the H$^+$ from the dissociation of H$_2$O. Therefore, for the purposes of the problem there is no H$^+$ initially. We know that at equilibrium some of the acetic acid in the solution will have dissociated. If we therefore let x equal the number of moles per dm^3 of HC$_2$H$_3$O$_2$ that dissociate, at equilibrium we will have produced x mol dm^{-3} of H$^+$, x mol dm^{-3} of C$_2$H$_3$O$_2^-$, and we will have lost x mol dm^{-3} of HC$_2$H$_3$O$_2$. At equilibrium we thus have $x\ M$ H$^+$, $x\ M$ C$_2$H$_3$O$_2^-$, and $(0.50 - x)\ M$ HC$_2$H$_3$O$_2$.

	Initial Molar Concentrations	Change	Equilibrium Molar Concentrations
H^+	0.0	$+x$	x
$C_2H_3O_2^-$	0.0	$+x$	x
$HC_2H_3O_2$	0.50	$-x$	$0.50 - x$

Substituting these values into the equilibrium expression gives us

$$\frac{(x)(x)}{(0.50 - x)} = 1.8 \times 10^{-5}$$

Without simplification this expression leads to a quadratic equation that can be solved using the quadratic formula. However, in Example 14.11 we saw that it is sometimes possible to make simplifying assumptions that greatly reduce the effort required to obtain solutions to problems of this type. Because K is small, very little $HC_2H_3O_2$ will have actually undergone dissociation, so x will be small. Let us assume that x will be negligible compared to 0.50; that is,

$$0.50 - x \approx 0.50$$

Our equation then becomes

$$\frac{x^2}{0.50} = 1.8 \times 10^{-5}$$

or

$$x = 3.0 \times 10^{-3}$$

If we look back on our assumption, we see that x is in fact small compared to 0.50 and that, when *rounded to the proper number of significant figures,*

$$0.50 - 0.0030 = 0.50$$

	Equilibrium Concentrations (M)
H^+	3.0×10^{-3}
$C_2H_3O_2^-$	3.0×10^{-3}
$HC_2H_3O_2$	0.50

Therefore, the equilibrium concentrations of the species involved in the dissociation of the acid are those given in the table at the left. Since the question asks for *all* concentrations, we must also calculate the concentration of OH^-, which comes from the dissociation of water. Here we use K_w.

$$[OH^-] = \frac{K_w}{[H^+]}$$

$$= \frac{1.0 \times 10^{-14}}{3.0 \times 10^{-3}}$$

$$= 3.3 \times 10^{-12} \ M$$

In the last example the only source of H^+ and $C_2H_3O_2^-$ was from the dissociation of the weak acid. Example 15.8 shows how we would handle a problem dealing with a solution for which there are two sources of one of the ions.

EXAMPLE 15.8 CALCULATING THE pH OF A SOLUTION THAT CONTAINS A WEAK ACID AND A SALT OF THE ACID

PROBLEM What are the concentrations of H^+, $C_2H_3O_2^-$, and $HC_2H_3O_2$ in a solution prepared by dissolving 0.10 mol of $NaC_2H_3O_2$ and 0.20 mol of $HC_2H_3O_2$ in enough water to give a total volume of 1.00 dm^3?

SOLUTION   There is only one equilibrium here that we must be concerned with,

$$HC_2H_3O_2 \rightleftharpoons H^+ + C_2H_3O_2^-$$

$$\frac{[H^+][C_2H_3O_2^-]}{[HC_2H_3O_2]} = 1.8 \times 10^{-5}$$

It is very important to remember that salts are strong electrolytes.

When $NaC_2H_3O_2$ dissolves, it is completely dissociated. It is important to remember that virtually all salts are 100% dissociated in solution. Therefore, 0.10 mol dm^{-3} of $NaC_2H_3O_2$ gives 0.10 mol dm^{-3} of Na^+ and 0.10 mol dm^{-3} of $C_2H_3O_2^-$. We are interested only in the $C_2H_3O_2^-$; the Na^+ is simply a *spectator ion* and we can ignore it. The initial concentrations of concern to us are found in the first column of our table. Since no H^+ is present, some $HC_2H_3O_2$ must ionize; so let's allow x to equal the number of moles per dm^3 of $HC_2H_3O_2$ that dissociates to give H^+ and $C_2H_3O_2^-$. This will increase $[H^+]$ and $[C_2H_3O_2^-]$ by x, and decrease $[HC_2H_3O_2]$ by x. The equilibrium concentrations are then found in the last column for our table.

	Initial Molar Concentrations	Change	Final Molar Concentrations
H^+	0.0	$+x$	x
$C_2H_3O_2^-$	0.10	$+x$	$0.10 + x \approx 0.10$
$HC_2H_3O_2$	0.20	$-x$	$0.20 - x \approx 0.20$

As before, we look at K_a and see that x will probably be small. We will therefore assume that $0.10 + x \approx 0.10$ and $0.20 - x \approx 0.20$. Substituting into the expression for K_a gives

$$\frac{(x)(0.10)}{(0.20)} = 1.8 \times 10^{-5}$$

$$x = 3.6 \times 10^{-5}$$

Note that x is small compared to both 0.10 and 0.20. This justifies our assumption. Finally, the equilibrium concentrations are

$$[H^+] = 3.6 \times 10^{-5}\ M$$

$$[C_2H_3O_2^-] = 0.10\ M$$

$$[HC_2H_3O_2] = 0.20\ M$$

EXAMPLE 15.9   CALCULATING THE pH OF A SOLUTION THAT CONTAINS A WEAK ACID AND A STRONG ACID

PROBLEM   What is the pH of a solution that contains 0.10 M HCl and 0.10 M $HC_2H_3O_2$? For acetic acid, $K_a = 1.8 \times 10^{-5}$.

SOLUTION   This kind of problem fools many students because they forget that HCl is a strong acid—it's completely ionized. This means that the solution contains 0.10 M H^+ just from the HCl, plus a little bit more from the weak acid, $HC_2H_3O_2$. If we let x be the number of moles per liter of $HC_2H_3O_2$ that ionizes by the reaction

$$HC_2H_3O_2 \rightleftharpoons H^+ + C_2H_3O_2^-$$

then we can construct the concentration table as follows:

	Initial Molar Concentrations	Change	Equilibrium Molar Concentrations
H^+	0.10	$+x$	$0.10 + x \approx 0.10$
$C_2H_3O_2^-$	0	$+x$	x
$HC_2H_3O_2$	0.10	$-x$	$0.10 - x \approx 0.10$

We expect x to be small, so we've assumed $0.10 \pm x \approx 0.10$. Substituting into the K_a expression, we get

$$1.8 \times 10^{-5} = K_a = \frac{[H^+][C_2H_3O_2^-]}{[HC_2H_3O_2]}$$

$$1.8 \times 10^{-5} = \frac{(0.10)(x)}{(0.10)}$$

$$x = 1.8 \times 10^{-5}$$

We see that x is indeed small compared to 0.10, so in the solution $[H^+] = 0.10\ M$. This gives a pH of 1.00.

15.3 DISSOCIATION OF POLYPROTIC ACIDS

Acids containing more than one atom of hydrogen that can be lost on dissociation are known as polyprotic acids. Some examples are H_2SO_4 and H_2S, both of which contain two ionizable hydrogens, and H_3PO_4, which contains three. These acids lose their hydrogens one at a time. Thus we write two steps for the dissociation of sulfuric acid, each with a corresponding equation for K_a,

$$H_2SO_4 \rightleftharpoons H^+ + HSO_4^- \qquad K_{a_1} = \frac{[H^+][HSO_4^-]}{[H_2SO_4]}$$

$$HSO_4^- \rightleftharpoons H^+ + SO_4^{2-} \qquad K_{a_2} = \frac{[H^+][SO_4^{2-}]}{[HSO_4^-]}$$

and for H_2S,

$$H_2S \rightleftharpoons H^+ + HS^- \qquad K_{a_1} = \frac{[H^+][HS^-]}{[H_2S]} \qquad [15.5]$$

$$HS^- \rightleftharpoons H^+ + S^{2-} \qquad K_{a_2} = \frac{[H^+][S^{2-}]}{[HS^-]} \qquad [15.6]$$

The three steps in the dissociation of H_3PO_4 are

$$H_3PO_4 \rightleftharpoons H^+ + H_2PO_4^- \qquad K_{a_1} = \frac{[H^+][H_2PO_4^-]}{[H_3PO_4]}$$

$$H_2PO_4^- \rightleftharpoons H^+ + HPO_4^{2-} \qquad K_{a_2} = \frac{[H^+][HPO_4^{2-}]}{[H_2PO_4^-]}$$

$$HPO_4^{2-} \rightleftharpoons H^+ + PO_4^{3-} \qquad K_{a_3} = \frac{[H^+][PO_4^{3-}]}{[HPO_4^{2-}]}$$

Table 15.2
Stepwise dissociation of some polyprotic acids at 25°C

Acid	Stepwise Dissociation	Dissociation Constant for Each Step	pK_a
Phosphoric	$H_3PO_4 \rightleftharpoons H^+ + H_2PO_4^-$	$K_{a_1} = 7.5 \times 10^{-3}$	2.13
	$H_2PO_4^- \rightleftharpoons H^+ + HPO_4^{2-}$	$K_{a_2} = 6.2 \times 10^{-8}$	7.21
	$HPO_4^{2-} \rightleftharpoons H^+ + PO_4^{3-}$	$K_{a_3} = 2.2 \times 10^{-12}$	11.66
Sulfuric	$H_2SO_4 \rightleftharpoons H^+ + HSO_4^-$	$K_{a_1} = $ very large	<0
	$HSO_4^- \rightleftharpoons H^+ + SO_4^{2-}$	$K_{a_2} = 1.2 \times 10^{-2}$	1.92
Sulfurous	$H_2SO_3 \rightleftharpoons H^+ + HSO_3^-$	$K_{a_1} = 1.5 \times 10^{-2}$	1.82
	$HSO_3^- \rightleftharpoons H^+ + SO_3^{2-}$	$K_{a_2} = 1.0 \times 10^{-7}$	7.00
Hydrosulfuric	$H_2S \rightleftharpoons H^+ + HS^-$	$K_{a_1} = 1.1 \times 10^{-7}$	6.96
	$HS^- \rightleftharpoons H^+ + S^{2-}$	$K_{a_2} = 1.0 \times 10^{-14}$	14.00
Carbonic	$H_2CO_3 \rightleftharpoons H^+ + HCO_3^-$	$K_{a_1} = 4.3 \times 10^{-7}$	6.37
	$HCO_3^- \rightleftharpoons H^+ + CO_3^{2-}$	$K_{a_2} = 5.6 \times 10^{-11}$	10.26
Acorbic (vitamin C)	$H_2C_6H_6O_6 \rightleftharpoons H^+ + HC_6H_6O_6^-$	$K_{a_1} = 7.9 \times 10^{-5}$	4.10
	$HC_6H_6O_6^- \rightleftharpoons H^+ + C_6H_6O_6^{2-}$	$K_{a_2} = 1.6 \times 10^{-12}$	11.79

Table 15.2 gives some polyprotic acids and their stepwise dissociation constants. We see from this table that the first dissociation goes essentially to completion for sulfuric acid, while the second occurs only to a relatively limited degree. Because of the virtual completion of its first dissociation, sulfuric acid is considered a strong acid. We also see that the first dissociation step of each of these acids occurs with the largest value of K_a, and that each successive step occurs with an ever-decreasing value of K_a. This trend in K_a is reasonable when we consider that it should be easiest to remove an H^+ ion from an uncharged species, and that is should become progressively more difficult to do so as the negative charge on the ion increases.

Because the equilibria involving polyprotic acids are more complex than those of monoprotic acids, equilibrium calculations are somewhat more complicated. The next example shows, however, that a number of approximations can be made that simplify the approach to these kinds of problems.

EXAMPLE 15.10

CALCULATING THE CONCENTRATIONS OF SPECIES IN A SOLUTION OF A WEAK DIPROTIC ACID

PROBLEM

Hydrogen sulfide, H_2S, is a gas produced by anaerobic (bacterial action in the absence of air) decomposition of organic compounds. Its disagreeable odor is responsible for the terrible smell of rotten eggs. In water, H_2S is a diprotic weak acid. What are the equilibrium concentrations of H^+, HS^-, S^{2-}, and H_2S in a saturated (0.10 M) aqueous solution of H_2S?

SOLUTION

The equilibria involved are shown in Equations 15.5 and 15.6. From Table 15.2, the equilibrium constants have the following values: $K_{a_1} = 1.1 \times 10^{-7}$ and $K_{a_2} = 1.0 \times 10^{-14}$.

Because K_{a_1} is *so much larger* than K_{a_2}, we can safely assume that nearly all the hydrogen ion in the solution is derived from the first step of the dissociation. In addition, only very little of the HS^- formed in the first step will undergo further dissociation. On the basis of this we can calculate the H^+ and HS^- concentrations using the expression for K_{a_1} alone.

$$K_{a_1} = \frac{[H^+][HS^-]}{[H_2S]}$$

If we let x equal the number of moles per dm^3 of H_2S that dissociate, we obtain, from the stoichiometry of the first step, x mol dm^{-3} of H^+ and x mol/L of HS^-. At equilibrium, there will be $(0.10 - x)$ mol dm^{-3} of H_2S remaining.

	Initial Molar Concentrations	Change	Equilibrium Molar Concentrations
H^+	0.0	$+x$	x
HS^-	0.0	$+x$	x
H_2S	0.10	$-x$	$0.10 - x \approx 0.10$

Note that as before, because K_{a_1} is very small, we may assume that x will be negligible compared to 0.10 and write

$$[H_2S] = 0.10 - x \approx 0.10 \; M$$

Substituting these equilibrium quantities into the expression for K_{a_1}, we have

$$\frac{(x)(x)}{0.10} = 1.1 \times 10^{-7}$$

$$x^2 = 1.1 \times 10^{-8}$$

$$x = 1.0 \times 10^{-4}$$

Therefore, the equilibrium concentrations from this first dissociation are

$$[H^+] = 1.0 \times 10^{-4} \; M$$

$$[HS^-] = 1.0 \times 10^{-4} \; M$$

$$[H_2S] = 0.10 - 1.0 \times 10^{-4} = 0.10 \; M$$

Notice that our approximation in the H_2S concentration is valid, because 1.0×10^{-4} is in fact negligible compared to 0.10.

By employing K_{a_2}, we can now calculate the equilibrium concentration of S^{2-},

$$HS^- \rightleftharpoons H^+ + S^{2-}$$

and

$$K_{a_2} = \frac{[H^+][S^{2-}]}{[HS^-]}$$

If we let y equal the number of moles per dm^3 of HS^- that dissociates, then, from the stoichiometry of this second dissociation step, the number of moles per dm^3 of H^+ and S^{2-} produced would also be y. Thus the total hydrogen ion concentration from both the first and second dissociations will be $[H^+] = (1.0 \times 10^{-4} + y)$, and the concentration of

HS$^-$ that remains at equilibrium will be $(1.0 \times 10^{-4} - y)$. For this second dissociation,

	Initial Concentrations	Change	Equilibrium Concentrations
H$^+$	1.0×10^{-4}	$+y$	$1.0 \times 10^{-4} + y$
S^{2-}	0.0	$+y$	y
HS$^-$	1.0×10^{-4}	$-y$	$1.0 \times 10^{-4} - y$

These quantities may be simplified by recognizing that because K_{a_2} is so very small, the amount of HS$^-$ that will dissociate will also be very small. We can therefore make the assumption that y will be negligible compared to 1.0×10^{-4}. Our equilibrium concentrations then become

$$[H^+] = 1.0 \times 10^{-4} + y \approx 1.0 \times 10^{-4} \ M$$

$$[S^{2-}] = y$$

$$[HS^-] = 1.0 \times 10^{-4} - y \approx 1.0 \times 10^{-4} \ M$$

Substituting, we obtain

$$K_{a_2} = \frac{(1.0 \times 10^{-4})(y)}{(1.0 \times 10^{-4})} = 1.0 \times 10^{-14}$$

$$y = 1.0 \times 10^{-14}$$

Therefore,

$$[S^{2-}] = 1.0 \times 10^{-14} \ M$$

Note that the value of y is very much smaller than the value of x obtained for the first step in the dissociation.

In summary, the concentrations of all solute species present at equilibrium in a 0.10 M H$_2$S solution are

$$[H^+] = 1.0 \times 10^{-4} \ M$$

$$[HS^-] = 1.0 \times 10^{-4} \ M$$

$$[S^{2-}] = 1.0 \times 10^{-14} \ M$$

$$[H_2S] = 0.10 \ M$$

HPO$_4^{2-}$ is the anion formed in the second step of the dissociation of H$_3$PO$_4$.

Reviewing this last example, we see that in any solution containing H$_2$S as the only solute, the concentration of the sulfide ion will be equal to K_{a_2}. In fact, for any polyprotic acid where $K_{a_2} \ll K_{a_1}$, the concentration of the anion formed in the second dissociation will always equal K_{a_2}, provided, of course, that the acid is the only solute. For example, K_{a_2} for H$_3$PO$_4$ has a value of 6.2×10^{-8}, and in a solution containing only H$_3$PO$_4$ and H$_2$O, the concentration of HPO$_4^{2-}$ is $6.2 \times 10^{-8} \ M$.

Saturated solutions of H$_2$S are sometimes used in chemical analysis to detect the presence of certain cations by the formation of an insoluble sulfide precipitate. In these analyses the concentration of the S^{2-} is critical and, therefore, must be controlled. By applying Le Châtelier's principle to the dissociation of H$_2$S, we see that any increase in the H$^+$ concentration (perhaps by the addition of a strong acid) will cause a shift in the equilibrium to the left, favoring the formation of more H$_2$S and decreasing the concentrations of both the S^{2-} and HS$^-$ species. Conversely, lowering the H$^+$ concentration will increase [HS$^-$] and

[S^{2-}]. Thus, we can control the concentration of the S^{2-} in a saturated H_2S solution by varying the H^+ concentration. A useful equation expressing the relationship that exists between H^+ and S^{2-} in an H_2S solution can be derived by multiplying K_{a_1} by K_{a_2}. Thus

$$K_a = K_{a_1} \times K_{a_2} = \frac{[H^+][\cancel{HS^-}]}{[H_2S]} \times \frac{[H^+][S^{2-}]}{[\cancel{HS^-}]}$$

$$= \frac{[H^+]^2[S^{2-}]}{[H_2S]} = 1.1 \times 10^{-21} \qquad [15.7]$$

Similar equations apply for other diprotic acids, and with similar restrictions.

A word of caution about the use of Equation 15.7 is in order. *This equation can be used only when two of the three equilibrium concentrations are given and we wish to calculate the third.* It cannot be used to determine, for example, both the H^+ and S^{2-} concentrations in solutions of known H_2S concentrations. You can verify this for yourself by using it to calculate the concentrations of H^+ and S^{2-} that are present in a 0.10 M H_2S solution and then comparing your answers with those that we calculated using the two dissociation constants. The proper use of Equation 15.7 is illustrated in Example 15.11.

EXAMPLE 15.11   USING THE pH TO CONTROL THE CONCENTRATION OF THE ANION FORMED IN THE SECOND STEP IN THE IONIZATION OF A DIPROTIC ACID

PROBLEM   Calculate the S^{2-} concentration in a saturated solution (0.10 M) of H_2S whose pH was adjusted to 2.00 by the addition of HCl.

SOLUTION   Since this is a saturated H_2S solution with a known H^+ concentration, we can use Equation 15.7.

$$K_a = \frac{[H^+]^2[S^{2-}]}{[H_2S]}$$

Rearranging the equation to solve for [S^{2-}], we have

$$[S^{2-}] = \frac{K_a[H_2S]}{[H^+]^2}$$

From the pH of this acidic solution we calculate that [H^+] = 1.0×10^{-2} M; since the solution is saturated with H_2S, we know that [H_2S] = 0.10 M. Substituting these values and the value of K_a into our equation, we have

$$[S^{2-}] = \frac{(1.1 \times 10^{-21})(1.0 \times 10^{-1})}{(1.0 \times 10^{-2})^2}$$

or

$$[S^{2-}] = 1.1 \times 10^{-18} \ M$$

15.4   BUFFERS

Any solution that contains both a weak acid and a weak base has the ability to absorb small amounts of either a strong acid or strong base with very little change in pH. When small quantities of a strong acid are added, they are neutralized by the weak base, while small quantities of a strong base are neutralized by the weak acid. Such solutions are said to be **buffers** because they resist significant changes in the pH.

A buffer whose pH is less than 7 generally can be prepared by mixing a weak acid with the salt of that weak acid—for example, acetic acid and sodium acetate. A buffer whose pH is greater than 7 generally can be prepared by mixing a weak base with the salt of that weak base—for example, ammonia and ammonium chloride. When H^+ or OH^- are added to an acetic acid-acetate buffer, the following neutralization reactions take place:

$$H^+ + C_2H_3O_2^- \longrightarrow HC_2H_3O_2$$

$$OH^- + HC_2H_3O_2 \longrightarrow H_2O + C_2H_3O_2^-$$

Similarly, for the NH_3, NH_4Cl, basic buffer we have

$$H^+ + NH_3 \longrightarrow NH_4^+$$

and

$$OH^- + NH_4^+ \longrightarrow H_2O + NH_3$$

In an acid buffer the H^+ concentration (and pH) is determined by the relative concentrations of the weak acid and its conjugate base—that is, the anion. For example, for acetic acid we have

$$K_a = \frac{[H^+][C_2H_3O_2^-]}{[HC_2H_3O_2]}$$

Solving for $[H^+]$ gives

$$[H^+] = K_a\left(\frac{[HC_2H_3O_2]}{[C_2H_3O_2^-]}\right) \qquad [15.8]$$

To calculate $[H^+]$, and then pH, we must know K_a for the weak acid (from Table 15.1) as well as the ratio of the concentrations of the weak acid and its anion.

Since salts completely dissociate in aqueous solution, the number of moles of the anion in the solution from this source is determined by the formula of the salt and the number of moles of salt dissolved. Thus a 1.0 M $NaC_2H_3O_2$ solution contains 1.0 mol dm^{-3} of $C_2H_3O_2^-$, while a 1.0 M $Ca(C_2H_3O_2)_2$ solution contains 2.0 mol dm^{-3} of $C_2H_3O_2^-$. In the buffer there is also an additional amount of acetate ion that comes from the dissociation of $HC_2H_3O_2$. The amount of H^+ and $C_2H_3O_2^-$ stemming from this source is very small even in solutions containing only acetic acid, and this small amount is reduced even further in the buffer because of the presence of the large concentration of $C_2H_3O_2^-$ from the salt.[2] The total anion concentration in the buffer is essentially determined by the salt concentration alone, since the contribution from the dissociation of the weak acid is negligible. As in our previous calculations on weak acids, the concentration of $HC_2H_3O_2$ will not be reduced appreciably by its dissociation and, in a mixture of 1.0 M $NaC_2H_3O_2$ and 1.0 M $HC_2H_3O_2$, for example, the concentrations of both the molecular acid and the anion are 1.0 M. The H^+ concentration in such a buffer could be found from Equation 15.8 using $K_a = 1.8 \times 10^{-5}$ (Table 15.1).

$$[H^+] = (1.8 \times 10^{-5})\frac{1.0}{1.0}$$

$$= 1.8 \times 10^{-5} \, M$$

The pH of this solution is 4.74.

Acetic acid is a weak acid and acetate ion is its weak conjugate base.

Ammonia is a weak base and ammonium ion is its weak conjugate acid.

[2] This can easily be seen by applying Le Châtelier's principle to the dissociation of $HC_2H_3O_2$. The presence of acetate ion from a salt causes the dissociation equilibrium of the acid to be shifted to the left and actually suppresses the dissociation.

If you are taking a biology or biochemistry course, you will probably encounter an equation derived (essentially) from Equation 15.8.

$$pH = pK_a + \log \frac{[\text{anion}]}{[\text{acid}]} \qquad [15.9]$$

It is not necessary to know the Henderson–Hasselbalch equation to do buffer problems. You can always start with the K_a or K_b expression itself.

This is called the Henderson–Hasselbalch equation.[3] For acetic acid, $pK_a = 4.74$, and for a buffer having $[HC_2H_3O_2] = [C_2H_3O_2^-] = 1.0\ M$.

$$pH = 4.74 + \log \left(\frac{1.0}{1.0}\right)$$

$$= 4.74 + \log 1$$

log 1 = 0

$$= 4.74$$

As before, we find the pH equal to 4.74.

Notice that when the concentrations of the acid and anion are the same in a buffer, the H^+ concentration in that solution is equal to the K_a of the weak acid, and the $pH = pK_a$. Thus, if a buffer was prepared by mixing 0.1 mol of formic acid ($K_a = 1.8 \times 10^{-4}$ from Table 15.1) and 0.1 mol of sodium formate into 1 dm^3 of solution, the resulting H^+ concentration would be

$$[H^+] = 1.8 \times 10^{-4}\ M$$

and

$$pH = 3.74$$

We can also use Equation 15.8 to calculate the concentrations of acid and salt that would be needed to achieve a certain pH buffer. This is illustrated in Example 15.12.

EXAMPLE 15.12 PREPARING A BUFFER WITH A SPECIFIED pH

PROBLEM   What ratio of acetic acid to sodium acetate concentration is needed to form a buffer whose pH is 5.70?

SOLUTION   To solve this problem we need to rearrange Equation 15.8 to solve for the ratio of concentrations. Thus

$$\frac{[HC_2H_3O_2]}{[C_2H_3O_2^-]} = \frac{[H^+]}{K_a}$$

The H^+ concentration when the pH is 5.70 is

$$[H^+] = 2.0 \times 10^{-6}\ M$$

Therefore,

$$\frac{[HC_2H_3O_2]}{[C_2H_3O_2^-]} = \frac{2.0 \times 10^{-6}}{1.8 \times 10^{-5}} = \frac{2.0 \times 10^{-6}}{18 \times 10^{-6}}$$

or

$$\frac{[HC_2H_3O_2]}{[C_2H_3O_2^-]} = \frac{1}{9}$$

---

[3] A similar equation that would apply to a basic buffer (e.g., NH_3, NH_4Cl) is

$$pOH = pK_b + \log \frac{[\text{cation, } BH^+]}{[\text{base, } B]}$$

As long as this ratio is maintained, the pH of an acetic acid–sodium acetate buffer is 5.70. For example, if 0.2 mol of $HC_2H_3O_2$ and 1.8 mol of $NaC_2H_3O_2$ are dissolved in 1 dm^3 of solution, the pH is 5.70. This same pH will result if 0.1 mol of $HC_2H_3O_2$ and 0.9 mol of $NaC_2H_3O_2$ are dissolved.

We have seen that by adjusting the ratio of concentrations of the weak acid to that of the salt, a buffer of almost any desired pH can be achieved. For example, a buffer composed of 0.010 mol of $HC_2H_3O_2$ and 1.0 mol of $NaC_2H_3O_2$ would have a pH of 6.74. However, when as little as 0.010 mol of base is added to this buffer, all of the acetic acid is neutralized and a large change in the pH of the buffer results. Therefore, *the most effective pH range for any buffer is at or near the pH where the acid and salt concentrations are equal (that is, pK_a)*. Also, in order to be most effective, the amounts of weak acid and base used to prepare the buffer must be considerably greater than the amounts of acid or base that may later be added to the buffer.

The larger the concentrations of the components of the buffer, the more effective it is at resisting changes in pH.

Let us now see how effective a buffer is at holding the pH nearly constant. Suppose we have an acetic acid-acetate buffer in which the concentrations of $HC_2H_3O_2$ and $C_2H_3O_2^-$ are each 1.00 M. We saw earlier that its pH will be 4.74 and $[H^+] = 1.8 \times 10^{-5}$ M. What happens to the pH if we add, say, 0.20 mol of HCl to 1 dm^3 of this buffer?

When a strong acid such as HCl is added, its H^+ reacts with acetate ion.

Any H^+ from a strong acid converts the conjugate weak base in the buffer to its corresponding conjugate acid.

$$H^+ + C_2H_3O_2^- \longrightarrow HC_2H_3O_2.$$

This decreases the $C_2H_3O_2^-$ concentration and increases the $HC_2H_3O_2$ concentration. Below are the number of moles per dm^3 of all species before and after the addition:

Initial	Final
$[H^+] = 1.8 \times 10^{-5}$ M	$[H^+] = x$
$[C_2H_3O_2^-] = 1.00$ M	$[C_2H_3O_2^-] = 1.00 - 0.20$ $M = 0.80$ M
$[HC_2H_3O_2] = 1.00$ M	$[HC_2H_3O_2] = 1.00 + 0.20$ $M = 1.20$ M

Substituting these final concentrations into Equation 15.8, we have

$$[H^+] = (1.8 \times 10^{-5}) \times \left(\frac{1.20}{0.80}\right)$$

$$= 2.7 \times 10^{-5} \ M$$

The pH of the buffer after the 0.20 mol of H^+ is added is 4.57—a change of only 0.17 pH units from its initial pH of 4.74.

Suppose now that we were to add 0.20 mol of H^+ to 1 dm^3 of a solution of HCl whose pH = 4.74 (that is, a 1.8×10^{-5} M HCl solution). Since Cl^- has virtually no tendency to react with H^+, the final H^+ concentration will be 0.20 M ($0.20 + 1.8 \times 10^{-5} = 0.20$), and the pH of the solution will be 0.70. The change in pH in this case is 4.04 pH units, as opposed to a change of only 0.17 pH units when the same quantity of H^+ is added to the buffer.

Additions of strong base are also absorbed by the buffer. When 0.20 mol of OH^- is added to the 1 dm^3 of our original buffer, it is neutralized according to the reaction

Any OH^- from a strong base converts the conjugate weak acid in the buffer to its corresponding conjugate base.

$$OH^- + HC_2H_3O_2 \longrightarrow H_2O + C_2H_3O_2^-$$

The number of moles per dm^3 before the addition are the same as above, but the number of moles per dm^3 after 0.20 mol of OH^- is added would be

Concentration After Addition of 0.20 mol of OH^-
$[H^+] = x$
$[C_2H_3O_2^-] = 1.00 + 0.20 = 1.20\ M$
$[HC_2H_3O_2] = 1.00 - 0.20 = 0.80\ M$

Substituting these values into Equation 15.8, we find that the H^+ concentration after the addition is $1.2 \times 10^{-5}\ M$ and the resulting pH is 4.92—once again, a small change of 0.18 pH units. Finally, note that the addition of an acid to the buffer lowers the pH, while the addition of a base raises the pH. Although the change in pH is small, the *direction* of the change is as expected.

EXAMPLE 15.13

CALCULATING THE pH OF A BUFFER AND THE CHANGE IN pH WHEN A STRONG ACID IS ADDED

PROBLEM A buffer was prepared by mixing exactly 200 cm^3 of a 0.60 M NH_3 solution and 300 cm^3 of a 0.30 M NH_4Cl solution. (a) What is the pH of this buffer, if we assume a final volume of 500 cm^3? (b) What will be the pH after 0.020 mol of H^+ is added?

SOLUTION The number of moles of NH_3 added to this solution is

$$0.60\ \frac{mol}{dm^3} \times 0.200\ dm^3 = 0.12\ mol$$

and the number of moles of NH_4^+ added is

$$0.30\ \frac{mol}{dm^3} \times 0.300\ dm^3 = 0.090\ mol$$

Therefore, the concentrations of these in the 500 cm^3 is

$$[NH_3] = \frac{0.12\ mol}{0.500\ dm^3} = 0.24\ M$$

$$[NH_4^+] = \frac{0.090\ mol}{0.500\ dm^3} = 0.18\ M$$

(a) The OH^- concentration for this buffer is found by using the K_b for NH_3:

$$NH_3 + H_2O \rightleftharpoons NH_4^+ + OH^-$$

$$K_b = \frac{[NH_4^+][OH^-]}{[NH_3]}$$

Rearranging and solving for $[OH^-]$, we have

$$[OH^-] = K_b \frac{[NH_3]}{[NH_4^+]}$$

Substituting K_b for NH_3 from Table 15.1 and the concentrations of NH_3 and NH_4^+ for this buffer into this equation gives

$$[OH^-] = (1.8 \times 10^{-5}) \times \left(\frac{0.24}{0.18}\right)$$

$$= 2.4 \times 10^{-5} M$$

$$pOH = 4.62$$

$$pH = 9.38$$

(b) The neutralization reaction for H^+ in this buffer is

$$H^+ + NH_3 \longrightarrow NH_4^+$$

We are adding 0.020 mol of H^+ to 500 cm^3, or 0.040 mol of H^+ per dm^3. Therefore, the concentrations before and after the addition of the acid are

Notice that we calculate the amount of H^+ added *per dm^3*.

Initial	Final
$[OH^-] = 2.4 \times 10^{-5} M$	$[OH^-] = x$
$[NH_3] = 0.24 M$	$[NH_3] = 0.24 - 0.040 M = 0.20 M$
$[NH_4^+] = 0.18 M$	$[NH_4^+] = 0.18 + 0.040 M = 0.22 M$

and the OH^- concentration is

$$[OH^-] = (1.8 \times 10^{-5}) \times \left(\frac{0.20}{0.22} \right)$$

$$= 1.6 \times 10^{-5} M$$

$$pOH = 4.80$$

The pH is therefore 9.20, a decrease of 0.18 pH units.

Buffers find many important applications. Living systems employ buffers to maintain nearly constant pH so that biochemical reactions can follow their correct paths. For example, blood contains, among other things, a H_2CO_3/HCO_3^- buffer system that helps maintain the pH at 7.4.

HCO_3^- is itself a buffer. In the photo at the beginning of the chapter, $NaHCO_3$ is being added to a swimming pool to control the pool's pH.

In the laboratory many inorganic and organic chemical reactions are performed in buffered solutions to minimize any adverse effects caused by acids or bases that might be consumed or produced during reaction.

15.5 HYDROLYSIS OF SALTS

In our previous discussions we have spoken of the reaction of an acid and a base as neutralization, and in Chapter 7 we pointed out that this reaction in aqueous solutions produces a salt plus water. From this you might come to the conclusion that when stoichiometric amounts of acid and base have been combined, the solution of the salt thus formed will be neutral in the sense of having a pH of 7. This is, in fact, true for the salt NaCl—sodium chloride solutions are neutral. But this generalization doesn't hold for all salts. For example, an aqueous solution of sodium acetate, $NaC_2H_3O_2$, is not neutral, but instead slightly basic. On the other hand, a solution of ammonium chloride, NH_4Cl, is slightly acidic. The origin of the basicity or acidity of some salt solutions is not difficult to understand, but to explain it we first must reexamine some notions about the relative strengths of Bronsted acids and bases.

In Chapter 7 we compared the strengths of HCl and HF. By now you know that HCl is a strong acid, so the position of equilibrium in the reaction

$$HCl(aq) + H_2O \rightleftharpoons H_3O^+(aq) + Cl^-(aq) \quad K_a \text{ very large}$$

lies very far to the right. The equilibrium constant for the ionization is so large that we usually don't even bother to write this as an equilibrium. On the other hand, HF is a weak acid that ionizes according to the equation

$$HF(aq) + H_2O \rightleftharpoons H_3O^+(aq) + F^-(aq) \quad K_a = 6.5 \times 10^{-4}$$

and the small value of K_a means that only a small degree of ionization actually occurs.

In Chapter 7 we commented that this information tells us something about the relative acidities of HCl and HF. HCl is a stronger acid than HF because it is better able to donate its protons to water molecules, as evidenced by the extent to which the reactions proceed toward completion. But these reactions also tell us something about the strengths of the conjugate bases of HCl and HF (Cl^- and F^-, respectively). The equilibrium constant for the ionization of HCl is so large because the reverse reaction, which is the capture of a proton by the chloride ion, has very little tendency to occur. This must mean that chloride ion is a very weak base. On the other hand, the fact that the ionization of HF does not go very far toward completion means that the fluoride ion must be a stronger base than Cl^-, because it is able to capture protons to some degree to reform molecules of HF.

We can make similar comparisons between the strengths of bases and their conjugate acids. Compare, for example the reactions of oxide ion and ammonia with water.

$$O^{2-}(aq) + H_2O \rightleftharpoons OH^-(aq) + OH^-(aq) \quad K_b \text{ is very large}$$

$$NH_3(aq) + H_2O \rightleftharpoons NH_4^+(aq) + OH^-(aq) \quad K_b = 1.8 \times 10^{-5}$$

Because K_b is much larger for O^{2-} than NH_3, oxide ion is a much stronger base than ammonia. In fact, it is such a strong base that the reaction proceeds essentially to completion and there is no detectable O^{2-} in the solution. The equilibrium constants for these reactions also allow us to compare the acidities of the conjugate acids, OH^- and NH_4^+. Because the first reaction proceeds so far toward completion, OH^- must be a very weak acid. But because the second reaction has a relatively small equilibrium constant, the ammonium ion must be at least a moderately weak acid. The NH_4^+ ion is acidic enough to cause some of the OH^- ions in the solution to accept protons so that it can become NH_3 again.

In general, there is a simple relationship between the relative strengths of acids and their conjugate bases—*the stronger an acid is, the weaker is its conjugate base.* Thus very strong acids such as HCl or HNO_3 have *very* weak conjugate bases, while acids that we have been classifying as weak have conjugate bases that are only *moderately* weak. And very weak acids, such as hydroxide ion, for example, have very strong conjugate bases. (OH^- is a very weak acid and its conjugate base, O^{2-}, is a strong base.) This inverse relationship between the strengths of conjugate acids and bases is illustrated in Figure 15.2.

Now that you know the way the strengths of conjugate acids and bases are related, let's study why solutions of some salts are neutral, while solutions of others are acidic or basic. In doing this, it is usually helpful to classify the salt according to the nature of the acid and base that would react to form it.

Salts of strong acids and strong bases

Sodium chloride is typical of this type of salt. It would be formed from the acid HCl and the base NaOH, both of which are strong. A solution of NaCl contains the ions Na^+ and Cl^-, and what we wish to know is whether either of these ions can cause the solution to be acidic or basic. What can we say about the strengths of these ions as acids or bases?

Figure 15.2

Relative strength of acid-base pairs.

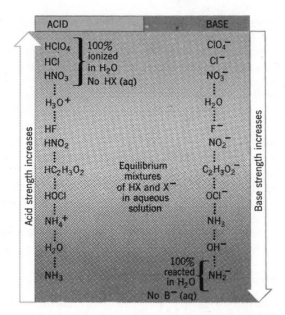

Metal cations with small positive charges generally do not affect the pH of a solution.

The sodium ion contains no hydrogen, so it can't be a Brønsted acid, and it certainly has no tendency to accept H^+ ions. Therefore, Na^+ is not able to either accept or donate protons and cannot affect the acidity of an aqueous solution. The chloride ion, Cl^-, is the conjugate base of HCl. Since HCl is a very strong acid, Cl^- is an extremely weak base and has virtually no tendency to accept protons. Chloride ion, therefore, has no effect on the acidity either. Because neither Na^+ nor Cl^- can affect the concentrations of H^+ or OH^- in water, solutions of sodium chloride are neutral and we can expect that their solutions would have a pH of 7.

Aqueous solutions in which the only solute is the salt of a strong acid and strong base are neutral.

A similar analysis for other salts of strong acids and strong bases will give the same result, so we can make the generalization: *Salts formed by the reaction of strong acids with strong bases have no effect on the pH of a solution in which they are dissolved.*

Salts of weak acids and strong bases

Sodium acetate is typical of this kind of salt. It is formed in the reaction between acetic acid, $HC_2H_3O_2$, and sodium hydroxide, and a solution of $NaC_2H_3O_2$ contains the ions Na^+ and $C_2H_3O_2^-$. What effect do these ions have on the acidity of a solution?

Sodium ion, as noted above, is neither an acid nor a base, so it doesn't affect the pH of the solution. Acetate ion, on the other hand, is the conjugate base of a weak acid, $HC_2H_3O_2$. This means that $C_2H_3O_2^-$ must be at least a moderately weak base. It reacts with water following the general equation given on page 544.

$C_2H_3O_2^-$ is a weak base.

$$C_2H_3O_2^- + H_2O \rightleftharpoons HC_2H_3O_2 + OH^-$$

Note the similarity between this reaction and the reaction of the base NH_3 with water

$$NH_3 + H_2O \rightleftharpoons NH_4^+ + OH^-$$

The only difference is that the base $C_2H_3O_2^-$ is an anion instead of a neutral molecule.

Because the reaction of $C_2H_3O_2^-$ produces some hydroxide ion, whereas there is no compensating reaction to produce H^+, a solution of $NaC_2H_3O_2$ will have a slight excess of OH^- and it will be slightly basic. Other analyses of salts of the same type will yield similar results, so we can say: *An aqueous solution of the salt of a weak acid and a strong base will be slightly basic because the anion of the salt is a weak Brønsted base.*

The title of this section contains the word "hydrolysis," and we haven't defined it up till now. **Hydrolysis** means *reaction with water*, and if we examine the ionic reaction in a solution of sodium acetate,

$$Na^+ + C_2H_3O_2^- + H_2O \rightleftharpoons Na^+ + HC_2H_3O_2 + OH^-$$

we see that there is indeed a reaction between the salt and water that affects the pH of the solution. In fact, in this solution we could even say that it is the anion that undergoes hydrolysis. This is another word that you should add to your vocabulary, but keep in mind that the reaction that takes place is no different than other reactions between acids and bases and water.

Salts of strong acids and weak bases

An example of this type of salt is ammonium chloride, which is formed in the reaction of the weak base ammonia, NH_3, and the strong acid HCl. A solution of ammonium chloride contains the ions NH_4^+ and Cl^-. As before, let's analyze how each of these affects the pH of the solution.

Ammonium ion is the conjugate acid of ammonia. Since NH_3 is a weak base, NH_4^+ must be at least a moderately weak acid. The general reaction of weak acids was given on page 543, and ammonium ion undergoes this same kind of reaction.

NH_4^+ is a weak acid in water.

$$NH_4^+ + H_2O \rightleftharpoons H_3O^+ + NH_3$$

or, if we abbreviate H_3O^+ as H^+ and leave out the water,

$$NH_4^+ \rightleftharpoons H^+ + NH_3$$

Notice the similarity between this reaction and the ionization of a weak acid such as $HC_2H_3O_2$.

$$HC_2H_3O_2 \rightleftharpoons H^+ + C_2H_3O_2^-$$

In this case, the only major difference is that the acid NH_4^+ is an ion, whereas $HC_2H_3O_2$ is a molecule.

In a solution of ammonium chloride, the ammonium ion undergoes a reaction with water. As we learned earlier, however, the chloride ion is such a weak base that it doesn't react. Therefore, in a solution of NH_4Cl, only the ammonium ion is able to affect the pH, and because its reaction adds H^+ to the solution, the solution becomes slightly acidic and has a pH less than 7. For solutions of salts of strong acids and weak bases, the following holds: *An aqueous solution of the salt of a strong acid and a weak base will be slightly acidic because the cation of the salt is a weak Brønsted acid.* In this case, it is the cation that undergoes hydrolysis.

Type of Salt	pH of Aqueous Solution
Strong acid—strong base	7
Weak acid—strong base	>7
Strong acid—weak base	<7
Weak acid—weak base	Depends on salt

Calculations involving the hydrolysis of salts

The equilibrium constants for the reactions of NH_4^+ and $C_2H_3O_2^-$ with water are sometimes called *hydrolysis constants* and given the symbol K_h.

To calculate the effect that acetate ion or ammonium ion has on the pH of a solution we obviously need the appropriate equilibrium constants. However, if you search through tables, you probably won't find the K_b for $C_2H_3O_2^-$ or the K_a for NH_4^+. This is because there is really no need to tabulate them—there is a very simple relationship between K_a and K_b for the members of an acid-base conjugate pair. This is

$$K_a \times K_b = K_w$$

To see why this is so, let's multiply the mass action expression corresponding to K_a for $HC_2H_3O_2$ with the mass action expression for K_b for the acetate ion. The ionization of $HC_2H_3O_2$ follows the equation

$$HC_2H_3O_2 \rightleftharpoons H^+ + C_2H_3O_2^-$$

for which we write

$$K_a = \frac{[H^+][C_2H_3O_2^-]}{[HC_2H_3O_2]}$$

The reaction of acetate ion as a base is

$$C_2H_3O_2^- + H_2O \rightleftharpoons HC_2H_3O_2 + OH^-$$

for which we write

$$K_b = \frac{[HC_2H_3O_2][OH^-]}{[C_2H_3O_2^-]}$$

Multiplying these expressions together gives

$$K_a \times K_b = \frac{[H^+][\cancel{C_2H_3O_2^-}]}{[\cancel{HC_2H_3O_2}]} \times \frac{[\cancel{HC_2H_3O_2}][OH^-]}{[\cancel{C_2H_3O_2^-}]} = [H^+][OH^-] = K_w$$

Thus, if we know K_a we can calculate K_b and vice versa. Let's tackle a few hydrolysis problems now, to see how these concepts are applied.

EXAMPLE 15.14 CALCULATING THE pH OF A SALT SOLTION

PROBLEM Calculate the pH of a 0.10 M solution of potassium nitrite, KNO_2.

SOLUTION Hydrolysis problems are easily recognized because they deal with a question involving the acidity of a solution of a salt. The first step in solving them is to write the correct equation for the equilibrium that is involved. The generalizations developed in the earlier parts of this section can help us do this.

The salt KNO_2 would be formed by the reaction of KOH and HNO_2. Potassium hydroxide is a metal hydroxide, which is ionic and fully dissociated in water; it is a strong base. However, HNO_2 (nitrous acid) is a weak acid. Thus KNO_2 is a salt of a weak acid and a strong base, so we can expect that it is the anion (NO_2^-) that hydrolyzes. The NO_2^- ion is the conjugate base of HNO_2, so the equation for its reaction is

$$NO_2^- + H_2O \rightleftharpoons HNO_2 + OH^-$$

for which we can write

$$K_b = \frac{[HNO_2][OH^-]}{[NO_2^-]}$$

If we examine Table 15.1, we do not find K_b for NO_2^-, but we are able to find K_a for HNO_2: $K_a = 4.5 \times 10^{-4}$. Therefore, we have to calculate K_b.

$$K_b = \frac{K_w}{K_a}$$

$$= \frac{1.0 \times 10^{-14}}{4.5 \times 10^{-4}}$$

$$= 2.2 \times 10^{-11}$$

Because K_b is so small, we know that the position of equilibrium lies far to the left in favor of NO_2^-. Therefore, if we let x be the number of moles per dm^3 of NO_2^- that react, we can set up our concentration table and expect that x will be small compared to 0.10.

	Initial Molar Concentration	Change	Equilibrium Molar Concentration
NO_2^-	0.10	$-x$	$0.10 - x \approx 0.10$
HNO_2	0.0	$+x$	x
OH^-	0.0	$+x$	x

Substituting values into the equilibrium expression gives

$$K_b = \frac{(x)(x)}{0.10} = 2.2 \times 10^{-11}$$

$$x^2 = 2.2 \times 10^{-12}$$

$$x = 1.5 \times 10^{-6}$$

1.5×10^{-6} is negligible compared to 0.10.

Thus

$$[OH^-] = 1.5 \times 10^{-6}\ M$$

$$pOH = 5.82$$

and therefore

$$pH = 8.18$$

Notice that the pH indicates that the solution is basic.

EXAMPLE 15.15   CALCULATING THE pH OF A SALT SOLUTION

PROBLEM   What is the pH of a 0.10 M solution of N_2H_5Cl?

SOLUTION   First we have to realize that this is a salt that gives the ions $N_2H_5^+$ and Cl^- in solution. We also must recognize that these ions come from a strong acid (HCl) and a weak base (N_2H_4). This means that only the cation, $N_2H_5^+$, affects the pH of the solution. Since $N_2H_5^+$ is the conjugate acid of N_2H_4, its reaction in water is

The only strong bases you have to worry about are metal hydroxides. Since N_2H_4 isn't a metal hydroxide, it must be a *weak base*.

$$N_2H_5^+ \rightleftharpoons H^+ + N_2H_4$$

for which we can write

$$K_a = \frac{[H^+][N_2H_4]}{[N_2H_5^+]}$$

Searching Table 15.1, we find K_b for N_2H_4, but not K_a for $N_2H_5^+$, so we have to calculate K_a. For N_2H_4, $K_b = 1.7 \times 10^{-6}$; therefore

$$K_a = \frac{K_w}{K_b}$$

$$K_a = \frac{1.0 \times 10^{-14}}{1.7 \times 10^{-6}} = 5.9 \times 10^{-9}$$

Once again, we can expect very little reaction to occur as the system approaches equilibrium. If we let x equal the number of moles of $N_2H_5^+$ that react, then our concentration table becomes:

	Initial Molar Concentrations	Change	Equilibrium Molar Concentrations
$N_2H_5^+$	0.10	$-x$	$0.10 - x \approx 0.10$
H^+	0.0	$+x$	x
N_2H_4	0.0	$+x$	x

Substituting equilibrium quantities into the expression for K_a gives

$$\frac{(x)(x)}{0.10} = 5.9 \times 10^{-9}$$

$$x^2 = 5.9 \times 10^{-10}$$

$$x = 2.4 \times 10^{-5}$$

2.4×10^{-5} is negligible compared to 0.10.

Therefore

$$[H^+] = 2.4 \times 10^{-5} \ M$$

and

$$pH = 4.62$$

Notice that the solution is acidic, which is what is expected.

Salts of weak acids and weak bases

Solutions of this type of salt can be either acidic, neutral, or basic, because both the cation and anion of the salt undergo hydrolysis. The pH of such a salt solution is determined by the relative extent of the reaction of each ion. By applying what we have learned earlier, we should be able to predict, at least qualitatively, whether the solution will be acidic or basic. If the K_a of the acid in the solution is larger than the K_b of the base, then the acid is stronger than the base and the solution will be acidic. If the situation is reversed ($K_b > K_a$), then the base is stronger than the acid and the solution should be basic. And if K_a and K_b are equal, then both ions react to the same extent and the solution should be neutral. Let's examine some examples.

Consider the salt NH_4CN. This is the salt of NH_3 and HCN, both of which are weak. The K_a for NH_4^+ is calculated from K_b for NH_3, and K_b for CN^- is calculated from K_a for HCN.

$$\text{For NH}_4{}^+, \qquad K_a = \frac{1.0 \times 10^{-14}}{1.8 \times 10^{-5}} = 5.6 \times 10^{-10}$$

$$\text{For CN}^- \qquad K_b = \frac{1.0 \times 10^{-14}}{4.9 \times 10^{-10}} = 2.0 \times 10^{-5}$$

These equilibrium constants tell us that CN^- as a base is stronger than $NH_4{}^+$ as an acid. Therefore, more OH^- will be produced than H^+, and the solution will be basic.

Following similar reasoning, we would conclude that a solution of $NH_4C_2H_3O_2$ is neutral. Ammonium ion is the conjugate acid of ammonia, whose value of K_b is 1.8×10^{-5}. Acetate ion is the conjugate base of acetic acid, whose value of K_a is also 1.8×10^{-5}. This means that the K_a for $NH_4{}^+$ will be equal to the K_b for $C_2H_3O_2{}^-$, and both hydrolysis reactions will proceed to the same degree. This will give equal amounts of H^+ and OH^-, so the solution will have to be neutral.

EXAMPLE 15.16

PREDICTING HYDROLYSIS OF SALTS OF WEAK ACIDS AND WEAK BASES

PROBLEM

Will a solution of ammonium formate, NH_4CHO_2, be acidic, neutral, or basic?

SOLUTION

The ions of this salt are $NH_4{}^+$ and $CHO_2{}^-$. Ammonium ion is the conjugate acid of NH_3, and we've calculated its K_a to be 5.6×10^{-10}. The formate ion, $CHO_2{}^-$, is the conjugate base of formic acid, $HCHO_2$, whose K_a is 1.8×10^{-4}. The K_b for $CHO_2{}^-$ is therefore

$$K_b = \frac{1.0 \times 10^{-14}}{1.8 \times 10^{-4}} = 5.6 \times 10^{-11}$$

In this case, the K_a for $NH_4{}^+$ is slightly larger than the K_b for $CHO_2{}^-$, so the solution will be slightly acidic.

Hydrolysis of salts of polyprotic acids

A typical example of this kind of salt is Na_2S, the salt of the weak acid H_2S and the strong base NaOH. As you learned in our earlier discussions, Na^+ has no effect on the pH of the solution, so we only have to consider the hydrolysis of the anion, S^{2-}. In this case, however, there are two stages to the hydrolysis, each of which has its own equilibrium constant.

$$\text{Step 1} \qquad S^{2-} + H_2O \rightleftharpoons HS^- + OH^-$$

$$\text{Step 2} \qquad HS^- + H_2O \rightleftharpoons H_2S + OH^-$$

$$K_{b_1} = \frac{[HS^-][OH^-]}{[S^{2-}]}$$

$$K_{b_2} = \frac{[H_2S][OH^-]}{[HS^-]}$$

The first step in the hydrolysis involves the reaction of S^{2-}, which is the conjugate base of HS^-. This means that in computing K_{b_1}, we must use the K_a for HS^-. This K_a is actually the *second* ionization constant for H_2S, K_{a_2}, which is found in Table 15.2. Similarly, to calculate the value of K_{b_2}, we must use the K_a for H_2S, which is K_{a_1}. Thus

$$K_{b_1} = \frac{K_w}{K_{a_2}}$$

$$= \frac{1.0 \times 10^{-14}}{1.0 \times 10^{-14}} = 1.0$$

and

$$K_{b_2} = \frac{K_w}{K_{a_1}}$$

$$= \frac{1.0 \times 10^{-14}}{1.1 \times 10^{-7}} = 9.1 \times 10^{-8}$$

This is similar to solutions of diprotic acids in which we only need to consider the first ionization step when we calculate the pH.

Notice that the value of K_{b_2} is much smaller than K_{b_1}. The relative magnitudes of these two equilibrium constants indicate that the second step in the hydrolysis occurs to a negligible extent compared to the first, so essentially all the OH^- in the solution comes from the first step. Therefore, if we are only interested in the pH of the solution, we only need to use K_{b_1}. This is shown in the next example.

EXAMPLE 15.17 CALCULATING THE pH OF A SOLUTION OF A SALT OF A POLYPROTIC ACID

PROBLEM What is the pH of a 0.20 M solution of sodium sulfide, Na_2S?

SOLUTION

Sodium sulfide is a chemical that is used in dehairing animal hides.

From our previous discussion we know that we only need to consider the first step in the hydrolysis. The equation for the reaction is

$$S^{2-} + H_2O \rightleftharpoons HS^- + OH^-$$

and the equilibrium expression is

$$K_{b_1} = \frac{[HS^-][OH^-]}{[S^{2-}]} = 1.0$$

As usual, we let x equal the number of moles per dm^3 of S^{2-} that hydrolyze. Then

	Initial Molar Concentrations	Change	Equilibrium Molar Concentrations
HS^-	0.0	$+x$	x
OH^-	0.0	$+x$	x
S^{2-}	0.20	$-x$	$0.20 - x$

Our first reaction here is to make the simplifying assumption that $0.20 - x \approx 0.20$. However, if we do so and then solve for x, we obtain $x = 0.45$. This is clearly impossible because the equilibrium sulfide concentration becomes $-0.25\ M$. Negative concentrations are absurd—we can't have less than nothing! Therefore, we cannot simplify the S^{2-} concentration term in our usual way, and when we substitute the quantities corresponding to the equilibrium concentrations into the mass action expression we obtain

$$\frac{(x)(x)}{(0.20 - x)} = 1.0$$

Multiplying both sides by $(0.20 - x)$ gives

$$x^2 = (0.20 - x)1.0$$

which can be rearranged as

$$x^2 + x - 0.20 = 0$$

This is a quadratic equation whose roots can be obtained using the quadratic formula. This gives

The roots of the quadratic equation, $ax^2 + bx + c = 0$, are given by the quadratic formula

$$x = \frac{-b \pm \sqrt{b^2 - 4ac}}{2a}$$

$$x = \frac{-1 \pm \sqrt{(1)^2 - 4(1)(-0.20)}}{(2)(1)}$$

$$x = \frac{-1 \pm \sqrt{1.8}}{2} = \frac{-1 \pm 1.34}{2}$$

Notice that two values of x are obtained.

$$x = \frac{-2.34}{2} = -1.17$$

$$x = \frac{0.34}{2} = 0.17$$

The first value of x makes no sense. It has no physical meaning because it tells us that the concentrations of HS^- and OH^- are negative. As we said before, we cannot have less than nothing. The second value of x is meaningful, and we conclude that

$$x = 0.17 \ M$$

and therefore,

$$[OH^-] = 0.17 \ M$$

from which we obtain

Sulfide ion is a pretty strong base.

$$pOH = 0.77$$

Thus, the pH of the solution is 13.23.

15.6 ACID-BASE TITRATIONS: THE EQUIVALENCE POINT

In Chapter 8 we saw that a titration is a useful and accurate way of determining the concentrations of acids and bases, provided that the equivalence point can be detected. The equivalence point, you should remember, occurs when equal numbers of equivalents of acid and base have been combined. In this section we will see how the pH of a solution changes during the course of typical acid-base titrations and what the pH is at the equivalence point.

Titration of a strong acid with a strong base

A typical example of a titration of a strong acid with a strong base occurs when 25.00 cm^3 of 0.10 M HCl is titrated with 0.10 M NaOH. We can mathematically determine the pH throughout the titration by calculating the H^+ concentration present in the flask each time a quantity of NaOH is added to the HCl. For example, the number of moles of H^+ present in the 25 cm^3 of a 0.10 M HCl solution is

$$\left(\frac{0.10 \text{ mol}}{1000 \text{ cm}^3} \right) \times 25 \text{ cm}^3 = 2.5 \times 10^{-3} \text{ mol of } H^+$$

Table 15.3
Titration of 25 cm³ of 0.10 M HCl with a 0.10 M NaOH solution

Volume of HCl (cm³)	Volume of NaOH (cm³)	Total Volume (cm³)	Moles of H⁺	Moles of OH⁻	Concentration of Ion in Excess (M)	pH
25.00	0.00	25.00	2.5×10^{-3}	0	0.10 (H⁺)	1.00
25.00	10.00	35.00	2.5×10^{-3}	1.0×10^{-3}	4.3×10^{-2} (H⁺)	1.37
25.00	24.99	49.99	2.5×10^{-3}	2.499×10^{-3}	2.0×10^{-5} (H⁺)	4.70
25.00	25.00	50.00	2.5×10^{-3}	2.50×10^{-3}	0	7.00
25.00	25.01	50.01	2.5×10^{-3}	2.501×10^{-3}	2.0×10^{-5} (OH⁻)	9.30
25.00	26.00	51.00	2.5×10^{-3}	2.60×10^{-3}	2.0×10^{-3} (OH⁻)	11.30
25.00	50.00	75.00	2.5×10^{-3}	5.0×10^{-3}	3.3×10^{-2} (OH⁻)	12.52

When 10 cm³ of the 0.10 M NaOH are added, we in fact have added

$$\left(\frac{0.10 \text{ mol}}{1000 \text{ cm}^3}\right) \times 10 \text{ cm}^3 = 1.0 \times 10^{-3} \text{ mol of OH}^-$$

The neutralization reaction,

$$\text{H}^+ + \text{OH}^- \longrightarrow \text{H}_2\text{O}$$

occurs, and the amount of H⁺ remaining is

$$(2.5 \times 10^{-3}) - (1.0 \times 10^{-3}) = 1.5 \times 10^{-3} \text{ mol of H}^+$$

The molar concentration of H⁺ is now

$$[\text{H}^+] = \frac{1.5 \times 10^{-3} \text{ mol}}{0.035 \text{ dm}^3} = 4.3 \times 10^{-2} M$$

The total volume

and the pH is calculated to be 1.37. The concentrations of H⁺ after further additions of NaOH have occurred are summarized in Table 15.3.

Our calculations show that the pH increases slowly at first, then rises rapidly near the equivalence point, and finally levels off gradually after the equivalence point is reached.

If a graph is drawn of pH versus the volume of base added, we obtain the plot shown in Figure 15.3. The equivalence point occurs, in this case, at a pH of 7. At the equivalence point the solution is neutral because neither of the ions of the salt that is left in solution (NaCl) undergoes hydrolysis.

Titration using a weak acid and a strong base

In an acid-base titration in which one substance is strong and the other weak, the solution is not neutral at the equivalence point because of the hydrolysis of the salt. For example, consider the titration of 25.0 cm³ of 0.10 M HC₂H₃O₂ with 0.10 M NaOH. Before any base is added the only solute is acetic acid. The pH of the solution is calculated as shown in Section 15.2—for 0.10 M HC₂H₃O₂, the pH = 2.89.

When we begin to add NaOH, acetic acid molecules are converted to acetate ions.

$$\text{HC}_2\text{H}_3\text{O}_2 + \text{OH}^- \longrightarrow \text{H}_2\text{O} + \text{C}_2\text{H}_3\text{O}_2{}^-$$

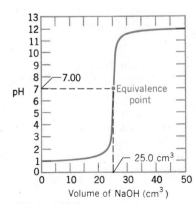

Figure 15.3
Titration of 0.10 M HCl with 0.10 M NaOH.

Since the solution then contains both $HC_2H_3O_2$ and $C_2H_3O_2^-$, it is a buffer, and we've also learned how to calculate the pH for this kind of mixture. For instance, when 10.0 cm³ of 0.10 M NaOH have been added, 1.0×10^{-3} mol of OH^- has been supplied. This neutralizes 1.0×10^{-3} mol of $HC_2H_3O_2$ and converts it to 1.0×10^{-3} mol of $C_2H_3O_2^-$. The original 25.0 cm³ of 0.10 M $HC_2H_3O_2$ contained 2.5×10^{-3} mol $HC_2H_3O_2$, so the amount that is left is

$$(2.5 \times 10^{-3} \text{ mol}) - (1.0 \times 10^{-3} \text{ mol}) = 1.5 \times 10^{-3} \text{ mol } HC_2H_3O_2$$

The concentrations of acetic acid and acetate ion in the total volume of 35.0 cm³ are therefore

$$[HC_2H_3O_2] = \frac{1.5 \times 10^{-3} \text{ mol}}{0.0350 \text{ dm}^3} = 4.3 \times 10^{-2} \, M$$

$$[C_2H_3O_2^-] = \frac{1.0 \times 10^{-3} \text{ mol}}{0.0350 \text{ dm}^3} = 2.9 \times 10^{-2} \, M$$

If we solve the K_a expression for acetic acid for the H^+ concentration and substitute these values for $[HC_2H_3O_2]$ and $[C_2H_3O_2^-]$, we obtain

$$[H^+] = K_a \times \frac{[HC_2H_3O_2]}{[C_2H_3O_2^-]}$$

$$= 1.8 \times 10^{-5} \left(\frac{4.3 \times 10^{-2}}{2.9 \times 10^{-2}} \right)$$

$$= 2.7 \times 10^{-5} \, M$$

Therefore,

$$pH = 4.57$$

From the time of the first addition of base until the equivalence point is reached, the solution contains both acetic acid and acetate ion, and the pH may be computed in this fashion.

When a total of 25.0 cm³ of NaOH are added, all the acetic acid is "neutralized," and we have produced 2.5×10^{-3} mol of $NaC_2H_3O_2$ in 50.0 cm³ of solution. The resulting 0.050 M $NaC_2H_3O_2$ solution undergoes hydrolysis because it contains the anion of a weak acid. We have seen that for this solute the equilibrium is

For the $NaC_2H_3O_2$,

$$\frac{2.5 \times 10^{-3} \text{ mol}}{0.050 \text{ dm}^3} = 0.050 \, M$$

$$C_2H_3O_2^- + H_2O \rightleftharpoons HC_2H_3O_2 + OH^-$$

From the last section we know that

$$K_b = \frac{[HC_2H_3O_2][OH^-]}{[C_2H_3O_2^-]} = \frac{K_w}{K_a} = 5.6 \times 10^{-10}$$

We can calculate the OH^- concentration in this solution as we did previously by letting x equal the number of moles per dm³ of $C_2H_3O_2^-$ that reacts. This allows us to construct our table.

	Initial Molar Concentrations	Change	Equilibrium Molar Concentrations
$HC_2H_3O_2$	0.0	$+x$	x
OH^-	0.0	$+x$	x
$C_2H_3O_2^-$	0.050	$-x$	$0.050 - x \approx 0.050$

Substituting into the K_b expression gives

$$K_b = \frac{(x)(x)}{0.050} = 5.6 \times 10^{-10}$$

$$x^2 = 2.8 \times 10^{-11}$$

$$x = 5.3 \times 10^{-6}$$

This means that $[OH^-] = 5.3 \times 10^{-6}$ M, from which we obtain

$$pOH = 5.28$$

and finally,

$$pH = 8.72$$

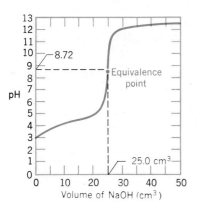

Figure 15.4

Titration of 25.0 cm³ of 0.10 M acetic acid with 0.10 M sodium hydroxide.

Thus the pH at which the equivalence point occurs is greater than 7. We find that this is true for any weak acid-strong base titration.

Thus far we have discussed only the first half of the titration (see Table 15.4). What takes place beyond the equivalence point? As soon as all the weak acid has been neutralized, any further addition of NaOH suppresses the hydrolysis of the anion and the pH is then solely dependent on the concentration of OH^- coming from the added NaOH. Thus we generate the last half of Table 15.4 in the same manner as we did Table 15.3 in the HCl/NaOH titration.

A graph of these data is shown in Figure 15.4, where we have plotted pH versus volume of base added. From both Table 15.4 and Figure 15.4, we can see that the change in pH near the equivalence point is not as drastic as in the case of the HCl/NaOH titration. This less rapid change near the equivalence point becomes even more pronounced for weaker acids such as HCN.

Titration of a weak base with a strong acid

When a weak base is titrated with a strong acid, the titration curve that is generated is very similar in shape to that obtained by reaction of a weak acid with a strong base. During the initial addition of acid the solution contains unreacted weak base and its salt; it therefore constitutes a buffer. At the equivalence point the solution contains the salt of the weak base, and the pH of the mixture is determined by the hydrolysis of the cation. Finally, beyond the equivalence point the pH of the solution is controlled by the excess hydrogen ion from the strong acid. The shape of the titration curve for such a titration is shown in

Table 15.4
Titration of 25.0 cm³ of 0.10 M HC$_2$H$_3$O$_2$ with 0.10 M NaOH

Volume of Base Added (cm³)	Molar Concentration of Species in Parentheses	pH
0.0	1.3×10^{-3} (H$^+$)	2.89
10.0	2.7×10^{-5} (H$^+$)	4.57
24.99	7.2×10^{-9} (H$^+$)	8.14
25.0	5.3×10^{-6} (OH$^-$)	8.72
25.01	2.0×10^{-5} (OH$^-$)	9.30
26.0	2.0×10^{-3} (OH$^-$)	11.30

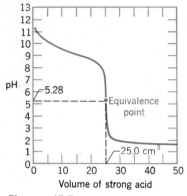

Figure 15.5
Titration of 25.0 cm³ of 0.10 M
NH₃ with 0.10 M HCl.

Figure 15.5 for the titration of 25.0 cm³ of 0.10 M NH_3 with 0.10 M HCl. We can show that the pH at the equivalence point is less than 7 by considering the hydrolysis of the NH_4Cl produced during the reaction.

From the last section we recall that the K_a for NH_4^+ is written as

$$K_a = \frac{[H^+][NH_3]}{[NH_4^+]} = \frac{K_w}{K_b} = 5.6 \times 10^{-10}$$

All the NH_3 is "neutralized" in this titration when exactly 25.0 cm³ of the 0.10 M HCl (2.5×10^{-3} mol HCl) have been added. At this point, the concentration of NH_4^+ is

$$\frac{2.5 \times 10^{-3} \text{ mol}}{0.0500 \text{ dm}^3} = 5.0 \times 10^{-2} \ M$$

If we let x equal the number of moles per dm³ of NH_4^+ that undergo hydrolysis

$$[H_3O^+] = x$$

$$[NH_3] = x$$

$$[NH_4^+] = 5.0 \times 10^{-2} - x = 5.0 \times 10^{-2} \ M$$

Substituting these concentrations into the equation for K_a gives

$$K_a = \frac{(x)(x)}{5.0 \times 10^{-2}} = 5.6 \times 10^{-10}$$

$$x^2 = 28.0 \times 10^{-12}$$

$$x = 5.3 \times 10^{-6}$$

$$[H^+] = 5.3 \times 10^{-6} \ M$$

and

$$pH = 5.28$$

The pH at the equivalence point of this titration is less than 7, which is typical for all weak base-strong acid titrations.

15.7 ACID-BASE INDICATORS

Indicators are often used, in very small amounts, to detect the equivalence point in an acid-base titration. They are usually weak organic acids or bases that change color on going from an acidic medium to a basic medium. Not all indicators change color at the same pH, however. The choice of indicator for a particular titration depends on the pH at which the equivalence point is expected to occur. A list of some common indicators, with their color changes and the pH ranges over which the color changes are observed, is found in Table 15.5. Let us examine briefly how these indicators work.

Some indicators, like thymol blue, have two color changes over separate pH ranges.

If we denote an indicator by the general formula HIn, we have the dissociation reaction

$$HIn \rightleftharpoons H^+ + In^-$$

Applying Le Châtelier's principle to this equilibrium, we see that in an acidic solution (excess H^+) the predominant species is HIn. On the other hand, in basic solutions the equilibrium is shifted to the right and the predominant species is In^-. Therefore, HIn is said to be the acid form and In^- the basic form of the indicator. The ability of HIn to function as an indicator is based on the difference

Phenolphthalein is often used as an indicator in the titration of a strong acid with a strong base.

Table 15.5
Some common indicators

Indicator	Color Change	pH Range in Which Color Change Occurs
Thymol blue	Red to yellow	1.2–2.8
Bromophenol blue	Yellow to blue	3.0–4.6
Congo red	Blue to red	3.0–5.0
Methyl orange	Red to yellow	3.2–4.4
Bromocresol green	Yellow to blue	3.8–5.4
Methyl red	Red to yellow	4.8–6.0
Bromocresol purple	Yellow to purple	5.2–6.8
Bromothymol blue	Yellow to blue	6.0–7.6
Cresol red	Yellow to red	7.0–8.8
Thymol blue	Yellow to blue	8.0–9.6
Phenolphthalein	Colorless to pink	8.2–10.0
Alizarin yellow	Yellow to red	10.1–12.0

in color between acid and basic forms. For example, with litmus the acid form (HIn) is pink while the basic form (In$^-$) is blue.

The dissociation constant, K_a, for an indicator is

$$K_a = \frac{[H^+][In^-]}{[HIn]}$$

Let's solve this for the ratio [In$^-$]/[HIn].

$$\frac{[In^-]}{[HIn]} = \frac{K_a}{[H^+]}$$

We have seen that as we pass through the equivalence point, the pH changes very rapidly. For example, in the NaOH/HCl titration described earlier—Table 15.3 and Figure 15.3—the pH changed from 4.7 to 9.3 with the addition of only 0.02 cm^3 of base, which is only about one-half drop of solution. This pH change corresponds to a change in [H$^+$] from 2×10^{-5} *M* to 5×10^{-10} *M*. How does this affect the [In$^-$]/[HIn] ratio?

Suppose that we were using an indicator whose $K_a = 1 \times 10^{-7}$. Then, before the equivalence point,

$$\frac{[In^-]}{[HIn]} = \frac{1 \times 10^{-7}}{2 \times 10^{-5}} = \frac{1}{200}$$

Visually, we observe the color of the species that is present in largest concentration.

This tells us that there is 200 times as much HIn as In$^-$, and the color observed is that resulting from HIn.

After the equivalence point,

$$\frac{[In^-]}{[HIn]} = \frac{1 \times 10^{-7}}{5 \times 10^{-10}} = \frac{200}{1}$$

Now there is 200 times as much In$^-$ as HIn, and the color that we see is due to In$^-$. Thus, as we pass through the equivalence point there is a sudden change in

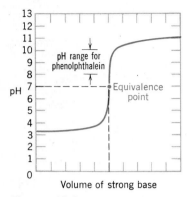

Figure 15.6

Titration curve for the titration of a strong acid with a strong base.

the relative amounts of the acid and basic forms of the indicator, which we notice as a change in color.

If the indicator changes color at the equivalence point, the end point in the titration—that point at which we observe the color change—occurs at the same pH as the equivalence point. Often, however, we find ourselves using an indicator whose color change takes place at a pH slightly different from that of the equivalence point. This is shown in Figure 15.6 for phenolphthalein. When the color change occurs we have actually gone slightly past the equivalence point.

In choosing an indicator, we wish to have it change color very close to the equivalence point. Phenolphthalein, for example, would be a poor choice of indicator for the titration depicted in Figure 15.5, because its color change would occur long before the equivalence point. We would find that we had stopped adding acid before the equivalence point had been reached, thereby defeating the purpose of using an indicator. A better choice would be an indicator such as methyl red, where the center of the color change range occurs very near the pH at the equivalence point.

REVIEW QUESTIONS AND PROBLEMS

(Problems whose numbers are in blue have their answers in Appendix C at the back of the book; the more difficult problems are marked with asterisks.)

Autoionization of Water

15.1 Write the equation for the autoionization of water. Why is this equilibrium important?

15.2 Why can we almost always ignore the H^+ contributed by the dissociation of water when we calculate the H^+ concentration in solutions containing an acid? Under what conditions would we have to consider the H^+ from the dissociation of H_2O, even though the solute was an acid?

15.3 Calculate the H^+ and OH^- concentrations and the pH of the following solutions of strong acids and bases:
(a) 0.0010 M HCl
(b) 0.125 M HNO_3
(c) 0.0031 M NaOH
(d) 0.012 M $Ba(OH)_2$
(e) 2.1×10^{-4} M $HClO_4$
(f) 1.3×10^{-5} M HCl
(g) 8.4×10^{-3} M NaOH
(h) 4.8×10^{-2} M KOH

***15.4** A dilute solution of hydrochloric acid was labeled 1.0×10^{-8} M HCl. Is this solution acidic or basic? What is its pH?

15.5 At 37°C (body temperature), $K_w = 2.42 \times 10^{-14}$. What are the concentrations of H^+ and OH^- in a neutral aqueous solution at 37°C?

pH and pOH

15.6 How is pH defined? How is pOH defined? Why does pH + pOH = 14?

15.7 Identify the following as representing acidic, basic, or neutral solutions:
(a) pH = 3.54   (d) pOH = 10.43
(b) pH = 8.25   (e) pOH = 2.25
(c) pOH = 7.00

15.8 Arrange the solutions in Problem 15.7 in order of increasing acidity.

15.9 What is the pH of a neutral solution at 37°C? (See Problem 15.5.)

15.10 Calculate the H^+ and OH^- concentrations in a solution having a pH equal to:
(a) 1.30   (d) 7.80
(b) 5.73   (e) 10.94
(c) 4.00   (f) 12.61

15.11 What is the pOH of each solution in Problem 15.10?

15.12 Calculate the pH and pOH of each of the solutions in Exercise 15.3.

pK

15.13 A weak acid has an equilibrium constant $K_a = 3.8 \times 10^{-9}$. What is the pK_a of the acid?

15.14 A base has $pK_b = 3.84$. What is K_b of the base?

Weak Acids and Bases

15.15 Refer to Table 15.1 and write the appropriate equilibrium constant expressions for the ionization of:
(a) benzoic acid
(b) hydrazine
(c) formic acid
(d) Veronal
(e) pyridine

15.16 What is the H^+ concentration in each of the following solutions:
(a) 0.30 M HNO_2
(b) 1.00 M HF
(c) 0.025 M HCN
(d) 0.10 M butyric acid
(e) 0.050 M barbituric acid

15.17 Calculate the OH^- concentration in the following solutions:
(a) 0.15 M NH_3
(b) 0.20 M N_2H_4
(c) 0.80 M CH_3NH_2
(d) 0.35 M hydroxylamine
(e) 0.010 M pyridine

15.18 What is the OH^- concentration in each solution in Problem 15.16?

15.19 What is the pH of each solution in Exercise 15.17?

15.20 A 0.25 M solution of a monoprotic weak acid was observed to have a pH = 1.35. What is K_a for this acid?

15.21 A 0.10 M solution of a weak monoprotic acid was found to have a pH = 5.37. What is K_a for the acid?

15.22 A weak base was found to give a solution with pH = 8.75 when its concentration was 0.10 M. What is K_b for the base?

15.23 What is the percent ionization of the acid in each of the following six solutions:
(a) 1.0 M formic acid
(b) 0.010 M propionic acid
(c) 0.025 M HCN
(d) 0.35 M nicotinic acid
(e) 0.50 M HOCl
(f) 0.25 M HNO_3

15.24 Calculate the percent ionization of each of the following acetic acid solutions. What conclusions can you draw? Can you explain on a molecular level why you obtain these results?
(a) 1.00 M $HC_2H_3O_2$
(b) 0.10 M $HC_2H_3O_2$
(c) 0.010 M $HC_2H_3O_2$

15.25 Calculate the hydrogen ion concentration in mol dm^{-3} for each of the following solutions of an acid or base and its salt:
(a) 0.25 M $HC_2H_3O_2$, 0.15 M $NaC_2H_3O_2$

(b) 0.50 M $HCHO_2$, 0.50 M $NaCHO_2$
(c) 0.30 M HNO_2, 0.40 M $NaNO_2$
(d) 0.25 M NH_3, 0.15 M NH_4Cl
(e) 0.30 M N_2H_4, 0.50 M $N_2H_5NO_3$

15.26 The pH of a 0.012 M solution of a weak base, B, was experimentally determined to be 11.40. Calculate K_b for the base.

15.27 How many grams of HCl gas would have to be dissolved in 500 cm^3 of 1.0 M $NaC_2H_3O_2$ to give a solution having a pH = 4.74?

15.28 If 10 mg of sodium barbituate is swallowed, what fraction is converted to barbituric acid if the pH of the stomach is 1.0 and there are 250 cm^3 of fluid in the stomach?

15.29 Nicotinic acid is another name for the important vitamin, niacin. What is the pH of a 0.010 M solution of nicotinic acid?

15.30 What is the molar concentration of a solution of acetic acid whose pH is 2.50?

15.31 What molar concentration of hydrazine, N_2H_4, yields a solution whose pH = 10.64?

15.32 A 0.010 M solution of a weak acid, HA, is found to have a pH of 4.55. What is the value of K_a for this acid?

15.33 Calculate the H^+ concentration in 0.0010 M $HC_2H_3O_2$.

15.34 Calculate the molar concentrations of all species in 0.010 M formic acid solution.

Polyprotic Acids

15.35 Write appropriate mass action expressions for K_{a_1} and K_{a_2} for ascorbic acid (vitamin C).

15.36 Citric acid, which is present in many fruits and vegetables, has the formula, $H_3C_6H_5O_7$. It is a triprotic acid. Write the three equilibria for the dissociation of the acid and the appropriate equilibrium constant expression for each step.

15.37 Selenious acid, H_2SeO_3, has $K_{a_1} = 3 \times 10^{-3}$ and $K_{a_2} = 5 \times 10^{-8}$. What is the pH of a 0.50 M solution of H_2SeO_3? What are the equilibrium molar concentrations of H_2SeO_3, $HSeO_3^-$, and SeO_3^{2-}?

15.38 Calculate the pH obtained by dissolving a 500-mg tablet of vitamin C in 250 cm^3 of H_2O.

15.39 Calculate the molar concentrations of all of the species in a 0.050 M solution of the diprotic acid, vitamin C.

15.40 Calculate the molar concentrations of all species present in a 1.0 M H_3PO_4 solution.

15.41 In the stomach the fluids have a pH ≈ 1.0 due to the strong acid, HCl. What fraction of the vitamin C in a

500-mg tablet will be dissociated if the volume of fluid in the stomach is 200 cm^3?

15.42 What is the HCO_3^- concentration (in mol dm^{-3}) in a 0.10 M solution of H_2CO_3 whose pH = 3.00? What is the CO_3^{2-} concentration in this solution?

15.43 Suppose that we wish the sulfide ion concentration to be 8.4×10^{-15} M in a saturated (0.10 M) solution of H_2S. What hydrogen ion concentration must be maintained by a buffer to give this S^{2-} concentration?

15.44 What is the sulfide ion concentration in a saturated H_2S solution (0.10 M H_2S) whose pH has been adjusted to a value of 4.60 by the addition of a buffer?

*15.45 A sample of arterial blood was found to contain 2.6×10^{-2} mol of dissolved CO_2 per dm^3. The pH of the sample was 7.43. If it is assumed that in solution the CO_2 forms H_2CO_3, what is the HCO_3^- concentration in this blood sample?

Buffers

15.46 What is a buffer? Explain how the following solutes function as buffers:
(a) $NaCHO_2$ and $HCHO_2$
(b) C_5H_5N and C_5H_5NHCl
(c) $NH_4C_2H_3O_2$
(d) $NaHCO_3$

15.47 Would a solution containing a mixture of NaCl and HCl be an effective buffer? Explain.

15.48 Blood contains, among others, the buffer system $HPO_4^{2-}/H_2PO_4^-$. Explain how this pair of ions can serve as a buffer.

15.49 What ratio of lactic acid to sodium lactate is required to give a solution having a pH = 4.25?

15.50 Calculate the pH of each of the following buffers prepared by placing, in 1.0 dm^3 of solution:
(a) 0.10 mol of NH_3 and 0.10 mol of NH_4Cl
(b) 0.20 mol of $HC_2H_3O_2$ and 0.40 mol of $NaC_2H_3O_2$
(c) 0.15 mol of N_2H_4 and 0.10 mol of N_2H_5Cl
(d) 0.20 mol of HCl and 0.30 mol of NaCl

15.51 How many grams of $NaC_2H_3O_2$ must be added to 1.00 mol of $HC_2H_3O_2$ in order to prepare 1.00 dm^3 of a buffer whose pH equals 5.15?

15.52 What must the ratio of NH_3 to NH_4^+ be to have a buffer with a pH of 10.00?

*15.53 How many moles of HCl must be added to 1.0 dm^3 of a mixture containing 0.010 M $HC_2H_3O_2$ and 0.010 M $NaC_2H_3O_2$ in order to give a solution whose pH = 3.00?

15.54 Calculate the pH change produced by adding 0.10 mol of solid NaOH to each of the following buffers:

(a) 500 cm^3 of 1.00 M $HC_2H_3O_2$ and 1.00 M $NaC_2H_3O_2$
(b) 500 cm^3 of 0.50 M $HC_2H_3O_2$ and 0.50 M $NaC_2H_3O_2$
(c) 500 cm^3 of 0.30 M $HC_2H_3O_2$ and 0.70 M $NaC_2H_3O_2$
(d) 500 cm^3 of 0.20 M $HC_2H_3O_2$ and 0.80 M $NaC_2H_3O_2$
(e) 500 cm^3 of 0.10 M $HC_2H_3O_2$ and 0.90 M $NaC_2H_3O_2$

15.55 How much would the pH change if 0.10 mol of HCl were added to 1.0 dm^3 of a formic acid–sodium formate buffer containing 0.45 mol of $HCHO_2$ and 0.55 mol of $NaCHO_2$?

15.56 How much would the pH change if 0.20 mol of NaOH were added to the original buffer in Exercise 15.55?

15.57 What happens to the pH of a buffer solution if it is diluted with water? Illustrate your answer using the buffer in part (a) of Exercise 15.50.

*15.58 Calculate the pH of 0.50 M $NaHCO_3$. How much will the pH change if 0.05 mol dm^{-3} of HCl is added?

*15.59 What volume of 6.0 M HCl is required to be added to 100 cm^3 of 0.10 M $NaC_2H_3O_2$ to give a solution having a pH = 4.25?

Relative Strengths of Conjugate Acids and Bases

15.60 From Figure 15.2, place the following reactions in order of increasing tendency to proceed toward completion:
(a) $H_2O + NH_3 \rightleftharpoons NH_4^+ + OH^-$
(b) $HClO_4 + NH_2^- \rightleftharpoons ClO_4^- + NH_3$
(c) $H_2O + NO_2^- \rightleftharpoons HNO_2 + OH^-$
(d) $NH_3 + Cl^- \rightleftharpoons NH_2^- + HCl$

15.61 Use Figure 15.2 to place the following reactions in order of increasing tendency to proceed toward completion:
(a) $OCl^- + HCl \rightleftharpoons HOCl + Cl^-$
(b) $HF + C_2H_3O_2^- \rightleftharpoons HC_2H_3O_2 + F^-$
(c) $NH_4^+ + ClO_4^- \rightleftharpoons NH_3 + HClO_4$
(d) $HNO_2 + F^- \rightleftharpoons HF + NO_2^-$

15.62 Hydrogen sulfide is a stronger acid than phosphine, PH_3. What may we conclude about the strengths of their conjugate bases, HS^- and PH_2^-?

15.63 Hydrogen cyanide, HCN, is a weaker acid in water than nitrous acid, HNO_2. What can we say about the relative strengths of CN^- and NO_2^- as bases?

15.64 Ammonia is a stronger base in water than hydrazine, N_2H_4. If we were to use these two substances as solvents for a very weak acid, in which of them would the acid be more fully ionized?

15.65 Given the following equilibria and equilibrium constants:

(a) $HOCl + H_2O \rightleftharpoons H_3O^+ + OCl^-$ $K_a = 3.2 \times 10^{-8}$
(b) $NH_4^+ + H_2O \rightleftharpoons H_3O^+ + NH_3$ $K_a = 5.6 \times 10^{-10}$
(c) $HC_2H_3O_2 + H_2O \rightleftharpoons H_3O^+ + C_2H_3O_2^-$ $K_a = 1.8 \times 10^{-5}$
(d) $H_2CO_3 + H_2O \rightleftharpoons H_3O^+ + HCO_3^-$ $K_a = 4.2 \times 10^{-7}$
(e) $HSO_4^- + H_2O \rightleftharpoons H_3O^+ + SO_4^{2-}$ $K_a = 1.3 \times 10^{-2}$

Arrange the conjugate bases in order of *increasing* strength.

15.66 For the members of an acid-base conjugate pair, what is the relationship between K_a and K_b in water?

Hydrolysis

15.67 What is hydrolysis? Without performing any computations, predict whether the following solutions will be acidic, basic, or neutral:

(a) KCl (c) $NaC_4H_7O_2$
(b) NH_4NO_3 (d) $C_6H_5NH_3NO_3$

15.68 If the concentration of each solute in Problem 15.67 were 0.10 M, which solution would be most basic? Which would be most acidic?

15.69 Why is it necessary to consider only the first step in the hydrolysis of a salt such as Na_2SO_3?

15.70 Determine the pH of each of the following salt solutions:

(a) 1.0×10^{-3} M $NaC_2H_3O_2$
(b) 0.125 M NH_4Cl
(c) 0.10 M Na_2CO_3
(d) 0.10 M NaCN
(e) 0.20 M NH_3OHCl

15.71 What percent of $C_5H_5NH^+$ reacts in a 0.10 M solution of pyridinium chloride, C_5H_5NHCl? For pyridine, C_5H_5N, the value of $K_b = 1.7 \times 10^{-9}$.

15.72 A 0.10 M solution of the sodium salt of a weak monoprotic acid has a pH of 9.35. What is the K_a of the weak acid?

15.73 Liquid chlorine bleach is really nothing more than a dilute solution of NaOCl, usually about 5% NaOCl by mass. A particular sample of bleach was found to contain 0.67 mol dm^{-3} NaOCl. Calculate the pH of the solution.

15.74 Veronal, a barbiturate drug, is generally administered as its sodium salt. What is the pH of a solution of $NaC_8H_{11}N_2O_3$ that contains 10 mg of the drug in 250 cm^3 of solution? For veronal, $HC_8H_{11}N_2O_3$, the value of $K_a = 3.7 \times 10^{-8}$.

15.75 What would be the concentration of barbituric acid in a 0.0010 M solution of sodium barbiturate?

15.76 What is the pH of a 0.20 M solution of sodium ascorbate?

15.77 What would be the pH of a 0.50 M solution of Na_3PO_4?

*15.78 What would be the pH of a 0.0010 M solution of potassium cyanide (a deadly poison)?

*15.79 Sodium benzoate is often used as a preservative in packaged food products. What would be the pH of a 0.020 M solution of sodium benzoate?

*15.80 Calculate the pH of 0.10 M NH_4NO_2.

Acid-Base Titrations

15.81 Is it possible to have a pH other than 7 at the equivalence point in an acid-base titration?

15.82 In a titration, the equivalence point and the end point are often not exactly the same. Justify this statement.

15.83 A 15.0-cm^3 portion of a solution of 0.0200 M HNO_3 is titrated with 0.0100 M KOH.

(a) What will be the pH at the equivalence point?
(b) What volume of the base will be required to reach the equivalence point?
(c) What will be the pH when 10.0 cm^3 of the KOH solution has been added?
(d) What will be the pH when 35.0 cm^3 of the KOH solution has been added?

15.84 What would be the pH at the equivalence point if 25.0 cm^3 of 0.010 M barbituric acid is titrated with 0.020 M NaOH?

*15.85 When 50.0 mL of 0.200 M HF is titrated with 0.100 M NaOH, what is the pH:

(a) after 5.0 cm^3 of base has been added?
(b) when half of the HF has been neutralized?
(c) at the equivalence point?

15.86 Plot a curve showing the pH of a solution of 100 cm^3 of 0.10 M butyric acid that is gradually neutralized by the addition of solid NaOH. Do this by calculating the pH after addition of 0.0, 0.0010, 0.0050, 0.0090, 0.010, and 0.011 mol of NaOH. Assume no change in volume. What is the pH at the equivalence point? What indicator in Table 15.5 could be used for this titration?

15.87 Determine the shape of the titration curve for the titration of 50.0 cm^3 of 0.10 M HCl with 0.10 M NaOH.

*15.88 Determine the shape of the titration curve when 100 cm^3 of 0.20 M H_2CO_3 is titrated with 0.10 M NaOH. Determine the pH at each equivalence point.

Indicators

15.89 Explain how an indicator works. Why do we want to use as little of the indicator as possible when we perform a titration?

15.90 What indicators might be acceptable for the titration depicted in Figure 15.4? Why would we not wish to use congo red as an indicator?

15.91 Would congo red be an acceptable indicator for the titration depicted in Figure 15.6? Explain your answer.

15.92 Using the data in Tables 15.1 and 15.5, choose an indicator that is suitable for the titration of:

(a) hydrocyanic acid with sodium hydroxide

(b) aniline with hydrochloric acid

15.93 An indicator, HIn, has an ionization constant, K_a, equal to 1×10^{-5}. If the molecular form of the indicator is yellow and the In$^-$ ion green, what is the color of a solution containing this indicator when its pH is 7.0?

CHAPTER 16

SOLUBILITY AND COMPLEX ION EQUILIBRIA

Oysters extract Ca^{2+} and CO_3^{2-} ions from sea water and deposit them on irritating grains of sand to form beautiful pearls that are composed of mostly insoluble $CaCO_3$. In this chapter we will study, quantitatively, the solubility of salts such as calcium carbonate and learn how it is possible to predict the conditions that are necessary for a precipitate to form from a solution.

In Chapter 15 we studied ionic equilibria involving acids and bases. These are not, however, the only dynamic equilibria that can take place in aqueous solutions. In this chapter we will turn our attention to the equilibria involved with salts that have very low solubilities—those we considered to be insoluble in water in our discussions in Chapter 8. We will also look at equilibria involving species that we call complex ions—ions composed of a metal atom surrounded by a number of ions, or neutral molecules, to which the metal is bound.

16.1 SOLUBILITY PRODUCT

In Chapter 8 you were presented with a list of solubility rules that described certain salts as soluble and others as quite insoluble. Even the most insoluble salts, however, dissolve in water to at least some degree, and their saturated solutions constitute dynamic equilibria that can be studied by the same principles that we applied to acid-base equilibria in the last chapter.

Nearly all salts are completely dissociated in water. There are some exceptions, such as $HgCl_2$ and $CdSO_4$, but they are rare. For simplicity, therefore, our discussions will not include them, and we will assume that in a saturated solution an equilibrium exists between the solid salt and its dissolved ions. For example, in a saturated solution of silver chloride we have the equilibrium

$$AgCl(s) \rightleftharpoons Ag^+(aq) + Cl^-(aq)$$

It is safe to assume that there is no molecular AgCl in the solution.

for which we can write

$$K = \frac{[Ag^+][Cl^-]}{[AgCl(s)]}$$

In Section 14.5 we saw that the *concentration* of a pure solid is independent of the amount of solid present. In other words, the concentration of the solid is a constant and can therefore be included with the constant K, so that

$$K[AgCl(s)] = K_{sp} = [Ag^+][Cl^-]$$

The equilibrium constant K multiplied by the concentration of solid AgCl is still another constant called the **solubility product constant,** given the label K_{sp}. Similarly, we can obtain the expression for the K_{sp} of silver acetate from its solubility equilibrium,

$$AgC_2H_3O_2(s) \rightleftharpoons Ag^+(aq) + C_2H_3O_2^-(aq)$$

The equilibrium expression is therefore,

$$K_{sp} = [Ag^+][C_2H_3O_2^-]$$

In the case of an insoluble solid such as $Mg(OH)_2$, the coefficients in the equilibrium are not all equal to one:

$$Mg(OH)_2(s) \rightleftharpoons Mg^{2+}(aq) + 2OH^-(aq)$$

The K_{sp} for $Mg(OH)_2$ is then given by

$$K_{sp} = [Mg^{2+}][OH^-]^2$$

Thus the solubility product constant is equal to the product of the concentrations of the ions produced in a saturated solution, each raised to a power equal to its coefficient in the balanced equation. A list of some ionic solids and their K_{sp}'s at temperatures ranging between 18 and 25°C is given in Table 16.1

A suspension of magnesium hydroxide in water.

Table 16.1
Solubility product constants

Compound	K_{sp}	Compound	K_{sp}
$Al(OH)_3$	2×10^{-33}	PbS	7×10^{-27}
$BaCO_3$	8.1×10^{-9}	$Mg(OH)_2$	1.2×10^{-11}
$BaCrO_4$	2.4×10^{-10}	MgC_2O_4	8.6×10^{-5}
BaF_2	1.7×10^{-6}	$Mn(OH)_2$	4.5×10^{-14}
$BaSO_4$	1.5×10^{-9}	MnS	7×10^{-16}
CdS	3.6×10^{-29}	Hg_2Cl_2	2×10^{-18}
$CaCO_3$	9×10^{-9}	HgS	1.6×10^{-54}
CaF_2	1.7×10^{-10}	NiS	2×10^{-21}
$CaSO_4$	2×10^{-4}	$AgC_2H_3O_2$	2.3×10^{-3}
CoS	7×10^{-23}	Ag_2CO_3	8.2×10^{-12}
CuS	8.5×10^{-36}	AgCl	1.7×10^{-10}
Cu_2S	2×10^{-47}	AgBr	5×10^{-13}
$Fe(OH)_2$	2×10^{-15}	AgI	8.5×10^{-17}
$Fe(OH)_3$	1.1×10^{-36}	Ag_2CrO_4	1.9×10^{-12}
FeC_2O_4	2.1×10^{-7}	AgCN	1.6×10^{-14}
FeS	3.7×10^{-19}	Ag_2S	2×10^{-49}
$PbCl_2$	1.6×10^{-5}	$Sn(OH)_2$	5×10^{-26}
$PbCrO_4$	1.8×10^{-14}	SnS	1×10^{-26}
PbC_2O_4	2.7×10^{-11}	$Zn(OH)_2$	4.5×10^{-17}
$PbSO_4$	2×10^{-8}	ZnS	1.2×10^{-23}

Calculations involving K_{sp} can be divided into three categories:

1. Calculating K_{sp} from solubility data.
2. Calculating solubility from K_{sp}.
3. Problems dealing with precipitation.

We will begin (as you might expect) with the first type.

EXAMPLE 16.1	CALCULATING K_{sp} FROM SOLUBILITY DATA

PROBLEM   It was experimentally determined that at 25°C the solubility of $BaSO_4$ in water is 0.0091 g dm^{-3}. What is the value of K_{sp} for barium sulfate?

SOLUTION   From the solubility we can calculate the number of moles of $BaSO_4$ that are dissolved in 1 dm^3 of solution.

The solubility in moles of solute per dm^3 of saturated solution is called the **molar solubility.**

$$\left(0.0091 \ \frac{g}{dm^3}\right) \times \left(\frac{1 \ mol}{233 \ g}\right) = 3.9 \times 10^{-5} \ mol \ dm^{-3}$$

The solubility equilibrium for $BaSO_4$ is

$$BaSO_4(s) \rightleftharpoons Ba^{2+}(aq) + SO_4{}^{2-}(aq)$$

so that for every mole of $BaSO_4$ that dissolves, 1 mol of Ba^{2+} and 1 mol of $SO_4{}^{2-}$ are produced. Therefore the molar concentrations of Ba^{2+} and SO_4^{2-} in this saturated solution at 25°C are

$$[Ba^{2+}] = 3.9 \times 10^{-5}\ M$$

$$[SO_4^{2-}] = 3.9 \times 10^{-5}\ M$$

and the K_{sp} would be

$$K_{sp} = [Ba^{2+}][SO_4^{2-}]$$
$$= (3.9 \times 10^{-5})(3.9 \times 10^{-5})$$
$$= 1.5 \times 10^{-9}$$

EXAMPLE 16.2 CALCULATING K_{sp} FROM SOLUBILITY DATA

PROBLEM The solubility of lead iodate, $Pb(IO_3)_2$, is 4.0×10^{-5} mol dm^{-3} at 25°C. What is the value of K_{sp} for this salt?

SOLUTION First we write the chemical equation and the K_{sp} expression.

$$Pb(IO_3)_2(s) \rightleftharpoons Pb^{2+}(aq) + 2IO_3^-(aq)$$

$$K_{sp} = [Pb^{2+}][IO_3^-]^2$$

When the $Pb(IO_3)_2$ dissolves, we get 1 mol of Pb^{2+} and 2 mol of IO_3^- for each mole of $Pb(IO_3)_2$. Therefore, when 4.0×10^{-5} mol of $Pb(IO_3)_2$ is dissolved in 1 dm^3, we obtain

$$[Pb^{2+}] = 4.0 \times 10^{-5}\ M$$

$$[IO_3^-] = 2(4.0 \times 10^{-5}) = 8.0 \times 10^{-5}\ M$$

These quantities are now substituted into the K_{sp} expression.

$$K_{sp} = (4.0 \times 10^{-5})(8.0 \times 10^{-5})^2$$
$$= 2.6 \times 10^{-13}$$

Let us now look at how we can determine solubility from a known value of K_{sp}—the second type of problem on our list.

EXAMPLE 16.3 CALCULATING MOLAR SOLUBILITY FROM K_{sp}

PROBLEM What is the molar solubility of AgCl in water at 25°C?

SOLUTION We are concerned with the equilibrium

$$AgCl(s) \rightleftharpoons Ag^+(aq) + Cl^-(aq)$$

for which

$$K_{sp} = [Ag^+][Cl^-] = 1.7 \times 10^{-10}$$

In solving solubility problems it is helpful to formulate a mental picture of what is happening. The statement of the problem indicates that the solvent into which the AgCl is dissolved is pure water, so there is no Ag^+ or Cl^- present initially.

In working problems of this type we will again set up a concentration table similar to those we've used in previous equilibrium calculations. In this particular example, the AgCl is dissolving into water that contains neither Ag^+ nor Cl^-, so the entries in the first column are both zero. Next, let's allow x to be the molar solubility—the number of moles of AgCl that dissolves per dm^3. Since one Ag^+ and one Cl^- are produced for each AgCl that dissolves, the concentrations of Ag^+ and Cl^- will each increase by x; both entries in the change column are therefore $+x$. Finally, the values in the equilibrium concentrations column are obtained by adding the change to the initial concentration—a rather trivial operation in this particular instance.

	Initial Molar Concentrations	Change	Equilibrium Molar Concentrations
Ag^+	0.0	$+x$	x
Cl^-	0.0	$+x$	x

Substituting the equilibrium quantities and the value of K_{sp} from Table 16.1 into the K_{sp} expression, we obtain

$$K_{sp} = (x)(x) = 1.7 \times 10^{-10}$$

$$x^2 = 1.7 \times 10^{-10}$$

$$x = 1.3 \times 10^{-5}$$

Therefore, the molar solubility of AgCl in water is 1.3×10^{-5} M.

EXAMPLE 16.4 CALCULATING MOLAR SOLUBILITY FROM K_{sp}

PROBLEM What are the concentrations of Ag^+ and CrO_4^{2-} in a saturated solution of Ag_2CrO_4 at 25°C?

SOLUTION Ag_2CrO_4 dissolves in water according to the equilibrium

$$Ag_2CrO_4(s) \rightleftharpoons 2Ag^+(aq) + CrO_4^{2-}(aq)$$

and the K_{sp} expression is

$$K_{sp} = [Ag^+]^2[CrO_4^{2-}]$$

Once again, neither of the ions involved in the equilibrium is present in the solvent before the salt is added, so the initial concentrations are zero. Next, we let x be the molar solubility. Since two Ag^+ and one CrO_4^{2-} are produced for each Ag_2CrO_4 that dissolves, then when x mol of Ag_2CrO_4 dissolves per dm³, the Ag^+ concentration increases by $2x$ and the CrO_4^{2-} concentration increases by x. Finally, we add the change to the initial concentration to obtain the equilibrium concentration in the last column.

By defining x as the molar solubility, the coefficients of x in the change column are the coefficients of the ions in the chemical equation for the equilibrium.

	Initial Molar Concentrations	Change	Equilibrium Molar Concentrations
Ag^+	0.0	$+2x$	$2x$
CrO_4^{2-}	0.0	$+x$	x

Substituting the value of K_{sp} from Table 16.1 and solving for x, we have

$$K_{sp} = (2x)^2(x) = 1.9 \times 10^{-12}$$

$$(4x^2)x = 4x^3 = 1.9 \times 10^{-12}$$

$$x^3 = 4.8 \times 10^{-13}$$

$(2x)^2 = 4x^2$

and

$$x = 7.8 \times 10^{-5}$$

Therefore,

$$[Ag^+] = 2(7.8 \times 10^{-5}) = 1.6 \times 10^{-4} \ M$$

$$[CrO_4^{2-}] = 7.8 \times 10^{-5} \ M$$

Determining when a precipitate will form in a solution

You should recall from earlier discussions that a saturated solution is one in which the undissolved solute is in dynamic equilibrium with the solution. This is precisely the situation to which we apply K_{sp}. In other words, a saturated solution exists *only* when the **ion product**, *the product of the concentrations of the dissolved ions each raised to its proper power*, is exactly equal to K_{sp}. When the ion product is less than K_{sp}, the solution is unsaturated, because more salt would have to dissolve in order to raise the concentrations to the point at which the ion product equals K_{sp}. On the other hand, when the ion product exceeds K_{sp}, a supersaturated solution exists because some of the salt would have to precipitate to lower the concentrations until the ion product is equal to K_{sp} once again.

In a solution, a precipitate will be formed only when the mixture is supersaturated. Therefore, we can use the value of the ion product in a solution to tell us whether or not precipitation will occur. In summary, we find that

Unsaturated:	Ion product $< K_{sp}$ $\Big\}$	No precipitate will form
Saturated:	Ion product $= K_{sp}$	
Supersaturated:	Ion product $> K_{sp}$	Precipitation will occur

---

EXAMPLE 16.5

DETERMINING WHETHER A PRECIPITATE WILL FORM IN A SOLUTION

PROBLEM Will a precipitate of $PbCl_2$ form in a solution having a $Pb(NO_3)_2$ concentration of 0.010 M and an HCl concentration of 0.010 M? For $PbCl_2$, $K_{sp} = 1.6 \times 10^{-5}$.

SOLUTION For lead chloride, we would write the following equilibrium

$$PbCl_2(s) \rightleftharpoons Pb^{2+}(aq) + 2Cl^-(aq)$$

We can ignore the H^+ and NO_3^- because they are simply spectator ions here.

so the ion product that we will use in our test for precipitation is $[Pb^{2+}][Cl^-]^2$. In a 0.010 M $Pb(NO_3)_2$ solution, $[Pb^{2+}] = 0.010$ M and in 0.010 M HCl, $[Cl^-] = 0.010$ M. Using these values, we now compute the ion product.

$$[Pb^{2+}][Cl^-]^2 = (0.010)(0.010)^2 = 1.0 \times 10^{-6}$$

Since this is less than the value of K_{sp} (1.6×10^{-5}), we conclude that no precipitate will form.

EXAMPLE 16.6

DETERMINING WHETHER A PRECIPITATE WILL FORM WHEN TWO SOLUTIONS ARE MIXED

PROBLEM Will a precipitate of $PbSO_4$ form when exactly 100 cm^3 of a 0.0030 M $Pb(NO_3)_2$ solution is mixed with exactly 400 cm^3 of 0.040 M Na_2SO_4?

SOLUTION For lead sulfate, the ion product that we must examine is

$$[Pb^{2+}][SO_4^{2-}]$$

but this time we have to take into account that one solution dilutes the other when they are mixed. We approach this by imagining that the solutions can be combined before any reaction can take place, and then we look at the value of the ion product in the final mixture. The first step is to calculate the concentrations of Pb^{2+} and SO_4^{2-} in our total volume of 500 cm^3. The original 100 cm^3 of the 0.0030 M $Pb(NO_3)_2$ solution contains

$$(0.100 \, \cancel{dm^3}) \times \left(0.0030 \, \frac{mol}{\cancel{dm^3}}\right) = 0.000\,30 \text{ mol of } Pb^{2+}$$

(This solution also contains 0.000 60 mol of NO_3^-, but this species is unimportant in this calculation—it is a spectator ion.)

The 400 cm^3 of Na$_2$SO$_4$ contains

$$(0.400 \ \cancel{dm^3}) \times \left(0.040 \ \frac{mol}{\cancel{dm^3}}\right) = 0.016 \ \text{mol of SO}_4^{2-}$$

(This solution also contains 0.032 mol of Na$^+$, but it also is unimportant in this problem.)

The concentration of the Pb^{2+} in the 500 cm^3 is then

$$\frac{0.00030 \ \text{mol}}{0.500 \ \text{dm}^3} = 0.00060 \ M = 6.0 \times 10^{-4} \ M$$

and the concentration of SO$_4^{2-}$ in the 500 cm^3 is

$$\frac{0.016 \ \text{mol}}{0.500 \ \text{dm}^3} = 0.032 \ M = 3.2 \times 10^{-2} \ M$$

The ion product in the final solution is therefore

$$[\text{Pb}^{2+}][\text{SO}_4^{2-}] = (6.0 \times 10^{-4})(3.2 \times 10^{-2}) = 1.9 \times 10^{-5}$$

When we compare the value of the ion product to the value of K_{sp} for PbSO$_4$ (from Table 16.1, $K_{sp} = 2 \times 10^{-8}$), we find that the ion product is *greater* than K_{sp} and therefore a precipitate will form.

Separation of ions by precipitation

From the solubility rules presented in Chapter 8 we know that it is possible to separate certain ions from each other when they are present together in solution. For instance, the addition of chloride ion to a solution containing both Na$^+$ and Ag$^+$ yields a precipitate of AgCl, thereby removing most of the Ag$^+$ from the mixture. In this case one possible product, NaCl, is soluble while the other, AgCl, is quite insoluble.

Even when both products are "insoluble," it is still frequently possible to achieve some degree of separation. Consider, for example, the salts CaSO$_4$ and BaSO$_4$. Although both have very low solubilities, as evidenced by their respective K_{sp}'s, we can compute that CaSO$_4$ is about 1000 times more soluble, on a mole basis, than BaSO$_4$. As a result, if we had a solution containing equal concentrations of Ca^{2+} and Ba^{2+}, we would find that as the SO$_4^{2-}$ concentration was increased in the solution, BaSO$_4$ would precipitate first. Conceivably, one could separate Ca^{2+} and Ba^{2+} by appropriately adjusting the SO$_4^{2-}$ concentration so that the Ca^{2+} would remain in solution while nearly all the Ba^{2+} would be removed as BaSO$_4$. This general concept is used often in the separation of ions in qualitative analysis.

When the anion employed in a separation is derived from a weak acid, it is possible to control its concentration by appropriately adjusting the hydrogen ion concentration. This is illustrated for the selective precipitation of metal sulfides in Example 16.7

EXAMPLE 16.7 SEPARATING IONS BY SELECTIVE PRECIPITATION

PROBLEM A solution containing 0.10 M Sn^{2+} and 0.10 M Zn^{2+} is kept saturated with H$_2$S ([H$_2$S] = 0.10 M). What range of hydrogen ion concentrations will permit the selective precipitation of one of these ions as its sulfide?

SOLUTION   From Table 16.1 we have

$$SnS \quad K_{sp} = 1 \times 10^{-26}$$

$$ZnS \quad K_{sp} = 1.2 \times 10^{-23}$$

In this problem the sulfide ion concentration must be controlled so that one ion will precipitate while the other remains in solution. Let us, therefore, calculate for each salt the value of $[S^{2-}]$ that will make the ion product equal to K_{sp}. For tin we have

$$K_{sp} = [Sn^{2+}][S^{2-}] = 1 \times 10^{-26}$$

Substituting the Sn^{2+} concentration into the expression gives

$$(0.10)[S^{2-}] = 1 \times 10^{-26}$$

$$[S^{2-}] = 1 \times 10^{-25} \, M$$

In a similar fashion for zinc, we obtain

$$[S^{2-}] = 1.2 \times 10^{-22} \, M$$

These numbers tell us that if the sulfide ion concentration is *greater* than 1×10^{-25} M but *less than or equal to* 1.2×10^{-22} M, only SnS will precipitate.

We saw in Equation 15.7 (page 554) that the sulfide ion concentration is directly related to the hydrogen ion concentration; that is,

$$\frac{[H^+]^2[S^{2-}]}{[H_2S]} = K_{a_1}K_{a_2} = 1.1 \times 10^{-21}$$

We can use this expression to calculate the $[H^+]$ that gives us our desired $[S^{2-}]$.

For the lower limit, $[S^{2-}] = 1 \times 10^{-25}$ M. Since the solution is saturated with H_2S, we have $[H_2S] = 0.10$ M. Substituting gives

$$\frac{[H^+]^2(1 \times 10^{-25})}{(0.10)} = 1.1 \times 10^{-21}$$

$$[H^+]^2 = \frac{(1.1 \times 10^{-21})(0.10)}{1 \times 10^{-25}}$$

$$= 1.1 \times 10^3$$

$$[H^+] = 3.3 \times 10^1 = 33 \, M$$

This calculation implies that in order to prevent SnS from precipitating, the H^+ concentration must be 33 M. This concentration is impossible to achieve; therefore, SnS *must* precipitate, no matter how acidic the solution is.

To prevent ZnS from forming, the S^{2-} concentration cannot be larger than 1.2×10^{-22} M. Using this value for $[S^{2-}]$ in Equation 15.7, we have

$$\frac{[H^+]^2(1.2 \times 10^{-22})}{0.10} = 1.1 \times 10^{-21}$$

$$[H^+]^2 = 0.92$$

$$[H^+] = 0.96 \, M$$

Thus, when $[H^+] = 0.96$ M, the S^{2-} concentration will be 1.2×10^{-22} M, the highest value it can have without causing the Zn^{2+} to precipitate. A hydrogen ion concentration *greater* than 0.96 M will produce a sulfide ion concentration *less* than 1.2×10^{-22} M. In summary, to achieve a separation,

$$[H^+] < 33 \, M \quad \text{and} \quad [H^+] \geq 0.96 \, M$$

Table 16.2
Metal ions that can be separated
according to the solubilities of their sulfides

Metal Ions with Acid-Insoluble Sulfides			Metal Ions with Basic-Insoluble Sulfides		
Metal Ion	Metal Sulfide	K_{sp}	Metal Ion	Metal Sulfide	K_{sp}
Cu^{2+}	CuS	8.5×10^{-36}	Zn^{2+}	ZnS	1.2×10^{-23}
Bi^{3+}	Bi_2S_3	2.9×10^{-70}	Co^{2+}	CoS	7.0×10^{-23}
Pb^{2+}	PbS	7×10^{-27}	Ni^{2+}	NiS	2×10^{-21}
Hg^{2+}	HgS	1.6×10^{-54}	Fe^{2+}	FeS	3.7×10^{-19}
Sn^{2+}	SnS	1×10^{-26}	Mn^{2+}	MnS	7×10^{-16}

The preceding example illustrates how two metal ions whose sulfides differ rather widely in solubility can be separated from one another by precipitating only one of them. This separation is really quite complete. For example, at a hydrogen ion concentration of 1 M, which will prevent zinc sulfide from forming, the solubility of SnS in the saturated H_2S solution is only 9×10^{-5} M, which means that 99.91% of the Sn^{2+} originally in solution would be precipitated as SnS! Table 16.2 lists some metal ions that can be separated from one another in this manner. They are divided into groups according to the solubilities of their sulfides in an acidic solution. Those in the first column of the table are often referred to as metals having acid-insoluble sulfides. They can be separated from the metal ions in the second column, which precipitate as insoluble sulfides at higher pH and are referred to as metals with basic-insoluble sulfides.

16.2 COMMON ION EFFECT AND SOLUBILITY

When a salt is dissolved in a solution that already contains one of its ions, its solubility is less than in pure water. Silver chloride, for example, is less soluble in a solution of NaCl than it is in pure water. In this case both solutes have an ion in common: chloride ion. The reduction in the solubility in the presence of a **common ion** is called the **common ion effect.**

The effect of a common ion on solubility can easily be understood on the basis of Le Châtelier's principle. Suppose that solid silver chloride is placed into pure water and allowed to come to equilibrium with its ions in solution.

$$AgCl(s) \rightleftharpoons Ag^+(aq) + Cl^-(aq)$$

If a soluble chloride salt such as NaCl is now added to this solution, the chloride ion concentration will increase and drive this equilibrium to the left, thereby causing some AgCl to precipitate. In other words, AgCl is less soluble in aqueous NaCl than in pure water.[1] Let's look at a few examples.

---

[1] A similar effect was seen in the section on buffers in Chapter 15 when a common ion was added to the equilibrium dissociation of a weak acid.

EXAMPLE 16.8 CALCULATING THE SOLUBILITY OF A SALT IN A SOLUTION THAT CONTAINS A COMMON ION

PROBLEM What is the molar solubility of AgCl in a 0.010 M solution of NaCl?

SOLUTION For this salt we have

$$K_{sp} = [Ag^+][Cl^-] = 1.7 \times 10^{-10}$$

NaCl is soluble and completely dissociated. Don't attempt to write an equilibrium equation for it.

Before any AgCl dissolves we have an initial Cl^- concentration of 0.010 M. We can ignore the Na^+ because it is not involved in the equilibrium. We now let x equal the number of moles per dm^3 of AgCl that dissolve. This increases both $[Cl^-]$ and $[Ag^+]$ by x. Thus,

	Initial Molar Concentrations	Change	Equilibrium Molar Concentrations
Ag^+	0.0	$+x$	x
Cl^-	0.010	$+x$	$0.010 + x \approx 0.010$

Note that we have assumed that we can neglect x in computing the equilibrium Cl^- concentration. We make this assumption because the value of K_{sp} is very small. In doing so we greatly simplify the algebra. Substituting the equilibrium concentrations into the expression for K_{sp} gives

$$(x)(0.010) = 1.7 \times 10^{-10}$$

Notice that 1.7×10^{-8} is negligible compared to 0.010, which justifies our approximation.

or

$$x = 1.7 \times 10^{-8} \ M$$

Thus, the molar solubility of AgCl is 1.7×10^{-8} M. We might compare this to the molar solubility of AgCl in pure water, which we found to be 1.3×10^{-5} M in Example 16.3. The solubility of AgCl is indeed much less in a solution containing a common ion.

EXAMPLE 16.9 CALCULATING THE SOLUBILITY OF A COMPOUND IN A SOLUTION THAT CONTAINS A COMMON ION

PROBLEM What is the molar solubility of $Mg(OH)_2$ in 0.10 M NaOH?

SOLUTION The K_{sp} for $Mg(OH)_2$ is

$$K_{sp} = [Mg^{2+}][OH^-]^2$$

From this point on, such problems seem to mystify some students, but if you approach your construction of the concentration table systematically, you should have no difficulty. First, ask yourself what the initial concentrations are before any $Mg(OH)_2$ is added. Since the solution initially contains 0.10 M NaOH, $[Na^+] = 0.10$ M and $[OH^-] = 0.10$ M. There is no magnesium ion, so $[Mg^{2+}] = 0.0$ M. We are only interested in the concentrations of OH^- and Mg^{2+}, so we enter their values in the first column.

Next, how do the concentrations change? If x mol dm^{-3} of $Mg(OH)_2$ dissolves, then $[Mg^{2+}]$ increases by x and $[OH^-]$ increases by $2x$. These are entered in the center column, and the equilibrium concentrations are obtained by adding the quantities in the first two columns (as usual).

	Initial Molar Concentrations	Change	Equilibrium Molar Concentrations
Mg^{2+}	0.0	$+x$	x
OH^-	0.10	$+2x$	$0.10 + 2x \approx 0.10$

Again, we simplify the algebra by assuming that $2x$ is negligible compared to 0.10. Substituting the equilibrium concentrations and the K_{sp} for $Mg(OH)_2$ from Table 16.1 into the solubility product expression gives

$$(x)(0.10)^2 = 1.2 \times 10^{-11}$$

or

Note that $2x$, which equals 2.4×10^{-9}, is negligible compared to 0.10.

$$x = \frac{1.2 \times 10^{-11}}{(0.10)^2}$$

$$= 1.2 \times 10^{-9} \ M$$

Thus 1.2×10^{-9} mol dm^{-3} of $Mg(OH)_2$ dissolves in a 0.10 M solution of NaOH.

EXAMPLE 16.10   CALCULATING THE SOLUBILITY OF A SALT IN A SOLUTION THAT CONTAINS A COMMON ION

PROBLEM   What is the molar solubility of PbI_2 in 0.10 M $Pb(NO_3)_2$ solution? For lead iodide, $K_{sp} = 1.4 \times 10^{-8}$.

SOLUTION   The K_{sp} expression is

$$K_{sp} = [Pb^{2+}][I^-]^2$$

To construct the solubility table, we begin by asking, "What are the initial concentrations of Pb^{2+} and I^-?" The solution initially contains 0.10 M $Pb(NO_3)_2$, so the initial Pb^{2+} concentration is 0.10 M. No iodide is present initially, so the initial I^- concentration is 0.0 M. These values go in the first column.

Next, we let x be the molar solubility of PbI_2. When x mol dm^{-3} of PbI_2 dissolves, $[Pb^{2+}]$ increases by x and $[I^-]$ increases by $2x$. These quantities are entered in the change column. Then the first and second columns are added to give the equilibrium concentrations.

	Initial Molar Concentrations	Change	Equilibrium Molar Concentrations
Pb^{2+}	0.10	$+x$	$0.10 + x \approx 0.10$
I^-	0.0	$+2x$	$2x$

After we make our usual simplification, the equilibrium quantities are substituted into the K_{sp} expression.

$$K_{sp} = (0.10)(2x)^2 = 1.4 \times 10^{-8}$$

$$4x^2 = 1.4 \times 10^{-7}$$

$$x^2 = 3.5 \times 10^{-8}$$

$$x = 1.9 \times 10^{-4}$$

The simplification was justified because 1.9×10^{-4} is negligible compared to 0.10.

The molar solubility is $1.9 \times 10^{-4} \ M$.

16.3 COMPLEX IONS AND THEIR EQUILIBRIA

Many metal ions, particularly those of the transition elements, are able to combine with one or more other molecules or ions to produce more complex species that are called **complex ions,** or simply **complexes.** The substances that combine with the metal ion are called **ligands** and are usually Lewis bases. They can be either (a) neutral molecules such as H_2O and NH_3, (b) monatomic anions such as Cl^- and Br^-, or (c) polyatomic anions such as CN^- and $C_2O_4^{2-}$. One example of a complex ion that we saw earlier is $Al(H_2O)_6^{3+}$. Another example, containing fewer ligands, is formed when NH_3 is added to a solution containing Ag^+. Its formula is $Ag(NH_3)_2^+$. The charge on a complex ion such as this is the algebraic sum of the charges of the metal ion and the ligands. Thus Ag^+ also forms a complex ion with CN^- having the formula $Ag(CN)_2^-$.

We will discuss the details of structure and bonding of complex ions in Chapter 21. For now we will focus our attention on their dissociation equilibria and the effect that their formation has on the solubility of salts.

There are two ways of dealing with the equilibria involving complex ions. One is to consider their dissociation equilibria. For example, the overall reaction for the equilibrium dissociation of $Ag(NH_3)_2^+$ can be written as

$$Ag(NH_3)_2^+(aq) \rightleftharpoons Ag^+(aq) + 2NH_3(aq)$$

The equilibrium constant for this reaction is called an **instability constant,** K_{inst}. This is because the larger the value of K_{inst}, the less stable the complex is, as reflected by its tendency to dissociate. For the $Ag(NH_3)_2^+$ ion the equilibrium expression is

$$K_{inst} = \frac{[Ag^+][NH_3]^2}{[Ag(NH_3)_2^+]}$$

The value of the instability constant for this complex has been found to be 6.0×10^{-8}. We can see by the size of this constant that this particular complex is quite stable and will readily form whenever Ag^+ and NH_3 are added to the same solution. Other examples of complex ions and their instability constants can be seen in Table 16.3.

Table 16.3
Instability constants
and formation constants at 25°C

Complex Ion	K_{inst}	K_{form}
AlF_6^{3-}	1.5×10^{-20}	6.7×10^{19}
$Cd(CN)_4^{2-}$	1.3×10^{-17}	7.7×10^{16}
$Co(NH_3)_6^{2+}$	1.3×10^{-5}	7.7×10^4
$Co(NH_3)_6^{3+}$	2.0×10^{-34}	5.0×10^{33}
$Cu(NH_3)_4^{2+}$	2.1×10^{-13}	4.8×10^{12}
$Cu(CN)_2^-$	1.0×10^{-16}	1.0×10^{16}
$Fe(CN)_6^{4-}$	1.0×10^{-35}	1.0×10^{35}
$Fe(CN)_6^{3-}$	1.1×10^{-42}	9.1×10^{41}
$Ni(NH_3)_4^{2+}$	1.1×10^{-8}	9.1×10^7
$Ni(NH_3)_6^{2+}$	2.0×10^{-9}	5.0×10^8
$Ag(NH_3)_2^+$	6.0×10^{-8}	1.7×10^7
$Ag(CN)_2^-$	1.9×10^{-19}	5.3×10^{18}
$Zn(OH)_4^{2-}$	3.6×10^{-16}	2.8×10^{15}

An alternative way of writing the equilibrium for a complex ion is as an equation representing its formation. For example,

$$Ag^+(aq) + 2NH_3(aq) \rightleftharpoons Ag(NH_3)_2^+(aq)$$

The equilibrium expression, of course, is simply the reciprocal of the K_{inst} expression. In this case the equilibrium constant (which equals the reciprocal of K_{inst}) is called a **formation constant**, K_{form}, or **stability constant.** These are also given in Table 16.3.

$$K_{form} = \frac{[Ag(NH_3)_2^+]}{[Ag^+][NH_3]^2}$$

$$K_{form} = \frac{1}{K_{inst}}$$

In the chemical literature the equilibrium constants for complex ions are sometimes tabulated as instability constants and at other times as formation constants or stability constants. You should know the difference between them.

16.4 COMPLEX IONS AND SOLUBILITY

When a complex ion is formed in a solution of an insoluble salt, it reduces the concentration of free metal ion. As a result, more solid must dissolve in order to replenish the amount of metal ion lost, until that concentration required by the K_{sp} of the salt is achieved. Thus the solubility of an insoluble salt generally increases when complex ions are formed. To see this more clearly, let us see what effect adding NH_3 has on a saturated solution of AgCl.

Before any NH_3 is added, we have the equilibrium

$$AgCl(s) \rightleftharpoons Ag^+ + Cl^- \qquad [16.1]$$

Because NH_3 forms such a stable complex with the free silver ion, when NH_3 is added to this system a second equilibrium is established, namely,

$$Ag^+ + 2NH_3 \rightleftharpoons Ag(NH_3)_2^+ \qquad [16.2]$$

The creation of this new equilibrium upsets the first by removing some of the Ag^+, thereby causing the first equilibrium to shift to the right. As a result, some of the solid AgCl dissolves.

The two equilibria represented by Equations 16.1 and 16.2 can be combined into one overall equilibrium by adding them together.

$$AgCl(s) \rightleftharpoons Ag^+ + Cl^-$$
$$\underline{Ag^+ + 2NH_3 \rightleftharpoons Ag(NH_3)_2^+}$$
$$AgCl(s) + 2NH_3 \rightleftharpoons Ag(NH_3)_2^+ + Cl^-$$

The equilibrium constant for this overall reaction is

$$K_c = \frac{[Ag(NH_3)_2^+][Cl^-]}{[NH_3]^2}$$

We can obtain this same expression by multiplying the K_{sp} of AgCl by the K_{form} of the complex ion.[2] Thus,

$$K_{sp} \times K_{form} = [\cancel{Ag^+}][Cl^-] \times \frac{[Ag(NH_3)_2^+]}{[\cancel{Ag^+}][NH_3]^2} = K_c$$

---

[2] In general, if some equilibrium equation is obtained as the sum of two or more equations, the K_c for the final equation is equal to the *product* of the K's of the equations that were added together.

Therefore, with a knowledge of K_{sp} of the salt, K_{form} of the complex ion (or K_{inst}), and the concentration of NH_3, it is possible to calculate the concentrations of Ag^+ and Cl^- present at equilibrium and thus determine the solubility of AgCl in NH_3, as shown by the next example.

EXAMPLE 16.11 CALCULATING THE SOLUBILITY OF A SALT IN A SOLUTION THAT CONTAINS A COMPLEXING AGENT

PROBLEM   What is the molar solubility of AgCl in 1.0 M NH_3 at 25°C?

SOLUTION   As we have seen, the overall equilibrium equation for this problem is

A *complexing agent* is a substance that can form a complex with a metal ion (e.g., NH_3).

$$AgCl(s) + 2NH_3(aq) \rightleftharpoons Ag(NH_3)_2^+(aq) + Cl^-(aq)$$

for which we write

$$K_c = \frac{[Ag(NH_3)_2^+][Cl^-]}{[NH_3]^2}$$

where

$$K_c = K_{sp} \times K_{form} = (1.7 \times 10^{-10}) \times (1.7 \times 10^7) = 2.9 \times 10^{-3}$$

If we let x equal the number of moles per dm^3 of AgCl that dissolves, then we have the following initial and equilibrium concentrations:

	Initial Molar Concentrations	Change	Equilibrium Molar Concentrations
NH_3	1.0	$-2x$	$(1.0 - 2x)$
$Ag(NH_3)_2^+$	0.0	$+x$	x
Cl^-	0.0	$+x$	x

Substituting the concentrations at equilibrium into the K_c equation, we have

$$K_c = \frac{(x)(x)}{(1.0 - 2x)^2} = \frac{x^2}{(1.0 - 2x)^2} = 2.9 \times 10^{-3}$$

Taking the square root of both sides, we have

$$\frac{x}{1.0 - 2x} = 5.4 \times 10^{-2}$$

from which we obtain

$$x = 0.049$$

Therefore, we find that 0.049 mol of AgCl will dissolve in 1.00 dm^3 of 1.0 M NH_3.

In Example 16.11 we assumed that when the AgCl dissolves in the ammonia solution, essentially all the Ag^+ becomes complexed by NH_3. In other words, we said that the chloride ion concentration was equal to the concentration of $Ag(NH_3)_2^+$. Note that this assumption is valid only if K_{form} is very large, indicating that the complex is very stable.

EXAMPLE 16.12 DISSOLVING AN INSOLUBLE COMPOUND BY
FORMING A COMPLEX ION

PROBLEM   How many moles of solid NaOH must be added to 1.00 dm^3 of H_2O in order to dissolve
0.10 mol of $Zn(OH)_2$ according to the reaction,

$Zn(OH)_2$ is amphoteric and
dissolves in base as well as
acid.

$$Zn(OH)_2(s) + 2OH^- \rightleftharpoons Zn(OH)_4^{2-}$$

SOLUTION   The two equilibria involved in this system are

$$Zn(OH)_2(s) \rightleftharpoons Zn^{2+} + 2OH^- \qquad K_{sp} = 4.5 \times 10^{-17}$$

$$Zn^{2+} + 4OH^- \rightleftharpoons Zn(OH)_4^{2-} \qquad K_{form} = 2.8 \times 10^{15}$$

As before, the overall reaction can be written as the sum of these two equilibria,

$$Zn(OH)_2(s) + 2OH^- \rightleftharpoons Zn(OH)_4^{2-}$$

for which

$$K_c = K_{sp} \times K_{form} = (4.5 \times 10^{-17})(2.8 \times 10^{15})$$

$$= 1.3 \times 10^{-1}$$

Therefore,

$$\frac{[Zn(OH)_4^{2-}]}{[OH^-]^2} = 1.3 \times 10^{-1}$$

In this problem we know that 0.10 mol of zinc goes into solution where it is present as
either free Zn^{2+} or $Zn(OH)_4^{2-}$. Because K_{form} is so very large, essentially all of the zinc
will be present as the complex ion; therefore we can write

$$[Zn(OH)_4^{2-}] = 0.10 \ M$$

Substituting this into the equilibrium expression gives

$$1.3 \times 10^{-1} = \frac{0.10}{[OH^-]^2}$$

Therefore,

$$[OH^-]^2 = \frac{0.10}{1.3 \times 10^{-1}} = 7.7 \times 10^{-1}$$

and

$$[OH^-] = 0.88 \ M$$

This corresponds to the equilibrium concentration of free OH^-. In this solution, however, we also have 0.10 mol of $Zn(OH)_4^{2-}$, which contains an additional 0.40 mol of
OH^-, 0.20 mol of which was contained in the original 0.10 mol of $Zn(OH)_2$ that dissolved. Therefore, the total number of moles of NaOH that must be *added* to the water is
0.88 + 0.20 = 1.08 mol.

REVIEW QUESTIONS AND PROBLEMS

(Problems whose numbers are in blue have their answers in Appendix C at the back of the book; the more difficult problems are marked with asterisks.)

Solubility Product

16.1 Why can the concentration of the solid be left out of the equilibrium expression for the solubility of a salt?

16.2 Write the K_{sp} expression for each of the following substances:
(a) Ag_2S (d) MgC_2O_4
(b) CaF_2 (e) Bi_2S_3
(c) $Fe(OH)_3$ (f) $BaCO_3$

16.3 Write the K_{sp} expression for these salts:
(a) PbF_2 (d) Li_2CO_3
(b) Cu_2S (e) $Ca(IO_3)_2$
(c) $Fe_3(PO_4)_2$ (f) $Ag_2Cr_2O_7$

Calculating K_{sp} from Solubility

16.4 The solubility of CuCl in water is 1.0×10^{-3} mol dm^{-3}. What is its value of K_{sp}?

16.5 The solubility of $PbCO_3$ is 1.8×10^{-7} mol dm^{-3}. What is K_{sp} for $PbCO_3$?

16.6 The solubility of barium oxalate, BaC_2O_4, is 0.0781 g dm^{-3}. Calculate K_{sp} for BaC_2O_4.

16.7 The solubility of $CaCrO_4$ is 1.0×10^{-2} mol dm^{-3}. What is K_{sp} for $CaCrO_4$?

16.8 The solubility of lead iodide, PbI_2, in water is 1.5×10^{-3} mol dm^{-3}. Calculate its K_{sp}.

16.9 A student determined that 0.0981 g of PbF_2 was dissolved in 200 cm^3 of saturated PbF_2 solution. What is K_{sp} for PbF_2?

16.10 The solubility of MgF_2 is 7.6×10^{-2} g dm^{-3}. Calculate K_{sp} for this salt.

16.11 The solubility of Bi_2S_3 is 2.5×10^{-12} g dm^{-3}. What is K_{sp} for Bi_2S_3?

16.12 The pH of a saturated solution of $Ni(OH)_2$ is 8.83. Calculate K_{sp} for $Ni(OH)_2$.

*16.13 A 500-cm^3 portion of 0.0020 M $Na_2C_2O_4$ (sodium oxalate) solution is able to dissolve 0.47 g of MgC_2O_4. What is K_{sp} for MgC_2O_4?

Calculating Solubility from K_{sp}

16.14 Using the data in Table 16.1, calculate the molar solubility in water of each of the following:
(a) PbS
(b) $Fe(OH)_2$
(c) $BaSO_4$
(d) Hg_2Cl_2 (which yields Hg_2^{2+} and $2Cl^-$)
(e) CaF_2
(f) MgC_2O_4

16.15 Milk of magnesia is a suspension of solid $Mg(OH)_2$ in water. Calculate the pH of the aqueous phase, assuming that it is composed of pure water saturated with $Mg(OH)_2(s)$.

16.16 How many grams of $CaSO_4$ will dissolve in 600 cm^3 of water?

16.17 What volume of saturated HgS solution contains a single Hg^{2+} ion?

*16.18 Plaster is composed of $CaSO_4$. Suppose that there was a leak above a ceiling through which water was seeping at the rate of 2.0 dm^3 per day. If the plaster in the ceiling is 1.50 cm thick, how long would it take to dissolve a circular hole 1 cm in diameter? Assume that the density of the plaster is 0.97 g cm^{-3}.

K_{sp} and Precipitation

16.19 What condition must be met to have a precipitate form in a solution?

16.20 Would a precipitate form in the following solutions?
(a) 5.0×10^{-2} mol of $AgNO_3$ and 1.0×10^{-3} mol of $NaC_2H_3O_2$ dissolved in 1.0 dm^3 of solution
(b) 1.0×10^{-2} mol of $Ba(NO_3)_2$ and 2.0×10^{-2} mol of NaF dissolved in 1.0 dm^3 of solution
(c) 500 cm^3 of 1.4×10^{-2} M $CaCl_2$ and 250 cm^3 of 0.25 M Na_2SO_4 mixed to give a final volume of 750 cm^3

16.21 What is the minimum pH necessary to cause a precipitate of $Fe(OH)_2$ to form in a 0.010 M $FeCl_2$ solution?

*16.22 A solution is prepared by mixing 100 cm^3 of 0.20 M $AgNO_3$ with 100 cm^3 of 0.10 M HCl. What are the molar concentrations of all species present in the solution when equilibrium is reached?

16.23 Will a precipitate form in a solution containing:
(a) 0.025 M $CaCl_2$ and 0.0050 M Na_2CO_3
(b) 0.010 M $Pb(NO_3)_2$ and 0.030 M $CaCl_2$
(c) 1.5×10^{-3} M $FeCl_2$ and 2.2×10^{-3} M $Na_2C_2O_4$

16.24 What would the H^+ concentration have to be in order to prevent the precipitation of HgS when a 0.0010 M $Hg(NO_3)_2$ solution is saturated with H_2S ([H_2S] = 0.10 M)? Can you explain why HgS is insoluble in concentrated (12 M) HCl?

16.25 Show that ZnS is soluble in concentrated (12 M) HCl.

Selective Precipitation

16.26 A solution is known to contain 0.010 M Pb^{2+} and 0.010 M Ni^{2+}. How must the pH be adjusted to achieve the maximum separation of these ions when the solution is saturated with H_2S ([H_2S] = 0.10 M)?

16.27 A solution containing 0.10 M Zn^{2+} and 0.10 M Fe^{2+} is saturated with H_2S ([H_2S] = 0.10 M). What must the H^+ concentration be to separate these ions by selectively precipitating ZnS? What is the smallest Zn^{2+} concentration that can be achieved without precipitating any of the Fe^{2+} as FeS?

16.28 What will precipitate first when $Na_2CrO_4(s)$ is gradually added to a solution containing 0.010 M Pb^{2+} and 0.010 M Ba^{2+}? What will be the molar concentration of the ion precipitated first when the other ion just begins to form a precipitate?

16.29 Magnesium oxalate, MgC_2O_4, has $K_{sp} = 8.6 \times 10^{-5}$ and calcium oxalate, CaC_2O_4, has $K_{sp} = 2.3 \times 10^{-9}$. What pH must be maintained to achieve *maximum* separation of Ca^{2+} from Mg^{2+} if both have a concentration of 0.10 M and the concentration of oxalic acid, $H_2C_2O_4$, is maintained at 0.10 M? For $H_2C_2O_4$, $K_{a_1} = 6.5 \times 10^{-2}$ and $K_{a_2} = 6.1 \times 10^{-5}$.

Common Ion Effect

16.30 What is the common ion effect?

16.31 What is the molar solubility of $CaCO_3$ in 0.50 M Na_2CO_3?

16.32 What is the molar solubility of AgCl in 0.020 M $AlCl_3$? Assume that $AlCl_3$ gives Al^{3+} and Cl^- in solution.

16.33 What is the molar solubility of $PbCl_2$ in 0.020 M $AlCl_3$? Assume that $AlCl_3$ gives Al^{3+} and Cl^- in solution.

16.34 How many moles of Ag_2CrO_4 will dissolve in 1.0 dm^3 of 0.10 M $AgNO_3$?

16.35 How many moles of Ag_2CrO_4 will dissolve in 1.0 dm^3 of 0.10 M Na_2CrO_4?

16.36 What is the molar solubility of CaF_2 in 0.010 M NaF?

16.37 How many grams of NaF must be added to 1.00 dm^3 of solution to reduce the molar solubility of BaF_2 to 6.8×10^{-4} mol dm^{-3}?

*16.38 25.0 mL of 0.10 M HCl is added to 1.000 dm^3 of saturated $Mg(OH)_2$ in contact with more than enough $Mg(OH)_2(s)$ to react with all the HCl. After reaction has ceased, what will be the molar concentration of Mg^{2+}? What will be the pH of the solution?

*16.39 2.20 g of NaOH(s) are added to 250 cm^3 of 0.10 M $FeCl_2$ solution. What mass of $Fe(OH)_2$ will be formed? What will be the molar concentration of Fe^{2+} in the final solution?

*16.40 1.75 g of NaOH(s) are added to 250 cm^3 of 0.10 M $NiCl_2$ solution. What mass, in grams, of $Ni(OH)_2$ will be formed? What will be the pH of the final solution? For $Ni(OH)_2$, $K_{sp} = 1.6 \times 10^{-14}$.

*16.41 Solid $Mn(OH)_2$ is added to a solution of 0.100 M $FeCl_2$. After reaction, what will be the molar concentrations of Mn^{2+} and Fe^{2+} in the solution? What will be the pH of the solution?

Complex Ions

16.42 What is a complex ion? What is a ligand? What kinds of substances serve as ligands?

16.43 Write equilibrium expressions corresponding to K_{form} for the complex ions:
(a) $AgCl_2^-$ (b) $Ag(S_2O_3)_2^{3-}$ (c) $Zn(NH_3)_4^{2+}$

16.44 Write equilibrium expressions corresponding to K_{inst} for the complex ions:
(a) $Fe(CN)_6^{4-}$ (b) $CuCl_4^{2-}$ (c) $Ni(NH_3)_6^{2+}$

Complex Ions and Solubility

16.45 Silver forms a relatively stable complex ion, AgI_2^-. When a solution containing this ion is diluted with water, AgI precipitates. Explain why this happens in terms of the equilibria that are involved.

16.46 Use the data in Tables 16.1 and 16.3 to determine the molar solubility of AgI in 0.010 M KCN solution.

16.47 The solubility of $Zn(OH)_2$ in 1.0 M NH_3 is 5.7×10^{-3} mol dm^{-3}. Determine the value of the instability constant of the complex ion, $Zn(NH_3)_4^{2+}$. Ignore the reaction, $NH_3 + H_2O \rightleftharpoons NH_4^+ + OH^-$.

Simultaneous Equilibria

16.48 On the basis of Le Châtelier's principle, explain why the addition of solid NH_4Cl to a beaker containing solid $Mg(OH)_2$ in contact with water causes the $Mg(OH)_2$ to dissolve.

*16.49 Will a precipitate form in a solution made by dissolving 1.0 mol of $AgNO_3$ and 1.0 mol $HC_2H_3O_2$ in 1.0 dm^3 of solution?

*16.50 How many moles of HCl must be added to 1.0 dm^3 of water to dissolve completely 0.20 mol of FeS? Remember that a saturated H_2S solution is 0.10 M.

*16.51 How many moles of solid NH_4Cl must be added to 1.0 dm^3 of water in order to dissolve 0.10 mol of solid $Mg(OH)_2$? *Hint:* Consider the simultaneous equilibria

$$Mg(OH)_2 \rightleftharpoons Mg^{2+} + 2OH^-$$

$$NH_3 + H_2O \rightleftharpoons NH_4^+ + OH^-$$

*16.52 How much solid sodium acetate would have to be added to 200 cm^3 of a solution containing 0.200 M $AgNO_3$ and 0.10 M nitric acid to cause silver acetate to begin to precipitate? For $HC_2H_3O_2$, $K_a = 1.8 \times 10^{-5}$ and for $AgC_2H_3O_2$, $K_{sp} = 2.3 \times 10^{-3}$.

*16.53 How many grams of solid potassium fluoride must be added to 200 cm^3 of a solution that contains 0.20 M $AgNO_3$ and 0.10 M acetic acid to cause silver acetate to begin to precipitate? For HF, $K_a = 6.5 \times 10^{-4}$; for $HC_2H_3O_2$, $K_a = 1.8 \times 10^{-5}$; for $AgC_2H_3O_2$, $K_{sp} = 2.3 \times 10^{-3}$.

*16.54 What is the molar solubility of $Mg(OH)_2$ in 0.10 M NH_3 solution? Remember that NH_3 is a weak base.

CHAPTER 17
ELECTROCHEMISTRY

We routinely rely on pocket calculators to solve chemistry problems, and we take for granted the chemical reactions occurring inside the batteries that provide the electricity to power them. In this chapter we will study how electrical energy can cause non-spontaneous chemical changes to occur, and how spontaneous reactions can serve as a source of electricity.

Electrochemistry is concerned with the conversion of electrical energy into chemical energy in **electrolytic cells,** as well as the conversion of chemical energy into electrical energy in **galvanic** or **voltaic cells.** In an electrolytic cell a process called electrolysis takes place in which the passage of electricity through a solution provides sufficient energy to cause an otherwise nonspontaneous oxidation-reduction reaction to take place. A galvanic cell, on the other hand, provides a source of electricity that results from a spontaneous oxidation-reduction reaction taking place in solution.

Electrochemical processes have a practical importance in chemistry and everyday life. Electrolytic cells can provide us with information about the chemical environment as well as the energy that is required for many important oxidation-reduction reactions to occur. In addition, electrolysis is used to make many important chemicals that find their way into our lives. Examples are lye, NaOH, which is used to make soap, paper, and many other chemicals, and liquid bleach, NaOCl. For years now, galvanic cells such as the dry cell and "nicad" battery have powered our flashlights, radios, electronic calculators, wristwatches, cameras, and children's toys. The familiar lead storage battery has widespread applications, especially in the automotive industry. More recently, fuel cells, in which the energy available from the combustion of fuels is converted directly into electricity, are finding many uses, especially in space vehicles. Electrochemical know-how has aided scientists in producing modern equipment for pollution analysis and biomedical research. With the aid of tiny electrochemical probes, scientists are beginning to study the chemical reactions taking place in living cells.

17.1 METALLIC AND ELECTROLYTIC CONDUCTION

For a substance to be classified as a conductor of electricity, it must be able to allow electrical charges within it to be moved from one point to another for the purpose of completing an electrical circuit. From our earlier discussion of solids, we know that most metals are conductors of electricity because of the relatively free movement of their *electrons* throughout the metallic lattice. This conduction is simply called **metallic conduction.** We also know from Chapter 7 that solutions containing electrolytes have the ability to conduct electricity. In this case, however, there are no free electrons in the solution to carry the current. How, then, do they conduct?

We can determine whether or not a solution is a conductor of electricity by using an apparatus similar to that shown in Figure 7.2 (p. 225). When the two electrodes are connected to a source of electricity and dipped into a solution, we can observe whether or not the bulb of the apparatus lights. When this is done, we find that the bulb burns brightly when the solution contains a salt such as NaCl, but not at all when the solution contains a molecular compound such as sugar. Only when there are mobile ions present is electrical conduction possible, and this condition is fulfilled only by solutions of electrolytes and by molten salts.

When the source of electricity to the electrodes of a conductivity apparatus is a battery or other direct current (D.C.) source, each electrode acquires an electrical charge as shown in Figure 17.1, and each ion in the liquid tends to move toward the electrode of opposite charge. Thus, when the electric potential is applied, the positive ions migrate toward the negative electrode and the nega-

Figure 17.1
Ion flow in an electrolytic cell.

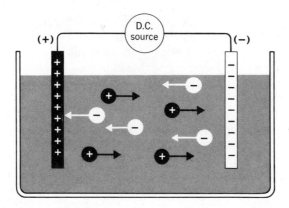

If redox didn't occur, the electrodes would become neutralized by the layer of oppositely charged ions and no further ion migration would tend to occur.

tive ions move toward the positive electrode. This movement of ionic charges through the liquid, brought about by the application of electricity, is called **electrolytic conduction** and the apparatus is called an **electrolytic cell.**

When electrolytic conduction occurs, chemical reactions take place as the ions in the liquid come into contact with the electrodes. At the positive electrode—where a deficiency of electrons exists—the negative ions are forced to give up electrons and are therefore oxidized. At the negative electrode—which has an excess of electrons—the positive ions pick up electrons and are reduced. Thus, during electrolytic conduction, oxidation is occurring at the positive electrode and reduction is taking place at the negative electrode. The liquid will continue to conduct electricity only as long as the oxidation-reduction reactions occurring at the electrodes continue.

The electrons that are deposited during the oxidation reaction are pumped out the electrode by the source of electric potential and transferred to the negative electrode. During electrolytic conduction we have electrons flowing through the exterior wire and ions flowing through the solution. This situation is illustrated in Figure 17.2a.

The ionic movement, as well as the reactions at the electrodes, must take place so that electrical neutrality is maintained. This means that even in the most minute part of the liquid, whenever a negative ion moves away, a positive ion must also leave, or another negative ion must immediately take its place (Figure 17.2b). In this way every portion of the liquid is electrically neutral at all times. During the reactions at the electrodes, electrical neutrality is assured by having equal numbers of electrons deposited and picked up. For example, whenever

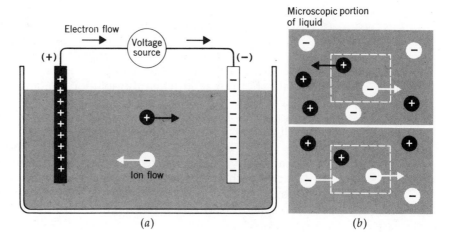

Figure 17.2
Electrolytic conduction
(a) *Electrolytic cell.* (b) *Maintaining electrical neutrality on a microscopic scale.*

one electron is deposited at the positive electrode, one electron must simultaneously be taken from the negative electrode. We focus our attention next on the chemical consequences of these last two processes.

17.2 ELECTROLYSIS

Molten means melted.

The chemical reactions that occur at the electrodes during electrolytic conduction constitute **electrolysis.** For example, when liquid (molten) sodium chloride, is *electrolyzed*, we find that the Na^+ ions move toward the negative electrode and the Cl^- ions move toward the positive electrode (Figure 17.3). The reactions that take place at the electrodes are

Positive electrode	$2Cl^-(l) \longrightarrow Cl_2(g) + 2e^-$	(Oxidation)
Negative electrode	$Na^+(l) + e^- \longrightarrow Na(l)$	(Reduction)

You might recognize these as half-reactions of the type that you learned to construct in Chapter 7. In fact, the methods that you learned to obtain balanced half-reactions in Chapter 7 give the electrode reactions that we will discuss in this chapter for reactions in aqueous solutions.

In electrochemistry we always assign the terms **cathode** and **anode** according to the chemical reaction that is taking place at the electrode. *Reduction always takes place at the cathode and oxidation always takes place at the anode.* Thus, in our electrolysis reactions above, we label the negative electrode the cathode and the positive electrode the anode.

The net chemical change that takes place in the electrolytic cell is called the **cell reaction.** It is obtained by adding together the anode and cathode half-reactions in such a way that the same number of electrons are gained and lost. This is the same procedure we used in the ion electron method of balancing oxidation-reduction reactions in Chapter 7. Thus, in this case, we must multiply the reduction half-reaction by 2 to obtain

In any redox reaction, the total electron gain must equal the total electron loss.

$$2Cl^-(l) \longrightarrow Cl_2(g) + 2e^-$$
$$\underline{2Na^+(l) + 2e^- \longrightarrow 2Na(l)}$$
$$2Na^+(l) + 2Cl^-(l) \longrightarrow Cl_2(g) + 2Na(l)$$

Therefore, in this electrolytic cell sodium is formed at the cathode and chlorine gas is produced at the anode. This is one of the major sources of pure sodium metal and chlorine gas in the United States (see Section 17.3).

The electrolysis of aqueous solutions of electrolytes is somewhat more com-

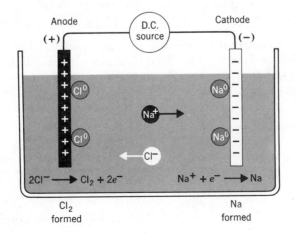

Figure 17.3
Electrolysis of molten NaCl.

plex because of the ability of water to be oxidized as well as reduced. The oxidation reaction for water is

$$2H_2O(l) \longrightarrow O_2(g) + 4H^+(aq) + 4e^-$$ [17.1]

and the reduction reaction takes the form

$$2H_2O(l) + 2e^- \longrightarrow H_2(g) + 2OH^-(aq)$$ [17.2]

In acidic solutions another reaction that may take place is the reduction of H^+, which is

$$2H^+(aq) + 2e^- \longrightarrow H_2(g)$$ [17.3]

Reaction 17.3 is not, however, a major reaction in most dilute aqueous solutions that we will consider.

It is not necessarily an easy matter to predict what reactions will actually occur at the electrodes.

In aqueous solution, we have the possible oxidation and reduction of the solvent in addition to the possible oxidation and reduction of the ions of the solute. Whether the solute anion or water is going to be oxidized, or whether the solute cation or water is going to be reduced, depends on the relative ease of the two competing reactions, as we will see in the next few examples.

Electrolysis of aqueous NaBr

In the electrolysis of aqueous NaBr, the following two anode reactions (oxidation reactions) are possible:

(1) $2Br^-(aq) \longrightarrow Br_2(aq) + 2e^-$

(2) $2H_2O(l) \longrightarrow O_2(g) + 4H^+(aq) + 4e^-$

and the following two cathode reactions (reduction reactions) are possible:

(3) $Na^+(aq) + e^- \longrightarrow Na(s)$

(4) $2H_2O(l) + 2e^- \longrightarrow H_2(g) + 2OH^-(aq)$

We can, of course, experimentally determine the outcome of the electrolysis simply by examining the products formed at the electrodes, and when we carry out the reaction we find that H_2 bubbles from the solution at the cathode and the solution around the anode takes on a red color due to the formation of Br_2 (Figure 17.4). Therefore, during the electrolysis of NaBr solutions, the two electrode half-reactions and the overall cell reaction must be

$$2Br^-(aq) \longrightarrow Br_2(aq) + 2e^- \qquad \text{(Anode)}$$
$$\underline{2H_2O(l) + 2e^- \longrightarrow H_2(g) + 2OH^-(aq) \qquad \text{(Cathode)}}$$
$$2Br^-(aq) + 2H_2O(l) \longrightarrow Br_2(aq) + H_2(g) + 2OH^-(aq) \quad \text{(Cell reaction)}$$

The results of this electrolysis experiment tell us that Na^+ is more difficult to reduce than H_2O and that H_2O is more difficult to oxidize than Br^-—that's why H_2O is reduced and Br^- is oxidized. In the competition, the reaction that takes place more easily is the reaction that we actually observe.

Electrolysis of aqueous CuSO₄

As in our last example, there are two possible oxidation and reduction reactions for the electrolysis of $CuSO_4$. These are

Oxidation (1) $2SO_4^{2-}(aq) \longrightarrow S_2O_8^{2-}(aq) + 2e^-$

(2) $2H_2O(l) \longrightarrow O_2(g) + 4H^+(aq) + 4e^-$

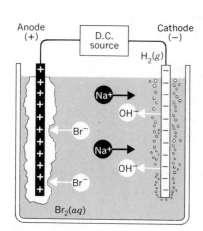

Anode (+)   D.C. source   Cathode (−) $H_2(g)$ Na+   OH−   Br−   Na+   OH−   Br− $Br_2(aq)$

Figure 17.4
The electrolysis of aqueous sodium bromide solution.

and

Reduction   (3) $Cu^{2+}(aq) + 2e^- \longrightarrow Cu(s)$

(4) $2H_2O(l) + 2e^- \longrightarrow H_2(g) + 2OH^-(aq)$

During this electrolysis we find experimentally that oxygen gas bubbles from the anode and a reddish coating of copper metal is deposited on the cathode. Therefore, we would write for the electrolysis of aqueous $CuSO_4$:

$$2H_2O(l) \longrightarrow O_2(g) + 4H^+(aq) + 4e^- \qquad \text{(Anode)}$$
$$\underline{2Cu^{2+}(aq) + 4e^- \longrightarrow 2\,Cu(s) \qquad\qquad\qquad \text{(Cathode)}}$$
$$2H_2O(l) + 2Cu^{2+}(aq) \longrightarrow O_2(g) + 4H^+(aq) + 2Cu(s) \quad \text{(Cell reaction)}$$

Notice that we multiplied the equation for the reduction of Cu^{2+} by 2 so that we have equal numbers of electrons lost and gained.

Because of the products formed, we can conclude that in the electrolysis of aqueous $CuSO_4$, the H_2O is more easily oxidized than the SO_4^{2-}, and the Cu^{2+} is more easily reduced than the H_2O.

Electrolysis of aqueous CuBr₂

At this point we should be able to apply what we have learned about the electrolysis of aqueous solutions of NaBr and $CuSO_4$ to the electrolysis of aqueous $CuBr_2$. In solution, this salt gives Cu^{2+} and Br^- ions. Therefore, at the anode we once again have a competition between the oxidation of water and the oxidation of Br^-. Earlier we learned that Br^- is easier to oxidize than H_2O, so we should expect the product of the electrolysis to be Br_2. Similarly, at the cathode we have a competition for reduction, and we have learned that Cu^{2+} is more easily reduced than H_2O. Therefore, at the cathode the reduction product should be metallic copper. As a result, we expect the following reactions:

$$2Br^-(aq) \longrightarrow Br_2(aq) + 2e^- \qquad \text{(Anode)}$$
$$\underline{Cu^{2+}(aq) + 2e^- \longrightarrow Cu(s) \qquad\quad \text{(Cathode)}}$$
$$Cu^{2+}(aq) + 2Br^-(aq) \longrightarrow Cu(s) + Br_2(aq) \quad \text{(Cell reaction)}$$

This is exactly what is found experimentally.

Electrolysis of aqueous Na₂SO₄

Once again we call on what we have learned previously. We know that water is more easily oxidized than SO_4^{2-} at the anode and that water is more easily reduced than Na^+ at the cathode. Therefore, in this solution H_2O is both oxidized and reduced, giving us

$$2H_2O(l) \longrightarrow O_2(g) + 4H^+(aq) + 4e^- \qquad\qquad \text{(Anode)}$$
$$\underline{4H_2O(l) + 4e^- \longrightarrow 2H_2(g) + 4OH^-(aq) \qquad\qquad \text{(Cathode)}}$$
$$6H_2O(l) \longrightarrow O_2(g) + 2H_2(g) + 4OH^-(aq) + 4H^+(aq) \quad \text{(Cell reaction)}$$

Notice that we had to multiply the reduction reaction (Equation 17.2) by 2 to have the same number of electrons as in the oxidation reaction.

During the electrolysis, H^+ ions are formed at the anode and OH^- ions are formed at the cathode. The solution in the immediate vicinity of the anode therefore becomes acidic, and around the cathode the solution becomes basic. We can see this if the electrolysis is performed in the presence of an acid-base indicator.

We can also see that if the solution is stirred afterwards, the H^+ and OH^- ions react to give water by our familiar neutralization reaction, and the solution becomes neutral again. Using the amounts of these ions in the equation above, we obtain

$$4OH^- + 4H^+ \longrightarrow 4H_2O$$

Thus the net overall reaction for the electrolysis of a stirred aqueous Na_2SO_4 solution is

$$2H_2O(l) \longrightarrow O_2(g) + 2H_2(g)$$

which is simply the reaction for the electrolysis of H_2O. Sodium sulfate does not participate in this electrolysis in the sense that it is not consumed at the electrodes. Yet we would find experimentally that it or some other similar salt is needed if electrolysis of water is to occur. What, then, is the role of the Na_2SO_4? The Na_2SO_4 is needed to maintain electrical neutrality (Figure 17.5). During the oxidation of H_2O, H^+ ions are produced in the immediate vicinity of the anode. A negative ion must also be present in that region to neutralize the positive charges. This is fulfilled by SO_4^{2-} ions. Likewise, at the cathode, where OH^- ions are produced, there must be positive ions present to neutralize the charges on the OH^- ions, and thereby keep the solution electrically neutral.

17.3 PRACTICAL APPLICATIONS OF ELECTROLYSIS

Each day our lives are touched either directly or indirectly by the products of electrolysis reactions. For example, the drinking water in most places is treated with chlorine to kill bacteria, and chlorine is used to manufacture many chemicals, from pesticides that protect crops to plastics such as polyvinyl chloride—usually just called vinyl. Yet elemental chlorine does not occur free in nature. It must be extracted from its compounds, and this is done most economically by electrolysis. In this section we will study how chlorine and some other commercially important substances are made.

Electrolysis of molten sodium chloride

The chemistry of this process was described in our introduction to the discussion of electrolysis reactions on page 599. The products—sodium and chlorine—are both commercially important.

Figure 17.5

Electrolysis of aqueous Na_2SO_4. The sodium ions and sulfate ions are needed to counter the charges of the ions formed at the electrodes and thereby maintain electrical neutrality.

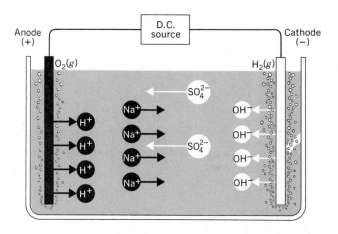

Cathode:

$$Na^+ + e^- \rightleftharpoons Na(l)$$

Anode:

$$2Cl^- \rightleftharpoons Cl_2(g) + 2e^-$$

Melting points:
NaCl 801°C
Na 98°C

Sodium and chlorine are very reactive chemicals. Therefore, when they are produced from NaCl they must be kept apart, otherwise they will react and reform sodium chloride. The Downs cell, illustrated in Figure 17.6, accomplishes this. The chlorine, of course, comes off as a gas. Because of the temperature at which the cell operates, the sodium is formed as a liquid and also easily removed. This allows the cell to operate continuously as fresh sodium chloride is added and the products are taken away.

Electrolysis of brine (NaCl solution)

In the electrolysis of a sodium chloride solution there is a competition at the anode between the oxidation of chloride ion and the oxidation of water.

$$(1) \quad 2Cl^-(aq) \longrightarrow Cl_2(g) + 2e^-$$

$$(2) \quad 2H_2O(l) \longrightarrow O_2(g) + 4H^+(aq) + 4e^-$$

As the NaCl becomes more dilute, Reaction (2) begins to compete and some O_2 is formed as well.

When a concentrated salt solution (**brine**) is used, the first reaction is the one that is observed. At the cathode, the reaction is the reduction of water (you've learned that H_2O is more easily reduced than Na^+).

$$2H_2O(l) + 2e^- \longrightarrow H_2(g) + 2OH^-(aq)$$

When we combine the anode and cathode reactions and include the Na^+ ion (which is actually a spectator ion), the overall reaction is

In 1984, production of NaOH by this reaction amounted to about 9×10^9 kg.

$$2Na^+ + 2Cl^- + 2H_2O \longrightarrow Cl_2(g) + H_2(g) + \underbrace{2Na^+ + 2OH^-}_{NaOH(aq)}$$

This is only a small sampling of the uses of Cl_2, H_2, and NaOH.

If we examine this equation, we can appreciate its commercial importance. All the products—chlorine, hydrogen, and sodium hydroxide—are important industrial chemicals. Chlorine is used to make plastics such as polyvinyl chloride (PVC) and to purify drinking water. Hydrogen is used to make ammonia and to manufacture hydrogenated vegetable oils such as Crisco®. And sodium hydroxide is used in huge quantities to neutralize acids in various chemical processes,

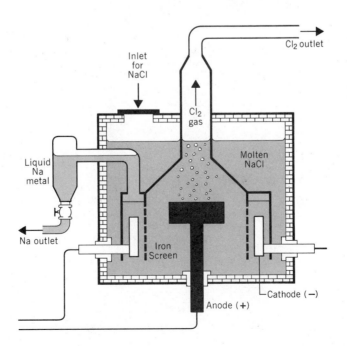

Figure 17.6

The Downs cell used for the electrolysis of molten sodium chloride. The cell is constructed to keep the metallic sodium and gaseous chlorine from reacting with each other after they are formed in the electrolysis reaction.

Figure 17.7

Electrolysis of aqueous sodium chloride.

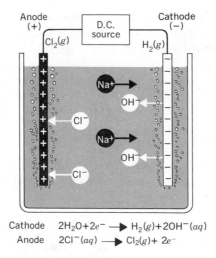

Cathode $2H_2O + 2e^- \longrightarrow H_2(g) + 2OH^-(aq)$
Anode $2Cl^-(aq) \longrightarrow Cl_2(g) + 2e^-$

process pulp and paper, purify aluminum ores, and in the manufacture of textiles and the refining of petroleum.

The electrolysis of NaCl can be carried out in an apparatus such as that illustrated in Figure 17.7, but doing so always leads to some contamination of the NaOH by unreacted NaCl and by hypochlorite ion, which is formed in a reaction between OH^- and the chlorine produced at the anode. (This reaction is discussed further later.) To prevent the formation of hypochlorite and avoid dangerously explosive mixtures of H_2 and Cl_2, the products formed at the cathode must be prevented from coming into contact with those formed at the anode. There have been several approaches to accomplishing this, and most of the NaOH made today is produced by an apparatus called a **diaphragm cell.**

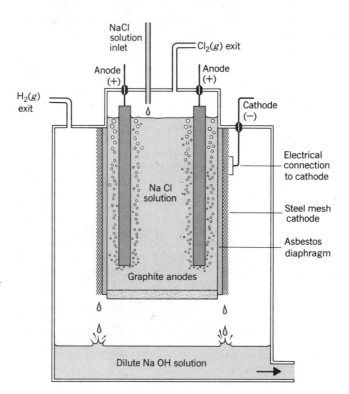

Figure 17.8

A diaphragm cell for the production of NaOH by the electrolysis of aqueous NaCl solutions. The diagram is a cross section of a cylindrical cell in which the NaCl solution is surrounded by an asbestos diaphragm held in place by the steel wire mesh cathode.

There are several variations in the design of this cell, but the drawing in Figure 17.8 illustrates its essential features.

The cell consists of a compartment into which a NaCl solution is fed slowly. Graphite anodes dip into this solution. The cylindrical wall of the compartment consists of a porous asbestos paper layer (the diaphragm) that is supported by an iron wire mesh that serves as the cathode. When the cell operates, the NaCl solution seeps slowly through the asbestos and comes into contact with the iron cathode where hydrogen is evolved and hydroxide ion is formed. This solution, now containing NaOH, drips to the bottom of the container that holds the cell and is removed for purification. Meanwhile, the anode reaction within the compartment generates Cl_2 that bubbles through the electrolyte and is removed at the top.

The NaOH produced in the diaphragm cell is always contaminated by a little unreacted NaCl. If very pure NaOH is needed, a mercury cell, illustrated in Figure 17.9, can be used. In this cell sodium ion is actually the substance that is reduced, and the free metal dissolves in the liquid mercury as it is formed. The solution of sodium in mercury—it's called sodium amalgam—is pumped to a separate vessel where the metallic sodium at the surface of the mercury can react with water. This reaction liberates hydrogen and leaves pure NaOH in solution.

$$2Na(\text{in Hg}) + 2H_2O \longrightarrow 2Na^+ + 2OH^- + H_2(g)$$

The mercury cell has a disadvantage of posing a serious threat of mercury water pollution and so must be carefully monitored.

If the electrolysis of brine is carried out in a vigorously stirred solution, the OH^- produced at the cathode reacts with the Cl_2 formed at the anode. The reaction is

$$Cl_2 + 2OH^- \longrightarrow Cl^- + OCl^- + H_2O$$

Continued electrolysis therefore gradually converts nearly all the chloride ion to hypochlorite ion, OCl^-, and the sodium chloride solution is changed to a solution of sodium hypochlorite. When diluted to about 5 to 6 percent by mass, this is sold as liquid laundry bleach (e.g., Clorox®).

Figure 17.9

Electrolysis of aqueous sodium chloride using a mercury cell. Chlorine gas is evolved at the anode where chloride ions are oxidized. At the cathode, sodium ions are reduced to sodium atoms, which dissolve in the mercury. The mercury is pumped to a tank where it comes into contact with water. Sodium atoms react there with water to liberate hydrogen and produce sodium hydroxide.

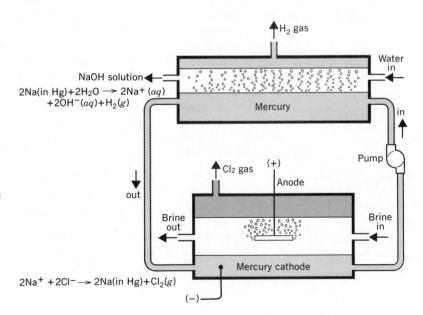

Graphite electrodes can be seen projecting from the tops of the electrolysis cells in this aluminum pot line at Alcan's Arvida Works in Quebec.

Aluminum

As you are undoubtedly aware, aluminum finds many important uses as a structural metal because of its strength and low density (light weight). Its commercial availability has been made possible through the application of electrochemical reduction.

If we were to electrolyze an aqueous solution of an aluminum salt, such as $AlCl_3$, we would find that H_2O is more easily reduced than the Al^{3+}, so an aqueous solution of an aluminum salt cannot be used to produce the metal. A 22-year-old graduate of Oberlin College, Charles Hall, invented a process, in 1886, whereby molten Al_2O_3 is used. He prepared a mixture of Al_2O_3 with the mineral *cryolite*, Na_3AlF_6, and electrolyzed it in the molten state. The cryolite, he found, reduced the melting temperature from 2000°C for Al_2O_3 to 1000°C for the mixture. A diagram of the electrolysis cell is shown in Figure 17.10. The vessel holding the molten mixture is made of iron lined with carbon and serves as the cathode. Carbon rods that serve as anodes are inserted into the melt. As the oxidation-reduction reactions proceed, pure aluminum is produced at the cathode and sinks to the bottom of the vessel. The reactions at the electrodes are

At the high temperature of the cell, the O_2 attacks the carbon electrodes and they must be replaced periodically.

$$3O^{2-}(l) \longrightarrow \tfrac{3}{2}O_2(g) + 6e^- \qquad \text{(Anode)}$$
$$\underline{2Al^{3+}(l) + 6e^- \longrightarrow 2Al(l) \qquad\qquad \text{(Cathode)}}$$
$$2Al^{3+}(l) + 3O^{2-}(l) \longrightarrow \tfrac{3}{2}O_2(g) + 2Al(l)$$

Today, cryolite has been replaced by a synthetic electrolyte composed of a mixture of NaF, CaF_2, and AlF_3. This mixture permits operation at still lower temperatures and is less dense than the cryolite used by Hall. This lower density of the electrolyte mix permits easier separation of the molten aluminum.

Figure 17.10

Production of aluminum by the Hall process.

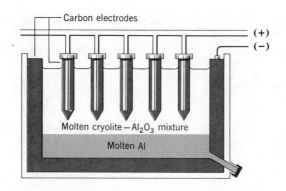

Magnesium

Mg²⁺ is the third most abundant ion in sea water.

Magnesium, another structural metal that is important because of its low density, occurs to an appreciable extent in sea water. Magnesium ions are precipitated from sea water as the hydroxide and the $Mg(OH)_2$ is then converted to the chloride by treatment with hydrochloric acid. After evaporation of the water, the $MgCl_2$ is melted and electrolyzed. Magnesium is produced at the cathode and chlorine is evolved at the anode. The overall net reaction is simply

$$MgCl_2(l) \longrightarrow Mg(l) + Cl_2(g)$$

Copper

An interesting application of electrolysis is the refining, or purification, of copper metal. When first separated from its ore, copper metal is about 99% pure, with iron, zinc, silver, gold, and platinum as major impurities. In the refining process the impure copper is used as the anode in an electrolytic cell containing aqueous copper(II) sulfate as the electrolyte. The cathode of the cell is constructed of high-purity copper (Figure 17.11).

When electrolysis is carried out, the electric potential across the cell is adjusted so that only copper and other more active metals, such as iron or zinc, are able to dissolve at the anode. The silver, gold, and platinum do not dissolve, and simply fall off and settle to the bottom of the electrolysis cell. At the cathode only the most easily reduced species, Cu^{2+}, is caused to pick up electrons; hence, only copper is deposited.

The net result of the operation of this cell is that copper is transferred from the anode to the cathode while the Fe and Zn impurities remain in solution as Fe^{2+} and Zn^{2+}. Afterwards the silver, gold, and platinum sludge is removed from the apparatus and sold for enough money to nearly pay for the cost of the electricity required in the electrolysis. As a result, the purification of copper (about 99.96% pure) costs nearly nothing! Nevertheless, the total production cost of copper is still considerable, because it includes the mining of the crude ore and its initial purification.

Electroplating

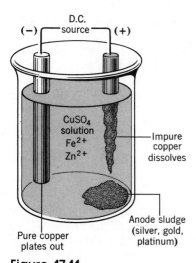

Figure 17.11

Purification of copper by electrolysis.

We have just seen how copper can be plated out on an electrode in an electrolysis cell. The plating out of a metal in this fashion is called **electroplating.** If we replaced the cathode in the cell in Figure 17.11 with another metal, the surface of that metal will also become covered with a layer of pure copper when the current is applied. Other metals can be electroplated as well as copper, which makes this process of great commercial importance. In the manufacturing of automobiles,

Copper cathodes, 99.96% pure, are pulled from the electrolytic refining tanks at Kennecott's Utah copper refinery. It takes about 28 days for the impure copper anodes to dissolve and deposit the pure metal on the cathodes.

for example, various parts, such as steel bumpers, are electroplated with chromium for beauty as well as protection against corrosion.

17.4 QUANTITATIVE ASPECTS OF ELECTROLYSIS

Michael Faraday was the first to describe, in a quantitative fashion, the relationship that exists between the amount of current used and the extent of the chemical change that takes place at the electrodes during electrolysis. He found that the amount of chemical change that occurs in an electrolysis reaction is related to the amount of electricity that passes through the cell. In modern terms, we say that the amount of change is related to the number of moles of electrons lost or gained in the oxidation-reduction reactions. For example, in the reduction of silver ion to silver metal,

$$Ag^+(aq) + e^- \longrightarrow Ag(s)$$

1 mol of electrons reacts with 1 mol of silver ions to give 1 mol, or 107.87 g, of solid silver. Thus, in this case, when 107.87 g of silver are deposited on the cathode, we know that 1 mol of electrons must have passed through the cell.

The quantity of electricity that must be supplied to a cell in order to deliver 1 mol *of electrons has been known historically as a* **faraday** *(ℱ).* In the above example, 1 ℱ must be supplied to produce the 107.87 g of silver, and it would take 2 ℱ to produce 215.74 g of silver, and so on. In other words, "faraday" is just another way of saying "a mole of electrons."

$$1 \ \mathscr{F} \equiv 1 \text{ mol of electrons}$$

Another important unit is the coulomb (C). This is the SI unit of electric charge. One coulomb is the quantity of charge that moves past any given point in a circuit when a current of 1 ampere (1 A) is supplied for 1 second (1 s). Thus

$$1 \text{ coulomb} = 1 \text{ ampere} \times 1 \text{ second}$$
$$1 \text{ C} = 1 \text{ A} \cdot \text{s}$$

Experimentally, it is found that 1 $\mathscr{F}$ is equivalent to 96 487 coulombs, or 96 500 when rounded to three significant figures. Thus

$$1 \text{ mol electrons} = 1 \mathscr{F} = 96 \text{ 500 C}$$

Let's look at some examples of how these concepts can be applied.

EXAMPLE 17.1   QUANTITATIVE PROBLEMS ON ELECTROLYSIS

PROBLEM   In an electrolytic cell like that in Figure 17.2, how many grams of Cu will be deposited from a solution of $CuSO_4$ by a current of 1.5 A flowing for 2.0 h?

SOLUTION   First, we must have an equation representing the reaction that takes place. Since metallic copper is being deposited from a solution containing Cu^{2+}, the copper ions must gain electrons and thereby undergo reduction. This lets us write the following half-reaction.

$$Cu^{2+}(aq) + 2e^- \longrightarrow Cu(s)$$

This equation is necessary because it gives the important relationship

$$1 \text{ mol Cu} \sim 2 \text{ mol } e^-$$

or simply

$$1 \text{ mol Cu} \sim 2 \mathscr{F}$$

Our procedure now is to use the current and time to calculate the number of coulombs delivered. Then we calculate the number of moles of electrons supplied. Finally we compute the number of moles of Cu, followed by the number of grams of Cu.

$$1.5 \text{ A} \times 2.0 \text{ h} \times \left(\frac{3600 \text{ s}}{1 \text{ h}}\right) = 11 \text{ 000 A} \cdot \text{s (rounded)}$$

$$11 \text{ 000 A} \cdot \text{s} \times \left(\frac{1 \text{ C}}{1 \text{ A} \cdot \text{s}}\right) = 11 \text{ 000 C}$$

$$11 \text{ 000 C} \times \left(\frac{1 \text{ mol } e^-}{96 \text{ 500 C}}\right) = 0.11 \text{ mol } e^-$$

$$0.11 \text{ mol } e^- \times \left(\frac{1 \text{ mol Cu}}{2 \text{ mol } e^-}\right) \times \left(\frac{63.55 \text{ g Cu}}{1 \text{ mol Cu}}\right) \sim 3.5 \text{ g Cu}$$

EXAMPLE 17.2   QUANTITATIVE PROBLEMS ON ELECTROLYSIS

PROBLEM   How long would it take to produce 25.0 g of Cr from a solution of $CrCl_3$ by a current of 2.75 A.

SOLUTION   Again, we begin by writing the equation for the reaction. It also is a reduction;

$$Cr^{3+}(aq) + 3e^- \longrightarrow Cr(s)$$

from which we can say that

$$1 \text{ mol Cr} \sim 3 \text{ mol } e^- \sim 3 \mathscr{F}$$

First we calculate the number of faradays (or moles of electrons) required.

$$25.0 \text{ g Cr} \times \left(\frac{1 \text{ mol Cr}}{52.0 \text{ g Cr}} \right) \times \left(\frac{3 \mathscr{F}}{1 \text{ mol Cr}} \right) \sim 1.44 \mathscr{F}$$

Next we calculate the number of coulombs required.

$$1.44 \mathscr{F} \times \left(\frac{96\,500 \text{ C}}{1 \mathscr{F}} \right) \sim 139\,000 \text{ C}$$

Since 1 C is equal to $1 \text{ A} \cdot \text{s}$,

$$139\,000 \text{ C} \times \left(\frac{1 \text{ A} \cdot \text{s}}{1 \text{ C}} \right) \times \left(\frac{1}{2.75 \text{ A}} \right) \sim 50\,500 \text{ s}$$

If we convert this to hours, we get

$$50\,500 \text{ s} \times \left(\frac{1 \text{ h}}{3600 \text{ s}} \right) = 14.0 \text{ h}$$

We can experimentally determine the mass of a substance that has been deposited on an electrode during electrolysis by weighing the electrode before and after the current has been supplied. The apparatus used in experiments of this kind is called a **coulometer**. In Figure 17.12 we see two such coulometers connected in series so that the same current, and thus the same number of moles of electrons, passes through both cells. With the aid of this apparatus it is possible to use a known oxidation-reduction reaction in one cell to provide an experimental measure of the molar mass of an unknown in the other cell. This type of analysis is illustrated in the next example.

EXAMPLE 17.3   CALCULATIONS INVOLVING A COULOMETER

PROBLEM   In the left cell of Figure 17.12 we place a solution containing Ag^+ ions, and in the right cell a solution containing X^{2+} ions of an unknown metal X. The same current is passed through both cells for the same total time. When the current is turned off and the

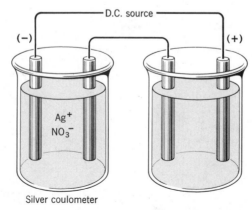

Figure 17.12

Two coulometers connected in series. The same current must pass through both cells. When 107.87 g of silver are deposited on the cathode in the cell at the left we know that 1 mol e^- has passed through both cells.

Silver coulometer

$\left[\begin{array}{c} 1 \text{ mol } e^- \text{ deposits} \\ 107.87 \text{ g Ag on a cathode} \end{array} \right]$

1 mol e^- passes through this cell when 107.87 g of Ag (1 mol Ag) is deposited in the other cell

electrodes are rinsed, dried, and weighed, it is found that 3.50 g of silver are deposited during the same period of time that 2.50 g of element X are deposited. What is the molar mass of element X?

SOLUTION

This rather long-winded problem has a relatively simple solution. Since the current and time are the same for both cells in this series circuit, the same number of electrons must pass through both. We can calculate the number of faradays (moles of electrons) supplied by using the information from the Ag^+ cell.

From our earlier discussion in this section, we know that the reduction reaction for Ag^+ is

$$Ag^+(aq) + e^- \longrightarrow Ag(s)$$

Therefore,

$$1 \text{ mol Ag} \sim 1 \text{ mol } e^- \ (1 \ \mathscr{F})$$

or

$$107.87 \text{ g Ag} \sim 1 \text{ mol } e^-$$

Total moles of electrons used in the experiment must be

$$3.50 \text{ g} \times \left(\frac{1 \text{ mol } e^-}{107.87 \text{ g}} \right) = 0.0324 \text{ mol } e^-$$

The reaction involving $X^{2+}(aq)$ in the cell is

$$X^{2+}(aq) + 2e^- \longrightarrow X(s)$$

Therefore,

$$1 \text{ mol X} \sim 2 \text{ mol } e^- \ (\text{or } 2 \ \mathscr{F})$$

Since 0.0324 mol of electrons passed through both cells, the total number of moles X deposited can be found:

$$0.0324 \text{ mol } e^- \times \frac{1 \text{ mol X}}{2 \text{ mol } e^-} = 0.0162 \text{ mol X}$$

Finally, recognizing that molar mass has units of g mol^{-1},

$$\text{molar mass X} = \frac{3.36 \text{ g X}}{0.0162 \text{ mol X}} = 207 \text{ g mol}^{-1}$$

17.5 GALVANIC CELLS

In our previous discussion of electrolytic cells, chemical changes occurred because an electric potential placed across electrodes forced an otherwise nonspontaneous chemical reaction to take place. We now turn to the opposite situation, in which electron flow is produced as a result of spontaneous oxidation-reduction reactions in **galvanic cells.**

Luigi Galvani (1737–1798) was an Italian anatomist who pioneered in the field of electrophysiology—the study of the relationship between electricity and living organisms.

An example of a spontaneous oxidation-reduction reaction taking place in a solution can be seen simply by placing a piece of metallic zinc into a solution of $CuSO_4$ (see Figure 7.13 on p. 263). A dark brown, spongelike layer begins to form on the piece of zinc and, at the same time, the blue color of the $CuSO_4$ begins to disappear. The brownish substance forming on the zinc is metallic copper, and if we were to analyze the solution, we would find that it now contains Zn^{2+}. The reaction that is taking place, therefore, is

$$Cu^{2+}(aq) + Zn(s) \longrightarrow Cu(s) + Zn^{2+}(aq)$$

This can be divided into a pair of redox half-reactions:

$$Cu^{2+}(aq) + 2e^- \longrightarrow Cu(s)$$
$$Zn(s) \longrightarrow Zn^{2+}(aq) + 2e^-$$

We see from these reactions that the Cu^{2+} ions are spontaneously removed from the solution and replaced by the colorless Zn^{2+} ions. Thus the blue color of the solution gradually disappears as more and more Zn^{2+} ions are formed.

As long as these reactions take place at the surface of the zinc, no useful flow of electrons can be obtained. The reaction simply generates heat. We can, however, take advantage of the spontaneous electron transfer by making the oxidation and reduction half-reactions occur in separate compartments of a cell, as shown in Figure 17.13. Each compartment is called a **half-cell,** and when they are properly connected the electrons produced by the oxidation of the zinc must travel through the wire and into the electrode in the $CuSO_4$ solution. The electrons are then picked up by the Cu^{2+} ions and reduction takes place. The electrons flowing through the external wire constitute an electric current, and the galvanic cell therefore serves as a source of electricity.

Although the zinc and copper have to be separated to obtain a useful flow of electrons, complete isolation of the two species would lead to an electrical imbalance at the electrodes, and the electron flow would soon cease. We can see how this would occur if we imagine that the two half-cells were completely isolated from each other and the oxidation-reduction reactions still continued to take place. On the left side of this hypothetical setup, Zn^{2+} ions entering the solution would give the solution an overall positive charge. This would prevent additional Zn^{2+} from entering. On the right we would find that when Cu^{2+} leaves the solution the SO_4^{2-} ions left behind would give the solution a negative charge and the electrode would become positively charged. This would cause the electrode to repel Cu^{2+} ions and prevent their further removal from the solution.

From the preceding discussion we see that a continuous electric current, accompanied by a continuous chemical activity, can only occur if the solution around each electrode is kept electrically neutral. For this to happen, however, ions must flow into or out of the cell compartments. In the zinc half-cell, for example, Zn^{2+} must leave the electrode compartment or anions must enter it. Similarly, in the copper half-cell, cations must enter to balance the charge of the SO_4^{2-}, or SO_4^{2-} ions must leave.

Although ions must be able to diffuse from one cell compartment to another, the two solutions cannot be allowed to mix freely, otherwise Cu^{2+} would

Figure 17.13

A galvanic cell that employs the reaction

$Zn(s) + Cu^{2+}(aq) \rightarrow$
$\qquad Zn^{2+}(aq) + Cu(s)$

The salt bridge in (a) serves the same purpose as the porous partition in (b)—they each allow electrical neutrality to be maintained as the reactions occur at the electrodes.

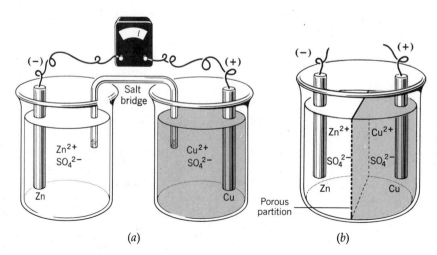

(a) (b)

react directly at the Zn electrode and no flow of electrons through the external circuit would have to occur.

The salt bridge in Figure 17.13*a* and the porous partition in Figure 17.13*b* allow for the slow mixing of the ions in the two solutions. A **salt bridge** is usually a tube filled with an electrolyte such as KNO_3 or KCl in gelatin. Cations from the salt bridge can move into one compartment to compensate for the excess negative charge, while the anions from the salt bridge diffuse into the other compartment to neutralize the excess positive charge. The porous partition in Figure 17.13*b* serves the same purpose as the salt bridge. With either the salt bridge or the porous partition in place, there is a continuous electron flow through the external wire and an ion flow through the solution as a result of the spontaneous oxidation-reduction reactions taking place at the electrodes.

The signs of the electrodes in galvanic cells

Earlier we defined the anode in electrochemistry at the electrode at which oxidation takes place and the cathode as the one at which reduction occurs. This applies regardless of whether the cell is an electrolytic cell or a galvanic cell. In the galvanic cell just described, oxidation takes place in the zinc compartment, so the zinc bar would be the anode and the copper electrode would be the cathode. As zinc ions leave the solid zinc anode and enter the solution, electrons are left behind and the zinc electrode acquires a negative charge. At the copper cathode, Cu^{2+} ions become attached to the electrode and seek electrons to become reduced This gives the copper electrode a positive charge. Thus we see that in a galvanic cell, the anode is negative and the cathode is positive, quite the opposite of what we found to be true in electrolytic cells.[1]

17.6 CELL POTENTIALS

The electric current obtained from a galvanic cell is a result of electrons being pushed or forced to flow from the negative electrode, through an external wire, to the positive electrode. The force with which these electrons move through the wire is called the **electromotive force,** or **emf,** and is measured in **volts** (V). Actually, the volt is a measure of the energy that is capable of being extracted from the flowing electric charge. If the emf is 1 V, the passage of 1 coulomb is able to accomplish 1 joule of work.

$$1 \text{ volt} = \frac{1 \text{ joule}}{\text{coulomb}}$$

$$1 \text{ V} = 1 \text{ J/C} \qquad [17.4]$$

The emf produced by a galvanic cell is called the **cell potential,** $\mathscr{E}_{cell}$. This emf depends on the concentrations of the ions in the cell, the temperature, and the partial pressures of any gases that might be involved in the cell reactions. When all ion concentrations are 1 M, all partial pressures of gases are 1 atm, and

---

[1] This labeling of electrodes in galvanic cells is, however, consistent with electrolytic cells when we consider the movement of the ions within the solution. The Zn^{2+} ions produced at the anode and the SO_4^{2-} ions freed at the cathode must mingle with each other if electrical neutrality is to prevail in the solution. To accomplish this, some of the Zn^{2+} ions must move toward the cathode and some of the SO_4^{2-} ions must move toward the anode. Thus we have cations moving toward the cathode and anions moving toward the anode, which is precisely the same situation as in electrolytic cells.

Notice that the standard conditions here are the same as those used in defining standard thermodynamic quantities.

the temperature of the cell is 25°C, the emf is called the **standard cell potential,**[2] designated as $\mathscr{E}^0_{cell}$.

To measure the cell potential accurately, care must be taken to avoid drawing current from the cell. This is because some of the cell's potential is required to overcome the cell's own internal resistance when current is drawn. The remaining electric potential that can be measured under these conditions is less than the maximum.

One device that can be used to measure the emf of a cell is called a **potentiometer.** In this instrument the potential generated by the cell is balanced by an opposing potential from within the potentiometer. When the two opposing potentials are equal, no current flows and the cell potential is equal to the opposing emf, which can be read directly from the instrument. The cell potential that is measured in this way is the maximum emf of the cell. Today modern advances in electronics have led to a variety of other devices that are able to measure the emf of a cell quickly and simply, without drawing significant amounts of current.

17.7 REDUCTION POTENTIALS

A very important and useful concept can be developed if we attempt to answer the question, "What is the origin of the cell potential?" To answer this question, we will use the Zn/Cu cell described earlier. In this cell we have a solution containing Zn^{2+} ions around one electrode and a solution containing Cu^{2+} ions around the other. Each of these ions has a certain tendency to acquire electrons from its respective electrode and become reduced. In other words, each reduction half-reaction,

$$Zn^{2+}(aq) + 2e^- \longrightarrow Zn(s)$$

and

$$Cu^{2+}(aq) + 2e^- \longrightarrow Cu(s)$$

has a certain intrinsic tendency to proceed from left to right that we can describe by its **reduction potential.** The larger the reduction potential for any half-reaction, the greater its tendency to undergo reduction.

When the cell reaction takes place what we are actually observing is a kind of "tug-of-war." Each of the species in solution attempts to pull electrons from its electrode so as to become reduced. The species with the greater tendency to acquire electrons—that is, the one with the larger reduction potential—wins the tug-of-war and does undergo reduction. The loser, on the other hand, must supply the electrons to the winner, so the loser is oxidized. In the zinc-copper cell, therefore, copper must have a larger reduction potential than zinc, because copper is reduced.

The potential that we measure for a cell corresponds to the *difference* in the tendencies of the two ions to become reduced, and is equal to the reduction potential for the substance that actually undergoes reduction minus the reduction potential for the substance that is forced to undergo oxidation. In terms of standard reduction potentials,

$$\mathscr{E}^0_{cell} = \mathscr{E}^0_{substance\ reduced} - \mathscr{E}^0_{substance\ oxidized} \qquad [17.5]$$

---

[2] In footnote 1 in Chapter 14, it was mentioned that activities (effective concentrations and pressures) should be used in the mass action expression when computing ΔG. This applies also to the effect of concentration on $\mathscr{E}$. The standard potential is obtained when all species are at unit activity. Only a small error is introduced, however, by using actual concentrations and pressures when solutions are relatively dilute.

In the zinc-copper cell, therefore,

$$\mathscr{E}^0_{cell} = \mathscr{E}^0_{Cu} - \mathscr{E}^0_{Zn}$$

where $\mathscr{E}^0_{Cu}$ is the standard reduction potential for copper and $\mathscr{E}^0_{Zn}$ is the standard reduction potential for zinc. Since $\mathscr{E}^0_{Cu}$ is larger than $\mathscr{E}^0_{Zn}$, $\mathscr{E}^0_{cell}$ is positive.

Experimentally, it is only possible to measure overall cell potentials. This means that we are only capable of obtaining differences between the reduction potentials for any two half-reactions. Then how can we obtain the reduction potential for any specific half-reaction? Clearly, if the cell potential and the $\mathscr{E}^0$ for one of the half-reactions are known, the $\mathscr{E}^0$ for the other half-reaction can be calculated. What has been done, therefore, is to choose a half-reaction arbitrarily and assign to it a standard reduction potential of zero volts. All other half-reactions can then be compared to this standard and a set of relative values of $\mathscr{E}^0$ obtained.

The electrode chosen to be the standard is called the **hydrogen electrode,** shown in Figure 17.14a. It consists of a platinum wire encased in a glass sleeve with hydrogen gas passing through it at a pressure of 1 atm (101.325 kPa). The platinum wire is attached to a platinum foil that is coated with a black velvety layer of finely divided platinum, which serves as a catalyst for the reaction,

$$2H^+(aq) + 2e^- \rightleftharpoons H_2(g)$$

This assembly is then immersed in an acid solution whose hydrogen ion concentration is 1.00 M.

By definition, the standard hydrogen electrode is assigned a reduction potential, $\mathscr{E}^0_{H_2}$, of exactly 0.000 V. Any substance that is more easily reduced than H^+ has a positive value of $\mathscr{E}^0$, and any substance that is more difficult to reduce has a negative $\mathscr{E}^0$.

When the hydrogen electrode is paired with another half-cell in a galvanic cell, it can undergo either oxidation or reduction, depending on the reduction potential of the other half-cell. For instance, if the reduction potential of the species in the other half-cell is greater than that for the hydrogen electrode—that is, if its $\mathscr{E}^0$ is positive—the hydrogen electrode is forced to undergo oxidation. The corresponding half-reaction for the oxidation of the hydrogen electrode is

$$H_2(g) \longrightarrow 2H^+(aq) + 2e^-$$

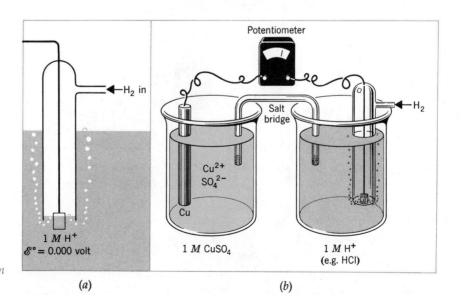

Figure 17.14
(a) *The hydrogen electrode.*
(b) *The hydrogen electrode used in a galvanic cell.*

On the other hand, if the reduction potential of the other half-reaction is less than 0.000 V, this species has a negative reduction potential and the hydrogen electrode undergoes reduction.

$$2H^+(aq) + 2e^- \longrightarrow H_2(g)$$

This causes the other species to become oxidized.

To illustrate how the hydrogen electrode is used, let's examine the galvanic cell in Figure 17.14b. When we connect a potentiometer to this cell to measure its potential, proper readings can only be obtained if we connect the terminal labeled $(+)$ to the positive electrode and the terminal labeled $(-)$ to the negative electrode. That's the way these instruments are designed. In this case, we find that to obtain a proper reading the $(+)$ terminal must be connected to the copper electrode and the $(-)$ terminal to the hydrogen electrode. From what we learned earlier, this tells us that the copper electrode is the cathode and the hydrogen electrode is the anode. The spontaneous half-reactions in this cell are therefore

> The cathode carries the positive charge in a galvanic cell.

$$Cu^{2+}(aq) + 2e^- \longrightarrow Cu(s) \qquad \text{(Cathode)}$$

$$H_2(g) \longrightarrow 2H^+(aq) + 2e^- \qquad \text{(Anode)}$$

Since copper is reduced, when we apply Equation 17.5 we must write

$$\mathscr{E}^0_{cell} = \mathscr{E}^0_{Cu} - \mathscr{E}^0_{H_2}$$

> A *measured* cell potential is always a positive number.

The measured cell potential (as read from the potentiometer) is 0.34 V. Therefore

$$0.34 \text{ V} = \mathscr{E}^0_{Cu} - 0.000 \text{ V}$$

or

$$\mathscr{E}^0_{Cu} = +0.34 \text{ V}$$

Now let's look at the cell in Figure 17.15. Here we have a zinc half-cell paired with the hydrogen electrode. To obtain a reading with the potentiometer we find, by experiment, that the positive terminal has to be connected to the hydrogen electrode, so this time it is the cathode. The spontaneous half-reactions in this cell are therefore

$$2H^+(aq) + 2e^- \longrightarrow H_2(g) \qquad \text{(Cathode)}$$

$$Zn(s) \longrightarrow Zn^{2+}(aq) + 2e^- \qquad \text{(Anode)}$$

In other words, zinc ion is more difficult to reduce than H^+.

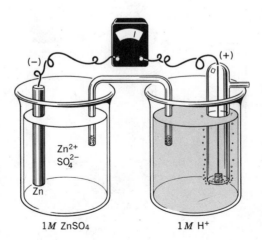

Figure 17.15

A galvanic cell that can be used to determine the standard reduction potential of Zn^{2+}.

The measured value of the cell potential, as read from the potentiometer, is 0.76 V. Using Equation 17.5 again, we obtain

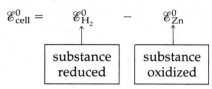

$$\mathscr{E}^0_{cell} = \mathscr{E}^0_{H_2} - \mathscr{E}^0_{Zn}$$

| substance reduced | substance oxidized |

Substituting yields

$$0.76 \text{ V} = 0.000 \text{ V} - \mathscr{E}^0_{Zn}$$

Solving for $\mathscr{E}^0_{Zn}$ gives

$$\mathscr{E}^0_{Zn} = -0.76 \text{ V}$$

$\mathscr{E}^0$ is negative for species that are more difficult to reduce than H^+.

The negative sign for $\mathscr{E}^0_{Zn}$ reflects the fact that Zn^{2+} is more difficult to reduce than H^+.

With a knowledge of the reduction potentials for the zinc and copper electrodes, we can now *predict* the cell potential and the spontaneous cell reaction for the Zn/Cu cell. This can be done even if we had no previous knowledge of the species undergoing oxidation or reduction.

First, simply by examining the reduction potentials,

$$Cu^{2+}(aq) + 2e^- \rightleftharpoons Cu(s) \quad \mathscr{E}^0_{Cu} = +0.34 \text{ V}$$

$$Zn^{2+}(aq) + 2e^- \rightleftharpoons Zn(s) \quad \mathscr{E}^0_{Zn} = -0.76 \text{ V}$$

we know immediately that Cu^{2+} is more easily reduced than Zn^{2+}. This is because Cu^{2+} has the higher (more positive) reduction potential. The cell reaction must therefore be

$$Zn(s) + Cu^{2+}(aq) \longrightarrow Cu(s) + Zn^{2+}(aq)$$

Next, the only way we can obtain a positive $\mathscr{E}^0_{cell}$ is to subtract $\mathscr{E}^0_{Zn}$ (the $\mathscr{E}^0$ of the substance oxidized) from $\mathscr{E}^0_{Cu}$ (the $\mathscr{E}^0$ of the substance reduced).

$$\mathscr{E}^0_{cell} = \mathscr{E}^0_{Cu} - \mathscr{E}^0_{Zn}$$

$$\mathscr{E}^0_{cell} = +0.34 \text{ V} - (-0.76 \text{ V})$$

$$\mathscr{E}^0_{cell} = +1.10 \text{ V}$$

This is precisely the value that we observe experimentally when we measure the potential of the cell.

We are now in a position to determine the reduction potentials for many different half-reactions, because all we have to do is construct galvanic cells in which the reduction potential of one half-cell is known, relative to the hydrogen electrode. Some standard reduction potentials, $\mathscr{E}^0$, determined in this manner are given in Table 17.1. In this table the reduction potential of the hydrogen electrode is placed in the middle, with the species more difficult to reduce than hydrogen listed below it and those more easily reduced placed above it. Such a table of reduction potentials serves many useful purposes.

The double arrows in the half-reactions in Table 17.1 signify that the reactions are reversible, not that they are equilibria.

1. From a table of reduction potentials we can, at a glance, pick out substances that are good oxidizing agents and those that are good reducing agents. Any species that appears on the left of the double arrow serves as an oxidizing agent if it undergoes reduction during the course of a chemical reaction. Since the substances at the top left side of the table are more easily reduced than those at the bottom, their ability to serve as oxidizing

Table 17.1
Standard reduction potentials at 25°C

$\mathscr{E}^0$ (volts)	Half-reaction
2.87	$F_2 + 2e^- \rightleftharpoons 2F^-$
2.00	$S_2O_8^{2-} + 2e^- \rightleftharpoons 2SO_4^{2-}$
1.78	$H_2O_2 + 2H^+ + 2e^- \rightleftharpoons 2H_2O$
1.69	$PbO_2 + SO_4^{2-} + 4H^+ + 2e^- \rightleftharpoons PbSO_4 + 2H_2O$
1.49	$8H^+ + MnO_4^- + 5e^- \rightleftharpoons Mn^{2+} + 4H_2O$
1.47	$2ClO_3^- + 12H^+ + 10e^- \rightleftharpoons Cl_2 + 6H_2O$
1.36	$Cl_2 \ (g) + 2e^- \rightleftharpoons 2Cl^-$
1.33	$Cr_2O_7^{2-} + 14H^+ + 6e^- \rightleftharpoons 2Cr^{3+} + 7H_2O$
1.28	$MnO_2 + 4H^+ + 2e^- \rightleftharpoons Mn^{2+} + 2H_2O$
1.23	$O_2 + 4H^+ + 4e^- \rightleftharpoons 2H_2O$
1.09	$Br_2 \ (aq) + 2e^- \rightleftharpoons 2Br^-$
0.80	$Ag^+ + e^- \rightleftharpoons Ag$
0.77	$Fe^{3+} + e^- \rightleftharpoons Fe^{2+}$
0.54	$I_2 \ (aq) + 2e^- \rightleftharpoons 2I^-$
0.52	$Cu^+ + e^- \rightleftharpoons Cu$
0.34	$Cu^{2+} + 2e^- \rightleftharpoons Cu$
0.27	$Hg_2Cl_2 + 2e^- \rightleftharpoons 2Hg + 2Cl^-$
0.22	$AgCl + e^- \rightleftharpoons Ag + Cl^-$
0.00	$2H^+ + 2e^- \rightleftharpoons H_2$
−0.04	$Fe^{3+} + 3e^- \rightleftharpoons Fe$
−0.13	$Pb^{2+} + 2e^- \rightleftharpoons Pb$
−0.14	$Sn^{2+} + 2e^- \rightleftharpoons Sn$
−0.25	$Ni^{2+} + 2e^- \rightleftharpoons Ni$
−0.36	$PbSO_4 + 2e^- \rightleftharpoons Pb + SO_4^{2-}$
−0.44	$Fe^{2+} + 2e^- \rightleftharpoons Fe$
−0.74	$Cr^{3+} + 3e^- \rightleftharpoons Cr$
−0.76	$Zn^{2+} + 2e^- \rightleftharpoons Zn$
−0.83	$2H_2O + 2e^- \rightleftharpoons H_2 + 2OH^-$
−1.03	$Mn^{2+} + 2e^- \rightleftharpoons Mn$
−1.67	$Al^{3+} + 3e^- \rightleftharpoons Al$
−2.38	$Mg^{2+} + 2e^- \rightleftharpoons Mg$
−2.71	$Na^+ + e^- \rightleftharpoons Na$
−2.76	$Ca^{2+} + 2e^- \rightleftharpoons Ca$
−2.90	$Ba^{2+} + 2e^- \rightleftharpoons Ba$
−2.92	$K^+ + e^- \rightleftharpoons K$
−3.05	$Li^+ + e^- \rightleftharpoons Li$

Notice that the order of increasing ease of oxidation of the metals (top to bottom) in Table 17.1 is the same as the order (bottom to top) in the activity series in Table 7.3 on p. 266.

agents decreases as we proceed down the table. Thus we could conclude, from their positions in this table, that H^+ is a better oxidizing agent than Zn^{2+}, and that F_2 is a better oxidizing agent than Cl_2. In brief, *good oxidizing agents are those species on the left of the double arrow at the top of the table.*

Each of the half-reactions listed in Table 17.1 is reversible. We saw, for example, that H_2 is oxidized to H^+ when placed in a cell with copper, and that H^+ is reduced to H_2 when placed against zinc. When the reactions in Table 17.1 are forced to proceed from right to left—that is, when they are caused to be the oxidation step in an overall reaction—then the species appearing at the right in Table 17.1 are functioning as reducing agents by being oxidized. *All the substances appearing on the right side of the reactions in Table 17.1 could behave as reducing agents; those species at the bottom right of the table, such as Li, are the best and those at the top right, such as F^-, are the poorest.*

2. Using Table 17.1 we can find rather quickly which combinations of reactants lead to spontaneous oxidation-reduction reactions (when the concentrations of the reactants and products are 1 M and the partial pressures of any gases involved are 1 atm). We can also determine whether or not a given reaction, as written, will proceed spontaneously in the forward direction.

Consider, for example, a mixture consisting of pieces of solid zinc and solid chromium in contact with a solution containing 1 M Zn^{2+} and 1 M Cr^{3+}. What reaction will occur in this mixture? To answer this question we look at the following half-reactions found in Table 17.1.

$$Cr^{3+}(aq) + 3e^- \rightleftharpoons Cr(s) \quad \mathscr{E}^0_{Cr} = -0.74 \text{ V}$$
$$Zn^{2+}(aq) + 2e^- \rightleftharpoons Zn(s) \quad \mathscr{E}^0_{Zn} = -0.76 \text{ V}$$

−0.74 V is more positive than −0.76 V.

The reduction potentials tell us that Cr^{3+} is more easily reduced than Zn^{2+}, so in this mixture reduction of Cr^{3+} will occur and the zinc half-reaction will be forced to occur as oxidation. To obtain the cell reaction we combine the half-reactions in a way that makes the total number of electrons gained equal the total number lost.

$$2 \times [Cr^{3+}(aq) + 3e^- \longrightarrow Cr(s)] \quad \text{(Reduction)}$$
$$\underline{3 \times [Zn(s) \longrightarrow Zn^{2+}(aq) + 2e^-]} \quad \text{(Oxidation)}$$
$$2Cr^{3+}(aq) + 3Zn(s) \longrightarrow 2Cr(s) + 3Zn^{2+}(aq) \quad \text{Cell reaction}$$

To calculate the $\mathscr{E}^0_{cell}$ for this reaction, we use Equation 17.5.

$$\mathscr{E}^0_{cell} = \mathscr{E}^0_{\text{substance reduced}} - \mathscr{E}^0_{\text{substance oxidized}}$$
$$= \mathscr{E}^0_{Cr} - \mathscr{E}^0_{Zn}$$
$$= (-0.74 \text{ V}) - (-0.76 \text{ V})$$
$$= +0.02 \text{ V}$$

Notice that $\mathscr{E}^0_{cell}$ is positive, as it must be for a spontaneous change. Also notice that even though the half-reactions are multiplied by factors before they are combined to give the net cell reaction, the reduction potentials are not. They are simply subtracted one from the other. This is because reduction potentials are intensive quantities and are therefore independent of the number of moles of reactants and products involved.

We have also said that we can determine whether a reaction, as written, will occur spontaneously. Let's consider the possible reaction

$$Fe^{2+}(aq) + Ni(s) \longrightarrow Fe(s) + Ni^{2+}(aq)$$

If the reaction is spontaneous, the calculated $\mathscr{E}^0_{cell}$ will be positive, as it was above. On the other hand, if the reaction is not spontaneous in the direction written, the calculated $\mathscr{E}^0_{cell}$ will be negative.

For the reaction we are considering, the first step is to divide the overall equation into two half-reactions.

$$Fe^{2+}(aq) + 2e^- \longrightarrow Fe(s)$$

$$Ni(s) \longrightarrow Ni^{2+}(aq) + 2e^-$$

We see that the first equation is a reduction and we can find its reduction potential, $\mathscr{E}^0_{Fe} = -0.44$ V, from Table 17.1. The second equation is an oxidation. If we were to rewrite it as a reduction, we would also be able to find its reduction potential, $\mathscr{E}^0_{Ni} = -0.25$ V. Since in our overall equation iron(II) is reduced and nickel is oxidized, when we substitute into Equation 17.5, we get

$$\mathscr{E}^0_{cell} = \mathscr{E}^0_{Fe} - \mathscr{E}^0_{Ni}$$

$$= -0.44 - (-0.25)$$

$$= -0.19 \text{ V}$$

Because $\mathscr{E}^0_{cell}$ is computed to be negative, we know the reaction of $Fe^{2+}(aq)$ with $Ni(s)$ is *not* spontaneous. In fact, it is the reverse reaction that is spontaneous.

EXAMPLE 17.4 **PREDICTING THE emf OF A CELL FROM STANDARD REDUCTION POTENTIALS**

PROBLEM What will be the spontaneous reaction between the following set of half-reactions? What is the value of $\mathscr{E}^0_{cell}$?

(1) $Cr^{3+}(aq) + 3e^- \rightleftharpoons Cr(s)$

(2) $MnO_2(s) + 4H^+(aq) + 2e^- \rightleftharpoons Mn^{2+}(aq) + 2H_2O(l)$

SOLUTION From Table 17.1, reaction (1) has $\mathscr{E}^0 = -0.74$ V and reaction (2) has $\mathscr{E}^0 = +1.28$ V. Since the reduction potential of Reaction (2) is larger (more positive) than that of Reaction (1), Reaction (2) will occur as reduction. Reaction (1) must be reversed and written as oxidation. To obtain the net overall reaction we multiply by appropriate coefficients so that electrons cancel.

$$3[MnO_2(s) + 4H^+(aq) + 2e^- \longrightarrow Mn^{2+}(aq) + 2H_2O(l)]$$
$$\underline{2[Cr(s) \longrightarrow Cr^{3+}(aq) + 3e^-]}$$
$$2Cr(s) + 3MnO_2(s) + 12H^+(aq) \longrightarrow 2Cr^{3+}(aq) + 3Mn^{2+}(aq) + 6H_2O(l)$$

To calculate $\mathscr{E}^0_{cell}$ we can subtract reduction potentials to obtain a positive value.

$$\mathscr{E}^0_{cell} = +1.28 - (-0.74)$$

$$= +2.02 \text{ V}$$

---

3. In the process of combining half-reactions from Table 17.1, we see that some of the reactants in the spontaneous reaction appear on the left side of one half-reaction while the rest of the reactants are found on the right side of another half-reaction. Among the reactants in Example 17.4, for instance, we have $MnO_2 + 4H^+$ from the left side of one half-reaction

and Cr(s) from the right side of the other. The order of these reactions in Table 17.1 is

$$MnO_2(s) + 4H^+(aq) + 2e^- \rightleftharpoons Mn^{2+}(aq) + 2H_2O(l) \quad \mathscr{E}^0 = 1.28 \text{ V}$$

$$Cr^{3+}(aq) + 3e^- \rightleftharpoons Cr(s) \qquad\qquad \mathscr{E}^0 = -0.74 \text{ V}$$

and we see that the reactants in the overall spontaneous reaction are those substances related by the diagonal line (colored arrow) running from upper left to lower right. As a general statement, *we can say that when comparing reactants and products having unit concentrations, any species on the left of a given half-reaction will react spontaneously with a substance that is found on the right of a half-reaction located below it in Table 17.1.* We could use this rule of thumb, for example, to tell us that Br_2 will react spontaneously with I^- to produce Br^- and I_2, while Br_2 will *not* react spontaneously with Cl^-. Our rule, therefore, permits us to determine the course of a reaction without having to worry about subtracting electrode potentials in the proper sequence.

4. A point worth noting is that a collection of half-reactions, such as that found in Table 17.1, enables us to predict the outcome of many chemical reactions when we know only a relatively few half-reactions and their corresponding reduction potentials. From the 36 half-reactions listed in Table 17.1, for example, we can predict the results of 630 different chemical reactions! A table of this type, therefore, provides us with a very compact way of storing chemical information.

5. With a knowledge of the standard reduction potentials listed in Table 17.1, we account for the course of electrolysis reactions. For example, we know from experiment that we can produce copper by electrolyzing an aqueous solution containing Cu^{2+}, but that we cannot obtain aluminum in this same fashion. From Table 17.1 we see that the reduction potential of copper is $+0.34$ V and that for H_2O it is -0.83 V Thus copper(II) ion is more readily reduced than H_2O and will plate out on the electrode according to the half-reaction,

$$Cu^{2+}(aq) + 2e^- \longrightarrow Cu(s)$$

In the case of aluminum, however, we find that the reduction potential for Al^{3+} is -1.66 V, which makes it more difficult to reduce than water. This means that when an aqueous solution containing Al^{3+} ions is electrolyzed, the H_2O will preferentially be reduced.

The reduction potentials of hundreds of half-reactions have been measured.

There are complicating factors we haven't discussed that make it difficult to accurately predict, based on emf data, what will or will not be formed at an electrode during electrolysis.

17.8 SPONTANEITY OF OXIDATION-REDUCTION REACTIONS

It was pointed out in Section 12.8 that the thermodynamic criterion for spontaneity of a chemical reaction taking place at constant temperature and pressure is that the change in free energy, ΔG, has to be a negative quantity. In Section 12.9 we saw that ΔG also represents the maximum quantity of useful work obtainable from a chemical reaction. The relationship between ΔG and maximum useful work (W_{max}) for any system takes the form

$$\Delta G = -W_{max}$$

But what is W_{max} for an electrochemical cell?

The work derived from an electrochemical cell might be compared to that

Figure 17.16
Work obtained from a waterwheel.

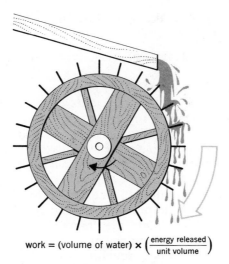

$$\text{work} = (\text{volume of water}) \times \left(\frac{\text{energy released}}{\text{unit volume}}\right)$$

obtained from a waterwheel, shown in Figure 17.16. The quantity of work that can be obtained from this waterwheel depends on two things: (1) the volume of water flowing over the blades of the wheel, and (2) the energy given to the wheel per unit volume of water as it drops to the lower level of the stream:

$$\text{work} = (\text{volume of water}) \times \left(\frac{\text{energy released}}{\text{unit volume}}\right)$$

Similarly, the work that can be done by an electrochemical cell is dependent on (1) the number of coulombs that flow and (2) the energy available per coulomb:

$$\text{work} = (\text{number of coulombs}) \times \left(\frac{\text{energy available}}{\text{coulomb}}\right)$$

The number of coulombs that flow is equal to the number of moles of electrons that are involved in the redox reaction, n, multiplied by the faraday (which is the number of coulombs per mole of electrons):

$$\text{number of coulombs} = n\mathscr{F}$$

The value of n depends on the nature of the half-reactions taking place in the cell and can be derived once the specific reactions are known. For example, in the Zn/Cu cell there are two electrons involved in both of the half-reactions; therefore, n for this cell is 2.

The energy available per coulomb is simply the emf of the cell, because the volt is equal to the energy per coulomb (Equation 17.4):

$$\frac{\text{energy available}}{\text{coulomb}} = \text{emf}$$

When the emf is a maximum, the work derived from the cell is also a maximum. In Section 17.6 we saw that the maximum emf is the cell potential, $\mathscr{E}_{\text{cell}}$.

Thus the equation for maximum work for an electrochemical cell is

$$W_{\text{max}} = \quad n \quad \times \quad \mathscr{F} \quad \times \quad \mathscr{E}$$
$$\updownarrow \qquad\qquad \updownarrow \qquad\qquad \updownarrow$$
$$\text{joules} = (\text{moles of electrons}) \times \left(\frac{\text{coulombs}}{\text{mole}}\right) \times \left(\frac{\text{joules}}{\text{coulomb}}\right)$$

Since $\Delta G = -W_{\text{max}}$, then

$$\Delta G = -n\mathscr{F}\mathscr{E}_{\text{cell}} \qquad\qquad\qquad [17.6]$$

When all species are at unit concentration as identified by the superscript zero in $\mathscr{E}^0_{cell}$, then ΔG becomes the standard free-energy change for the reaction, ΔG^0. Thus, Equation 17.6 becomes

$$\Delta G^0 = -n\mathscr{F}\mathscr{E}^0_{cell} \qquad [17.7]$$

With the aid of Equation 17.7 we can calculate the standard free-energy change for an oxidation-reduction reaction from a knowledge of its standard cell potential. Consider, for example, then Zn/Cu cell.

$$\begin{array}{r} Zn(s) \longrightarrow Zn^{2+}(aq) + 2e^- \\ Cu^{2+}(aq) + 2e^- \longrightarrow Cu(s) \\ \hline Zn(s) + Cu^{2+}(aq) \longrightarrow Zn^{2+}(aq) + Cu(s) \end{array}$$

The n for this reaction is 2 (because two electrons are transferred), $\mathscr{F} = 96\,500$ C/mol e^-, and $\mathscr{E}^0_{cell}$, either derived from Table 17.1 or determined experimentally, is $+1.10$ V; therefore,

$$\Delta G^0 = -2 \, \cancel{\text{mol } e^-} \times \left(\frac{96\,500\,\cancel{C}}{\cancel{\text{mol } e^-}} \right) \times \left(\frac{+1.10\,\text{J}}{\cancel{C}} \right)$$
$$= -212\,000\,\text{J}$$

This relationship between the standard cell potential and ΔG^0 is extremely important, because it ties together two different aspects of spontaneity while at the same time it gives us a readily accessible pathway for calculating standard free-energy changes. Experimentally, we have only to measure the standard cell emf, from which we can then compute the value of ΔG^0. But still more important, perhaps, is that through this equation we can derive even more useful thermodynamic quantities.

17.9 THERMODYNAMIC EQUILIBRIUM CONSTANTS FROM STANDARD CELL POTENTIALS

In Section 14.3 we saw that for reactions in solution,

$$\Delta G^0 = -RT \ln K_c$$

Scientists who have studied electrochemistry have traditionally worked with common logarithms, so this equation is expressed as

$$\Delta G^0 = -2.303RT \log K_c \qquad [17.8]$$

Combining Equations 17.7 and 17.8 gives us

$$\Delta G^0 = -n\mathscr{F}\mathscr{E}^0 = -2.303RT \log K_c$$

or simply

$$n\mathscr{F}\mathscr{E}^0 = 2.303RT \log K_c$$

Solving for $\mathscr{E}^0$, we have

$$\mathscr{E}^0 = \frac{2.303RT}{n\mathscr{F}} \log K_c \qquad [17.9]$$

If we choose to restrict ourselves to discussing reactions that take place at 25°C (298 K), the quantity $2.303RT/\mathscr{F}$ becomes a constant,

$$\frac{2.303RT}{\mathscr{F}} = \frac{2.303(8.314\,\text{J mol}^{-1}\,\text{K}^{-1})(298\,\text{K})}{96\,500\,\text{C mol}^{-1}} = 0.0592\,\text{J C}^{-1}$$

Since 1 V = 1 J/C,

$$\frac{2.303RT}{\mathscr{F}} = 0.0592 \text{ V}$$

Thus, at 25°C, Equation 17.9 becomes

$$\mathscr{E}^0 = \frac{0.0592}{n} \log K_c$$

Solving for K_c gives

$$\log K_c = \frac{n\mathscr{E}^0}{0.0592} \qquad [17.10]$$

We see, therefore, that from a knowledge of the standard cell potential, the equilibrium constant for the cell reaction can be calculated. For the Zn/Cu cell, we have

$$\log K_c = \frac{n\mathscr{E}^0}{0.0592} = \frac{2(+1.10)}{0.0592} = 37.2$$

Hence,

$$K_c \approx 2 \times 10^{37}$$

Standard cell potentials are an important source of equilibrium constants.

From the magnitude of this equilibrium constant, we could certainly say that the spontaneous Zn/Cu cell reaction will go very nearly to completion.

EXAMPLE 17.5   **DETERMINING THE SPONTANEITY OF A REDOX REACTION AND CALCULATING ITS EQUILIBRIUM CONSTANT**

PROBLEM   Using Table 17.1, determine whether the oxidation-reduction reaction

$$\text{Sn}(s) + \text{Ni}^{2+} \longrightarrow \text{Sn}^{2+} + \text{Ni}(s)$$

is spontaneous, and calculate its equilibrium constant at 25°C.

SOLUTION   The two half-reactions for this overall reaction are

$$\text{Sn}(s) \longrightarrow \text{Sn}^{2+}(aq) + 2e^- \qquad \text{(Oxidation)}$$

$$\text{Ni}^{2+}(aq) + 2e^- \longrightarrow \text{Ni}(s) \qquad \text{(Reduction)}$$

From Table 17.1 we find the reduction potential for the half-reaction involving Sn and Sn^{2+} to be $\mathscr{E}^0_{\text{Sn}} = -0.14$ V; for the nickel half-reaction $\mathscr{E}^0_{\text{Ni}} = -0.25$ V. Applying Equation 17.5, we get

$$\mathscr{E}^0_{\text{cell}} = -0.25 - (-0.14)$$

$$= -0.11 \text{ V}$$

This means that under *standard conditions* (unit concentrations), the reaction in this direction is nonspontaneous. We can still calculate the equilibrium constant in the same fashion as outlined above. Since two electrons are transferred during the reaction,

$$\log K_c = \frac{2(-0.11)}{0.0592} = -3.7$$

Taking the antilogarithm gives

$$K_c = 2 \times 10^{-4}$$

From the size of this equilibrium constant, we can say that this reaction will not occur to an appreciable extent in the forward direction.

17.10 THE EFFECTS OF CONCENTRATION ON CELL POTENTIALS: THE NERNST EQUATION

Until now, we have only discussed cells containing reactants at unit concentration. In the laboratory, however, we usually do not restrict ourselves to only this one set of conditions, and it is found that the cell emf, and in fact the direction of the cell reaction, can be controlled by the concentrations of the species taking part in the reaction. Let us examine this now from a quantitative point of view.

An equation that summarizes how the free energy of the reactants and products of a given reaction varies with temperature and concentration was given in Section 14.3. For the generalized reaction,

$$aA + bB \longrightarrow eE + fF$$

this equation takes the form (using common logarithms)

$$\Delta G = \Delta G^0 + 2.303RT \log \left(\frac{[E]^e[F]^f}{[A]^a[B]^b} \right)$$

The quantity

$$\frac{[E]^e[F]^f}{[A]^a[B]^b}$$

is the mass action expression for the reaction.

where the quantity within parentheses is the mass action expression (reaction quotient) for the reaction.

Equations 17.6 and 17.7 (in Section 17.8) show the relationship between ΔG and $\mathscr{E}$, and ΔG^0 and $\mathscr{E}^0$, respectively. Substituting these expressions for ΔG and ΔG^0 into the preceding equation we have

$$-n\mathscr{F}\mathscr{E} = -n\mathscr{F}\mathscr{E}^0 + 2.303RT \log \left(\frac{[E]^e[F]^f}{[A]^a[B]^b} \right)$$

which can be rearranged to give

$$\mathscr{E} = \mathscr{E}^0 - \frac{2.303RT}{n\mathscr{F}} \log \left(\frac{[E]^e[F]^f}{[A]^a[B]^b} \right) \qquad [17.11]$$

It was Walter Nernst who discovered the third law of thermodynamics.

This equation, first developed by Walter Nernst in 1889, now bears his name and is called the **Nernst equation.**

At 25°C we have seen that the numerical value of $2.303RT/\mathscr{F}$ is 0.0592. Therefore, at 25°C the Nernst equation becomes

$$\mathscr{E} = \mathscr{E}^0 - \frac{0.0592}{n} \log \left(\frac{[E]^e[F]^f}{[A]^a[B]^b} \right) \qquad [17.12]$$

$\log 1 = 0$

We can see from Equation 17.12 that when all ionic species are present at unit concentration, the log term becomes zero and the emf of the cell becomes $\mathscr{E}^0$—that is, at unit concentration $\mathscr{E} = \mathscr{E}^0$. This, of course, must be true in light of our basic definition of $\mathscr{E}^0$. When the species in a cell are not present at unit concentration, $\mathscr{E}$ is generally not equal to $\mathscr{E}^0$ and the Nernst equation can be used to calculate $\mathscr{E}$. For example, in the case of the Zn/Cu cell, whose cell reaction is

$$Zn(s) + Cu^{2+}(aq) \longrightarrow Cu(s) + Zn^{2+}(aq)$$

the Nernst equation takes the form

$$\mathscr{E} = \mathscr{E}^0 - \frac{0.0592}{n} \log \frac{[Zn^{2+}]}{[Cu^{2+}]}$$

Note that as usual we omit the concentrations of pure solids from the mass action expression.[3] Since two electrons are transferred in the reaction, $n = 2$, and

$$\mathscr{E} = \mathscr{E}^0 - 0.0296 \log \frac{[Zn^{2+}]}{[Cu^{2+}]}$$

Thus, $\mathscr{E}$ can be calculated for any particular set of concentrations of Zn^{2+} and Cu^{2+}. This use of the Nernst equation is illustrated in the next example.

EXAMPLE 17.6

USING THE NERNST EQUATION TO CALCULATE THE POTENTIAL OF A CELL UNDER NONSTANDARD CONDITIONS

PROBLEM   Calculate the emf of the Zn/Cu cell at 25°C under the following conditions:

$$Zn(s) + Cu^{2+}(0.020\ M) \longrightarrow Cu(s) + Zn^{2+}(0.40\ M)$$

SOLUTION   We have just seen that for this system the Nernst equation is

$$\mathscr{E} = \mathscr{E}^0 - 0.0296 \log \frac{[Zn^{2+}]}{[Cu^{2+}]}$$

We can also calculate $\mathscr{E}^0_{cell} = +1.10$ V for the equation as written. Substituting this $\mathscr{E}^0$ and the concentrations of the Zn^{2+} and Cu^{2+} into the Nernst equation, we have

$$\mathscr{E} = 1.10 - 0.0296 \log \frac{(0.40)}{(0.020)}$$

$$= 1.10 - 0.0296 \log 20$$

$$= 1.10 - 0.0296(1.30)$$

$$= 1.10 - 0.0385$$

$$= 1.06\ V$$

Thus, we see that under these conditions of concentration the emf obtained from this cell is slightly less than that obtained at unit concentration.

17.11   APPLICATIONS OF THE NERNST EQUATION

Just as the cell emf is dependent on the concentrations of the ions involved in the half-reactions, so we find that the reduction potential of the individual half-reactions is also determined by the concentrations of the ions involved. This effect of the concentration on the reduction potential can also be given by the Nernst equation. For example, if we consider the half-reaction,

$$Zn^{2+}(aq) + 2e^- \rightleftharpoons Zn(s)$$

[3] The Nernst equation applies exactly only if we use activities. The activity of any pure solid or liquid is equal to 1. Errors introduced by using the concentrations of the ions instead of their activities are small, as mentioned before, provided that all the solutions are relatively dilute.

the Nernst equation at 25°C takes the form

$$\mathcal{E}_{Zn} = \mathcal{E}_{Zn}^0 - \frac{0.0592}{2} \log \frac{1}{[Zn^{2+}]}$$

As usual, we have omitted the concentration of the solid from the mass action expression.

Because the reduction potential of an electrode depends on the concentrations of the ions in solution, it is possible to construct a cell in which the cathode and anode compartments contain the same electrode materials but different concentrations of the ions. Such a cell is called a **concentration cell** and is illustrated in Figure 17.17.

In Figure 17.17 we have a cell composed of two zinc electrodes dipping into separate solutions of $ZnSO_4$ whose Zn^{2+} concentrations are different. The concentration of the Zn^{2+} on the left (1.0 M) is 100 times greater than the Zn^{2+} concentration in the right compartment, and when the circuit is completed a spontaneous reaction takes place in a direction that tends to make the two Zn^{2+} concentrations become equal. Thus, in the more concentrated side Zn^{2+} ions disappear forming $Zn(s)$, in order to decrease the Zn^{2+} concentration, and in the more dilute side, more Zn^{2+} will be produced. Thus, in the more concentrated compartment, we have

$$Zn^{2+} (1\ M) + 2e^- \longrightarrow Zn(s) \quad \text{(Reduction)}$$

and on the more dilute side,

$$Zn(s) \longrightarrow Zn^{2+} (0.10\ M) + 2e^- \quad \text{(Oxidation)}$$

From our earlier discussions we know that the potential of a cell is found by subtracting the reduction potential of the half-cell in which oxidation occurs from the reduction potential of the half-cell that undergoes reduction. For this concentration cell

$$\mathcal{E}_{cell} = \mathcal{E}_{conc} - \mathcal{E}_{dil}$$

where $\mathcal{E}_{conc}$ and $\mathcal{E}_{dil}$ are the electrode potentials of the concentrated and dilute half-cells, respectively. These are given as

$$\mathcal{E}_{conc} = \mathcal{E}_{Zn}^0 - \frac{0.0592}{2} \log \frac{1}{[Zn^{2+}]_{conc}}$$

Figure 17.17

A concentration cell.

and

$$\mathcal{E}_{dil} = \mathcal{E}_{Zn}^0 - \frac{0.0592}{2} \log \frac{1}{[Zn^{2+}]_{dil}}$$

Therefore,

$$\mathcal{E}_{cell} = (\mathcal{E}_{Zn}^0 - \mathcal{E}_{Zn}^0) - \frac{0.0592}{2} \left(\log \frac{1}{[Zn^{2+}]_{conc}} - \log \frac{1}{[Zn^{2+}]_{dil}} \right)$$

or

$$\mathcal{E}_{cell} = -\frac{0.0592}{2} \log \frac{[Zn^{2+}]_{dil}}{[Zn^{2+}]_{conc}}$$

Substituting the concentrations of Zn^{2+} into this expression allows us to compute the cell potential.

$$\mathcal{E}_{cell} = -\frac{0.0592}{2} \log \frac{(0.01)}{(1)}$$

$$= 0.0592 \text{ V}$$

In general, for any concentration cell we could write

$$\mathcal{E}_{cell} = -\frac{0.0592}{n} \log \frac{[M^{n+}]_{dil}}{[M^{n+}]_{conc}}$$

The electric potential obtained from this type of cell is usually small and will continually decrease as the concentrations in the two compartments approach each other. The emf becomes zero when the concentrations of these ions in each compartment of the cell are the same.

Solubility product constant

The Nernst equation can also be useful in determining the solubility product constant of an insoluble salt. To find the K_{sp} of $PbSO_4$, for example, an experiment might be designed in the following fashion: A galvanic cell is prepared consisting of Pb/Pb^{2+} versus an Sn/Sn^{2+} electrode with a salt bridge connecting them. In the tin compartment the Sn^{2+} concentration is held constant at $1 \, M$. In the lead compartment SO_4^{2-} is added to precipitate $PbSO_4$ and thereby establish the equilibrium

$$PbSO_4(s) \rightleftharpoons Pb^{2+}(aq) + SO_4^{2-}(aq)$$

The SO_4^{2-} concentration in the lead compartment is then adjusted until it is $1 \, M$, and the emf of the cell is found to be $+0.22$ V. It is also observed that the Pb electrode is negative with respect to the Sn electrode, thereby indicating that the Pb is undergoing oxidation while the Sn^{2+} is reduced. The cell reaction must therefore be

$$Pb(s) + Sn^{2+}(1 \, M) \longrightarrow Pb^{2+}(?) + Sn(s)$$

and the calculated $\mathcal{E}^0$ is

$$\mathcal{E}^0 = \mathcal{E}_{Sn} - \mathcal{E}_{Pb}$$

$$= (-0.14 \text{ V}) - (-0.13 \text{ V})$$

$$= -0.01 \text{ V}$$

(Note that *if* the Sn^{2+} and Pb^{2+} concentrations were both $1 \, M$, this calculation tells us that the reaction would be spontaneous from right to left, rather than from left to right.)

We can calculate the concentration of Pb^{2+} by using the Nernst equation for this cell reaction which takes the form

$$\mathscr{E} = \mathscr{E}^0 - \frac{0.0592}{n} \log \frac{[Pb^{2+}]}{[Sn^{2+}]}$$

We know $\mathscr{E}$, $\mathscr{E}^0$, and $[Sn^{2+}]$ and, because there are two electrons transferred in this reaction, the above equation becomes, after substitution,

$$0.22 \text{ V} = -0.01 \text{ V} - \frac{0.0592}{2} \log \frac{[Pb^{2+}]}{(1)}$$

Next we solve for $\log[Pb^{2+}]$.

$$-0.22 \text{ V} = 0.01 \text{ V} + 0.0296 \log[Pb^{2+}]$$

or

$$\log[Pb^{2+}] = \frac{-0.22 \text{ V} - 0.01 \text{ V}}{0.0296 \text{ V}}$$

and

$$\log[Pb^{2+}] = -7.8$$

Taking the antilogarithm, we find that the concentration of Pb^{2+} in this cell is

$$[Pb^{2+}] = 2 \times 10^{-8} \ M$$

$10^{-7.8} = 2 \times 10^{-8}$

The expression for K_{sp} of $PbSO_4$ is

$$K_{sp} = [Pb^{2+}][SO_4{}^{2-}]$$

Since

$$[Pb^{2+}] = 2 \times 10^{-8}$$
$$[SO_4{}^{2-}] = 1 \ M$$

then

$$K_{sp} = (2 \times 10^{-8})(1) = 2 \times 10^{-8}$$

which is the value that was given in Table 16.1.

Determination of pH

An extremely important application of the Nernst equation is its use in calculating the concentration of a single ionic species from the experimentally measured potential of a carefully designed cell. We have already seen one example of this in the determination of K_{sp}. If we were to use the Cu/H_2 cell, discussed in Section 17.7, we could determine the H^+ concentration of a solution and then calculate its pH. The cell reaction for the Cu/H_2 cell is

$$Cu^{2+}(aq) + H_2(g) \longrightarrow Cu(s) + 2H^+(aq)$$

and the corresponding form of the Nernst equation at 25°C is

p_{H_2} is the partial pressure of H_2.

$$\mathscr{E} = \mathscr{E}^0 - \frac{0.0592}{n} \log \frac{[H^+]^2}{[Cu^{2+}]p_{H_2}}$$

If the concentration of Cu^{2+} is 1 M and the pressure of H_2 is 1 atm, this equation reduces to

$$\mathscr{E} = \mathscr{E}^0 - \frac{0.0592}{n} \log[H^+]^2$$

which is the same as

$$\mathscr{E} = \mathscr{E}^0 - \frac{(0.0592)(2)}{n} \log[H^+]$$

$\log x^2 = 2 \log x$

Let's rewrite this equation as

$$\mathscr{E} = \mathscr{E}^0 + \frac{(0.0592)(2)}{n} (-\log[H^+])$$

We see that because $\mathscr{E}^0$ and the quantity $0.0592(2)/n$ are both constant for a specific reaction, then

$$\mathscr{E} \propto -\log[H^+]$$

By definition, $pH = -\log[H^+]$ and, therefore, we have

$$\mathscr{E} \propto pH$$

Thus, by measuring the emf of a galvanic cell containing a reference electrode (such as the Cu, Cu^{2+} electrode here) and the hydrogen electrode, the pH of a solution can be calculated. One application is shown in the next example.

EXAMPLE 17.7   USING THE NERNST EQUATION TO CALCULATE THE pH OF A SOLUTION

PROBLEM   A galvanic cell consisting of a Cu versus a hydrogen electrode was used to determine the pH of an unknown solution. The unknown was placed in the hydrogen electrode compartment and the pressure of the hydrogen gas was controlled at 1 atm. The concentration of Cu^{2+} was $1\ M$ and the emf of the cell at 25°C was determined to be +0.48 V. Calculate the pH of this unknown solution.

SOLUTION   The cell reaction for the Cu/H_2 cell is

$$Cu^{2+}(1\ M) + H_2(g)(1\ atm) \longrightarrow Cu(s) + 2H^+(?M)$$

for which we write the Nernst equation as

$$\mathscr{E} = \mathscr{E}^0 - \frac{0.0592}{n} \log \frac{[H^+]^2}{[Cu^{2+}]p_{H_2}}$$

Because $[Cu^{2+}] = 1\ M$ and $p_{H_2} = 1$ atm,

$$\mathscr{E} = \mathscr{E}^0 - \frac{(0.0592)(2)}{n} \log[H^+]$$

The $\mathscr{E}^0$ for this cell is +0.34 V; the value of n is 2. Substituting these values as well as the measured value of $\mathscr{E}$ into the equation gives

$$-0.48 = +0.34 - 0.0592 \log[H^+]$$

and hence

$$-\log[H^+] = pH = \frac{0.48 - 0.34}{0.0592} = 2.4$$

Therefore the pH of this solution is 2.4.

The convenience of modern pH meters, such as the one illustrated in Chapter 15 on page 540, has been made possible by the development of specialized electrodes that are rugged, easily used, and are sensitive to the concentration of

Figure 17.18
(a) *A typical glass electrode.*
(b) *Cut-away view of the construction of a glass electrode.*

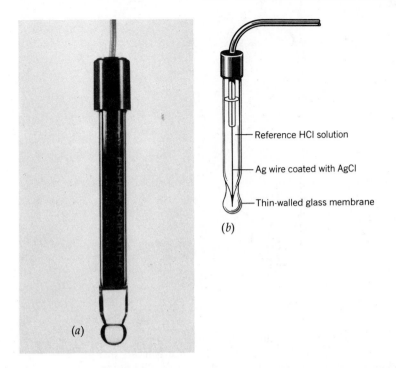

— Reference HCl solution

— Ag wire coated with AgCl

— Thin-walled glass membrane

(b)

(a)

hydrogen ion in a solution. An example is the **glass electrode,** which is shown in Figure 17.18. This electrode consists of a silver wire covered with silver chloride that dips into a dilute solution of HCl. The HCl solution is enclosed within a very thin walled glass membrane sealed into the end of a hollow glass tube. The emf of this electrode is sensitive to the difference between the concentrations of H^+ inside and outside. Since the concentration inside the electrode is constant, the electrode in effect becomes sensitive to the hydrogen ion concentration in the solution in which it is dipped. The glass electrode is always used with another reference electrode to generate a potential difference that can be measured. The measurement of this potential difference, and translating it to pH, are the jobs of the pH meter.

17.12 PRACTICAL APPLICATIONS OF GALVANIC CELLS

The use of galvanic cells for the production of electricity has a long history. There is evidence, for example, that the Parthians (early inhabitants of the land now known as Iran) may have used primitive galvanic cells for gold-plating jewelry as early as 250 B.C. Today we enjoy the choice of a variety of galvanic cells, or batteries as they are better known, and they have become an intimate part of our modern society. The following discussion describes the chemistry of some of the more important and common types.

Zinc-carbon dry cell

The common dry cell is technically known as a Leclanché cell.

This type of cell is used in flashlights, portable radios, toys, and the like. A cut-away diagram of a typical zinc-carbon dry cell is shown in Figure 17.19. It has an exterior layer of either cardboard or metal that serves only as a seal against the atmosphere. Inside this outer shell is a zinc cup that serves as the anode. The zinc cup is filled with a moist paste consisting of ammonium chloride, manganese dioxide, and finely divided carbon. Immersed in this paste is a

Figure 17.19
The common dry cell.

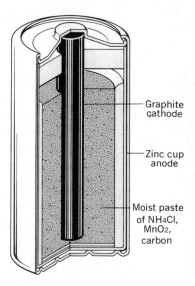

Graphite cathode

Zinc cup anode

Moist paste of NH4Cl, MnO2, carbon

The physical size of a cell doesn't affect its electric potential, but it does affect the amount of current that can be delivered. A small AA battery doesn't produce as much current as a larger D-size cell.

graphite rod, which serves as the cathode. The chemical reactions that take place when the circuit is completed are actually quite complex and, in fact, are not completely understood. The following, however, is perhaps a reasonable estimate of what occurs.

At the anode zinc is oxidized,

$$Zn(s) \longrightarrow Zn^{2+}(aq) + 2e^- \quad \text{(Anode)}$$

while at the carbon cathode the MnO_2/NH_4Cl mixture undergoes reduction to give a complex mixture of products. One of these reactions appears to be

$$2MnO_2(s) + 2NH_4^+(aq) + 2e^- \longrightarrow Mn_2O_3(s) + 2NH_3(aq) + H_2O(l) \quad \text{(Cathode)}$$

The ammonia produced at the cathode reacts with part of the Zn^{2+} formed at the anode to give the complex ion, $Zn(NH_3)_4^{2+}$. Because of the complex nature of the dry cell, no simple overall cell reaction can be written.

The common zinc-carbon dry cell suffers from the disadvantage that under heavy use it rather quickly appears to become "dead." Yet, if allowed to rest for a while, it appears to come back to life and can deliver additional current. When the cell delivers current the reaction products can't diffuse away from electrodes very quickly. As they accumulate it becomes more difficult for the electrode reactions to occur, and the cell potential drops. After sitting idle for a while, however, these products diffuse away from the electrodes and the cell regains the ability to function.

We say the cell has become polarized.

Dry cells cannot be effectively recharged and, therefore, have a relatively short lifetime (as compared to the rechargeable lead storage and nickel-cadmium batteries, for example).

Alkaline battery

Another type of dry cell that uses zinc and manganese dioxide as reactants is the alkaline dry cell. Once again, zinc serves as the anode and manganese dioxide functions as the cathode. The electrolyte, however, contains potassium hydroxide and is therefore basic (alkaline). The zinc anode is also slightly porous, giving it a larger effective area. This allows the cell to deliver more current than the common zinc cell. As you've probably learned from TV commercials, these batteries are able to stand up better under heavy use and have a longer shelf life.

The reactions in the alkaline battery are

$$Zn(s) + 2OH^-(aq) \longrightarrow Zn(OH)_2(s) + 2e^- \qquad \text{(Anode)}$$

$$2MnO_2(s) + 2H_2O + 2e^- \longrightarrow 2MnO(OH)(s) + 2OH^-(aq) \quad \text{(Cathode)}$$

The cell produces an emf of about 1.5 V.

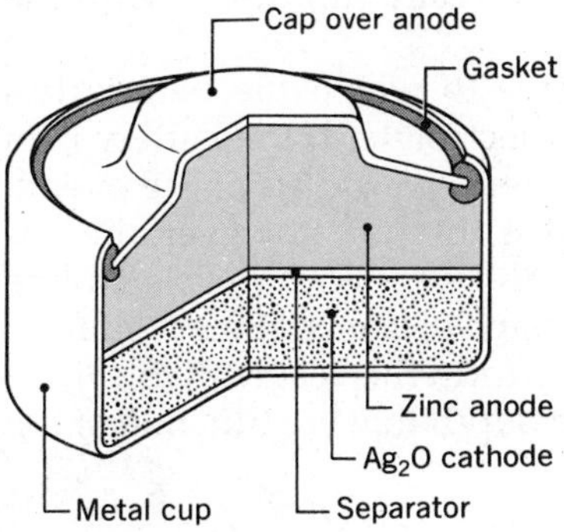

Silver oxide battery

These tiny and rather expensive batteries (Figure 17.20) have become popular as power sources in electronic wristwatches, auto exposure cameras, and electronic calculators. The cathode reactant is silver oxide, Ag_2O, and the anode once again is zinc. The electrode reactions occur in a basic electrolyte.

$$Zn(s) + 2OH^-(aq) \longrightarrow Zn(OH)_2(s) + 2e^- \qquad \text{(Anode)}$$

$$Ag_2O(s) + H_2O + 2e^- \longrightarrow 2Ag(s) + 2OH^-(aq) \quad \text{(Cathode)}$$

The emf of this battery is about 1.5 V.

Lead storage battery

The common automobile battery is a lead storage battery that usually delivers either 6 or 12 V, depending on the number of cells used in its construction. The inside of the battery consists of a number of galvanic cells connected to each other in series.

To increase the current output, each of the individual cells (Figure 17.21) contains a number of lead anodes connected together, plus a number of cathodes, composed of PbO_2, also joined together. These electrodes are immersed in an electrolyte composed of dilute sulfuric acid (actually about 30% by mass in a fully charged cell). A single lead storage cell delivers 2 V, so that a 12-V battery contains six such cells connected in series.

When the external circuit is complete and the battery is in operation, the following oxidation-reduction reactions take place:

$$Pb(s) + SO_4^{2-}(aq) \longrightarrow PbSO_4(s) + 2e^- \qquad \text{(Anode)}$$

$$PbO_2(s) + 4H^+(aq) + SO_4^{2-}(aq) + 2e^- \longrightarrow PbSO_4(s) + 2H_2O \quad \text{(Cathode)}$$

and the overall reaction is

$$Pb(s) + PbO_2(s) + 4H^+(aq) + 2SO_4^{2-}(aq) \longrightarrow 2PbSO_4(s) + 2H_2O$$

Figure 17.20

A silver oxide battery.

Figure 17.21

A 12 V lead storage battery, such as those used in most automobiles, consists of six cells like the one illustrated here. Notice that the anode and cathode each consist of several plates connected together. This allows the cell to produce the large currents needed to start a car.

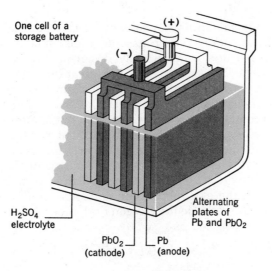

- Cap over anode
- Gasket
- Zinc anode
- Ag₂O cathode
- Separator
- Metal cup

One cell of a storage battery
(+)
(−)
H₂SO₄ electrolyte
Alternating plates of Pb and PbO₂
PbO₂ (cathode)
Pb (anode)

These batteries have the advantage that the electrode reactions can be reversed by placing a potential across the electrodes that is slightly larger than that which the battery can deliver. The recharging operation is performed in such a way that the negative external potential is applied to the negative pole and the positive potential to the positive pole. In doing this, the H_2SO_4 that is used up while the battery is in operation is restored. Recharging is accomplished by the generator or alternator of the car, or, if the battery is really run down, with the aid of a battery charger.

A convenient method of estimating the degree to which the battery has been discharged is by checking the density of the electrolyte. If the battery is in a weakened state, the electrolyte will be mostly water—a product of the overall reaction—and will have a density somewhere near 1 g cm^{-3}. If, however, the battery is in good operating order, with a full charge, the density of the electrolyte will be somewhat higher than 1 g cm^{-3} (the density of concentrated sulfuric acid is 1.8 g cm^{-3}). The mechanic in a garage can perform this test with the aid of a **hydrometer,** a device having a float that sinks to a depth that is a function of the density of the liquid in which it is immersed.

A hydrometer being used to check the state of charge of a lead storage battery.

Nickel-cadmium cell

A storage cell that has acquired widespread use in recent years is the *nicad*, or nickel-cadmium battery. The anode in the cell is composed of cadmium, which undergoes oxidation in an alkaline (basic) electrolyte.

$$Cd(s) + 2OH^-(aq) \longrightarrow Cd(OH)_2(s) + 2e^- \quad \text{(Anode)}$$

The cathode is composed of NiO_2, which undergoes reduction.

$$NiO_2(s) + 2H_2O + 2e^- \longrightarrow Ni(OH)_2(s) + 2OH^-(aq) \quad \text{(Cathode)}$$

The net cell reaction during discharge is therefore

$$Cd(s) + NiO_2(s) + 2H_2O \longrightarrow Cd(OH)_2(s) + Ni(OH)_2(s)$$

The voltage of the cell is about 1.4 V, somewhat less than the dry cell.

The nicad battery has some appealing features. First, it has a longer life than a lead storage battery. Second, it can be packaged in a sealed unit, much like the common dry cell. These advantages have made the nicad the choice among manufacturers of such devices as rechargeable calculators and electronic flash units in photography.

Fuel cells

Fuel cells are another means by which chemical energy may be converted into electrical energy. When gaseous fuels, such as H_2 and O_2, are allowed to undergo reaction in a carefully designed environment, electrical energy can be obtained. This type of cell finds great importance in space vehicles, where the fuels used in such cells can be the same as those used to power the rockets.

A diagram of a H_2/O_2 fuel cell is shown in Figure 17.22. In this cell there are three compartments separated from one another by porous electrodes. The hydrogen gas is fed into one compartment and the oxygen gas into another. These gases then diffuse (not bubble) slowly through the electrodes and react with an electrolyte that is in the center compartment. The electrodes are made of a conducting material, such as carbon, with a sprinkling of platinum to act as a catalyst, and the electrolyte is an aqueous solution of a base.

At the cathode the oxygen undergoes reduction, producing OH^- ions. This can be expressed as

$$O_2(g) + 2H_2O(l) + 4e^- \longrightarrow 4OH^-(aq)$$

Figure 17.22

Hydrogen-oxygen fuel cell.

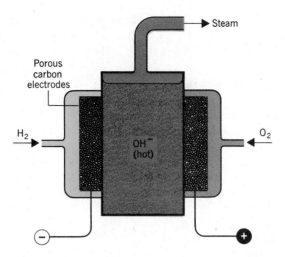

These OH^- ions travel to the anode where they undergo reaction with H_2.

$$H_2(g) + 2OH^-(aq) \longrightarrow 2H_2O(l) + 2e^-$$

The net reaction in the cell is

$$2H_2(g) + O_2(g) \longrightarrow 2H_2O(l)$$

The fuel cell is operated at a high temperature so that the water that is formed as a product of the cell reaction evaporates and may be condensed and used as drinking water for an astronaut. A number of these cells are usually connected together so that several kilowatts of power can be produced.

Fuel cells offer several advantages over other sources of energy. Unlike the dry cell or storage battery, the cathode and anode reactants may be continually supplied so that, in principle, energy can be withdrawn indefinitely from a fuel cell as long as the outside supply of fuel is maintained. Another advantage of the fuel cell is that the energy is extracted from the reactants under more nearly reversible conditions. Therefore, the thermodynamic efficiency of the reaction, in terms of producing useful work, is higher than when the reactants such as H_2 and O_2 are burned to produce heat that must be subsequently harnessed to produce work. These two advantages suggest that the development of fuel cells will probably continue at an accelerated pace in the future, particularly in light of predicted future energy shortages caused by greater demands for energy and dwindling supplies of fossil fuels.

Fuel cells have efficiencies that can approach 75%, whereas power plants that burn fuels have efficiencies of only about 40%.

REVIEW QUESTIONS AND PROBLEMS

(Problems whose numbers are in blue have their answers in Appendix C at the back of the book; the more difficult problems are marked with asterisks.)

General

17.1 Distinguish between: (a) electrolytic and galvanic cells, (b) metallic and electrolytic conduction, (c) oxidation and reduction.

17.2 Why must oxidation-reduction occur to maintain a steady flow of electricity during electrolytic conduction?

17.3 How do we define anode and cathode?

17.4 What is a volt? What is an ampere?

Electrolysis

17.5 Sketch an electrolysis cell in which the net cell reaction is

$$MgCl_2(l) \longrightarrow Mg(l) + Cl_2(g)$$

(the electrolysis of molten $MgCl_2$).
(a) Identify the anode and cathode.
(b) Write the half-reactions that occur at each electrode.

(c) Indicate the direction of ion flow in the molten $MgCl_2$.

(d) Indicate the direction of electron flow in the external circuit.

17.6 Write equations for the half-reactions for the oxidation and reduction of water.

17.7 From the reactions discussed in Section 17.2, predict the products that you would obtain in the electrolysis of an aqueous solution of H_2SO_4.

17.8 What is the function of an electrolyte such as Na_2SO_4 or H_2SO_4 during the electrolysis of water? Why can't we carry out electrolysis on pure H_2O in the absence of an electrolyte?

17.9 The oxidation of I^- to I_2 occurs more easily than the oxidation of water, and the reduction of Ni^{2+} occurs more easily than the reduction of water. Write the anode and cathode half-reactions and the cell reaction for the electrolysis of aqueous NiI_2.

17.10 The reduction of Ni^{2+} to Ni occurs more easily than the reduction of water. Write the anode and cathode half-reactions and the cell reaction for the electrolysis of an aqueous solution of $NiSO_4$.

Practical Applications of Electrolysis

17.11 Write equations for the electrode reactions and the net reaction for (a) electrolysis of an unstirred brine solution; (b) electrolysis of a stirred brine solution.

17.12 What are the advantages and disadvantages of the diaphragm cell used to produce NaOH and Cl_2?

17.13 What are the advantages and disadvantages of the mercury electrolysis cell used to produce NaOH and Cl_2?

17.14 What function does the design of the Downs cell serve?

17.15 Why was cryolite mixed with the Al_2O_3 prior to its electrolysis to produce Al?

17.16 Why can Al not be produced by electrolysis of an aqueous solution containing a salt such as $Al_2(SO_4)_3$?

17.17 Write a series of chemical equations representing the reactions involved with the recovery of Mg from sea water.

17.18 Describe the electrolytic purification of metallic copper. Why is the process economically feasible?

17.19 What is electroplating? If an object were to be plated with nickel from a solution of $NiSO_4$, to which electrode (anode or cathode) would the object be connected? Why?

Quantitative Aspects of Electrolysis

17.20 What is a faraday? What relationships connect faradays to measurements of current and time in the laboratory?

17.21 How many moles of electrons would be required to reduce 1 mol of each of the following to the indicated product?
(a) Cu^{2+} to Cu   (d) F_2 to $2F^-$
(b) Fe^{3+} to Fe^{2+} (e) NO_3^- to NH_3
(c) MnO_4^- to Mn^{2+}

17.22 Calculate the number of electrons that corresponds to 1 coulomb of charge.

17.23 How many moles of electrons would be required to oxidize 1 mol of each of the following to give the indicated product?
(a) Cu^+ to Cu^{2+}
(b) Pb to PbO_2
(c) Cl_2 to $2ClO_3^-$
(d) O_2 to H_2O_2 (hydrogen peroxide)
(e) NH_3 to NO_3^-

17.24 How many moles of electrons are given by
(a) 8950 C
(b) a current of 1.5 A for 30 s
(c) a current of 14.7 A for 10 min

17.25 State how many minutes it would take to
(a) deliver 10 500 C using a current of 25.0 A
(b) deliver 0.65 $\mathscr{F}$ using a current of 15 A
(c) reduce 0.20 mol of Cu^{2+} to Cu using a current of 12 A

17.26 State how many minutes it would take to
(a) deliver 84 200 C using a current of 6.30 A
(b) deliver 1.25 $\mathscr{F}$ using current of 8.40 A
(c) produce 0.500 mol of Al from molten $AlCl_3$ using a current of 18.3 A.

17.27 How many moles of electrons are required to produce the following?
(a) 10.0 cm^3 of O_2 gas (at STP) from aqueous Na_2SO_4
(b) 10.0 g of Al from molten Al_2O_3 (in cryolite)
(c) 5.00 g of Na from molten NaCl
(d) 5.00 g of Mg from molten $MgCl_2$

17.28 How many grams of Na and Cl_2 would be produced if a current of 25 A was applied for 8.0 h to the cell shown in Figure 17.3?

17.29 How many grams of O_2 and H_2 are produced in 1.0 h when water is electrolyzed at a current of 0.50 A? What would be the volume, at STP, of O_2 and H_2?

17.30 How many grams of copper could be purified by a current of 115 A for 8.00 h? Refer to Figure 17.12.

17.31 How many grams of silver could be plated out on a serving tray by electrolysis of a solution containing Ag in the +1 oxidation state for a period of 8.00 h at a current of 8.46 A? What area would this cover, if we assume that the density of Ag is 10.5 g cm^{-3} and the thickness of the silver plate is 0.002 54 cm?

17.32 How many seconds would it take to deposit 21.4 g of Ag from a solution of $AgNO_3$ by a current of 10.0 A?

17.33 How many hours would it take to deposit 35.3 g of Cr from a solution of $CrCl_3$ at a current of 6.00 A?

17.34 How many minutes would it take to plate out 5.00 g of copper from a solution of $CuSO_4$ at a current of 5.00 A?

17.35 What current is required to deposit 0.225 g of Ni from a solution of $NiSO_4$ in 10.0 min?

17.36 What current is required to produce 1.33 g of Cl_2 from a solution of NaCl in 45.0 min?

17.37 What current is required to produce 50.0 cm^3 of O_2 gas, measured at STP, by the electrolysis of H_2O for a period of 3.00 h?

17.38 How does a coulometer work? What advantages are there to the use of a coulometer?

17.39 How many minutes would it take to remove all the Cr from 500 cm^3 of 0.270 M $Cr_2(SO_4)_3$ by a current of 3.00 A?

17.40 In an experiment two coulometers were connected in series, one containing $CuSO_4$, the other an unknown salt. It was found that 1.25 g of copper was plated out during the same period of time that 3.42 g of the unknown metal was plated out.
(a) How many moles of electrons passed through this coulometer?
(b) If the oxidation state of the unknown metal ion in the solution was 2+, what is the molar mass of the unknown?

17.41 Two coulometers were connected in series so that the same current passes through each of them. In an experiment, 0.125 mol of Cu was deposited from a solution of $CuSO_4$ in one of the coulometers How many moles of Cr were deposited at the same time from a $Cr_2(SO_4)_3$ solution in the other?

*17.42 A current of 0.250 A is passed through 400 cm^3 of a 2.00 M solution of NaCl for 35.0 min. What will be the pH of the solution after the current is turned off?

*17.43 An unstirred solution 2.00 M of NaCl was electrolyzed for a period of 25.0 min and then titrated with 0.250 M HCl. The titration required 15.5 cm^3 of the acid. What was the average current during the electrolysis?

*17.44 A student set up an electrolysis apparatus and passed a current of 1.22 A through a 3 M H_2SO_4 solution for 30.0 min. He collected the H_2 evolved and found that it occupied a volume, over water at 27°C, of 288 cm^3 at a total pressure of 102 kPa. Use these data to calculate the charge on the electron, expressed in the units, coulombs.

*17.45 What current would be required to deposit 1 m^2 of chrome plate having a thickness of 0.050 mm in 25 min from a solution containing H_2CrO_4? The density of Cr is 7.19 g cm^{-3}.

Galvanic Cells

17.46 In Section 17.5 we saw that electrical energy can be extracted from a Zn/Cu cell. If Cu^{2+} is brought into contact with metallic Zn, it is reduced to Cu while the Zn is oxidized, without the generation of electricity. In this case, what happens to the energy that is *not* being extracted as electrical energy?

17.47 What is the function of a salt bridge in a galvanic cell?

17.48 What are the signs of the anode and cathode in galvanic and electrolytic cells?

17.49 Sketch a galvanic cell in which the following net reaction occurs: $Ni^{2+}(aq) + Fe(s) \longrightarrow Ni(s) + Fe^{2+}(aq)$.
(a) Label the cathode and anode.
(b) Indicate the charges on the electrodes.
(c) Indicate the direction of electron flow.
(d) Indicate the direction of the flow of cations and anions.
(e) If the concentrations of the ions are each 1 M, what is the potential of the cell?

17.50 How can one identify *experimentally* which electrode in a galvanic cell is the anode and which is the cathode?

Reduction Potentials

17.51 Sketch a hydrogen electrode. Under what conditions is its reduction potential defined as exactly 0.000 V?

17.52 A galvanic cell was set up with a zinc electrode dipping into 1.00 M $ZnSO_4$ as one half-cell and a gallium electrode dipping into 1.00 M $GaCl_3$ solution as the other. The emf of the cell was measured to be 0.23 V. The zinc electrode was found to be negatively charged.
(a) Write the anode and cathode half-reactions that are taking place in this cell.
(b) Write the cell reaction.
(c) Calculate $\mathscr{E}^0_{Ga}$ for $Ga^{3+} + 3e^- \rightleftharpoons Ga(s)$.
(d) What would be the standard potential of a gallium-copper galvanic cell?

17.53 Which is the better oxidizing agent? (Refer to Table 17.1.)
(a) Li^+ or Ca^{2+}
(b) Cl_2 or F_2
(c) H_2O or Al^{3+}
(d) $S_2O_8{}^{2-}$ or Cl_2
(e) Br_2 or H_2O

17.54 Which is the better oxidizing agent? (Refer to Table 17.1.)
(a) Cl_2 or $ClO_3{}^-$
(b) O_2 or $Cr_2O_7{}^{2-}$
(c) $MnO_4{}^-$ or $Cr_2O_7{}^{2-}$
(d) PbO_2 or Hg_2Cl_2

17.55 Which is the better reducing agent? (Refer to Table 17.1.)
(a) Ni or Fe
(b) H_2 or Mg
(c) Br^- or I^-
(d) $SO_4{}^{2-}$ or F^-
(e) Sn or Mn

17.56 Which is the better reducing agent? (Refer to Table 17.1.)
(a) Na or Cr
(b) $PbSO_4$ or Cl_2
(c) Ag or Cu
(d) I^- or Sn
(e) H_2 or H_2O

Cell Potentials and Spontaneity of Reactions

17.57 Without computing $\mathscr{E}^0$, determine what reactions will occur spontaneously among the following sets of reactants in aqueous solution:
   (a) Al(s), Ni(s), $NiSO_4(aq)$, $Al_2(SO_4)_3(aq)$
   (b) $PbO_2(s)$, $K_2Cr_2O_7(aq)$, $H_2SO_4(aq)$, $PbSO_4(s)$, $Cr_2(SO_4)_3(aq)$
   (c) Ag(s), $AgNO_3(aq)$, Pb(s), $Pb(NO_3)_2(aq)$
   (d) $MnO_2(s)$, HCl(aq), $Cl_2(g)$, $MnCl_2(aq)$
   (e) Mn(s), HCl(aq), $MnCl_2(aq)$, $H_2(g)$

17.58 Without computing $\mathscr{E}^0$, determine whether the following reactions will occur spontaneously:
   (a) $2Fe^{3+} + Sn \rightarrow 2Fe^{2+} + Sn^{2+}$
   (b) $Cu + 2H^+ \rightarrow Cu^{2+} + H_2$
   (c) $3Mg^{2+} + 2Al \rightarrow 3Mg + 2Al^{3+}$
   (d) $Mn + Zn^{2+} \rightarrow Mn^{2+} + Zn$
   (e) $PbO_2 + SO_4^{2-} + 4H^+ + 2Hg + 2Cl^- \rightarrow$
   $$Hg_2Cl_2 + PbSO_4 + 2H_2O$$

17.59 Compute $\mathscr{E}^0$ and use its value to determine whether the following reactions will occur spontaneously:
   (a) $Ca^{2+} + Mg \rightarrow Ca + Mg^{2+}$
   (b) $Pb^{2+} + 2Cl^- \rightarrow Pb + Cl_2$
   (c) $2Cl^- + S_2O_8^{2-} \rightarrow Cl_2 + 2SO_4^{2-}$
   (d) $6Mn^{2+} + 5Cr_2O_7^{2-} + 22H^+ \rightarrow$
   $$6MnO_4^- + 10Cr^{3+} + 11H_2O$$
   (e) $O_2 + 4Cl^- + 4H^+ \rightarrow 2H_2O + 2Cl_2$

17.60 Given the following sets of half-reactions, write the net cell reaction and calculate $\mathscr{E}^0$ for the spontaneous changes that will occur.
   (a) $Hg_2Cl_2 + 2e^- \rightleftharpoons 2Hg + 2Cl^-$
   $PbSO_4 + 2e^- \rightleftharpoons Pb + SO_4^{2-}$
   (b) $AgCl + e^- \rightleftharpoons Ag + Cl^-$
   $Cu^{2+} + 2e^- \rightleftharpoons Cu$
   (c) $Mn^{2+} + 2e^- \rightleftharpoons Mn$
   $Cl_2(g) + 2e^- \rightleftharpoons 2Cl^-$
   (d) $Al^{3+} + 3e^- \rightleftharpoons Al$
   $Br_2(aq) + 2e^- \rightleftharpoons 2Br^-$

17.61 Determine the value of $\mathscr{E}^0$ for each of the spontaneous reactions in Problem 17.57.

17.62 Determine the value of $\mathscr{E}^0$ for each of the reactions as written from left to right in Problem 17.58.

Calculating Equilibrium Constants from $\mathscr{E}^0$

17.63 Calculate the equilibrium constants for the following cell reactions:
   (a) $Ni(s) + Sn^{2+}(aq) \rightleftharpoons Ni^{2+}(aq) + Sn(s)$
   (b) $Cl_2(g) + 2Br^-(aq) \rightleftharpoons Br_2(aq) + 2Cl^-(aq)$
   (c) $Fe^{2+}(aq) + Ag^+(aq) \rightleftharpoons Ag(s) + Fe^{3+}(aq)$

17.64 Calculate the equilibrium constants for the reactions in Problem 17.58.

17.65 Calculate the equilibrium constants for the reactions in Problem 17.59.

Standard Cell Potentials and ΔG^0

17.66 Calculate ΔG_{298}^0 in kilojoules for each reaction in Problem 17.58.

17.67 Calculate ΔG_{298}^0 in kilojoules for each reaction in Problem 17.59.

Nernst Equation

17.68 Write the Nernst equation and calculate $\mathscr{E}^0$ and $\mathscr{E}$ for the following reactions:
   (a) $Cu^{2+}(0.1\ M) + Zn(s) \rightarrow Cu(s) + Zn^{2+}(1.0\ M)$
   (b) $Sn^{2+}(0.5\ M) + Ni(s) \rightarrow Sn(s) + Ni^{2+}(0.01\ M)$
   (c) $F_2(g,\ 1\ atm) + 2Li(s) \rightarrow 2Li^+(1\ M) + 2F^-(0.5\ M)$
   (d) $Zn(s) + 2H^+(0.01\ M) \rightarrow Zn^{2+}(1\ M) + H_2(1\ atm)$
   (e) $2H^+(1.0\ M) + Fe(s) \rightarrow H_2(1\ atm) + Fe^{2+}(0.2\ M)$

17.69 Calculate $\mathscr{E}^0$, $\mathscr{E}$ and ΔG (in kilojoules) for the following cell reactions (not balanced):
   (a) $Al(s) + Ni^{2+}(0.80\ M) \rightarrow Al^{3+}(0.020\ M) + Ni(s)$
   (b) $Ni(s) + Sn^{2+}(1.10\ M) \rightarrow Sn(s) + Ni^{2+}(0.010\ M)$
   (c) $Ag^+(0.050\ M) + Zn(s) \rightarrow Ag(s) + Zn^{2+}(0.010\ M)$

17.70 Calculate the cell potential for the following:
   (a) $Sn(s) + Pb^{2+}(0.050\ M) \rightarrow Sn^{2+}(1.50\ M) + Pb(s)$
   (b) $3Zn(s) + 2Cr^{3+}(0.010\ M) \rightarrow 3Zn^{2+}(0.020\ M) + Cr(s)$
   (c) $PbO_2(s) + SO_4^{2-}(0.010\ M) + 4H^+(0.10\ M) +$
   $Cu(s) \rightarrow PbSO_4(s) + 2H_2O + Cu^{2+}(0.0010\ M)$

17.71 What is the reduction potential of a half-cell composed of a copper wire dipping into $2 \times 10^{-4}\ M$ $CuSO_4$?

*17.72 Calculate the value of ΔG (in kilojoules) for a system containing the following species: $Mn^{2+}(0.10\ M)$, $Cr_2O_7^{2-}(0.010\ M)$, $MnO_4^-(0.0010\ M)$, Cr^{3+} $(0.0010\ M)$. The pH of the solution is 6.00. The reaction that you should consider is

$$6Mn^{2+}(aq) + 5Cr_2O_7^{2-}(aq) + 22H^+(aq) \rightleftharpoons$$
$$6MnO_4^-(aq) + 10Cr^{3+}(aq) + 11H_2O(l)$$

In which direction will this reaction proceed to get to equilibrium from the starting conditions given above?

*17.73 The standard reduction potential for Ag^+ is 0.80 V. Compute the standard reduction potential for the half-reaction.

$$Ag_2S(s) + 2e^- \rightleftharpoons 2Ag(s) + S^{2-}(aq)$$

in a solution buffered to a pH of 3.00.

*17.74 A hydrogen electrode is immersed in a 0.10 M solution of acetic acid. This electrode is connected to another consisting of an iron nail dipping into 0.10 M $FeCl_2$. What will be the measured emf of this cell? Assume $p_{H_2} = 1$ atm.

*17.75 An Ag/AgCl electrode dipping into 1 M HCl has a standard reduction potential of +0.22 V [AgCl(s) + e^- $\rightleftharpoons$ Ag(s) + Cl$^-$(aq)]. A second Ag/AgCl electrode is dipped into a solution containing Cl$^-$ at an unknown concentration. The cell generates a potential of 0.0435 V, with the electrode in the unknown solution serving as the anode. What is the molar concentration of Cl$^-$ in the unknown?

*17.76 A galvanic cell was set up using silver as one electrode dipping into 200 cm^3 of 0.100 M AgNO$_3$ and magnesium as the other electrode dipping into 250 cm^3 of 0.100 M Mg(NO$_3$)$_2$ solution.
(a) What is the potential of the cell?
(b) Suppose that current was drawn from the cell for a period of time until 1.00 g of silver plated out on the silver electrode. What is the potential of the cell now?
(c) Suppose that the original magnesium electrode had a mass of 0.080 g (it consisted of 0.080 g of magnesium deposited on an inert platinum electrode). What would the potential of the cell be just before the last tiny bit of magnesium dissolved?

$\mathscr{E}^0$ and Work

*17.77 How many hours will a 25-watt (W) light bulb burn if it is powered by a lead storage battery that has available 25.0 g of Pb that can react as an anode. Assume a constant voltage of 1.5 V. (1 W = 1 J s^{-1})

*17.78 What masses of H$_2$ and O$_2$ in grams would have to react each second in a fuel cell at 110°C to provide 1.0 kilowatt (kW) of power, if we assume a thermodynamic efficiency of 70%. (*Hint:* Use the data in Chapter 12 to compute ΔG^0 for the reaction, H$_2$(g) + $\frac{1}{2}$O$_2$(g) → H$_2$O(g) at 110°C. 1 W = 1 J s^{-1}.)

*17.79 How much work, expressed in kilojoules, is able to be accomplished by a 5.00-min flow of electricity having a potential of 110 V and a current of 1.00 A?

*17.80 If we assume that the typical electric generating plant has an efficiency of only about 30%, what volume (in dm^3) of fuel having an average formula of C$_{12}$H$_{26}$ must be burned, giving H$_2$O(g) and CO$_2$(g), to produce 3.6 MJ of electricity? Assume ΔH_f^0 of C$_{12}$H$_{26}$(l) = 291 kJ mol^{-1} and a density of 0.74 g cm^{-3}. 1 W = J s^{-1}.

Concentration Cells

17.81 What is a concentration cell?

17.82 Calculate the potential generated by a concentration cell consisting of a pair of iron electrodes dipping into two solutions, one containing 0.10 M Fe^{2+} and the other containing 0.0010 M Fe^{2+}

17.83 Calculate the potential of a concentration cell containing 0.0020 M Cr^{3+} in one compartment and 0.10 M Cr^{3+} in the other compartment with Cr(s) electrodes dipping into each solution.

Solubility Product

17.84 The solubility product constant of AgBr is 5×10^{-13}. What will be the potential of a cell constructed using the H$_2$ electrode ([H$^+$] = 1.0 M, p_{H_2} = 1 atm) versus a half-cell containing a silver wire coated with AgBr immersed in 0.010 M HBr?

17.85 A cell was constructed using the standard hydrogen electrode ([H$^+$] = 1.0 M, p_{H_2} = 1 atm) in one compartment and a lead electrode in a 0.10 M K$_2$CrO$_4$ solution in contact with undissolved PbCrO$_4$. The potential of the cell was measured to be 0.51 V with the Pb electrode as the anode. Determine the K_{sp} of PbCrO$_4$ from these data.

*17.86 A student set up a galvanic cell to measure the K_{sp} of CuS. On one side of the cell she had a copper electrode dipping into a 0.10 M Cu^{2+} solution and on the other side a zinc electrode in a Zn^{2+} solution. The Zn^{2+} concentration was held constant at 1.0 M and the [Cu^{2+}] brought to a minimum by saturating the Cu^{2+} solution with H$_2$S. The emf of the cell was read as +0.67 V, with the Cu electrode serving as the cathode. Calculate the Cu^{2+} concentration and the K_{sp} of CuS. Compare your answer to the K_{sp} reported in Table 16.1. In a saturated solution the concentration of H$_2$S is 0.10 M. The solution in which the CuS was formed was not buffered.

Practical Galvanic Cells

17.87 Describe the anode and cathode reactions in the ordinary dry cell (the Leclanché cell).

17.88 Why does a dead dry cell seem to come back to life if left idle for a while?

17.89 What are the anode, cathode, and net cell reactions in the alkaline zinc-manganese dioxide battery? What is the electrolyte?

17.90 What are the cathode, anode, and net cell reactions in the silver oxide battery?

17.91 What are the reactions that take place during the discharge of the lead storage battery? What reactions occur when this battery is being charged?

17.92 Write the anode, cathode, and overall cell reaction for the discharge of the nickel-cadmium battery.

17.93 What is a fuel cell? What advantages does a fuel cell offer over the lead storage battery?

17.94 What possible advantages do fuel cells offer over current electrical power plants?

CHAPTER 18

METALS AND THEIR COMPOUNDS; THE REPRESENTATIVE METALS

A novel use for a metal—as a rocket propellant. White clouds of aluminum oxide produced by burning aluminum powder billow from the solid booster rockets that lift the space shuttle Discovery from its launch pad at the Kennedy Space Center in Florida. The representative metals and their compounds find many practical uses, as we will discover in this chapter.

At various times throughout this book we have described some of the physical and chemical properties of the elements. In Chapter 3, for example, we studied some of the physical and chemical properties of the metals and nonmetals, and in Chapter 7 we examined some of the redox reactions that typical metals and nonmetals undergo. Now that you have acquired a broader understanding of basic chemical principles—thermodynamics, kinetics, and equilibrium, for example—we again turn our attention toward the properties of the elements. Our goal will be not only to learn about some of their specific chemical properties, but also to see how intimately our daily lives are tied to chemistry.

In this chapter we begin by studying how metals are obtained from natural sources and how the properties of metal compounds are related to bonding and structure. Then we will discuss some of the specific chemical properties of the representative metals. These include the elements in Group IA (excluding hydrogen) and IIA, as well as the heavier members of Groups IIIA, IVA, and VA.

18.1 METALLURGY

Metallurgy is the science and technology of extracting metals from their ores and bringing them to the point at which they can be put to practical use. It is a subject with a long and rich history, dating from the time when humans first fashioned tools and weapons from natural deposits of metals such as copper. Modern metallurgy touches our lives at every turn, from special alloys in automobile engines to stainless steels used in surgical implants.

An **ore** is a substance that contains a particular desirable constituent in a high enough concentration that its extraction from the ore is economically worthwhile. For example, many minerals may contain small amounts of iron, but only those that are rich in iron would be considered iron ores. As the earth's supply of rich ores is consumed, it will be the job of chemists to devise new and efficient ways of extracting metals such as iron from less rich ores.

The sources of metals range from deposits within the earth to the sea itself. The oceans provide a huge storehouse of minerals in which the metals occur primarily as soluble sulfates and halides. The major source of magnesium, for example, is the oceans, and in the future greater attention will no doubt be focused on the sea as a source of raw materials as supplies of ore deposits on land are depleted. There has already been some interest in mining *manganese nodules*—lumps about the size of an orange that are relatively rich in manganese

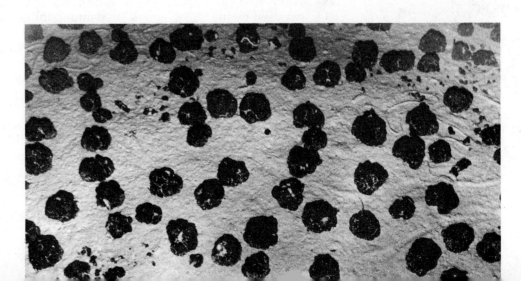

Manganese nodules lying on the floor of the north Pacific Ocean. The depth here is about 5 km.

($\sim 25\%$) and iron ($\sim 15\%$)—that seem to line the ocean floor in certain places.

On land, some metals occur as deposits of their carbonates or sulfates. Limestone, for example, is primarily $CaCO_3$, and gypsum is a hydrate of calcium sulfate. Oxides are also important sources of metals. Two examples are aluminum (Al_2O_3) and iron (Fe_2O_3 and Fe_3O_4). Sulfides are the primary sources of lead (as PbS) and copper (as Cu_2S).

Because of the wide variety of sources of metals, there obviously is no single method that can be used in obtaining them all. Nevertheless, it is possible to divide metallurgical processes into three categories.

1. *Concentration.* Ores that contain substantial amounts of impurities, such as rock, must often be treated to concentrate the metal-bearing constituent. Pretreatment of an ore is also carried out to convert some metal compounds into substances that can be more easily reduced.
2. *Reduction.* In their compounds, metals nearly always exist in positive oxidation states. Therefore, to obtain a metal from its ore it must be reduced. The particular procedure employed for a given metal depends on its ease of reduction to the free state.
3. *Refining.* Often, during reduction, substantial amounts of impurities become introduced into the metal. Refining is the process whereby these impurities are removed and the composition of the metal adjusted (alloys formed) to meet specific applications.

Let us take a brief look at each of these as they apply to some important metals.

Concentration

Not all ores have to be subjected to a pretreatment step prior to reduction, although most of them must. These pretreatment procedures involve the separation of the metal-bearing component of the ore from unwanted or interfering impurities. This is particularly important for low-grade ores in which the desired metal is present only in small amounts.

As expected, different methods are applied to different ores, depending on the specific properties of the impurities and the metal compounds. We can divide these procedures into two classes: *physical separations*, in which the chemical compositions of the constituents are not altered, and *chemical separations*, which use the chemical properties of the different substances in the ore.

Some metals, such as silver and gold, are found in deposits as the free element, and their recovery simply involves removing them from the rock and sand with which they are mixed. One of the earliest forms of physical separation was used by the "forty-niners" in panning for gold. A mixture of sand containing (it was hoped!) particles of metallic gold was placed in a shallow pan with water. The mixture was swirled about and the sand was washed over the rim, leaving the gold dust in the bottom of the pan. The success of this procedure is based on the fact that gold is about nine times as dense as the sand and gravel impurities. As a result, the lighter impurities are more easily washed away than the more dense metal.

Another way of removing metallic gold and silver from their ores is to treat the mixture with metallic mercury—a liquid in which silver and gold dissolve to form an alloy called an **amalgam.** The silver and gold are later recovered by distilling away the mercury, which is reclaimed and used again. You are probably familiar with silver and gold amalgams as the material used by dentists to fill teeth.

A physical separation technique that can be applied to the sulfide ores of

A rare photograph of a nineteenth century miner panning for gold. The technique is still used occasionally, especially by amateur prospectors.

An *amalgam* is a solution of a metal in mercury.

Figure 18.1
The flotation process.

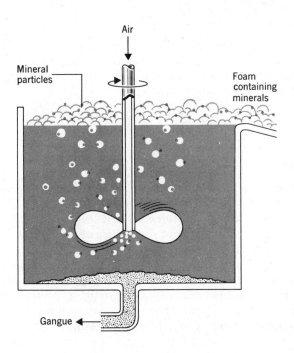

Air

Mineral
particles

Foam
containing
minerals

Gangue

You are familiar with the reddish-brown color of Fe_2O_3 that is characteristic of rust.

zinc, copper, and lead is called **flotation.** In this process, illustrated in Figure 18.1, the ore is pulverized and added to a mixture of water and oil containing suitable additives. The metal-bearing component of the ore becomes coated by the oil while the unwanted material, called the **gangue,** is wetted by the water. A stream of air is then blown through the mixture and the oil-covered mineral is carried to the surface by bubbles where it is trapped in a froth that can be removed to recover the metal compound. The gangue, on the other hand, simply settles to the bottom of the apparatus and is later discarded.

Iron, as you are probably aware, is society's most important metal, and its ores have traditionally been rich deposits of reddish-brown **hematite,** composed primarily of Fe_2O_3. Through decades of mining, hematite deposits in regions such as the Mesabi range of Minnesota have been nearly depleted and other sources of iron ore have been sought. One such ore being mined today is *taconite.* Taconites have Fe_3O_4 as their iron-containing compound, and the better of them contain 30 to 40% iron by mass. The compound Fe_3O_4 is called **magnetite,** and as its name suggests, it is magnetic. This provides the basis for the enrichment of taconite ores by a physical separation process. The rock-hard ore is first ground to give particles of very small size and then the Fe_3O_4 is pulled out using a powerful electromagnet.

Chemical methods of concentrating the metal-bearing component of an ore vary considerably because of the variety of chemical properties exhibited by the metals and their compounds. For example, aluminum, whose electrolytic reduction was described in Chapter 17, occurs in deposits of *bauxite,* a form of Al_2O_3. In this case the ore is concentrated by taking advantage of the amphoteric behavior of aluminum. The bauxite is treated with concentrated base, which dissolves the Al_2O_3 to produce aluminate ion, AlO_2^-.

$$Al_2O_3 + 2OH^- \longrightarrow 2AlO_2^- + H_2O$$

After the solution is removed from the gangue, it is acidified. This precipitates $Al(OH)_3$, which yields pure Al_2O_3 when heated.

$$2Al(OH)_3 \xrightarrow{\text{heat}} Al_2O_3 + 3H_2O$$

This purified aluminum oxide serves as the raw material in the Hall process discussed previously.

Another chemical pretreatment, often given to a sulfide ore, is called **roasting.** Here the ore is heated in air, converting the metal sulfide to an oxide that is more conveniently reduced.

$$2PbS + 3O_2 \longrightarrow 2PbO + 2SO_2$$

$$2ZnS + 3O_2 \longrightarrow 2ZnO + 2SO_2$$

Reduction

Most metals, including those of both the representative and transition elements, are always found in nature in the combined state where they exist in positive oxidation states. Therefore, recovering them from their compounds involves a chemical reduction, and the nature of the reduction process depends on the ease with which the metal can be reduced.

Some metals are so easily reduced that many of their compounds can be decomposed just by heating them at relatively low temperatures. Priestley, for example, in his experiments on oxygen, produced metallic mercury and oxygen from mercuric oxide by simply heating it with sunlight focused on the HgO by means of a magnifying glass. In this case, HgO decomposes quite spontaneously at elevated temperatures according to the equation

$$2HgO(s) \longrightarrow 2Hg(g) + O_2(g)$$

The practicality of using a thermal decomposition reaction of this type to produce a free metal depends on the extent to which the reaction proceeds to completion at a given temperature. In Chapter 12 we saw that at 25°C the position of equilibrium in a reaction is governed by the sign and magnitude of ΔG^0. If we take $\Delta G'$ to be the equivalent of ΔG^0, but at some other temperature, we have the relationship

$$\Delta G' = \Delta H' - T\,\Delta S'$$

where $\Delta H'$ and $\Delta S'$ are the heat and entropy changes that accompany the reaction. As discussed in Chapter 14, for most systems ΔH and ΔS do not change much with temperature, so $\Delta H'$ and $\Delta S'$ can reasonably be approximated by ΔH^0 and ΔS^0. We have seen that when ΔG^0 for a reaction is negative, the reaction is feasible from a practical standpoint because significant amounts of products will be formed. Extending this idea to temperatures other than 25°C, we can say that a thermal decomposition reaction will be feasible when $\Delta G'$ is negative, since under these conditions an appreciable amount of product will be formed. We must now look at the magnitudes of $\Delta H'$ and $\Delta S'$, because they control the sign and magnitude of $\Delta G'$.

Since a gas (O_2) and sometimes the metal vapor is produced in the decomposition of an oxide, the process occurs with a sizable increase in entropy, so $\Delta S'$ will be positive. The enthalpy change for the decomposition, $\Delta H'$, is simply the negative of the heat of formation of the oxide. Since ΔH_f is generally negative for metal oxides, $\Delta H'$ for the decomposition reaction will be positive. As a result, the sign of $\Delta G'$ is determined by the difference between two positive quantities, $\Delta H'$ and $T\,\Delta S'$.

If the metal oxide has a large negative heat of formation—that is, if a great deal of energy is evolved when the oxide is formed—then $\Delta H'$ for the decomposition will have a large positive value. As a result, the value of $\Delta G'$, which is given by the difference $\Delta H' - T\,\Delta S'$, will be negative *only* at very high temperatures, where $T\,\Delta S'$ is larger than $\Delta H'$. We express this by saying that the metal oxide is very stable with respect to thermal decomposition. On the other hand, if

Remember, T is always positive.

The low thermal stabilities of compounds of silver and gold is one of the reasons these metals are found in the free state in nature.

the $\Delta H_f'$ of the metal oxide is relatively small, as with HgO and certain other oxides (for example, Ag_2O, CuO, and Au_2O_3), then $\Delta H'$ for the decomposition reaction is a small positive quantity, and $\Delta G'$ for the reaction becomes negative at relatively low temperatures. These oxides, therefore, are said to have relatively low thermal stabilities.

EXAMPLE 18.1 CALCULATING THE TEMPERATURE NEEDED TO CAUSE THERMAL DECOMPOSITION OF A METAL COMPOUND

PROBLEM Above what temperature would the decomposition of Ag_2O be expected to proceed to an appreciable extent toward completion? At 25°C, ΔH_f^0 for Ag_2O is -30.5 kJ mol^{-1}, $\Delta S_f^0 = -66.1$ J mol^{-1} K^{-1}.

SOLUTION Since the decomposition of Ag_2O,

$$Ag_2O(s) \longrightarrow 2Ag(s) + \tfrac{1}{2}O_2(g)$$

is the reverse of formation, we have for this reaction

$$\Delta H^0 = -\Delta H_f^0 = +30.5 \text{ kJ mol}^{-1}$$

$$\Delta S^0 = -\Delta S_f^0 = +66.1 \text{ J mol}^{-1} \text{ K}^{-1}$$

Now let's calculate the temperature at which $\Delta G' = 0$, because above that temperature $\Delta G'$ will be negative. We will assume, as stated in the text, that $\Delta H'$ and $\Delta S'$ are approximately independent of temperature so that we can use ΔH^0 and ΔS^0 in the equation for $\Delta G'$.

$$\Delta G' = \Delta H^0 - T\,\Delta S^0$$

When $\Delta G' = 0$

$$0 = \Delta H^0 - T\,\Delta S^0$$

Solving for T, we obtain

$$T = \frac{\Delta H^0}{\Delta S^0}$$

Notice that we converted kJ to J so that the units cancel properly.

$$= \frac{30\,500 \text{ J mol}^{-1}}{66.1 \text{ J mol}^{-1} \text{ K}^{-1}}$$

$$= 461 \text{ K}$$

Because ΔH^0 and ΔS^0 are both positive, $\Delta G'$ will become negative at temperatures above 461 K (188°C). This means that above 461 K the reaction should become feasible, with much of the Ag_2O undergoing decomposition.

EXAMPLE 18.2 CALCULATING THE TEMPERATURE NEEDED TO CAUSE THERMAL DECOMPOSITION OF A METAL COMPOUND

PROBLEM Above what temperature would $\Delta G'$ be negative for the reaction

$$Au_2O_3(s) \longrightarrow 2Au(s) + \tfrac{3}{2}O_2(g)$$

At 25°C, $\Delta H_f^0 = +80.8$ kJ mol^{-1} for Au_2O_3. Also for Au_2O_3, $S^0 = 125$ J mol^{-1} K^{-1}; for Au, $S^0 = 47.7$ J mol^{-1} K^{-1}; for O_2, $S^0 = 205$ J mol^{-1} K^{-1}.

SOLUTION First, let's calculate ΔH^0 and ΔS^0 for the decomposition reaction. ΔH^0 is simply the negative of ΔH_f^0—that is, $\Delta H^0 = -80.8$ kJ for the decomposition of 1 mol of Au_2O_3. The value of ΔS^0 is

$$\Delta S^0 = (2S_{Au}^0 + \tfrac{3}{2}S_{O_2}^0) - (S_{Au_2O_3}^0)$$

$$= (2 \text{ mol}) \times \left(\frac{47.7 \text{ J}}{\text{mol K}}\right) + \tfrac{3}{2}\text{mol} \times \left(\frac{205 \text{ J}}{\text{mol K}}\right) - 1 \text{ mol} \left(\frac{125 \text{ J}}{\text{mol K}}\right)$$

$$= 278 \text{ J K}^{-1} = 0.278 \text{ kJ K}^{-1}$$

Again, we obtain $\Delta G'$ by assuming that $\Delta H'$ and $\Delta S'$ are the same as ΔH^0 and ΔS^0. Therefore,

$$\Delta G' = \Delta H^0 - T \, \Delta S^0$$

Substituting yields,

$$\Delta G' = -80.8 \text{ kJ} - T(0.278 \text{ kJ K}^{-1})$$

Here is an example in which kinetics, not thermodynamics, controls what we are able to observe.

Note that regardless of the temperature $\Delta G'$ will be negative because the absolute temperature is always a positive quantity. What this tells us is that Au_2O_3 is unstable with respect to decomposition at any temperature. It exists only because at low temperatures its rate of decomposition is very slow.

Except in a few cases, thermal decomposition is not a practical way of extracting a metal from its compounds. Usually, the metal compound is allowed to react with some other substance that is a better reducing agent than the metal being sought. Titanium, for example, is obtained from $TiCl_4$ by its reaction with the more active metal magnesium. The $TiCl_4$ is obtained from the mineral rutile, a fairly pure source of TiO_2, by reaction with chlorine gas and carbon.

Titanium is used in making many aircraft parts. It is very strong, but only about 60% as dense as iron. This means that a part made of Ti will have only about 60% as much mass as the same size part made of steel.

$$TiO_2 + 2C + 2Cl_2 \longrightarrow TiCl_4 + 2CO$$

The $TiCl_4$ is a volatile liquid (b.p. = 136°C) and can be separated easily from less volatile impurities by distillation. The final reduction to the metal then follows the equation

$$TiCl_4 + 2Mg \longrightarrow Ti + 2MgCl_2$$

Very reactive metals, such as sodium and magnesium, are produced by electrolysis, as described in Chapter 17.

Active metals such as magnesium are very expensive reducing agents because they themselves are difficult or costly to prepare. Therefore, less expensive reducing agents are employed whenever possible. One such reducing agent is hydrogen, which can be used to liberate metals of moderate chemical activity from their compounds. For example, tin and lead oxides can be reduced by heating them in the presence of H_2.

$$SnO + H_2 \xrightarrow{\text{heat}} Sn + H_2O$$

$$PbO + H_2 \xrightarrow{\text{heat}} Pb + H_2O$$

Although hydrogen is sometimes used to reduce metal oxides, it is still a rather expensive reducing agent compared to the least expensive of all—carbon. Carbon is generally used in the form of **coke,** which is made from coal by heating it at high temperatures in the absence of air. This treatment drives off the volatile components of the coal (from which other important chemicals are derived) and leaves nearly pure carbon behind.

Tin and lead are normally prepared by heating their oxides with carbon.

$$2SnO + C \xrightarrow{\text{heat}} Sn + CO_2$$

$$2PbO + C \xrightarrow{\text{heat}} Pb + CO_2$$

Undoubtedly the most important chemical reduction brought about by carbon is that of iron oxide, either Fe_2O_3 or Fe_3O_4. This is accomplished in an apparatus called a **blast furnace,** developed in about 1300 A.D. and which in modern times has taken the form shown in Figure 18.2.

The mixture of ingredients added to the blast furnace is called the **charge,** and its composition depends on the type of ore being reduced. For a hematite ore, which contains Fe_2O_3 as the iron-bearing compound, a typical charge consists of a mixture of limestone, coke, and the iron ore. The ore is normally composed of mostly Fe_2O_3 with impurities of sand (SiO_2, about 10%) and smaller amounts of compounds containing sulfur, phosphorus, aluminum, and manganese.

During the operation of the blast furnace, heated air is forced in at the bottom of the furnace where it reacts with carbon in a very exothermic reaction to produce carbon dioxide.

This hot blast of air gives the blast furnace its name.

$$C + O_2 \longrightarrow CO_2 \qquad \Delta H = -394 \text{ kJ}$$

The large quantity of heat generated in this region of the furnace raises the temperature to nearly 1900°C. As the hot gases rise, the CO_2 reacts with addi-

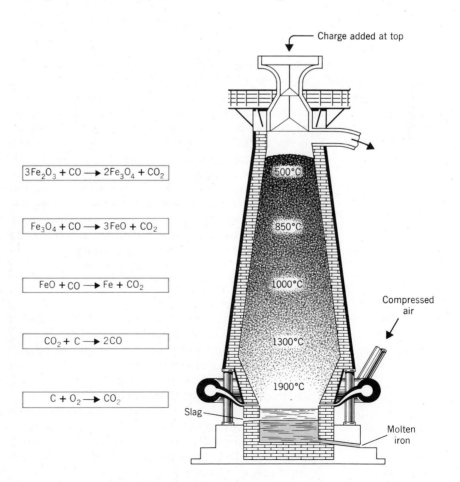

Figure 18.2
The blast furnace.

tional carbon in an endothermic reaction to form carbon monoxide, the active reducing agent in the furnace.

$$CO_2 + C \longrightarrow 2CO \qquad \Delta H = +173 \text{ kJ}$$

The reduction of the iron oxide takes place in a series of steps. Near the top of the furnace, Fe_2O_3 is reduced to Fe_3O_4.

$$3Fe_2O_3 + CO \longrightarrow 2Fe_3O_4 + CO_2$$

Farther down, in a hotter region of the furnace, this is reduced to FeO.

$$Fe_3O_4 + CO \longrightarrow 3FeO + CO_2$$

Finally, still farther down the FeO is reduced to the metal that, at these high temperatures, is a liquid and trickles down to form a pool of molten metal at the base of the tower.

$$FeO + CO \longrightarrow Fe + CO_2$$

The function of the limestone in the furnace is to provide a basic medium with which acidic oxides, such as SiO_2 and P_4O_{10}, or amphoteric oxides such as Al_2O_3 can react. At elevated temperatures limestone, $CaCO_3$, decomposes to form lime (CaO) and CO_2 according to the equation

$$CaCO_3 \longrightarrow CaO + CO_2$$

The lime then reacts as follows:

$$CaO + SiO_2 \longrightarrow CaSiO_3$$

$$6CaO + P_4O_{10} \longrightarrow 2Ca_3(PO_4)_2$$

$$CaO + Al_2O_3 \longrightarrow Ca(AlO_2)_2$$

The products of these reactions have relatively low melting points and are liquids when they are formed. The mixture, called **slag,** also runs to the base of the furnace, where it floats atop the molten iron. As these two layers are formed, the charge in the furnace settles and additional limestone-coke-ore mixture is added at the top. In this way the blast furnace operates continuously, with fresh charge being added at the top and molten iron and slag being tapped off at the bottom. These furnaces are often run for months at a time before they are shut down for routine maintenance.

Some blast furnaces are as high as a 15-story building and produce up to 2.2×10^6 kg of iron each day.

The liquid iron, when it is withdrawn from the blast furnace, is called **pig iron** and consists of about 95% Fe and approximately 4% carbon, with small amounts of silicon, manganese, phosphorus, and sulfur. This somewhat impure iron is very hard and can be poured into molds as **cast iron.** The slag that comes from the furnace can be used in making cement.

Refining

In the process of separating a metal from its ore, impurities are often introduced that impart undesirable properties to the final product. Therefore, it is generally necessary to purify the metal before it can be put to practical use. This purification process is called **refining.**

The specific procedure employed for refining a given metal depends on the chemical and physical properties of the metal as well as the properties of the impurities. As a result, there is no single method applicable to a very large number of different metals. We saw in Chapter 17 that copper can be economically refined electrolytically. This occurs, however, primarily because the silver and other precious metals recovered from the electrolytic cell offset the generally high cost of electricity.

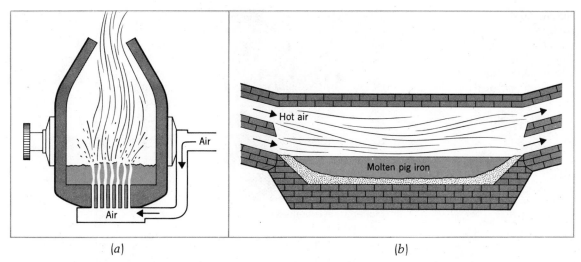

Figure 18.3
(a) *Bessemer converter.* (b) *Open hearth furnace.*

An interesting process for refining nickel, called the **Mond process,** makes use of the relative ease of formation of nickel carbonyl—a compound formed between nickel and carbon monoxide.

$$Ni + 4CO \longrightarrow Ni(CO)_4$$
nickel carbonyl

Besides being easily formed, nickel carbonyl is also very volatile (and very poisonous). The impure nickel is therefore treated with CO at a moderately low temperature of 60°C, where the $Ni(CO)_4$ that is formed exists as a gas. This is circulated to another portion of the apparatus, where it is heated to about 200°C and decomposes to give pure nickel plus CO, which can be recycled through the process.[1]

The most important commercial refining processes involve the conversion of pig iron into steel. This requires the removal of impurities such as silicon, sulfur, and phosphorus and lowering the carbon content significantly from the approximately 4% introduced into the pig iron in the blast furnace.

Modern steel making began with the introduction of the **Bessemer converter** in England in 1856. A batch of molten pig iron from the blast furnace, weighing about 23 megagrams, is transferred to a tapered cylindrical vessel containing a refractory lining (Figure 18.3). The composition of the lining is determined in part by the nature of the impurities in the iron. Since these impurities are usually silicon, phosphorus, and sulfur, whose oxides are acidic, a basic lining of dolomite (a $MgCO_3$, $CaCO_3$ mineral) is generally used. A blast of air (or oxygen) is blown through the metal from a set of small holes at the bottom of the vessel. The oxygen passing through the molten metal converts the silicon, phosphorus, and sulfur to oxides that then react with the lining to form a slag. The carbon in the pig iron is also oxidized to CO, so its concentration is reduced, too. The conversion of the pig iron to steel by this process is rapid, requiring about 15 minutes, and gives rise to a spectacular display of fire and showers of sparks. The reaction is difficult to control, however, and the quality of the steel produced in the Bessemer converter can be quite variable.

A somewhat newer method that virtually replaced the Bessemer process employs an **open hearth furnace,** a large, shallow hearth usually lined with a basic oxide refractory (for example, MgO, CaO). The furnace is charged with a mixture of pig iron, Fe_2O_3, scrap iron, and limestone. A mixture of burning

A *refractory substance* is one that is difficult to melt.

[1] For a time it was believed that extremely toxic $Ni(CO)_4$ was responsible for the so-called Legionnaires' disease that killed a group of people attending an American Legion Convention in Philadelphia in 1976. Later, however, this idea was abandoned.

Figure 18.4

Basic oxygen furnace used for the production of steel. Oxygen is forced through a tube called an oxygen lance. This burns away the impurities that combine with CaO, formed by the decomposing limestone, to give a slag.

Molten steel is poured from a basic oxygen furnace.

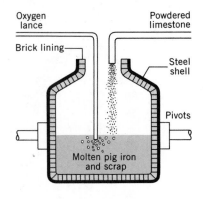

gases and hot air is played over the surface of the charge to maintain it in a molten state while a series of chemical reactions take place. Impurities in the steel are oxidized by the Fe_2O_3 and air. Carbon dioxide, formed by oxidation of the carbon in the pig iron, bubbles out of the mixture, keeping it stirred, while the SiO_2 and other acidic oxides combine with CaO (from the limestone) and the refractory lining to form a slag. This entire process takes much longer than the Bessemer process, requiring 8 to 10 hours to complete. However, the quality of the steel is much more easily controlled because chemical analyses can be constantly carried out on samples of the mixture. The increased length of time required to process a batch of steel is also offset by the fact that much larger quantities (about 2×10^5 kg) can be handled at one time. In addition, prior to being poured from the furnace, other metals (e.g., cobalt, chromium, nickel, vanadium, and tungsten) can be added to the steel to form alloys with special properties. A typical stainless steel, for instance, is composed of approximately 72% iron, 19% chromium, and 9% nickel.

Modern methods of chemical analysis, making use of high-speed computers, have enabled a return to a modified form of the Bessemer process called the **basic oxygen process.** This newer procedure, which has largely replaced the open hearth furnace because of its speed, involves forcing a mixture of powdered $CaCO_3$ and oxygen gas into the molten pig iron (Figure 18.4). This rapidly burns away the impurities, which form a slag. The characteristic emission spectra of the elements in the steel permit rapid chemical analysis, and additives can be incorporated into the steel in the proper proportions to give a product with the desired properties. This process takes only about 20 to 25 minutes to complete, thereby yielding very substantial savings in time (and, of course, money) over the open hearth process.

18.2 TRENDS IN METALLIC BEHAVIOR

In the earlier chapters we described some of the physical and chemical properties normally associated with metals, such as their tendencies to form ionic compounds with nonmetals and the basicity of their oxides. All metals are not equally metallic, however, especially in their chemical properties. For example, we commonly think of compounds formed from a metal and a halogen, such as chlorine or bromine, as being ionic. Beryllium chloride, however, does not fit this mold. When melted, $BeCl_2$ is a poor conductor of electricity, which suggests that molecules are present instead of ions. Similarly, aluminum forms molecular species such as Al_2Cl_6 and Al_2Br_6 in which the atoms are held together by covalent rather than ionic bonds.

Within the periodic table there are clear trends in the tendency of atoms to form covalent bonds. In Chapter 5 we learned that ionic bonds are preferred when a metal from the far left of the table combines with a nonmetal from the upper right corner. Covalent bonds are preferred, however, if two nonmetals combine. The deciding factor in each case is the difference in electronegativity between the bonded atoms. Metals have low electronegativities, which reflect their low ionization energies and electron affinities. Nonmetals, on the other hand, have high electronegativities. An atom's electronegativity, therefore, is a measure of how metallic it is chemically.

Metals with low electronegativities are said to be **electropositive.**

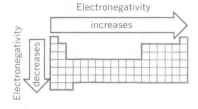

You learned in Chapter 5 that electronegativity increases from left to right across a period. If we compare the bonds formed by a metal with a given nonmetal—chlorine, for example—they should become less ionic moving from left to right across the table. This is because the difference in electronegativity between the metal and nonmetal becomes progressively smaller. In Period 2 we see this by comparing LiCl, $BeCl_2$, and BCl_3. Lithium chloride is distinctly ionic—it conducts electricity well in the molten state. As we've noted, $BeCl_2$ is a poor conductor when molten, which suggests that Be—Cl bonds are covalent. Boron trichloride is also covalent and exists as distinct molecules of BCl_3.

You also learned that electronegativity decreases going down a group. Metal–nonmetal compounds should therefore become more ionic as we descend a group, which they do. In Group IIA, for example, $BeCl_2$ is covalent, but $MgCl_2$ is ionic. In Group IIIA, BCl_3 is molecular and boron shows no evidence of forming B^{3+} ions. Aluminum chloride, on the other hand, appears to be ionic in the solid, although it exists in the vapor as molecules of Al_2Cl_6.

The elements of Group IVA, perhaps more than any others, illustrate the transition toward increasing metallic character as we descend a group. Carbon, at the head of the group, is nonmetallic in virtually every way.[2] Below carbon are the metalloids, silicon and germanium, and below these the metals, tin and lead.

Tin is unusual. At high temperatures it forms crystals that are metallic in appearance. At low temperatures (below 13.2°C) these very gradually change to a nonmetallic, powdery form in which tin has a diamond-type lattice. Tin articles left for long periods in the cold therefore gradually crumble. At one time it was thought that some pest attacked the tin, and this disintegration of tin articles was called tin disease.

Lead, at the bottom of the group, demonstrates only metallic properties in the elemental state.

The trends in the metallic character of the elements that we have just seen—decreasing from left to right across a period and increasing from top to bottom in a group—can also be demonstrated by considering acid-base properties. Recall that metal oxides such as Na_2O are typically basic. They react with water to form hydroxides.

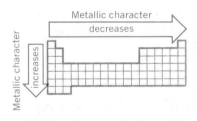

$$Na_2O + H_2O \longrightarrow 2NaOH$$

Nonmetal oxides are acidic. They form acids if they are able to react with water.

$$O^{2-} + H_2O \longrightarrow 2OH^-$$

$$SO_3 + H_2O \longrightarrow H_2SO_4$$

The acid-base properties of the oxide of an element can therefore be used as an indicator of its metallic character.

If we look again at Period 2, the trend in metallic character is clearly seen.

---

[2] One form of carbon, graphite, does conduct electricity. This, however, is more a result of the bonding in graphite than evidence for metallic behavior. The structure of graphite and the mechanism of its electrical conductivity are discussed in Section 19.2.

Lithium oxide reacts with water to form LiOH. It also reacts with acids to neutralize them.

$$Li_2O(s) + 2HCl(aq) \longrightarrow 2LiCl(aq) + H_2O$$

Lithium hydroxide reacts with acids, but it does not react with bases. This purely basic behavior of Li_2O and LiOH identifies lithium as a distinctly metallic element.

When we examine beryllium oxide, we find quite a different state of affairs. Although BeO is rather inert (resistant to chemical attack), under appropriate conditions it will react with *either acids or bases*. The hydroxide, $Be(OH)_2$, is also able to react with either acids or bases. Substances that behave this way are said to be **amphoteric**—they show both acidic and basic characteristics. Because beryllium compounds show some acidic characteristics, while the corresponding lithium compounds are purely basic, beryllium is less metallic than lithium. Moving farther to the right, boron is even less metallic than beryllium. Its oxide shows only acidic properties.

B_2O_3 reacts with water to give $B(OH)_3$, boric acid.

The same trend toward decreasing metallic character is also seen in moving from left to right in Period 3. The hydroxides of sodium and magnesium are only basic—they react with acids but not with other bases. Aluminum, on the other hand, is amphoteric, indicating that it is less metallic than either sodium or magnesium. The free element as well as its oxide and hydroxide reacts with both acids and bases.

$$2Al(s) + 6H^+(aq) \longrightarrow 2Al^{3+}(aq) + 3H_2(g)$$

$$2Al(s) + 2OH^-(aq) + 2H_2O \longrightarrow 2AlO_2^-(aq) + 3H_2(g)$$

The ion AlO_2^- is called aluminate ion.

The reaction of metallic aluminum with a base explains why you will find warnings on many oven cleaners not to use them on aluminum pots and pans. These oven cleaners contain lye (NaOH), which would quickly attack aluminum cookware. The popular drain cleaner Drano® also makes use of this reaction, as mentioned in Example 9.11. It consist of crystals of lye mixed with bits of metallic aluminum. When the Drano® is placed into a plugged drain, the reaction of the NaOH with the aluminum releases bubbles of H_2 that stir the mixture and help the cleaner dissolve any hair and grease that might be causing the stoppage.

18.3 IONIC–COVALENT CHARACTER OF METAL–NONMETAL BONDS

In the last section we saw that there is a gradual transition in the properties of the elements from metallic to nonmetallic as we move from left to right across a period. We also saw that the elements become more metallic going down a group. Although the changes in the covalence of metal–nonmetal compounds can be related to the variations in electronegativity, there is another way of looking at this that is useful in accounting for many properties.

Metal cations are almost always smaller than nonmetal anions. The positive charge on the cation is therefore concentrated in a rather small volume, while the negative charge on the anion is spread over a much larger volume. When a cation is placed near an anion, it can distort the electron cloud of the anion, pulling part of the electron density into the region between the two nuclei as shown in Figure 18.5. This causes the anion to become something of a dipole (besides being an anion) and we say that the electron cloud of the anion has been *polarized* by the cation. Since a covalent bond consists of electrons shared *between* nuclei, polarization of the anion results in the partial formation of a covalent bond. Furthermore, *the greater the degree of polarization, the greater the extent of covalent character of the bond.*

Figure 18.5

Polarization of an anion by the positive charge on the cation.
(a) Unpolarized anion and cation.
(b) Electrical charge on the anion is distorted by the cation.

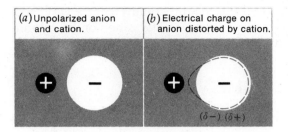

There are two major factors that contribute to a cation's ability to polarize a given anion. One of these is the number of positive charges on the cation. All other things being equal, a cation with a 2+ charge will distort the electron cloud of an anion more than one with a 1+ charge. The second factor is the size of the cation. In a small cation, the positive charge is highly concentrated and has a strong effect on the anion. The same positive charge on a large cation will be more spread out and diffuse, so it won't distort the anion's electron cloud as well. A cation's effect on an anion is therefore directly proportional to its positive charge and inversely proportional to its size. We could also say that the effect of the cation is proportional to the *ratio* of its charge to its size. This ratio is called the **ionic potential,** ϕ,

$$\phi = \frac{q}{r}$$

where q is the cation's charge and r is its radius.

To see how this relates to the covalent character of metal–nonmetal bonds, let's look once more at the chlorides of the Period 2 elements: LiCl, $BeCl_2$, and BCl_3. Suppose, for the moment, that these compounds could all be made ionic. The cations would then be Li^+, Be^{2+}, and B^{3+}, and each would be surrounded by one or more chloride ions. Suppose we now ask the question, "How would the ionic potentials of these cations vary?" Studying their electronic structures, we find that each has just a filled $1s$ subshell. All the valence electrons have been stripped away. As the nuclear charge increases from Li ($Z = 3$) to boron ($Z = 5$), this $1s$ subshell is drawn more and more closely to the nucleus, so the size of the cations should decrease from Li^+ to B^{3+}. At the same time, the charge increases. The ionic potentials, therefore, increase very rapidly from Li^+ to B^{3+}, which means that their polarizing effects on the Cl^- ions surrounding them should also increase dramatically. This would cause a rapid increase in the degree of covalent bonding going from Li^+—Cl^- to B^{3+}—Cl^-, which, of course, is exactly what we've found. In LiCl the bonding is ionic, in $BeCl_2$ it is largely covalent, and in BCl_3 the bonds are even more covalent.

Going down a group, the charge on the cations stays the same (for example, Be^{2+}, Mg^{2+}, Ca^{2+}, etc.). The sizes of the cations increase, however, so their ionic potentials decrease. This means that if we compare $BeCl_2$ with $MgCl_2$, we would expect Mg^{2+} to have less of a polarizing effect on Cl^- than Be^{2+}. The result is that $MgCl_2$ should be more ionic than $BeCl_2$, which is also what we've seen before.

The usefulness of this approach toward discussing the covalent character of metal–nonmetal bonds is that it lets us easily understand why certain metal compounds are so clearly covalent. Consider, for instance, tin(IV) chloride, $SnCl_4$. This substance is a clear colorless liquid at room temperature. It boils at 114°C, and its colorless cubic crystals are soft and melt at -33°C. These are properties we learned to identify with molecular substances, so the Sn—Cl bonds must be covalent. This is really not surprising in light of our previous discussions. If $SnCl_4$ were initially ionic, the cation would be Sn^{4+}. The large

True ions with large positive charges are rare; their ionic potentials are too large.

Of course, we can also explain the bonding in $SnCl_4$ by valence bond or molecular orbital theory.

charge on this cation would give it a very large ionic potential and we would expect very extensive polarization of the chloride ions to occur. This would lead to quite covalent Sn—Cl bonds.

18.4 COLORS OF METAL COMPOUNDS

Many inorganic compounds, such as common table salt or sodium bicarbonate, appear white, which means that when they are bathed in visible white light, their crystals reflect all the wavelengths of visible light equally well. You've only to look around, of course, to also appreciate that there is a huge number of very brightly colored compounds. Although many of these are organic substances, many also are inorganic compounds formed from a metal and a nonmetal.

A substance appears colored if it absorbs some wavelengths (or a band of wavelengths) from white light. The light that is reflected has these wavelengths missing and the color we perceive results from the wavelengths that remain.[3] Figure 18.6 illustrates a color wheel that we can use to anticipate the colors resulting when selected wavelength regions are absorbed. Across from each other are complementary colors. Green is the complementary color to red; yellow is the complementary color to violet. If a substance absorbs a particular color when bathed in white light, the reflected light is perceived as its complement. A substance that absorbs yellow light, for example, will appear violet. Chlorophyll, the substance that gives plants their green color, absorbs both red and blue light as shown by its absorption spectrum, Figure 18.7. The light that is reflected is yellow-green, and that's the color we see.

When a substance absorbs light of a particular color, the energy of the ab-

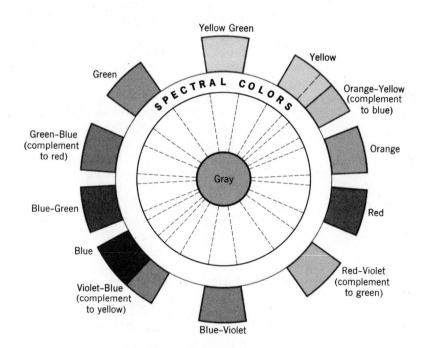

Figure 18.6

A color wheel. Colors that are across from each other are called complementary colors. When a substance absorbs a particular color, light that is reflected or transmitted has the color of its complement. Thus, something that absorbs red light appears green-blue, and vice versa.

[3] The perception of color is actually somewhat more complex than this because of the varying sensitivity of the human eye to various wavelengths. For example, the eye is much more sensitive to green than to red. If a compound reflects both of these colors with equal intensity, it will appear greenish simply because the eye sees green better than it sees red.

sorbed photon raises an electron from one energy level to another. Thus the energy levels in substances affect the wavelengths (colors) of light absorbed, just as we saw in Chapter 4 that they affect the wavelengths of light emitted. In molecules or polyatomic ions, these are molecular orbital energy levels somewhat like those discussed in Section 6.9.

Typical ionic compounds such as NaCl or Na_2S don't absorb visible light. They have no electronic transitions with energies corresponding to visible wavelengths, so they appear white or colorless. The absorption that does take place occurs in the shorter-wavelength (higher-frequency) ultraviolet region of the spectrum. Here the energy that is absorbed is used to shift an electron from the anion to the cation.

Recall that the energy of a photon is $E = h\nu$.

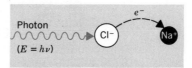

This is a **charge transfer process,** and the band of wavelengths absorbed from the ultraviolet rainbow is called a **charge transfer absorption band.**

As the bond between a metal and a nonmetal becomes somewhat covalent (that is, as electron density shifts away from the anion *toward* the cation), less energy is required to produce the charge transfer. This means that light of lower frequency (longer wavelength) is required, and the absorption band shifts from the high-frequency ultraviolet region toward the violet and blue end of the visible spectrum. As a portion of the absorption band moves into the visible region, some violet light is absorbed and the compound takes on a pale yellow color. As the bond becomes even more covalent, the band shifts farther into the visible region and even more violet light is absorbed, so the compound appears even more intensely yellow. The depth or intensity of the color, therefore, is related to the degree of covalent character of the metal–nonmetal bonds.

Using color as a way to gauge the covalent character of bonds, we can now look at two other factors related to the ease with which a cation is able to polarize an anion. One is the size of the anion. For anions of the same charge the ease of polarization increases with size. This occurs because as the size gets larger, the outer electrons are farther from the nucleus and are not held as tightly, so the cloud becomes more easily deformed. We can see this effect by looking at the colors of the silver halides (Table 18.1). Comparing AgCl, AgBr, and AgI, we see that as the anion becomes larger, the colors of the compounds become progres-

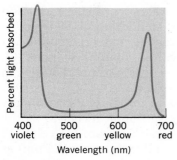

Figure 18.7

The absorption spectrum of chlorophyll. The light that is not absorbed is in the green and yellow parts of the spectrum—which is why the leaves of plants appear green.

The cream color of unexposed photographic film results from AgBr that is used as the light sensitive agent.

Table 18.1

Colors of the silver halides

Compound	Color	Anion Radius (pm)
AgF	White	136
AgCl	White	181
AgBr	Cream	195
AgI	Yellow	216

Table 18.2
Colors of some metal oxides and sulfides

Oxides	Color	Sulfides	Color
Al_2O_3	White	Al_2S_3	Yellow
Ga_2O_3	White	Ga_2S_3	Yellow
Sb_2O_3	White	Sb_2S_3	Yellow-red
Bi_2O_3	Yellow	Bi_2S_3	Brown-black
SnO_2	White	SnS_2	Yellow

sively deeper. This indicates a progressive increase in the covalent character of the Ag—X bonds as the halide ions become easier to polarize.[4]

We also find a similar relationship between covalent character and anion size if we compare metal oxides ($r_{O^{2-}} = 140$ pm) and sulfides ($r_{S^{2-}} = 184$ pm), as shown in Table 18.2. Both of these anions are colorless as evidenced by the fact that both Na_2O and Na_2S are colorless. However, with aluminum we find Al_2O_3 to be white while Al_2S_3 is yellow, suggesting a greater degree of ionic bonding in the oxide.

A second factor related to the ease of anion polarization is the charge on the anion. For anions of the same size, the ease of polarization increases as the negative charge becomes greater. For example, we can compare the colors of some compounds containing chloride ($r = 181$ pm) and sulfide ($r = 184$ pm) as shown in Table 18.3. In each case the sulfide has a deeper color than the corresponding chloride salt, indicating that the compounds containing the more highly charged sulfide ion are more covalent. We also see that, in general, most

Table 18.3
Colors of metal chlorides and sulfides

Chloride ($r = 181$ pm)		Sulfide ($r = 184$ pm)	
AgCl	White	Ag_2S	Black
CuCl	White	Cu_2S	Black
AuCl	Yellow	Au_2S	Brown-black
$CdCl_2$	White	CdS	Yellow
$HgCl_2$	White	HgS	Black or red, depending on crystal structure
$PbCl_2$	White	PbS	Black
$SnCl_2$	White	SnS	Black
$AlCl_3$	White	Al_2S_3	Yellow
$GaCl_3$	White	Ga_2S_3	Yellow
$BiCl_3$	White	Bi_2S_3	Brown-black

[4] The salts NaF, NaCl, NaBr, and NaI, which are predominantly ionic, are colorless. This indicates that the halide ions themselves are colorless. The colors of the AgX compounds are therefore a reflection of covalent bonding.

metal sulfides, except those of the alkali and alkaline earth metals, are deeply colored and possess quite a substantial degree of covalent bonding.

The presence of these colored compounds can often be seen. For instance, when silver tarnishes, it reacts with traces of H_2S in the air to give a dull film of black Ag_2S. This hydrogen sulfide, produced generally from decomposing organic matter, also darkens lead-based paint. The pigment, $Pb_3(OH)_2(CO_3)_2$, called *white lead*, reacts with H_2S to form black PbS.

The fact that many of these compounds possess rather striking colors has also been put to use throughout history. For example, the brilliant yellow of natural CdS and the vermilion red of HgS have led them to be employed as pigments for the oil paints used by artists. Not long ago, several sticks of black PbS that had been used as a type of mascara were recovered from an ancient Egyptian burial ground.

18.5 GROUP IA: THE ALKALI METALS

The elements of Group IA consist of hydrogen, lithium, sodium, potassium, rubidium, cesium, and francium. Except for hydrogen, all of them are very reactive metals. They are called alkali metals because their oxides are quite soluble in water and produce very basic (alkaline) solutions. For example, sodium oxide—the basic anhydride of sodium hydroxide—reacts as follows:

$$Na_2O(s) + H_2O(l) \longrightarrow 2Na^+(aq) + 2OH^-(aq)$$

The oxides of the other alkali metals react in a similar fashion.

Each of the alkali metals has only a single, rather loosely held electron in its outer shell. Loss of this electron gives an ion with a charge of 1+. This is the only oxidation state (other than zero, of course) that these elements exhibit, which leads to some rather simple chemistry. In fact, the chemical and physical similarities among the metals of Group IA illustrate in a very striking way the empirical basis of the periodic table—elements having similar properties are placed in the same vertical column.

Occurrence and preparation

Sodium is not required at all by plants, except for some salt marsh species.

The most abundant of the alkali metals are sodium and potassium. It is not surprising, therefore, that they are also the most biologically important. In animals, for instance, both sodium and potassium ions are needed, and the proper balance between them must be maintained in order for the organism to function properly. In plants, potassium is significantly more important than sodium.

Lithium, rubidium, and cesium are present in the earth's crust in much smaller amounts than sodium and potassium. They and their compounds are therefore more difficult to come by and, as a result, they are more expensive. Because of this, they have few practical applications.

Francium is radioactive, and even its longest-lived isotope, ^{223}Fr, has a half-life of only 21 minutes. (It is produced in the radioactive decay of another radioactive isotope, ^{227}Ac.) Because of francium's short half-life it is estimated that at any given time there is less than 30 g of it in the entire Earth's crust.

The alkali metals never occur free in nature. They always exist in compounds, which are found in both the Earth's crust as well as the ocean. For example, gigantic salt deposits (Figure 18.8a) are located in many places below the surface of the Earth—in Louisiana, New York, Michigan, Oklahoma, California, and Texas, to name just some. Where the climate is arid, salt deposits

Molar Concentrations in Sea Water	
Li$^+$	6×10^{-5}
Na$^+$	0.47
K$^+$	0.010
Rb$^+$	$\approx 10^{-6}$
Cs$^+$	$\approx 10^{-8}$

Clays, mica, and silicate ores also contain alkali metal ions.

even occur on the surface (Figure 18.8b). Nearly all the compounds of the alkali metals are soluble in water, and where rainfall is plentiful they have been largely washed away into the oceans and salt lakes, or leached into subterranean waters. The recovery of the alkali metals therefore occurs from both land and water—from salt mines like that shown in Figure 18.8, by the evaporation of sea water in large ponds (Figure 18.9), and from brine wells.

The separation of the alkali metals from their compounds can be accomplished by the electrolytic reduction of a molten compound. Normally, halides are chosen because they have relatively low melting points (compared to oxides, for instance). An example is the electrolysis of molten sodium chloride using the Downs cell, which was described in Chapter 17.

The Downs cell does not work well for the electrolysis of molten KCl, RbCl, and CsCl because at the temperatures required to melt these salts, the metals are very volatile. This makes their collection somewhat difficult. Instead, the molten chlorides are exposed to sodium vapor. The equilibrium

$$Na(g) + MCl(l) \rightleftharpoons NaCl(l) + M(g)$$

is shifted to the right because potassium, rubidium, and cesium are considerably more volatile than sodium—sodium condenses and the other metal evaporates. What makes this reaction especially interesting is that it is driven to the right even though sodium is not as strong a reducing agent as potassium, rubidium, or cesium when they are all compared under identical conditions.

One of the principal uses of sodium is the manufacture of tetraethyllead, a gasoline octane booster.

(a)

(b)

Figure 18.8
Sodium chloride occurs in large deposits both below and above ground. (a) The interior of a salt mine located in Texas. (b) The Bonneville Salt Flats in Utah.

Figure 18.9

Salt is harvested from the sea in many parts of the world, including the United States. Here we see brine being agitated by hand in seawater evaporation ponds in the Canary Islands. Most of the salt collected by these people is used in fish factories.

Physical properties

The alkali metals exhibit many of the typical properties that we've come to expect of metals: high luster and good thermal and electrical conductivity. Nevertheless, applications of the free metals rarely exploit these properties because the metals are so reactive. An exception has been the use of sodium in cooling nuclear reactors, which takes advantage of its low melting point, its relatively high boiling point, and its good thermal conductivity. These characteristics allow the metal to be easily melted and pumped through pipes that pass through the hot core of the reactor where the sodium quickly absorbs heat. The sodium is then pumped through the pipes of a heat exchanger outside the reactor where the heat is transferred to water, which can be made into steam to generate electricity.

The operation of a nuclear reactor is described further in Chapter 22.

The softness and low melting points of the alkali metals (Table 18.4) reflect the existence in the metallic lattice of singly charged cations that attract the surrounding "electron sea" weakly. These cations are formed by the loss of the single *s* electrons from the outer shells of the atoms. Because these outer-shell electrons are only weakly held, the alkali metals also have low ionization energies, electron affinities, and electronegativities.

An important physical property of the alkali metals is their emission spectra, which can be produced by passing an electric discharge through their vapors

Table 18.4
Some physical properties of the alkali metals

Element	Ionization Energy (kJ mol^{-1})	$\mathscr{E}^0$ (V)	M^+ Radius (pm)	Melting Point (°C)	Boiling Point (°C)
Lithium	520.1	−3.05	60	180.5	1326
Sodium	495.8	−2.71	95	97.8	883
Potassium	418.8	−2.92	133	63.7	756
Rubidium	402.9	−2.99	148	38.98	688
Cesium	375.6	−3.02	169	28.59	690

or introducing one of their salts into a Bunsen burner flame. Lithium salts, for example, impart a beautiful red color to a flame, sodium salts give a brilliant yellow color, whereas potassium salts produce a violet colored flame. These colors are intense enough to serve as useful qualitative tests called **flame tests** that can be used in analyzing mixtures of unknown composition. For example, if a drop of a solution of an unknown is placed into a flame and the yellow color is observed, sodium ions are in the unknown. If no yellow color is seen, then sodium is absent. The violet color of the potassium flame test is not as bright as the yellow sodium flame and is easily masked, even by traces of sodium in the unknown. Viewing the flame through blue glass—called *cobalt glass*—filters out the yellow and allows the violet potassium flame to be seen.

The brilliant yellow emission by sodium accounts for one of this element's growing commercial uses: in sodium vapor lamps. These are becoming more and more widely used throughout the country for street lighting because they are much more economical to operate than incandescent lamps. An incandescent lamp, such as an ordinary light bulb, gives off much of its energy in the form of invisible infrared radiation, so much of the electrical energy used to operate it is wasted. However, in a sodium vapor lamp, which is really nothing more than a gas discharge tube containing sodium vapor as the gas, most of the energy of the electrical discharge appears as visible yellow light ($\lambda = 589$ nm). This actually corresponds to a pair of closely spaced lines in the emission spectrum of sodium.

Chemical properties and important compounds

The alkali metals are the most reactive metals. They are very powerful reducing agents and they are able to reduce water with the evolution of hydrogen. The general reaction is

$$2M(s) + 2H_2O(l) \longrightarrow 2M^+(aq) + 2OH^-(aq) + H_2(g)$$

The effectiveness of the alkali metals as reducing agents is reflected in the extremely negative reduction potentials of their ions (Table 18.4). These suggest that the half-reactions

$$M^+(aq) + e^- \longrightarrow M(s)$$

occur with great difficulty, and therefore that the oxidation half-reactions

$$M(s) \longrightarrow M^+(aq) + e^-$$

take place easily.

Figure 18.10

Enthalpy diagram for the reaction:
M(s) → M⁺(aq) + e⁻.

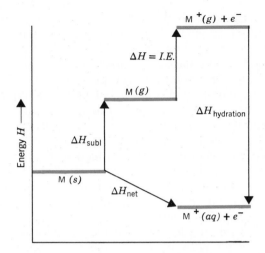

Figure 18.11

Solvation of a cation by water dipoles.

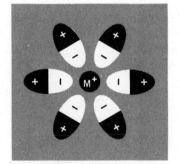

Recall that the ionic potential is the ratio of an ion's charge to its radius.

A close examination of the trends in the ionization energy and in $\mathscr{E}^0$ reveals an apparent contradiction, however. Note that the ionization energy (IE) decreases as we proceed down within the group, suggesting that it becomes progressively easier to strip an electron from the atom as we go from Li to Cs. In Chapter 4 we saw that this is, in fact, expected. We would also anticipate that the reduction potentials should become more negative as the IE becomes smaller since the elements should become more easily oxidized. This trend is indeed followed from Na downward; however, Li has an $\mathscr{E}^0$ that is more negative than Na (or any of the other alkali metals for that matter). Why is this so?

The ionization energy, remember, is a measure of the ease with which a *gaseous* atom loses electrons to produce a *gaseous* cation. The reduction potential, on the other hand, is concerned with the transfer of electrons between the *solid* metal and the corresponding cation in *aqueous solution*, where it is hydrated by the water molecules surrounding it. This latter process is more complex than simply removing an electron from the isolated metal atom. To understand the trends in the reduction potentials, we must break down the overall reaction into several steps. If we concentrate on the enthalpy changes involved in the reaction, we can construct the diagram in Figure 18.10. We see that the net enthalpy change is the sum of three energy terms. Two of these are endothermic—the sublimation energy, ΔH_{subl}, which is the energy needed to convert the solid into gaseous atoms, and the ionization energy, which we have already examined. The third quantity, called the hydration energy, is strongly exothermic. It corresponds to the energy *released* when the cation is placed into the solvent cage where it is surrounded by the water dipoles oriented in such a way that their negative ends are directed at the positive ion (Figure 18.11).

Among the alkali metals the sublimation energy remains approximately constant as we descend the group while the ionization energy decreases. To reach the peak on the energy diagram, we require the greatest quantity of energy for Li and the least quantity for Cs. However, because of its small size and high ionic potential, upon hydration Li⁺ interacts much more strongly with the water dipoles than any of the other Group IA ions do. As a result, the hydration energy of Li⁺ is unusually large, and much greater than for the other M^+ ions in the group. Therefore, the net overall enthalpy change is most exothermic for Li. This in turn causes Li to be more easily oxidized in an aqueous medium than any other alkali metal. In other words, the extraordinarily high hydration energy of the small Li⁺ ion more than compensates for its relatively high ionization energy, and causes Li to have an unexpectedly large negative $\mathscr{E}^0$.

Associated with the easy loss of electrons from the alkali metals is their interesting behavior in liquid ammonia. We have already seen that these metals are capable of reducing water to liberate H_2. Ammonia is not as easily reduced as water and, when placed into this solvent, alkali metals dissolve without reaction to form deep blue solutions. It is generally agreed that this color, which is identical for liquid ammonia solutions of all the alkali metals (as well as for Ca, Sr, and Ba from Group IIA), is a result of the presence of free electrons that have become solvated by ammonia molecules. Apparently, when the metal dissolves in NH_3, it loses its valence electron to become a cation. This electron becomes surrounded by NH_3 molecules arranged so that the positive ends of their dipoles are directed at the negatively charged electron, as shown in Figure 18.12, thereby stabilizing it through solvation. Solutions containing alkali metals in liquid ammonia are, as we would expect from the presence of readily available electrons, excellent reducing agents.

The strong tendency of the alkali metals to undergo oxidation permits them to react readily with most of the elemental nonmetals. The design of the Downs cell, you recall, is based on the need to keep Cl_2 and Na apart after they are formed by electrolysis. All the halogens (F_2, Cl_2, Br_2, and I_2) react with all the alkali metals to form the corresponding salts.

For example,

$$2Na(s) + Cl_2(g) \longrightarrow 2NaCl(s)$$

Among the most interesting reactions of the alkali metals is their behavior toward elemental oxygen. Only lithium burns in air to form the normal oxide, Li_2O.

$$4Li(s) + O_2(g) \longrightarrow 2Li_2O(s)$$

Na$_2$O has a strong affinity for moisture and is an effective drying agent.

Sodium undergoes a similar reaction, but only if the supply of oxygen is limited. In the presence of excess oxygen sodium forms the pale yellow peroxide.

$$2Na(s) + O_2(g) \longrightarrow Na_2O_2(s)$$
sodium peroxide

This solid contains the peroxide ion, O_2^{2-}, which itself is an effective oxidizing agent. Sodium peroxide is therefore used commercially as an oxidizing agent. It is also used as a bleaching agent, because the oxidation of intensely colored molecules often gives colorless reaction products. When dissolved in water, the peroxide ion hydrolyzes extensively

$$Na_2O_2(s) + 2H_2O(l) \longrightarrow 2Na^+(aq) + 2OH^-(aq) + H_2O_2(aq)$$

This causes solutions of Na_2O_2 to be very alkaline.

Potassium, rubidium, and cesium react with oxygen to form *superoxides*. For example,

$$K(s) + O_2(g) \longrightarrow KO_2(s)$$
**potassium
superoxide**

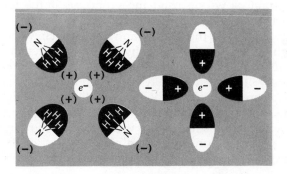

Figure 18.12
Solvated electron in liquid ammonia.

These compounds contain the paramagnetic superoxide ion, O_2^-. When the superoxides are dissolved in water, oxygen is evolved.

$$2MO_2(s) + 2H_2O(l) \longrightarrow 2M^+(aq) + 2OH^-(aq) + O_2(g) + H_2O_2(aq)$$

A recirculating breathing apparatus like this allows the user to avoid breathing toxic fumes.

Because of this reaction, potassium superoxide has found applications in breathing equipment designed to recirculate the air of the user. When air is exhaled, it contains both moisture and carbon dioxide. This is circulated through a canister containing KO_2. Moisture reacts as above, and carbon dioxide is removed from the air and replaced with oxygen by the overall reaction

$$4KO_2(s) + 2CO_2(g) \longrightarrow 2K_2CO_3(s) + 3O_2(g)$$

This allows the user to continue to breathe without having to draw in possibly contaminated air from outside the apparatus.

Because of their vigorous reactions with both moisture and oxygen, the alkali metals must be stored under an inert (nonreactive) liquid—often an oil or kerosene. Use is made of their affinity for H_2O in drying solvents. Sodium, for instance, is often employed to effectively remove traces of moisture from laboratory solvents. Commercially, the alkali metals have been used as *getters* in the production of electronic vacuum tubes. Getters combine with the last traces of H_2O and O_2 that would otherwise interfere with the operation of the tube.

Besides combining with the halogens and oxygen, the alkali metals react directly with virtually all the other nonmetals as well. Lithium, however, is the only one that combines directly with gaseous nitrogen to form a *nitride*.

$$6Li(s) + N_2(g) \longrightarrow 2Li_3N(s)$$
lithium nitride

Most of the important compounds of the alkali metals are those of sodium and potassium. This is simply because among the Group IA metals these two elements have the largest abundance. Sodium chloride is by far the most readily available and easily recoverable of all the alkali metal compounds, which makes it an inexpensive raw material for preparing other sodium compounds. For this reason, sodium compounds are considerably cheaper than those of potassium or the other alkali metals. Since there are such close similarities among the chemical properties of alkali metal compounds of a given type, when given a choice industry will almost always select a sodium compound simply because of economics.

The primary raw material for the preparation of potassium compounds is potassium chloride. It is obtained from *sylvite,* a mineral form of KCl, and *carnallite,* $KCl \cdot MgCl_2 \cdot 6H_2O$. These are important as fertilizers because of the need of plants for both potassium and magnesium.

Over 10^{10} kg of Cl_2 and NaOH are made annually from NaCl by electrolysis.

Sodium hydroxide—made from salt by the electrolysis of brine—is industry's most important strong base. Its common name is *lye* or *caustic soda.* Around the home it is found in oven and drain cleaners. It is used industrially to make soap and detergents, pulp and paper, textiles, in removing sulfur from petroleum, and in the manufacture of myriad other chemicals. In the majority of its applications, NaOH is used to neutralize acids.

Potassium hydroxide, which is produced by electrolysis of KCl solution, is more expensive than NaOH, so its uses have been limited. It serves as the electrolyte in alkaline Zn/MnO_2 batteries, as we learned in Chapter 17.

The carbonates of the alkali metals constitute another class of important compounds. Once again, the sodium salts are used most widely because of their low cost. The principal source of sodium carbonate is *Trona ore,* a mixture of sodium carbonate and sodium bicarbonate with the composition,

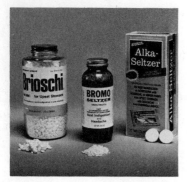

Sodium bicarbonate is an ingredient in these products that also contain a weak acid. When they dissolve in water CO_2 is produced, which is why they fizz.

$Na_2CO_3 \cdot NaHCO_3 \cdot 2H_2O$. Large amounts of this ore are mined from deposits in Wyoming.

Sodium carbonate is also made chemically from salt and limestone ($CaCO_3$) by the **Solvay process,** which takes advantage of the fact that $NaHCO_3$ is less soluble in cold water than $NaCl$. The process begins by thermally decomposing limestone to give CaO and CO_2.

$$CaCO_3 \xrightarrow{\text{heat}} CaO(s) + CO_2(g)$$

The carbon dioxide, along with ammonia, is bubbled into a concentrated solution of salt. The ammonia partially neutralizes carbonic acid that is formed from the carbon dioxide.

$$CO_2(g) + H_2O(l) \longrightarrow H_2CO_3(aq)$$
$$H_2CO_3(aq) + NH_3(aq) \longrightarrow NH_4^+(aq) + HCO_3^-(aq)$$

As the concentration of HCO_3^- builds up, the less soluble $NaHCO_3$ begins to precipitate. The overall reaction can be written

$$Na^+(aq) + Cl^-(aq) + NH_4^+(aq) + HCO_3^-(aq) \longrightarrow NaHCO_3(s) + NH_4^+(aq) + Cl^-(aq)$$

After the solid $NaHCO_3$ is separated by filtration it is heated to convert it to sodium carbonate.

$$2NaHCO_3(s) \xrightarrow{\text{heat}} Na_2CO_3(s) + H_2O(g) + CO_2(g)$$

Meanwhile, the solution containing NH_4Cl is treated with the calcium oxide left over from the decomposition of the limestone. This regenerates ammonia, which is recycled and used again. The reactions are

$$CaO + H_2O \longrightarrow Ca(OH)_2$$
$$Ca(OH)_2 + 2NH_4Cl \longrightarrow CaCl_2 + 2H_2O + 2NH_3(g)$$

If care is taken not to lose any ammonia, the process consumes $NaCl$ and $CaCO_3$ and produces Na_2CO_3 and $CaCl_2$. The net overall change is

$$2NaCl + CaCO_3 \longrightarrow Na_2CO_3 + CaCl_2$$

The hydrate, $Na_2CO_3 \cdot 10H_2O$, is called washing soda. Its solutions are basic because of hydrolysis of the CO_3^{2-} ion.

Sodium carbonate is used in huge amounts by industry (about 8×10^9 kg of it are produced each year). About half is used to manufacture glass; the rest is used to manufacture other chemicals, process pulp and paper, and make soap and detergents.

Sodium bicarbonate, the intermediate product in the Solvay process is also a useful chemical. A solution of it is mildly basic and serves as a buffer because of the ability of the HCO_3^- ion to react with both acids and bases.

The bicarbonate ion is simultaneously a weak Brønsted acid and a weak Brønsted base.

$$HCO_3^- + H^+ \longrightarrow H_2CO_3$$
$$HCO_3^- + OH^- \longrightarrow H_2O + CO_3^{2-}$$

For this reason, sodium bicarbonate is recommended as an additive to swimming pools because it helps control the pH of the water when other chemicals are added to destroy bacteria.

The common name for sodium bicarbonate is *baking soda*. When added to dough it decomposes during baking to release carbon dioxide.

$$2NaHCO_3(s) \xrightarrow{\text{heat}} Na_2CO_3(s) + CO_2(g) + H_2O(g)$$

The CO_2 produces tiny bubbles throughout the dough and makes the baked product rise and become light and appealing.

The carbonates and bicarbonates of the other alkali metals have properties similar to the sodium compounds. Potassium carbonate, K_2CO_3, is called *potash* and is a major constituent of wood ashes. When needed in quantity it is prepared by first reacting potassium hydroxide with carbon dioxide to give potassium bicarbonate.

$$KOH + CO_2 \longrightarrow KHCO_3$$

Thermal decomposition gives K_2CO_3.

Lithium carbonate, Li_2CO_3, has been found to be useful as a drug in the treatment of manic depression.

18.6 GROUP IIA: THE ALKALINE EARTH METALS

The elements of Group IIA consist of beryllium, magnesium, calcium, strontium, barium, and radium. All are very reactive metals, although not as reactive as the metals of Group IA. They are called *alkaline earth* metals because their oxides are basic, and because many of their compounds are of low solubility in water and are therefore found in mineral deposits in the Earth's crust.

Each of the alkaline earth metals has two electrons in an *s* subshell that lies outside a noble gas core. Each exhibits only a 2+ oxidation state in its compounds and, with the exception of the compounds of beryllium and some of those of magnesium, their compounds are largely ionic. As in Group IA, there are strong similarities among all members of this group, but among the heavier ones—calcium through radium—the similarities are particularly close.

Occurrence and preparation

Calcium is the third most abundant *metal* in the Earth's crust.

Calcium and magnesium are among the most abundant of the elements in the Earth's crust, ranking fifth and eighth in terms of percent by mass. They are found in many locations in large mineral deposits of various compositions. Examples are gypsum ($CaSO_4 \cdot 2H_2O$), limestone ($CaCO_3$), dolomite ($CaCO_3 \cdot MgCO_3$), and carnallite ($MgCl_2 \cdot KCl \cdot H_2O$). In sea water, Ca^{2+} and Mg^{2+} are both major constituents among the dissolved ions. On a mole basis, Mg^{2+} is the third most abundant ion in the sea, and Ca^{2+} places sixth. Calcium and magnesium are also the most biologically important alkaline earth metals. Calcium is found in the bones of animals, and shellfish extract Ca^{2+} from sea water to form their $CaCO_3$ shells. Magnesium is vital to plants where it is found at the center of the chlorophyll molecule—the substance that captures solar energy and begins the biological food chain.

In sea water, the concentration of Mg^{2+} is 0.056 M and that of Ca^{2+} is 0.011 M.

The major source of beryllium is the mineral *beryl*, $Be_3Al_2(SiO_3)_6$. Sometimes beautiful large pure crystals of this substance are found and, when polished, become the gems emerald and aquamarine.

Radium was discovered by Pierre and Marie Curie.

Strontium and barium are recovered from deposits of their insoluble sulfates and carbonates. Radium, however, occurs principally as an impurity in *pitchblende*—a mineral from which uranium is extracted. Radium, which itself is radioactive, is formed as a product of the radioactive decay of heavier elements. For example, ^{226}Ra is radium's longest-lived isotope, with a half-life of 1600 years. It is one of a chain of isotopes produced from ^{238}U when it decays (see Chapter 22).

Of the metals in Group IIA, only beryllium and magnesium are produced and used in significant quantities. This is because they are the only alkaline earth metals that do not react rapidly with air and moisture at room temperature.

Beryllium is obtained by the electrolysis of molten beryllium chloride. However, sodium chloride must be added to the melt as an electrolyte because $BeCl_2$ is primarily covalent and therefore is a very poor electrical conductor. During the electrolysis, the less active metal, Be, is produced at the cathode and Cl_2 is evolved at the anode.

$$BeCl_2(l) \xrightarrow[\text{(NaCl)}]{\text{electrolysis}} Be(l) + Cl_2(g)$$

Beryllium-copper alloy tools can be used in situations like this because they are nonsparking if accidentally struck against other metal objects.

In recent times beryllium has found a variety of practical uses. It is a very low-density, strong metal that has structural applications in missiles and spacecraft. It absorbs X rays less than any other metal that's stable toward air, and is used as windows for X-ray tubes. Beryllium-copper alloy is fashioned into springs, electrical contacts, and welding rods. It is also made into tools—hammers, for example—that will not create sparks when used. This is particularly important when workers must labor in an explosive atmosphere. A major drawback of beryllium, however, is that its compounds are highly toxic, and stringent safety standards must be maintained when it or its alloys are machined.

Magnesium is extracted both from its land-based ores and the sea. On land its principal sources are its chloride, found in carnallite, $MgCl_2 \cdot KCl \cdot 6H_2O$, and its carbonate in dolomite, $CaCO_3 \cdot MgCO_3$. Extraction from dolomite involves first heating the ore strongly—a process called *calcining*—which decomposes the carbonates and forms the oxides.

$$CaCO_3 \cdot MgCO_3(s) \xrightarrow{\text{heat}} CaO \cdot MgO(s) + 2CO_2(g)$$

The mixed oxides are then treated with an excess of sea water, which contains appreciable amounts of dissolved Mg^{2+}. The water converts the oxides to the hydroxides.

$$CaO(s) + H_2O \longrightarrow Ca^{2+}(aq) + 2OH^-(aq)$$

$$MgO(s) + H_2O \longrightarrow Mg(OH)_2(s)$$

Calcium hydroxide, although not extremely soluble, is appreciably more soluble than $Mg(OH)_2$. As it dissolves, it makes the water basic, causing the Mg^{2+} that's in the sea water to precipitate as $Mg(OH)_2$. The combined precipitate of $Mg(OH)_2$ is then filtered and dissolved in hydrochloric acid to convert it to the chloride. This solution is evaporated and the solid $MgCl_2$ is melted and electrolyzed, giving magnesium and chlorine. The chlorine produced is made into HCl again and recycled.

In the absence of dolomite, the magnesium in sea water can be recovered by making the water basic with calcium oxide. This is obtained by calcining limestone, $CaCO_3$, or even sea shells, which are also composed of $CaCO_3$.

$$CaCO_3(s) \xrightarrow{\text{heat}} CaO(s) + CO_2(g)$$

$$CaO(s) + H_2O + Mg^{2+}(aq) \longrightarrow Ca^{2+}(aq) + Mg(OH)_2(s)$$

The precipitate of $Mg(OH)_2$ is treated as described previously.

A modern plant that produces magnesium from seawater. In the foreground, we see the magnesium hydroxide settling ponds.

Magnesium metal has a number of practical uses. Its low density and moderate strength when alloyed with aluminum make it a useful structural metal. (Perhaps you or your family owns a magnesium alloy stepladder.) Presently magnesium is more expensive than aluminum, but its virtually inexhaustible supply in the sea is likely to make it comparatively less expensive as land-bases supplies of aluminum ore are depleted.

Another application of magnesium—one you've surely seen—is in flash-bulbs and signal flares. The combustion of magnesium

$$2 \ Mg(s) + O_2(g) \longrightarrow 2 \ MgO(s)$$

produces not only a great deal of heat, but also intense light. As noted in Chapter 7, a flashbulb (Figure 7.15 on p. 268) contains a fine magnesium wire in an atmosphere of pure oxygen. The flashbulb is fired by passing a small electrical current through the wire, which heats it and sets off the combustion reaction. Afterwards, the interior of the bulb is coated with a thin deposit of the white powder MgO.

Calcium, strontium, and barium have very few commercial applications as free metals because of their reactivity toward oxygen and moisture. As a result, they are prepared only in small quantities, usually by electrolysis of their molten chlorides.

Barium is used as a *getter* to remove traces of oxygen in the manufacture of vacuum tubes.

Physical properties

Table 18.5 lists some physical properties of the alkaline earth metals. We see that in general they have higher melting points and higher densities than their neighbors in Group IA. This is not difficult to explain. The greater effective nuclear charges experienced by their outer electrons make the atoms of the alkaline earth metals smaller than those of the alkali metals alongside them in the periodic table, so more mass is packed into a smaller volume, which leads to higher densities. The 2+ charge on the cations in the metallic lattice of an alkaline earth metal causes them to be attracted more strongly to the "sea of electrons" and makes it more difficult to pull them apart than the 1+ cations in the metallic lattice of an alkali metal. The Group IIA metals therefore have the higher melting points. The Group IIA metals are also harder than the Group IA metals for the same reason, and they have larger ionization energies.

Salts of the heavier alkaline earth metals, like those of the metals of Group IA, produce striking colors when introduced into a Bunsen burner flame. These colors serve as flame tests for them. Calcium salts, for example, give a brick-red color; strontium salts produce a crimson flame; and barium salts give a yellowish-green flame. Salts of these metals are often used to give spectacular colors to fireworks displays.

Table 18.5
Some properties of the alkaline earth metals

| Element | Ionization Energy (kJ mol^{-1}) | | $\mathscr{E}^0$ (V)[a] | M^{2+} Radius (pm) | Melting Point (°C) |
	First	Second			
Beryllium	900	1757	−1.70	31[b]	1278
Magnesium	737.6	1450	−2.38	65	651
Calcium	589.5	1146	−2.76	99	843
Strontium	549	1064	−2.89	113	769
Barium	503	965	−2.90	135	725
Radium	509	979	−2.92	140	700

[a] For $M^{2+}(aq) + 2e^- \longrightarrow M(s)$.
[b] Estimated.

Chemical properties and important compounds

When $\mathscr{E}^0$ has a large negative value, the metal ion is difficult to reduce and the metal itself is easily oxidized.

Ionic Radii (pm)			
Li$^+$	60	Be^{2+}	31*
Na$^+$	95	Mg^{2+}	65
K$^+$	133	Ca^{2+}	99
Rb$^+$	148	Sr^{2+}	113
Cs$^+$	169	Ba^{2+}	135

*Estimated.

Their protective oxide coatings permit beryllium and magnesium to serve as useful structural metals.

As we've already noted, the alkaline earth metals are all very reactive elements. They are easily oxidized and therefore serve as excellent reducing agents, as evidenced by their very negative reduction potentials. The heavier of them, in fact, have reduction potentials comparable to those of the alkali metals. At first glance, this may appear somewhat surprising, because the sum of the first two ionization energies of an alkaline earth metal is considerably greater than just the first ionization energy of an alkali metal. In other words, removing two electrons from an isolated alkaline earth metal atom is much more difficult than removing a single electron from an isolated metal atom of Group IA. However, as we learned earlier in our discussion of lithium's unusually negative $\mathscr{E}^0$, the size of the reduction potential is controlled by more than simply the ionization energy. The hydration energy of the ion also plays a very important role.

The magnitude of the hydration energy of an ion depends both on its size and charge. A small ion can get closer to the water dipoles than a large ion can, so the smaller ion interacts more strongly with the solvent and its hydration energy is larger. A highly charged ion attracts the water dipoles more strongly than one of low charge, so the higher the ion's charge, the larger its hydration energy. Therefore, because the ions of the alkaline earth metals are both smaller *and* more highly charged than those of their neighbors in Group IA, their hydration energies are much larger. This serves to offset their much larger ionization energies. As a result, the alkaline earth metals have reduction potentials almost as negative as those of the alkali metals.

As reducing agents, the Group IIA metals are all powerful enough to reduce water, at least in principle. However, beryllium and magnesium both form insoluble oxide coatings that protect them from attack by water. Beryllium, in particular, is quite resistant to oxidation, even by acids, because of its BeO coating. Magnesium is more reactive than beryllium. Even though it is not attacked by cold water, magnesium reacts slowly with boiling water and quite rapidly with steam to liberate hydrogen.

The remaining alkaline earth metals—calcium through radium—form oxides that are at least moderately soluble in water, so their oxides are unable to protect them. As a result, they reduce water and liberate hydrogen. The general reaction is

$$M(s) + 2H_2O \longrightarrow M(OH)_2(aq) + H_2(g)$$

The vigor with which this reaction occurs increases from calcium to strontium to barium.

Like the alkali metals, the Group IIA elements react directly with most elemental nonmetals. For example, magnesium, we learned, reacts directly with oxygen to form MgO. It also is able to react with nitrogen to give magnesium nitride.

$$3Mg(s) + N_2(g) \longrightarrow Mg_3N_2(s)$$

With sulfur it gives MgS, and with the halogens it yields MgX$_2$ (X = halide ion). The other Group IIA metals react similarly (except with N$_2$).

An interesting phenomenon among the Group IIA elements is the variation in their metallic character. Although physically they all exhibit metallic characteristics, chemically we find that beryllium, and, to some slight extent, magnesium exhibit a degree of nonmetallic character as well.

Beryllium and its oxide are amphoteric—they dissolve in both acids and bases. For example, in base they react as follows:

The ion Be(OH)$_4{}^{2-}$ is called the beryllate ion.

$$Be + 2H_2O + 2OH^- \longrightarrow Be(OH)_4{}^{2-} + H_2(g)$$

$$BeO + H_2O + 2OH^- \longrightarrow Be(OH)_4{}^{2-}$$

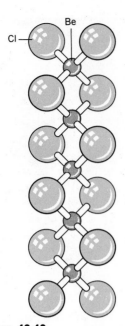

Figure 18.13
The structure of $(BeCl_2)_x$.

The remainder of the alkaline earth metals and their oxides are not amphoteric, so beryllium is chemically less metallic than the other members of its group.

Another way that beryllium is less metallic than the other Group IIA elements is in the degree of covalence of its compounds. There is no evidence that beryllium exists as Be^{2+} in any of its compounds—they all show a significant degree of covalent character. Beryllium chloride, for example, is a poor conductor when melted, and we saw that an electrolyte—sodium chloride—has to be added to molten $BeCl_2$ so that it can be electrolyzed. In the solid state, $BeCl_2$ exists as a covalently linked chain of $BeCl_2$ units in which Be completes its octet by forming coordinate covalent bonds with chlorine atoms on adjacent $BeCl_2$ molecules.

(Arrows indicate coordinate covalent bonds.)

Since each Be atom has four separate electron pairs around it, the arrangement of the chlorines around the Be is tetrahedral, as shown in Figure 18.13.

The covalent character of bonds to beryllium can be related to the small size and high charge that a true Be^{2+} ion would have. We saw in Section 18.3, that this would give the Be^{2+} a very large ionic potential—sufficiently large to polarize other ions next to it and cause the bonds to become mostly covalent.

Although most magnesium compounds are ionic, magnesium does form a variety of covalently bonded compounds in which portions of organic molecules—molecules derived from methane (CH_4), ethane (C_2H_6), and other hydrocarbons—are bonded to magnesium. Examples are C_2H_5MgBr and $Mg(C_2H_5)_2$. These are called *organomagnesium compounds* and are important reagents in organic chemistry. The compounds of the rest of the alkaline earth metals are nearly all predominantly ionic.

The most important compounds of the alkaline earth metals are their carbonates, oxides, hydroxides, and sulfates. The carbonates are all quite insoluble in water, but there are trends in the solubilities of the oxides, hydroxides, and sulfates that influence their practical applications.

The solubilities of the oxides increase going down the group. BeO and MgO are insoluble, but CaO, SrO, and BaO react with water to form the corresponding hydroxides.

$$MO + H_2O \longrightarrow M(OH)_2 \quad (M = Ca, Sr, Ba)$$

$$O^{2-} + H_2O \longrightarrow 2\,OH^-$$

The solubilities of the hydroxides also increase going down the group. Magnesium hydroxide is insoluble, calcium hydroxide is slightly soluble, and the hydroxides of strontium and barium are even more soluble. In contrast, the solubilities of the sulfates vary in the opposite direction: $BaSO_4$, $SrSO_4$ and $CaSO_4$ are insoluble, but their K_{sp}'s increase from Ba to Ca. Magnesium sulfate, on the other hand, is quite soluble.

	K_{sp}
$CaSO_4$	2×10^{-4}
$SrSO_4$	2.9×10^{-7}
$BaSO_4$	1.5×10^{-9}

We learned earlier that calcium and magnesium occur in large deposits of their carbonates—$CaCO_3$ (limestone and marble) and $MgCO_3 \cdot CaCO_3$ (dolomite). Calcium carbonate is a very common chemical. Seashells are composed almost entirely of $CaCO_3$, as is the chalk your teacher uses to write on the blackboard. Calcium carbonate is also used as a mild abrasive in toothpaste and household cleansers, and as an antacid. Limestone is an extremely important raw material in many industrial reactions. As we saw earlier, its thermal decomposition produces calcium oxide and carbon dioxide.

Some things that are composed of or contain calcium carbonate.

$$CaCO_3 \xrightarrow[900°C]{\text{heat}} CaO + CO_2$$

Crushed limestone or dolomite is often spread on lawns and gardens to make the soil less acidic.

Calcium oxide is called *lime*, or sometimes *quicklime*, and its annual production ranks second among industrial chemicals because it is an inexpensive, relatively strong base. When treated with water—a process called *slaking*—calcium hydroxide (*slaked lime*) is formed with considerable evolution of heat.

$$CaO + H_2O \longrightarrow Ca(OH)_2$$

About 10^{10} kg of CaO are produced each year.

Calcium oxide is an ingredient in Portland cement, and this reaction is among the first to occur when water is added to the cement.

Magnesium oxide can also be formed by decomposing its carbonate. However, unlike CaO, magnesium oxide is relatively unreactive toward water, especially if heated to very high temperatures. It is used as a component in refractory bricks—those used to line the interiors of high temperature furnaces. It is also used in the manufacture of paper, and medicinally as an antacid.

MgO melts at about 2800°C.

Magnesium hydroxide, $Mg(OH)_2$, is much less soluble in water than $Ca(OH)_2$. It also is used as an antacid and a laxative; it is the creamy white substance in milk of magnesia.

Important sulfates of the alkaline earth metals are those of magnesium, calcium, and barium. Magnesium sulfate in the form of its hydrate, $MgSO_4 \cdot 7H_2O$, is called *epsom salts*, and is used to treat fabrics so that they readily accept dyes, to fireproof fabrics, as a fertilizer, and medicinally.

The mineral gypsum is $CaSO_4 \cdot 2H_2O$ and is used to make plaster and plasterboard—a building material commonly called sheet rock. When gypsum is partially dehydrated by heating, *plaster of paris* is formed.

In plaster of paris there is 1 mol of H_2O for each 2 mol of $CaSO_4$.

$$2CaSO_4 \cdot 2H_2O \xrightarrow{\text{heat}} (CaSO_4)_2 \cdot H_2O$$
gypsum   **plaster of paris**

The formula for plaster of paris is normally written $CaSO_4 \cdot \frac{1}{2}H_2O$. When water is added to $CaSO_4 \cdot \frac{1}{2}H_2O$ the crystals absorb it and reform the dihydrate. This is an exothermic reaction, and anyone who has had a broken limb set in a plaster cast has probably noticed how warm the cast became as it hardened.

Barium sulfate has a number of uses that are based on its whiteness and low solubility in water. It is used as a whitener in photographic papers and a filler in papers and polymeric fibers. Medicinally it is used for X-ray diagnosis of intesti-

Figure 18.14

An X-ray photograph of a patient's large intestine that has been filled with a suspension of barium sulfate. The BaSO$_4$ is opaque to X rays and allows the shape of the intestine to be seen.

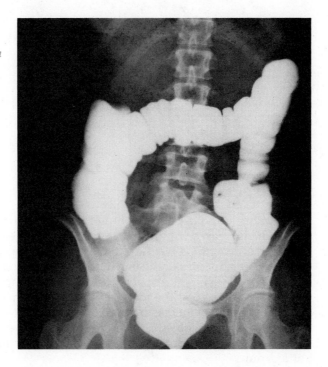

nal tract disorders because BaSO$_4$ is quite opaque to X rays. Even though barium salts are normally quite poisonous, BaSO$_4$ is so insoluble that a suspension of it can be swallowed without harm. This is because so little Ba^{2+} is in solution. As the suspension of BaSO$_4$ passes through a patient's intestines, its path can be followed by X-ray photographs like that in Figure 18.14.

18.7 METALS OF GROUPS IIIA, IVA, AND VA

In Section 18.2, you learned that the metallic character of the elements decreases from left to right in a period and increases from top to bottom in a group. The impact of these trends is especially apparent in Groups IIIA, IVA, and VA, where fewer metallic elements are found as the group number increases (Figure 18.15). In Group IIIA, for example, the metals are aluminum, gallium, indium, and thallium. In Group IVA, only tin and lead are metals, and in Group VA the only metal is bismuth. In general, the metals in these three groups tend to be less reactive and less metallic in their chemical behavior than the metals of Groups IA and IIA. For instance, many are amphoteric and many form covalent compounds.

There are no genuine metals in Groups VIA or VIIA.

Aluminum has three electrons in its valence shell outside a neon core ([Ne]$3s^2 3p^1$), and in its compounds aluminum always occurs in the +3 oxidation state. The metals in Periods 4 (Ga, In, Tl), 5 (Sn, Pb), and 6 (Bi) occur after a row of transition elements and are called *post-transition metals*. They have pseudonoble gas cores beneath their valence shells, and their chemistry is characterized by the occurrence of two oxidation states. In each case, the lower oxidation state corresponds to the loss of the outer *p* electron(s) and the higher one corresponds to the further loss of the pair of *s* electrons. Thus, in Group IIIA the two oxidation states are +1 and +3; in Group IVA they are +2 and +4; and in Group VA they are +3 and +5.

The pseudonoble gas core is $ns^2 np^6 nd^{10}$.

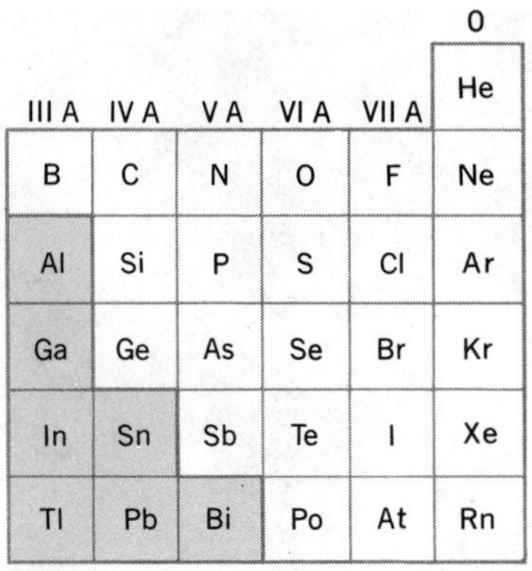

Figure 18.15

The metals of Groups IIIA, IVA, and VA.

IIIA	IVA	VA	VIA	VIIA	0
					He
B	C	N	O	F	Ne
Al	Si	P	S	Cl	Ar
Ga	Ge	As	Se	Br	Kr
In	Sn	Sb	Te	I	Xe
Tl	Pb	Bi	Po	At	Rn

Among the post-transition metals the relative stability of the lower oxidation state increases with increasing atomic number within a group. For instance, Ga^{3+} is more stable than Ga^+, while Tl^+ is more stable than Tl^{3+}. This trend in the relative stabilities of the high and low oxidation states persists in Groups IVA and VA, and results, it appears, from a decreasing stability of $M{-}X$ bonds with the increasing size of the metal atom. Since atoms become larger going down a group, it becomes increasingly more difficult to recover, by bond formation, the extra energy that must be expended to remove the pair of s electrons. This causes the lower oxidation state to become much more stable toward the bottom of a group.

Occurrence and preparation

Aluminum Of the metals in Group IIIA, aluminum is the only one of much practical importance and the only one whose chemistry we will examine in much detail. On a mass basis, aluminum is the third most abundant element in the Earth's crust. For example, it is found combined with silicon and oxygen in *aluminosilicates*, which occur in rock such as granite and various clays. Unfortunately, no practical method has yet been found to extract aluminum from these sources.

The major ore of aluminum is *bauxite*, which contains its oxide Al_2O_3. However, before the metal can be obtained electrolytically by the Hall process, which was described in Chapter 17, it must first be purified. The method of purification takes advantage of the amphoteric nature of aluminum and its compounds. The ore is first treated with a concentrated NaOH solution, which dissolves the Al_2O_3 and leaves most of the impurities behind.

The Earth's crust is composed of about 8.8% aluminum by mass, which makes aluminum the most abundant metal *in the crust.*

Purification of bauxite consumes about 5.4×10^8 kg of NaOH each year.

$$Al_2O_3(s) + 2OH^- \longrightarrow 2AlO_2^- + H_2O$$
$$\text{aluminate ion}$$

Then the basic aluminum-containing solution is acidified, which precipitates the insoluble hydroxide.

$$AlO_2^- + H_3O^+ \longrightarrow Al(OH)_3(s)$$

After filtration the aluminum hydroxide is heated. This drives off water and gives the oxide again, which is now pure.

$$2Al(OH)_3(s) \xrightarrow{heat} Al_2O_3(s) + 3H_2O(g)$$

The Al_2O_3 is then dissolved in molten cryolite and electrolyzed as described in Chapter 17.

Aluminum finds many uses in modern society. Large amounts of it serve as a structural metal in kitchen utensils, automobiles, aircraft, beverage cans, aluminum foil, and other consumer products. Aluminum is a good electrical conductor and is used in electrical wiring. Alloyed with magnesium, it is employed in structural applications, and an alloy called *alnico*—50% Fe, 20% Ni, 20% Al, 10% Co—forms powerful magnets.

Tin and Lead Tin occurs as SnO_2 in an ore called *cassiterite*. To recover the metal, the ore is first heated strongly in air to drive off volatile oxides of some of its impurities—for example, arsenic and sulfur. Afterwards, the SnO_2 is reduced with carbon.

$$SnO_2 + C \longrightarrow Sn + CO_2$$

The metal can be further purified in a manner similar to the electrolytic purification of copper. The impure tin is made the anode in an electrolytic cell and pure tin is made the cathode. Operation of the cell causes the impure tin to dissolve and pure tin to plate out on the cathode.

Elemental tin occurs in three different physical forms. This phenomenon, you recall, is termed allotropism, and the various forms of the element are called allotropes. The most common allotrope of tin is called *white tin* or *malleable tin*— you've seen it as the shiny coating on tin cans. Below 13.2°C, the white form very gradually changes to a powdery, nonmetallic form called *gray tin*. A third form, called *brittle tin*, is obtained when white tin is heated. Its properties are obvious from its name.

One of the principal uses of tin is as a protective coating over steel—the familiar tin can, for example. The coating is thin and usually applied electrolytically. It protects the steel simply by excluding air and moisture. However, once the coating is scratched and the steel below is exposed, corrosion occurs very rapidly. Iron is more easily oxidized than tin, so when both metals are exposed to moisture a galvanic cell is established in which iron is the anode and is therefore oxidized in preference to the tin. Other applications of tin are in alloys such as bronze (copper and tin) and solder (tin and lead).

Lead is found in nature as its sulfate, $PbSO_4$, its carbonate, $PbCO_3$, and its sulfide, PbS. The principal ore of lead is called *galena* and contains PbS. The metal is obtained from the ore by first heating it strongly in air, which converts it to the oxide.

$$2PbS + 3O_2 \longrightarrow 2PbO + 2SO_2$$

The oxide is then reduced with carbon.

$$2PbO + C \longrightarrow 2Pb + CO_2$$

The metallic lead that comes from this process contains impurities of silver, gold, and other metals. It also can be purified by electrolysis in the same manner as copper and tin. The silver and gold, of course, are recovered and help offset the cost of the electricity used in the purification process.

Metallic lead is used to make lead storage batteries and as a raw material in the manufacture of tetraethyl lead, $Pb(C_2H_5)_4$—the additive in leaded gasoline. It is also used to make compounds found in lead-based paints. Lead's ability to absorb high-energy radiation makes it a useful shielding material for X rays and high-energy radiation produced in nuclear reactors.

Bismuth Bismuth is found as its oxide, Bi_2O_3, and sulfide, Bi_2S_3. The sulfide must first be heated in air to convert it to the oxide before reduction.

$$2Bi_2S_3 + 9O_2 \longrightarrow 2Bi_2O_3 + 6SO_2$$

$$2Bi_2O_3 + 3C \longrightarrow 4Bi + 3CO_2$$

Bismuth is also obtained as a byproduct in the production of lead.

The allotropes of tin differ in their crystal structures.

Bismuth is a hard, brittle, slightly reddish-colored metal. It is fairly resistant to corrosion and is one of only a few known substances that expand when they freeze. It is used to make alloys whose volumes stay very nearly constant when they solidify.

In general, alloys of bismuth have low melting points. One of them, called *Wood's metal,* is composed of 50% Bi, 25% Pb, and 12½% each of Sn and Cd. It has a melting point of about 70°C, well below the boiling point of water, and is used in the triggering mechanism of automatic sprinkler systems. Similar alloys are used in fuses designed to protect electrical circuits. When too much current is drawn through the alloy wire in the fuse, it becomes hot and quickly melts. This breaks the electrical circuit and prevents damage to the device that it's meant to protect.

Chemical properties and compounds

Aluminum Aluminum is a very reactive element, a fact that prevented the simple recovery of the metal from its compounds until the invention of the Hall process. It is a good reducing agent, and its reduction potential ($\mathscr{E}^0 = -1.67$ V) is sufficiently negative that aluminum should react with water and liberate hydrogen. However, it forms a tough oxide coating that adheres to the metal and protects it from attack both by moisture and oxygen.[5] This allows aluminum to be a useful structural metal. In fact, this oxide coating is sometimes deliberately made especially thick by having aluminum be the anode in an electrolytic cell—the product is called *anodized aluminum.*

Earlier, you learned that aluminum metal is amphoteric and is able to dissolve in both dilute acids and in base. Both reactions liberate hydrogen.

$$2Al(s) + 6H^+(aq) \longrightarrow 2Al^{3+}(aq) + 3H_2(g)$$

$$2Al(s) + 2OH^-(aq) + 2H_2O \longrightarrow 2AlO_2^-(aq) + 3H_2(g)$$

However, toward concentrated HNO_3 aluminum appears passive—that is, it doesn't react. The strong oxidizing power of HNO_3 creates an oxide coating that protects the metal from further attack.

Besides protecting the metal from oxidation, aluminum oxide is an important compound in its own right. It occurs in two principal forms, which are designated as γ-Al_2O_3 and α-Al_2O_3. The γ-Al_2O_3 is formed by dehydrating aluminum hydroxide at a relatively low temperature.

$$2Al(OH)_3(s) \xrightarrow{<450°C} \gamma\text{-}Al_2O_3(s) + 3H_2O(g)$$

This form of the oxide is quite reactive, dissolving readily in both acids and bases.

If γ-Al_2O_3 is heated strongly, its crystal structure changes to that of the α-Al_2O_3 form, which is called *corundum.* This substance has a very high melting point (about 2050°C), is very hard and quite inert, especially toward acids. Corundum from deposits that are found in nature is used as an abrasive in sandpaper and in making refractory bricks that line furnace interiors.

A number of familiar gemstones are also composed of almost pure α-Al_2O_3. For example, ruby consists of α-Al_2O_3 with small amounts of dissolved Cr^{3+}.

The thermite reaction being used to weld reinforcing rods for concrete at a construction site.

[5] A freshly exposed aluminum surface can be amalgamated—coated with a film of mercury in which the metallic aluminum dissolves. The Al_2O_3 that's formed on exposure to air does not adhere to the amalgamated surface and the aluminum reacts easily with oxygen, corroding very rapidly. This very exothermic reaction can make the aluminum hot.

Other impurities in the α-Al_2O_3 produce gems with other colors. Sapphire, for instance, contains traces of Fe^{2+} and Ti^{4+}, oriental topaz contains Fe^{3+}, and oriental amethyst contains Mn^{3+}.

Aluminum oxide has a very large exothermic heat of formation ($\Delta H_f^0 = -1676$ kJ mol^{-1}). This allows aluminum to extract oxygen from other metal oxides with the simultaneous release of large quantities of heat—enough to melt the products of the reaction. For example, the reaction of Fe_2O_3 with aluminum

$$2Al(s) + Fe_2O_3(s) \longrightarrow Al_2O_3(l) + Fe(l)$$

produces temperatures approaching 3000°C. This is called the **thermite reaction** and is used to weld large masses of iron and steel. Thermite bombs have also been used by the military as incendiary devices because of the intense heat of the reaction.

The large exothermic heat of formation of Al_2O_3 has had another interesting application in recent times: It provides the thrust for the booster rockets that enable the space shuttle to take off. The solid propellant in these booster rockets is a mixture of powdered aluminum (the fuel) and ammonium perchlorate, NH_4ClO_4 (the oxidizer). They are mixed with a small amount of iron oxide catalyst, and the entire mixture is held together in a solid mass by an epoxy plastic. When the rocket is ignited, the aluminum is oxidized, and the formation of Al_2O_3 liberates large quantities of heat that cause the gases also formed to expand with great force. That is what lifts the rocket.

The anhydrous aluminum halides are interesting compounds because of their tendency to form *dimeric species*—molecules formed by the pairing of two AlX_3 units. This is especially so in the vapor and solutions of these compounds in nonpolar solvents, such as benzene and carbon tetrachloride. The pairing occurs in a way similar to the linking together of $BeCl_2$ molecules in solid $BeCl_2$—by the formation of a coordinate covalent bond from the halogen on one AlX_3 unit to the aluminum atom of another. For example, with $AlCl_3$ the species Al_2Cl_6 is formed.

When aluminum chloride or other aluminum salts are dissolved in water, their solutions are acidic. In Chapter 7 we saw that this could be accounted for by considering how the Al^{3+} ion polarizes the water molecules that surround it in solution, making it easier for the hydrogens to be removed as H^+.

$$H_2O + Al(H_2O)_6^{3+} \longrightarrow Al(H_2O)_5(OH)^{2+} + H_3O^+$$

As base is added to a solution containing Al^{3+}, it neutralizes H_3O^+ and gradually strips protons from the water molecules that surround the Al^{3+} until the insoluble hydroxide is produced.

$$Al(H_2O)_6^{3+} + 3\,OH^- \longrightarrow Al(H_2O)_3(OH)_3(s) + 3\,HOH$$

Because of all the water contained within it, the aluminum hydroxide formed in this way is gelatinous (gelatinlike), rather than crystalline.

When base is added to the aluminum hydroxide precipitate, it dissolves, presumably because the OH^- extracts a proton from yet another water molecule that is attached to the Al^{3+}.

$$Al(H_2O)_3(OH)_3 + OH^- \longrightarrow Al(H_2O)_2(OH)_4^- + HOH$$

The white cloud seen billowing beneath the rising space shuttle in the photograph at the beginning of this chapter is composed of fine particles of Al_2O_3.

The chemistries of beryllium and aluminum are similar in many ways. This has been attributed to their similar ionic potentials.

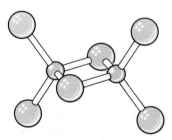

Structure of Al_2X_6 (small spheres, Al; large spheres, X).

The species $Al(H_2O)_2(OH)_4^-$ is called the aluminate ion, although its precise composition is uncertain. Notice, for example, the equivalence

$$Al(H_2O)_2(OH)_4^- \text{ and } AlO_2^- + 4H_2O$$

The reactions described above can be reversed by the addition of acid. For instance, when a basic solution containing aluminate ion is gradually neutralized the hydroxide precipitates and then redissolves as more acid is added,

$$Al(H_2O)_2(OH)_4^- + H_3O^+ \longrightarrow Al(H_2O)_3(OH)_3(s) + H_2O$$

$$Al(H_2O)_3(OH)_3(s) + H_3O^+ \longrightarrow Al(H_2O)_4(OH)_2^+ + H_2O$$

and ultimately, in solutions that are sufficiently acidic,

$$Al(H_2O)_5OH^{2+} + H_3O^+ \longrightarrow Al(H_2O)_6^{3+} + H_2O$$

A compound of aluminum produced in large quantities, much of which is destined to be changed to the hydroxide, is aluminum sulfate. It is usually made from bauxite and sulfuric acid.

> $Al_2(SO_4)_3$ ranks about 37th in total production among industrial chemicals.

$$Al_2O_3(s) + 3H_2SO_4(aq) \longrightarrow Al_2(SO_4)_3(aq) + 3H_2O$$

When crystallized it forms hydrates with as many as 18 water molecules ($Al_2(SO_4)_3 \cdot 18H_2O$).

The principal use of $Al_2(SO_4)_3$ is in the paper industry, where it is used to adjust acidity and treat paper to make it water resistant. Aluminum sulfate is also employed in municipal water treatment. It is added to the water, which is then made basic by the addition of lime (CaO). This causes a gelatinous aluminum hydroxide precipitate to be formed that settles slowly to the bottom, taking fine sediment and bacteria with it.

When an aqueous mixture of aluminum sulfate and sodium sulfate is evaporated, crystals having the composition $NaAl(SO_4)_2 \cdot 12H_2O$ are formed. Similar crystals are also formed if $(NH_4)_2SO_4$ or K_2SO_4 are substituted for the Na_2SO_4 and if sulfates of Cr^{3+} or Fe^{3+} are substituted for $Al_2(SO_4)_3$. These crystals are called **alums.** They are characterized by the general formula $M^+M^{3+}(SO_4)_2 \cdot 12H_2O$, and are examples of **double salts.** If conditions are right, large well-formed octahedrally shaped crystals can be produced, as shown in Figure 18.16.

> $M^+ = Na^+, K^+, NH_4^+$
> $M^{3+} = Al^{3+}, Cr^{3+}, Fe^{3+}$

One use of sodium alum is in baking powder, where it is combined with sodium bicarbonate. When added to moist dough, the acidity of the aluminum ion in water causes carbon dioxide to be released by reaction of hydronium ion with the bicarbonate ion,

$$HCO_3^- + H_3O^+ \longrightarrow 2H_2O + CO_2(g)$$

As mentioned earlier, when the dough is baked the small bubbles of CO_2 give the finished product a light and porous texture.

Figure 18.16
An octahedrally shaped crystal of $KAl(SO_4)_2 \cdot 12H_2O$, potassium alum, shown in actual size.

Tin, Lead, and Bismuth The chemistries of these elements are characterized by the existence of two oxidation states. Tin and lead, for example, form compounds in both the +2 and +4 states, while bismuth forms compounds in the +3 and +5 states. The relative stabilities of their higher and lower oxidation states are reflected in the way the metals react with various substances. For example, tin reacts with nonoxidizing acids such as HCl to form Sn^{2+}, but with oxidizing acids the +4 state is produced.

$$Sn + 2HCl \longrightarrow SnCl_2 + H_2$$

$$Sn + 4HNO_3 \longrightarrow SnO_2 + 4NO_2 + 2H_2O$$

Lead, however, forms only lead(II) compounds, even when strong oxidizing acids such as concentrated HNO_3 are used.

$$3Pb + 8HNO_3 \longrightarrow 3Pb(NO_3)_2 + 2NO + 4H_2O$$

This tells us that when tin and lead are compared, the higher oxidation state is relatively more stable for tin than lead. This is also revealed in their reaction products with chlorine. Tin combines with chlorine to form $SnCl_4$, but lead reacts to produce $PbCl_2$.

$$Sn + 2Cl_2 \longrightarrow SnCl_4$$

$$Pb + Cl_2 \longrightarrow PbCl_2$$

In the case of bismuth, the +3 state is much more stable than the +5 one. Direct combination of bismuth with oxygen or chlorine, for example, produces Bi_2O_3 and $BiCl_3$, respectively. The oxidation of Bi_2O_3 to Bi_2O_5 requires very severe oxidizing conditions, and the compound $BiCl_5$ does not exist at all because the +5 oxidation state of bismuth is such a powerful oxidizing agent—that is, it has such a strong tendency to be reduced—that it would oxidize Cl^- to Cl_2.

Besides dissolving in acids, tin and lead also dissolve in base with the evolution of hydrogen.

$$Sn + 2OH^- + 2H_2O \longrightarrow Sn(OH)_4{}^{2-} + H_2(g)$$

$$Pb + 2OH^- + 2H_2O \longrightarrow Pb(OH)_4{}^{2-} + H_2(g)$$

The *stannite ion*, $Sn(OH)_4{}^{2-}$, is a very powerful reducing agent and is easily oxidized. On the other hand, the *plumbite ion*, $Pb(OH)_4{}^{2-}$, is less easily oxidized, which again reveals that the +2 state is more stable for lead than it is for tin.

Compounds of tin, lead, and bismuth have a variety of practical uses. For example, tin(II) fluoride, SnF_2, also called stannous fluoride, is the decay-inhibiting ingredient in some fluoride toothpastes. (The way fluoride ion helps prevent tooth decay was discussed in Chapter 14, page 523). Some tin compounds are also useful fungicides. An example is $(C_4H_9)_3SnO$, whose chemical name is tributyltin oxide. It is used in some antifouling paints that are applied to boat hulls to prevent marine growth, and in preparations that are applied to wood to prevent rotting.

Among the most useful compounds of lead are its oxides. Lead(II) oxide, PbO, a yellow powder that is also called litharge, is used in pottery glazes and in making fine lead crystal. If PbO is heated carefully in air, it can be oxidized to Pb_3O_4. This is a mixed oxide containing both lead(II) and lead(IV). Its common name is *red lead*, and it is used in corrosion-inhibiting paints that are applied to structural steel.

If solutions of plumbite ion are subjected to strong oxidizing conditions, PbO_2 can be prepared. For example,

$$Pb(OH)_4{}^{2-} + OCl^- \longrightarrow PbO_2 + H_2O + 2OH^- + Cl^-$$

Unlike SnO_2, PbO_2 is a strong oxidizing agent. Its most common use is as the cathode material in lead storage batteries. Recall that during the discharge of this battery the net reaction is

$$Pb + PbO_2 + 2H_2SO_4 \longrightarrow 2PbSO_4 + 2H_2O$$

As you are probably aware, in relatively recent times there has been a great deal of concern over the past use of lead-based paints, particularly on surfaces in living areas children may be exposed to. This is because lead compounds are very toxic. The pigment most used in lead-based paints is a white basic carbonate, $Pb_3(OH)_2(CO_3)_2$, and its use in interior paints is now restricted. Although it

Tin and lead are both amphoteric.

OCl^- is a strong oxidizing agent.

provides excellent covering power as a pigment, it suffers from the disadvantage of being darkened on contact with H_2S produced in the environment by decaying vegetation. The darkening is caused by the production of black PbS. Another lead-based pigment that is used in artists' oil colors is lead chromate, $PbCrO_4$. This compound has a bright yellow color.

Bismuth compounds are utilized frequently in the cosmetic and pharmaceutical industry. In fact, these industries account for about 30 percent of the bismuth produced each year. For example, a substance known as bismuth subnitrate, produced by partial hydrolysis of $Bi(NO_3)_3$, is used medicinally as an antacid in the treatment of gastric ulcers. Its exact composition varies according to how it is prepared, but it can be approximately formulated as $BiO(NO_3)$.

The hydrolysis of bismuth(III) compounds is not unusual. When $BiCl_3$ is dissolved in water it hydrolyzes, producing the bismuthyl ion, BiO^+. As the solution is diluted an insoluble precipitate of BiOCl is formed that redissolves if the solution is made more acidic by the addition of hydrochloric acid.

The oxide of bismuth, Bi_2O_3, is formed when bismuth is heated in air. As mentioned earlier, under extreme oxidizing conditions this can be oxidized to give bismuth in the +5 oxidation state. Compounds containing bismuth(V), such as $NaBiO_3$ (*sodium bismuthate*), are extremely powerful oxidizing agents as a result of bismuth's strong tendency to revert to the +3 oxidation state by acquiring electrons from other substances.

REVIEW QUESTIONS AND PROBLEMS

(Problems whose numbers are in blue have their answers in Appendix C at the back of the book; the more difficult problems are marked with asterisks.)

Metallurgy

18.1 How are most metals found in nature? What are some important sources of metals?

18.2 What is an *ore*?

18.3 What are the three steps involved in extracting a metal from its ore and making it ready for practical use?

18.4 Why is "panning" able to separate sand and mud from tiny particles of gold?

18.5 What is an amalgam? How is it used in the recovery of gold from gold ores?

18.6 Describe the process called flotation. What is meant by *roasting* as applied to metallurgy?

18.7 Write chemical equations for the purification of bauxite, Al_2O_3.

18.8 Why must metal compounds always be reduced to extract the metals from them?

18.9 What property must a metal compound possess to be easily decomposed thermally?

18.10 What is implied thermodynamically when we say that a particular compound is thermally stable?

18.11 From the data in Tables 12.1 and 12.4, calculate the temperature (in °C) above which the thermal decomposition of ZnO should become feasible.

18.12 Given the data below, determine the temperature in °C at which $K_P = 1$ for the reaction,

$$CuO(s) \longrightarrow Cu(s) + \tfrac{1}{2}O_2(g)$$

For CuO(s), $\Delta H_f^0 = -155$ kJ mol^{-1}. Absolute entropies; CuO(s), 43.5 J mol^{-1} K^{-1}; Cu(s), 33.3 J mol^{-1} K^{-1}; $O_2(g)$, 205.0 J mol^{-1} K^{-1}.

18.13 Calculate K_P at 100, 500, and at 2000°C for the reaction,

$$MoO_3(s) \longrightarrow Mo(s) + \tfrac{3}{2}O_2(g)$$

given the following data:

	ΔH_f^0 (kJ mol^{-1})	S^0 (J mol^{-1} K^{-1})
$MoO_3(s)$	−754.4	78.2
$Mo(s)$	0.0	28.6
$O_2(g)$	0.0	205.0

18.14 Write chemical equations for (a) the reduction of Fe_2O_3 in the blast furnace and (b) the production of slag from SiO_2 and $CaCO_3$.

18.15 Why is carbon a preferred reducing agent in commercial metallurgy?

18.16 Write equations showing the use of carbon and hydrogen as reducing agents in the extraction of a metal from one of its compounds.

18.17 Why isn't sodium produced commercially by reduction of its compounds with a reducing agent that is stronger than sodium?

18.18 Why are halide salts often used when preparing metals by electrolysis?

18.19 Write equations for the commercial electrolytic production of sodium and aluminum.

18.20 Why must pig iron be refined to be useful as a strong structural metal? Describe the Bessemer converter; the open hearth furnace; the basic oxygen process. Which of these is the principal method used today to make steel?

18.21 Describe the Mond process.

Trends in Metallic Behavior

18.22 How does the metallic character of the elements depend on electronegativity? What vertical and horizontal trends in metallic character exist in the periodic table? Illustrate these trends for the elements in the second period and in Group IVA.

18.23 In each pair below, choose the element expected to have the more metallic character.
(a) Li or Be   (d) Sn or P
(b) B or Al   (e) Ga or I
(c) Al or Cs

18.24 Which oxide should be more basic, Al_2O_3 or Ga_2O_3?

18.25 Which should be more acidic, MgO or Al_2O_3?

18.26 What is meant by amphoteric? Write chemical equations to illustrate the amphoteric behavior of beryllium and aluminum.

18.27 In what way are the metals in Groups IIIA, IVA, and VA less metallic than those in Groups IA and IIA?

Covalent Character of Metal Compounds

18.28 Define ionic potential. How is it related to the degree of covalent character in a metal–nonmetal bond?

18.29 Which should have the greater degree of covalent bonding?
(a) $GaCl_3$ or $GeCl_4$ (d) $LiCl$ or Li_2S
(b) Bi_2O_3 or Bi_2O_5 (e) Na_2S or MgS
(c) PbO or PbS

18.30 Which should be more ionic?
(a) SnO or SnS (d) SnS or PbS
(b) $AlCl_3$ or $AlBr_3$ (e) SnS or SnS_2
(c) $BeCl_2$ or BeF_2

Colors of Metal Compounds

18.31 What is responsible for the color in compounds such as SnS_2 or PbS?

18.32 Which compound would you expect to be more deeply colored?
(a) Ag_2O or Ag_2S
(b) $CuCl$ or $CuBr$
(c) SnS or SnS_2
(d) MgS or Al_2S_3

18.33 If a particular compound absorbs green light, what color will it appear to be?

Group IA Metals: General

18.34 Why are the Group IA elements called *alkali metals*? What oxidation states are observed for the Group IA metals?

18.35 Why do compounds of lithium, rubidium, and cesium have little commercial importance?

18.36 Why are sodium compounds more important, commercially, than compounds of the other alkali metals?

18.37 Given the following thermodynamic data, calculate the hydration energy for the Na^+ ion in $kJ\ mol^{-1}$:

$$\Delta H_f^0 \text{ of } Na^+(aq) = -239.7\ kJ\ mol^{-1}$$
$$\Delta H_{atom}^0 \text{ of } Na(s) = 108.7\ kJ\ mol^{-1}$$
$$IE \text{ of } Na(g) = 493.7\ kJ\ mol^{-1}$$

18.38 Using the data for the atomization energy of Na (Problem 18.37) and the first and second ionization energies for Na in Table 4.5, compute the value of the hydration energy (in $kJ\ mol^{-1}$) required to produce a negative ΔH_f for $Na^{2+}(aq)$.

*18.39 The standard reduction potential of potassium is -2.92 V. ΔH_{hyd}^0 of K^+ in 1 M aqueous solution is $-759\ kJ\ mol^{-1}$. The atomization energy of K is $+90.0\ kJ\ mol^{-1}$ and the ionization energy of $K(g)$ is $+418\ kJ\ mol^{-1}$. Calculate ΔS^0 (in $J\ mol^{-1}\ K^{-1}$) for the process,

$$K(s) \longrightarrow K^+(1\ M) + e^-$$

18.40 Which alkali metals are most abundant? Which one is least abundant? Why?

18.41 Where do the alkali metals occur in nature?

18.42 Write a chemical equation to show how elemental potassium, rubidium, and cesium are usually made.

Group IA Metals: Physical and Chemical Properties

18.43 Describe two applications of metallic sodium related to its physical properties.

18.44 What colors are given to a Bunsen burner flame by compounds of (a) sodium, (b) lithium, and (c) potassium?

18.45 How can potassium ion in a mixture be detected if the mixture also contains sodium ion?

18.46 Write a chemical equation for the reaction of rubidium with water.

18.47 Why is the reduction potential of lithium more negative than the reduction potential of sodium?

18.48 What happens when an alkali metal is added to liquid ammonia? Why are these solutions such good reducing agents?

18.49 Write chemical equations to show the reaction (if any) of lithium and sodium with (a) bromine, (b) sulfur, and (c) nitrogen.

18.50 Write equations that illustrate the reactions of each of the alkali metals with oxygen (present in excess).

18.51 What compound of the alkali metals is used in a recirculating breathing apparatus? Write chemical equations to show how it functions.

18.52 Why are solutions of Na_2O_2 basic?

18.53 What is a *getter*?

18.54 What are other common names for sodium hydroxide? What are some of its uses?

18.55 What is trona ore? Give the chemical reactions involved in the Solvay process. What is the net reaction in the Solvay process?

18.56 How does sodium bicarbonate serve as a buffer? What is the common name for sodium bicarbonate? How can $NaHCO_3$ serve as a fire extinguisher?

18.57 What is potash? How is it made?

18.58 What alkali metal compound is used to treat manic depression?

Group IIA Metals: General

18.59 Why are the Group IIA elements called alkaline earth metals? How do their densities, hardness, melting points, and ionization energies compare to their neighbors in Group IA?

18.60 Where are calcium and magnesium found? What is the source of radium?

18.61 How is magnesium recovered from dolomite? How is magnesium recovered from sea water?

Group IIA Metals: Physical and Chemical Properties

18.62 Why do the alkaline earth metals have reduction potentials that are nearly as negative as the metals in Group IA?

18.63 Define *calcining*. What is lime? What happens when water is added to lime? Why is lime such an important industrial chemical?

18.64 What reaction takes place inside a flashbulb when it is fired?

18.65 What color is given to a Bunsen burner flame by compounds of (a) calcium, (b) strontium, and (c) barium?

18.66 What chemical fact is responsible for the structural uses of metallic beryllium and magnesium? Why are metallic calcium, strontium, and barium not used as structural metals?

18.67 Write chemical equations for the reaction of magnesium with elemental oxygen, sulfur, and nitrogen.

18.68 Write chemical equations that illustrate the amphoteric behavior of metallic beryllium.

18.69 What is the structure of solid $BeCl_2$? How does this support the statements in earlier chapters that molecules of $BeCl_2$ have less than an octet in the valence shell of beryllium?

18.70 Why are beryllium compounds covalent? What are organomagnesium compounds?

18.71 Write an equation for the reaction of water with (a) calcium and (b) potassium.

18.72 How do the solubilities of the alkaline earth hydroxides vary from top to bottom in the group? How do the solubilities of the sulfates vary?

18.73 What are some uses of calcium carbonate?

18.74 What is milk of magnesia composed of?

18.75 What is gypsum? Write chemical equations showing how plaster of paris is made and what happens when it combines with water during hardening.

18.76 What property of barium sulfate allows it to be used for X-ray diagnosis of intestinal tract disorders?

18.77 What is *epsom salts*? What are some uses of magnesium oxide?

Metals in Group IIIA, IVA, and VA

18.78 What oxidation states are observed for the metals in Groups IIIA, IVA, and VA?

18.79 Which are the post-transition metals? Why do their lower oxidation states become more stable relative to the higher ones going down a group?

Aluminum

18.80 What is the ore of aluminum? Write chemical equations to show how it is purified.

18.81 What are some commercial uses of metallic aluminum?

18.82 Why doesn't aluminum corrode rapidly in the presence of air and moisture? How can it be made to corrode very quickly?

18.83 Write chemical equations showing the amphoteric behavior of metallic aluminum.

18.84 How do the properties of γ-Al_2O_3 and α-Al_2O_3 differ? What gems are composed primarily of aluminum oxide?

18.85 What is the thermite reaction?

18.86 What is the structure of dimeric aluminum chloride?

18.87 Why are solutions of aluminum salts acidic?

18.88 Write chemical equations for the gradual neutralization of $Al(H_2O)_6^{3+}$, and the dissolving in base of the gelatinous aluminum hydroxide precipitate.

18.89 What are two ways of writing the formula of the aluminate ion?

18.90 How is aluminum sulfate used in water treatment plants?

18.91 What is an alum? Give an example. Which alum is used in baking powders and how does it function?

18.92 When aluminum oxide dissolves in base, the aluminate ion, which can be written as AlO_2^-, is formed. Write an equation for this reaction.

Tin, Lead, and Bismuth

18.93 Write chemical equations showing how tin, lead, and bismuth are recovered from their ores.

18.94 Why does a tin can rust so rapidly if the tin coating is scratched, exposing the steel beneath it?

18.95 What are allotropes? Which of the representative metals exhibits allotropism?

18.96 What unusual property is possessed by metallic bismuth? What is Wood's metal? What are its properties and uses?

18.97 How do the elements tin and lead differ in their behavior toward nitric acid and chlorine? Illustrate using chemical equations.

18.98 Why doesn't $BiCl_5$ exist?

18.99 Write chemical equations for the reactions of metallic tin and lead with base.

18.100 What are the oxides of lead and what are some of their uses?

18.101 Why do lead-based paints gradually darken over a period of time?

18.102 What are some uses of bismuth compounds? What is the formula of the bismuthyl ion?

THE CHEMISTRY OF SELECTED NONMETALS PART I: HYDROGEN, CARBON, OXYGEN, AND NITROGEN

Carbon, hydrogen, oxygen, and nitrogen touch our lives closely. Not only are they found in nearly all the molecules in our bodies, they are also involved, one way or another, in the production of photochemical smog, seen here hanging over a western city. How photochemical smog is formed is one of the topics discussed in this chapter.

In Chapter 18 we discussed the chemical and physical properties of the representative metals. We now turn our attention, in this chapter and the next, to the remainder of the representative elements: the nonmetals and metalloids. In many ways, the chemistry of these nonmetallic elements is more interesting than that of the A-group metals because of the large variety of compounds that they form. Not only do they combine with metals to form substances that are often predominantly ionic, they also combine with each other by way of covalent bonds to form molecules and polyatomic ions that range from simple clusters of atoms to gigantic molecules, such as the DNA that controls heredity and guides the chemical functions of our cells.

In this chapter we begin by studying the properties of hydrogen, carbon, oxygen, and nitrogen. Although our discussions will not be biased toward biology, these four nonmetals are uniformly important to all life as we know it, and much of our interest in their chemical properties is tied to how they and their compounds influence our lives and affect the quality of our environment.

19.1 HYDROGEN

How stars produce energy is discussed further in Chapter 22.

Hydrogen, first recognized as an element by the English chemist, Henry Cavendish (1731–1810), is the most abundant of all the elements in the universe—approximately 93% if counted by atoms. It is the principal element in the solar atmosphere, and it is the nuclear fuel that stars consume in their generation of energy. Here on Earth, hydrogen is much less abundant—about 3% by atoms, or about 0.14% by mass—presumably because the Earth's gravity was insufficient to hold onto most of the hydrogen that was present when the planet was formed.

Virtually all the Earth's hydrogen exists in the combined state. Its reactivity, especially toward oxygen, is too great to allow it to be present as the free element in the atmosphere except in trace amounts. Water, of course, is two-thirds hydrogen on an atom basis (about 11% by mass) and the oceans represent a huge storehouse of this element. Hydrogen is also a principal element in all organic material. This includes living things, both animal and vegetable, as well as their fossils—petroleum and natural gas.

The official name for H_2 is dihydrogen.

Only helium has a lower boiling point than hydrogen.

In its elemental state hydrogen exists as diatomic molecules of H_2. Its boiling point is $-253°C$ and its freezing point (melting point) is $-259°C$. At room temperature, of course, hydrogen is a gas, and its very small molar mass makes it the least dense gas of all (it is only half as dense as helium). Hydrogen therefore has great lifting power in balloons, but as we learned in Chapter 1 (Figure 1.1, page 4), its extreme flammability poses a threat of disaster!

H $1s^1$

In the periodic table, hydrogen is placed in Group IA because its valence shell has the same electron configuration as the other members of this group. Here any similarities cease, however. Hydrogen has no electrons below its valence shell, and for this reason its chemistry does not resemble that of the Group IA metals at all. In fact, hydrogen really does not fit well into any group within the periodic table.

Isotopes of hydrogen

Many elements occur in nature as mixtures of isotopes, but hydrogen is the only one whose isotopes have their own names. The nucleus of ordinary hydrogen, $_1^1H$, consists of a single proton. It is the most abundant of hydrogen's three isotopes, and on rare occasions is called **protium.**

Deuterium is called heavy hydrogen.

Deuterium, $_1^2H$, has a nucleus consisting of one proton and one neutron, and is often given the symbol D. For example, the formula for deuterium oxide—

also known as **heavy water**—is usually written D_2O. Only about 1 of every 5000 atoms of naturally occurring hydrogen is deuterium.

Only 1 of every 10 million atoms of naturally occurring hydrogen is ^3_1H.

The third isotope of hydrogen, ^3_1H is called **tritium** (symbol T). It is radioactive, and because of its relatively short half-life of 12.3 years, it is found in only very minute amounts in naturally occurring hydrogen. However, it can be made in nuclear reactions—for example, by bombarding lithium or boron with neutrons. In fact, tritium is a by-product of the operation of nuclear power plants, where it is produced by a variety of nuclear reactions including ones involving lithium and boron additives in the reactor cooling system.

Replacing H with D doesn't alter the chemical properties of a compound.

Chemically, the isotopes of hydrogen are identical except for small differences in the rates at which they react. This is both a blessing and a curse. One beneficial use of hydrogen's isotopes is in the study of reaction mechanisms. For example, a compound can be labeled by replacing one or more of the hydrogens in its structure with atoms of deuterium. After this compound is allowed to react, the locations of the labels in the product molecules can be determined using a mass spectrometer, and the information gathered in this way can help a chemist deduce the mechanism of the reaction.

A danger also exists in the chemical similarities of isotopes. A large-scale use of nuclear power plants will almost surely increase the concentration of tritium in the environment. This isotope can easily be incorporated into biological molecules because it behaves chemically just like ordinary hydrogen, and the radiation that it would give off within an organism could cause many problems, including cancer and other radiation-related maladies.

Preparation and uses

Hydrogen is an important industrial chemical. Its principal source is natural gas, from which it is extracted by reactions with steam at high temperature in the presence of a catalyst.

Similar reactions can occur with other hydrocarbons.

$$CH_4(g) + H_2O(g) \xrightarrow[\text{catalyst}]{\text{heat}} CO(g) + 3H_2(g)$$

methane
(natural gas)

$$CO(g) + H_2O(g) \xrightarrow[\text{catalyst}]{\text{heat}} CO_2(g) + H_2(g)$$

The hydrogen and carbon dioxide can be easily separated from each other by bubbling the gas mixture through water, in which CO_2 is fairly soluble and H_2 virtually insoluble.

Hydrogen can also be extracted from water by allowing steam to react with carbon (from coal, for instance) at temperatures of about 1000°C.

$$C(s) + H_2O(g) \xrightarrow{1000°C} CO(g) + H_2(g)$$

The mixture of CO and H_2 is called *water gas* and the reaction is referred to as the *water gas reaction*. It is used by industry because it changes a solid fuel (coal), which is awkward to handle in large quantities, into a gaseous combustible mixture that is easily piped to where it is needed.

Combustion of water gas:

$2CO + O_2 \longrightarrow 2CO_2$

$2H_2 + O_2 \longrightarrow 2H_2O$

Another way of obtaining hydrogen from water is by electrolysis. In Chapter 17 we saw that the electrolysis of brine is used to produce huge quantities of caustic soda (sodium hydroxide). A second product of this electrolysis is hydrogen, which also becomes part of the industrial supply of this element. The overall cell reaction is

$$2NaCl(aq) + 2H_2O \longrightarrow 2NaOH(aq) + Cl_2(g) + H_2(g)$$

Figure 19.1

Laboratory preparation of hydrogen by the reaction of a metal such as zinc with an acid.

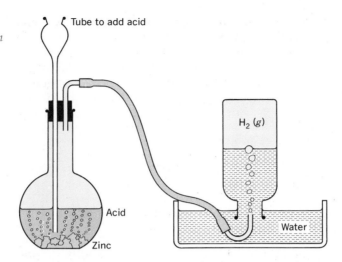

Zinc metal readily dissolves in hydrochloric acid with the evolution of bubbles of hydrogen.

By the use of appropriate catalysts, CO and H$_2$ can be made to form other alcohols and hydrocarbons. Industrially, mixtures of CO and H$_2$ are also known as **synthesis gas**.

If hydrogen is needed in small quantities in the laboratory, it can be conveniently made by reacting an active metal with an acid. Any metal that is below hydrogen in Table 17.1 is a better reducing agent than hydrogen, and, in principle, should cause H$^+$ to be reduced. However, those metals having reduction potentials near zero (lead and tin, for example) react with acids very slowly. At the other extreme, those metals near the bottom of Table 17.1 are such active reducing agents that they react very vigorously with water (as we saw in the last chapter) and almost explosively with an acid. Metals having intermediate reduction potentials react smoothly with acids and serve as practical sources of hydrogen. A metal frequently used for this purpose is zinc. With dilute sulfuric acid it reacts as follows.

$$H_2SO_4(aq) + Zn(s) \longrightarrow ZnSO_4(aq) + H_2(g)$$

Figure 19.1 shows the type of apparatus that can be used to prepare H$_2$ in the laboratory. The gas is collected by allowing it to displace water from the inverted bottle, which has its mouth below the surface of the water in the tray. This method of collection works because H$_2$ has a very low solubility in water.

The greatest single use of hydrogen is in the production of ammonia from nitrogen.

$$3H_2(g) + N_2(g) \longrightarrow 2NH_3(g)$$

This reaction, which consumes approximately two-thirds of the annual world production of hydrogen, is discussed further in Section 19.4.

Hydrogen is also used in large quantities in the manufacture of methanol (also called methyl alcohol or wood alcohol[1]). The reaction combines carbon monoxide and hydrogen at high pressure and temperature over a catalyst.

$$CO + 2H_2 \xrightarrow[\substack{20-30 \text{ MPa} \\ \text{catalyst}}]{300-400°C} CH_3OH$$

This reaction is important because it provides a simple route by which coal can be converted to a liquid fuel. First the coal can be converted to a mixture of CO and H$_2$ by the water gas reaction. Then CO and H$_2$ can be combined to produce methanol, which is itself a potentially useful fuel. What makes matters even

---

[1] At one time methanol was obtained as one of the products formed by heating wood in the absence of air—a process called *destructive distillation*. Although methanol is still called wood alcohol in many places, very little of it is currently produced by this older method.

A methanol to gasoline conversion plant has begun commercial operation in New Zealand.

more interesting is a catalytic process discovered by Mobil Oil Corporation that is able to change CH_3OH into high octane gasoline.[2] At the present time, however, in the United States gasoline produced from methanol is more expensive than that obtained directly from petroleum.

Still another commercial use of hydrogen is the **hydrogenation** of vegetable oils, in which hydrogen is added chemically to carbon-carbon double bonds. For example, ethylene can be converted to ethane by hydrogenation.

Because organic molecules having double bonds possess the ability to take on additional hydrogen, they are said to be *unsaturated*. On the other hand, molecules with only carbon-carbon single bonds cannot react with more hydrogen, so they are termed *saturated*. Hydrogenation therefore converts unsaturated vegetable oils into saturated fats. We will say more about this in Chapter 23.

Compounds of hydrogen

Hydrogen is found in more compounds than any other element. Virtually all organic compounds contain hydrogen—they are either hydrocarbons or are formed from hydrocarbons by replacing some of their hydrogen atoms with other elements. Hydrogen is found in both binary acids and oxoacids, and hydrogen even combines directly with active metals.

Binary compounds with hydrogen are called **hydrides,** and can be divided into two main types: ionic or saltlike hydrides and covalent hydrides. The ionic hydrides are formed from hydrogen and an active metal. Because of hydrogen's rather high electronegativity, it combines directly with the active metals in Groups IA and IIA by acquiring an electron and forming the **hydride ion,** H^-, thereby completing its valence shell.

$$2Li(s) + H_2(g) \longrightarrow 2LiH(s)$$

$$Ca(s) + H_2(g) \longrightarrow CaH_2(s)$$

These compounds are very sensitive toward moisture because of the basicity of the hydride ion. In water their overall reactions are as follows.

$$LiH(s) + H_2O \longrightarrow LiOH(aq) + H_2(g)$$

$$CaH_2(s) + 2H_2O \longrightarrow Ca(OH)_2(aq) + 2H_2(g)$$

but the reaction is really one between H^- and H_2O.

$$H^- + H_2O \longrightarrow H_2 + OH^-$$

This can viewed as either an acid-base reaction or a redox reaction.

Hydrogen, of course, can also complete its $1s$ subshell by electron sharing. You no doubt recall drawing many Lewis structures showing hydrogen covalently bonded to another atom. Except for the very special case of boron and the hydrogen difluoride ion HF_2^-, in which a H atom lies equidistant between two F atoms (that is, $[F—H—F]^-$), hydrogen is capable of binding to only one other atom by way of electron sharing, because it seeks only one additional electron. As a result, the structural chemistry of the simple hydrides is rather straightforward and, perhaps, even somewhat mundane. With the halogens it forms com-

---

[2] Direct conversion of coal to liquid hydrocarbon fuels by catalytic reactions of carbon in the form of coal dust with hydrogen has also been under study, and some success has been achieved.

pounds with the general formula HX (e.g., HF, HCl). Similarly, we find, quite expectedly, that the Group VIA elements form molecules of general formula H$_2X$ (e.g., H$_2$O, H$_2$S), the elements of Group VA form H$_3X$ (usually written XH$_3$, for example, NH$_3$, PH$_3$), and those in Group IVA form H$_4X$ (or XH$_4$, for example, CH$_4$, SiH$_4$). The geometries of these molecules are all readily predicted by the VSEPR theory discussed in Section 6.2.

In addition to these simple hydrides, which contain a single atom of the nonmetal, there are others that possess two or more nonmetal atoms. Some examples are given in Table 19.1. All these compounds are characterized by nonmetal atoms of the same element linked directly to one another, a phenomenon called **catenation.** Thus hydrogen peroxide, H$_2$O$_2$, has the Lewis structure

$$:\overset{\displaystyle H}{\underset{\displaystyle H}{\ddot{O}}}\!-\!\ddot{O}:$$

Similarly, there are others such as

$$H\!-\!\overset{H}{\underset{\cdot\cdot}{N}}\!-\!\overset{H}{\underset{\cdot\cdot}{N}}\!-\!H \qquad H\!-\!\overset{\displaystyle H}{\underset{\displaystyle H}{Si}}\!-\!\overset{\displaystyle H}{\underset{\displaystyle H}{Si}}\!-\!H \qquad H\!-\!\overset{\displaystyle H}{\underset{\displaystyle H}{C}}\!-\!\overset{\displaystyle H}{\underset{\displaystyle H}{C}}\!-\!\overset{\displaystyle H}{\underset{\displaystyle H}{C}}\!-\!H$$

hydrazine   disilane   propane

The ability of nonmetals to form compounds in which they bond to other like atoms varies greatly. You will notice, for example, that in Group VIA only oxygen and sulfur form such compounds. In Group VA we find that both nitrogen and phosphorus catenate, but the chain length seems to be limited to two atoms. When we proceed to Group IVA, all the elements, down to and including tin, exhibit this property and here we find chains containing three, four, and even more atoms. We also see that the tendency toward catenation generally decreases downward in a group, as evidenced by the trend toward shorter chains demonstrated by the heavier elements in Group IVA, Ge and Sn.

Of all the elements, carbon has the greatest capacity to form bonds to itself. In fact, the broad area of organic chemistry is concerned entirely with hydrocarbons and compounds that are derived from them by substituting other elements

We will see later that catenation is not restricted to nonmetal hydrides.

Table 19.1
Catenation among nonmetal hydrides

Group IVA	CH$_4$	SiH$_4$	GeH$_4$	SnH$_4$	
	C$_2$H$_6$	Si$_2$H$_6$	Ge$_2$H$_6$	Sn$_2$H$_6$	
	C$_3$H$_8$	Si$_3$H$_8$	Ge$_3$H$_8$		
	⋮	⋮			
	C$_n$H$_{2n+2}$	Si$_6$H$_{14}$			
	+				
	many others				
Group VA	NH$_3$	PH$_3$	AsH$_3$	SbH$_3$	BiH$_3$
	N$_2$H$_4$	P$_2$H$_4$			
Group VIA	H$_2$O	H$_2$S	H$_2$Se	H$_2$Te	H$_2$Po
	H$_2$O$_2$	H$_2$S$_2$			
		H$_2$S$_n$ ($n = 1 - 6$)			

Table 19.2
Standard free energies and
enthalpies of formation of nonmetal hydrides

	XH_n ΔG_f^0(kJ mol^{-1}) ΔH_f^0(kJ mol^{-1})			
BH$_3$ Not stable, simplest hydride is B$_2$H$_6$	CH$_4$ −50.6 −74.9	NH$_3$ −17 −46.0	H$_2$O −228 −242	HF −273 −271
	SiH$_4$ +55.2 −34	PH$_3$ +12.9 +5.4	H$_2$S −33.6 −20.6	HCl −95.4 −92.5
	GeH$_4$ +117 +90.4	AsH$_3$ +68.9 +66.4	H$_2$Se +62.3 +76	HBr −53.1 −36
	SbH$_3$ +148 +145	H$_2$Te +138 +154	HI +1.3 +26

for hydrogen. Organic compounds are compounds in which the molecular framework consists primarily of carbon-carbon chains. The unique ability of carbon to form such diverse compounds containing these long stable carbon chains is undoubtedly the reason why life has evolved around the element carbon instead of around another element such as silicon.

Preparation of nonmetal hydrides

Nonmetal hydrides are formed as products of many different chemical reactions; however, we will consider only two general methods of preparation here. One of these is the direct combination of the elements, as illustrated, for example, by the reaction of hydrogen with either chlorine,

$$H_2 + Cl_2 \longrightarrow 2HCl$$

or with oxygen

$$2H_2 + O_2 \longrightarrow 2H_2O$$

However, this method is not applicable to all the hydrides, as we can see by examining some of their thermodynamic properties shown in Table 19.2. Here we see that only the hydrides of the more active nonmetals possess negative free energies of formation. Those lying below the heavy colored line in the table have positive free energies of formation and from a practical standpoint cannot be prepared directly from the free elements. Instead an indirect procedure must be employed.

The rates of reaction toward hydrogen vary substantially among the nonmetals. In period 2, for instance, fluorine reacts immediately with hydrogen when they are placed in contact. On the other hand, H$_2$ and O$_2$ mixtures are stable virtually indefinitely, unless the reaction is initiated in some way—for example, by applying heat or introducing a catalyst.

Nitrogen is even less reactive than oxygen, not only toward hydrogen, but toward nearly all other chemical reagents as well. Presumably this is because of the high stability of the N$_2$ molecule that arises as a consequence of its strong triple bond (the bond energy of N_2 is 946 kJ mol^{-1}, compared to 502 and 159 kJ mol^{-1} for O_2 and F_2, respectively).

The second method of preparation of nonmetal hydrides involves the addition of protons, from a Brønsted acid, to the conjugate base of a nonmetal hydride, a reaction that we might depict as

$$X^{n-} + nHA \longrightarrow H_nX + nA^-$$

where X^{n-} is the conjugate base of the hydride H_nX, and HA is the Brønsted acid. Let's look at some examples.

The hydrogen halides are commonly prepared in the laboratory by treating a halide salt with a nonvolatile acid such as sulfuric or phosphoric acid.

$$NaCl(s) + H_2SO_4(l) \longrightarrow HCl(g) + NaHSO_4(s)$$

$$NaCl(s) + H_3PO_4(l) \longrightarrow HCl(g) + NaH_2PO_4(s)$$

In these examples HCl is removed as a gas, which causes the reaction to proceed to completion.

With the heavier halogens—bromine and iodine—sulfuric acid cannot be used because it is a sufficiently strong oxidizing agent to oxidize the halide ion to the free halogen. For example, when treated with concentrated H_2SO_4, I^- reacts as follows:

$$2I^- + HSO_4^- + 3H^+ \longrightarrow I_2 + SO_2 + 2H_2O$$

Phosphoric acid, being a much weaker oxidizing agent than H_2SO_4, simply supplies protons to I^-, and HI can therefore be produced in a reaction analogous to the production of HCl above, that is,

$$NaI(s) + H_3PO_4(l) \longrightarrow HI(g) + NaH_2PO_4(s)$$

The reaction is slow and the reaction mixture must be warmed to expel the HI.

As we proceed from right to left across a period—for example, from fluorine toward carbon—we have seen that the acid strengths of the H_nX compounds decrease. Thus HF is a stronger acid than H_2O which, in turn, is stronger than NH_3, and so forth. This means that the strengths of their corresponding conjugate bases *increase* from right to left ($C^{4-} > N^{3-} > O^{2-} > F^-$). As a result, the strength of the Brønsted acid required to react with the anion of the nonmetal to produce the hydride decreases. For example, the production of HF, whose conjugate base, F^-, is relatively weak, requires a strong acid such as H_2SO_4. Oxide ion, on the other hand, is a much stronger base than F^- and when treated with even a relatively weak source of protons—for example, acetic acid—the oxide ion gobbles them up to produce water.

$$O^{2-} + 2HC_2H_3O_2 \longrightarrow H_2O + 2C_2H_3O_2^-$$

This is a reaction we have seen before in Chapter 7.

Nitride ion, N^{3-}, is expected to be even a stronger base than O^{2-}. Therefore it is not surprising to find that Mg_3N_2 reacts with an acid as weak as H_2O to produce NH_3 in a reaction that we can interpret as a hydrolysis of the N^{3-} ion.

$$Mg_3N_2 + 6H_2O \longrightarrow 3Mg(OH)_2 + 2NH_3$$

Metal carbides, which can be prepared by heating an active metal with carbon, also react with water in the same fashion. Aluminum carbide, for instance, which contains C^{4-} ions, hydrolyzes according to the reaction

$$Al_4C_3 + 12H_2O \longrightarrow 4Al(OH)_3 + 3CH_4$$

This general method of preparation also extends to the third, fourth, and fifth periods, with the same trends in the strength of the Brønsted acid required to liberate the hydride. In period 3 we have these anions:

Group	IV	V	VI	VII
Anion	Si^{4-}	P^{3-}	S^{2-}	Cl^-

We again expect the anions to become increasingly basic as we move from right to left (from Cl^- to Si^{4-}); therefore the strength of the Brønsted acid needed to protonate the anion decreases. To form HCl from NaCl, a strong acid is required. Sulfide ion, on the other hand, is sufficiently basic to be extensively hydrolyzed in aqueous solution (as we found in Chapter 15), and solutions containing a soluble sulfide such as Na_2S always are very basic and have a strong odor of H_2S because of the reaction[3]

$$S^{2-} + 2H_2O \longrightarrow H_2S + 2OH^-$$

The ability of S^{2-} to pick up protons also explains why many insoluble metal sulfides dissolve in acid with the evolution of H_2S.

Phosphides, like sulfides, also hydrolyze on contact with water. However, because the P^{3-} ion is more basic than the S^{2-} ion, the hydrolysis proceeds essentially to completion. Thus aluminum phosphide, AlP, reacts with water to produce phosphine, PH_3.

PH_3 = phosphine

$$AlP + 3H_2O \longrightarrow Al(OH)_3 + PH_3$$

Moving left to Group IVA, we again find that a hydrolysis reaction serves to prepare silicon hydrides. A metal silicide such as Mg_2Si (which can be formed by simply heating Mg and Si together) reacts with water to generate a mixture of silanes; SiH_4, Si_3H_8, and so on, up to Si_6H_{14}.

SiH_4 = silane

The heavier nonmetals behave in much the same fashion as those above them. Thus H_2S and H_2Te, like H_2S, can be prepared by adding an acid to the metal selenide or telluride. Arsine, AsH_3, like phosphine, PH_3, is made by the hydrolysis of a metal arsenide such as Na_3As or AlAs, and the germanes, GeH_4, Ge_2H_6, and Ge_3H_8 are produced by the action of dilute HCl on Mg_2Ge.

AsH_3 = arsine

GeH_4 = germane

A hydrogen economy

As the world's supplies of petroleum dwindle and become increasingly more expensive, the search for alternative fuels grows more urgent. Coal, and its conversion to synthetic petroleumlike products, has been mentioned as a means of providing substitutes, as we've noted earlier. Another interesting alternative is suggested by proponents of the large-scale use of hydrogen as a fuel, around which virtually our entire energy economy could be built.

Hydrogen has some very attractive features as a fuel. Its reaction with oxygen

$$2H_2(g) + O_2(g) \longrightarrow 2H_2O(l) \quad \Delta H^0 = -572 \text{ kJ}$$

is highly exothermic and produces no pollutants. The gas could be either burned according to this reaction to produce heat or it could be used as a fuel in hydrogen-oxygen fuel cells to generate electricity directly. Since hydrogen is a gas, it could be pumped through the already existing network of pipelines that crisscross the nation carrying natural gas. Hydrogen is also available in virtually infinite amounts from water.

Despite all these advantages, however, there are problems. One of them is that hydrogen is not a very convenient portable fuel for use in an automobile.

[3] Actually, there is a two-step equilibrium involving both HS^- and H_2S. See Chapter 15.

High pressure tanks of gaseous hydrogen are heavy and can carry only relatively small amounts of this fuel. Other approaches are being tried, including combining hydrogen with certain metals to form metal hydrides that can be decomposed when the hydrogen is needed, but many technical difficulties still exist.

The main problem with hydrogen as a fuel is that it is not a primary energy source like petroleum and natural gas, which already exist in a state ready for use. Hydrogen can only be obtained by first investing energy to extract it from water, so hydrogen will only become a viable fuel if some inexpensive method can be found to produce it. A number of approaches to this problem have been proposed that make use of solar or nuclear energy.

Solar energy can be harnessed in several ways. One is to employ large arrays of solar cells to generate electricity that can be used in the electrolysis of water. Another is to focus solar energy by mirrors into solar furnaces that would heat water vapor to very high temperatures where it would decompose into H_2 and O_2. The mixture of gases would then be rapidly cooled before they could recombine to form H_2O. Nuclear reactors could also be used either to generate the electricity needed to decompose water by electrolysis, or to produce the high temperatures necessary to crack water into H_2 and O_2.

Recently, advances have been made in harnessing solar energy directly to split water into H_2 and O_2. Using special catalysts, scientists in France have been able to achieve the efficient decomposition of water under visible and ultraviolet illumination. If this process can be made industrially practicable, a convenient means of converting solar energy directly to a useful form of stored chemical energy will be available.

It makes no sense, of course, to burn petroleum to make electricity to produce hydrogen by electrolysis!

19.2 CARBON

The element carbon is found in every living thing; all life as we know it is based on carbon-containing compounds. It is this fact that is responsible for the term *organic* that is used to describe hydrocarbons, and those compounds that come from hydrocarbons by substituting other atoms for some of the hydrogen atoms in their molecules. At one time it was believed that such substances could only be made within *living* organisms. This has since been shown to be false, but the name organic is still used in discussing these kinds of compounds, and the name *inorganic* is used to describe all the rest. This section deals with the inorganic compounds of carbon and the properties of the element itself. Organic chemistry is discussed in Chapter 23.

Carbon is found in period 2 at the head of Group IVA. Its atoms each have four valence electrons that they tend to share with other atoms in the formation of four covalent bonds. Although in most of its compounds carbon forms four bonds, we will see some exceptions in this section.

Carbon occurs in nature as both the free element and in compounds. Coal contains elemental carbon, and when it is heated strongly in the absence of air, volatile substances are driven off and the material that remains—called **coke**—is almost all carbon. Diamonds are, for all practical purposes, pure carbon. In the combined state, we find carbon in all living things and in fossil fuels such as methane, CH_4, and petroleum. Carbon also occurs in large amounts in carbonates such as limestone.

The free element

In Chapter 6 we saw that carbon can exist in two different allotropic forms. One of them is graphite, a soft black slippery solid that is a reasonably good conductor of electricity. The other is diamond, an extremely hard, nonconducting substance well known for its gem quality crystals. Their different properties can be

At room temperature and pressure, graphite is the most thermodynamically stable form of carbon.

traced to the differences in the way the carbon atoms are bonded to each other.

Graphite's loosely stacked layer structure (Figure 6.27 on p. 212) explains both its slippery feel and electrical conductivity. Because the layers are only weakly attracted to each other by London forces, they can slide over each other easily. Gas molecules trapped between the layers also help considerably by acting as molecular "ball bearings." The electrical conductivity of graphite is also a consequence of the solid's structure. Within each layer there is extensive delocalization of the pi-electron cloud. An electron forced into one end of a layer can produce a flow of charge through the pi cloud and cause an electron to exit from the other end. (Interestingly, it has been found by studing single crystals of graphite that the electrical conductivity is very good in a direction parallel to the layers, but very poor in a direction perpendicular to the layers.) Because of these properties, graphite is used commercially as a lubricant and to construct electrodes—for example, in ordinary dry cells.

In diamond, the interlocking network of covalent bond produces large crystals that are actually single molecules! As we learned earlier, this makes diamond extremely hard. Aside from its decorative use in jewelry, diamond is one of industry's most important abrasives, and is used to make cutting and grinding tools.

The thermodynamics of the conversion of graphite to diamond were discussed in Chapter 12.

Slippery graphite and brilliant diamonds represent quite pure forms of elemental carbon. Less pure, predominantly graphitic forms of carbon are also known. One of these—charcoal—is formed by heating wood strongly in the absence of air. Charcoal has a particularly open structure with an enormous surface area for a given mass of carbon. *Activated charcoal*—a finely pulverized form having a surface are of about 1000 m^2 per gram—has many commercial uses. Its large surface area permits small amounts of it to adsorb large numbers of molecules, a property that allows it to remove molecules from the air that have offensive odors and to remove toxic impurities from water. In fact, several municipalities located in areas polluted by chemical spills have had to install activated carbon filtration systems to make contaminated well water pure enough to drink.

When a hydrocarbon is burned in a very limited supply of oxygen the hydrogen combines with the oxygen to form water, and finely divided carbon—called *carbon black*—remains.

$$CH_4(g) + O_2(g) \longrightarrow C(s) + 2H_2O(g)$$

Carbon black is used as a pigment in black inks, and large amounts of it are used in making automobile tires.

A slurry of granular activated carbon (activated charcoal) is pumped into a water treatment facility where it will be used to remove offensive odors and toxic impurities from municipal drinking water.

Oxides and oxoacids of carbon

Carbon forms two principal oxides: carbon monoxide and carbon dioxide. Carbon monoxide has the Lewis structure

$$:C\equiv O:$$

and is one of a rather small number of species in which carbon has only three bonds. Carbon monoxide is a nearly nonpolar substance with a low melting point and boiling point (mp = −199°C, bp = −192°C). It is unreactive toward water and has a low solubility. It is also flammable and burns with a hot flame, which is why the mixture of CO and H$_2$ from the water gas reaction (Section 19.1) is a useful industrial fuel.

Carbon monoxide is particularly dangerous because it is both colorless and odorless.

One way that carbon monoxide can be formed is by burning carbon or a hydrocarbon in a limited supply of oxygen. These are exactly the conditions that exist in the fire in a charcoal barbecue or gasoline engine, and both produce some carbon monoxide. As you are probably aware, carbon monoxide is quite toxic—it binds to hemoglobin in the blood, thereby preventing it from carrying oxygen—and that is why it is dangerous to be in an enclosed space with either a charcoal fire or a running automobile engine.

In the laboratory, small amounts of carbon monoxide can be made by treating formic acid, $HCHO_2$, with concentrated sulfuric acid. Sulfuric acid has a strong affinity for water and actually extracts the components of water (two hydrogens and an oxygen) from a formic acid molecule.

$$HCHO_2(l) \xrightarrow{H_2SO_4} H_2O(l) + CO(g)$$

Industrially, carbon monoxide is proving to be an extremely important chemical. More and more interest is being shown in reactions like those discussed in the last section in which CO and H_2 are combined catalytically to form hydrocarbons—a replacement for petroleum. We saw in the last section that CO can be made by reacting steam with white-hot carbon in the form of coke.

$$C(s) + H_2O(g) \xrightarrow{1000°C} CO(g) + H_2(g)$$

At high temperatures, carbon monoxide is an effective reducing agent. For example,

$$Fe_2O_3(s) + 3CO(g) \xrightarrow{heat} 2Fe(s) + 3CO_2(g)$$

This overall reaction is used to extract iron from its ore and was discussed in greater detail in Chapter 18.

An interesting chemical property of carbon monoxide is its ability to form covalent compounds with transition metals in which the metal is in a low (or zero) oxidation state. These are called **metal carbonyl compounds.** An example is $Ni(CO)_4$, nickel carbonyl, which is formed by simply warming metallic nickel in the presence of CO. The compound is quite volatile and very toxic.

$$Ni(s) + 4CO(g) \longrightarrow Ni(CO)_4(g)$$

This reaction, you may recall, is used in the Mond process for purifying nickel.

Some other examples of metal carbonyls are $Cr(CO)_6$, $Mn_2(CO)_{10}$, $Fe(CO)_5$, and $Co_2(CO)_8$.

The second major oxide of carbon is carbon dioxide.

$$\ddot{O}=C=\ddot{O}$$

As predicted by VSEPR theory, CO_2 is a linear molecule. It is also nonpolar, because of its symmetrical structure. The triple point of CO_2 is above 1 atm, and when CO_2 is cooled at atmospheric pressure it condenses to a solid rather than to a liquid. Solid carbon dioxide is called **dry ice** and its temperature is −78°C, the temperature at which CO_2 sublimes at 1 atm. Because gaseous CO_2 is more dense ("heavier") than air and because it doesn't support combustion, CO_2 is used to smother fires, especially those that are difficult to extinguish using water.

Carbon dioxide can be formed in a number of ways. Complete combustion of carbon, a hydrocarbon, or carbon monoxide gives CO_2

$$C + O_2 \longrightarrow CO_2$$

$$CH_4 + 2O_2 \longrightarrow CO_2 + 2H_2O$$

$$2CO + O_2 \longrightarrow 2CO_2$$

Industrially, it is often made by the thermal decomposition of limestone.

$$CaCO_3(s) \xrightarrow{heat} CaO(s) + CO_2(g)$$
limestone

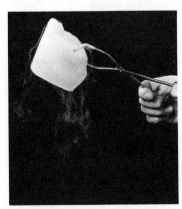

Dry ice—solid carbon dioxide at a temperature of −78°C—sublimes at atmospheric pressure.

Carbon dioxide is also formed naturally by decomposing organic matter.

In the laboratory, CO_2 can be conveniently prepared by the reaction of an acid with a carbonate—for example, calcium carbonate.

$$CaCO_3(s) + 2HCl(aq) \longrightarrow CaCl_2(aq) + H_2O(l) + CO_2(g)$$

Similar reactions with bicarbonates (especially $NaHCO_3$) were mentioned in Chapter 18 (page 664).

Carbon dioxide is consumed in large quantities by industry. You learned in Chapter 18 that it is used to make sodium carbonate by the Solvay process. Major uses also include refrigeration (including dry ice) and beverage carbonation. The process of carbonation consumes about 35% of a total annual production of about 2.3×10^9 kg of CO_2.

In nature, green plants consume carbon dioxide in photosynthesis and produce glucose, $C_6H_{12}O_6$, a sugar from which they manufacture starch, cellulose, and other chemicals. In the process, oxygen is released to the atmosphere.

$$6CO_2(g) + 6H_2O(l) \xrightarrow[\text{chlorophyll}]{\text{light}} C_6H_{12}O_6(aq) + 6O_2(g)$$

Carbon dioxide is moderately soluble in water, and its solutions are slightly acidic due to the formation of carbonic acid.

$$CO_2(aq) + H_2O(l) \rightleftharpoons H_2CO_3(aq)$$
carbonic acid

Carbonic acid, as we've seen earlier, is a weak diprotic acid that ionizes in two steps. Neutralization with a base is able to give two types of salts: carbonates, formed by complete neutralization, and bicarbonates (hydrogen carbonates), formed by partial neutralization.

$$NaOH + H_2CO_3 \longrightarrow NaHCO_3 + H_2O$$

$$2NaOH + H_2CO_3 \longrightarrow N_2CO_3 + 2H_2O$$

Bicarbonates such as $NaHCO_3$, you recall, are easily decomposed by heating to give carbonates ($2NaHCO_3 \rightarrow Na_2CO_3 + H_2O + CO_2$).

In Chapter 7 it was pointed out that when groundwater containing dissolved CO_2 trickles slowly through large deposits of limestone it gradually dissolves the $CaCO_3$ and forms huge limestone caverns. The equilibrium equations are

$$H_2O + CO_2(g) \rightleftharpoons H_2CO_3(aq)$$

$$H_2CO_3(aq) + CaCO_3(s) \rightleftharpoons Ca(HCO_3)_2(aq)$$

When a drop of water containing a dilute solution of $Ca(HCO_3)_2$ collects on the ceiling of a limestone cave, it gradually evaporates. This causes the first of these equilibria to be shifted to the left as H_2CO_3 decomposes and gaseous CO_2 is released. The loss of H_2CO_3 also affects the second reaction. That equilibrium is also shifted to the left, causing $CaCO_3$ to precipitate. Over a period of many years, drop after drop evaporates and a stalactite composed of calcium carbonate is slowly formed. Drops that fall to the floor evaporate, too, and give rise to stalagmites. Limestone caves are not the only places to observe the growth of stalactites. Calcium carbonate is one of the substances formed during the curing of concrete, and when water seeps through the concrete of a highway overpass, stalactites looking a bit like whiskers are formed that hang from the overpass above the roadway, as shown in Figure 19.2.

Because of reactions like those just described, the groundwater in many parts of the country contains dissolved calcium as Ca^{2+}, as well as other ions such as Mg^{2+} and Fe^{3+}. If their concentrations are sufficiently large, these ions can interfere with the action of ordinary soap by forming a precipitate with it.

CO_2 has replaced chlorofluorocarbons as a propellant in many aerosol products.

The oceans represent a huge reservoir of dissolved CO_2.

Figure 19.2

Photograph of stalagtites formed on the ceiling of a highway overpass.

Figure 19.3

Photograph of boiler scale.

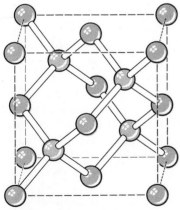

Figure 19.4

The structure of carborundum. Gray spheres represent silicon and the colored spheres represent carbon. In diamond all the spheres would be the same.

The water is then described as **hard water,** and the ions causing the problem are called **hardness ions.** Hardness ions can be removed from water in a number of ways. If the solution also contains bicarbonate ion—as it would if the ions in it came from the dissolving of limestone—heating the water will drive off CO_2 and cause a precipitate to form. For example,

$$Ca(HCO_3)_2(aq) \xrightarrow{\text{heat}} CaCO_3(s) + H_2O(l) + CO_2(g)$$

This type of hard water causes serious problems because the precipitate (called *boiler scale*) can clog hot water pipes and make it more difficult to heat water in a boiler (see Figure 19.3).

If the hard water doesn't contain HCO_3^-, the ions can still be precipitated by adding *washing soda*—a hydrate of sodium carbonate, $Na_2CO_3 \cdot 10H_2O$. The carbonate ion precipitates the hardness ions,

$$Ca^{2+}(aq) + CO_3^{2-}(aq) \longrightarrow CaCO_3(s)$$

and, by removing them, allows the soap to do its job.

Other inorganic compounds of carbon

Carbides Binary compounds formed between carbon and a metal or metalloid are generally referred to as carbides. They are of three types. In *covalent carbides*, carbon is bonded covalently to the other element. An example is silicon carbide, SiC, known commercially as **carborundum.** It is made by heating silicon dioxide (sand) and carbon to very high temperatures.

$$SiO_2(s) + 3C(s) \xrightarrow{\text{heat}} SiC(s) + 2CO(g)$$

This solid has the same structure as diamond, except that silicon atoms alternate with carbon atoms (Figure 19.4). Like diamond, silicon carbide is very hard and is used as an abrasive—for example, in sandpaper.

Ionic carbides (also called *saltlike carbides*) are formed from carbon and active metals, and contain more or less discrete ions of carbon in their solids. An example mentioned earlier (page 689) is Al_4C_3, which contains the anion C^{4-}. When placed into water it hydrolyzes and methane is formed. A particularly important saltlike carbide is calcium carbide, CaC_2, which contains the ion,

$[:C≡C:]^{2-}$, called acetylide ion. It is made by reacting calcium oxide (lime) with carbon at high temperatures.

$$CaO(s) + 3C(s) \longrightarrow CaC_2(s) + CO(g)$$

Calcium carbide reacts with water to form acetylene, the gas used in welding torches.

$$CaC_2(s) + 2H_2O(l) \longrightarrow \underset{\textbf{acetylene}}{C_2H_2(g)} + Ca(OH)_2(s)$$

The third type of carbide is called an *interstitial carbide*. In these substances, carbon atoms fit into voids (empty spaces called *interstices*) between the metal atoms in a crystal. An example is tungsten carbide, WC, which is extremely hard and brittle and is used to make cutting tools used to machine other metals.

Cyanides Hydrogen cyanide, HCN, can be made by a number of reactions, but one that is especially important commercially is the reaction of ammonia with methane. It is carried out at high temperatures over a platinum catalyst.

$$NH_3(g) + CH_4(g) \xrightarrow[\text{Pt}]{1200°C} HCN(g) + 3H_2(g)$$

Hydrogen cyanide is an intermediate in the synthesis of a number of well-known plastics, including nylon. It is also a very potent, fast-acting poison that both deactivates critical enzymes in the body and binds irreversibly to iron in hemoglobin in the blood. In water, HCN is a weak acid and salts such as NaCN can be formed by neutralization. Like HCN, NaCN is a deadly poison. It is used in certain electroplating baths and to extract gold from its ores.

CN^- forms many stable complex ions.

Carbon Disulfide Carbon reacts directly with sulfur to form carbon disulfide, CS_2.

$$C + 2S \xrightarrow[\text{temp}]{\text{high}} CS_2$$

$:\overset{..}{S}=C=\overset{..}{S}:$

carbon disulfide

It is a useful solvent, but it is extremely flammable—boiling water is hot enough to ignite it! When it burns, CO_2 and SO_2 are formed. Besides being a solvent, carbon disulfide is a useful chemical reagent. It is used commercially to manufacture carbon tetrachloride.

$$CS_2 + 3Cl_2 \longrightarrow \underset{\substack{\textbf{carbon} \\ \textbf{tetrachloride}}}{CCl_4} + S_2Cl_2$$

19.3 OXYGEN

The atmosphere is 20.9% oxygen by volume.

There are very few people who are unaware of the importance of oxygen to our existence. Breathing oxygen keeps us alive, and fuels couldn't be burned without it. Oxygen is literally everywhere—not only as a free element in the atmosphere, but also in all living creatures as well as in most of the substances that surround them. Water, which covers much of the Earth, is one-third oxygen on an atom basis, but about 89% oxygen by mass. In the Earth's crust, which is composed mostly of silicate minerals, oxygen is the *most* abundant element—46.6% by mass, 62.6% if counted by atoms, and an amazing 93.8% if measured by volume!

In its ground state, the element oxygen has the electron configuration $1s^2 2s^2 2p^4$, and therefore needs only two electrons to complete its octet. As you know, it is able to accomplish this either by acquiring electrons or by electron sharing. In the free state, oxygen exists as diatomic molecules, O_2 (called dioxygen). The molecular orbital energy diagram for O_2 was discussed in Chap-

The bond length and bond energy both suggest a double bond in O_2.

ter 6 and accounts for both the double bond between the oxygen atoms as well as the fact that O_2 is paramagnetic, with two unpaired electrons. When cooled, oxygen forms a pale blue liquid that boils at $-183°C$. Oxygen freezes at $-219°C$, giving a pale blue solid.

Preparation and uses

Because it is so freely available, O_2 is normally produced at the plant site where it will be used.

The most obvious source of oxygen when it is needed in large quantities is the atmosphere, and the commercial preparation of oxygen involves separating it from liquefied air. Nitrogen, which has a lower boiling point than O_2, is removed from the liquid air by allowing it to boil off, thereby leaving behind liquid oxygen contaminated with small amounts of N_2 and argon (another component of air). Warming the liquid, of course, converts the oxygen to a gas. Most of the oxygen prepared this way (about 85%) is used in the steel industry and metal fabrication. Some is also used to make chemical intermediates in the manufacture of plastics, in waste water treatment, for life support in hospitals, and in bleaching pulp and paper.

In the laboratory, small amounts of pure oxygen can be prepared in a number of ways. Joseph Priestley (1773–1804), credited with the discovery of oxygen, obtained the gas by thermally decomposing mercury(II) oxide. Usually, however, oxygen is made in the laboratory either by electrolysis of water or thermally decomposing potassium chlorate, $KClO_3$, using manganese dioxide as a catalyst.

$$2KClO_3(s) \xrightarrow[\text{heat}]{MnO_2} 2KCl(s) + 3O_2(g)$$

Like hydrogen, oxygen can be collected by the displacement of water because of its relatively low solubility (Figure 19.5).

In nature, oxygen is generated by green plants during photosynthesis. The chemical equation for the conversion of CO_2 and H_2O into glucose and O_2 was described in the last section. Large forests and the plankton in the sea are responsible for maintaining the balance of oxygen in the Earth's atmosphere.

Ozone

When an electric discharge is passed through molecular oxygen a second allotrope of the element is formed that is called **ozone,** O_3.

$$3O_2(g) \longrightarrow 2O_3(g) \qquad \Delta H = +284 \text{ kJ}$$

Its pungent odor can sometimes be detected after a severe thunderstorm or near electrical machinery. In high concentrations it is poisonous. In Chapter 9 we saw

Liquid oxygen was used to oxidize the fuel in this modified Saturn V rocket, which put the Skylab space station into orbit in 1973. More recently, it has also been used along with liquid hydrogen to power the space shuttle.

Figure 19.5

The laboratory preparation of oxygen by the thermal decomposition of potassium chlorate, $KClO_3$, using manganese dioxide, MnO_2, as a catalyst. Care must be taken not to heat the rubber stopper or allow the hot $KClO_3$ to come in contact with it.

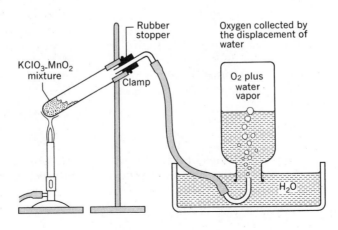

that the structure of ozone can be represented by the resonance formulas

$$\underset{\ddot{O}}{\overset{O}{\diagup}}\overset{\diagdown}{\underset{\ddot{O}}{}} \longleftrightarrow \underset{\ddot{O}}{\overset{O}{\diagdown}}\overset{\diagup}{\underset{\ddot{O}}{}}$$

VSEPR theory predicts that O_3 will be nonlinear.

As we would expect from the number of groups of electrons around the central oxygen atom, the molecule is nonlinear.

When ozone is formed, a large quantity of energy must be supplied, and an equally large quantity of energy, of course, is released when the ozone decomposes. Therefore, since highly exothermic reactions tend to be quite spontaneous, ozone decomposes very easily. Ozone is also a very powerful oxidizing agent—considerably more powerful than ordinary O_2. This has both advantages and disadvantages. On the positive side, the oxidizing power of ozone shows promise as an alternative to chlorine in the treatment of municipal drinking water supplies. It has been found that Cl_2 in drinking water is able to form chlorine compounds with some of the organic substances that are also present in small amounts. Some of these products, such as chloroform, $CHCl_3$, have been

Recall that carcinogenic means cancer-causing.

shown to be carcinogenic, and the long-range toxic effects of others are open to question. All these problems are avoided if bacteria in the water are killed by O_3 instead of Cl_2. Another problem arises, however, because little residual O_3 remains after treatment, so any bacteria that get into the water after the treatment process are not destroyed.

On the negative side, the powerful oxidizing ability of ozone can cause extensive damage to plants and articles made of natural rubber. As we will see in the next section, ozone is one of the major constituents of smog, so in locations that experience severe smog episodes, damage caused by ozone can be particularly troublesome.

The ozone concentration at these altitudes approaches 27% by mass.

In the Earth's upper atmosphere, at altitudes ranging from about 15 to 24 km, ozone is formed in appreciable amounts from O_2 by absorption of ultraviolet radiation from the sun. The light energy first splits oxygen molecules into oxygen atoms.

$$O_2 \xrightarrow{h\nu} 2O$$

Reaction of oxygen atoms with oxygen molecules produces O_3.

$$O_2 + O \longrightarrow O_3$$

Ozone also absorbs ultraviolet light, especially at wavelengths that prove harmful to living organisms. This causes the O_3 to decompose and form O_2 again. The absorption of UV radiation by ozone converts the energy of the UV light into heat and also protects the inhabitants of our planet from the radiation's harmful effects.

You are probably aware of the controversy that has arisen in recent years concerning the effects of certain human activity on this ozone shield. For a time there was worry that high-flying supersonic airliners such as the Concorde would be emitting nitrogen oxide pollutants in quantities that would have a significant effect on the ozone concentration. Even small amounts of NO could have damaging effects because of such reactions as

Destruction of stratospheric ozone could lead to increased incidents of skin cancer and a rise in the Earth's average temperature.

$$NO + O_3 \longrightarrow O_2 + NO_2$$

$$NO_2 + O \longrightarrow O_2 + NO$$

Thus, molecules of NO remove not only O_3, but also oxygen atoms needed to reform the O_3. Since the product of the second reaction is the reactant in the first, the cycle can be repeated many times by each NO molecule. Despite these early concerns it appears that the Concorde, as presently operated, has not caused any noticeable change in the stratospheric ozone concentration.

Another danger to the ozone shield has been the release into the atmosphere of Freons. These substances are composed of carbon, fluorine, and chlorine, and are called chlorofluorocarbons. They have been widely used as propellants in aerosol cans and as refrigerants. An example is Freon-11, $CFCl_3$, which is colorless, virtually odorless, and unreactive under ordinary conditions. However, as these chlorofluorocarbons diffuse into the upper atmosphere they can absorb ultraviolet radiation, which ruptures carbon-chlorine bonds to give chlorine atoms.

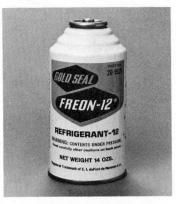

Freon-12, CCl_2F_2, is the refrigerant used in the air conditioners in automobiles. Containers of it can be purchased in any automotive store.

$$CFCl_3 \xrightarrow{h\nu} CFCl_2 + Cl$$

Both products are reactive, but the chlorine atoms are believed to be involved in the destruction of ozone. Possible reactions are

$$Cl + O_3 \longrightarrow ClO + O_2$$

$$ClO + O \longrightarrow Cl + O_2$$

Once again we have a cycle by which a single species—a chlorine atom—destroys not only O_3, but oxygen atoms needed to replace O_3. Concern over the threat posed by Freons has led to a ban on their use as aerosol propellants in the United States.

Compounds of oxygen

Oxygen forms compounds with every element except helium, neon, and argon. These compounds are called **oxides** and are generally of two types: ionic or covalent. *Ionic oxides* are formed with many metals and can usually be made by direct combination of the elements, for example,

$$4Li(s) + O_2(g) \longrightarrow 2Li_2O(s)$$

$$2Ca(s) + O_2(g) \longrightarrow 2CaO(s)$$

Remember, metal oxides are basic anhydrides.

Metal oxides are basic and, if soluble, give solutions containing hydroxide ion caused by the virtual complete hydrolysis of O^{2-}.

$$Li_2O(s) + H_2O(l) \longrightarrow 2LiOH(aq) \qquad \text{(Molecular equation)}$$

$$O^{2-}(aq) + H_2O(l) \xrightarrow{100\%} 2OH^-(aq) \qquad \text{(Net ionic equation)}$$

Even insoluble metal oxides are basic because they neutralize acids. Iron(III) oxide, for example, dissolves in acids such as HCl or H_2SO_4.

$$Fe_2O_3(s) + 6H^+(aq) \longrightarrow 2Fe^{3+}(aq) + 3H_2O$$

This reaction, in fact, is often used to remove rust (Fe_2O_3) from iron or steel prior to being given a protective coating of zinc or tin. The acid treatment is called **pickling.**

Some oxides of metals show acidic properties as well as basic ones, and are said to be amphoteric. Examples that we have discussed earlier are the oxides of beryllium and aluminum, which dissolve in both acids and bases For instance,

$$Al_2O_3 + 6H^+ \longrightarrow 2Al^{3+} + 3H_2O$$

$$Al_2O_3 + 2OH^- \longrightarrow 2AlO_2^- + H_2O$$

Covalent oxides are generally formed by the nonmetals. In them, oxygen completes its octet by electron sharing—normally by forming either two single bonds (as in H_2O) or one double bond (as in CO_2). Exceptions are CO and NO,

Table 19.3
Typical oxides of the nonmetallic elements

Group III	B_2O_3			
Group IV	CO	SiO_2	GeO_2	
	CO_2			
Group V	N_2O	P_4O_6	As_4O_6	Sb_4O_6
	NO	P_4O_{10}	$As_2O_5{}^a$	$Sb_2O_5{}^a$
	N_2O_3			
	NO_2; (N_2O_4)			
	N_2O_5			
Group VI	O_2	SO_2	SeO_2	TeO_2
	O_3	SO_3	SeO_3	TeO_3
Group VII	OF_2	Cl_2O	Br_2O	I_2O_5
	O_2F_2	ClO_2	BrO_2	I_2O_7
		Cl_2O_7		

^{a} Molecular structure unknown.

which contain triple bonds. Table 19.3 contains a list of many oxides of the nonmetallic elements.

There are several ways that oxides of the nonmetallic elements can be made. Except for the halogens and the noble gases most oxides can be prepared simply by the direct union of the elements, as typified by the reactions

$$S + O_2 \longrightarrow SO_2$$

$$C + O_2 \longrightarrow CO_2 \qquad \text{(Excess oxygen)}$$

$$2C + O_2 \longrightarrow 2CO \qquad \text{(Limited supply of oxygen)}$$

$$2H_2 + O_2 \longrightarrow 2H_2O$$

Not all oxides can be prepared effectively in this manner, however. For example, in Table 19.4 we see that many oxides of nitrogen have positive free energies of formation and, therefore, from what we know of thermodynamics, they cannot be synthesized in significant amounts directly from the elements.[4] In these instances, and in others too, indirect methods of preparation are employed.

The indirect procedures are, expectedly, many in number. However, a few

Table 19.4
Thermodynamic properties
of some nitrogen oxides

Oxide	ΔH_f^0(kJ mol^{-1})	ΔG_f^0(kJ mol^{-1})
$N_2O(g)$	+81.5	+104
$NO(g)$	90.4	86.8
$NO_2(g)$	38	51.9
$N_2O_4(g)$	9.7	98.3
$N_2O_5(g)$	11	115

[4] Since they also have positive ΔS_f^0, it should be possible to prepare them at very high temperatures where $\Delta H - T\,\Delta S$, and hence ΔG, is negative. This is the case with NO, which is formed in small amounts near 3000°C. The high-temperature production of NO in motor vehicle engines is a major source of urban air pollution.

generalizations can be made. In some cases an oxide can be prepared from a lower oxide by further reaction with oxygen. The synthesis of SO_3, for example, consists of the catalytic oxidation of SO_2,

$$2SO_2 + O_2 \longrightarrow 2SO_3$$

Earlier (page 503) it was mentioned that this reaction is promoted in catalytic mufflers originally designed to speed up the conversion of CO to CO_2.

$$2CO + O_2 \longrightarrow 2CO_2$$

The combustion of CO, you recall, is an important industrial reaction because CO is often used as a fuel.

Another technique that serves to produce oxides is the combustion of non-metal hydrides. Methane, the chief constituent of natural gas, and other hydro-carbons burn to produce CO_2 and H_2O when an excess of O_2 is present.

$$CH_4 + 2O_2 \longrightarrow CO_2 + 2H_2O$$
methane

$$2C_8H_{18} + 25O_2 \longrightarrow 16CO_2 + 18H_2O$$
octane
(gasoline)

When insufficient O_2 is available, as in an automobile engine, CO may be produced instead of CO_2.

A reaction of this general type that is of great commercial importance is the oxidation of ammonia. In this case a platinum catalyst is used and the reaction is

$$4NH_3 + 5O_2 \xrightarrow{Pt} 4NO + 6H_2O$$

The NO formed in this reaction is readily oxidized further to produce NO_2,

$$2NO + O_2 \longrightarrow 2NO_2$$

In the absence of a catalyst, combustion of ammonia in air gives N_2 and H_2O.

$$4NH_3(g) + 3O_2(g) \longrightarrow 2N_2(g) + 6H_2O(g)$$

These reactions are essential to the production of nitric acid and nitrates, which are used in manufacturing many chemicals from explosives to fertilizers. We will discuss them in the next section.

Finally, another indirect method of obtaining nonmetal oxides makes use of oxidation-reduction reactions. For instance, when nitric acid serves as an oxidizing agent, the nitrate ion is reduced and, depending on conditions, nitrogen in various oxidation states may be produced. When concentrated nitric acid is used, the reduction product is frequently NO_2. Dilute solutions of HNO_3 often yield NO as the reduction product.

$$4HNO_3 + Cu \longrightarrow Cu(NO_3)_2 + 2NO_2 + 2H_2O \quad \text{(Concentrated)}$$

$$8HNO_3 + 3Cu \longrightarrow 3Cu(NO_3)_2 + 2NO + 4H_2O \quad \text{(Dilute)}$$

Similarly, hot concentrated sulfuric acid is a fairly potent oxidizing agent, and the reduction product is usually SO_2; for example,

$$Cu + 2H_2SO_4 \longrightarrow CuSO_4 + SO_2 + 2H_2O$$

In Chapter 7 we saw that nonmetal oxides are acid anhydrides; when they react with water they produce oxoacids. For example,

$$SO_2 + H_2O \longrightarrow H_2SO_3$$

Neutralization gives the corresponding oxoanion. Thus, sulfurous acid (H_2SO_3) gives sulfite ion (SO_3^{2-}). Table 19.5 provides a summary of the simple oxoacids and oxoanions of the nonmetallic elements. As indicated in the table, some of the oxoacids cannot actually be isolated, although their corresponding anions can be.

Table 19.5
Simple oxoacids and oxoanions of the nonmetallic elements

Group IIIA	H_3BO_3 (no simple borates)			
Group IVA	H_2CO_3 (CO_3^{2-})	$H_4SiO_4{}^a$ (SiO_4^{4-})	$H_4GeO_4{}^a$ (GeO_4^{4-})	
Group VA	HNO_2 (NO_2^-) HNO_3 (NO_3^-)	H_3PO_3 (HPO_3^{2-}) H_3PO_4 (PO_4^{3-})	H_3AsO_4 (AsO_4^{3-})	
Group VIA		H_2SO_3 (SO_3^{2-}) H_2SO_4 (SO_4^{2-})	H_2SeO_3 (SeO_3^{2-}) H_2SeO_4 (SeO_4^{2-})	$H_2TeO_3{}^a$ (TeO_3^{2-}) $Te(OH)_6$ ($TeO(OH)_5^-$)
Group VIIA	HOF	$HOCl$ (OCl^-) $HClO_2$ (ClO_2^-) $HClO_3$ (ClO_3^-) $HClO_4$ (ClO_4^-)	$HOBr$ (OBr^-) $HBrO_2$ (BrO_2^-) $HBrO_3$ (BrO_3^-) $HBrO_4$ (BrO_4^-)	HOI (OI^-) HIO_3 (IO_3^-) H_5IO_6 ($H_2IO_6^{3-}$) HIO_4 (IO_4^-)

^{a} Not observed.

Other compounds of oxygen

Peroxides and superoxides of the alkali metals were discussed on page 663.

Peroxides and Superoxides In addition to forming normal oxides with metals (i.e., compounds that contain the O^{2-} ion), oxygen combines with the more active alkali metals to form peroxides and superoxides, which contain the O_2^{2-} and O_2^- ions, respectively. Sodium forms the yellow peroxide.

$$2Na(s) + O_2(g) \longrightarrow Na_2O_2(s)$$
sodium peroxide

Potassium, rubidium, and cesium form yellow to orange superoxides.

$$K(s) + O_2 \longrightarrow KO_2(s)$$
potassium superoxide

Hydrogen Peroxide In Chapter 18 we saw that when Na_2O_2 is placed in water the peroxide ion hydrolyzes to form hydrogen peroxide, H_2O_2.

$$Na_2O_2(s) + 2H_2O(l) \xrightarrow{\text{cold}} 2NaOH(aq) + H_2O_2(aq)$$

Hydrogen peroxide is a molecular substance having an oxygen-oxygen bond.

The structure of the molecule is shown in Figure 19.6. In water, H_2O_2 is a weak acid,

$$H_2O_2 \rightleftharpoons H^+ + HO_2^- \qquad K_a = 1.5 \times 10^{-12}$$

Pure H_2O_2 is a colorless liquid with a boiling point of 150°C, but in its pure

Figure 19.6
The structure of hydrogen peroxide.

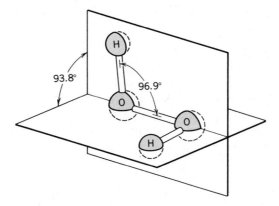

state it is an extremely hazardous substance to work with. Its decomposition to water and oxygen can occur explosively.

$$2H_2O_2(l) \longrightarrow 2H_2O(l) + O_2(g)$$

This decomposition is promoted by heat or by traces of many heavy metal ions.

Hydrogen peroxide is usually purchased as a solution in water. Solutions of 3% H_2O_2 by mass can be purchased in drug stores for use as an antiseptic. Its oxidizing power destroys bacteria, while blood catalyzes its decomposition. The fizzing action caused by escaping O_2 dislodges dirt and other foreign matter from a wound. More concentrated solutions of H_2O_2 are effective oxidizing agents and are used as bleaches for hair and industrially in the bleaching of cotton fabrics.

19.4 NITROGEN

The nitrogen in organic matter represents a small fraction of the Earth's total.

Nitrogen is another of the principal elements found in all living creatures. It is an essential constituent of amino acids—the primary building blocks of proteins—and is the key element in a precise lock-and-key fit between molecules that cells use to decipher the genetic code incorporated in the DNA residing in their nuclei. Because of the importance of nitrogen to the growth of plants and animals, it is not surprising that nitrogen compounds—particularly fertilizers—rank very high in total annual commercial production.

The element nitrogen has as its ground state electron configuration, $1s^2 2s^2 2p^3$. It achieves a completed octet either by gaining electrons to form ionic nitrides that contain the N^{3-} ion or by covalent bonding. In many of its compounds, nitrogen forms three bonds.

Very little nitrogen is present in a combined state in the Earth's crust. Instead, most occurs as diatomic molecules of N_2 (dinitrogen) in the atmosphere. This is a colorless, odorless, quite unreactive gas that boils at $-196°C$ and freezes at $-210°C$. Its lack of reactivity is attributed to the strength of the nitrogen-nitrogen triple bond

$$:N{\equiv}N:$$

The bond energy of N_2 (946 kJ mol^{-1}) is very high, so the molecule is very stable. For example, at 25°C (room temperature), N_2 reacts directly only with lithium to form Li_3N, although the nitrogen-fixing bacteria in the root nodules of clover, beans, peas, and certain other plants are also able to convert N_2 into usable nitrogen compounds. The high bond energy of N_2 also causes many nitrogen compounds to have positive enthalpies and free energies of formation.

Preparation and uses of elemental nitrogen

The Earth's atmosphere is composed of about 78% N_2 by volume and, as with oxygen, the commercial source of N_2 is from the liquefaction of air. As liquid air boils, the lower boiling nitrogen escapes and is collected. Generally, commercial nitrogen from this source consists of about 99% N_2, with a small amount of argon and traces of oxygen.

In the laboratory, nitrogen can be made chemically by warming a solution that contains an ammonium salt (such as NH_4Cl) and a nitrite salt (such as $NaNO_2$). The net ionic reaction corresponds to the decomposition of NH_4NO_2.

Solid NH_4NO_2 explodes if warmed and forms the same products.

$$NH_4^+(aq) + NO_2^-(aq) \xrightarrow{\text{warm}} N_2(g) + 2H_2O(l)$$

Commercially, the leading use of nitrogen is in the manufacture of ammonia, which is described later in this section. Because of nitrogen's low reactivity, it is also used in large quantities as an inert gaseous blanket to exclude oxygen during the manufacture of chemicals, the fabrication of metals, and in the production of electronic devices. Large amounts of liquid nitrogen are employed in the food industry where its low temperature ($-196°C$) is used to rapidly freeze foods.[5]

Gaseous N_2 is often used in chemical apparatus in the laboratory to provide an inert atmosphere.

Ionic compounds of nitrogen

Nitrides Elemental nitrogen combines directly with some of the very reactive metals on the left side of the periodic table. Earlier it was mentioned that lithium reacts with N_2 at room temperature to form an ionic nitride.

$$6Li(s) + N_2(g) \longrightarrow 2Li_3N$$

Similar nitrides are formed at higher temperatures by magnesium, calcium, strontium, and barium, and have the general formula M_3N_2. When placed into water the nitrides immediately hydrolyze, liberating ammonia.

The N^{3-} ion is a very strong Brønsted–Lowry base.

$$Li_3N(s) + 3H_2O(l) \longrightarrow 3LiOH(aq) + NH_3(g)$$

$$Mg_3N_2(s) + 6H_2O(l) \longrightarrow 3Mg(OH)_2(s) + 2NH_3(g)$$

Covalent compounds of nitrogen

Nitrogen forms covalent compounds with many nonmetals. The most important are those with hydrogen and oxygen, and among them we find nitrogen in every oxidation state from -3 to $+5$, as shown in Table 19.6. Actually, the oxidation numbers have no real physical significance for these substances—for example, the nitrogen in NH_3 certainly doesn't carry a charge of $3-$. Nevertheless, the oxidation numbers are useful in balancing redox equations and organizing the discussion of the chemistry of nitrogen.

About 1.6×10^{10} kg of ammonia are produced each year.

Ammonia (-3 Oxidation State) Ammonia (NH_3) is by far the most important compound of nitrogen. It serves as a route to virtually all other nitrogen compounds and is itself a useful fertilizer. Ammonia is prepared industrially by the **Haber process,** which combines hydrogen and nitrogen on a catalytic iron surface.

$$N_2(g) + 3H_2(g) \xrightarrow[\text{catalyst}]{\text{iron}} 2NH_3(g) \qquad \Delta H^0 = -92 \text{ kJ}$$

[5] Businesses have been established that, for a fee, offer to freeze your body in liquid nitrogen after you've died, and maintain it in a frozen state until (hopefully) a cure is found for whatever killed you. No one has yet been returned to life, however, and most scientists question whether the cell damage caused by rapid freezing can ever be reversed in humans.

Table 19.6
Oxidation states of nitrogen

Oxidation State	Example	Preparative Reaction
−3	NH_3 (ammonia)	$N_2 + 3H_2 \rightarrow \mathbf{2NH_3}$
−2	N_2H_4 (hydrazine)	$2NH_3 + NaOCl \rightarrow \mathbf{N_2H_4} + NaCl + H_2O$
−1	NH_2OH (hydroxylamine)	$NaNO_2 + NaHSO_3 + SO_2 + 2H_2O \rightarrow$ $2NaHSO_4 + \mathbf{NH_2OH}$
0	N_2 (dinitrogen)	$NH_4NO_2 \xrightarrow{\text{heat}} \mathbf{N_2} + 2H_2O$
+1	N_2O (nitrous oxide)	$NH_4NO_3 \xrightarrow{\text{heat}} \mathbf{N_2O} + 2H_2O$
+2	NO (nitric oxide)	$4NH_3 + 5O_2 \rightarrow \mathbf{4NO} + 6H_2O$
+3	N_2O_3 (dinitrogen trioxide)	$NO + NO_2 \xrightarrow{-30°C} \mathbf{N_2O_3}$
+4	NO_2 (nitrogen dioxide)	$2NO + O_2 \rightarrow \mathbf{2NO_2} \rightleftharpoons N_2O_4$
	N_2O_4 (dinitrogen tetroxide)	
+5	HNO_3 (nitric acid)	$3NO_2 + H_2O \rightarrow \mathbf{2HNO_3} + NO$

The reaction is run at pressures of several megapascals, which favors the production of NH_3. It is also run at temperatures between 400 and 500°C in order to cause the reaction to proceed at a reasonable rate, although in principle the position of equilibrium is less favorable than at lower temperatures.

Ammonia is a colorless gas with a powerful irritating odor that most people would recognize immediately. It has a boiling point of −33.35°C and freezes at −77.7°C. As a liquid it has many solvent properties similar to those of water.

The Lewis structure of ammonia is

$$H—\overset{\displaystyle ..}{N}—H$$
$$\underset{\displaystyle H}{|}$$

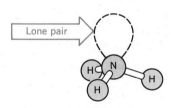

Lone pair

The pyramidal ammonia molecule.

and, as we would expect from VSEPR theory, NH_3 is a pyramidal molecule. It is quite polar and is extremely soluble in water. A saturated solution at room temperature contains about 28% by mass of NH_3 and is about 15 M. Its high solubility in water is the result of its ability to form hydrogen bonds with water molecules, both of the type $O—H\cdots N$ as well as $N—H\cdots O$.

Aqueous solutions of ammonia are basic, as you recall from Chapter 15. The reaction between NH_3 and H_2O is

$$NH_3 + H_2O \rightleftharpoons NH_4^+ + OH^- \qquad K_b = 1.8 \times 10^{-5}$$

Bottles of concentrated ammonia purchased from chemical supply companies are almost always labeled "ammonium hydroxide"; however, there is no evidence that the species of NH_4OH actually exists. These solutions simply consist of molecules of NH_3 dissolved in water (along with small amounts of NH_4^+ and OH^- produced by the ionization of NH_3).

Ammonia can serve as either a proton donor or a proton acceptor, depending on conditions, and therefore forms two types of salts. When acting as a proton donor, NH_3 forms *amide salts* such as $NaNH_2$ that contains the NH_2^- ion. These can only be made in nonaqueous media, because NH_2^- is a very strong base and is completely hydrolyzed in water to give NH_3.

$$\underset{\textbf{amide ion}}{NH_2^-} + H_2O \longrightarrow NH_3 + OH^-$$

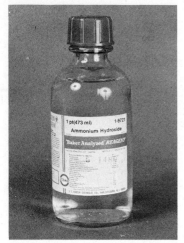

Reagent bottles of concentrated aqueous ammonia are usually labeled ammonium hydroxide.

When acting as a proton acceptor, ammonia forms ammonium salts—for example, NH_4Cl.

$$NH_3 + H_3O^+ \longrightarrow NH_4^+ + H_2O \qquad K = 1.8 \times 10^9$$

As we learned in Chapter 15, NH_4^+ also hydrolyzes slightly, so solutions of NH_4^+ tend to be slightly acidic (providing the anion present is that of a strong acid).

$$NH_4^+ + H_2O \rightleftharpoons NH_3 + H_3O \qquad K = 5.5 \times 10^{-10}$$

In the laboratory, ammonia can be prepared by reacting an ammonium salt with a strong base such as oxide ion or hydroxide ion.

$$CaO + 2NH_4Cl \longrightarrow 2NH_3 + CaCl_2 + H_2O$$

$$NaOH + NH_4Cl \longrightarrow NH_3 + NaCl + H_2O$$

These strong bases displace the weaker base, NH_3, from the NH_4^+ ion. The reaction of ammonium ion with a base serves as part of the qualitative test for ammonium ion in a mixture. If a sample of the mixture is warmed with base, an odor of ammonia serves to confirm the presence of NH_4^+ in the mixture. Alternatively, the NH_3 can be detected by the basic reaction of its vapors with moist litmus paper (pink → blue).

One of the principal reactions of ammonia is its catalytic oxidation to nitric oxide in the **Ostwald process,** shown in Figure 19.7.

This is a very exothermic reaction.

$$4NH_3(g) + 5O_2(g) \xrightarrow[\substack{catalyst \\ 750-900°C}]{Pt} 4NO(g) + 6H_2O(g)$$

The NO is quickly oxidized to NO_2 in the presence of excess oxygen,

$$2NO(g) + O_2(g) \longrightarrow 2NO_2(g)$$

and when the NO_2 is dissolved in water it **disproportionates**—that is, it enters into a redox reaction with itself so that some of it is oxidized while the rest is reduced.

N is oxidized from +4 to +5

$$3NO_2(g) + H_2O(l) \longrightarrow \underbrace{2H^+(aq) + 2NO_3^-(aq)}_{2HNO_3} + NO(g)$$

N is reduced from +4 to +2

Figure 19.7

In this illustration of the Ostwald process, ammonia escaping from a concentrated aqueous solution reacts with oxygen on a catalytic platinum surface, which glows from the heat of reaction. The colorless NO produced in this reaction is further oxidized to give reddish-brown gas, NO_2.

The commercial application of this sequence of reactions accounts for the major source of nitric acid and nitrates used in the manufacture of explosives, fertilizers, plastics, and many other useful substances. In fact, the development of this process in Germany by Wilhelm Ostwald, accompanied by the successful preparation of NH_3 from N_2 and H_2 by Haber, is said to have prolonged World War I, since the Allied blockade of Germany was unable to halt the German manufacture of munitions that had depended, prior to these processes, on the importation of nitrates from other countries.

Hydrazine (−2 Oxidation State) Hydrazine has the formula N_2H_4 and can be considered the nitrogen analog of hydrogen peroxide (although hydrazine's reactions are vastly different than those of H_2O_2). Its Lewis structure is

$$\begin{array}{cc} H & H \\ | & | \\ H-N-N-H \\ \cdot\cdot & \cdot\cdot \end{array}$$

and its molecular geometry is shown in Figure 19.8. The staggered arrangement of the hydrogens probably arises because of the tendency of the lone pairs of the nitrogens to be as far apart as possible so that the repulsions between them are a minimum.

Hydrazine is prepared by the reaction of ammonia with hypochlorite ion in aqueous solutions.

$$2NH_3 + NaOCl \longrightarrow N_2H_4 + NaCl + H_2O$$

Hydrazine is a violent poison, and this reaction is one of the reasons that warnings are given about the dangers of mixing household cleaning agents—specifically household ammonia and liquid bleach, which contains NaOCl. The danger of such a mixture is somewhat lessened fortunately, because an intermediate in the reaction (NH_2Cl) has a tendency to react with N_2H_4 yielding harmless NH_4Cl, especially if traces of Cu^{2+} are present.

Pure hydrazine is a liquid that freezes at 2°C and boils at 114°C. It has a strong affinity for water and its solutions are weakly basic because of the reaction,

$$N_2H_4 + H_2O \rightleftharpoons N_2H_5^+ + OH^- \qquad K_b = 1.7 \times 10^{-6}$$
$$\text{hydrazinium ion}$$

In basic solutions, hydrazine is a powerful reducing agent. For example, it reacts with iodine to liberate nitrogen.

$$N_2H_4 + 2I_2 + 4OH^- \longrightarrow N_2 + 4H_2O + 4I^-$$

The combustion of hydrazine is very exothermic.

$$N_2H_4(l) + O_2(g) \longrightarrow N_2(g) + 2H_2O(g) \qquad \Delta H^0 = -534 \text{ kJ}$$

Because of this, hydrazine as well as some compounds derived from it have been used as rocket fuels.

Hydroxylamine (−1 Oxidation State) Hydroxylamine can be thought of as an ammonia molecule in which one hydrogen has been replaced by an —O—H group.

$$\begin{array}{c} \cdot\cdot \\ H-N-OH \\ | \\ H \end{array}$$

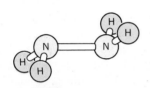

Figure 19.8
The structure of hydrazine, N_2H_4.

Traces of Cu^{2+} are common in water that passes through the copper water pipes found in many homes.

A major industrial use of hydrazine is as an oxygen scavenger in high-temperature boilers.

$$\begin{array}{cc} H & CH_3 \\ \diagdown \cdot\cdot \; \cdot\cdot \diagup \\ N-N \\ \diagup \qquad \diagdown \\ H \qquad\quad H \end{array}$$

Monomethylhydrazine—one of the fuels used by the space shuttle.

It can be made by reduction of nitrites with SO_2.

$$NaNO_2 + NaHSO_3 + SO_2 + 2H_2O \longrightarrow 2NaHSO_4 + NH_2OH$$

Pure hydroxylamine is a white solid that melts at 33°C, but it decomposes very easily. In water it is a weak base

NH_3OH^+ is called the *hydroxylammonium ion.*

$$NH_2OH + H_2O \rightleftharpoons NH_3OH^+ + OH^- \qquad K_b = 1.1 \times 10^{-8}$$

and forms salts such as $[NH_3OH]Cl$ and $[NH_3OH]_2SO_4$. The salts are stable and are used as mild reducing agents in photography. The sulfate salt is used for removing hair from animal hides.

Nitrous Oxide (+1 Oxidation State) Nitrous oxide (N_2O) is made by decomposing molten ammonium nitrate, NH_4NO_3,

NH_4NO_3 is used as a high explosive.

$$NH_4NO_3(l) \xrightarrow{\text{heat}} N_2O(g) + 2H_2O(g)$$

Although this reaction proceeds smoothly under most circumstances, NH_4NO_3 can be made to explode if detonated by another explosive.

The structure of N_2O can be described by the resonance formulas. As the VSEPR theory would predict, it is a linear molecule.

$$:\ddot{\text{N}}\!\!=\!\!\text{N}\!\!=\!\!\ddot{\text{O}}: \longleftrightarrow :\text{N}\!\!\equiv\!\!\text{N}\!\!-\!\!\ddot{\ddot{\text{O}}}:$$

Thermodynamically, N_2O is unstable with respect to decomposition to the elements. This is because its heat of formation and free energy of formation are both positive ($\Delta H_f^0 = +81.5$ kJ mol^{-1}, $\Delta G_f^0 = +104$ kJ mol^{-1}). Nitrous oxide is stable at room temperature only because its *rate* of decomposition is extremely slow, but at elevated temperatures it decomposes easily to N_2 and O_2 with the *release* of heat. Oxygen, of course, supports combustion, so burning a fuel with N_2O releases more heat than with just O_2 because of the added heat released by the decomposition of the N_2O. This is the reason that race car drivers often use N_2O as an oxidizer for their fuel—it gives their engines extra power.

Nitrous oxide is also used as an anesthetic. It produces a mild intoxicating effect and is commonly called laughing gas. Another application is as a propellant in aerosol cans.

Nitrogen Oxide (+2 Oxidation State) and Nitrogen Dioxide (+4 Oxidation State) Nitrogen oxide (also called nitric oxide), NO, and nitrogen dioxide, NO_2, are the two most important oxides of nitrogen. They are intermediates in the conversion of ammonia to nitric acid, and both play a major role in the formation of a type of air pollution called photochemical smog. In this context, they are generally referred to together as NO_x.

The path from ammonia to nitric acid was discussed earlier in this section. Oxidation of ammonia on a platinum catalyst gives NO, which is rapidly oxidized to NO_2 by excess oxygen. Dissolving the NO_2 in water gives nitric acid plus NO (which is oxidized again to NO_2). This can be represented schematically as

$$NH_3(g) \xrightarrow[\text{catalyst}]{O_2} NO(g) \xrightarrow{O_2} NO_2(g) \xrightarrow{H_2O} HNO_3(aq) + NO(g)$$

Nitrogen oxide is a fairly reactive, colorless gas. It has an odd number of electrons, so they can't all be paired. NO is therefore paramagnetic. The bonding in nitric oxide, like that in O_2, is perhaps best described by molecular orbital theory. The molecular orbital energy level diagram for NO is given in Figure 19.9. The net bond order in the molecule is 2.5 because the electron in the π^*

Figure 19.9

Molecular orbital energy level diagram for NO. Note that the energies of the atomic orbitals on the separate nitrogen and oxygen atoms are not the same.

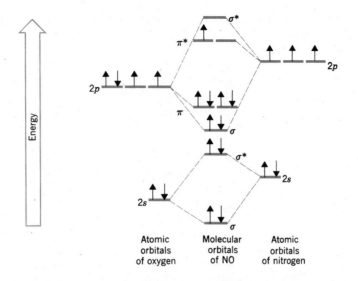

antibonding MO cancels the effect of one of the electrons in the π bonding orbitals. It is interesting that NO loses an electron rather easily to form the NO^+ (nitrosonium) ion. This is because the electron that is lost is from a high-energy π^* orbital. Removal of the antibonding electron raises the bond order to 3.0, and the NO^+ ion actually has a shorter, stronger bond than NO.

Compounds containing the NO^+ ion can be isolated from nonaqueous solutions.

Nitrogen dioxide is a toxic, reddish-brown gas. Its structure is given by the resonance formulas

and in the gas phase there is an equilibrium between NO_2 and its dimer, dinitrogen tetroxide, N_2O_4. (A *dimer* is a molecule formed by joining two simpler ones.)

$$2NO_2(g) \rightleftharpoons N_2O_4(g) + 57 \text{ kJ}$$
reddish-brown   colorless

The formation of N_2O_4 from NO_2 can be explained by the tendency of the unpaired electron in NO_2, which spends most of its time on the nitrogen, to become paired with another electron from a neighboring NO_2 molecule.

NO₂ is paramagnetic, but N₂O₄ is not.

Dinitrogen tetroxide is a low-boiling liquid (bp = 21°C) that is deep brown at its boiling point, caused by some NO_2 produced by its dissociation (pure N_2O_4 is colorless). It is a good oxidizing agent and has been used as an oxidizer in liquid rockets. In fact, the space shuttle burns monomethylhydrazine (see page 707) with N_2O_4 in the rocket engine that it uses to drop out of orbit on its return to Earth.

Photochemical Smog Within the last several decades, the growing number of automobiles in urban areas has produced a kind of air pollution never before experienced by civilization. It is characterized by a reddish-brown haze that contains substances irritating to the eyes, nose, and lungs, and that cause extensive damage to vegetation and rubber products not containing antioxidants. This

Nitrogen oxides, along with sulfur oxides, have been implicated as the source of acid rain. This problem is discussed further in the next chapter.

haze has come to be known as **photochemical smog** (or often just smog) and has been traced to abnormally high levels of ozone and oxides of nitrogen in the atmosphere.

A typical smog episode begins in the early morning when urban rush-hour traffic spews out the primary pollutant, nitrogen oxide. This is formed in the gasoline engine during combustion by the direct combination of N_2 and O_2, which are present in the air drawn into the engine to burn the fuel. Even though the reaction

$$N_2(g) + O_2(g) \rightleftharpoons 2NO(g)$$

has an extremely small equilibrium constant at ordinary temperatures, the K increases with increasing temperature, so at the high temperature inside the engine small amounts of NO are produced. When the exhaust gases leave the engine, they cool so rapidly that the NO doesn't have an opportunity to decompose back to N_2 and O_2. As a result, it is released into the atmosphere. In Figure 19.10 we see that during the early hours the NO concentration rises.

$2NO + O_2 \longrightarrow 2NO_2$

The next step in the sequence of reactions is the oxidation of NO to NO_2. This occurs slowly as the morning wears on, and in Figure 19.10 we see that the NO concentration starts to drop as the NO_2 concentration rises. The presence of the NO_2 in smog is what gives this form of air pollution its characteristic reddish-brown color.

As the sun climbs higher in the sky, the NO_2 begins to undergo a photochemical reaction—a reaction brought on by the absorption of light. In this case, ultraviolet light from the sun's rays causes NO_2 to decompose.

$h\nu$ is the energy absorbed from a photon of frequency ν.

$$NO_2 \xrightarrow{h\nu} NO + O$$

The oxygen atoms produced in this reaction combine with oxygen molecules and ozone is formed.

Over 99% of the oxygen atoms react with O_2 to form O_3.

$$O_2 + O \longrightarrow O_3$$

In Figure 19.10 we see that the ozone concentration peaks about noon as the NO_2 concentration begins to decline.

As we learned in Section 19.3, ozone is an extremely reactive substance. It reacts with NO to form NO_2, and this reaction helps moderate the rate of buildup of O_3.

$$O_3 + NO \longrightarrow NO_2 + O_2$$

Ozone also attacks other substances and is particularly hard on vegetation because it reacts with chlorophyll. It is especially reactive toward molecules that contain carbon-carbon double bonds. Until antipollution devices were installed on automobiles, considerable amounts of hydrocarbons (including those with

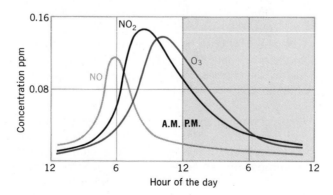

Figure 19.10

The concentrations of pollutants during a typical photochemical smog episode.

double bonds) were released into the atmosphere by the evaporation of gasoline and as unburned hydrocarbons in auto exhausts. These substances react with ozone in a series of complex reactions to give final products with the formula

$$
\begin{array}{c}
: \ddot{O} : \\
\parallel \\
R - C - \ddot{O} - \ddot{O} - NO_2
\end{array}
$$

where R is a portion of a hydrocarbon molecule—for example, $-CH_3$, $-C_2H_5$, and so on. They are collectively called *peroxyacylnitrates*, or PAN for short. Molecules of PAN are oxidizing agents—a property that is typical of compounds having an oxygen-oxygen bond (e.g., H_2O_2)—and contribute to the oxidizing nature of photochemical smog. Molecules of PAN are also eye irritants, even at concentration levels of parts per billion. This is one of the reasons that smog is so unpleasant.

$$
\begin{array}{c}
: \ddot{O} : \\
\parallel \\
CH_3 - C - \ddot{O} - \ddot{O} - NO_2
\end{array}
$$
peroxyacetylnitrate

Finally, as late afternoon approaches, the O_3 concentration begins to fall off as it reacts with airborn hydrocarbons and other substances, and the smog attack gradually subsides—until daybreak the next day.

Dinitrogen Trioxide, Nitrous Acid, and Nitrites (the +3 Oxidation State) Dinitrogen trioxide, N_2O_3, is produced by condensing an equimolar mixture of NO and NO_2 at very low temperatures (below $-30°C$).

$$NO + NO_2 \longrightarrow N_2O_3$$

It is a blue liquid in which there are molecules having the structure

$$
\begin{array}{ccc}
\cdot \ddot{O} \cdot & & \cdot \ddot{O} \cdot \\
 \diagdown & & \diagup \\
 & N - N & \\
 & & \diagdown \\
 & & \ddot{O} \cdot
\end{array}
$$

$N_2O_3 + H_2O \longrightarrow 2HNO_2$

At least in a formal sense, N_2O_3 is the acid anhydride of nitrous acid, HNO_2. In fact, if an equimolar mixture of NO and NO_2 is bubbled into water, nitrous acid is produced.

Nitrous acid is a weak acid ($K_a = 4.5 \times 10^{-4}$) and is only stable in solution and in the gas phase. It cannot be isolated in pure liquid form because it decomposes by a disproportionation reaction.

$$3HNO_2 \longrightarrow HNO_3 + H_2O + 2NO$$

When HNO_2 is needed for a chemical reaction it is usually generated in aqueous solution by adding a strong acid, such as HCl, to a salt of HNO_2 such as $NaNO_2$ (sodium nitrite). Hydrogen ion and nitrite ion combine in the net reaction

$$H^+ + NO_2^- \longrightarrow HNO_2$$

The formation of the weak electrolyte HNO_2 drives this reaction to the right.

Nitrous acid is able to function as either an oxidizing agent or a reducing agent, depending on the ease of oxidation or reduction of the other reactant. For example, it is oxidized to HNO_3 by permanganate ion,

Here HNO_2 is oxidized and is a reducing agent.

$$H^+ + 5HNO_2 + 2MnO_4^- \longrightarrow 5NO_3^- + 2Mn^{2+} + 3H_2O$$

but, in the presence of iodide ion and excess H^+, HNO_2 is reduced.

Here HNO_2 is reduced and is an oxidizing agent.

$$2H^+ + 2HNO_2 + 2I^- \longrightarrow I_2 + 2NO + 2H_2O$$

Neutralization of nitrous acid gives nitrite ion, although nitrite salts are normally made by reducing a metal nitrate. For example,

$$NaNO_3 + Pb \xrightarrow{\text{heat}} NaNO_2 + PbO$$

The nitrite ion has an angular structure that is predicted by VSEPR theory from either of its resonance formulas

$$\left[\begin{array}{c} \ddot{N} \\ \ddot{O} \quad \ddot{O} \end{array}\right]^{-} \longleftrightarrow \left[\begin{array}{c} \ddot{N} \\ \ddot{O} \quad \ddot{O} \end{array}\right]^{-}$$

A major but controversial use of nitrites is in preserving meats such as ham, frankfurters, bologna, and bacon. The nitrite serves two functions. The most important one is that it inhibits the growth of bacteria, especially *clostridium botulinum*, which produces the very poisonous botulinus toxin that causes fatal food poisoning. The second is that it preserves the red color of the meat and thereby maintains the food's appetizing appearance.

The controversy over the use of nitrites in cured meat products stems from the effect that nitrous acid has on organic compounds called amines. The reaction of HNO_2 with amines produces other compounds called nitrosoamines, which have been shown to cause cancer, mutations, and birth defects in experimental animals. Meat and body fluids contain many amines, and it is feared by some scientists that when the NO_2^- contacts the acidic condition in the stomach some of the HNO_2 that is formed might react with amines that are there and thereby give nitrosoamines. Although no direct evidence has been found linking nitrites in meat to cancer in either animals or humans, the FDA has set limits on the maximum allowable NO_2^- concentrations in these food products.

Some scientists feel that the risk of food poisoning from meats untreated with nitrites is greater than the risk of cancer from meats that are treated.

Dinitrogen Pentoxide, Nitric Acid, and Nitrates (the +5 Oxidation State) Dinitrogen pentoxide, N_2O_5, exists in molecular form in the vapor. Its structure is

$$\begin{array}{ccc} \ddot{O} & & \ddot{O} \\ \diagdown & & \diagup \\ N-\ddot{O}-N & \\ \diagup & & \diagdown \\ \ddot{O} & & \ddot{O} \end{array}$$

In the solid state, N_2O_5 dissociates into ions and exists as $NO_2^+NO_3^-$. When allowed to react with water, N_2O_5 gives nitric acid, HNO_3,

N_2O_5 is the acid anhydride of HNO_3.

$$N_2O_5 + H_2O \longrightarrow 2HNO_3$$

This reaction can be reversed, and N_2O_5 is produced by removing the components of water from a pair of HNO_3 molecules. This requires a very powerful dehydrating agent such as P_4O_{10}. (The reaction of P_4O_{10} with water is discussed in the next chapter.)

Nitric acid is one of the world's most vital chemicals, and huge amounts of it are produced each year. Much of it is made into ammonium nitrate to be used as a nitrogen fertilizer, because plants can utilize both NH_4^+ and NO_3^- ion. The manufacture of explosives such as TNT and nitroglycerine also require nitric acid, and nitrates are used in curing meats along with nitrites.

About 8.5 million tons of HNO_3 are made annually.

The commercial production of nitric acid by dissolving NO_2 in water was described earlier in our discussion of the Ostwald process. In the laboratory, nitric acid can be made by heating a mixture of KNO_3 and concentrated sulfuric acid.

$$KNO_3(s) + H_2SO_4(l) \xrightarrow{\text{heat}} KHSO_4(s) + HNO_3(g)$$

The nitric acid vapors condense to a liquid when they are cooled.

Pure nitric acid is a colorless liquid that decomposes easily above 0°C into NO_2, H_2O, and O_2. The concentrated laboratory reagent is about 70% HNO_3 and is often slightly yellow in color from the presence of small amounts of NO_2 formed by a photochemical decomposition.

$$4HNO_3 \xrightarrow{h\nu} 4NO_2 + O_2 + 2H_2O$$

colorless   red-brown
(appears yellow
when dilute)

Nitric acid is a strong acid, and the presence of nitrate ion in its solutions makes it an especially powerful oxidizing agent. It is therefore able to dissolve many metals that fail to dissolve in an acid such as HCl, which contains H^+ as its strongest (and only) oxidizing agent. For example, copper dissolves in both concentrated and dilute HNO_3, but the reduction products differ. These reactions were described previously.

$$Cu + 2NO_3^- + 4H^+ \longrightarrow Cu^{2+} + 2NO_2 + 2H_2O \quad \text{(Concentrated)}$$

$$3Cu + 2NO_3^- + 8H^+ \longrightarrow 3Cu^{2+} + 2NO + 4H_2O \quad \text{(Dilute)}$$

With stronger reducing agents and more dilute solutions, the nitrogen can be reduced all the way to the -3 oxidation state. For example,

$$4Zn + NO_3^- + 10H^+ \longrightarrow 4Zn^{2+} + NH_4^+ + 3H_2O$$

A mixture of one part by volume of concentrated HNO_3 and three parts by volume of concentrated HCl is called aqua regia. As discussed earlier, this mixture is able to dissolve the noble metals such as gold and platinum that fail to dissolve in concentrated HNO_3 alone. The chloride ion in aqua regia forms complex ions with the ions of these metals and that helps draw them into solution. For instance, gold dissolves as follows.

$$4H^+(aq) + 4Cl^-(aq) + NO_3^-(aq) + Au(s) \longrightarrow AuCl_4^-(aq) + NO(g) + 2H_2O$$

REVIEW QUESTIONS AND PROBLEMS

(Problems whose numbers are in blue have their answers in Appendix C at the back of the book.)

Hydrogen: General

19.1 What is the correct name for H_2?

19.2 Which is the most abundant element in the universe?

19.3 What is the probable reason that the abundance of hydrogen on Earth is less than it is in the rest of the universe?

19.4 Why is there little free hydrogen in the Earth's atmosphere?

19.5 What advantage does hydrogen have over helium for use in lighter-than-air balloons? What is a disadvantage?

19.6 Why are the properties of hydrogen so different from the properties of the other elements in Group IA? Why is hydrogen placed in Group IA?

19.7 Give the names and symbols of the three isotopes of hydrogen. Which one is radioactive? Why does it potentially pose an environmental problem?

19.8 What are the advantages of hydrogen as a fuel? Give two disadvantages (see Exercise 12.56 on page 474).

Hydrogen: Preparation and Uses

19.9 Give chemical reactions for the preparation of hydrogen from water and (a) methane (natural gas); (b) coal.

19.10 Write chemical equations for the combustion of water gas.

19.11 In what way is the commercial production of caustic soda related to the commercial production of hydrogen?

19.12 What is the greatest industrial use for hydrogen? Give the appropriate chemical equation.

19.13 Write a chemical equation for the laboratory preparation of hydrogen. Sketch the apparatus you would use.

19.14 How is hydrogen used in the synthesis of methanol? What importance is this commercially?

19.15 Use Lewis structures to illustrate the hydrogenation of acetylene, C_2H_2. What effect does hydrogenation have on vegetable oils?

Hydrogen Compounds

19.16 What general name is given to binary compounds of hydrogen? What would be the name for the compound NaH? How could it be formed? What is its reaction with water?

19.17 How many covalent bonds does hydrogen normally form? Why?

19.18 Predict the molecular structures of (a) H_2S; (b) PH_3; (c) SiH_4.

19.19 Explain how the reaction $H^- + H_2O \rightarrow H_2 + OH^-$ can be interpreted as both an acid-base reaction and a redox reaction.

19.20 Define *catenation*. Which element has the greatest tendency to catenate? How does the tendency to catenate vary within a group in the periodic table?

19.21 Why can't all the nonmetal hydrides be made by direct combination of the elements? Which nonmetal hydrides can be made by this method?

19.22 How do the basicities of the conjugate bases of the nonmetal hydrides vary within periods and groups in the periodic table? Give two examples.

Carbon: General

19.23 What is an *organic* compound?

19.24 How is coke made?

19.25 How do the structures of diamond and graphite differ?

19.26 What are two uses of graphite?

19.27 What is activated charcoal? What property does it possess that makes it commercially useful?

19.28 How is carbon black made? What are two of its uses?

Oxides and Oxoacids of Carbon

19.29 Give the structures of carbon monoxide and carbon dioxide.

19.30 How can carbon monoxide be prepared in the laboratory?

19.31 Write a chemical equation showing the reducing properties of carbon monoxide toward Fe_2O_3.

19.32 What is a metal carbonyl compound? Give an example.

19.33 How is CO_2 usually made industrially? How can it be made conveniently in the laboratory?

19.34 Give three major industrial uses of CO_2.

19.35 How do green plants use CO_2?

19.36 Give the equations for the dissolving of limestone in water by dissolved carbon dioxide.

19.37 How are stalagmites and stalactites formed in limestone caves?

19.38 What is hard water? How can hardness ions be removed from hard water?

19.39 Can you suggest a chemical method for removing boiler scale from the insides of pipes and water boilers?

Other Inorganic Compounds of Carbon

19.40 What is carborundum? How is its structure related to that of diamond? How is carborundum made?

19.41 Write an equation for the reaction of aluminum carbide with water.

19.42 How is acetylene made?

19.43 What is an interstitial carbide? Give an example.

19.44 How is hydrogen cyanide synthesized? How does it function as a poison?

19.45 Give the structure of carbon disulfide. What makes CS_2 a hazardous solvent with which to work?

Oxygen: General

19.46 What are the commercial sources of oxygen and nitrogen?

19.47 What is the correct chemical name for O_2?

19.48 How is oxygen conveniently prepared in the laboratory?

19.49 What is the largest use for O_2 industrially?

19.50 Give the structure for ozone. How can it be made in the laboratory? What advantage does ozone have over Cl_2 for the purification of drinking water?

19.51 How is ozone formed in the Earth's upper atmosphere? How does it protect the Earth's inhabitants from ultraviolet radiation from the sun? How could nitrogen oxide pollutants affect the Earth's ozone layer (give equations)? How could Freons affect the ozone layer (give equations)?

Oxygen Compounds

19.52 Write a chemical equation for the hydrolysis of oxide ion.

19.53 What does pickling mean with respect to the treatment of iron and steel objects?

19.54 Write equations that show the amphoteric behavior of a metal oxide.

19.55 Give three ways that the oxide, CO_2, can be made. Illustrate each with a chemical equation.

19.56 Why can N_2O not be synthesized from its elements?

19.57 Write an equation for the oxidation of ammonia with oxygen.

19.58 Write chemical equations for the reaction of copper with (a) concentrated HNO_3; (b) dilute HNO_3.

19.59 Give equations showing reactions of oxygen to form (a) a metal peroxide; (b) a metal superoxide.

19.60 What is the structure of hydrogen peroxide?

19.61 Write an equation for the decomposition of hydrogen peroxide.

Nitrogen: General

19.62 Why is dinitrogen so unreactive?

19.63 Give three commercial uses of N_2.

19.64 What are *nitrogen-fixing bacteria*?

19.65 How can small amounts of N_2 be made in the laboratory?

19.66 Which is the only element N_2 reacts with at room temperature?

19.67 What happens when an ionic nitride such as Mg_3N_2 is placed into water?

Ammonia

19.68 Why is ammonia so soluble in water?

19.69 Give the formula for the principal nitrogen-containing species in a solution of *ammonium hydroxide*.

19.70 How can ammonia be made in the laboratory? Could it be collected by displacement of water in the same manner as hydrogen?

19.71 How could you determine whether a particular unknown solid was an ammonium salt?

Nitric Acid

19.72 Give the chemical reactions in the Ostwald process.

19.73 Why does concentrated HNO_3 usually have a pale yellow color?

19.74 How could pure HNO_3 be prepared in the laboratory?

19.75 Give three uses of nitric acid and nitrates.

19.76 What nitrogen-containing product is formed when zinc reacts with HNO_3?

19.77 Why does nitric acid dissolve metals such as silver and copper which are unaffected by hydrochloric acid?

19.78 What is aqua regia? How does it function in dissolving noble metals such as gold?

Other Compounds and Reactions of Nitrogen

19.79 Give an example of a nitrogen compound in which the oxidation number of nitrogen is (a) -3, (b) -1, (c) $+1$, (d) $+3$, and (e) $+5$.

19.80 Give the chemical equation and reaction conditions in the Haber process. Why are those particular conditions chosen?

19.81 Give the formula for potassium amide. Why can potassium amide not be made in an aqueous solution?

19.82 What is a disproportionation reaction?

19.83 What influence did the Haber and Ostwald processes have on the waging of World War I?

19.84 Give the structure of hydrazine.

19.85 How is hydrazine prepared? Why has hydrazine been used as a rocket fuel?

19.86 Give the structure of hydroxylamine. Compare the basicities of ammonia, hydrazine, and hydroxylamine.

19.87 Why is it potentially dangerous to mix household ammonia with liquid laundry bleach?

19.88 Give a chemical equation for the preparation of nitrous oxide. What are the resonance structures for nitrous oxide?

19.89 Look up the standard free energies of formation of N_2O, NO and NO_2. What accounts for the apparent stability of these oxides of nitrogen?

19.90 Construct molecular orbital energy level diagrams for NO and for O_2. (If necessary, refer to Figure 19.9 and Figure 6.33.) Compare expected bond orders, bond lengths, and bond energies for (a) the series NO^-, NO, NO^+, (b) the series O_2^+, O_2, O_2^-, O_2^{2-}.

19.91 Give the Lewis structures for NO_2 and N_2O_4. Write the equilibrium that exists between these two molecules.

19.92 What is the structure of N_2O_3? Under what conditions can it be formed? Of which acid is N_2O_3 the acid anhydride?

19.93 Write a chemical equation for the decomposition of nitrous acid.

19.94 Write Lewis formulas for NO_2 and NO_2^-. On the basis of the principles of VSEPR theory, which one should have the larger bond angle?

19.95 Write equations that show nitrous acid functioning as (a) a reducing agent, (b) an oxidizing agent.

19.96 How is $NaNO_2$ normally made? What are the functions of $NaNO_2$ in meat products such as bologna? What are the pros and cons concerning its use in such products?

19.97 Give the Lewis structure for N_2O_5 as it exists in the vapor. How does N_2O_5 exist in the solid state?

19.98 How can N_2O_5 be made?

19.99 Write a chemical equation for the reaction of N_2O_5 with water.

19.100 For octane, $\Delta H_f^0 = -255.1$ kJ mol^{-1}. Use this and the data in Table 12.1 to calculate the standard heats of reaction for the following:

$$C_8H_{18}(l) + 12\tfrac{1}{2}O_2(g) \longrightarrow 8CO_2(g) + 9H_2O(g)$$

$$C_8H_{18}(l) + 25N_2O(g) \longrightarrow 8CO_2(g) + 9H_2O(g) + 25N_2(g)$$

Smog

19.101 Describe the series of chemical reactions responsible for photochemical smog.

19.102 What substance causes the reddish-brown haze of photochemical smog? What is PAN?

CHAPTER 20

THE CHEMISTRY OF SELECTED NONMETALS PART II: PHOSPHORUS, SULFUR, THE HALOGENS, THE NOBLE GASES, AND SILICON

Sulfur, the solid seen here piled in railroad freight cars, is one of the world's important raw materials. Its principal use is in the manufacture of sulfuric acid—industry's number one chemical. More sulfuric acid is made each year than any other single chemical compound.

The nonmetallic elements discussed in this chapter cover a broad range of chemical reactivity and commercial importance. They range from fluorine—the most reactive element—to helium—the least reactive. Here we find sulfuric and phosphoric acids, chemicals that consistently rank among the top 10 in total mass manufactured each year. At the other extreme we have compounds of the heavier noble gases, substances that have so far found no practical uses. (In fact, the ability of any of the noble gases to react at all was a bombshell in the chemical world only two decades ago.)

We also see in this chapter a range of biological importance. Compounds of phosphorus and sulfur occur in all living systems, but many of those containing the halogens are lethal to living organisms. On the other hand, living things are almost indifferent toward the noble gases.

20.1 PHOSPHORUS

The Earth's crust contains about 0.1% phosphorus by mass.

Phosphorus is a relatively abundant element, ranking 12th in the Earth as a whole. It is always found in the combined state, normally in *phosphate rock*, which consists primarily of minerals that contain calcium phosphate, $Ca_3(PO_4)_2$. In living systems, phosphorus serves a number of critical functions: Phosphate units play a role in the structure of DNA, which directs the chemistry of our cells and transmits genetic information from one generation to another. Phosphate units are also present in phospholipids—substances that make up cell membranes—and phosphorus-oxygen bonds store the energy that we derive from the metabolism of foods.

About 85% of the phosphate rock mined in the United States now comes from Florida and North Carolina.

Because phosphorus is so important to biological systems of all kinds, it is not surprising that most industrial applications of phosphorus compounds, especially phosphates, are ultimately related in one way or another to providing sufficient phosphorus for the growth of agricultural products. For this reason also, deposits of phosphate rock are an important national resource. Until recently, some rich U.S. deposits of phosphate rock appeared on the verge of depletion during the 1990s. However, discoveries of huge phosphate deposits covering hundreds of square kilometres below the ocean floor within 100 km of North Carolina have relieved fears of imminent shortages.

The free element

Phosphorus atoms have the electron configuration, [Ne] $3s^23p^3$, in their ground state and therefore require three electrons to complete an octet. In the free element this is accomplished by linking each phosphorus atom to three others by single covalent bonds. As we learned in Chapter 6, this gives rise to three principal allotropes.

White phosphorus (Figure 20.1) consists of individual tetrahedral P_4 molecules. It is a waxy solid that melts at 44.1°C and boils at 280°C. Because of the small, highly strained 60° P—P—P bond angle in P_4, this allotrope is extremely reactive, especially toward oxygen, and must be stored under water. When exposed to air, white phosphorus ignites spontaneously, and contact with the skin produces painful slow-healing burns. White phosphorus is also very toxic, and long-term exposure to even low levels of P_4 vapor produces a gradual deterioration and softening of the bones of the jaw.

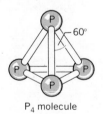

P$_4$ molecule

About 0.05 g of white phosphorus is a fatal dose.

White phosphorus was first prepared (unintentionally) in 1669 by an alchemist named Hennig Brand, who heated a mixture of dried urine and sand and condensed the vapors that were given off by passing them through water. The new element was named phosphorus, from the Greek *phosphoros* = light bringer, because the dry solid in a sealed bottle glows in the dark. The glow is

Figure 20.1

White phosphorus (left) is a yellowish-white waxy solid that is extremely reactive toward oxygen. Red phosphorus (right) is much less reactive than the white allotrope.

actually caused by a slow oxidation of the phosphorus surface by residual oxygen in the container.

Modern methods of producing white phosphorus differ only slightly from Brand's. A mixture of phosphate rock, $Ca_3(PO_4)_2$, silica, SiO_2, and coke is heated to about 1300°C in an **electric furnace**—a furnace in which the contents are heated to very high temperatures by passing an electric current through them. The reaction is

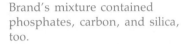

Brand's mixture contained phosphates, carbon, and silica, too.

$$2Ca_3(PO_4)_2(s) + 6SiO_2(s) + 10C(s) \longrightarrow 6CaSiO_3(l) + 10CO(g) + P_4(g)$$

The CO and P_4 vapor is passed through water, which causes the P_4 to condense. Most of the phosphorus produced in this way is ultimately used to make phosphoric acid.

Red phosphorus (also shown in Figure 20.1) is formed when white phosphorus is heated or exposed to ultraviolet light. It is amorphous and as we learned earlier, probably consists of P_4 tetrahedra joined at their corners. It is also relatively nonpoisonous and is considerably less reactive than white phosphorus. Red phosphorus is used to make incendiary devices (bombs, fireworks, flares, etc.) and, when mixed with fine sand, is used on the striking surfaces of matchbooks.

Black phosphorus is the least reactive form of phosphorus. It consists of layers of phosphorus atoms. Within each layer the phosphorus atoms are covalently bonded to each other, but the attractions between layers are the result of much weaker London forces. This gives it a flaky, graphitelike appearance.

Compounds of phosphorus

Phosphorus combines with most of the nonmetals, as well as with active metals. With the metals of Groups IA and IIA, for example, binary *phosphides* are formed that are predominantly ionic and contain the *phosphide ion*, P^{3-}. In water, this ion hydrolyzes to give *phosphine*, PH_3, a very bad-smelling poisonous gas.

$$Na_3P(s) + 3H_2O(l) \longrightarrow 3NaOH(aq) + PH_3(g)$$

The most important compounds of phosphorus are formed with nonmetals, particularly oxygen and the halogens.

Oxides of Phosphorus When any of the allotropes of phosphorus are burned in an excess supply of oxygen, white, powdery phosphorus(V) oxide, P_4O_{10}, is formed. For example,

$$P_4(s) + 5O_2(g) \longrightarrow P_4O_{10}(s)$$

This oxide is often called *phosphorus pentoxide* because its empirical formula,

The systematic name of P_4O_{10} is *tetraphosphorus decaoxide.*

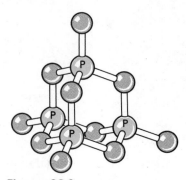

Figure 20.2
The structure of P_4O_{10}.

The systematic name of P_4O_6 is *tetraphosphorus hexaoxide.*

Note the spelling: Phosphorus is the name of the element; phosphorous implies the lower oxidation state of phosphorus in its oxoacids.

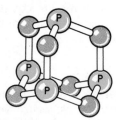

Figure 20.3
The structure of P_4O_6.

85% of the elemental phosphorus made each year is used to manufacture pure H_3PO_4.

P_2O_5, was recognized long before its molecular formula and structure were discovered. The structure of P_4O_{10} is shown in Figure 20.2. Notice that it is related to the basic P_4 tetrahedron by insertion of an oxygen bridge between phosphorus atoms along each edge of the tetrahedron, plus an additional oxygen at each vertex.

The most striking property of P_4O_{10} is its strong affinity for water. When placed into water it reacts vigorously and exothermically to form phosphoric acid, H_3PO_4.

$$P_4O_{10}(s) + 6H_2O(l) \longrightarrow 4H_3PO_4(l)$$

This oxide is such a strong dehydrating agent that it is even able to extract the components of water—two hydrogens and an oxygen—from other molecules. For example, in Chapter 19 we saw that P_4O_{10} is able to remove hydrogen and oxygen from HNO_3 molecules to form N_2O_5.

$$P_4O_{10} + 12HNO_3 \longrightarrow 4H_3PO_4 + 6N_2O_5$$

In the laboratory, P_4O_{10} is often used as a **desiccant,** an agent that removes moisture from air or other gases. When the moist air or gas mixture comes in contact with the P_4O_{10}, the water vapor reacts to form H_3PO_4, thereby leaving the gas virtually moisture-free.

If white phosphorus is burned in a limited supply of oxygen, not all of it is oxidized to P_4O_{10}. About half is oxidized to phosphorus(III) oxide, P_4O_6 (often called *phosphorus trioxide* because its empirical formula is P_2O_3). The structure of P_4O_6, shown in Figure 20.3, is similar to that of P_4O_{10}, except that the oxygen at each apex of the tetrahedron is missing in P_4O_6.

Phosphorus(III) oxide is a crystalline solid that melts at 23.8°C and boils at 175°C. Its vapors are quite poisonous. When stirred with cold water, P_4O_6 reacts to form phosphorous acid, H_3PO_3.

$$P_4O_6(s) + 6H_2O(l) \longrightarrow 4H_3PO_3(aq)$$

Phosphoric Acid and Phosphates The most important oxoacid of phosphorus, by far, is **phosphoric acid,** H_3PO_4 (also called **orthophosphoric acid**). It is produced in huge quantities, about 10^{10} kg annually, for use in the manufacture of fertilizers such as ammonium phosphate, as a food additive, and for the production of detergents.

Most phosphoric acid is made directly from phosphate rock by reaction with sulfuric acid.

$$Ca_3(PO_4)_2(s) + 3H_2SO_4(aq) + 6H_2O \longrightarrow 3CaSO_4 \cdot 2H_2O(s) + 2H_3PO_4(aq)$$

After reaction, the insoluble calcium sulfate is separated from the mixture by filtration and the phosphoric acid is then concentrated by evaporating water from the solution. This gives a concentrated solution that is about 85% H_3PO_4 by mass. It is known as *syrupy phosphoric acid* because of its viscous nature.

A purer form of phosphoric acid is obtained from elemental phosphorus by burning the element to give P_4O_{10} and then dissolving the oxide in water. As we saw earlier, P_4O_{10} reacts with water to give H_3PO_4. The phosphoric acid produced in this way is used to make detergents and in food preparations. For example, H_3PO_4 is added to many carbonated beverages to give them tartness. The acid is also used to make the acid salt $Ca(H_2PO_4)_2$, which is used in baking powders.

Pure phosphoric acid is a solid at room temperature; its clear, colorless crystals melt at 42.4°C. In water, H_3PO_4 is a weak triprotic acid and is a very poor oxidizing agent. Its structure is

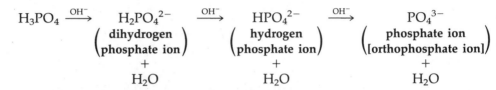

Neutralization of H_3PO_4 with a base produces three different ions, depending on the amount of base that is added.

$$H_3PO_4 \xrightarrow{OH^-} \begin{matrix} H_2PO_4^{2-} \\ \left(\begin{matrix} \text{dihydrogen} \\ \text{phosphate ion} \end{matrix}\right) \\ + \\ H_2O \end{matrix} \xrightarrow{OH^-} \begin{matrix} HPO_4^{2-} \\ \left(\begin{matrix} \text{hydrogen} \\ \text{phosphate ion} \end{matrix}\right) \\ + \\ H_2O \end{matrix} \xrightarrow{OH^-} \begin{matrix} PO_4^{3-} \\ \left(\begin{matrix} \text{phosphate ion} \\ \text{[orthophosphate ion]} \end{matrix}\right) \\ + \\ H_2O \end{matrix}$$

It is possible to isolate salts of each of the three anions.

Salts of phosphoric acid have a variety of applications, and one of the most common is in fertilizers. Phosphate rock itself can be pulverized and used as a phosphate fertilizer, but because $Ca_3(PO_4)_2$ has a very low solubility it is able to deliver phosphate only slowly and in small amounts. However, treatment of $Ca_3(PO_4)_2$ with dilute sulfuric acid gives a fertilizer known as *superphosphate*, a mixture of $CaSO_4$ and $Ca(H_2PO_4)_2$.

$$Ca_3(PO_4)_2 + 2H_2SO_4 + 4H_2O \longrightarrow \underbrace{2CaSO_4 \cdot 2H_2O + Ca(H_2PO_4)_2}_{\textbf{superphosphate}}$$

Because $Ca(H_2PO_4)_2$ is water soluble, this mixture is a more effective fertilizer than phosphate rock—hence the name superphosphate.

The anions of phosphoric acid are also found in the blood where they serve as one of the buffer systems. This system consists of the ions $H_2PO_4^-$ and HPO_4^{2-}. If base is added to the blood it is able to react with $H_2PO_4^-$.

$$H_2PO_4^- + OH^- \longrightarrow HPO_4^{2-}$$

and if acid is added, it reacts with HPO_4^{2-}.

$$HPO_4^{2-} + H^+ \longrightarrow H_2PO_4^-$$

These reactions help prevent large changes in the pH of the blood.

Trisodium phosphate, Na_3PO_4, often called TSP, is an effective water softener and cleansing agent. In hard water it reacts with hardness ions such as Ca^{2+}, Mg^{2+}, and Fe^{3+} to form precipitates or complex ions, which removes the ions from solution so they can't interfere with the action of the soap. Solutions of Na_3PO_4 are basic because of the hydrolysis of the PO_4^{3-} ion.

$$PO_4^{3-} + H_2O \rightleftharpoons HPO_4^{2-} + OH^-$$

A 1.0 M Na_3PO_4 solution has a pH of about 12.8. This is very basic, so you should always wear rubber gloves when working with concentrated TSP solutions.

Polymeric Phosphoric Acids and Their Anions Orthophosphoric acid is the parent of a host of more complex acids and anions that contain more than one phosphorus atom. They are formed by eliminating the components of water from −OH groups on neighboring acid molecules and linking together adjacent PO_4 tetra-

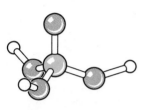

Phosphoric acid

○ = Hydrogen

◉ = Oxygen

◯ = Phosphorus

hedra by the mutual sharing of an oxygen atom. For example, when H_3PO_4 is heated to 250°C, **pyrophosphoric acid,** $H_4P_2O_7$, is formed.

$H_4P_2O_7$ is also called *diphosphoric acid.*

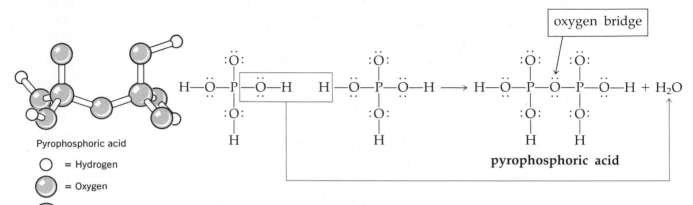

Pyrophosphoric acid

○ = Hydrogen

◐ = Oxygen

● = Phosphorus

Hydrolysis adds water to the P—O—P bridge to regenerate the pair of P—OH units.

Pyrophosphoric acid is a colorless, glassy solid that is very soluble in water. It forms salts such as $Na_4P_2O_7$ and $Na_2H_2P_2O_7$. Both the acid and its salts are very slowly hydrolyzed to phosphoric acid or its anions. The reaction is slow enough, however, so that $Na_4P_2O_7$ is used as the phosphate ingredient in many liquid detergents.

If orthophosphoric acid is heated above 400°C, extensive polymerization occurs by elimination of water, as shown below. The product is called **metaphosphoric acid.** Its formula is $(HPO_3)_n$ where n is a large number.

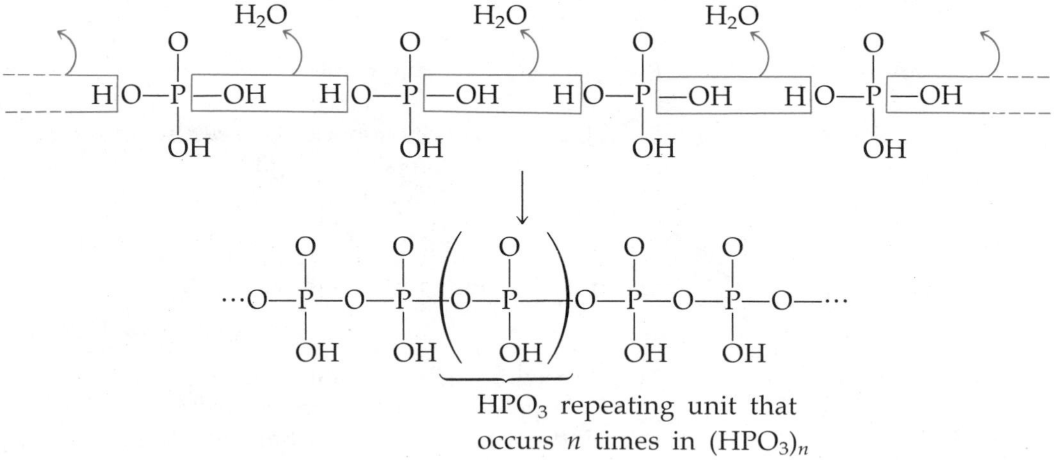

HPO_3 repeating unit that occurs n times in $(HPO_3)_n$

The sodium salt of metaphosphoric acid, which contains the $(PO_3^-)_n$ ion, is normally made by heating sodium dihydrogen phosphate.

$$n \ NaH_2PO_4 \xrightarrow{heat} (NaPO_3)_n + n \ H_2O$$

Polyphosphates of intermediate chain length can also be made. For example,

$$NaH_2PO_4 + 2Na_2HPO_4 \xrightarrow{heat} Na_5P_3O_{10} + 2H_2O$$

sodium
tripolyphosphate

Sodium tripolyphosphate is used in many solid detergent mixtures and contains the ion

$$(-):\overset{..}{\underset{..}{O}}-\overset{\overset{\displaystyle :\overset{..}{O}:}{|}}{\underset{\underset{(-)}{\displaystyle :\overset{..}{O}:}}{P}}-\overset{..}{\underset{..}{O}}-\overset{\overset{\displaystyle :\overset{..}{O}:}{|}}{\underset{\underset{(-)}{\displaystyle :\overset{..}{O}:}}{P}}-\overset{..}{\underset{..}{O}}-\overset{\overset{\displaystyle :\overset{..}{O}:}{|}}{\underset{\underset{(-)}{\displaystyle :\overset{..}{O}:}}{P}}-\overset{..}{\underset{..}{O}}:(-)$$

In many parts of the United States the use of phosphate-based detergents has been banned or severely restricted because of effects that high phosphate concentrations have on lakes. In general, the rate of algae growth in lakes is determined by the nutrient present in the most limited amount. Normally, this limiting nutrient is phosphate, and when soluble phosphates from fertilizers and detergents wash into lakes, episodes of rapid growth of algae—*algae blooms*— can occur. When the algae die and settle to the bottom they begin to decompose. This causes the waters to be depleted of oxygen, so other marine organisms— fish, for example—begin to die. These processes speed the natural aging or *eutrophication* of the lake and gradually reduce the oxygen level to the point where no aquatic life can survive in the waters.

Phosphorous Acid We learned earlier that **phosphorous acid,** H_3PO_3, is formed when phosphorus(III) oxide is dissolved in water.

$$P_4O_6(s) + 6H_2O(l) \longrightarrow 4H_3PO_3(aq)$$

Despite the way its formula is written, H_3PO_3 is only a diprotic acid. Its structure is

$$H-\overset{..}{\underset{..}{O}}-\overset{\overset{\displaystyle H}{|}}{\underset{\underset{..}{\displaystyle :\overset{..}{O}:}}{P}}-\overset{..}{\underset{..}{O}}-H$$

Only the O—H bonds break to give H^+; the hydrogen attached directly to the phosphorus is not acidic. This means that only two kinds of salts can be prepared—for example, NaH_2PO_3 and Na_2HPO_3. Phosphorous acid and phosphites are reasonably good reducing agents. For example, they reduce silver compounds to metallic silver.

Halogen Compounds Phosphorus forms two kinds of binary compounds with the halogens: the trihalides, PX_3 ($X =$ F, Cl, Br, and I) and the pentahalides, PX_5 ($X =$ F, Cl, and Br). The compound PI_5 is not known, presumably because iodine atoms are so large that five of them simply cannot be packed around a phosphorus atom.

The most important halogen compounds of phosphorus are the chlorides, PCl_3 and PCl_5. Their structures are shown in Figure 20.4. Phosphorus trichloride

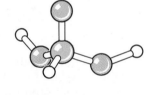

○ = Hydrogen

◐ = Oxygen

◓ = Phosphorus

Figure 20.4

The structures of phosphorus trichloride and phosphorus pentachloride.

PCl_3

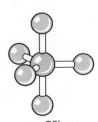

PCl_5

is made by reacting molten phosphorus with chlorine. If an excess of chlorine is present, the pentachloride can also be formed.

$$P_4(l) + 6Cl_2(g) \longrightarrow 4PCl_3(g)$$

$$PCl_3(g) + Cl_2(g) \rightleftharpoons PCl_5(g)$$

The second reaction is an equilibrium, and PCl_5 decomposes if heated.

Phosphorus trichloride is a volatile liquid that boils at 76°C. It is used as a starting material for the preparation of many other phosphorus compounds. When exposed to water, PCl_3 hydrolyzes to give phosphorous acid.

Many phosphorus-containing pesticides are made using PCl_3.

$$PCl_3 + 3H_2O \longrightarrow H_3PO_3 + 3HCl$$

In fact, this is the method usually used to make phosphorous acid.

Phosphorus trichloride reacts with oxygen to give **phosphoryl chloride,** $POCl_3$ (also called **phosphorus oxychloride**). About half of the PCl_3 manufactured each year is oxidized to $POCl_3$ and much of that is used to make compounds that are employed as flame retardants.

Phosphorus pentachloride in the vapor or liquid has the trigonal bipyramidal structure shown in Figure 20.4. In the solid, however, PCl_5 appears to exist as an ionic compound composed of tetrahedral PCl_4^+ ions and octahedral PCl_6^- ions (i.e., in the solid, PCl_5 is really $PCl_4^+PCl_6^-$). One of the principal uses of PCl_5 is in the manufacture of $POCl_3$ by the reaction

$$P_4O_{10} + 6PCl_5 \longrightarrow 10POCl_3$$

20.2 SULFUR

Brimstone is mentioned in the Bible.

The element sulfur is widely distributed in nature, although it is only about half as abundant as phosphorus. It occurs as the free element, and long before sulfur was recognized as an element it was known as *brimstone*, meaning stone that burns. In the combined state it is found in mineral sulfides (those of lead and copper, for example), in sulfates such as gypsum and epsom salts, and as sulfate ion in the ocean. Sulfur compounds also occur as impurities in natural gas, petroleum, and coal.

Sulfur is also found in Spain, Mexico, Japan, and Italy.

Somewhat more than half of the sulfur used by industry each year is mined from large underground deposits of the free element, chiefly in Texas and Louisiana. A rather clever method to accomplish this was invented in 1890 by an American engineer, Herman Frasch, and has come to bear his name. It involves pumping superheated water under pressure into the sulfur deposit, which causes the sulfur to melt (Figure 20.5). The hot sulfur-water mixture is then foamed to the surface with compressed air and sprayed into huge piles to dry (Figure 20.6).

The rest of the industrial supply of sulfur is recovered during the purification of natural gas and petroleum, which contain sulfur compounds such as H_2S that would create pollution problems if burned along with these fuels.

The free element

Sulfur atoms in their ground state have the electron configuration [Ne] $3s^23p^4$. As we saw in Chapter 6, these combine with each other by forming two covalent bonds to separate sulfur atoms, and eight member rings are formed. The most stable allotrope of sulfur at room temperature is known as rhombic sulfur in which the S_8 rings are stacked in a way that gives a rhombic crystal structure.

Figure 20.5

Sulfur is extracted from deposits of the free element deep below the surface by the Frasch process.

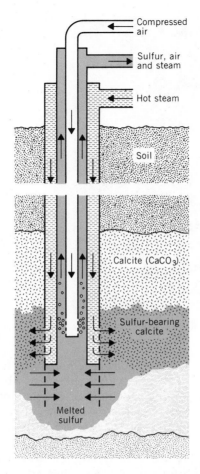

Compressed air

Sulfur, air and steam

Hot steam

Soil

Calcite (CaCO₃)

Sulfur-bearing calcite

Melted sulfur

Figure 20.6

Liquid sulfur is pumped into large ponds to cool and solidify after being extracted from deep below the earth by the Frasch process. In the background is an enormous pile of elemental sulfur.

Figure 20.7

Crystals of naturally occurring rhombic sulfur are seen here surrounded by crystals of quartz.

Crystals of yellow rhombic sulfur are shown in Figure 20.7. A second allotrope of sulfur is known as monoclinic sulfur in which S_8 rings are stacked in a monoclinic crystal structure. Rhombic sulfur melts at about 112°C, and when heated above 120°C and then allowed to cool slowly, needlelike crystals of the monoclinic allotrope (mp 119°C) are formed (Figure 20.8).

The behavior of elemental sulfur as it is heated from its melting point to its boiling point is quite interesting. When it first melts, a straw-colored, relatively nonviscous liquid is formed. As this is heated it gradually darkens and becomes molasseslike. At still higher temperatures it thins out again and the dark red liquid finally boils at 445°C. These changes are shown in Figure 20.9.

The changes in the sulfur as it is heated are explained as follows. When the sulfur first melts, the liquid is composed of S_8 rings in the usual jumbled arrangement characteristic of liquids in general. As the temperature is raised, ther-

Figure 20.8

Needlelike crystals of monoclinic sulfur.

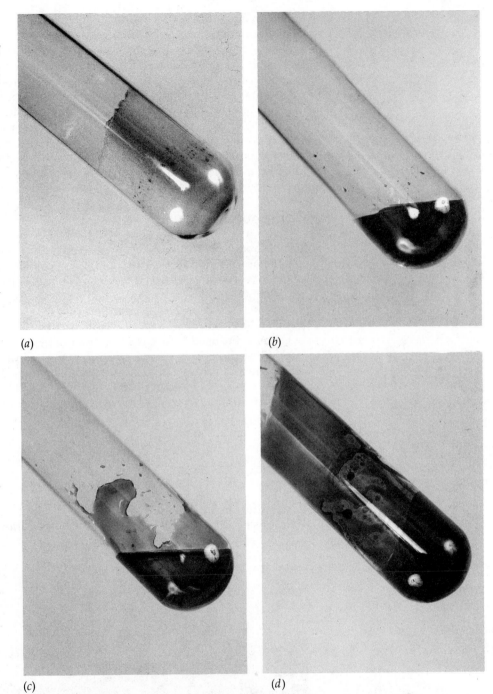

Figure 20.9

What happens when sulfur is heated. (a) *Solid sulfur.* (b) *Liquid sulfur just above its melting point.* (c) *Liquid sulfur becomes dark and very viscous as its temperature is raised.* (d) *At its boiling point, the sulfur is a dark, nonviscous liquid.*

(a)

(b)

(c)

(d)

mal energy increases the vibrational motion of the sulfur atoms in the rings, and sulfur-sulfur bonds begin to break. This gives chains of sulfur atoms that have an unpaired electron at each end.

When an end sulfur atom of one chain encounters another from a different chain a covalent bond is formed and an S_{16} chain is produced. Coupling can continue and long chains of S_{24}, S_{32}, S_{40}, and so on are formed that intertwine and cause the liquid to become very viscous. At still higher temperatures the more violent motions of the sulfur atoms cause the long chains to start to break into smaller fragments, and the liquid becomes relatively nonviscous again.

It is interesting to note that if the thickened liquid sulfur is cooled rapidly—for example, by being poured into cold water—the sulfur atoms do not have a chance to rearrange into S_8 rings. As a result, a supercooled liquid called *amorphous sulfur* is produced that has many of the elastic properties of rubber. When allowed to stand, the S_x chains in the amorphous sulfur gradually revert to the thermodynamically more stable S_8 rings of the rhombic form.

Another name for amorphous sulfur is *plastic sulfur*.

Compounds of Sulfur

Sulfur Dioxide and Sulfurous Acid Elemental sulfur is easily ignited and burns with a blue flame to give **sulfur dioxide,** SO_2, a colorless gas with a choking, irritating odor. If you've ever gotten a whiff of the fumes from an igniting match, you've smelled SO_2—sulfur is one of the substances used in matches. Sulfur dioxide is a nonlinear molecule and therefore is polar. This gives it a relatively high boiling point ($-10°C$) and allows it to be easily liquefied under pressure. For this reason, SO_2 has been used as a gas in refrigeration systems.

$$S(s) + O_2(g) \longrightarrow SO_2(g)$$

Sulfur dioxide is also produced when sulfur compounds are burned and when metal sulfides are heated in air—an important step in the extraction of metals such as lead and copper from their ores. The SO_2 formed in metallurgical process is now recovered, but the SO_2 released into the atmosphere by combustion of high sulfur fuels, both petroleum based and coal, presents a major pollution problem in some areas.

resonance structures of SO_2

When sulfur dioxide is dissolved in water, in which it is quite soluble, it forms sulfurous acid, a weak diprotic acid.

$$SO_2(aq) + H_2O \rightleftharpoons H_2SO_3(aq)$$

Pure sulfurous acid cannot be isolated. It decomposes into SO_2 and H_2O as its solutions become more concentrated by evaporation of the water. Neutralization of H_2SO_3 by base produces two kinds of salts—for example, Na_2SO_3 (sodium

Heating metal sulfides in air to convert them to their oxides is called smelting and is a common metallurgical process, but the SO_2 that is released must not be allowed to escape into the atmosphere. Here we see the effects of a longtime smelting operation at Copper Hill, Tennessee where SO_2 was released into the atmosphere.

NaHSO₃ is used in the manufacture of paper and pulp.

sulfite) and $NaHSO_3$ (sodium hydrogen sulfite or sodium bisulfite). These neutralization reactions are easily reversed, and the most convenient method of preparation of SO_2 in the laboratory is by reacting a sulfite or bisulfite with a concentrated strong acid such as H_2SO_4. For example,

$$Na_2SO_3(s) + H_2SO_4(aq) \longrightarrow Na_2SO_4(aq) + H_2O + SO_2(g)$$

Sulfurous acid and sulfites are both rather good reducing agents. On exposure to air they are slowly oxidized to sulfuric acid and sulfates.

Acid Rain The emissions of sulfur dioxide by industries and power plants that burn high-sulfur fuels have produced a major environmental problem affecting many regions that lie down wind from industrialized areas. As mentioned, the combustion of sulfur and sulfur compounds gives SO_2 as the major sulfur-containing product. This SO_2 is carried by the wind and is slowly oxidized to sulfur trioxide, SO_3. When it rains, both of these sulfur oxides are washed from the atmosphere as they dissolve in the rain drops. Once dissolved, the oxides react to form dilute solutions of acids—sulfurous acid from the dissolved SO_2 and sulfuric acid from the dissolved SO_3. The rain is made even more acidic by reactions of nitrogen oxides. In the last chapter, you learned that NO_2 dissolves in water where it disproportionates to give a solution of nitric acid.

Rain as acidic as vinegar once fell during a rainstorm in Pitlochry, Scotland, in April 1974.

As you have probably learned from the news media, this acidified rain has been called *acid rain*, and it is not unusual to find precipitation in northwestern Europe and in large areas of North America with a pH between 4 and 4.5. In cities this can cause structural damage to vehicles, buildings, and statues (Figure 20.10), and in the countryside it has damaged trees and caused lakes to become too acidic to support fish (Figure 20.11).

One method of removing SO_2 from the exhaust gases of industrial furnaces is to allow it to pass over moist limestone. It is absorbed by the moisture and reacts as follows

$$CaCO_3(s) + SO_2(g) \longrightarrow CaSO_3(s) + CO_2(g)$$

This is essentially the same reaction that occurs when acid rain falls on limestone and marble building materials.

Figure 20.10

How polluted air and acid rain cause decay is seen in this statue made of Baumberg sandstone at the Herten Castle in Westphalia, West Germany. On the left is its appearance in 1908 after 206 years of exposure to the atmosphere. On the right is its appearance 60 years later after exposure to air pollution produced by European heavy industry.

(a)   *(b)*

Figure 20.11
(a) *The effects of acid rain on fir trees in the forests of Bavaria, Germany.* (b) *In Boksjo, Sweden, "lime" (actually, pulverized limestone, $CaCO_3$) is pumped into a lake to neutralize acidity caused by acid rain. The reaction is*

$$CaCO_3(s) + 2H^+(aq) \longrightarrow Ca^{2+}(aq) + CO_2(g) + H_2O$$

Sulfur Trioxide and Sulfuric Acid   At room temperature sulfur trioxide is a solid consisting of SO_3 units linked together in various complex chain structures. The solid is easily vaporized and in the gas phase SO_3 exists as discrete planar triangular molecules.

Sulfur trioxide is made by the oxidation of sulfur dioxide with oxygen. As noted earlier, this is a highly favorable reaction thermodynamically, as can be seen by its large negative ΔG^0.

$$2SO_2(g) + O_2(g) \longrightarrow 2SO_3(g) \qquad \Delta G^0 = -140 \text{ kJ mol}^{-1}$$

The reaction is slow in the absence of a catalyst, which explains the apparent stability of SO_2 in air. However, in the presence of an appropriate catalyst the oxidation of SO_2 occurs quickly. As mentioned in Chapter 13, catalysts used in automobile catalytic converters accelerate this reaction and the SO_3 produced combines with moisture in the exhaust gases to form a mist of H_2SO_4.

$$SO_3(g) + H_2O(g) \longrightarrow H_2SO_4(l)$$

The oxidation of SO_2 to SO_3 and its subsequent conversion to sulfuric acid are, by far, industry's most important chemical reactions. The amount of sulfuric acid produced annually—approximately 3.8×10^{10} kg—is more than *twice* that of any other single substance. About 60% of it is used to treat phosphate rock in the production of phosphate fertilizers and in making ammonium sulfate (also a fertilizer). Sulfuric acid is also used in the refining of petroleum, the steel industry, lead storage batteries, and chemical reactions involved in the manufacture of paints, explosives, plastics, drugs, and many other chemicals.

Most sulfuric acid is made by the **contact process.** The raw material is sulfur, which is first burned in air to form sulfur dioxide. The SO_2 and additional oxygen (in air) are then passed over a vanadium pentoxide (V_2O_5) catalyst, which oxidizes the SO_2 to SO_3. Next, the SO_3 is absorbed into concentrated H_2SO_4, with which it reacts to form **pyrosulfuric acid,** $H_2S_2O_7$.

$$H_2SO_4(l) + SO_3(g) \longrightarrow H_2S_2O_7(l)$$

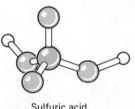

Sulfuric acid

○ = Hydrogen

◑ = Oxygen

◐ = Sulfur

Figure 20.12

Effects of H_2SO_4 on sugar.
(a) *Before H_2SO_4 is added.*
(b) *After H_2SO_4 has attacked the sugar.*

(a)

(b)

The H_2SO_4 is used to trap the SO_3 because it is more effective than water. Finally, the $H_2S_2O_7$ is diluted with water, which converts it to sulfuric acid.

$$H_2S_2O_7(l) + H_2O(l) \longrightarrow 2H_2SO_4(l)$$

Pure H_2SO_4 is a dense, colorless, viscous liquid that tends to decompose into SO_3 and H_2O when heated. The concentrated H_2SO_4 found in bottles on laboratory shelves is 96% H_2SO_4 by mass (18 M). Dilution of the concentrated acid with water is *very* exothermic because of the reaction

$$H_2SO_4 \text{ (conc.)} + H_2O(l) \longrightarrow H_3O^+(aq) + HSO_4^-(aq)$$

Care should always be taken to add concentrated H_2SO_4 *to the water*—never the other way around. If water is added to the concentrated acid, the heat generated causes the water to boil and the expanding steam spatters the neighborhood with H_2SO_4.

The strong affinity that H_2SO_4 has for water makes the concentrated acid an effective dehydrating agent. For example, if concentrated H_2SO_4 is poured onto sugar, $C_{12}H_{22}O_{11}$, it extracts the components of water and leaves a blackened, charred mass behind (Figure 20.12).

$$C_{12}H_{22}O_{11} \xrightarrow{\text{conc. } H_2SO_4} 12C + 11H_2O$$

Sulfuric acid is a strong diprotic acid. The first step in its dissociation is complete in aqueous solutions, but the second step only occurs to about 10% in a 1 M solution. This gives a 1.0 M solution of H_2SO_4 an H_3O^+ concentration of approximately 1.1 M.

$$HSO_4^- \rightleftharpoons H^+ + SO_4^{2-}$$
$$K_a = 1.2 \times 10^{-2}$$

By controlling the amount of base during the neutralization of H_2SO_4, two series of salts can be formed: sulfates such as Na_2SO_4 and hydrogen sulfates or bisulfates such as $NaHSO_4$. Solutions of bisulfates are quite acidic because the HSO_4^- ion is a relatively strong acid.

A 0.10 M solution of $NaHSO_4$ has a pH of 1.4.

Concentrated sulfuric acid is a mild oxidizing agent when cold, but a fairly strong oxidizing agent when hot. For example, cold concentrated H_2SO_4 oxidizes iodide ion to iodine and, to some extent, bromide ion to bromine.

$$2X^- + 3H_2SO_4 \longrightarrow X_2 + SO_2 + 2H_2O + 2HSO_4^- \qquad (X = I \text{ or } Br)$$

Cold H_2SO_4 doesn't attack metallic copper, but the hot concentrated acid does.

The reaction, which causes the sulfur to be reduced from the +6 state to the +4 state, is

$$Cu + 2H_2SO_4 \xrightarrow{heat} CuSO_4 + SO_2 + 2H_2O$$

A stronger reducing agent such as zinc reduces the sulfuric acid even further to free sulfur (oxidation number = zero) or even hydrogen sulfide (oxidation number of sulfur = −2).

Other Compounds of Sulfur Sulfur combines with most metals and nonmetals to form a large variety of different compounds. We will look only briefly at some of the more important types.

Binary compounds of metals with sulfur are sulfides and contain the sulfide ion, S^{2-}. They can often be prepared by direct reaction of the metal with sulfur. For example, zinc and sulfur react vigorously to produce zinc sulfide.

$$Zn(s) + S(s) \longrightarrow ZnS(s)$$

Metal sulfides tend to be less ionic than their oxides.

Since most metal sulfides are insoluble in water, they can also be formed by precipitation reactions using hydrogen sulfide, H_2S. The use of hydrogen sulfide in separating metal ions by selective precipitation was discussed in Chapter 16.

Hydrogen sulfide itself is a poisonous gas with an odor of rotten eggs. It is more poisonous than carbon monoxide, but its bad odor allows it to be detected at low concentrations. In nature, hydrogen sulfide is released into the air during volcanic eruptions and by the decomposition of organic matter in the absence of air—that's why rotten eggs smell of H_2S. Airborn H_2S is responsible for the gradual tarnishing of silver and the darkening of lead-based paints. In each case, a black metal sulfide (Ag_2S and PbS) is formed.

In the laboratory, H_2S can be prepared by reacting a metal sulfide with a strong nonoxidizing acid—for example, HCl.

We learned in Chapter 15 that H_2S is a weak diprotic acid.

$$FeS(s) + 2H^+(aq) \longrightarrow Fe^{2+}(aq) + H_2S(g)$$

Another method, commonly used when one wants to generate the H_2S in an aqueous solution, is the hydrolysis of an organic compound called thioacetamide.

$$\underset{\textbf{thioacetamide}}{CH_3-\overset{\overset{S}{\|}}{C}-NH_2(aq)} + 2H_2O \longrightarrow H_2S(aq) + NH_4^+(aq) + \underset{\textbf{acetate ion}}{\left(CH_3\overset{\overset{O}{\|}}{C}-O\right)^-(aq)}$$

The advantage of this reaction is that it avoids the release of significant amounts of toxic H_2S into the atmosphere.

The prefix *thio* in a chemical name implies that sulfur replaces oxygen in a compound. For example,

$$\underset{\textbf{acetamide}}{CH_3-\overset{\overset{O}{\|}}{C}-NH_2} \qquad \underset{\textbf{thioacetamide}}{CH_3-\overset{\overset{S}{\|}}{C}-NH_2}$$

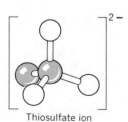
Thiosulfate ion

○ Oxygen
◐ Sulfur

An important oxoanion that fits this nomenclature pattern is the thiosulfate ion, $S_2O_3^{2-}$. It is formed by boiling sulfur with a solution containing sulfite ion.

$$S(s) + SO_3^{2-}(aq) \longrightarrow S_2O_3^{2-}(aq)$$

The Lewis structure for the $S_2O_3^{2-}$ ion is

$$\left[\begin{array}{c} :\overset{\displaystyle ..}{\overset{\displaystyle O}{|}}: \\ :\overset{\displaystyle ..}{\underset{\displaystyle ..}{S}}-\overset{\displaystyle ..}{\underset{\displaystyle ..}{S}}-\overset{\displaystyle ..}{O}: \\ :\overset{\displaystyle ..}{\underset{\displaystyle ..}{O}}: \end{array}\right]^{2-}$$

Notice that it can be viewed as a sulfate ion in which one oxygen is replaced by sulfur—hence, *thio*sulfate.

Thiosulfate ion forms quite stable complex ions with metal ions. The one with silver is particularly important in photography. Silver bromide is the usual light-sensitive substance found in photographic films and paper. After an image is developed, the photosensitive emulsion still contains unexposed silver bromide that must be removed to prevent the picture from darkening gradually when viewed in the light. Solutions of sodium thiosulfate—known as *hypo* to photographers—are used to remove the unexposed silver bromide. The equilibria involved are

$$AgBr(s) \rightleftharpoons Ag^+(aq) + Br^-(aq)$$

$$Ag^+(aq) + 2S_2O_3^{2-}(aq) \rightleftharpoons Ag(S_2O_3)_2^{3-}(aq)$$

Le Châtelier's principle explains how the thiosulfate works. As $S_2O_3^{2-}$ is added, the second equilibrium shifts to the right and the Ag^+ concentration drops. This causes the AgBr to dissolve in an attempt to restore the Ag^+ concentration to its former value. If sufficient thiosulfate is present, equilibrium is never reestablished and all the AgBr dissolves. Afterwards, the $Ag(S_2O_3)_2^{3-}$ complex ion and residual thiosulfate are washed from the film or paper, leaving only the desired image.

Thiosulfate ion is also a good reducing agent. It reacts with strong oxidizing agents such as chlorine to give sulfate ion.

$$4Cl_2(g) + S_2O_3^{2-} + 5H_2O \longrightarrow 8Cl^- + 2SO_4^{2-} + 10H^+$$

Thiosulfate ion is often used in titrations of the weaker oxidizing agent iodine. Its reaction with iodine is

$$2S_2O_3^{2-} + I_2 \longrightarrow S_4O_6^{2-} + 2I^-$$

thiosulfate   **tetrathionate**
ion   **ion**

$$\left[\begin{array}{c} :\overset{\displaystyle ..}{O}: \quad\quad :\overset{\displaystyle ..}{O}: \\ :\overset{\displaystyle ..}{\underset{\displaystyle ..}{O}}-\overset{\displaystyle ..}{\underset{\displaystyle ..}{S}}-\overset{\displaystyle ..}{\underset{\displaystyle ..}{S}}-\overset{\displaystyle ..}{\underset{\displaystyle ..}{S}}-\overset{\displaystyle ..}{\underset{\displaystyle ..}{S}}-\overset{\displaystyle ..}{O}: \\ :\overset{\displaystyle ..}{\underset{\displaystyle ..}{O}}: \quad\quad :\overset{\displaystyle ..}{\underset{\displaystyle ..}{O}}: \end{array}\right]^{2-}$$

tetrathionate ion

20.3 THE HALOGENS

Astatine, At, is also a halogen, but it is rare and radioactive. We will not include it in our discussions.

Until now we've discussed the nonmetallic elements one at a time because the variations in their properties are rather substantial. For example, although oxygen and sulfur are in the same group and similar in some respects, they also have many differences, and the same applies to nitrogen and phosphorus. Among the halogens—the Group VIIA elements—many close similarities and trends in properties exist that are easy to follow, so we will discuss them as a group.

The halogens—fluorine, chlorine, bromine, and iodine—are never found free in nature because of their high reactivity. Until humans began their study of chemistry and began to produce the free halogens, these elements occurred almost exclusively in inorganic salts. In fact, the name halogen comes from the Greek *halos*, meaning salt. Today, halogens have been incorporated into many useful organic compounds, ranging from Teflon and polyvinyl chloride plastics to Freon refrigerants and aerosol propellants to insecticides such as DDT. In recent times it has been found that some of these halogenated organic compounds (DDT, chloroform, and polychlorinated biphenyls (PCB's), for example) can have very harmful effects on our environment and its inhabitants—us!

Cryolite, you recall, was used by Hall in his process to recover aluminum from Al_2O_3 by electrolysis.

The principal sources of the halogens are their salts. Fluorine is found in the earth's crust principally in deposits of fluorspar, CaF_2, cryolite, Na_3AlF_6, and fluoroapatite, $Ca_5(PO_4)_3F$. The main source of chlorine is NaCl which, as we learned earlier, is recovered from the sea and large underground deposits formed, presumably, by evaporation of ancient seas. Bromine and iodine are also found in the sea, but their concentrations as Br^- and I^- are *much* less than that of Cl^-. (Sea water contains these concentrations of the halide ions: 0.53 M Cl^-, $8.1 \times 10^{-4}\,M\,Br^-$, and $5 \times 10^{-7}\,M\,I^-$). Bromine and iodine are also found in the waters of brine (salt) wells, and iodine is obtained from sodium iodate, $NaIO_3$, which is recovered from deposits of saltpeter, $NaNO_3$, imported from Chile.

Properties of the free elements

Halogen atoms each have seven valence electrons and therefore need only one more to achieve a noble gas configuration. As a result, in the elemental state they form singly bonded diatomic molecules, X_2.

X = F, Cl, Br, or I.

$$:\!\overset{..}{\underset{..}{X}}\!:\!\overset{..}{\underset{..}{X}}\!:$$

Some physical properties of the halogens are given in Table 20.1. We see that as their molecules become larger, the melting points and boiling points increase, corresponding to an increase in the strengths of the London forces.

At room temperature fluorine is a pale yellow gas. It is the most reactive of all the elements because of the low F—F bond energy, that is, little energy is needed to split the molecule into very reactive fluorine atoms. This low bond energy is believed to be caused by electron-electron repulsions between the small, compact, electron-rich valence shells of the bonded fluorine atoms.

Chlorine is a pale green gas at room temperature. Chlorine molecules have a slightly larger bond energy than F_2 molecules, and chlorine atoms are less electronegative than fluorine. As a result, Cl_2 is somewhat less reactive than F_2.

At room temperature, bromine is a volatile, nonviscous red liquid, and iodine forms dark metallic-looking crystals that easily sublime to give a purple vapor. Chemical reactivity decreases from chlorine to bromine to iodine, following the decrease in their electronegativities.

Because of the high electronegativities of the halogens compared to other elements, they tend to gain electrons from other substances and thereby serve as oxidizing agents. The ability of the halogens to serve as oxidizing agents decreases going down the group. As a result we find that a given halogen is able to oxidize the anions of the halogens below it in Group VIIA. Thus F_2 will oxidize

Table 20.1
Some physical properties of the halogens

Halogen	Melting Point (°C)	Boiling Point (°C)	Bond Energy (kJ mol^{-1})	Electronegativity
Fluorine (F_2)	−233	−188	157	4.0
Chlorine (Cl_2)	−103	−34.6	242	3.0
Bromine (Br_2)	−7.2	58.8	193	2.8
Iodine (I_2)	113.5	184.4	150	2.5

Cl^-, Br^-, and I^-, while Cl_2 will oxidize only Br^- and I^- but not F^-. This is illustrated by the following typical reactions:

Fluorine

$$F_2 + \left\{\begin{array}{l} 2NaCl \\ 2NaBr \\ 2NaI \end{array}\right\} \longrightarrow 2NaF + \left\{\begin{array}{l} Cl_2 \\ Br_2 \\ I_2 \end{array}\right\}$$

Chlorine

$$Cl_2 + NaF \longrightarrow \text{No reaction}$$

$$Cl_2 + \left\{\begin{array}{l} 2NaBr \\ 2NaI \end{array}\right\} \longrightarrow 2NaCl + \left\{\begin{array}{l} Br_2 \\ I_2 \end{array}\right\}$$

Bromine

$$Br_2 + \left\{\begin{array}{l} NaF \\ NaCl \end{array}\right\} \longrightarrow \text{No reaction}$$

$$Br_2 + 2NaI \longrightarrow 2NaBr + I_2$$

Iodine

$$I_2 + \left\{\begin{array}{l} NaF \\ NaCl \\ NaBr \end{array}\right\} \longrightarrow \text{No reaction}$$

Preparation of the free elements

Fluorine Elemental fluorine is such an active oxidizing agent that it can only be made by electrolysis. The raw material is hydrogen fluoride, which is dissolved in molten KF. Electrolysis produces fluorine gas at the anode and hydrogen gas at the cathode. The KF in the mixture serves as an electrolyte because pure hydrogen fluoride is molecular and therefore is nonconducting.

$$2HF \xrightarrow[\text{KF}]{\text{electrolysis}} H_2(g) + F_2(g)$$

Fluorine is used to make a variety of fluorine-containing organic compounds. Examples are Teflon and the Freons. As we learned in Chapter 19, Freons are chlorofluorocarbons that are used as refrigerants and aerosol propellants, and they may have harmful effects on the Earth's ozone layer as they diffuse into the stratosphere.

Chlorine The production of chlorine by electrolysis both of molten NaCl and aqueous NaCl (brine) has been described previously. The reactions are

$$2NaCl(l) \xrightarrow{\text{electrolysis}} 2Na(l) + Cl_2(g)$$

$$2Na^+(aq) + 2Cl^-(aq) + 2H_2O \xrightarrow{\text{electrolysis}} 2Na^+(aq) + 2OH^-(aq) + Cl_2(g) + H_2(g)$$

Industrially, chlorine ranks about eighth in total annual tons produced.

Over 10^{10} kg of chlorine are produced each year. Its major use has been in the manufacture of chemical intermediates—chemicals that are used to make other chemicals. Chlorine is also used to treat municipal drinking water, make solvents and plastics such as polyvinyl chloride (vinyl plastics), and to manufacture pesticides.

In the laboratory chlorine can be made by oxidation of chloride ion in an acidic solution by a strong oxidizing agent such as manganese dioxide, MnO_2, or potassium permanganate, $KMnO_4$. The simplest method is to allow concentrated hydrochloric acid to react with MnO_2.

$$MnO_2(s) + 2Cl^-(aq) + 4H^+(aq) \longrightarrow Mn^{2+}(aq) + Cl_2(g) + 2H_2O$$

Bromine Although bromide ion occurs in a low concentration in sea water, bromine can be recovered from it by taking advantage of the ease of oxidation of Br^- by chlorine and of the volatility of Br_2. First, chlorine is dissolved in the water and oxidizes bromide ion to bromine.

$$2Br^-(aq) + Cl_2(aq) \longrightarrow Br_2(aq) + 2Cl^-(aq)$$

Air is then blown through the water and the volatile bromine and residual unreacted chlorine are flushed out. Cooling the air causes the Br_2 to condense to a liquid. Today, most bromine is extracted from brines obtained from wells in Arkansas and Michigan. These contain bromide ion in concentrations that are 50 to 60 times greater than in sea water.

About half the bromine produced each year is used to make ethylene dibromide, $C_2H_4Br_2$, which is an additive in leaded gasoline. Its purpose is to prevent deposits of lead compounds from forming inside the engine. During combustion lead bromide, $PbBr_2$, is produced, which is volatile at temperatures that exist in the cylinders and therefore leaves the engine as a gas in the exhaust. Leaded gasoline cannot be used in cars equipped with catalytic converters because the lead compounds would be adsorbed on the catalyst surface and destroy its catalytic activity. Bromine is also used to make silver bromide, the principal ingredient in the light sensitive emulsions on photographic film and paper.

In the laboratory, Br_2 can be made by oxidation of a bromide salt by MnO_2 in an acidic solution (e.g., a solution containing H_2SO_4).

$$MnO_2(s) + 2Br^- + 4H^+ \longrightarrow Mn^{2+} + Br_2 + 2H_2O$$

It can also be made by oxidation of bromide ion by chlorine.

$$Cl_2 + 2Br^- \longrightarrow Br_2 + 2Cl^-$$

Iodine Iodide ion is present in very low concentrations in sea water, but seaweed extracts it from the water and concentrates it. Commercial quantities of iodine are recovered from the ashes of burned seaweed, in which the iodide concentrations approach 1%. The I^- is oxidized to I_2 using chlorine or other oxidizing agents.

Another commercial source of iodine is Chilean saltpeter, which contains sodium iodate, $NaIO_3$. The iodate is reduced to free iodine using sodium bisulfite, $NaHSO_3$, as the reducing agent.

$$2IO_3^- + 5HSO_3^- \longrightarrow I_2 + 5SO_4^{2-} + H_2O + 3H^+$$

The recovery of iodine is expensive and its applications are limited. It is used to make various medicinal products (tincture of iodine, for example) and silver iodide, which is also employed in photographic film.

Compounds of the halogens

Binary Halides of Metals Halogen atoms easily gain an electron to form the halide ions—fluoride, chloride, bromide, and iodide—and their compounds with metals are quite common. Most metal halides are ionic, providing the metal is in a low oxidation state. When the metal is in a high oxidation state, polarization of the anion often produces covalently bonded species. For example, in the vapor, aluminum chloride exists as Al_2Cl_6 molecules. Tin(IV) chloride ($SnCl_4$) and titanium(IV) chloride ($TiCl_4$) are both covalent liquids at room temperature.

Hydrogen Halides The binary hydrogen compounds of the halogens have the general formula, HX. They can be prepared by direct combination of the elements

$$H_2 + X_2 \longrightarrow 2HX$$

HCl is sometimes made commercially by the reaction

$$H_2 + Cl_2 \longrightarrow 2HCl$$

but the vigor of the reaction varies substantially from fluorine to iodine.

Fluorine reacts violently with hydrogen as soon as the two gases are mixed. Chlorine and hydrogen react at a much slower rate, however, provided their mixtures are not heated or exposed to ultraviolet light. Light or heat splits Cl_2 molecules into Cl atoms and initiates a chain reaction that is explosively fast. Chain mechanisms are also involved in the reaction of H_2 with Br_2 and I_2, but the reactions are less vigorous than with chlorine.

The chain mechanism for the reaction of H_2 with Br_2 was discussed in Chapter 13.

The hydrogen halides can also be made from their binary salts by reaction with a nonvolatile acid. For example, hydrogen fluoride is prepared by reacting CaF_2 with sulfuric acid.

$$CaF_2(s) + H_2SO_4(l) \longrightarrow CaSO_4(s) + 2HF(g)$$

Similarly, HCl is evolved if concentrated H_2SO_4 is added to NaCl.

$$NaCl(s) + H_2SO_4(l) \longrightarrow HCl(g) + NaHSO_4(s)$$

Additional HCl can be produced by adding more salt to the $NaHSO_4$ and heating the mixture.

$$NaCl(s) + NaHSO_4(s) \xrightarrow{\text{heat}} Na_2SO_4(s) + HCl(g)$$

Concentrated sulfuric acid is too powerful an oxidizing agent, even when cold, to be used to generate HBr and HI from their salts. Oxidation of the halide ion to free Br_2 or I_2 occurs. Phosphoric acid—a very poor oxidizing agent—can be used in place of sulfuric acid, but the mixtures must be warmed to expel the hydrogen bromide or hydrogen iodide.

Hydrogen fluoride has a substantially higher boiling point than the other hydrogen halides because of extensive hydrogen bonding that produces long staggered chains of HF molecules in the liquid.

(Dots indicate hydrogen bonds.)

	Boiling Points (°C)
HF	+19.7
HCl	−85.1
HBr	−66.8
HI	−35.4

Water solutions of the hydrogen halides are the *hydrohalic acids*—hydrofluoric acid, hydrochloric acid, hydrobromic acid, and hydriodic acid. In water, HF is a weak acid, whereas the others are all 100% ionized. Despite being a weak acid, hydrofluoric acid attacks glass and sand. In this case the reaction produces silicon tetrafluoride, SiF_4, a volatile substance that can escape as a gas.

Even though HF is a weak acid, it causes very severe skin burns.

$$\underset{\substack{\text{(in sand} \\ \text{and glass)}}}{SiO_2(s)} + 4HF(aq) \longrightarrow SiF_4(g) + 2H_2O(l)$$

Concentrated solutions of HF contain the HF_2^- ion in which a hydrogen ion is shared between two fluoride ions.

$$[F\cdots H\cdots F]^-$$

Solutions of hydrochloric acid can be purchased in hardware stores under the name *muriatic acid.*

Hydrochloric acid is the most important of the hydrohalic acids because it is relatively inexpensive. It is used to remove rust from iron and steel before the metal is coated with zinc (galvanizing) and in the manufacture of many other chemicals. Hydrochloric acid is also produced in the stomach and helps us digest our foods.

Table 20.2
Oxoacids and oxoanions of the halogens

Fluorine	Chlorine	Bromine	Iodine
HOF	HOCl	HOBr	HOI
	(OCl^-)	(OBr^-)	(OI^-)
	$HClO_2$	$HBrO_2$	HIO_3
	(ClO_2^-)	(BrO_2^-)	(IO_3^-)
	$HClO_3$	$HBrO_3$	H_5IO_6
	(ClO_3^-)	(BrO_3^-)	$(H_2IO_6^{3-})$
	$HClO_4$	$HBrO_4$	HIO_4
	(ClO_4^-)	(BrO_4^-)	(IO_4^-)

HOCl OCl^-	hypochlorous acid hypochlorite ion	H—O—Cl:
$HClO_2$ (or HOClO) ClO_2^-	chlorous acid chlorite ion	H—O—Cl—O:
$HClO_3$ (or $HOClO_2$) ClO_3^-	chloric acid chlorate ion	H—O—Cl—O: (with :O: above)
$HClO_4$ (or $HOClO_3$) ClO_4^-	perchloric acid perchlorate ion	H—O—Cl—O: (with :O: above and :O: below)

Oxoacids and Oxoanions The halogens form four kinds of oxoacids and anions (Table 20.2). Those of chlorine are the most familiar and the most important.

As we learned earlier, the strengths of these acids increase from HOCl, which is a weak acid, to $HClO_4$, which is an extremely powerful acid. Recall that the explanation given is that the electron withdrawing effect of the lone oxygens leads to an increased polarization of the O—H bond as the number of lone oxygens becomes larger.

Hypochlorous acid is formed by a disproportionation reaction—a reaction in which the same chemical is both oxidized and reduced—when chlorine is dissolved in cold water.

$$Cl_2(aq) + H_2O \rightleftharpoons HOCl(aq) + H^+(aq) + Cl^-(aq)$$

Similar reactions of Br_2 and I_2 occur to much lesser extents.

Approximately 30% of the dissolved chlorine exists in the form of HOCl and Cl^-. When chlorine is dissolved in base, the equilibrium is shifted to the right because the HOCl is neutralized. The net reaction is

$$Cl_2 + 2OH^- \longrightarrow OCl^- + Cl^- + H_2O$$

Electrolysis of a stirred NaCl solution, you recall, produces Cl_2, which reacts with the OH^- formed at the cathode. In this way, the aqueous NaCl is gradually changed to aqueous NaOCl, which is diluted and sold as liquid laundry bleach (e.g., Clorox®).

Reaction of Cl_2 with lime, CaO, produces a calcium salt that is sold as a solid laundry bleach and in antimildew preparations.

$$CaO(s) + Cl_2(g) \longrightarrow CaCl(OCl)(s)$$

A similar compound, calcium hypochlorite, formed by the neutralization of hypochlorous acid with calcium oxide, is sold as an algicide for backyard swimming pools. It goes by the commercial name HTH, meaning "high test hypochlorite"

$$CaO(s) + 2HOCl(aq) \longrightarrow Ca(OCl)_2(aq) + H_2O$$

HOCl is too unstable to be isolated in pure form.

Hypochlorous acid and its salts are powerful oxidizing agents. This is why hypochlorites are used as bleaches—the oxidation of colored compounds often produces colorless oxidation products. This same strong oxidizing ability also kills bacteria and fungi.

The hypohalites (OCl^-, OBr^-, OI^-) all tend to disproportionate to form the corresponding halides (X^-) and halates (ClO_3^-, BrO_3^-, IO_3^-).

$$3OX^- \longrightarrow XO_3^- + 2X^- \qquad (X = Cl, Br, I)$$

The equilibrium constants for this reaction are large for all three of these halogens; however, the *rates* of disproportionation differ greatly. The reaction of OI^- is very rapid at all temperatures, while OBr^- reacts moderately fast at room temperature. (Solutions of OBr^- ion can be prepared only if they are kept cold.) At room temperature the disproportionation of OCl^- is very slow, so its solutions can be stored for reasonable periods, which explains why they can be sold as liquid bleaches. This is an interesting example of stability being determined by the slow rate of reaction rather than thermodynamics.

Chlorous acid and the chlorite ion are among the lesser important compounds of chlorine. Reaction of chlorine dioxide with base produces ClO_2^- ion.

$$2ClO_2(g) + 2OH^-(aq) \longrightarrow ClO_2^-(aq) + ClO_3^-(aq) + H_2O$$

Chlorous acid is made from its barium salt by a metathesis reaction using H_2SO_4.

$$2H^+ + SO_4^{2-} + Ba(ClO_2)_2(s) \longrightarrow BaSO_4(s) + 2HClO_2(aq)$$

The reaction proceeds because of the very low solubility of $BaSO_4$. Like HOCl, $HClO_2$ is unstable and decomposes when attempts are made to isolate it in pure form.

The halate ions, XO_3^-, are obtained when Cl_2, Br_2, and I_2 are dissolved in hot basic solutions.

$$3X_2 + 6OH^-(aq) \longrightarrow 5X^-(aq) + XO_3^-(aq) + 3H_2O$$

Iodic acid, HIO_3, is the only halic acid that can be isolated in the pure state.

Chloric acid, like chlorous acid, can be made in solution from its barium salt.

$$H_2SO_4(aq) + Ba(ClO_3)_2(aq) \longrightarrow BaSO_4(s) + 2HClO_3(aq)$$

or

$$2H^+(aq) + SO_4^{2-}(aq) + Ba^{2+}(aq) + 2ClO_3^-(aq) \longrightarrow BaSO_4(s) + 2H^+(aq) + 2ClO_3^-(aq)$$

Chloric acid is a very powerful oxidizing agent. It is a strong acid, and it too cannot be isolated in pure form.

Perchlorates are commonly prepared by electrolytic oxidation of chlorates. Disproportionation of $KClO_3$ at moderate temperatures in the absence of catalysts also produces $KClO_4$.

In the presence of a catalyst (MnO_2) or at high temperatures, $KClO_3$ decomposes to KCl and O_2.

$$4KClO_3 \longrightarrow 3KClO_4 + KCl$$

Table 20.3

Halogen compounds of the nonmetals

Group IIIA	BX_3 (X = F, Cl, Br, I) BF_4^-			
Group IVA	CX_4 (X = F, Cl, Br, I)	SiF_4 SiF_6^{2-} $SiCl_4$	GeF_4 GeF_6^{2-} $GeCl_4$	
Group VA	NX_3 (X = F, Cl, Br, I) N_2F_4	PX_3 (X = F, Cl, Br, I) PF_5 PCl_5 PBr_5	AsF_3 AsF_5	SbF_3 SbF_5
Group VIA	OF_2 (O_2F_2) OCl_2 OBr_2	SF_2 SCl_2 S_2F_2 S_2Cl_2 SF_4 SCl_4 SF_6	SeF_4 SeF_6 $SeCl_2$ $SeCl_4$ $SeBr_4$	TeF_4 TeF_6 $TeCl_4$ $TeBr_4$ TeI_4
Group VIIA	ICl IBr BrF BrCl ClF	ClF_3 BrF_3 ICl_3 IF_3	ClF_5 BrF_5 IF_5	IF_7

Perchloric acid itself can be obtained pure, but it is unstable and tends to explode. In concentrated solutions it is a very strong oxidizing agent, although its dilute solutions, for unknown reasons, have very little oxidizing power.

Other halogen compounds of the nonmetals

In Table 20.3 you will find a list (although not an exhaustive one) of many of the compounds that are formed between the nonmetals and the halogens. The structures of the substances found in this table can, without exception, be predicted on the basis of the valence shell electron-pair repulsion theory.

In Table 20.3 we see that most of the nonmetals form more than one compound with a given halogen. The number of halogen atoms that can become bound to any particular nonmetal can be related to two factors. One is the electronic structures of the elements that are combined together and the other has to do with the sizes of the atoms.

Each halogen atom contains seven electrons in its valence shell and requires only one more to achieve the stable noble gas configuration. As a result, there is little tendency for them to form multiple bonds with other nonmetals. Furthermore, the halogens ordinarily do not accept electrons in the formation of coordinate covalent bonds because this would mean the addition of two electrons to a valence shell that already contains seven, thereby exceeding the stable octet by one electron.[1]

On this basis, we can divide the halogen compounds into two groups: those that obey the octet rule and those that do not. The compositions of the compounds in the first category are determined by the number of electrons that a given nonmetal requires to reach an octet, since each halogen atom furnishes one electron. For example, in Group VIIA—the halogens themselves—only one

---

[1] The halide ions (for example, Cl⁻) do *furnish* electron pairs toward the formation of coordinate covalent bonds and are common ligands in complex ions.

electron is needed and only one bond is formed. Thus the halogens are diatomic, and substitution of one halogen for another is possible, as we see for substances such as ClF, BrF, BrCl, BrI, and ICl.

$$:\ddot{C}l \times \ddot{\underset{\times\times}{\overset{\times\times}{F}}}\times$$

In Group VIA each element requires two electrons to reach an octet, so compounds such as OF_2 and SCl_2 are formed, and in Group VA, three electrons are given to the central atom by three halogens in molecules such as NF_3, PF_3, AsF_3, and so on. Similarly, the Group IVA elements pick up four electrons from four halogen atoms in compounds such as CCl_4 and $SiCl_4$.

Boron, in Group IIA, is a special case. Since the boron atom has only three valence electrons, it forms only three ordinary covalent bonds to the halogens. With fluorine, however, BF_3 can add on an additional F^- ion to form the BF_4^- (tetrafluoroborate) anion.

$$:\ddot{F}\times B\overset{..}{\underset{..}{\overset{:\ddot{F}:}{\times}}}\ \ +\ \ \left[:\ddot{F}:\right]^- \longrightarrow \left[\begin{array}{c}:\ddot{F}:\\ :\ddot{F}\times B:\ddot{F}:\\ :\ddot{F}:\end{array}\right]^-$$

We see that the fourth B—F bond can be considered a coordinate covalent bond (although by now you know that we really cannot distinguish the source of the electrons once the bond has been formed). Since the boron halides have less than an octet of electrons in the valence shell of boron, they are all powerful Lewis acids and the formation of the BF_4^- ion is a typical Lewis acid-base reaction.

In the second category of halogen compounds we have substances in which more than four pairs of electrons surround the central atom. These are limited to those nonmetals beyond the second period, because the second-period elements have a valence shell that can contain a maximum of only eight electrons, corresponding to the completion of the $2s$ and $2p$ subshells. The elements below the second period, however, also have in their valence shell a low-energy set of vacant d orbitals as well as the s and p subshells. These d orbitals may be used, through hybridization, to make additional electrons available for bonding.

The second factor influencing the number of halogen atoms that can become bound to a nonmetal is the relative sizes of the different atoms. In Table 20.3 we see that the compounds containing a large number of halogen atoms are formed from nonmetals found toward the bottom of the periodic table. This makes sense because these nonmetals are large and therefore able to accommodate a relatively large number of bonded atoms with a minimum of crowding. On the other hand, an element near the top of a group is much smaller, so only a relatively few halogen atoms could be expected to be packed about it.

An interesting facet of the chemistry of the nonmetal halides is their reactivity toward compounds containing an —OH group, the most familiar of which is water. Here once again we find that both thermodynamics and kinetics are involved in determining the course of reactions.

The kind of reaction we will focus our attention on is the hydrolysis of the nonmetal halide to produce either the oxoacid, or an oxide, plus the corresponding hydrogen halide. Some examples are the reactions of PCl_5, $SiCl_4$, and SF_4 with water.

$$PCl_5 + 4H_2O \longrightarrow H_3PO_4 + 5HCl$$

$$SiCl_4 + 2H_2O \longrightarrow SiO_2 + 4HCl$$

$$SF_4 + 2H_2O \longrightarrow SO_2 + 4HF$$

These reactions occur very rapidly and proceed to completion with the evolution of considerable quantities of heat. In fact, it is quite common for many halogen compounds of the elements below period 2 to react very rapidly in this same way. For example, the tin(IV) and lead(IV) halides, which are covalent, also hydrolyze in this manner with the formation of a mixture of species including complexes of Sn^{4+} with the halide ion.

There are also some nonmetal halogen compounds that are quite *unreactive* toward water, for example, CCl_4, SF_6, and NF_3. In these cases it is unfavorable kinetics, instead of thermodynamics, that prevents the hydrolysis from taking place.

Consider, for example, the potential hydrolysis reactions of CCl_4 and $SiCl_4$. Calculations based on thermodynamics imply that both should proceed very nearly to completion.

$$SiCl_4(l) + 2H_2O(l) \longrightarrow SiO_2(s) + 4HCl(aq) \qquad \Delta G^0 = -282 \text{ kJ}$$

$$CCl_4(l) + 2H_2O(l) \longrightarrow CO_2(g) + 4HCl(aq) \qquad \Delta G^0 = -377 \text{ kJ}$$

In fact, we see from the values of ΔG^0 that the hydrolysis of CCl_4 is even more "spontaneous" than $SiCl_4$. Kinetically, however, the hydrolysis of CCl_4 is essentially prohibited. This is attributed to the absence of a low-energy path for the hydrolysis of CCl_4. Attack by water on the carbon atom of CCl_4 is prevented by the crowding of the Cl atoms. In $SiCl_4$, on the other hand, the larger Si atom provides a greater opportunity for attack and, in addition, the presence of low-energy $3d$ orbitals in the valence shell of the Si atom permits a temporary bonding of the water molecule to the Si atom prior to the expulsion of a molecule of HCl. The mechanism of this hydrolysis is believed to be

Repetition of this process eventually yields $Si(OH)_4$ (orthosilicic acid), which loses water spontaneously to give a hydrated SiO_2.

SiO_2 is a complex network solid.

$$Si(OH)_4 \longrightarrow SiO_2 + 2H_2O$$

The stability of SF_6 and NF_3 toward hydrolysis can also be attributed to the absence of a low-energy reaction mechanism. Like CCl_4, SF_6 should also undergo hydrolysis quite spontaneously, with the value of ΔG^0 for the reaction being tremendous.

$$SF_6(g) + 4H_2O(l) \longrightarrow H_2SO_4(aq) + 6HF(g) \qquad \Delta G^0 = -423 \text{ kJ}$$

However, the crowding of the fluorine atoms around the sulfur atom apparently prevents attack by water (even up to 500°C), as well as most other reagents. This crowding is absent with SF_4, and hydrolysis by water is instantaneous.

The resistance of NF_3 toward attack by water cannot be attributed to interference by the fluorine atoms as in SF_6, since the NF_3 molecule is pyramidal, with the nitrogen atom quite openly exposed to an attacking water molecule. We might compare NF_3 with NCl_3, which *does* hydrolyze (if it doesn't explode first—NCl_3 is extremely unstable). The mechanism for this reaction appears to involve the initial formation of a hydrogen bond from water to the lone pair on the nitrogen atom, followed by expulsion of hypochlorous acid.

The ultimate products of the hydrolysis are ammonia and HOCl. This mechanism is not favorable for NF_3 because of its very low basicity. In this case the highly electronegative fluorine atoms draw electron density from the nitrogen atom. As a result, it has been suggested that the lone pair of electrons on nitrogen, in NF_3, may not be available to serve as a point of attachment for the H_2O molecule which, as we see above, is a necessary step in the mechanism proposed for the hydrolysis of NCl_3.

20.4 NOBLE GAS COMPOUNDS

In our discussion of the nonmetals we have not mentioned compounds of the noble gases. These are rather unusual substances because, on the basis of the electronic structure of the noble gases, we would perhaps not have predicted their existence. In fact, until 1962 most chemists firmly believed that these elements were totally incapable of forming compounds (other than several **clathrates,** in which the noble gas atoms are trapped in cagelike sites within a crystalline lattice). For this reason, chemists had referred to them as the *inert gases.* Today they are spoken of as the noble gases in recognition of the fact that although some do react, they nevertheless possess a very low degree of reactivity.

The first real chemistry of the noble gases was discovered in 1962 by Neil Bartlett at the University of British Columbia. He had found that molecular oxygen, O_2, reacts with PtF_6 to form an orange-red compound, O_2PtF_6, containing the ion, O_2^+. Since the ionization energies of O_2 and Xe are nearly the same (1210 and 1170 kJ mol^{-1}, respectively), he reasoned that Xe should react in the same way O_2 does. When he reacted Xe with PtF_6 he isolated a yellow compound, containing Xe, that was formulated as $XePtF_6$.

After the initial report by Bartlett, it was not long before chemists at Argonne National Laboratory found that Xe also reacts directly with fluorine at elevated temperatures. This reaction yields a series of fluorides: XeF_2, XeF_4, and XeF_6. Other reactions and compounds were soon discovered and a partial list of the known Xe compounds is given in Table 20.4. The oxides and oxofluorides result from the hydrolysis of the fluorides

$$XeF_6 + 3H_2O \longrightarrow XeO_3 + 6HF$$

$$XeF_6 + H_2O \longrightarrow XeOF_4 + 2HF$$

Some of these compounds are quite unstable and tend to decompose. This is particularly true for the oxides XeO_3 and XeO_4, which explode (XeO_3 has $\Delta H_f^0 = +400$ kJ mol^{-1}). Others, on the other hand, appear quite stable. For example, Cs_2XeF_8 does not decompose even when heated to 400°C, and the fluorides have moderately high melting points, suggesting a modest degree of thermal stability.

Table 20.4
Some compounds of xenon

	Melting Point (°C)	Physical Form
Fluorides		
XeF_2	140	Colorless crystals
XeF_4	114	Colorless crystals
XeF_6	47.7	Colorless crystals
Oxides		
XeO_3	Explodes	Colorless crystals
XeO_4	Explodes	Colorless gas
Salts		
$XePtF_6{}^a$	—	Red-orange crystals
$CsXeF_7$	Decomp > 50°C	Colorless solid
Cs_2XeF_8	Decomp > 400°C	Yellow solid

[a] Since shown to be more complex; $Xe(PtF_6)_x$, where x lies between 1 and 2.

The structure and bonding in these compounds is quite interesting. Since Xe has four pairs of electrons in its valence shell, corresponding to a completed 5s and 5p subshell, unpairing of electrons and expansion of the octet must occur to provide unpaired electrons for bonding. Let us consider XeF_2 and XeF_4.

The electronic structure of Xe can be represented as

Xe   ⇅   ⇅ ⇅ ⇅   __ __ __ __ __
     5s   5p   5d

In order to form XeF_2, one electron must be promoted to the 5d subshell, followed by hybrid orbital formation. The smallest hybrid set that will accommodate all the electrons is sp^3d.

Xe   ⇅   ⇅ ⇅ ↑   ↑ __ __ __ __
     5s   5p   5d

gives

Xe   ⇅ ⇅ ⇅ ↑ ↑   __ __ __ __
          sp^3d unhybridized 5d

The two unpaired electrons can now be used in bonding to fluorine to give

Xe (in XeF_2)   ⇅ ⇅ ⇅ ⇅ ⇅   __ __ __ __
                    sp^3d unhybridized 5d
(Colored arrows represent fluorine electrons)

In Chapter 6 we saw that the sp^3d hybrids point to the vertices of a trigonal bipyramid. In terms of the valence shell electron-pair repulsion theory, these five electron pairs will also be situated in this fashion, and from our rules (presented on pp. 186–187), we expect that the three lone pairs will locate themselves in the triangular plane with the fluorine atoms above and below (Figure 20.13). The XeF_2 molecule is therefore linear.

In the case of XeF_4 we must provide four unpaired electrons for bonding to fluorine. This requires promotion of two electrons to the 5d subshell and the

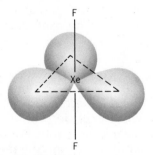

Figure 20.13
Molecular structure of XeF_2.

formation of sp^3d^2 hybrid orbitals.

Xe $\underset{5s}{\uparrow\downarrow}$ $\underset{5p}{\uparrow\downarrow\ \uparrow\ \uparrow}$ $\underset{5d}{\uparrow\ \uparrow\ \_\ \_\ \_}$

gives

Xe $\underset{sp^3d^2}{\uparrow\downarrow\ \uparrow\downarrow\ \uparrow\ \uparrow\ \uparrow\ \uparrow}$ $\underset{\text{unhybridized } 5d}{\_\ \_\ \_}$

Finally, bonding with fluorine gives

Xe (in XeF$_4$) $\underset{sp^3d^2}{\uparrow\downarrow\ \uparrow\downarrow\ \uparrow\downarrow\ \uparrow\downarrow\ \uparrow\downarrow\ \uparrow\downarrow}$ $\underset{\text{unhybridized } 5d}{\_\ \_\ \_}$

In XeF$_4$ there are six electron pairs about the Xe. Both valence bond theory, with its sp^3d^2 hybrids, as well as the VSEPR theory predict that these are directed towards the corners of an octahedron. As we have seen before, the two lone pairs occupy positions on opposite sides of a square plane containing the four ligand atoms, so that XeF$_4$ has a square planar structure (Figure 20.14).

Since the initial discovery of noble gas compounds by Bartlett, three noble gases have been demonstrated to form compounds, Rn, Xe, and Kr. The lighter ones—helium, neon, and argon—do not appear able to form chemical compounds because of their much higher ionization energies.

Bartlett's work on Xe has taught chemists an important lesson. So firmly convinced were they that the noble gases were totally unreactive that after some initial experiments attempting to react Xe with fluorine had failed in the 1930s, no further efforts were made to explore the possibility that they were not inert. It is interesting that the noble gas compounds obtained do not present any particular problem in bonding. In fact, many of the interhalogen compounds that had already been found (for example, BrF$_5$, ICl$_4^-$, IF$_7$) have the same number of electrons in the valence shell of the central atom as do the noble gas compounds. The same concepts that we applied to other compounds in which the octet was exceeded, therefore, work with the noble gas compounds, too. The failure by chemists to recognize the possibility that the noble gases might react reflects a blind spot in their thinking that was probably founded in an overzealous acceptance of the stability and inertness of the ns^2np^6 octet of electrons.

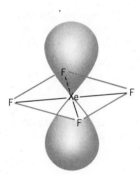

Figure 20.14
Molecular structure of XeF$_4$.

20.5 SILICON

The second most abundant element in the Earth's crust is silicon. It constitutes approximately 27.7% of the crust, by mass, where it occurs in rocks of various kinds in the form of silicates—silicon-oxygen compounds. The free element is normally obtained by reduction of silicon dioxide (found in many kinds of sand) using carbon in an electric furnace. The reaction is

$$SiO_2(s) + 2C(s) \xrightarrow{\text{heat}} Si(s) + 2CO(g)$$

Silicon is a dark, metallic looking solid that melts at 1410°C. In Chapter 3 we learned that it is a semiconductor, a fact that has made possible the fantastic world of miniature electronic devices such as pocket calculators and microcomputers. The silicon used in these devices must be extremely pure. It is usually made from a tetrahalide such as silicon tetrachloride (SiCl$_4$) by high temperature reduction with hydrogen.

$$SiCl_4(g) + 2H_2(g) \longrightarrow Si(s) + 4HCl(g)$$

Silica rock (top). Silicon (bottom). The silicon is reduced from the mineral silica through reaction with carbon in an electric arc furnace.

The silicon is then further purified by an interesting process called **zone refining** (Figure 20.15). This method is based on the fact that when a solution freezes, the crystal lattice of the solid formed does not accommodate the impurities very well, so the impurities tend to become concentrated in the remaining solution. In zone refining, a bar of silicon is placed in a device that melts a thin cross-sectional wafer of the solid near one end. The heat source is then gradually moved toward the other end of the bar, and as it moves the melted zone follows along. Behind this molten section, the silicon solidifies in a very pure crystalline state, while the impurities collect in the molten band. In this way the impurities are brought to one end of the bar. After the procedure is repeated several times the impure end is cut off and discarded. The rest of the bar consists of very pure silicon with impurity levels ranging from a few parts per million to as little as a few parts per billion.

Chemical properties

Silicon is found in Group IVA of the periodic table, beneath carbon. Its atoms in their ground state have the electron configuration [Ne] $3s^2 3p^2$. Unlike carbon, silicon has virtually no tendency to form π bonds by overlap of p atomic orbitals ($p\pi$—$p\pi$ bonds). As a result, silicon only forms single bonds. In the free element, each silicon atom completes its octet by bonding to four separate silicon atoms located at the corners of a tetrahedron, and, as we saw in Chapter 6, elemental silicon crystallizes in a diamond-type lattice.

Silicon is not a very reactive element at room temperature. It is unaffected by acids, but it does dissolve in hot basic solutions (NaOH, for example) to give hydrogen plus a mixture of silicates.

$$\text{Si}(s) + 4\text{OH}^-(aq) \longrightarrow \text{SiO}_4^{4-}(aq) + 2\text{H}_2(g)$$
$$\text{(plus other silicates)}$$

At high temperatures silicon also combines with halogens to produce the tetrahalides (for example, SiCl_4) and with hydrogen to form hydrides called silanes (SiH_4, Si_2H_6, etc.). The hydrides are not very stable, and the ability of silicon atoms to link to each other in these compounds is limited. The longest chain to be observed is in Si_6H_{14}, which tends to decompose to SiH_4.

Compounds with silicon oxygen bonds

Silicon has a strong affinity for oxygen because it forms very stable silicon-oxygen bonds. In fact, all the naturally occurring silicon compounds are silicates in which the basic structural unit is the SiO_4 tetrahedron. Although some are rather

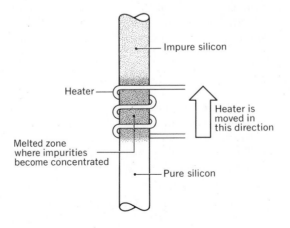

Figure 20.15
Zone refining.

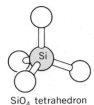

SiO$_4$ tetrahedron

complex, the structures of these naturally occurring silicates can be understood by considering them as polymers of more basic silicate units.

The simplest of the silicates contain the *orthosilicate ion*, SiO_4^{4-}, which is the anion of orthosilicic acid.

$$
\begin{array}{c}
OH \\
| \\
HO{-}Si{-}OH \\
| \\
OH
\end{array}
\qquad
\left[
\begin{array}{c}
:\!\overset{..}{O}\!: \\
| \\
:\!\overset{..}{\underset{..}{O}}\!{-}Si{-}\overset{..}{\underset{..}{O}}\!: \\
| \\
:\!\overset{..}{\underset{..}{O}}\!:
\end{array}
\right]^{4-}
$$

orthosilicic acid   **orthosilicate ion**
(cannot be isolated)

An example of a substance containing the SiO_4^{4-} ion is the gem *zircon*, which consists of crystals of $ZrSiO_4$.

In Section 20.1 we saw that phosphoric acid units are able to be polymerized by the removal of the components of water from a pair of —OH groups on neighboring molecules and the formation of an oxygen bridge between the phosphorus atoms. Among the silicates this kind of polymerization also occurs, and as the pH of a solution of sodium silicate, Na_4SiO_4, is lowered the SiO_4 units become joined by similar oxygen bridges. For example, the pyrosilicate ion, $Si_2O_7^{6-}$, can be considered to be formed as follows:

$$
\begin{array}{ccc}
& H_2O & \\
(-)\;:\!\overset{..}{O}\!: & & :\!\overset{..}{O}\!:\;(-) \\
(-)\;:\!\overset{..}{O}\!{-}Si{-}\boxed{\overset{..}{O}\!{-}H \quad H{-}\overset{..}{O}}{-}Si{-}\overset{..}{O}\!:\;(-) \\
:\!\overset{..}{O}\!: & & :\!\overset{..}{O}\!: \\
(-) & & (-)
\end{array}
$$

$$
\downarrow
$$

$$
\left[
\begin{array}{ccc}
:\!\overset{..}{O}\!: & & :\!\overset{..}{O}\!: \\
:\!\overset{..}{O}\!{-}Si{-}\overset{..}{O}\!{-}Si{-}\overset{..}{O}\!: \\
:\!\overset{..}{O}\!: & & :\!\overset{..}{O}\!:
\end{array}
\right]^{6-}
$$

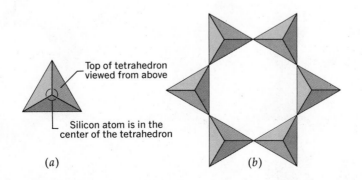

$Si_2O_7^{6-}$ ion

Figure 20.16

The structure of the pyrosilicate anion, $Si_2O_7^{6-}$. Colored spheres are silicon atoms. Each nonbridging oxygen atom carries a negative charge.

Notice that in the various silicates discussed in this section each nonbridging oxygen carries a negative charge.

In this anion a pair of SiO_4 tetrahedra share a corner in common (Figure 20.16). In nature, the $Si_2O_7^{6-}$ ion is found in the mineral $Sc_2Si_2O_7$, called thortveitite.

When one SiO_4 tetrahedron shares *two* of its corners with other SiO_4 tetrahedra, rings and long chain structures are able to be formed. For example, the polymeric anion $Si_6O_{18}^{12-}$, shown in Figure 20.17, is present in beryl,

Figure 20.17

(a) *A representation of the SiO$_4$ tetrahedron as viewed from above.* (b) *The structure of the $Si_6O_{18}^{12-}$ ion showing the SiO$_4$ tetrahedra linked by oxygen bridges at their corners.*

Top of tetrahedron
viewed from above

Silicon atom is in the
center of the tetrahedron

(a)   (b)

Figure 20.18

The staggered arrangement of SiO_4 tetrahedra in a linear $(SiO_3^{2-})_x$ metasilicate chain.

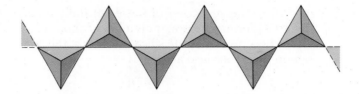

$Be_3Al_2(Si_6O_{18})$, which forms crystals of gem quality that are known as emeralds.

Linking the SiO_4 tetrahedra in long chains gives huge polymeric anions having the empirical formula SiO_3^{2-}.

repeating

SiO_3^{2-} unit

The staggered structure of an $(SiO_3)_x^{2x-}$ chain is shown in Figure 20.18. As you might expect, minerals such as $MgSiO_3$ and $LiAl(SiO_3)_2$, which contain these "linear" SiO_3^{2-} chains, have a fibrous appearance. In addition to the simple strands of SiO_4 tetrahedra, double strands, formed by the sharing of three corners by *every other* SiO_4 unit, are also found. This time, an infinite *double* chain results, a small segment of which is illustrated in Figure 20.19. Once again, each unshared oxygen atom carries a negative charge, and the repeating unit along the chain is $Si_4O_{11}^{6-}$. This anion is found in asbestos, $[Ca_2Mg_5(Si_4O_{11})_2(OH)_2]_x$, and, as we might expect, asbestos has a fiberlike nature because of the presence of the long $(Si_4O_{11}^{6-})_x$ chains that line up more or less parallel to one another.

Asbestos has many useful properties. Its fibrous nature makes it excellent as a reinforcing filler in brake and clutch linings. Because it is fireproof it has been spun and woven into fabrics used in fireproof curtains and ironing board covers. It has been used in insulation and sealers in cars, trucks, and planes. Once again, however, nature takes as well as gives. Asbestos fibers breathed into the lungs have been implicated in many cases of lung cancer. There is also evidence that if consumed in food it can produce stomach cancer. Because of these problems, the uses of asbestos have been sharply curtailed.

Still another type of silicate is formed if *each* SiO_4 tetrahedron shares three of its corners so that each Si is attached to three others by oxygen bridges. When this occurs a planar sheet of SiO_4 units results. A portion of one of these is

Figure 20.19

Linear double chain of the $(Si_4O_{11}^{6-})_x$ anion found in asbestos.

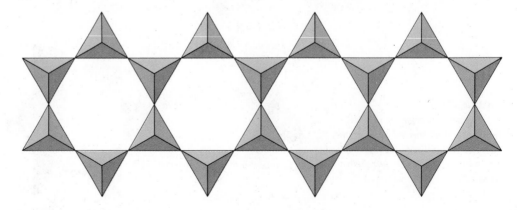

Figure 20.20

Planar sheet silicate formed by sharing three corners of every SiO_4 tetrahedron.

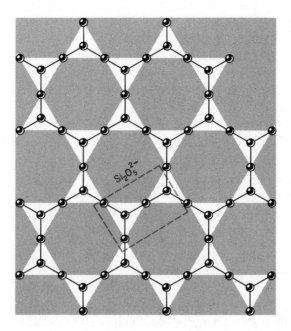

shown in Figure 20.20 with the repeating unit, $Si_2O_5^{2-}$, outlined by the rectangle.

A number of different minerals are known to contain these $(Si_2O_5^{2-})_x$ sheets. They differ in the way the silicate layers are stacked, and the nature of the cations and other anions that are also present in the structure: However, all have certain similarities to each other. Some examples are talc (used in bath powder) and soapstone, in which the $(Si_2O_5^{2-})_x$ sheets are packed together with cations in such a way that there is a minimum of attractive forces between successive layers. These layers therefore slide over each other easily and both these minerals feel slippery.

In other related minerals there is substitution of another element for Si. In mica, for instance every fourth Si atom is replaced by an Al^{3+} ion. The properties of mica are therefore different from talc; however, the layer structure is still apparent. A mica-type material you may be familiar with is *vermiculite*, used in place of soil in propagating house plants and as a cushioning filler in packaging items for shipment. This solid flakes into thin flat layers characteristic of mica.

When Si finally shares all four of its oxygen atoms with other Si atoms, a three-dimensional framework is produced that has the empirical formula SiO_2. Silicon dioxide, also called silica, sometimes occurs in large beautiful crystals of quartz (Figure 20.21), although most people are more familiar with quartz in a finer state of subdivision—sand.

Figure 20.21

Quartz crystals.

Figure 20.22
Left- and right-handed helixes in quartz. The spiral chains of SiO_2 tetrahedra can twist in either of two directions. This gives two kinds of quartz crystals known as d-quartz and l-quartz.

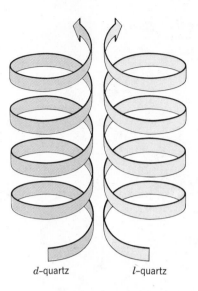

d-quartz   *l*-quartz

Quartz is composed of spiral chains of SiO_4 tetrahedra that are linked to each other. Since the spiral can twist either clockwise or counterclockwise (Figure 20.22), two types of crystals are produced that are exactly alike in all but one respect. They differ in the same way that your left and right hands differ. If you examine your two hands with their palms facing you, the thumbs point in opposite directions—otherwise they are essentially the same. That's why a left-handed glove doesn't fit your right hand. The two kinds of quartz crystals have a similar "handedness," and we will see later that this phenomenon extends to certain individual molecules as well. The topic of molecular "handedness" is discussed further in the next chapter.

Silicones

Organic polymers, such as polyethylene, nylon, and the polyesters, have many desirable physical properties that make them ideally suited for packaging materials and the production of synthetic fibers used in clothing. There are drawbacks, however. They are generally flammable and they do not stand up well to high temperatures.

Inorganic polymers—polymers in which atoms other than carbon make up the primary chain or backbone of the molecules—often do not suffer from these disadvantages. A type of inorganic polymer that has achieved widespread use are the silicones. These are polymers in which the backbone is made up of alternating silicon and oxygen atoms. Their thermal stability arises from the strength and inertness of Si—O bonds. They are formed by hydrolysis of compounds such as $(CH_3)_2SiCl_2$. The formation of silicone polymers can be thought

Silicones have many medical applications. For example silicone oils are used to lubricate the skin of burn victims.

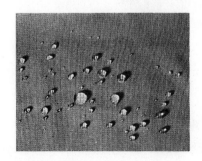

Water forms drops on a fabric treated with a silicone water repellent.

of as proceeding by formation of a hydroxy intermediate that then eliminates water.

$$\underset{\underset{CH_3}{|}}{\overset{\overset{CH_3}{|}}{Cl-Si-Cl}} + 2H_2O \longrightarrow \underset{\underset{CH_3}{|}}{\overset{\overset{CH_3}{|}}{HO-Si-OH}} + 2HCl$$

This silicone is known medicinally as *simethicone.* It is a common antigas agent found in many antacid preparations.

Depending on the chain length, and the degree to which chains may be cross-linked to one another, the silicones may be oils, greases, or rubbery solids. They are useful in waterproofing garments. In low-temperature applications the oils remain fluid (hydrocarbon oils become very viscous), and the rubbers retain their elastic properties (ordinary rubber becomes brittle). Silicones are also unaffected by hydrocarbon solvents and greases that soften ordinary rubber and they are not attacked by ozone in the air, which causes ordinary rubber to crack.

REVIEW QUESTIONS AND PROBLEMS

(Problems whose numbers are in blue have their answers in Appendix C at the back of the book.)

Phosphorus: General

20.1 What is the formula for the phosphorus-containing compound in phosphate rock?

20.2 What kinds of roles does phosphorus play in living systems?

20.3 Give the electron configuration of phosphorus.

20.4 What is the structure of the molecules found in white phosphorus? Why is white phosphorus so reactive?

20.5 How is white phosphorus made? Give a chemical equation.

20.6 What is an *electric furnace?*

20.7 Describe red phosphorus and black phosphorus. How do their reactivities compare to white phosphorus?

Phosphorus Oxides

20.8 Write chemical equations for the reaction of white phosphorus with oxygen (a) when the O_2 is present in excess and (b) when there is a limited amount of O_2.

20.9 Sketch the structures of the molecules found in (a) phosphorus(V) oxide and (b) phosphorus(III) oxide.

20.10 Write equations for the reaction of water with (a) phosphorus(V) oxide and (b) phosphorus(III) oxide.

20.11 What is a *desiccant?* Why is P_4O_{10} a good desiccant?

Phosphoric Acid and Phosphate

20.12 Give a chemical equation showing how phosphoric acid is made directly from phosphate rock.

20.13 What is *syrupy phosphoric acid?*

20.14 How is the phosphoric acid used in food products manufactured?

20.15 Give three commercial uses of phosphoric acid.

20.16 Give the formulas and names of three salts that can be formed from $Mg(OH)_2$ and phosphoric acid.

20.17 What is orthophosphoric acid?

20.18 What ions make up the phosphate buffer in the blood? How do they serve to help maintain a nearly constant blood pH?

20.19 What are some uses of TSP? Give a chemical equation to explain why its solutions are basic.

20.20 Give Lewis structures for H_3PO_4 and H_3PO_3.

20.21 What is superphosphate fertilizer? Give a chemical equation showing how it is made.

20.22 Why is phosphate rock itself a poor phosphate fertilizer?

20.23 What is the name of (a) Na_2HPO_4 and (b) NaH_2PO_4?

Phosphorus: Other Compounds

20.24 Write a chemical equation for the reaction of sodium with phosphorus? What happens when the product of this reaction is placed in water?

20.25 Use Lewis structures to illustrate how pyrophosphoric acid is formed from phosphoric acid.

20.26 What is the basic structural unit in the polymeric phosphoric acids and polyphosphates?

20.27 What is the empirical formula for the metaphosphate ion? Give its Lewis structure and show how it can be considered to be formed from phosphoric acid.

20.28 Give the structure of the tripolyphosphate ion. What uses does it have?

20.29 What mole ratio of NaH_2PO_4 and Na_2HPO_4 should be chosen to obtain a linear polyphosphate containing five phosphorus atoms?

20.30 Explain why phosphate pollution can be very harmful to lakes.

20.31 What is the name of the acid, H_3PO_3? What salts can be formed by reacting it with $Mg(OH)_2$?

20.32 How do H_3PO_4 and H_3PO_3 compare as oxidizing and/or reducing agents?

20.33 Sketch the structures of PCl_3 and PCl_5. How does PCl_5 exist in the solid state? What kinds of hybrid orbitals does phosphorus use in PCl_3 and PCl_5?

Sulfur: General

20.34 Give the electron configuration of sulfur.

20.35 What is *brimstone?* What does this name mean?

20.36 How does the element sulfur occur in nature?

20.37 What are the two allotropic forms of sulfur? How do they differ? Outline the physical changes that take place when sulfur is gradually heated to its boiling point and relate them to the structural changes that occur.

20.38 What is plastic sulfur?

20.39 Describe the Frasch process for mining sulfur.

Sulfur: Oxides and Oxoacids

20.40 Which air pollutant is produced by the combustion of high-sulfur fuels?

20.41 Write chemical equations for the production of H_2SO_4 using O_2, S, and H_2O as starting materials.

20.42 Why is SO_2 stable toward oxidation to SO_3 in the air?

20.43 How can SO_2 be conveniently prepared in the laboratory?

20.44 Write chemical equations for the reaction of water with (a) sulfur dioxide and (b) sulfur trioxide.

20.45 Describe *acid rain*—how it is formed and the damage it does.

20.46 How do the acidities of H_2SO_3 and H_2SO_4 compare? How can the differences be explained?

20.47 List four uses for sulfuric acid.

20.48 Draw the resonance structures for and describe the shapes of SO_2 and SO_3.

20.49 What is the proper way to prepare dilute H_2SO_4 from concentrated H_2SO_4?

20.50 Even though H_2SO_4 isn't fully dissociated into H^+ and SO_4^{2-}, it is considered a strong acid. Why?

20.51 Write an equation showing the dehydrating action of concentrated H_2SO_4 on the sugar, glucose, $C_6H_{12}O_6$.

20.52 Name these salts: (a) Na_2SO_3 and (b) $NaHSO_4$.

Sulfur: Other Compounds

20.53 What is the structure of pyrosulfuric acid? (*Hint:* What is the structure of pyrophosphoric acid?)

20.54 The cyanate ion has the Lewis structure

$$\left[:N \equiv C - \ddot{\underset{\displaystyle ..}{O}} : \right]^-$$

What would be the structure of the thiocyanate ion?

20.55 Write an equation for the generation of H_2S by the hydrolysis of thioacetamide.

20.56 Why are hydrogen sulfide fumes to be avoided in the laboratory?

20.57 What is the structure of the thiosulfate ion? Write a chemical equation showing how thiosulfate ion is made.

20.58 Explain the function of sodium thiosulfate, $Na_2S_2O_3$, in photography.

20.59 Write balanced chemical equations for the oxidation of $S_2O_3^{2-}$ by
(a) Cl_2
(b) I_2

20.60 What products, if any, are formed in the following reactions?
(a) cold concentrated H_2SO_4 and NaI
(b) hot concentrated H_2SO_4 and copper
(c) hot concentrated H_2SO_4 and zinc
(d) cold dilute H_2SO_4 and zinc
(e) cold dilute H_2SO_4 and copper

Halogens: General

20.61 Write the electron configuration of each of the halogens.

20.62 What is the origin of the name *halogen?*

20.63 How are the halogens normally found in nature?

20.64 What are the principal sources of each of the halogens?

20.65 How is fluorine prepared?

20.66 Describe how you could prepare chlorine in the laboratory. How is chlorine made industrially? What are three commercial uses for chlorine?

20.67 Describe the physical characteristics of the halogens.

20.68 Complete and balance the following equations. If no reaction occurs, write N.R.
(a) $Cl_2 + KI \rightarrow$
(b) $F_2 + KBr \rightarrow$
(c) $I_2 + NaCl \rightarrow$
(d) $Br_2 + NaI \rightarrow$

20.69 Why is fluorine so reactive?

20.70 How is bromine obtained industrially? Give two principal uses of Br_2. How can Br_2 be made in the laboratory? (Give a balanced equation.)

20.71 What are the commercial sources of iodine?

Halogens: Binary Compounds

20.72 How do the reactivities of the halogens toward hydrogen compare?

20.73 How is HF made?

20.74 How would you prepare HCl in the laboratory? What are two commercial uses of hydrochloric acid?

20.75 How would you prepare HBr and HI in the laboratory?

20.76 Why is glass attacked by hydrofluoric acid?

20.77 Compare the boiling points of the hydrogen halides. How do HF molecules exist in liquid HF?

20.78 HF is a weak acid, yet it is very dangerous to work with. Why?

Halogens: Oxoacids and Oxoanions

20.79 Name these compounds.
(a) HOBr   (e) HIO_4
(b) NaOCl   (f) $HBrO_3$
(c) $KBrO_3$ (g) $NaIO_3$
(d) $Mg(ClO_4)_2$ (h) $KClO_2$

20.80 What is a disproportionation reaction?

20.81 Write a chemical equation for the reaction that takes place when Cl_2 is dissolved in
(a) cold water
(b) cold NaOH solution

20.82 CaCl(OCl) is a solid water-soluble bleach. What reaction would occur if acid were added to this solid?

20.83 Compare the stabilities of the ions OCl^-, OBr^-, and OI^- in cold aqueous solutions. What accounts for their relative stabilities?

20.84 What reaction occurs if $KClO_3$ is heated at moderate temperatures? What happens if MnO_2 is present while the $KClO_3$ is heated?

20.85 What is the chemical formula for HTH, which is sold as a swimming pool chlorinating agent?

Halogens: Other Compounds

20.86 Use the VSEPR theory to predict the molecular shapes of
(a) SF_2 (e) BrF_3
(b) PBr_5 (f) $TeBr_4$
(c) SiF_6^{2-} (g) BrF_5
(d) AsF_3 (h) $GeCl_4$

20.87 What is the likely reason that IF_7 can be made but ClF_7 cannot?

20.88 Suggest two reasons why oxygen doesn't form OF_4 even though sulfur forms SF_4.

20.89 Predict what products would be formed in the reaction between $GeCl_4$ and H_2O.

20.90 Why does $SiCl_4$ hydrolyze but CCl_4 does not?

20.91 Compare the effectiveness of NH_3 and NF_3 as Lewis bases.

Noble Gas Compounds

20.92 What is a *clathrate?*

20.93 Predict the molecular structures of XeF_4 and XeF_2 using the VSEPR theory.

20.94 Why were chemists so surprised when it was found that the noble gases were not inert?

Silicon and Silicates

20.95 How is elemental silicon prepared? Give an equation.

20.96 Describe *zone refining*.

20.97 Why doesn't silicon form graphitelike crystals?

20.98 Carbon forms CO_2, which is a gas, whereas silicon forms SiO_2, which is a complex solid. Why doesn't Si form an SiO_2 molecule with the same structure as CO_2?

20.99 What is the structure of the orthosilicate ion?

20.100 Use Lewis structures to describe the linking together of SiO_4 tetrahedra in the formation of the pyrosilicate ion.

20.101 Sketch the structure of the cyclic $Si_6O_{18}^{12-}$ anion. Which mineral contains this anion?

20.102 Sketch a portion of the anion found in asbestos. What is the repeating unit in the structure?

20.103 Sketch a portion of the polymeric metasilicate anion, $(SiO_3^{2-})_x$.

20.104 What kind of structure is formed when each SiO_4 tetrahedron shares three of its corners with neighboring SiO_4 tetrahedra? What are some common minerals containing this structure?

20.105 What is the empirical formula for quartz? Why do quartz crystals occur in left- and right-handed forms?

20.106 What structure is found along the backbone of the silicone polymers?

20.107 What compound would be formed by hydrolysis of

$$CH_3-\underset{\underset{CH_3}{|}}{\overset{\overset{CH_3}{|}}{Si}}-Cl$$

20.108 Describe the structure of simethicone.

CHAPTER 21

THE TRANSITION ELEMENTS

As nearly everyone knows, chlorophyll is the substance responsible for the green glow of sunlight filtering through leaves such as these. It is used by plants to harness solar energy in the production of sugars and ultimately other materials that they need. Chlorophyll is an example of a "complex ion"—a substance composed of a metal ion (in this case Mg^{2+}) bonded covalently to atoms of the ligands that surround it. One of the properties often associated with the transition metals is their great tendency to form complex ions, and the study of them has become an important branch of chemistry.

In this chapter we consider the collection of elements, generally called the transition elements, that fit into the periodic table between Groups IIA and IIIA. We saw in Chapter 4 that these elements arise as a consequence of the gradual filling of d and f subshells and, as a rule, a transition element is usually considered to be one that possesses an incompletely filled d or f subshell in either the free state or one of its compounds. We will also include in this discussion the elements in Group IIB, zinc, cadmium, and mercury, which are found at the extreme right of the transition elements and which complete a *transition series* (a horizontal row) by having their outer d subshells filled.

21.1 GENERAL PROPERTIES

In discussing the transition elements it is convenient to divide them into two categories: the **d-block elements** (or **main transition elements**), which in our condensed version of the periodic table are located in the main body of the table between Groups IIA and IIIA, and the **inner transition elements,** which correspond to the two long rows of 14 elements each that are placed just below the table (Figure 21.1). The d-block elements themselves consist of three rows frequently referred to as the first, second, and third transition series.

Like the representative elements, most of the d-block transition elements possess certain vertical similarities in chemical and physical properties and are therefore divided into groups, designated as B groups. They begin with Group IIIB on the left and proceed through Group VIIB. Next is a set of nine elements collectively termed Group VIII, and finally, on the right, we find Groups IB and IIB. The group numbers are chosen to correspond to the highest positive oxidation state that their elements normally exhibit.

The division of the periodic table into A and B groups—for example, IIIA and IIIB—suggests that there may be certain parallels between the two, and to a limited degree this is true. The similarities, however, are restricted primarily to likenesses of composition, structure, and maximum positive oxidation state, instead of chemical reactivity. Some examples of these similarities are found in Table 21.1.

The Group VIII elements, which lie between Groups VIIB and IB, are classed differently from the other d-block elements because they have no counterparts among the representative elements. Within this group there are greater *horizontal similarities* than vertical ones, and the description of the behavior of these elements is usually organized on the basis of horizontal groups of three elements each, called **triads.** Each triad is named after the best-known element

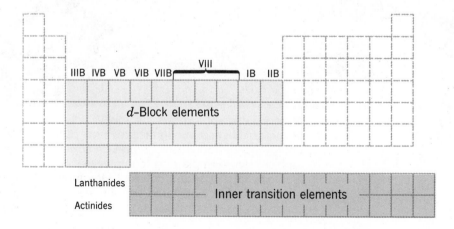

Figure 21.1
The transition elements.

Table 21.1
Some similarities of chemical composition
between A and B group compounds

Group	Compounds	Group	Compounds
IVA	CCl_4, $SnCl_4$, CO_2	IVB	$TiCl_4$, TiO_2
VA	PO_4^{3-}, $POCl_3$	VB	VO_4^{3-}, $VOCl_3$
VIA	SO_4^{2-}, $S_2O_7^{2-}$	VIB	CrO_4^{2-}, $Cr_2O_7^{2-}$
VIIA	ClO_4^-, Cl_2O_7	VIIB	MnO_4^-, Mn_2O_7
IA	NaCl	IB	CuCl, AgCl
IIA	$CaCl_2$	IIB	$ZnCl_2$

About one-quarter of the silver consumed by industry each year is used to make electronic equipment.

within it. Thus we have the iron triad, the palladium triad, and the platinum triad.

As a class, the transition elements are all typical metals; they possess a characteristic metallic luster and are good conductors of heat and electricity. Silver has the highest electrical and thermal conductivity of any metal, followed closely by copper. Copper, of course, is used in vast quantities in electrical wiring. Silver is the preferred coating for mirrors because of its high reflectivity.

The chemical and physical properties of the transition elements cover a wide range and account for the large variety of uses to which they are applied. Some of these metals are very hard and strong and are used as structural metals, either in the pure state or as alloys. Iron is the prime example; steels with a variety of different properties are formed by incorporating iron with other transition elements such as chromium, cobalt, and nickel. Even copper, which is very soft when pure, can be made very strong by forming an alloy with beryllium. Beryllium-copper alloys are used in place of steel in nonsparking tools for use in explosive atmospheres and as high-quality springs in cameras and other precision instruments.

Tungsten has the highest melting point of any metal.

The melting points of the transition elements also vary over a wide range (Figure 21.2). Most are high-melting. Tungsten, with a melting point of approximately 3400°C, is used for filaments in light bulbs. At the other extreme is mercury, which is a liquid at room temperature and is used in thermometers.

The chemical reactivity of the free elements varies greatly too. Most react

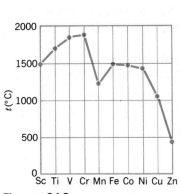

Figure 21.2

Variation among the melting points of the transition elements.

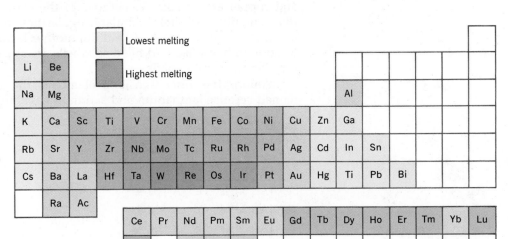

Filament

The filament in an electric light bulb is made of tungsten.

directly with nonmetals such as oxygen and the halogens to produce the corresponding oxides and halides. In fact, some transition elements are so easily oxidized that they react with water to liberate hydrogen. This is true for scandium (Sc), yttrium (Y), lanthanum (La), and the lanthanide elements (atomic numbers 58 to 71). They have very negative reduction potentials and react according to the equation

$$2M + 6H_2O \longrightarrow 3H_2 + 2M(OH)_3$$

Some other transition elements—such as platinum and gold—are very resistant to oxidation. They are insoluble in both nonoxidizing acids such as HCl as well as in oxidizing acids such as HNO_3, although they do dissolve slowly in *aqua regia*.

Despite some rather marked differences in behavior, the transition elements have several characteristics in common with each other:

1. *Multiple oxidation states.* With only a few exceptions, the transition elements tend to exhibit more than one oxidation state.
2. *Many of their compounds are paramagnetic.* Because the transition elements tend to have partially completed *d* or *f* subshells in both the free state and in their compounds, the metal atoms often possess unpaired electrons. These impart the property of paramagnetism.
3. *Many (if not most) of their compounds are colored.* The origin of the colors of complex ions of the transition elements will be discussed in Section 21.11.
4. *They have a strong tendency to form complex ions.* As a group, these elements form a huge number of complex ions of varying degrees of complexity. The last six sections of this chapter are devoted to a discussion of their structures and bonding.

21.2 ELECTRONIC STRUCTURE AND OXIDATION STATES

In Chapter 4 we saw that as we proceed from left to right across a period through the main transition elements, there is a gradual filling of the *d* subshell that lies just below the outer shell. In Period 4, for example, this corresponds to the 3*d* subshell, as seen for the electronic structures of the first row elements given in Table 21.2. Each of these elements possesses a completed argon core with additional electrons in the 3*d* and the 4*s* subshells. Notice once again that chromium and copper are anomalous because of the extra stability associated with half-filled and filled subshells. Similar irregularities are found in the second and third transition series (Table 21.3), although other factors in addition to those having to do with half-filled and filled subshells are apparently involved, so no simple correlations can be made.

Among the inner transition elements, the lanthanides and actinides—so named because lanthanum and actinium have properties more or less typical of

Table 21.2
Electronic structures of the elements in the first transition series

Sc	[Ar] $3d^14s^2$	Fe	[Ar] $3d^64s^2$
Ti	[Ar] $3d^24s^2$	Co	[Ar] $3d^74s^2$
V	[Ar] $3d^34s^2$	Ni	[Ar] $3d^84s^2$
Cr	[Ar] $3d^54s^1$	Cu	[Ar] $3d^{10}4s^1$
Mn	[Ar] $3d^54s^2$	Zn	[Ar] $3d^{10}4s^2$

Table 21.3
Electronic structures of elements of the
second and third transition series

Period 5 Second Transition Series		Period 6 Third Transition Series	
Y	[Kr] $4d^1 5s^2$	La	[Xe] $5d^1 6s^2$
Zr	[Kr] $4d^2 5s^2$	Hf	[Xe,$4f^{14}$] $5d^2 6s^2$
Nb	[Kr] $4d^4 5s^1$	Ta	[Xe,$4f^{14}$] $5d^3 6s^2$
Mo	[Kr] $4d^5 5s^1$	W	[Xe,$4f^{14}$] $5d^4 6s^2$
Tc	[Kr] $4d^6 5s^1$	Re	[Xe,$4f^{14}$] $5d^5 6s^2$
Ru	[Kr] $4d^7 5s^1$	Os	[Xe,$4f^{14}$] $5d^6 6s^2$
Rh	[Kr] $4d^8 5s^1$	Ir	[Xe,$4f^{14}$] $5d^7 6s^2$
Pd	[Kr] $4d^{10} 5s^0$	Pt	[Xe,$4f^{14}$] $5d^9 6s^1$
Ag	[Kr] $4d^{10} 5s^1$	Au	[Xe,$4f^{14}$] $5d^{10} 6s^1$
Cd	[Kr] $4d^{10} 5s^2$	Hg	[Xe,$4f^{14}$] $5d^{10} 6s^2$

their respective series—there is a gradual filling of an f subshell that lies *two* shells below the outer shell. Thus, in Table 21.4 we see that as we pass through the lanthanides in Period 6, the $4f$ subshell is completed. In the following period, as we pass through the actinides, the $5f$ subshell becomes populated.

The chemical and physical properties of the transition elements are controlled, of course, by their electronic structures. With the d-block elements the outer s and underlying d subshells are of nearly equal energy. Therefore, when these elements react, the d electrons are able to participate in bonding. The importance of the d electrons in determining the chemistry of these elements accounts for their varied chemical properties, including the multiplicity of oxidation states.

Table 21.4
Electronic structures of the
lanthanide and actinide elements

Lanthanides		Actinides	
La	[Xe] $5d^1 6s^2$	Ac	[Rn] $6d^1 7s^2$
Ce	[Xe] $4f^1 5d^1$	Th	[Rn] $6d^2 7s^2$
Pr	[Xe] $4f^3 6s^2$	Pa	[Rn] $5f^2 6d^1 7s^2$
Nd	[Xe] $4f^4 6s^2$	U	[Rn] $5f^3 6d^1 7s^2$
Pm	[Xe] $4f^5 6s^2$	Np	[Rn] $5f^5 7s^2$
Sm	[Xe] $4f^6 6s^2$	Pu	[Rn] $5f^6 7s^2$
Eu	[Xe] $4f^7 6s^2$	Am	[Rn] $5f^7 7s^2$
Gd	[Xe] $4f^7 5d^1 6s^2$	Cm	[Rn] $5f^7 6d^1 7s^2$
Tb	[Xe] $4f^9 6s^2$	Bk	[Rn] $5f^8 6d^1 7s^2$
Dy	[Xe] $4f^{10} 6s^2$	Cf	[Rn] $5f^{10} 7s^2$
Ho	[Xe] $4f^{11} 6s^2$	Es	[Rn] $5f^{11} 7s^2$
Er	[Xe] $4f^{12} 6s^2$	Fm	[Rn] $5f^{12} 7s^2$
Tm	[Xe] $4f^{13} 6s^2$	Md	[Rn] $5f^{13} 7s^2$
Yb	[Xe] $4f^{14} 6s^2$	No	[Rn] $5f^{14} 7s^2$
Lu	[Xe] $4f^{14} 5d^1 6s^2$	Lr	[Rn] $5f^{14} 6d^1 7s^2$

The +2 oxidation state is common because many transition elements have a pair of s electrons that are the first to be lost.

The wide variety of different oxidation states found for the transition elements is illustrated in Figure 21.3. Don't be overwhelmed—we are only interested here in looking at trends in the relative stabilities of oxidation states. When we compare these stabilities, we can see two important trends.

1. Elements at the left of a row of transition elements prefer the highest oxidation states. As we proceed to the right, the lower oxidation states become increasingly more stable relative to the higher ones. (Remember, the *highest* oxidation state is usually equal to the group number.)
2. Going down a group of transition elements, the higher oxidation states become increasingly more stable than the lower ones.

Knowing these trends is useful because it helps us make comparisons of the strengths of oxidizing agents. For example, in the first row of transition elements (the first *transition series*), we see an increasing tendency toward a stable +2 oxidation state and fewer and fewer high oxidation states. Suppose we use this to compare the relative stabilities of Fe^{2+} and Fe^{3+} with Ni^{2+} and Ni^{3+}. The horizontal trend suggests that Ni^{3+} has a greater tendency to become Ni^{2+} than Fe^{3+} has to become Fe^{2+}. In other words, Ni^{3+} has a greater tendency to gain an electron than Fe^{3+}. Since these species are functioning as oxidizing agents when they gain electrons, we conclude that Ni^{3+} is a stronger oxidizing agent than Fe^{3+}. In fact, this is what is found experimentally.

Remember, an oxidizing agent becomes reduced in a redox reaction.

We can also make vertical comparisons. For example, suppose we consider the ions MnO_4^- and ReO_4^-. Since high oxidation states become more stable going down, ReO_4^- should be a weaker oxidizing agent than MnO_4^-. This is also found to be true experimentally.

MnO_4^- is *permanganate* ion.
ReO_4^- is *perrhenate* ion.

IIIB	IVB	VB	VIB	VIIB	VIII			IB	IIB
Sc	Ti	V	Cr	Mn	Fe	Co	Ni	Cu	Zn
+3	+2	+1	+2	+2	+2	+2	+2	+1	+2
	+3	+2	+3	+3	+3	+3	+3	+2	
	+4	+3	+6	+4	+4				
		+4		+6	+6				
		+5		+7					
Y	Zr	Nb	Mo	Tc	Ru	Rh	Pd	Ag	Cd
+3	+2	+2	+2	+2	+2	+1	+2	+1	+2
	+3	+3	+3	+3	+3	+2	+3	+2	
	+4	+4	+4	+4	+4	+3	+4	+3	
		+5	+5	+5	+5	+4			
			+6	+6	+6	+5			
			+8	+7	+7	+6			
					+8				
La	Hf	Ta	W	Re	Os	Ir	Pt	Au	Hg
+3	+3	+2	+2	+3	+2	+1	+2	+1	+1
	+4	+3	+3	+4	+3	+2	+3	+3	+2
		+4	+4	+5	+4	+3	+4		
		+5	+5	+6	+5	+4	+5		
			+6	+7	+6	+5	+6		
					+8	+6			

Figure 21.3
Oxidation states of the transition metals. The most stable oxidation states are shown in color.

EXAMPLE 21.1 PREDICTING RELATIVE STABILITIES OF
OXIDATION STATES BASED ON AN ELEMENT'S
POSITION IN THE PERIODIC TABLE

PROBLEM Which would you expect to be a more powerful oxidizing agent, TiO_2 or MnO_2?

SOLUTION First we locate Ti and Mn in the periodic table. Since the higher oxidation states become less stable relative to the lower ones going from left to right, Mn(IV) should be relatively less stable than Ti(IV) toward reduction—Mn(IV) should have a greater tendency to be reduced than Ti(IV). Therefore, MnO_2 should be the stronger oxidizing agent. (Actually, TiO_2 is a stable white pigment used in paint, and MnO_2 is the oxidizing agent in the common dry cell.)

In contrast to the wide range of chemical properties of the *d*-block elements, the lanthanides exhibit a remarkable sameness of properties. The 4*f* subshell, which is only partially filled for most of these elements, is buried beneath the outer 5*d* and 6*s* subshells and does not interact to an appreciable extent with the surrounding chemical environment. Consequently, the chemistry of the lanthanides, like that of lanthanum itself, is predominantly that of the 3+ ion, and differences in behavior depend primarily on differences in ionic size.

The actinide elements exhibit a greater variation in oxidation numbers than the lanthanides (e.g., uranium forms compounds in the +3, +4, +5, and +6 oxidation states). An explanation sometimes given for this is that the 5*f* orbitals of the actinides project farther toward the outer parts of the atoms than the 4*f* orbitals of the lanthanides. As a result, the 5*f* orbitals are able to become involved to a greater degree in chemical bonding, so more complex chemistry is observed.

21.3 ATOMIC AND IONIC RADII

We have seen before that many trends in properties can be correlated with variations that occur in atomic and ionic radii. This is true among the transition elements as well as the representative elements.

In Chapter 4 the horizontal and vertical trends in atomic size were discussed. Let's briefly review them here. As we move across a given transition series there is only a gradual decrease in atomic radius. This is because the 3*d* electrons that are added to the atom shield the outer 4*s* electrons quite well from the increasing nuclear charge and, as a result, the effective nuclear charge experienced by the outer electrons rises only slowly. Therefore, only a small size decrease occurs (Table 21.5).

Vertically, we find a rather large increase in size among the *d*-block elements going from Period 4 to Period 5. However, between Periods 5 and 6 there is only a very small size increase and, in some cases, none at all. As we learned earlier, this is a consequence of the lanthanide contraction—the gradual decrease in size that occurs across the lanthanide series. Apparently this just cancels the size increase that would be otherwise expected as we go from Period 5 to Period 6, so the Period 6 elements that follow the lanthanides are essentially the same size as those above them in Period 5.

These variations in size have some very pronounced chemical and physical consequences. They can be correlated, for example, with variations in ionization

Table 21.5
Atomic Radii (pm)

Sc	162	Y	180	La	187
Ti	147	Zr	160	Hf	158
V	134	Nb	146	Ta	146
Cr	127	Mo	139	W	139
Mn	126	Tc	136	Re	137
Fe	126	Ru	134	Os	135
Co	125	Rh	134	Ir	136
Ni	124	Pd	137	Pt	138
Cu	128	Ag	144	Au	144
Zn	138	Cd	154	Hg	157

energies (IE), shown in Table 21.6. Here we see that the gradual decrease in size along a period that is associated with an increase in effective nuclear charge is also accompanied by an increase in IE. As we might expect, this increasing difficulty encountered in removing an outer electron from the isolated atoms is reflected in a gradual, although not altogether uniform, rise in their standard reduction potentials. In other words, as we proceed across the table from left to right, it generally becomes more difficult to oxidize the elements.

The effect of the lanthanide contraction is demonstrated in these properties as well. Among the representative elements the IE generally decreases as we proceed down a group, paralleling the increase in size. This phenomenon is also observed among the transition elements on going from Period 4 to 5. However, from Period 5 to 6 there is an increase in nuclear charge without an accompanying increase in size, so the IE increases. This in turn manifests itself in reduction potentials that tend to be quite high for the third transition series elements, thereby accounting for their virtually inert behavior toward many oxidizing agents.

Trends in the ease of oxidation of the metals were summarized in Figure 7.10 on page 261.

21.4 MAGNETISM

In Chapter 4 we saw that the presence of unpaired electrons in an atom or molecule imparts the property called paramagnetism to the substance. The tiny

Table 21.6
Ionization energy (kJ mol^{-1})

Sc	632	Y	616	La	540
Ti	660	Zr	672	Hf	675
V	651	Nb	665	Ta	763
Cr	653	Mo	694	W	771
Mn	718	Tc	720	Re	761
Fe	763	Ru	711	Os	842
Co	760	Rh	720	Ir	868
Ni	737	Pd	805	Pt	866
Cu	746	Ag	732	Au	891
Zn	907	Cd	869	Hg	1008

electron magnets cause the atom or molecule as a whole to behave as a small magnet. When these are placed into a magnetic field, the microscopic magnets tend to align themselves with and be attracted toward the field. However, thermal motion operates to randomize the orientations of the little magnets. The net result is that only a relatively small fraction of the number of tiny magnets are aligned with the field at any particular instant, so an ordinary paramagnetic substance is drawn only weakly into an external magnetic field.

Characteristically, the transition elements and their compounds possess a partially filled d or f subshell, so many of them exhibit the phenomenon of paramagnetism. The prediction of magnetic properties of transition metal compounds is therefore one requisite of a theory of bonding that is applicable to these substances. We will explore this a little further when we discuss complex ions later in this chapter.

Related to the property of paramagnetism is the phenomenon called **ferromagnetism,** observed for the three pure elements, iron, cobalt, and nickel. Ferromagnetic materials, like paramagnetic ones, are also attracted to a magnetic field; however, the magnitude of the interaction for a ferromagnetic substance is approximately a million times stronger than it is with paramagnetic materials. How does this occur?

The origin of ferromagnetism is the same as paramagnetism—that is, the existence of unpaired electrons in the ferromagnetic material. In these substances it is believed that regions exist, called **domains,** that contain very large numbers of paramagnetic atoms with their atomic magnets all lined up in the same direction, as illustrated in Figure 21.4. Ordinarily, these domains are randomly oriented in a ferromagnetic solid, so even though each domain behaves as a relatively large magnet, their combined effects cancel. When the ferromagnetic material is placed in a magnetic field, the domains tend to become aligned, much the same as the atomic magnets of a paramagnetic substance. In this case, however, each time one domain becomes aligned with the field, millions of tiny atomic magnets become aligned all at once. As a result, the interaction between the ferromagnetic solid and the magnetic field is very much larger than that experienced by paramagnetic substances.

When the magnetic field is removed from a paramagnetic substance, its atomic magnets very quickly become randomly oriented and no permanent magnetism is induced. For a ferromagnetic substance, however, the domains tend to remain in the orientation in which they found themselves when the

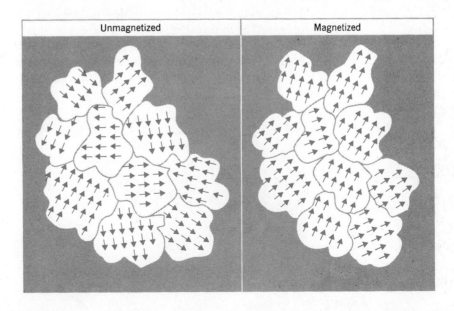

Figure 21.4

Domains in a ferromagnetic solid.

external magnetic field was present. This alignment of domains in the absence of an external field causes the substance to possess a residual magnetism, and we say that it has become *permanently magnetized*. Any piece of iron (e.g., a pin) can be magnetized simply by stroking it with another permanent magnet.

A permanent magnet is not really permanent, because the magnetism may be destroyed either by heating the solid or pounding it. In the first case the increased thermal motion causes the domains to become randomly oriented, while in the second instance violent vibrational motions cause the domains to twist and turn and become disoriented.

The phenomenon of ferromagnetism is associated only with the solid state. Iron, for example, is no longer ferromagnetic when it is melted. Instead it exhibits only paramagnetism. Melting of the solid thus appears to destroy the domains, and each individual atom in the liquid behaves more or less independently of the others nearby.

Even in the solid state not all elements containing unpaired electrons are ferromagnetic. Manganese, for example, possesses five unpaired electrons compared to only four for iron; yet pure iron is ferromagnetic but pure manganese is not. Apparently a requirement for ferromagnetism is for the spacings between paramagnetic ions to be just right so that they may lock onto each other to form a domain. Nonferromagnetic metals, in which the ions are too close together, can sometimes be made ferromagnetic by forming an alloy. This is the case with manganese, where the addition of the proper amount of copper permits the Mn^{2+} ions in the metallic lattice to interact strongly and form domains, thereby producing a ferromagnetic alloy.

21.5 PROPERTIES OF SOME IMPORTANT TRANSITION METALS

Many transition metals have useful applications that depend on both their chemical and physical properties. We find them almost anywhere we look—iron in the many steel products that are all around us, chromium on automobile bumpers, the zinc coating on galvanized steel, and the titanium dioxide pigment that is in nearly all paints. In this section we will examine the chemical and physical properties of some of the more important transition elements.

Titanium

Titanium is a strong corrosion-resistant metal with a density only about 60% that of iron. These properties make it especially useful in the aircraft industry where it is used often in place of steel and aluminum. It is also used in aircraft jet engines because it doesn't lose its strength at high temperatures.

Titanium is difficult to extract from its compounds because the metal reacts

TiCl₄ was used by the Navy during WWII to make smoke screens. When the liquid comes into contact with moisture in the air, it reacts as follows:

TiCl₄ + 2H₂O → TiO₂ + 4HCl.

The dense smoke is caused by the white TiO₂.

with carbon, oxygen, and even nitrogen at very high temperatures. As described in Chapter 18, the metal is made by first converting its oxide, TiO_2 (found in the ore *rutile*), to $TiCl_4$. This is accomplished by heating the oxide with carbon and chlorine.

$$TiO_2 + 2C + 2Cl_2 \longrightarrow TiCl_4 + 2CO_2$$

The low-boiling $TiCl_4$ is vaporized to separate it from impurities in the ore and then reduced using magnesium.

$$TiCl_4 + 2Mg \longrightarrow Ti + MgCl_2$$

The most important compound of titanium is its oxide, TiO_2, which is usually called titanium dioxide. It is a very white solid and is a common pigment in paint, where it possesses good hiding power. It is better than white lead, $Pb_3(OH)_2(CO_3)_2$, because it appears to be of low toxicity and because it doesn't darken on exposure to H_2S as lead-based pigments do.

Chromium

Chromium is a hard, brittle, lustrous metal that is very resistant to corrosion. That's why it is used as a protective coating over steel for automobile bumpers. Thin layers of chromium are also deposited by electroplating on brass or bronze objects for decorative purposes.

One of the principal uses of chromium is in stainless steel—a type of steel that is very resistant to corrosion. A typical stainless steel contains about 19% chromium, 9% nickel, with the rest iron. Unlike ordinary iron and steel, a high-quality stainless steel is not ferromagnetic. A small magnet is therefore a handy tool to check whether or not a metal object claimed to be made of stainless steel is, in fact, composed of this alloy.

The principal oxidation states of chromium are +2, +3, and +6. The +2 state, characterized by the blue Cr^{2+} ion in aqueous solutions, is very easily oxidized to the +3 state, which is the most stable oxidation state. Chromium(III) ion forms many stable complex ions, and in aqueous solutions it actually exists as the violet complex ion, $Cr(H_2O)_6^{3+}$. This ion is what gives many chromium(III) salts their violet color.

Like the $Al(H_2O)_6^{3+}$ ion, the $Cr(H_2O)_6^{3+}$ ion is slightly acidic. When base is added to a solution of the ion, protons are extracted from the water molecules attached to the chromium and a pale blue-violet precipitate is formed. The net reaction is

$$Cr(H_2O)_6^{3+}(aq) + 3OH^-(aq) \longrightarrow Cr(H_2O)_3(OH)_3(s) + 3H_2O$$

This precipitate redissolves in additional base *or* in acid, so chromium(III) is amphoteric.

In base

$$Cr(H_2O)_3(OH)_3(s) + OH^-(aq) \longrightarrow Cr(H_2O)_2(OH)_4^-(aq) + H_2O$$

In acid

$$Cr(H_2O)_3(OH)_3(s) + H^+(aq) \longrightarrow Cr(H_2O)_4(OH)_2^+(aq) + H_2O$$

$$Cr(H_2O)_4(OH)_2^+(aq) + H^+(aq) \longrightarrow Cr(H_2O)_5OH^{2+}(aq) + H_2O$$

$$Cr(H_2O)_5OH^{2+}(aq) + H^+(aq) \longrightarrow Cr(H_2O)_6^{3+}(aq) + H_2O$$

In the +6 oxidation state, chromium forms the red oxide, CrO_3. Chromium (VI) oxide is a strong oxidizing agent and the acid anhydride of *chromic acid*, H_2CrO_4. In very acidic solutions, H_2CrO_4 is the principal species. As the pH is raised, however, two other species are formed. One is the yellow *chromate ion*,

Cr^{2+} is also called *chromous ion*; Cr^{3+} is called *chromic ion*.

The $Cr(H_2O)_6^{3+}$ ion

In general, oxides of metals in high oxidation states tend to be acidic instead of basic.

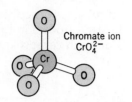

Chromate ion
CrO_4^{2-}

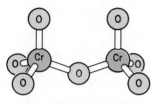

Dichromate ion
$Cr_2O_7^{2-}$

CrO_4^{2-}, and the other is the red-orange *dichromate ion*, $Cr_2O_7^{2-}$. Both are strong oxidizing agents, and exist in equilibrium with each other.

$$2CrO_4^{2-} + 2H^+ \rightleftharpoons Cr_2O_7^{2-} + H_2O$$

By Le Châtelier's principle, we see that $Cr_2O_7^{2-}$ predominates at low pH and CrO_4^{2-} is the major species at high pH.

Manganese

Manganese is much less corrosion-resistant than its neighbor chromium. It corrodes in moist air, and dilute acids dissolve it, much like iron. Manganese is used mostly in making alloys such as the ferromagnetic manganese-copper alloy mentioned in the last section.

The most stable oxidation state of manganese is +2. In solution it exists as the very pale pink $Mn(H_2O)_6^{2+}$ ion. An important compound of manganese in the +4 oxidation state is MnO_2, commonly called manganese dioxide. This is the substance, you recall, that undergoes reduction at the cathode in the ordinary dry cell (see Section 17.12).

When MnO_2 is added to molten KOH and oxidized with O_2 or KNO_3, the green manganate ion, MnO_4^{2-}, is formed. It is stable only in very alkaline (basic) solutions, and when they are acidified the MnO_4^{2-} disproportionates—some of it is oxidized to MnO_4^- (permanganate ion) while the rest is reduced to MnO_2.

$$3MnO_4^{2-} + 4H^+(aq) \longrightarrow 2MnO_4^-(aq) + MnO_2(s) + 2H_2O$$

Solutions of permanganate ion have a deep violet color, and are strong oxidizing agents. When the MnO_4^- is reduced in an acidic solution, the pale pink Mn^{2+} ion is formed.

$$\underset{\text{deep violet}}{MnO_4^-(aq)} + 8H^+(aq) + 5e^- \longrightarrow \underset{\text{very pale pink}}{Mn^{2+}(aq)} + 4H_2O$$

Permanganate titrations were discussed in Chapter 8.

This makes MnO_4^- a useful analytical reagent for performing redox titrations. As the aqueous MnO_4^- is added from a buret to a solution of a reducing agent, reduction of the MnO_4^- takes place. This produces Mn^{2+} ion whose pale pink color is invisible at the concentrations used in the titration. Therefore, when the MnO_4^- is reduced, its violet color disappears and the solution appears colorless. When all the reducing agent has finally been consumed, the next drop of MnO_4^- solution gives an excess of this ion and that makes the solution appear pink, signaling the endpoint.

Reduction of MnO_4^- in neutral or somewhat basic solutions produces a dark brown precipitate of MnO_2.

$$MnO_4^-(aq) + 4H^+(aq) + 3e^- \longrightarrow MnO_2(s) + 2H_2O$$

Because a precipitate is formed, MnO_4^- is not used for redox titrations in neutral or basic solutions. The dark brown color of the precipitate obscures the endpoint.

Iron

Iron is the second most abundant metal and the fourth most abundant element in the Earth's crust.

Iron is the most used of the transition metals because it is relatively abundant and easily extracted from its ores. In the pure state, iron is not very hard, but when small amounts of carbon and other metals are added to it, strong steel alloys are formed.

Iron is a fairly reactive metal. It forms compounds principally in two oxidation states: +2 and +3. Generally, iron(II) compounds are easily oxidized to give the corresponding iron(III) compounds. Three oxides of iron are formed: FeO, Fe_2O_3, and Fe_3O_4. Iron(II) oxide is difficult to prepare and disproportionates into

Fe and Fe_2O_3 when heated.

$$3FeO(s) \longrightarrow Fe(s) + Fe_2O_3(s)$$

Iron(III) oxide is the principal component of many iron ores, and its hydrated form is produced when iron rusts (see below). The oxide Fe_3O_4, known as *magnetite*, contains iron in two different oxidation states, and can be formulated as $Fe^{II}Fe_2^{III}O_4$ (Roman numerals are used here to express the oxidation states of the iron atoms). As its name suggests, magnetite is magnetic. It is an important iron ore because a strong magnet can easily separate it from useless rock and because it is rich in iron.

Metallic iron dissolves in nonoxidizing acids such as HCl or H_2SO_4 with the release of hydrogen (see photo on page 262).

$$Fe(s) + 2H^+(aq) \longrightarrow Fe^{2+}(aq) + H_2(g)$$

One of the most important reactions of iron, at least from an economic point of view, is its corrosion in the presence of air and moisture. The mechanism for the reaction appears to be electrochemical, as shown in Figure 21.5. The iron—

Figure 21.5

Corrosion of iron. Iron dissolves at anode sites and diffuses through the water as Fe^{2+} to cathodic sites where it precipitates as $Fe(OH)_2$. Iron(II) hydroxide is easily oxidized by O_2 in the presence of water to give the hydrated iron(III) oxide, $Fe_2O_3 \cdot xH_2O$, called rust.

$$Fe^{2+}(aq) + 2OH^-(aq) \longrightarrow Fe(OH)_2(s)$$

$$Fe(OH)_2(s) \xrightarrow{O_2, H_2O} Fe_2O_3 \cdot xH_2O(s)$$

Rust ($Fe_2O_3 \cdot xH_2O$)

Water

Fe^{2+} Fe^{2+}

Iron

Cathode

Anode e^-

Anode $Fe(s) \longrightarrow Fe^{2+}(aq) + 2e^-$

Cathode $\frac{1}{2}O_2(aq) + H_2O + 2e^- \longrightarrow 2\,OH^-(aq)$

Iron is acting as the anode in a galvanic cell. That's why it is oxidized.

in contact with water—is oxidized to the +2 state.

$$Fe(s) \longrightarrow Fe^{2+}(aq) + 2e^-$$

The electrons released by the iron during this oxidation are transmitted to sites on the iron that are in contact with oxygen and moisture, where reduction of O_2 to hydroxide ions occurs.

$$\tfrac{1}{2}O_2(aq) + H_2O + 2e^- \longrightarrow 2\ OH^-(aq)$$

Notice that complete dehydration of $Fe(OH)_3$ would give Fe_2O_3:

$$2Fe(OH)_3 \rightarrow Fe_2O_3 + 3H_2O$$

The iron(II) ions diffuse through the water and when they contact the OH^- a precipitate of $Fe(OH)_2$ is formed, which is very easily oxidized to $Fe(OH)_3$ by oxygen.

$$4Fe(OH)_2(s) + O_2(aq) + 2H_2O \longrightarrow 4Fe(OH)_3(s)$$

Iron(III) hydroxide is easily dehydrated and is better represented as a hydrated oxide, $Fe_2O_3 \cdot xH_2O$, in which the proportion of water is somewhat variable. This hydrated oxide is **rust**.

This mechanism for the rusting of iron explains some interesting observations. First, rusting only occurs if *both* oxygen and water are present. Iron won't rust in dry air or oxygen-free water. Second, the formation of rust often occurs at a site somewhat removed from the location where the iron is pitting. Have you noticed, for example, that a car rusts out *under* the paint around places where the paint has been scratched. The iron that dissolves migrates under the paint to the place of the scratch where it can be oxidized by O_2 to give the hydrated oxide.

Rusting has taken place beneath the paint on this very badly corroded car.

Cobalt

Cobalt is an important metal because it is used in many alloys having special properties. *Stellite*, for example, is an alloy containing cobalt, chromium, and tungsten that retains its hardness even when very hot. This property allows it to be used to make cutting tools such as drill bits for high-speed machining of steel parts. Cobalt is also used in catalysts.

Blue glass is made using cobalt salts. This is the *cobalt glass* used in the flame test for potassium.

The principle oxidation states of cobalt are +2 and +3. In the absence of complex ion forming substances other than water, the +2 state is most stable. In most complex ions, the +3 state is very easily formed and is the most stable.

Nickel

Nickel is very useful because it resists corrosion, and its alloys have desirable properties. Electroplating of steel parts with nickel gives them a thin protective coating. Nickel, along with chromium, is added to iron to produce stainless steel, and iron-nickel alloys are used for armor plating because they are impact-resistant. Even the familiar five-cent piece is a copper-nickel alloy.

The most stable oxidation state of nickel is +2. Like the other transition elements, nickel forms many complex ions and many nickel salts are green because they contain the $Ni(H_2O)_6{}^{2+}$ ion. An important compound of nickel in the +4 oxidation state is NiO_2—the cathode material in the nickel-cadmium battery. When the battery discharges, the nickel is reduced to the +2 state.

$$NiO_2(s) + Cd(s) + 2H_2O \longrightarrow Ni(OH)_2(s) + Cd(OH)_2(s)$$

The coinage metals—copper, silver, and gold

The name "coinage metals" given to the Group IB elements is not surprising since these metals have been used for many centuries to make coins as well as jewelry. They are relatively unreactive and their reactivity decreases going down the group. As a result, gold is found free in nature, as are silver and copper, but silver and especially copper also are found in deposits of their compounds.

All the coinage metals have practical uses. As mentioned earlier, copper has the second highest electrical conductivity of any metal and is used extensively for electrical wiring. Silver is an even better conductor than copper, but its lower abundance and its value as a decorative metal generally make it too expensive to use in place of copper. Nevertheless, U.S. industries use about 5×10^6 kg of silver each year—about 24 percent is consumed by the electronics industry. Large amounts of silver are also used to make photographic film and paper. Even gold has practical applications. Low-potential electrical contacts are often gold plated because of gold's resistance to corrosion. Even a thin film of corrosion on contacts made of other metals would seriously impede the flow of electricity in low electric potential circuits.

Copper, silver, and gold have more positive reduction potentials than hydrogen, which means that they cannot displace hydrogen from acids such as HCl or H_2SO_4 in which the strongest oxidizing agent is H^+. Copper and silver do dissolve in HNO_3, however, because it contains $NO_3{}^-$, which serves as the oxidizing agent.

$$3Cu(s) + 8H^+(aq) + 2NO_3{}^-(aq) \longrightarrow 3Cu^{2+}(aq) + 2NO(g) + 4H_2O$$

$$3Ag(s) + 4H^+(aq) + NO_3{}^-(aq) \longrightarrow 3Ag^+(aq) + NO(g) + 2H_2O$$

Gold is much more difficult to oxidize than copper or silver, and even concentrated HNO_3 is not a strong enough oxidizing agent to do the job. However, aqua regia—a 3-to-1 mixture of concentrated HCl and NHO_3—does dissolve gold slowly because the chloride ion stabilizes the Au^{3+} by forming a complex ion with it.

$$Au(s) + 6H^+(aq) + 3NO_3{}^-(aq) + 4Cl^-(aq) \longrightarrow AuCl_4{}^-(aq) + 3H_2O + 3NO_2(g)$$

Each of the Group IB metals forms compounds in the +1 oxidation state. Examples are CuCl, AgCl, and AuCl, all of which are insoluble in water. For copper and gold, the +1 oxidation states are not especially stable and tend to disproportionate. In aqueous solution, for example, copper(I) ion spontaneously gives metallic copper and copper(II).

$$2Cu^+(aq) \longrightarrow Cu(s) + Cu^{2+}(aq)$$

The most stable oxidation state of gold is +3, so gold(I) compounds tend to disproportionate to give the free metal and gold(III).

By far, the most stable oxidation state for silver is the +1 state. The most important compound of silver is silver nitrate, $AgNO_3$—made by dissolving silver in nitric acid. It serves as the starting material for nearly all other silver compounds—for instance, the halides AgCl, AgBr, and AgI that are used in photographic films and papers.

The ions of the Group IB metals form many complex ions. Copper salts are pale blue in aqueous solutions because of the pale blue ion $Cu(H_2O)_4^{2+}$. Addition of ammonia produces the deep blue $Cu(NH_3)_4^{2+}$ ion. The qualitative test for silver ion in a solution uses a similar complex between Ag^+ and NH_3. First the solution to be tested is made acidic with HCl, which precipitates white AgCl, as well as the chlorides of lead $(PbCl_2)$ and mercury(I), (Hg_2Cl_2) if any of these metal ions are in the solution. To determine whether the precipitate contains any AgCl, it is treated with aqueous NH_3. This causes AgCl to dissolve, but not $PbCl_2$ or Hg_2Cl_2 because lead and mercury do not form soluble complex ions with ammonia.

The precipitate might be just $PbCl_2$ and/or Hg_2Cl_2.

$$AgCl(s) + 2NH_3(aq) \rightleftharpoons Ag(NH_3)_2^+(aq) + Cl^-(aq)$$

The solution, which would contain $Ag(NH_3)_2^+$ and Cl^-, is then acidified. If any AgCl had dissolved when the aqueous NH_3 was added, it is now reprecipitated because the equilibrium is shifted to the left as NH_3 is converted to NH_4^+ by the acid.

$NH_3 + H^+ \rightarrow NH_4^+$

Zinc, cadmium, and mercury

These metals are often not considered true transition elements because their d subshells are completed. Each has a valence shell consisting of only two electrons in an s orbital, so their highest oxidation state is $+2$. In fact, zinc and cadmium show only a $+2$ oxidation state (other than zero, of course), while mercury has oxidation states of $+2$ and $+1$.

Zinc and cadmium are both silvery, rather reactive metals. They dissolve readily in nonoxidizing acids such as HCl and H_2SO_4 with the evolution of hydrogen. In fact, the reaction of zinc with dilute H_2SO_4 is a common way of preparing hydrogen in the laboratory.

$$Zn(s) + H_2SO_4(aq) \longrightarrow ZnSO_4(aq) + H_2(g)$$

The reactivities of zinc and cadmium account for one of their principal uses as free metals—providing corrosion protection for iron and steel. Coating an iron or steel object with zinc is called **galvanizing.** The zinc protects the steel in two ways. First, it reacts with moisture and CO_2 to form a film of $Zn_2(OH)_2CO_3$ that prevents oxygen and moisture from coming into contact with and reacting with the zinc below or with the iron. But even if this zinc coating is scratched through to the iron, the zinc still protects the iron by electrolytic action. Zinc is more easily oxidized than iron, so when the metals are in contact the zinc becomes the anode of a galvanic cell and iron becomes the cathode. Since oxidation always occurs at the anode, zinc is oxidized, but iron is protected by being the cathode. The phenomenon is called **cathodic protection.** Cadmium plating on steel functions in a similar way, but cadmium is used less often than zinc for a number of reasons. One is that cadmium is less abundant than zinc, and is therefore more expensive. Another is that cadmium compounds are very toxic— they cause high blood pressure, heart disease, and can even lead to painful death.

The shiny zinc coating on a galvanized steel object such as a garbage pail becomes dull as the zinc reacts with air and moisture to form $Zn_2(OH)_2CO_3$.

Cadmium generally occurs in nature as an impurity in zinc ores.

Zinc and cadmium have uses other than as protective coatings over other metals. Zinc is the metal used as the anode in the common dry cell, and it is alloyed with copper to produce *brass* and with copper and tin to produce *bronze*. Cadmium is used to make rechargeable nickel-cadmium batteries.

Mercury is the only metal that is a liquid at ordinary room temperatures. It freezes at $-38.9°C$ and boils at $357°C$. This liquid range has led to its widespread use as the fluid in thermometers. As we learned earlier, mercury has the ability to dissolve many metals to form solutions called amalgams.

Not *all* metals dissolve in mercury. Iron is an example of one that doesn't.

Zinc wire being placed beside the Alaskan pipeline. When connected electrically to the pipeline, it prevents corrosion by providing cathodic protection.

Cadmium and zinc have many similar chemical properties, which differ considerably from those of mercury. For instance, we have seen that both Zn and Cd dissolve in dilute acids with the evolution of hydrogen. Mercury, however, is considerably less reactive. It doesn't dissolve in acids such as HCl or H_2SO_4, although it does react with nitric acid.

$$3Hg(l) + 8H^+(aq) + 2NO_3^-(aq) \longrightarrow 3Hg^{2+}(aq) + 2NO(g) + 4H_2O$$

One important difference between zinc and cadmium is that zinc is amphoteric but cadmium is not. For example, zinc dissolves in base.

$$Zn(s) + 2OH^-(aq) + 2H_2O \longrightarrow Zn(OH)_4^{2-}(aq) + H_2(g)$$

Its hydroxide is also amphoteric. When base is added to Zn^{2+} in aqueous solution, a precipitate of $Zn(OH)_2$ is formed.

$$Zn^{2+}(aq) + 2OH^-(aq) \longrightarrow Zn(OH)_2(s)$$

The zinc hydroxide dissolves either in acid or additional base.

$$Zn(OH)_2(s) + 2H^+(aq) \longrightarrow Zn^{2+}(aq) + 2H_2O$$

$$Zn(OH)_2(s) + 2OH^-(aq) \longrightarrow Zn(OH)_4^{2-}(aq)$$

Tubes of zinc oxide ointment can be purchased for use as sun screens.

This chemical difference between zinc and cadmium explains why metals are sometimes cadmium plated rather than zinc plated. Cadmium is used if the metal is to be exposed to an alkaline environment, because the base will destroy a zinc coating but not a cadmium one.

Many zinc compounds have important commercial uses. Zinc oxide, which forms when zinc is heated in air, is used as a paint pigment, in creams that are spread on the skin as sun screens, and in fast-setting dental cements. Zinc chloride is used in many different ways, including deodorants, embalming, and fireproofing lumber.

As mentioned earlier, mercury forms compounds in both the +1 and +2 oxidation states. In the +1 state, two mercury(I) ions are joined by a covalent bond to give Hg_2^{2+}. The existence of this ion is supported by equilibrium studies (see Problem 21.66), X-ray determination of crystal structures, and the fact that mercury(I) compounds are diamagnetic. If there were a simple Hg^+ ion, it would have an odd number of electrons, so all the electrons couldn't be paired and Hg^+ would be paramagnetic. Pairing of the odd electron with an electron from another Hg^+ gives a covalent bond and a diamagnetic Hg_2^{2+} species.

Mercury compounds have quite different properties from those of zinc and cadmium. They are considerably less ionic, for example, and $HgCl_2$ in aqueous solution exists primarily (99%) as $HgCl_2$ molecules.

> $HgCl_2$ is a weak electrolyte.

$$HgCl_2(aq) + H_2O \rightleftharpoons Hg(OH)Cl + H^+ + Cl^-$$

As you probably know, mercury and its compounds are very toxic. Mercury spills should be avoided in the laboratory because the vapor pressure of mercury (about 10^{-1} Pa at room temperature) is sufficient to cause mercury poisoning if the vapors are breathed for long periods of time. Soluble mercury compounds, such as mercury(II) chloride, are especially poisonous because they can quickly provide sufficient mercury to the body to cause death. By contrast, mercury(I) chloride, Hg_2Cl_2, is very insoluble in water and at one time—before the discovery of penicillin—it was used medicinally as a treatment for syphilis. Its low solubility prevents the body from absorbing lethal doses of mercury. Mercury is a cumulative poison, however, so even small amounts absorbed over extended periods can lead to serious medical problems.

> One or two grams of $HgCl_2$ is a fatal dose.

Because Hg_2Cl_2 is insoluble in water, the presence of Hg_2^{2+} ion in a solution can be tested by treating the solution with HCl, which causes the Hg_2Cl_2 to precipitate. However, since AgCl and $PbCl_2$ are also white and insoluble, and would also be formed if Ag^+ and Pb^{2+} were present, the precipitate must be tested further. Addition of aqueous ammonia—which is part of the test for silver—causes the Hg_2Cl_2 to disproportionate.

> $Hg(NH_2)Cl$ is called *mercury(II) amido chloride.*

$$Hg_2Cl_2 + 2NH_3 \longrightarrow Hg(NH_2)Cl + Hg + NH_4^+ + Cl^-$$

The mixture of Hg and $Hg(NH_2)Cl$ appears black or gray because the black color of the finely divided mercury masks the white of the $Hg(NH_2)Cl$.

21.6 COORDINATION COMPOUNDS

As we have noted before, the transition elements are known for their ability to form many complex ions—substances in which the metal cation is surrounded by and bonded to one or more other ions or molecules, which we refer to as ligands. Complex ions have many interesting properties that have been extensively studied. Among them are their vivid colors. Complex ions also possess unusual magnetic properties, undergo many kinds of chemical reactions, and have the ability to exist in a variety of structural forms. These and other factors have caused the study of complex ions to become a major specialty within the field of chemistry.

So far, we have been using the term *complex ion* to describe the species that are formed from metal ions and ligands. Actually this is too restrictive, because a number of these complex species are electrically neutral. Therefore, to avoid problems of description they are often simply called *complexes*. They are also frequently called **coordination compounds,** because from the point of view of the valence bond theory—the first bonding theory to be applied to them—they are considered to be held together by *coordinate* covalent bonds between the ligands and the metal.

Coordination compounds have a number of important uses. You've learned that unexposed silver salts in photographic film and paper are removed by dissolving these salts in a solution containing thiosulfate ion, with which Ag^+ forms a complex ion. Complex ions are used in water softening (phosphates binding to iron and manganese ions) and as catalysts in a variety of industrial processes. The formation of complex ions has also been used to alleviate poisoning produced by beryllium and lead and to retard the spoilage of foods. As the study of biochemistry has progressed, it has also become evident that many biologically important molecules owe their biological activity to a metal ion held in a complex ion within the molecule. Hemoglobin, containing iron(II) atoms, is a well-known example. The importance of metal ions in biosystems (not to mention the increased availability of research funding in biochemistry), has turned many inorganic chemists into bioinorganic chemists. We will take a closer look at metal-containing biomolecules in Chapter 24.

The father of modern coordination chemistry was Alfred Werner, who received the Nobel Prize in chemistry in 1913 for his work on these compounds. Werner was the first to recognize that metal ions could combine with other molecules or ions through more than one type of "valence" to produce relatively stable complex species. He was also the first to propose structures of complex ions that were consistent with their properties.

Types of ligands

Metal complexes are formed with many kinds of ligands, including ions such as Cl^-, CN^-, and NO_2^-, as well as neutral molecules such as H_2O or NH_3. Nearly all ligands, however, have one thing in common: They possess a lone pair of electrons that may be shared with the metal cation in coordinate covalent bonds. In this sense the formation of a complex can be viewed as a Lewis acid-base reaction; in general, we can expect that the ligands in coordination compounds will all be Lewis bases with few exceptions.

In a complex, the ligands attached to the metal are considered in a **first coordination sphere,** and in solution they are held tightly by the metal ion, compared to other ions and molecules that might also be present nearby in the mixture. When the formula of a metal complex is written, we usually indicate the species that are bonded to the metal in the first coordination sphere by enclosing them and the metal ion within square brackets. An example is the ion

$$[CoCl_6]^{3-}$$

These brackets are *not* to be confused with those that we used earlier when we wished to denote molar concentration.[1] Note that the charge on the complex is indicated *outside* the brackets, showing that the entire complex ion carries, in this example, a charge 3−.

Ligands such as Cl^- or NH_3, which have one atom that can bond to a metal cation, are said to be **monodentate** (one "tooth") ligands. There are also many molecules and ions that are able to attach themselves to a metal ion through more than one donor atom to produce a cyclic *ring*-type of arrangement. Two very common examples that have been much studied are oxalate ion, $C_2O_4^{2-}$,

One can usually tell from the context of a discussion whether or not square brackets mean molar concentration.

[1] In our earlier discussions of complex ion equilibria, we avoided this notation specifically to prevent such confusion.

and ethylenediamine, $H_2N—CH_2—CH_2—NH_2$,

$$H—\underset{\underset{H}{|}}{\overset{\overset{H}{|}}{N}}—\underset{\underset{H}{|}}{\overset{\overset{H}{|}}{C}}—\underset{\underset{H}{|}}{\overset{\overset{H}{|}}{C}}—\underset{\underset{H}{|}}{\overset{\overset{H}{|}}{N}}—H$$

They bond to a metal ion as shown below.[2]

With cobalt(III), for example, these two ligands form complexes such as $[Co(C_2O_4)_3]^{3-}$ and $[Co(H_2NCH_2CH_2NH_2)_3]^{3+}$, whose structures are illustrated in Figure 21.6. Ethylenediamine is such a common ligand in coordination chemistry that in writing formulas containing it, the abbreviation, **en,** is almost always used. The latter ion is therefore normally written as simply $[Co(en)_3]^{3+}$.

Molecules or ions such as ethylenediamine or oxalate ion, which have two atoms that may coordinate to a metal ion, are said to be **bidentate** ligands. There are also more complex **polydentate** ligands containing three, four, or even more donor atoms. Table 21.7 contains a list of some common mono- and bidentate ligands, as well as a few examples of polydentate ligands.

A particularly important polydentate ligand is the anion of ethylenediaminetetraacetic acid (EDTA). Its six coordinating atoms firmly attach themselves to free metal ions, and EDTA has been used as an antidote in lead poisoning because it binds to Pb^{2+}, thereby preventing it from inhibiting certain important enzyme functions. EDTA also binds to iron and calcium and is used as a water softener in products such as shampoos. EDTA is added to foods to tie up metal ions that catalyze oxidation (and hence deterioration) of the food product. It has also been found that EDTA increases the storage life of whole blood by removing free Ca^{2+}, which promotes clotting.

To prevent EDTA from extracting calcium from the body, $CaNa_2EDTA$ is the salt that's added to food products such as salad dressings.

Figure 21.6

Structures of $[Co(C_2O_4)_3]^{3-}$ and $[Co(en)_3]^{3+}$. (a) $[Co(C_2O_4)_3]^{3-}$. The colored atom is cobalt, solid gray atoms are carbon, and white atoms are oxygen. (b) $[Co(en)_3]^{3+}$. The colored atom is cobalt, solid gray atoms are nitrogen, the large white atoms are carbon, and the small white atoms are hydrogen.

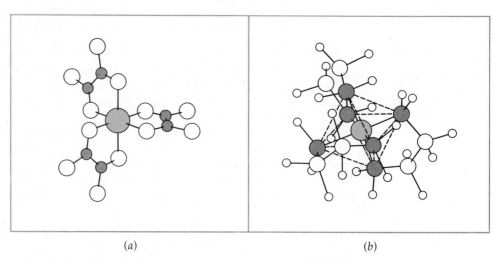

(a)   (b)

[2] The origin of chemical terminology is sometimes rather colorful. Complexes of this general type are often called **chelates,** from the Greek *chele,* meaning claw. The ligand in this case bites the metal with two claws (donor atoms) much like a crab.

Table 21.7
Common ligands found in complex ions

Monodentate

H_2O	Water	Br^-	Bromide
NH_3	Ammonia	I^-	Iodide
CN^-	Cyanide	NO_2^-	Nitrite
OH^-	Hydroxide	SCN^-	Thiocyanate
F^-	Fluoride	$S_2O_3^{2-}$	Thiosulfate
Cl^-	Chloride		

Bidentate

oxalate

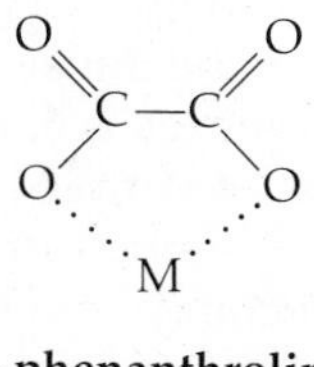

ethylenediamine

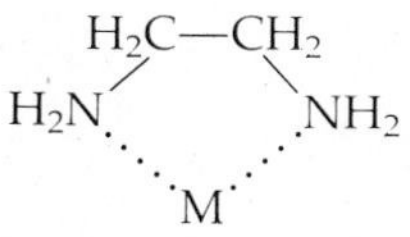

o-phenanthroline

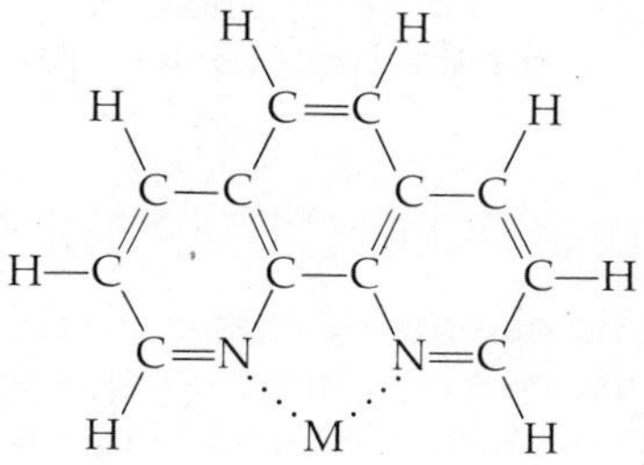

dipyridyl

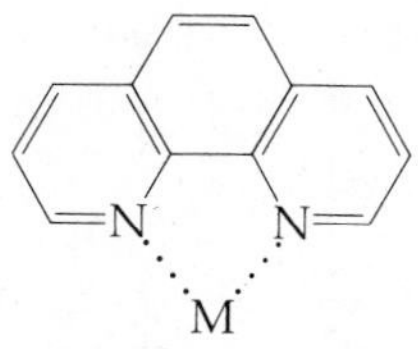

or

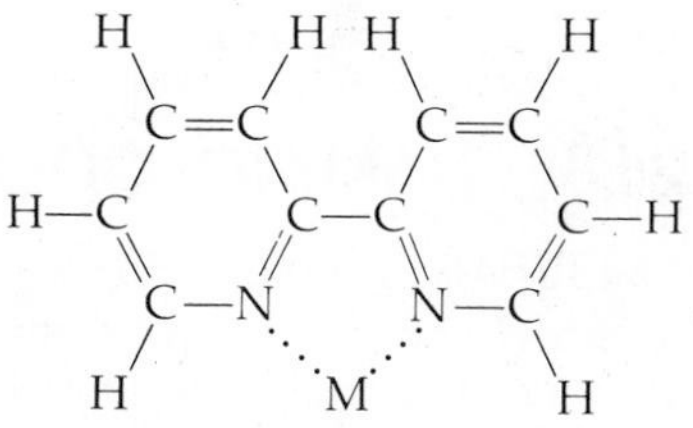

or

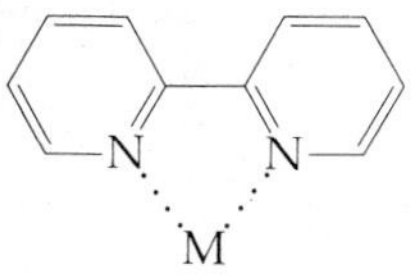

Polydentate

(coordinating atoms indicated with asterisks)

Diethylenetriamine (three coordinating atoms)

$$\overset{*}{H_2N}-CH_2-CH_2-\overset{*}{NH}-CH_2-CH_2-\overset{*}{NH_2}$$

Ethylenediaminetetraacetate ion (six coordinating atoms)

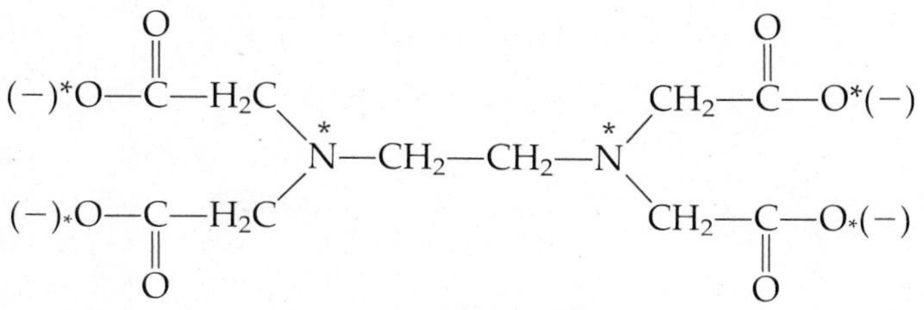

also called EDTA

21.7 COORDINATION NUMBER AND STRUCTURE

The term **coordination number (C.N.)** refers to the total number of ligand atoms that are bound to a given metal ion in a complex. These atoms may be supplied by either monodentate or polydentate ligands, or both. Thus, in the three complexes, $[CoCl_6]^{3-}$, $[Co(en)_2Cl_2]^+$, and $[Co(en)_3]^{3+}$, the C.N. of cobalt is the same, since in each case there are *six* donor atoms about the Co^{3+} ion.

Coordination numbers ranging from 2 to more than 8 are observed in various coordination compounds, with the C.N. in any given instance determined by the nature of the metal ion, its oxidation state, and to some extent, the ligands and the environment surrounding the complex. The most common coordination numbers are observed to be 2, 4, and 6. The basic structural types found for them are shown in Figure 21.7.

By far the most frequently occurring coordination number in transition metal complexes is 6, and the geometry that is observed in nearly all instances is octahedral. A simple two-dimensional way of representing the octahedral geometry was described in Chapter 6 on page 181 and is shown in Figure 21.8. The dashed rectangle represents the square plane that joins the upper and lower pyramids in the octahedron. The six solid lines connect the center of the metal cation to the coordinated ligand atoms. This arrangement is illustrated in Figure 21.8 for the complex ion, $[CoCl_6]^{3-}$.

21.8 NAMING COORDINATION COMPOUNDS

The IUPAC is composed of a group of chemists drawn from all over the world. One of their tasks is to meet periodically to discuss current problems in nomenclature.

The naming of chemical compounds was introduced in Chapter 5 where we discussed the nomenclature system for simple inorganic compounds. This system, developed and kept up to date by the International Union of Pure and

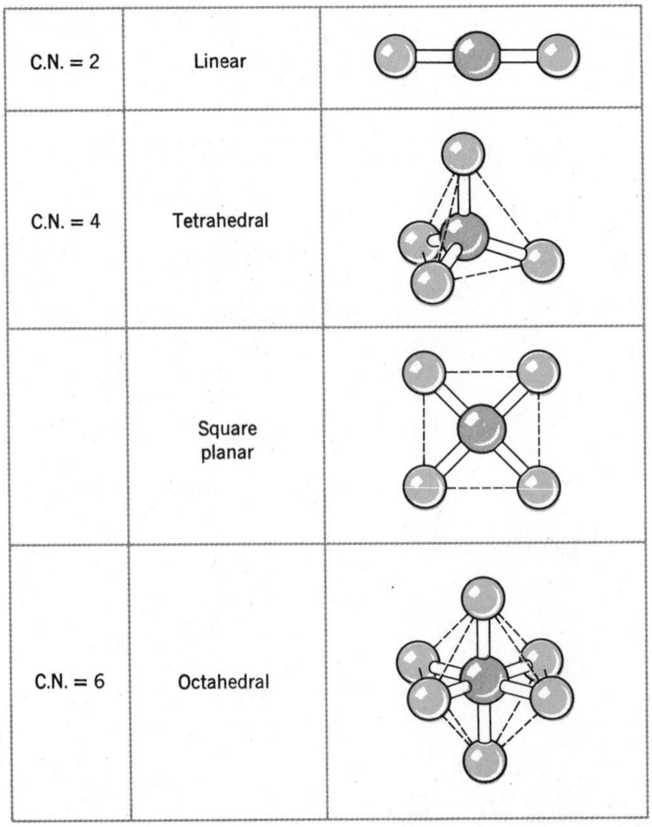

C.N. = 2	Linear	
C.N. = 4	Tetrahedral	
	Square planar	
C.N. = 6	Octahedral	

Figure 21.7

Structure types for coordination numbers of 2, 4, and 6.

Figure 21.8

Two-dimensional representation of octahedral coordination.

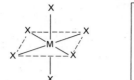
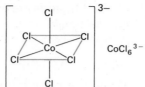

$CoCl_6{}^{3-}$

Applied Chemistry (IUPAC) has been extended to cover names for complexes, too. Below are some rules that have been developed by the IUPAC to name coordination complexes. A few of the names assigned following these rules may sound odd, and even funny. Remember, however, that we are primarily interested in formulating a name that is able to transmit the maximum quantity of information in the shortest possible name. The end result is therefore sometimes difficult to pronounce.

Rules of nomenclature of coordination compounds

At one time, negative ligands were named first, followed by neutral ligands.

1. **Cationic species are named before anionic species.** This is just as in other cases of ionic compounds, such as NaCl, which is named as sodium chloride (cation, anion).
2. **Within a complex ion, the ligands are named first in alphabetical order followed by the metal ion.** This is opposite to the sequence in which they appear in the formula. For example, the complex $[Co(NH_3)_6]^{3+}$ is named by specifying the ammonia first, then the cobalt.
3. **The names of anionic ligands end in the suffix -o.**
   (a) Ligands whose names end in *-ide* have this suffix replaced by *-o*.

Anion		Ligand
chloride	Cl^-	chloro
bromide	Br^-	bromo
cyanide	CN^-	cyano
oxide	O^{2-}	oxo

   (b) Ligands whose names end in *-ite* or *-ate* become *-ito* and *-ato*, respectively.

Anion		Ligand
carbonate	$CO_3{}^{2-}$	carbonato
thiosulfate	$S_2O_3{}^{2-}$	thiosulfato
thiocyanate	SCN^-	thiocyanato (bonded through sulfur) isothiocyanato (bonded through nitrogen)
oxalate	$C_2O_4{}^{2-}$	oxalato
nitrite (bonded through oxygen, ONO^-)[a]	$NO_2{}^-$	nitrito

[a] An exception to this is $NO_2{}^-$ when bonded through nitrogen, in which case it is named as *nitro*.

4. **Neutral ligands are given the same names as the neutral molecule.** Thus ethylenediamine as a ligand is called ethylenediamine in the name of the complex. Two very important exceptions to this, however, are

In all but the most recent literature, water is named as *aquo*.

H_2O Aqua NH_3 Ammine (note double *m*)

5. **When there are more than one of a particular ligand, their number is specified by di = 2, tri = 3, tetra = 4, penta = 5, hexa = 6, and so forth. When confusion might result, the prefixes bis = 2, tris = 3, tetrakis = 4, and so forth, are employed.** Thus the presence of two chloride ions is specified as *dichloro.* however, because ethylenediamine already contains the term *di*, two of these molecules are indicated by placing the name of the ligand in parentheses preceded by the term *bis*; that is, *bis(ethylenediamine)*.

6. **Negative (anionic) complex ions always end in the suffix -ate.** This suffix is appended to the English name of the metal atom in most cases.

Element	Metal as Named in Anionic Complex
aluminum	aluminate
chromium	chromate
manganese	manganate
nickel	nickelate
cobalt	cobaltate
zinc	zincate
molybdenum	molybdate
tungsten	tungstate

For some metals, the *-ate* is appended to the Latin stem.

Except for mercury, metals whose symbols are derived from their Latin names are specified in anionic complexes by using the Latin stem.

Element	Stem	Metal as Named in Anionic Complex
iron	ferr-	ferrate
copper	cupra-	cuprate
lead	plumb-	plumbate
silver	argent-	argentate
gold	aur-	aurate
tin	stann-	stannate

In neutral or positively charged complexes the metal *always* appears with the common English name for the element.

7. **The oxidation number of the metal in the complex is written in Roman numerals within parentheses following the name of the metal.** For example,

$[Co(H_2O)_6]^{3+}$ is the hexaaquacobalt(III) ion.

$[CoCl_6]^{3-}$ is the hexachlorocobaltate(III) ion.

Note that the charge on the complex is obtained as the *algebraic sum* of the oxidation number of the metal and the charges on the ligands.

Some additional examples illustrate these rules.

$[Ni(CN)_4]^{2-}$	tetracyanonickelate(II) ion
$[Co(NH_3)_4Cl_2]^+$	tetraamminedichlorocobalt(III) ion
$Na_3[Cr(NO_2)_6]$	sodium hexanitrochromate(III)
$[Ag(NH_3)_2]^+$	diamminesilver(I) ion
$[Ag(CN)_2]^-$	dicyanoargentate(I) ion
$[Co(en)_3]Cl_3$	tris(ethylenediamine)cobalt(III) chloride
$[Cr(NH_3)_3Cl_3]$	triamminetrichlorochromium(III)

21.9 ISOMERISM AND COORDINATION COMPOUNDS

When two different compounds have the same molecular formula, but differ in the way that their atoms are arranged, they are said to be **isomers** of one another. For example, there are two compounds with the general formula

$$Cr(NH_3)_5SO_4Br$$

One of these we should formulate as

$$[Cr(NH_3)_5SO_4]Br$$

because it yields a precipitate of AgBr when treated in aqueous solution with $AgNO_3$ but does not give a precipitate of $BaSO_4$ when treated with $Ba(NO_3)_2$. This last observation means that the SO_4^{2-} is not free in the solution and, hence, must be bound to the chromium.

The second compound is written as

$$[Cr(NH_3)_5Br]SO_4$$

and produces $BaSO_4$ when treated with $Ba(NO_3)_2$. On the other hand, addition of $AgNO_3$ to a solution of the compound does not yield AgBr.

The two compounds just described have different chemical properties and are clearly different chemical substances, even though they are composed of the same number of the same kinds of atoms. This particular type of isomerism is not uncommon among coordination compounds and is called **ionization isomerism.**

Another type of isomerism that is very important is called **stereoisomerism,** and results when a given molecule or ion can exist in more than one structural form in which the same atoms are bonded to one another but find themselves oriented differently in space. To illustrate this, we will focus our attention on octahedral complexes because they represent the most common structural type.

The simplest form of stereoisomerism results when a complex has the general formula Ma_4b_2 in which a and b represent monodentate ligands. An example would be the ion, $[Co(NH_3)_4Cl_2]^+$. (How would you name it?) This complex can exist in two different isomeric forms, called **geometrical isomers,** as shown in Figure 21.9. As you can see, in one of these isomers the two b ligands are located across from one another on opposite sides of the metal ion. Such an isomer is given the designation **trans** (Latin *trans* means across). The other isomer has the two b ligands adjacent to one another and is referred to as the **cis** isomer (L. *cis* = on the same side). Thus the two isomers would be specified as

$$trans\text{-}[Co(NH_3)_4Cl_2]^+$$

Figure 21.9
Cis-trans *isomers for complexes* Ma_4b_2.

cis—Ma_4b_2 $trans$—Ma_4b_2

and

$$cis\text{-}[Co(NH_3)_4Cl_2]^+$$

Because *cis*- and *trans*-isomers possess different structures, they are different chemical species, each with its own set of chemical and physical properties. While these properties may often be similar, the fact that they are not identical clearly tells us that the two structures represent truly different compounds.

Geometrical isomers also occur when there are bidentate ligands in a complex, as illustrated by the *cis* and *trans* forms of the ion $[Cr(en)_2Cl_2]^+$ shown in Figure 21.10. Once again, in the *trans* form we see that the chloride ligands are on opposite sides of the metal, while in the *cis* form they are alongside each other.

A second form of stereoisomerism is called **optical isomerism.** As we will see, optical isomers affect polarized light differently and bear the same structural relationship to each other as do your left and right hands—that is, they are *nonsuperimposable mirror images* of one another. To see what this means, try this simple experiment. Place your right hand in front of a mirror, with the palm toward the mirror, and hold your left hand alongside with the palm facing you (Figure 21.11). Notice that the image of your *right* hand in the mirror looks the same as your left hand. That is why we say that your left and right hands are mirror images of each other. The nonsuperimposable aspect arises because your left and right hands, while similar in appearance, do not match exactly when one is placed over the other, both with palms down; the thumbs point in different directions. This difference is perhaps seen even more clearly if you attempt to place you right hand into a left-hand glove; it doesn't fit properly. Thus your left and right hand, and optical isomers too, cannot be superimposed on each other.

Molecules or ions that can have two structures related to each other in the same way that your left and right hands are related—that is, as nonsuperimposable mirror images—are said to be **chiral** (from the Greek *cheir*, meaning "hand"). An example of a chiral complex is the $[Co(en)_3]^{3+}$ ion. The mirror

Bidentate ligands always span adjacent or *cis* positions in a complex.

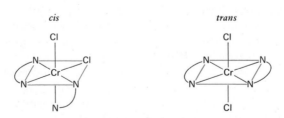

cis   *trans*

Figure 21.10
Cis-trans *isomerism in* $[Cr(en)_2Cl_2]^+$.

N———N represents the bidentate ethylenediamine ligand

Figure 21.11

Illustration of nonsuperimposable mirror images.

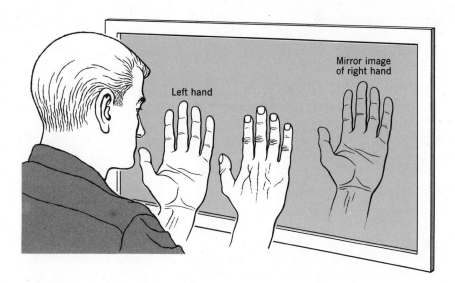

If you make a model of a molecule and use its reflection as a guide in constructing its mirror image, you can test for superimposability.

image relationship between the two isomers is shown in Figure 21.12 in which the hydrogen atoms have been omitted for simplicity. A pair of isomers related in this manner are called **enantiomers.**

In the complex $[Co(en)_3]^{3+}$ it is the arrangement of the metal-ligand rings (called *chelate rings*) that gives rise to the chirality, or optical isomerism. In fact, any octahedral complex containing three bidentate ligands is chiral and can exist as two optical isomers.

Optical isomerism is also important for the *cis* form of complex ions containing two bidentate ligands and two monodentate ligands—for example, *cis*-$[Co(en)_2Cl_2]^+$ (Figure 21.13). The *trans* form of this complex is not chiral and does not exhibit optical isomerism, however, because it and its mirror image are identical. They are superimposable.

In general, the properties of optical isomers are identical except for the way in which they interact with outside influences that are able to distinguish between left and right handedness. The situation here is analogous to having a group of baseball players, some of whom are left-handed and some right-handed. Since they are all able to toss a baseball with equal ease, a baseball will not differentiate between left- and right-handed players. The same applies to a

Figure 21.12

The mirror image relationship between the isomers of the chiral $[Co(en)_3]^{3+}$ ion. Isomer II is the same as the mirror image of Isomer I, but I and II are not identical. They are nonsuperimposable isomers.

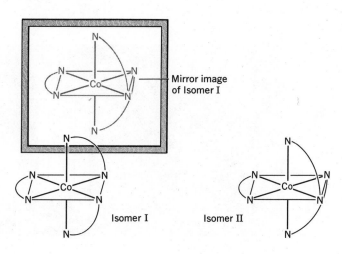

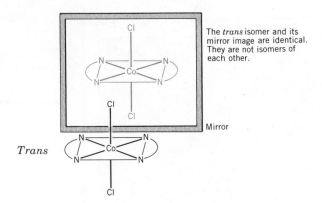

The *trans* isomer and its mirror image are identical. They are not isomers of each other.

Mirror

Trans

Isomer II cannot be superimposed exactly on Isomer I. They are not identical structures.

Mirror image of isomer I

Cis

(Isomer I)   (Isomer II)

Isomer II has the same structure as this mirror image of isomer I

Figure 21.13

Isomers of $[Co(en)_2Cl_2]^+$. The mirror image of the trans isomer is identical to the original, so the trans isomer is not chiral. The mirror image of the cis isomer is not superimposable on the original, so the cis isomer exists as two nonidentical optical isomers.

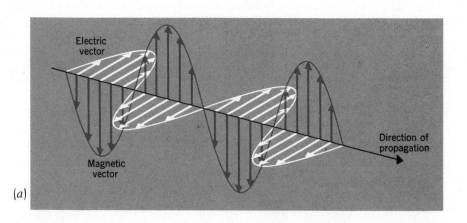

Electric vector

Magnetic vector

Direction of propagation

(a)

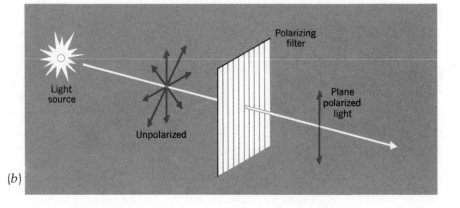

Light source

Unpolarized

Polarizing filter

Plane polarized light

(b)

Figure 21.14

Polarized light. (a) Electromagnetic radiation composed of electric and magnetic vectors. (b) Orientation of electric vectors in unpolarized and polarized light.

bat, since there are no left- or right-handed baseball bats. A fielder's glove, however, will fit only one hand. A glove designed to be worn on the left hand cannot be used by a player who catches the ball in the right hand. In this case, the glove differentiates between these two kinds of players because it also has a left or right handedness to it. In this same fashion, optical isomers interact in an identical way with most chemical reagents and physical probes. They do differ, however, in the way they react toward other optically active chemicals and polarized light.

Light, in general, is composed of electromagnetic radiation that possesses both electric and magnetic components that behave like vectors. These vectors oscillate in a sinusoidal fashion perpendicular to the direction in which the light wave is traveling (Figure 21.14). If we examine the electric vectors, all different orientations are observed in an unpolarized beam. However, when such a beam is passed through a polarizing medium, only the vibrations in one plane remain. The result is called **plane polarized light.** A unique feature of optical isomers is that when plane polarized light is passed through them (or their solutions) the plane of polarization is rotated through some angle, θ, as shown in Figure 21.15. Substances that affect polarized light this way are said to be optically active. One enantiomer (optical isomer) causes the light to be rotated to the right—that is, clockwise when viewed down the axis of the oncoming light beam—and is said to be **dextrorotatory.** The other enantiomer causes the polarized beam to be rotated to the left and is described as **levorotatory.** The two isomers are therefore designated as *d* or *l* depending on the direction of rotation of the polarized light.

An equal mixture of two enantiomers tends to rotate polarized light to both the left and right simultaneously. These effects therefore cancel each other, and such a mixture shows no optical activity; it is said to be **racemic.** In almost all cases, when enantiomers are produced in a chemical reaction carried out in the laboratory, they are formed in equal numbers so that a racemic mixture results. One of the arts in chemistry is the separation of optical isomers from one another.

Quartz crystals occur in two nonsuperimposable forms and they are optically active.

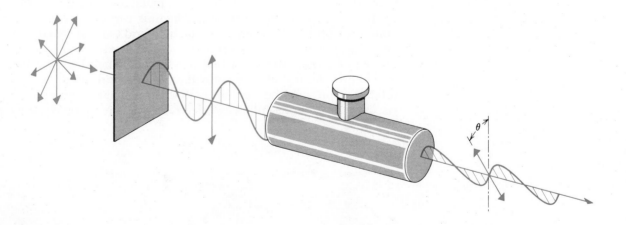

Figure 21.15

Rotation of plane polarized light by an optical isomer in solution. The plane of polarization is rotated by an angle θ as it passes through the solution containing the optically active com- *pound. In this example the light is rotated to the left; the substance in the solution is said to be levorotatory.*

21.10 BONDING IN COORDINATION COMPOUNDS: VALENCE BOND THEORY

There are three important properties of transition metal complexes that must be explained by a bonding theory: (1) structure, (2) magnetic properties, and (3) color.

In the earliest theories of coordination complexes the metal was considered to be attached to the ligands by way of coordinate covalent bonds, and the first serious attempt to explain the structures and magnetic properties of complexes was made by applying the concepts of valence bond theory. Much of what follows is simply an extension of some of the ideas that were developed in Chapter 6.

In valence bond theory, you recall, a bond is formed by the overlap of two orbitals and the subsequent sharing of a pair of electrons between the two atoms in the region of overlap. Ligands, as a rule, do not possess unpaired electrons, so the bonding in the complex must result from the overlap of ligand orbitals containing lone pairs of electrons with vacant orbitals on the metal ion, thereby giving rise to a coordinate covalent bond as illustrated in Figure 21.16. To see how this occurs, let's consider an octahedral ion, such as the blue-violet $[Cr(H_2O)_6]^{3+}$, characteristic of simple Cr(III) salts in aqueous solution. The electronic structure of a free chromium atom is

$$\text{Cr} \quad [\text{Ar}] \quad \underbrace{\uparrow \ \uparrow \ \uparrow \ \uparrow \ \uparrow}_{3d} \quad \underbrace{\uparrow}_{4s} \quad \underbrace{\underline{\ \ }\ \underline{\ \ }\ \underline{\ \ }}_{4p} \quad \underbrace{\underline{\ \ }\ \underline{\ \ }\ \underline{\ \ }\ \underline{\ \ }\ \underline{\ \ }}_{4d}$$

where we have shown the empty 4p and 4d subshells as well as those that are occupied by electrons. The central Cr^{3+} ion in our complex results from the loss of three electrons to give

The 4s electrons are lost before any of the 3d electrons.

$$\text{Cr}^{3+} \quad [\text{Ar}] \quad \underbrace{\uparrow \ \uparrow \ \uparrow \ \underline{\ \ }\ \underline{\ \ }}_{3d} \quad \underbrace{\underline{\ \ }}_{4s} \quad \underbrace{\underline{\ \ }\ \underline{\ \ }\ \underline{\ \ }}_{4p} \quad \underbrace{\underline{\ \ }\ \underline{\ \ }\ \underline{\ \ }\ \underline{\ \ }\ \underline{\ \ }}_{4d}$$

In Chapter 6 we saw that an octahedral structure is formed when the central atom uses hybrid orbitals derived from two atomic d orbitals, an s orbital, and three p orbitals. We called them sp^3d^2 hybrids. We have the additional requirement here that the orbitals used by the metal to create the hybrid must be empty so that the electron pairs from the ligands can be placed into them when the bonds are formed. In this case, suitable hybrids can be constructed using the two vacant 3d orbitals. The 3d orbitals are preferred over the 4d because they are lower in energy; stronger bonds are formed when the 3d orbitals are used instead of the 4d. In this case we will call the hybrids d^2sp^3, recognizing that the d orbitals come from the shell beneath the one that supplies the s and p orbitals. If we use dots to represent ligand electrons, we can now write the electronic struc-

Figure 21.16

Formation of a coordinate covalent bond by the overlap of a filled ligand orbital and an empty hybrid orbital of the metal ion.

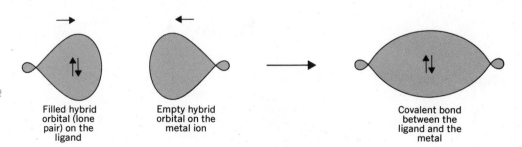

| Filled hybrid orbital (lone pair) on the ligand | Empty hybrid orbital on the metal ion | | Covalent bond between the ligand and the metal |

ture for the complex as

$$[Cr(H_2O)_6]^{3+} \quad \underset{3d}{\uparrow \;\; \uparrow \;\; \uparrow} \;\; \overbrace{\underset{4s}{\cdot\cdot \;\; \cdot\cdot} \;\; \underset{4p}{\cdot\cdot \;\; \cdot\cdot \;\; \cdot\cdot}}^{d^2sp^3} \quad \underset{4d}{—\;\;—\;\;—\;\;—\;\;—}$$

Experimental measurements demonstrate the presence of three unpaired electrons in this ion, as suggested by our bonding picture.

Let's consider next the emerald-green complex ion, $[Ni(H_2O)_6]^{2+}$. The electron configuration of the Ni^{2+} ion, obtained in the same manner as Cr^{3+} above, is

$$Ni^{2+} \quad [Ar] \quad \underset{3d}{\uparrow\downarrow \;\; \uparrow\downarrow \;\; \uparrow\downarrow \;\; \uparrow \;\; \uparrow} \quad \underset{4s}{—} \quad \underset{4p}{—\;\;—\;\;—} \quad \underset{4d}{—\;\;—\;\;—\;\;—\;\;—}$$

Once again we must have a set of six vacant hybrids in order to obtain the octahedral geometry. However, this time we cannot use a pair of $3d$ orbitals to form the hybrid set. At best, we could obtain only one empty $3d$ orbital if we paired all the electrons in the $3d$ subshell. When the hybrids are created, both d orbitals must come from the *same* subshell. This means that two $4d$ orbitals must be used. The electronic structure of the complex then becomes

$$[Ni(H_2O)_6]^{2+} \quad \underset{3d}{\uparrow\downarrow \;\; \uparrow\downarrow \;\; \uparrow\downarrow \;\; \uparrow \;\; \uparrow} \;\; \overbrace{\underset{4s}{\cdot\cdot} \;\; \underset{4p}{\cdot\cdot \;\; \cdot\cdot \;\; \cdot\cdot} \;\; \underset{4d}{\cdot\cdot \;\; \cdot\cdot}}^{sp^3d^2} \quad —\;\;—\;\;—$$

Note that the complex contains two unpaired electrons. This is also in agreement with experiment.

We have now seen two complex ions that can be considered to employ an octahedral set of hybrids for bonding. In valence bond language, when $3d$ orbitals are used to form the hybrids, an **inner orbital complex** results. On the other hand, when the $4d$ orbitals are used an **outer orbital complex** is formed.

In these last two examples there really was no choice about which type of bonding (inner or outer orbital) would occur. Let's look at a situation now in which we do have a choice. An example is Co(III). The electronic structure of the Co^{3+} ion is

$$Co^{3+} \quad [Ar] \quad \underset{3d}{\uparrow\downarrow \;\; \uparrow \;\; \uparrow \;\; \uparrow \;\; \uparrow} \quad \underset{4s}{—} \quad \underset{4p}{—\;\;—\;\;—} \quad \underset{4d}{—\;\;—\;\;—\;\;—\;\;—}$$

In this case the hybrids can be formed in either of two ways. One is to make use of two $4d$ orbitals, thereby giving an outer orbital complex. The second is to pair the electrons together to produce two vacant $3d$ orbitals that can be used in the hybrids. This gives rise to an inner orbital complex.

Outer orbital

$$Co^{III}X_6 \quad \underset{3d}{\uparrow\downarrow \;\; \uparrow \;\; \uparrow \;\; \uparrow \;\; \uparrow} \;\; \overbrace{\underset{4s}{\cdot\cdot} \;\; \underset{4p}{\cdot\cdot \;\; \cdot\cdot \;\; \cdot\cdot} \;\; \underset{4d}{\cdot\cdot \;\; \cdot\cdot}}^{sp^3d^2} \quad —\;\;—\;\;—$$

Inner orbital

$$Co^{III}X_6 \quad \underset{3d}{\uparrow\downarrow \;\; \uparrow\downarrow \;\; \uparrow\downarrow} \;\; \overbrace{\underset{4s}{\cdot\cdot \;\; \cdot\cdot} \;\; \underset{4p}{\cdot\cdot} \;\; \underset{}{\cdot\cdot \;\; \cdot\cdot \;\; \cdot\cdot}}^{d^2sp^3} \quad \underset{4d}{—\;\;—\;\;—\;\;—\;\;—}$$

The Roman numeral superscript indicates the oxidation state of the cobalt.

Notice that we can distinguish experimentally between these two possibilities by examining the number of unpaired electrons in the complex. The outer orbital complex has four unpaired electrons and is paramagnetic; the inner orbital complex has none and is diamagnetic.

Both inner and outer orbital complexes are possible whenever the metal ion contains either four, five, or six d electrons. In each case, two empty $3d$ orbitals can be created by the pairing of electrons. Failure to pair them, however, leads to the formation of outer orbital complexes.

When there is a choice, what determines whether inner or outer orbital bonding will occur? To answer this question we must consider two opposing factors:

1. As stated earlier, when $3d$ orbitals are used to form the hybrid orbitals, stronger metal-ligand bonds result than when $4d$ orbitals are used. This favors the pairing of electrons and the production of inner orbital complexes.
2. The pairing of electrons required to produce the necessary vacant $3d$ orbitals for inner orbital bonding also requires an input of energy. This is because electrons, which repel each other, are being forced to occupy the same orbital. Since this pairing energy doesn't have to be invested if the $4d$ orbitals are used to form the hybrids, this factor favors the production of outer orbital complexes.

The way in which these two factors come into play is illustrated in Figure 21.17. In the first drawing we see that the energy released when the bonds are formed using hybrids composed of $3d$ orbitals is so great that it more than compensates for the pairing energy, and the resulting inner orbital complex is of lower energy than the outer orbital complex. In this case the preferred complex would be the inner orbital one. On the right in Figure 21.17 we find the other situation—the energy released on inner orbital bond formation does not lead to an overall lower energy than that achieved with the formation of an outer orbital complex. In this case, outer orbital bonding would occur in preference to inner orbital bonding.

As a general rule most ligands tend to give inner orbital complexes with first-row transition elements having either d^4 or d^6 configurations. Exceptions are the ligands H_2O and F^-, which usually produce outer orbital complexes.

With a $3d^5$ configuration the subshell is half-filled, and we have noted earlier that a half-filled subshell in which all electrons have the same spin possesses a certain extra stability. As a result, these electron configurations are difficult to disturb and electron pairing is difficult to accomplish. Consequently, metal ions with a d^5 structure tend to keep their electrons spread out with parallel spins and thus tend to form outer orbital complexes with most ligands. An exception to this occurs with cyanide ion. This particular ligand forms very stable metal-ligand bonds,[3] sufficiently stable that electron pairing of the d^5 configuration can

Lower energy means greater stability.

Figure 21.17

Energy changes in the production of inner orbital and outer orbital complexes. (a) inner orbital favored; $E_{outer} > E_{inner}$. (b) Outer orbital favored; $E_{outer} < E_{inner}$. (E_b = Energy released upon bond formation. P = pairing energy.)

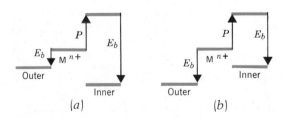

[3] Cyanides, such as KCN or HCN, are very poisonous because of the ability of CN^- to irreversibly bind to iron atoms in hemoglobin and because CN^- is able to inactivate certain enzymes.

The $[Fe(CN)_6]^{3-}$ ion is also called the *ferricyanide ion*.

occur. This is illustrated by the hexacyanoferrate(III) ion, $[Fe(CN)_6]^{3-}$. For iron(III) we have

Fe^{3+} [Ar]   ↑ ↑ ↑ ↑ ↑   __   __ __ __   __ __ __ __ __
                        3d   4s   4p   4d

and for the $[Fe(CN)_6]^{3-}$ inner orbital complex,

$$d^2sp^3$$

$[Fe(CN)_6]^{3-}$ ⇅ ⇅ ↑ ⤬⤬ ⤬⤬ ⤬ ⤬⤬ ⤬⤬ ⤬⤬   __ __ __ __ __
                       3d   4s   4p   4d

Other geometries

Valence bond theory can also be applied to geometries other than octahedral. For example, the complex ion, $[Ni(CN)_4]^{2-}$, is found to have a square planar shape and is diamagnetic. To have a square planar geometry requires a set of hybrid orbitals that point to the corners of a square. These can be obtained by mixing one *d* orbital, one *s* orbital, and two *p* orbitals to give a set of four dsp^2 hybrids. Once again, we have nickel(II) and therefore a d^8 electron configuration for the metal ion.

dsp^2 hybrids were not included in our discussion of hybrid orbitals in Chapter 6 because they are not needed to account for bonding in molecules formed by the representative elements.

Ni^{2+} [Ar]   ⇅ ⇅ ⇅ ↑ ↑   __   __ __ __
                        3d   4s   4p

We can create the one empty *d* orbital needed for the hybrids by pairing the electrons in the 3d subshell, so that for the complex we have

$$dsp^2$$

$[Ni(CN)_4]^{2-}$ ⇅ ⇅ ⇅ ⇅ ⤬⤬   ⤬⤬   ⤬⤬ ⤬⤬ __
                        3d   4s   4p

Since we have paired all the electrons, the complex should be diamagnetic, which it is.

Tetrahedral complexes such as $[CoCl_4]^{2-}$ can also be accounted for. This ion, containing cobalt(II) with a d^7 configuration, makes use of tetrahedral sp^3 hybrids and has the electronic structure

$$sp^3$$

$[CoCl_4]^{2-}$ ⇅ ⇅ ↑ ↑ ↑   ⤬⤬   ⤬⤬ ⤬⤬ ⤬⤬
                     3d   4s   4p

From our discussion, valence bond theory appears to be quite effective at accounting for the structures and magnetic properties of complex ions. However, there are some problems with the theory. One very serious drawback is that it does not allow us to explain why complex ions exist in such a wide profusion of colors, even when they contain the same metal ion in the same oxidation state. There are also certain complex ions that are quite difficult to account for in a reasonable and satisfying way. The ion $[Co(NO_2)_6]^{4-}$ is one. This ion contains Co(II), a d^7 ion,

Co^{2+} [Ar]   ⇅ ⇅ ↑ ↑ ↑   __   __ __ __   __ __ __ __ __
                        3d   4s   4p   4d

With water we saw that cobalt(II) forms an outer orbital complex containing three unpaired electrons. The $[Co(NO_2)_6]^{4-}$ ion, however, contains only one unpaired electron; therefore, it must be postulated that two of the three unpaired electrons in Co^{2+} become paired and that the third is promoted to the 4d subshell to make two $3d$ orbitals available for inner orbital complex formation. The final result looks like this:

$$[Co(NO_2)_6]^{4-} \quad \overset{\uparrow\downarrow}{\underset{3d}{}} \overset{\uparrow\downarrow}{} \overset{\uparrow\downarrow}{} \quad \overbrace{\underset{4s}{\overset{\cdot\cdot\quad\cdot\cdot}{}} \underset{}{\overset{\cdot\cdot}{}} \underset{4p}{\overset{\cdot\cdot\quad\cdot\cdot\quad\cdot\cdot}{}}}^{d^2sp^3} \quad \overset{\uparrow}{\underset{}{}} \underset{4d}{} $$

Thus with valence bond theory we can account for the magnetic properties of this ion, but they certainly are not what we would have predicted. Let us now look at another bonding theory that manages to avoid some of the pitfalls of valence bond theory as applied to these transition metal complexes.

21.11 CRYSTAL FIELD THEORY

A second theory of bonding in transition metal complexes, that has been extensively applied over the past 30 years, is called **crystal field theory** (CFT). It was developed by physicists in the early 1930s to deal with metal ions trapped in crystalline lattices, but it was not until the early 1950s that chemists realized that it could be applied to coordination complexes in general. It differs from valence bond theory in that it views the complex as held together by purely electrostatic attractions, that is, *in its simplest form, CFT ignores covalent bonding.* The most significant aspect of the theory, however, is its concern with the effect that the ligands have on the energies of the d orbitals of the metal.

Generally, the ligands in a transition metal complex are either anions, or they are polar molecules. When they are polar molecules the negative ends of the ligand dipoles point in the direction of the metal cation. Let's examine how these ligands affect the d orbitals. One of the simplest complex ions that we can consider for this purpose is the $[Ti(H_2O)_6]^{3+}$ cation, consisting of a Ti^{3+} ion surrounded octahedrally by six water molecules. Titanium(III) has a single $3d$ electron,

$$Ti^{3+} \quad \overset{\uparrow}{\underset{3d}{}} \quad \underset{4s}{} \quad \underset{4p}{}$$

Which of the five $3d$ orbitals will this electron prefer to occupy? Before we can answer this question, we must examine the shapes and directional properties of d orbitals.

The labels for the d orbitals have their origins in the mathematics of quantum mechanics.

In Chapter 4 we discussed the shapes and directional properties of the d orbitals. That description is repeated here in Figure 21.18. Notice that four of the

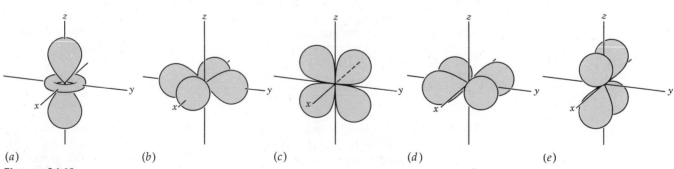

Figure 21.18

Directional properties of the d *orbitals.* (a) d_{z^2}. (b) $d_{x^2-y^2}$. (c) d_{yz}. (d) d_{xy}. (e) d_{xz}.

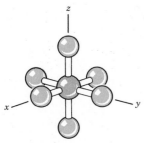

Figure 21.19

An octahedral arrangement of ligands about a central metal ion produced by placing the ligands along the x, y, and z axes.

d orbitals, labeled d_{xy}, d_{xz}, d_{yz}, and $d_{x^2-y^2}$, have the same shape, and are composed of four lobes each. The fifth, the d_{z^2}, consists of two large lobes directed along the positive and negative z axis plus a donut of charge in the xy plane. For our purposes here, it is important to notice that two of these d orbitals have lobes that are pointed along the coordinate axes—the $d_{x^2-y^2}$ and d_{z^2} orbitals. The other three—the d_{xy}, d_{xz}, and d_{yz}—have lobes that point between the axes at 45° angles to them.

Now let's consider constructing an octahedral complex by placing the six ligands along the xyz axes as shown in Figure 21.19. Since the ligands are negatively charged, or have the negative ends of their dipoles pointing at the metal ion, when the complex is formed, electrons in the d orbitals will feel an electrostatic repulsion and their energies will be raised. What is especially important, however, is that an electron in a $d_{x^2-y^2}$ or d_{z^2} orbital will be repelled *more* than an electron in one of the d_{xy}, d_{xz}, or d_{yz} orbitals because the $d_{x^2-y^2}$ and d_{z^2} orbitals point directly at the ligands (Figure 21.20). As a result, the energies of the $d_{x^2-y^2}$ and d_{z^2} are raised more than the energies of the d_{xy}, d_{xz}, and d_{yz} orbitals, as shown in Figure 21.21. This splits the d subshell into two energy levels. The lower one consists of the d_{xy}, d_{xz}, and d_{yz} orbitals and the higher one has the $d_{x^2-y^2}$ and d_{z^2} orbitals. For reasons beyond the scope of this book, in an octahedral complex the lower level is labeled t_{2g} and the upper level is labeled e_g. The

Figure 21.20

Interaction of the ligands with the d orbitals of the metal. (a) d_{z^2}. (b) $d_{x^2-y^2}$. (c) d_{yz^2}. (d) d_{xz}. (e) d_{xy}.

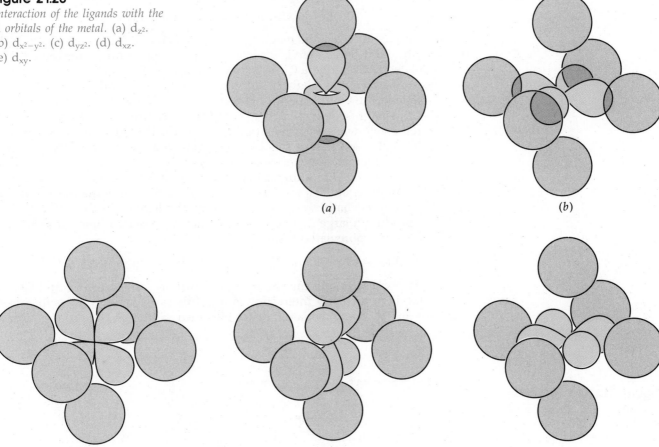

(a)

(b)

(c)

(d)

(e)

Figure 21.21

Repulsions caused by the ligands along the xyz axes raises the energy of the $d_{x^2-y^2}$ and d_{z^2} orbitals more than the energies of the d_{xy}, d_{xz}, and d_{yz} orbitals.

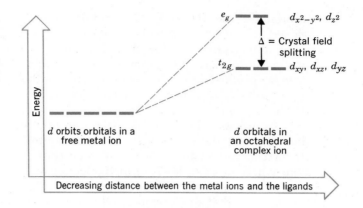

energy difference between the t_{2g} and e_g levels is called the **crystal field splitting** and is usually indicated by the symbol Δ.

Returning to the $[Ti(H_2O)_6]^{3+}$ ion, we see that its single d electron will have the lowest energy if it occupies one of the orbitals of the t_{2g} level. We can also understand what happens when the complex absorbs light (Figure 21.22). If the energy of the light that strikes the complex is equal to Δ, the light can be absorbed and the electron raised from the t_{2g} level to the e_g level. The energy of this absorbed light, of course, depends on the magnitude of Δ and, as you learned in Chapter 4, the energy of a light wave, E, is also related to its frequency, ν.

$$E = h\nu \qquad (h \text{ is Planck's constant})$$

For most complexes, the magnitude of Δ is such that the frequencies of light that are absorbed lie in the visible portion of the spectrum. Since the color of light is related to its frequency, the color of the complex depends on the frequencies absorbed when white light is reflected from it or passes through it. In other words, we see the complement of the color of the light that is absorbed. For example, the magnitude of Δ for the $[Ti(H_2O)_6]^{3+}$ ion corresponds to the energy of light in the yellow-green portion of the spectrum. Therefore, when white light passes through a solution of this complex, yellow-green light is absorbed and the light that emerges appears violet.

For a given metal ion, different ligands have different effects on the magnitude of the splitting of the d orbitals—that is, on Δ. By examining the absorption spectra of various complexes, we can arrange the ligands in the order of their ability to produce a large Δ. This series is called the **spectrochemical series** and can be given in abbreviated form as

$$I^- < Br^- < Cl^- < F^- < OH^- < H_2O < NH_3 < en < NO_2^- < CN^-$$

Thus I^- is poorest at splitting the energies of the t_{2g} and e_g levels, and CN^- is best. What is particularly interesting is that this same series applies for essentially any metal in any oxidation state. However, although the order is usually

Figure 21.22

Absorption of light ($h\nu$) by the $[Ti(H_2O)_6]^{3+}$ ion promotes the electron from the t_{2g} to the e_g level.

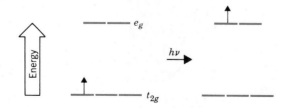

the same, the actual magnitude of Δ for a given complex in a given geometry depends on the ligand, the metal, and its oxidation state.

As with nearly any generalization we might make in attempting to describe chemical properties, there are exceptions. This is true here with the order of ligands in the spectrochemical series, because in some instances the relative positions of neighboring ligands in the series is reversed. With cobalt(III), for example, Cl^- appears to produce a greater crystal field splitting than F^-. Nevertheless, the spectrochemical series often serves as a useful guide in understanding, and sometimes even predicting, the properties of complexes. For instance, we have just seen how CFT accounts for the colors of complexes. We can explain their magnetic properties as well. Consider, for example, the cobalt(III) complexes of F^- and Cl^-. The metal ion here contains six d electrons and, in a weak crystal field—one that produces a small Δ—they will be unpaired as much as possible, as shown in Figure 21.23a, to give a complex with four unpaired electrons. This is what occurs with F^- in the $[CoF_6]^{3-}$ ion.

When the ligand produces a large crystal field splitting we have the possibility of pairing all of the d electrons in the t_{2g} level (Figure 21.23b) to produce a diamagnetic complex. This will occur if the magnitude of Δ is greater than the energy needed to pair the electrons in a given orbital. In other words, when the pairing energy (let's call it P) is less than Δ, more energy is required to place the electron in the e_g orbital than to pair them and place them in the t_{2g} level. This happens when the ligand is Cl^-.

For Co(III) complexes (or for that matter, any d^6 system), a paramagnetic complex with four unpaired electrons will occur whenever $\Delta < P$; diamagnetic complexes will be formed when $\Delta > P$. In general, we speak of these two possibilities as **high-spin** complexes (minimum pairing of electrons) and **low-spin** complexes (maximum pairing of electrons from the e_g into the t_{2g}). Comparing them to the valence bond treatment, we find that low-spin complexes correspond to inner orbital complexes whereas high-spin complexes correspond to outer orbital complexes.

The possibility of both low-spin and high-spin complexes exist when the central metal ion contains four, five, six, or seven d electrons. The electron configurations of the t_{2g} and e_g levels in these species are left to you as a problem (Problem 21.96). For d^1, d^2, and d^3 systems the electrons will naturally prefer the three low-energy t_{2g} orbitals because no pairing is required; therefore, only one type of electron configuration will be found for them. Likewise, with a d^8 or d^9 configuration, six electrons will be forced to occupy the t_{2g} (thereby filling it) and

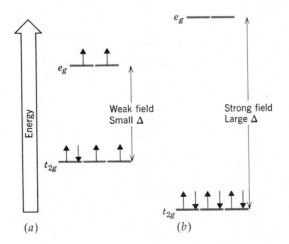

Figure 21.23

Electron configuration of Co(III) in weak and strong crystal fields.

Figure 21.24

Pairing of electrons in the t$_{2g}$ level in [Co(NO$_2$)$_6$]$^{4-}$.

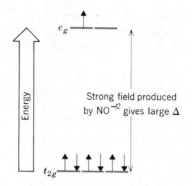

the e_g level will contain either two or three electrons, respectively. Once again we see that d^8 and d^9 ions will each have only one type of electron configuration in an octahedral complex.

With this as background, we note that the magnetic properties of the [Co(NO$_2$)$_6$]$^{4-}$ ion, which presented such a problem with the valence bond theory, are easily explained in terms of the CFT. Recall that this complex contains the Co^{2+} ion, a d^7 system. From the spectrochemical series we also note that NO$_2^-$ produces a very strong crystal field; therefore, we would expect that Δ is probably quite large, larger in fact than the pairing energy. Under these circumstances there will be pairing of electrons in the t_{2g} level, as shown in Figure

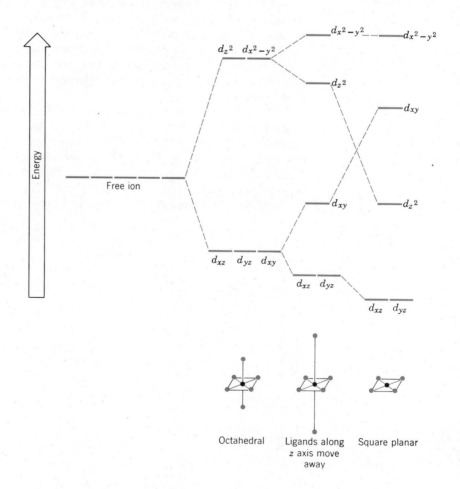

Figure 21.25

The splitting pattern of the d orbitals changes as the geometry of the complex changes.

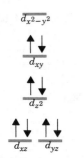

Figure 21.26

Electron distribution among the d *orbitals of the diamagnetic* $[Ni(CN)_4]^{2-}$ *complex.*

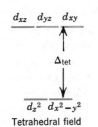

Tetrahedral field

Figure 21.27

Splitting pattern for a tetrahedral field. $\Delta_{tet} \approx \frac{4}{9}\Delta_{oct}$.

21.24. Since the t_{2g} level can accommodate only six electrons (three pairs), the seventh electron is forced to occupy the e_g level. The complex is therefore low-spin (analogous to inner orbital of the valence bond theory) and will contain a single unpaired electron, in agreement with experiment. Thus we see that there is really nothing unusual about the $[Co(NO_2)_6]^{4-}$ ion, quite opposite to what we would have concluded based on the valence bond theory.

Crystal field theory can be extended to other geometries besides octahedral; the difference is that other splitting patterns are observed. For example, a square planar complex can be thought of as being derived from an octahedral complex by removing the ligands that lie along the z axis. As shown in Figure 21.25, when this occurs the energies of the d_{z^2}, d_{xz} and d_{yz} orbitals decrease because an electron placed into them experiences less repulsion than in an octahedral complex. Also, by removing the ligands along the z axis, those along the x and y axes can move in slightly and therefore the energies of the $d_{x^2-y^2}$ and d_{xy} orbitals rise somewhat. In the $[Ni(CN)_4]^{2-}$ ion, the energy separation between the d_{xy} and $d_{x^2-y^2}$ is large enough so that the eight d electrons of the Ni^{2+} ion can exist as four pairs (Figure 21.26).

Finally, in tetrahedral complexes the splitting pattern of the d orbitals is that shown in Figure 21.27. Notice that the order of the energy levels is exactly opposite to that found in octahedral complexes. The magnitude of Δ is also considerably smaller (actually, $\Delta_{tet} \approx \frac{4}{9}\Delta_{oct}$ for the same ligands and metal ion). The small Δ observed for tetrahedral complexes is always less than the pairing energy, and tetrahedral complexes are always of the high-spin variety. Note that this agrees with the valence bond theory in which sp^3 hybrids were used by the metal thereby making all the d orbitals available for spreading out the d electrons.

REVIEW QUESTIONS AND PROBLEMS

(Problems whose numbers are in blue have their answers in Appendix C at the back of the book.)

General Properties of Transition Metals

21.1 In general, what distinguishes a transition element from a representative element?

21.2 Which elements are called *inner transition elements*?

21.3 What similarities exist between elements in the A and B groups?

21.4 Why do the Group VIII elements consist of *three* columns?

21.5 Which elements are in the iron triad?

21.6 Make a list of the d-block elements you are familiar with and give as many applications of each as you can.

21.7 What are the four general properties we normally associate with the transition metals?

21.8 What oxoanions of sulfur are analogous to CrO_4^{2-} and $Cr_2O_7^{2-}$?

21.9 What manganese compound is analogous to $KClO_4$?

Electron Configurations and Oxidation States

21.10 Write the electron configurations of the members of the first transition series.

21.11 Why do many transition metals exhibit a +2 oxidation state in their compounds?

21.12 Why do many of the transition metals exhibit multiple oxidation states?

21.13 How do the relative stabilities of high and low oxidation states vary within the d-block elements?

21.14 Which is probably the better oxidizing agent, CrO_4^{2-} or WO_4^{2-}?

21.15 Which would you expect to be the stronger oxidizing agent, Cr^{3+} or Ni^{3+}?

21.16 Which would you expect to be more easily oxidized, Cr^{2+} or Fe^{2+}?

21.17 Which would you expect to be a better oxidizing agent, Cu^{3+} or Au^{3+}?

Atomic and Ionic Radii

21.18 What is the lanthanide contraction? Why does it occur?

21.19 Why are the chemical properties of the lanthanide elements so similar?

21.20 Construct a graph showing how atomic radius varies among the *d*-block elements. Plot atomic radius vertically and atomic number, from 21 to 30, horizontally. On the graph, plot the radii of the elements for each of the three transition series. Do you see evidence of the lanthanide contraction? What general relationship is there between atomic radius and the number of unpaired *d* electrons?

21.21 The chemistries of Zr and Hf are very similar; the elements are always found together in nature and are difficult to separate from one another. What explanation can be offered to account for the similar properties of Zr and Hf compounds?

21.22 The density of molybdenum (Mo) is 10.2 g cm^{-3}. Calculate the density of tungsten and compare your value to the experimental value of 19.3 g cm^{-3}.

21.23 In general, there are only small changes in atomic size crossing a row of transition elements. Among the representative elements, the size-changes across a period are much larger. Why?

Magnetism

21.24 Which transition elements are ferromagnetic in the pure state?

21.25 Compare paramagnetism and ferromagnetism. Why can a ferromagnetic material become permanently magnetized?

21.26 What happens to the ferromagnetism of iron when the metal is melted? Why?

Properties of Transition Metals and Their Compounds

21.27 Which *d*-block elements react with water with the evolution of hydrogen? Give a chemical equation for the reaction.

21.28 CrO_3 is the anhydride of $CrO_2(OH)_2$, which was written as H_2CrO_4 in the text because it is acidic. On the other hand, Cr_2O_3 is the anhydride (at least in a formal sense) of $Cr(OH)_3$, which exhibits both basic and acidic properties. On the basis of what you learned in Chapter 7 about the factors that affect the acidity of X—O—H bonds, explain why $CrO_2(OH)_2$ is more acidic than $Cr(OH)_3$.

21.29 Based on your answer to Question 21.28, explain why the oxides of metals in high oxidation states tend to be acid anhydrides instead of basic anhydrides.

21.30 Why is TiO_2 a better paint pigment that white lead? Why is titanium metal useful in the manufacture of aircraft?

21.31 $TiCl_4$ is a low-boiling molecular substance. Why isn't $TiCl_4$ ionic?

21.32 Write the equation for the reaction of $TiCl_4$ with water.

21.33 Why is chromium used to coat other metals such as steel? What are the most important oxidation states of chromium?

21.34 What is stainless steel? How does it differ from ordinary steel?

21.35 Why is chromium(III) ion in aqueous solution acidic?

21.36 Write chemical equations showing the amphoteric behavior of chromium(III) hydroxide.

21.37 When solutions containing chromate ion (CrO_4^{2-}) are acidified, dichromate ion ($Cr_2O_7^{2-}$) is produced. Use structural formulas to indicate how this polymerization occurs. (*Hint:* How does polymerization of H_3PO_4 occur? See Section 20.1 if necessary.)

21.38 What are the principal oxidation states of manganese? Which one is most stable with respect to redox in aqueous solution?

21.39 Why is $KMnO_4$ a useful titrant for redox reactions in acidic solutions? Why is it not used for reactions in neutral or basic solutions?

21.40 Write an equation for the reaction of a strong acid, such as HCl, with manganese.

21.41 Write an equation for the disproportionation of manganate ion in an acidic solution.

21.42 Why does iron have so many practical uses?

21.43 What are the oxides of iron? Which one is magnetic?

21.44 What is the apparent mechanism for the rusting of iron in moist air?

21.45 Write a chemical equation for the reaction of hydrochloric acid with iron.

21.46 Why is cobalt an important metal? What are the principal oxidation states of cobalt?

21.47 Why is nickel such a useful metal?

21.48 Why are aqueous solutions of many nickel salts green? What practical application is there for the compound, NiO_2?

21.49 How does the ease of oxidation of the coinage metals vary? How is this related to how they are found in nature?

21.50 Give a practical application for each of the metals: copper, silver, and gold.

21.51 What oxidation states are observed for each of the coinage metals?

21.52 Why can't hydrochloric acid be used to dissolve the coinage metals? Which of them react with nitric acid? Give chemical equations.

21.53 Write a chemical equation for the reaction of gold with aqua regia.

21.54 Which compounds of silver are used in photography?

21.55 Describe how a solution can be tested for the presence of Ag^+. Give all the important chemical equations.

21.56 What happens when concentrated aqueous ammonia is added to a solution containing copper(II) ion?

21.57 What reactions occur, if any, when dilute sulfuric acid is added to (a) zinc (b) cadmium (c) mercury?

21.58 What is *cathodic protection*?

21.59 What is galvanizing? How does it protect steel?

21.60 Why is cadmium sometimes used in place of zinc as a protective coating over steel? Why is cadmium not used more often as a protective coating over other metals?

21.61 What are two common alloys that contain zinc?

21.62 What are some common uses for zinc oxide?

21.63 Aqueous solutions of $HgCl_2$ are poor conductors of electricity. Why? What equilibrium accounts for the small degree of conductivity that is observed?

21.64 Describe how a solution can be tested for the presence of Hg_2^{2+}.

21.65 Salts of which metals are added to glass to give it (a) a blue color and (b) a green color?

21.66 Saturated solutions of mercurous chloride were prepared by adding the solid to solutions having various chloride ion concentrations. The total concentration of mercury in each solution was then determined. Use the concepts that you learned having to do with equilibrium and solubility product to show that the data below are only consistent with mercury(I) having the formula Hg_2^{2+}, [and hence mercury(I) chloride being Hg_2Cl_2], and not with Hg^+ (and therefore HgCl).

Chloride Ion Concentration	Moles of Mercury per dm^3
1.0 M	2.2×10^{-18}
0.5 M	8.8×10^{-18}
0.2 M	5.5×10^{-17}
0.1 M	2.2×10^{-16}

Coordination Compounds

21.67 Define ligand, first coordination sphere, coordination compound, monodentate ligand, polydentate ligand, chelate, and coordination number.

21.68 Sketch the structures of the following ligands and circle the atoms that each ligand uses when it becomes attached to a metal ion:
(a) oxalate ion
(b) *o*-phenathroline
(c) ethylenediamine
(d) ethylenediaminetetraacetate ion

21.69 What are the full names of these ligands: (a) en and (b) EDTA?

21.70 What are some applications of coordination compounds?

21.71 What are some uses of EDTA?

Nomenclature of Coordination Compounds

21.72 Give IUPAC names for each of the following:
(a) $[Ni(NH_3)_6]^{2+}$
(b) $[CrCl_3(NH_3)_3]$
(c) $[Co(NO_2)_6]^{3-}$
(d) $[Mn(C_2O_4)_3]^{3-}$
(e) MnO_4^-

21.73 What are the IUPAC names for the following:
(a) $[AgI_2]^-$
(b) $[Cr(NH_3)_5Cl]^{2+}$
(c) $[Co(H_2O)_4(NH_3)_2]Cl_2$
(d) $[Co(en)_2(H_2O)_2]_2(SO_4)_3$
(e) $[Cr(NH_3)_4Cl_2]Cl$

21.74 Write chemical formulas for the following:
(a) tetraaquadicyanoiron(III) ion
(b) tetraammineoxalatonickel(II)
(c) potassium hexacyanomanganate(III)
(d) tetrachlorocuprate(II) ion
(e) tetraoxochromate(VI) ion

21.75 Write chemical formulas for the following:
(a) tetrachloroaurate(III) ion
(b) bis(ethylenediamine)dinitroiron(III) sulfate
(c) tetraamminecarbonatocobalt(III) nitrate
(d) ethylenediaminetetracetatoferrate(II) ion
(e) dithiosulfatoargentate(I) ion

Structure and Isomerism

21.76 Sketch the common structures found for complex ions with coordination number 4 and coordination number 6.

21.77 Sketch the structure of an octahedral EDTA complex.

21.78 Nitrilotriacetic acid, NTA (structure shown on the next page), was used by detergent manufacturers for a while in place of phosphates because it is biodegradable and does not promote the growth of algae. However, it was found to increase the solubility of some heavy metals that are poisonous to a variety of life forms, and its use has been discontinued. Can you suggest, using appropriate chemical equations and structural formulas, how NTA dissolves metal compounds?

$$
\begin{array}{c}
\text{CH}_2-\overset{\displaystyle \overset{O}{\|}}{\text{C}}-\text{O}^{(-)} \\[2ex]
\text{:N}-\text{CH}_2-\overset{\displaystyle \overset{O}{\|}}{\text{C}}-\text{O}^{(-)} \\[2ex]
\text{CH}_2-\overset{\displaystyle \overset{O}{\|}}{\text{C}}-\text{O}^{(-)}
\end{array}
$$

nitrilotriacetate ion

21.79 What is meant by *isomer*? What are stereoisomers?

21.80 Sketch the isomers of $[Co(NH_3)_2Cl_4]^-$. Identify *cis* and *trans* isomers. How many isomers are there for the complex, $[Co(NH_3)_3Cl_3]$? Sketch them.

21.81 What condition must be met for a molecule or ion to be *chiral*?

21.82 Why are chiral substances said to be optically active?

21.83 Sketch the isomers of $[Cr(en)_2Cl_2]^+$. Identify *cis* and *trans* isomers and indicate any isomers that exhibit optical isomerism.

21.84 Draw the two optical isomers of $[Co(EDTA)]^-$.

21.85 What are enantiomers? What is meant by *racemic*?

Bonding: Valence Bond Theory

21.86 What is the difference between an inner orbital complex and an outer orbital complex?

21.87 Use valence bond theory to predict the electron configuration, the type of bonding (inner orbital or outer orbital), and the number of unpaired electrons for

each of the following:
(a) $[VCl_6]^{3-}$
(b) $[Ni(NH_3)_6]^{2+}$
(c) $[Fe(NH_3)_6]^{3+}$
(d) $[Co(CN)_6]^{3-}$
(e) $[CrCl_6]^{3-}$

21.88 What magnetic properties would you predict for the square planar complex $[Cu(NH_3)_4]^{2+}$?

Bonding: Crystal Field Theory

21.89 Predict the number of unpaired electrons in (a) $[Cr(H_2O)_6]^{2+}$ and (b) $[Cr(CN)_6]^{4-}$.

21.90 Sketch on appropriate coordinate axes the shapes of the five *d* orbitals.

21.91 Diagram the crystal field splitting of the *d* orbitals, and indicate the electron population of each energy level, in the paramagnetic complex $[Mn(H_2O)_6]^{3+}$. Label the energy levels.

21.92 How does crystal field theory account for the colors of complex ions?

21.93 What relationship exists between Δ (the crystal field splitting) and the pairing energy in determining whether a given complex will be paramagnetic or diamagnetic?

21.94 What are meant by high-spin and low-spin complexes? How do these compare with inner orbital and outer orbital complexes in the valence bond theory?

21.95 Sketch the CFT splitting patterns of the *d* orbitals for
(a) square planar complexes
(b) tetrahedral complexes

21.96 Using the CFT splitting pattern for an octahedral complex, indicate the high-spin and low-spin distribution of electrons among the t_{2g} and e_g levels for the configurations: (a) d^4, (b) d^5, (c) d^6, (d) d^7.

CHAPTER 22
NUCLEAR CHEMISTRY

The Mammoth Site in Hot Springs, South Dakota is an archaeological "dig" that is open to the public. The site contains the remains of at least 34 mammoths. They died in a pond which was located here 26,000 years ago. The dating of a site such as this is often accomplished by measuring in once-living things the ratio of radioactive carbon-14 to ordinary nonradioactive carbon-12. This technique, described further in this chapter, is just one of the many practical uses of radioactivity.

In our discussions of chemistry until now we have paid little attention to the nuclei of atoms, other than to note that it is the number of positive charges on the nucleus that determine the number of electrons that a neutral atom must have. There are nuclear phenomena, however, that have applications in chemistry and one of the goals of this chapter is to explore some of them. Equally important in this nuclear age is the production of nuclear energy for both peaceful and defense applications. No matter how you feel about this topic politically, it is one that we all must face, now and in the future. Therefore, in this chapter we will also describe the origin of the tremendous quantities of energy available from nuclear reactions and attempts, both presently successful and promising for the future, at harnessing this energy for peaceful purposes.

22.1 SPONTANEOUS RADIOACTIVE DECAY

Natural radioactivity was discovered, quite by accident, by a French physicist named Antoine Henri Becquerel (1852–1908). Becquerel found that when uranium salts were left in contact with photographic plates, the plates were darkened in the same manner as when they were exposed to X rays. He reasoned correctly that uranium spontaneously emits a penetrating form of radiation that was responsible for blackening the photographic plates. Two colleagues of Becquerel, Pierre and Marie Curie, were successful in isolating two other radioactive elements from uranium ore—polonium (Po) and radium (Ra)—and found that they were even more intensely radioactive. For their discoveries, Becquerel and the two Curies were awarded the 1903 Nobel Prize for Physics.

We now know that the radiation emitted from radioactive substances is of three main types: **alpha (α) particles, beta (β) particles,** and **gamma (γ) rays.** Experimentally it is found that the alpha particle carries a positive charge, the beta particle is negatively charged, and the gamma rays carry no charge. The actual charges and masses of these and other particles with which we will be concerned are listed in Table 22.1.

When a substance spontaneously emits either an alpha or a beta particle there is a change in the charge on the nucleus, and therefore a change in atomic number. For example, nuclei of the most abundant isotope of uranium, $^{238}_{92}U$, spontaneously emit alpha particles. This removes two units of charge and four units of mass from the nucleus, and gives the isotope $^{234}_{90}Th$. Thus the emission of an alpha particle changes a uranium atom into a thorium atom—we say that $^{238}_{92}U$ *decays* to $^{234}_{90}Th$.

The changes that occur during a nuclear reaction such as the decay of $^{238}_{92}U$

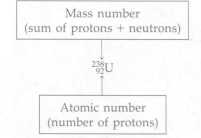

Mass number (sum of protons + neutrons)

$^{238}_{92}U$

Atomic number (number of protons)

Table 22.1
Principal types of radiation emitted by radioactive nuclei

Radiation	Approximate Mass (u)	Charge	Symbol	Type
Alpha	4	2+	^4_2He	Particle
Beta	0	1−	$^0_{-1}e$	Particle
Gamma	0	0	γ	Electromagnetic radiation
Neutron	1	0	^1_0n	Particle
Proton	1	1+	^1_1p (^1_1H)	Particle
Positron	0	1+	^0_1e	Particle

can be represented by a nuclear equation. For example,

$$^{238}_{92}U \longrightarrow \, ^4_2He + \, ^{234}_{90}Th$$
$$\alpha\text{-particle}$$

The algebraic sum of the subscripts must be the same on both sides, and the algebraic sum of the superscripts must match.

In balancing nuclear equations, notice that both the total charge and the total mass must balance.

The thorium produced in the decay of $^{238}_{92}U$ is itself radioactive and decays by beta emission. The nuclear equation for the change is

$$^{234}_{90}Th \longrightarrow \, ^{\;0}_{-1}e + \, ^{234}_{91}Pa$$
$$\beta\text{-particle}$$

Thus beta emission increases the atomic number by one unit, but has (essentially) no effect on the mass.

γ-rays are emitted by nearly all radioactive substances.

Gamma radiation, as we learned in Chapter 4, is really nothing more than a very energetic form of electromagnetic radiation. Its emission from a nucleus doesn't change the charge or mass number, so gamma radiation is often omitted from nuclear equations.

In discussions of nuclear reactions and radioactive decay, certain terms are used frequently, so you should become familiar with them. The most common is **nuclide**—a general term used when referring to the nucleus of a particular isotope. Radioactive nuclei are called **radionuclides** and the atoms having these nuclei are called **radioisotopes.** In a radioactive decay, the isotope that decays is often called the **parent isotope** and the isotope formed is referred to as the **daughter isotope.** Thus in the decay of uranium-238, the nuclide $^{238}_{92}U$ is the parent and $^{234}_{90}Th$ is the daughter.

EXAMPLE 22.1 WRITING EQUATIONS FOR NUCLEAR REACTIONS

PROBLEM (a) ^{234}U decays by alpha emissions. What is its daughter isotope?

(b) ^{214}Pb decays to ^{214}Bi. By what type of radiation is this accomplished?

SOLUTION (a) We begin by writing a balanced nuclear equation, keeping in mind that the total mass and charge must be the same on both sides. Using X to represent the symbol for the daughter isotope, we obtain

$$^{234}_{92}U \longrightarrow \, ^4_2He + \, ^{230}_{90}X$$

The element with $Z = 90$ is thorium. The daughter isotope is therefore $^{230}_{90}Th$.

(b) Again, we write a balanced nuclear equation. This time we will let X be the symbol for the type of radiation.

$$^{214}_{82}Pb \longrightarrow \, ^{214}_{83}Bi + \, ^{\;0}_{-1}X$$

From Table 22.1 we see that $^{\;0}_{-1}X$ corresponds to $^{\;0}_{-1}e$; therefore the decay occurs by beta emission.

Radioactive decay series

We have seen that ^{238}U decays to ^{234}Th, which is also radioactive and decays to ^{234}Pa. This isotope is also unstable and decays to ^{234}U (how?), which is also radioactive, and so on. This decay process continues until a stable (nonradioactive) isotope of an element is formed. This entire scheme, in which one isotope decays to another and that to another, and so on, is called a **radioactive series** or

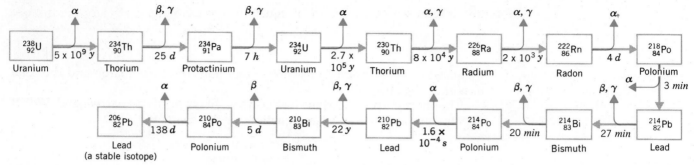

Figure 22.1

The uranium-238 radioactive disintegration series. The time given beneath each arrow is the half-life period of the preceding isotope (y = year; m = month; d = day; h = hour; min = minute; and s = second). Source: J. E. Brady and J. R. Holum, Fundamentals of Chemistry, *2nd ed., John Wiley & Sons, 1984, used by permission.*

decay series. ^{238}U, for example, decays by some 14 steps to stable ^{206}Pb, as shown in Figure 22.1.

Kinetics of radioactive decay

In Chapter 13 we saw that the rate of a chemical reaction is given by the change in the concentration of a reactant per unit time. For a radioactive substance, its concentration is directly proportional to the number of particles or rays that it emits per unit time. These emissions can be detected and counted by a suitable device such as the Geiger-Müller counter, which we will describe later. Experimentally, the concentration of the radioactive isotope is proportional to the number of counts per second (cps) or per minute (cpm) recorded by the counter.

Each *count* is a record of a nuclear event that is recorded by some measuring device that counts the individual radiations during the radioactive decay process.

If, for a particular radioactive isotope, we construct a graph in which the number of counts per minute is plotted against time, we obtain a curve similar to that shown in Figure 22.2a. If the log of the cpm is graphed versus time, a straight line is obtained as shown in Figure 22.2b. Both of these plots are indicative of a first-order process and, as it turns out, all radioactive decay reactions obey first-order kinetics.

In Chapter 13, we saw that for a first-order process the concentration of a reactant (for example, a radioactive isotope) at any time *t* is given by the equation

We could also use the equation

$$\log \frac{[A]_0}{[A]_t} = \frac{kt}{2.303}$$

$$\ln \frac{[A]_0}{[A]_t} = kt \qquad [22.1]$$

Figure 22.2

Kinetics of radioactive decay. (a) A graph of concentration expressed in counts per minute versus time. (b) A logarithmic plot of concentration versus time. The straight line indicates a first-order process.

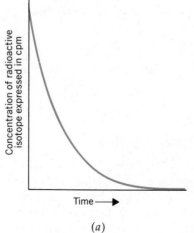

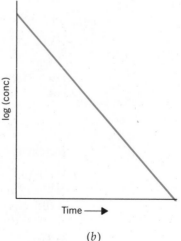

(a)

(b)

In this equation $[A]_0$ is the initial concentration of the isotope (concentration at time zero) and $[A]_t$ is its concentration at any time t. Equation 22.1 can be employed to calculate rate constants as shown in the following example.

EXAMPLE 22.2   CALCULATING THE RATE CONSTANT FOR A RADIOACTIVE DECAY

PROBLEM A chemist determined that after exactly one week an initial 10.0 μg (1 μg $= 10^{-6}$ g) of ^{222}Rn had decayed, and there was now 2.82 μg of the radon. What is the rate constant for the alpha decay of ^{222}Rn in the units d^{-1}?

SOLUTION The rate constant can easily be obtained by using Equation 22.1, where $[A]_0 = 10.0$ μg, $[A]_t = 2.82$ μg, and $t = 7.00$ d. Solving Equation 22.1 for k, we have

Since we are taking a ratio of concentrations in Equation 22.1, we can use any quantity that is proportional to the concentration in place of the concentration itself.

$$k = \frac{1}{t} \ln \frac{[A]_0}{[A]_t}$$

Substituting the values given in the problem yields

$$k = \left(\frac{1}{7.00 \text{ d}}\right) \ln \left(\frac{10.0}{2.82}\right)$$

$$k = 0.181 \text{ d}^{-1}$$

In Chapter 13 we also learned that the *half-life*, $t_{1/2}$, of a reactant in a first-order process is independent of the initial concentration of the reactant. In other words, the time required for half of a radioactive isotope to decay is the same regardless of the amount of the isotope that is present. Thus, if we were to begin with 10.0 g of a particular isotope, after one half-life only 5.0 g would remain. In a time equal to another half-life this would decay to 2.5 g, and so on. This is represented graphically in Figure 22.3. In this figure we see in general how an isotope decays through many half-lives. We have seen that the half-life of any substance is inversely proportional to the rate constant for its decay. The relationship is

$$t_{1/2} = \frac{0.693}{k} \tag{22.2}$$

From this equation we see that if the rate constant for decay is large, its half-life will be small and vice versa. Also, we see that if either $t_{1/2}$ or k is known, the other can be calculated.

Figure 22.3

A graph of a first-order radioactive decay illustrating the concept of half-life. The initial concentration is a, which drops to $\frac{a}{2}$ after one half-life, to $\frac{a}{4}$ after the second half-life, and so forth.

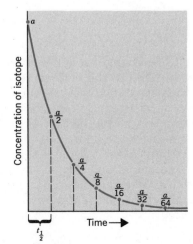

EXAMPLE 22.3 USING THE RATE CONSTANT FOR RADIOACTIVE DECAY

PROBLEM
The rate constant for the decay of ^{222}Rn is 0.181 d^{-1}.

(a) What is the half-life for ^{222}Rn expressed in days?

(b) What fraction of a sample of ^{222}Rn decays in a period of one week?

SOLUTION
(a) Substituting the value of k into Equation 22.2, we obtain

$$t_{1/2} = \frac{0.693}{0.181 \text{ d}^{-1}}$$

which gives

$$t_{1/2} = 3.83 \text{ d}$$

(b) We first have to calculate the fraction left after decay. By this we mean the ratio $[A]_t/[A]_0$. Therefore we employ Equation 22.1 for this purpose and solve it for $[A]_0/[A]_t$.

$$\ln \frac{[A]_0}{[A]_t} = kt$$

Substituting in $k = 0.181$ d^{-1} and $t = 7.00$ d, we have

$$\ln \frac{[A]_0}{[A]_t} = (0.181 \text{ d}^{-1})(7.00 \text{ d})$$

or

$$\ln \frac{[A]_0}{[A]_t} = 1.267$$

Taking the antilog gives

$$\frac{[A]_0}{[A]_t} = 3.55$$

The reciprocal, therefore, is

$$\frac{[A]_t}{[A]_0} = 0.282$$

This is the fraction left after one week, so the fraction that decayed is simply $1 - 0.282$, or 0.718.

Measurement of radioactivity

In order to study nuclear changes it is necessary to measure radiation and express it quantitatively. One of the earliest devices used to detect radioactive emissions was the Geiger–Müller counter, shown schematically in Figure 22.4. In this device, alpha or beta particles enter a chamber filled with low-pressure argon through the thin window at the left. As they pass through the argon they knock electrons off the gaseous atoms, leaving behind a trail of positive argon ions and electrons. The presence of these charged particles causes the gas to suddenly become conducting, and this allows a discharge—a spark—to jump between the electrodes. Sensitive electronics (not shown in the figure) detect this momentary small burst of current and register it on some sort of meter or other counting device.

Figure 22.4

The Geiger–Müller counter. An alpha or beta particle enters the Geiger tube through the thin window shown at the left of the apparatus. As the particle passes through the gas inside the tube, it ionizes argon atoms along its path. These ions cause an electrical breakdown (discharge) between the wire and the wall of the tube, thereby producing a current pulse. This current pulse is readily amplified and counted electronically.

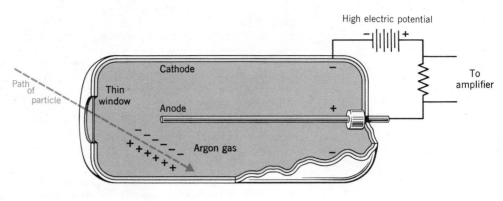

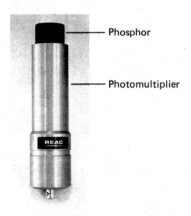

Figure 22.5

A typical scintillation probe used to measure radioactive emissions. Pulses of light are amplified first by a photomultiplier and then detected and counted electronically.

Newer methods of counting radioactive emissions often make use of **scintillation counters.** In these devices, the radiation is absorbed and causes the emission of a photon of visible light. For example, if beta particles are being studied, a zinc sulfide phosphor can be used. It is this production of visible light that is called *scintillation.* Once the light is generated, it can be detected by photo-optical methods and the signal produced in these light detectors is then amplified. Figure 22.5 shows a typical scintillation probe.

In working with radioactive substances, it is desirable (and usually necessary) to be concerned with the quantity of radiation emitted by the material. We express this by referring to its *activity,* which is the number of nuclear disintegrations or other nuclear changes that occur per second in the sample under study. The SI unit of activity is called the **becquerel (Bq),** named after Henri Becquerel, the discoverer of radioactivity. A becquerel is defined as one nuclear disintegration (or other nuclear change that produces radiation) per second.

$$1 \text{ Bq} = 1 \text{ disintegration/s}$$

An older non-SI unit that is still widely used is the **curie (Ci),** which is the number of disintegrations per second in one gram of radium. (In other words, the activity in Ci is an activity relative to radium as a standard.) In one gram of radium there are 3.7×10^{10} disintegrations per second, so the relationship between the curie and the becquerel is

$$1 \text{ Ci} = 3.7 \times 10^{10} \text{ Bq}$$

Another measure of activity that is sometimes convenient to use is **specific activity,** which is the number of disintegrations per second per gram of sample. In SI units it would be Bq g^{-1}.

We are often interested in knowing not only the number of emissions from a radioactive source, but also the quantity of radiation absorbed by some object in the path of the radiation. We refer to this as the **absorbed dose** or simply the **dose.** In the SI, absorbed dose is expressed in terms of the quantity of energy (in joules) absorbed per kilogram of absorbing material. The unit is called the **gray (Gy)** and is named after British radiologist Harold Gray.

$$1 \text{ Gy} = 1 \text{ J kg}^{-1}$$

The gray does not differentiate between the effectiveness of various forms of radiation at causing damage to animal tissue, and that is certainly an important aspect of our concern about the quantity of radiation that we are exposed to. Alpha, beta, and gamma radiation are absorbed to different degrees by tissue and therefore penetrate our bodies to different depths. Table 22.2 gives the

Table 22.2
Penetrating ability (approximate) of various kinds of radiation

Radiation Type	Approximate Depth of Penetration[a]		
	Dry Air	Animal Tissue	Lead
Alpha	4 cm	0.05 mm	0
Beta	6 cm to 300 cm	0.06 mm to 4 mm	0.005 mm to 0.3 mm
Gamma (Data for reduction of intensity by 10%)	400 m	50 cm	30 mm

[a] Depth of penetration depends also on the initial energy of the radiation.

depth of penetration of these forms of radiation in some familiar materials. From this table, we can see that alpha radiation is far less dangerous to us than gamma radiation. Whatever damage is done by alpha radiation occurs at the very surface of our skin and only in rare cases does this extend below our outer layer of dead skin cells. Gamma radiation, however, passes through this protective layer of skin and can cause damage to cells within our inner organs. This damage involves the breaking of chemical bonds and the production of free radicals (these were discussed in Chapter 13).

A unit that has been defined to take into account the *kind* of radiation that is absorbed by animal tissue is the **rem,** which stands for **r**adiation **e**quivalent for **m**an. To obtain a dose in rem, the absorbed dose is multiplied by an appropriate factor for the kind of radiation absorbed.

22.2 APPLICATIONS OF NUCLEAR REACTIONS

One of the most interesting applications of radioactive decay is in the dating of ancient objects of either natural origin, such as ancient rocks, or of human origin, such as objects made by prehistoric people. In these applications use is made of the known and constant rate of decay of radioactive isotopes. For example, the age of rocks containing uranium can be determined by measuring the ratio of ^{238}U to ^{206}Pb. Remember that ^{206}Pb is the stable isotope to which ^{238}U eventually decays. Using a $^{238}U/^{206}Pb$ ratio of 1 to 0 as corresponding to the ratio at zero time and a $^{238}U/^{206}Pb$ ratio of 1 to 1 as corresponding to the ratio after one half-life—that is, 4.5×10^9 years—we can closely approximate the age of these rocks. The oldest rocks that have been found on earth have an age of 3.9×10^9 years using this method.

Rocks that do not contain uranium may be dated using a potassium-argon method. This makes use of the reaction

$$^{40}_{19}K + {}^{0}_{-1}e \longrightarrow {}^{40}_{18}Ar \quad t_{1/2} = 1.3 \times 10^9 \text{ y}$$

The electron in this case comes from the 1s orbital of the potassium and is captured by the unstable $^{40}_{19}K$ nucleus. In the dating procedure the ratio of ^{40}K to ^{40}Ar is measured, and the age of the rock is determined in the same manner as the uranium dating described above.

The age of materials that were once living, such as bones and wood, can be estimated quite accurately by measuring their ratio of ^{14}C to ^{12}C. Carbon-14 is radioactive and is constantly being produced in the upper atmosphere by the

^{14}C dating is reliable provided the object isn't more than 70,000 years old.

bombardment of cosmic neutrons upon $^{14}_{7}N$, which is present there in large amounts. The equation for this reaction is

$$^{14}_{7}N + ^{1}_{0}n \longrightarrow ^{14}_{6}C + ^{1}_{1}p$$

The carbon-14 thus produced immediately begins to decay.

$$^{14}_{6}C \longrightarrow ^{14}_{7}N + ^{0}_{-1}e \quad t_{1/2} = 5770 \text{ y}$$

The steady-state concentration of ^{14}C is 15 cpm per gram of carbon. This is an activity of 0.25 Bq g^{-1}.

Because ^{14}C is being both formed in the atmosphere and removed by its decay, a constant concentration is maintained. We say that a *steady-state* concentration is achieved. This ^{14}C becomes incorporated in carbon dioxide in the atmosphere where it can be taken in by plants through the process of photosynthesis. The intake of ^{14}C into animals is by the consumption of such plants or by the consumption of plant-eating animals. While they are alive, plants and animals consume and excrete carbon so that they also maintain a steady-state concentration of ^{14}C and are thus in equilibrium with their surroundings. Once they die, however, the ^{14}C that they possess is not replaced as the organisms decay, so the ^{14}C concentration begins to decrease. The half-life of the ^{14}C is 5770 y; therefore, if we find that the carbon-14 concentration in an object that had once been living has dropped to half its initial value, we could conclude that the object is 5770 years old.

Atmospheric testing of nuclear weapons has made it impossible for our current history to be dated this way by future archaeologists.

EXAMPLE 22.4 — USING THE KINETICS OF RADIOACTIVE DECAY FOR ARCHAEOLOGICAL DATING

PROBLEM A piece of charcoal from the ruins of a settlement in Japan was found to have a $^{14}C/^{12}C$ ratio that was 0.617 times that found in living organisms. How old is this piece of charcoal?

SOLUTION The answer to this problem can be obtained once again by employing Equations 22.2 and 22.1. From Equation 22.2 we can obtain the k for the decay of ^{14}C.

$$k = \frac{0.693}{5770 \text{ y}} = 1.20 \times 10^{-4} \text{ y}^{-1}$$

Next, we need values to substitute for $[A]_0$ and $[A]_t$. Fortunately, we don't need actual values for them. If we *arbitrarily* take the $^{14}C/^{12}C$ ratio in a living organism to be equal to 1.000, then according to the data in the problem, the charcoal has a $^{14}C/^{12}C$ ratio of $0.617 \times 1.000 = 0.617$. Taking the ratio of these numbers is the *same* as taking the ratio of the concentrations. Therefore, substituting into Equation 22.1, we have

$$\ln \frac{[A]_0}{[A]_t} = kt$$

$$\ln \left(\frac{1.000}{0.617} \right) = \ln 1.62 = 0.482 = (1.20 \times 10^{-4} \text{ y}^{-1})t$$

or

$$t = 4020 \text{ years}$$

Chemical applications

Ordinarily, nuclear changes such as those involved in radioactive decay have very little direct effect on chemical reactions, although in some cases the high-energy radiation emitted by radioactive nuclei can influence the products of

DNA is the genetic material in cells. Its functions are described in Chapter 24.

reaction. This radiation is generally capable of disrupting chemical bonds. If the cleavage of a chemical bond occurs in DNA, for example, mutations of the DNA strand can be brought about. Mutational changes can also take place by reactions between DNA and the products of other cleavage reactions. For instance, free radicals are produced by splitting an H—O bond in water. This takes place through a series of steps, the first of which is the absorption of radiation by the water molecule and the subsequent ejection of an electron

$$\underbrace{H_2O + h\nu}_{\substack{\text{energy from}\\\text{radiation}}} \longrightarrow H_2O^+ + e^-$$

The final products of the reaction are a hydrogen atom, H· (the dot indicates the unpaired electron in the radical) and a *hydroxyl radical,* ·OH. The net overall change is

$$H_2O + h\nu \longrightarrow H· + ·OH$$

Free radicals are extremely reactive because of the presence of their unpaired electrons, and their interactions with DNA can cause mutations or otherwise disrupt the replication of the DNA strands.

It is only in rare cases that the chemist makes direct use of the energy emitted in nuclear transformations. Most chemical applications of radioactive nuclides stem from their ease of identification and detection, even when they are present in very small amounts. Hence radioactive isotopes are usually employed in **tracer studies,** where they may be added in very small amounts and used to follow, or trace, the course of a chemical reaction. The range of applications of these tracer techniques is limited only by the imagination and ingenuity of the experimenter. Let's take a brief look at some examples that demonstrate the scope of these applications.

Analytical chemistry

There are many examples of analytical uses for radioactive isotopes. One of these techniques, called **isotope dilution,** can be used when it is impossible to completely separate a desired substance from a mixture. In this case, a small measured amount of the substance containing a known quantity of a radioactive isotope is *added* to the mixture. After making sure that complete mixing has occurred, a small amount of the *pure* desired substance is separated from the mixture. This sample will contain some of the added radioactive isotope, and from the proportion of the labeled isotope present in the sample, the total quantity of the substance in the original mixture can be computed.

Consider, for instance, a mixture of salts of similar solubilities, such as a mixture of KNO_3 and $NaCl$. By fractional crystallization only a portion of the KNO_3 can be separated from the mixture. As a result, we cannot determine, in a simple fashion, how much of this salt is in the mixture.

Suppose, now, that 1.0 g of KNO_3 containing a small amount of radioactive ^{40}K is added to the salt mixture and then some KNO_3 (now containing K from the original mixture as well as from the added tagged KNO_3) is separated by fractional crystallization. If the specific activity of this KNO_3—that is, the number of counts per minute per gram (cpm/g)—has dropped to 1% of the specific activity of the added KNO_3, then we know that only 1% of the added solid has been recovered in our KNO_3 sample and the other 99% of the KNO_3 must have been present in the original mixture. In other words, after we had added the 1 g of labeled KNO_3 there was a 99-to-1 ratio of unlabeled to labeled salt. Therefore the original mixture must have contained 99 g of KNO_3.

Isotope dilution methods are also used when the volume of a liquid in an irregular container must be measured. A small known volume of radioactive material is added and after mixing is complete, the extent of dilution allows one to calculate backward to find the initial liquid volume. This method has been used to measure blood volumes in living animals and the volumes of underground reservoirs of water.

Another technique applicable to analytical chemistry is called **neutron activation analysis.** When nonradioactive isotopes are bombarded by neutrons, heavy isotopes of these elements can be produced. The product of this reaction may be unstable and hence radioactive. Even if another nonradioactive nucleus is produced, however, the absorption of these neutrons generally gives nuclei that are excited and emit gamma radiation in much the same way that an excited atom emits light when it returns to the ground state.

$$\ce{^{A}_{Z}X} + \ce{^{1}_{0}}n \longrightarrow \ce{^{A+1}_{Z}X^*} \qquad \text{(the asterisk indicates an excited nucleus)}$$

$$\ce{^{A+1}_{Z}X^*} \longrightarrow \ce{^{A+1}_{Z}X} + h\nu \qquad (h\nu = \gamma \text{ photon})$$

Since each element has its own characteristic gamma-emission spectrum, an analysis of the energies of the gamma emissions from the activated sample allows its composition to be determined. In addition, from the intensity of the emitted gamma radiation, the concentration of each element can be computed.

This technique has some very useful advantages. First, it is nondestructive. Since the number of nuclei that must be activated to perform the analysis is small, most of the sample is unaffected. Second, as implied in the preceding sentence, the method is very sensitive and is, therefore, well suited to the analysis of trace amounts of impurities. In some cases, sensitivities of the order of 10^{-12} g can be achieved.

Descriptive chemistry

Many elements having atomic numbers greater than $Z = 83$ (bismuth) have short half-lives and, therefore, are not observed to occur naturally. Instead, they must be synthesized in particle accelerators. As a result, only extremely small quantities of these elements have ever been prepared. A question then arises: How can we study their chemistry if we cannot even obtain enough to be able to see them?

To arrive at the solution to this problem, let us consider the element astatine. Astatine was first produced in a device called a cyclotron by the reaction

$$\ce{^{209}_{83}Bi} + \ce{^{4}_{2}He} \longrightarrow \ce{^{211}_{85}At} + 2\ce{^{1}_{0}}n$$

in which the ^{211}At produced has a half-life of only about 7.5 h. The most stable isotope, ^{210}At, has a half-life of only 8.3 h, so large quantities of the element cannot be accumulated.

Since astatine occurs in Group VIIA, we expect the element to be similar in some of its properties to iodine. To verify this the astatine is added as a tracer in reactions involving iodine, and the fate of the At is followed as the iodine undergoes reactions. If in a given reaction the At occurs in the products along with the iodine, we conclude that, in this reaction, At behaves just as I does. We have discovered something about the chemical behavior of an element that we cannot even see. For instance, it is observed that, like iodine, elemental astatine is rather volatile, since it is carried with the iodine when I_2 is sublimed. In solution At^- is carried from solution along with I^- upon the addition of Ag^+. Thus we conclude that AgAt is insoluble just as is AgI.

How would you study the chemical properties of francium, Fr ($Z = 87$)?

Reaction mechanisms

In Chapter 13 we saw that a study of the effect of the concentrations of the reactants on the rate of a chemical reaction can often give some insight into the mechanism of the reaction. Such studies, however, seldom answer all the questions that we might ask about the reaction mechanism. Consider, for example, the reaction of an alcohol and an organic acid to produce a type of compound called an ester and water.

$$C_2H_5\text{—OH} + \underset{\text{acid}}{HO\text{—}\overset{\displaystyle O}{\overset{\|}{C}}\text{—}CH_3} \longrightarrow \underset{\text{ester}}{C_2H_5\text{—O}\text{—}\overset{\displaystyle O}{\overset{\|}{C}}\text{—}CH_3} + H_2O$$

Upon the formation of the ester molecule, two hydrogen atoms and one oxygen atom are removed from the alcohol and acid to become a molecule of water. There seems little doubt about the origin of the two hydrogen atoms; however, there is a question about which one of the —OH oxygen atoms is removed and finds its way into the H_2O molecule.

This question can be resolved by carrying out the reaction with a labeled oxygen (for example, ^{18}O) incorporated into the OH group of either the alcohol or the acid. For instance, if the alcohol is labeled with ^{18}O, it is found that all the labeled oxygen becomes incorporated into the ester. On the other hand, if the acid contains ^{18}O in the OH group, all the labeled oxygen ends up in the water with none in the ester. It is clear, therefore, that the reaction involves the removal of the OH from the acid and the H from the alcohol.

Reaction using labeled alcohol

$$C_2H_5\text{—}\overset{*}{O}\text{—H} + H\text{—O}\text{—}\overset{\displaystyle O}{\overset{\|}{C}}\text{—}CH_3 \longrightarrow C_2H_5\text{—}\overset{*}{O}\text{—}\overset{\displaystyle O}{\overset{\|}{C}}\text{—}CH_3 + H_2O$$

Reaction using labeled acid

$$C_2H_5\text{—O}\text{—H} + H\text{—}\overset{*}{O}\text{—}\overset{\displaystyle O}{\overset{\|}{C}}\text{—}CH_3 \longrightarrow C_2H_5\text{—O}\text{—}\overset{\displaystyle O}{\overset{\|}{C}}\text{—}CH_3 + H_2O^*$$

Many similar experiments using tagged atoms have been employed to aid in the elucidation of a large number of reaction mechanisms, including biological and biochemical processes. For instance, labeled water can be added to the root system of a plant, and its progression into the stem—and ultimately into the leaves—can be traced. Experiments using ^{14}C-labeled CO_2 have been used to follow the course of carbon in photosynthesis in plants. In this case plants are exposed to CO_2 and, at various intervals, are killed and their cellular components separated to determine which compounds have had ^{14}C built into them. In this way the sequence of reactions in photosynthesis can be unraveled.

22.3 NUCLEAR STABILITY

Experimentally it is observed that all the elements with atomic numbers greater than 83 (bismuth) are radioactive and possess no known stable isotopes. On the other hand, all the lighter elements, with the exception of technetium (Z = 43) and promethium (Z = 61), have one or more stable, nonradioactive isotopes. In addition, radioactive isotopes undergo nuclear transformations that lead ultimately to stable nuclei. Sometimes this is accomplished by a simple one-step process, while in other cases a series of nuclear reactions occur before a stable isotope is reached. A question that naturally arises from these observations is, what factors give rise to stable or unstable nuclei?

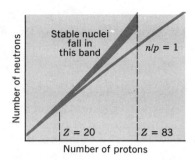

Figure 22.6
Band of stability.

There are some interesting facts about nuclear stability that emerge if we study the numbers of protons and neutrons that are found in stable nuclei. For example, if a graph is made of the numbers of neutrons (vertical axis) versus the numbers of protons (horizontal axis) that are observed for stable nuclei, we find that all the stable nuclei fall in a relatively narrow band, as shown in Figure 22.6. This band has been called a **band of stability**. In Figure 22.6, we see that at low atomic numbers stable nuclei possess approximately equal numbers of protons and neutrons. Above about $Z = 20$, however, the number of neutrons always exceeds the number of protons, and the neutron-to-proton ratio gradually increases to about 1.5 at the upper end of the band of stability. Apparently, as the number of protons in the nucleus increases, there must be more and more neutrons present to help overcome the strong repulsive forces between the protons. It also seems that there is an upper limit to the number of protons that can exist in a stable nucleus, with that number being reached at bismuth.

Nuclei that lie outside the band of stability are unstable and decay in a manner that tends to give them a stable neutron-to-proton (n/p) ratio. On this basis, we can understand why certain nuclei undergo the type of radioactive decay that they do. For instance, a nucleus that lies above the band of stability must either lose neutrons or gain protons to achieve stability. Thus we can understand why elements such as ^{14}C (which lies above the band) decay by β-emission, because this process converts a neutron into a proton (^1_1p).

$$^1_0n \longrightarrow {}^1_1p + {}^0_{-1}e$$

For ^{14}C we have

$$^{14}_6C \longrightarrow {}^{14}_7N + {}^0_{-1}e$$

Another way that an element located above the band can achieve a stable n/p ratio is by emitting a neutron, although this particular mode of decay is rare. An example is the decay of ^{137}I.

$$^{137}_{53}I \longrightarrow {}^{136}_{53}I + {}^1_0n$$

Elements located *below* the band of stability must increase their n/p ratio to achieve stability. This is accomplished generally in either of two ways. One involves the emission of a positron—a particle having the same mass as the electron but with a unit positive charge. The positron is symbolized as ^0_1e. The ejection of a positron by an unstable nucleus converts a proton into a neutron

$$^1_1p \longrightarrow {}^1_0n + {}^0_1e$$

An example is the decay of ^{11}C.

$$^{11}_6C \longrightarrow {}^{11}_5B + {}^0_1e$$

The second mode of decay that results in an increased n/p ratio is called **electron capture.** In this case the unstable nucleus captures an electron, usually from its own $1s$ orbital. Since the captured electron most often originates in the K shell, the process is also called **K-capture**. The addition of this electron to the nucleus transforms a proton into a neutron.

$$^1_1p + {}^0_{-1}e \longrightarrow {}^1_0n$$

Two examples of decay by K-capture are

$$^7_4Be + {}^0_{-1}e \xrightarrow{\text{K-capture}} {}^7_3Li$$

and

$$^{40}_{19}K + {}^0_{-1}e \xrightarrow{\text{K-capture}} {}^{40}_{18}Ar$$

The *K*-shell has $n = 1$.

The vacancy created in the 1*s* subshell as a result of *K*-capture is only temporary, and electrons from higher energy levels quickly drop to fill the 1*s* orbital. Since electrons are falling from higher energy levels to lower ones, energy is emitted in the form of electromagnetic radiation (light)—in this instance in the X-ray region of the spectrum.

Elements having atomic numbers higher than 83—that is, those beyond the end of the band of stability—cannot find their way to a stable *n/p* ratio by any of the decay modes we just discussed. In these cases the unstable nuclei must lose both protons *and* neutrons. As a result, their decay usually involves emission of α-particles, since each α-emission removes two protons and two neutrons simultaneously. Earlier, for example, we saw this type of decay process for uranium,

$$^{238}_{92}\text{U} \longrightarrow {}^{4}_{2}\text{He} + {}^{234}_{90}\text{Th}$$

Another type of nuclear transformation that is available to the heavy elements is **fission,** in which a heavy nucleus splits into several much lighter fragments, many of which may also lie outside the band of stability and hence may be radioactive. The smaller nuclei that are produced by fission, if they are unstable, are able to undergo the simpler types of decay in order to produce a stable nucleus. We will take a closer look at nuclear fission in Section 22.7.

We may also observe that nuclei with even numbers of protons and neutrons are apparently more stable than those containing an odd number of these particles. For example, there are 157 stable isotopes in which there are even numbers of both protons and neutrons, 52 isotopes having an even number of protons and an odd number of neutrons, and 50 with an even number of neutrons but an odd number of protons. By contrast, there are only five stable nuclides in which there are odd numbers of both protons and neutrons.

Protons	Even	Even	Odd	Odd
Neutrons	Even	Odd	Even	Odd
Stable nuclei	157	52	50	5

This phenomenon suggests that in stable nuclei, protons and neutrons each tend to be paired, in much the same way that electrons become paired in the outer region of an atom. Apparently, extra stability, as evidenced by the number of stable nuclides, results when pairing takes place with both protons and neutrons. On the other hand, when pairing cannot occur, as must be true when the numbers of protons and neutrons are both odd, very few stable isotopes occur (most isotopes having odd numbers of both protons and neutrons are radioactive).

A final observation on nuclear stability is that nuclei that contain certain specific numbers of protons and neutrons possess a degree of extra stability. These so-called *magic numbers* for protons and neutrons are 2, 8, 20, 28, 50, and 82, with an additional magic number of 126 for neutrons. When nuclei contain a magic number of both protons and neutrons, they are said to be *doubly magic* and are extremely stable. Examples are ${}^{4}_{2}\text{He}$, ${}^{16}_{8}\text{O}$, ${}^{40}_{20}\text{Ca}$, and ${}^{208}_{82}\text{Pb}$.

The occurrence of these magic numbers suggests a shell structure for the nucleus somewhat akin to the shell structure exhibited by electrons. For example, we have seen that very stable (unreactive) electron configurations occur when an atom contains magic numbers of 2, 8, 18, 36, or 54 electrons, corresponding to the noble gases, He through Kr. In the nucleus, it seems that nuclear shells of either protons or neutrons become completed when the nuclear magic numbers are reached and that a particularly stable nucleus occurs whenever there is a completed shell of either neutrons or protons. Exceptionally stable nuclei result when the nucleus contains filled shells of protons and neutrons simultaneously.

22.4   NUCLEAR TRANSFORMATIONS

A **nuclear transformation** is a nuclear reaction in which a bombarding particle is absorbed and causes the absorbing nucleus to change into a nucleus of another element. Lord Rutherford was the first scientist to cause a nuclear transformation. He found that by bombarding $^{14}_{7}\text{N}$ with alpha particles he produced the isotope of oxygen, $^{17}_{8}\text{O}$. The equation for this reaction is written as

$$^{14}_{7}\text{N} + ^{4}_{2}\text{He} \longrightarrow [^{18}_{9}\text{F}] \longrightarrow ^{17}_{8}\text{O} + ^{1}_{1}\text{H}$$

The square brackets indicate that the isotope $^{18}_{9}\text{F}$ is a very unstable intermediate that rapidly decays to the products above.

In 1933 Irène Curie and her husband Frédéric Joliot showed that other light elements could similarly be transformed by bombardment with α-particles. For example, $^{27}_{13}\text{Al}$ can be transformed into $^{30}_{15}\text{P}$ by this process.

$$^{27}_{13}\text{Al} + ^{4}_{2}\text{He} \longrightarrow ^{30}_{15}\text{P} + ^{1}_{0}n$$

Reactions of this type, where an α-particle is used for bombardment and a neutron is one of the products, is known as an *alpha, neutron reaction*—symbolized by (α,n). A shorthand notation for the reaction is $^{27}_{13}\text{Al}(\alpha,n)^{30}_{15}\text{P}$.

Since 1933 many isotopes have been produced by bombardment reactions, some in which particles other than α-particles have been used. One of the main problems with such experiments is that a positively charged nucleus is being bombarded with positively charged particles. Heavy elements, with their very highly positive nuclei, repel particles with positive charges like the α-particle. One way to circumvent this problem is to use neutrons as the bombarding particles.

Neutrons, which have no charge, are not repelled by the nucleus and, therefore, are excellent for bombardment reactions. The supply of neutrons for these reactions can be obtained from either of two sources—a transformation reaction in which neutrons are produced or a nuclear reactor in which fission reactions (which will be examined in Section 22.7) occur at a controlled rate. Examples of neutron-producing reactions are the $^{27}_{13}\text{Al}(\alpha,n)^{30}_{15}\text{P}$ reaction mentioned above and

$$^{9}_{4}\text{Be} + ^{4}_{2}\text{He} \longrightarrow ^{12}_{6}\text{C} + ^{1}_{0}n$$

in which the beryllium isotope undergoes an α,n reaction.

A second way that nuclear transformations can be brought about is through the use of particle accelerators. Particle accelerators, such as the cyclotron illustrated in Figure 22.7, speed up particles to extremely high velocities and then

Irène Joliot-Curie was the daughter of Marie and Pierre Curie, the discoverers of polonium and radium.

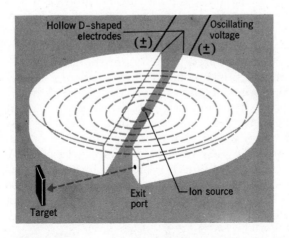

Figure 22.7

Diagram of a cyclotron.

Table 22.3
Some elements produced in particle accelerators

$$^{238}_{92}U + ^4_2He \longrightarrow ^{239}_{94}Pu + 3\,^1_0n$$

$$^{239}_{94}Pu + ^4_2He \longrightarrow ^{240}_{95}Am + ^1_1H + 2\,^1_0n$$

$$^{239}_{94}Pu + ^4_2He \longrightarrow ^{242}_{96}Cm + ^1_0n$$

$$^{244}_{96}Cm + ^4_2He \longrightarrow ^{245}_{97}Bk + ^1_1H + 2\,^1_0n$$

$$^{238}_{92}U + ^{12}_6C \longrightarrow ^{246}_{98}Cf + 4\,^1_0n$$

$$^{238}_{92}U + ^{14}_7N \longrightarrow ^{247}_{99}Es + 5\,^1_0n$$

$$^{238}_{92}U + ^{16}_8O \longrightarrow ^{249}_{100}Fm + 5\,^1_0n$$

$$^{253}_{99}Es + ^4_2He \longrightarrow ^{256}_{101}Md + ^1_0n$$

$$^{246}_{96}Cm + ^{13}_6C \longrightarrow ^{254}_{102}No + 5\,^1_0n$$

$$^{252}_{98}Cf + ^{10}_5B \longrightarrow ^{257}_{103}Lw + 5\,^1_0n$$

$$^{249}_{98}Cf + ^{12}_6C \longrightarrow ^{257}_{104}Unq + 4\,^1_0n$$

direct them at target nuclei. At these speeds the positive particles are able to overcome the coulombic repulsion of a nucleus and collide with it. Accelerators have been used extensively by Dr. Glenn Seaborg and his colleagues at the University of California in producing many of the *transuranium elements*—that is, elements 93 to 106. In these accelerators positive ions of such isotopes as ^2_1H (deuterium), $^{12}_6C$, $^{13}_6C$, $^{16}_8O$, $^{14}_7N$, $^{10}_6B$, as well as ^4_2He have been used in producing new, artificial elements. A list of these bombardment reactions and the elements they produce is shown in Table 22.3. Many of these elements are extremely short-lived and, as a result, only a few atoms, especially of the high-atomic-numbered isotopes, have ever been formed.

22.5 EXTENSION OF THE PERIODIC TABLE

The heaviest naturally occurring element is uranium and, as we saw in the previous section, elements beyond $Z = 92$ are all prepared artificially by bombarding lighter nuclei with protons, alpha particles, and the positive ions of some of the second-period elements. The discovery of these new elements quite expectedly prompted chemists to begin to think about a whole host of new elements with new and interesting properties to be studied. However, it soon became apparent that as the atomic number of the artificial element became higher, its half-life became shorter, so the prospects for stable elements of very high atomic numbers became dim.

Calculations by many nuclear physicists, based on the nuclear shell model, now suggest that a closed nuclear shell for protons exists at $Z = 114$ and that one for neutrons occurs at 184. As a result, chemists have once again begun to speculate about the possibilities of new stable elements.

One proposed extension of the periodic table to include these heavy and superheavy elements is shown in Figure 22.8. Recall that the actinide series, which occurs as the result of the filling of the $5f$ subshell, ends at lawrencium, $Z = 103$. The next element, 104, therefore lies under hafnium if we follow the scheme for the filling of subshells developed in Chapter 3. Elements 104 to 112, therefore, would correspond to the filling of the $6d$ subshell. Next we have six elements in the p-block (113 to 118), which would have their $7p$ subshell gradually filled. Elements 119 and 120, in Period 8, correspond to the completion of the $8s$ subshell, and following element 121 a sequence of 32 inner transition elements occurs, from $Z = 122$ to 153. These *superactinides* would be accounted

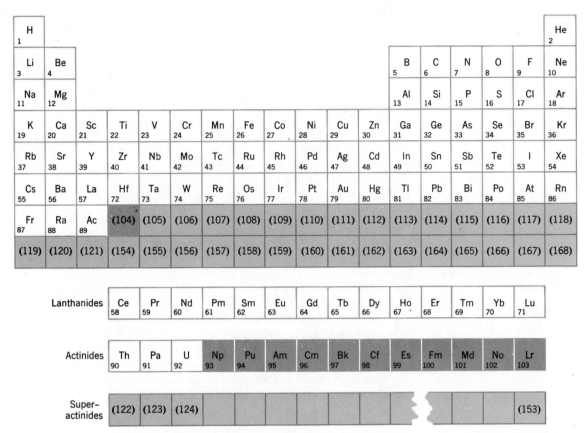

Figure 22.8

Extended periodic table. From G. Seaborg, Journal of Chemical
Education, *Vol. 46, p. 626, October 1969, used by permission.*

for by the completion of first the *6f* subshell (14 elements) followed by the filling
of a *5g* subshell (a *g* subshell would contain nine orbitals that can accommodate
18 electrons; therefore we would have 18 more elements, 136 to 153). After the
superactinides we would fill the *7d* subshell (elements 154 to 162) and then the
8p subshell (163 to 168).

In the search for new elements it is expected that nuclides that differ much
from $Z = 114$ would be extremely unstable and decompose by fission with very
short half-lives. However, in the vicinity of $Z = 114$ it has been suggested that
fission should not occur and the half-lives of these elements with respect to
alpha and beta decay should be long enough so that it should be possible to
detect them and perhaps even investigate their chemical properties. There is
even the possibility that these superheavy elements will not be radioactive at all.

The relative stabilities of nuclides containing differing numbers of protons
and neutrons have been dramatized in a drawing (Figure 22.9) published by Dr.
Glenn T. Seaborg, formerly head of the U. S. Atomic Energy Commission. Here
the stable nuclei of our band of stability are shown as a long peninsula extending
out into a sea of instability. Stable nuclei correspond to points above sea level,
whereas submerged regions constitute unstable nuclei. Notice that nuclei with
magic numbers of either protons or neutrons are shown as higher, more stable
ridges, while doubly magic nuclei are shown as mountains of stability.

The superheavy elements with approximately a magic number of 114 protons and either 184 or 196 neutrons are depicted as an island of stability separated from the peninsula by a region of high nuclear instability. As a result, in

Figure 22.9

Known and predicted regions of nuclear stability, surrounded by a sea of instability. From G. Seaborg, Journal of Chemical Education, *Vol. 46, p. 626, October 1969, used by permission.*

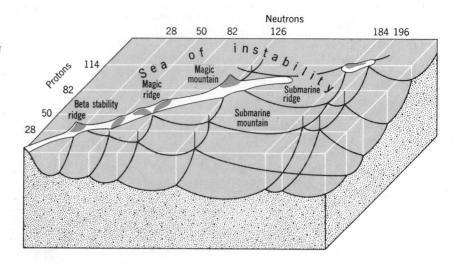

order to reach this island we cannot bombard stable (or relatively stable) nuclei with light particles such as ^4_2He, because this simply places the product nuclei into the sea of instability where they decompose rapidly before any additional mass and charge can be added. Consequently, the jump to the island must be made in one step. At the present time this presents substantial problems, because bombarding nuclei must contain a n/p ratio of at least 1.6 (184/114 = 1.61). However, light nuclei such as $^{40}_{18}Ar$, while possessing sufficient protons to give the desired atomic number by a reaction such as

$$^{248}_{96}Cm + ^{40}_{18}Ar \longrightarrow ^{284}_{114}X + 4^1_0n$$

do not contain enough neutrons to place the product isotope within the island of stability. For example, $^{284}_{114}X$ contains only 284 − 114 = 170 neutrons, 14 less than the 184 that we would want to achieve a doubly magic nucleus. Research today is directed at obtaining suitable target and projectile nuclei that will give not only Z = 114 but also a sufficient number of neutrons to place the product nucleus within the bounds of the predicted island of stability.

While physicists continue their search for ways to synthesize these superheavy elements, other scientists are looking for evidence of their current or past existence in the universe. In 1975 evidence was discovered by Dr. Edward Anders at the University of Chicago's Fermi Institute that suggested the one-time existence of element 114 (or perhaps 115 or 113) in a meteorite that fell in Mexico in 1969. Other scientists at Oak Ridge National Laboratories have found crystals that may once have contained elements 116 and 126.

Naming elements with large atomic numbers

It has been historical practice to allow the discoverer of an element to assign the element's name. In recent times this had led to some controversy, because elements with very high atomic numbers are so unstable that only minute quantities of them (sometimes only one or two atoms) are prepared before scientists claim credit for their discovery. This has led on occasion to questions of the reliability of the data and whether the sought-after element had in fact been made. For example, both American and Soviet scientists claim credit for discovering element 104. The Americans named it rutherfordium and the Soviets named it kurchatovium.

Because of this problem, the International Union of Pure and Applied Chemistry (IUPAC) has made the official recommendation that until a new element's discovery has been proven, a systematic nomenclature be applied. The rules of this system are as follows:

1. All elements will end in the letters *-ium*.
2. The name will be constructed from the following numerical roots:
   0 = nil   1 = un   2 = bi   3 = tri   4 = quad
   5 = pent   6 = hex   7 = sept   8 = oct   9 = enn
3. The symbol will consist of three letters derived from the numerical roots above.

The following illustrates how this works for element 104:

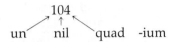

The name for the element is *unnilquadium* and its symbol is Unq.

EXAMPLE 22.5   NAMING SUPER-HEAVY ELEMENTS

PROBLEM   What would be the systematic name and symbol for element 114?

SOLUTION   The numerical root for 1 is un, so it has to be repeated; the root for 4 is quad. Therefore, the name is ununquadium, and its symbol is Uuq.

22.6 NUCLEAR BINDING ENERGY

Nuclei are composed of protons and neutrons. Naturally we would expect that if we added up the mass of all the protons that go into forming a nucleus, and then added in the mass of all the neutrons, we would obtain the mass of the nucleus. Actually, however, the mass of the nucleus is always somewhat *less* than the total mass of the individual protons and neutrons. How can this be?

Within the nucleus the nuclear particles are bound together by very strong forces, the nature of which is not understood very well. Nevertheless, enormous quantities of energy have to be supplied to separate the nucleus into its component protons and neutrons. It follows that if we were to form a nucleus, this same large quantity of energy would be released.

Einstein showed that mass and energy are related by his now famous equation, $E = mc^2$, where c is the speed of light. The energy liberated when the nucleus is formed comes at the expense of some of the mass of the nucleons. In other words, some mass is converted to the energy that is liberated as the nucleus is formed. Consequently, the final mass of the nucleus is less than we might have expected it to be.

Nucleons are the individual particles found in the nucleus.

The energy needed to decompose the nucleus (or the energy released when it is formed) is called the **binding energy.** The difference between the actual mass of a nucleus and the sum of the masses of its individual protons and neutrons is termed the **mass defect.** Let's look at an example.

A $^4_2$He atom is composed of two protons, two neutrons, and two electrons. These individual particles have the following masses:

$$p \quad\quad 1.007\ 277\ u$$
$$n \quad\quad 1.008\ 665\ u$$
$$e^- \quad\quad 0.000\ 548\ 6\ u$$

The calculated mass of a $^4_2$He atom is therefore

$$(2 \times 1.007\ 277\ u) + (2 \times 1.008\ 665\ u) + (2 \times 0.000\ 548\ 6\ u) = 4.032\ 981\ u$$

$$\text{calculated mass } ^4_2\text{He} = 4.032\ 981\ u$$

The actual mass of $^4_2$He, as measured with a mass spectrometer, is 4.002 603 u. The mass defect is the difference between the computed and measured mass; that is

$$4.032\ 981\ u - 4.002\ 603\ u = 0.030\ 378\ u$$

Thus when a helium nucleus is formed from two protons and two neutrons, 0.030 378 u of mass is converted to energy and released. How much energy does this represent?

Suppose that we were to form 1 mol of He atoms. The total mass lost would then be 0.030 378 g or 3.0378×10^{-5} kg. We can use Einstein's equation to calculate the energy equivalent. Using $c = 2.9979 \times 10^8$ m s^{-1}, we have

$$E = (3.0378 \times 10^{-5}\ \text{kg}) \times (2.9979 \times 10^8\ \text{m s}^{-1})^2$$

$$= 2.7302 \times 10^{12}\ \text{kg m}^2\ \text{s}^{-2}$$

Since 1 J = 1 kg m^2 s^{-2}, for 1 mol of He,

$$E = 2.73 \times 10^{12}\ \text{J mol}^{-1}$$

$$= 2.73 \times 10^9\ \text{kJ mol}^{-1}$$

The binding energy of a mole of helium is about three million times larger than the energy liberated by burning a mole of methane.

For comparison, combustion of 1 mol of CH$_4$ liberates only 8.9×10^2 kJ. The binding energy, therefore, represents a huge quantity of energy.

Let us now look at the average binding energy per nucleon that occurs for various atoms. This is usually expressed in energy units of *MeV per nucleon*. Nuclear physicists generally deal in energy units of MeV. One MeV is 1 million **electron volts,** where the electron volt is the kinetic energy that an electron would acquire if it were accelerated from one electrode to another across a potential difference of 1 V.

1 MeV per molecule is equivalent to 9.65×10^7 kJ mol^{-1}.

$$1\ \text{MeV} = 10^6\ \text{eV}$$

From Einstein's equation, the energy equivalence of 1 u can be calculated. Expressed in MeV, this is

$$1\ u \sim 931\ \text{MeV}$$

For $^4_2$He the binding energy is therefore

$$0.030\ 378\ u \times \left(\frac{931\ \text{MeV}}{1\ u}\right) = 28.3\ \text{MeV}$$

Since $^4_2$He is composed of four nucleons ($2p$, $2n$),

$$\text{average binding energy per nucleon} = \frac{28.3\ \text{MeV}}{4\ \text{nucleons}} = 7.07\ \text{MeV/nucleon}$$

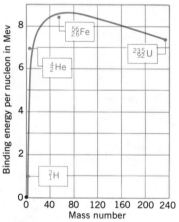

Figure 22.10

Average nuclear binding energy per nucleon for atoms of different mass number.

The average binding energy per nucleon varies, of course, for different atoms. Figure 22.10 is a plot of this average binding energy versus mass number.

22.7 FISSION, FUSION, AND NUCLEAR ENERGY

In the late 1930s in Germany, two chemists, Otto Hahn and Fritz Strassmann, and a physicist, Lise Meitner, found that when ^{235}U was bombarded with neutrons the unexpected products were the isotopes ^{139}Ba and ^{94}Kr, as well as three neutrons:

$$^{235}_{92}U + {}^{1}_{0}n \longrightarrow {}^{139}_{56}Ba + {}^{94}_{36}Kr + 3\,{}^{1}_{0}n$$

The significance of this accidental discovery was explained by Meitner and her nephew, Otto Frisch, who pointed out that fragmentation (splitting) of the ^{235}U was taking place and a very large quantity of energy was emitted during the process. The splitting of an atom into two approximately equal parts is known as **fission.** The results of the fission of ^{235}U were tested and substantiated by Enrico Fermi at Columbia University in New York City and physicists at Berkeley in California. Scientists saw military applications of the fission process, and it was Albert Einstein who alerted President Roosevelt to its possibilities. Roosevelt responded by establishing the Manhattan Project, whose research efforts led to the two bombs that were dropped on Hiroshima and Nagasaki, and ultimately to the production of energy by controlled nuclear fission.

Since three neutrons are produced during fission of each ^{235}U nucleus and since neutrons are required as a reactant for each fission process, then, potentially at least, the initial reaction is capable of triggering several additional reactions, as we can see in Figure 22.11. These reactions can in turn trigger many

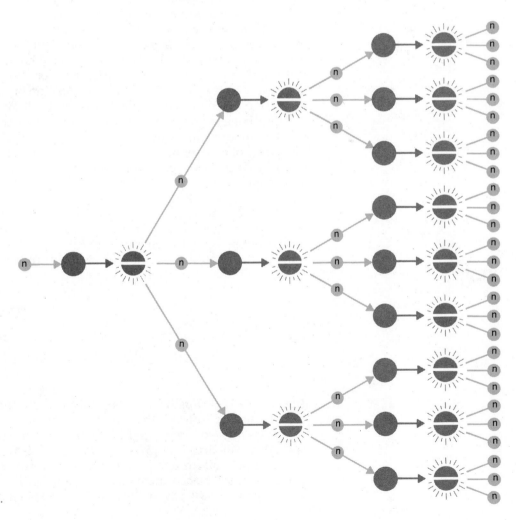

Figure 22.11
A nuclear chain reaction for ^{235}U.

more and so on, permitting a nuclear chain reaction to take place. This is indeed what occurs if enough pure ^{235}U is present. Each fission reaction causes several others to take place, with the evolution of a tremendous quantity of energy.

The origin of the energy released in nuclear fission can be seen by examining Figure 22.10. Here we see that nuclei of intermediate mass have larger binding energies per nucleon than nuclei that are either very light or very heavy. When a very heavy nucleus is split to give lighter fragments, the fission products have a greater binding energy per nucleon. Energy equal in quantity to the increase in binding energy is therefore released.

We mentioned earlier that the most abundant isotope of uranium found in its naturally occurring ores is ^{238}U. This isotope is nonfissionable and can, in fact, prohibit the fission chain reaction of the ^{235}U from occurring. The ^{238}U absorbs the neutrons emitted during the fission reaction, thus preventing the chain from continuing. There is, then, a minimum quantity of ^{235}U that must be present in order for the fission reaction to sustain itself. This minimum quantity needed for fission is called the isotope's **critical mass.**

Nuclear reactors

In a nuclear reactor, fission reactions take place, but at a controlled rate: slow enough to avoid a chain explosion but fast enough to produce usable heat. The fission reactions are controlled by the use of control rods made of such materials as cadmium, which absorb neutrons and thus prohibit the chain from occurring too rapidly. When these rods are extended all the way into the reactor (or pile), fission occurs very slowly. The rate can be increased by withdrawing the rods. When they are pulled out, they absorb fewer and fewer neutrons and the reaction occurs faster and faster. The ideal position of the rods is at that point where the fission reaction is just able to sustain itself at a desired level.

The large quantity of energy generated in the controlled nuclear fission reaction appears primarily as heat, and hence nuclear reactors must be cooled. There are, of course, many current applications of this released thermal energy in the production of electrical power. The heat removed from the reactor is used to convert water to steam, which is then used to drive turbines that generate electricity (Figure 22.12).

As with our supply of fossil fuels, there is only a limited supply of the fissionable ^{235}U, and nuclear reactors would face a somewhat uncertain future were it not possible to generate other fissionable isotopes. In a **breeder reactor** some of the control rods are replaced with rods containing ^{238}U. Some of the neutrons produced in the fission reaction are absorbed by the ^{238}U and give the reaction

^{235}U is the only naturally occurring fissionable isotope.

$$^{238}_{92}U + ^{1}_{0}n \longrightarrow ^{239}_{92}U$$

The $^{239}_{92}U$ decays rapidly to yield ultimately $^{239}_{94}Pu$ that, like $^{235}_{92}U$, is fissionable.

$$^{239}_{92}U \longrightarrow ^{239}_{93}Np + ^{0}_{-1}e$$

$$^{239}_{93}Np \longrightarrow ^{239}_{94}Pu + ^{0}_{-1}e$$

Breeder reactors thus have the useful property that they produce as much or *more* fissionable isotopes than they consume.

Nuclear fission reactors, in general, have the undesirable effect of producing highly radioactive by-products, some with extremely long half-lives, that are nearly impossible to dispose of safely. Nevertheless, the energy crisis of the recent past has placed increasing pressure on the need to develop additional sources of electrical power, including the exploitation of nuclear energy.

Figure 22.12

Application of nuclear fission to the production of electricity.

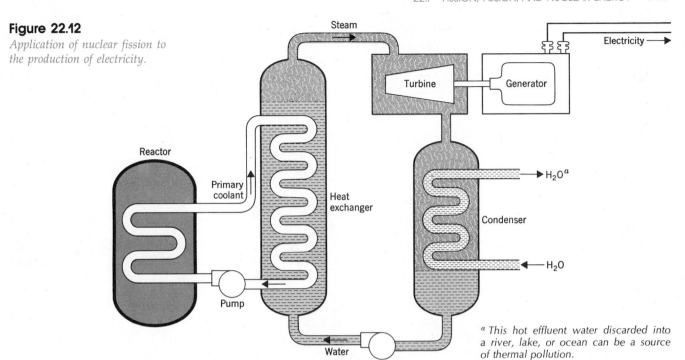

^a This hot effluent water discarded into a river, lake, or ocean can be a source of thermal pollution.

The Big Rock Nuclear Power Plant on the shore of Lake Michigan at Charlevoix, a town on the northwestern tip of Michigan's lower peninsula. The reactor itself is located in the spherically shaped containment building at the right.

Nuclear fusion

Quite the opposite of nuclear fission (fragmentation) is nuclear fusion. In this process two isotopes, usually very light ones, are brought together to form a heavier one. In so doing, a very large quantity of energy is released. The fusion reaction known as the hydrogen bomb is the reaction

$$^2_1H + {}^3_1H \longrightarrow {}^4_2He + {}^1_0n + \text{energy}$$

This is also one of the reactions taking place on the sun and accounts for the production of a good deal of its energy.

The origin of the energy released in nuclear fusion can also be seen from an examination of Figure 22.10. As we noted earlier, nuclei of intermediate mass have a greater binding energy per nucleon than light nuclei. When light nuclei such as ^2_1H combine to form heavier ones, there is an increase in the binding energy. As in fission, energy equal to the change in binding energy is released. However, if we study Figure 22.10 we see that the relative change in binding energy is much larger when very light nuclei are fused than during the fission of very heavy nuclei such as uranium. Thus, for reactions involving the same number of nuclei, the total quantity of energy released by fusion is considerably larger than that given off during fission.

Fusion reactions—not surprisingly—possess a high energy of activation mainly because of the electrostatic repulsion between the two nuclei being joined. As a result, they occur only at extremely high temperatures where the kinetic energies of the nuclei that are being joined are sufficient to overcome this repulsion. In fact, it is estimated that temperatures of approximately 40 million kelvins are needed for fusion to occur. The temperatures required to initiate such fusion reactions can be supplied by using an atomic (fission) bomb as a sort of nuclear match. The energy obtained from one fusion reaction is sufficient to cause other reactions to occur, so a chain reaction is established that results in a thermonuclear explosion. In controlled fusion applications, such as the generation of electrical power, the use of an atomic bomb to initiate the fusion process is, to say the least, unacceptable. Currently work is centering on the use of multiple high-energy lasers to provide the high temperatures required to get the fusion process started (Figure 22.13).

As a potential source of commercial power, nuclear fusion has some clear advantages over nuclear fission. First, the quantity of energy that is liberated in a fusion reaction is much greater than in a fission reaction. A second advantage is that the supply of fuel for nuclear fusion is virtually inexhaustible. One sequence of reactions that is being studied employs deuterium as the primary fuel.

$$^2_1H + {}^2_1H \longrightarrow {}^1_1H + {}^3_1H$$

$$^2_1H + {}^3_1H \longrightarrow {}^4_2He + {}^1_0n + \text{energy}$$

Deuterium atoms account for about 1 percent of the hydrogen atoms found in natural samples of water. Even if only a small fraction of all the deuterium present in the oceans could be separated and recovered, the energy that it could supply by fusion reactions such as those above is hundreds of thousands of times larger than all the energy contained in all the fossil fuels in the world! Still another advantage is that fusion reactions are relatively clean, in the sense that the products are generally not radioactive, so the waste disposal problem posed by fission reactors is avoided. In addition, the dangers of a runaway reactor are very small with fusion because all the nuclear fuel will not be present at the same time as it is in a fission reactor. Instead, it will be added a bit at a time as it is needed.

Figure 22.13

Fusion by laser implosion.

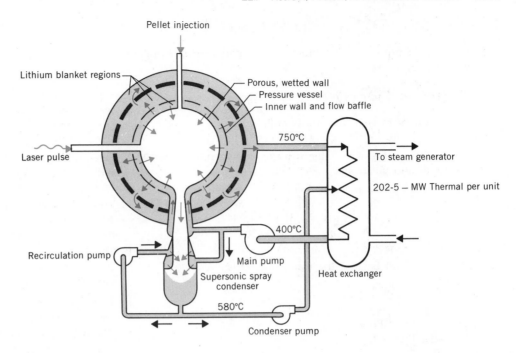

Pellet injection

Lithium blanket regions

Porous, wetted wall
Pressure vessel
Inner wall and flow baffle

750°C

Laser pulse

To steam generator

202-5 — MW Thermal per unit

400°C

Recirculation pump

Main pump

Supersonic spray condenser

Heat exchanger

580°C

Condenser pump

Despite these advantages, there are some major technical obstacles that must be overcome before fusion energy becomes a commercial reality. One of these is getting the reaction started. As mentioned above, high energy lasers are being developed to provide the necessary activation energy. An equally important problem is the lack of a container that is able to hold a reaction that is at a temperature of several million kelvins. One approach that is being tested is to suspend the reacting mass of ions, called a **plasma,** within a powerful magnetic field (see Figure 22.14).

Figure 22.14

Fusion by magnetic confinement of a plasma. This is an artist's conception of a full-scale Ormak reactor nearing the end of its assembly. In the foreground are two partially assembled sectors of the donut-shaped magnetic confinement apparatus. During operation, the plasma will circulate through the circular tunnel under the influence of powerful magnetic forces.

REVIEW QUESTIONS AND PROBLEMS

(Problems whose numbers are in blue have their answers in Appendix C at the back of the book; the more difficult problems are marked with asterisks.)

Radioactive Decay

22.1 What are the three main types of radiation emitted by radioactive nuclei? What are their properties?

22.2 What differences and similarities exist between beta particles and positrons?

22.3 Complete and balance the following nuclear equations:
(a) $^{81}_{36}Kr + ^{0}_{-1}e \rightarrow$?   (d) $^{104}_{48}Cd \rightarrow ^{104}_{47}Ag +$?
(b) $^{104}_{47}Ag \rightarrow ^{0}_{1}e +$?   (e) ? $+ ^{0}_{-1}e \rightarrow ^{54}_{24}Cr$
(c) $^{73}_{31}Ga \rightarrow ^{0}_{-1}e +$?

22.4 Complete and balance the following nuclear equations:
(a) $^{47}_{20}Ca \rightarrow ^{47}_{21}Sc +$?
(b) $^{55}_{27}Co \rightarrow ^{55}_{26}Fe +$?
(c) $^{220}_{86}Rn \rightarrow ^{216}_{84}Po +$?
(d) $^{54}_{26}Fe + ^{1}_{0}n \rightarrow ^{1}_{1}H +$?
(e) $^{46}_{20}Ca + ^{1}_{0}n \rightarrow$?

22.5 Complete and balance the following nuclear equations:
(a) $^{135}_{53}I \rightarrow ^{135}_{54}Xe +$?
(b) $^{245}_{97}Bk \rightarrow ^{4}_{2}He +$?
(c) $^{238}_{92}U + ^{12}_{6}C \rightarrow ^{246}_{98}Cf +$?
(d) $^{96}_{42}Mo + ^{2}_{1}H \rightarrow ^{1}_{0}n +$?
(e) $^{20}_{8}O \rightarrow ^{20}_{9}F +$?

22.6 Complete and balance the following nuclear equations:
(a) $^{35}_{17}Cl + ^{1}_{0}n \rightarrow ^{35}_{16}S +$?
(b) $^{40}_{19}K \rightarrow ^{0}_{-1}e +$?
(c) $^{98}_{42}Mo + ^{1}_{0}n \rightarrow ^{0}_{-1}e +$?
(d) $^{229}_{90}Th \rightarrow ^{4}_{2}He +$?
(e) $^{184}_{80}Hg \rightarrow ^{184}_{79}Au +$?

22.7 Write balanced equations for the nuclear decay reactions below.
(a) alpha emission by $^{11}_{5}B$
(b) beta emission by $^{98}_{38}Sr$
(c) neutron absorption by $^{107}_{47}Ag$
(d) neutron emission by $^{88}_{35}Br$
(e) electron absorption by $^{116}_{51}Sb$
(f) positron emission by $^{70}_{33}As$
(g) proton emission by $^{41}_{19}K$

22.8 What is a radioactive decay series? For ^{238}U, why does it stop at ^{206}Pb?

22.9 What is a parent isotope? What is a daughter isotope?

22.10 In the nucleus of $^{41}_{19}K$, how many nucleons are there?

Kinetics of Radioactive Decay

22.11 Show that Equation 22.1 reduces to Equation 22.2 if we take $[A]_t = \frac{1}{2}[A]_0$.

22.12 Cobalt-60 has a half-life of 5.26 y. If 1.00 g of ^{60}Co was allowed to decay, how many grams would be present after (a) one half-life, (b) three half-lives, (c) five half-lives?

22.13 Selenium-75 has a half-life of 120.0 d. If we began with 8.00 g of ^{75}Se, how many grams would remain after (a) 240 d, (b) 480 d, (c) 960 d?

22.14 The rate constant for the decay of ^{45}Ca is 4.23×10^{-3} d^{-1}. What is the half-life of ^{45}Ca expressed in days.

22.15 The rate constant for the decay of ^{36}Cl is 2.30×10^{-6} y^{-1}; what is the half-life of ^{36}Cl (expressed in years)?

22.16 The half-life of ^{51}Cr is 27.72 d. What is the rate constant for decay of ^{51}Cr in units of s^{-1}?

22.17 The half-life for the decay of ^{109}Cd is 470 d. What is the value for the rate constant for this decay in units of d^{-1}?

*__22.18__ The following data were obtained for the decay of ^{47}Ca:

Time (h)	cpm
0.0	4720
8.0	4485
12.0	4372
24.0	4050
48.0	3475
72.0	2983
96.0	2560

Determine (a) the rate constant for the decay, (b) the half-life of ^{47}Ca.

Measurement of Radioactivity

22.19 How does a Geiger–Müller counter work? How does a scintillation counter work?

22.20 Define the following units: (a) becquerel, (b) curie, (c) specific activity, (d) gray.

22.21 What is the difference between a gray and a rem?

22.22 Suppose a hospital owns a 150-g radioactive source. It has an activity of 1.24 Ci.
(a) What is its activity in Bq?
(b) What is the specific activity of the sample?

22.23 The isotope ^{145}Pr decays by emission of beta particles with an energy of 1.80 MeV each. Suppose a person swallowed, by accident, 1.0 mg of Pr having a specific activity of 140 Bq g^{-1}. What would be the absorbed dose in Gy from this isotope over a period of 10 min? (1 MeV = 1.6×10^{-13} J) Assume all the beta particles are absorbed by the person's body.

22.24 Which kind of radiation is most dangerous to humans?
(a) alpha or beta
(b) beta or gamma

Applications of Radioactivity

22.25 A sample of rock was found to contain 2.07×10^{-5} mol of ^{40}K and 1.15×10^{-5} mol of ^{40}Ar. If we assume that all of the ^{40}Ar came from the decay of ^{40}K, what is the age of the rock in years ($t_{1/2} = 1.3 \times 10^9$ y for ^{40}K)?

22.26 The ^{14}C content of an ancient piece of wood was found to be one-eighth of that in living trees. How many years old is this piece of wood ($t_{1/2} = 5770$ y for ^{14}C)?

22.27 Dinitrogen trioxide, N_2O_3, is largely dissociated into NO and NO_2 in the gas phase where there exists the equilibrium, $N_2O_3 \rightleftharpoons NO + NO_2$. In an effort to determine the structure of N_2O_3, a mixture of NO and *NO_2 was prepared containing isotopically labeled N in the NO_2. After a period of time the mixture was analyzed and found to contain substantial amounts of both *NO and *NO_2. Explain how this is consistent with the structure for N_2O_3 being ONONO.

22.28 The reaction $(CH_3)_2Hg + HgI_2 \rightarrow 2CH_3HgI$ is believed to occur through a transition state with the structure

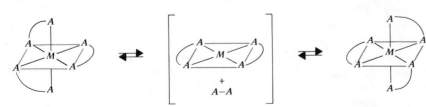

If this is so, what should be observed if CH_3HgI and *HgI_2 are mixed? Explain your answer.

22.29 *Racemization* is a chemical reaction in which one optical isomer of a compound is converted into its mirror image. One possible mechanism for the racemization of octahedral complex ions containing three bidentate ligands involves the temporary loss of one of the ligands,

$$d\text{-}[M(AA)_3] \longrightarrow \begin{pmatrix} M(AA)_2 \\ + \\ AA \end{pmatrix} \longrightarrow l\text{-}[M(AA)_3]$$

This can be pictured as shown in Figure 22.15. Can you suggest a simple experiment, making use of radioisotopes, that would be able to confirm whether or not this mechanism is operative in the racemization of the $[Co(C_2O_4)_3]^{3-}$ ion?

*22.30 A large, complex piece of apparatus has built into it a cooling system containing an unknown volume of cooling liquid. It is desired to measure the volume of the coolant without draining the lines. To the coolant was added 10.0 cm^3 of methanol labeled with ^{14}C and having a specific activity of 580 cpm per gram. The coolant was permitted to circulate to assure complete mixing before a sample was withdrawn that was found to have a specific activity of 29 cpm per gram. Calculate the volume of coolant in the system in cubic centimetres. The density of methanol is 0.792 g cm^{-3} and the density of the coolant is 0.884 g cm^{-3}.

*22.31 A complex ion of chromium(III) with oxalate ion was prepared from ^{51}Cr-labeled $K_2Cr_2O_7$, having a specific activity of 843 cpm/g, and ^{14}C-labeled oxalic acid, $H_2C_2O_4$, having a specific activity of 345 cpm/g. Chromium-51 decays by electron capture with the emission of a gamma-ray, whereas ^{14}C is a pure beta-emitter. Because of the characteristics of the beta and gamma detectors, each of these isotopes may be counted independently. A sample of the complex ion was observed to give a gamma count of 165 cpm and a beta count of 83 cpm. From these data, determine the number of oxalate ions bound to each Cr(III) in the complex ion. (*Hint:* For the starting materials calculate the cpm per mole of Cr and oxalate, respectively.)

Nuclear Stability

22.32 What is the significance of the *band of stability*? What decay processes are likely to occur for nuclides that have n/p ratios that place them above the band of stability?

Figure 22.15
A possible mechanism for the racemization of an octahedral [M(AA)$_3$] complex (AA = bidentate ligand).

22.33 Elements with atomic numbers greater than 83 generally decay by either alpha emission or fission. Why are the other forms of decay less likely for these nuclides?

22.34 What is a *magic number*? What magic numbers occur for protons? For neutrons? What magic numbers do we observe for orbital electrons?

22.35 In the absence of any specific information about their actual stability, rank the following nuclides in their expected order of decreasing stability: $_2^4He$, $_{20}^{39}Ca$, $_5^{10}B$, $_{32}^{71}Ge$, $_{28}^{58}Ni$.

22.36 What would you anticipate for the order of increasing nuclear stability for the following nuclides? $_2^3He$, $_{20}^{40}Ca$, $_{50}^{116}Sn$, $_6^{13}C$, $_{77}^{192}Ir$

22.37 The element $_{17}^{34}Cl$ emits gamma radiation with energies of 0.14, 1.15, 2.27, 3.22, and 4.80 MeV. How does this observation support the nuclear shell theory?

22.38 Technetium and promethium do not possess any stable isotopes. In light of the discussion in Section 22.3, comment on this observation.

Nuclear Transformation

22.39 What is a nuclear transformation?

22.40 How does a cyclotron work?

22.41 What are the transuranium elements? Where do they occur naturally?

22.42 Write nuclear equations for the following processes:
(a) $_{13}^{27}Al(\alpha,n)_{15}^{30}P$
(b) $_{83}^{209}Bi(d,n)_{84}^{210}Po$; ($d$ = deuteron, $_1^2H$)
(c) $_7^{15}N(p,\alpha)_6^{12}C$
(d) $_6^{12}C(p,\gamma)_7^{13}N$
(e) $_7^{14}N(\alpha,p)_8^{17}O$

22.43 Write nuclear equations for the following:
(a) $_{96}^{242}Cm(\alpha,n)_{98}^{245}Cf$
(b) $_{48}^{108}Cd(n,\gamma)_{48}^{109}Cd$
(c) $_7^{14}N(n,p)_6^{14}C$
(d) $_{13}^{27}Al(d,\alpha)_{12}^{25}Mg$
(e) $_{98}^{249}Cf(_8^{18}O,4n)_{106}^{263}Xe$

Extension of the Periodic Table

22.44 What chemical and physical properties would you predict for element number 114? If any of this element were formed at the time the universe came into being, where among earthly minerals would be a likely place to search for it?

22.45 If element 116 is found, what would be the expected formula of (a) its sodium salt, (b) its simple hydride, (c) its oxide? Would 116 be a metal or a nonmetal?

22.46 What would be the probable mass numbers of the most stable isotopes of element 114?

22.47 Explain the difficulties that must be overcome if a stable element with Z = 114 is to be made by nuclear bombardment.

22.48 What are the IUPAC names and symbols for these elements?
(a) 115 (b) 127

22.49 Specify the atomic numbers of
(a) unquadoctium
(b) unbipentium
(c) Unt

Nuclear Binding Energy and Nuclear Energy

22.50 What is the origin of the mass defect for a given nucleus?

22.51 What is the binding energy of $_{26}^{56}Fe$ expressed in MeV (atomic mass, 55.9349 u)? Where is this on the binding energy versus mass number curve? Why don't we have to worry about an enemy that claims they have developed the iron bomb?

22.52 Calculate the energy (in kilojoules) liberated in the fusion reaction to produce 1 mol of helium from deuterium.

$$_1^2H + _1^2H \longrightarrow _2^4He$$

The accurate atomic masses are

$$_1^2H = 2.014\ 102\ u$$

$$_2^4He = 4.002\ 603\ u$$

22.53 When an electron and a positron (a positively charged electron) encounter each other, they destroy each other with the production of energy. How much energy, in joules, results from such an encounter? The rest mass of the electron is 9.1096×10^{-28} g.

*22.54 Calculate the binding energy in kJ mol^{-1} and in MeV for the following isotopes, given the following masses: $p = 1.007\ 277$ u, $n = 1.008\ 665$ u, $e = 5.4859 \times 10^{-4}$ u.

Isotope	Actual Atomic Mass (u)
$_3^7Li$	7.016 00
$_7^{14}N$	14.003 074
$_9^{19}F$	18.998 40

22.55 Define nuclear fission. Why is it possible to have a nuclear explosion with ^{235}U?

22.56 Why doesn't a nuclear reactor explode?

22.57 Define nuclear fusion. How does the energy obtained in fusion reactions compare with that obtained in fission?

22.58 Why is it so difficult to build a fusion reactor for the controlled production of nuclear energy?

22.59 What is a plasma? Why is it difficult to contain?

22.60 Give two reasons why nuclear fusion would be better than nuclear fission as a source of energy for our society.

*22.61 Compare the energies liberated in the following fusion reactions:
(a) $2{}^2_1H \rightarrow {}^4_2He$
(b) $2{}^{12}_6C \rightarrow {}^{24}_{12}Mg$

Which reaction produces the most energy per mole of product? If each were equally feasible from an engineering standpoint for producing energy by controlled fusion, which would be preferable? Actual atomic masses: ${}^2_1H = 2.014\ 102$, ${}^4_2He = 4.002\ 603$, ${}^{24}_{12}Mg = 23.985\ 04$.

*22.62 How many litres (dm³) of gasoline, C_8H_{18}, having a density of 0.703 g cm⁻³, would have to be burned [to give $H_2O(l) + CO_2(g)$] to produce the same quantity of energy as in the production of 1 mol of ${}^4_2He$ by the fusion reaction described in Problem 22.52? ΔH_f^0 of $C_8H_{18}(l)$ is -208.4 kJ mol⁻¹.

CHAPTER 23

ORGANIC CHEMISTRY

The search for crude oil, carried on by "roughnecks" such as these who operate the oil drilling rigs, is a search for organic chemicals. Besides being our most important source of fuel, petroleum is the major source of raw materials for the synthesis of organic chemicals that range from exotic pharmaceuticals to plastic wrap.

As far back as the eighteenth century chemists were able to distinguish between two types of compounds: those derived from plants and animals and those that come from the mineral constituents of the earth. The second kind, called inorganic substances, have received much of our attention in this book. This chapter and the next are devoted to a discussion of organic compounds, many of which are found in nature but most of which have been synthesized in the laboratory.

One does not have to delve very far into the chemistry of life before observing that the element carbon is present in all the molecules in life's makeup. Of the 100 odd elements, carbon is the only one found universally in these substances. Thus, organic chemistry has become known as the study of carbon and its compounds. Besides carbon, organic molecules contain relatively few other elements; among the most prevalent are hydrogen, oxygen, nitrogen, and, to a lesser extent, phosphorus and sulfur.

Until 1828 chemists believed that the only source of these organic compounds was nature itself. It was thought to be impossible to synthesize them in the laboratory because nature's "vital force" was missing. In 1828 Friedrich Wöhler first synthesized urea—a compound found in urine—from inorganic materials. During one of his experiments, Wöhler evaporated a solution of the inorganic salt ammonium cyanate, $NH_4^+CNO^-$. When he heated the solid he discovered that another substance was formed, which he analyzed and found to be urea.

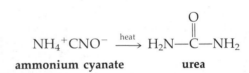

$$NH_4^+CNO^- \xrightarrow{\text{heat}} H_2N-\overset{\overset{\textstyle O}{\|}}{C}-NH_2$$

ammonium cyanate   **urea**

Once the results of Wöhler's synthesis were known, many scientists began to attempt to prepare other organic materials with a good deal of success. Gradually, the vital force theory was abandoned, and during the next several years a great profusion of organic compounds were made. Today there are well over six million known organic compounds and new ones are being discovered every day. The great abundance of organic compounds is a result of carbon's ability to bond to itself to form long chains, rings, and complex combinations of both.

In organic chemistry we are fortunate to be able to systematically categorize a gigantic number of compounds into a relatively small number of groups quite successfully. In this chapter we will survey a number of these groups and look at some sample organic reactions. The goal here will be simply to provide an overview of some topics that are of importance in this vast field of organic chemistry.

Friedrich Wöhler, the father of organic chemistry.

23.1   HYDROCARBONS

In Section 19.1 we saw that unsaturated hydrocarbons can combine with additional hydrogen until they have formed saturated hydrocarbons.

We begin our classification with the **hydrocarbons,** which are compounds containing only carbon and hydrogen. All the remaining types of organic compounds can then be looked on as being derived from the hydrocarbons. Hydrocarbons can be divided into two main categories: **aliphatic hydrocarbons,** which include straight-chain, branched-chain, and cyclic compounds, and **aromatic hydrocarbons,** which contain highly stable rings of carbon atoms. The aliphatic hydrocarbons can be further subdivided into two groups based on the multiplicity of the carbon–carbon bond: **saturated hydrocarbons** that contain only carbon–carbon single bonds; and **unsaturated hydrocarbons** that possess at least one carbon–carbon double bond or triple bond.

Saturated hydrocarbons

The compounds that constitute the saturated hydrocarbons are collectively called the **alkanes** or **paraffins.** Except for methane, they contain chains of carbon atoms linked by single bonds. The simplest ones have all the carbons in one continuous chain and are known as *normal* or *straight chain* alkanes. More complex alkanes have chains that branch. As you might expect, they are called *branched chain* alkanes. The first ten members of the straight-chain alkanes are listed in Table 23.1 in order of an increasing number of carbon atoms in the chain.

The alkanes are important as fuels and raw materials in the synthesis of other organic compounds. They occur in abundance in petroleum from which they may be separated to a large degree by fractional distillation. Methane, a gas at room temperature, is the principal component in natural gas used for cooking and as a heating fuel. Propane, which liquefies under high pressure, is used as a fuel in many rural areas where tanks of LPG (liquefied petroleum gas) are delivered to homes. Butane liquefies more easily than propane and is used in cigarette lighters. Octane, of course, has a boiling point that places it in the range of gasoline fuel. Heavier alkanes are found in kerosene, diesel fuel, lubricating oils, and the paraffins used to make candles.

The names of each of the members of the alkanes are composed of two parts. The first part, *meth-*, *eth-*, *prop-*, and so on in Table 23.1, reflects the number of carbon atoms in the chain. The second part, which is the same for all the members, is *ane* after the parent name alk*ane*. Thus we have methane, an alkane with one carbon, ethane having two carbons, propane having three carbons, and so on. You should become familiar with the names of all ten of these simple alkanes, because they serve as the basis for naming many of the remaining organic compounds.

In listing the alkanes by increasing number of carbon atoms, two things become apparent. First, the molecular formula for each member of this series can be represented by a single general formula, C_nH_{2n+2}, where n is the number of carbons in the molecular chain. For example, the formula for the alkane with

Ordinary candle wax—paraffin—is composed of alkanes.

Remember: *meth* = 1, *eth* = 2, *prop* = 3, *but* = 4. When there are more than four carbon atoms, the stems of the Greek number prefixes are used—for example, *pent* = 5, *hex* = 6, *hept* = 7, *oct* = 8, *non* = 9, *dec* = 10.

"Straight chain" doesn't mean the carbons are in a straight line—only that they are in one continuous sequence.

Table 23.1
The first ten members of the straight-chain alkanes

Formula	Name	Boiling Point (°C) at 1 atm
CH_4	Methane	−161
C_2H_6	Ethane	−89
C_3H_8	Propane	−44
C_4H_{10}	Butane	−0.5
C_5H_{12}	Pentane	36
C_6H_{14}	Hexane	68
C_7H_{16}	Heptane	98
C_8H_{18}	Octane	125
C_9H_{20}	Nonane	151
$C_{10}H_{22}$	Decane	174

four carbon atoms would be C_4H_{10} and, according to Table 23.1, would be called butane. Second, we see that any two successive members of the series differ from each other by a single CH_2 group (this becomes quite apparent after ethane). Such a series, in which one member differs from the next by the same repeating cluster of atoms, is called a **homologous series.** The alkanes form such a series.

In organic chemistry it is important for us to be able to write molecular formulas, which indicate the number of each of the various atoms present in a molecule, as well as structural formulas, which show the relative positions of each of the atoms in a molecule. Since carbon forms the backbone of all organic compounds, it is mainly the shape of the carbon skeleton that is responsible for the overall shape of the various molecules. Carbon normally forms four covalent bonds, and in the alkanes these are single bonds. Each carbon atom is sp^3 hybridized, with the hybrid orbitals pointing to the corners of a tetrahedron.

$$\underset{1s}{\underline{\uparrow\downarrow}} \quad \underset{2s}{\underline{\uparrow\downarrow}} \quad \underset{2p}{\underline{\uparrow}\;\underline{\uparrow}\;\underline{}} \qquad \underset{1s}{\underline{\uparrow\downarrow}} \quad \underset{sp^3 \text{ hybrid}}{\underline{\uparrow}\;\underline{\uparrow}\;\underline{\uparrow}\;\underline{\uparrow}}$$

unhybridized carbon

Thus in the straight- and branched-chain alkanes, carbon is tetrahedrally surrounded by bonded atoms. There are several ways of representing tetrahedral carbon in two dimensions, as illustrated in Figure 23.1 for methane. The most common, and the simplest, two-dimensional representation used for most organic compounds is the structural formula shown as Figure 23.1a. The remaining members of the straight-chain alkanes are drawn simply by connecting several tetrahedral carbons along with their respective hydrogens.

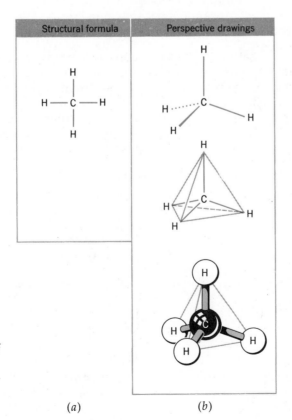

Structural formula	Perspective drawings

(a) (b)

Figure 23.1

Two-dimensional representations of tetrahedral carbon. Methane is chosen as an example. (a) *A structural formula.* (b) *Perspective drawings.*

Figure 23.2

Two- and three-dimensional illustrations of the structure of butane. (a) A two-dimensional structural formula. (b) A three-dimensional drawing. Note that the carbon atoms are not actually in a straight line.

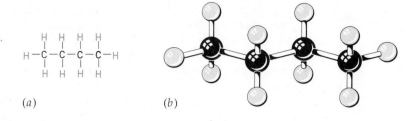

(a)   (b)

In Figure 23.2 we see two representations of the four-carbon alkane, butane. Although the two-dimensional representation drawn in Figure 23.2*a* suggests a flat molecule with a straight carbon chain, you should remember that such molecules are not planar. The tetrahedral geometry about the carbon atoms prevents them from being in a straight line, as shown in Figure 23.2*b*.

Very often we find it convenient to condense structural formulas somewhat by not drawing all the C—H bonds. For instance, the butane formula can be condensed to either

$$CH_3—CH_2—CH_2—CH_3$$

or

$$CH_3CH_2CH_2CH_3$$

We will often use condensed formulas in this chapter. Their advantage is that they save time (and space), yet they still convey structural information.

Unsaturated hydrocarbons

The unsaturated hydrocarbons can be divided into two groups: **alkenes** or **olefins,** which contain at least one carbon–carbon double bond and **alkynes,** which contain at least a carbon–carbon triple bond.

Following a procedure similar to that used for the alkanes, we can introduce the alkenes by considering a series of compounds with one carbon–carbon double bond and an increasing number of carbon atoms in the chain. The simplest alkene contains two carbons:

$CH_2=CH_2$

$$\underset{H}{\overset{H}{>}}C=C\underset{H}{\overset{H}{<}} \quad (C_2H_4)$$

**ethene
(ethylene)**

which is then followed by the three-carbon alkene,

$CH_3—CH=CH_2$

$$H—\overset{H}{\underset{H}{C}}—\overset{H}{C}=C\underset{H}{\overset{H}{<}} \quad (C_3H_6)$$

**propene
(propylene)**

Next come the alkenes with the formulas C_4H_8, C_5H_{10}, C_6H_{12}, and so on. Thus we see that the alkenes, like the alkanes, form a homologous series, with a CH_2 group as the difference between any two successive members. We also see that, like the alkanes, the alkenes can collectively be represented by one general formula, C_nH_{2n}, where n is the number of carbons in the molecule.

The names of the straight-chain alkenes also consist of two parts. The first part, which is the same as that used in the naming of the alkanes, indicates the number of carbons in the chain: *eth-*, two; *prop-*, three; *but-*, four; and so on. To this stem we add *ene*, which tells us that there is a carbon–carbon double bond present. The first two members of the alkenes are thus ethene and propene. These are also called by their older names, ethylene and propylene, and are probably the best known of the alkenes because of their use in making the plastics, polyethylene and polypropylene. We will discuss the nonmenclature of these and other hydrocarbons in greater detail in Section 23.3.

The bonding in ethene was discussed earlier in Section 6.6, so it will be reviewed only briefly here. The carbon atoms participating in the double bond are each sp^2 hybridized. The overlap of the sp^2 hybrid orbitals between the two carbon atoms produces a σ bond along the C—C axis. Each carbon also possesses an unhybridized pure p orbital. These overlap sideways to produce a π-bond with electron density concentrated above and below the bond axis. Because of the geometry of the sp^2 hybrids and the restriction that the unhybridized p orbitals must overlap in the π-bond, a planar configuration of atoms with approximately 120° bond angles is formed.

Besides the straight-chain alkenes there also exist alkenes with more than one double bond, alkenes that have other groups attached to the straight chain, alkenes with branched chains, and alkenes that are cyclic in structure.

The second type of unsaturated hydrocarbon is the **alkynes**—a series of compounds containing a carbon–carbon triple bond. The simplest, but truly one of the most important alkynes is HC≡CH, which is called ethyne or, more commonly, acetylene. This is the fuel used in welding torches. The next two members of this group are CH_3—C≡CH, propyne, and CH_3—CH_2—C≡CH, butyne. In naming these compounds we once again use the stems *eth-*, *prop-*, and *but-* to mean two, three, and four carbons, respectively. To this stem is added *yne* to denote the existence of the carbon–carbon triple bond.

The bonding in the carbon–carbon triple bond was also discussed earlier (Section 6.6). The hybridization used by the two carbons in the triple bond is sp. Thus one of the bonds—a σ-bond—is formed by sp–sp overlap. The remaining two bonds are both π-bonds that are formed by p_x–p_x and p_y–p_y overlap.

23.2 ISOMERS OF ORGANIC COMPOUNDS

Beginning with hydrocarbons containing four carbon atoms, we find that besides the normal straight-chain structures, there also exist branched-chain structures bearing the same molecular formula. For example, we find that there are two compounds with the formula C_4H_{10}.

Recall that isomers are compounds with the same molecular formula but with distinctly different properties.

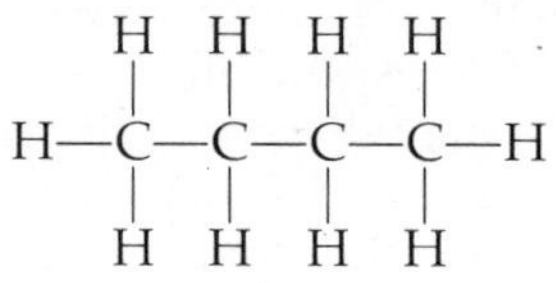

(1)
straight chain
m.p. = −138.3°C
b.p. = −0.5°C

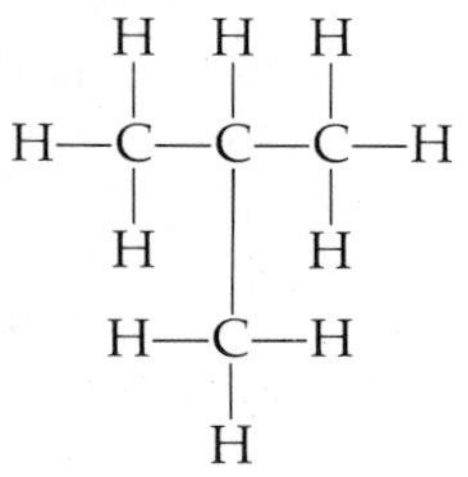

(2)
branched chain
m.p. = −159°C
b.p. = −12°C

Two or more compounds that have the same molecular formula but that differ in the sequence in which the atoms are joined together are said to be **structural isomers.** Thus there are two structural isomers of butane, each with its own chemical and physical properties. This is an important point to keep in mind. Each structural isomer is a unique chemical compound.

Each of the remaining members of the alkane series shows an even greater number of isomers. With C_5H_{12}, for example, we can write three structures:

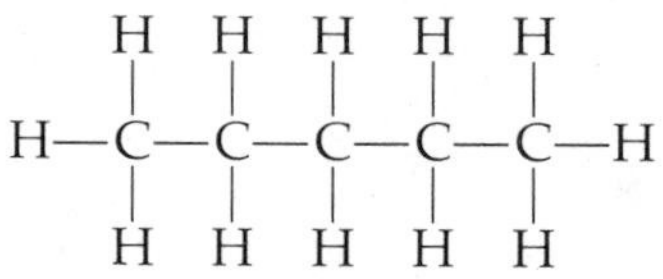

(1)

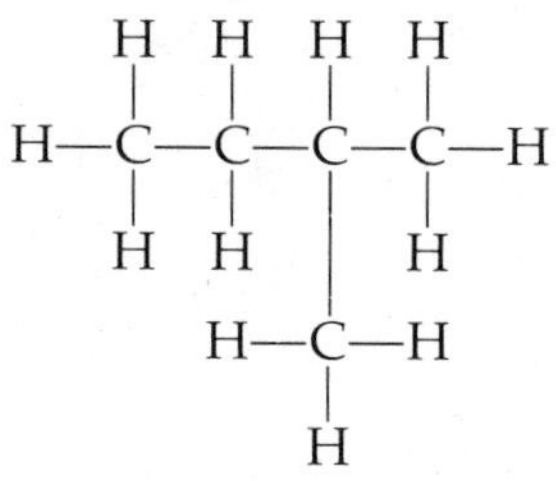

(2)

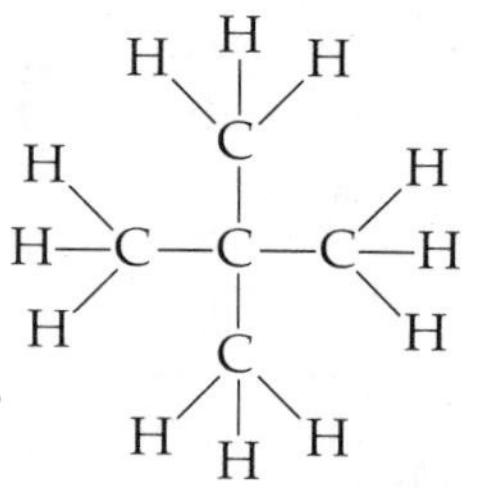

(3)

In the case of C_6H_{14} we find five isomers, which are listed in Table 23.2. In this table we have left room for you to add the names of each of the isomers following our discussion of nomenclature in the next section.

Structural isomerism also exists in the alkenes. Butene (C_4H_8), for example, can be written as a straight chain with the double bond between the first and second carbon atoms.

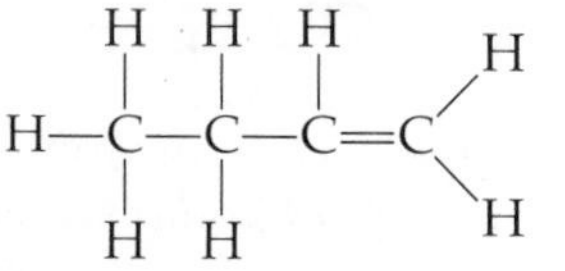

$$CH_3—CH_2—CH=CH_2$$ or

isomer 1

Table 23.2
The five isomers of hexane

Isomer	Name (fill in this column after reading Section 23.3)
| ```
 H  H  H  H  H  H
 |  |  |  |  |  |
H—C—C—C—C—C—C—H
 |  |  |  |  |  |
 H  H  H  H  H  H
``` | |
| ```
 H  H  H  H  H
 |  |  |  |  |
H—C—C—C—C—C—H
 |  |  |     |
 H  H  H     H
          |
       H—C—H
          |
          H
``` | |
| ```
 H  H  H  H  H
 |  |  |  |  |
H—C—C—C—C—C—H
 |  |     |  |
 H  H     H  H
       |
    H—C—H
       |
       H
``` | |
| ```
 H        H         H        H
 |        |         |        |
H—C———————C—————————C————————C—H
 |        |         |        |
 H    H—C—H     H—C—H        H
          |         |
          H         H
``` | |
| ```
            H
            |
         H—C—H
 H  H       |   H
 |  |       |   |
H—C—C———————C———C—H
 |  |       |   |
 H  H       |   H
         H—C—H
            |
            H
``` | |

There is also an isomer with the double bond between the middle two carbons,

```
 H  H  H  H
 |  |  |  |
H—C—C=C—C—H      or      CH₃—CH=CH—CH₃
 |        |
 H        H
```

$$H-C-C=C-C-H \quad\text{or}\quad CH_3-CH=CH-CH_3$$

isomer 2
```

and another is a branched isomer in which two CH_3 groups are attached to one of the carbon atoms participating in the carbon–carbon double bond.

$$
\underset{\textbf{isomer 3}}{
H-\overset{\overset{\displaystyle H}{|}}{\underset{\underset{\displaystyle H-\overset{\overset{\displaystyle H}{|}}{\underset{\underset{\displaystyle H}{|}}{C}}-H}{|}}{C}}-\overset{\overset{\displaystyle H}{|}}{C}=\overset{}{C}-H}
\qquad \text{or} \qquad CH_3-\overset{\overset{\displaystyle }{\underset{\underset{\displaystyle CH_3}{|}}{C}}}{}=CH_2
$$

Notice that the arrangement of atoms in the molecules is different in all three cases even though the molecular formula, C_4H_8, is the same. Thus in alkenes we have isomers that result from the branching of the carbon chains as well as isomers that result from a difference in the relative position of the double bond.

Another type of isomerism exists in organic compounds in which the sequence of atoms in the molecule is the same but in which the relative positions of the atoms or groups of atoms are different. In general, this type of isomerism is called **stereoisomerism,** and is found in two forms: geometrical isomerism and optical isomerism (see also Section 21.9).

Geometrical isomerism

In organic compounds *cis* and *trans* geometrical isomers can occur in molecules that possess one or more carbon–carbon double bonds. This is illustrated by one of the isomers of butene,

cis **isomer**   *trans* **isomer**

These are not identical molecules because rotation of the atoms about the axis of the C=C double bond is restricted. When one of the $CH(CH_3)$ groups begins to rotate, the p orbitals that overlap sideways to form the π-bond become misaligned—the overlap is lost. This destroys the bond, so in effect, conversion of *cis* to *trans* in this manner would require the breaking of a bond. The energy available through molecular collisions is generally not sufficient to accomplish this, so the rate of interconversion between *cis* and *trans* isomers is usually so slow that the individual isomers can be isolated.

Optical isomerism

Optical isomers, as we saw in Section 21.9, are molecules that have the same formula but possess structures that are nonsuperimposable mirror images of each other. In organic compounds the presence of a **chiral carbon atom** is responsible for optical isomerism. A chiral carbon atom occurs when the carbon is bonded to four *different* atoms or groups of atoms; for example,

$$
A-\overset{\overset{\displaystyle B}{|}}{\underset{\underset{\displaystyle D}{|}}{C}}-E
$$

Rotation about a C—C single bond is relatively unrestricted, so geometrical isomers of

can't be isolated.

A chiral carbon atom is often called an **asymmetric carbon atom.**

where *A*, *B*, *D*, and *E* represent different groups. For instance, the isomer of heptane with the structural formula

$$H-\overset{\displaystyle H}{\underset{\displaystyle H}{C}}-\overset{\displaystyle H}{\underset{\displaystyle H}{C}}-\overset{\displaystyle H}{\underset{\displaystyle |}{\overset{\displaystyle *}{C}}}-\overset{\displaystyle H}{\underset{\displaystyle H}{C}}-\overset{\displaystyle H}{\underset{\displaystyle H}{C}}-\overset{\displaystyle H}{\underset{\displaystyle H}{C}}-H$$

$$H-\overset{\displaystyle }{\underset{\displaystyle H}{C}}-H$$

has one chiral carbon atom—the one marked with an asterisk. This can be seen more readily if we rewrite the structural formula for this isomer in a slightly more condensed fashion.

$$H_5C_2-\overset{\displaystyle H}{\underset{\displaystyle CH_3}{\overset{\displaystyle *}{C}}}-C_3H_7$$

Let's see how compounds with a chiral carbon atom give rise to optical isomers and why molecules without one cannot. To show this, we compare the structures of two different isomers of a compound to see if one is the nonsuperimposable mirror image of the other. If this condition is fulfilled, then they are optical isomers. If not, then they are identical and the compound is not optically active.

In the case of our general molecule, the two structures shown in Figure 23.3

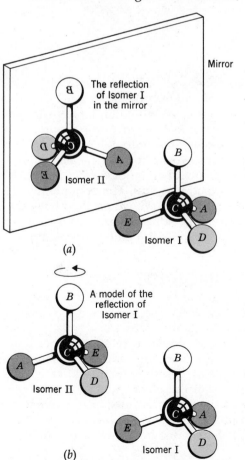

Figure 23.3

Optical isomers. (a) Isomer II is a reflection of Isomer I in the mirror; that is, it is the mirror image of Isomer I. (b) When Isomer II is rotated about the B—C bond so that atoms D in both I and II are in same relative position, Isomers I and II are not superimposable. Atoms B, C, and D match, but A and E do not.

Figure 23.4

Lack of optical isomers when two groups attached to carbon are the same. (a) "Isomer I" is formed as the mirror image of II. (b) "Isomer I" rotated about the B—C bond so that atoms D are matched. These two are identical, so they are not isomers of each other.

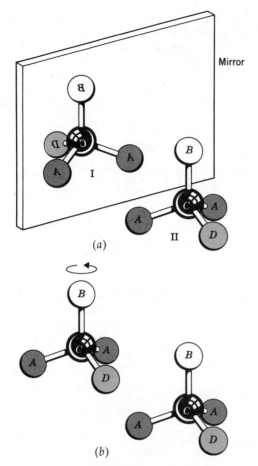

are mirror images of each other with nonsuperimposable structures. Note that when one is placed over the other, the D's and B's line up, but the E's and A's do not. Furthermore, we would find that, regardless of the manipulation of the two structures, we could never bring about the situation in which one structure would be exactly superimposable on the other.

If there are two groups that are alike in each structure, the asymmetry of the central carbon is lost and the two structures no longer represent optical isomers; they are, in fact, identical. For example, by replacing the E in our general formula with another A, we would then have the two structures shown in Figure 23.4, which can be superimposed on each other. Thus we see that optical isomerism will occur only if a chiral carbon atom is present.

23.3 NAMING ORGANIC COMPOUNDS

We have already seen some of the basic elements of naming organic compounds in our discussion of the hydrocarbons. The complexities produced by isomerism, and the introduction of atoms other than carbon and hydrogen, create a need for a systematic procedure for naming compounds. The guidelines presented below are currently followed in the modern chemical literature. Unfortunately, many of the older, nonsystematic *common names* are still often used in less formal situations. For instance, ethyne, C_2H_2, is also called acetylene; ethene and propene are also known as ethylene and propylene. A student of organic chemistry must be aware of this dual nomenclature.

The systematic nomenclature of organic compounds, like that of coordination compounds, has been established by the IUPAC. The application of the IUPAC system to the saturated and unsaturated hydrocarbons is defined by the following rules.

Rules of nomenclature of organic compounds

1. **The longest unbroken chain of carbon atoms in a molecule serves as the parent name for any hydrocarbon or its derivative.** With the alkenes this longest chain must contain the double bond; in the alkynes the triple bond must be included. For example, each of the following three compounds would be named as a derivative of pentane, because in each case the longest carbon chain consists of five carbon atoms.

$$CH_3-\underset{\underset{CH_3}{|}}{\overset{\overset{CH_3}{|}}{C}}-CH_2-CH_2-CH_3 \qquad CH_3-CH_2-\underset{}{\overset{\overset{CH_3}{|}}{CH}}-\overset{\overset{CH_3}{|}}{CH}-CH_3$$

$$CH_3-\underset{\underset{CH_3}{|}}{\overset{\overset{CH_3}{|}}{C}}-CH_2-\underset{\underset{CH_3}{|}}{\overset{\overset{CH_3}{|}}{C}}-CH_3$$

Sometimes, when a structural formula is drawn, the longest carbon chain might not be written in a straight line. For instance, the compound below has an eight-carbon chain (can you find it?) and would be named as a derivative of octene (note the double bond).

$$CH_3-\overset{\overset{CH_3}{|}}{CH}-\underset{\underset{CH_2}{|}}{CH}-CH_2-CH_2-CH_3$$
$$CH_3-\overset{}{\underset{\underset{CH_3}{|}}{C}}-CH=CH_2$$

2. **The number of carbon atoms in the chain is indicated by the first part of the parent name.** We've already seen that *meth* implies one carbon atom, *eth* implies two, *prop* implies three, and so forth. (See page 828.)
3. **Saturation or unsaturation is indicated by the second part of the parent name.** The ending *-ane* means that only single bonds between carbon atoms occur in the chain, *-ene* means there is a carbon–carbon double bond, and *-yne* means there is a carbon–carbon triple bond. When two double bonds are present the ending is **-diene,** and when three are present the ending is **-triene.** Thus we would have

$$CH_3-CH_2-CH_2-CH_3 \quad \text{butane}$$
$$CH_3-CH_2-CH=CH_2 \quad \text{butene}$$
$$CH_3-CH_2-C\equiv CH \quad \text{butyne}$$
$$CH_2=CH-CH=CH_2 \quad \text{butadiene}$$

4. **Branched isomers are named as derivatives of straight-chain hydrocarbons in which one or more hydrogen atoms are replaced by hydrocarbon fragments.** For example, isomer 2 of butane (on page 832) can be considered to be derived from propane by replacing one of the middle hydrogen atoms with a CH_3 group.

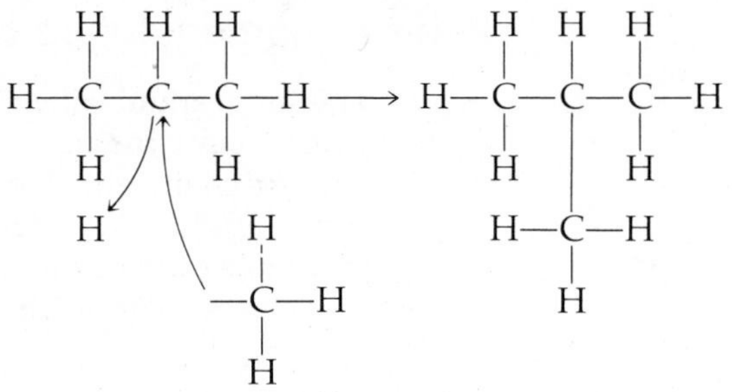

Hydrocarbon groups that are derived from members of the alkane series are called **alkyl groups.** The CH_3 group is derived from methane and is called a *methyl group.*

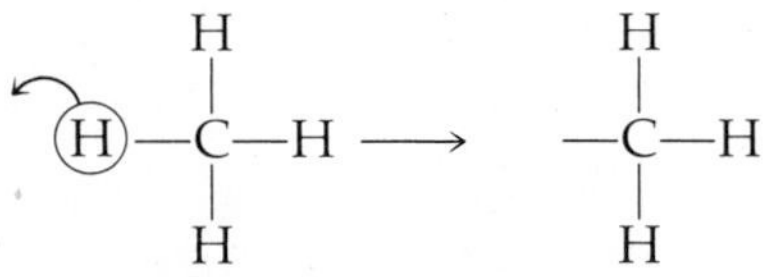

methane   methyl group

Similarly, the C_2H_5 group is derived from ethane and is called an *ethyl group.*

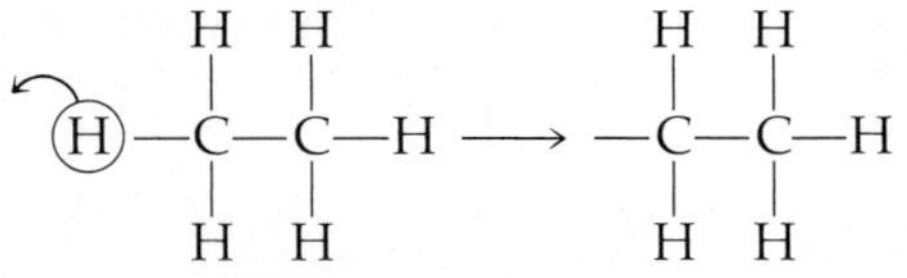

ethane   ethyl group

Thus, in naming an alkyl group the ending **-yl** is added to the alkane stem that indicates the number of carbon atoms in the group. Some additional alkyl groups are listed in Table 23.3.

Referring back to the branched isomer of butane, we see that it has a three-carbon chain as its longest chain and a CH_3 group attached to the middle carbon; therefore it would be called

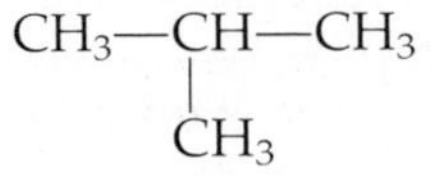

$$CH_3—\underset{\underset{CH_3}{|}}{CH}—CH_3 \qquad \text{methylpropane}$$

Likewise, we have

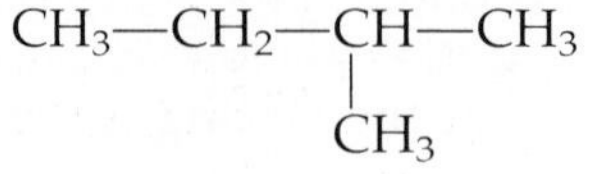

$$CH_3—CH_2—\underset{\underset{CH_3}{|}}{CH}—CH_3 \qquad \text{methylbutane}$$

and

$$CH_3—\underset{\underset{CH_3}{|}}{C}=CH_2 \qquad \text{methylpropene}$$

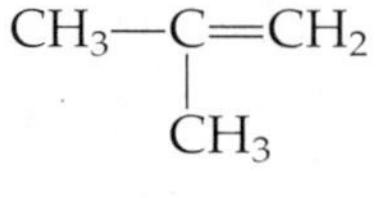

Table 23.3
The names of some alkyl groups

Alkyl Group	Condensed Formula	Name
H—C— with H above and H below	CH_3—	Methyl
H—C—C— with H's	CH_3—CH_2—	Ethyl
H—C—C—C— with H's	CH_3—CH_2—CH_2—	Propyl
H—C—C—C—C— with H's	CH_3—CH_2—CH_2—CH_2—	Butyl

5. **To denote the positions of alkyl groups attached to the parent chain, as well as the positions of double and triple bonds, a numbering system is used.** In alkanes the numbering starts from the end of the molecule that gives the lowest numbers to the alkyl groups. When multiple bonds are present the numbering begins from the end of the molecule that gives the lowest numbers to the multiple bonds. In the case of an alkyl group, the number identifying its position immediately precedes its name, while the numbers identifying a double or triple bond precede the name of the parent chain. Examples of this include

CH_3—CH_2—CH_2—CH—CH_2—CH_3 3-methylhexane
with CH_3 below the CH

CH_3—CH=CH—CH_3 2-butene

CH_3—CH=CH—CH_2—CH=CH_2 1,4-hexadiene

CH_3—CH—CH=CH—CH_3 4-methyl-2-pentene
with CH_3 below the CH

6. **When two or more identical alkyl groups are attached to a carbon chain, their number is specified by the Greek prefixes di- (two), tri- (three), tetra- (four), and so on. The position of each is also specified.** For example,

$$CH_3-\overset{\overset{\displaystyle CH_3}{|}}{CH}-\overset{\overset{\displaystyle CH_3}{|}}{CH}-CH_3 \qquad \text{2,3-dimethylbutane}$$

$$CH_3-\overset{\overset{\displaystyle CH_3}{|}}{CH}-\overset{\overset{\displaystyle CH_3}{|}}{\underset{\underset{\displaystyle CH_3}{|}}{C}}-CH_3 \qquad \text{2,2,3-trimethylbutane}$$

7. **When different alkyl groups are present along the parent chain, they are given in alphabetical order.** For example,

$$CH_3-CH_2-\overset{\overset{\displaystyle CH_2}{|}}{\underset{\underset{\displaystyle CH_3}{|}}{C}}-CH_2-CH_2-CH_2-\overset{\overset{\displaystyle CH_3}{|}}{CH}-CH_3 \qquad \text{6-ethyl-2-methyloctane}$$

$$CH_3-\overset{\overset{\displaystyle CH_3}{|}}{\underset{\underset{\displaystyle CH_3}{|}}{C}}-\overset{\overset{\displaystyle }{}}{\underset{\underset{\displaystyle CH_2-CH_3}{|}}{CH}}-CH_2-CH_3 \qquad \text{3-ethyl-2,2-dimethylpentane}$$

EXAMPLE 23.1   APPLYING THE IUPAC NOMENCLATURE RULES

PROBLEM   (a) Give the IUPAC name for the following compounds:

(1)
$$\begin{matrix} H_3C \\ \\ H_3C \end{matrix}\!\!\!\!\diagup\!\!\!\diagdown\!\!\!\! CH-CH_2-CH_2-CH_3$$

(2)
$$CH_3-\overset{\overset{\displaystyle CH_3}{|}}{\underset{\underset{\displaystyle CH_3}{|}}{C}}-CH_2-\overset{\overset{\displaystyle CH_2-CH_3}{|}}{CH}-CH_2-\overset{\overset{\displaystyle }{}}{\underset{\underset{\displaystyle CH_3}{|}}{CH}}-CH_3$$

(3)
$$\begin{matrix} H_3C \\ \\ H_3C \end{matrix}\!\!\!\diagup\!\!\!\diagdown C=C \diagup\!\!\!\diagdown\!\!\!\begin{matrix} CH_3 \\ \\ CH_3 \end{matrix}$$

(4)
$$CH_3-\overset{\overset{\displaystyle }{}}{\underset{\underset{\displaystyle CH_3}{|}}{CH}}-CH=CH-CH=CH_2$$

(5) $CH_3-CH_2-C\equiv C-CH_2-CH_3$

(b) Write structural formulas for the following:
   (1) 2,3-dimethylbutane
   (2) 2-pentyne
   (3) 2-ethyl-1-butene
   (4) 1,5-octadiene
   (5) 2-ethyl-3-methyl-1-pentene

SOLUTION

(a) (1) 2-methylpentane
(2) 4-ethyl-2,2,6-trimethylheptane
(3) 2,3-dimethyl-2-butene
(4) 5-methyl-1,3-hexadiene
(5) 3-hexyne

(b) (1) 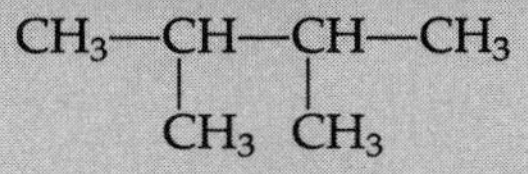 $CH_3—CH—CH—CH_3$ with CH_3 CH_3

(2) $CH_3—CH_2—C≡C—CH_3$

(3) 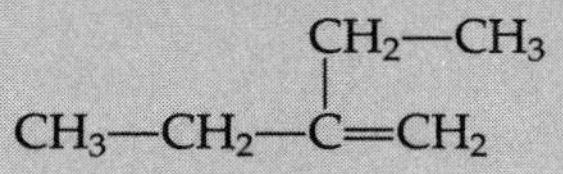 $CH_2—CH_3$ / $CH_3—CH_2—C=CH_2$

(4) $CH_3—CH_2—CH=CH—CH_2—CH_2—CH=CH_2$

(5) 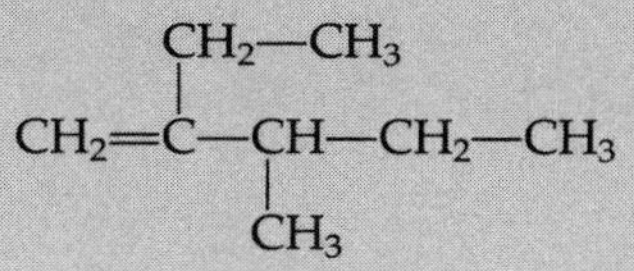 $CH_2—CH_3$ / $CH_2=C—CH—CH_2—CH_3$ / CH_3

As an exercise, you should now go back and fill in the names of all the isomers of hexane in Table 23.2.

Common names

As mentioned earlier, quite a large number of organic compounds were known prior to the introduction of the IUPAC rules. As a result, many had already been named using other systems of nomenclature, and some of these names are carried over into current usage. For example, methylpropane,

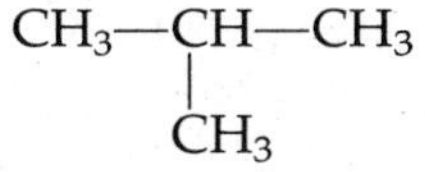

$$CH_3—CH—CH_3$$
$$CH_3$$

is more commonly referred to as isobutane. This name indicates that there are four carbons present (*but*ane), but that they are not in straight chain and the molecule is an isomer of butane, hence *iso*. In general, alkanes that possess the arrangement

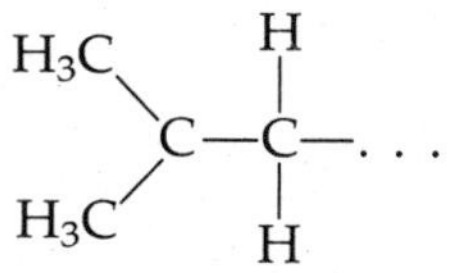

$$H_3C \quad H$$
$$C—C—...$$
$$H_3C \quad H$$

are called *iso* compounds. Thus we have

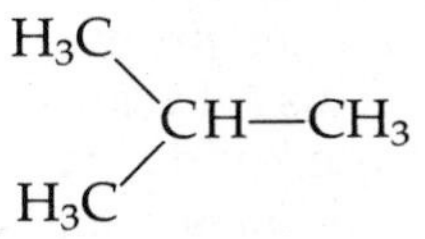

isobutane 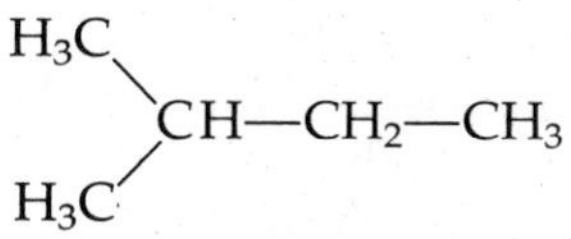 isopentane 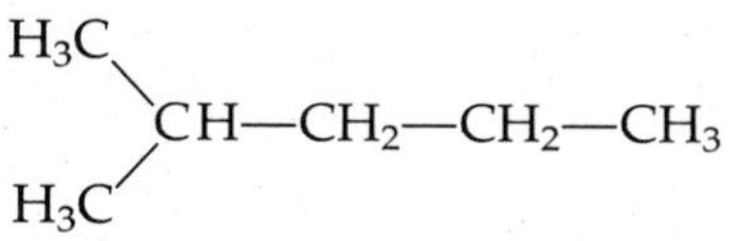 isohexane

In the case of the alkynes, the first member, acetylene, is so important that its name is sometimes used in the common nomenclature of all the remaining members. Thus we have

Alkyne	Common	IUPAC
$CH_3-C{\equiv}CH$	Methylacetylene	Propyne
$CH_3-CH_2-C{\equiv}CH$	Ethylacetylene	1-Butyne
$CH_3-CH_2-CH_2-C{\equiv}CH$	Propylacetylene	1-Pentyne

We will see that the use of common names carries over to organic compounds that also contain elements in addition to carbon and hydrogen.

23.4 CYCLIC HYDROCARBONS

Cyclic alkanes

With alkanes containing three or more carbon atoms it is possible for the atoms to be arranged in a ring. An example is

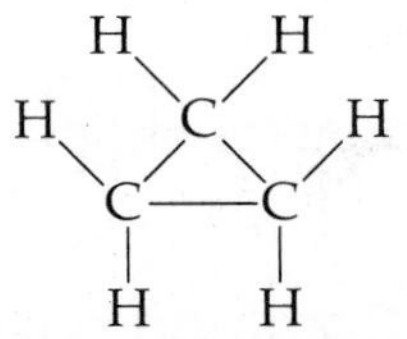

This compound, as well as the other members of this group, contains only carbon–carbon single bonds, and is therefore saturated. These hydrocarbons are named in the same manner as the straight-chain alkanes, but with the prefix *cyclo-* added. The compound above would therefore be named cyclopropane (*cyclo* meaning ringed, *prop* for three carbon atoms, and *ane* because it is saturated). Some other cycloalkanes, with their molecular and structural formulas, are listed in Table 23.4. Notice that the general formula for the cycloalkanes is C_nH_{2n}. This is the same as the general formula for the noncyclic alkenes and demonstrates that caution must be exercised when writing structures from molecular formulas alone.

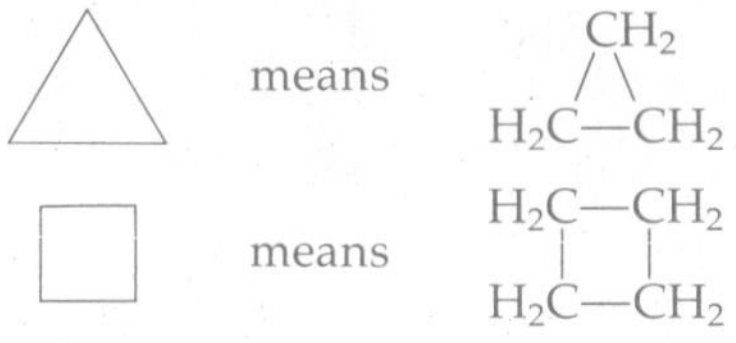

In cyclopropane, shown above, the C—C—C bond angle is 60°, while in cyclobutane,

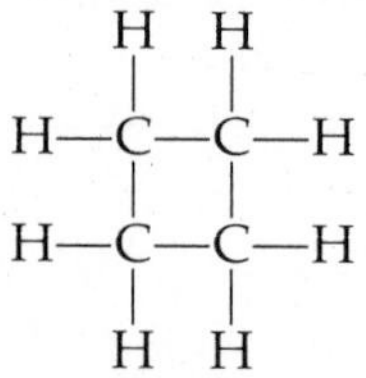

the C—C—C bond angle is about 90°. In both cases the bond angle is much less than the stable tetrahedral bond angle of 109° exhibited by carbon in the noncyclic alkanes. These small bond angles produce a relatively poor overlap of the orbitals used to form the bonds. This is because the orbitals do not point directly at one another and, as a result, the bonds are weak. In the language of organic chemistry, the bonds are said to be *strained*, and as a rule such molecules possess

Cyclic hydrocarbons are often represented by a polygon, where it is assumed that at each vertex there is a carbon plus a sufficient number of hydrogens to give a total of four bonds to the carbon.

Table 23.4
Some cycloalkanes

Structural Formula	Molecular Formula	Name
	C_3H_6	Cyclopropane
	C_5H_{10}	Cyclopentane
	C_6H_{12}	Cyclohexane
	C_7H_{14}	Cycloheptane

a high degree of reactivity. Cyclopropane, for example, which is used as a very fast-acting anesthetic, forms extremely flammable mixtures with air. The straight-chain propane, on the other hand, is much less reactive.

The next two members that follow cyclobutane (i.e., cyclopentane and cyclohexane) are quite stable. In cyclopentane, which has a nonplanar pentagonal structure, the C—C—C bond angle is very nearly 109° and consequently we expect it to be stable. In cyclohexane we find that in order to achieve more closely the tetrahedral angle of 109°, the ring is warped or puckered. In a planar or flat hexagon,

cyclohexane

the C—C—C bond angle would have to be 120° and, as a result, the structure

Figure 23.5

Two structures of cyclohexane that are free of "angle strain."

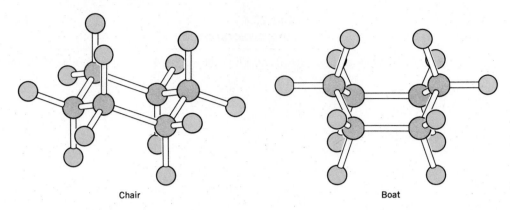

Chair   Boat

would be strained. Two structures of cyclohexane that are free of this angle strain are shown in Figure 23.5.

Cyclic alkenes

Cyclic alkenes are formed containing three or more carbon atoms and are named in a fashion similar to that for the cycloalkanes. For example, cyclopentene would be

cyclopentene

Other examples include

1,3-cyclopentadiene   **1,3-cyclohexadiene**

1,3-cyclopentadiene

In these compounds the numbering begins at the carbon with the first double bond and continues in the direction leading to the smallest numbers for the remaining double bonds. Following this idea, the compound would be called 1,3,5-cyclohexatriene. However, the orientation of the orbitals that would be involved in the bonding in this molecule gives rise to some special properties. This compound is called benzene and forms the basis of another entire series of organic compounds.

23.5 AROMATIC HYDROCARBONS

Benzene has properties that are unlike those of the other cycloalkenes and is placed in a separate class. Benzene, and the host of benzenelike compounds (those containing a benzene ring), are collectively called the *aromatic compounds* because many of them have pleasing aromatic odors. In spite of this, however, benzene and many other aromatic compounds are quite toxic.

The main structural feature in the aromatic compounds, which is responsible for their distinctive chemical properties, is the benzene ring. We saw in the last section that the stoichiometry of benzene corresponds to that of the cyclotriene,

Physical and chemical evidence, however, reveals that this does not give an accurate representation of the molecule. For instance, it is found experimentally that all the C—C bond distances in benzene are the same. If there were, in fact, alternating double and single bonds, some bond distances (C=C double bonds) would be shorter than others. In addition, the benzene molecule does not readily undergo chemical reactions typical of molecules containing double bonds.

The uniform bond distances in benzene can be accounted for by resonance. The two resonance structures (also called Kekulé structures[1]) that are usually written are

These are usually represented simply as

and

It is understood that at each vertex of the hexagon there is a carbon atom bound to a hydrogen atom.

[1] These structures are named after the German chemist, August Kekulé, who first proposed them in 1865.

Figure 23.6
Sigma bond framework in benzene. Each carbon uses sp² hybrid orbitals.

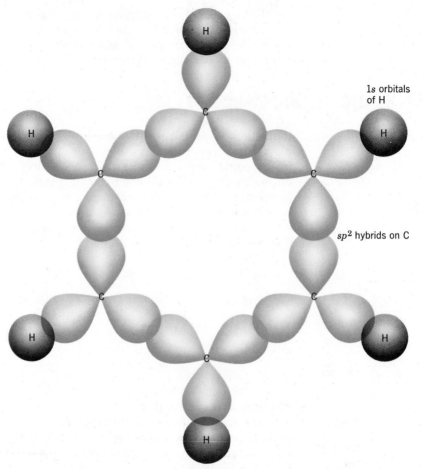

1s orbitals of H

sp² hybrids on C

According to *valence bond theory*, to achieve a C—C—C bond angle of 120°, the carbon atoms must use sp^2 hybrid orbitals to form the σ-bond framework of the ring. Thus on each of the carbons in the benzene ring we have three of the valence electrons in sp^2 hybrids and a fourth in a pure p orbital. The carbon skeleton showing only the hybridization can be seen in Figure 23.6. This same skeleton showing the usual dash for the σ-bonds as well as the electrons in the pure p orbitals is illustrated in Figure 23.7. The two resonance structures of benzene would be formed by the overlap of pairs of adjacent p orbitals, as shown in Figure 23.8.

In Section 6.9 it was pointed out that the *molecular orbital theory* can quite successfully explain the bonding in polyatomic molecules without resorting to resonance. According to this theory, the six unhybridized p atomic orbitals in benzene overlap to form one continuous molecular orbital that extends over the entire molecule. This is illustrated in Figure 23.9, where we see the σ-bonds as the solid lines between the carbons and the π-bonds as a donut-shaped, cloud with electron density above and below the carbon ring. You may recall that electrons belonging to a molecular orbital that extends over several nuclei (such as the six π electrons in benzene) are said to be delocalized. The benzene ring is frequently drawn as

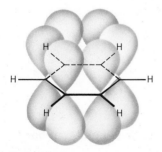

Figure 23.7
Electrons in unhybridized p orbitals on C atoms in benzene.

to emphasize the delocalized nature of the π electrons.

Figure 23.8

Valence bond resonance structures of benzene.

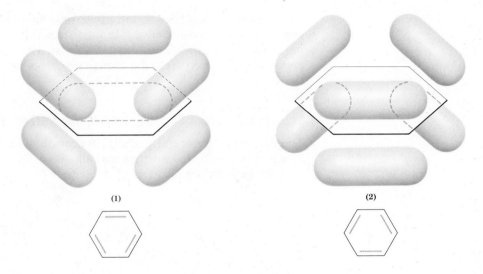

This extra stability is called the resonance energy or delocalization energy.

The delocalization of the π-electron cloud in benzene leads to a very stable ring structure. In fact, thermochemical calculations show that benzene is more stable than the hypothetical 1,3,5-cyclohexatriene by approximately 150 kJ mol^{-1}.

Aromatic compounds are important in many ways. We often find them in solvents of various kinds as well as many plastics. The benzene ring is also in many biologically important compounds such as vitamins, proteins, and hormones. Unfortunately, the human body does not possess the ability to synthesize the benzene ring and we must therefore obtain them from an outside source. Plants have the ability to synthesize some of the needed aromatics and they are therefore essential to our diets.

Benzene itself is poisonous.

Not all aromatics are beneficial. In fact, some of them are extremely harmful. One example is the compound 1,2-benzopyrene, composed of five benzene rings fused together.

1,2-benzopyrene

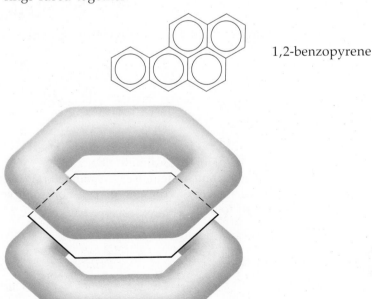

Figure 23.9

The delocalized molecular orbital electron cloud in benzene sandwiches the sigma bond framework of the hexagonal ring.

This compound has been found in cigarette smoke, in the exhaust from gasoline engines, and even on charcoal-broiled steaks. It is one of the most potent carcinogens (cancer-producing agents) known. For example, even small amounts applied to a shaved area of a mouse produces skin cancer very nearly 100% of the time.

Nomenclature of some benzene derivatives

The nomenclature of the aromatic hydrocarbons follows much the same pattern as we outlined for the alkanes and alkenes. Here, *benzene is generally taken to be the parent and the attached groups are named as before.* For example, the compound

is called methylbenzene. Its common name is *toluene.* Another example is ethylbenzene,

In a compound having more than one substituent attached to the ring, two systems may be used to identify their positions. In one of them the carbon atoms in the ring are numbered 1 to 6, beginning with the carbon that is bonded to the first group and continuing in such a direction as to lead to the lowest numbers for the remaining ones. For example, the compound,

would be called 1,2-dimethylbenzene, whereas the compound

would be named 1,3-dimethylbenzene.

The second system uses the names *ortho, meta,* and *para* to identify positions that are adjacent to the first group. For example, the *ortho, meta,* and *para* positions on toluene are

Thus the compounds

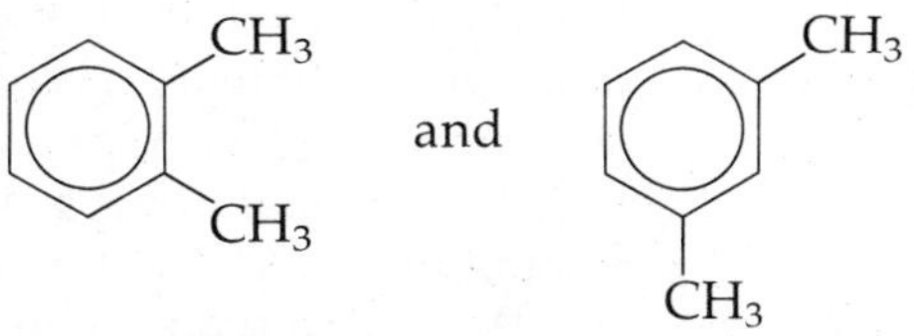

and

would be called *ortho*-methyltoluene and *meta*-methyltoluene, respectively. These compounds also have common names, which are *ortho*-xylene and *meta*-xylene (or simply *o*-xylene and *m*-xylene).

In some compounds it is sometimes easier to consider the benzene ring as the attached group and the longest carbon chain as the parent. In this case the benzene ring (C_6H_5) is called **phenyl** (rhymes with kennel), and the same rules developed earlier for the alkanes and alkenes are employed. For example, we have

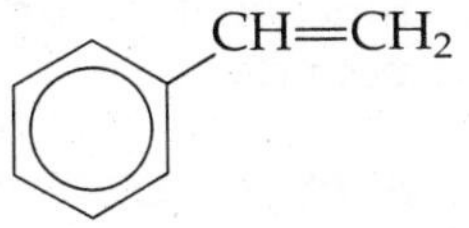

which we would call phenylethene. Its more common name is *styrene*. This compound is used to manufacture the plastic *polystyrene*, which is easily molded into various forms (e.g., model airplanes) and foamed to give the thermal insulating product, *styrofoam*.

Some other examples are

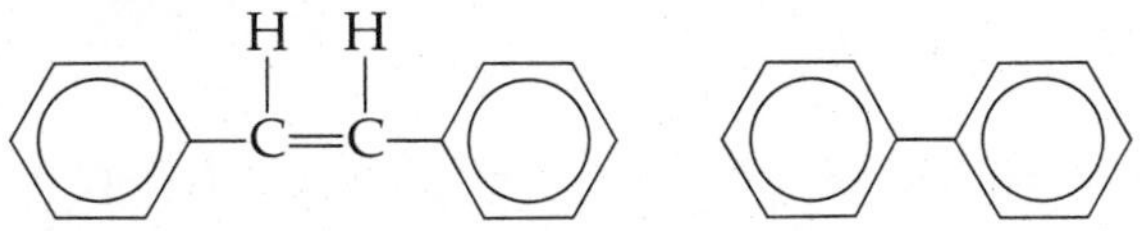

1,2-diphenylethene   **biphenyl**

The second compound, biphenyl, has become the villain in a very serious pollution problem. When the hydrogen atoms in biphenyl are replaced by chlorine atoms, polychlorinated biphenyls (PCBs) are produced. These have many desirable physical and chemical properties that have led to their widespread use in many consumer products. For example, they are excellent flame retardants and, as oils, have been used in many types of electric motors; for example, in refrigerators, air conditioners, clothes washers and dryers, and furnace blowers. They have also been used in electrical transformers such as those seen on utility poles that feed electricity into homes. Over the past five decades large quantities of these very stable PCBs have found their way into the environment where they are just now beginning to be seen as a potential hazard of immense proportions. It has been discovered that they produce severe acne and loss of hair in humans, while in test animals, PCBs have caused birth defects, cancer, and death. Their presence in high concentrations in sediments in the Hudson River has already led to a ban on commercial fishing in this river. At the same time, PCBs are beginning to show up in fish caught in other areas too.

PCB's are nonpolar and become concentrated in the fatty tissue of animals.

23.6 HYDROCARBON DERIVATIVES

Only a small fraction of all organic compounds contain just carbon and hydrogen; most contain other elements as well. In attempting to organize these compounds it is most convenient to view them as being derived from a hydrocarbon by replacing one or more of the hydrogens of the parent molecule by other atoms or groups of atoms. That is why we call them *hydrocarbon derivatives.* Usually the attached atoms or groups bestow some characteristic property to the molecule so that any molecule with the same grouping will react chemically in a similar fashion. Groups bestowing such properties on organic compounds are called **functional groups.** The study of organic chemistry is greatly simplified by examining the properties and reactions of various functional groups.

We have already seen two functional groups within the hydrocarbons—the carbon–carbon double bond and the carbon–carbon triple bond. Some other important functional groups, such as the halogens (X), the hydroxyl group (—OH, in alcohols), and so on, are listed in Table 23.5.

For convenience, organic compounds can be viewed as composed of two parts: A hydrocarbon fragment that is generally denoted as R (aliphatic) or Ar (aromatic) plus one or more functional groups, such as those listed in Table 23.5. Thus we have R—X or Ar—X, an alkyl halide or aryl halide, respectively, R—OH, an alcohol, and so forth. Let's now take a brief look at some of these functional groups and some of the properties they impart to organic compounds.

Alkenes and alkynes

The double and triple bonds in alkenes and alkynes show a characteristic tendency to undergo addition reactions, that is, reactions in which the components of a reactant molecule become incorporated into the alkene or alkyne. For example, we have already seen that molecules having carbon–carbon double or triple bonds can be made to react with hydrogen.

The most common catalysts for hydrogenation reactions are finely divided nickel, paladium, platinum, ruthenium, and rhodium.

$$
\begin{array}{c}
\mathrm{H} \\
\diagdown \\
\mathrm{C}=\mathrm{C} \\
\diagup \\
\mathrm{H}
\end{array}
\begin{array}{c}
\mathrm{H} \\
\diagup \\
\diagdown \\
\mathrm{H}
\end{array}
+ \mathrm{H_2} \xrightarrow[\text{pressure}]{\text{catalyst}}
\begin{array}{c}
\mathrm{H}\ \ \mathrm{H} \\
|\ \ \ | \\
\mathrm{H{-}C{-}C{-}H} \\
|\ \ \ | \\
\mathrm{H}\ \ \mathrm{H}
\end{array}
$$

$$
\mathrm{H{-}C{\equiv}C{-}H} + 2\mathrm{H_2} \xrightarrow[\text{pressure}]{\text{catalyst}}
\begin{array}{c}
\mathrm{H}\ \ \mathrm{H} \\
|\ \ \ | \\
\mathrm{H{-}C{-}C{-}H} \\
|\ \ \ | \\
\mathrm{H}\ \ \mathrm{H}
\end{array}
$$

Hydrogen is not the only substance that can be added to a double bond. Some other examples are

$$
\mathrm{CH_3{-}CH{=}CH{-}CH_3} + \mathrm{HBr} \longrightarrow
\begin{array}{c}
\mathrm{CH_3{-}CH{-}CH{-}CH_3} \\
|\quad\ \ | \\
\mathrm{H}\quad\ \mathrm{Br}
\end{array}
$$

$$
\mathrm{H{-}C{\equiv}C{-}H} + \mathrm{Br_2} \longrightarrow
\begin{array}{c}
\mathrm{H{-}C{=}C{-}H} \\
|\quad\ | \\
\mathrm{Br}\ \ \mathrm{Br}
\end{array}
$$

$$
\mathrm{H{-}C{\equiv}C{-}H} + 2\mathrm{Br_2} \longrightarrow
\begin{array}{c}
\mathrm{Br}\ \ \mathrm{Br} \\
|\ \ \ | \\
\mathrm{H{-}C{-}C{-}H} \\
|\ \ \ | \\
\mathrm{Br}\ \ \mathrm{Br}
\end{array}
$$

$$
\mathrm{CH_2{=}CH_2} + \mathrm{H_2O} \xrightarrow{\mathrm{H^+}} \mathrm{CH_3CH_2OH}
$$

In this last case $\mathrm{H^+}$ serves as a catalyst and is written over the arrow.

Table 23.5
Some functional groups in organic compounds

Functional Group	Compound Class	Example	Name
—C=C—	Alkenes	$CH_2=CH_2$	Ethene
—C≡C—	Alkynes	$HC≡CH$	Ethyne
F, Cl, Br, I	Halides	CH_3Cl	Chloromethane
—OH	Alcohols	CH_3OH	Methanol
—C(=O)H	Aldehydes	$CH_3—C(=O)—H$ or CH_3CHO	Ethanal (acetaldehyde)
—C(=O)—	Ketones	$CH_3—C(=O)—CH_3$	Propanone (acetone)
—C(=O)OH	Carboxylic acids	$CH_3—C(=O)—OH$	Ethanoic acid (acetic acid)
—C—N(H)(H)	Amines	CH_3NH_2	Methylamine
—C(=O)—N(H)(H)	Amides	$CH_3—C(=O)—NH_2$	Ethanamide (acetamide)
—C—O—C—	Ethers	$CH_3—O—CH_3$	Dimethyl ether
—C—C(=O)—O—C—	Esters	$CH_3—C(=O)—O—CH_3$	Methyl acetate

When a substance such as HBr is added to an unsymmetrical alkene there are two possibilities for reaction.

$$CH_3—CH=CH_2 + HBr \longrightarrow CH_3—CHBr—CH_3$$
2-bromopropane

$$CH_3—CH=CH_2 + HBr \longrightarrow CH_3—CH_2—CH_2Br$$
1-bromopropane

In this situation it happens that very little of the 1-bromo product is formed. Nearly 90% of the product is 2-bromopropane. The products of such reactions can be predicted by *Markovnikov's*[2] *rule*, which states that during an addition to

[2] The publication stating Vladimir W. Markovnikov's rule was published in 1905, one year after his death.

ordinary alkenes the hydrogen of the reactant becomes attached to that carbon of the alkene already bonded to the most hydrogen atoms. In propene, the end carbon has two hydrogen atoms while the middle carbon has only one. The H of the HBr therefore becomes attached to the end carbon, while the Br becomes bonded to the middle carbon. Other unsymmetrical reagents that follow this rule are H_2O, HCN, and the other hydrogen halides. For example, we find that the addition of water to 2-methylpropene gives the following:

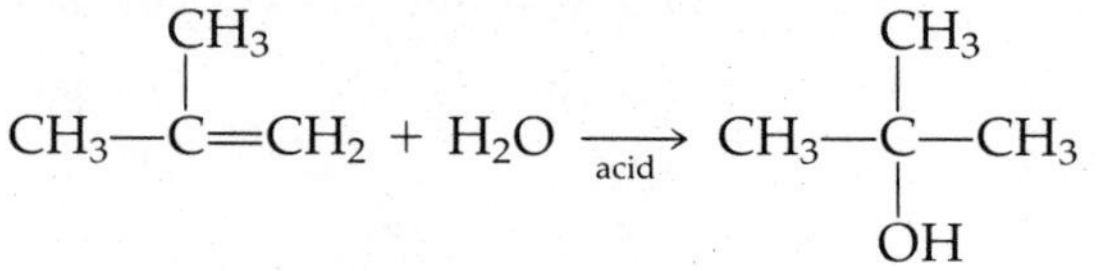

$$CH_3-\underset{\underset{}{\overset{\overset{CH_3}{|}}{C}}}{}=CH_2 + H_2O \xrightarrow[\text{acid}]{} CH_3-\underset{\underset{OH}{|}}{\overset{\overset{CH_3}{|}}{C}}-CH_3$$

The tendency of the unsaturated hydrocarbons to undergo addition reactions can be contrasted with the tendency of the alkanes to react through substitution. For example, methane will react with Cl_2 to give a variety of products in which Cl replaces H.

$$CH_4 + Cl_2 \longrightarrow \left\{\begin{array}{l} CH_3Cl \\ CH_2Cl_2 \\ CHCl_3 \\ CCl_4 \end{array}\right\} + HCl$$

Under similar conditions the alkenes and alkynes have little tendency to undergo substitution.

23.7 HALOGEN DERIVATIVES

A *halogen derivative* is a molecule in which a halogen atom has replaced a hydrogen in the parent hydrocarbon. Some examples of these were seen in the last section as products of reactions of hydrocarbons with halogens or hydrogen halides. In naming these compounds, the halogen is specified as *fluoro*, *chloro*, *bromo*, or *iodo*.

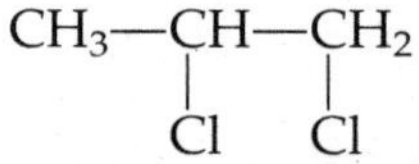

1,2-dichloropropane

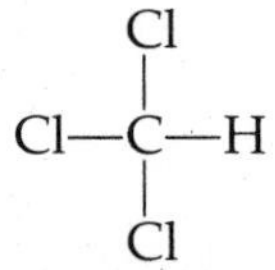

trichloromethane
(chloroform)

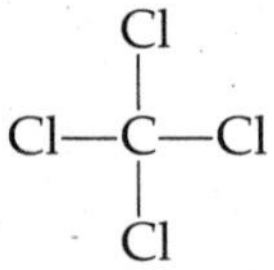

tetrachloromethane
(carbon tetrachloride)

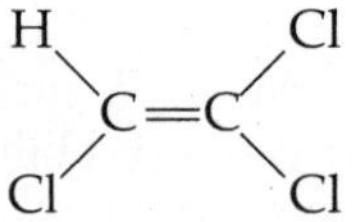

trichloroethene
(trichloroethylene)

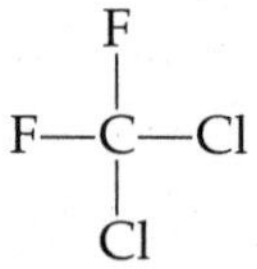

dichlorodifluoromethane
(*Freon-12*)

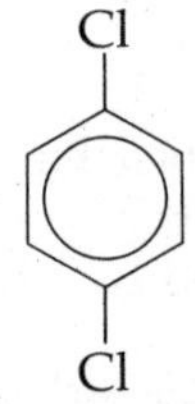

p-dichlorobenzene

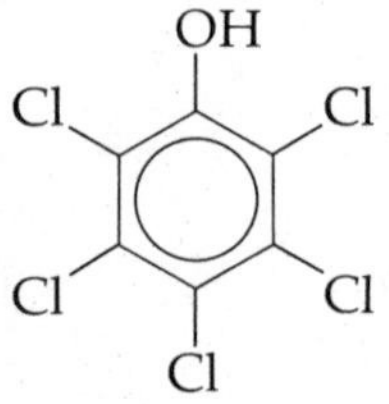

pentachlorophenol

Many halogenated hydrocarbons have important commercial applications. Trichloroethene, for example, is a common dry cleaning solvent. The *freons* are used as refrigerants and propellants in aerosol products. As we discussed earlier, there is considerable controversy over whether these materials are seriously depleting the ozone shield in the earth's upper atmosphere (see page 699).

Many halogenated compounds are toxic. Carbon tetrachloride, for example, is no longer used as a dry cleaning solvent because it is a cumulative poison. Many insecticides contain halogenated compounds. For example, *p*-dichlorobenzene (above) has been used in moth balls. Pentachlorophenol, also shown above, is used as a wood preservative because it is toxic to creatures that attack wood.

The alkyl halides can be prepared by addition of the halogen or hydrogen halides to alkenes, as we have seen, as well as by substitution of a halogen for a hydrogen in an alkane.

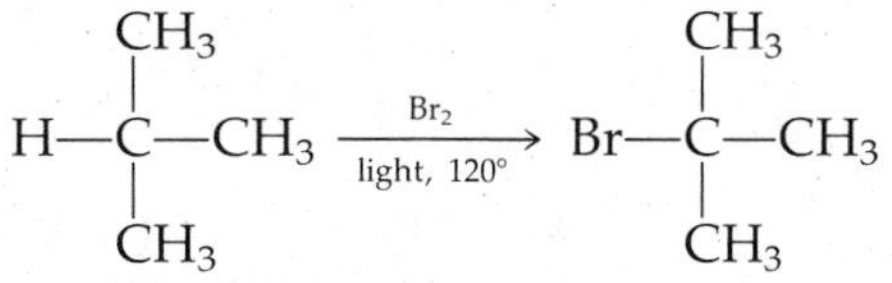

$$
\underset{\underset{CH_3}{|}}{\overset{\overset{CH_3}{|}}{H-C-CH_3}} \xrightarrow[\text{light, }120°]{Br_2} \underset{\underset{CH_3}{|}}{\overset{\overset{CH_3}{|}}{Br-C-CH_3}}
$$

Here the Br replaces the hydrogen bonded least strongly to the carbon in the molecule.

Perhaps the most important method of preparing alkyl halides is from alcohols. During this reaction the OH of the alcohol is displaced by a halide ion. Examples of this are

$$CH_3-CH_2-CH_2-CH_2-OH + HBr \longrightarrow CH_3-CH_2-CH_2-CH_2-Br + H_2O$$

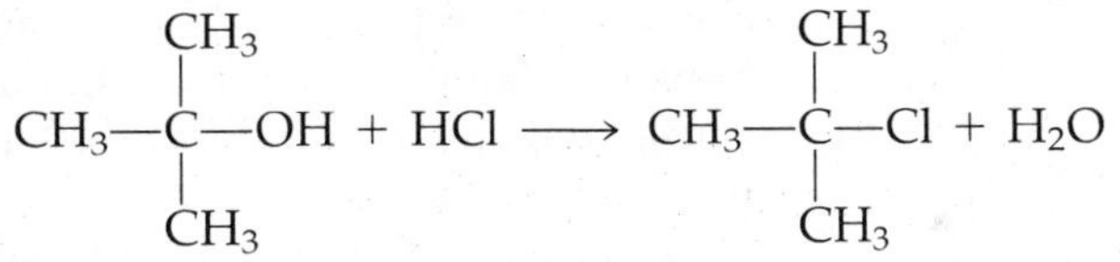

$$
\underset{\underset{CH_3}{|}}{\overset{\overset{CH_3}{|}}{CH_3-C-OH}} + HCl \longrightarrow \underset{\underset{CH_3}{|}}{\overset{\overset{CH_3}{|}}{CH_3-C-Cl}} + H_2O
$$

23.8 ORGANIC COMPOUNDS THAT CONTAIN OXYGEN

Many common organic compounds have functional groups that contain oxygen— for example, alcohols, aldehydes and ketones, carboxylic acids and esters, and ethers. Many are related to each other and therefore we discuss them together in this section.

Alcohols

Alcohols are characterized by the functional group, —OH. Some typical alcohols with their IUPAC and common names are

Formula	IUPAC Name	Common Name
CH_3-OH	Methanol	Methyl alcohol
CH_3-CH_2-OH	Ethanol	Ethyl alcohol
$CH_3-CH_2-CH_2-OH$	1-Propanol	*n*-Propyl alcohol

As you can see, the name of an alcohol is obtained from the name of the parent alkane by replacing the *e* by *ol.* Also, when necessary, the position of the OH group is identified by number. Thus we have

$$
\begin{array}{ll}
CH_3-\underset{\underset{OH}{|}}{CH}-CH_3 & \text{2-propanol} \\[2em]
CH_3-\underset{\underset{OH}{|}}{CH}-CH_2-CH_3 & \text{2-butanol} \\[2em]
CH_3-CH_2-\underset{\underset{OH}{|}}{CH}-CH_2-CH_3 & \text{3-pentanol} \\[2em]
\underset{H_3C}{\overset{H_3C}{\diagup}}CH-CH_2-OH & \text{2-methyl-1-propanol}
\end{array}
$$

There are many important alcohols. Ethanol, of course, is found in alcoholic beverages. Compared with most other alcohols, ethanol is relatively nontoxic. When it is consumed even in small quantities, however, it causes the blood vessels to dilate, resulting in a lowering of the blood pressure followed by a general feeling of relaxation. In larger amounts ethanol causes intoxication, and excessively prolonged use can permanently damage the liver and eventually lead to death.

Methanol, also known as wood alcohol, is a deadly poison, while the remaining alcohols, other than ethanol, are somewhat milder poisons. Methanol can cause blindness and eventually total loss of motor control and death. Perhaps one familiar use of methanol is in dry gas, which is added to gasoline to cause suspended water droplets in the fuel to dissolve. In this way water that may have condensed in a fuel tank can pass harmlessly through the engine.

Methanol is prepared in large quantities from the reaction between carbon monoxide and hydrogen in the presence of a metal oxide catalyst at high temperature and pressure. The balanced equation for this preparation is

$$CO + 2H_2 \xrightarrow[\substack{20\text{ MPa}\\350-400°C}]{ZnO/Cr_2O_3} CH_3OH$$

This reaction was discussed in Section 19.1.

The common method of preparing ethanol is through the fermentation of carbohydrates (sugars or starch). Sugars are converted into ethanol and carbon dioxide by the action of yeast in the absence of oxygen. For example,

$$\underset{\substack{\text{glucose}\\\text{(a carbohydrate)}}}{C_6H_{12}O_6} \xrightarrow{\text{yeast}} \underset{\text{ethanol}}{2C_2H_5OH} + 2CO_2$$

Another important alcohol is 2-propanol, commonly known as isopropyl alcohol, which is used widely as rubbing alcohol.

The number of groups attached to the carbon to which the OH is bonded aids us in classifying the alcohols. The C—OH grouping is called the **carbinol group,** and the carbon of this group is referred to as the **carbinol carbon.** Compounds in which there is one hydrocarbon group (R) attached to the carbinol carbon are known as **primary alcohols.** Alcohols that contain two such R groups are known as **secondary alcohols. Tertiary alcohols,** on the other hand, have three R groups bonded to the carbinol carbon.

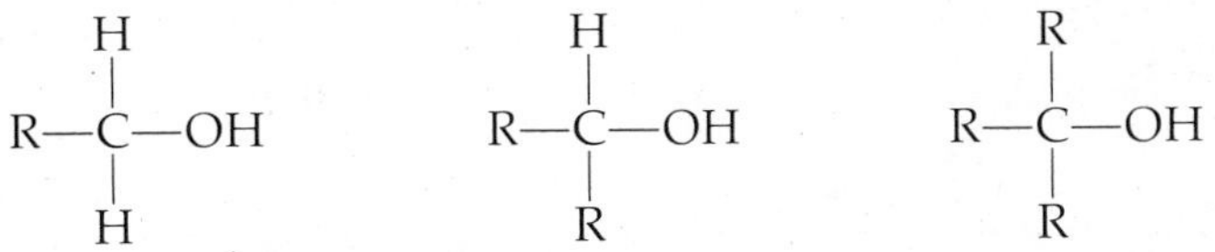

$$
\begin{array}{ccc}
\text{H} & \text{H} & \text{R} \\
| & | & | \\
\text{R}-\text{C}-\text{OH} & \text{R}-\text{C}-\text{OH} & \text{R}-\text{C}-\text{OH} \\
| & | & | \\
\text{H} & \text{R} & \text{R}
\end{array}
$$

primary alcohol secondary alcohol tertiary alcohol

Some specific examples are

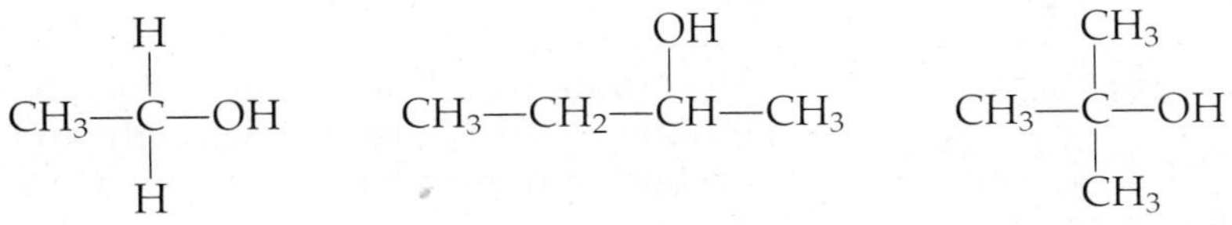

$$
\begin{array}{ccc}
\text{H} & \text{OH} & \text{CH}_3 \\
| & | & | \\
\text{CH}_3-\text{C}-\text{OH} & \text{CH}_3-\text{CH}_2-\text{CH}-\text{CH}_3 & \text{CH}_3-\text{C}-\text{OH} \\
| & & | \\
\text{H} & & \text{CH}_3
\end{array}
$$

a primary alcohol a secondary alcohol a tertiary alcohol

Following the IUPAC nomenclature, these alcohols are called ethanol, 2-butanol, and 2-methyl-2-propanol. Their common names are ethyl alcohol, *sec*-butyl alcohol, and *tert*-butyl alcohol.

Alcohols containing more than one hydroxyl group are also possible and are called polyhydroxy alcohols. Important examples are ethylene glycol (1,2-ethanediol),

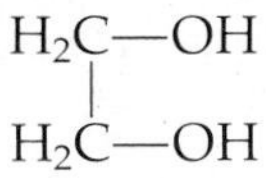

$$
\begin{array}{l}
\text{H}_2\text{C}-\text{OH} \\
| \\
\text{H}_2\text{C}-\text{OH}
\end{array}
\quad \text{ethylene glycol}
$$

propylene glycol (1,2-propanediol)

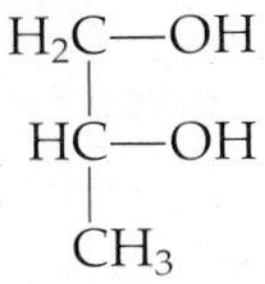

$$
\begin{array}{l}
\text{H}_2\text{C}-\text{OH} \\
| \\
\text{HC}-\text{OH} \\
| \\
\text{CH}_3
\end{array}
\quad \text{propylene glycol}
$$

and glycerol (1,2,3-propanetriol)

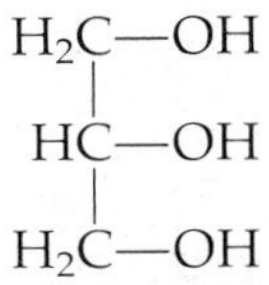

$$
\begin{array}{l}
\text{H}_2\text{C}-\text{OH} \\
| \\
\text{HC}-\text{OH} \\
| \\
\text{H}_2\text{C}-\text{OH}
\end{array}
\quad \text{glycerol (also, glycerin)}
$$

Ethylene glycol is toxic, but propylene glycol and glycerol are not.

These three compounds are soluble in water in all proportions. Ethylene glycol is used widely as an automotive antifreeze and in the manufacture of polyesters such as *Dacron* and *Mylar*. Propylene glycol is used as a nontoxic antifreeze and as a food additive to prevent the growth of mold. Glycerol occurs in fats, as we will see in the next chapter, and is also used in antifreeze applications when a nontoxic antifreeze is necessary. Glycerol is also sometimes added to alcoholic beverages to promote smoothness.

Polyhydroxy alcohols can be *nitrated* by their cautious addition to a mixture of nitric and sulfuric acids.

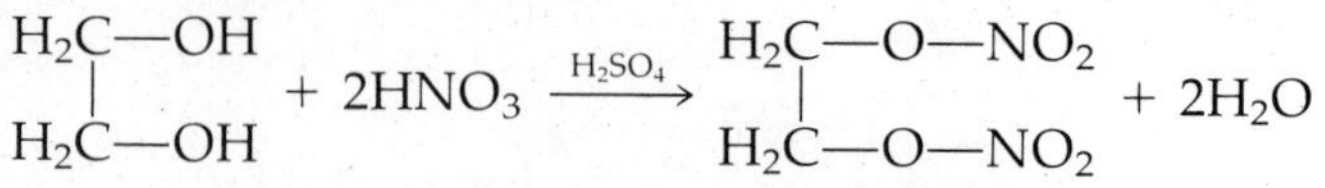

$$
\begin{array}{c}
H_2C\!-\!OH \\
| \\
H_2C\!-\!OH
\end{array}
+ 2HNO_3 \xrightarrow{\ H_2SO_4\ }
\begin{array}{c}
H_2C\!-\!O\!-\!NO_2 \\
| \\
H_2C\!-\!O\!-\!NO_2
\end{array}
+ 2H_2O
$$

glycol dinitrate

$$
\begin{array}{c}
H_2C\!-\!OH \\
| \\
HC\!-\!OH \\
| \\
H_2C\!-\!OH
\end{array}
+ 3HNO_3 \xrightarrow{\ H_2SO_4\ }
\begin{array}{c}
H_2C\!-\!O\!-\!NO_2 \\
| \\
HC\!-\!O\!-\!NO_2 \\
| \\
H_2C\!-\!O\!-\!NO_2
\end{array}
+ 3H_2O
$$

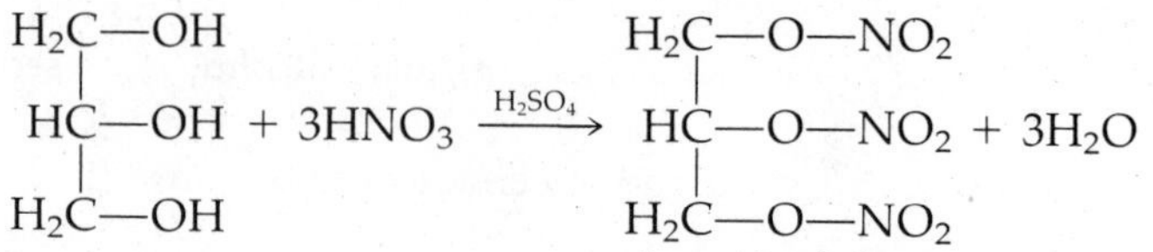

glyceryl trinitrate

The products of these reactions must be handled with extreme caution. Glyceryl trinitrate is also called nitroglycerin and is used with glycol dinitrate in the production of dynamite.

Aldehydes and ketones

Aldehydes and ketones are characterized by the presence of a **carbonyl group,** $\diagup\!\!\!\!C\!=\!O$. In aldehydes this occurs on an end carbon,

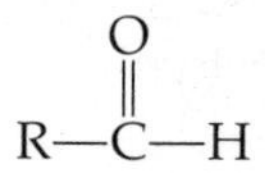

$$
\begin{array}{c}
O \\
\parallel \\
R\!-\!C\!-\!H
\end{array}
\quad \text{aldehyde}
$$

while in ketones it occurs on one of the middle carbon atoms,

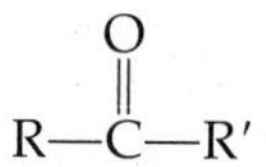

$$
\begin{array}{c}
O \\
\parallel \\
R\!-\!C\!-\!R'
\end{array}
\quad \text{ketone}
$$

(R and R' indicate the possibility of having different alkyl groups attached to the carbonyl group.)

Many aldehydes have pleasant odors, particularly those in which the R group is aromatic.

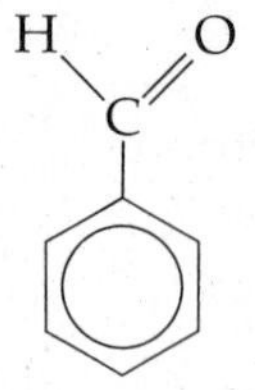

benzaldehyde   **vanillin**   **cinnamaldehyde**

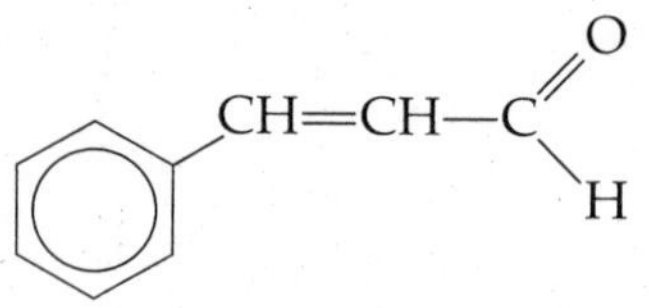

(bitter almonds)   **(vanilla bean)**   **(cinnamon)**

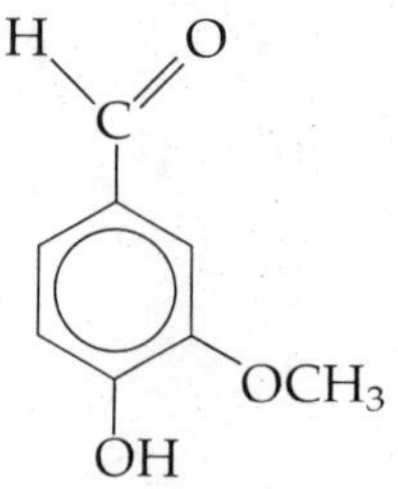

Ketones often have very desirable solvent properties. Acetone, for example, is found in nail polish remover; methyl ethyl ketone is a solvent in airplane glue.

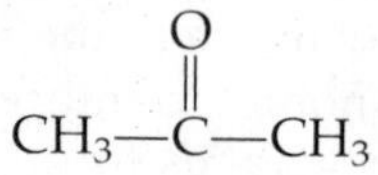

$$
\begin{array}{cc}
O & O \\
\parallel & \parallel \\
CH_3\!-\!C\!-\!CH_3 \qquad & CH_3\!-\!C\!-\!CH_2\!-\!CH_3 \\
\text{acetone} & \text{methyl ethyl ketone}
\end{array}
$$

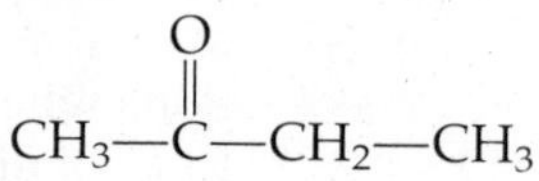

In naming aldehydes and ketones, the $-e$ of the corresponding hydrocarbon parent is dropped and replaced by **-al** for aldehydes or **-one** for ketones. Thus we have

$$CH_3-\overset{\overset{\displaystyle O}{\|}}{C}-H \qquad \text{ethanal (common: acetaldehyde)}$$

The IUPAC names of the ketones, acetone and methyl ethyl ketone, would be propanone and butanone, respectively. When there are possible alternative locations of the carbonyl group, its position is indicated by number.

$$CH_3-\overset{\overset{\displaystyle O}{\|}}{C}-CH_2-CH_2-CH_3 \qquad CH_3-CH_2-\overset{\overset{\displaystyle O}{\|}}{C}-CH_2-CH_3$$

2-pentanone   **3-pentanone**

Aldehydes and ketones are related to alcohols through oxidation and reduction. Oxidation of a primary alcohol can yield an aldehyde (although these are usually difficult to isolate because they are generally easily oxidized further).

$$CH_3-\overset{\overset{\displaystyle H}{|}}{\underset{\underset{\displaystyle H}{|}}{C}}-OH \xrightarrow{(O)} CH_3-\overset{\overset{\displaystyle H}{|}}{C}=O$$

ethanol   **ethanal**

Secondary alcohols are oxidized to ketones.

$$CH_3-\overset{\overset{\displaystyle OH}{|}}{CH}-CH_3 \xrightarrow{(O)} CH_3-\overset{\overset{\displaystyle O}{\|}}{C}-CH_3$$

2-propanol   **propanone**
                                         (acetone)

The symbol (O) is used here to indicate oxidation without specifying the oxidizing agent. Oxidation of tertiary alcohols, which is considerably more difficult, breaks down the carbon chain.

The reason for the various products seems to be the number of hydrogens attached to the carbinol carbon. During the oxidation reaction one hydrogen is eliminated from the carbinol carbon as well as the one on the OH. With a primary alcohol, therefore, one hydrogen remains on the carbon after oxidation, whereas with a secondary alcohol none remain. Since no hydrogens are attached to the carbinol carbon in a tertiary alcohol, no reaction occurs when H_2CrO_4 or other similar oxidizing materials are added.

The oxidation of alcohols to aldehydes and ketones can be reversed through reduction. Often a metal hydride is used (for example, $NaBH_4$, sodium borohydride) because of the availability, at least in principle, of electron-rich H^- ions.

$$CH_3-\overset{\overset{\displaystyle O}{\|}}{C}-H \xrightarrow{NaBH_4} CH_3-\overset{\overset{\displaystyle H}{|}}{\underset{\underset{\displaystyle H}{|}}{C}}-OH$$

$$CH_3-\overset{\overset{\displaystyle O}{\|}}{C}-CH_3 \xrightarrow{NaBH_4} CH_3-\overset{\overset{\displaystyle OH}{|}}{CH}-CH_3$$

Organic acids and esters

The formula for a carboxylic acid is sometimes written RCO_2H.

Organic acids are characterized by the presence of the **carboxyl group,** —COOH. Structurally this is

$$\begin{array}{c} O \\ \| \\ R-C-O-H \end{array}$$

The presence of the lone oxygen bonded to the carbon that is attached to the —OH group polarizes the O—H bond and permits the H to be lost as a proton.

$$\begin{array}{c} :O: \\ \| \\ R-C-\ddot{O}-H \end{array} + H_2O \longrightarrow \left[\begin{array}{c} :O: \\ \| \\ R-C-\ddot{O}: \end{array}\right]^{-} + H_3O^{+}$$

Many important organic compounds are acids or their salts. Some examples are,

$$\begin{array}{c} COOH \\ | \\ CH_2 \\ | \\ HO-C-COOH \\ | \\ CH_2 \\ | \\ COOH \end{array}$$

citric acid
(citrus fruits)

acetylsalicylic acid
(aspirin)

$$\left[\begin{array}{c} COO^{-} \end{array}\right] Na^{+}$$

sodium benzoate
(a food preservative)

$$\left[\begin{array}{c} O \quad H \qquad\qquad O \\ \| \quad | \qquad\qquad \| \\ HO-C-C-CH_2-CH_2-C-O \\ \quad\quad | \\ \quad\quad NH_2 \end{array}\right]^{-} Na^{+}$$

monosodium glutamate, MSG
(*Accent,* **a flavor enhancer**)

Organic acids derived from hydrocarbons are named by dropping the $-e$ from the end of the name of the parent and adding **-oic acid.**

$$\begin{array}{c} O \\ \| \\ CH_2-C-OH \end{array} \quad \text{ethanoic acid (acetic acid)}$$

$$\begin{array}{c} O \\ \| \\ CH_3-CH_2-CH_2-C-OH \end{array} \quad \text{butanoic acid (butyric acid, from rancid butter)}$$

One method of preparing an acid is by oxidation of an aldehyde (or by thorough oxidation of a primary alcohol).

$$CH_3CH_2OH \xrightarrow{(O)} CH_3CHO \xrightarrow{(O)} CH_3COOH$$

ethanol   **ethanal**   **ethanoic acid**

Esters are characterized by the presence of the functional group,

$$\begin{array}{c} O \\ \| \\ R-C-O-R' \end{array}$$

They are products of the acid-catalyzed elimination of water from between a carboxylic acid and an alcohol, a process called **esterification.**

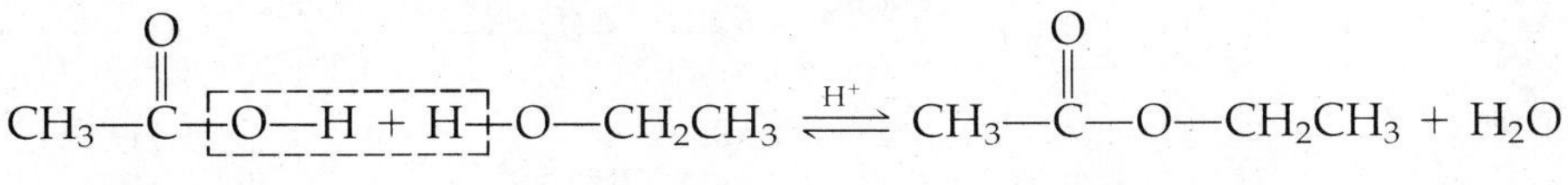

$$CH_3—\overset{\overset{\textstyle O}{\|}}{C}\underset{}{\overline{—O—H + H—}}O—CH_2CH_3 \overset{H^+}{\rightleftharpoons} CH_3—\overset{\overset{\textstyle O}{\|}}{C}—O—CH_2CH_3 + H_2O$$

acetic acid　　　　　**ethyl alcohol**　　　　　**ethyl acetate**

This reaction leads to an equilibrium. The position of equilibrium can be shifted to the right by employing a dehydrating agent to remove H_2O from the reaction mixture as it is formed. This type of reaction, where molecules are joined with the simultaneous elimination of another smaller molecule, is also called a **condensation reaction.**

The polymerization of H_3PO_4 discussed in Chapter 20 is also an example of a condensation reaction.

When an alcohol is treated with an inorganic acid an inorganic ester is produced. For example, when nitric acid is *cautiously* added to ethyl alcohol,

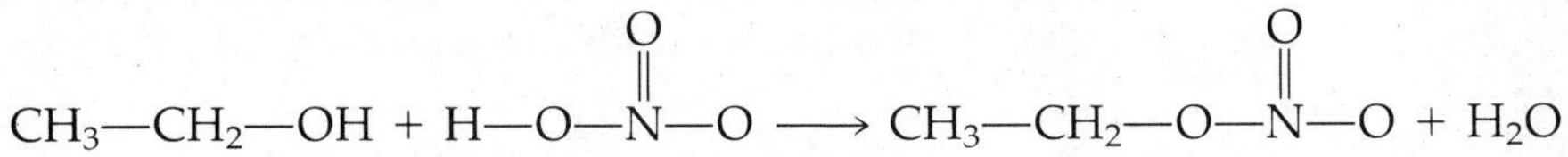

$$CH_3—CH_2—OH + H—O—\overset{\overset{\textstyle O}{\|}}{N}—O \longrightarrow CH_3—CH_2—O—\overset{\overset{\textstyle O}{\|}}{N}—O + H_2O$$

the product is ethyl nitrate (quite explosive). Organic phosphate esters of the type

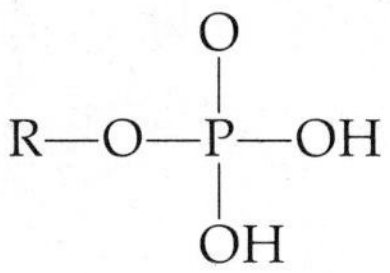

$$R—O—\overset{\overset{\textstyle O}{\|}}{\underset{\underset{\textstyle OH}{|}}{P}}—OH$$

are very important in biological systems as are esters of the trihydroxy alcohol, glycerol.

The reaction to produce an ester is reversible and the insertion of an H_2O molecule into the ester to give the acid and alcohol is termed hydrolysis. An example is the hydrolysis of methyl acetate.

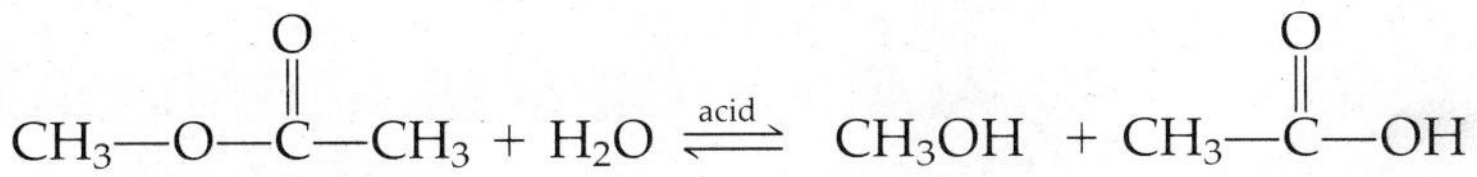

$$CH_3—O—\overset{\overset{\textstyle O}{\|}}{C}—CH_3 + H_2O \overset{acid}{\rightleftharpoons} CH_3OH + CH_3—\overset{\overset{\textstyle O}{\|}}{C}—OH$$

methyl acetate　　　　　**methanol**　　　**acetic acid**

This equilibrium is established rapidly, and in order to drive the reaction to the right, the alcohol can be removed by distillation.

In the base-catalyzed hydrolysis, also called **saponification,** the acid produced in the forward reaction is neutralized, thereby shifting the position of equilibrium to the right. For example, using molecular formulas,

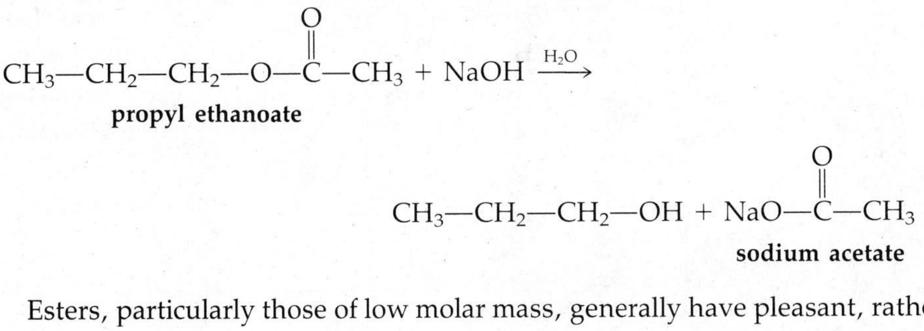

Sodium acetate is really $Na^+CH_3CO_2^-$.

$$CH_3—CH_2—CH_2—O—\overset{\overset{\textstyle O}{\|}}{C}—CH_3 + NaOH \overset{H_2O}{\longrightarrow}$$

propyl ethanoate

$$CH_3—CH_2—CH_2—OH + NaO—\overset{\overset{\textstyle O}{\|}}{C}—CH_3$$

sodium acetate

Esters, particularly those of low molar mass, generally have pleasant, rather

Table 23.6
Odors of some common esters

Name	Formula	Odor
n-Amyl acetate	$CH_3COOCH_2(CH_2)_3CH_3$	Banana
n-Octyl acetate	$CH_3COOCH_2(CH_2)_6CH_3$	Orange
iso-Amyl butyrate	$CH_3(CH_2)_2COOCH(CH_3)CH_2CH_2CH_3$	Pear

agreeable odors, as seen in Table 23.6. Some of them have very desirable solvent properties and are used in paints and varnishes.

Ethers

We saw that esters are produced by a condensation reaction between an alcohol and an acid. Alcohols can also undergo self-condensation to give **ethers** in which two hydrocarbon units are joined by an oxygen bridge.

$$R{-}OH + H{-}O{-}R \longrightarrow R{-}O{-}R + H_2O$$
ether

A specific example is

$$2CH_3{-}CH_2{-}OH \xrightarrow[H_2SO_4]{conc} CH_3{-}CH_2{-}O{-}CH_2{-}CH_3 + H_2O$$
diethyl ether

The concentrated sulfuric acid is used as a dehydrating agent to help in the removal of the H_2O as it is formed. Diethyl ether has been used as an anesthetic. It must be used with care because it is extremely flammable. Most ethers are used primarily as solvents.

23.9 AMINES AND AMIDES: ORGANIC DERIVATIVES OF AMMONIA

The **amines** are a group of compounds that can be viewed as derivatives of ammonia. Below are some typical amines:

Formula	IUPAC Name	Common Name
$CH_3{-}NH_2$	Aminomethane	Methylamine
$CH_3{-}CH_2{-}NH_2$	Aminoethane	Ethylamine
$CH_3{-}CH_2{-}NH{-}CH_2{-}CH_3$	Ethylaminoethane	Diethylamine
$(CH_3)_3C{-}NH_2$	2-Amino-2-methylpropane	*t*-Butylamine
$H_2N{-}CH_2{-}CH_2{-}NH_2$	1,2-Diaminoethane	Ethylenediamine
⬡—NH_2	Aminobenzene	Aniline

Amines can be classified as being primary, secondary, or tertiary, depending on the number of R groups attached to the nitrogen.

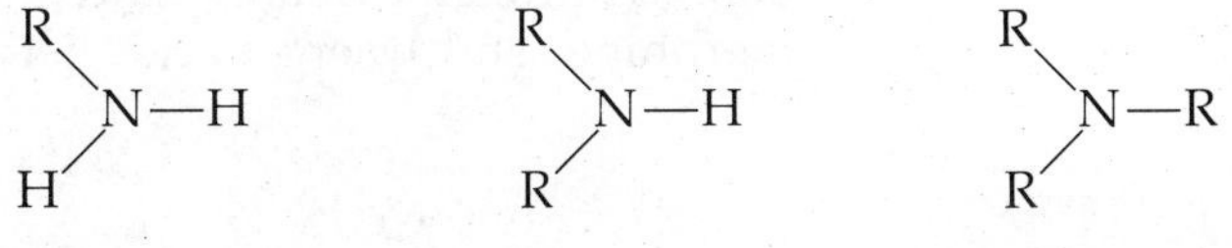

primary amine   secondary amine   tertiary amine

All the amines in our list are primary amines except diethylamine, which is a secondary amine.

Amines also exist in which the nitrogen is a member of a ring. Examples of this type are

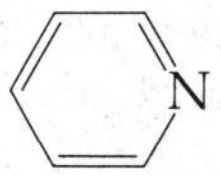

pyridine

piperidine

These compounds are referred to as **heterocycles** because not all the atoms in the ring are identical.

Amines, like ammonia, are weak bases. Most of them also have very unpleasant odors. The stench of decaying protein, for instance, can be traced to compounds like those below.

$H_2N-CH_2-CH_2-CH_2-CH_2-NH_2$ putrescine

$H_2N-CH_2-CH_2-CH_2-CH_2-CH_2-NH_2$ cadaverine

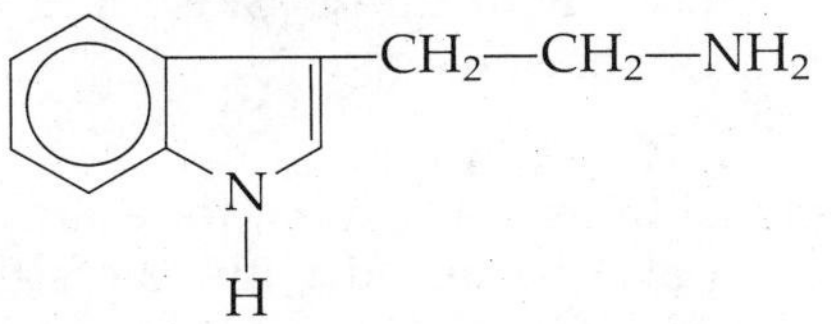

skatole (in feces)

Amides are identified by the functional group $-\overset{\overset{\displaystyle O}{\|}}{C}-NH_2$. Some examples of this type of compound are

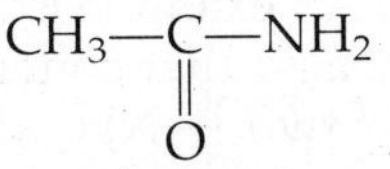

acetamide 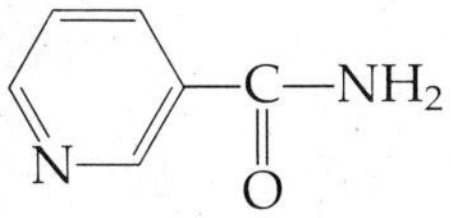   nicotinamide

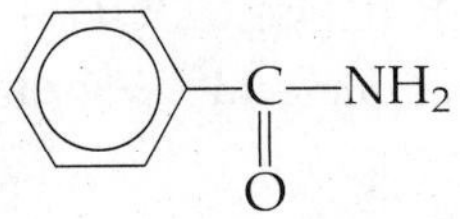

benzamide 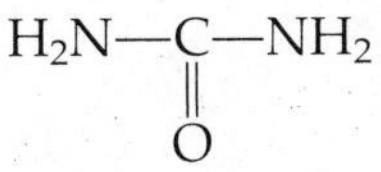   urea

The functional groups of the amines and amides are found in many important biological compounds that will be discussed in the next chapter. These compounds include the nucleic acids, the amino acids, thiamin, riboflavin, and biotin. The heterocyclic amines are also found as a basic unit of a group of compounds known as alkaloids. Alkaloids are rather complex compounds containing nitrogen that are found in plants. Compounds such as nicotine, codeine, morphine, and lysergic acid diethylamide (LSD) are all alkaloids.

23.10 POLYMERS

Polymers are very large molecules made by bonding together many smaller molecules called **monomers.** For example, polyethylene (plastic food wrap) is prepared by linking together a large number of $CH_2{=}CH_2$ monomer units to give a hydrocarbon having the general formula, $({-}CH_2{-}CH_2{-})_n$. Many naturally occurring substances are polymers, for example, rubber, starch, proteins, and the nucleic acids. Artificial polymers include such familiar materials as Bakelite®, Melmac®, nylon, Dacron®, Plexiglass®, Teflon®, and polyvinyl chloride (PVC).

Polymers can be classified as addition polymers or condensation polymers. **Addition polymers** are formed by the direct linking together of the monomer units. An example is polyvinyl chloride (PVC), which is formed by polymerizing vinyl chloride.

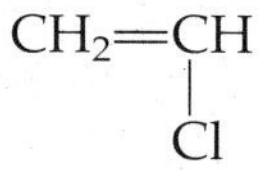

$$CH_2{=}CH{-}Cl$$

vinyl chloride

In the presence of a suitable initiator (generally, a peroxide) the double bond opens and the monomer units link end-to-end.

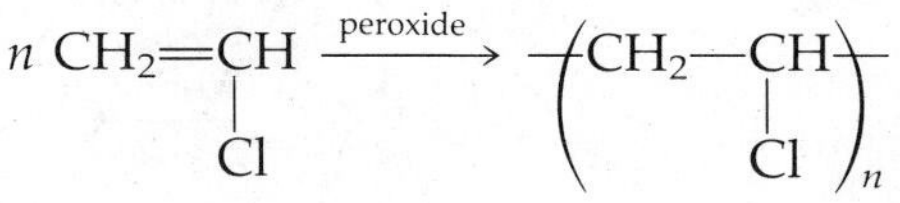

$$n\ CH_2{=}CH{-}Cl \xrightarrow{\text{peroxide}} (CH_2{-}CH{-}Cl)_n$$

This material finds many uses, for example, in phonograph records and plastic pipe. It can also be mixed with esters that soften the polymer. The softened material finds uses in such products as plastic garden hoses, tablecloths, raincoats, and vinyl leather products.

When a **condensation polymer** is formed, two monomers are joined with the simultaneous elimination of a small molecule such as water or methanol. Usually two different monomers are joined and the polymer is said to be a **copolymer.** An example is nylon, which is formed when a dicarboxylic acid (i.e., a carboxylic acid that contains two —COOH groups) reacts with a diamine (an amine with two —NH$_2$ groups). The overall reaction for the production of nylon is

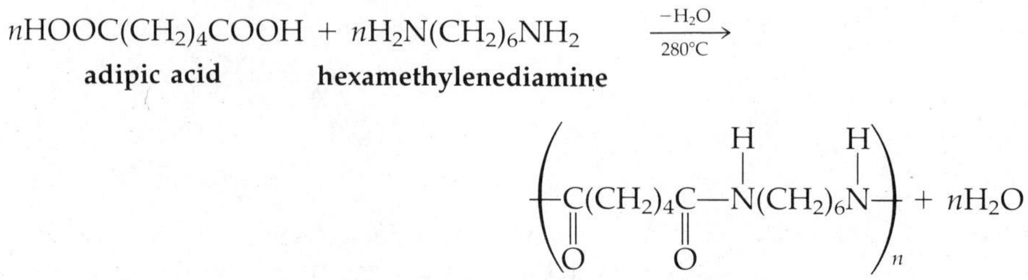

$$n\text{HOOC(CH}_2)_4\text{COOH} + n\text{H}_2\text{N(CH}_2)_6\text{NH}_2 \xrightarrow[280°C]{-H_2O}$$

adipic acid **hexamethylenediamine**

$$(C(CH_2)_4C-N(CH_2)_6N)_n + n\text{H}_2\text{O}$$

nylon

Dacron, a **polyester,** is prepared by the reaction of methyl terephthalate (a diester) with ethylene glycol in the presence of an acid or base.

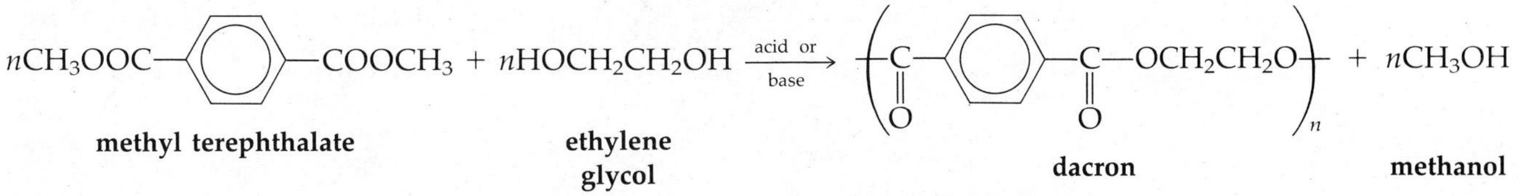

| methyl terephthalate | ethylene glycol | dacron | methanol |

In the production of nylon, H_2O is eliminated; and with dacron, methanol is eliminated. Table 23.7 contains a number of important polymers, the reactants needed for their production, and indicates whether they are addition or condensation polymers.

If we use appropriate monomers—ones that are able to link to more than two other species—bonds can be formed between adjacent polymer strands. This binds the strands in place, so the greater the degree of this **cross-linking** between parallel rows of polymer molecules, the stronger the material. Bakelite, for example, owes its strength and hardness to a three-dimensional network of covalent bonds throughout the entire polymer, as illustrated in Figure 23.10.

Natural rubber, too, can be made harder and stronger by a process known as **vulcanization.** In this reaction sulfur bridges between different chains create cross-links that lead to a tougher material.

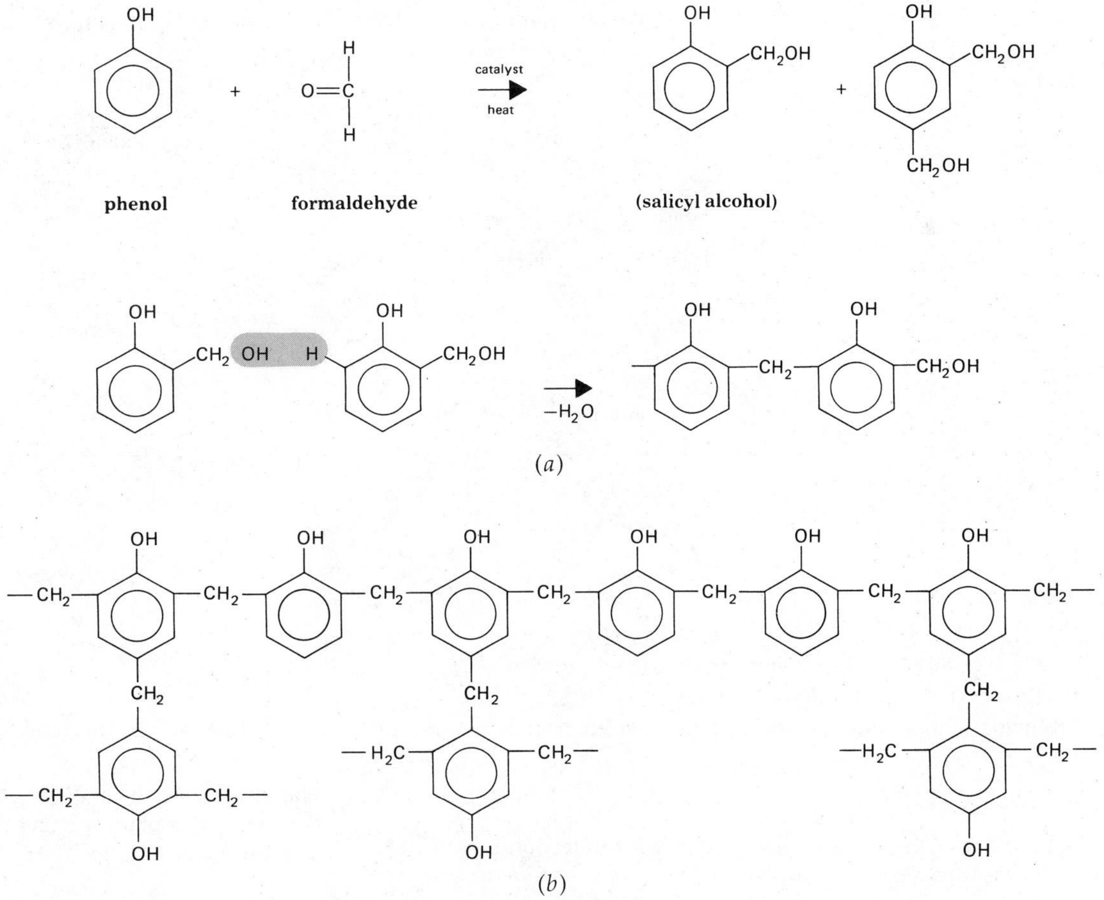

Figure 23.10

Bakelite®. (a) Polymerization of salicyl alcohol gives a linear polymer. (b) Continued polymerization, with cross-linking, gives a rigid three-dimensional structure called Bakelite®.

phenol   formaldehyde   (salicyl alcohol)

(a)

(b)

Table 23.7
Compositions of some common polymers

Monomer	Polymer	Type
$CH_2{=}CH_2$ ethylene	Polyethylene	Addition
$CH_2{=}CHCl$ vinyl chloride	Polyvinyl chloride (PVC)	Addition
$F_2C{=}CF_2$ tetrafluoroethylene	Teflon®	Addition
⬡—CH=CH$_2$ styrene	Polystyrene	Addition
$CH_2{=}C(CH_3)COOCH_3$ methyl methacrylate	Plexiglass®	Addition
$HOOC{-}(CH_2)_4{-}COOH + NH_2{-}(CH_2)_6{-}NH_2$ adipic acid hexamethylenediamine	Nylon	Condensation
$HOOC{-}C_6H_4{-}COOH + HO{-}CH_2CH_2{-}OH$ terephthalic acid ethylene glycol	Dacron®	Condensation
$C_6H_5OH + HCHO$ phenol formaldehyde	Bakelite®	Condensation
C_6H_5OH + furfural ring phenol furfural	Durite®	Condensation
melamine + HCHO melamine formaldehyde	Melmac®	Condensation

REVIEW QUESTIONS AND PROBLEMS

(Problems whose numbers are in blue have their answers in Appendix C at the back of the book.)

Hydrocarbons

23.1 What is the difference between a saturated and an unsaturated hydrocarbon?

23.2 The straight-chain alkanes are nonpolar molecules. How do we explain the fact that their boiling points increase from CH_4 to $C_{10}H_{22}$?

23.3 What would be the molecular formulas for: (a) an alkane having 30 carbon atoms, (b) an alkene having 27 carbon atoms, (c) an alkyne having 33 carbon atoms?

23.4 What would be the molecular formula for a straight-chain hydrocarbon having 17 carbon atoms and (a) all C—C single bonds, (b) one C—C double bond, (c) one C—C triple bond, (d) three C—C double bonds, (e) two C—C triple bonds?

23.5 What is a homologous series?

Isomerism

23.6 Draw the remaining isomers of the compound

$$CH_3-\underset{\underset{CH_3}{|}}{C}=\underset{\underset{CH_3}{|}}{C}-CH_3$$

23.7 How does optical isomerism arise in organic compounds? Draw the optical isomers of

$$Br-\underset{\underset{Cl}{|}}{\overset{\overset{H}{|}}{C}}-I$$

23.8 Sketch the *cis* and *trans* isomers of 2-pentene.

23.9 Draw the structural formulas for and name the nine isomers of heptane. Which of these isomers would give rise to optical isomerism?

23.10 Draw all possible isomers of hexene and show geometric isomers wherever possible.

23.11 Draw *all* the structural isomers, including cyclic structures, of $C_3H_4Cl_2$.

Structure and Bonding

23.12 What geometry do we expect for the molecules described in Question 23.7?

23.13 The molecule C_2Cl_4 is planar. Why?

23.14 The carbon atoms in 2-butyne lie in a straight line while those in butane do not. Explain why this is so.

23.15 What would be the C—C—C bond angles in *planar* cyclopropane, cyclobutane, cyclopentane, and cyclohexane?

23.16 The chair form of cyclohexane is more stable (of lower energy) than the boat form by several kilocalories per mole. By examining these two structures, can you suggest why the boat form has a higher energy?

23.17 Among the alcohols,

$$CH_3-(CH_2)_x-CH_2OH$$

as x increases the molar solubility in water decreases. Why does this occur?

23.18 Describe the bonding in (a) C_2H_4, (b) C_2H_2.

23.19 Why is it possible to obtain isomers of 2-butene, but not of *n*-butane?

23.20 Why is cyclopropane so chemically reactive?

23.21 It is possible to prepare the molecule cyclopentene, but not the molecule cyclopentyne. Why?

23.22 Describe the bonding in benzene. How does it compare to the bonding in graphite?

Nomenclature

23.23 Name the following compounds:

(a)
$$\underset{H_3C}{\overset{H_3C}{>}}CH-CH_2-\underset{\underset{CH_3}{|}}{CH}-CH_2-CH_3$$

(b)
$$CH_3-\underset{\underset{\underset{CH_3}{|}}{CH_2}}{CH}-CH_2-\underset{\underset{CH_3}{|}}{CH}-CH_2-CH_3$$

(c)
$$CH_3-CH_2-\underset{\underset{\underset{CH_2-CH_2-CH_3}{|}}{CH}}{CH}-CH_2-\underset{\overset{\overset{CH_3}{|}}{}}{CH}-CH_2-CH_3$$

(d)
$$CH_3-CH_2-CH=CH-\underset{\underset{CH_2-CH_3}{|}}{CH}-CH_3$$

(e)
$$CH_3-CH_2-\underset{\underset{CH_3}{|}}{CH}-CH_2-\underset{\overset{\overset{CH_3}{|}}{}}{CH}-CH_3$$

23.24 Name the following compounds:

(a)
$$CH_3-\underset{\underset{CH_3-C=CH-CH_3}{|}}{\overset{\overset{CH_2-CH_3}{|}}{C}}=C-CH_2-CH_3$$

(b)
$$CH_3-CH_2-C\equiv C-\underset{\underset{CH_2-CH_3}{|}}{CH}-CH_3$$

(c)
$$CH_3-\underset{\underset{CH_3}{|}}{\overset{\overset{CH_3}{|}}{CH}}-\underset{\underset{CH_3}{|}}{\overset{\overset{CH_3}{|}}{C}}-C-CH_2-CH_3$$

(d)
$$CH_3-C\equiv C-\underset{\underset{CH_3}{|}}{\overset{\overset{CH_3}{|}}{CH}}$$

(e)

$$H_3C, \quad CH=CH$$
$$C=C \quad CH_2-CH_3$$
$$CH_3-CH_2 \quad CH_3$$

23.25 Name the following compounds using IUPAC rules:

(a)

(b)

$$CH_3$$

$$Cl$$

(c)

$$\overset{O}{\overset{\|}{CH_3-CH_2-C-CH_3}}$$

(d)

$$CH_3$$
$$CH_3-CH-CH_2-NH_2$$

23.26 Write structural formulas for the following:
(a) 2-methylpentane
(b) 2,3-dimethylbutane
(c) 2,3-dimethyl-2-butene
(d) 1,3,5-octatriene
(e) 3,3,4-trimethyl-1-pentyne

23.27 What is the proper IUPAC name for 2,4-diethyl-3,3,4-trimethylpentane?

23.28 Write the structural formula for (a) *cis*-1,2-dichloropropene, (b) isohexane, (c) isopropyl alcohol, (d) *m*-dichlorobenzene.

23.29 Write structural formulas for the following compounds:
(a) 2-methyl-1-butene
(b) 2,3-dimethyl-2-butanol
(c) 2-bromo-1-phenylpropane
(d) 3-methyl-2-pentanone
(e) 1,3,5-tribromo-2,4,6-trichlorobenzene

Familiar Properties and Applications

23.30 Name some uses of the alkanes.

23.31 What is a carcinogen? In what way is the structure of 1,2-benzopyrene similar to graphite? Can you suggest why incomplete combustion of many hydrocarbons produces products similar to 1,2-benzopyrene?

23.32 What are some uses of halogenated hydrocarbons?

23.33 Examine the labels of a number of common household insecticides and make a list of their active ingredients. Identify the kinds of functional groups present in these ingredients.

23.34 What kinds of compounds often have fruity odors?

23.35 The odor of slightly spoiled fish is characteristic of which kind of compound?

Functional Groups

23.36 What is a functional group? Give the structural formula of
(a) an aldehyde
(b) a ketone
(c) a carboxylic acid
(d) an amine
(e) an alcohol
(f) an ester
(g) an ether

23.37 Vanillin (page 856) contains three types of functional groups. What are they?

23.38 Draw the structures of the esters formed from
(a) acetic acid and 2-propanol
(b) acetic acid and 1-pentanol
(c) benzoic acid and methanol
(d) formic acid (methanoic acid) and methanol

23.39 Diethylamine gives a basic solution in water. Why?

Reactions of Organic Compounds

23.40 What type of reaction is characteristic of the alkenes and alkynes? What type of reaction is characteristic of the alkanes?

23.41 Using Markovnikov's rule, complete the following addition reactions. Name the reactants and products.
(a) $CH_3-CH_2-CH=CH_2 + HI \longrightarrow$
(b) $CH_3-CH=CH_2 + H_2O \xrightarrow{H_2SO_4}$
(c)

$$CH_3-CH_2-CH=C\overset{CH_3}{\underset{CH_3}{}} + H_2O \xrightarrow{H_2SO_4}$$

23.42 What products (if any) are obtained by the *mild* oxidation, using $K_2Cr_2O_7$ in acid solution, of each of the following:

(a) CH_3-CH_2-OH
(b)
$$CH_3$$
$$CH_3-CH-CH_2-OH$$
(c)
$$OH$$
$$CH_3-CH-CH_3$$
(d) CH_3-CH_2-CHO

(e)
$$CH_3$$
$$CH_3-C-CH_3$$
$$OH$$
(f) CH_3-COOH
(g)
$$\overset{O}{\overset{\|}{CH_3-C-CH_3}}$$

23.43 What chemical reactions discussed in this chapter could be used to distinguish between the following:
(a) 2-propanol and 2-methyl-2-propanol
(b) 1-butanol and 2-butanol
(c) *n*-butane and 1-butene
(d) ethanal and 2-propanone

23.44 Write equations for the saponification of:

(a)

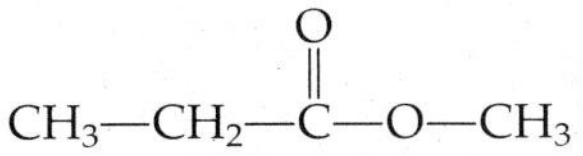

$$CH_3-CH_2-\overset{\overset{\displaystyle O}{\|}}{C}-O-CH_3$$

(b)

$$CH_3-CH_2-O-\overset{\overset{\displaystyle O}{\|}}{C}-CH_2-CH_2-\overset{\overset{\displaystyle O}{\|}}{C}-O-CH_2-CH_3$$

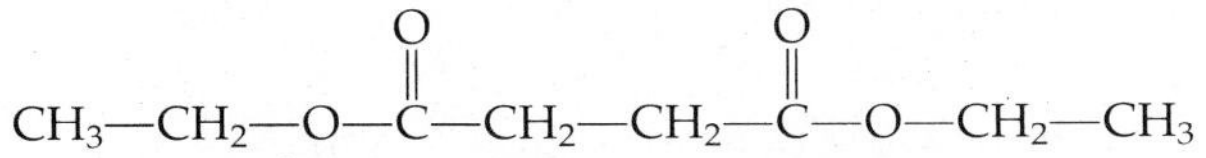

23.45 The synthesis of organic compounds from simple starting materials is an important aspect of organic chemistry. From the reactions described in this chapter, describe how you could prepare
(a) dichloroethane from ethene
(b) propanoic acid from 1-propanol
(c) 2-propanol from 1-chloropropane
(d) ethyl acetate from ethanal
(e) methyl ethyl ketone from 1-bromobutane

23.46 Predict the results of the following reactions:

(a)

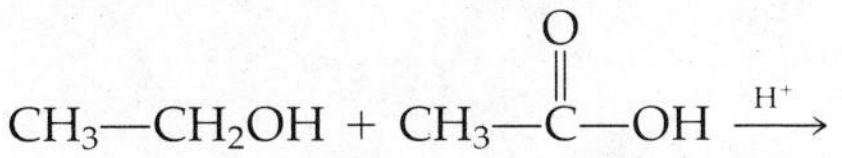

$$CH_3-CH_2OH + CH_3-\overset{\overset{\displaystyle O}{\|}}{C}-OH \xrightarrow{H^+}$$

(b) $CH_3-CH_2-CH_2-OH \xrightarrow[\Delta]{KMnO_4}$

(c)

$$CH_3-CH_2-\overset{\overset{\displaystyle OH}{|}}{CH}-CH_3 \xrightarrow{KMnO_4}$$

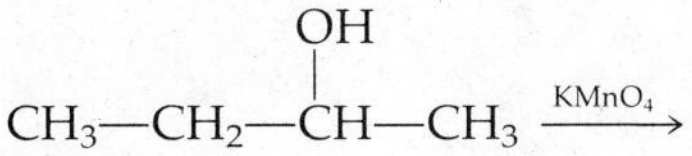

(d) $CH_3-CH_2-CH_2-CH_2-OH \xrightarrow[conc]{H_2SO_4}$

23.47 Write a chemical equation showing the saponification of propyl acetate. What drives this reaction toward completion?

23.48 Compare the products obtained by reduction of an aldehyde and a ketone with hydrogen.

Polymers

23.49 Styrene (phenylethene) forms an addition polymer. Sketch the structure of the repeating unit in polystyrene.

23.50 What is the difference between addition polymerization and condensation polymerization?

23.51 *Saran* is an addition copolymer of vinyl chloride and vinylidene chloride, $CH_2{=}CCl_2$. Sketch a portion of the polymer chain showing several monomer units.

23.52 If ethylene glycol and dimethylmalonate were to form a condensation polymer by the elimination of methanol, what would be the structure of the repeating unit in the polymer chain?

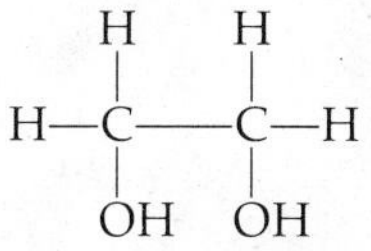

$$H-\overset{\overset{\displaystyle H}{|}}{\underset{\underset{\displaystyle OH}{|}}{C}}-\overset{\overset{\displaystyle H}{|}}{\underset{\underset{\displaystyle OH}{|}}{C}}-H$$

ethylene glycol

$$CH_3-O-\overset{\overset{\displaystyle O}{\|}}{C}-CH_2-\overset{\overset{\displaystyle O}{\|}}{C}-O-CH_3$$

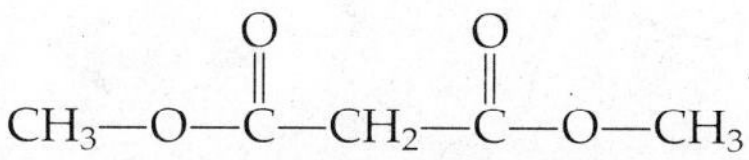

dimethylmalonate

23.53 Nylon stockings appear practically to disintegrate when hydrochloric or sulfuric acid is accidentally spilled on them. Actually, they dissolve very rapidly. Suggest what occurs chemically when nylon is dissolved by acid?

23.54 What is cross-linking? What effect does it have on the physical properties of a polymer?

CHAPTER 24
BIOCHEMISTRY

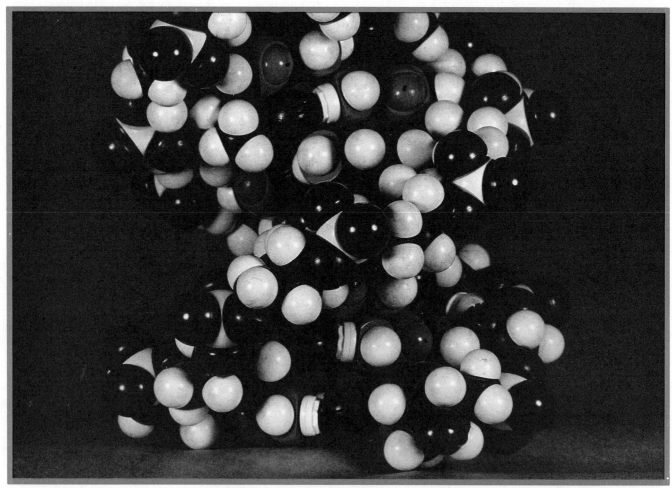

One of the major triumphs of science in general, and of biochemistry in particular, was solving the mystery of the structure of DNA, the substance responsible for heredity and the synthesis of the various proteins that each of our cells need to function. The twisted double-helix structure of just a short segment of DNA is illustrated in this molecular model, which actually reveals only half a turn of the helix. The structure of DNA and how its function is related to its structure is one of the topics described in this chapter.

Without question, one of the most active areas of chemical research today is the field of biochemistry which, as its name implies, is concerned with the chemistry that takes place in living systems. The modern biochemist views a living organism as a collection of organic molecules that interact with each other and their environment in a very unique and special way. When isolated from a living system these biomolecules are themselves lifeless. They obey all of the laws of chemistry and thermodynamics that we have examined up to now, and it is the goal of the biochemist to understand the functions and intricate interactions of these molecules that give rise to the phenomenon we call life.

The field of biochemistry is very large and quite complex, and we certainly cannot hope to explore it fully in a single chapter. Instead, we will be content to examine some of the types of biomolecules and their apparent functions in the operations of a living cell.

Nearly all the compounds found in a living system are composed primarily of carbon, hydrogen, oxygen, nitrogen, and some sulfur and, for the most part, they depend on carbon for their molecular backbone. In general these molecules are very large, with molecular weights ranging up to a million or more. We will see, however, that in many cases these **macromolecules** are constructed using a relatively small number of different simple molecules. For the purpose of discussing the various kinds of biomolecules we can place nearly all of them into one or another of four classes: **proteins, carbohydrates, lipids,** and **nucleic acids.**

Before we proceed, remember that you are not expected to memorize the names and formulas of all the different compounds that we will discuss in this chapter. Instead, concentrate on the types of molecules that are involved, the way that they combine with each other, and the general features of the resulting structures.

24.1 PROTEINS

Proteins are very large molecules having molecular weights ranging from about 6000 to approximately 1 000 000. They constitute nearly 50% of the dry mass of cells and, depending on the individual protein, serve a variety of different functions within a living organism. Some of them are hormones, like insulin, that serve as chemical messengers that coordinate certain biochemical activities. Insulin, for example, controls the level of sugar in the bloodstream by promoting its absorption into cells. Others are enzymes that act as catalysts for biochemical reactions. We will discuss these further in the next section. Some proteins serve to transport substances through the organism. Hemoglobin, for instance, carries oxygen in the bloodstream and delivers it to different parts of the body. There are long fibrous proteins, such as *actin* and *myosin*, that are found in muscle. Another fibrous protein, *α-keratin*, serves as the major constituent of hair, nails, and skin, while *collagen* is the prime constituent of tendons. Proteins are also found in toxins (poisonous materials) such as botulinus toxin as well as in antibodies.

Despite their wide range of functions, all proteins have something in common with one another. They are polymers made up by linking together, in various combinations, a number of different simple monomeric units called **α-amino acids.**

An amino acid is a bifunctional organic molecule that contains both a carboxyl group, —COOH, as well as an amine group, —NH_2. In an α-amino acid the amine group is located on the carbon atom adjacent to the carboxyl group (the α-carbon atom). This gives a structure that we can generalize as

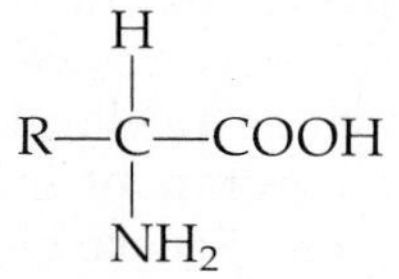

$$R-\underset{\underset{NH_2}{|}}{\overset{\overset{H}{|}}{C}}-COOH$$

Since the —NH$_2$ group (like ammonia) is basic and the —COOH group is acidic, in neutral solution the amino acid exists in an internal ionic form called a *zwitterion* where the proton of the —COOH group is transferred to the NH$_2$ to give

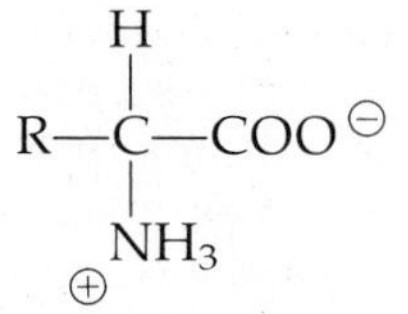

$$R-\underset{\underset{\overset{\oplus}{NH_3}}{|}}{\overset{\overset{H}{|}}{C}}-COO^{\ominus}$$

A very interesting and important fact is that in all organisms nearly all proteins are constructed using as building blocks a set of only 20 α-amino acids. These are shown in Figure 24.1. Note that except for proline, all fit the general

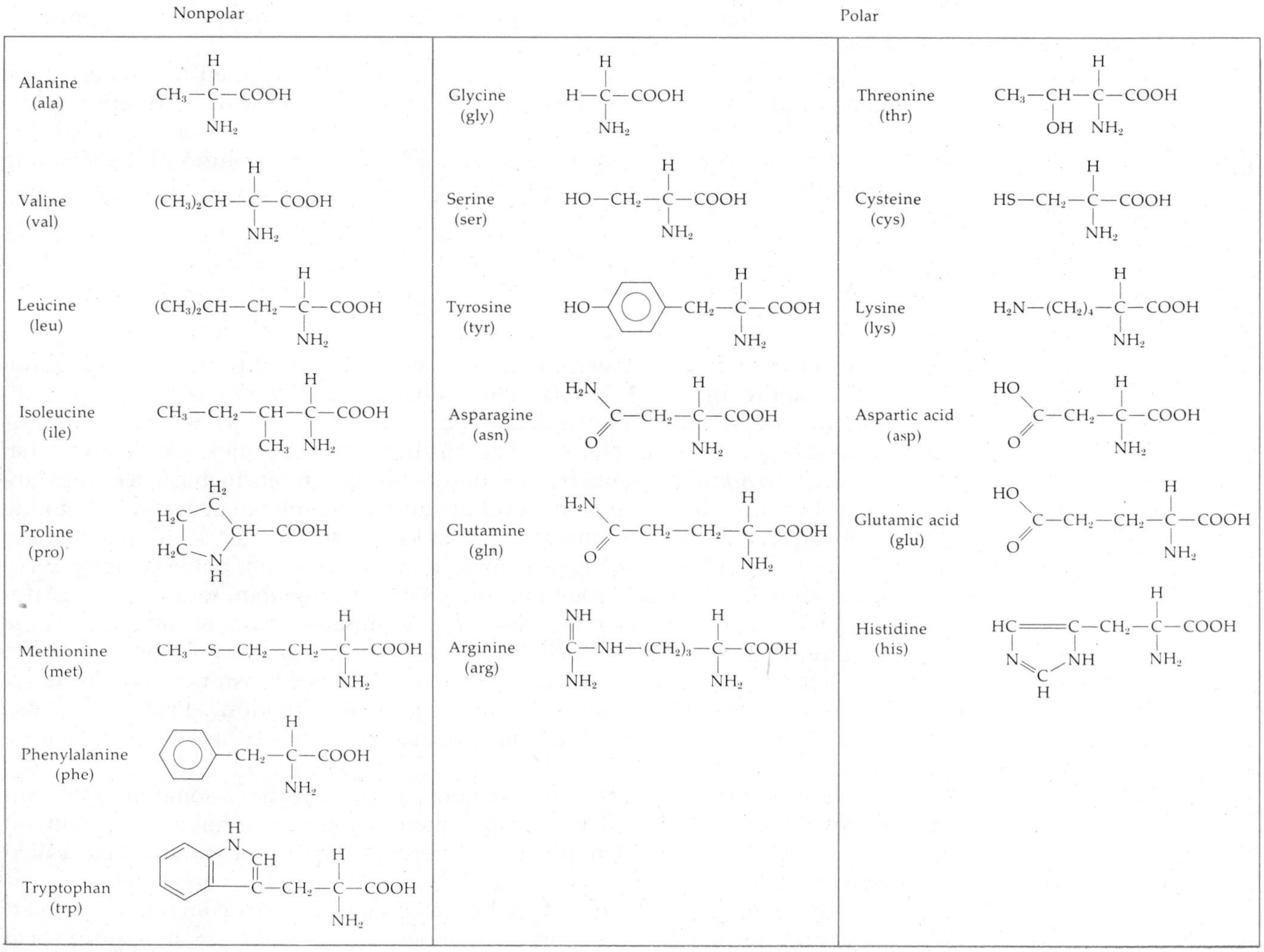

Figure 24.1

The 20 amino acids found in most proteins.

formula. Another point of interest is that except for glycine, all these amino acids have four different groups attached to the α-carbon atom, which is therefore an asymmetric (chiral) carbon atom. Each of these amino acids can therefore exist in two different isomeric forms (optical isomers). In proteins, however, only one isomer of each is commonly observed to occur. Apparently, substitution of one isomer for another destroys the biological activity of the protein molecule.

All the α-amino acids except glycine are chiral.

In a protein molecule, amino acids are linked together to form a long chain. This can be viewed as the result of the elimination of a water molecule from between the —NH$_2$ group of one amino acid molecule and the —COOH group of another,

$$\text{H}_2\text{N}-\overset{\overset{\text{H}}{|}}{\underset{\underset{\text{R}_1}{|}}{\text{C}}}-\overset{\overset{\text{O}}{\|}}{\text{C}}-\text{O}-\text{H} \quad \text{H}-\overset{\overset{\text{H}}{|}}{\underset{\underset{\text{R}_2}{|}}{\text{N}}}-\overset{}{\underset{}{\text{C}}}-\text{COOH} \xrightarrow{-\text{H}_2\text{O}} \text{H}_2\text{N}-\overset{\overset{\text{H}}{|}}{\underset{\underset{\text{R}_1}{|}}{\text{C}}}-\overset{\overset{\text{O}}{\|}}{\text{C}}-\overset{\overset{}{}}{\underset{\underset{\text{H}}{|}}{\text{N}}}-\overset{\overset{\text{H}}{|}}{\underset{\underset{\text{R}_2}{|}}{\text{C}}}-\text{COOH}$$

peptide bond or peptide linkage

Notice that the amide bond that connects monomers in polypeptides is the same as the linkage that connects monomers in the polymer nylon (page 862).

The molecule that results is called a **peptide** and the group of atoms within the dotted line constitutes a **peptide bond** (amide bond), or **peptide linkage.** In the particular example above, the peptide is composed of two amino acids and is said to be a **dipeptide.** Since one end of this molecule contains a carboxyl group and the other a free —NH$_2$ group, additional amino acids may be joined to give ultimately a **polypeptide,** a long chain composed of many amino acid molecules linked by peptide bonds. A segment of such a chain could be indicated as

$$\cdots-\overset{\overset{\text{H}}{|}}{\underset{\underset{\text{H}}{|}}{\text{N}}}-\overset{\overset{\text{O}}{\|}}{\underset{\underset{\text{R}_1}{|}}{\text{C}}}-\overset{\overset{\text{H}}{|}}{\underset{\underset{\text{H}}{|}}{\text{N}}}-\overset{\overset{\text{O}}{\|}}{\underset{\underset{\text{R}_2}{|}}{\text{C}}}-\overset{\overset{\text{H}}{|}}{\underset{\underset{\text{H}}{|}}{\text{N}}}-\overset{\overset{\text{O}}{\|}}{\underset{\underset{\text{R}_3}{|}}{\text{C}}}-\overset{\overset{\text{H}}{|}}{\underset{\underset{\text{H}}{|}}{\text{N}}}-\overset{\overset{\text{O}}{\|}}{\underset{\underset{\text{R}_4}{|}}{\text{C}}}-\cdots$$

The backbone of the chain is thus the same series of atoms repeated over and over again, with only the R groups changing as we move along the chain.

Before we continue, we should note that in the formation of a protein the linking together of the different amino acids is not a random process. Each molecule of a given protein has the same sequence of amino acids along its polypeptide chain. In fact, it is this very sequence that imparts to a particular protein its own specific properties.

The amino acid sequence that exists in a polypeptide is called its **primary structure.** In addition to this, the polypeptide chain twists and turns and assumes a **secondary structure** that is determined by hydrogen bonding that occurs between different groups along the chain. An example of this is found in the fibrous protein α-keratin (the major component of hair), in which the polypeptide chains coil themselves into the **α-helix,** shown in Figure 24.2. The hydrogen bonding, in this case, takes place between the oxygen atom in a carbonyl group ($>$C=O) and a hydrogen atom attached to a nitrogen atom that lies in an adjacent loop of the helix (Figure 24.3). This therefore serves to hold the chain in its coiled shape.

Altering the temperature or changing the properties of the solvent (e.g., by changing the pH) can alter the shape of the protein and destroy its biological activity. We say the protein is *denatured.*

In globular proteins, so named because of their overall shape, coiled polypeptide chains are also folded to give a complex three-dimensional structure,

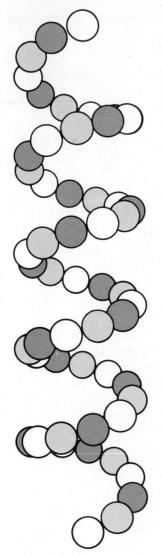

Figure 24.2

The α-helix, composed of a repetition of amino acid units.

$$-\text{C}-\overset{\text{O}}{\underset{\text{R}}{\text{C}}}-\overset{\text{H}}{\text{N}}-\left(\overset{\text{O}}{\text{C}}-\overset{\text{O H}}{\underset{\underbrace{\text{R} \ \ \text{C} \ \ \text{N}}_{n}}{\text{C}-\text{N}}}\right)-\overset{\text{O H}}{\underset{\text{R}}{\text{C}-\text{C}-\text{N}}}-$$

In the illustration, N is white, C is pale blue, and R is dark blue.

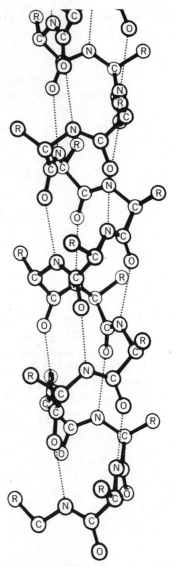

Figure 24.3

Hydrogen bonding (dotted lines) in the α-helix. Courtesy of Carroll K. Johnson, Oak Ridge National Laboratory, Oak Ridge, Tennessee.

referred to as its **tertiary structure.** This is shown in Figure 24.4 for the protein myoglobin, a substance that stores oxygen in muscle tissue until it is needed in metabolic oxidation. It is the presence of large amounts of myoglobin in the leg and thigh muscles of birds, for instance, that gives this meat a darker color than the breast meat. Myoglobin is also responsible for the red color of beef steak.

The tertiary structure of a protein is controlled by several different kinds of interactions that serve to hold the folded segments of the chain in place. For

Figure 24.4

The tertiary structure of myoglobin. The polypeptide helix folds and turns to produce a globular protein.

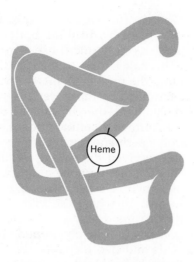

example, besides hydrogen bonding there are also ionic attractions that occur between a negatively charged deprotonated carboxyl group (like that found in the R group of glutamic acid) and a positively charged protonated amine group (like that found in lysine). This is shown in Figure 24.5.

The solvent is also important in determining the shape of the protein molecule. In the presence of the polar solvent water, nonpolar R groups such as the phenyl ring in phenylalanine (see Figure 24.1) are forced toward the center of the folded polypeptide chain, away from the solvent. This is the same phenomenon, you may remember, that leads to the low solubility of nonpolar substances in polar solvents. It also helps determine the tertiary structure of proteins because the polypeptide chain tends to fold in such a way that nonpolar groups do not contact the solvent.

Still another type of interaction that maintains the folded conformation of the protein molecule is the formation of covalent bonds between cysteine molecules located at different points along the chain. This occurs by partial oxidation of the —SH (thiol) group,

$$R{-}SH + HS{-}R \xrightarrow{\text{oxidation}} H_2O + R{-}S{-}S{-}R$$

The resulting linkage is called a **disulfide bridge.**

Disulfide bridges not only help to keep the polypeptide chain folded but can also bind two such chains together. For example, beef insulin (Figure 24.6) consists of two polypeptide chains that are cross-linked at two points by these disulfide bridges.[1]

There is an interesting sidelight to the subject of the disulfide bridge in

Nonpolar substances that repel water are said to be *hydrophobic* (water hating). Biochemists call interactions involving nonpolar groups *hydrophobic* interactions.

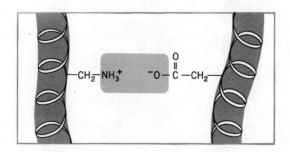

Figure 24.5

Ionic interactions hold portions of the polypeptide chains together in the tertiary structure of proteins.

[1] The elucidation of the primary structure of this protein by Frederick Sanger won him the Nobel Prize in 1958.

protein chemistry. The curl (or lack of curl) in hair is determined by protein conformation locked in place by disulfide bridges. The permanent wave treatment that women (and, recently, men) use to produce curly hair involves two chemical reactions. The hair is first treated with a chemical able to break the disulfide bridges so that the protein chains in the hair are free to twist into any desired shape, as determined by the curlers. A second solution is then applied that produces a mild oxidation, thereby causing the disulfide bridges to be reestablished. These newly formed disulfide bridges set the hair in the newly curled shape. The only reason a permanent wave isn't really permanent is because new hair grows that hasn't received the setting treatment.

Proteins that contain more than one independent polypeptide chain exhibit still another degree of structural sophistication, called **quaternary structure.** This is determined by the way in which the folded chains orient themselves with respect to one another. A good example of this occurs in hemoglobin (Figure 24.7). This protein consists of four polypeptide chains; two α-chains each containing 141 amino acids and two β-chains each with 146 amino acids.

In addition to the polypeptide chains, this protein also contains groupings of atoms, called **heme groups,** that serve to bind oxygen so that it can be transported through the bloodstream and be deposited at oxygen-poor cells. As depicted in Figure 24.7, the four folded polypeptide chains with their heme groups are packed together in a roughly tetrahedral fashion.

The heme group in hemoglobin is also found in myoglobin and accounts for the red color of blood and muscle tissue. The structure of heme is

Figure 24.7
Quaternary structure of hemoglobin. Four globular protein molecules containing heme groups are packed into the hemoglobin structure. Adapted from R. E. Dickerson and I. Geis, The Structure and Action of Proteins, *W. A. Benjamin, Inc., Menlo Park, Calif., 1969. Original illustration copyright 1969 by R. E. Dickerson and I. Geis.*

Amino—terminal ends

A chain:
Gly
Ile
Val
Glu
Gln
Cys
Cys —— S—S —— Cys
Ala
Ser
Val
Cys
Ser
Leu
Tyr
Gln
Leu
Glu
Asn
Tyr
Cys —— S—S —— Cys
Asn

B chain:
Phe
Val
Asn
Gln
His
Leu
Cys
Gly
Ser
His
Leu
Val
Glu
Ala
Leu
Tyr
Leu
Val
Cys
Gly
Glu
Arg
Gly
Phe
Phe
Tyr
Thr
Pro
Lys
Ala

Figure 24.6
Amino acid sequence in beef insulin.

Figure 24.8

Attachment of a heme group to the hemoglobin protein by coordination of histidine to the iron atom of heme. (a) Sixth coordination site is vacant. (b) When hemoglobin carries oxygen, the O_2 is bound to the Fe of heme at the sixth coordination site.

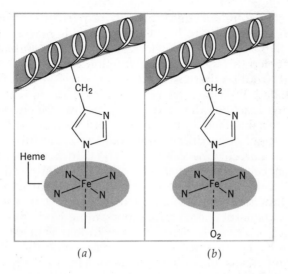

(a) (b)

It is, in fact, a complex ion containing iron(II) enclosed within a square planar grouping of nitrogen atoms. In hemoglobin each heme group is attached to its polypeptide by additional coordination to the nitrogen atom of a histidine, as shown in Figure 24.8. The sixth coordination site about the iron(II) is empty and is used to bind an oxygen molecule.

The basic square planar ligand structure in heme is called a **porphyrin** and forms very stable complexes with several different metal ions. For example, a structure very similar to heme containing Mg^{2+} instead of Fe^{2+} is found in chlorophyll, the green pigment in plants that is used in photosynthesis. Still another porphyrin structure, containing Co^{2+} in the center, exists in a substance called vitamin B_{12} coenzyme (Figure 24.9). Thus, even though most of the structures of biomolecules are made up of carbon, hydrogen, nitrogen, and oxygen, some metals are also critically important to the well-being of a living organism.

Normally, metals needed by the body in trace amounts are present in our cells in the form of complex ions.

24.2 ENZYMES

Enzymes are globular proteins that serve to catalyze specific biochemical reactions with what can only be judged as amazing effectiveness. In some cases

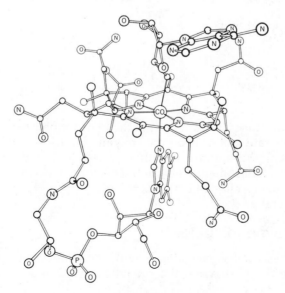

Figure 24.9

Vitamin B-12 coenzyme. Courtesy of Carroll K. Johnson, Oak Ridge National Laboratory, Oak Ridge, Tennessee.

reactions are speeded up, with respect to their uncatalyzed paths, by factors ranging from 10^9 to 10^{20}! Competing side reactions are not affected and are very slow by comparison. The result is that essentially 100% of the reactants are funneled through the same reaction path. In this way a buildup of by-products that would otherwise cause a waste-disposal problem for the organism is avoided. Enzymes also provide the organism with a way of controlling the rates of the reactions that take place, because biochemical reactions do not occur at appreciable rates in the absence of the catalyst. Removal, or at least temporary blockage, of a critically important enzyme turns off the chemistry which that particular enzyme catalyzes. Thus, in a very real sense, enzymes direct the chemical reactions that take place in a living cell.

Some enzymes require an additional substance called a **coenzyme** in order to function. Many vitamins that we must ingest to maintain good health are precursors of coenzymes. An example is vitamin B_{12}, whose absence from the diet leads to a deficiency disease known as pernicious anemia. In the body vitamin B_{12} is converted to its coenzyme, with the structure we saw in Figure 24.9. Many metals, like cobalt, are needed in small amounts to promote enzyme activity.

The mechanism by which an enzyme acts has long been the subject of intense research. It appears to depend on the ability of the enzyme to bind very selectively to a reactant molecule (called the **substrate** of the enzyme). There thus seems to be a lock-and-key relationship between an enzyme and its substrate, in which the substrate molecule just precisely fits into (or onto) the folded globular protein. There is evidence that when this occurs, there is a slight alteration in the shape of the enzyme that strains certain key bonds in the substrate, thereby making them more susceptible to chemical attack.

Some enzymes are very specific in their activity, affecting the rate of reaction of a single compound. Others are less choosy and simply promote a certain kind of chemical reaction on a whole class of compounds having similar structures. This behavior, also, is the direct result of the lock-and-key relationship between the enzyme and its substrate. An example is the enzyme *chymotrypsin*, which accelerates the hydrolysis of the dotted bond in the compounds,

where X = N or O

An example is

In this case it is thought that the hydrophobic benzene ring serves to position the substrate molecule on the enzyme (Figure 24.10), which then interacts with the $\diagup C{=}O$ group in a way that makes the dotted bond more susceptible to hydrolysis. Since the function of the enzyme depends on the hydrophobic tail and the proper location of the $\diagup C{=}O$ group, a family of similar compounds are affected by the enzyme.

Enzyme inhibition

When a substance other than the enzyme substrate becomes bound to the active site of an enzyme, the catalytic activity is lost and the enzyme is said to be

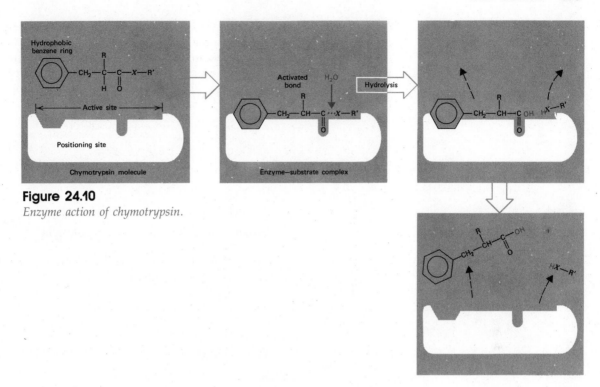

Figure 24.10

Enzyme action of chymotrypsin.

inhibited. In some cases this inhibition is irreversible. This occurs when the inhibitor becomes permanently bound to the enzyme by covalent bond formation and cannot be displaced. An example is diisopropylfluorophosphate,

$$\begin{array}{ccccc} CH_3 & & F & & CH_3 \\ | & & | & & | \\ H-C-O-P-O-C-H \\ | & & | & & | \\ CH_3 & & O & & CH_3 \end{array}$$

Botulinus toxin, the most powerful poison yet discovered, blocks the production of acetylcholine, so it is also a nerve poison.

a highly toxic nerve poison and an ingredient in some nerve gases. This molecule reacts with and poisons (inhibits) an enzyme called *acetylcholine esterase*, which is required for the transport of impulses along nerve tissue.

$$\text{enzyme}-O-(H+F)-\underset{\underset{\displaystyle CH(CH_3)_2}{\overset{\displaystyle O}{|}}}{\overset{\overset{\displaystyle CH(CH_3)_2}{|}}{\overset{\displaystyle O}{|}}}P-O \longrightarrow HF + \text{enzyme}-O-\underset{\underset{\displaystyle CH(CH_3)_2}{\overset{\displaystyle O}{|}}}{\overset{\overset{\displaystyle CH(CH_3)_2}{|}}{\overset{\displaystyle O}{|}}}P-O$$

A second type of enzyme inhibition is called **competitive inhibition,** in which there is a competition between the inhibitor and the substrate for the enzyme active site. This is a system in which there are two simultaneous equilibria, one between the enzyme (E) and the substrate (S),

$$E + S \rightleftharpoons ES$$

and one between the enzyme and the inhibitor (I),

$$E + I \rightleftharpoons EI$$

As we would predict from Le Châtelier's principle, increasing the substrate concentration displaces the inhibitor.

An example of competitive inhibition is the action of the sulfa drug sulfanilamide. This molecule, shown below, bears a very close similarity to *p*-aminobenzoic acid, which is acted on by an enzyme to produce an important coenzyme required by bacteria.

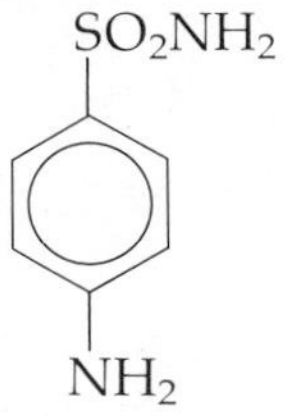

sulfanilamide 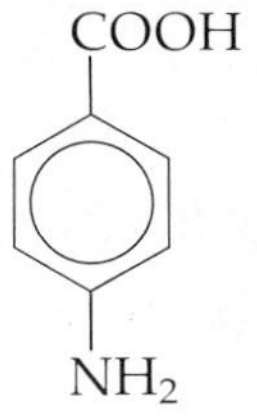   *p*-**aminobenzoic acid**

The sulfanilamide, by occupying the active site of the enzyme that works on the *p*-aminobenzoic acid, prevents the production of the required coenzyme and hence leads to the demise of the bacterium.

As a final note, enzyme activity is also affected by temperature and pH. These factors alter the conformation and shape of the globular protein structure; therefore, changes in temperature or pH can destroy the precise fit that must exist between the enzyme and its substrate in order to obtain the desired catalytic activity.

24.3   CARBOHYDRATES

The carbohydrates form an important class of compounds that are used by living organisms in a variety of ways—as a source of energy, a source of carbon to be used in the synthesis of other biomolecules, and a structural element in cells and tissues. Historically the name carbohydrate arose as a consequence of the empirical formula exhibited by many of them, $C_n(H_2O)_m$, which suggested that they were *hydrates of carbon*. Examples are glucose, $C_6H_{12}O_6$, having the empirical formula CH_2O, and sucrose (ordinary cane sugar), $C_{12}H_{22}O_{11}$, with the empirical formula $C_{12}(H_2O)_{11}$. The name carbohydrate has remained with these substances, even though it is now known that they do not contain intact water molecules.

Most carbohydrates, such as starch and cellulose, are very large molecules having enormous molecular weights. However, like the proteins, they are composed of many relatively simple units arranged in long polymeric chains. The simplest of these units are called **monosaccharides,** and constitute the **simple sugars.** As a class, the monosaccharides are polyhydroxy aldehydes or ketones, the simplest of which is glyceraldehyde,

The Latin *saccharum* means sugar.

$$
\begin{array}{c}
CHO \\
| \\
H-C-OH \\
| \\
CH_2OH
\end{array}
$$

Glyceraldehyde is a **triose,** *tri* denoting three carbon atoms and *ose*, the characteristic ending used in naming the sugars (for example, gluc*ose*, sucr*ose*, and fruct*ose*).

Glyceraldehyde, like the other saccharides, contains an asymmetric carbon atom and exhibits optical isomerism. In the two-dimensional structural formulas written for the saccharides, the H and OH units attached to the chiral carbon

Figure 24.11

Optical isomers of glyceraldehyde.

$$
\begin{array}{c}
\text{CHO} \\
| \\
\text{H—C—OH} \\
| \\
\text{CH}_2\text{OH}
\end{array}
\qquad
\begin{array}{c}
\text{CHO} \\
\vdots \\
\text{H—C—OH} \\
\vdots \\
\text{CH}_2\text{OH}
\end{array}
$$

$$
\begin{array}{c}
\text{CHO} \\
| \\
\text{HO—C—H} \\
| \\
\text{CH}_2\text{OH}
\end{array}
\qquad
\begin{array}{c}
\text{CHO} \\
| \\
\text{HO—C—H} \\
| \\
\text{CH}_2\text{OH}
\end{array}
$$

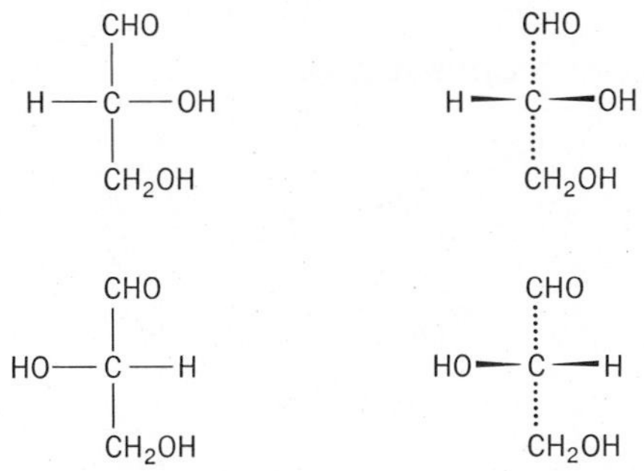

atoms project upward from the paper while the bonds to other carbon atoms project downward. For example, the two optical isomers of glyceraldehyde are shown in Figure 24.11.

Among the saccharides it is generally observed that one optical isomer is significantly more important than the others. Glucose, for example, contains four chiral carbon atoms (indicated by asterisks).

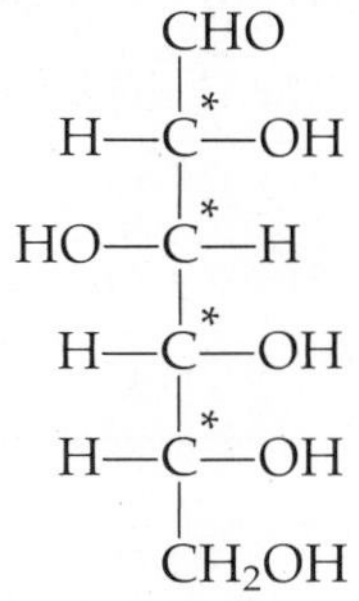

$$
\begin{array}{c}
\text{CHO} \\
| \\
\text{H—}\overset{*}{\text{C}}\text{—OH} \\
| \\
\text{HO—}\overset{*}{\text{C}}\text{—H} \\
| \\
\text{H—}\overset{*}{\text{C}}\text{—OH} \\
| \\
\text{H—}\overset{*}{\text{C}}\text{—OH} \\
| \\
\text{CH}_2\text{OH}
\end{array}
$$

D-**glucose**

There are 16 possible optical isomers of glucose; and the most important one is D-glucose, shown above.

By far the most important monosaccharides are those containing five and six carbon atoms, the pentoses and hexoses, respectively. Some of the more prominent ones are found in Table 24.1. The most common hexose is glucose, whose structure is shown above. Glucose, however, like most of the other pentoses and hexoses, exists predominantly in a cyclic structure in which the molecule turns on itself as shown in Figure 24.12. When the ring is closed, the —OH group that is created from the aldehyde functional group can point either up or down (this is the —OH group on the rightmost carbon atom in the structures drawn in Figure 24.12). Two isomers are thus created, α-D-glucose and β-D-glucose. As we will see, the orientation of this —OH group is quite significant in the polysaccharides, starch and cellulose.

Another important six-carbon sugar is fructose. In its open-chain structure the molecule is a ketone.

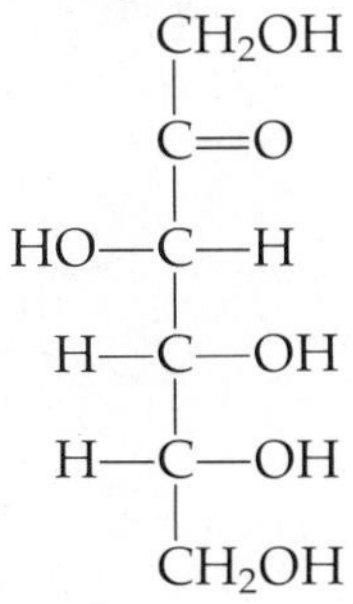

Fructose is found along with glucose and sucrose in honey and fruit juices.

$$
\begin{array}{c}
\text{CH}_2\text{OH} \\
| \\
\text{C}=\text{O} \\
| \\
\text{HO—C—H} \\
| \\
\text{H—C—OH} \\
| \\
\text{H—C—OH} \\
| \\
\text{CH}_2\text{OH}
\end{array}
$$

Table 24.1
Some important monosaccharides

Pentoses

D-Ribose	D-Arabinose	D-Ribulose
CHO	CHO	CH₂OH
H—C—OH	HO—C—H	C=O
H—C—OH	H—C—OH	H—C—OH
H—C—OH	H—C—OH	H—C—OH
CH₂OH	CH₂OH	CH₂OH

Hexoses

D-Glucose	D-Mannose	D-Galactose	D-Fructose
CHO	CHO	CHO	CH₂OH
H—C—OH	HO—C—H	H—C—OH	C=O
HO—C—H	HO—C—H	HO—C—H	HO—C—H
H—C—OH	H—C—OH	HO—C—H	H—C—OH
H—C—OH	H—C—OH	H—C—OH	H—C—OH
CH₂OH	CH₂OH	CH₂OH	CH₂OH

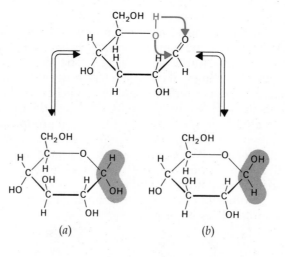

Figure 24.12
Cyclic structures for glucose. (a) α-D-glucose. (b) β-D-glucose. (c) Puckered ring. Note orientation of H and OH on the rightmost carbon.

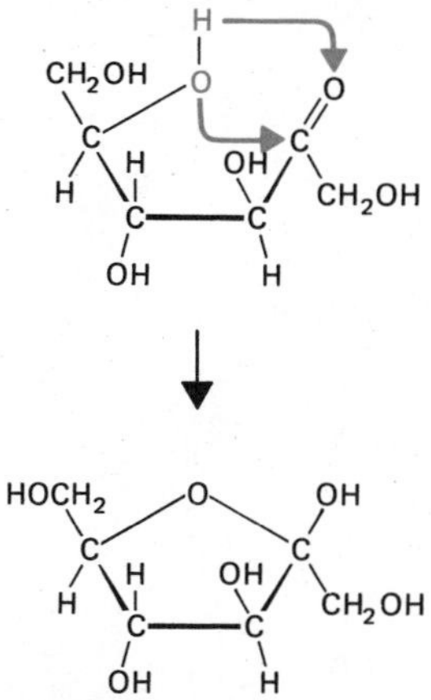

Figure 24.13

Cyclic structure for fructose.

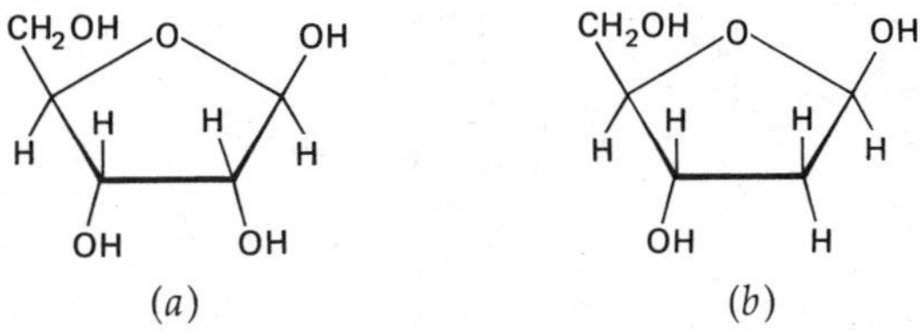

Figure 24.14

(a) *Ribose and* (b) *deoxyribose.*

Like glucose, however, it also prefers a cyclic structure, as shown in Figure 24.13. In this case a five-membered ring is formed.

The five-membered ring also occurs in two very important pentoses, **ribose** and **deoxyribose** (Figure 24.14), sugars that are part of the backbone of RNA and DNA, respectively. We will examine the structures of these in Section 24.5.

In the more complex sugars and the polysaccharides, monosaccharide units are condensed together by way of C—O—C bridges called **glycoside linkages.** Sucrose, for example, is a disaccharide consisting of a glucose and a fructose unit joined by eliminating H_2O from an —OH group on the glucose and an —OH group on the fructose. This is illustrated in Figure 24.15. As indicated, addition of H_2O to the glycoside linkage (hydrolysis) splits the sucrose molecule into the simple monosaccharides from which it is formed. This hydrolysis reaction is accelerated by the presence of dilute acid and, in general, polysaccharides can be broken down into their simple sugars by this reaction. Special enzymes in saliva start digesting carbohydrates in the mouth by this hydrolysis reaction.

The formation of two glycoside linkages by a single monosaccharide unit permits the formation of long polymeric chains called polysaccharides, the two most important of which are starch and cellulose. Starch (amylose) is composed

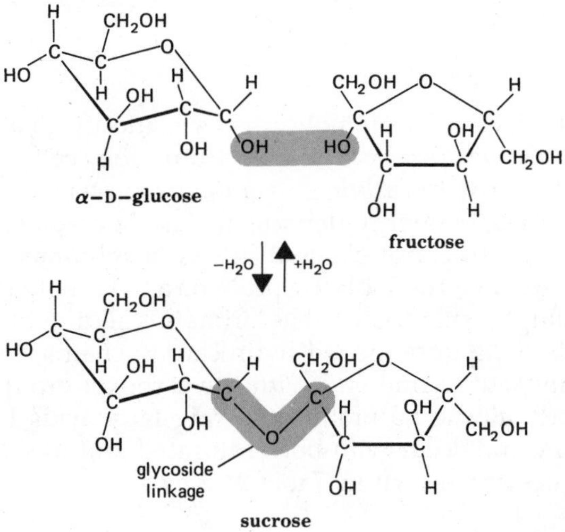

Figure 24.15

Sucrose.

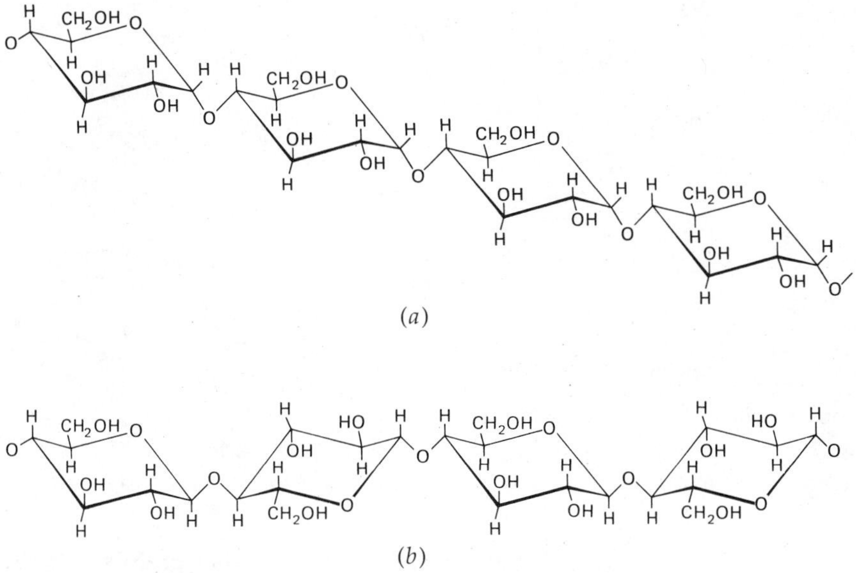

Figure 24.16
Structures of the polysaccharides, starch and cellulose. (a) *Amylose (starch).* (b) *Cellulose.*

(a)

(b)

of α-D-glucose units strung together, while cellulose is composed of β-D-glucose units, as shown in Figure 24.16.

The difference between these two structures is rather subtle, but nevertheless has very profound effects. In starch the polysaccharide chains tend to coil in a helical structure with the polar —OH groups pointing outward. When placed into water these —OH groups on the starch molecule interact strongly with the polar solvent and cause the starch to be slightly water-soluble. Cellulose, on the other hand, forms linear chains that interact with each other via hydrogen bonding. This phenomenon gives wood—which is composed of approximately 50% cellulose—its structural strength.

The relatively minor structural differences between starch and cellulose also account for the fact that starch can be digested by humans but cellulose cannot. In the digestive tract the starch molecule is hydrolyzed enzymatically, which requires a certain fit between the carbohydrate molecule and the enzyme. With cellulose this necessary fit is not achieved and, hence, cellulose is unaffected. In termites, cows, and many other animals, however, cellulose is hydrolyzed and digested with the aid of bacteria in their digestive tract.

24.4 LIPIDS

A third class of biomolecules is made up of the **lipids**—water-insoluble substances that can be extracted from other cell components by nonpolar organic solvents (hydrocarbon solvents, carbon tetrachloride, etc.). Lipids serve mainly as storage of energy-rich fuel for use in metabolism (for example, in fats) and as a major structural element in cell membranes.

As was true with the proteins and carbohydrates, most lipids are composed of simpler substances. The primary building blocks of the lipids are called **fatty acids**, long unbranched hydrocarbon chains, from 12 to 28 carbon atoms long, terminated at one end with the carboxyl group characteristic of organic acids. Nearly all the naturally occurring fatty acids have an even number of carbon atoms and occur with both saturated and unsaturated chains. Some typical examples are shown in Table 24.2.

Table 24.2
Some naturally occurring fatty acids

Fatty Acid		Melting Point (°C)
Saturated		
Lauric acid (coconut or palm kernal oil)	$CH_3(CH_2)_{10}COOH$	44
Myristic acid (nutmeg fat)	$CH_3(CH_2)_{12}COOH$	54
Palmitic acid (palm oil, animal fats)	$CH_3(CH_2)_{14}COOH$	63
Stearic acid (animal fats)	$CH_3(CH_2)_{16}COOH$	70
Unsaturated		
Palmitoleic acid (butter fat)	$CH_3(CH_2)_5CH{=}CH(CH_2)_7COOH$	−1
Oleic acid (olive oil, animal fats)	$CH_3(CH_2)_7CH{=}CH(CH_2)_7COOH$	13.4
Linoleic acid (linseed oil)	$CH_3(CH_2)_4CH{=}CHCH_2CH{=}CH(CH_2)_7COOH$	−5
Linolenic acid (linseed oil)	$CH_3(CH_2)CH{=}CHCH_2CH{=}CHCH_2CH{=}CH(CH_2)_7COOH$	−11

Most lipids can be classed as either **neutral lipids** or **polar lipids.** Fats, for example, are neutral lipids and are esters of the fatty acids with the alcohol, glycerol.

glycerol

The resulting triester is called a **triglyceride.**

As you might expect, many different triglycerides are found to occur, as determined by the nature and location of the fatty acids attached to the glycerol molecule. Lipids containing saturated fatty acids, such as tristearin (glycerol esterified with three stearic acid molecules),

$$
\begin{array}{c}
\quad\quad\; H \quad\quad\; O \\
\quad\quad\; | \quad\quad\;\; \| \\
H-C-O-C-C_{17}H_{35} \\
\quad\quad\; | \quad\quad\quad\; O \\
\quad\quad\quad\quad\quad\;\; \| \\
H-C-O-C-C_{17}H_{35} \\
\quad\quad\; | \quad\quad\quad\; O \\
\quad\quad\quad\quad\quad\;\; \| \\
H-C-O-C-C_{17}H_{35} \\
\quad\quad\; | \\
\quad\quad\; H
\end{array}
$$

are solids, while those containing three unsaturated fatty acids are liquids at room temperature. An example of this liquid type is *triolein*, in which glycerol is esterified with three oleic acid molecules. This substance is the major constituent of olive oil. The liquid unsaturated triglycerides that are found in vegetable oils, such as olive oil and corn oil, serve as the basis of oleomargarine. Addition of hydrogen to the double bonds of unsaturated vegetable oils produces saturated chains and hence solid fats.

Recall that one of the principal commercial uses of hydrogen is the hydrogenation of vegetable oils.

In recent years the growth of the processed food industry has led to wide-spread use, as food additives, of monoglycerides and diglycerides, in which only one or two of the —OH groups of glycerol are esterified. These are added to foods as emulsifiers to improve texture and keep oils suspended.

Fats, like other esters, can be **saponified** on treatment with aqueous base. The products of this reaction are glycerol plus the anions of the fatty acids that were bound to the glycerol in the fat.

$$
\begin{array}{ccc}
H_2COOCC_{17}H_{35} & & H_2COH \\
| & \xrightarrow[H_2O]{OH^-} & | \\
HCOOCC_{17}H_{35} & & HCOH \; + \; 3C_{17}H_{35}COO^- \\
| & & | \\
H_2COOCC_{17}H_{35} & & H_2COH
\end{array}
$$

Hydrophilic means "water loving."

These anions constitute a soap and have rather peculiar properties that result from having a polar hydrophilic "head" and a nonpolar hydrophobic "tail."

$$
CH_3-CH_2-\cdots-CH_2-CH_2-CH_2-C \begin{array}{c} {\diagup} O \\ {\diagdown} O^- \end{array}
$$

$$\underbrace{\qquad\qquad\qquad\qquad}_{\text{tail}} \quad \underbrace{\qquad\qquad}_{\text{head}}$$

The polar end of the anion tends to be water-soluble while the other end, the nonpolar hydrocarbon tail, tends to be insoluble in water but soluble in nonpolar solvents. As a result, in water these anions group themselves into small globules called **micelles** in which the nonpolar tails dissolve in each other, leaving the polar heads facing outward toward the aqueous surroundings. This is illustrated in Figure 24.17. The same properties that lead to micelle formation are also responsible for the ability of soap to dissolve grease. In this case the nonpolar tails dissolve in the grease particle and the polar heads dissolve in water (Figure 24.18). This keeps the grease particle suspended in water so that it can be rinsed away.

Recall that micelle formation and the function of soaps and detergents were discussed in Chapter 11.

Another very important class of lipids are the **phospholipids**. These are polar lipids and, like the fats, are esters of glycerol. In this case, however, only

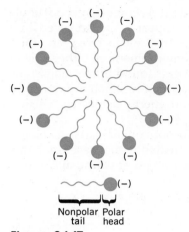

Figure 24.17

Micelle formation with soap.

Nonpolar tail Polar head

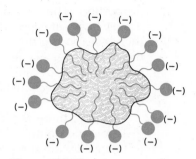

Figure 24.18

The dissolving of a grease globule by soap.

two fatty acid molecules are esterified to glycerol, at the first and second carbon atom. The remaining end position of the glycerol is esterified to a molecule of phosphoric acid, which in turn is also esterified to another alcohol. This gives a general structure,

$$
\begin{array}{c}
O \\
\parallel \\
O-P-O-R'' \\
\mid \\
O \\
\mid \\
H_2C-CH-CH_2 \\
\mid \quad\quad \mid \\
O \quad\quad O \\
\mid \quad\quad \mid \\
O=C \quad C=O \\
\mid \quad\quad \mid \\
R \quad\quad R'
\end{array}
$$

An example is the phospholipid *phosphatidyl ethanolamine,*

$$
\begin{array}{l}
CH_3-CH_2-CH_2-CH_2-CH_2-CH_2-CH_2-(CH_2)_8\overset{\displaystyle O}{\overset{\displaystyle \parallel}{C}}-O-\overset{\displaystyle H}{\underset{\displaystyle |}{C}}-H \\[2em]
CH_3-CH_2-CH_2-CH_2-CH_2-CH_2-CH=CH-(CH_2)_7\overset{\displaystyle O}{\overset{\displaystyle \parallel}{C}}-O-\overset{\displaystyle |}{\underset{\displaystyle |}{C}}-H \quad O \\[2em]
 H-\overset{\displaystyle |}{\underset{\displaystyle |}{C}}-O-P-O^{(-)} \\
 H \quad\quad O \\
 CH_2 \\
 CH_2 \\
 NH_2^{(+)}
\end{array}
$$

nonpolar tail   **polar head**

As indicated, this type of lipid contains a polar head and nonpolar tail, much the same as the anions of the fatty acids.

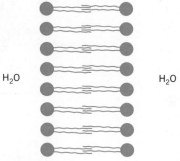

Figure 24.19

Formation of bilayers by phospholipids. Nonpolar tails dissolve in each other. Polar heads are exposed to the aqueous environment.

Cell membranes are composed of phospholipids and proteins in about equal proportion. The phospholipids in the membrane appear to be arranged in a double layer, or *bilayer* (Figure 24.19) in which the nonpolar tails face each other, thereby exposing the polar heads to the aqueous environment on either side of the membrane. As shown in Figure 24.20, the proteins found in the membrane are embedded in the mosaic formed by the lipids. Much research today is centered on the mechanism of transport of matter and energy across such membranes.

Finally, there is a third class of lipids that do not contain fatty acids and glycerol, and do not undergo saponification when treated with a base. Included in this group are the **steroids,** complex substances that possess unusually high biological activity.

The steroids have in common a fused ring structure.

Examples of some important steroids, whose names you may have come across before, are the following:

estrone   **estradiol**

(Female sex hormones: Note similarity in structures.)

Figure 24.20

Bilayer structure of a cell membrane. Lipid mosaic model: Irregularly shaped proteins float randomly in a lipid sea. Source: S. J. Singer, Annals of the New York Academy of Sciences, *Vol. 195, p. 21, 1982. Used by permission of the publisher and author.*

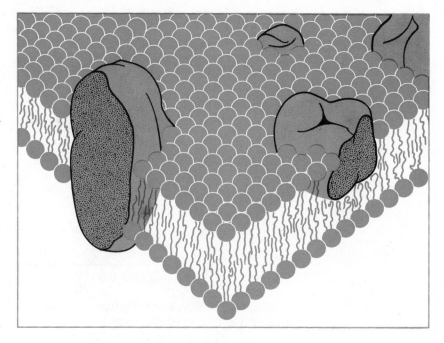

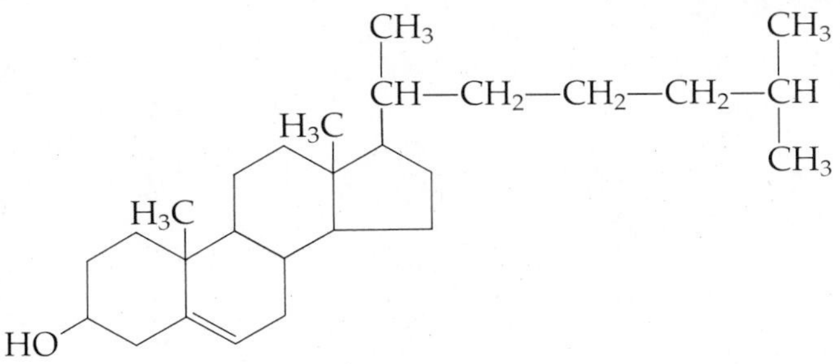

cholesterol

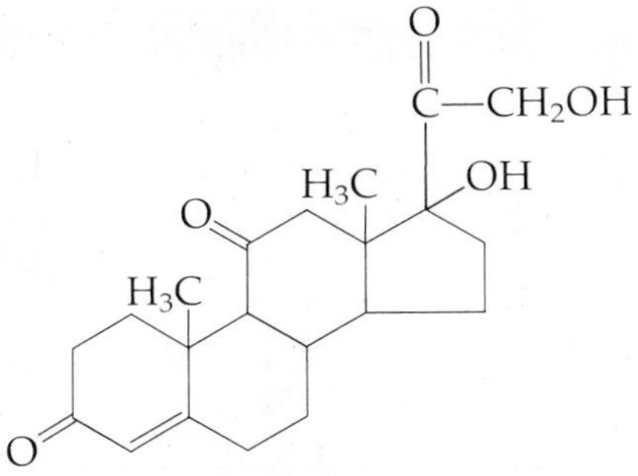

cortisone
(affects protein metabolism)

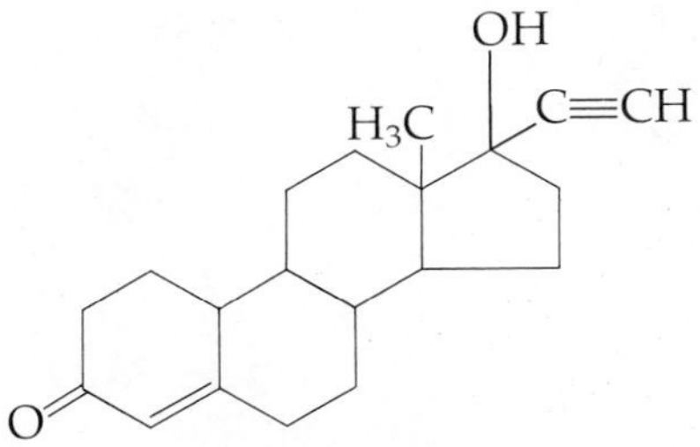

norethindrone: an
oral contraceptive
(the pill)

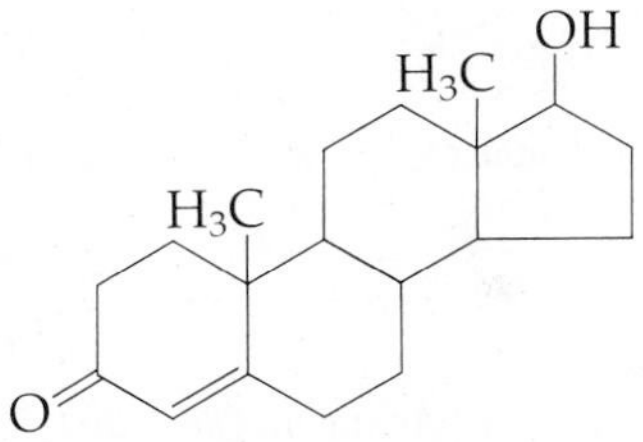

testosterone
(male sex hormone)

24.5 NUCLEIC ACIDS

One of the most intriguing aspects of biochemistry has been, from its very beginnings, the mechanism whereby an organism transmits its genetic information from one generation to the next during cell division. It is now believed, with a good deal of confidence, that this process is controlled by a substance found in the nucleus of the cell—the **nucleic acid, DNA** (deoxyribonucleic acid). Furthermore, DNA in conjunction with **RNA** (ribonucleic acid, another type of nucleic acid) is responsible for the synthesis of the proteins that are characteristic of a given organism.

Nucleic acids, like the proteins and carbohydrates that we looked at earlier, are polymers. The simpler units that make up the nucleic acid are called **nucleotides,** and are themselves composed of three even simpler molecules. They include the following:

1. **A nitrogenous base.** These are heterocyclic organic compounds with two or more nitrogen atoms in the ring skeleton. They are called bases because the lone pairs of electrons on the nitrogen atoms make them Lewis bases. These substances are shown in Figure 24.21.
2. **A five-carbon sugar (pentose).** In RNA this sugar is ribose, whereas in DNA the sugar is deoxyribose. These two are shown in Figure 24.22. Notice that they differ only at carbon atom number 2 in the ring.
3. **Phosphoric acid.** H_3PO_4, as we will see, forms esters to —OH groups of the sugar to bind nucleotide segments together.

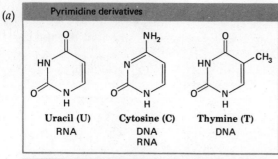

Figure 24.21

Nitrogeneous bases found in DNA and RNA.
(a) *Pyrimidine derivatives.* (b) *Purine derivatives.*

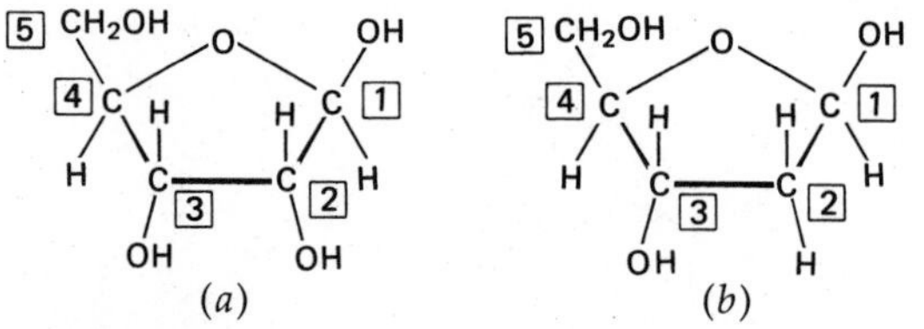

Figure 24.22

(a) *Ribose and* (b) *deoxyribose.*

A molecule called a **nucleoside** is formed from these components by condensing a molecule of the base with the appropriate pentose. For example, adenine combines with ribose and deoxyribose at carbon number 1 to give the compounds shown in Figure 24.23a. Adenosine is an important constituent of ATP

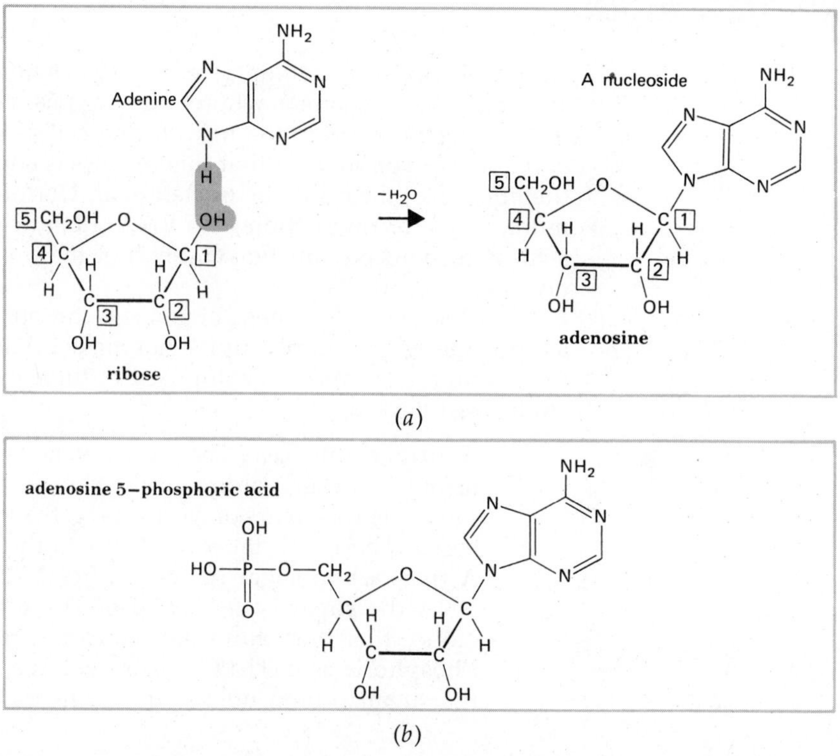

(a)

(b)

Figure 24.23

Nucleosides and nucleotides.
(a) *Adenine combines with ribose to form a nucleoside.* (b) *Linkage of phosphoric acid to carbon 5 gives a nucleotide.*

(adenosine triphosphate) and ADP (adenosine diphosphate), which are both involved in energy-transfer processes in a cell. Finally, linkage of phosphoric acid to carbon atom number 5 (Figure 24.23b) produces a **nucleotide,** the basic building block of both DNA and RNA.

The nucleic acids are condensation polymers of the nucleotide monomers and are formed by the creation of an ester linkage from the phosphoric acid residue on one nucleotide to the hydroxy group on carbon number 3 in the pentose of the second nucleotide, as illustrated in Figure 24.24. The result is a very long polymeric chain, possessing up to a billion or so nucleotide units in DNA.

In Figure 24.21 it was indicated that the base compositions of DNA and RNA are not the same. In DNA the organic bases adenine (A), guanine (G), cytosine (C), and thymine (T) occur bound to the deoxyribose ring, whereas in RNA the bases adenine, guanine, cytosine, and uracil (U) are found. As in the proteins, the sequence of bases along the DNA or RNA chain establishes its

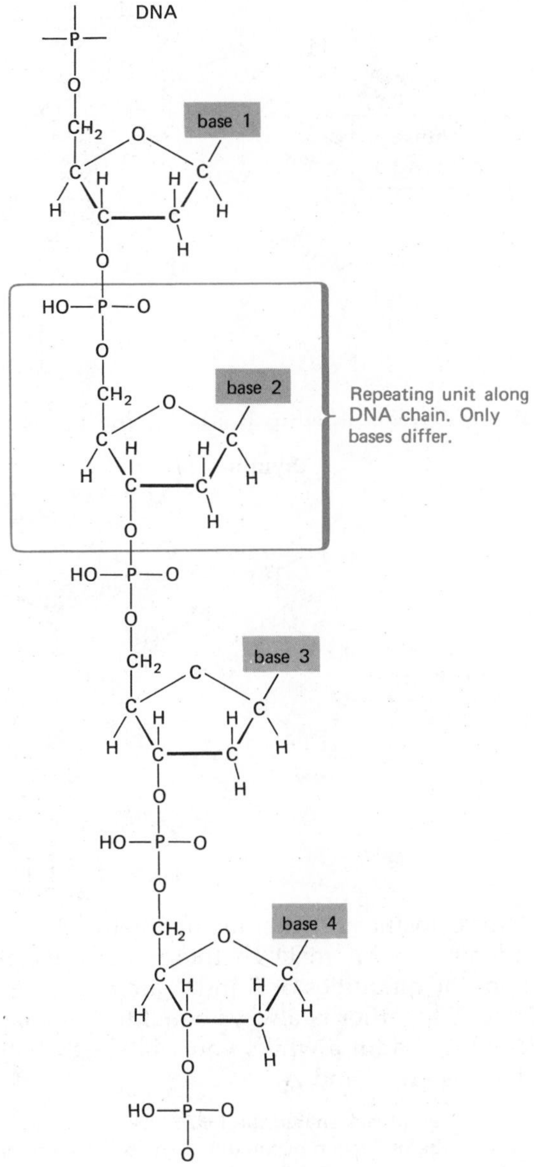

Figure 24.24

Polymerization of nucleotides gives nucleic acids.

primary structure, which controls the specific properties of the nucleic acid. For instance, the base sequence in DNA contains coded information that the cell utilizes in the synthesis of its own characteristic proteins. We will look at this more closely in the next section.

The truly remarkable properties of DNA that account for its ability to reproduce itself exactly are dictated by its secondary structure, in which two strands of DNA intertwine into the now much celebrated **double helix** proposed by Watson and Crick, for which they received the Nobel Prize in 1962.[2] The key to the formation of the double helix, as well as the function of DNA and RNA in protein synthesis, lies in the interaction of the nitrogenous bases by way of hydrogen bonding.

Consider, for example, the bases cytosine and guanine, situated across from one another on separate DNA strands. The structure of these bases, as shown below, allows them to be ideally suited to interact with each other through the formation of hydrogen bonds.

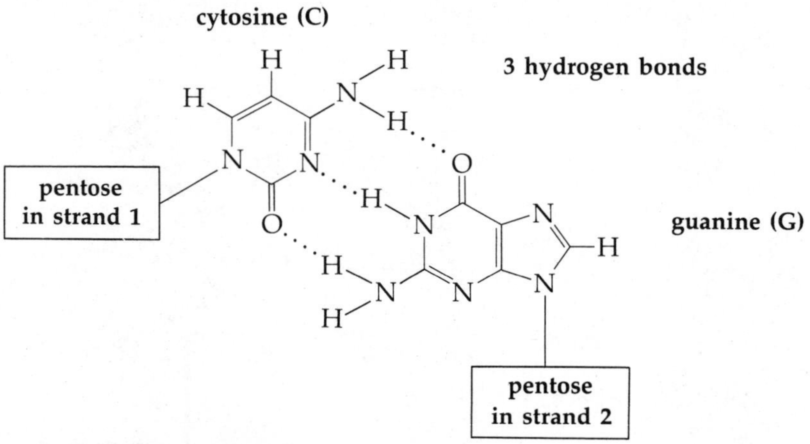

A similar relationship holds for the bases thymine and adenine.

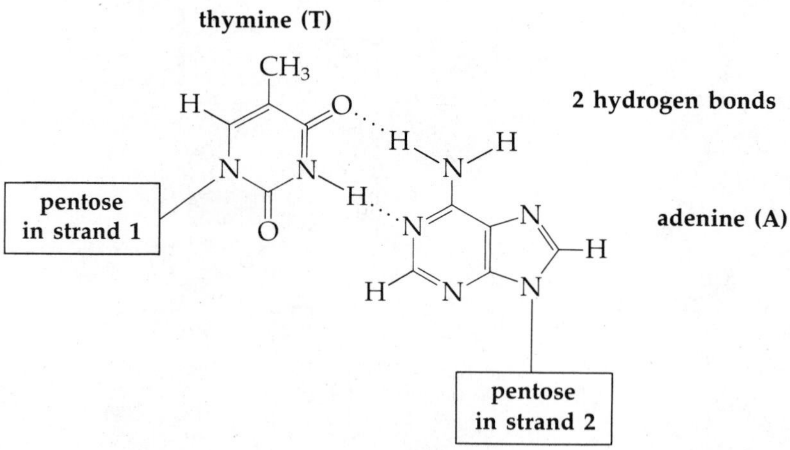

Thus cytosine and guanine fit together like a hand in a glove, as do thymine and adenine. Now, in DNA there is always the same amount of C as G. In addition, the quantities of T and A are the same. Furthermore, the total amount of C and T together is always equal to the combined total of G and A. If you think about this for a while, you will see this suggests that C and G are paired together, as are T and A.

[2] Watson and Crick shared the 1962 Nobel Prize with Wilkins. They used Rosalind Franklin's X-ray data in the deduction of the double-helix structure of DNA.

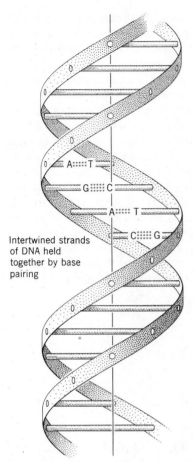

Figure 24.25

The double helix. Interwined strands of DNA held together by base pairing.

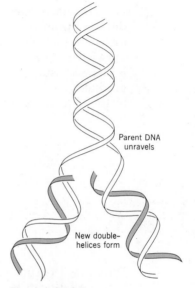

Figure 24.26

Replication of DNA.

These observations, in conjunction with X-ray diffraction data, led Watson and Crick to propose the double-helical structure of DNA, shown schematically in Figure 24.25. In this structure we see that for each G on one strand there is a C opposite it, across the axis of the helix, on the other strand. A similar relationship also holds for T and A. The two DNA strands are not identical, but instead complement one another, and it is this property that accounts for the replication of the DNA on cell division.

It is believed that during cell division the two DNA strands begin to unravel, as shown in Figure 24.26, giving the two complementary chains that serve as templates for the construction of two new daughter chains. The restrictions on the base pairing (that is, T with A and C with G) causes the newly formed strands to be identical to the departing complementary parent chains and, as a result, a pair of DNA double helices are produced that are exact copies of the original.

24.6 PROTEIN SYNTHESIS

The DNA found in the nucleus of a cell serves indirectly to determine the makeup of the proteins that are synthesized at *ribosomes* located outside the nucleus. The genetic information that determines the amino acid sequence in each of the enzymes is stored in the DNA in a genetic code that is made up of the sequence of bases along a DNA strand. The transcription of this code to the site of protein synthesis, as well as the decoding and construction of the polypeptide chain of the protein, is accomplished by the other nucleic acids—the RNA.

Unlike DNA, RNA occurs only in single strands. In addition, there are several types of RNA. One of these is called **messenger RNA, mRNA,** and serves to carry the genetic code from the DNA template within the nucleus to the ribosomes outside. Another type of RNA molecule, called **transfer RNA, tRNA,** is much smaller than mRNA. It acts as an amino acid carrier and through a decoding mechanism that we will examine momentarily, adds the amino acid to the growing polypeptide chain at just the right place at just the right time.

The mechanism by which this process is believed to occur is not really too different from that involved in the duplication of DNA itself. It is known, for instance, that the production of mRNA takes place within the nucleus on an untwisted segment of a DNA chain. This segment corresponds to the gene that is characteristic of the particular protein to be synthesized. The mRNA strand produced contains a sequence of bases that is determined, through base pairing, by the sequence of bases in the DNA; however, the pairing scheme in this case is

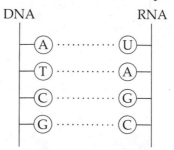

Thus uracil occurs in RNA instead of thymine. Once formed, the mRNA is transported from the nucleus to the active site of protein synthesis.

It is now known that the genetic code that directs the insertion of amino acids in the proper sequence in the growing polypeptide chain consists of sets of three bases (called **codons**). The amino acids are brought to the proper place by tRNA, which is able to decipher the code.

A molecule of tRNA contains from 75 to 90 nucleotide units containing the bases U, A, G, C, and, in addition, many minor bases. What is interesting is that although the tRNAs corresponding to different amino acids have different minor bases and base sequences, they can all be brought into a characteristic cloverleaf shape if maximum base pairing is assumed. For example, the yeast tRNA for the amino acid alanine has the structure shown in Figure 24.27. The molecule is folded in such a way as to give maximum base pairing (colored lines). On the stem of this RNA molecule, alanine becomes attached by the action of the proper enzyme. The two arms to either side of the site of attachment of the amino acid appear to be important in the interaction of the tRNA with this enzyme. The bottom loop contains the three-unit anticodon that attaches itself to the complementary codon on the mRNA chain.

The sequence of operations that leads to the synthesis of a polypeptide is summarized in Figure 24.28. In the bacterium, *E. coli,* for example, it is known that the amino acid sequence is initiated by a derivative of methionine. The tRNA containing this derivative becomes attached at the head of the mRNA through its codon-anticodon pairing. Once this is in place the next amino acid in the sequence is brought into place by its tRNA, which becomes bound to the mRNA at the second codon site. This is followed by an enzyme-induced forma-

Figure 24.27

The base sequence in a yeast alanine tRNA.

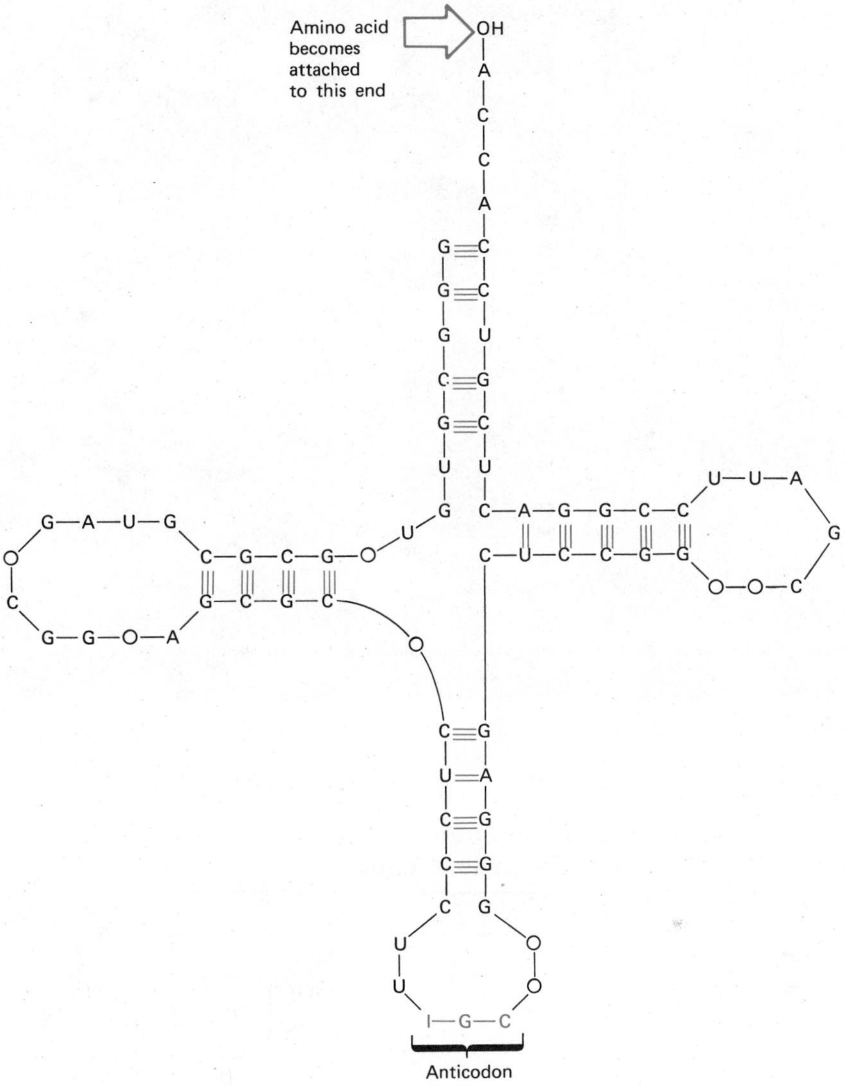

tion of the peptide linkage and the subsequent departure of the tRNA from the first codon. Next, the third amino acid is delivered by its tRNA, which attaches itself to the third codon. Again a peptide linkage is formed and the second tRNA leaves. This process is repeated over and over along the mRNA as the polypeptide chain grows in size. The chain is finally terminated when a nonsense codon (one that cannot be recognized by any tRNA molecule) is encountered.

The genetic code

Through a series of very clever experiments, which earned Nirenberg, Holley, and Khorana a Nobel Prize in 1968, the base sequences in the genetic code triplets were determined. These are given in Table 24.3. Notice that most amino acids are specified by more than one code word. For these, apparently, the attachment of the tRNA to the mRNA is determined primarily by the first two bases in the sequence (which are usually the same for a given amino acid). The

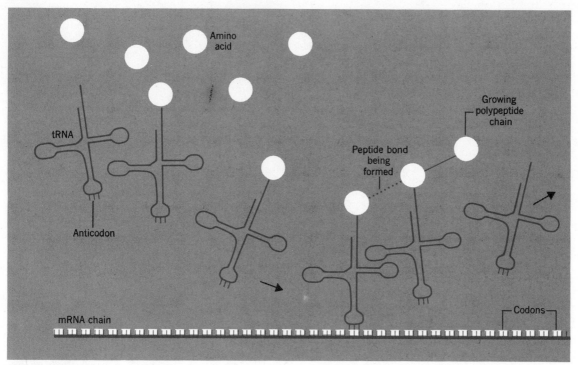

Figure 24.28

Synthesis of a polypeptide.

Table 24.3
The genetic code

		Second Base in Codon			
		U	C	A	G
First Base in Codon	U	UUU $\}$ Phe UUC UUA $\}$ Leu UUG	UCU $\}$ Ser UCC UCA UCG	UAU $\}$ Tyr UAC UAA^b UAG^b	UGU $\}$ Cys UGC UGA^b UGG Trp
	C	CUU CUC $\}$ Leu CUA CUG	CCU CCC $\}$ Pro CCA CCG	CAU $\}$ His CAC CAA $\}$ Gln CAG	CGU CGC $\}$ Arg CGA CGG
	A	AUU $\}$ Ile AUC AUA AUG Met^a	ACU ACC $\}$ Thr ACA ACG	AAU $\}$ Asn AAC AAA $\}$ Lys AAG	AGU $\}$ Ser AGC AGA $\}$ Arg AGG
	G	GUU GUC $\}$ Val GUA GUG	GCU GCC $\}$ Ala GCA GCG	GAU $\}$ Asp GAC GAA $\}$ Glu GAG	GGU GGC $\}$ Gly GGA GGG

^{a} AUG appears to initiate an amino acid sequence with methionine.
^{b} Nonsense codons: these do not code for any amino acid. They serve to terminate peptide chains.

interaction between the third base in the mRNA codon and the tRNA anticodon thus does not appear to be as critical as the first two in determining specificity.

With this code we can now see how the amino acid sequence in a polypeptide is fixed by the primary structure of the DNA. Consider, for example, a segment of DNA having the base sequence

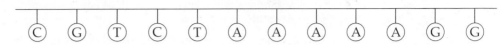

The mRNA formed from it will have the base sequence

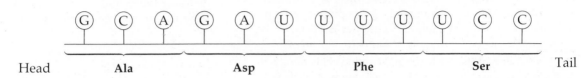

which, as you can see, is composed of the code words for the polypeptide

Ala—Asp—Phe—Ser

The triplet code presented in Table 24.3 has been shown to apply to such diverse species as the bacterium, *E. coli,* the tobacco plant, the guinea pig, and even humans. It is widely believed that this code is universal for all species.

Mutations

There are many chemicals that have been shown to be mutagenic.

Any factor that will have the effect of altering the base sequence in a DNA molecule will, in effect, alter the mRNA transcribed from it and thereby cause a change in the amino acid sequence produced using that mRNA as a template. Provided that this does not prove fatal to the cell, a mutation will have occurred that will be transmitted from one generation to another as the DNA reproduces itself on cell division.

Often these mutations prove harmful, although not immediately lethal to the organism in which they occur, and give rise to symptoms that cause them to be referred to as diseases. Since the origin of the diseases is in the genetic material of the cell, they are said to be genetic diseases. Many such diseases are recognized. Some of the more well known are cystic fibrosis, hemophilia, and sickle-cell anemia. Some evidence even suggests that schizophrenia may be of genetic origin.

In sickle-cell anemia, for example, the red blood cells assume a crescent shape instead of the flat disklike shape of normal cells. This is caused by a mutated DNA in the gene that is responsible for the synthesis of hemoglobin. It has been found that in sickle-cell hemoglobin a *glutamic acid* unit in one of the hemoglobin chains is replaced by *valine.* This alters the secondary and tertiary structure of the protein and reduces its ability to carry oxygen.

Sickle-cell anemia is thus a disease with its origin in the DNA of the cell nucleus and is therefore passed from one generation to the next because of its genetic nature.

REVIEW QUESTIONS AND PROBLEMS

(Problems whose numbers are in blue have their answers in Appendix C at the back of the book.)

Amino Acids

24.1 What is an α-amino acid? Indicate how amino acids are linked together to give a dipeptide, a polypeptide.

24.2 What is a zwitterion? Why do amino acids form zwitterions?

24.3 Monosodium glutamate (MSG), which is the monosodium salt of glutamic acid, is a popular flavor enhancer. However, only the L-isomer is effective. Is this particularly surprising? Explain your answer.

24.4 Only one of the basic set of 20 amino acids does not exhibit optical isomerism. Which one is it? Why is it not optically active?

Proteins

24.5 Lye, NaOH, is able to dissolve proteins such as hair that are lodged in a sink drain. What chemical reaction is involved?

24.6 Describe what is meant by the primary structure of a protein. What is meant by secondary structure, tertiary structure, and quaternary structure?

24.7 Indicate three functions served by proteins in a living organism.

24.8 Describe the α-helix structure found in many polypeptides. What holds the polypeptide chain in this helical conformation?

24.9 How do the properties of the solvent affect the tertiary structures of proteins? What interactions in addition to hydrogen bonding determine tertiary structure?

24.10 What is meant by a quaternary structure? What proteins exhibit this kind of structural sophistication?

24.11 How do the roles played by hemoglobin and myoglobin differ? In what way are they similar?

24.12 What is a porphyrin? Name two biologically important porphyrin structures.

Enzymes

24.13 In what sense do enzymes guide the chemistry that takes place in living organisms? What problems would a living cell encounter without the existence of enzymes?

24.14 Describe in qualitative terms how an enzyme operates. What is meant by *enzyme substrate?* What is enzyme inhibition? How do the sulfa drugs function?

24.15 How does diisopropylfluorophosphate function as a nerve gas? Many insecticides contain organic phosphates. Can you guess how they function?

Carbohydrates

24.16 What is a monosaccharide? Give an example of (a) a pentose and (b) a hexose.

24.17 How many optical isomers of fructose are there?

24.18 Compare the structures of starch and cellulose. What is the major difference between them?

24.19 Using the method described on page 879, draw formulas for the remaining optical isomers of

$$
\begin{array}{c}
\text{CHO} \\
|\\
\text{HO—C—H} \\
|\\
\text{H—C—OH} \\
|\\
\text{HO—C—H} \\
|\\
\text{CH}_2\text{OH}
\end{array}
$$

24.20 Sucrose, cane sugar, is used in large amounts in food products of all kinds. What monosaccharides are produced by hydrolysis of a sucrose molecule?

24.21 What functions do the carbohydrates serve in living organisms?

Lipids

24.22 What is a lipid? What functions do they serve in living systems?

24.23 What is a fatty acid? Give an example of a triglyceride. What difference in physical properties exists between the saturated and unsaturated triglycerides?

24.24 Write a chemical equation for the saponification of tristearin.

24.25 Give the structure of triolein. What product is formed on addition of H_2 to the double bonds in triolein?

24.26 What is a soap? How does it form micelles? What is responsible for the ability of soap to dissolve grease?

24.27 How does a phospholipid differ from a triglyceride?

24.28 How are phospholipids believed to contribute to the structure of cellular membranes?

24.29 What characteristic structural feature is found among the steroids?

Nucleic Acids

24.30 Describe the structure of a *nucleotide.* What is the difference between the nucleotide units in DNA and RNA?

24.31 Imagine that a single DNA strand contained a segment with the composition

What would be the base sequence in the complementary strand?

24.32 What holds the two DNA strands together in the double helix? Illustrate the base pairing between C and G, T and A.

24.33 What occurs with DNA during cell division? How is the DNA replicated?

Protein Synthesis

24.34 What is meant by the *genetic code?* Describe the functions of mRNA and tRNA.

24.35 Give a base sequence that would have to exist on mRNA to give a polypeptide with the amino acid sequence,

$$Arg—Leu—Lys—Gly—Cys$$

24.36 If a DNA strand contains the base sequence

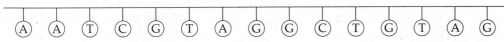

what will be the base sequence transcribed onto the mRNA?

24.37 What amino acid sequence is specified by the base sequence (from left to right) in the following mRNA strand?

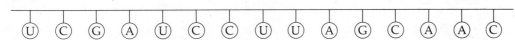

24.38 What base sequence on a DNA strand would give a mRNA that would produce a polypeptide having the amino acid sequence,

$$Ala—Pro—Asp—Tyr—Ile—Gly$$

24.39 What amino acid sequence is specified by the base sequence (starting at the left) in the DNA strand in Question 24.36?

Mutations

24.40 Suppose that through some mutational change the first base was removed from the sequence in the DNA strand depicted in Question 24.36. What amino acid sequence would result in the polypeptide synthesized using this modified (mutant) DNA template?

24.41 What is a genetic disease? Must it always be fatal to the organism?

24.42 How could X rays or γ-rays cause mutations in cells?

MATHEMATICS FOR GENERAL CHEMISTRY

For many students, solving numerical problems is often the most difficult part of any chemistry course. In this appendix we review some of the mathematical concepts that you will find useful in your study of chemistry.

A.1 THE FACTOR-LABEL METHOD OF PROBLEM SOLVING

Even after learning the principles of chemistry, students sometimes have difficulty in correctly setting up the arithmetic to give the proper numerical answer to a problem. The "factor-label" method uses the units associated with numbers as a guide in working out the arithmetic. The method is based on the fact that the units associated with numbers undergo the same kinds of mathematical operations of multiplication and division as the numbers themselves. For example, if two numbers are multiplied to give some desired quantity, the units belonging to those numbers are also multiplied together. As an illustration, suppose we wished to determine the area of a rectangular carpet whose dimensions had been measured to be 4.0 m (the length) and 3.0 m (the width). Area is calculated as length × width, so the area is

$$\text{area} = 4.0 \text{ m} \times 3.0 \text{ m} = 12 \text{ m}^2$$

We've obtained the numerical value of the area by multiplying the numbers, and we've obtained the units of the answer by multiplying the units for the length and the width (i.e., meter × meter = meter^2, or m × m = m^2).

An especially useful application of the properties of units is based on the mathematical operation of division: *Units that are the same in the numerator and denominator of a fraction cancel.* For instance, if the units *seconds* appeared in numerator and denominator, they would cancel.

$$\frac{3 \text{ s}}{2 \text{ s}} = \frac{3}{2}$$

We use this in solving numerical problems by constructing **conversion factors** from relationships between units. A conversion factor is a fraction that we use to convert a given quantity to a desired quantity by multiplication.

$$(\text{given quantity}) \times (\text{conversion factor}) = (\text{desired quantity})$$

A conversion factor has the property of changing the *units* of the given quantity to the appropriate *units* of the desired quantity. It is easiest to see how this happens if we examine a specific example. Suppose we wished to convert 4 minutes into the equivalent time measured in seconds. To do this, we must use the relationship between seconds and minutes

$$60 \text{ s} = 1 \text{ min}$$

This can be used to construct *two* conversion factors, which are

$$\frac{1 \text{ min}}{60 \text{ s}} \quad \text{and} \quad \frac{60 \text{ s}}{1 \text{ min}}$$

Each of these has a numerical equivalence of 1. For instance, since 60 s equals 1 min, we can substitute 1 min wherever we find 60 s. The first factor becomes

$$\frac{1 \text{ min}}{60 \text{ s}} = \frac{1 \text{ min}}{1 \text{ min}} = \frac{1}{1} = 1$$

Of course, if we multiply anything by 1, we don't change its size. Therefore, if we multiply any quantity by a conversion factor, its magnitude doesn't change—all we do is change the units.

To convert 4 min to s, we must multiply by one of the conversion factors that we've constructed. To make the choice, we simply examine the units. We know that the unit min must not be in the answer, so we choose the factor that allows us to cancel min. This is the second one.

$$4 \text{ min} \times \frac{60 \text{ s}}{1 \text{ min}} = 240 \text{ s}$$

Notice that we have used the units to guide us in the arithmetic. The units tell us that to make the conversion, we must multiply 4 min by 60 to obtain the time in s. Of course, you probably knew that, so the factor-label method just prolonged a simple problem. But in solving chemistry problems, the solution is not always so obvious. Furthermore, the units can often help you detect mistakes, because if you do the *wrong* arithmetic, you get the *wrong* units. For example, suppose we had used the first conversion factor by mistake.

$$4 \text{ min} \times \frac{1 \text{ min}}{60 \text{ s}} = \frac{4 \text{ min}^2}{60 \text{ s}}$$

The units are $\text{min}^2 \text{ s}^{-1}$ (square minutes per second). These are not the units we want, so we know that we made a mistake. In solving chemistry problems, it is very helpful to know when you've made a mistake.

It should be obvious that the key to using the factor-label method is choosing *correct* relationships between the units. For instance, if you didn't know how many seconds equal 1 min and decided to use the relationship

$$90 \text{ s} = 1 \text{ min}$$

you would certainly obtain the wrong answer, no matter how clever you are at arithmetic or applying the factor-label method. You will have many opportunities to use the factor-label method throughout your chemistry course. *Pay special attention to the correctness of the relationships that you use to make your conversion factors.*

Before concluding this section, we would like to point out that in many places throughout the book, we use the concept of an *equivalence* between quantities, instead of an equality. For example, if you have a job that pays 5 dollars per hour, there is an equivalence between the time you work and the dollars you earn: Each hour that you work is equivalent to 5 dollars pay. We express this in a mathematical way by using the symbol ~ to mean "is equivalent to." Thus

$$1 \text{ h} \sim 5 \text{ dollars}$$

We can't really say that one hour *equals* five dollars, so we express the relationship as an equivalence.

Equivalencies such as this are also used to make conversion factors. In this case, we can make these two factors

$$\frac{1 \text{ h}}{5 \text{ dollars}} \quad \text{and} \quad \frac{5 \text{ dollars}}{1 \text{ h}}$$

If you worked for 12 h, you could use the second factor to calculate your earnings.

$$12 \, \cancel{hr} \times \frac{5 \text{ dollars}}{1 \, \cancel{hr}} = 60 \text{ dollars}$$

A.2 EXPONENTIAL NOTATION (SCIENTIFIC NOTATION)

Quite often in science it is necessary to deal with numbers that are very large, such as Avogadro's number,

602 200 000 000 000 000 000 000

or numbers that are very small, such as the mass of a single molecule of water,

0.000 000 000 000 000 000 000 03 g

These numbers are very cumbersome and difficult to work with without making mistakes in arithmetic computations. To aid us in handling these large and small numbers, a system called either **exponential notation** or **scientific notation** is employed. In this system, a number is expressed as a decimal part multiplied by 10 raised to an appropriate power. Thus

$$200 = 2 \times 10 \times 10 = 2 \times 10^2$$

$$205 \, 000 = 2.05 \times 100{,}000 = 2.05 \times (10 \times 10 \times 10 \times 10 \times 10)$$

$$= 2.05 \times 10^5$$

To determine the exponent on the 10, we can simply count the number of places the decimal must be moved to produce the number that precedes the 10 when the number is expressed in the scientific notation

$$205 \quad 000 \quad_{\circ} \quad = 2.05 \times 10^5$$

5 places

Note that the exponent on the 10 is positive when the decimal is moved to the left. When it is moved to the right, the exponent is negative.

$$0 \quad_{\circ} \quad 000 \, 000 \, 315 = 3.15 \times 10^{-7}$$

7 places

Most students today perform their computations using an electronic calculator, and virtually every "scientific calculator" is designed to handle calculations involving numbers expressed in scientific notation. As simple as these calculators are to manipulate, many students make mistakes in entering numbers in scientific notation. If you plan to use one of these calculators, take it out now and turn it on, so we can review the procedure.

Most scientific calculators have a key labeled EE or EXP. These are the keys that are used to activate the scientific notation function of the calculator. When you press this key, you should say to yourself, "times ten to the ...." To enter the number 4.5×10^3, the key sequence is

| 4 | | . | | 5 | | EE | (or | EXP |) | | 3 |

4 means the key labeled 4, . means the key with the decimal point, and so forth.

Try it on your calculator. As you press the keys, say to yourself, "four point five *times ten to the* third." On some calculators, the display will show a small 10 with the exponent 3; on other calculators, there will be a space after the 4.5, which is followed by 03.

To enter a negative exponent, you *do not* use the subtraction key, which appears as $\boxed{-}$. Instead, after entering the decimal part of the number and the EE (or EXP) key, press the change-sign key, which is usually labeled $\boxed{+/-}$. For example, to enter the number 4.5×10^{-3}, the key sequence is

$$\boxed{4} \quad \boxed{.} \quad \boxed{5} \quad \boxed{EE} \quad \boxed{+/-} \quad \boxed{3}$$

Once again, try it on your calculator. Notice that the exponent appears with a negative sign on the display.

Once you know how to enter numbers expressed in scientific notation, arithmetic operations are performed on them just as on other kinds of numbers. Refer to the directions that accompany your calculator and read Section A.5 of this appendix for additional help, if you need it.

Even though you may have learned to work with numbers in scientific notation using your calculator, someday the batteries may fail and you might find it necessary to do arithmetic the "old-fashioned" way. Therefore, it is a good idea to know the following rules for arithmetic.

Multiplication

Try these on your calculator to be sure you obtain the correct answers.

In multiplication, the decimal portions of the numbers are multiplied and the exponents on the 10 are *added* algebraically.

$$(2.0 \times 10^4) \times (3.0 \times 10^3) = (2.0 \times 3.0) \times 10^{(4+3)} = 6.0 \times 10^7$$

$$(4.0 \times 10^8) \times (-2.0 \times 10^{-5}) = (4.0 \times (-2.0)) \times 10^{(8+(-5))} = -8.0 \times 10^3$$

Division

The decimal portions are divided, and the exponent on 10 in the denominator is *subtracted algebraically* from the exponent on 10 in the numerator.

$$\frac{8.0 \times 10^7}{4.0 \times 10^3} = \left(\frac{8.0}{4.0}\right) \times 10^{(7-3)} = 2.0 \times 10^4$$

$$\frac{6.0 \times 10^5}{2.0 \times 10^{-3}} = \left(\frac{6.0}{2.0}\right) \times 10^{(5-(-3))} = 3.0 \times 10^8$$

$$\frac{9.0 \times 10^{-4}}{3.0 \times 10^{-6}} = \left(\frac{9.0}{3.0}\right) \times 10^{(-4-(-6))} = 3.0 \times 10^2$$

You have probably noticed that the usual practice is to express a number with the decimal point located between the first and second digit. There are, of course, other ways that these numbers can be written that are all equivalent, and you will undoubtedly find occasions when it is convenient to use a number in other than its standard form. An example of a few equivalent expressions of the same number are

$$3.15 \times 10^{-7} = 315 \times 10^{-9} = 0.0315 \times 10^{-5}$$

Notice that in converting from one to another, one part of the number is increased while the other is decreased. For instance, to change 8.25×10^6 to 825×10^4, multiply *and* divide by 100 (or 10^2)

$$8.25 \times 10^6\left(\frac{100}{100}\right) = (8.25 \times 100) \times \left(\frac{10^6}{10^2}\right) = 825 \times 10^4$$

Addition and subtraction

When carrying out addition and subtraction, each quantity must first be written with the same power of 10. Then addition or subtraction is performed on the decimal parts; the power of 10 remains the same. For example,

$$(2.17 \times 10^5) + (3.0 \times 10^4) = ?$$

If we express both numbers with the same power of 10, we have

$$
\begin{array}{ll}
2.17 \times 10^5 & \text{or} \quad\ 21.7 \times 10^4 \\
\underline{+0.30 \times 10^5} & \quad\ \ \underline{+3.0 \times 10^4} \\
2.47 \times 10^5 & \quad\ 24.7 \times 10^4
\end{array}
$$

Taking a root

To extract a root (e.g., the square root), the exponent on the 10 is made to be divisible by the desired root. For instance, to take the square root of 3.7×10^7, we first change the number so that the power of 10 is divisible by 2. Then we take the square root of the decimal part and divide the exponent by 2.

$$\sqrt{3.7 \times 10^7} = \sqrt{37 \times 10^6} = \sqrt{37} \times 10^3 = 6.1 \times 10^3$$

A.3   LOGARITHMS

A logarithm is an exponent. **Common logarithms** are exponents to which 10 must be raised to give a specified number. For instance, the log (100) = 2 because $10^2 = 100$. Similarly, log (1000) = log (10^3) = 3.

Since logarithms are exponents, when we perform mathematical operations the same rules that apply to exponents also apply to logarithms. Thus we have

Multiplication $\begin{cases} \text{add exponents} \\ \text{add logarithms} \end{cases}$

Division $\begin{cases} \text{subtract exponents} \\ \text{subtract logarithms} \end{cases}$

Your scientific calculator will handle logarithms with ease. See Section A.5 of this appendix.

For example,

$$\boxed{10^3 \times 10^4} = 10^{3+4} = \boxed{10^7}$$
$$\log(10^3 \times 10^4) = \log(10^3) + \log(10^4) = 3 + 4 = 7 = \log(10^7)$$

Similarly, for division

$$\boxed{\frac{10^8}{10^6}} = 10^{8-6} = \boxed{10^2}$$
$$\log\left(\frac{10^8}{10^6}\right) = \log(10^8) - \log(10^6) = 8 - 6 = 2 = \log(10^2)$$

For decimal numbers between 1 and 10 their logarithms lie between 0 and 1, since

$$\log(1) = 0 \quad (1 = 10^0)$$
$$\log(10) = 1 \quad (10 = 10^1)$$

For example, log 2 = 0.3010 or

$$10^{0.3010} = 2$$

The common logarithm of 2 and other numbers between 1 and 10 can be obtained from the table of logarithms in Appendix B.

To use this table to find the logarithm of a number, we use the extreme left column to locate the first two digits of the number, and the top horizontal row to locate the third digit. The value in the table corresponding to these is the logarithm of our number. For example, if we want log(4.61), we would locate 46 in the left column and proceed to the right until we were in the column headed by 1. The answer is

$$\log(4.61) = 0.6637$$

This table is extremely easy to use as long as our numbers are expressed in this fashion, that is, as a decimal number between 1 and 10. If the number whose logarithm we seek does not appear this way, we can first express the number in exponential notation and then take its logarithm. For example, what is log(728)?

$$\log(728) = \log(7.28 \times 10^2)$$

$$\log(7.28 \times 10^2) = \log(7.28) + \log(10^2)$$

$$\log(7.28) = 0.8621 \qquad \text{(from table)}$$

$$+\log(10^2) = \underline{2.0000}$$
$$2.8621$$

therefore,

$$\log(728) = 2.8621$$

What would be the value of log(0.005 83)? Once again we first express the number in exponential notation

$$\log(0.005\ 83) = \log(5.83 \times 10^{-3})$$

$$\log(5.83 \times 10^{-3}) = \log(5.83) + \log(10^{-3})$$

$$\log(5.83) = +0.7657$$

$$+\log(10^{-3}) = \underline{-3.0000}$$

Adding these algebraically, we get

$$\log(0.005\ 83) = -2.2343$$

Sometimes it is necessary to obtain the number whose logarithm is known. This is called taking the **antilogarithm.** The procedure is simply the reverse of that given above. For example, suppose that we wish to find the number whose logarithm is 3.253.

$$\log x = 3.253$$

First, we divide the number into two parts, a positive integer and a decimal.

$$3.253 = 3 + 0.253 = 0.253 + 3$$

$$\log x = (0.253 + 3)$$

We locate 0.253 in the body of the log table and find that it is the log of 1.79; we also know that 3 is the log of 10^3. Therefore,

$$\log x = \log(1.79) + \log(10^3) = \log(1.79 \times 10^3)$$

$$x = 1.79 \times 10^3$$

If the logarithm of the number is negative, the procedure for taking the antilogarithm is just slightly different. For example, suppose we have the problem,

$$\log x = -8.475$$

Once again, we divide the logarithm into two parts—a *positive decimal* and a *negative integer*.

$$-8.475 = (+0.525) + (-9)$$

Therefore,

$$\log x = (+0.525) + (-9)$$

Next, we locate 0.525 in the body of the log table and find that is in the log of 3.35. The -9 becomes the exponent on the 10, so our answer is

$$x = 3.35 \times 10^{-9}$$

What you've learned about logarithms here can be used in working problems dealing with pH in Chapter 15. The pH of an aqueous solution is defined as

$$\text{pH} = -\log[\text{H}^+]$$

where $[\text{H}^+]$ stands for the molar concentration of hydrogen ion in the solution. As promised in the text (page 541), here are the solutions to the logarithm calculations in Examples 15.2, 15.3, and 15.4.

EXAMPLE 15.2

SOLUTION

We are given that

$$[\text{H}^+] = 2.0 \times 10^{-3}$$

Therefore

$$\text{pH} = -\log(2.0 \times 10^{-3})$$
$$= -(\log 2.0 + \log 10^{-3})$$
$$= -[0.30 + (-3)]$$
$$= -(-2.70)$$
$$= 2.70$$

EXAMPLE 15.3

SOLUTION

Method 1

$$[\text{H}^+] = 2.0 \times 10^{-11}$$
$$\text{pH} = -\log(2.0 \times 10^{-11})$$
$$= -(\log 2.0 + \log 10^{-11})$$
$$= -[0.30 + (-11)]$$
$$= -(-10.70)$$
$$= 10.70$$

Method 2

$$[OH^-] = 5.0 \times 10^{-4}$$

$$pOH = -\log[OH^-]$$

$$= -\log(5.0 \times 10^{-4})$$

$$= -[0.70 + (-4)]$$

$$= 3.30$$

Then

$$pH = 14.00 - pOH$$

$$= 10.70$$

EXAMPLE 15.4

SOLUTION   In this case, pH = 3.80 and we must compute $[H^+]$

$$pH = -\log[H^+] = 3.80$$

Therefore

$$\log[H^+] = -3.80$$

To take the antilog, we have to express −3.80 as the sum of a negative whole number and a positive decimal.

$$\log[H^+] = (0.20) + (-4.00)$$

The antilog of 0.20, obtained from the log table, is 1.6; the antilog of −4 is 10^{-4}. Therefore

$$\log[H^+] = \log(1.6) + \log(10^{-4})$$

$$= \log(1.6 \times 10^{-4})$$

This gives

$$[H^+] = 1.6 \times 10^{-4}$$

Also, the calculated pOH = 10.20. Therefore

$$-\log[OH^-] = 10.20$$

$$\log[OH^-] = -10.20$$

$$= (0.80) + (-11)$$

$$= \log(6.3) + \log(10^{-11})$$

$$= \log(6.3 \times 10^{-11})$$

Hence,

$$[OH^-] = 6.3 \times 10^{-11}$$

Natural logarithms

A system of logarithms encountered frequently in the sciences, known as natural logarithms, has as its base $e = 2.718\ 28.\ \ldots$. In other words, natural logarithms are exponents to which e must be raised to give a number. The relationship between common logs and natural logs is seen below

$$\log_{10}(10) = 1 \qquad \text{or} \qquad 10^1 = 10$$
$$\log_e(10) = 2.303 \qquad \text{or} \qquad e^{2.303} = 10$$

With common logarithms we usually omit the base and write $\log 10 = 1$. With natural logarithms the base e is omitted, and they are written

$$\ln 10 = 2.303$$

The conversion from base e to base 10 logarithm is accomplished by the equation

$$\ln x = 2.303 \log x$$

A.4 THE QUADRATIC EQUATION

When an equation can be written in the form

$$ax^2 + bx + c = 0$$

in which the coefficients a, b, and c are known, two values (called roots) of the variable x can be obtained by substituting the values of a, b, and c into the expression

$$x = \frac{-b \pm \sqrt{b^2 - 4ac}}{2a}$$

For example, given the equation

$$x^2 - 5x + 4 = 0$$

what is the value of x? In this equation $a = 1$, $b = -5$, and $c = 4$. Thus

$$x = \frac{-(-5) \pm \sqrt{(-5)^2 - 4(1)(4)}}{2(1)} = \frac{5 \pm \sqrt{25 - 16}}{2}$$

$$= \frac{5 \pm \sqrt{9}}{2} = \frac{5 \pm 3}{2}$$

Therefore,

$$x = \frac{2}{2} = 1 \qquad \text{and} \qquad x = \frac{8}{2} = 4$$

Both values of x are mathematically correct. Usually when a quadratic equation is encountered in a chemical problem, only one of the roots has any real significance. Generally, the other root will be clearly meaningless—for instance, a negative concentration, which is impossible (you can't have a smaller amount of matter than no matter at all!).

A.5 ELECTRONIC CALCULATORS

Today, much of the tiresome work of arithmetic is relieved by the use of small hand-held electronic calculators. For example, we have already described how they can be useful for calculations that involve scientific notation. To obtain the most out of these electronic marvels, you should be aware of some simple mathematical relationships, and those that are most useful to you in chemistry are discussed below. Since operational procedures differ on various calculators, you will have to refer to the direction booklet that accompanies your calculator for specific instructions about how to apply these relationships.

Successive multiplication and division

One of the most common kinds of computations you will encounter, especially if you apply the factor-label method, involves evaluating expressions that have several numbers in the numerator and several in the denominator. An example might be something like this:

$$\frac{14 \times 92 \times 32}{73 \times 43 \times 51} = ?$$

There are several ways that this expression can be evaluated and they all give the same result. For example, you could evaluate the numerator and write down the answer; then evaluate the denominator and write down its value; and then finally enter the value of the numerator and divide it by the value of the denominator. This will certainly give the correct answer, but it involves a lot of unnecessary writing and the error-prone reentering of numbers. The expression can be evaluated in one step without writing down any intermediate values as follows:

1. Enter the first value in the numerator (in this case, 14).
2. Each time you enter another value from the numerator, perform multiplication.
3. Each time you enter a value from the denominator, perform a division.

To evaluate the expression above, a suitable sequence of operations would be

$$14 \times 92 \times 32 \div 73 \div 43 \div 51 =$$

Actually, the order in which the operations are performed doesn't matter, so an equally suitable sequence is

$$14 \times 92 \div 73 \div 43 \times 32 \div 52 =$$

Obviously, there are other sequences, too. Try some of them, just for practice.

Logarithms and antilogarithms

We've seen that a logarithm is an exponent and that there are two systems of logarithms generally encountered, base e and base 10. If your calculator possesses logarithm capabilities, you will probably find a key labeled LN for base e (natural) logarithms and a key labeled LOG for base 10 (common) logarithms. Generally, if a number is entered and the LN key depressed, the display will show the natural log of that number. The common log will appear if you depress the LOG key.

Useful relationships among logarithms are

$$10^{\log X} = X$$
$$e^{\ln X} = X$$

For example, log 2 = 0.3010, ln 2 = 0.6931.

$$10^{\log 2} = 10^{0.3010} = 2$$
$$e^{\ln 2} = e^{0.6931} = 2$$

Some calculators don't have keys labeled e^x and 10^x. They usually do antilogarithms by combining an inverse key, INV , with the appropriate logarithm key. For example, e^x is accomplished by the sequence INV LN .

These provide a means to obtain the antilogarithms. If you have the natural log of a number, enter it and depress the e^x key. If you have the common log, enter it and depress the 10^x key. In each case you will obtain the antilogarithm.

If your calculator does not have a 10^x key, but does have an x^y (or y^x) key, enter *10* and raise it to an exponent that corresponds to the common log. The result will be the antilogarithm.

Exponents and roots

Most calculators have X^2 and $\sqrt{X}$ keys, and these operations are simple. For higher powers and roots you can use either of two methods.

1. **Using the x^y key.** To compute, $X = a^b$, enter a and raise it to the power b. To compute $X = \sqrt[b]{a}$, enter a and raise it to the power, $1/b$. For example,

$$X = 2^3 = 8$$
$$X = \sqrt[3]{2} = 2^{1/3} = 2^{0.3333...3} = 1.259\ 92$$

2. **Using logarithms.** To compute $X = a^b$ with natural logarithms, we use the relationship that

$$\ln X = \ln a^b = b \ln a$$

Therefore, $$X = e^{\ln X}$$
$$X = e^{b \ln a}$$

Let's suppose that we wished to compute 3^5. On a typical calculator we would perform this computation in the following sequence:

1. Take ln 3   ln 3 = 1.098 612

2. Multiply ln 3 by 5   5 ln 3 = 5.493 061

3. Raise e to this exponent $e^{5 \ln 3} = 243$

$$3^5 = 243$$

These calculations, of course, can also be done using common logarithms, in which case

$$X = 10^{b \log a}$$

To compute a root, $X = \sqrt[b]{a}$, we find that

$$X = e^{(\ln a)/b}$$

For example, suppose that we wished to calculate $\sqrt[5]{12}$. The sequence of operations is

1. Take ln 12   ln 12 = 2.484 907

2. Divide ln 12 by 5 $\cdot \dfrac{(\ln 12)}{5} = 0.496\ 981$

3. Raise e to this exponent

$$e^{(\ln 12)/5} = 1.643\ 752$$

$$\sqrt[5]{12} = 1.643\ 752$$

COMMON LOGARITHMS

	0	1	2	3	4	5	6	7	8	9
10	0000	0043	0086	0128	0170	0212	0253	0294	0334	0374
11	0414	0453	0492	0531	0569	0607	0645	0682	0719	0755
12	0792	0828	0864	0899	0934	0969	1004	1038	1072	1106
13	1139	1173	1206	1239	1271	1303	1335	1367	1399	1430
14	1461	1492	1523	1553	1584	1614	1644	1673	1703	1732
15	1761	1790	1818	1847	1875	1903	1931	1959	1987	2014
16	2041	2068	2095	2122	2148	2175	2201	2227	2253	2279
17	2304	2330	2355	2380	2405	2430	2455	2480	2504	2529
18	2553	2577	2601	2625	2648	2672	2695	2718	2742	2765
19	2788	2810	2833	2856	2878	2900	2923	2945	2967	2989
20	3010	3032	3054	3075	3096	3118	3139	3160	3181	3201
21	3222	3243	3263	3284	3304	3324	3345	3365	3385	3404
22	3424	3444	3464	3483	3502	3522	3541	3560	3579	3598
23	3617	3636	3655	3674	3692	3711	3729	3747	3766	3784
24	3802	3820	3838	3856	3874	3892	3909	3927	3945	3962
25	3979	3997	4014	4031	4048	4065	4082	4099	4116	4133
26	4150	4166	4183	4200	4216	4232	4249	4265	4281	4298
27	4314	4330	4346	4362	4378	4393	4409	4425	4440	4456
28	4472	4487	4502	4518	4533	4548	4564	4579	4594	4609
29	4624	4639	4654	4669	4683	4698	4713	4728	4742	4757
30	4771	4786	4800	4814	4829	4843	4857	4871	4886	4900
31	4914	4928	4942	4955	4969	4983	4997	5011	5024	5038
32	5051	5065	5079	5092	5105	5119	5132	5145	5159	5172
33	5185	5198	5211	5224	5237	5250	5263	5276	5289	5302
34	5315	5328	5340	5353	5366	5378	5391	5403	5416	5428
35	5441	5453	5465	5478	5490	5502	5514	5527	5539	5551
36	5563	5575	5587	5599	5611	5623	5635	5647	5658	5670
37	5682	5694	5705	5717	5729	5740	5752	5763	5775	5786
38	5798	5809	5821	5832	5843	5855	5866	5877	5888	5899
39	5911	5922	5933	5944	5955	5966	5977	5988	5999	6010
40	6021	6031	6042	6053	6064	6075	6085	6096	6107	6117
41	6128	6138	6149	6160	6170	6180	6191	6201	6212	6222
42	6232	6243	6253	6263	6274	6284	6294	6304	6314	6325
43	6335	6345	6355	6365	6375	6385	6395	6405	6415	6425
44	6435	6444	6454	6464	6474	6484	6493	6503	6513	6522
45	6532	6542	6551	6561	6571	6580	6590	6599	6609	6618
46	6628	6637	6646	6656	6665	6675	6684	6693	6702	6712
47	6721	6730	6739	6749	6758	6767	6776	6785	6794	6803
48	6812	6821	6830	6839	6848	6857	6866	6875	6884	6893
49	6902	6911	6920	6928	6937	6946	6955	6964	6972	6981
50	6990	6998	7007	7016	7024	7033	7042	7050	7059	7067
51	7076	7084	7093	7101	7110	7118	7126	7135	7143	7152
52	7160	7168	7177	7185	7193	7202	7210	7218	7226	7235
53	7243	7251	7259	7267	7275	7284	7292	7300	7308	7316
54	7324	7332	7340	7348	7356	7364	7372	7380	7388	7396

	0	1	2	3	4	5	6	7	8	9
55	7404	7412	7419	7427	7435	7443	7451	7459	7466	7474
56	7482	7490	7497	7505	7513	7520	7528	7536	7543	7551
57	7559	7566	7574	7582	7589	7597	7604	7612	7619	7627
58	7634	7642	7649	7657	7664	7672	7679	7686	7694	7701
59	7709	7716	7723	7731	7738	7745	7752	7760	7767	7774
60	7782	7789	7796	7803	7810	7818	7825	7832	7839	7846
61	7853	7860	7868	7875	7882	7889	7896	7903	7910	7917
62	7924	7931	7938	7945	7952	7959	7966	7973	7980	7987
63	7993	8000	8007	8014	8021	8028	8035	8041	8048	8055
64	8062	8069	8075	8082	8089	8096	8102	8109	8116	8122
65	8129	8136	8142	8149	8156	8162	8169	8176	8182	8189
66	8195	8202	8209	8215	8222	8228	8235	8241	8248	8254
67	8261	8267	8274	8280	8287	8293	8299	8306	8312	8319
68	8325	8331	8338	8344	8351	8357	8363	8370	8376	8382
69	8388	8395	8401	8407	8414	8420	8426	8432	8439	8445
70	8451	8457	8463	8470	8476	8482	8488	8494	8500	8506
71	8513	8519	8525	8531	8537	8543	8549	8555	8561	8567
72	8573	8579	8585	8591	8597	8603	8609	8615	8621	8627
73	8633	8639	8645	8651	8657	8663	8669	8675	8681	8686
74	8692	8698	8704	8710	8716	8722	8727	8733	8739	8745
75	8751	8756	8762	8768	8774	8779	8785	8791	8797	8802
76	8808	8814	8820	8825	8831	8837	8842	8848	8854	8859
77	8865	8871	8876	8882	8887	8893	8899	8904	8910	8915
78	8921	8927	8932	8938	8943	8949	8954	8960	8965	8971
79	8976	8982	8987	8993	8998	9004	9009	9015	9020	9025
80	9031	9036	9042	9047	9053	9058	9063	9069	9074	9079
81	9085	9090	9096	9101	9106	9112	9117	9122	9128	9133
82	9138	9143	9149	9154	9159	9165	9170	9175	9180	9186
83	9191	9196	9201	9206	9212	9217	9222	9227	9232	9238
84	9243	9248	9253	9258	9263	9269	9274	9279	9284	9289
85	9294	9299	9304	9309	9315	9320	9325	9330	9335	9340
86	9345	9350	9355	9360	9365	9370	9375	9380	9385	9390
87	9395	9400	9405	9410	9415	9420	9425	9430	9435	9440
88	9445	9450	9455	9460	9465	9469	9474	9479	9484	9489
89	9494	9499	9504	9509	9513	9518	9523	9528	9533	9538
90	9542	9547	9552	9557	9562	9566	9571	9576	9581	9586
91	9590	9595	9600	9605	9609	9614	9619	9624	9628	9633
92	9638	9643	9647	9652	9657	9661	9666	9671	9675	9680
93	9685	9689	9694	9699	9703	9708	9713	9717	9722	9727
94	9731	9736	9741	9745	9750	9754	9759	9763	9768	9773
95	9777	9782	9786	9791	9795	9800	9805	9809	9814	9818
96	9823	9827	9832	9836	9841	9845	9850	9854	9859	9863
97	9868	9872	9877	9881	9886	9890	9894	9899	9903	9908
98	9912	9917	9921	9926	9930	9934	9939	9943	9948	9952
99	9956	9961	9965	9969	9974	9978	9983	9987	9991	9996

ANSWERS TO SELECTED QUESTIONS AND PROBLEMS

CHAPTER 1

1.9 (a) 1.25×10^3 (b) 1.3×10^7 (c) 6.023×10^{22} (d) 2.1457×10^5 (e) 3.147×10

1.11 (a) 30 000 000 000 (b) 0.000 025 4 (c) 1.22 (d) 0.000 000 34 (e) 32 500

1.13 (a) 7.7 (b) 73.3 (c) 0.785 (d) 3.478 (e) 81.4

1.15 (a) 2.14×10^3 (b) 1.21×10^4 (c) 41 (d) 5.9 (e) 261 (f) 0.690

1.16 (a) 7.32×10^{14} (b) 1.52×10^4 (c) 1.22×10^{20} (d) 1.7×10^{-4} (e) 2.70×10^{-25}

1.17 (a) 5.56×10^3 (b) 2.9×10^4 (c) 1.49×10^{10} (d) 3.8×10^{-6} (e) 9.0×10^{-31}

1.18 (a) 6.3×10^{-1} (b) 3.66×10^4 (c) 3.03×10^{-2} (d) -4.02×10^{-1} (e) 5.06×10^{-2}

1.21 (a) 1.40×10^2 cm (b) 2.8 m (c) 0.185 dm³ (d) 0.018 kg (e) 1.5×10^{-3} m² (f) 3.22×10^7 cm (g) 1.6×10^2 km h⁻¹ (h) 1×10^9 m³ (i) 2.0×10^4 mm s⁻¹ (j) 2.533×10^4 J

1.24 (a) nm (b) mm (c) mg (d) cm (e) mm³ (f) dm³

1.26 For the Pferdburper, 11 km L⁻¹; for the Smokebelcher, 12 km L⁻¹. The Smokebelcher is more economical.

1.28 6.9 km

1.30 303 K, 2256 K

1.32 $-78°C$

1.33 °N = (°C -80) $\times$ (100 °N/138°C). 0°C = -58 °N and 100°C = 14 °N

1.41 1.69 g cm⁻³

1.43 4.55 g cm⁻³

1.45 (a) 6.702 cm³ (b) 14.92 g

1.48 (a) 0.787 (b) 0.787 g cm⁻³

1.50 The first beaker, because density = 1.00 g cm⁻³.

1.56 7.21×10^3 J

1.58 6.30×10^4 J or 63.0 kJ

1.60 2.7×10^6 J

1.62 Temperature decreases by 0.0558°C; temperature increases by 0.0637°C.

1.64 27.1°C

1.68 Compound. Both samples contain 46.8% silicon.

1.78 Carbon, 0.632 22 u; hydrogen, 0.053 053 u.

1.80 35.5 u

1.82 1.4×10^2 g C

1.84 Sample 1: 0.158 g C/1 g F, 0.0857 g C/1 g Cl, 0.542 g F/1 g Cl
Sample 2: 0.158 g C/1 g F, 0.0857 g C/1 g Cl, 0.542 g F/1 g Cl
Both samples contain the same elements in the same ratios by mass.

1.86 Sample 1: 1.29 g P/1 g O. Sample 2: 0.773 g P/1 g O.
Ratio of phosphorus masses is 5-to-3.

1.88 4.58 g oxygen

1.91 (a) K, 2 atoms; S, 1 atom
(b) Na, 2 atoms; C, 1 atom; O, 3 atoms
(c) K, 4 atoms; Fe, 1 atom; C and N, 6 atoms each
(d) N, 3 atoms; H, 12 atoms; P, 1 atom; O, 4 atoms
(e) Na, 3 atoms; Ag, 1 atom; S, 4 atoms; O, 6 atoms

1.93 Al, 1 atom; H, 24 atoms; O, 20 atoms; K, 1 atom; S, 2 atoms

CHAPTER 2

2.4 (a) 1-to-1 (b) 1-to-1 (c) 1-to-3 (d) 1-to-1 (e) 1-to-3 (f) 1-to-3

2.6 (a) 2.00 (b) 0.720 (c) 6.00 (d) 3.00

2.8 3.00 mol S

2.10 1.00 mol CO_2

2.13 (a) 2.17 mol Na (b) 0.668 mol As (c) 0.962 mol Cr (d) 1.85 mol Al (e) 1.28 mol K (f) 0.463 mol Ag

2.15 (a) 40.304 (b) 110.99 (c) 208.239 (d) 135.03 (e) 163.9407

2.17 262 g caffeine

2.18 779 g penicillin

2.21 2.88 mol $NaHCO_3$

2.23 0.870 mol H_2SO_4

2.25 1.68 mol K

2.27 7.81 mol FeS_2

2.28 9.274×10^{-23} g/atom Fe; 1.064×10^{-22} g/molecule SO_2

2.30 5.684×10^{-22}, 28.50 times more massive, 4.40×10^{22} molecules, 1.98×10^{24} atoms

2.33 0.844 g Cu
2.34 (a) 34.43% Fe, 65.57% Cl
      (b) 42.07% Na, 18.89% P, 39.04% O
      (c) 28.71% K, 0.74% H, 23.55% S, 47.00% O
      (d) 21.21% N, 6.87% H, 23.46% P, 48.46% O
      (e) 84.98% Hg, 15.02% Cl
2.36 5.60 g N
2.39 (a) 2.55 g C, 0.569 g H
      (b) 1.13 g O
      (c) 60.0% C, 13.4% H, 26.6% O
2.41 (a) NH_4SO_4 (b) Fe_2O_3 (c) $AlCl_3$ (d) CH (e) $C_3H_8O_3$
      (f) CH_2O (g) Hg_2So_4
2.44 SO_3
2.46 P_4S_3
2.48 CCl_2
2.51 $NaBH_4$
2.53 C_2H_6O
2.55 (a) 25.1% C, 4.23% H, 9.79% N, 49.7% Cl,
      11.2% O (b) $C_3H_6NCl_2O$
2.57 Empirical formula = $C_6H_8O_7$; Molecular
      formula = $C_6H_8O_7$
2.60 Coefficients are: (a) 1,2,1,1 (b) 2,1,1,1 (c) 8,3,4,9
      (d) 1,1,2 (e) 2,1,2,1
2.62 Coefficients are: (a) 2,13,8,10 (b) 2,17,14,6 (c) 1,6,4
      (d) 4,11,2,8 (e) 4,5,4,6
2.64 (a) 2.50 mol C_2H_2 (b) 13.0 g C_2H_2 (c) 6.40 mol H_2O
      (d) 79.8 g $Ca(OH)_2$
2.66 (a) $P_4 + 5O_2 \rightarrow P_4O_{10}$ (b) 0.100 mol P_4O_{10} (c) 21.8 g P_4
      (d) 19.4 g P_4
2.68 (a) 52.5 mol CO
      (b) 1.50 mol Fe_2O_3
      (c) 45.5 g Fe_2O_3
      (d) 0.911 mol CO
      (e) 24.7 g Fe
2.70 1.574×10^3 kg DDT
2.72 (a) $(CH_3)_2NNH_2 + 2N_2O_4 \rightarrow 4H_2O + 2CO_2 + 3N_2$
      (b) 153 kg N_2O_4
2.74 (a) 12.5 g white lead (b) 12.8 g CO_2
2.77 (a) HCl (b) 0.38 mol H_2 (c) 0.02 mol Fe remains.
2.79 (a) C_2H_2 (b) 83.8 g C_2H_3Cl (c) 2 g HCl remains.
2.81 1.02 g Ag_2S
2.83 20 g $C_2H_2Br_4$, 24.9 g $C_2H_2Br_2$
2.85 (a) 30.1 g C_6H_5Br (b) 0.828 g C_6H_6 (c) 28.5 g C_6H_5Br
      (d) 94.7%
2.87 0.590 g of product, 508 g of reactant
2.92 (a) 0.625 M NaCl (b) 4.20 M sucrose (c) 2.27 M
      H_2SO_4 (d) 7.13 M KOH
2.95 (a) 0.0375 mol Li_2CO_3 (b) 6.98 g Li_2CO_3 (c) 66.7 cm^3
      soln (d) 7.22 cm^3 soln
2.97 79.1 g $Ca(C_2H_3O_2)_2$
2.99 18.5 g $MgSO_4 \cdot 7H_2O$
2.102 38 cm^3
2.103 8.3 cm^3
2.106 46 cm^3
2.108 12 cm^3
2.109 (a) 973 cm^3 (b) 8.00×10^{-2} mol H_2
2.111 (a) 25.0 cm^3 (b) 270 cm^3 (c) 0.005 25 mol Na_2SO_4
2.113 0.1000 M $AgNO_3$

CHAPTER 3

3.46 -8.0×10^{-20} C
3.51 (a) 9.59×10^4 C g^{-1} (b) e/m for proton is
      5.45×10^{-4} as large as e/m for the electron.
3.60 3.19×10^{15} g cm^{-3}
3.62 0.153 km
3.67

	Protons	Neutrons	Electrons
^{132}Cs	55	77	55
$^{115}Cd^{2+}$	48	67	46
^{194}Tl	81	113	81
$^{105}Ag^+$	47	58	46
$^{78}Se^{2-}$	34	44	36

3.69 (a) $^{55}_{26}Fe$ (b) $^{86}_{37}Rb$ (c) $^{204}_{81}Tl$ (d) $^{170}_{71}Lu$ (e) $^{169}_{70}Yb$
3.72 151.91
3.74 207 g
3.76 51.82% ^{107}Ag, 48.18% ^{109}Ag

CHAPTER 4

4.7 (a) 3.8×10^{-8} m (38 nm)
      (b) 1.50×10^{15} Hz
4.9 (a) 2.97 m (b) 341 m
4.14 $\nu = 5.49 \times 10^{14}$ Hz
4.16 $\nu = 5.09 \times 10^{14}$ Hz
4.18 $\lambda = 7459.85$ nm, $\lambda = 4653.77$ nm
4.22 2×10^{-18} J, 10 μm (1×10^{-7} m)
4.27 (a) 486.273 nm (b) 1094.11 nm
4.31 2.2×10^8 kg
4.33 3.0×10^{19} s
4.42 4
4.44 f
4.47 (a) $s = 2$, $p = 6$, $d = 10$, $f = 14$, $g = 18$, $h = 22$
      (b) $n = 6$ (c) $-5, -4, -3, -2, -1, 0, 1, 2, 3, 4, 5$
4.59 Rb: $1s^2 2s^2 2p^6 3s^2 3p^6 4s^2 3d^{10} 4p^6 5s^1$
      Sn: $1s^2 2s^2 2p^6 3s^2 3p^6 4s^2 3d^{10} 4p^6 5s^2 4d^{10} 5p^2$
      Br: $1s^2 2s^2 2p^6 3s^2 3p^6 4s^2 3d^{10} 4p^5$
      Cr: $1s^2 2s^2 2p^6 3s^2 3p^6 4s^1 3d^5$
      Cu: $1s^2 2s^2 2p^6 3s^2 3p^6 4s^1 3d^{10}$
4.60 K $4s^1$; Al $3s^2 3p^1$; F $2s^2 2p^5$; S $3s^2 3p^4$; Tl $6s^2 6p^1$; Bi $6s^2 6p^3$
4.64 (a) P

      (b) Ca

4.67 Cd, Sr, and Kr
4.77 Sn
4.78 (a) Se (b) C (c) Fe^{2+} (d) O^- (e) S^{2-}
4.84 (a) Be (b) Be (c) N (d) N (e) Ne (f) S^+ (g) Na^+
4.86 (a) Cl (b) S (c) P (d) S
4.89 1.3×10^4 g H_2O

CHAPTER 5

5.13 (a) Ba^{2+} $[Kr]4d^{10}5s^25p^6$ or $[Xe]$ (b) Se^{2-} $[Ar]3d^{10}4s^24p^6$ or $[Kr]$ (c) Al^{3+} $[He]2s^22p^6$ or $[Ne]$ (d) Na^+ $[He]2s^22p^6$ or $[Ne]$ (e) Br^- $[Ar]3d^{10}4s^24p^6$ or $[Kr]$

5.16 (a) $Ba^{2+}[:\ddot{O}:]^{2-}$ (b) $2Na^+,[:\ddot{O}:]^{2-}$ (c) $K^+[:\ddot{F}:]^-$ (d) $Ca^{2+}[:\ddot{S}:]^{2-}$ (e) $2Mg^{2+},[:\ddot{C}:]^{4-}$

5.21 (a), (d) and (e)

5.26 See Table 5.3 (a) $CrCl_2$, $CrCl_3$, CrS and Cr_2S_3 (b) $MnCl_2$, $MnCl_3$, MnS and Mn_2S_3 (c) $FeCl_2$, $FeCl_3$, FeS and Fe_2S_3 (d) $CoCl_2$, $CoCl_3$, CoS and Co_2S_3 (e) $NiCl_2$ and NiS (f) $CuCl$, $CuCl_2$, Cu_2S and CuS (g) $AgCl$ and Ag_2S

5.30 (a) Na_2CO_3 (b) $Ca(ClO_3)_2$ (c) SrS (d) $CrCl_3$ (e) $Ti(ClO_4)_4$

5.38

$$[1] + [2] + [3] + [4] + [5] = [6]$$

[1] =	90.0	kJ
[2] =	119	kJ
[3] =	419	kJ
[4] =	−348	kJ
[5] =	−704.2	kJ
[6] =	−424	kJ (exothermic)

5.40 Electron affinity = -342 kJ mol^{-1}

5.45

5.47

5.48

5.66 154 pm, 146 pm, 140 pm, and 137 pm. Bond energies also increase in this direction.

5.69

5.72 The S–O bond order in SO_2 is approximately 1.5 and in SO_3 approximately 1.3. Therefore, the bond length should be a little shorter and the bond energy a little higher in SO_2. The average vibrational frequency will be higher in SO_2.

5.75

Molecule	Bond Order	Bond Length	Bond Energy	Vib. Frequency
CO	3			
CO_2	2			
CH_3COO^-	1.5	increases	decreases	decreases
CO_3^{2-}	1.3			
CH_3CH_2OH	1			

5.77

5.83 (a) P—F (b) Al—Cl (c) Se—Cl

5.84 MgO, Al_2O_3, CsF

5.86 F_2 < H_2Se < H_2S < OF_2 < SO_2 = ClF_3 < SF_2

5.90 The difference between an experimental bond energy and a calculated bond energy is proportional to the electronegativity difference between the bonded atoms.

Compounds	HF	HCl	HBr	HI
Difference between calc. and exp. bond energies (kJ mol^{-1})	270	94	50	10

Since the electronegativity of hydrogen is constant in each compound, the differences must be due to the presence of the other elements. The change in bond energy indicates a decrease in electronegativity from F to I.

5.92 NaBr sodium bromide CaO calcium oxide FeCl$_3$ ferric chloride; iron(III) chloride AsCl$_5$ arsenic pentachloride CuCO$_3$ cupric carbonate; copper(II) carbonate CBr$_4$ carbon tetrabromide P$_4$O$_6$ tetraphosphorus hexoxide Mn(HCO$_3$)$_2$ manganous hydrogen carbonate; manganous bicarbonate; manganese(II) hydrogen carbonate; manganese(II) bicarbonate $NaMnO_4$ sodium permanganate O_2F_2 dioxygen difluoride.

5.93 $Al(NO_3)_3$, $FeSO_4$, $NH_4H_2PO_4$, IF_5, PCl_3, N_2O_4, $KMnO_4$, $Mg(OH)_2$, H_2Se, NaH

5.96 (a) $Fe_2(SO_4)_3$ (b) $FeCl_2$ (c) $Hg_2(NO_3)_2$ (d) $CuCl$ (e) $SnCl_4$ (f) $Co(OH)_2$ (g) $AuCl_3$ (from Table 5.3) (h) $Cr(C_2H_3O_2)_3$

CHAPTER 6

6.7 (a) AX_3E; pyramidal (b) AX_4; tetrahedral (c) AX_3; planar triangular (d) AX_2E; V-shaped (e) AX_4E; distorted tetrahedral (f) AX_2E_3; linear (g) AX_5E; square pyramidal (h) AX_4; tetrahedral (i) AX_6; octahedral (j) AX_5; trigonal bipyramidal (k) AX_2E; V-shaped

6.9

	(1)	(2)
(a)	tetrahedral	bent
(b)	tetrahedral	bent
(c)	octahedral	octahedral
(d)	tetrahedral	tetrahedral
(e)	octahedral	square planar
(f)	tetrahedral	tetrahedral
(g)	planar triangular	planar triangular
(h)	trigonal bipyramidal	T-shaped
(i)	planar triangular	planar triangular
(j)	tetrahedral	pyramidal
(k)	tetrahedral	tetrahedral

6.15 (a), (c), (d), (g), (h), and (j)

6.27 The overlap of the unpaired electrons in the p orbitals of As with the unpaired electron in the s orbital of the hydrogens will yield bonding with a bond angle of close to 90°.

6.29 P $\underline{\uparrow\downarrow}_{3s}$ $\underline{\uparrow}\,\underline{\uparrow}\,\underline{\uparrow}_{3p}$ F $\underline{\uparrow\downarrow}_{2s}$ $\underline{\uparrow\downarrow}\,\underline{\uparrow\downarrow}\,\underline{\uparrow}_{2p}$

$\underbrace{\underline{\uparrow\downarrow}\,\underline{\uparrow}\,\underline{\uparrow}\,\underline{\uparrow}}_{sp^3}$

P(in PF_3) $\underline{\uparrow\downarrow}\,\underline{\uparrow\downarrow}\,\underline{\uparrow\downarrow}\,\underline{\uparrow\downarrow}$ (Colored arrows are F electrons) sp^3

F(in PF_3) $\underline{\uparrow\downarrow}\,\underline{\uparrow\downarrow}\,\underline{\uparrow\downarrow}\,\underline{\uparrow\downarrow}$ (Colored arrow is P electron) $2s$ $2p$

6.32 (a) planar triangular, sp^2 bonding
(b) tetrahedral, sp^3 bonding
(c) trigonal bipyramidal, sp^3d bonding
(d) octahedral, sp^3d^2 bonding
(e) linear, sp bonding
(f) octahedral, sp^3d^2 bonding
(g) pyramidal, sp^3 bonding
(h) distorted tetrahedral, sp^3d bonding
(i) tetrahedral, sp^3 bonding

6.44 Valence bond theory shows that the CN^- triple bond consists of one sp-sp σ and two p-p π-bonds.

6.60 N_2^+ and N_2^- are both less stable than N_2. Bond orders: N_2, 3; N_2^+ and N_2^-, 2.5. Both N_2^+ and N_2^- would have longer bonds than N_2.

6.62 Bond energy of O_2 would be greater, its bond length less, and its vibrational frequency greater than O_2^{2-}

CHAPTER 7

7.11 $KCl(s) \rightarrow K^+(aq) + Cl^-(aq)$
$(NH_4)_2SO_4(s) \rightarrow 2NH_4^+(aq) + SO_4^{2-}(aq)$
$Na_3PO_4(s) \rightarrow 3Na^+(aq) + PO_4^{3-}(aq)$
$NaOH(s) \rightarrow Na^+(aq) + OH^-(aq)$
$HCl(g) \rightarrow H^+(aq) + Cl^-(aq)$

7.14 $CdSO_4(aq) \rightleftharpoons Cd^{2+}(aq) + SO_4^{2-}(aq)$

7.26 $Ca^{2+}(aq) + 2Cl^-(aq) + 2K^+(aq) + CO_3^{2-}(aq) \rightarrow CaCO_3(s) + 2K^+(aq) + 2Cl^-(aq)$
Net ionic $Ca^{2+}(aq) + CO_3^{2-}(aq) \rightarrow CaCO_3(s)$

7.28 (a) $Cu^{2+} + 2NO_3^- + 2Na^+ + 2OH^- \rightarrow Cu(OH)_2(s) + 2Na^+ + 2NO_3^-$
$Cu^{2+}(aq) + 2OH^-(aq) \rightarrow Cu(OH)_2(s)$
(b) $3Ba^{2+} + 6Cl^- + 2Al^{3+} + 3SO_4^{2-} \rightarrow 3BaSO_4(s) + 2Al^{3+} + 6Cl^-$ $Ba^{2+}(aq) + SO_4^{2-}(aq) \rightarrow BaSO_4(s)$
(c) $Hg_2^{2+} + 2NO_3^- + 2H^+ + 2Cl^- \rightarrow Hg_2Cl_2(s) + 2H^+ + 2NO_3^-$ $Hg_2^{2+}(aq) + 2Cl^-(aq) \rightarrow Hg_2Cl_2(s)$
(d) $2Bi^{3+} + 6NO_3^- + 6Na^+ + 3S^{2-} \rightarrow Bi_2S_3(s) + 6Na^+ + 6NO_3^-$ $2Bi^{3+}(aq) + 3S^{2-}(aq) \rightarrow Bi_2S_3(s)$
(e) $Ca^{2+} + 2Cl^- + 2Na^+ + SO_4^{2-} \rightarrow CaSO_4(s) + 2Na^+ + 2Cl^-$ $Ca^{2+}(aq) + SO_4^{2-}(aq) \rightarrow CaSO_4(s)$

7.37 (a) acidic (b) basic (c) acidic (d) acidic (e) basic

7.45 $N_2H_4(aq) + H_2O \rightleftharpoons N_2H_5^+(aq) + OH^-(aq)$

7.51 (a) NH_2^- (b) NH_3 (c) $C_2H_3O_2^-$ (d) $H_2PO_4^-$ (e) NO_3^-

7.54 Acid-base conjugate pairs (acid written first in each pair)
(a) $HC_2H_3O_2$, $C_2H_3O_2^-$ and H_2O, OH^-
(b) HF, F^- and NH_4^+, NH_3
(c) $Zn(OH)_2$, ZnO_2^- and H_2O, OH^-
(d) $Al(H_2O)_6^{3+}$, $Al(H_2O)_5OH^{2+}$ and H_2O, OH^-
(e) $N_2H_5^+$, N_2H_4 and H_2O, OH^-
(f) NH_3OH^+, NH_2OH and HCl, CL^-
(g) OH^-, O^{2-} and H_2O, OH^-
(h) H_2, H^- and H_2O, OH^-
(i) NH_3, NH_2^- and N_2H_4, $N_2H_3^-$
(j) HNO_3, NO_3^- and $H_3SO_4^+$, H_2SO_4

7.57 acid, $(CH_3)_2NH_2^+$; base, $(CH_3)_2N^-$

7.59 The highly charged Cr^{3+} ion polarizes the O—H bonds of the water molecules attached to it, thereby making it easier for the H^+ to be transferred to neighboring H_2O molecules.

7.60 (a) $HClO_3$ (b) HNO_3 (c) H_3PO_4 (d) $CHCl_2COOH$ (e) CH_2FCOOH (f) H_2SO_4 (g) $HClO_3$ (h) $HBrO_3$

7.61 (a) H_2Se (b) HBr (c) PH_3

7.67 Ammonia acts as a Lewis base on donating a pair of electrons to form a covalent bond with H^+ which acts as a Lewis acid.

7.72 (a) $K = +1$, $Cl = +3$, $O = -2$
(b) $Ba = +2$, $Mn = +6$, $O = -2$

(c) $Fe = +8/3$, $O = -2$
(d) $O = +1$, $F = -1$
(e) $I = +5$, $F = -1$
(f) $H = +1$, $O = -2$, $Cl = +1$
(g) $Ca = +2$, $S = +6$, $O = -2$
(h) $Cr = +3$, $S = +6$, $O = -2$
(i) $O = 0$
(j) $Hg = +1$, $Cl = -1$

7.76 (a) $10HNO_3 + 4Zn \rightarrow 4Zn(NO_3)_2 + 3H_2O +$
NH_4NO_3
(b) $10K + 2KNO_3 \rightarrow N_2 + 6K_2O$
(c) $3C_3H_7OH + 2Na_2Cr_2O_7 + 8H_2SO_4 \rightarrow$
$2Cr_2(SO_4)_3 + 2Na_2SO_4 + 11H_2O + 3HC_3H_5O_2$
(d) $3H_2S + 2HNO_3 \rightarrow 3S + 2NO + 4H_2O$
(e) $4Fe(OH)_2 + O_2 + 2H_2O \rightarrow 4Fe(OH)_3$

7.80 (a) $8H^+ + 2NO_3^- + 3Cu \rightarrow 2NO + 3Cu^{2+} + 4H_2O$
(b) $10H^+ + NO_3^- + 4Zn \rightarrow NH_4^+ + 4Zn^{2+} +$
$3H_2O$
(c) $2Cr + 6H^+ \rightarrow 2Cr^3 + 3H_2$
(d) $8H^+ + Cr_2O_7^{2-} + 3H_3AsO_3 \rightarrow 2Cr^{3+} + 4H_2O +$
$3H_3AsO_4$
(e) $10H^+ + SO_4^{2-} + 8I^- \rightarrow 4I_2 + H_2S + 4H_2O$
(f) $4H_2O + 8Ag^+ + AsH_3 \rightarrow H_3AsO_4 + 8Ag +$
$8H^+$
(g) $H_2O + S_2O_8^{2-} + HNO_2 \rightarrow NO_3^- + 2SO_4^{2-} +$
$3H^+$
(h) $4H^+ + MnO_2 + 2Br^- \rightarrow Mn^{2+} + Br_2 + 2H_2O$
(i) $2S_2O_3^{2-} + I_2 \rightarrow 2I^- + S_4O_6^{2-}$
(j) $IO_3^- + 3HSO_3^- \rightarrow I^- + 3SO_4^{2-} + 3H^+$

7.82 (a) $H_2O + CN^- + AsO_4^{3-} \rightarrow AsO_2^- + CNO^- +$
$2OH^-$
(b) $2CrO_2^- + 3HO_2^- \rightarrow 2CRO_4^{2-} + H_2O + OH^-$
(c) $7OH^- + 4Zn + NO_3^- + 6H_2O \rightarrow$
$4Zn(OH)_4^{2-} + NH_3$
(d) $4OH^- + Cu(NH_3)_4^{2+} + S_2O_4^{2-} \rightarrow 2SO_3^{2-} +$
$Cu + 4NH_3 + 2H_2O$
(e) $N_2H_4 + 2Mn(OH)_3 \rightarrow 2Mn(OH)_2 + 2NH_2OH$
(f) $4OH + 2MnO_4^- + 3C_2O_4^{2-} \rightarrow 2MnO_2 +$
$6CO_3^{2-} + 2H_2O$
(g) $6OH^- + 7ClO_3^- + 3N_2H_4 \rightarrow 6NO_3^- + 7Cl^- +$
$9H_2O$

7.84 (a) $Zn + 2H_2O + 2OH^- \rightarrow Zn(OH)_4^{2-} + H_2$
(b) $2CrO_2^- + 3HO_2^- \rightarrow 2CrO_4^{2-} + H_2O + OH^-$

7.88 $Mn(s) + 2HCl(aq) \rightarrow H_2(g) + MnCl_2(aq)$

7.90 $2Al(s) + 6HBr(aq) \rightarrow 2\,AlBr_3(aq) + 3H_2(g)$

7.94 $4Zn(s) + NO_3^-(aq) + 10H^+(aq) \rightarrow 4Zn^{2+}(aq) +$
$NH_4^+(aq) + 3H_2O$

7.97 (a) Rb (b) Rb (c) Na (d) Ca

7.101 (a) $2Cr + 6HCl \rightarrow 2CrCl_3 + 3H_2$
(b) $Ni + H_2SO_4 \rightarrow NiSO_4 + H_2$

7.106 (a) $2Al(s) + 3Zn^{2+}(aq) \rightarrow 2Al^{3+}(aq) + 3Zn(s)$
(b) $Sn(s) + Cu^{2+}(aq) \rightarrow Sn^{2+}(aq) + Cu(s)$
(c) $Ag(s) + Co^{2+}(aq) \rightarrow$ No reaction
(d) $Mn(s) + Pb^{2+}(aq) \rightarrow Mn^{2+}(aq) + Pb(s)$
(e) $Cu(s) + Mg^{2+}(aq) \rightarrow$ No reaction
(f) $Hg(s) + H^+(aq) \rightarrow$ No reaction
(g) $Ni(s) + 2H^+(aq) \rightarrow Ni^{2+}(aq) + H_2(g)$
(h) $Cd(s) + H_2O \rightarrow$ No reaction
(i) $Ba(s) + 2H_2O \rightarrow Ba(OH)_2(aq) + H_2(g)$

(j) $H_2(g) + Pt^{2+} \rightarrow Pt(s) + 2H^+(aq)$

7.107 (a) $C < N < O < F$
(b) $I < Br < Cl < F$

7.114 (a) $2Fe(s) + 3O_2(g) \rightarrow 2Fe_2O_3(s)$
(b) $4Li(s) + O_2(g) \rightarrow 2Li_2O(s)$
(c) $2Ca(s) + O_2(g) \rightarrow 2CaO(s)$
(d) $2Mg(s) + O_2(g) \rightarrow 2MgO(s)$
(e) $4Al(s) + 3O_2(g) \rightarrow 2Al_2O_3(s)$

7.117 (a) $C_9H_{20} + 14O_2 \rightarrow 9CO_2 + 10H_2O$
(b) $2C_2H_4(OH)_2 + 5O_2 \rightarrow 4CO_2 + 6H_2O$
(c) $2(CH_3)_2S + 9O_2 \rightarrow 4CO_2 + 6H_2O + 2SO_2$

CHAPTER 8

8.2

Soluble	Insoluble
KCl	$PbSO_4$
$(NH_4)_2SO_4$	$Mn(OH)_2$
$AgNO_3$	$FePO_4$
$Zn(ClO_4)_2$	$CaCO_3$
$Ba(C_2H_3O_2)_2$	NiO

8.5 (a) $Ag^+(aq) + Br^-(aq) \rightarrow AgBr(s)$
(b) $CoCO_3(s) + 2H^+(aq) \rightarrow Co^{2+}(aq) + CO_2(g) +$
H_2O
(c) $C_2H_3O_2^-(aq) + H^+(aq) \rightarrow HC_2H_3O_2(aq)$
(d) $Pb^+(aq) + SO_4^{2-}(aq) \rightarrow PbSO_4(s)$
(e) $H_2S(aq) + Cu^{2+}(aq) \rightarrow 2H^+(aq) + CuS(s)$
(f) $NH_4^+(aq) + OH^-(aq) \rightarrow NH_3(g) + H_2O$

8.7 (a) $Na_2SO_4(aq) + BaCl_2(aq) \rightarrow BaSO_4(s) + 2NaCl(aq)$
$2Na^+(aq) + SO_4^{2-}(aq) + Ba^{2+}(aq) + 2Cl^- \rightarrow$
$BaSO_4(s) + 2Na^+ + 2Cl^-$
$Ba^{2+}(aq) + SO_4^{2-}(aq) \rightarrow BaSO_4(s)$
(b) $Ca(NO_3)_2(aq) + (NH_4)_2CO_3(aq) \rightarrow CaCO_3(s) +$
$2NH_4NO_3(aq)$
$Ca^{2+}(aq) + 2NO_3^-(aq) + 2NH_4^+ + CO_3^{2-} \rightarrow$
$CaCO_3 + 2NH_4^+ + 2NO_3^-$
$Ca^{2+}(aq) + CO_3^{2-}(aq) \rightarrow CaCO_3(s)$
(c) $NaC_2H_3O_2(aq) + HNO_3(aq) \rightarrow NaNO_3(aq) +$
$HC_2H_3O_2(aq)$
$Na^+ + C_2H_3O_2^- + H^+ + NO_3^- \rightarrow Na^+ +$
$NO_3^- + HC_2H_3O_2$
$H^+(aq) + C_2H_3O_2^-(aq) \rightarrow HC_2H_3O_2(aq)$
(d) $2NaOH(aq) + CuCl_2(aq) \rightarrow 2NaCl(aq) +$
$Cu(OH)_2(s)$
$2Na^+ + 2OH^-(aq) + Cu^{2+}(aq) + 2Cl^- \rightarrow$
$2Na^+(aq) + 2Cl^-(aq) + Cu(OH)_2(s)$
$Cu^{2+}(aq) + 2OH^-(aq) \rightarrow Cu(OH)_2(s)$
(e) $(NH_4)_2CO_3(aq) + 2HNO_3(aq) \rightarrow 2NH_4NO_3(aq) +$
$H_2O + CO_2(g)$
$2NH_4^+ + CO_3^{2-} + 2H^+ + 2NO_3^- \rightarrow 2NH_4^+ +$
$2NO_3^- + H_2O + CO_2(g)$
$2H^+(aq) + CO_3^{2-}(aq) \rightarrow H_2O + CO_2(g)$

8.12 One possible set of reactions is:
(a) See (d) above
(b) $MnCl_2(aq) + Ba(OH)_2(aq) \rightarrow Mn(OH)_2(s) +$
$BaCl_2(aq)$

(c) $FeCl_2(aq) + Pb(C_2H_3O_2)_2(aq) \rightarrow Fe(C_2H_3O_2)_2(aq) + PbCl_2(s)$

(d) $NiCl_2(aq) + Pb(ClO_4)_2(aq) \rightarrow PbCl_2(s) + Ni(ClO_4)_2(aq)$

(e) $(NH_4)SO_3(aq) + BaCl_2(aq) \rightarrow 2NH_4Cl(aq) + BaSO_3(s)$

8.14 (a) $Ca(OH)_2(aq) + 2HNO_3(aq) \rightarrow Ca(NO_3)_2(aq) + 2H_2O$

(b) $2NaOH(aq) + H_2C_2O_4(aq) \rightarrow Na_2C_2O_4(aq) + 2H_2O$

(c) $Fe(OH)_2(s) + 2H_2SO_4(aq) \rightarrow 2H_2O + Fe(HSO_4)_2(aq)$

(d) $Al_2O_3(s) + 6HClO_4(aq) \rightarrow 2Al(ClO_4)_3(aq) + 3H_2O$

(e) $NiO(s) + 2HBr(aq) \rightarrow NiBr_2(aq) + H_2O$

8.19 (a) 1×10^{-4} % F^- by mass (b) 1.0 ppm F^- (c) 1.0×10^3 ppb F^-

8.21 (a) 0.750 M (b) 0.992 M (c) 0.556 M (d) 1.36 M (e) 0.962 M

8.23 4.77 g Na_2CO_3

8.25 24.01 M

8.27 0.0825%, 0.0106 M

8.29 (a) 0.0250 M Ba^{2+} and 0.0500 M OH^-
(b) 0.300 M Cd^{2+} and 0.600 M NO_3^-
(c) 0.800 M Na^+ and 0.400 M HPO_4^{2-}
(d) 0.200 M Cr^{3+} and 0.300 M SO_4^{2-}
(e) 0.0450 M Hg_2^{2+} and 0.0900 M NO_3^-

8.31 0.0533 M $FeCl_3$

8.33 (a) 0.0100 mol Na^+ and 0.0100 mol Cl^-
(b) 0.004 80 mol Ca^{2+} and 0.009 60 mol Cl^-
(c) 0.0351 mol Na^+ and 0.0176 mol SO_4^{2-}
(d) 0.221 mol NH_4^+ and 0.111 mol SO_4^{2-}
(e) 0.0375 mol Al^{3+} and 0.0562 mol SO_4^{2-}

8.35 100 cm^3

8.37 (a) $2H^+(aq) + CuCO_3(s) \rightarrow CO_2(g) + H_2O + Cu^{2+}(aq)$ (b) 29.6 cm^3 (c) 2.47 g $CuCO_3$

8.39 50.0 cm^3 $BaCl_2$ soln

8.41 122 g cm^{-3}

8.43 0.18 g AgCl

8.45 19.1 cm^3 HNO_3 soln

8.47 (a) HCl in excess; acidic (b) 0.059 M H^+

8.51 62.40% $PbCO_3$

8.53 50.0% NaCl, 20% NaBr, 30% NaI

8.57 21.7% $CaCO_3$

8.59 26.8%

8.61 64% $CaCO_3$ and 36% $MgCO_3$

8.72 31.4%

8.73 (a) 0.009 820 mol $CaCO_3$ ppt. (b) 45.70% $CaCl_2$

CHAPTER 9

9.6 21.1 kPa

9.8 239 mm

9.10 117.6 kPa

9.13 288 cm^3

9.15 107 kPa

9.17 61.4 cm

9.22 1.88 dm^3

9.24 2.49 dm^3

9.26 375°C

9.28 50.2 kPa

9.30 243 kPa

9.32 (a) 66 kPa (b) 68 kPa

9.34 350°C

9.36 26.2 cm^3

9.38 102 kPa

9.40 90.7 kPa

9.42 81.5 cm^3

9.44 $x_{N_2} = 0.749$, $x_{O_2} = 0.153$, $x_{CO_2} = 0.037$, $x_{H_2O} = 0.061$

9.46 19.6 cm^3

9.48 208 kPa

9.50 256 cm^3

9.52 (a) 4.48 dm^3 (STP) (b) 3.92 dm^3 (STP) (c) 3.36 dm^3 (Total at STP)

9.54 2.59 g dm^{-3} (STP)

9.55 43.9 g mol^{-1}

9.59 16.1 g mol^{-1}

9.61 45.0 g mol^{-1}

9.63 (a) CH_3 (b) 30.0 g mol^{-1} (c) C_2H_6

9.65 0.198 g N_2

9.67 200 cm^3 N_2, 600 cm^3 H_2

9.69 (a) 0.144 g O_2 (b) 0.368 g $KClO_3$

9.71 130 kPa

9.73 (a) 2.10×10^{-6} mol I_2 (b) 2.10×10^{-6} mol I_2
(c) 2.10×10^{-6} mol O_3 (d) 4.70×10^{-2} cm^3
(e) 0.000 235 ppm

9.75 (a) 1.36×10^{-3} mol O_2 (b) 0.111 g $KClO_3$ (c) 44.4%

9.78 He effuses 2.25 times faster than Ne.

9.80 1.99 g mol^{-1}

9.93 $P = 101.34$ kPa. An ideal gas would exert a pressure of 101.325 kPa. He acts very nearly like an ideal gas.

9.95 22.38 dm^3 mol^{-1} at STP

CHAPTER 10

10.35 46.1 kJ

10.37 59.3 kJ mol^{-1}

10.39 2.50 kJ of energy absorbed

10.41 40.3°C

10.71 (a) $\theta = 6.57°$ (b) $\theta = 34.9°$

10.73 20.5°, 44.8°

10.74 $r = 124.9$ pm

10.76 $r = 144.20$ pm

10.78 $r = 176$ pm

10.80 (a) 7.50 g cm^{-3} (b) 9.74 g cm^{-3} (c) 10.6 g cm^{-3}. The density calculated for face-centered cubic is closest to the actual density.

10.82 (a) 48% vacant (b) 32% vacant (c) 26% vacant

10.84 For fcc, 15.96 g cm^{-3}; for bcc, 7.98 g cm^{-3}. Neither is close to known density of 3.99 g cm^{-3}.

10.86 4 formula units per unit cell (rounded from 3.998).

10.89 (a) molecular (b) molecular (c) metallic (d) ionic (e) covalent (f) ionic (g) molecular

10.91 molecular

10.93 molecular

10.95 ionic

CHAPTER 11

11.20 $x_{glycerin} = 0.0810$, $w_{glycerin} = 0.310$, mass% glycerin = 31.0%

11.22 (a) 11.00% (b) 0.011 62 $x_{2n(NO_3)_2} = 0.011\ 62$ (c) 0.6431 M

11.24 (a) $1.24 \times 10^{-3}\%$ (b) $1.04 \times 10^{-4}\ M$

11.26 $x_{NaHCO_3} = 0.019$

11.28 $x_{Na_2CO_3} = 0.0269$

11.30 $x_{H_2SO_4} = 0.815$, $x_{H_2O} = 0.185$

11.32 (a) 76.0 mol% $CHCl_3$ (b) mass% = 17.1% C_6H_6

11.34 23.9% ethylene glycol, $x_{C_2H_4(OH)_2} = 0.0837$

11.36 13.5 g $Na_2CO_3 \cdot 10\ H_2O$

11.51 (a) Na^+ (b) F^- (c) Ca^{2+} (d) Fe^{3+} (e) S^{2-}

11.56 24.1 kJ liberated

11.61 88 g

11.66 76.3 kPa

11.68 0.213 g dm^{-3}

11.73 12.3 kPa

11.75 74 kPa

11.77 150 g mol^{-1}

11.79 55°C

11.81 Approximately three times.

11.86 B.P. = 101.17°C, F.P. = −4.25°C

11.88 $C_8H_8O_4$ (167 g mol^{-1})

11.90 B.P. = 100.682°C

11.92 F.P. = −25.2°C

11.98 2.0×10^3 g mol^{-1}

11.99 −0.372°C

11.102 50 kPa

11.104 Π = 2.89 MPa; a pressure greater than 2.89 MPa is needed for reverse osmosis.

11.106 −0.225°C

11.109 $Al_2(SO_4)_3$

CHAPTER 12

12.17 First step: $\Delta T = 0$, $\Delta E_{system} = \Delta E_{surroundings} = 0$, $q = w = 7500$ J
Second step: $\Delta T = 0$, $\Delta E_{system} = \Delta E_{surroundings} = 0$, $q = w = 13\ 000$ J

12.18 $w = q = 5.0 \times 10^3$ kJ

12.20 (a) $q = 35$ J, $w = -40$ J, $\Delta E_{system} = 75$ J
(b) $q_{surr.} = -35$ J, $w_{surr} = 40$ J, $\Delta E_{surr} = -75$ J

12.23 heat cap. = 1187 J °C^{-1}

12.26 (a) 2.22×10^5 J (b) $\Delta E = -2.22 \times 10^3$ kJ mol^{-1}

12.28 48.8 kJ mol^{-1}

12.31 −2230 kJ

12.33 −3940 kJ

12.35 $q = 43.9$ kJ mol^{-1}, $w = 2.48$ kJ mol^{-1}, $\Delta E = 41.4$ kJ mol^{-1}

12.41 −276 kJ mol^{-1}

12.43 (a) −854 kJ (b) −1427 kJ (c) −402 kJ (d) −87 kJ (e) −136 kJ

12.45 −394 kJ

12.47 −17 kJ

12.49 225 kJ mol^{-1}

12.51 In order, they are: −113 kJ, 305 kJ, −106 kJ.

12.53 2140 kJ

12.55 24 kJ

12.56 3.4×10^4 kJ liberated; 240 g H_2; V = 17 dm^3; H_2 would occupy 17 times as much space for storage as would octane.

12.58 630 g

12.60 785 kJ

12.63 203 kJ

12.65 8 kJ mol^{-1}

12.71 (a), (b), and (d)

12.73 (a) positive (b) negative (c) positive (d) positive (e) negative

12.83 (a) -237.3 J mol^{-1} K^{-1} (b) -305.0 J mol^{-1} K^{-1} (c) -358 J mol^{-1} K^{-1} (d) -274 J mol^{-1} K^{-1} (e) -99.2 J mol^{-1} K^{-1}

12.84 (a) -65.9 J K^{-1} (b) -231 J K^{-1} (c) -205 J K^{-1} (d) $+51$ J K^{-1} (e) -128 J K^{-1}

12.86 (a) −836 kJ (b) −1364 kJ (c) −346 kJ (d) −101 kJ (e) −101 kJ

12.88 (a) −498 kJ (b) −396 kJ (c) -3.11×10^3 kJ

12.93 ΔS_{vap} = 109 J mol^{-1} K^{-1}, ΔS_{fus} = 22.1 J mol^{-1} K^{-1}

12.100 (a) $\Delta G° = +51.9$ kJ (No) (b) $\Delta G° = -57$ kJ (Yes) (c) $\Delta G° = -511$ kJ (Yes) (d) $\Delta G° = +92$ kJ (No) (e) $\Delta G° = -34$ kJ (Yes) (f) $\Delta G° = +1637$ kJ (No)

CHAPTER 13

13.9 Rate (for CO_2) = 0.16 mol CO_2 dm^{-3} s^{-1}
Rate (for H_2O) = 0.32 mol H_2O dm^{-3} s^{-1}

13.10 For H_2O, Rate = 2.0 mol H_2O dm^{-3} s^{-1}. For NH_3, Rate = −1.4 mol NH_3 dm^{-3} s^{-1}. For O_2, Rate = −1.0 mol O_2 dm^{-3} s^{-1}

13.15 Zero-order

13.17 (a) dm^3 mol^{-1} s^{-1} (b) dm^3 mol^{-1} s^{-1} (c) dm^9 mol^{-3} s^{-1}

13.19 −1 (Rate = $k[A]^{-1}$)

13.21 (a) 2.35×10^{-6} mol dm^{-3} s^{-1} (b) 1.91×10^{-7} mol dm^{-3} s^{-1}

13.23 (a) Rate = $k[NO_2][O_3]$ (b) $k = 4.4 \times 10^7$ dm^3 mol^{-1} s^{-1}

13.24 (a) Rate of formation of $C = k[A][B]^2$ (b) $k = 1.2 \times 10^3$ dm^6 mol^{-2} s^{-1} (c) 1.7×10^{-1} mol D dm^{-3} s^{-1}

13.26 (a) Rate = $k[NO]^2[Cl_2]$ (b) $k = 2.5 \times 10^{-3}$ dm^6 mol^{-2} s^{-1}

13.31 (a) 2.43×10^{-4} mol dm^{-3} (b) 4.75 s (c) $5.62 \times 10^{-4}\ M$

13.32 21 min

13.34 (a) 21 s (b) 0.068 M

13.39 (a) Rate = $k[NO][Br_2]$ (b) Rate = $k[NO]^2[Br_2]$ (c) Step 2 must be rate determining. (d) A termolecular collision is unlikely for a fairly rapid reaction. (e) No.

13.41 Step 1: $(CH_3)_3CBr \rightarrow (CH_3)_3C^+ + Br^-$ (slow)
Step 2: $(CH_3)_3C^+ + OH^- \rightarrow (CH_3)_3COH$ (fast) for which Rate = $k[(CH_3)_3CBr]$

13.43 $NO_2 + O_3 \rightarrow NO_3 + O_2$ (slow)
$NO_3 + NO_2 \rightarrow N_2O_5$ (fast)
13.47 $NO + O_2 \rightleftharpoons NO_3$ (fast)
$NO_3 + NO \rightarrow 2 NO_2$ (slow)
13.53 $E_a = 162$ kJ mol^{-1}, $A = 1.28 \times 10^{16}$ dm^3 mol^{-1} s^{-1}
13.55 1.19×10^{-4} dm^3 mol^{-1} s^{-1}
13.57 18.6 kJ mol^{-1}
13.59 1.6×10^2 kJ
13.61 100 h
13.67 344 nm

CHAPTER 14

14.4 (a) $K_P = \dfrac{p_{CH_3OH}}{p_{CO}p^2_{H_2}}$, $K_c = \dfrac{[CH_3OH]}{[CO][H_2]^2}$

(b) $K_P = \dfrac{p_{CO_2}p_{H_2}}{p_{CO}p_{H_2O}}$, $K_c = \dfrac{[CO_2][H_2]}{[CO][H_2O]}$

(c) $K_P = \dfrac{p_{PCl_5}}{p_{PCl_3}p_{Cl_2}}$, $K_c = \dfrac{[PCl_5]}{[PCl_3][Cl_2]}$

(d) $K_P = \dfrac{p_{N_2}p^4_{H_2O}}{p^2_{NO_2}p^4_{H_2}}$, $K_c = \dfrac{[N_2][H_2O]^4}{[NO_2]^2[H_2]^4}$

(e) $K_P = \dfrac{p^2_{H_2O}p^2_{SO_2}}{p^2_{H_2S}p^3_{O_2}}$, $K_c = \dfrac{[H_2O]^2[SO_2]^2}{[H_2S]^2[O_2]^3}$

14.5 (a) $\dfrac{[HCl]^2}{[H_2][Cl_2]}$ (b) $\dfrac{[HCl]}{[H_2]^{1/2}[Cl_2]^{1/2}}$ $K_{(b)}$ would be the square root of $K_{(a)}$.
14.7 All 4 experiments give K_c very close to 5.5.
14.13 $K_P = 1.82 \times 10^{62}$
14.14 1.4×10^6
14.16 10.2
14.18 -10.8 kJ
14.20 1.13×10^3 K, or 860°C
14.23 4.05
14.24 1.20×10^5 atm^2 (1.23×10^{15} Pa^2)
14.28 (a) $K_c = [CO_2(g)]$ $K_P = p_{CO_2(g)}$

(b) $K_c = \dfrac{[Ni(CO)_4(g)]}{[CO(g)]^4}$ $K_P = \dfrac{p_{Ni(CO)_4(g)}}{p^4_{CO(g)}}$

(c) $K_c = \dfrac{[I_2(g)][CO_2(g)]^5}{[CO(g)]^5}$ $K_P = \dfrac{p_{I_2(g)}\,p^5_{CO_2(g)}}{p^5_{CO(g)}}$

(d) $K_c = \dfrac{[CO_2(g)]}{[Ca(HCO_3)_2(aq)]}$ $K_P = \dfrac{p_{CO_2}}{[Ca(HCO_3)_2]}$

(e) $K_c = [Ag^+(aq)][Cl^-(aq)]$
14.29 (a) Shift to the right. (b) Shift to the left. (c) Shift to the right. (d) Shift to the right. (e) No effect.
14.31 (a) decreased (b) increased (c) no change (d) decreased (e) no change
14.33 (a) increased (b) decreased (c) increased (d) no change
14.36 (a) no change (b) increased (c) decreased (d) decreased
14.37 $K_c = 0.18$

14.39 The system is not at equilibrium. The reaction must proceed to the right to reach equilibrium.
14.40 2.30×10^{-1} M
14.42 $K_c = 5.2 \times 10^2$
14.43 $K_p = 0.30$ atm^{-1}
14.44 (a) $[N_2O] = 0.01$ M, $[O_2] = 0.041$ M (b) 23
14.46 $[H_2] = [CO_2] = 0.106$ M, $[CO] = [H_2O] = 0.0935$ M
14.48 $[CO_2] = 8.9 \times 10^{-4}$, $[CO] = 1.1 \times 10^{-4}$
$[O_2] = 5.4 \times 10^{-5}$
14.51 $[HI] = 0.113$ M, $[H_2] = [I_2] = 0.0153$ M
14.53 $p_{NO} = 8.2 \times 10^{-4}$ MPa (0.82 kPa)
14.55 At equilibrium $[NO_2] = 0.23$ M $[N_2O_4] = 0.39$ M. When container volume is doubled, $[NO_2] = 0.15$ M, $[N_2O_4] = 0.17$ M. Yes.

CHAPTER 15

15.3

	$[H^+]$	$[OH^-]$	pH
(a)	1.0×10^{-3} M	1.0×10^{-11} M	3.00
(b)	1.25×10^{-1} M	8.00×10^{-14} M	0.90
(c)	3.2×10^{-12} M	3.1×10^{-3} M	11.49
(d)	4.2×10^{-13} M	2.4×10^{-2} M	12.38
(e)	2.1×10^{-4} M	4.8×10^{-11} M	3.68
(f)	1.3×10^{-5} M	7.7×10^{-10} M	4.89
(g)	1.2×10^{-12} M	8.4×10^{-3} M	11.92
(h)	2.1×10^{-13} M	4.8×10^{-2} M	12.68

15.5 $[H^+] = [OH^-] = 1.56 \times 10^{-7}$ M
15.7 (a) acidic (b) basic (c) neutral (d) acidic (e) basic
15.9 6.81
15.10 (a) $[H^+] = 0.050$ M, $[OH^-] = 2.0 \times 10^{-13}$ M
(b) $[H^+] = 1.9 \times 10^{-6}$ M, $[OH^-] = 5.3 \times 10^{-9}$ M
(c) $[H^+] = 1.0 \times 10^{-4}$ M, $[OH^-] = 1.0 \times 10^{-10}$ M
(d) $[H^+] = 1.6 \times 10^{-8}$ M, $[OH^-] = 6.3 \times 10^{-7}$ M
(e) $[H^+] = 1.1 \times 10^{-11}$ M, $[OH^-] = 9.1 \times 10^{-4}$ M
(f) $[H^+] = 2.5 \times 10^{-13}$ M, $[OH^-] = 4.0 \times 10^{-2}$ M
15.11 (a) 12.70 (b) 8.27 (c) 10.00 (d) 6.20 (e) 3.06 (f) 1.39
15.13 8.42
15.16 (a) 1.2×10^{-2} M (b) 2.5×10^{-2} M (c) 3.5×10^{-6} M (d) 1.2×10^{-3} M (e) 7.1×10^{-4} M
15.18 (a) 8.3×10^{-13} M (b) 4.0×10^{-13} M (c) 2.9×10^{-9} M (d) 8.3×10^{-12} M (e) 1.4×10^{-11} M
15.20 9.7×10^{-3}
15.22 3.2×10^{-10}
15.23 (a) 1.3% (b) 3.7% (c) 0.014% (d) 0.63% (e) 0.025% (f) 100% (strong acid)
15.25 (a) 3.0×10^{-5} M (b) 1.8×10^{-4} M (c) 3.4×10^{-4} M (d) 3.3×10^{-10} M (e) 9.8×10^{-9} M
15.27 9.1 g HCl
15.28 Essentially 100%
15.30 0.55 M
15.32 7.9×10^{-8}
15.34 $[H^+] = 1.3 \times 10^{-3}$ M, $[CHO_2^-] = 1.3 \times 10^{-3}$ M, $[HCHO_2] = 0.0087$ M, $[OH^-] = 7.7 \times 10^{-12}$ M
15.37 $[H^+] = 0.04$ M, pH = 1.4, $[H_2SeO_3] = 0.46$ M, $[HSeO_3^-] = 0.04$ M, $[SeO_3^{2-}] = 5.0 \times 10^{-8}$ M

15.39 $[H^+] = [HC_6H_6O_7^-] = 2.0 \times 10^{-3}\,M$, pH $= 2.7$, $[OH^-] = 5.0 \times 10^{-12}\,M$ $[H_2C_6H_6O_6] = 0.048\,M$, $[C_6H_6O_6^{2-}] = 1.6 \times 10^{-12}\,M$
15.41 0.079% dissociated
15.43 $1.1 \times 10^{-4}\,M$
15.45 $2.4 \times 10^{-2}\,M$
15.49 4.1-to-1
15.51 209 g
15.53 1.1×10^{-2} mol HCl
15.54 (a) ΔpH $= 0.18$ (b) ΔpH $= 0.37$ (c) ΔpH $= 0.59$ (d) ΔpH $= 4.02$ (e) ΔpH $= 7.30$
15.56 ΔpH $= -0.39$
15.58 ΔpH $= 2.7$
15.60 (d) < (c) < (a) < (b)
15.62 CN^- is a stronger base than NO_2^-.
15.64 Ammonia
15.67 (a) neutral (b) acidic (c) basic (d) acidic
15.70 (a) 7.87 (b) 5.08 (c) 11.63 (d) 11.15 (e) 3.37
15.72 2.0×10^{-6}
15.74 8.86
15.76 12.51
15.78 10.12
15.80 6.30
15.84 8.41
15.85 (a) 2.16 (b) 3.19 (c) 8.01
15.93 Green

CHAPTER 16

16.3 (a) $K_{sp} = [Pb^{2+}][F^-]^2$ (b) $K_{sp} = [Cu^+]^2[S^{2-}]$ (c) $K_{sp} = [Fe^{2+}]^3[PO_4^{3-}]^2$ (d) $K_{sp} = [Li^+]^2[CO_3^{2-}]$ (e) $K_{sp} = [Ca^{2+}][IO_3^-]$ (f) $K_{sp} = [Ag^+]^2[Cr_2O_7^{2-}]$
16.4 1.0×10^{-6}
16.6 1.20×10^{-7}
16.8 1.4×10^{-8}
16.10 7.3×10^{-9}
16.12 1.5×10^{-16}
16.14 (a) $8 \times 10^{-14}\,M$ (b) $8 \times 10^{-6}\,M$ (c) $3.9 \times 10^{-5}\,M$ (d) $8 \times 10^{-7}\,M$ (e) $3.5 \times 10^{-4}\,M$ (f) $9.3 \times 10^{-3}\,M$
16.16 1 g $CaSO_4$
16.18 0.3 day
16.20 (a) No precipitate. (b) BaF_2 will precipitate. (c) $CaSO_4$ will precipitate.
16.22 $[Ag^+] = 0.050\,M$, $[NO_3^-] = 0.10\,M$, $[H^+] = 0.050\,M$, $[Cl^-] = 3.4 \times 10^{-9}\,M$
16.24 $[H^+] = 2.6 \times 10^{14}\,M$, an impossible value.
16.26 Choose a pH slightly below 1.63.
16.28 $PbCrO_4$ will precipitate first. When $BaCrO_4$ begins to precipitate, $[Pb^{2+}] = 7.5 \times 10^{-7}\,M$.
16.31 $2 \times 10^{-8}\,M$
16.33 $3.4 \times 10^{-3}\,M$
16.35 2.2×10^{-6} mol
16.37 2.1 g NaF
16.39 2.2 g $Fe(OH)_2$ precipitated, $[Fe^{2+}] = 5 \times 10^{-12}\,M$
16.41 $[Fe^{2+}] = 4.5 \times 10^{-3}\,M$, $[Mn^{2+}] = 0.10\,M$, pH $= 7.83$

16.43 (a) $K_{form} = \dfrac{[AgCl_2^-]}{[Ag^+][Cl^-]^2}$ (b) $K_{form} = \dfrac{[Ag(S_2O_3)_2^{3-}]}{[Ag^+][S_2O_3^{2-}]^2}$ (c) $K_{form} = \dfrac{[Zn(NH_3)_4^{2+}]}{[Zn^{2+}][NH_3]^4}$
16.46 $4.9 \times 10^3\,M$
16.49 A precipitate of $AgC_2H_3O_2$ will form.
16.51 0.53 mol NH_4Cl
16.53 0.75 g KF (or more)

CHAPTER 17

17.9 anode: $2I^- \rightarrow I_2 + 2e^-$
cathode: $Ni^{2+} + 2e^- \rightarrow Ni$
cell: $Ni^{2+} + 2I^- \rightarrow Ni + I_2$
17.21 (a) 2 (b) 1 (c) 5 (d) 2 (e) 8
17.23 (a) 1 (b) 4 (c) 10 (d) 2 (e) 8
17.25 (a) 7.00 min (b) 70 min (c) 54 min
17.27 (a) 1.79×10^{-3} (b) 1.11 (c) 0.217 (d) 0.411
17.29 0.15 g O_2 and 0.019 g H_2; 0.11 dm^3 O_2 and 0.21 dm^3 H_2 (at STP)
17.31 (a) 272 g Ag (b) 1.02×10^4 cm^2
17.33 9.10 h
17.35 1.23 A
17.37 0.0798 A
17.39 434 min
17.41 0.0833 mol Cr
17.43 0.249 A
17.45 3×10^3 A (1 significant figure)
17.52 (a) anode: $Zn(s) \rightarrow Zn^{2+} + 2e^-$
cathode: $Ga^{3+} + 3e^- \rightarrow Ga(s)$
(b) $3\,Zn(s) + 2\,Ga^{3+} \rightarrow 3\,Zn^{2+} + 2\,Ga(s)$
(c) -0.53 V
(d) 0.87 V
17.54 (a) ClO_3^- (b) $Cr_2O_7^{2-}$ (c) MnO_4^- (d) PbO_2
17.56 (a) Na (b) Cl_2 (c) Cu (d) Sn (e) H_2
17.57 (a) $2Al(s) + 3Ni^{2+} \rightarrow 3Ni(s) + 2Al^{3+}$
(b) $3PbO_2(s) + 2Cr^{3+} + 3SO_4^{2-} + H_2O \rightarrow 3PbSO_4(s) + 2H^+ + Cr_2O_7^{2-}$
(c) $2Ag^+ + Pb(s) \rightarrow 2Ag(s) + Pb^{2+}$
(d) $Cl_2(g) + Mn^2 + 2H_2O \rightarrow 2Cl^- + 4H^+ + MnO_2(s)$
(e) $2H^+ + Mn(s) \rightarrow Mn^{2+} + H_2(g)$
17.59 (a) -0.38 V (b) -1.49 V (c) $+0.64$ V (d) -0.16 V (e) -0.13 V
17.61 (a) 1.42 V (b) 0.36 V (c) 0.93 V (d) 0.08 V (e) 1.03 V
17.62 (a) 0.91 V (b) -0.34 V (c) -0.71 V (d) 0.27 V (e) 1.42 V
17.63 (a) 5×10^3 (b) 1×10^9 (c) 3
17.65 (a) 1×10^{-13} (b) 5×10^{-51} (c) 4×10^{21} (d) 8×10^{-82} (e) 2×10^{-9}
17.66 (a) -180 kJ (b) 66 kJ (c) 410 kJ (d) -52 kJ (e) -274 kJ
17.69 (a) $\mathscr{E}° = 1.42$ V, $\mathscr{E} = 1.45$ V, $\Delta G = -840$ kJ
(b) $\mathscr{E}° = 0.11$ V, $\mathscr{E} = 0.17$ V, $\Delta G = -33$ kJ
(c) $\mathscr{E}° = 1.56$ V, $\mathscr{E} = 1.54$ V, $\Delta G = -297$ kJ
17.71 0.23 V
17.73 -0.17 V

17.75   5.43 M
17.77   0.39 h
17.79   33.0 kJ
17.82   0.059 V
17.84   0.19 V
17.86 8.4×10^{-36}

CHAPTER 18

18.11   3180°C
18.13 $K_{p(373)} = 6.7 \times 10^{-93}$, $K_{p(773)} = 3.1 \times 10^{-38}$,
        $K_{p(2273)} = 1.4 \times 10^{-4}$
18.23   (a) Li (b) Al (c) Cs (d) Sn (e) Ga
18.25 Al_2O_3
18.29   (a) $GeCl_4$ (b) Bi_2O_5 (c) PbS (d) Li_2S (e) MgS
18.32   (a) Ag_2S (b) CuBr (c) SnS_2 (d) Al_2S_3
18.37 $\Delta H° = -842.1$ kJ mol^{-1}
18.39   103 J K^{-1}

CHAPTER 19

19.16   Hydrides. NaH is sodium hydride.
        $2Na(s) + H_2(g) \rightarrow 2NaH(s)$
        $NaH + H_2O \rightarrow NaOH + H_2(g)$
19.18   (a) nonlinear (b) trigonal pyramidal (c) tetrahedral
19.39   Wash with dilute acid, which will dissolve $CaCO_3$.
19.55   From elemental carbon: $C + O_2 \rightarrow CO_2$
        From lower oxide: $2CO + O_2 \rightarrow 2CO_2$
        From hydride: $CH_4 + 2O_2 \rightarrow CO_2 + 2H_2O$
19.81 KNH_2. In water NH_2^- hydrolyzes immediately to
        give NH_3 because NH_2^- is a very strong base.
19.89   Their rates of decomposition are very slow.
19.94 NO_2^- has the larger bond angle.
19.100 (a) -5075 kJ (b) -7112 kJ

CHAPTER 20

20.16 $Mg(H_2PO_4)_2$ magnesium dihydrogen phosphate
        $MgHPO_4$ magnesium hydrogen phosphate
        $Mg_3(PO_4)_2$ magnesium phosphate
20.23   (a) sodium hydrogen phosphate (b) sodium
        dihydrogen phosphate
20.24 $12Na + P_4 \rightarrow 4Na_3P$
        $Na_3P + 3H_2O \rightarrow 3NaOH + PH_3$
20.29   3 mol NaH_2PO_4 to 2 mol Na_2HPO_4
20.31   Phosphorous acid. $Mg(H_2PO_3)_2$ and $MgHPO_3$
20.33   See Figure 20.4. PCl_5 exists as $PCl_4^+ PCl_6^-$; sp^3 in
        PCl_3 and sp^3d in PCl_5.
20.52   (a) sodium sulfite (b) sodium hydrogen sulfate
20.54 $[:N{\equiv}C{-}\ddot{S}:]$
20.68   (a) $Cl_2 + 2 KI \rightarrow I_2 + 2 KCl$ (b) $F_2 +$
        $2 KBr \rightarrow Br_2 + 2 KF$ (c) $I_2 + NaCl \rightarrow$ N.R.
        (d) $Br_2 + 2 NaI \rightarrow I_2 + 2 NaBr$

20.79   (a) hypobromous acid (b) sodium hypochlorite
        (c) potassium bromate (d) magnesium perchlorate
        (e) periodic acid (f) bromic acid (g) sodium iodate
        (h) potassium chlorite
20.82 $CaCl(OCl) + 2H^+ \rightarrow Ca^{2+} + H_2O + Cl_2$
20.86   (a) nonlinear (b) trigonal bipyramidal (c) octahedral
        (d) trigonal pyramidal (e) T-shaped (f) distorted
        tetrahedral (g) square pyramidal (h) tetrahedral
20.88   Oxygen is too small to accommodate four flourine
        atoms in OF_4, and the valence shell of oxygen can
        only have an octet of electrons.
20.98   Because Si doesn't form stable pi bonds to other
        atoms.
20.107
        $$CH_3{-}\underset{\underset{CH_3}{|}}{\overset{\overset{CH_3}{|}}{Si}}{-}O{-}\underset{\underset{CH_3}{|}}{\overset{\overset{CH_3}{|}}{Si}}{-}CH_3$$

CHAPTER 21

21.14 CrO_4^{2-}
21.16 Cr^{2+}
21.22   Calculated: 19.5 g cm^{-3}; Measured 19.3 g cm^{-3}.
21.29   Oxides of metals in high oxidation states are
        anhydrides of oxoacids that have one or more lone
        oxygens and are therefore acidic.
21.31   The energy needed to remove electrons from Ti is
        so large that Ti^{4+} has no real existence. If it were
        to form, it would polarize the Cl^- ions so much
        that the bonds would become covalent.
21.40 $2 HCl + Mn \rightarrow MnCl_2 + H_2$
21.45 $2 H^+(aq) + Fe(s) \rightarrow Fe^{2+}(aq) + H_2(g)$
21.72   (a) hexaamminenickel(II) ion
        (b) triamminetrichlorochromium(III)
        (c) hexanitrocobaltate(III) ion
        (d) trioxalatomanganate(III) ion
        (e) tetraoxomanganate(VII) ion
21.74   (a) $[Fe(H_2O)_4(CN)_2]^+$ (b) $[Ni(NH_3)_4(C_2O_4)]$
        (c) $K_3[Mn(CN)_6]$ (d) $[CuCl_4]^{2-}$ (e) $[CrO_4]^{2-}$
21.78   NTA can coordinate to four sites in an octahedral
        complex in much the same manner as EDTA.

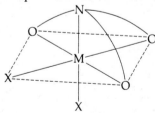

        The remaining two sites, X, can be occupied by other
        ligands or by H_2O molecules.
            The NTA would increase the solubility of metal
        salts by shifting equilibria such as:
        $MX_n(s) + NTA \rightleftharpoons M(NTA)(aq) + nX(aq)$ to the
        right.

21.80 Isomers of $[Co(NH_3)_2Cl_4]^-$

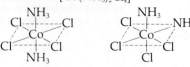

trans   cis

Isomers of $[Co(NH_3)_3Cl_3]$

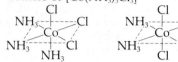

21.84

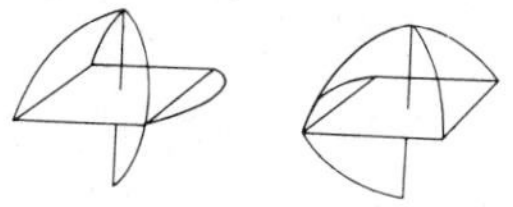

21.87 $3d$ $4s$ $4p$ $4d$

(a) inner

$\uparrow \quad \uparrow$ ___ xx xx xx xx xx xx _ _ _ _ _

(b) outer

$\downarrow\uparrow \ \downarrow\uparrow \ \downarrow\uparrow \ \uparrow \quad \uparrow$ xx xx xx xx xx xx _ _ _ _

(c) outer

$\uparrow \quad \uparrow \quad \uparrow \quad \uparrow \quad \uparrow$ xx xx xx xx xx xx _ _ _ _

(d) inner

$\downarrow\uparrow \ \downarrow\uparrow \ \downarrow\uparrow$ xx xx xx xx xx xx _ _ _ _ _

(e) inner

$\uparrow \quad \uparrow \quad \uparrow$ xx xx xx xx xx xx _ _ _ _

An inner orbital with three unpaired electrons.

21.89 (a) four (b) two

CHAPTER 22

22.3 (a) $^{81}_{36}Kr + ^{0}_{-1}e \rightarrow ^{81}_{35}Br$

(b) $^{104}_{47}Ag \rightarrow ^{0}_{1}e + ^{104}_{46}Pd$

(c) $^{73}_{31}Ga \rightarrow ^{0}_{-1}e + ^{73}_{32}Ge$

(d) $^{104}_{48}Cd \rightarrow ^{104}_{47}Ag + ^{0}_{1}e$

(e) $^{54}_{25}Mn + ^{0}_{-1}e \rightarrow ^{54}_{24}Cr$

22.5 (a) $^{135}_{53}I \rightarrow ^{135}_{54}Xe + ^{0}_{-1}e$

(b) $^{245}_{97}Bk \rightarrow ^{4}_{2}He + ^{241}_{95}Am$

(c) $^{238}_{92}U + ^{12}_{6}C \rightarrow ^{246}_{98}Cf + 4\ ^{1}_{0}n$

(d) $^{96}_{42}Mo + ^{2}_{1}H \rightarrow ^{1}_{0}n + ^{97}_{43}Tc$

(e) $^{20}_{8}O \rightarrow ^{20}_{9}F + ^{0}_{-1}e$

22.7 (a) $^{11}_{5}B \rightarrow ^{4}_{2}He + ^{7}_{3}Li$

(b) $^{90}_{38}Sr \rightarrow ^{0}_{-1}e + ^{90}_{39}Y$

(c) $^{107}_{47}Ag + ^{1}_{0}n \rightarrow ^{108}_{47}Ag$

(d) $^{88}_{35}Br \rightarrow ^{1}_{0}n + ^{87}_{35}Br$

(e) $^{116}_{51}Sb + ^{0}_{-1}e \rightarrow ^{116}_{50}Sn$

(f) $^{70}_{33}As \rightarrow ^{0}_{1}e + ^{70}_{32}Ge$

(g) $^{41}_{19}K \rightarrow ^{1}_{1}H + ^{40}_{18}Ar$

22.12 (a) 0.500 g (b) 0.125 g (d) 0.0313 g

22.14 164 days

22.16 2.89×10^{-7} s^{-1}

22.18 (a) 6.37×10^{-3} h^{-1} (b) $t_{1/2} = 109$ h

22.22 (a) 4.6×10^{10} Bq (b) 3.1×10^{8} Bq g^{-1}

22.25 8.3×10^8 y Bq

22.27 ON + O*NO   (bond breaking at 1)

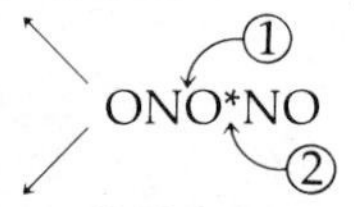

ONO*NO ① ②

ONO + *NO   (bond breaking at 2)

22.30 180 cm^3

22.35 $^{4}_{2}He > ^{58}_{28}Ni > ^{39}_{20}Ca > ^{71}_{32}Ge > ^{10}_{5}B$

22.42 (a) $^{27}_{13}Al + ^{4}_{2}He \rightarrow ^{1}_{0}n + ^{30}_{15}P$

(b) $^{209}_{83}Bi + ^{2}_{1}H \rightarrow ^{1}_{0}n + ^{210}_{84}Po$

(c) $^{15}_{7}N + ^{1}_{1}H \rightarrow ^{4}_{2}He + ^{12}_{6}C$

(d) $^{12}_{6}C + ^{1}_{1}H \rightarrow ^{13}_{7}N + \gamma$

(e) $^{14}_{7}N + ^{4}_{2}He \rightarrow ^{1}_{1}H + ^{17}_{8}O$

22.45 (a) Na_2X (b) H_2X (c) XO_2 (d) metallic

22.48 (a) ununpentium (b) unbiseptium

22.52 2.3009×10^9 kJ mol^{-1}

22.54 ^7Li 3.7870×10^9 kJ mol^{-1}, 39.250 MeV

^{19}F 1.4261×10^{10} kJ mol^{-1}, 147.73 MeV

^{14}N 1.0098×10^{10} kJ mol^{-1}, 104.61 MeV

CHAPTER 23

23.3 (a) $C_{30}H_{62}$ (b) $C_{27}H_{54}$ (c) $C_{33}H_{64}$

23.6 All the isomers of this compound are:

23.6

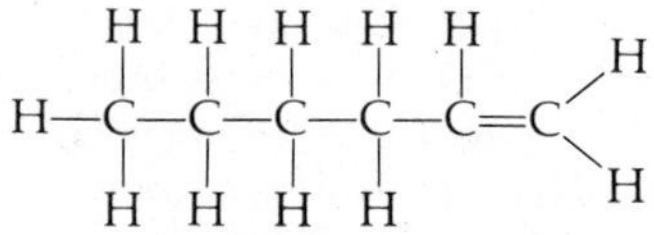

H H H H H
H—C—C—C—C—C=C<H/H   1-hexene
H H H H

H H H H H H
H—C—C—C—C=C—C—H   2-hexene
H H H H

H H   H H
H—C—C—C=C—C—C—H   3-hexene
H H H H H H

H H H
H—C—C—C—C=C<H   2-methyl-1-pentene
H H H   H
H—C—H
H

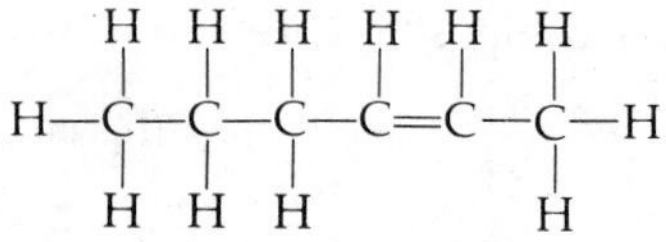

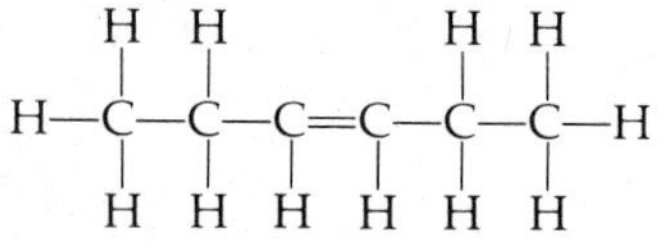

H H H H   H
H—C—C—C—C=C<H   3-methyl-1-pentene
H H   H
H—C—H
H

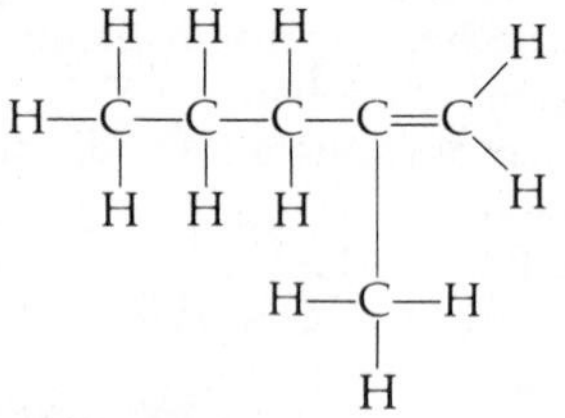

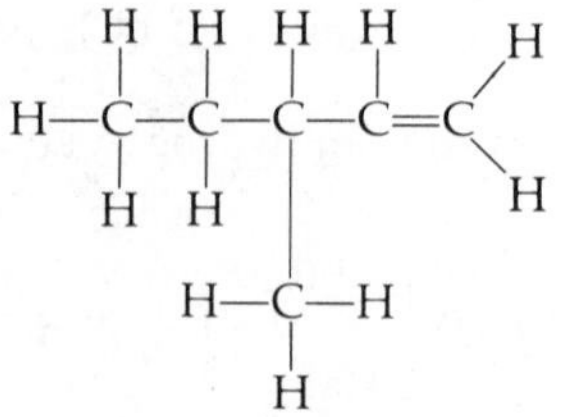

23.6 (cont'd)

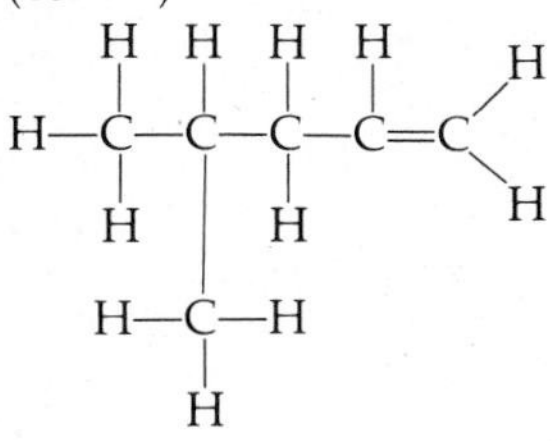

4-methyl-1-pentene

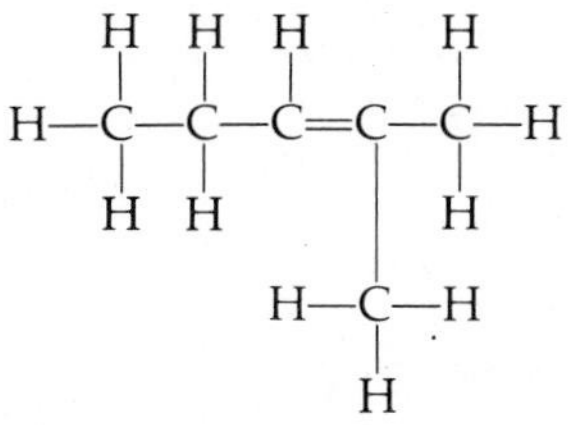

2-methyl-2-pentene

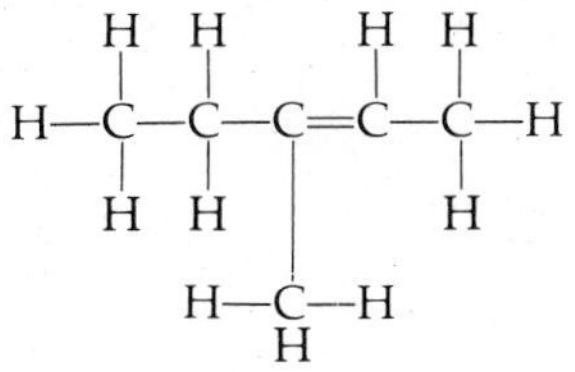

3-methyl-2-pentene isomers

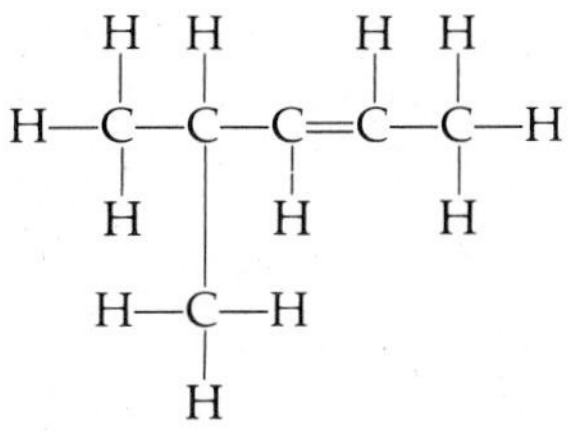

4-methyl-2-pentene isomers

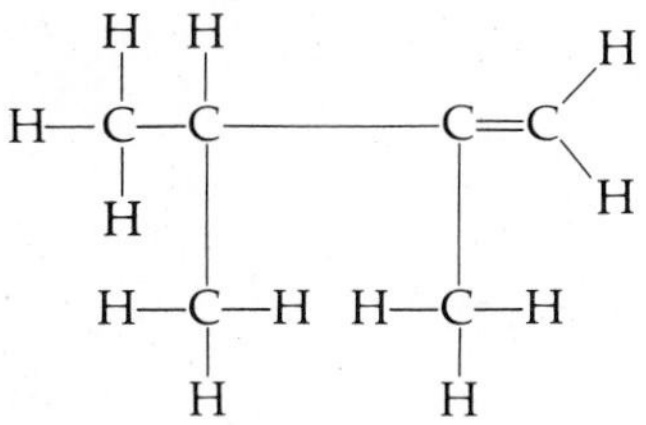

2,3-dimethyl-1-butene

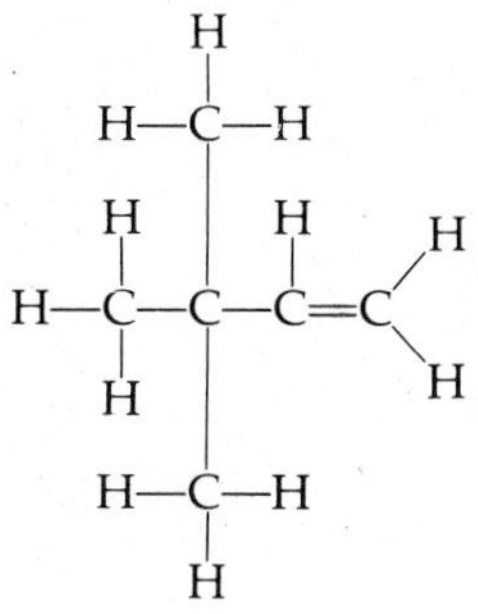

3,3-dimethyl-1-butene

23.6 (cont'd)

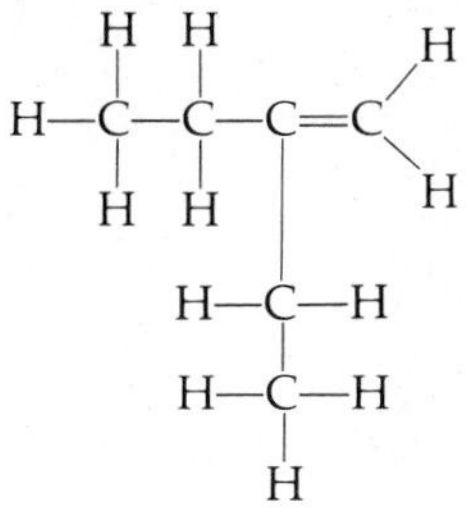

2-ethyl-1-butene

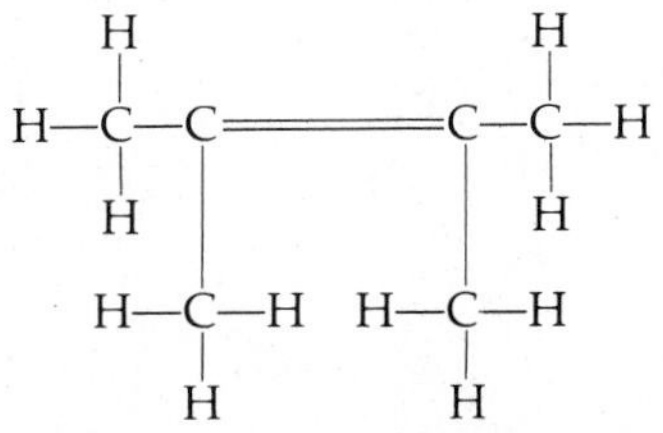

2,3-dimethyl-2-butene

23.19 There is restricted rotation around a double bond, but not around a single bond.

23.21 The C—C≡C—C arrangement must be linear. A closed ring of 5 carbon atoms can't be formed.

23.24 (a) 4-ethyl-3,5-dimethyl-2,4-heptadiene
(b) 5-methyl-3-heptyne
(c) 2,3,3,4,4-pentamethylhexane
(d) 4-methyl-2-pentyne
(e) 3,4-dimethyl-3,5-octadiene

23.26 (a) 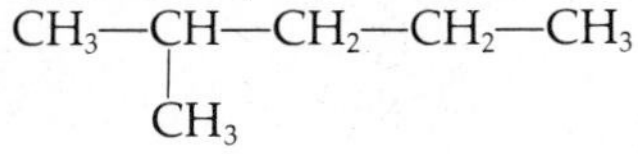 CH_3—CH—CH_2—CH_2—CH_3
$\qquad\qquad$ $\overset{|}{CH_3}$

(b) 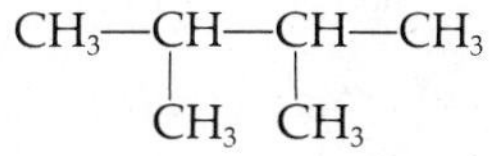 CH_3—CH—CH—CH_3
$\qquad\qquad$ $\overset{|}{CH_3}$ $\overset{|}{CH_3}$

(c) 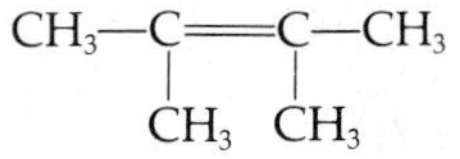 CH_3—C=C—CH_3
$\qquad\qquad$ $\overset{|}{CH_3}$ $\overset{|}{CH_3}$

(d) CH_2=CH—CH=CH—CH=CH—CH_2—CH_3

(e) 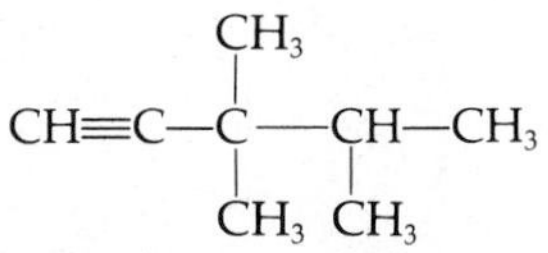
$\qquad\qquad$ CH_3
$\qquad\qquad$ $\overset{|}{}$
CH≡C—C—CH—CH_3
$\qquad\qquad$ $\overset{|}{CH_3}$ $\overset{|}{CH_3}$

23.27 3,3,4,4,5-pentamethylheptane

23.77 C=O, OH, ether

CHAPTER 24

24.31 T G A G C T A G T A C

24.35 CGU-UUA-AAA-GGU-UGU (There are other base sequences that could be chosen from Table 24.3.)

24.37 Ser-Ile-Leu-Ser-Asn

24.39 Leu-Ala-Ser-Asp-Ile

24.40 His-Pro-Thr

GLOSSARY

A number that appears in parentheses after an entry refers to the section in the text where that term is discussed.

Absolute temperature (9.3) A temperature measured in kelvins (K).

Absolute zero (9.3) $0 \text{ K} = -273.15 \text{ °C}$.

Absorption (1.6) The act of being swallowed up, much as a sponge swallows up water.

Accuracy (1.2) How closely an experimental observation lies to the true value.

Acid (7.2, 7.3, 7.5) In aqueous solutions, a substance that gives H_3O^+. Under the Brønsted-Lowry definition, a proton donor. Under the Lewis definition, an electron-pair acceptor.

Acid ionization constant (Acid dissociation constant) (15.2) The equilibrium constant for the ionization of a weak acid.

Acid rain (20.2) Rain made acidic by dissolved sulfur oxides and nitrogen oxides, which come from burning high-sulfur fuels.

Acid salt (5.11, 7.2) A salt of a partially neutralized polyprotic acid: for example, $NaHSO_4$ and NaH_2PO_4.

Acidic anhydride (7.2) A nonmetal oxide that reacts with water to form an acid.

Actinide elements (3.3) Elements 90 through 103.

Activated complex (13.7) The chemical species that exists with partly broken and partly formed bonds in the transition state.

Activation energy, E_a (13.7, 13.8) The minimum kinetic energy that must be possessed by reactant molecules in order to give an effective collision (one that forms the products).

Activity (14.3) The effective concentration of a solute or the effective pressure of a gas in an equilibrium system. At low concentrations or low pressures, activities approach molar concentrations or pressures.

Activity series (7.9) A listing of metals in order of decreasing ease of oxidation.

Actual yield (2.12) The actual amount of products obtained in a particular chemical reaction when the experiment is performed in the laboratory.

Addition compound (5.9) A compound formed by the joining of two molecules with a coordinate covalent bond.

Addition polymer (23.10) A polymer formed simply by the joining together, or addition, of monomer units.

Addition reaction (23.6) A reaction in which a molecule such as H_2 is added across a double or triple bond in an organic molecule.

Adiabatic (12.1) A change that occurs without heat transfer between the system and its surroundings.

Adsorption (1.6) A process whereby a substance sticks to the surface of some other substance.

Alcohol (23.8) An organic compound with an —OH group attached to a hydrocarbon in place of a hydrogen, for example, CH_3OH.

Aldehyde (23.8) An organic molecule with the structure

$$\begin{array}{c} O \\ \parallel \\ R-C-H \end{array}$$

Aliphatic hydrocarbons (23.1) Hydrocarbons that lack the benzene ring structure.

Alkaline battery (17.12) A dry cell in which zinc serves as the anode and MnO_2 serves as the cathode in an alkaline KOH electrolyte.

Alkaline earth metals (3.3) The elements in Group IIA.

Alkali metals (3.3) The elements of Group IA, except hydrogen.

Alkane (23.1) A saturated hydrocarbon of general formula C_nH_{2n+2}.

Alkene (23.1) A hydrocarbon with one carbon-carbon double bond and having the general formula C_nH_{2n}.

Alkyl group (23.3) A group of atoms, derived from an alkane by loss of a hydrogen, that replaces a hydrogen in another molecule.

Alkyne (23.1) A hydrocarbon with one carbon-carbon triple bond and having the general formula C_nH_{2n-2}.

Allotropism (6.8) The existence of an element in two or more different forms.

Alloy (11.2) A solid solution of two or more metals.

Alpha amino acid (24.1) A molecule with the structure

$$R-\underset{\underset{NH_2}{|}}{CH}-\overset{\overset{O}{\|}}{C}-OH$$

Alpha helix (α-helix) (24.1) The coil structure assumed by a polypeptide chain.

Alpha particle (α-particle) (22.1) The nucleus of a helium atom, ^4_2He. One of the types of radiation given off by radioactive substances.

Alum (18.7) A double salt with the general formula $M^+M^{3+}(SO_4)_2 \cdot 12H_2O$, for example, $KA1(SO_4)_2 \cdot 12H_2O$.

Amalgam (18.1) A solution of a metal in mercury.

Amide (23.9) An organic molecule with the structure

$$R-\overset{\overset{O}{\|}}{C}-NH_2$$

Amine (23.9) An organic molecule derived from ammonia by replacing hydrogen atoms in NH_3 with organic groups.

Amorphous solid (10.12) A noncrystalline solid. It lacks the long-range order found in crystals. It is also called a supercooled liquid. Glass is an example.

Ampere (A) (3.4) The SI unit of electric current. One coulomb per second: $1\ A = 1\ C\ s^{-1}$.

Amphiprotic (7.3) The ability of a substance to either accept or donate a proton.

Amphoteric (7.3) The ability of a substance to behave as either an acid or a base.

Amplitude (4.1) The intensity of a wave. The maximum height of a wave as measured from the average height of peak and trough.

Analysis (8.4) See *Chemical analysis*.

Anion (5.3) A negative ion.

Anode (3.4) In a gas discharge tube, it is positive electrode. In electrochemistry, it is the electrode at which oxidation takes place. The anode carries a positive charge in an electrolytic cell and a negative charge in a galvanic cell.

Antibonding orbital (6.8) A molecular orbital that places electron density outside of the region between the nuclei. When filled, an antibonding orbital destabilizes a molecule.

Aqua regia (7.9) One part concentrated HNO_3 and three parts concentrated HCl, by volume. It is a mixture that is able to dissolve the very unreactive metals gold and platinum.

Aromatic hydrocarbon (23.5) A hydrocarbon that contains a benzene ring structure.

Arrhenius concept of acids and bases (7.2) Acids produce H_3O^+ in water and bases produce OH^- in water.

Arrhenius equation (13.8) The equation that relates the rate constant for a reaction, k, to the activation energy, E_a.

$$k = Ae^{-E_a/RT}$$

Association (11.12) The joining together of two or more molecules in a solution to produce a particle of higher apparent molecular weight. The forces of attraction between the particles are usually of dipole-dipole type, including hydrogen bonding.

Asymmetric carbon atom (23.2) A carbon atom that is bonded to four different groups and which is a center of chirality.

Atmosphere (9.1) See *Standard atmosphere*.

Atomic emission spectrum (4.1) The spectrum emitted by atoms that have been energized (*excited*). This kind of spectrum consists of a relatively small number of different wavelengths of light. See also, *Line spectrum*.

Atomic number (3.3) The number of protons in the nucleus of an atom.

Atomic radius (4.8) The effective radius of an atom.

Atomic spectrum (4.1) See *Atomic emission spectrum*.

Atomic theory (1.8) The theory proposed by Dalton that matter consists of tiny indestructible particles (atoms). The theory explained the laws of conservation of mass and definite proportions, and predicted the law of multiple proportions.

Atomic weight (1.9) The relative average atomic mass of the atoms of an element expressed in unified atomic mass units.

Atomization energy (12.6) The quantity of energy needed to convert one mole of a compound into neutral gaseous atoms. It is the total energy needed to break all the bonds in a molecule to give gaseous atoms.

Autoionization reaction (7.3) The reaction of a substance with itself to produce ions. For example

$$H_2O + H_2O \rightleftharpoons H_3O^+ + OH^-$$

Avogadro's number (2.2) The number of things (atoms, molecules, or whatever) in one mole. 6.022×10^{23} things = 1 mole of things.

Avogadro's principle (9.7) Equal volumes of gases at the same temperature and pressure contain equal numbers of molecules.

Azeotrope (11.9) A mixture that has either a higher boiling point or a lower boiling point than either of the two pure components in a liquid mixture.

Azimuthal quantum number (l) (4.3) The secondary quantum number whose values can be 0, 1, 2, ... $(n-1)$ where n is the principal quantum number.

Balance (1.4) A device used to measure mass by comparison of the weights of two objects, one of known mass and the other of unknown mass.

Balanced equation (1.10) A chemical equation that has the same number of atoms of each kind, and the same net electrical charge on both sides of the arrow.

Band of stability (22.3) The collection of stable nuclei with various numbers of protons and neutrons that fall within a narrow band on a plot of numbers of neutrons versus numbers of protons.

Bar (12.5) A unit pressure, $1\ bar = 10^5\ Pa$.

Barometer (9.1) A tube in which a column of mercury is supported by the atmospheric pressure and that is used to measure the pressure exerted by the atmosphere by measuring the height of the mercury column.

Base (7.2, 7.3, 7.5) In aqueous solutions, a substance that gives OH⁻. According to the Brønsted-Lowry definition, a proton acceptor. According to the Lewis definition, an electron-pair donor.

Base ionization constant, K_b (15.2) The equilibrium constant for the ionization of a weak base.

Base units (1.3) The seven primary measurement standards in SI.

Basic anhydride (7.2) A metal oxide that reacts with water to give a metal hydroxide.

Basic oxygen process (18.1) A modern relatively fast method that is used to convert pig iron into steel. It involves blowing pure oxygen and powdered limestone through the molten pig iron, which lowers the carbon content and converts other impurities into slag.

Becquerel (Bq) (22.1) A unit of radioactive activity: 1 Bq = 1 disintegration/second.

Bessemer converter (18.1) An outdated method for converting pig iron into steel, which involves blowing air through the molten pig iron to lower its carbon content.

Beta particle (β-particle) (22.1) A particle given off by a radioactive nucleus. It is actually an electron. Its symbol is $_{-1}^{0}e$.

Bidentate ligand (21.6) A ligand that has two atoms that can become simultaneously attached to the same metal ion in a complex.

Bilayer structure (24.4) The structure of a cell membrane in which the nonpolar tails of two layers of phospholipids face each other while the polar heads face the aqueous environment on either side of the bilayer.

Bimolecular collision (13.4) A collision between two molecules.

Binary acid (7.5) A substance with the general formula H_nX which produces acidic aqueous solutions (e.g., HCl, H_2S).

Binary compound (5.11) A compound that consists of two different elements (e.g., HCl, Na_2S, $FeCl_3$).

Binding energy (22.6) The energy equivalent of the mass defect of a nuclide. This energy would have been released on formation of the nuclide from its protons and neutrons.

Black phosphorus (6.8) An allotrope of phosphorus having a layer structure, which imparts to it some properties similar to graphite.

Blast furnace (18.1) An apparatus that is used to reduce metal ores such as iron oxides to the free metal. The reducing agent is carbon.

Body-centered cubic unit cell (10.8) A cubic unit cell that has identical atoms, molecules, or ions at the corners, plus one in the center of the cell.

Boiling point (10.5) The temperature at which the vapor pressure of a liquid equals the prevailing external pressure.

Boiling point diagram (11.9) A graph on which is plotted the boiling points of mixtures of varying composition. Also plotted is a curve showing the compositions of the vapor given off when solutions of different compositions boil. A vapor composition can be obtained from a liquid composition by a tie line running horizontally between the two curves.

Bomb calorimeter (12.3) A constant-volume calorimeter. Heats of reaction measured with this apparatus correspond to ΔE for the reaction.

Bond angle (6.1) When one atom forms a bond to each of two other atoms, that angle between the two bonds is the bond angle.

Bond dipole (6.3) The dipoles within a molecule caused by the unequal sharing of electrons in the bonds.

Bond distance (5.5) See *Bond length*.

Bond energy (5.5) The quantity of energy needed to separate two atoms that are joined by a chemical bond and produce electrically neutral particles. It is also the quantity of energy released when such a bond is formed.

Bond length (5.5) Also called the bond distance. The distance between the nuclei of two atoms that are joined by a chemical bond.

Bond order (5.7) The *net* number of pairs of bonding electrons:

$$\frac{\text{(no. of bonding electrons} - \text{no. of antibonding electrons)}}{2}$$

Bonding molecular orbital (6.9) A molecular orbital that gives a buildup of electron density between nuclei and helps stabilize a molecule.

Born–Haber cycle (5.4) A method of examining the contributing energy factors in an overall energy change. It involves constructing an alternative path from reactants to products and analyzing the individual energy changes accompanying each step along the alternative path.

Boyle's law (9.2) At constant temperature, for a fixed quantity of gas, PV = constant.

Bragg equation (10.7) $n\lambda = 2d \sin \theta$. The equation is used to analyze X-ray diffraction data obtained from crystals.

Branching step (13.10) A step in a chain reaction mechanism that produces more free radicals than it consumes.

Breeder reactor (22.7) A nuclear reactor that produces plutonium from ^{238}U. The amount of fissionable Pu produced is greater than the amount of nuclear fuel consumed.

Brine (17.3) A concentrated aqueous solution of sodium chloride.

Brønsted acid (7.3) A proton (H^+) donor.

Brønsted base (7.3) A proton (H^+) acceptor.

Buffer (15.4) A mixture that contains both a weak acid and a weak base. It is capable of absorbing small additions of either a strong acid or strong base with little change in pH.

Buret (8.4) A graduated glass tube fitted with a valve (stopcock) at one end. It is used to dispense measured volumes of solutions.

Calcining (18.6) Heating a substance strongly in air.

Calorie (cal) (1.4) 1 cal = 4.184 J (exactly). One calorie will raise the temperature of 1 g of water by 1 °C.

Calorimeter (12.3) A device used for the measurement of heats of reaction.

Carbide (19.2) An inorganic compound of carbon (e.g., Mg_2C).

Carbinol group (23.8) The group of atoms, C—OH, in an organic molecule.

Carbohydrate (24.3) An organic compound that contains carbon, hydrogen, and oxygen and in which the ratio of hydrogen to oxygen is 2 to 1. This gives it a general formula $C_n(H_2O)_m$, although there are no intact water molecules in the structure. An example is sucrose (table sugar) $C_{12}H_{22}O_{11}$.

Carbonyl group (23.8) The group

$$\overset{\displaystyle O}{\underset{\displaystyle ---C---}{\|}}$$

Carborundum (19.2) Silicon carbide, SiC, which is a common abrasive.

Carboxyl group (23.8) The group

$$\overset{\displaystyle O}{\underset{\displaystyle ---C---OH}{\|}}$$

Cast iron (18.1) Pig iron that has been cast into shapes by pouring the liquid metal into molds.

Catalyst (13.9) A substance that alters the rate of a chemical reaction by providing a lower energy path (mechanism) that leads from reactants to products. The catalyst is not used up as the reaction progresses.

Catenation (19.1) Linking together of atoms of the same element to give chains of atoms.

Cathode (3.4) In a gas discharge tube, it is the negative electrode from which cathode rays are emitted. In electrochemistry, it is the electrode at which reduction takes place. The cathode carries a negative charge in an electrolytic cell and a positive charge in a galvanic cell.

Cathode rays (3.4) The stream of electrons that are emitted by the cathode in a gas discharge tube and that move through the tube to the anode.

Cathodic protection (21.5) Protecting a metal from corrosion by coating it with another metal that is more easily oxidized, and which thereby causes the protected metal to be the cathode in a galvanic cell.

Cation (5.3) A positive ion.

Cell potential, ε_{cell} (17.6) The emf that can be produced by a galvanic cell when no current is drawn from the cell.

Cell reaction (17.2) The overall chemical change that takes place in a galvanic or electrolytic cell.

Celsius scale (1.3) A temperature scale on which water boils at 100 °C and freezes at 0 °C.

Chain reaction (13.10) A reaction in which a product of one step is the reactant in another. Usually, they involve free radicals and are very rapid once initiated.

Change of state (10.11) Transformation of matter from one state to another, for example, from liquid to solid.

Charge cloud (4.7) Electron cloud.

Charge-to-mass ratio (3.4) The ratio of a particle's charge to its mass, expressed in units of coulombs per gram.

Charge transfer absorption band (18.4) A band of wavelengths that a substance absorbs from the electromagnetic spectrum. The photons that are absorbed cause a charge transfer process to occur.

Charge transfer process (18.4) The transfer of an electron from one atom to another brought about by absorption of a photon that has the energy required for the process.

Charles' law (9.3) At constant pressure and for a fixed quantity of a gas, V/T = constant, where T is the temperature in kelvins.

Chelate (21.6) A complex that contains rings formed by polydentate ligands.

Chemical analysis (8.4) The experimental determination of the composition of a substance.

Chemical bond (5.1) Forces of attraction that link atoms together in compounds.

Chemical change (1.6) A chemical reaction. An event that causes the chemical and physical properties of substances to change.

Chemical equation (1.10) A representation using chemical formulas of the changes that occur during a chemical reaction. It is a sort of before-and-after view of the chemical system.

Chemical equivalence (1.8) A relationship between the amounts of two chemicals in a compound or a reaction. For example, in H_2O, 2.0 g H $\sim$ 16.0 g O.

Chemical formula (1.10) A shorthand way of representing the composition of a substance using chemical symbols.

Chemical kinetics (13, Introduction) The study of the speeds of chemical reactions.

Chemical property (1.5) A statement that describes how a substance reacts chemically with another substance.

Chemical reaction (1.6) An event that produces changes in the chemical and physical properties of the substances involved.

Chemical symbol (1.10) The symbol that represents the name of an element. In formulas and equations it is used to represent an atom of the element.

Chemistry (1.1) The study of the composition of substances, the way their properties depend on their compositions, and the way substances interact with one another to form new materials.

Chiral (21.9) Possessing a "handedness." A term applied to optical isomers, which are not superimposable on their mirror images.

Cholesteric liquid crystal (10.10) Rodlike molecules, similar in structure to cholesterol, that are arranged in layers in which the parallel rods in one layer are oriented in a different direction than the parallel rods in an adjoining layer.

Chromatography (1.6) A method of separating mixtures that relies on the different tendencies of substances to be adsorbed onto the surfaces of certain solids.

Cis- (21.9) A term applied to a geometrical isomer in which two groups are located on the same side of some central reference line in the molecule or ion.

Clathrate (20.4) Crystals in which noble gas atoms are trapped in a cagelike lattice.

Codon (24.6) A sequence of three bases along an *m*-RNA strand that serves as a code for a particular amino acid in a growing polypeptide chain.

Coefficients (2.9) Numbers that precede chemical formulas in a chemical equation.

Coenzyme (24.2) A substance needed by an enzyme in order to function.

Coinage metals (21.5) Copper, silver, and gold.

Coke (18.1) Coal that has had its volatile components driven off at high temperature. It is mostly carbon.

Colligative property (11.10) A property that depends only on the number of particles in a solution, and not on their chemical identity. Examples are freezing point depression, boiling point elevation, vapor pressure lowering, and osmotic pressure.

Collision theory (13.4) A theory of reaction rates that postulates that the rate of a reaction is proportional to the number of collisions that occur each second between the reactant molecules.

Colloidal dispersion (11.1) A mixture in which particles of one of the components are of a size intermediate between those in a true solution and those in a suspension.

Combined gas law (9.5) For a fixed quantity of gas, $P_1V_1/T_1 = P_2V_2/T_2$.

Combustion (7.11) A reaction with oxygen that produces heat and light.

Common ion (16.2) An ion that is common to more than one salt in a solution. Na^+ is the common ion between $NaCl$ and $NaNO_3$.

Common ion effect (16.2) The solubility of a salt is less in a solution that already contains one of its ions than it is in pure water.

Complex ion (or simply a **complex**) (21.6) A substance formed when one or more anions or neutral molecules become bonded to a metal ion.

Compound (1.6) A substance consisting of two or more elements combined in fixed proportions by mass (and by atoms).

Compressibility (10.1) The ease with which a gas, a liquid, or a solid can be compressed to a smaller volume.

Concentrated (2.13) A large proportion of solute to solvent in a solution.

Concentration (2.13) A quantitative statement of the proportion of solute to solvent, or of solute to the total quantity of solution.

Concentration cell (17.11) An electrochemical cell in which both electrodes are composed of the same substance, but the ion concentrations in the two half-cells are different.

Condensation polymer (23.10) A polymer formed by linking together monomers, with the simultaneous elimination of small molecules such as water.

Condensation reaction (23.8) A reaction that joins together two molecules with the simultaneous elimination of a small molecule such as water.

Conjugate acid (7.3) A Brønsted acid, which is formed by the addition of a proton to a base. (NH_4^+ is the conjugate acid of the base NH_3).

Conjugate acid-base pair (7.3) Two substances related to each other by the gain or loss of a single proton, for example, NH_4^+ and NH_3.

Conjugate base (7.3) A Brønsted base, which is formed by the removal of a proton from a Brønsted acid. (NH_3 is the conjugate base of the acid NH_4^+).

Constructive interference (4.3) The addition of the intensities (amplitudes) of waves that are in phase so that the resultant wave has a greater amplitude.

Contact process (20.2) The commercial method of manufacturing sulfuric acid from sulfur

$$S + O_2 \longrightarrow SO_2$$
$$2SO_2 + O_2 \longrightarrow SO_3$$
$$SO_3 + H_2SO_4 \longrightarrow H_2S_2O_7$$
$$H_2S_2O_7 + H_2O \longrightarrow 2H_2SO_4$$

Continuous spectrum (4.1) An electromagnetic spectrum that contains all wavelengths.

Contributing structures (5.8) See *Resonance structure*.

Conversion factor (Appendix A) A fraction formed from a valid relationship between two quantities that is used in problem-solving to convert the units of the given quantity to the units of the desired quantity.

Cooling curve (10.11) For a particular substance, a graph of temperature versus quantity of heat removed. Temperatures corresponding to condensation of the vapor and freezing of the liquid can be read from the graph.

Coordinate covalent bond (5.9) A term used to describe a covalent bond in which both of the shared electrons are contributed by the same atom. Once formed, a coordinate covalent bond is no different than any other covalent bond.

Coordination compound (21.6) A compound that contains a metal ion bonded to one or more neutral molecules or anions (ligands).

Coordination number (21.7) The number of donor atoms that surround a metal ion in a complex.

Copolymer (23.10) A polymer formed from two or more different monomer units.

Core electrons (4.5) The electrons in shells below an atoms's outer shell.

Corrosion (7.9) The oxidation of a metal, which gives products that lack desirable metallic properties.

Coulomb (3.4) The SI unit of electrical charge. It is the quantity of charge that passes a given point in a wire when a current of 1 ampere flows for 1 second.

Coulometer (17.4) An electrolysis cell in which the amount of chemical change that takes place is used to compute the number of coulombs that have passed through the cell.

Covalent bond (5.3, 5.5) A chemical bond formed by the sharing of electrons between two atoms.

Covalent crystal (10.9) A crystal in which lattice positions are occupied by atoms that are covalently bonded to other atoms at neighboring lattice sites.

Critical mass (22.7) The minimum amount of fissionable material that must be present to sustain a nuclear chain reaction.

Critical pressure, P_c (10.4) The vapor pressure of a substance at its critical temperature.

Critical temperature, T_c (10.4) The temperature above which a substance cannot exist as a separate liquid phase, regardless of the pressure.

Cross linking (23.10) The joining of adjacent polymer strands to give a three-dimensional rigid structure.

Cryolite (17.3) Na_3AlF_6, the solvent used in the original Hall process for the production of aluminum by electrolysis.

Crystal field splitting, Δ (21.11) The energy difference between sets of d orbitals in a complex.

Crystal field theory (21.11) A theory that considers the effects of the polar or ionic ligands of a complex on the energies of the d orbitals of the central metal ion.

Crystal lattice (10.8) The repeating symmetrical pattern of atoms, molecules, or ions that occurs in a crystal.

Crystalline solid (10.7) A solid in which the particles are arranged in an orderly, repeating pattern.

Curie (Ci) (22.1) A measure of radioactive activity related to radium as a standard

$$1 \text{ Ci} = 3.7 \times 10^{10} \text{ Bq}$$

Cyclo- (23.4) A prefix that means that the carbon chain in an organic molecule exists in the form of a ring.

Cyclotron (22.4) A type of particle accelerator that is used to study nuclear reactions.

d-block elements (21.1) The transition elements found in the main part of the periodic table between groups IIA and IIIA.

Dalton's law of partial pressures (9.6) For a mixture of gases, A, B, C, etc., the total pressure P_T is given by

$$P_T = p_A + p_B + p_C + \cdots$$

where p_A, p_B, etc., are the partial pressures of each of the gases.

Data (1.1) The information obtained in an experiment.

Dative bond (5.9) A coordinate covalent bond.

Daughter isotope (22.1) An isotope produced in a radioactive decay.

Decay series (22.1) See *Radioactive series*.

Delocalized molecular orbital (6.9) A molecular orbital that spreads over more than two nuclei.

$\Delta G°$ (12.10) See *Standard free energy change*.

$\Delta H°$ (12.5) See *Standard heat of reaction*.

ΔH_{cryst} (10.6) See *Molar heat of crystallization*.

ΔH_{fus} (10.6) See *Molar heat of fusion*.

ΔH_{soln} (11.5) See *Heat of solution*.

$\Delta H_{sublimation}$ (10.3) See *Molar heat of sublimation*.

$\Delta H_{vaporization}$ (10.3) See *Molar heat of vaporization*.

Density (1.5) The ratio of an object's mass to its volume.

Deoxyribose (24.3) A five-carbon sugar that is one of the building blocks of DNA.

Derived units (1.3) Units for quantities that are derived by appropriately combining the SI base units. For example,

the unit for area is m^2, which is derived from the SI base unit for length.

Desiccant (20.1) A drying agent.

Destructive interference (4.3) The cancellation of intensity (amplitude) that occurs when two waves are out of phase. This produces a wave of diminished (or even zero) amplitude.

Deuterium (19.1) An isotope of hydrogen. ^2_1H. Sometimes represented by the symbol D.

Dextrorotatory (21.9) An optical isomer that causes a rotation of plane polarized light in a clockwise direction, when looking toward the source, as the light passes through a solution of the isomer.

Dialysis (11.11) The passage of water and small molecules and ions, but not large molecules, through a membrane.

Diamagnetism (4.4) A magnetic property associated with the absence of unpaired electrons in a substance. It gives rise to a very slight repulsion away from a magnetic field.

Diamond (6.8) An allotrope of carbon. The hardest substance known.

Diatomic molecule (3.1) A molecule that is composed of only two atoms, for example, H_2 and HCl.

Diffraction (4.3) The scattering of light as it passes through a tiny pinhole or through a very narrow slit.

Diffraction pattern (4.3) The pattern that is produced by constructive and destructive interference of diffracted waves.

Diffusion (9.9, 10.1) The mixing of two fluids, one into the other.

Difunctional molecule (24.1) A molecule having two functional groups.

Dilute (2.13) Very little solute dissolved in a solution. A low ratio of solute to solvent.

Dimer (11.12) A particle formed by the joining together of two smaller particles.

Dipole (5.10) A molecule having partial positive and negative charges on opposite ends.

Dipole–dipole attraction (10.2) Attractions between molecules that are dipoles.

Dipole moment (5.10) The product of the partial charge on either end of a dipole multiplied by the distance between the partial charges. It is a measure of the extent of polarity of a molecule.

Diprotic acid (7.2) An acid that can furnish two H^+ ions per molecule of the acid.

Dispersed phase (11.1) In a colloidal dispersion, the substance analogous to the solute in a solution. It is the substance dispersed in the dispersing medium.

Dispersing medium (11.1) In a colloidal dispersion, the substance analogous to the solvent in a solution. It is the substance into which the colloidal substance is dispersed.

Disproportionation (19.4) A redox reaction in which a portion of a substance is oxidized while the rest is reduced. The same chemical substance undergoes both oxidation and reduction.

Dissociation (7.1) In general, the breaking apart of a substance into simpler substances. For aqueous solutions, the separation of the ions of an ionic compound as it dissolves. The term is also applied to the ionization of molecular compounds in water, which gives ions in solution.

Dissociation constant (15.1) See *Ionization constant*.

Distillation (1.6) A means of separating liquid mixtures into their individual components. The method involves boiling the mixture and then condensing the more volatile component, which becomes more concentrated in the vapor.

Disulfide bridge (24.1) A bridge between portions of polypeptide chains. It is created by the formation of —S—S— between adjacent chains, and this holds the polypeptide chains in a particular configuration.

DNA (24.5) Deoxyribonucleic acid, the carrier of genetic information in a cell nucleus.

Domain (21.4) A region in a ferromagnetic substance in which the individual paramagnetic atoms are all aligned in the same direction.

Double bond (5.5) A covalent bond in which two pairs of electrons are shared.

Double helix (24.5) The intertwining of complementary DNA strands.

Double replacement reaction (7.1) Metathesis. A reaction between two salts in which cations exchange partners, for example

$$AgNO_3(aq) + NaCl(aq) \longrightarrow AgCl(s) + NaNO_3(aq)$$

Double salt (18.7) Crystals that contain the components of two different salts in a definite ratio.

Dry ice (19.2) Solid CO_2.

Ductility (3.1) A metal's ability to be stretched (drawn) into wire.

Dynamic equilibrium (7.1) An equilibrium in which two opposing processes are occurring at equal rates.

Effective collisions (13.6) Collisions between reactant molecules that lead to a net chemical change.

Effective nuclear charge (4.8) The effective charge experienced by a particular electron in an atom, which is a composite of the positive charge on the nucleus and the offsetting negative charge of the electrons in inner shells (and to some extent, by other electrons in the same shell).

Effusion (9.9) The escape of a gas under pressure through a very small opening into a region of low pressure.

Electric furnace (20.1) A furnace in which heat is generated by the passage of a large electric current through the contents of the furnace.

Electrochemical change (17.1) A chemical change that is caused by or that produces electricity.

Electrochemistry (17, Introduction) The study of electrochemical changes.

Electrode (17.1) An electrically conducting substance that carries an electrical charge, either given to it by an ex-

ternal electric potential or acquired as a result of a chemical reaction such as one that occurs in a battery.

Electrolysis (17.2) A chemical change caused by the passage of electricity through a molten ionic compound or through a solution that contains ions.

Electrolysis cell (17.2) An apparatus for carrying out electrolysis.

Electrolyte (7.1) A substance that gives ions in an aqueous solution and thereby gives a solution that conducts electricity.

Electrolytic cell (17.1) An electrolysis apparatus.

Electrolytic conduction (17.1) The transport of charge through a solution by the movement of ions.

Electromagnetic radiation (4.1) General term used to describe light waves in all their various forms—e.g., X rays, ultraviolet and infrared radiation, visible light, TV waves, microwaves, and radio waves.

Electromagnetic spectrum (4.1) The entire range of electromagnetic radiation, from short wavelength gamma rays to long wavelength radio waves.

Electromotive force (emf) (17.6) The electric potential produced by a galvanic cell.

Electron (3.4, 3.5) A subatomic particle that is found outside the nucleus of an atom. It carries one unit of negative charge (-1.60×10^{-16} C) and has a mass (9.1×10^{-28} g) that is about 1/1800 of the mass of a proton. Its symbol is e.

Electron affinity (EA) (4.8) The energy change that occurs when an electron is added to an isolated gaseous atom or ion (usually expressed in kJ mol^{-1}).

Electron capture (22.3) Capture of an electron from an atom's 1s orbital (K shell) by an unstable nucleus. It converts a proton to a neutron in the nucleus.

Electron cloud (4.7) Because of its wave properties, an electron is spread out like a cloud around the nucleus.

Electron configuration (4.5) The distribution of electrons in an atom's orbitals.

Electron density (4.7) The concentration of the electronic charge within a given volume.

Electron-dot formula (5.6) See *Lewis structure*.

Electron-pair bond (5.5) A covalent bond.

Electron spin (4.4) A property that the electron appears to have because it behaves like a tiny magnet.

Electronic structure (4.1, 4.5) The distribution of an atom's electrons among the atom's orbitals.

Electronegativity (5.10) The relative attraction that an atom has for the electrons in a bond.

Electroplating (17.3) The deposition of a thin layer of a metal on an object by electrolysis.

Electropositive (18.2) Having a low electronegativity.

Electrovalent bond (5.3) An ionic bond.

Element (1.6) The simplest forms of matter that can exist under conditions normally encountered in a chemistry laboratory. Elements cannot be decomposed into simpler substances by chemical reactions.

Elementary process (13.5) One of the individual steps in a reaction mechanism.

Emf (17.6) See *Electromotive force.*

Empirical formula (2.6) The simple whole-number ratio of atoms in a compound.

Enantiomers (21.9) Optical isomers.

Endothermic change (1.4) A change that absorbs energy from the surroundings.

Endpoint (8.4) That point in a titration when delivery of the titrant is halted because the indicator changes color or some other event signals the completion of the reaction.

Energy (1.4) The capacity to do work.

Energy level (4.2) A particular energy that an electron can have in an atom or molecule.

Enthalpy, H (12.3) Also called heat content. Defined by the equation, $H = E + PV$. At constant T and P, $\Delta H = \Delta E + P\Delta V$. ΔH is also the heat of reaction at constant pressure.

Enthalpy diagram (12.4) A diagram that displays in graphical form the enthalpy changes that accompany the various thermochemical equations that combine to give some net change.

Enthalpy of formation (12.4) See *Heat of formation.*

Entropy, S (12.7) The thermodynamic quantity that describes the degree of randomness of a system. The greater the disorder, the higher the statistical probability of the state, and the higher the entropy.

Enzyme (24.2) A biological catalyst that is very effective and highly specific for a particular reaction.

Equation of state (12.1) An equation relating state variables.

Equation of state for an ideal gas (12.1) The ideal gas law, $PV = nRT$.

Equilibrium constant (14.1) The value that the mass action expression has when a chemical system is at equilibrium. It is K_p when partial pressures are used in the mass action expression, and it is K_c when molar concentrations are used.

Equilibrium law (14.1) An equation that sets the mass action expression equal to the equilibrium constant. A condition that must be fulfilled for a reaction to be at equilibrium.

Equilibrium vapor pressure of a liquid (10.4) The pressure exerted by a vapor that is in dynamic equilibrium with its liquid.

Equilibrium vapor pressure of a solid (10.12) The pressure exerted by a vapor that is in dynamic equilibrium with its solid.

Ester (23.8) An organic molecule with the structure

$$R—\overset{\overset{\textstyle O}{\|}}{C}—O—R$$

Esterification (23.8) The reaction of an organic acid with an alcohol to form an ester.

Ether (23.8) An organic molecule with the structure R—O—R.

Eutrophication (20.1) The natural aging process of a lake.

Evaporation (10.1) The conversion of a liquid to a gas.

Exact numbers (1.2) Numbers that come from definitions or a direct count of objects. They contain an infinite number of significant figures because they contain no uncertainty.

Excluded volume (9.11) The volume that one molecule in a gas prevents other molecules from occupying.

Exothermic change (1.4) A change that releases energy into the surroundings.

Exponential notation (1.2) See *Scientific notation.*

Extensive property (1.5) A property that depends on the size of a sample.

Face-centered cubic (fcc) unit cell (10.8) A cubic unit cell that has atoms, molecules, or ions at the corners and in the center of each face.

Factor-label method (1.2, Appendix A) The use of the cancellation of units that are associated with numbers in the solving of numerical problems.

Family of elements (3.2) A column of elements (a group) in the periodic table.

Faraday ($\mathscr{F}$) (17.4) One mole of electrons. 96 500 coulombs.

Fatty acid (24.4) An organic acid having a long hydrocarbon-chain tail.

Ferromagnetism (21.4) The strong magnetism associated with iron, cobalt, and nickel, which results from the alignment of the magnetic poles (in the solid state) of large numbers of paramagnetic atoms.

Filtrate (7.1) The liquid that passes through a filter.

First coordination sphere (21.6) The set of ligands that surround a metal ion in a complex.

First law of thermodynamics (12.2) When a system undergoes a series of changes that ultimately brings it back to its original state, the net energy change for the system is zero. This is a formal statement of the law of conservation of energy, and is the basis of Hess's law. Often stated mathematically as $\Delta E = q - w$, where ΔE is the change in the internal energy, q is the heat added to the system, and w is the work done by the system.

Fission (22.3) Splitting of a heavy unstable nucleus into several pieces, usually with the emission of large quantities of energy.

Flame test (18.5) Identification of an ion by the color produced when the ion is introduced into a flame. For example, sodium ion gives a flame a yellow color.

Flotation (18.1) A method for concentrating sulfide ores of copper and lead. Air is bubbled through a slurry of oil-coated ore particles, which stick to rising air bubbles and collect in a foam at the surface.

Formation constant (16.3) The equilibrium constant for the formation of a complex ion.

Formula unit (2.4) The collection of atoms specified by the chemical formula. For example, a formula unit of $CaCl_2$ contains one calcium atom and two chlorine atoms.

Formula weight (2.4) The sum of the atomic weights of all the atoms in one formula unit.

Fractional crystallization (11.6) A procedure for purifying substances in which the impure solid is dissolved in a minimum amount of hot solvent. The solution is then gradually cooled, which causes crystals of the pure substance to precipitate. These are collected by filtration, while the impurities remain in the solution.

Fractional distillation (11.9) A method used to separate mixtures of volatile liquids into their components. It involves boiling the mixture, condensing the vapor, then boiling this vapor, then condensing the new vapor, and so on. In each step, the vapor becomes richer in the more volatile component.

Frasch process (20.2) A method of mining sulfur from deep wells. Compressed air and superheated water are forced into a sulfur deposit where they melt the surface and bring it to the surface.

Free radical (13.10) An extremely reactive chemical species that contains one or more unpaired electrons.

Freezing point (10.6) At a particular pressure, the temperature at which an equilibrium can exist between the liquid and solid forms of a substance.

Frequency (4.1) The number of peaks in a wave that pass a given point per second. The SI unit of frequency is the hertz. $1 \text{ Hz} = 1 \text{ s}^{-1}$.

Fuel cell (17.12) A galvanic cell in which the anode and cathode reactants can be fed continuously, so power can be drawn from the cell continuously.

Functional group (23.6) A group of atoms in an organic molecule that gives the molecule certain characteristic properties.

Fundamental particle (3.4) A basic building block of all matter. Examples are the electron, proton, and neutron.

Fusion (22.7) Melting. In nuclear reactions, it is the joining of low-mass nuclei to produce more massive nuclei with the simultaneous emission of large quantities of energy.

Galvanic cell (17.5) An electrochemical cell in which a spontaneous redox reaction produces electricity.

Galvanizing (21.5) Coating a steel object with zinc to protect it from corrosion.

Gamma rays (γ-rays) (22.1) High energy, short wavelength (high frequency) radiation similar to X rays that is given off by radioactive substances.

Gangue (18.1) The unwanted rock and sand that is separated from an ore.

Gas constant (9.8) See *Universal gas constant*.

Gas discharge tube (3.4) A glass tube fitted with metal electrodes at either end and containing a gas at a low pressure. When a high electric potential is applied across the electrodes, electric current passes through the tube and the gas glows.

Gay–Lussac's law (9.4) For a fixed quantity of gas at a constant volume, P/T = constant, where T is the temperature in kelvins.

Gay–Lussac's law of combining volumes (9.7) When measured at the same temperature and pressure, the volumes of gases consumed or produced in a chemical reaction are in ratios of small whole numbers.

Geiger–Müller counter (22.1) A device used to detect radioactive emissions.

Genetic code (24.6) The base sequences found in *m*-RNA that specify the various amino acids used in building protein molecules.

Genetic disease (24.6) A "disease" or malfunction of cells that is caused by a fault in the DNA.

Geometrical isomers (21.9) Isomers that differ in the relative orientations of the atoms.

Gibbs free energy, G (12.8) A thermodynamic quantity that relates energy (enthalpy, H), and entropy, S. It is defined as $G = H - TS$.

Glass electrode (17.11) A special electrode that is sensitive to the concentration of hydrogen ion in the solution that surrounds the electrode. It consists of a silver wire coated with AgCl immersed in an HCl solution, which is separated by a thin glass membrane from the bulk of the solution being tested.

Glycoside linkage (24.3) The —C—O—C— linkage that holds sugar molecules together in a polysaccharide.

Graham's law (9.9) The rate of effusion of a gas is inversely proportional to the square root of the molecular weight of the gas. Comparing two gases,

$$\text{Rate(1)}/\text{Rate(2)} = \sqrt{M_2/M_1}$$

where M is the molecular weight.

Gram (1.3) 0.001 of the SI base unit of mass, the kilogram.

Graphite (6.8) The common allotrope of carbon. It is a black slippery solid that has a layer structure in which the carbon atoms in each layer are arranged in hexagonal rings fused together. Graphite is a conductor of electricity.

Gray (Gy) (22.1) A unit used to express the absorbed dose of radiation. One gray (1 Gy) is a dose of 1 joule per kilogram of absorbing material.

Ground state (4.3) The lowest-energy electron configuration for an atom or a molecule.

Group (3.2, 3.3) A column of elements in the periodic table.

Haber process (19.4) The process used in the commercial preparation of ammonia.

Half-cell (17.5) One of the two electrode/electrolyte components of a galvanic cell.

Half-life, $t_{1/2}$ (13.3) The time required in a chemical reaction for the concentration of a given reactant to be reduced to half its initial value.

Half-reaction (7.8) An individual oxidation or reduction reaction that includes the correct formulas for all species taking part in the reaction as well as the electrons that are lost or gained. An example is

$$2H_2O \longrightarrow O_2(g) + 2H^+(aq) + 4e^-$$

Hall process (17.3) The method for producing aluminum by electrolysis of Al_2O_3 dissolved in a cryolite, Na_3AlF_6.

Halogens (3.3) The elements of Group VIIA.

Hard water (19.2) Water containing the ions Ca^{2+}, Mg^{2+}, and Fe^{3+}, which interfere with the action of soap by forming insoluble precipitates with the anions in the soap.

Hardness ions (19.2) The ions found in hard water: Ca^{2+}, Mg^{2+}, and Fe^{3+}.

Heat capacity (12.1) The quantity of heat energy needed to raise the temperature of a system by 1 °C.

Heat of crystallization (10.6) See *Molar heat of crystallization*.

Heat of formation (12.4) The quantity of heat liberated or absorbed when one mole of a compound is formed from its elements.

Heat of fusion (10.6) See *Molar heat of fusion*.

Heat of vaporization (10.3) See *Molar heat of vaporization*.

Heat content (12.3) Enthalpy.

Heat of solution (11.5) The quantity of heat absorbed or evolved when a given amount of solute dissolves in a solvent to form a solution.

Heating curve (10.11) For a particular substance, a graph of temperature versus quantity of head added. Temperatures corresponding to the melting and boiling points of the substance can be read from the graph.

Heavy water (19.1) Water in which the hydrogen atoms are $^2_1$H (deuterium, D). Also called deuterium oxide, D_2O.

Heme group (24.1) A porphyrin structure having an Fe^{2+} ion in the center. It is responsible for hemoglobin's ability to carry oxygen and for myoglobin's ability to hold oxygen until needed for metabolism.

Henry's law (24.1) $C_g = k_g p_g$, where C_g is the concentration of a gas dissolved in a solvent at a particular temperature, p_g is the partial pressure of the gas over the solution, and k_g is a proportionality constant called the Henry's law constant.

Hertz (Hz) (4.1) The SI unit of frequency: 1 Hz $= 1$ s^{-1}.

Hess's law of heat summation (12.4) Also known simply as Hess's law. When thermochemical equations are added to give the net equation for a reaction, the corresponding heats of reaction are added to give the net heat of reaction. Also,

$$\Delta H_{reaction} = (\text{sum } \Delta H^0_f \text{ products}) - (\text{sum } \Delta H^0_f \text{ reactants})$$

Heterocycle (23.9) An organic ring structure in which one or more atoms in the ring is an element other than carbon.

Heterogeneous catalyst (13.9) A catalyst that is in a different phase than the reactants. The reactants are adsorbed on the surface of the catalyst, which is where the reaction occurs.

Heterogeneous equilibrium (1.6) An equilibrium in which the reactants and products are not all in the same phase.

Heterogeneous mixture (1.6) A mixture that consists of two or more phases.

Heterogeneous reaction (14.5) A chemical reaction in which the reactants and products are not all in the same phase.

High spin complex (21.11) A complex in which there is minimum pairing of electrons in the d orbitals of the central metal ion.

Homogeneous catalyst (13.9) A catalyst that is in the same phase as the reactants.

Homogeneous mixture (1.6) A mixture that has the same properties and composition throughout. A solution.

Homogeneous reaction (14.5) A reaction in which the reactants and products are in the same phase.

Homologous series (23.1) A series of hydrocarbons in which each member differs from the preceding member by the same grouping of atoms.

Hund's rule (4.6) The lowest-energy electron configuration results when electrons that occupy orbitals of equal energy are spread out over the orbitals as much as possible with spins unpaired.

Hybrid atomic orbitals (6.5) Orbitals formed by mixing the basic atomic orbitals of an atom. They are more effective at overlapping with other orbitals than are ordinary unhybridized atomic orbitals.

Hydrate (1.10) A crystal that contains molecules of water in a fixed proportion relative to other substances present.

Hydrated ion (7.1) An ion that has become surrounded by molecules of water to which it is attracted.

Hydration (11.4) The act of a molecule or ion becoming surrounded by water molecules.

Hydration energy (11.5) The quantity of energy liberated when an ion or other solute particle becomes surrounded by water molecules. It is usually expressed as kJ mol^{-1} of solute hydrated.

Hydride ion (19.1) The ion, H^-.

Hydrides (19.1) Binary compounds that contain hydrogen.

Hydrocarbon (23.1) A compound composed of only carbon and hydrogen.

Hydrogen bond (10.2) An extra strong dipole-dipole attraction that occurs between molecules in which hydrogen is covalently bonded to nitrogen, oxygen, or fluorine.

Hydrogen economy (19.1) An economy built around the extensive use of hydrogen as a fuel.

Hydrogen electrode (standard hydrogen electrode) (17.7) The standard of comparison for reduction potentials

$$2H^+(aq) + 2e^- \rightleftharpoons H_2(g), \quad E^\circ_{H^+} = 0.0 \text{ V}$$

at 25 °C, 1 atom, and 1 M H^+.

Hydrogenation (19.1) The addition of H_2 to organic molecules having double or triple bonds.

Hydrolysis (15.5) The reaction of a substance with water. Hydrolysis of anions produces basic solutions and hydrolysis of cations produces acidic solutions.

Hydronium ion (7.1) The ion H_3O^+.

Hydrophilic (11.4) A hydrophilic substance is attracted strongly to water molecules and tends to be soluble in water.

Hydrophobic (11.4) A hydrophobic substance is attracted very weakly to water molecules and tends to be insoluble in water.

Hypothesis (1.1) A tentative explanation of the results of a series of experiments.

Ideal gas (9.2) A hypothetical gas that would obey the gas laws perfectly under all conditions.

Ideal gas law (9.8) $PV = nRT$.

Ideal solution (11.5) A solution for which $\Delta H_{soln} = 0$, and in which solute-solute, solvent-solvent, and solute-solvent attractions are all equal.

In phase (4.3) A condition that is met when the peaks and troughs of waves coincide.

Indicator (7.1, 15.7) A substance that changes color to signal the completion of a reaction during a titration. For acid-base reactions, an indicator is a weak acid or base whose molecular form differs in color from its ionic form.

Induced dipole (10.2) A dipole created when the electron cloud of an atom, molecule, or ion is distorted by a neighboring dipole or by an ion.

Inhibition (24.2) The blocking of enzyme activity by blockage of the active enzyme site.

Inhibition step (13.10) In a chain reaction, a step in the mechanism that removes product molecules and thereby inhibits or slows the overall rate of production of the products.

Inhibitor (13.10) A substance that blocks the action of a catalyst.

Initiation step (13.11) The first step in a chain reaction in which a reactant molecule is converted into one or more free radicals.

Inner orbital complex (3.3) A complex in which the metal ion uses *d* oribitals below its outer shell in forming hybrid orbitals used for bonding to the ligands.

Inner transition element (3.3) A member of the two long rows of elements below the main body of the periodic table. ($Z = 58 - 71$ and $90 - 103$).

Inorganic compound (5.11) A compound whose structure is not primarily determined by the linking together of carbon atoms.

Instability constant (16.3) The equilibrium constant for the decomposition of a complex into its components.

Instantaneous dipole (10.2) A momentary dipole caused by the erratic movement of electrons in an atom, molecule, or ion.

Intensive property (1.5) A property whose value is independent of the size of the sample under consideration.

Intermolecular forces of attraction (10.2) The attractive forces that occur *between* neighboring molecules.

Internal energy, *E* (12.2) The total kinetic and potential energy of a system.

International system of units (1.3) Known as SI. It is the modernized metric system adopted for worldwide use in 1960 by the General Conference of Weights and Measures.

Interstitial solid solution (11.2) A solid solution in which the solute particles fit into empty spaces between particles of the solvent.

Ion (2.4) An electrically charged atom or group of atoms.

Ion-electron method (7.4) A method for balancing redox reactions. It divides the overall reaction into half-reactions that are balanced separately and then combined to give the net ionic equation for the redox reaction.

Ion product (16.1) The product of ion concentrations. For a salt, the product of the concentrations of the ions, each raised to a power that is equal to the number of ions of that kind produced by one formula unit of the salt. (For example, the ion product for Ag_2S is $[Ag^+]^2[S^{2-}]$.).

Ion product constant (15.1) An equilibrium constant that is equal to an ion product, for example, $K_w = [H^+][OH^-]$.

Ionic bond (5.3) The electrostatic attraction that holds ions together in an ionic compound.

Ionic compound (2.4, 5.1) A compound composed of positive and negative ions.

Ionic crystal (10.9) A crystal in which the particles are positive and negative ions.

Ionic equation (7.1) A chemical equation in which all water-soluble electrolytes are written in ionic form, while solids and weak electrolytes are written in molecular form.

Ionic potential (18.3) The ratio of an ion's charge (q) to its radius (r). $\phi = q/r$.

Ionic radius (4.8) The effective radius of an ion, which is nearly constant from one compound to another.

Ionization constant (15.1) The equilibrium constant for the ionization of a weak electrolyte.

Ionization energy (IE) (4.8) The energy needed to remove an electron from an isolated gaseous atom, ion, or molecule (usually expressed in kJ mol^{-1}).

Ionization isomers (21.9) Coordination compounds that have the same molecular formulas, but in which different anions serve as ligands.

Ionization reaction (7.1) A reaction that produces ions. A reaction of a molecular substance with water in which ions are formed.

Isomers (21.9) Compounds that have the same formula but differ in the way their atoms are arranged.

Isothermal (12.1) A change that occurs at constant temperature.

Isotonic solutions (11.11) Solutions having the same osmotic pressure.

Isotope dilution (22.2) An analytical method in which a small amount of a radioactive isotope is added to a sample and in which the extent of dilution is used to compute the amount of nonradioactive isotope originally present in the sample.

Isotopes (3.10) Atoms of the same element that differ slightly in their masses. The nuclei of all isotopes of a given element have the same number of protons, but they differ in the number of neutrons.

Joule (1.4) The SI unit of energy. $1 J = 1$ kg m^2 s^{-2}. It is the kinetic energy possessed by an object with a mass of 2 kg moving at a speed of 1 m s^{-1}.

***K*-capture** (22.3) See *Electon capture*.

Kelvin (1.3) The SI unit of temperature, equal in size to the degree Celsius.

Kelvin temperature (1.3) The temperature expressed in kelvins: $K = °C + 273.15$.

Ketone (23.8) An organic compound with the structure

$$\overset{\displaystyle O}{\underset{\displaystyle R-C-R}{\|}}$$

Kilojoule (kJ) (1.4) $1 kJ = 1000 J$.

Kinetic energy (1.4) Energy that an object possesses because of its motion. K.E. $= 1/2\ mv^2$, where m is the object's mass and v is the object's velocity.

Kinetic molecular theory (9.10) The theoretical model that explains the properties of gases and accounts for the gas laws. It states that an ideal gas is composed of point-sized particles in rapid random motion and that the temperature of the gas is directly proportional to the average kinetic energy of its particles.

Lanthanide contraction (4.8) The gradual decrease in size that occurs from element 58 to 71 which causes the elements that follow the lanthanides to have unusually small sizes and large ionization energies.

Lanthanide elements (3.3) Elements with atomic numbers 58 through 71.

Lattice (10.8) A repeating pattern of points. In a crystal, it corresponds to the repeating pattern formed by the particles of the substance.

Lattice energy (5.4) The quantity of energy that would be released by the imaginary process in which isolated particles of a substance (atoms, molecules, or ions) come together to form one mole of a crystal of that substance.

Law (1.1) A statement of behavior based on the results of many experiments. Laws are often expressed in equation form.

Law of conservation of energy (1.4) Energy is neither created nor destroyed but, instead, can only be transformed from one kind of energy to another.

Law of conservation of mass (1.7) Mass is neither created nor destroyed during a chemical reaction.

Law of definite proportions (Law of definite composition) (1.7) In a pure chemical substance, the elements are always present in the same definite proportions by mass.

Law of multiple proportions (1.8) When two compounds are formed from the same two elements, the masses of one element that combine with the same mass of the other element are in a ratio of small whole numbers.

Le Chatelier's principle (10.4, 14.6) When a system that is in dynamic equilibrium is subjected to a disturbance that upsets the equilibrium, the system undergoes a change in a direction that counteracts the disturbance and restores equilibrium.

Lead storage battery (17.12) The common automobile battery in which during discharge Pb serves as the anode and PbO_2 serves as the cathode in an H_2SO_4 electrolyte.

Levorotatory (21.9) An optical isomer that causes a rotation of plane polarized light in a counterclockwise direction, when looking toward the source, as the light passes through a solution of the isomer.

Lewis acid (7.5) An electron pair acceptor during the formation of a coordinate covalent bond.

Lewis base (7.5) An electron pair donor during the formation of a coordinate covalent bond.

Lewis structure (5.6) Also called Lewis formula or electron-dot formula. A structural formula drawn with Lewis symbols, which represents the valence electrons as dots (or as dashes for pairs of electrons).

Lewis symbol (5.2) The chemical symbol of an element surrounded by dots that represent the valence electrons of an atom of the element.

Ligand (6.2, 21.6) An atom or a group of atoms bonded to a central atom in a molecule or polyatomic ion. A molecule or anion that can bind to a metal ion to form a complex.

Limiting reactant (2.11) The reactant that is completely consumed in a chemical reaction. It is the reactant that limits the amount of products that can be formed in a particular experiment.

Line spectrum (4.1) A spectrum that consists of only a relatively few wavelengths that is produced when the light emitted by energized or excited atoms is passed first through a thin slit, then through a prism, and then allowed to fall on a screen or piece of photographic film. It is also called an atomic spectrum.

Linear molecule (6.1) A molecule in which all the atoms lie in a straight line.

Lipid (24.4) A water-insoluble substance that can be extracted from cells by nonpolar solvents.

Liquid crystal (10.10) A substance that is able to flow like a liquid but which has some physical properties normally associated with crystals.

Liter (L) (1.3) A unit of volume. By definition, $1 \text{ L} = 1 \text{ dm}^3$.

London forces (10.2) Weak attractive forces caused by instantaneous dipole-induced dipole attractions.

Lone pair (6.1) An unshared pair of electrons in the valence shell of an atom.

Low spin complex (21.11) A complex in which there is a maximum pairing of electrons in the orbitals of the central metal atom.

Magic numbers (22.3) Numbers of protons and neutrons found in especially stable nuclei.

Magnetic quantum number (4.3) The quantum number ml which can have values from $-l$ to $+l$.

Main transition elements (21.1) See *d-block elements*.

Malleability (3.1) A metal's ability to be hammered or rolled into thin sheets.

Manganese nodules (18.1) Lumps about the size of an orange that contain large concentrations of manganese and iron. They are found on the ocean floor.

Manometer (9.1) A U-shaped tube, filled partially with a liquid, that is used to measure pressures of trapped gases.

Markonikov's rule (23.6) During an addition reaction, the hydrogen of the molecule being added becomes attached to the carbon that is already bonded to the larger number of hydrogen atoms.

Mass (1.4) A measure of the quantity of matter in an object. A measure of an object's resistance to a change in velocity.

Mass action expression (14.1) A fraction that can be constructed from the overall balanced equation for a reaction. For the general reaction

$$aA + bB \rightleftharpoons mM + nN$$

the mass action expression is

$$\frac{[M]^m[N]^n}{[A]^a[B]^b}$$

For reactions involving gases, partial pressures can be used in place of molar concentrations.

Mass defect (22.6) The differences between the actual mass of a nuclide and the mass computed by adding the masses of the corresponding number of protons and neutrons.

Mass fraction (11.3) The number of grams of solute divided by the total number of grams of solution.

Mass number (3.10) The sum of the number of protons and the number of neutrons in a particular nucleus.

Mass percent (11.3) Mass percent = mass fraction × 100.

Mass spectrometer (3.6) A device that allows the determination of the charge-to-mass ratio of positive particles.

Mass spectrum (3.6) The distribution of particles that is found when a mass spectrometer separates a mixture of ions of different masses.

Matter (1.4) Anything that has mass and occupies space.

Mean free path (10.1) The average distance traveled by a molecule between collisions.

Mechanism of a reaction (13.5) The series of individual steps in a chemical reaction that gives the net overall change.

Melting point (10.6) The temperature at which a substance melts. It is the same temperature as the freezing point of the substance.

Messenger RNA (m-RNA) (24.6) A form of RNA that carries the code for the sequence of amino acids in proteins from the DNA inside the cell nucleus to locations outside the nucleus where protein synthesis takes place.

Meta- (23.5) In a benzene ring, positions separated by one intervening carbon atom.

Metal carbonyl compound (19.2) A compound formed from a metal and carbon monoxide, for example, $Ni(CO)_4$.

Metallic conduction (17.1) The transport of electrical charge through a metal by the movement of electrons.

Metallic crystal (10.9) A solid having only positive ions at the lattice points, which are attracted to a 'sea of electrons' that extends throughout the entire crystal.

Metalloids (3.1) Elements with properties that lie between those of metals and nonmetals. They are located around the diagonal line running from boron to astatine in the periodic table.

Metallurgy (18.1) The science and technology of metals. It is concerned with the procedures and chemical reactions that are used to separate metals from their ores and make them ready for practical uses.

Metals (3.1) Elements that are lustrous, have high thermal and electrical conductivity, and that are generally ductile and malleable. In compounds they nearly always have positive oxidation states. In the periodic table, they are located to the left of the diagonal line running from boron to astatine.

Metathesis reaction (7.1) See *Double replacement reaction*.

Meter (1.3) The SI base unit for length.

MeV (million electron volts) (22.6) An energy unit used to express binding energies and energy changes that accompany nuclear reactions. 1 MeV per nuclide corresponds to an energy of 9.65×10^7 kJ mol^{-1}.

Micelle (11.4) A particle formed by the grouping together of fatty acid anions (soap anions) with their nonpolar tails intermingling and their anionic heads facing outward toward the aqueous environment.

Miscible (11.4) Two liquids are miscible if they are soluble in each other in all proportions.

Mixture (1.6) Two or more substances combined in no particular proportions.

Molar (2.13) A term that describes the molar concentration of a solute in a solution. It means *moles of solute per dm^3 of solution*.

Molar concentration (2.13, 11.3) A ratio of moles of solute to cubic decimetres of solution. It is the number of moles of solute per dm^3 (L) of solution.

Molar heat capacity (12.1) The quantity of heat required to raise the temperature of one mole of a substance by 1 °C.

Molar heat of crystallization, ΔH_{cryst} (10.6) The heat evolved when one mole of a liquid freezes. It is equal in magnitude, but opposite in sign to the molar heat of fusion.

Molar heat of fusion, ΔH_{fus} (10.6) The quantity of heat absorbed when one mole of solid melts to give a mole of liquid at constant temperature and pressure.

Molar heat of sublimation, $\Delta H_{sublimation}$ (10.3) The quantity of heat absorbed when one mole of solid sublimes to give one mole of vapor at constant temperature and pressure.

Molar heat of vaporization, ΔH_{vap} (10.3) The quantity of heat absorbed when one mole of liquid is converted to a vapor at constant temperature and pressure. Usually, it is measured at the boiling point of the substance.

Molar solubility (16.1) The number of moles of solute dissolved in one cubic decimetre (litre) of its saturated solution.

Molarity (2.13) See *Molar concentration*.

Mole (mol) (2.2) The SI unit for amount of substance. 6.022×10^{23} things. A quantity of a substance whose mass in grams is numerically equal to the substance's formula weight.

Mole fraction (9.10, 11.3) A concentration unit that is the ratio of the number of moles of a given component to the total number of moles in the solution.

Mole percent (mol %) (11.3) Mole percent = 100 × mole fraction.

Molecular compound (5.1) A compound whose particles are molecules.

Molecular crystal (10.9) A crystal composed of molecules (such as water) or individual atoms (such as the noble gases).

Molecular equation (7.1) A chemical equation for a reaction in solution in which the formulas of strong electrolytes are written as if the substances were molecular.

Molecular formula (2.6) A chemical formula that specifies the number of atoms of each kind that are present in a molecule of the substance.

Molecular orbital (6.9) An orbital in a molecule that is able to extend over all the nuclei of the molecule.

Molecular orbital theory (6.4, 6.9) A theory of covalent bonding that views a molecule as a collection of positive nuclei surrounded by a set of orbitals that belong to the molecule as a whole and extend over all the positive centers in the molecule.

Molecular weight (2.4) The sum of the atomic weights of all the atoms in a molecule.

Molecule (1.8, 5.1) An electrically neutral group of atoms bound tightly enough together that they behave as and can be recognized as a single particle.

Monodentate ligand (21.6) A ligand that can attach itself to a metal atom by only one of its atoms. Examples are ammonia and water.

Monomer (23.10) A small molecule that combines with others to form polymers.

Monoprotic acid (7.2) An acid that is capable of furnishing only one H^+ per molecule.

Monosaccharide (24.3) A simple sugar unit that combines with others to form polysaccharide chains in more complex sugars such as sucrose, starch, and cellulose.

Negative deviations (from Raoult's law) (11.8) Solutions that exhibit negative deviations from Raoult's law have vapor pressures that are lower than predicted by Raoult's law.

Nematic liquid crystal (10.10) A liquid crystal composed of long rodlike molecules packed like short pieces of uncooked spaghetti.

Nernst equation (17.10)

$$\mathscr{E}_{cell} = \mathscr{E}_{cell}^{\circ} - \frac{0.0592}{n} \log Q$$

where Q is the mass action expression for the cell reaction as written.

Net bond order (6.9)

$$\frac{\text{(no. of bonding electrons)} - \text{(no. of antibonding electrons)}}{2}$$

Net ionic equation (7.1) A chemical equation obtained by omitting spectator ions from an ionic equation. It shows the net chemical change that occurs.

Neutralization (7.2) The reaction of an acid with a base. In aqueous solutions, the products are a salt and water.

Neutron (3.9) A subatomic particle found in the nuclei of atoms. It is electrically neutral and has a mass $(1.67 \times 10^{-24}$ g) that is just slightly larger than that of a proton.

Neutron activation analysis (22.2) An analytical procedure in which nonradioactive isotopes are bombarded by neutrons, making some of their atoms radioactive. From the frequencies of the rays emitted by the bombarded sample, the concentrations of the various elements in the sample can be determined.

Newton (9.1) The SI unit of force, $1 N = kg\ m\ s^{-2}$. A force of $1 N$ applied to an area of $1\ m^2$ produces a pressure of 1 pascal (Pa).

Nickel-cadmium cell (17.12) The nicad battery, in which during discharge Cd serves as the anode and NiO_2 serves as the cathode in an alkaline electrolyte. The nicad battery is rechargeable.

Noble gases (3.3) The elements in Group 0 of the periodic table: helium, neon, argon, krypton, xenon, and radon.

Noble metal (7.9) A metal that is very unreactive, for example, gold and platinum.

Node (4.3) In a wave, a place where the amplitude or intensity is zero.

Nomenclature (5.11) A system of names used in a particular branch of knowledge, such as chemistry.

Nonelectrolyte (7.1) A substance that does not dissociate into ions in an aqueous solution.

Nonmetals (3.1) Elements that are poor conductors of heat and electricity and lack the other properties normally associated with metals. In the periodic table, they are located to the right of the diagonal stair-step line running from boron to astatine.

Nonoxidizing acid (7.9) In a solution of a nonoxidizing acid, the strongest oxidizing agent is H_3O^+.

Normal boiling point (10.5) The temperature at which the vapor pressure of a liquid equals 101.325 kPa (1 atm).

Nuclear transformations (22.4) Changes in nuclei brought about by bombarding them with high energy particles such as neutrons.

Nucleic acid (24.5) A polymer of nucleotide units. Varieties include DNA and RNA.

Nucleotide (24.5) The units that form the building blocks of nucleic acids. They are composed of even simpler units—a five-carbon sugar, phosphoric acid, and a nitrogenous base.

Nucleus (3.8) The very tiny, massive particle found at the center of an atom. It contains all of the atom's positive charge and nearly all of its mass. Protons and neutrons are found in the nucleus.

Nuclide (22.1) A general term used to describe the nucleus of a particular isotope.

Octahedral molecule (6.1) A molecule in which the central atom is bonded to six others that are located at the vertices of an octahedron (a figure consisting of two square pyramids that share a common base).

Octet rule (5.3) An atom tends to gain or lose electrons until its outer shell contains eight electrons.

Olefin (23.1) An alkene.

Open hearth furnace (18.1) A now-obsolete furnace used to convert pig iron into steel.

Optical activity (21.9) The rotation of the plane of polarized light as it passes through a chiral substance, either in its pure state or in a solution.

Optical isomers (21.9) Isomers that are nonsuperimposable mirror images of each other.

Orbital (4.3) A particular electron waveform with a particular energy. In an atom, each orbital has a specific set of values of its quantum numbers, n, l, and m_l.

Orbital diagram (4.5) A diagram that represents an atom's orbitals by dashes (or some other suitable device) and that represents electrons that populate the orbitals by arrows. For example, the orbital diagram of carbon is

$$C\ \underset{1s}{\underline{\uparrow\downarrow}}\ \underset{2s}{\underline{\uparrow\downarrow}}\ \underset{2p}{\underline{\uparrow}\ \underline{\uparrow}\ \underline{}}$$

Order (of reaction) (13.2) The sum of the exponents in the rate law is the overall order of the reaction. Each exponent gives the order with respect to a certain reactant.

Ore (18.1) A substance that contains a desirable constituent in concentrations large enough to make its recovery economically worthwhile.

Organic acid (23.8) An organic molecule with the structure

$$R{-}{-}{-}\overset{\overset{\displaystyle O}{\|}}{C}{-}{-}{-}O{-}{-}{-}H$$

Organic compounds (5.11) Compounds whose structures are determined primarily by the linking together of carbon atoms. They are hydrocarbons or can be considered to be derived from hydrocarbons.

Ortho- (23.5) In the benzene ring, positions that are adjacent to each other.

Osmosis (11.11) The selective passage of solvent through a semipermeable membrane from a solution of low solute concentration to a solution of high solute concentration.

Osmotic pressure (11.11) The pressure that must be exerted on the more concentrated solution to prevent osmosis from occurring.

Ostwald process (19.4) The commercial process for the manufacture of nitric acid from ammonia.

$$4NH_3 + 5O_2 \longrightarrow 4NO + 6H_2O$$
$$2NO + O_2 \longrightarrow 2NO_2$$
$$3NO_2 + H_2O \longrightarrow 2HNO_3 + NO$$

Out of phase (4.3) A condition that is met when the peak of one wave coincides with the trough of another wave.

Outer orbital complex (21.10) A complex in which the metal ion uses d orbitals in its outer shell to form the hybrid orbitals that are used in bonding to the ligands.

Overlap of orbitals (6.4) Portions of two orbitals from different atoms share the same space.

Oxidation (7.6) Loss of electrons. Increase in oxidation number.

Oxidation number (7.6) The charge an atom would have if all the electrons in each of its bonds belonged to the more electronegative atom. Normally, oxidation numbers are assigned following the rules given in the text.

Oxidation-number-change method (7.7) A method for balancing redox equations that makes the total increase in oxidation number equal the total decrease.

Oxidation-reduction reaction (7.6) A reaction that involves the transfer of electrons from one substance to another. A reaction that involves a change in oxidation numbers. Also called a redox reaction.

Oxidation state (7.6) The same as oxidation number.

Oxidizing acid (7.9) An acid whose aqueous solutions contain an oxidizing agent that is stronger than H_3O^+. An example is HNO_3, which gives the strong oxidizing agent NO_3^- in aqueous solutions.

Oxidizing agent (7.6) In a redox reaction, the substance that is reduced, thereby causing the oxidation.

Oxoacid (5.11) An acid that contains hydrogen, oxygen, and one other element (for example, HNO_3, H_3PO_4, H_2SO_4).

Ozone (19.3) An allotrope of oxygen, O_3.

Paired electrons (4.5) Two electrons, one with $m_s = +1/2$ and the other with $m_s = -1/2$. Two electrons with opposite spins.

Pairing energy (21.11) The quantity of energy that must be absorbed to cause two electrons to occupy the same orbital with their spins paired.

Para- (23.5) In the benzene ring, positions separated by two intervening carbon atoms.

Paraffin (23.1) The general name given to the alkane series of hydrocarbons. Also, high molecular weight hydrocarbons such as $C_{20}H_{42}$ that exist as waxes at room temperature.

Paramagnetism (4.4) A weak attraction toward a magnetic field. It is a property possessed by substances having unpaired electrons.

Parent isotope (22.1) The isotope that changes into some other isotope in a radioactive decay.

Partial charge (5.10) A charge found on either end of a dipole that is less than full $+1$ or -1 charges.

Partial pressure (9.6) The pressure that is exerted by a particular gas in a mixture of gases.

Particle accelerator (22.4) A device that accelerates charged particles to very high speeds (high energies) before allowing them to bombard target nuclei.

Parts per billion (ppb) (8.3) Usually expressed on a mass basis. The number of grams of the component in question per billion grams of the solution. It is the mass fraction multiplied by 10^9.

$$\text{ppb } X = \frac{\text{grams } X}{\text{grams solution}} \times 10^9$$

Parts per million (8.3) Usually expressed on a mass basis. The number of grams of the component in question per million grams of solution. It is mass fraction multiplied by 10^6.

$$\text{ppm } X = \frac{\text{grams } X}{\text{grams solution}} \times 10^6$$

Pascal (9.1) The SI unit for pressure. A pressure of 1 newton per square meter. $1 \text{ Pa} = 1 \text{ N m}^{-2}$.

Pauli exclusion principle (4.4) No two electrons in the same atom can have all four of their quantum numbers the same.

Peptide (24.1) A polymer of α-amino acids. Proteins are composed of peptides.

Peptide bond (24.1) The structure

$$\cdots {-}{-}{-}\overset{\overset{\displaystyle O}{\|}}{C}{-}{-}{-}\underset{\underset{\displaystyle H}{|}}{N}{-}{-}{-} \cdots$$

Peptide linkage (24.1) The same as a peptide bond.

Percent by mass (11.3) The number of grams of the component in question per 100 grams of sample (or solution).

$$\% \ X = \frac{\text{grams } X}{\text{grams sample}} \times 100$$

Percent composition (2.5) The percents by mass of the elements in a compound.

Percent dissociation (15.2)

$$\% \text{ dissociation} = \frac{\text{amount dissociated}}{\text{total amount available}} \times 100$$

Percent yield (2.12)

$$\% \text{ yield} = \frac{\text{actual yield}}{\text{theoretical yield}} \times 100$$

Period (3.2) A horizontal row of elements in the periodic table.

Periodic law (3.3) When the elements are arranged in order of increasing atomic number, there is a periodic recurrence of properties.

pH (15.1) A logarithmic measure of acidity (pH = $-\log$ [H$^+$]). A solution is acidic if its pH < 7, neutral if pH = 7, and basic if pH > 7 (at 25°C).

Phase (1.6) Any part of a system that has the same uniform properties and composition.

Phase change (1.6) Transition of a sample from solid to liquid, liquid to gas, solid to gas, or any of the reverse of these changes.

Phase diagram (10.13) A pressure-temperature graph on which are plotted temperatures and pressures at which equilibrium exists between the states of a substance. It defines temperature-pressure regions in which the solid, liquid, and gaseous states of the substance can exist.

Phenyl group (23.5) Benzene, with one hydrogen removed, that becomes attached to some other organic molecule in place of a hydrogen.

Phlogiston (1.7) An imaginary substance once thought to be lost by an object when it burned.

Phospholipid (24.4) An ester of glycerol, two fatty acids, and phosphoric acid, which in turn is esterified to another alcohol. Phospholipids are found in cell membranes.

Photochemical smog (19.4) A type of urban air pollution caused by the interaction of sunlight and nitrogen oxides, which produces ozone and the unpleasant products of the reaction of ozone with airborne hydrocarbons.

Photon (3.7, 4.2) A tiny packet of electromagnet energy (light energy) whose energy is given by the equation $E = h\nu$.

Physical change (1.6) A change that does not alter the chemical properties of the substance involved.

Physical process (1.6) A process or event in which the chemical properties of the substances involved do not change. For example, the chemical properties of sugar and water do not change when they are mixed to form a solution. Formation of a solution is a physical process.

Physical property (1.5) A property (e.g., color) that can be specified without reference to any other chemical substance.

Pi bond (π bond) (6.6) A bond formed by the sideways overlap of a pair of p orbitals. Electron density in the bond is concentrated in two separate regions that lie on opposite sides of an imaginary plane that contains both nuclei.

Pickling (19.3) The removal of rust from iron or steel by reaction with acid.

Pig iron (18.1) The impure iron that comes from a blast furnace.

pK_w (15.1) $-\log[K_w] = 14.00$ (at 25°C).

Planar triangular molecule (6.1) A molecule in which three atoms surround the central atom at the corners of a triangle. All four atoms are in the same plane.

Planck's constant (4.2) A constant, h, that permits us to calculate the energy of a photon of frequency ν by the equation $E = h\nu$. $h = 6.6262 \times 10^{-34}$ J s.

Plasma (22.7) A very hot, high energy gas composed of ions.

pOH (15.1) $-\log[OH^-]$.

Polar bond (5.10) A polar covalent bond.

Polar covalent bond (5.10) A covalent bond in which more than half of the bond's negative charge is concentrated around one of the two atoms joined by the bond. This atom acquires a partial negative charge, while the other atom acquires a partial positive charge.

Polarizability (10.3) The ease with which the electron cloud of an atom, molecule, or ion is distorted, thereby causing the particle to become a dipole.

Polarized light (21.9) Light in which the oscillations of the light waves are all in the same plane.

Polyatomic ion (5.3) An ion composed of two or more atoms.

Polydentate ligand (21.6) A ligand that has two or more donor atoms that can become simultaneously attached to a metal ion.

Polyester (23.10) A copolymer of a difunctional alcohol with a difunctional organic acid.

Polymer (23.10) A large molecule formed by linking together many smaller molecules.

Polypeptide (24.1) A peptide composed of many α-amino acid units. Proteins are polypeptides.

Polyprotic acid (7.2) An acid that is capable of furnishing more than one H$^+$ per molecule of the acid.

Polysaccharide (24.3) A polymer formed from simple, monomeric sugar molecules.

Porphyrin (24.1) A square planar ligand structure that is found in a number of biologically important molecules such as heme and chlorophyll.

Position of equilibrium (7.1) The relative proportions of reactants and products in an equilibrium system.

Positive deviations (from Raoult's law) (11.8) Solutions that exhibit positive deviations have vapor pressures that are higher than predicted by Raoult's law.

Post-transition element (5.3) In the periodic table, a representative element in Group IIIA through Group VIIA that occurs after a series of transition elements.

Potential energy (1.4) Energy that an object possesses because of the attractions or repulsions that it experiences. It is stored energy, and the quantity of this stored energy changes when there are changes in the attractions or repulsions.

Potentiometer (17.6) A device that permits the measurement of the emf of a galvanic cell without drawing current from the cell.

Precipitate (2.15, 7.1) A solid that forms in a solution, often as the result of a chemical reaction.

Precision (1.2) How closely repeated measurements of the same quantity come to each other.

Pressure (9.1) Force per unit area. In SI, its units are pascals, $1 \text{ Pa} = 1 \text{ N m}^{-2}$.

Primary alcohol (23.8) An alcohol with the formula, $R—CH_2OH$.

Primary structure (24.1) The sequence of the various amino acids in a polypeptide or protein.

Primitive lattice (10.8) A lattice in which the unit cell has lattice points only at the corners.

Principle quantum number (4.3) The quantum number n, which can have values of $1,2,3,…,\infty$. The value of n specifies the electrons' shell and determines the size of the orbital.

Probability distribution (4.7) How the probability of finding a particular electron varies from place to place around the nucleus of an atom, or around the nuclei of a molecule or polyatomic ion.

Products (1.10) The substances that are formed in a chemical reaction. The substances that appear on the right side of a chemical equation.

Propagation step (13.10) A step in a chain reaction in which a free radical reacts with a reactant molecule to give a product molecule plus another free radical.

Properties (1.5) The characteristics of a sample of matter that serve to identify it.

Protein (24.1) A polymer of α-amino acids that has a specific biological function.

Protium (19.1) ¦H, the most abundant isotope of hydrogen.

Proton (3.6) A subatomic particle found in the nuclei of all atoms. It carries one unit of positive charge $(+1.60 \times 10^{-19} \text{ C})$ and its mass $(1.67 \times 10^{-24} \text{ g})$ is very nearly 1 u (about the same as the mass of a hydrogen atom). It is often symbolized as H^+ in chemical equations.

Pseudonoble gas configuration (5.3) For a given n, the configuration $ns^2np^6nd^{10}$.

Qualitative observation (1.1) An observation that does not involve numerical measurements.

Quantitative observation (1.1) An observation involving measurements that result in numerical data.

Quantum (4.2) A tiny packet of light energy having energy $E = h\nu$. Also called a photon.

Quantum mechanics (4.3) See *Wave mechanics*.

Quantum number (4.2) A number related to the energy, shape, or orientation of an orbital. Also a number related to the spin of an electron.

Quaternary structure (24.1) The way certain folded polypeptide chains pack together in a complex protein.

Racemic (21.9) A mixture that contains equal numbers of the two optical isomers of a substance, and which does not rotate the plane of polarized light in either direction.

Radioactive decay (22.1) The gradual transformation of a collection of unstable nuclei into a collection of stable nuclei by the emission of various forms of radiation (α, β, or γ).

Radioactive series (22.1) The series of radioactive isotopes produced one after another during the decay of some particular radioactive element. The series terminates when a stable nonradioactive nucleus is formed.

Radioactivity (3.7) The spontaneous emission of radiation by certain unstable atomic nuclei.

Radioisotope (22.1) A radioactive isotope.

Radionuclide (22.1) The nucleus of a radioactive isotope.

Raoult's law (11.8) $p_A = x_A P_A^\circ$, where P_A° is the vapor pressure of pure substance A, x_A is the mole fraction of A in the solution, and p_A is the vapor pressure of A over the solution.

Rare earth metals (3.3) The lanthanide elements (atomic numbers $58 - 71$).

Rate (13.1) A ratio in which units of time appear in the denominator, for example, 100 km h^{-1} or 3.0 mol L^{-1} s^{-1}.

Rate constant (13.2) The proportionality constant in the rate law for a reaction. Its value is the rate of the reaction when all the reactant concentrations are 1 M.

Rate-determining step (13.5) The slow step in a reaction mechanism, which determines how fast the products appear.

Rate law (13.2) An equation that relates the rate of a reaction to the molar concentrations of the reactants, each raised to some appropriate power.

Rate of reaction (13.1) How quickly the reactants disappear and the products form, expressed as mol L^{-1} s^{-1}.

Reactants (1.10) The substances that react during a chemical reaction. The substances that appear on the left side of a chemical equation.

Reaction coordinate (13.7) In an analysis of the energy changes that take place during a collision, it is the path followed as reactant molecules approach each in a collision, followed by bonds being broken as new ones are formed, and then the product species moving apart.

Reaction mechanism (13.5) See *Mechanism of a reaction*.

Reaction quotient (14.1) See *Mass action expression*. The numerical value of the mass action expression.

Reaction rate (13.1) See *Rate of reaction*.

Reagent (2.14) A term often used to refer to common chemicals that are stocked in the laboratory.

Red phosphorus (6.8) A relatively unreactive allotrope of phosphorus whose molecular structure is presently unknown.

Redox (7.6) A term meaning *oxidation-reduction*.

Reducing agent (7.6) A substance that causes reduction by supplying electrons. It is the substance that is oxidized in a redox reaction.

Reduction (7.6) A gain of electrons, or a decrease in oxidation number. In metallurgy, it is a process whereby a metal is extracted from its ore—a process that requires the chemical reduction of the metal ion.

Reduction potential (17.7) A measure of the tendency of a given reduction half-reaction to occur, measured in units of volts.

Refining (18.1) Purification and treatment of a metal to give it properties that are appropriate for specific applications.

Rem (22.1) *R*adiation *e*quivalent in *m*an. A unit of absorbed dose that takes into account the kind of radiation and the degree to which it is absorbed by animal tissue.

Representative element (3.3) An element in one of the A-groups of the periodic table.

Resonance (5.8) When the actual electronic structure of a molecule or polyatomic ion cannot be adequately represented by a single Lewis structure, it is instead represented as a composite of two or more Lewis structures that are called resonance structures. None of the individual resonance structures describes an actual electronic structure of a molecule or ion.

Resonance hybrid (5.8) The actual structure of a molecule or polyatomic ion which is represented by two or more resonance structures.

Resonance structure (5.8) One of two or more Lewis structures that can be drawn for a molecule or ion, none of which adequately describes the bonding in the species.

Reversible process (12.2) A process that occurs by an infinite number of steps during which the driving force for the change is just barely greater than the force that resists the change. Any slight increase in the resisting force causes the change to reverse direction.

Ribose (24.3) A five-carbon sugar that is one of the building blocks of RNA.

RNA (24.5) Ribonucleic acid. A type of nucleic acid involved in protein synthesis. (See *transfer RNA* and *messenger RNA*).

Roasting (18.1) The heating of an ore in air, which converts sulfides to oxides.

Rust (21.5) Hydrated iron(III) oxide, $Fe_2O_3 \cdot xH_2O$.

Rydberg equation (4.1) An empirical equation that allows the computation of the wavelengths of the lines in the emission spectrum of hydrogen.

Salt (5.11) Any ionic compound that does not contain O^{2-} or OH^-. The common name for NaCl.

Salt-bridge (17.5) A tube containing an electrolyte that connects the two half-cells of a galvanic cell.

Saponification (23.8) The base-catalyzed hydrolysis of an ester.

Saturated hydrocarbon (23.1) A hydrocarbon in which all the carbon-carbon bonds are single bonds.

Saturated solution (7.1) A solution that contains as much dissolved solute as it can hold while in equilibrium with undissolved solute.

Scientific method (1.1) Observation, explanation, and the testing of the explanation by further observation.

Scientific notation (1.2) Numbers expressed as the product of a decimal number between 1 and 10 multiplied by 10 raised to an appropriate power, for example, $0.035 = 3.5 \times 10^{-2}$.

Scintillation counter A device used for counting emissions from radioactive substances. The radiation is absorbed by the device, which causes the emission of a burst of light that is then detected electronically.

Second law of thermodynamics (12.8) Whenever a spontaneous event takes place, it is accompanied by an increase in the total entropy of the universe.

Secondary alcohol (23.8) An alcohol with the structure

$$R\text{---}\underset{\underset{H}{|}}{\overset{\overset{R}{|}}{C}}\text{---}OH$$

Secondary structure (24.1) The coiling that takes place in a polypeptide chain.

Semiconductor (3.1) A substance that conducts electricity weakly.

Semipermeable membrane (11.11) A membrane that permits the passage of solvent molecules but not solute molecules.

Shell (4.3) A term used to describe all the electrons (or orbitals) in an atom that have a given value of the principal quantum number, n.

SI (1.3) System International d'Unités. The International System of Units.

Sigma bond (σ bond) (6.6) A bond formed by the head-to-head overlap of two orbitals. The electron density in the bond is concentrated along an imaginary straight line that connects the two nuclei.

Significant figures (1.2) Digits obtained in a measurement such that only the rightmost digit in the measured value contains any uncertainty.

Silicone (20.5) A type of polymer in which the 'backbone' consists of alternating silicon and oxygen atoms.

Silver oxide battery (17.12) A battery in which zinc serves as the anode and Ag_2O serves as the cathode in an alkaline electrolyte.

Simple cubic unit cell (10.8) A cubic unit cell with identical atoms, molecules, or ions located only at the corners.

Simple sugar (24.3) See *Monosaccharide*.

Simplest formula (2.6) See *Empirical formula*.

Single bond (5.5) A covalent bond in which a single pair of electrons is shared between two atoms.

Single displacement reaction (7.9) A reaction in which one element displaces another from a compound. For example: $Zn(s) + CuSO_4(aq) \longrightarrow Cu(s) + ZnSO_4(aq)$.

Slag (18.1) A relatively low-melting mixture of impurities that forms in the blast furnace and other furnaces used in refining metals.

Slaking (18.6) Treating lime (CaO) with water, which produces $Ca(OH)_2$.

Smectic liquid crystal (10.10) A liquid crystal that consists of rodlike molecules arranged in layers of parallel rods.

Smog (19.4) See *Photochemical smog*.

Soap (11.4) A solution of fatty acid anions.

Solubility (7.1) The amount of solute required to give a saturated solution in a given quantity of solvent or solution.

Solubility product constant, K_{sp} (16.1) The equilibrium constant for the solubility of a salt. For a saturated solution, K_{sp} is equal to the product of the molar concentrations of the ions each raised to appropriate powers.

Solubility rules See page 277.

Solute (2.13) A substance dissolved in a solvent.

Solution (1.6) A homogeneous mixture.

Solvation (11.4) The surrounding of a solute particle by molecules of the solvent.

Solvay process (18.5) A commercial process used to prepare Na_2CO_3 from NaCl and CO_2 in an aqueous solution made basic by ammonia.

Solvent (2.13) Generally the substance in a solution that is present in largest quantity. If one substance is a liquid, it is normally considered to be the solvent. When water is present, it is taken to be the solvent.

Specific activity (22.1) The number of nuclear disintegrations per second per gram of sample. In the SI, its units are Bq g^{-1}.

Specific gravity (1.5) The ratio of the density of a substance to the density of water. It has no units.

Specific heat capacity (1.5) The quantity of heat needed to raise the temperature of one gram of a substance by one degree Celsius. Common units are J g^{-1} °C^{-1}. Also called specific heat.

Spectator ion (7.1) An ion that does not participate in a particular chemical reaction.

Spectrochemical series (21.11) A list of ligands arranged in order of their ability to produce a large crystal field splitting.

Spin quantum number (4.4) The quantum number m_s, which determines the direction in which the electron appears to be spinning. Its values are $+1/2$ and $-1/2$.

Spontaneous change (12.7) A change that occurs by itself without outside assistance.

Stability constant (16.3) See *Formation constant*.

Stainless steel (21.5) A type of steel alloy that is resistant to corrosion. It generally contains chromium and nickel alloyed with iron.

Standard atmosphere (9.1) A unit of pressure sufficient to support a column of mercury 760 mm high. It corresponds to exactly 101 325 Pa.

Standard cell potential, $\mathscr{E}°_{cell}$ (17.6) The potential of a galvanic cell at 25°C and 1 atm when all ionic concentrations are 1 M and the partial pressures of all gases involved in the cell reaction are 1 atm.

Standard entropy, $S°$ (12.10) The entropy that one mole of a substance has at 25°C and 1 atm.

Standard entropy change, $\Delta S°$ (12.10) Formally, $\Delta S° = S°_{final} - S°_{initial}$. $\Delta S° = $ (Sum of $S°$ of products) $-$ (Sum of $S°$ of reactants).

Standard free energy change, $\Delta G°$ (12.10) $\Delta G° = \Delta H° - T\Delta S°$. The magnitude and sign of $\Delta G°$ determines the size of the equilibrium constant for a reaction. $\Delta G° = $ (Sum of $\Delta G°_f$ of the products) $-$ (Sum of $\Delta G°_f$ of the reactants).

Standard free energy of formation, $\Delta G°_f$ (12.10) The change in free energy when one mole of a compound in its standard state is formed from its elements in their standard states.

Standard heat of formation, $\Delta H°_f$ (12.5) The enthalpy change associated with the formation of one mole of a substance in its standard state from its elements in their standard states.

Standard heat of reaction, $\Delta H°$ (12.5) $\Delta H° = $ (Sum of $\Delta H°_f$ of products) $-$ (Sum of $\Delta H°_f$ of reactants).

Standard reduction potential (17.7) The reduction potential of a half-reaction at 25°C when all ion concentrations are 1 M and all partial pressures of gases involved in the half-reaction are 1 atm.

Standard state (12.5) The natural state of a substance at 25°C and 1 atm pressure.

Standard temperature and pressure (STP) (9.5) The reference conditions for problems involving gases: 0°C and 101.325 kPa (1 atm).

Standardizing a solution (8.4) Measuring accurately the concentration of solute in a solution. A standard solution is one whose solute concentration is accurately known.

Standing wave (4.3) A wave whose peaks and nodes do not change position with time.

State (12.1) A particular set of conditions of pressure, temperature, volume, and amount of each component of a system.

State function (12.1) A quantity whose value depends only on the current state of the system and not on the system's prior history. The magnitude of the change in a state function depends only on the initial and final states of the system and is independent of the path followed between these states.

State variable (12.1) A state function.

States of matter (10.1) Solid, liquid, or gas.

Stereoisomers (21.9) Isomers that have the same atoms bonded to each other but differ in the way the atoms are arranged in space.

Steroid (24.4) A lipid with a complex ring structure that possesses very high biological activity.

Stock system (5.11) A system of nomenclature that uses Roman numerals within parentheses to indicate oxidation states.

Stoichiometry (2.1) A term that applies to the quantitative aspects of chemical composition and chemical reactions.

Stopcock (8.4) A valve (for example, at the end of a buret) that is used to control the flow of a liquid or sometimes a gas.

STP (9.5) Standard temperature and pressure: 0°C and 101.325 kPa (1 atm).

Strong acid (7.2) An acid that is 100 percent ionized in water.

Strong base (7.2) A base that is 100 percent dissociated in water.

Strong electrolyte (7.1) An electrolyte that is 100 percent dissociated in water.

Structural formula (2.6) A chemical formula that shows which atoms are bonded together in a molecule or polyatomic ion.

Structural isomers (23.2) Isomers that differ in the sequence in which their atoms are bonded together.

Sublimation (10.1) Conversion of a solid directly to a vapor without passing through the liquid state.

Subshell (4.3) All orbitals of a given shell that have the same value of the quantum number l.

Substitution reaction (23.6) A reaction in which one atom replaces another in a molecule.

Substitutional solid solution (11.2) A solid solution in which a particle of the solute (e.g., an atom or a molecule) replaces a particle of the solvent in the lattice of the solvent.

Substrate (24.2) The substance acted upon by an enzyme.

Supercooled liquid (10.12) A liquid at a temperature below its freezing point. An amorphous solid. A glass.

Supercooling (10.12) Cooling a liquid to a temperature below its freezing point.

Supercritical fluid (10.4) A substance as it exists at a temperature above its critical temperature.

Superoxide (18.5) A compound that contains the O_2^- ion.

Supersaturated (7.1) A term that describes a solution in which there is more dissolved solute than could normally exist in the solution if the solution were in contact with excess solute. Supersaturated solutions are unstable with respect to the spontaneous crystallization of excess solute.

Surface tension (10.1) A measure of the quantity of energy needed to expand the surface area of a liquid.

Surfactant (10.1) A substance that lowers the surface tension of a liquid and promotes wetting.

Surroundings (12.1) That which exists outside a system.

Suspension (11.1) Relatively large particles of one substance are suspended in another. For example, fine sand suspended in the wind, or mud suspended in water. Suspensions settle and separate upon standing.

Synthesis (8.2) The planned preparation of a compound from other materials.

Synthesis gas (19.1) A mixture of CO and H_2.

System (12.1) That particular portion of the universe upon which we wish to focus our attention.

$t_{1/2}$ (13.3) See *Half-life*.

Temperature (1.3) A quantity that is a measure of the hotness of a body. Temperature determines the direction of spontaneous heat flow (always from hot to cold).

Termination step (13.10) In a chain reaction, a step in the mechanism that removes free radicals and thereby terminates the chain.

Tertiary alcohol (23.8) An alcohol with the structure

$$R-\overset{\overset{\displaystyle R}{|}}{\underset{\underset{\displaystyle R}{|}}{C}}-OH$$

Tertiary structure (24.1) The way a coiled polypeptide chain folds to give a globular protein.

Tetrahedron (6.1) A four-sided pyramid with triangular faces. A tetrahedral molecule has an atom in the center of the tetrahedron with other atoms joined to it that are located at the four vertices.

Theoretical yield (2.12) The maximum amount of product(s) that could be formed from a particular mixture of reactants.

Theory (1.1) A tested explanation of the results obtained in many experiments.

Thermite reaction (18.7) The very exothermic reaction,

$$2Al + Fe_2O_3 \rightarrow 2Fe + Al_2O_3$$

Thermochemical equation (12.4) An equation whose coefficients are interpreted as representing numbers of moles of reactants and products, and which is accompanied by the energy change for the reaction.

Thermodynamic equilibrium constant (14.3) An equilibrium constant computed from the equation $\Delta G° = -RT \ln K$. For reactions involving gases, $K = K_p$.

Thermodynamics (12.1) The study of energy changes and the flow of energy from one substance to another.

Thio (20.2) In a chemical name, thio means sulfur in place of oxygen.

Third law of thermodynamics (12.10) For a pure crytalline substance, $S = 0$ at 0 K.

Tie line (11.9) A horizontal line on a boiling point diagram that connects the boiling point curve to the vapor composition curve. It allows one to read the boiling point of a solution having a particular composition and to determine the composition of the vapor that is in equilibrium with the boiling solution.

Titrant (8.4) The solution dispensed from a buret.

Titration (8.4) An analytical procedure in which a solution, generally of known concentration, is added gradually from a buret to another solution where the solutes react. When the completion of the reaction is signaled by an indicator, the volume of the solution added from the buret is read and recorded.

Tracer study (22.2) The use of radioisotopes to follow the fate of certain atoms during chemical reactions.

Trans- (21.9) A termed applied to a geometrical isomer in which two groups are located on opposite sides of some central imaginary line in the molecule or ion.

Transfer RNA (t-RNA) (24.6) A small RNA unit that carries amino acids to their proper location along m-RNA during protein synthesis.

Transition elements (3.3) Elements located between Groups IIA and IIIA in the periodic table.

Transition state (13.7) The brief moment during a reaction when the reactants have collided and are at the high point on the potential energy diagram for the reaction.

Transuranium elements (22.4) All the elements after uranium (Z = 92) in the periodic table. They are not naturally occuring.

Triad (21.1) Any one of the three horizontal sets of elements in Group VIII (e.g., Fe, Co, and Ni).

Triglyceride (24.4) An ester of glycerol and three fatty acid molecules.

Trigonal bipyramid (6.1) A geometrical figure composed of two trigonal pyramids (pyramids with three-sided faces) sharing a common triangular face. A trigonal bipyramidal molecule has an atom in the center of this triangular plane and is joined to five others that are located at the vertices of the trigonal bipyramid.

Triple bond (5.5) A covalent bond in which three pairs of electrons are shared between two atoms.

Triple point (10.13) The temperature and pressure at which the liquid, solid, and vapor states of a substance can coexist in equilibrium.

Triprotic acid (7.2) An acid that can furnish three H^+ per molecule.

Tritium (19.1) The radioactive isotope of hydrogen, ^3_1H, sometimes represented by the symbol T.

Trivial name (23.3) The common name for a compound.

Tyndall effect (11.1) The scattering of light at large angles by colloidal particles. This makes a light beam passing through a colloid visible when viewed from the side.

Uncertainty (1.2) The estimated quantity by which a measured quantity may be in error. For example, a reported measured length of 3.14 cm suggests an estimated uncertainty of ±0.01 cm.

Uncertainty principle (4.7) There are limits to our ability to simultaneously measure a particle's speed (or momentum) and its position.

Unified atomic mass unit (1.9) A unit of mass equal to one-twelfth of the mass of one atom of carbon-12.

Unit cell (10.8) The smallest portion of a lattice that can be used to generate the entire lattice by repeatedly moving the unit cell in directions parallel to its edges by distances equal to the lengths of those edges. In a less formal sense, it is the smallest portion of a crystal that can be repeated over and over in all directions to give the entire crystal lattice.

Universal gas constant (9.8)

$$R = 8.314 \text{ kPa dm}^3 \text{ mol}^{-1} \text{ K}^{-1}$$
$$= 8.314 \text{ J mol}^{-1} \text{ K}^{-1}$$

Unsaturated (7.1) A term applied to a solution that is capable of dissolving more solute.

Unsaturated hydrocarbon (23.1) A hydrocarbon having one or more carbon-carbon double or triple bonds in its structure.

Valence bond theory (6.4) A theory of covalent bonding that views a bond as being formed by the sharing of one pair of electrons between two overlapping atomic or hybrid orbitals.

Valence electrons (5.2) The electrons in an atom's outer shell.

Valence shell (4.6) The shell with highest n that is occupied by electrons.

Valence Shell Electron Pair Repulsion Theory (6.2) VSEPR theory. A theory used to predict molecular structure. It is based on the idea that electron pairs in the valence shell of an atom stay as far apart as possible.

Van der Waals equation of state (9.11) A modified form of $PV = nRT$ in which corrections for the finite volume of the gas molecules and intermolecular attractions are applied to the measured volume and pressure of a real gas.

van't Hoff factor (11.12) The ratio of the measures ΔT_f to the ΔT_f calculated assuming the solute in a solution to be a nonelectrolyte.

Vapor pressure (10.4) The pressure exerted by the vapor above a liquid. Usually, this term means the *equilibrium vapor pressure*, which is the vapor pressure when the liquid is in equilibrium with its vapor.

Vapor pressure curve (10.4) A graph of equilibrium vapor pressure versus temperature.

Vibrational frequency (5.7) The frequency with which atoms joined by a bond vibrate back and forth, toward and away from each other.

Volt (V) (17.6) The SI unit of electric potential or emf. $1 \text{ V} = 1 \text{ J C}^{-1}$.

Voltaic cell (17.5) Another term used to describe a *galvanic cell*.

Volume The space occupied by a sample of matter.

Volumetric analysis (8.4) An analytical procedure that makes use of reactions in solution, where volumes and concentrations of solutions are carefully measured.

Volumetric flask (11.3) A special flask having a long thin neck that has a mark etched around it. When filled to this mark with water, it contains precisely the volume specified. It is used to prepare solutions of accurately known concentration.

VSEPR Theory (6.2) See *Valence Shell Electron Pair Repulsion Theory*.

Vulcanization (23.10) Treatment of natural rubber with sulfur, which forms sulfur bridges between adjacent polymer strands.

Wave function (4.3) A mathematical function represented by the symbol ψ and obtained by solution of a wave equation. It describes the shape and size of an orbital and can be used to calculate the energy of an electron in that orbital.

Wave mechanics (4.3) A theory of atomic structure based on the wave properties of matter.

Wavelength (4.1) The distance between successive peaks in a wave.

Weak acid (7.2) An acid that is less than 100 percent dissociated in water.

Weak base (7.2) A base that is less than 100 percent dissociated in water.

Weak electrolyte (7.1) An electrolyte that is less than 100 percent dissociated in water.

Weight (1.4) The force with which an object having a given mass is attracted to the earth or some other object that it may be near. Sometimes the term weight is used to mean mass, even though they are not the same.

Wetting (10.1) The spreading of a liquid across a solid surface.

White phosphorus (6.8) A very reactive form of phosphorus that consists of P_4 molecules.

Work (12.2) The energy involved in moving an opposing force a given distance.

X-ray diffraction (10.7) The diffraction of X rays by the atoms in a crystalline solid. A technique for studying the structures of crystalline solids.

Zinc carbon dry cell (17.12) The ordinary dry cell in which zinc serves as the anode and MnO_2 serves as the cathode reactant.

Zone refining (20.5) A method for producing very high purity solids. A thin cross section of a bar of the solid is melted and the molten zone is moved slowly from one end to the other. Impurities collect in the molten zone.

INDEX

Absolute temperature, 314
Absolute zero, 17, 314
Absorbed dose, 803
Absorption, 27
Accuracy, 7
Acetic acid, structure of, 53
Acetone, 856
Acetylene:
  bonding in, 207
  preparation of, 76, 696
Acetylide ion, 696
Acidic anhydrides, 234
Acidic hydrogen, 233
Acid ionization constant, 543
Acid rain, 234, 729
Acids:
  in aqueous solution, 232
  Brønsted-Lowry, 238
  ionization of, 543
  Lewis, 246
  organic, 858
  strong, 233
  table of, 235
  weak, 233
Acid salts:
  formation of, 236
  nomenclature of, 173
Acids and bases, strengths of, 241
Actinides, 89
Activated charcoal, 692
Activated complex, 497
Activation energy, 495
  measurement of, 498
  potential energy diagram, 496
Activities, 415
  in Nernst equation, 626
Activity series, 265
Actual yield, 63
Addition compound, 166
Addition polymers, 862

Adiabatic processes, 436
Adsorption, 27, 391
Alcohols, 853
  oxidation of, 857, 858
Aldehydes, 856
  oxidation of, 858
  reduction of, 857
Algae blooms, 723
Aliphatic hydrocarbons, 827
Alkali, word origin, 89
Alkali metals, 89, 657
  biological importance, 657
  hydrides of, 686
  ionic radii, 668
  in liquid ammonia, 662
  physical properties of, 660
  reaction with oxygen, 662
  reaction with water, 660
  reduction potentials of, 660, 661
  in sea water, 658
  sources of, 658
Alkaline battery, 632
Alkaline earth metals, 89, 665
  biological importance, 665
  compounds, solubilities of, 669
  ionic radii, 668
  physical properties of, 667
  reaction with water, 668
  reduction potentials, 668
Alkanes, 828
  bonding in, 829
  reactions of, 852
  structures, 829
  table of, 828
Alkenes, 830
  hydrogenation of, 850
  reactions of, 850
Alkyl groups, 838
  table of, 839
Alkyl halides, 853

Alkynes, 830, 831
  reactions of, 850
Allotropism, 210
Alloys, 393
Alnico, 673
Alpha helix, 871
Alpha particles, 96, 798
Alpha radiation, 96, 798
Alum, 676
Aluminate ion, 672
Aluminosilicates, 672
Aluminum, 671
  amphoteric behavior, 652, 674
  Hall process, 606
  ore purification, 643, 672
  production of, 672
Aluminum carbide, hydrolysis of, 689
Aluminum halides, structure of, 675
Aluminum hydroxide, 675
Aluminum ion:
  as Brønsted acid, 241
  hydrated, 401
Aluminum oxide, 674
  electrolysis of, 606
Aluminum sulfate, 676
Amalgam, 642
Amide ion, 705
Amides, 861
Amines, 860
Amino acid, 869
  table of, 870
Ammonia:
  acid/base properties, 705
  complex ion with copper, 770
  laboratory preparation, 706
  Le Châtelier's principle and synthesis of, 525
  as Lewis base, 247
  oxidation of, 701, 706

Ammonia (*Continued*)
  reaction with bleach, 707
  structure of, 705
  synthesis of, 704
  valence bond description, 196, 201
Ammonium ion, 706
  bonding in, 205
Ammonium nitrate, 708
Amorphous solids, 376
Ampere, 12, 92, 609
Amphiprotic, definition of, 240
Amphoteric, definition of, 240, 652
Amplitude, 104
Analysis, chemical, 291
Analytical balance, 18
Anion, definition of, 146
Anode, 90
  charge on:
    in electrolytic cell, 599
    in galvanic cell, 613
  chemical reaction at, 599
Anodized aluminum, 674
Antibonding orbitals, 214
Antimony, 84
  molecular structure, 213
Aquamarine, 665
Aqua regia, 263, 713
Aromatic hydrocarbons, 827, 845
Arrhenius, Svante, 225
Arrhenius acids and bases, 232
Arrhenius equation, 498
Arsenic, 84
  molecular structure, 213
Arsine, 690
Asbestos, 748
Aspirin, 858
Association in solutions, 428
Astatine, 733. *See also* Halogens
Atmosphere, standard, 307
Atomic emission spectrum, 106
Atomic number, 87, 95, 98
  discovery of, 110
Atomic radii, from crystal structures, 371
Atomic size, 131
  table, 134
  variation in periodic table, 132
Atomic spectrum, 106
Atomic structure, definition of, 90
Atomic theory, 30
Atomic weight, 33
  calculation from isotopic abundances, 99

Atomization energy, 455
Attractive forces, *see* Intermolecular attractive forces
Autoionization reactions, 240
Autoionization of water, 537
Avogadro's number, 43
  in calculations, 49
Avogadro's principle, 321
Azeotropes, 417
Azimuthal quantum number, 117

Baking powder, 676
Baking soda, 73, 665
Balance, 18
Balancing equations:
  by inspection, 57
  ion-electron method, 253
  oxidation number change method, 251
Balmer, Johann, 107
Balmer equation, 107
Balmer series in hydrogen spectrum, 107
Band of stability, 809
Barium, 665
  flame test, 667
  see also Alkaline earth metals
Barium sulfate, 670
Barometer, 307
Bartlett, Neil, 743
Base ionization constant, 544
Bases:
  in aqueous solution, 234
  Brønsted-Lowry, 238
  ionization of, 544
  Lewis, 246
  strong, 234
  table of, 235
  weak, 235
Basic anhydrides, 235
Basic oxygen process, 650
Bauxite, 672
Becquerel, Antoine Henri, 798
Becquerel (unit), 803
Bends, 409
Benzene, 845
  bonding in, 218, 846
Benzopyrene, 847
Beryl, 665
Beryllium, 665
  amphoteric behavior, 668
  preparation of, 666
  see also Alkaline earth metals
Beryllium chloride, structure of, 669

Beryllium copper alloy, 666
Bessemer converter, 649
Beta particles, 96, 798
Beta radiation, 96, 798
Bidentate ligands, 774
Bimolecular collisions, 489
Binary acids:
  nomenclature of, 171
  strengths of, 246
Binary compounds, nomenclature of, 169
Binding energy, 815
Biochemistry, definition, 869
Biphenyl, 849
Bismuth, 671
  compounds of, 677
  hydrolysis of compounds, 678
  oxidation states of, 676
  production of, 673
Bismuthate ion, 678
Bisulfite ion, 729
  as reducing agent, 298
Black phosphorus, 719
Blast furnace, 647
Bleach, 739
  production by electrolysis, 605
  reaction with ammonia, 707
Body-centered cubic unit cell, 369
Bohr, Neils, 111
  model of hydrogen atom, 111
Boiler scale, 695
Boiling point, 360
  effect of hydrogen bonding on, 361
Boiling point diagram, 417
Boiling point elevation, 419
  table of constants for, 420
Bomb calorimeter, 440
Bond angle, 179
  in octahedral molecule, 181
  in planar triangular molecule, 180
  in tetrahedral molecule, 180
  in trigonal bipyramidal molecule, 180
Bond dipoles, 191
Bond distance, 155
Bond energy, 154, 155
  average, table, 456
  calculation from ΔH_f°, 455
  dependence on bond order, 161
Bonding molecular orbital, 214
Bond length, 154, 155
  dependence on bond order, 160
Bond order, 160

bond energy and, 161
bond length and, 160
bond vibrational frequency and, 162
Bond properties, dependence on bond order, 161
Bond vibrational frequency, 161
measurement of, 162
Boric acid, 652
Born-Haber cycle, 152
Boron trichloride:
molecular shap, 183
reaction with ammonia, 166
Botulinus toxin, 877
Boyle, Robert, 311
Boyle's law, 311
Bragg, Lawrence, 365
Bragg, William, 365
Bragg equation, 365
Brand, Hennig, 718
Brass, 771
Breeder reactor, 818
Brimstone, 724
Brine, 603
electolysis of, 604, 605
Bromine, 84, 733
preparation, 736
see also Halogens
Brønsted, J. N., 238
Brønsted-Lowry acids and bases, 237
Bronze, 771
Buffers, 555
effectiveness of, 557
preparation of, 556
Buret, 292
Butane, structure, 830

Cadmium, 770
plating with, 772
Calcining, 666
Calcium, 665
flame test, 667
ores of, 665
reaction with water, 262
see also Alkaline earth metals
Calcium carbide, 76, 696
Calcium carbonate, 669
Calcium hydride, 686
Calcium sulfate, 670
Calibration of instruments, 7
Calorie, 21
Canal rays, 94
Candela, 12

Canned heat, 79
Carbides, 695
Carbinol carbon, 854
Carbohydrates, 878
Carbon, 691
abundance, 691
allotropes of, 210, 211, 691
in metallurgy, 647
Carbonates, reaction with acid, 694
Carbon black, 692
Carbon dioxide, 693
as Lewis acid, 248
phase diagram, 382
uses of, 694
Carbon disulfide, 696
Carbonic acid, 694
Carbon monoxide, 692
Carbon-12, 98
Carbonyl compounds, 693
Carbonyl group, 856
Carborundum, 695
Carboxyl group, 858
Carnallite, 663, 665
Cassiterite, 673
Catalysts, 501
heterogeneous and homogeneous, 502
Catalytic converter, 503
oxidation of SO_2 in, 517
Catenation, 687
Cathode, 90
charge on:
in electrolytic cell, 599
in galvanic cell, 613
chemical reaction at, 599
Cathode rays, 90
Cathode ray tube, 91
Cathodic protection, 771
Cation, 146
Caustic soda, 663
Cell membranes, 886
Cell potential, 603
effect of concentration on, 635
Cell reaction, 599
Cellulose, 882
Celsius temperature scale, 16
Centrifuge, 388
Chadwick, J., 97
Chain reaction, 504
nuclear, 817
Changes of state, 375
Charge cloud, 128
Charge–to–mass–ratio, 91
of electron, 92

of positive particles, 94
Charge transfer process, 655
Charles, Jacques, 313
Charles' law, 313
Chelate, 774
Chemical analysis, 291
Chemical bond, definition, 143
Chemical change, 26
Chemical equation, 34
balanced, 35
balancing by inspection, 57
balancing by ion–electron method, 253
balancing by oxidation number change, 251
calculations from, 58
coefficients in, 35
ionic, 230
limiting reactant calculations, 61
molecular, 229
net ionic, 231
reactants and products in, 34
Chemical formula, 34
interpretation of, 42
types of, 52
Chemical kinetics, 478
Chemical property, 24
Chemical reaction, 26
Chemical symbol, 33
Chemistry, definition of, 2
Chiral isomers, 780
Chlorate ion, 739
Chloric acid, 739
Chlorine, 733
disproportionation in water, 738
laboratory preparation, 735
preparation, 602, 735
reaction with CaO, 739
uses of, 603
see also Halogens
Chlorites, 739
Chloroacetic acid, 244
Chlorofluorocarbons, effect on ozone layer, 699
Chloroform, 852
Chlorophyll, 875
absorption spectrum of, 655
Chlorous acid, 739
Cholesterol, 887
Chromate ion, 765
color of, 151
as oxidizing agent, 296
Chromatography, 27
Chromic acid, 765

Chromium, 765
Chromium(III) ion, amphoteric
     properties, 765
Chymotrypsin, 876
Cis isomers, 779
Citric acid, 858
Clathrates, 743
Cobalt, 768
Cobalt glass, 660, 768
Coefficients in chemical equation, 35
Coenzymes, 876
Coinage metals, 769
Coke, 646, 691
Colligative properties, 418
     solutions of electrolytes, 426, 427
     vapor pressure lowering, 411
Collision theory, 489
     effective collisions, 493
Colloidal dispersions, 389
Colloids, 389
     destabilization of, 392
     stabilization of, 391
     types of, 390
Color wheel, 654
Combined gas law, 316
Combustion, 267
Common ion effect, 587
     solubility and, 587
Complexes, 590, 773. *See also*
          Coordination compounds
Complex ions, 590, 773
     equilibria, 590
     formation constants, table, 590
     instability constants, table, 590
     solubility and, 591
     see also Coordination compounds
Complexing agent, 592
Compound, definition of, 25
Compressibility of states of matter,
     343
Concentrated solution, 64
Concentration, 64
Concentration cell, 627
Concentration units:
     conversions among, 396
     mass fraction, 395
     mass percent, 395
     molar concentration, 64, 288, 395
     mole fraction, 395
     mole percent, 395
     parts per million, 287, 395
     percent by mass, 287
Condensation polymers, 862
Condensation reaction, 859
Condensed formulas, 830

Conductivity apparatus, 143
Conjugate acids and bases, 238, 239
     relative strengths of, 560
Constructive interference, 115
Contact process, 730
Continuous spectrum, 105
Contributing structures, 164
Conversion factors, 31, 32
     in chemical calculations, 45
Cooling curves, 376
Coordinate covalent bond, 165
     valence bond theory and, 204
Coordination compounds, 773
     colors of, 770, 773, 790
     crystal field theory, 788
     geometrical isomers, 779
     isomers of, 779
     nomenclature, 776
     optical isomers, 780
     stereoisomers, 779
     structures of, 776
     valence bond theory, 784
Coordination number, 776
Copolymer, 862
Copper:
     complex ions, 770
     electrolytic refining of, 607
     reaction with nitric acid, 260, 769
     sulfide ore, 643, 644
Copper(II) sulfate, 34
Core electrons, 121
Corrosion of metals, 260
Cortisone, 887
Corundum, 674
Coulomb, 92
Coulomb's law, 4
Coulometer, 610
Covalent bond, 154
     energy diagram for formation of,
          154
     molecular orbital theory of, 213
     valence bond theory of, 194
Covalent crystals, 373
Crick, Francis, 366
Critical mass, 818
Critical point, 357
Critical temperature and pressure,
     357
Cryolite, 606
Crystal field splitting, 790
Crystal field theory, 788
     splitting patterns for various
          geometries, 791
Crystal lattice, 367
Crystalline solids, 365

Crystal systems, table, 368
Crystal types, 371
     table, 373
Curie (Ci), 803
Curie, Marie, 698
Curie, Pierre, 798
Cyanides, 696
Cycloalkanes, 842
Cyclohexane, structures of, 844
Cyclopropane, 842
Cyclotron, 811

Dacron, 855
Dalton, John, 30
Dalton's law of partial pressures,
     318, 333
Data, 3
Dating of archaeological objects, 804
Daughter isotope, 799
d-block elements, 756
de Broglie, Louis, 114
Decompression sickness, 409
Delocalized molecular orbitals, 218
Denaturation of protein, 871
Density, 21
Deoxyribose, 881
Derived units, 11
Desalination of sea water, 425
Desiccant, 720
Destructive distillation, 685
Destructive interference, 115
Detergents, 347
     action of, 401, 402
     phosphates in, 722, 723
     synthetic, 402
Deuterium, 683
Dextrorotatory, 783
Dialysis, 422
Diamagnetism, 120
Diamond, 84, 692
     from graphite, thermodynamics
          of, 464
     structure of, 210
Diaphragm cell, 604, 605
Diatomic molecules, 84
Dichromate ion, 764
     color of, 151
     as oxidizing agent, 296
Diffraction, 114
Diffraction pattern, 114
Diffusion, 328
     of gases, 328
     rates in liquids and solids, 344,
          345
Dihydrogen, 683

Dilute solution, 64
Dilution, 67
Dinitrogen, 703
Dinitrogen pentoxide, 712
Dinitrogen tetroxide, 709
Dinitrogen trioxide, 711
Dioxygen, 696
Dipeptide, 871
Diphosphoric acid, 722
Dipole, definition, 167
Dipole-dipole attractions, 349
Dipole moment, 167
    molecular structure and, 190
Diprotic acids, 233
Dirac, Paul, 116
Disilane, 687
Dispersed phase, 390
Dispersing medium, 390
Disproportionation, 706
Dissociation constant, 543
    for water, 537
    see also Ionization constant
Distillation, 27
Distribution of molecular speeds,
        331
Disulfide bridge, 873
DNA, 887, 888, 890, 891, 892
Dolomite, 665, 669
Domaine, 763
d orbitals, shapes of, 130
Double bonds, 155
    valence bond description of, 206
Double replacement reactions, 229.
        See also Metathesis
Double salt, 676
Downs cell, 603
Dry cell, 631
Dry gas, 431
Dry ice, 347, 693
Ductility, 82
Dynamic equilibrium, 227, 512. See
        also Equilibrium
Dynamite, 856

EDTA, 774
Effective collisions, 493
Effective nuclear charge, 132
Effusion, 328
Eicosane, 143
Einstein, Albert, 111, 817
Eka-silicon, 87
Electric current, unit of, 12
Electric furnace, 719
Electrodes, in gas discharge tubes,
        90

Electrolysis, 599
    aqueous copper sulfate, 600
    aqueous copper(II) bromide, 601
    aqueous sodium bromide, 600
    aqueous sodium chloride, 604
    aqueous sodium sulfate, 601
    molten sodium chloride, 599, 602
    need for electrolyte in, 602
    quantitative aspects of, 608
Electrolytes, 225
    interionic attractions in solutions,
        429
    properties of solutions of, 425
Electrolytic cells, 597, 598
Electrolytic conduction, 598
Electromagnetic radiation, 104
Electromagnetic spectrum, 105
Electromotive force, 613
Electron:
    charge, 93
    charge–to–mass ratio, 92
Electron affinity, 136
    of representative elements, table,
        137
Electron capture, 809
Electron cloud, 128
Electron configuration:
    definition, 120
    of elements, table, 122
    periodic table and, 124
    unexpected for Cu and Cr, 124
    valence shell, 127
Electron density, 128
Electron–dot formulas, 145
Electronegativity, 166
    table of, 168
Electronic structure, 104, 120
Electron pair bond, 155
Electrons, 50
    paired, 121
    spin of, 120
Electroplating, 607
Electropositive, 167
Electrostatic attractions, 146
Electrovalent bonds, 146
Element, definition of, 25
Elementary process, 400
Emeralds, 748
Emf, 613
Emission spectrum, 106
Empirical formula, 52
    calculation of, 53
    from combustion analysis, 55
Emulsifying agent, 391
Enantiomers, 781

Endothermic change, 20
Endpoint, 292
Energy:
    definition of, 18, 19
    kinetic, 19
    potential, 19
    units of, 20
Energy level diagram for
        multielectron atom, 119
Energy levels, 111
Enthalpy, defined, 446
Enthalpy change, 446
    relationship of ΔH to ΔE, 446
Enthalpy diagram, 449
Enthalpy of formation, 450. See also
        Heat of formation
Entropy, 460
    changes in, 460
    spontaneity and, 460
    units of, 461
Enzymes, 875
    inhibition, 876
    mechanism of action, 876
Epsom salts, 670
Equation of state, 437
    for ideal gas, 324
Equilibrium:
    chemical, 512
    heterogeneous, 520
    weak electrolytes, 227
Equilibrium calculations, 525
    approximations in, 531
Equilibrium constant, 512
    in calculations, 525
    from cell potential, 623
    conversion between K_p and K_c,
        518
    and $\Delta G°$, 516
    interpretation of, 514
    K_p and K_c, 514
    K_w, 537
Equilibrium law, 512
Equilibrium vapor pressure, 355. See
        also Vapor pressure
Equivalence, chemical, 31
Equivalence point, pH:
    strong acid-strong base titration,
        569
    weak acid-strong base titration,
        571
    weak base-strong acid titration,
        572
Ester formation, mechanism, 808
Esterification, 859
Esters, 858

Esters (*Continued*)
  formation of, 859
  saponification of, 859
Estradiol, 886
Estrone, 886
Ethanol, 854
Ethers, 860
Ethyl alcohol, azeotropic
    composition, 418
Ethylene, bonding in, 206
Ethylenediamine, 774
Ethylene glycol, 855
Ethyl group, 838
Eutrophication, 723
Evaporation, 347
  cooling effect from, 347
  factors that affect rate of, 347, 348
Exact numbers, 10
Excluded volume, 335
Exothermic change, 20
Exponential notation, 7
Extensive properties, 21

Face-centered cubic unit cell, 367
Factor-label method, 10
Family of elements, 88
Faraday, Michael, 90, 608
Faraday ($\mathcal{F}$), 608
Fats, 884
Fatty acids, 882
  table of, 883
Fermentation of glucose, 854
Fermi, Enrico, 817
Ferricyanide ion, 787
Ferromagnetism, 120, 763
Filtrate, definition of, 229
Filtration, 230
  separation of suspensions, 388
First coordination sphere, 774
First law of thermodynamics, 438
Fission, 810, 817
Flotation, 643
Fluorescent lamps, 90
Fluorine, 733
  preparation, 735
  see also Halogens
Formation constant, 591
Formula unit, 50
Formula weight, 50
Fractional crystallization, 408
Fractional distillation, 416, 417
Franklin, Rosalind, 366
Frasch, Herman, 724
Frasch process, 724
Free energy, 463

and equilibrium, 469
and spontaneity, 463
Free energy of formation, 466, 468
Free radical, 503
Freezing point, 363
Freezing point depression, 419
  table of constants for, 420
Freons, effect on ozone layer, 699
Freon-12, 77, 852
Frequency, 104
  SI unit, 105
Frisch, Otto, 817
Fructose, 879
Fuel cell, 634
Functional groups, 850
  table of, 851
Fundamental particles, 92
Fusion:
  defined, 363
  nuclear, 820

Galena, 673
Gallium, 671
Galvanic cells, 597, 611
Galvanizing, 770
Gamma radiation, 96, 798
Gamma rays, 96, 798
  position in electromagnetic
      spectrum, 105
Gangue, 643
Gas constant, 324
Gas discharge tubes, 90
Gases:
  collection over water, 319
  kinetic theory of, 329
  liquefaction of, 356
  real, 334
Gas laws:
  combined, 316
  combining volumes, 321
  ideal gas law, 324
  partial pressures, 318, 333
  pressure–temperature, 315
  pressure–volume, 311
  rates of effusion, 328
  volume-temperature, 313
Gasohol, 418
Gasoline, from methanol, 686
Gay-Lussac, Joseph, 315
Gay-Lussac's law, 315
  of combining volumes, 321
Geiger–Müller counter, 801, 802, 803
General Conference of Weights and
      Measures, 11
Genetic code, 893

table, 894
Genetic disease, 895
Geometrical isomers, 779
  organic, 834
Germane, 690
Germanium, properties of, 87
Gibbs free energy, 463. *See also* Free
    energy
Glass, 376
  etching with HF, 737
Glass electrode, 631
Glucose, 879
Glycerol, 855
Glycoside linkage, 881
Gold:
  ore treatment, 642
  oxide, thermal decomposition, 645
  reaction with aqua regia, 713, 769
  from refining of copper, 607
Gold leaf, 82
Graham, Thomas, 328
Graham's law of effusion, 328
Gram, 14
Graphite, 84, 692
  electrical conductivity, 219
  structure, 211
  thermodynamics of conversion to
      diamond, 464
Gray (Gy), 803
Gound state, 117, 120
Groups in periodic table, 88
Gypsum, 665, 670

Haber process, 704
Halate ions, 739
Half-cell, 612
Half-life:
  defined, 487
  first-order reactions, 487
  radioactive, 801
  second-order reactions, 488
Half-reactions, balancing equations
    with, 253
Hall, Charles, 606
Hall process, 607
Halogens, 733
  compounds of hydrogen, 737
  compounds with metals, 736
  compounds with nonmetals,
      table, 740
  hydrocarbon derivatives, 852
  hydrogen compounds,
      preparation, 689
  hydrolysis of nonmetal halides,
      742

oxidizing abilities, 735
as oxidizing agents, 267
physical properties of, 734
sources of, 734
uses, 733
Hardness ions, 695
Hard water, 35, 695
Heat:
  of formation, 450
    from bond energies, 457
    standard, 451
    standard, of gaseous atoms, table, 455
    standard, table, 452
  of reaction:
    combining thermochemical equations, 448
    constant pressure, 446
    constant volume, 444
    Hess's law, 448
    standard, from ΔH_f°, 452
  of solution, 403
    hydration energy, 406
    liquids in liquids, 403, 404
    solids in liquids, 405
  of vaporization, 351
    table, 353
Heat capacity, 437
Heat content, 446
Heating curves, 375
Heavy water, 684
Heisenberg uncertainty principle, 128
Hematite, 643
Heme group, 874
Henderson-Hasselbalch equation, 556
Henry's law, 410
Hertz, 105
Hess's law of heat summation, 448
Heterocycles, 861
Heterogeneous mixtures, 25
High spin complexes, 791
Hindenburg disaster, 4
Hodgkin, Dorothy, 366
Homogeneous mixtures, 25
Homologous series, 829
HTH, 739
Hund's rule, 123
Hybrid orbitals, 196
  orientations of, table, 198
Hydrates, 34
Hydration:
  of ions, 225
  of solute in aqueous solutions, 401

Hydration energy, 406
  alkali halides, table, 407
Hydrazine, 687, 707
Hydride ion, 686
Hydrides, 686
  catenation among, table, 687
  combustion of, 701
  nonmetal:
    preparation of, 688
    thermodynamic properties, 688
Hydro acids, 172
Hydrocarbons, 827
  aliphatic, 827
  alkanes, 828
  alkenes, 830
  alkynes, 830
  aromatic, 827, 845
  combustion of, 269
  cyclic, 842
  cycloalkanes, 842
  cycloalkenes, 844
  derivatives, 849
  saturated, 827, 828
  unsaturated, 830
  uses, 828
Hydrochloric acid, 737
Hydrofluoric acid, 735
Hydrogen, 683
  abundance, 683
  atomic spectrum, table, 109
  Bohr theory of, 111
  compounds of, 686
  energy diagram for H_2, 154
  isotopes of, 683
  laboratory preparation, 685
  in metallurgy, 646
  molecular orbital description of H_2, 216
  physical properties, 683
  preparation of, 684
  uses of, 603, 685
  valence bond description of H_2, 195
Hydrogenation, 686
Hydrogen bonding, 350
  effect on boiling point, 361
  effect on heat of vaporization, 353
  in HF, 362
  in ice, 350
  in proteins, 872
Hydrogen chloride, preparation of, 689, 737
Hydrogen cyanide, 696
Hydrogen difluoride ion, 686
Hydrogen economy, 690
Hydrogen electrode, 615

Hydrogen fluoride, 737
  bonding in, 195
Hydrogen halides, 737
Hydrogen peroxide, 702
Hydrogen sulfide:
  by hydrolysis of thioacetamide, 732
  separation of metal ions, 585
Hydrolysis:
  of metal ions, 241
  of salts, 559
    calculations, 563
    of polyprotic acids, 566
    of strong acids and strong bases, 560
    of strong acids and weak bases, 562
    of weak acids and strong bases, 561
    of weak acids and weak bases, 565
Hydrolysis constant, 563
Hydrometer, 634
Hydronium ion, 226
Hydrophilic molecules, 402
Hydrophobic molecules, 402
Hydroxide ion, as Lewis base, 248
  in basic solutions, 234
Hydroxyapatite, 523
Hydroxylamine, 707
Hypochlorite ion, 739
  from electrolysis of brine, 605
Hypochlorous acid, 739
Hypohalites, stability of, 739
Hypothesis, 4

Ice, hydrogen bonding in, 350
Ideal gas, 312
Ideal gas law, 324
Ideal solution, 404
  Raoult's law and, 414
Indicators, 232, 572
  how they function, 573
  table of, 573
  in titrations, 292
Indium, 671
Induced dipole, 351
Inert gases, 89. See also Noble gases
Infrared absorption spectrum, 162
Infrared radiation, 105
Inhibitor, 503
Inner orbital complexes, 785
Inner transition elements, 88, 756
Inorganic compound, 169
Instability constant, 590

Instantaneous dipole, 351
Intensive properties, 21
Intermolecular attractive forces, 306, 349
  dipole-dipole forces, 349
  heat of vaporization and, 352
  hydrogen bonding, 350
  London forces, 351
  in real gases, 335, 337
Internal energy:
  defined, 438
  relationship of ΔE to ΔH, 446
International System of Units:
  base units, 11
    table of, 12
  derived units, 11
  SI prefixes, table of, 12
Interstitial solid solutions, 394
Iodic acid, 739
Iodine, 84, 733
  preparation, 736
  thyroid cancer treatment, 98
  see also Halogens
Ionic bonding, 146
Ionic compounds, 50, 144
  dissociation of, in water, 226
  factors influencing formation of, 151
  properties of, 143
  writing formulas of, 148
Ionic crystals, 372
Ionic equation, 230
Ionic potential, 653
Ionic radii, from crystal structures, 371
Ionic reactions, 229
Ionization constant:
  polyprotic acids, table, 551
  for water, 537
  weak acids, 543
  weak acids and bases, table, 545
Ionization energy, 134
  of first 20 elements, table, 135
  variation in periodic table, 136
Ionization isomerism, 779
Ionization reaction, 226
Ion product, value of, in controlling precipitation, 584
Ion product constant, 537
Ions, 50
  of representative elements, table, 148
  sizes of, 133
    table of, 134

Iron, 767
  cast, 648
  corrosion of, 767
  ferromagnetisms of, 763
  ores of, 643
  oxides of, 767
  pickling of, 699
  pig, 648
  production in blast furnace, 647
  reaction with acids, 259, 767
  refining of, 649, 650
Iron pyrite, 73
Isomers, 779
Isothermal processes, 437
Isotonic solutions, 425
Isotope dilution analysis, 806
Isotopes, 98
IUPAC, 776

Joule, 20

K-capture, 819
kelvin, 16
Kelvin temperature scale, 16
Ketones, 856
  reduction of, 857
Kilogram, 12
Kinetic energy, 19
Kinetic molecular theory, 309
Kinetics of reactions, 478

Laboratory reagents, table of, 68
Lanthanide contraction, 133
  effects of, 762
Lanthanides, 88
Latin prefixes, 171
Lattice, defined, 367
Lattice energy, 153
  alkali halides, table, 407
Laughing gas, 708
Lavoisier, Antoine, 29
Law:
  of conservation of energy, 20
  of conservation of mass, 29
  of definite composition, 29
  of definite proportions, 29
  definition of, 3
  of multiple proportions, 30
Lead, 671
  amphoteric behavior, 677
  oxidation states of, 676
  in paint, 677, 678
  production of, 673
  reaction with chlorine, 677

Lead dioxide, 677
Lead oxide, reduction of, 646
Lead storage battery, 633
Le Châtelier, Henry, 359
Le Châtelier's principle, 359
  chemical equilibrium, 522
  phase changes, 359
  pressure and solubility of gases, 410
  solubility and temperature, 408
Leclanché cell, 631
Levorotatory, 783
Lewis, Gilbert N., 145
Lewis acids and bases, 246
Lewis structures, 145
  rules for constructing, 157
Lewis symbols, 145
  for A-group elements, 146
Ligand, 182, 590
Ligands:
  bidentate, 774
  monodentate, 774
  polydentate, 774
  table of, 775
Lime, 670
Limestone, 665, 669
  decomposition of, 693
Limestone caverns, formation of, 234, 694
Limiting reactants, 61
  in ionic reactions, 290
Linear molecular shape, 179
Line spectrum, 106
Lipids, 882
Liquid crystals, 374
Liquid nitrogen, 704
Liquids:
  compressibility of, 343
  diffusion in, 344
  evaporation of, 347
  heats of vaporization, 352
    table, 353
  surface tension of, 345
  vapor pressures of, 354
  volume and shape, 345
  wetting ability, 347
Litre, 14, 15
Lithium, 657
  flame test for, 660
  reaction with oxygen, 662
  see also Alkali metals
Lithium carbonate, 665
Lithium fluoride, crystal structure, 149

Lithium nitride, 663
London, Fritz, 351
London forces, 351
    effect of molecular structure, 352
Lone pair, 184
    effect on molecular structure, 185–188
Low spin complexes, 791
Lowry, T. M., 238
Luminous intensity, unit of, 12
Lye, 663
Lyman series, 108

Macromolecules, 869
Magic numbers, 810
Magnesium, 665
    production of, 607, 666
    reaction with nitrogen, 668
    reaction with oxygen, 667
    see also Alkaline earth metals
Magnesium nitride, hydrolysis of, 689
Magnesium sulfate, 670
Magnetite, 643, 767
Main transition elements, 756
Malleability, 82
Manganate ion, 766
Manganese, 766
Manganese dioxide, 766
Manganese nodules, 641
Manometer fluids, effect of density, 310
Manometers, 308
Markovnikov's rule, 851
Mass:
    definition of, 17
    measurement of, 18
    units of, 14
Mass action expression, 513
Mass defect, 815
Mass fraction, 395
Mass number, 98
Mass percent, 395
Mass spectrometer, 94
Mass spectrum, 95
Matter, 17
    classification of, 27
Maxwell-Boltzmann distribution, 330
Mean free path, 345
Mean square speed, 332
Measurement, 5
    accuracy of, 7
    laboratory units, 14
    precision of, 6

significant figures and, 5
    uncertainty in, 6
    units of, 11. See also International System of Units; Units
Mechanism of reaction, 490
    rate determining step in, 491
    using labeled isotopes, 808
Melting point, 363
Mendeleev, Dmitri, 85
Mercury, 770, 771
    test for, 772
Mercury cell, 605
Mercury(I) chloride, 772
Mercury(II) amido chloride, 772
Mercury(II) chloride, 772
Mercury(II) oxide, thermal decomposition, 644
Mercury vapor lamps, 107
Meitner, Lise, 816
Messenger RNA, 892
Meta, 848
Metal compounds:
    colors of, 654
    thermal decomposition of, 644
Metal halides, 736
Metal ions, reaction with water, 240
    tables of, 148
Metallic character, periodic trends, 650
Metallic conduction, 597
Metallic crystals, 373
Metallic luster, 82
Metallurgy, 641
Metal-nonmetal bonds, ionic character of, 652
Metals, 82
    activity series for, 266
    ease of oxidation, periodic trends, 261
    naming compounds of, 169
    reactions with acids, 259
    reactions with oxygen, 267
    as reducing agents, 259
Metal sulfides, 732
Metaphosphoric acid, 722
Metathesis, 229
Metathesis reactions:
    gases formed in, table, 283
    predicting, 277, 279, 281
    in synthesis, 283
    why they occur, 276
Meter, 14
Methane, bonding in, 200
Methanol, 854

synthesis of, 685
Methyl group, 838
Metric system, 11
Meyer, Julius Lothar, 85
Mica, 749
Micelles, 884
    in soap solutions, 402
Microwaves, 105
Millikan, R. A., 92
Millilitre, 15
Millimole, 68
Miscible, 399
Mixture, 25
Molar, 64
Molar concentration, 64, 395
    preparing solutions, 67
Molar heat:
    of crystallization, 363
    of fusion, 363
    of vaporization, 351
Molar heat capacity, 437
Molar mass, 50
    calculation from colligative properties, 421
    of gases, calculation of, 325
Molarity, see Molar concentration
Molar solubility, 581
Molar volume, 322
    of real gases, table, 322
Mole (mol), 43
    mass of, 46
Molecular compounds:
    definition, 144
    properties of, 143
Molecular crystals, 372
Molecular equations, 229
Molecular formula, 52
    calculation of, 56
Molecular orbital theory, 194, 213
    bonding in NO, 709
Molecular shapes, 179
Molecular weight, 49
Molecule, 30
Mole fraction, 395
    Dalton's law of partial pressures, 333
Mole percent (mol %), 395
Mole ratios, 44
Mond process, 649
Monoclinic sulfur, 726
Monodentate ligands, 774
Monomers, 862
Monomethylhydrazine, 707
Monoprotic acids, 233

Monosaccharides, 878
    table of, 880
Moseley, Henry, 110
Mulliken, R. S., 167
Muriatic acid, 737
Mutations, 895
Mylar, 855
Myoglobin, 874

Naming compounds, 169. *See also*
    Nomenclature
Natural gas, 828
    liquid, 358
Natural law, 3
Neon signs, 90
Nernst, Walter, 625
Nernst equation, 625
Nerve poison, 876
Net bond order, 216
Net ionic equation, 231
    stoichiometry of, 289
Neutralization:
    in aqueous solution, 236
    synthesis of compounds by, 285
Neutron activation analysis, 807
Neutrons, 97
Newton, 37
Nickel, 768
    refining of, 649
Nickel cadmium cell, 634
Nitric acid, 712
    as oxidizing acid, 713
    photochemical decomposition of,
        713
    uses of, 712
Nitric oxide, 708
Nitrides, 704
    hydrolysis of, 689
Nitrite ion, 712
    in foods, 712
Nitrogen, 703
    laboratory preparation, 704
    low reactivity of N_2, 703
    molecular orbital description of
        N_2, 217
    oxidation states of, table, 705
    preparation and uses, 704
    reactions with metals, 703, 704
Nitrogen dioxide, 708, 709
Nitrogen oxide, 708
Nitrogen oxides:
    in acid rain, 729
    effect on ozone layer, 698
    in smog, 710

thermodynamic properties, table,
    700
Nitroglycerin, 712, 856
Nitrosoamines, 712
Nitrosonium ion, 709
Nitrous acid, 711
Nitrous oxide, 708
Noble gases, 89
    compounds of, 743
        table, 744
Noble metals, 263
Nodes, 116
    in atomic orbitals, 129, 130
Nomenclature:
    benzene derivatives, 848
    coordination compounds, 777
    of inorganic compounds, 169
    organic compounds, 836
    super-heavy elements, 814
Nonelectrolytes, 227
Nonideal solutions, 414
Nonmetals, 84
    names of anions of, 169
    as oxidizing agents, 265
    reaction with oxygen, 268
Nonoxidizing acids, 259
Norethindrone, 887
Normal boiling point, 360
Nuclear energy, 818
Nuclear equations, 799
Nuclear fission, 810, 817
Nuclear fusion, 820
Nuclear reactors, 818
Nuclear stability, 808
Nuclear transformation, 811
Nucleic acid, 887
Nucleon, 815
Nucleosides, 888
Nucleotides, 887, 889
Nucleus, 96
    density of, 97
Nuclide, 799
Nylon, 862

Observation, 3
Octahedral molecular shapes, 181
Octet rule, 147
Open hearth furnace, 649
Optical isomerism, 780
    organic, 834
Orbital, definition of, 117
Orbital diagram, 121
Orbitals, shapes of in atoms, 129,
    130

Order of reaction, 481
Ores, 641
    concentration of, 642
Organic acids, 858
    reactions of, 858, 859
Organic compounds, 827
    common names, 841
    geometrical isomers, 834
    isomers, 831
    naming, 836
    optical isomerism, 834
    stereoisomerism, 834
    structural isomers, 832
Organomagnesium compounds, 669
Oriental amethyst, 675
Oriental topaz, 675
Ortho, 848
Orthosilicate ion, 747
Orthosilicic acid, 747
Osmosis, 422
    reverse, desalination of sea water,
        425
Osmotic pressure, 424
    molecular weights from, 425
Ostwald, Wilhelm, 707
Ostwald process, 706, 712
Outer orbital complexes, 785
Overlap of orbitals, 194
Oxidation, defined, 249
Oxidation number change method,
    251
Oxidation numbers:
    assignment of, 249
    balancing equations with, 251
Oxidation–reduction reactions, 248
Oxidation state, defined, 249
Oxide ion, as Lewis base, 247
Oxides, 708
    metal, colors of, 665
    of nonmetals, 709
    preparation of, 709
Oxidizing agent:
    defined, 249
    in laboratory, 295
Oxoacids:
    of halogens, table, 738
    nomenclature of, 172
    of nonmetals, table, 702
    strengths of, 242
Oxoanions:
    of halogens, 738
    of nonmetals, table, 702
Oxygen, 696
    abundance, 696

allotropes of, 210, 697
laboratory preparation, 697
molecular orbital description of O_2, 217
as oxidizing agent, 267
preparation, 697
reaction with nonmetals, 700
reactions with organic compounds, 269
Ozone, 210, 697
in smog, 710
Ozone layer, 698

Paired electrons, 121
Paired energy, 786, 791
PAN, 711
Panning for gold, 642
Para, 848
Paraffins, 828
properties of, 143
Paramagnetism, 120
Parent isotope, 799
Partial pressure, 318
Particle accelerators, 811
Parts per million, defined, 287
Pascal, defined, 308
Pauli exclusion principle, 120
Pauling, Linus, 168
PCB, 849
Peptide, 871
Peptide bond, 871
resonance structures of, 165
Peptide linkage, 871
Percent composition, 51
Percent dissociation, 546
Percent yield, 63
Perchlorates, 739
Perchloric acid, 740
Periodic law, 88
Periodic table:
extension of, 812
fully extended version, 89
Mendeleev's, 85, 86
modern version, 88
Periodic trends:
atomic size, 131
ease of oxidation of metals, 261
electron affinity, 136
electronegativity, 168
ionization energy, 134
melting points of transition metals, 757
metallic character, 650

nonmetals as oxidizing agents, 266
strengths of acids, 241
Periods in periodic table, 88
Permanent magnet, 764
Permanganate ion, 766
as oxidizing agent, 295
titrations with, 296
Peroxides, 702
Peroxyacylnitrates, 771
Perpetual motion machine, 438
pH:
acidity and, 540
from cell potentials, 629
of common materials, 541
control with buffers, 555
defined, 539
of solutions of strong acids and bases, 541
Phase, definition of, 25
Phase change, 26
Phase diagram, 379
interpretation of, 380
Phenolphthalein, in acid-base titrations, 293
Phenyl group, 849
Phlogiston theory, 29
pH meter, 540
Phosphate fertilizer, 77
Phosphine, 719
preparation of, 690
Phospholipid, 885
Phosphoric acid, 720
polymeric acids and anions, 721
salts of, 721
structure of, 721
Phosphorous acid, 723
Phosphorus, 718
abundance, 718
allotropes, 212, 718
black, 213
halogen compounds, 723
oxides of, 719, 720
preparation of, 718
red, 212
white, structure of, 212
Phosphorus oxychloride, 724
Phosphorus pentachloride, 723
Phosphorus trichloride, 723
Phosphoryl chloride, 714
Photochemical smog, 709
Photography, thiosulfate ion in, 733
Photons, 111
Physical process, 26

Physical property, 24
Pi bonds, 206
importance of, 209
Pickling, 699
Pig iron, 648
Pitchblende, 665
pK_a, 544
pK_b, 544
Planar triangular molecular shape, 180
Planck, Max, 111
Planck's constant, 111
Plaster of paris, 670
Platinum, from refining of copper, 607
Plumbite ion, 677
Plutonium, 818
pOH, defined, 539
Polar bonds, 167
Polarizability, 353
Polarized light, 783
Polar molecules, effect of molecular structure, 167, 190
Pollution, thermodynamics and, 465
Polyatomic ion, 150
table of, 150
Polydentate ligands, 774
Polyester, 863
Polyethylene, 864
Polymers:
organic, 862
table of, 864
Polypeptide, 871
Polyprotic acids, 233
control of anion concentration with pH, 553
dissociation of, 550
dissociation constants, table, 551
hydrolysis of anions of, 566
Polysaccharides, 881
Polystyrene, 75
Polyvinyl chloride, 862
p orbitals, shapes of, 129
Porphyrin structure, 875
Position of equilibrium, 228
effect of catalyst on, 525
effect of concentration changes on, 522
effect of pressure/volume changes on, 524
effect of temperature on, 524
Post-transition elements, 147
cations of, table, 148
Post-transition metals, 671

Potash, 665
Potassium, 657
    flame test for, 660
    in plants, 657
    preparation of, 659
    reaction with oxygen, 663
    sources of, 663
    see also Alkali metals
Potassium chlorate, decomposition
        of, 502
Potassium hydroxide, 664
Potassium superoxide, 663
Potential energy, 19
Potentiometer, 614
Precipitate, 70, 230
Precipitation, K_{sp} and, 584
Precipitation reactions, 277
    synthesis of compounds by, 283
Precision, 6
Pressure, 306
Pressure units:
    atmosphere, 307
    bar, 451
    millimetre of mercury, 307
    pascal, 308
Primary alcohol, 855
Primary structure, 871
Primitive cubic unit cell, 368
Principal quantum number, 117
Probability distribution, electron,
        128
Products in chemical equation, 34
Properties of matter, 21
Propylene glycol, 855
Proteins, 869
    primary, secondary, and tertiary
        structures, 871
    quaternary structure, 874
Protein synthesis, 891, 894
Protium, 683
Proton, 95
Pseudonoble gas configuration, 148
Pycnometer, 38
Pyrophosphoric acid, 722
Pyrosilicate ion, 747
Pyrosulfuric acid, 730

Qualitative observation, 3
Quanta, 111
Quantitative observation, 3
Quantum mechanics, 116
Quantum number:
    azimuthal, 117
    in Bohr theory, 112
    magnetic, 118

principal, 117
    spin, 110
    summary of, 109
Quartz, 749
    optical isomers of, 750
Quicklime, 670

Racemic, 783
Radiation, penetrating ability of,
        table, 804
Radioactive decay, 98, 798
    kinetics of, 800
    modes of, 809
Radioactive nuclei, particles emitted
        by, 798
Radioactivity, 96
    biological damage from, 806
    discovery of, 798
    measurement of, 802
    units of, 803
Radionuclides, 799
Radio waves, 105
Radium, 665
Raoult's law, 411
    deviations from, 414, 417
    two or more volatile components,
        413
Rare earth elements, 88
Rate constant, 482
Rate determining step, 491
Rate law, 482
    determination of, 483
Rate of reaction, 479
    effect of temperature, 497
    factors that influence, 478
    measurement of, 479, 480
    variation with time, 486
Reactants, 34
Reagent, 68
Reaction coordinate, 495
Reaction quotient, 513
    and $\Delta G°$, 515
Reaction rate, 479. See also Rate of
        reaction
Red lead, 677
Redox, defined, 249
Redox equations:
    balancing by ion-electron method,
        253
    balancing with oxidation
        numbers, 251
Red phosphorus, 719
Reducing agent:
    defined, 249
    laboratory, 297

metals as, 259
Reduction, defined, 249
Reduction potential, 614
    cell potentials from, 614
    standard, table, 618
    uses of, 617–621
Refining of metals, 648
Rem, 804
Representative elements, 88
    ions of, 148
    location in periodic table, 88
Resonance, 163
    valence bond theory and, 208
Resonance hybrid, 164
Resonance structures, 164
Reverse osmosis, 425
Reversible process, 443
Rhombic sulfur, 726
Ribose, 881
RNA, 887, 888
Roasting of ores, 644
Rock salt structure, 369
Roentgen, Wilhelm, 110
Rounding off numbers, 9
Ruby, 674
Rust, 768
    removal by pickling, 699
Rusting of iron, 267, 768
Rutherford, Ernest, 96
Rutile, 646
Rydberg equation, 108

Salt:
    definition, 172
    harvesting from sea water, 658
Salt bridge, 613
Salts, nomenclature of, 172
Salt solutions, ion concentrations in,
        288
Saponification, 859
Sapphire, 675
Saturated hydrocarbons, 827, 828
Saturated solution, 224
Scintillation counter, 801
Sea water:
    bromine from, 736
    calcium and magnesium in, 665
    desalination of, 425
Seaborg, Glenn, 812
Schrödinger, Erwin, 116
Scientific method, 3
Scientific notation, 7
    significant figures and, 7, 8
Secondary alcohol, 855
Secondary structure, 871

Second law of thermodynamics, 462
Selenium, molecular structure, 212
Semiconductors, 85
Semipermeable membranes, 422
Separation of ions, sulfide
    precipitations, 585
Shapes of molecules, predicting, 182
SI, *see* International System of Units
Sickle–cell anemia, 895
Sigma bonds, 206
Significant figures, 5
  in calculated quantities, 9, 10
  counting zeros, 8
  electronic calculators, 8, 10
  exact numbers and, 10
Silane, preparation of, 690
Silica, 746, 749
Silicates, 747
  containing $(SiO_3^{2-})_x$, 748
  containing $(Si_2O_5^{2-})_x$, 749
  containing $(Si_4O_{11}^{6-})_x$, 748
  containing $(Si_6O_{18}^{12-})$, 747
Silicon, 745
  chemical properties, 746
  molecular structure, 213
  prepration of, 745
  properties of, 87
Silicon carbide, 695
Silicones, 750
Silver:
  ore treatment, 642
  in photography, 769
  reaction with nitric acid, 769
  from refining of copper, 607
  tarnishing of, 78, 657
  test for, 770
  uses, 757
Silver halides, 769
  colors of, 655
  relative solubilities, 302
Silver oxide, thermal decomposition,
  645
Silver oxide battery, 633
Simethicone, 751
Simple cubic unit cell, 368
Simplest formula, 52. *See also*
  Empirical formula
Single bonds, 155
Single replacement reactions, 263
SI prefixes, table of, 12
Skeletal structure, 157
Slag, 648
Slaked lime, 670
Slaking, 670
Smog, 709

Soaps, 401
Soapstone, 749
Sodium, 657
  flame test for, 660
  in plants, 657
  production by electrolysis, 603,
    604
  properties of, 83
  reaction with oxygen, 663
  reaction with water, 262
  street lighting, 660
  see also Alkali metals
Sodium bicarbonate, 664
  as buffer, 664
Sodium carbonate, 664
Sodium chloride:
  crystal structure, 369
  molten, electrolysis of, 599
Sodium hydroxide, 663
  uses of, 603, 604
Sodium peroxide, 663
Sodium vapor lamps, 107
Solids:
  compressibility of, 343
  diffusion in, 345
  sublimation of, 347
  vapor pressures of, 377
  volume and shape, 345
Solid solutions, 393
Solubility:
  of alcohols in water, table, 400
  definition, 224
  effect of pressure on, 409
  effect of temperature on, 408
Solubility curves, 409
Solubility product, 580
Solubility product constant, 580
  from cell potentials, 627
  table of, 581
Solubility rules, 277
Solute, 64
Solution, definition of, 25
Solution process:
  energy changes during, 403, 405
  liquids in liquids, 399
  solids in liquids, 400
Solutions, 389
  freezing and boiling points of, 419
  interionic attractions in, 429
  osmotic pressure, 424
  reactions in, 223
  types of, 393
  vapor pressure of, 411
Solvation, 401
Solvay process, 664

Solvent, 64
s orbitals, shapes of, 129
Space shuttle, fuel used, 675
Specific activity, 803
Specific gravity, 23
Specific heat, 24, 437
Spectator ions, 230
Spectrochemical series, 790
Spontaneity of reaction, prediction
  from cell potential, 619
Spontaneous change, 458
Stability constant, 591
Stainless steel, 765, 768
Standard cell potential, 614
Standard entropies:
  of reaction, 467
  table, 467
Standard free energy:
  from cell potentials, 622
  of formation, 466
    table, 468
  of reaction, 468
    and equilibrium, 470, 515
    as guide to spontaneity, 470
Standard heat of reaction, 451. *See
  also* Heats of reaction
Standardizing solution, 294
Standard states, 451
Standard temperature and pressure,
  316
Standing waves, 116
  in atoms, 129
Stannite ion, 677
Starch, 882
  complex with iodine, 299
State function, 437
States of matter, 306, 343
State variable, 437
Stellite, 768
Stereoisomerism, 779
  organic, 834
Steroids, 886
Stock system of nomenclature, 170
Stoichiometry, 42
  gaseous reactions, 321
  ionic reactions, 289
  of reactions in solution, 70
Stoney, G. J., 90
Stopcock, 292
STP, 316
Straight chain compounds, 828
Strong acids, 226
  concentrations of ions in, 538
Strong bases, 234
Strong electrolytes, 227

Strontium, 665
   flame test, 667
   see also Alkaline earth metals
Structural formula, 53
Structural isomers, organic, 832
Styrene, 849
Styrofoam, 849
Subatomic particles, table of, 97
Sublimation, 347
Substance, definition of, 21
Substitutional solid solution, 393
Sucrose, 881
Sulfa drug, 876
Sulfide ores:
   concentration by flotation, 643
   roasting of, 644
Sulfide precipitations, 585
   ions separated by, table, 587
Sulfides, metal, colors of, 656
Sulfite ion, 728
   as reducing agent, 298
Sulfur, 724
   allotropes, 724, 726
   amorphous, 728
   changes when heated, 727
   compounds with metals, 732
   mining of, 724
   monoclinic, 726
   plastic, 728
   reaction with copper, 26
   reaction with zinc, 34
   rhombic, 726
   structure of S_8, 212
Sulfur dioxide, 728
   air pollution from, 728, 729
   oxidation of, 730
Sulfuric acid:
   concentrated, 731
   as dehydrating agent, 731
   dilution of concentrated acid, 69
   manufacture of, 730
   as oxidizing agent, 260
Sulfur oxides, in smog, 710
Sulfur trioxide, 730
   as Lewis acid, 247
Sulfurous acid, 728
Superactinides, 812
Supercooling, 376
Superoxides, 702
Superphosphate, 721
Supersaturated solution, 224
Surface tension, 345
   intermolecular attractions and, 346
Surfactants, 347

Surroundings, 436
Suspension, properties of, 388
Sylvite, 663
Synthesis gas, 685
Synthesis of compounds, 283
System, 436

Taconite, 643
Talc, 749
Teflon, 864
Tellurium, molecular structure, 212
Temperature:
   Celsius scale, 16
   definition of, 16
   Kelvin scale, 16
   kinetic theory and, 330
Tertiary alcohol, 855
Tertiary structure, 871
Testosterone, 887
Tetraethyl lead, 673
Tetrahedral molecular shape, 180
Tetraphosphorus decaoxide, 720
Tetraphosphorus hexaoxide, 720
Tetrathionate ion, 733
Thallium, 671
Theoretical yield, 63
Theory, definition of, 4
Thermal pollution, 465
Thermite reaction, 675
Thermochemical equations:
   combining by Hess' law, 448
   defined, 448
Thermochemistry, 444
Thermodynamic efficiency, 465
Thermodynamic equilibrium
      constant, 516
   from cell potentials, 623
Thermodynamics:
   defined, 436
   first law of, 438
   second law of, 462
   thermal decomposition of metal
      compounds, 644
   third law of, 466
Thermometer, 15, 16
Thin-layer chromatography, 27
Thio, 732
Thioacetamide, 732
Thiosulfate ion, 732
   as reducing agent, 298
   in titrations of I_3^-, 299
Third law of thermodynamics, 466
Thomson, J. J., 91
Thortveitite, 747

Tie line, 417
Tin, 671
   allotropes of, 673
   amphoteric behavior, 677
   oxidation states of, 676
   production of, 673
   properties of, 87
   reaction with chlorine, 677
Tin(IV) chloride, 653
Tin(II), as reducing agent, 298
Tin oxide, reduction of, 646
Tin oxides, 677
Titanium, 764, 765
   ore purification, 646
   preparation of, 646
   uses of, 646
Titanium tetrachloride, 646
   hydrolysis of, 764
Titrant, 292
Titration curves:
   strong acid–strong base, 569
   weak acid–strong base, 571
   weak base–strong acid, 572
Titrations, 292
   endpoint in, 292
   iodine–thiosulfate, 299
   strong acid–strong base, 568
      using $KMnO_4$, 296
   weak acid–strong base, 569
   weak base–strong acid, 571
Toluene, 848
Tracer studies, 806
Transfer RNA, 892
Trans isomers, 779
Transition elements:
   atomic and ionic radii, 761, 762
   cations of, table, 148
   electronic structures, 758, 759
   general properties, 756
   location in periodic table, 88
   melting points, 757
   stability of oxidation states, 760
Transition state, 497
Transition state theory, 494
Transuranium elements, 812
Tributyltin oxide, 677
Trichloroethylene, 852
Triglyceride, 883
Trigonal bipyramidal molecular
      shape, 180
Triiodide ion, 298
Triple bonds, 156
   valence bond description of, 207
Triple point, 378

Tripolyphosphate ion, 722
Triprotic acids, 233
Tritium, 684
Trona ore, 664
Trouton's rule, 383
TSP, 721
Tungsten, 757
    in light bulbs, 83
Tungsten carbide, 696
    use in cutting tools, 394
Tyndall effect, 391

Ultraviolet radiation, 105
Uncertainty in measurement, 6
Uncertainty principle, 128
Unified atomic mass unit, 33, 34
Unit cell, 367
Units:
    amount of substance, 12
    atomic mass, 33
    base, SI, 12
    concentration, 64, 394
    density, 22
    derived, 11
    electric charge, 609
    electric current, 12, 609
    emf, 613
    energy, 20
    entropy, 461
    heat capacity, 437
    length, 12, 14
    luminous intensity, 12
    mass, 12, 14
    metric, 11
    molar concentration, 64
    molar heat capacity, 437
    molarity, 64
    pressure, 307, 308
    radioactivity, 803
    rate constant, 482
    rate of reaction, 479
    specific heat, 24, 437
    surface tension, 345
    temperature, 12, 15, 16
    time, 12
    volume, 14, 15

Units of Measurement, 11. *See also* International System of Units
Unsaturated hydrocarbons, 830
Unsaturated solution, 224
Uranium, decay series for, 800
Urea, synthesis of, 827

Valence bond theory, 194
    coordination compounds, 784
Valence shell, 127
Valence shell electron-pair repulsion theory, 182
van der Waals, J. D., 335
van der Waals constants, table, 336
van der Waals equation, 336
van't Hoff, Jacobus Henricus, 424
van't Hoff equation, 424
van't Hoff factor, 429
Vapor pressure, 354
    factors influencing, 356
    of solids, 377
    of solutions, 411
    solutions of electrolytes, 426
    of water, table, 320
Vapor pressure curves, 357
Vegetable oils, 884
Vermiculite, 749
Visible spectrum, 105
Voltaic cells, 597
Volumetric analyses, 292
Volumetric flask, 67
von Laue, Max, 365
VSEPR theory, 182
    use with valence bond theory, 202
Vulcanization, 863

Washing soda, 664, 695
Water:
    autoionization of, 537
    critical T and P, 357
    phase diagram, 379
    valence bond description, 195, 201
Water gas, 684
Water treatment, 676
Watson, James, 366
Wave functions, 117

Wavelength, 104
Wave mechanics, 114
Waves, properties of, 104
Weak acids, 233
    dissociation of, 543
Weak bases, 235
    dissociation of, 544
Weak electrolytes, 227
    table of, 228
Weight, 17
Werner, Alfred, 773
White lead, 77, 657
White phosphorus, 718
Wilkins, Maurice, 366
Wire, manufacture of, 82
Wohler, Friedrich, 827
Woods metal, 674
Work, 440
    in chemical systems, 442
    first law of thermodynamics, 439
    free energy and, 465
    from galvanic cells, 622
    maximum, 443
    pressure-volume, 441

Xenon:
    compounds of, 743
    compounds of, table, 744
X-ray diffraction, 365
X rays:
    atomic numbers and, 110
    position in electromagnetic spectrum, 105
Xylene, 849

Zinc, 770
    alloys of, 771
    amphoteric behavior, 771
    galvanizing, 770
    reaction with sulfur, 34
    sulfide ore, 643, 644
Zinc-carbon dry cell, 631
Zinc chloride, 772
Zinc oxide, 772
Zircon, 747
Zone refining, 746
Zwitterion, 870